Multiple Abiotic Stresses in Plants

Mechanisms and Management Strategies

Multiple Abiotic Stresses in Plants

Mechanisms and Management Strategies

Edited by

Koushik Chakraborty
Division of Crop Physiology and Biochemistry, ICAR-Central Rice Research Institute, Cuttack, Odisha, India

Md. Hasanuzzaman
Department of Agronomy, Sher-e-Bangla Agricultural University, Dhaka, Bangladesh

Debarati Bhaduri
Division of Crop Production, ICAR-Central Rice Research Institute, Cuttack, Odisha, India

Somnath Roy
Central Rainfed Upland Rice Research Station, ICAR-Central Rice Research Institute, Hazaribag, Jharkhand, India

Honghong Wu
College of Plant Science and Technology, Huazhong Agricultural University, Wuhan, Hubei, P.R. China

ELSEVIER

ACADEMIC PRESS
An imprint of Elsevier

Academic Press is an imprint of Elsevier
125 London Wall, London EC2Y 5AS, United Kingdom
50 Hampshire Street, 5th Floor, Cambridge, MA 02139, United States

ISBN: 978-0-443-27424-4

For Information on all Academic Press publications
visit our website at https://www.elsevier.com/books-and-journals

Publisher: Andre Wolff
Acquisitions Editor: Chiara Giglio
Editorial Project Manager: Sweta Bharti
Production Project Manager: Aishwarya R.
Cover Designer: Greg Harris

Typeset by MPS Limited, Chennai, India

Contents

Section B
Physiological and molecular bases of multiple abiotic stress tolerance in plants

12. Genomic breeding empowered by next-generation sequencing and generative artificial intelligence for multiple abiotic stress tolerance in plants

Vanavichit A., Poncheewin W., Wanchana S., Ruenjaichol V., Ruengphayak S. and Arikit S.

13. Role of phenomics and precision phenotyping in breeding strategies to develop multiple abiotic stress-tolerant crops

H.G.K. Chanumolu, Siddartha Kumar Neteti, Mahesh Kumar, Biswabiplab Singh, Elangovan A., Sudhir Kumar, Madhurima Das, Jagadish Rane and Viswanathan C.

14. Use of nanobiotechnology to improve multiple abiotic stress tolerance in plants

Mohammad Faizan, Mohammad Nauman Khan, Honghong Wu, Lixiao Nie and Pravej Alam

15. Transgenic and genome editing-based approaches for improving multiple abiotic stress tolerance in plants

Tanmoy Halder, Sonali Panda, Chandana Ghosh, M.J. Baig and Kutubuddin A. Molla

List of contributors

Pravej Alam Department of Biology, College of Science and Humanities, Prince Sattam bin Abdulaziz University, Alkharj, Riyadh Province, Saudi Arabia

Abida Anjum Plant Breeding and Biotechnology Laboratory, Department of Botany, University of Dhaka, Dhaka, Bangladesh

Arikit S. Rice Science Center, Kasetsart University, Kamphang Saen, Nakhon Pathom, Thailand

M.J. Baig ICAR-Central Rice Research Institute, Cuttack, Odisha, India

Amrita Banerjee Central Rainfed Upland Rice Research Station, ICAR-Central Rice Research Institute, Hazaribag, Jharkhand, India

K.C. Bansal National Academy of Agricultural Sciences, New Delhi, India

Nirmalendu Basak ICAR-National Bureau of Soil Survey and Land Use Planning, Regional Centre Kolkata, Salt Lake city Kolkata, West Bengal, India; ICAR-Central Soil Salinity Research Institute, Karnal, Haryana, India

Piu Basak Department of Agricultural Chemistry and Soil Science, Bidhan Chandra Krishi Viswavidyalaya, Nadia, West Bengal, India

Debarati Bhaduri Division of Crop Production, ICAR-Central Rice Research Institute, Cuttack, Odisha, India

Ruchika Bhagat Dr. Panjabrao Deshmukh Krishi Vidyapeeth, Akola, Maharashtra, India

Arnab Bhowmik Department of Natural Resources and Environmental Design, North Carolina Agricultural and Technical State University, Greensboro, NC, United States

Dhiman Burman Regional Research Station (CSSRI, RRS), ICAR-Central Soil Salinity Research Institute, Canning Town, West Bengal, India

Koushik Chakraborty Division of Crop Physiology and Biochemistry, ICAR-Central Rice Research Institute, Cuttack, Odisha, India

Rakhi Chandoliya CSIR-Institute of Himalayan Bioresource Technology, Palampur, Himachal Pradesh, India; Department of Biotechnology, School of Bioengineering and Biosciences, Lovely Professional University, Phagwara, Punjab, India

Priyanka Chandra ICAR-Central Soil Salinity Research Institute, Karnal, Haryana, India

H.G.K. Chanumolu Division of Plant Physiology, ICAR-Indian Agricultural Research Institute, Pusa Campus, New Delhi, India

Shreyasi Gupta Choudhury ICAR-National Bureau of Soil Survey and Land Use Planning, Regional Centre Kolkata, Salt Lake city Kolkata, West Bengal, India

Raviprakash Dani Gen Scan Inc. Global Consultancy, Houston, TX, United States; Denau Institute of Entrepreneurship and Pedagogy, Surkhandarya Region, Viloyat, Uzbekistan

Ankan Das ICAR-Indian Institute of Horticultural Research, Bengaluru, Karnataka, India

Madhurima Das Division of Plant Physiology, ICAR-Indian Agricultural Research Institute, Pusa Campus, New Delhi, India

Chandan K. Deb ICAR Indian Agricultural Statistical Research Institute, New Delhi, Delhi, India

Sandip Debnath Department of Genetics and Plant Breeding, Institute of Agriculture, Visva-Bharati, West Bengal, India

Rupesh Deshmukh Department of Biotechnology, Central University of Haryana, Mahendergarh, Haryana, India

Devika S. Regional Research Station (CSSRI, RRS), ICAR-Central Soil Salinity Research Institute, Canning Town, West Bengal, India

Hena Dhar Faculty of Applied Sciences and Biotechnology, School of Biotechnology, Shoolini University, Solan, Himachal Pradesh, India

Ahmad Diyan Hubei Hongshan Laboratory, MOA Key Laboratory of Crop Ecophysiology and Farming System in the Middle Reaches of the Yangtze River, The Center of Crop Nanobiotechnology, College of Plant Science and Technology, Huazhong Agricultural University, Wuhan, Hubei, P.R. China

Zihao Dong Hubei Hongshan Laboratory, MOA Key Laboratory of Crop Ecophysiology and Farming System in the Middle Reaches of the Yangtze River, The Center of Crop Nanobiotechnology, College of Plant Science and Technology, Huazhong Agricultural University, Wuhan, Hubei, P.R. China

Elangovan A. Division of Plant Physiology, ICAR-Indian Agricultural Research Institute, Pusa Campus, New Delhi, India

Mohammad Faizan Botany Section, School of Sciences, Maulana Azad National Urdu University, Hyderabad, Telangana, India

Chandana Ghosh ICAR-Central Rice Research Institute, Cuttack, Odisha, India

Rajendran Gobinath ICAR-Indian Institute of Rice Research, Hyderabad, Telangana, India

B. Raghavendra Goud ICAR-Central Rice Research Institute, Cuttack, Odisha, India

Suman Gusain CSIR-Institute of Himalayan Bioresource Technology, Palampur, Himachal Pradesh, India; Academy of Scientific and Innovative Research (AcSIR), Ghaziabad, India

Tanmoy Halder ICAR-Central Rice Research Institute, Cuttack, Odisha, India

Md. Hasanuzzaman Department of Agronomy, Sher-e-Bangla Agricultural University, Dhaka, Bangladesh

Md Arif Hussain Department of Biochemistry and Molecular Biology, Sher-e-Bangla Agricultural University, Dhaka, Bangladesh; Institute of Plant Nutrition and Soil Science, Kiel University, Kiel, Schleswig-Holstein, Germany

Muhammad Ikram Faculty of Agricultural Sciences and Technology, Institute of Agronomy, Bahauddin Zakariya University, Multan, Punjab, Pakistan

Tahmina Islam Plant Breeding and Biotechnology Laboratory, Department of Botany, University of Dhaka, Dhaka, Bangladesh

Shyam Jadhao Dr. Panjabrao Deshmukh Krishi Vidyapeeth, Akola, Maharashtra, India

Balaji Jadhav Dr. Panjabrao Deshmukh Krishi Vidyapeeth, Akola, Maharashtra, India

Pravin Jadhav Dr. Panjabrao Deshmukh Krishi Vidyapeeth, Akola, Maharashtra, India

Mohammad Shah Jahan Key Laboratory of Stress Physiology and Molecular Biology, Department of Horticulture, Sher-e-Bangla Agricultural University, Dhaka, Bangladesh

Uday Chand Jha ICAR - Indian Institute of Pulses Research (IIPR), Kanpur, Uttar Pradesh, India

Rohit Joshi CSIR-Institute of Himalayan Bioresource Technology, Palampur, Himachal Pradesh, India; Academy of Scientific and Innovative Research (AcSIR), Ghaziabad, India

Shrushti Joshi Department of Biotechnology, Modern College of Arts, Science and Commission, Savitribai, Phule Pune University, Ganeshkhind, Pune, Maharashtra, India

Shubham Joshi CSIR-Institute of Himalayan Bioresource Technology, Palampur, Himachal Pradesh, India; Academy of Scientific and Innovative Research (AcSIR), Ghaziabad, India

Prashant Kale Dr. Panjabrao Deshmukh Krishi Vidyapeeth, Akola, Maharashtra, India

Hamroeva Marguba Kamilovna Gen Scan Inc. Global Consultancy, Houston, TX, United States; Denau Institute of Entrepreneurship and Pedagogy, Surkhandarya Region, Viloyat, Uzbekistan

Pandian Kannan Tamil Nadu Agricultural University, Coimbatore, Tamil Nadu, India

Ehtisham Hassan Khan Hubei Hongshan Laboratory, MOA Key Laboratory of Crop Ecophysiology and Farming System in the Middle Reaches of the Yangtze River, The Center of Crop Nanobiotechnology, College of Plant Science and Technology, Huazhong Agricultural University, Wuhan, Hubei, P.R. China; Department of Agronomy, Faculty of Crop Production Sciences, The University of Agriculture, Peshawar, Pakistan

Md. Rakibul Hasan Khan Department of Agronomy, Sher-e-Bangla Agricultural University, Dhaka, Bangladesh

Mohammad Nauman Khan National Key Laboratory for Germplasm Innovation & Utilization of Horticultural Crops, MOA Key Laboratory of Crop Ecophysiology and Farming System in the Middle Reaches of the Yangtze River, The Center of Crop Nanobiotechnology, Huazhong Agricultural University, Wuhan, Hubei, P.R. China; School of Breeding and Multiplication (Sanya Institute of Breeding and Multiplication), Hainan University, Sanya, Hainan, P.R. China

Vilas Kharche Dr. Panjabrao Deshmukh Krishi Vidyapeeth, Akola, Maharashtra, India

Ashok Kumar Division of Plant Physiology, ICAR-Indian Agricultural Research Institute, Pusa Campus, New Delhi, India

Mahesh Kumar Division of Plant Physiology, ICAR-Indian Agricultural Research Institute, Pusa Campus, New Delhi, India

Ranjeet R. Kumar Division of Plant Physiology, ICAR-Indian Agricultural Research Institute, Pusa Campus, New Delhi, India

Sudhir Kumar Division of Plant Physiology, ICAR-Indian Agricultural Research Institute, Pusa Campus, New Delhi, India

Vinay Kumar Department of Biotechnology, Modern College of Arts, Science and Commission, Savitribai, Phule Pune University, Ganeshkhind, Pune, Maharashtra, India

Virender Kumar International Rice Research Institute, Los Banos, Laguna, Philippines

Anita Kumari CSIR-Institute of Himalayan Bioresource Technology, Palampur, Himachal Pradesh, India; Academy of Scientific and Innovative Research (AcSIR), Ghaziabad, India

Khushbu Kumari CSIR-Institute of Himalayan Bioresource Technology, Palampur, Himachal Pradesh, India; Academy of Scientific and Innovative Research (AcSIR), Ghaziabad, India

Tashi Dorjee Lama Regional Research Station (CSSRI, RRS), ICAR-Central Soil Salinity Research Institute, Canning Town, West Bengal, India

Guangjing Li Hubei Hongshan Laboratory, MOA Key Laboratory of Crop Ecophysiology and Farming System in the Middle Reaches of the Yangtze River, The Center of Crop Nanobiotechnology, College of Plant Science and Technology, Huazhong Agricultural University, Wuhan, Hubei, P.R. China

Yanhui Li Hubei Hongshan Laboratory, MOA Key Laboratory of Crop Ecophysiology and Farming System in the Middle Reaches of the Yangtze River, The Center of Crop Nanobiotechnology, College of Plant Science and Technology, Huazhong Agricultural University, Wuhan, Hubei, P.R. China

Tianyang Liu Key Laboratory of Oasis Eco-agriculture, Xinjiang Production and Construction Corps/College of Agriculture, Shihezi University, Shihezi, Xinjiang, P.R. China

Badal Mahakalkar Department of Biotechnology, Central University of Haryana, Mahendergarh, Haryana, India

Arpita Mahobia Dr. Panjabrao Deshmukh Krishi Vidyapeeth, Akola, Maharashtra, India

N.P. Mandal Central Rainfed Upland Rice Research Station, ICAR-Central Rice Research Institute, Hazaribag, Jharkhand, India

Uttam Kumar Mandal Regional Research Station (CSSRI, RRS), ICAR-Central Soil Salinity Research Institute, Canning Town, West Bengal, India

Shyamsunder Mane Dr. Panjabrao Deshmukh Krishi Vidyapeeth, Akola, Maharashtra, India

Sheikh Muhammad Masum Department of Agronomy, Sher-e-Bangla Agricultural University, Dhaka, Bangladesh

Ibrahim A.A. Mohamed National Key Laboratory of Cotton Bio-breeding and Integrated Utilization, Institute of Cotton Research, Chinese Academy of Agricultural Sciences, Anyang, Henan, P.R. China; Hubei Hongshan Laboratory, MOA Key Laboratory of Crop Ecophysiology and Farming System in the Middle Reaches of the Yangtze River, The Center of Crop Nanobiotechnology, College of Plant Science and Technology, Huazhong Agricultural University, Wuhan, Hubei, P.R. China; Botany Department, Faculty of Agriculture, Fayoum University, Fayoum, Egypt

Mangesh Moharil Dr. Panjabrao Deshmukh Krishi Vidyapeeth, Akola, Maharashtra, India

Kutubuddin A. Molla ICAR-Central Rice Research Institute, Cuttack, Odisha, India

Gopika Mote Dr. Panjabrao Deshmukh Krishi Vidyapeeth, Akola, Maharashtra, India

Nuram Mubina Plant Breeding and Biotechnology Laboratory, Department of Botany, University of Dhaka, Dhaka, Bangladesh

Karl H. Mühling Institute of Plant Nutrition and Soil Science, Kiel University, Kiel, Schleswig-Holstein, Germany

Balwant Munde Dr. Panjabrao Deshmukh Krishi Vidyapeeth, Akola, Maharashtra, India

Giverson Mupambi Umass Cranberry Research Station, East Wareham, MA, United States

Shivani Nagar Division of Plant Physiology, ICAR-Indian Agricultural Research Institute, Pusa Campus, New Delhi, India

Nazima Nasrullah Division of Plant Physiology, ICAR-Indian Agricultural Research Institute, Pusa Campus, New Delhi, India

Jhilmil Nath CSIR-Institute of Himalayan Bioresource Technology, Palampur, Himachal Pradesh, India; Academy of Scientific and Innovative Research (AcSIR), Ghaziabad, India

Harsh Nayyar Department of Botany, Panjab University, Chandigarh, Panjab, India

Siddartha Kumar Neteti Division of Plant Physiology, ICAR-Indian Agricultural Research Institute, Pusa Campus, New Delhi, India

Lixiao Nie School of Breeding and Multiplication (Sanya Institute of Breeding and Multiplication), Hainan University, Sanya, Hainan, P.R. China

Ashadu Nyande Hubei Hongshan Laboratory, MOA Key Laboratory of Crop Ecophysiology and Farming System in the Middle Reaches of the Yangtze River, The Center of Crop Nanobiotechnology, College of Plant Science and Technology, Huazhong Agricultural University, Wuhan, Hubei, P.R. China

Suraj K. Padhi Department of Biodiversity and Conservation of Natural Resources, Central University of Odisha, Koraput, Odisha, India

Debabrata Panda Department of Biodiversity and Conservation of Natural Resources, Central University of Odisha, Koraput, Odisha, India

Sonali Panda ICAR-Central Rice Research Institute, Cuttack, Odisha, India; Ravenshaw University, Cuttack, Odisha, India

Ankesh Pandey Division of Plant Physiology, ICAR-Indian Agricultural Research Institute, Pusa Campus, New Delhi, India

Akshita Patial CSIR-Institute of Himalayan Bioresource Technology, Palampur, Himachal Pradesh, India; Department of Molecular Biology & Genetic Engineering, School of Bioengineering and Biosciences, Lovely Professional University, Phagwara, Punjab, India

Suraj Patil Department of Biotechnology, Modern College of Arts, Science and Commission, Savitribai, Phule Pune University, Ganeshkhind, Pune, Maharashtra, India

Poncheewin W. National Center for Genetic Engineering and Biotechnology (BIOTEC), National Science and Technology Development Agency (NSTDA), Khlong Luang, Pathum Thani, Thailand

Manu Priya Umass Cranberry Research Station, East Wareham, MA, United States

Arvind Kumar Rai ICAR-Central Soil Salinity Research Institute, Karnal, Haryana, India

William Randle Department of Natural Resources and Environmental Design, North Carolina Agricultural and Technical State University, Greensboro, NC, United States

Jagadish Rane ICAR-CIAH, Bikaner, Rajasthan, India

Aditya Rathod Dr. Panjabrao Deshmukh Krishi Vidyapeeth, Akola, Maharashtra, India

Tarasingh Rathod Dr. Panjabrao Deshmukh Krishi Vidyapeeth, Akola, Maharashtra, India

Meenakshi Rawat CSIR-Institute of Himalayan Bioresource Technology, Palampur, Himachal Pradesh, India; Academy of Scientific and Innovative Research (AcSIR), Ghaziabad, India

Soham Ray Division of Plant Physiology, ICAR-Indian Agricultural Research Institute, Pusa Campus, New Delhi, India

Somnath Roy Central Rainfed Upland Rice Research Station, ICAR-Central Rice Research Institute, Hazaribag, Jharkhand, India

Ruengphayak S. Rice Science Center, Kasetsart University, Kamphang Saen, Nakhon Pathom, Thailand

Ruenjaichol V. National Center for Genetic Engineering and Biotechnology (BIOTEC), National Science and Technology Development Agency (NSTDA), Khlong Luang, Pathum Thani, Thailand

Chinmay Saha Department of Agricultural Chemistry, Sher-e-Bangla Agricultural University, Dhaka, Bangladesh

Md. Nazmus Sakib Department of Agronomy, Sher-e-Bangla Agricultural University, Dhaka, Bangladesh

Puranjoy Sar Central Rainfed Upland Rice Research Station, ICAR-Central Rice Research Institute, Hazaribag, Jharkhand, India

R.K. Sarkar ICAR-Central Rice Research Institute, Cuttack, Odisha, India

Harmandeep Sharma Department of Natural Resources and Environmental Design, North Carolina Agricultural and Technical State University, Greensboro, NC, United States

Laxmi Sharma ICAR- National Bureau of Plant Genetic Resources, New Delhi, Delhi, India

Umesh Shinde Dr. Panjabrao Deshmukh Krishi Vidyapeeth, Akola, Maharashtra, India

Shajneen Jahan Shoily Plant Breeding and Biotechnology Laboratory, Department of Botany, University of Dhaka, Dhaka, Bangladesh

Abu Bakar Siddique Tasmanian Institute of Agriculture, University of Tasmania, Launceston, Tasmania, Australia

Biswabiplab Singh Division of Plant Physiology, ICAR-Indian Agricultural Research Institute, Pusa Campus, New Delhi, India

Sudhanshu Singh International Rice Research Institute, Los Banos, Laguna, Philippines

Humira Sonah Department of Biotechnology, Central University of Haryana, Mahendergarh, Haryana, India

Puja Gupta Soni Forage Section, Department of Genetics and Plant Breeding, Chaudhary Charan Singh Haryana Agricultural University, Hisar, Haryana, India

Tanvir Ahmad Sourav Department of Agronomy, Sher-e-Bangla Agricultural University, Dhaka, Bangladesh

Ashish Srivastava International Rice Research Institute, Los Banos, Laguna, Philippines

Tamima Sultana Plant Breeding and Biotechnology Laboratory, Department of Botany, University of Dhaka, Dhaka, Bangladesh

Parul Sundha ICAR-Central Soil Salinity Research Institute, Karnal, Haryana, India

Abida Lotif Tamanna Plant Breeding and Biotechnology Laboratory, Department of Botany, University of Dhaka, Dhaka, Bangladesh

Pranita Thakur School of Biotechnology and Bioinformatics, Dr. D.Y. Patil Deemed to be University, Navi Mumbai, Maharashtra, India

Rishu Thakur Menzies School of Health Research, Charles Darwin University, Alice Springs, NT, Australia

Chibuzor Uche Department of Natural Resources and Environmental Design, North Carolina Agricultural and Technical State University, Greensboro, NC, United States

Vanavichit A. Rice Science Center, Kasetsart University, Kamphang Saen, Nakhon Pathom, Thailand

Shanmugam Vijayakumar ICAR-Indian Institute of Rice Research, Hyderabad, Telangana, India; International Rice Research Institute, Los Banos, Laguna, Philippines

Joshitha Vijayan ICAR- National Institute for Plant Biotechnology, New Delhi, Delhi, India

Viswanathan C. Division of Plant Physiology, ICAR-Indian Agricultural Research Institute, Pusa Campus, New Delhi, India

Wanchana S. National Center for Genetic Engineering and Biotechnology (BIOTEC), National Science and Technology Development Agency (NSTDA), Khlong Luang, Pathum Thani, Thailand

Jie Wang National Key Laboratory of Cotton Bio-breeding and Integrated Utilization, Institute of Cotton Research, Chinese Academy of Agricultural Sciences, Anyang, Henan, P.R. China

Archana Watts Division of Plant Physiology, ICAR-Indian Agricultural Research Institute, Pusa Campus, New Delhi, India

Honghong Wu College of Plant Science and Technology, Huazhong Agricultural University, Wuhan, Hubei, P.R. China

Amit Yadav Department of Biotechnology, Central University of Haryana, Mahendergarh, Haryana, India

Pranjal Yadava Division of Plant Physiology, ICAR-Indian Agricultural Research Institute, Pusa Campus, New Delhi, India

Zujun Yin National Key Laboratory of Cotton Bio-breeding and Integrated Utilization, Institute of Cotton Research, Chinese Academy of Agricultural Sciences, Anyang, Henan, P.R. China

Yali Zhang Key Laboratory of Oasis Eco-agriculture, Xinjiang Production and Construction Corps/College of Agriculture, Shihezi University, Shihezi, Xinjiang, P.R. China

Lan Zhu Key Laboratory of Oasis Eco-agriculture, Xinjiang Production and Construction Corps/College of Agriculture, Shihezi University, Shihezi, Xinjiang, P.R. China

Section A

Multiple abiotic stresses in crop plants: A global climate change perspective

Chapter 1

Incidence of multiple abiotic stresses in crop plants and their implications for global food security

Koushik Chakraborty[1,*,**], Debarati Bhaduri[2,**], Md. Hasanuzzaman[3], Somnath Roy[4], and R.K. Sarkar[5]

[1]*Division of Crop Physiology and Biochemistry, ICAR-Central Rice Research Institute, Cuttack, Odisha, India,* [2]*Division of Crop Production, ICAR-Central Rice Research Institute, Cuttack, Odisha, India,* [3]*Department of Agronomy, Sher-e-Bangla Agricultural University, Dhaka, Bangladesh,* [4]*Central Rainfed Upland Rice Research Station, ICAR-Central Rice Research Institute, Hazaribag, Jharkhand, India,* [5]*ICAR-Central Rice Research Institute, Cuttack, Odisha, India*

[*]*Corresponding author. e-mail address: Koushik.Chakraborty@icar.org.in*

1.1 Introduction

Crop plants in natural and agricultural ecosystems are increasingly exposed to multiple abiotic stresses simultaneously rather than individually. The concurrent occurrence of stresses such as drought, salinity, heat, flooding, and nutrient imbalance profoundly alters plant growth, development, and productivity, posing a serious challenge to global food security under changing climatic conditions. In recent decades, the incidence of multiple abiotic stresses in crop plants has emerged as a critical concern for agricultural sustainability. Environmental factors such as drought, heat, salinity, and nutrient deficiencies often occur concurrently, and their combined effects impose greater damage than individual stresses alone, complicating stress mitigation strategies and crop improvement efforts (Sarkar et al., 2019). Among the abiotic stresses, drought is the most dominant. Drought is the number one abiotic stress, which impacts crop productivity in almost all continents. The second most important stress is salinity, whereas flooding/submergence also influences to a great extent, especially in South and Southeast Asia, where a large proportion of the world's poor reside. In Africa, the second most important stress is iron toxicity, whereas in East Asia, temperature-related stress is the most dominant stress, which affects rice growth and productivity. Overall, a bunch of abiotic stresses such as drought, salinity, submergence/flooding, heat and cold, iron and other heavy metal toxicity, and soil infertility are major challenges to global food production (Fig. 1.1).

According to the first major global assessment of salt-affected soils by the FAO in 50 years, approximately 1.4 billion hectares of land covering 10% of the global land area are currently affected by salinity, with an additional 1 billion hectares at risk due to climate change and unsustainable land management practices. Salinity severely threatens soil fertility and environmental sustainability, particularly in ten countries: Afghanistan, Australia, Argentina, China, Kazakhstan, Russia, the United States, Iran, Sudan, and Uzbekistan, which together account for 70% of the world's salt-affected soils. Yield losses of up to 70% have been observed in salinity-sensitive crops such as rice and beans, and around 10% of both irrigated and rainfed cropland is already affected (FAO, 2024). Floods also contribute significantly to global crop losses. A meta-analysis covering 1982–2016 estimates that flood events adversely affected 13 million hectares of rice, 8 million hectares each of maize and wheat, and 4 million hectares of soy, with yield reductions observed over nearly two-thirds of rice, maize, and soy areas, and half of wheat areas (Kim et al., 2023). Furthermore, projections indicate that a sea-level rise of 1 m by 2100 could inundate up to 15% of low-lying land globally, including 20% of farmland, potentially causing up to 21% loss in agricultural production, especially in coastal and delta regions (Chauhan et al., 2014). Within events between 1983 and 2009, drought affected 161 Mha of wheat, 124 Mha of maize, 102 Mha of rice, and 67 Mha of soybean, representing 75%, 82%, 62%, and

** Both authors have equally contributed.

Multiple Abiotic Stresses in Plants. DOI: https://doi.org/10.1016/B978-0-443-27424-4.00002-1

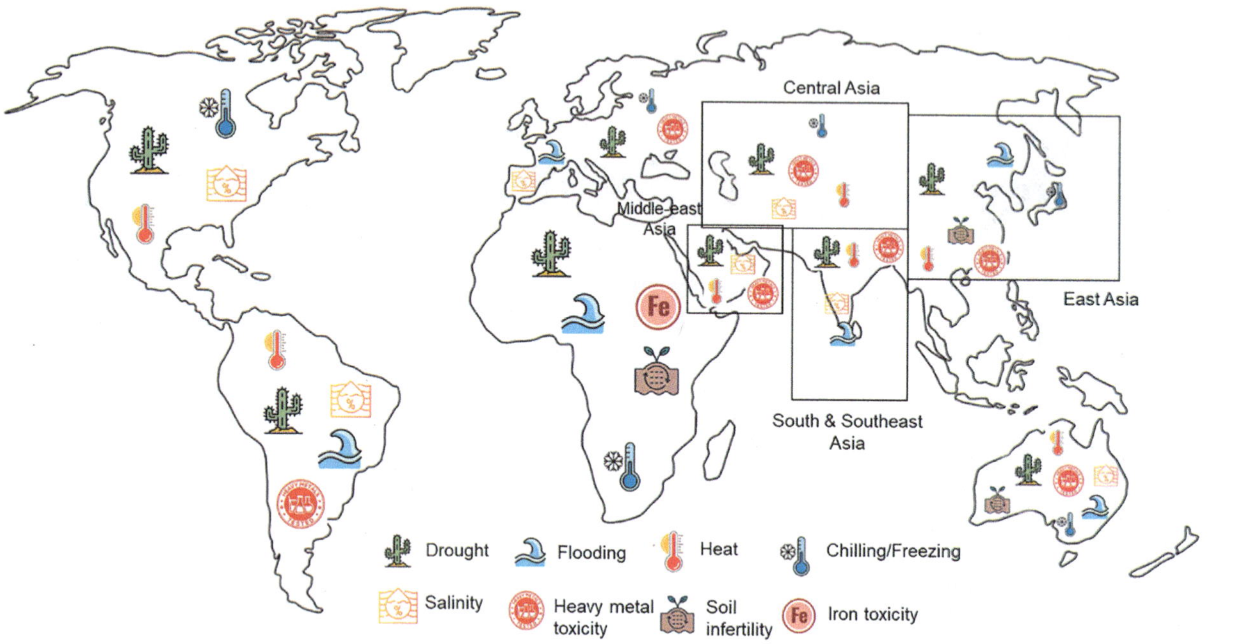

Depending on the topography and cultivation condition any combination of major abiotic stresses is possible in tandem or in sequence

FIGURE 1.1 Prevalence of major abiotic stresses in crop plants across the world, with special reference to rice. Map lines delineate study areas and do not necessarily depict accepted national boundaries.

91% of the respective global harvested areas. Yield losses during individual droughts averaged 8% for wheat, 7% for maize and soybean, and 3% for rice (Kim et al., 2019). According to OECD estimates, in the 10% driest years, average yield losses ranged from 12% to 21.9%, depending on crop type (OECD Publishing, 2025).

Between 2015 and 2019 alone, droughts and floods damaged 9.42 and 3.72 Mha of cropland, respectively, resulting in an estimated loss of 20.64 M tonnes of crop production. Rainfed systems in Thailand, Cambodia, and Myanmar were particularly affected by drought, while Indonesia, the Philippines, and Malaysia suffered more from floods (Venkatappa et al., 2021). Soil acidity also poses a serious threat to crop productivity. In China, continued cropland acidification could result in yield losses of 4%–24% for wheat, maize, and rice by 2050 if fertilizer practices are not improved (Zhu et al., 2020). In sub-Saharan Africa, acidic soils cover 10.8 million km^2 (~35% of land area), with especially severe conditions in East and Central Africa (Agegnehu et al., 2021). These findings underscore the widespread and compounding impacts of abiotic stresses on global food production and emphasize the urgent need for climate-resilient agricultural strategies.

1.1.1 Single versus multiple/combined abiotic stresses—the challenges

Crop cultivation in a stress-free environment is more or less impossible even with assured irrigation, pesticide, and fertilizer management due to weather aberrations. For example, rice cultivation under an assured irrigation system during the Winter/Boro season of Bangladesh and eastern India can encounter both cold during the seedling stage and heat at the reproductive stage. Improving tolerance of a particular stress can boost production. It was observed that rice with submergence-tolerant QTL *SUB1*, which produced 45%–90% more grain yield in a flooded year, as compared to Swarna without *SUB1*. Farmers got 2%–10% more yield in a nonflooded year. The increase in yield was due to better fertilizer management. Farmers in flood-prone areas generally apply a low dose of fertilizer. They become more confident with *SUB1* variety and applied proper doses of fertilizer (Sarkar et al., 2009; Dar et al., 2013; Emerick and Ronald, 2019). Improving the single tolerance level is always beneficial. The benefit we got by cultivating the *SUB1* variety became insignificant when drought occurred following a flash flood. Climate-driven temperature rise or overall climate change has made the situation worse (Fig. 1.2). Greater evaporative demand due to the intense and increasing heat waves, sea level rise, and alterations of precipitation patterns created a situation where single abiotic stress tolerance characteristics became insignificant for sustainable agriculture. Extreme weather events are now more frequent in many parts of crop-growing areas (Pathak et al., 2021). These threats are intensifying daily, making crop cultivation increasingly risky. Farmers face challenges such as both excessive and insufficient rainfall occurring within the same growing season. Additionally, salt

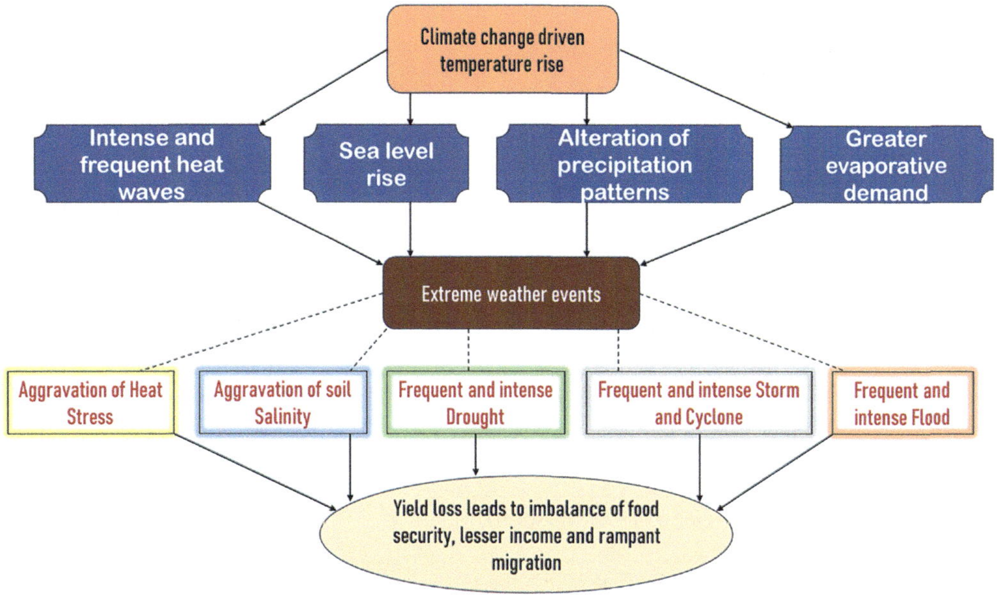

FIGURE 1.2 A flowchart illustrating the impact of climate change on extreme weather events and subsequent effects on food security, lesser income, and rampant migration.

stress is becoming more prevalent in coastal areas, exacerbated by factors such as excessive irrigation, salt deposition in the soil, seawater intrusion, and other climatic anomalies, including strong winds and high temperatures (Pathak et al., 2021).

Crop cultivation, particularly rice cultivated in river deltas and coastal areas during wet seasons, encounters frequent flooding. However, the deficit of rainfall, which has become increasingly common, introduces other abiotic stresses, such as drought and salinity. Tropical cyclones contribute to heavy downpours and strong winds. Flooding with saline water is frequent due to extreme climatic events (Cantelon et al., 2022). A rainfall deficit, combined with rising temperatures, leads to moisture scarcity and soil salinization, which significantly impact crop productivity and the livelihoods of farmers. According to recent data from the United Nations Development Programme (UNDP) and the Climate Impact Lab, coastal floodplains may expand up to five times over this century, jeopardizing the lives of many people and overall crop productivity (UNDP, 2024). Flooding can occur at any stage of crop growth. Some crops, such as rice, can adopt different strategies at various growth stages to thwart flooding/submergence. Varieties with multiple tolerances, such as resistance to moisture deficits, salinity, and excess water stresses, are best suited for conditions where drought and salt stress precede or follow periods of flooding.

In areas where drought and salinity are most dominant, plants may experience both stresses together. Non-availability of sufficient water induces moisture deficit conditions, which is likely to increase the salinity level of the soil. Increased temperature also impacts plant growth and productivity to a greater extent. Here, plants could overcome a more successful condition if they are tolerant to multiple stresses, such as drought, salinity, and heat simultaneously. Plants grown in the winter season, as noticed in Bangladesh and India, encounter cold stress at the seedling stage, whereas heat stress occurs at the flowering stage. In East Asia, rice growth can be jeopardized due to both cold and heat stresses. It shows that developing multiple abiotic stress-tolerant crops could be more rewarding in today's climatic conditions. Understanding weather patterns and zone-specific problems can help to identify how different abiotic stresses impact rice growth and productivity individually, in combination, or sequentially, one after another.

1.1.2 Greater fatality due to multiple abiotic stresses—a global concern

Abiotic stress reduces plant growth, leading to stunted development and a substantial decline in biomass production. Prolonged stress increases plant mortality, severely reducing crop productivity, and in extreme cases, resulting in complete yield loss (Fig. 1.3). The impact of different abiotic stresses is more or less similar, with little variation, depending on the nature and intensity of the stresses. The severity of submergence/stagnant flooding is significantly higher in weaker plants than in stronger plants (Sarkar and Ray, 2016). Due to climatic aberrations, nowadays, flooding as well as drought in the same growing season are more common. During post-submergence regeneration, if plants experience water scarcity,

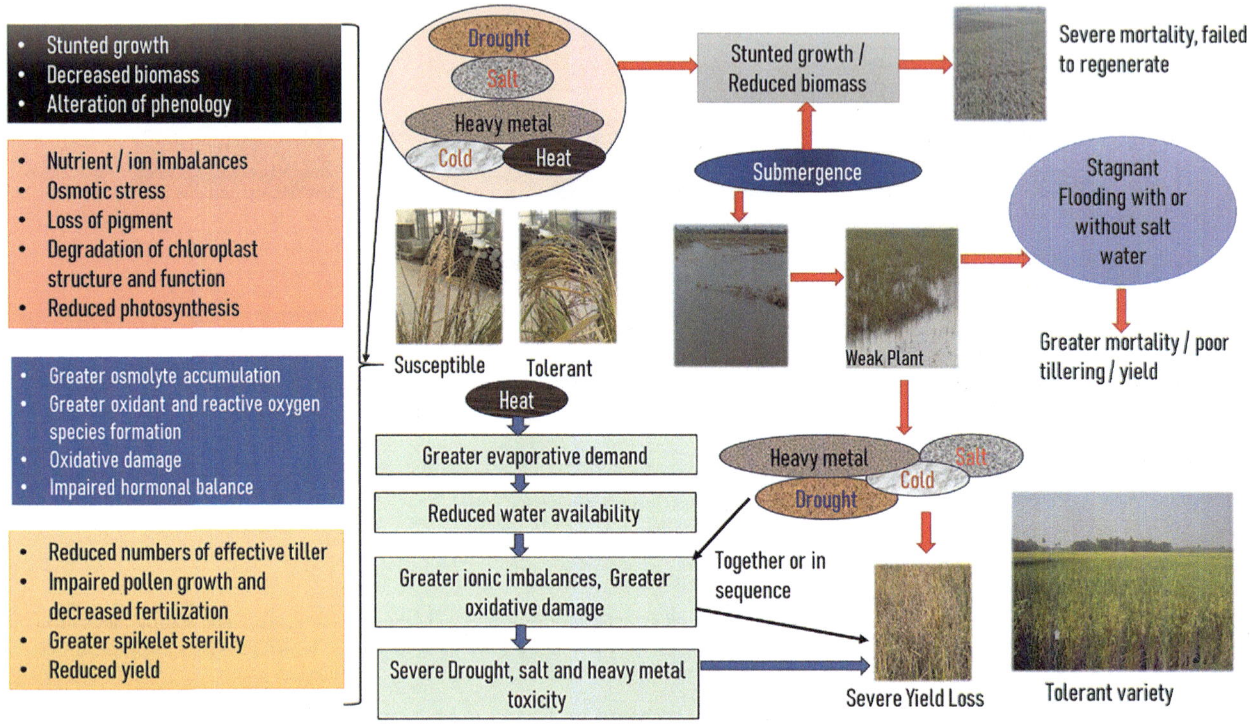

FIGURE 1.3 An infographic illustrating the effects of environmental stressors like drought, salt, heavy metal, heat and cold on plant growth, yield and tolerance.

the impact of drought becomes more severe than when drought occurs alone in singly stressed plants. Similarly, drought can intensify salt stress in coastal areas, further aggravating its negative effects on plant growth and productivity. In such a situation, getting an economic yield is highly insecure. The combined effect of salt and drought is much more severe than drought or salt alone (Li et al., 2024). Heat stress, when combined with drought, salinity, or heavy metal toxicity, becomes more severe and causes greater damage to plants due to reduced pollen viability and enhanced oxidative stress (Costa et al., 2021; Yadav et al., 2022; Mondal et al., 2024). Combined stresses strongly impair photosynthesis, disrupt ion homeostasis, and reduce overall plant productivity. Although heat, drought, and salinity individually exert similar effects on plants, their simultaneous or sequential occurrence leads to substantially greater damage (dos Santos et al., 2022; Salgotra and Chauhan, 2023; Sethuraman et al., 2024).

Climate change has increased rainfall uncertainty, leading to more frequent episodes of intense flooding. Flooding can damage crops at all growth stages, from germination to reproduction. Moreover, the nature of flooding differs in duration, depth, and timing. Flooding may occur as flash flooding, involving short-term submergence, or as stagnant flooding, characterized by partial submergence at varying water depths for extended periods. Flooding may also occur at the germination stage. Although excess water accumulation is the primary cause of growth inhibition in all cases, the mechanisms of plant tolerance vary greatly depending on the flooding scenario (Sarkar et al., 2019). Soil flooding during the germination stage creates near-anoxic conditions around germinating seeds, severely inhibiting crop establishment (Ray et al., 2016; Vijayan et al., 2018). Under such conditions, successful germination in anoxia and rapid elongation of the shoot to emerge above the water surface are the primary strategies enabling proper crop establishment (Castano-Duque et al., 2021). Submergence tolerance, on the other hand, involves a quiescence strategy in which plant elongation is strongly suppressed under water, conserving carbohydrate reserves for postsubmergence regrowth and stress recovery (Xu et al., 2006). In contrast, under stagnant flooding, plants adopt an elongation-based strategy, adjusting shoot growth according to the depth of standing water (Kuanar et al., 2017). Greater shoot elongation is required under deep-water conditions, where water depth exceeds 100 cm, whereas substantially less elongation is needed under medium-depth (~50 cm) stagnant flooding. In contrast, no elongation is required under shallow (~30 cm) stagnant flooding. Notably, submergence-tolerant cultivars may exhibit high susceptibility to stagnant flooding (Kuanar et al., 2019). Therefore, to ensure yield stability in flood-prone areas, the development of plants with diverse tolerance mechanisms capable of withstanding complete submergence, stagnant flooding of varying depths, and/or flooding stress at the germination stage is critically important.

1.1.3 From single to multiple abiotic stress tolerance

Large genetic diversity exists among natural populations. Identification of unique stress-tolerant genotypes and their utilization in developing stress-tolerant commercial cultivars is the oldest and time-tested method. For example, the discovery of submergence-tolerant QTL "*SUB1*" from the flood-tolerant rice genotype FR13A (vernacular name Dhallaputia, Odisha, India) can be termed as a leap forward in utilizing the natural stress-tolerant genotypes to develop stress-tolerant commercial varieties (Xu and Mackill, 1996). Several submergence-tolerant rice varieties have been developed utilizing this QTL/gene pertaining to this QTL in different submergence-prone areas of South and Southeast Asia and Africa (Toojinda et al., 2005; Dar et al., 2013; Iftekharuddaula et al., 2016; Septiningsih et al., 2015; Dixit et al., 2017; Singh et al., 2017; AfricaRice, 2020; Luu et al., 2012; Linh et al., 2013; Azarin et al., 2017; Sarkar et al., 2019). Among the different QTLs/genes identified so far for various abiotic stresses, QTLs for drought (*qDTYs*), salinity (*Saltol*), and anaerobic germination potential/submergence at germination stage (*qAGs*) are significant. Several high-yielding varieties have been developed and released for commercial cultivation utilizing these QTLs through marker-assisted backcrossing (MAB)/selection (MAS). The development of multiple abiotic stress-tolerant varieties is achieved through target breeding. Varieties tolerant to both submergence and salinity have been developed by the incorporation of both *SUB1* and *Saltol* (Nair and Shylaraj, 2021), submergence tolerance with greater anaerobic germination potential by incorporating both *SUB1* and *AG1* (Shin et al., 2022), submergence, salinity, and drought tolerance by incorporating *qDTY*s, *Saltol,* and *SUB1* (Muthu et al., 2020; Sazali et al., 2023). WITA 4, a mega rice variety in Africa, is tolerant to both drought and iron toxicity. Following the introgression of the *SUB1* QTL (released as FARO 66), it gained wider acceptance and surpassed the yield performance of its parent cultivar, WITA 4. Similarly, NERICA-L19, which is tolerant to iron toxicity, became more popular after the incorporation of the *SUB1* QTL and its subsequent release as FARO 67 (Africa Rice Center, 2020). The utilization of genotypes tolerant to multiple abiotic stresses, together with the fine-tuning of key QTLs and genes through MAS and MAB, significantly enhances the prospects for developing rice cultivars with combined abiotic stress tolerance.

1.2 Managing multiple abiotic stresses to meet Sustainable Development Goals

Multiple abiotic stresses—such as drought, salinity, heat, submergence, acidity, and heavy metals—frequently cooccur in field conditions, causing severe physiological disruptions and yield losses in crops (Zandalinas et al., 2021; Suzuki et al., 2014; Pandey et al., 2017). Stress combinations, such as salinity × submergence, impair ion and oxygen balance in rice, drought × salinity reduces water uptake and photosynthesis, and heat × drought leads to pollen sterility in maize and wheat (Xiong et al., 2025; Li et al., 2025). Other interactions, such as cold × drought in pea or cucumber, drought × nutrient deficiency in canola or maize, and salinity × heat in mustard, also limit growth and productivity (Zhang et al., 2024a; Khan et al., 2025). Root systems are particularly sensitive: heat, drought, and salinity individually or in combination reduce root growth and alter structure (Wu et al., 2020; Rivero et al., 2014; Vescio et al., 2021), whereas nutrient deficiencies further restrict length and surface area (Luo et al., 2012; Tindall et al., 1990). Elevated CO_2 may partially offset damage by enhancing root biomass under drought or salinity (Uddin et al., 2018; Li et al., 2020). These findings emphasize the need for integrated strategies and multistress-tolerant cultivars to sustain crop productivity under complex stress environments. Multiple stresses could be synergistic, antagonistic, or additive, impairing morphological, physiological, and biochemical damage; for example, drought stress reduces photosynthesis due to the closure of stomata, whereas concurrent heat stress accelerates leaf senescence, leading to severe yield penalties. Similarly, waterlogging and salinity not only impose osmotic stress but also play a significant role in nutrient imbalances, resulting in severe plant metabolic dysfunctions.

Developing and deploying climate-resilient crop varieties is a key strategy to mitigate the adverse impacts of abiotic stresses across diverse agroecosystems (Boraiah et al., 2023). However, the large-scale adoption of varieties with traits such as the *SUB1* gene for submergence and QTLs for salinity tolerance remains limited to specific regions. Integrating advanced technologies such as omics (transcriptomic, proteomic, metabolomic approaches), CRISPR/Cas9, speed breeding, and systems biology has accelerated the identification of stress-responsive pathways and the development of resilient genotypes (Khan et al., 2025). Complementary strategies, such as optimized nutrient management, particularly silicon and micronutrient supplementation, along with sustainable practices such as reduced tillage, cover cropping, and the application of plant growth-promoting rhizobacteria (PGPR), further enhance crop stress tolerance. In addition, agronomic adaptations, including adjusted sowing dates, efficient irrigation, and conservation tillage, play a crucial role in mitigating climate-induced stresses like drought and heat (Smith and Almaraz, 2004).

Among the 17 Sustainable Development Goals (SDGs) proposed by the United Nations 2030 agenda, combating abiotic stresses (single or multiple) is directly linked to four goals: (1) SDG 1 (No Poverty): Improving crop yields and rural livelihoods; (2) SDG 2 (Zero Hunger): Ensuring food security by maintaining and increasing agricultural productivity; (3) SDG 3 (Good Health and Well-being): Supporting a sustainable food supply and improving nutrition; (4) SDG 6 (Clean Water and Sanitation): Improving water use efficiency and managing water resources sustainably.

1.3 Multiple abiotic stresses—understanding crop physiology, molecular mechanisms, and crop improvement strategies

1.3.1 Plant genetic resources and their potential use to combat multiple abiotic stresses

Plant genetic resources (PGR) comprise a broad spectrum of genetic material, including wild relatives, traditional landraces, and improved cultivars, and are fundamental to sustaining agricultural productivity and enhancing resilience under changing environmental conditions. PGRs represent a critical reservoir of adaptive traits for mitigating the impacts of multiple abiotic stresses, such as drought, salinity, submergence, and extreme temperature extremes. The extensive genetic diversity conserved within PGR enables the identification and utilization of stress-tolerant alleles through conventional breeding, MAS, and modern biotechnological approaches. Exploitation of these resources has facilitated the development of crop varieties with improved stress tolerance, yield stability, and adaptability across diverse agroecological regions. Consequently, the strategic conservation and effective utilization of PGR are essential for developing climate-resilient cultivars, strengthening food security, and promoting long-term agricultural sustainability under increasing climatic variability (FAO, 2010; Khoury et al., 2014; Warschefsky et al., 2014).

1.3.2 Physiological and anatomical adaptations in plants under multiple abiotic stresses

Plants respond to diverse environmental stresses through coordinated modifications in their morphological, physiological, and biochemical traits. Understanding the extent of natural variation in plant stress responses is essential for identifying key adaptive mechanisms and engineering specific traits that confer enhanced stress tolerance. In this context, elucidating how plants undergo morphological and anatomical adaptations to cope with multiple abiotic stresses represents a critical area of research for the development of climate-resilient crops with broad-spectrum stress tolerance (Mittler, 2006; Zhu, 2016).

Photosynthetic regulation and stomatal behavior are fundamental processes through which plants perceive and respond to abiotic stress conditions, thereby maintaining survival and productivity under adverse environments. Under stresses such as drought, salinity, elevated temperatures, and nutrient deficiency, plants dynamically adjust their photosynthetic machinery to optimize resource use efficiency and reduce stress-induced damage (Flexas et al., 2014; Chaves et al., 2009). These adjustments often involve alterations in the activity of key enzymes associated with carbon fixation, particularly ribulose-1,5-bisphosphate carboxylase/oxygenase (Rubisco), as well as changes in the expression of genes related to photosynthesis and energy metabolism (Lawlor and Tezara, 2009).

In parallel, plants regulate stomatal aperture to balance carbon dioxide uptake with transpirational water loss. Stomatal closure under conditions such as drought and high temperature limits excessive water loss while protecting the photosynthetic apparatus from oxidative damage, albeit at the cost of reduced carbon assimilation (Hetherington and Woodward, 2003; Flexas et al., 2004). Together, the coordinated regulation of photosynthesis and stomatal conductance constitutes a central adaptive strategy enabling plants to withstand multiple abiotic stresses.

1.3.3 Phytohormones and plant growth regulators for multiple abiotic stress tolerance

Phytohormones and plant growth regulators play pivotal roles in coordinating plant growth, development, and adaptive responses to environmental stresses. Among these, abscisic acid (ABA) functions as a central regulator of abiotic stress signaling. Stress-induced accumulation of ABA promotes stomatal closure, thereby limiting transpirational water loss and maintaining cellular hydration under drought and osmotic stress conditions. In addition, ABA regulates ion homeostasis, activates antioxidant defense systems, and modulates the transcription of stress-responsive genes, collectively enhancing plant tolerance to adverse environments. Ethylene is another key phytohormone involved in growth regulation and stress adaptation, particularly under flooding, heat, and mechanical stress. Ethylene signaling stimulates root elongation and aerenchyma formation, facilitating internal oxygen diffusion in waterlogged soils, and induces the expression of genes associated with stress acclimation. Other phytohormones, including jasmonic acid, cytokinins, and brassinosteroids, also contribute to stress tolerance through the regulation of reactive oxygen species (ROS) scavenging, stomatal behavior, and complex hormonal crosstalk. The coordinated interaction among these hormonal pathways enables plants to fine-tune physiological and molecular responses, thereby improving resilience to multiple abiotic stresses (Peleg and Blumwald, 2011; Voesenek and Bailey-Serres, 2015; Verma et al., 2016).

1.3.4 Crop improvement strategies for multiple abiotic stress-tolerant crops

Conventional breeding and molecular marker-assisted breeding (MAB) are complementary strategies widely employed for the development of stress-resilient crop cultivars. Conventional breeding exploits naturally occurring genetic variation within plant populations to select genotypes exhibiting superior performance and economic yield under stress conditions, followed by hybridization to transfer desirable traits into elite genetic backgrounds. Numerous stress-tolerant cultivars have been successfully developed through phenotypic selection and recurrent breeding. However, this approach is often time-consuming and largely dependent on environmental conditions, which may limit the accuracy of stress tolerance assessment. In contrast, MAB enhances breeding efficiency by enabling indirect selection using molecular markers tightly linked to genes or quantitative trait loci (QTLs) associated with abiotic stress tolerance. Marker-assisted introgression of QTLs and major genes has facilitated the development of improved cultivars with enhanced tolerance to drought, submergence, and other abiotic stresses. Together, conventional and MAB constitute essential tools for developing cultivars capable of withstanding multiple environmental stresses (Collard and Mackill, 2008; Varshney et al., 2014).

Genomics-assisted breeding approaches have further transformed efforts to improve abiotic stress tolerance by elucidating the genetic architecture of stress responses and enabling targeted genetic improvement. Advances in high-throughput sequencing and bioinformatics have enabled genome-wide identification of stress-responsive genes and regulatory networks. Approaches such as genome-wide association studies, QTL mapping, and transcriptomic analyses facilitate the discovery of candidate genes underlying complex stress tolerance traits. Moreover, genome editing technologies, particularly CRISPR–Cas systems, allow precise modification of stress-responsive genes, offering new opportunities to enhance climate resilience in crops (Varshney et al., 2021; Zhang et al., 2018).

Precision phenotyping and phenomics further strengthen modern breeding pipelines by enabling high-throughput, accurate assessment of plant responses to abiotic stresses using advanced imaging, sensor technologies, and data analytics. These approaches allow comprehensive evaluation of natural variation in large and diverse populations. Integration of phenomics with genomic data improves selection efficiency, accelerates breeding cycles, and supports the development of predictive models and phenotypic databases for rapid germplasm screening. Consequently, precision phenotyping plays a pivotal role in enhancing the efficiency and accuracy of breeding strategies for stress-resilient cultivars (Furbank and Tester, 2011; Yang et al., 2020).

1.3.5 Genome editing and multi-omics-based approaches for improving multiple abiotic stress tolerance in plants

Transgenic approaches have been widely employed to enhance abiotic stress tolerance in plants through the introduction of foreign genes encoding stress-responsive proteins, osmoprotectants, transcription factors, or enzymes involved in stress-mitigation pathways. Although several transgenic events have demonstrated improved stress tolerance under controlled conditions, their large-scale field deployment remains constrained by stringent regulatory frameworks, biosafety concerns, and public acceptance issues. In contrast, genome editing technologies, particularly CRISPR–Cas systems, enable precise and targeted modification of endogenous genes associated with stress tolerance without necessarily introducing foreign DNA. Consequently, genome-edited crops face comparatively fewer regulatory hurdles and exhibit greater potential for field-level applications than conventional transgenics. CRISPR-based approaches allow targeted manipulation of genes that are difficult to modify using traditional breeding, thereby offering tailored improvements in plant stress-response mechanisms (Bortesi and Fischer, 2015; Chen et al., 2019).

In parallel, multiomics approaches integrating genomics, transcriptomics, proteomics, and metabolomics have emerged as powerful tools for dissecting complex abiotic stress responses in plants. These integrative analyses provide a holistic understanding of molecular, biochemical, and physiological processes governing stress tolerance by identifying key genes, regulatory networks, and metabolic pathways. Coupling multiomics datasets with advanced bioinformatics and statistical tools facilitates the discovery of novel biomarkers, allelic variants, and metabolic signatures associated with stress-adaptive traits. Furthermore, multiomics-driven insights enable the prioritization of molecular targets for genetic improvement using transgenic or genome editing strategies, thereby accelerating the development of climate-resilient crop varieties (Weckwerth et al., 2020; Varshney et al., 2021).

1.3.6 Use of nanobiotechnology to improve multiple abiotic stress tolerance in plants

Plant nanobiotechnology involves the application of engineered nanomaterials, including nanoparticles, nanotubes, and nanofibres, to plant systems to modulate physiological processes and enhance tolerance to abiotic stresses. Functionalized nanoparticles containing essential nutrients, plant growth regulators, or antioxidants can be designed to improve nutrient-use efficiency, promote root development, and scavenge ROS, thereby alleviating oxidative damage under stress conditions. Such nano-enabled inputs have demonstrated potential to improve plant growth and productivity under drought, salinity, and thermal stress. In addition, nanomaterial-based sensors and nanoprobes enable real-time, high-resolution monitoring of plant physiological parameters and stress responses, facilitating early stress detection and precision management of cultivation practices. Nanomaterials can also function as efficient carriers for the targeted delivery of genetic materials, including RNA interference constructs and genome-editing components, allowing precise modulation of gene expression and offering novel avenues for stress tolerance enhancement. Collectively, nanobiotechnology provides promising tools for developing climate-resilient crops and advancing sustainable agricultural systems (Wang et al., 2020; Tripathi et al., 2022; Kah et al., 2019).

1.4 Overcoming multiple abiotic stresses—management strategies

Abiotic stresses such as salinity, drought, heat, cold, waterlogging, nutrient imbalance, and heavy metal toxicity are major threats to crop growth, productivity, and global food security. Abiotic stresses, such as salinity + waterlogging, salinity + drought, drought + heat, waterlogging + ion toxicity, and salinity + nutrient deficiency, often occur simultaneously, creating compound effects on crop growth and causing significant yield losses (Zandalinas and Mittler, 2022; Priya et al., 2023). Therefore agricultural production is increasingly constrained by multiple abiotic stresses globally, which collectively reduce crop growth, yield stability, and food security across the globe (Ramegowda et al., 2024). Most of the research and management approaches have focused on individual stress scenarios, such as single stress-tolerant breeding or remediation. However, a single stress rarely occurs under field conditions. Instead, plants frequently experience combinations of different stresses instantaneously under field conditions, such as salinity combined with waterlogging or drought coupled with heat stress, salinity or waterlogging accompanied by nutrient imbalance. The complexity of multiple abiotic stresses can be partially managed through multidimensional and integrated stress management approaches, as briefly described below, which address multiple stress factors concurrently rather than focusing on a single solution.

1.4.1 Soil management strategies for mitigating multiple abiotic stresses

Soil fertility and productivity are the key factors of resilient crop production under abiotic stress conditions. Soil physiochemical properties, such as soil texture, structure, bulk density, electrical conductivity, pH, organic matter content, nutrient content, and microbial diversity, are directly involved in how plants respond to multiple abiotic stresses. Therefore soil management practices that increase organic matter content, nutrient status, nutrient mineralization, and nutrient recycling and optimize water-holding capacity are essential for mitigating multiple stresses simultaneously, as has been highlighted in different segments in the relevant section of the compilation (Fig. 1.4).

1.4.1.1 Organic amendments

Organic amendments, such as compost, farmyard manures (FYMs), vermicompost, tricho-compost, humic acid, and biochar, are rich in humic substances, enzymes, plant growth regulators, and nutrients. Previous studies have suggested that these organic amendments enhance soil porosity and cation exchange capacity; improve soil physical structure; increase water retention and adsorbed toxic ions; contribute to long-term carbon sequestration; and facilitate microbial colonization, thereby improving root permeability; reducing membrane injury; increasing relative water content; increasing stomatal conductance; maintaining higher chlorophyll content; and improving photosynthetic efficiency, resulting in greater growth, root and shoot biomass and yield of crops in stress-prone environments (Lehmann and Joseph, 2015; Zhang et al., 2022; Becagli et al., 2021; Chaudhari et al., 2021). Biochar application improved rice and tomato yield under combined salinity and drought stress conditions (Murtaza et al., 2024; Zhang et al., 2019a).

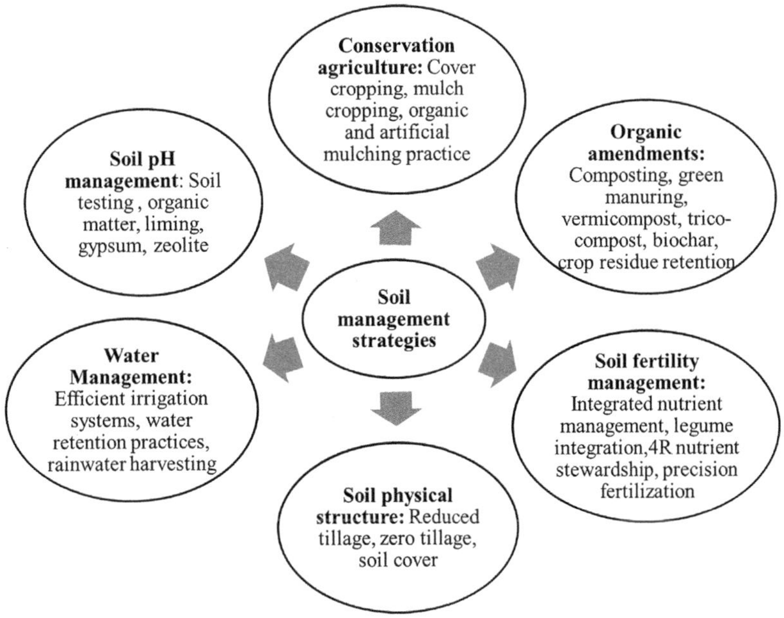

FIGURE 1.4 A cyclic infographic detailing six soil management strategies for mitigating multiple abiotic stresses.

1.4.1.2 Conservation agriculture

Conservation agriculture practices, including reduced or zero tillage, mulching, cover crops, residue retention, intercropping, crop rotation, and legume incorporation, play a crucial role in mitigating abiotic stress impacts (Fig. 1.4). Reduced or zero tillage combined with residue retention minimizes soil disturbance, conserves soil organic carbon, enhances water-holding capacity, reduces evaporation losses, stabilizes soil pH, and moderates soil temperature under drought- and heat-prone conditions (Wittwer et al., 2023; Gaevaya et al., 2023; Nurbekov et al., 2024). Compared with conventional tillage, zero tillage with residue retention significantly improved wheat yield under combined drought and heat stress (Nurbekov et al., 2024; Chaudhary et al., 2024). Mulching moderates soil microclimate, reduces evaporation, and improves water-use efficiency, while organic matter decomposition supplies nutrients and carbon. Plastic, straw, and film mulching further reduce net evaporation and enhance drought tolerance, resulting in higher yields (Zambrano et al., 2024; Zhao et al., 2024; Zhang et al., 2025). Cover crops, particularly legumes, reduce soil erosion, enhance nitrogen fixation, and improve resilience to drought and nutrient stress; cowpea cover cropping increased soil organic matter by 20% (Nelson et al., 2022; Whippo et al., 2024). Relay intercropping with transplanted aman rice also alleviates moisture stress and shortens crop duration (Ghorai, 2022).

1.4.1.3 Nutrient management strategies for mitigating multiple abiotic stresses

Strategic nutrient management plays a vital role in enhancing plant tolerance to abiotic stresses. Balanced and timely nutrient management confers resilience against multiple abiotic stresses by optimizing photosynthesis efficiency, maintaining osmotic adjustment and ion homeostasis, improving antioxidant defense systems against ROS, enhancing nutrient-use efficiency, and improving growth and development (Kumari et al., 2022; Singhal et al., 2023). Recent advances in strategic nutrient management, including integrated plant nutrient systems, site-specific nutrient management (SSNM), and 4R (right source, right time, right dose, and right location) stewardship, nanofertilizer, slow-releasing fertilizers, foliar application of nutrients, and biofertilizer consortia, have provided promising solutions to improve resilience in multiple stress-prone environments (Fig. 1.5). This overall document presents an overview of soil nutrient management strategies for mitigating multiple abiotic stresses.

1.4.1.3.1 Balanced fertilizer application

The balanced application of primary, secondary, and micronutrients is crucial for stress management. Nitrogen (N) increased chlorophyll synthesis and photosynthesis. Phosphorus (P) increases root proliferation for better water acquisition and energy transfer, which are vital for drought and salinity stress conditions. Potassium (K) plays an important role in stomatal regulation and osmotic adjustment, enabling plants to maintain water balance and sustain

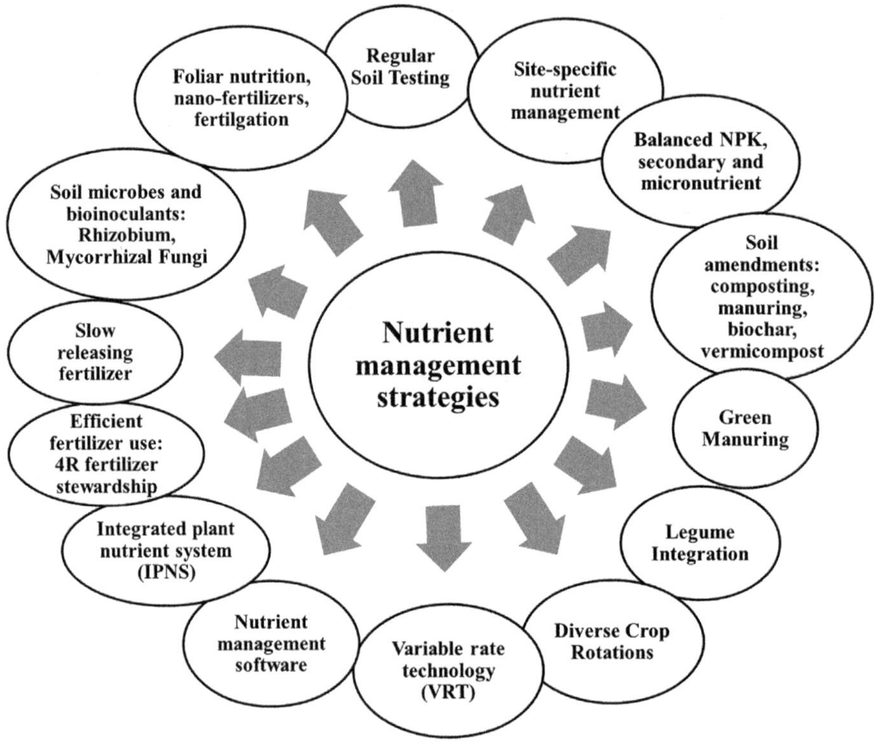

FIGURE 1.5 A cyclic infographic illustrating 14 different nutrient management strategies for mitigating multiple abiotic stresses.

photosynthesis under drought and extreme temperature stress conditions (Marschner, 2011). Therefore balanced NPK fertilizer management not only increases crop productivity but also reduces physiological stress markers such as electrolyte leakage and ROS accumulation. The balanced application of NPK improved barley and canola water use efficiency and productivity under water-deficit conditions (Aslam et al., 2021; Bagues et al., 2024). A previous study reported that foliar application of Si and B in rice improved grain yield by enhancing enzyme function and improving antioxidant defense under salinity and heat stress conditions (Farooq et al., 2019).

1.4.1.3.2 Foliar nutrition and nanofertilizers

Foliar application of macro- and micronutrients provides an immediate nutrient supply directly to plant tissues, such as salinity-induced nutrient antagonism and drought-induced limited root activity. The foliar fertilization approach mitigates multiple abiotic stresses; foliar application of KNO_3 improved the relative water content and increased yield under drought, heat, and their combined stress conditions (Bolat et al., 2024).

Recently, nanofertilizers, having large surface areas, facilitated the controlled release of nutrients, improved nutrient solubility, targeted delivery to plant tissue, and enhanced nutrient use efficiency under stress conditions. Nanofertilizers, such as nano-ZnO, enhance chlorophyll synthesis and strengthen the antioxidant defense system under drought and salinity stress (Abbas et al., 2023; Heikal et al., 2023); nano-Si reduces transpiration, strengthens cell integrity, and mitigates heavy metal stress (Wang et al., 2016; Mousavi, 2022), and nano-Fe improves electron transport and reduces chlorosis under different stress conditions (Farajollahi et al., 2023; Seyed Sharifi et al., 2024).

1.4.1.3.3 Integrated nutrient management approaches and nutrient stewardship

The integrated nutrient management (INM) approach involves the combined application of chemical and organic amendments (compost, vermicompost, FYM, biochar, humic acid, and tricho-compost), and biofertilizers (PGPR and mycorrhizae) can mitigate abiotic stresses by improving soil fertility and stress resistance. In the INM approach, chemical fertilizers provide immediate nutrient availability; organic amendments improve water and nutrient retention and microbial activity; and biofertilizers increase nutrient uptake and stress signaling. Additionally, the adoption of the 4R concept of "nutrient stewardship" with the right fertilizer source, the right time—when crops need, the right rate matches the amount of fertilizer type as the crop needs and the right place where crops can use them to match the fertilizer type to the crop need—can maintain the optimum nutrient content in the soil for sustainable crop production under multiple stress conditions (Fixen, 2020; Reetz et al., 2015).

1.4.1.4 *Beneficial microbes and bioinoculants for mitigating multiple abiotic stresses*

Beneficial microbes and bioinoculants, including plant growth-promoting rhizobacteria (PGPR), arbuscular mycorrhizal fungi (AMF), cyanobacteria, and *Trichoderma* spp., play a critical role in mitigating abiotic stresses in crops. These microorganisms enhance nutrient and water uptake, regulate phytohormones and ethylene levels, strengthen antioxidant defense systems against ROS, maintain osmotic balance, and improve soil physicochemical properties. Their integration with organic amendments and mineral fertilizers provides synergistic benefits, positioning microbial bioinoculants as a key component of climate-smart agriculture (CSA).

PGPR alleviate abiotic stress through nutrient solubilization, phytohormone production, and reduction of stress-induced ethylene via 1-aminocyclopropane-1-carboxylic acid (ACC) deaminase activity, thereby promoting root growth and stress resilience under drought and salinity (Misra and Chauhan, 2020; Danish et al., 2020; Orozco-Mosqueda et al., 2022). In maize, *Azospirillum brasilense* improved water status and photosynthetic efficiency (Sivasakthivelan et al., 2021; Checchio et al., 2021), whereas ACC deaminase-producing PGPR enhanced salinity tolerance in wheat by improving root architecture and ion homeostasis (Arshadullah et al., 2017).

AMF mitigate stress by enhancing water uptake, phosphorus and micronutrient mobilization, osmolyte accumulation, and soil structure. They reduce Na^+ uptake, increase the K^+/Na^+ ratio under salinity (Evelin et al., 2019; Wang et al., 2021), and improve stomatal conductance and chlorophyll content under drought (Begum et al., 2019; Ye et al., 2022). Endophytes, cyanobacteria (*Anabaena*, *Nostoc*), and *Trichoderma* spp. further enhance stress tolerance via nitrogen fixation, antioxidant activity, and bioactive metabolite production (Fouda, 2021; Sood et al., 2020). Microbial consortia often outperform single strains under field conditions, improving water-use efficiency, nutrient uptake, and oxidative stress tolerance (Tahiri et al., 2025). However, large-scale adoption requires optimized strain selection, formulations, and delivery systems supported by multiomics approaches.

1.4.2 Climate-smart agronomic practices for mitigating multiple abiotic stresses

Climate-smart agronomic (CSA) practices encompass soil, water, crop, and other agricultural resource management strategies that enable crops to cope with multiple abiotic stresses. In multiple abiotic stress-prone areas, CSA practices form a cornerstone of adaptive integrated pathways for overcoming multiple abiotic stresses by improving soil fertility and physio chemical structure through the judicial application of soil amendments and mineral nutrients, increasing water-use efficiency through efficient irrigation techniques, increasing nutrient-use efficiency via precision nutrient management and nanotechnologies, reducing the risk of crop failure through crop diversification and stress-tolerant crop varieties, and contributing to climate change mitigation via reduced GHG emissions and increased carbon sequestration. This compilation presents an inclusive summary of CSA strategies for mitigating multiple abiotic stresses (Fig. 1.6).

1.4.2.1 *Climate-smart soil and water management practices*

The adoption of climate-smart soil and water management practices is essential for mitigating the impacts of multiple abiotic stresses. Key strategies include conservation tillage (minimum and zero tillage), which preserves soil organic matter and microbial activity, and crop residue retention, which enhances soil moisture retention and moderates soil surface temperature. Crop rotation with diverse rooting systems and legume inclusion further improves resilience to drought and nutrient depletion. Conservation agriculture practices have been shown to increase maize yield under drought by improving soil water-holding capacity (Deng et al., 2023). Mulching, another core CSA practice, reduces soil temperature and evaporation, suppresses weeds, and enhances water infiltration. Organic mulches such as straw, leaves, and compost additionally enrich soil organic matter, nutrients, and microbial communities, thereby improving crop yield under heat stress by maintaining a cooler rhizosphere (Li et al., 2024; Khanzada et al., 2025). Climate-smart irrigation practices, including drip and microirrigation, alternate wetting and drying, deficit irrigation, and remote sensing- and IoT-based precision irrigation, optimize water use by reducing evaporation, nutrient leaching, and waterlogging while improving water-use efficiency. For instance, drip irrigation increased tomato yield by 50% under limited water supply (Wu et al., 2021). Rainwater harvesting structures further enhance yield stability in drought-prone and coastal regions.

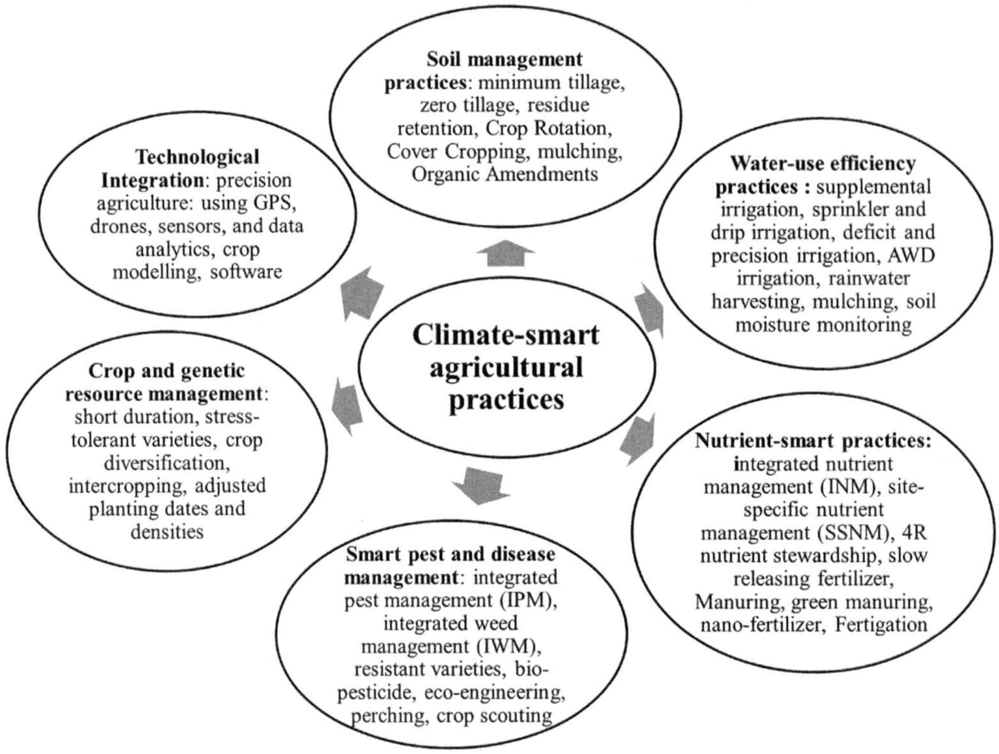

FIGURE 1.6 A cyclic infographic detailing six climate-smart agricultural practices for mitigating multiple abiotic stresses.

1.4.2.2 *Climate-smart nutrient management practices*

Climate-smart nutrient management practices—including soil test–based fertilization, INM, SSNM, need-based foliar application, and use of controlled-release and nanofertilizers—are critical for mitigating multiple abiotic stresses. INM integrates organic amendments with mineral fertilizers to optimize nutrient supply, whereas SSNM employs soil testing and sensors to deliver precise, site-specific fertilizer recommendations. Controlled-release fertilizers, such as urea supergranules, neem-coated urea, and sulfur-coated urea, enhance nutrient-use efficiency by minimizing leaching and ensuring sustained nutrient availability under stress conditions. Nanofertilizers, including nano-urea, nano-Diammonium Phosphate (DAP), and metal-based nanoparticles (e.g., ZnO, Si, Fe, and nanozeolite), further reduce volatilization and leaching losses in stress-prone environments. Additionally, soil acidity correction through liming and alkalinity management using gypsum improves nutrient availability and soil fertility. Adoption of the 4R nutrient stewardship framework—applying nutrients from the right source, at the right rate, time, and place—offers an effective strategy to enhance fertilizer efficiency, reduce environmental losses, and improve crop resilience to multiple abiotic stresses.

1.4.2.3 *Crop and genetic management in climate-smart agriculture*

Breeding programs have released drought-, salt-, waterlogging-, and heat-tolerant varieties of different crops worldwide. Cultivating these stress-tolerant crop varieties in different stress-prone environments enhances crop productivity, resilience, and sustainability under multiple stress conditions. Crop diversification, intercropping, relay cropping, legume intercropping, and deep-rooted crop intercropping with shallow-rooted crops may reduce the risk of drought stress impact under field conditions. Timely planting/sowing, transplanting of older seedlings, increasing the seedling density, altering the planting geometry, and double transplanting to avoid peak stress periods and improve the canopy microclimate and water use under different stress conditions are recommended. Climate-smart planting techniques, such as raised-bed planting/sowing, floating-bed planting/sowing, and sorjan cultivation, are recommended to grow upland crops in waterlogged, flood-prone, and saline-prone areas to maintain yield stability.

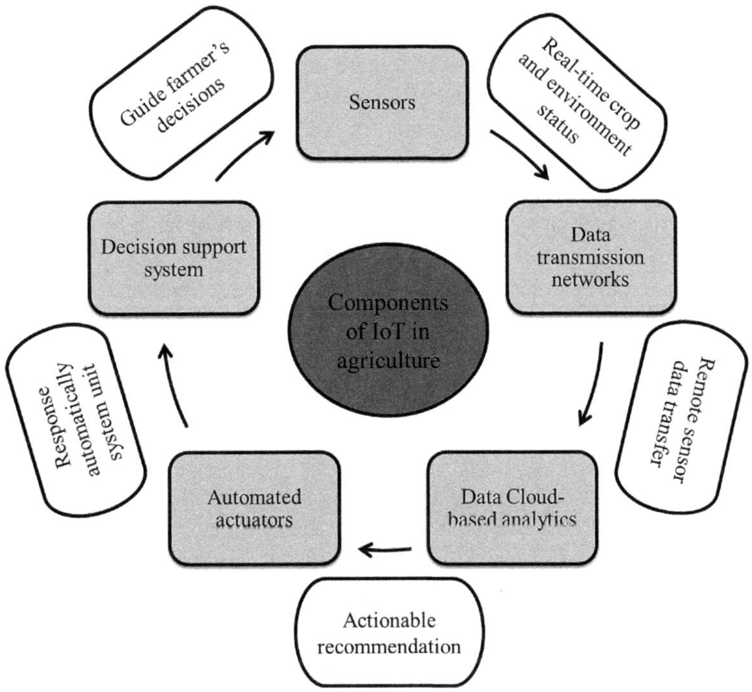

FIGURE 1.7 A cyclic diagram illustrating five major components of IoT-based agronomy for mitigating multiple abiotic stresses. *IoT*, Internet of Things.

1.4.3 Internet of Things-based new-generation agronomic practices for the management of multiple abiotic stresses

The fourth Industrial Revolution (IR 4.0) has transformed agriculture, enabling precision, efficiency, and real-time monitoring of crop growth under different stress conditions. Internet of Things (IoT)-based agronomic practices integrate sensors, remote sensing, geographic information systems (GIS), drones, unmanned aerial vehicles (UAVs), robots, data analytics, and automation technologies to optimize water, nutrient, and climate management; enhance resource-use efficiency; improve yield stability and crop quality; and increase crop resilience to different abiotic stresses. IoT-based agronomy relies on interconnected devices such as sensors of soil moisture, temperature, humidity, nutrients, pH, and EC, which are optical for monitoring real-time environmental and crop status; data transmission networks such as LoRa and ZigBee NB-IoT enabling wireless remote data transfer; cloud-based analytics big data platforms for real-time data processing to provide actionable recommendations; automated actuators for control devices automatically based on sensor data; and decision support systems to guide farmer decisions by integrating weather forecasts, crop models, and sensor data (Fig. 1.7). This compilation explores new insights into the latest IoT-based technological applications for monitoring specific abiotic stresses and combinations of stresses at different spatial scales, including greenhouse and field studies using imaging technologies with and without UAVs, contributing to improved crop input efficiency, yield, and quality of both field and horticultural crops.

1.4.3.1 Stress monitoring through Internet of Things-based technologies

IoT-based remote sensing technologies, such as UAVs equipped with drones and thermal sensors, can detect early stress signals such as leaf temperature anomalies, chlorophyll degradation, key vegetation indices such as the normalized difference vegetation index and normalized difference water index derived from visible-near-infrared data, and monitor plant health and nutrient and water status effectively (Asaari et al., 2019; Awais et al., 2023). UAV-based additional indices, such as the enhanced vegetation index, photochemical reflectance index, and water band index, also contribute to the assessment of water stress (Zhang et al., 2019b, 2021). IoT-connected satellites that provide high-resolution imaging data analysis through artificial intelligence (AI) analytics can be used for large-scale monitoring to predict stress onset and severity, enabling preemptive stress interventions.

1.4.3.2 Internet of Things-based application of irrigation and fertilizer management

IoT systems can incorporate local weather forecasts to anticipate upcoming stress events and enable automated irrigation before heatwaves or drought stress, resulting in reduced plant water deficit and improved soil plant ananlysis development (SPAD) values in leaves. Similarly, IoT platforms monitor soil and plant water status continuously and enable deficit irrigation strategies. A previous study reported that real-time soil moisture data-guided irrigation schedules improved the yields of wheat and maize under drought stress conditions (Zhang et al., 2024b). Additionally, soil moisture sensors coupled with drip irrigation systems deliver water to the root zone by monitoring drought stress (Seyar and Ahamed, 2024).

IoT-based nutrient and fertilizer management strategies could be very effective ways to deliver soil nutrients and fertilizers through real-time leaf and soil status monitoring. Real-time leaf and soil monitoring sensors that detect leaf chlorophyll or soil nutrients can trigger automated foliar sprays or fertigation events, thereby improving photosynthesis and stress resistance. Additionally, IoT-based fertigation systems enable the adjustment of nutrient delivery on the basis of the soil nutrient status and critical crop growth-stage requirements.

1.5 Future research needs and conclusion

Recent advances in biotechnology, nanotechnology, and AI are reshaping agriculture and offer new avenues for mitigating multiple abiotic stresses (Fig. 1.8). Genome editing, particularly CRISPR–Cas systems, enables precise modification of stress-responsive genes without introducing foreign DNA (Erdoğan et al., 2023; Wang et al., 2022). Targeted editing of genes regulating stomatal density, ABA signaling, osmolyte biosynthesis, Na^+ transporters (e.g., *HKT1*), heat shock proteins, and antifreeze proteins has shown potential for improving tolerance to drought, salinity, heat, and cold. For instance, CRISPR knockout of *OsRR22* enhanced salinity tolerance in rice without yield penalties (Zhang et al., 2019c), whereas editing *HKT1;1* in wheat improved Na^+ exclusion and salinity tolerance (Ahmad et al., 2021).

Synthetic biology and metabolic engineering further enable the design of novel pathways to enhance osmotic adjustment and antioxidant capacity, such as proline biosynthesis and ROS-scavenging systems (SOD, CAT). Bioengineered microbial consortia can simultaneously promote phytohormone production, nutrient solubilization, and heavy metal detoxification. Nanotechnology-based inputs, including nano-Zn, Fe, and Si, improve nutrient-use efficiency and reduce leaching under stress conditions. Integration of high-throughput omics with AI, IoT, robotics, and automated climate-control systems can accelerate stress phenotyping, identify tolerant alleles, enable early stress detection, and optimize adaptive management under multiple abiotic stresses.

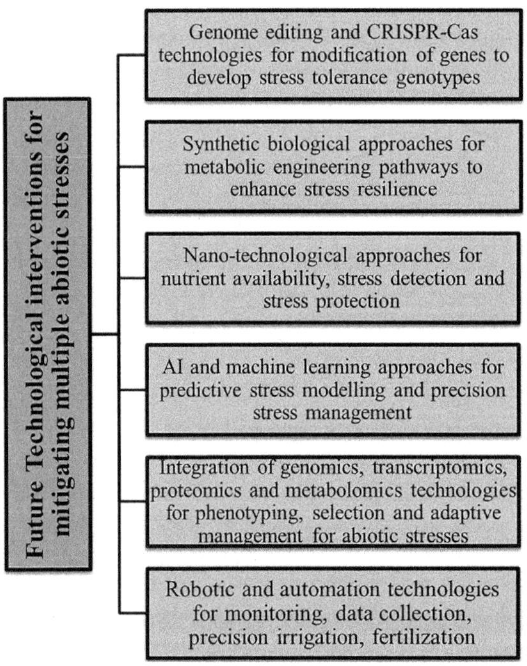

FIGURE 1.8 Summary of future technological interventions for mitigating multiple abiotic stresses.

Despite significant advances in understanding plant responses to individual abiotic stresses, knowledge of plant adaptation to multiple and combined abiotic stresses remains limited. Most physiological, molecular, and genetic studies have focused on single-stress conditions, which poorly represent natural field environments where stresses such as drought, heat, salinity, flooding, and nutrient toxicity frequently occur simultaneously or sequentially. Increasing evidence indicates that plant responses to combined stresses are distinct rather than additive, involving unique signaling networks, transcriptional regulation, and metabolic adjustments not observed under individual stress conditions. A major research gap also exists in the identification, validation, and deployment of genes and QTLs that confer tolerance to multiple abiotic stresses across different growth stages. Although numerous stress-responsive genes and QTLs have been characterized under controlled single-stress experiments, their effectiveness and stability under combined or sequential stress scenarios remain inadequately explored. Moreover, breeding strategies largely rely on trait pyramiding without sufficient field-based validation under realistic stress combinations. Another underexplored area is the impact of stress order and stress memory, where prior exposure to one stress modifies plant responses to subsequent stresses. This phenomenon is increasingly relevant under climate change, yet the underlying mechanisms involving hormonal crosstalk, redox regulation, and energy allocation remain poorly understood. Finally, the translation of combined-stress biology into crop improvement programs is constrained by limited integration of high-throughput phenotyping, multiomics tools, and marker-assisted or genomic selection strategies specifically tailored for multiple abiotic stress tolerance. Addressing these gaps is essential for developing climate-resilient crop varieties with stable productivity under complex and fluctuating stress environments.

References

Abbas, S.F., Bukhari, M.A., Raza, M.A.S., Abbasi, G.H., Ahmad, Z., Alqahtani, M.D., et al., 2023. Enhancing drought tolerance in wheat cultivars through nano-ZnO priming by improving leaf pigments and antioxidant activity. Sustainability (Switzerland) 15 (7). Available at https://doi.org/10.3390/su15075835.

Africa Rice Center, 2020. Africa Rice Center (AfricaRice) Annual Report 2019: Toward rice-based food systems transformation in Africa. Africa Rice Center, Abidjan, Côte d'Ivoire:, pp. 28. Available at https://hdl.handle.net/10568/110844.

Agegnehu, G., Amede, T., Erkossa, T., Yirga, C., Henry, C., Tyler, R., et al., 2021. Extent and management of acid soils for sustainable crop production system in the tropical agroecosystems: a review. Acta Agriculturae Scandinavica Section B: Soil and Plant Science 71 (9), 852–869. Available at https://doi.org/10.1080/09064710.2021.1954239.

Ahmad, A., Ashraf, S., Munawar, N., Jamil, A., Ghaffar, A., Shahbaz, M., 2021. CRISPR/Cas mediated abiotic stress tolerance in crops. CRISPR Crops: The Future of Food Security. Springer Singapore, Pakistan, pp. 177–211. Available at https://link.springer.com/book/10.1007/978-981-15-7142-8, 10.1007/978-981-15-7142-8_6.

Arshadullah, M., Hyder, S.I., Mahmood, I.A., Sultan, T., Naveed, S., 2017. Mitigation of salt stress in wheat plant (Triticum aestivum) by plant growth promoting rhizobacteria for ACC deaminase. International Journal of Advanced Research in Biological Sciences 4 (6), 41–46.

Asaari, M.S.M., Mertens, S., Dhondt, S., Inzé, D., Wuyts, N., Scheunders, P., 2019. Analysis of hyperspectral images for detection of drought stress and recovery in maize plants in a high-throughput phenotyping platform. Computers and Electronics in Agriculture 162, 749–758. Available at https://doi.org/10.1016/j.compag.2019.05.018, www.elsevier.com/inca/publications/store/5/0/3/3/0/4.

Aslam, M.M., Farhat, F., Siddiqui, M.A., Yasmeen, S., Khan, M.T., Sial, M.A., et al., 2021. Exploration of physiological and biochemical processes of canola with exogenously applied fertilizers and plant growth regulators under drought stress. PLoS One 16 (12). Available at https://doi.org/10.1371/journal.pone.0260960, https://journals.plos.org/plosone/article?id=10.1371/journal.pone.0260960.

Awais, M., Li, W., Cheema, M.J.M., Zaman, Q.U., Shaheen, A., Aslam, B., et al., 2023. UAV-based remote sensing in plant stress imagine using high-resolution thermal sensor for digital agriculture practices: a meta-review. International Journal of Environmental Science and Technology 20 (1), 1135–1152. Available at https://doi.org/10.1007/s13762-021-03801-5.

Azarin, K.V., Usatov, A.V., Kostylev, P.I., 2017. Molecular breeding of submergence-tolerant rice. Annual Research & Review in Biology 18 (1), 1–10.

Bagues, M., Neji, M., Karbout, N., Boussora, F., Triki, T., Guasmi, F., et al., 2024. Mitigating salinity stress in barley (Hordeum vulgare L.) through biochar and NPK fertilizers: impacts on physio-biochemical behavior and grain yield. Agronomy 14 (2), 317. Available at https://doi.org/10.3390/agronomy14020317.

Becagli, M., Guglielminetti, L., Cardelli, R., 2021. Effects of combined biochar and vermicompost solution on leachate characterization and nitrogen balance from a greenhouse tomato (Solanum lycopersicum) cultivation soil. Communications in Soil Science and Plant Analysis 52 (16), 1879–1893.

Begum, N., Ahanger, M.A., Su, Y., Lei, Y., Mustafa, N.S.A., Ahmad, P., et al., 2019. Improved drought tolerance by AMF inoculation in maize (Zea mays) involves physiological and biochemical implications. Plants 8 (12). Available at https://doi.org/10.3390/plants8120579, https://www.mdpi.com/2223-7747/8/12/579/pdf.

Bolat, I., Korkmaz, K., Dogan, M., Turan, M., Kaya, C., Seyed Hajizadeh, H., et al., 2024. Enhancing drought, heat shock, and combined stress tolerance in Myrobalan 29C rootstocks with foliar application of potassium nitrate. BMC Plant Biology 24 (1). Available at https://doi.org/10.1186/s12870-024-04811-4.

Boraiah, K.M., Basavaraj, P.S., Vijaysinha, D.K., Harisha, C.B., Halagundegowda, G.R., Krishnamurthy, D., et al., 2023. Abiotic stress-tolerant crop varieties in India: status and a way forward. Recent Trends in Plant Breeding and Genetic Improvement. IntechOpen.

Bortesi, L., Fischer, R., 2015. The CRISPR/Cas9 system for plant genome editing and beyond. Biotechnology Advances 33 (1), 41–52. Available at https://doi.org/10.1016/j.biotechadv.2014.12.006, www.elsevier.com/inca/publications/store/5/2/5/4/5/5/index.htt.

Cantelon, J.A., Guimond, J.A., Robinson, C.E., Michael, H.A., Kurylyk, B.L., 2022. Vertical saltwater intrusion in coastal aquifers driven by episodic flooding: a review. Water Resources Research 58 (11). Available at https://doi.org/10.1029/2022WR032614, http://agupubs.onlinelibrary.wiley.com/hub/journal/10.1002/(ISSN)1944-7973/.

Castano-Duque, L., Ghosal, S., Quilloy, F.A., Mitchell-Olds, T., Dixit, S., 2021. An epigenetic pathway in rice connects genetic variation to anaerobic germination and seedling establishment. Plant Physiology 186 (2), 1042–1059. Available at https://doi.org/10.1093/PLPHYS/KIAB100, https://academic.oup.com/plphys/article/186/2/1042/6151751.

Chaudhari, S., Upadhyay, A., Kulshreshtha, S., 2021. Influence of organic amendments on soil properties, microflora and plant growth. Sustainable Agriculture Review. Springer International Publishing, Cham, pp. 147–191.

Chaudhary, A., Yadav, D.B., Poonia, T.M., Roohi, Sihag, N., 2024. Conservation tillage and in-situ rice residue driven soil temperature moderation and wheat productivity under terminal heat stress in North-Western Indo Gangetic Plains of India. International Journal of Plant Production 18 (4), 485–496. Available at https://doi.org/10.1007/s42106-024-00305-3, https://www.springer.com/journal/42106.

Chauhan, B.S., Prabhjyot, K., Mahajan, G., Randhawa, R.K., Singh, H., Kang, M.S., 2014. Global warming and its possible impact on agriculture in India. Advances in Agronomy 123, 65–121. Available at https://doi.org/10.1016/B978-0-12-420225-2.00002-9, http://www.elsevier.com/wps/find/bookdescription.cws_home/675816/description#description.

Chaves, M.M., Flexas, J., Pinheiro, C., 2009. Photosynthesis under drought and salt stress: regulation mechanisms from whole plant to cell. Annals of Botany 103 (4), 551–560.

Checchio, M.V., de Cássia Alves, R., de Oliveira, K.R., Moro, G.V., Santos, D.M.Md, Gratão, P.L., 2021. Enhancement of salt tolerance in corn using Azospirillum brasilense: an approach on antioxidant systems. Journal of Plant Research 134 (6), 1279–1289. Available at https://doi.org/10.1007/s10265-021-01332-1, www.springer.com/journal/10265.

Chen, K., Wang, Y., Zhang, R., Zhang, H., Gao, C., 2019. CRISPR/Cas genome editing and precision plant breeding in agriculture. Annual Review of Plant Biology 70, 667–697. Available at https://doi.org/10.1146/annurev-arplant-050718-100049, http://www.annualreviews.org/journal/arplant.

Collard, B.C.Y., Mackill, D.J., 2008. Marker-assisted selection: an approach for precision plant breeding in the twenty-first century. Philosophical Transactions of the Royal Society B: Biological Sciences 363 (1491), 557–572. Available at https://doi.org/10.1098/rstb.2007.2170, http://rstb.royalsocietypublishing.org/.

Costa, M.V.J.D., Ramegowda, Y., Ramegowda, V., Karaba, N.N., Sreeman, S.M., Udayakumar, M., 2021. Combined drought and heat stress in rice: responses, phenotyping and strategies to improve tolerance. Rice Science 28 (3), 233–242. Available at https://doi.org/10.1016/j.rsci.2021.04.003, http://www.elsevier.com/wps/find/journaldescription.cws_home/713745/description#description.

Danish, S., Zafar-ul-Hye, M., Mohsin, F., Hussain, M., Farooq, S., 2020. ACC-deaminase producing plant growth promoting rhizobacteria and biochar mitigate adverse effects of drought stress on maize growth. PLoS ONE 15 (4), e0230615. Available at https://doi.org/10.1371/journal.pone.0230615.

Dar, M.H., de Janvry, A., Emerick, K., Raitzer, D., Sadoulet, E., 2013. Flood-tolerant rice reduces yield variability and raises expected yield, differentially benefitting socially disadvantaged groups. Scientific Reports 3 (1). Available at https://doi.org/10.1038/srep03315.

Deng, Z., Huang, M., Zhang, W., Wang, G., Huang, X., Liang, G., et al., 2023. Effects of five years conservation tillage for hedging against drought, stabilizing maize yield, and improving soil environment in the drylands of northern China. PLoS One 18 (3), e0282359. Available at https://doi.org/10.1371/journal.pone.0282359.

Dixit, S., Singh, A., Sandhu, N., Bhandari, A., Vikram, P., Kumar, A., 2017. Combining drought and submergence tolerance in rice: marker-assisted breeding and QTL combination effects. Molecular Breeding 37 (12). Available at https://doi.org/10.1007/s11032-017-0737-2.

Emerick, K., Ronald, P.C., 2019. Sub1 rice: engineering rice for climate change. Cold Spring Harbor Perspectives in Biology 11 (12), a034637. Available at https://doi.org/10.1101/cshperspect.a034637.

Erdoğan, İ., Cevher-Keskin, B., Bilir, Ö., Hong, Y., Tör, M., 2023. Recent developments in CRISPR/Cas9 genome-editing technology related to plant disease resistance and abiotic stress tolerance. Biology 12 (7). Available at https://doi.org/10.3390/biology12071037, www.mdpi.com/journal/biology.

Evelin, H., Devi, T.S., Gupta, S., Kapoor, R., 2019. Mitigation of salinity stress in plants by arbuscular mycorrhizal symbiosis: current understanding and new challenges. Frontiers in Plant Science 10. Available at https://doi.org/10.3389/fpls.2019.00470, https://www.frontiersin.org/articles/10.3389/fpls.2019.00470/pdf.

FAO, 2010. The second report on the state of the world's plant genetic resources for food and agriculture. Food and Agriculture Organization of the United Nations.

FAO, 2024. Global status of salt-affected soils – Main report. Available at https://doi.org/10.4060/cd3044en.

Farajollahi, Z., Eisvand, H.R., Nazarian-Firouzabadi, F., Nasrollahi, A.H., 2023. Nano-Fe nutrition improves soybean physiological characteristics, yield, root features and water productivity in different planting dates under drought stress conditions. Industrial Crops and Products 198. Available at https://doi.org/10.1016/j.indcrop.2023.116698, www.elsevier.com/inca/publications/store/5/2/2/8/2/5.

Farooq, M.A., Saqib, Z.A., Akhtar, J., Bakhat, H.F., Pasala, R.K., Dietz, K.J., 2019. Protective role of silicon (Si) against combined stress of salinity and boron (B) toxicity by improving antioxidant enzymes activity in rice. Silicon 11 (4), 2193–2197. Available at https://doi.org/10.1007/s12633-015-9346-z.

Fixen, P.E., 2020. A brief account of the genesis of 4R nutrient stewardship. Agronomy Journal 112 (5), 4511–4518. Available at https://doi.org/10.1002/agj2.20315.

Flexas, J., Bota, J., Loreto, F., Cornic, G., Sharkey, T.D., 2004. Diffusive and metabolic limitations to photosynthesis under drought and salinity in C3 plants. Plant Biology 6 (3), 269–279.

Flexas, J., Diaz-Espejo, A., Conesa, M.A., Coopman, R.E., Douthe, C., Gago, J., et al., 2014. Mesophyll conductance to CO_2 and Rubisco as targets for improving intrinsic water-use efficiency in C3 plants. Plant, Cell & Environment 37 (5), 965–982.

Fouda, S.E.E., 2021. Cyanobacteria and N-fertilization enhance the efficiency of rice plants grown under saline soil conditions. Egyptian Journal of Soil Science 61 (1), 63–77. Available at https://doi.org/10.21608/ejss.2021.56333.1420.

Furbank, R.T., Tester, M., 2011. Phenomics - technologies to relieve the phenotyping bottleneck. Trends in Plant Science 16 (12), 635–644. Available at https://doi.org/10.1016/j.tplants.2011.09.005.

Gaevaya, E., Ilyinskaya, I., Bezuglova, O., Klimenko, A., Taradin, S., Nezhinskaya, E., et al., 2023. Influence of heat stress and water availability on productivity of silage maize (Zea mays L.) under different tillage and fertilizer management practices in rostov region of Russia. Agronomy 13 (2), 320. Available at https://doi.org/10.3390/agronomy13020320.

Ghorai, A.K., 2022. Climate resilient rice production, intercropping in rice, alternate crops, insitu water harvestinig and ground water recharge to combat deficit rainfall. International Journal of Environment and Climate Change 829–839. Available at https://doi.org/10.9734/ijecc/2022/v12i121521.

OECD Publishing, 2025. Global Drought Outlook: Trends, Impacts and Policies to Adapt to a Drier World. OECD Publishing, ParisISBN. Available at https://doi.org/10.1787/d492583a-en.

Heikal, Y.M., El-Esawi, M.A., El-Ballat, E.M., Abdel-Aziz, H.M.M., 2023. Applications of nanoparticles for mitigating salinity and drought stress in plants: an overview on the physiological, biochemical and molecular genetic aspects. New Zealand Journal of Crop and Horticultural Science 51 (3), 297–327. Available at https://doi.org/10.1080/01140671.2021.2016870.

Hetherington, A.M., Woodward, F.I., 2003. The role of stomata in sensing and driving environmental change. Nature 424 (6951), 901–908. Available at https://doi.org/10.1038/nature01843.

Iftekharuddaula, K.M., Ghosal, S., Gonzaga, Z.J., Amin, A., Barman, H.N., Yasmeen, R., et al., 2016. Allelic diversity of newly characterized submergence-tolerant rice (Oryza sativa L.) germplasm from Bangladesh. Genetic Resources and Crop Evolution 63 (5), 859–867. Available at https://doi.org/10.1007/s10722-015-0289-4.

Kah, M., Tufenkji, N., White, J.C., 2019. Nano-enabled strategies to enhance crop nutrition and protection. Nature Nanotechnology 14 (6), 532–540.

Khanzada, A., Yan, K., Hu, W., Malko, M., Khan, K.A., Bao, Y., et al., 2025. Heat stress response mechanisms and resilience strategies in wheat. Journal of Agronomy and Crop Science 211 (1). Available at https://doi.org/10.1111/jac.70023.

Khan, N.A., Owens, L., Nuñez, M.A., Khan, A.L., 2025. Complexity of combined abiotic stresses to crop plants. Plant Stress 17. Available at https://doi.org/10.1016/j.stress.2025.100926.

Khoury, C.K., Bjorkman, A.D., Dempewolf, H., Ramirez-Villegas, J., Guarino, L., Jarvis, A., et al., 2014. Increasing homogeneity in global food supplies and the implications for food security. Proceedings of the National Academy of Sciences 111 (11), 4001–4006. Available at https://doi.org/10.1073/pnas.1313490111.

Kim, W., Iizumi, T., Hosokawa, N., Tanoue, M., Hirabayashi, Y., 2023. Flood impacts on global crop production: advances and limitations. Environmental Research Letters 18 (5), 054007. Available at https://doi.org/10.1088/1748-9326/accd85.

Kim, W., Iizumi, T., Nishimori, M., 2019. Global patterns of crop production losses associated with droughts from 1983 to 2009. Journal of Applied Meteorology and Climatology 58 (6), 1233–1244.

Kuanar, S.R., Molla, K.A., Chattopadhyay, K., Sarkar, R.K., Mohapatra, P.K., 2019. Introgression of Sub1 (SUB1) QTL in mega rice cultivars increases ethylene production to the detriment of grain- filling under stagnant flooding. Scientific Reports 9 (1). Available at https://doi.org/10.1038/s41598-019-54908-2.

Kuanar, S.R., Ray, A., Sethi, S.K., Chattopadhyay, K., Sarkar, R.K., 2017. Physiological basis of stagnant flooding tolerance in rice. Rice Science 24 (2), 73–84. Available at https://doi.org/10.1016/j.rsci.2016.08.008.

Kumari, V.V., Banerjee, P., Verma, V.C., Sukumaran, S., Chandran, M.A.S., Gopinath, K.A., et al., 2022. Plant nutrition: an effective way to alleviate abiotic stress in agricultural crops. International Journal of Molecular Sciences 23 (15). Available at https://doi.org/10.3390/ijms23158519.

Lawlor, D.W., Tezara, W., 2009. Causes of decreased photosynthetic rate and metabolic capacity in water-deficient leaf cells: a critical evaluation of mechanisms and integration of processes. Annals of Botany 103 (4), 561–579. Available at https://doi.org/10.1093/aob/mcn244.

Lehmann, J., Joseph, S., 2015. Biochar for environmental management: an introduction. Biochar for Environmental Management. Routledge, pp. 1–13.

Linh, T.-H., T.H.L., Le-Hung Linh, L.H.L., Cuc, D.T.K., Le-Huy Ham, L.H.H., Tran-Dang Khanh, T.D.K., 2013. Improving submergence tolerance of Vietnamese rice cultivar by molecular breeding. Journal of Plant Breeding and Genetics 1 (3), 157–168.

Li, Y., Li, S., He, X., Jiang, W., Zhang, D., Liu, B., et al., 2020. CO2 enrichment enhanced drought resistance by regulating growth, hydraulic conductivity and phytohormone contents in the root of cucumber seedlings. Plant Physiology and Biochemistry 152, 62–71. Available at https://doi.org/10.1016/j.plaphy.2020.04.037.

Li, Y., Zeleke, K., Wang, B., Liu, D.-L., 2025. A review of data for compound drought and heatwave stress impacts on crops: current progress, knowledge gaps, and future pathways. Plants 14 (14), 2158. Available at https://doi.org/10.3390/plants14142158.

Li, Q., Zhu, P., Yu, X., Xu, J., Liu, G., 2024. Physiological and molecular mechanisms of rice tolerance to salt and drought stress: advances and future directions. International Journal of Molecular Sciences 25 (17), 9404. Available at https://doi.org/10.3390/ijms25179404.

Luo, X., Bai, X., Zhu, D., Li, Y., Ji, W., Cai, H., et al., 2012. GsZFP1, a new Cys2/His2-type zinc-finger protein, is a positive regulator of plant tolerance to cold and drought stress. Planta 235 (6), 1141–1155. Available at https://doi.org/10.1007/s00425-011-1563-0.

Luu, T.N.H., Luu, M.C., Le Huy, H., 2012. Introgression of the Saltol into as 996, the elite variety of Vietnam, using marker assisted backcrossing. VNU Journal of Science: Natural Sciences and Technology 28 (1).

Marschner, H. (Ed.), 2011. Marschner's Mineral Nutrition of Higher Plants. Academic Press.

Misra, S., Chauhan, P.S., 2020. ACC deaminase-producing rhizosphere competent Bacillus spp. mitigate salt stress and promote Zea mays growth by modulating ethylene metabolism. 3 Biotech 10 (3). Available at https://doi.org/10.1007/s13205-020-2104-y.

Mittler, R., 2006. Abiotic stress, the field environment and stress combination. Trends in Plant Science 11 (1), 15–19. Available at https://doi.org/10.1016/j.tplants.2005.11.002.

Mondal, S., Paul, A., Mitra, D., Pradhan, C., Seth, C.S., Chattopadhyay, K., et al., 2024. The multifaceted role of different SnRK gene family members in regulating multiple abiotic stresses in plants. Physiologia Plantarum 176 (5). Available at https://doi.org/10.1111/ppl.14543.

Mousavi, S.M., 2022. Silicon and nano-silicon mediated heavy metal stress tolerance in plants. Silicon and Nano-Silicon in Environmental Stress Management and Crop Quality Improvement. Academic Press, pp. 181–191.

Murtaza, G., Usman, M., Iqbal, J., Tahir, M.N., Elshikh, M.S., Alkahtani, J., et al., 2024. The impact of biochar addition on morpho-physiological characteristics, yield and water use efficiency of tomato plants under drought and salinity stress. BMC Plant Biology 24 (1). Available at https://doi.org/10.1186/s12870-024-05058-9.

Muthu, V., Abbai, R., Nallathambi, J., Rahman, H., Ramasamy, S., Kambale, R., et al., 2020. Pyramiding QTLs controlling tolerance against drought, salinity, and submergence in rice through marker assisted breeding. PLoS One 15 (1), e0227421. Available at https://doi.org/10.1371/journal.pone.0227421.

Nair, M.M., Shylaraj, K.S., 2021. Introgression of dual abiotic stress tolerance QTLs (Saltol QTL and Sub1 gene) into Rice (Oryza sativa L.) variety Aiswarya through marker assisted backcross breeding. Physiology and Molecular Biology of Plants 27 (3), 497–514. Available at https://doi.org/10.1007/s12298-020-00893-0.

Nelson, W.C.D., Hoffmann, M.P., Vadez, V., Rötter, R.P., Koch, M., Whitbread, A.M., 2022. Can intercropping be an adaptation to drought? A model-based analysis for pearl millet–cowpea. Journal of Agronomy and Crop Science 208 (6), 910–927. Available at https://doi.org/10.1111/jac.12552.

Nurbekov, A., Kosimov, M., Islamov, S., Khaitov, B., Qodirova, D., Yuldasheva, Z., et al., 2024. No-till, crop residue management and winter wheat-based crop rotation strategies under rainfed environment. Frontiers in Agronomy 6. Available at https://doi.org/10.3389/fagro.2024.1453976.

Orozco-Mosqueda, Md.C., Kumar, A., Glick, B.R., Santoyo, G., 2022. The role of bacterial ACC deaminase and trehalose in increasing salt and drought tolerance in plants. Mitigation of Plant Abiotic Stress by Microorganisms: Applicability and Future Directions. Elsevier, Mexico, pp. 41–52. Available at 10.1016/B978-0-323-90568-8.00003-1.

Pandey, P., Irulappan, V., Bagavathiannan, M.V., Senthil-Kumar, M., 2017. Impact of combined abiotic and biotic stresses on plant growth and avenues for crop improvement by exploiting physio-morphological traits. Frontiers in Plant Science 8. Available at https://doi.org/10.3389/fpls.2017.00537.

Pathak, H., Kumar, M., A Molla, K., Chakraborty, K., 2021. Abiotic stresses in rice production: impacts and management. Oryza-An International Journal on Rice 58 (Special), 103–125. Available at https://doi.org/10.35709/ory.2021.58.spl.4.

Peleg, Z., Blumwald, E., 2011. Hormone balance and abiotic stress tolerance in crop plants. Current Opinion in Plant Biology 14 (3), 290–295. Available at https://doi.org/10.1016/j.pbi.2011.02.001.

Priya, P., Patil, M., Pandey, P., Singh, A., Babu, V.S., Senthil-Kumar, M., 2023. Stress combinations and their interactions in plants database: a one-stop resource on combined stress responses in plants. Plant Journal 116 (4), 1097–1117. Available at https://doi.org/10.1111/tpj.16497.

Ramegowda, V., Senthil, A., Senthil-Kumar, M., 2024. Stress combinations and their interactions in crop plants. Plant Physiology Reports 29 (1), 1–5.

Ray, S., Vijayan, J., Sarkar, R.K., 2016. Germination stage oxygen deficiency (GSOD): an emerging stress in the era of changing trends in climate and rice cultivation practice. Frontiers in Plant Science 7, 671.

Reetz, H.F., Heffer, P., Bruulsema, T.W., 2015. 4R nutrient stewardship: a global framework for sustainable fertilizer management. Managing Water and Fertilizer for Sustainable Agricultural Intensification 65.

Rivero, R.M., Mestre, T.C., Mittler, R., Rubio, F., Garcia-Sanchez, F., Martinez, V., 2014. The combined effect of salinity and heat reveals a specific physiological, biochemical and molecular response in tomato plants. Plant, Cell and Environment 37 (5), 1059–1073. Available at https://doi.org/10.1111/pce.12199.

Salgotra, R.K., Chauhan, B.S., 2023. Ecophysiological responses of rice (Oryza sativa L.) to drought and high temperature. Agronomy 13 (7). Available at https://doi.org/10.3390/agronomy13071877.

dos Santos, T.B., Ribas, A.F., de Souza, S.G.H., Budzinski, I.G.F., Domingues, D.S., 2022. Physiological responses to drought, salinity, and heat stress in plants: a review. Stresses 2 (1), 113–135. Available at https://doi.org/10.3390/stresses2010009.

Sarkar, R.K., Chakraborty, K., Chattopadhyay, K., Ray, S., Panda, D., Ismail, A.M., 2019. Responses of rice to individual and combined stresses of flooding and salinity. Advances in Rice Research for Abiotic Stress Tolerance. Woodhead Publishing, pp. 281–297.

Sarkar, R.K., Panda, D., Reddy, J.N., Patnaik, S.S.C., Mackill, D.J., Ismail, A.M., 2009. Performance of submergence tolerant rice (oryza sativa) genotypes carrying the Sub1 quantitative trait locus under stressed and non-stressed natural field conditions. Indian Journal of Agricultural Sciences 79 (11), 876–883.

Sarkar, R.K., Ray, A., 2016. Submergence-tolerant rice withstands complete submergence even in saline water: probing through chlorophyll a fluorescence induction O-J-I-P transients. Photosynthetica 54 (2), 275–287. Available at https://doi.org/10.1007/s11099-016-0082-4.

Sazali, S.A., Shamsudin, N.A.A., Rafii, M.Y., Ab Razak, M.S.F., Yunus, M.F., Yah, F.N.C., et al., 2023. Genetics, physiological mechanism and breeding for tolerance against submergence, salinity, and saline-submergence stress in rice (Oryza sativa L.). Plant Science Today 10, 235–242. Available at https://doi.org/10.14719/pst.2536.

Septiningsih, E.M., Hidayatun, N., Sanchez, D.L., Nugraha, Y., Carandang, J., Pamplona, A.M., et al., 2015. Accelerating the development of new submergence tolerant rice varieties: the case of Ciherang-Sub1 and PSB Rc18-Sub1. Euphytica 202 (2), 259–268. Available at https://doi.org/10.1007/s10681-014-1287-x.

Sethuraman, G., Zain, N.A.M., Osman, N., Ismail, M.R., Mispan, M.S., Hanafiah, N.M., et al., 2024. Individual and concurrent effects of heat and drought stress on the growth and yield of two Malaysian rice cultivars. Chilean Journal of Agricultural Research 84 (5), 709–717. Available at https://doi.org/10.4067/s0718-58392024000500709.

Seyar, M.H., Ahamed, T., 2024. Optimization of soil-based irrigation scheduling through the integration of machine learning, remote sensing, and soil moisture sensor technology. IoT and AI in Agriculture: Smart Automation Systems for Increasing Agricultural Productivity to Achieve SDGs and Society 5.0. Springer, Singapore., pp. 275–300. Available at https://doi.org/10.1007/978-981-97-1263-2_18.

Seyed Sharifi, R., Khalilzadeh, R., Dadashzadeh, S., 2024. Effect of biofertilizer and nano-Fe oxide foliar application on alleviation of water deficit in yield and some physico-chemical properties of barley. Russian Journal of Plant Physiology 71 (1), 46.

Shin, N.H., Han, J.H., Vo, K.T.X., Seo, J., Navea, I.P., Yoo, S.C., et al., 2022. Development of a temperate climate-adapted indica multi-stress tolerant rice variety by pyramiding quantitative trait loci. Rice 15 (1). Available at https://doi.org/10.1186/s12284-022-00568-2.

Singhal, R.K., Fahad, S., Kumar, P., Choyal, P., Javed, T., Jinger, D., et al., 2023. Beneficial elements: new players in improving nutrient use efficiency and abiotic stress tolerance. Plant Growth Regulation 100 (2), 237–265. Available at https://doi.org/10.1007/s10725-022-00843-8.

Singh, A., Septiningsih, E.M., Balyan, H.S., Singh, N.K., Rai, V., 2017. Genetics, physiological mechanisms and breeding of flood-tolerant rice (Oryza sativa L.). Plant and Cell Physiology 58 (2), 185–197. Available at https://doi.org/10.1093/pcp/pcw206.

Sivasakthivelan, P., Saranraj, P., Al-Tawaha, A.R.M., Amala, K., Al Tawaha, A.R., Thangadurai, D., et al., 2021. Adaptation of Azospirillum to stress conditions: a review. Advances in Environmental Biology 15 (4), 1–6.

Smith, D.L., Almaraz, J.J., 2004. Climate change and crop production: contributions, impacts, and adaptations. Canadian Journal of Plant Pathology 26 (3), 253–266. Available at https://doi.org/10.1080/07060660409507142.

Sood, G., Kaushal, R., Sharma, M., 2020. Significance of inoculation with Bacillus subtilis to alleviate drought stress in wheat (*Triticum aestivum* L.). Vegetos 33 (4), 782–792.

Suzuki, N., Rivero, R.M., Shulaev, V., Blumwald, E., Mittler, R., 2014. Abiotic and biotic stress combinations. New Phytologist 203 (1), 32–43. Available at https://doi.org/10.1111/nph.12797.

Tahiri, A.I., Slimani, A., Raklami, A., Meddich, A., Oufdou, K., 2025. Drought stress mitigation in tomato through combined use of plant growth promoting Rhizobacteria, Arbuscular Mycorrhizal fungi, and organic-mineral amendments. Journal of Soil Science and Plant Nutrition 25 (3), 7909–7928. Available at https://doi.org/10.1007/s42729-025-02641-3.

Tindall, J.A., Mills, H.A., Radcliffe, D.E., 1990. The effect of root zone temperature on nutrient uptake of tomato. Journal of Plant Nutrition 13, 939–956. Available at https://doi.org/10.1080/01904169009364127.

Toojinda, T., Tragoonrung, S., Vanavichit, A., Siangliw, J.L., Pa-In, N., Jantaboon, J., et al., 2005. Molecular breeding for rainfed lowland rice in the Mekong region. Plant Production Science 8 (3), 330–333. Available at https://doi.org/10.1626/pps.8.330.

Tripathi, D.K., Singh, S., Singh, S., Pandey, R., Singh, V.P., Sharma, N.C., et al., 2022. 'An overview on manufactured nanoparticles in plants: uptake, translocation, accumulation and phytotoxicity. Plant Physiology and Biochemistry 170, 12–27.

Uddin, S., Löw, M., Parvin, S., Fitzgerald, G.J., Tausz-Posch, S., Armstrong, R., et al., 2018. Elevated [CO2] mitigates the effect of surface drought by stimulating root growth to access sub-soil water. PLoS One 13 (6). Available at https://doi.org/10.1371/journal.pone.0198928.

UNDP (2024). Annual Report 2023. New York: United Nations Development Programme.

Varshney, R.K., Bohra, A., Yu, J., Graner, A., Zhang, Q., Sorrells, M.E., 2021. Designing future crops: genomics-assisted breeding comes of age. Trends in Plant Science 26 (6), 631–649. Available at https://doi.org/10.1016/j.tplants.2021.03.010.

Varshney, R.K., Terauchi, R., McCouch, S.R., 2014. Harvesting the promising fruits of genomics: applying genome sequencing technologies to crop breeding. PLoS Biology 12 (6), e1001883.

Venkatappa, M., Sasaki, N., Han, P., Abe, I., 2021. Impacts of droughts and floods on croplands and crop production in Southeast Asia – an application of Google Earth Engine. Science of The Total Environment 795, 148829. Available at https://doi.org/10.1016/j.scitotenv.2021.148829.

Verma, V., Ravindran, P., Kumar, P.P., 2016. Plant hormone-mediated regulation of stress responses. BMC Plant Biology 16, 86.

Vescio, R., Abenavoli, M.R., Sorgonà, A., 2021. Single and combined abiotic stress in maize root morphology. Plants 10, 5. Available at https://doi.org/10.3390/plants10010005.

Vijayan, J., Senapati, S., Ray, S., Chakraborty, K., Molla, K.A., Basak, N., et al., 2018. Transcriptomic and physiological studies identify cues for germination stage oxygen deficiency tolerance in rice. Environmental and Experimental Botany 147, 234–248.

Voesenek, L.A.C.J., Bailey-Serres, J., 2015. Flood adaptive traits and processes: an overview. New Phytologist 206 (1), 57–73. Available at https://doi.org/10.1111/nph.13209.

Wang, H., An, T., Huang, D., Liu, R., Xu, B., Zhang, S., et al., 2021. Arbuscular mycorrhizal symbioses alleviating salt stress in maize is associated with a decline in root-to-leaf gradient of Na+/K+ ratio. BMC Plant Biology 21 (1), 457. Available at https://doi.org/10.1186/s12870-021-03237-6.

Wang, P., Lombi, E., Zhao, F.J., Kopittke, P.M., 2020. Nanotechnology: a new opportunity in plant sciences. Trends in Plant Science 25 (6), 537–550.

Wang, S., Wang, F., Gao, S., Wang, X., 2016. Heavy metal accumulation in different rice cultivars as influenced by foliar application of nano-silicon. Water, Air, & Soil Pollution 227 (7). Available at https://doi.org/10.1007/s11270-016-2928-6.

Wang, Y., Zafar, N., Ali, Q., Manghwar, H., Wang, G., Yu, L., et al., 2022. CRISPR/Cas genome editing technologies for plant improvement against biotic and abiotic stresses: advances, limitations, and future perspectives. Cells 11 (23), 3928. Available at https://doi.org/10.3390/cells11233928.

Warschefsky, E., Penmetsa, R.V., Cook, D.R., Von Wettberg, E.J.B., 2014. Back to the wilds: tapping evolutionary adaptations for resilient crops through systematic hybridization with crop wild relatives. American Journal of Botany 101 (10), 1791–1800. Available at https://doi.org/10.3732/ajb.1400116.

Weckwerth, W., Ghatak, A., Bellaire, A., Chaturvedi, P., Varshney, R.K., 2020. PANOMICS meets germplasm. Plant Biotechnology Journal 18 (7), 1507–1525. Available at https://doi.org/10.1111/pbi.13372.

Whippo, C.W., Saliendra, N.Z., Liebig, M.A., 2024. Cover crop inclusion and residue retention improves soybean production and physiology in drought conditions. Heliyon 10 (8), e29838. Available at https://doi.org/10.1016/j.heliyon.2024.e29838.

Wittwer, R.A., Klaus, V.H., Miranda Oliveira, E., Sun, Q., Liu, Y., Gilgen, A.K., et al., 2023. Limited capability of organic farming and conservation tillage to enhance agroecosystem resilience to severe drought. Agricultural Systems 211, 103721. Available at https://doi.org/10.1016/j.agsy.2023.103721.

Wu, W., Shah, F., Duncan, R.W., Ma, B.L., 2020. Grain yield, root growth habit and lodging of eight oilseed rape genotypes in response to a short period of heat stress during flowering. Agricultural and Forest Meteorology 287, 107954. Available at https://doi.org/10.1016/j.agrformet.2020.107954.

Wu, Y., Yan, S., Fan, J., Zhang, F., Xiang, Y., Zheng, J., et al., 2021. Responses of growth, fruit yield, quality and water productivity of greenhouse tomato to deficit drip irrigation. Scientia Horticulturae 275, 109710. Available at https://doi.org/10.1016/j.scienta.2020.109710.

Xiong, H., He, H., Chang, Y., Miao, B., Liu, Z., Wang, Q., et al., 2025. Multiple roles of NAC transcription factors in plant development and stress responses. Journal of Integrative Plant Biology 67 (3), 510–538. Available at https://doi.org/10.1111/jipb.13854.

Xu, K., Mackill, D.J., 1996. A major locus for submergence tolerance mapped on rice chromosome 9. Molecular Breeding 2 (3), 219–224. Available at https://doi.org/10.1007/BF00564199.

Xu, K., Xu, X., Fukao, T., Canlas, P., Maghirang-Rodriguez, R., Heuer, S., et al., 2006. Sub1A is an ethylene-response-factor-like gene that confers submergence tolerance to rice. Nature 442 (7103), 705–708. Available at https://doi.org/10.1038/nature04920.

Yadav, C., Bahuguna, R.N., Dhankher, O.P., Singla-Pareek, S.L., Pareek, A., 2022. Physiological and molecular signatures reveal differential response of rice genotypes to drought and drought combination with heat and salinity stress. Physiology and Molecular Biology of Plants 28 (4), 899–910. Available at https://doi.org/10.1007/s12298-022-01162-y.

Yang, W., Feng, H., Zhang, X., Zhang, J., Doonan, J.H., Batchelor, W.D., et al., 2020. Crop phenomics and high-throughput phenotyping: past decades, current challenges, and future perspectives. Molecular Plant 13 (2), 187–214. Available at https://doi.org/10.1016/j.molp.2020.01.008.

Ye, Q., Wang, H., Li, H., 2022. Arbuscular mycorrhizal fungi improve growth, photosynthetic activity, and chlorophyll fluorescence of Vitis vinifera L. cv. Ecolly under drought stress. Agronomy 12 (7), 1563.

Zambrano, J.L., Cartagena, Y., Sangoquiza, C., Pincay, A., Parra, A.R., Maiguashca, J., et al., 2024. Exploring plastic mulching as a strategy for mitigating drought stress and boosting maize yield in the Ecuadorian Andes. Water (Switzerland) 16 (7), 1033. Available at https://doi.org/10.3390/w16071033.

Zandalinas, S.I., Mittler, R., 2022. Plant responses to multifactorial stress combination. New Phytologist 234 (4), 1161–1167. Available at https://doi.org/10.1111/nph.18087.

Zandalinas, S.I., Mittler, R., Balfagón, D., Arbona, V., Gómez-Cadenas, A., 2021. Plant adaptations to the combination of drought and high temperatures. Physiologia Plantarum 171 (1), 6–18.

Zhang, J., Bai, Z., Huang, J., Hussain, S., Zhao, F., Zhu, C., et al., 2019a. Biochar alleviated the salt stress of induced saline paddy soil and improved the biochemical characteristics of rice seedlings differing in salt tolerance. Soil and Tillage Research 195, 104372.

Zhang, L., Han, W., Niu, Y., Chávez, J.L., Shao, G., Zhang, H., 2021. Evaluating the sensitivity of water stressed maize chlorophyll and structure based on UAV derived vegetation indices. Computers and Electronics in Agriculture 185, 106174. Available at https://doi.org/10.1016/j.compag.2021.106174.

Zhang, A., Liu, Y., Wang, F., Li, T., Chen, Z., Kong, D., et al., 2019b. Enhanced rice salinity tolerance via CRISPR/Cas9-targeted mutagenesis of the OsRR22 gene. Molecular Breeding 39 (3), 47.

Zhang, X., Li, Y., Sun, Y., Yu, L., Xu, J., Gu, X., et al., 2025. Effects of straw mulching and plastic mulching on maize yield and crop water productivity in China: a meta-analysis. Agricultural Water Management 315, 109549. Available at https://doi.org/10.1016/j.agwat.2025.109549.

Zhang, F., Rosental, L., Ji, B., Brotman, Y., Dai, M., 2024a. Metabolite-mediated adaptation of crops to drought and the acquisition of tolerance. The Plant Journal 118 (3), 626–644.

Zhang, X., Wang, K., Sun, C., Yang, K., Zheng, J., 2022. Differences in soil physical properties caused by applying three organic amendments to loamy clay soil under field conditions. Journal of Soils and Sediments 22 (1), 43–55. Available at https://doi.org/10.1007/s11368-021-03049-z.

Zhang, Q., Yang, X., Liu, C., Yang, N., Yu, G., Zhang, Z., et al., 2024b. Monitoring soil moisture in winter wheat with crop water stress index based on canopy-air temperature time lag effect. International Journal of Biometeorology 68 (4), 647–659.

Zhang, L., Zhang, H., Niu, Y., Han, W., 2019c. Mapping maize water stress based on UAV multispectral remote sensing. Remote Sensing 11 (6), 605.

Zhang, H., Zhang, J., Wei, P., Zhang, B., Gou, F., Feng, Z., et al., 2018. The CRISPR/Cas9 system produces specific and homozygous targeted gene editing in rice in one generation. Plant Biotechnology Journal 12 (6), 797–807.

Zhao, H., Liu, G., Dou, Y., Yang, H., Wang, T., Wang, Z., et al., 2024. Plastic mulch increases dryland wheat yield and water-use productivity, while straw mulch increases soil water storage. Journal of Integrative Agriculture 23 (9), 3174–3185. Available at https://doi.org/10.1016/j.jia.2024.01.008.

Zhu, Q., Liu, X., Hao, T., Zeng, M., Shen, J., Zhang, F., et al., 2020. Cropland acidification increases risk of yield losses and food insecurity in China. Environmental Pollution 256, 113145.

Zhu, J.K., 2016. Abiotic stress signaling and responses in plants. Cell 167 (2), 313–324. Available at https://doi.org/10.1016/j.cell.2016.08.029.

Chapter 2

Plant genetic resources and their potential use to combat multiple abiotic stresses

Somnath Roy[1,*], Ankan Das[2], Sandip Debnath[3], Koushik Chakraborty[4], Puranjoy Sar[1], Amrita Banerjee[1], N.P. Mandal[1], and K.C. Bansal[5]

[1]*Central Rainfed Upland Rice Research Station, ICAR-Central Rice Research Institute, Hazaribag, Jharkhand, India,* [2]*ICAR-Indian Institute of Horticultural Research, Bengaluru, Karnataka, India,* [3]*Department of Genetics and Plant Breeding, Institute of Agriculture, Visva-Bharati, West Bengal, India,* [4]*Division of Crop Physiology and Biochemistry, ICAR-Central Rice Research Institute, Cuttack, Odisha, India,* [5]*National Academy of Agricultural Sciences, New Delhi, India*

**Corresponding author. e-mail address: Somnath.roy@icar.org.in*

2.1 Introduction

Global agriculture faces mounting pressures from climate change, environmental degradation, and population growth. Among the most urgent challenges is the prevalence of multifactorial stresses (MFSs) characterized by simultaneous or sequential exposures to drought, heat, salinity, submergence, nutrient imbalance, and flooding. Unlike single stresses, MFSs represent the realistic field conditions in which crops grow, often producing synergistic damage that far exceeds additive expectations (Khan et al., 2025; Mittler, 2006; Zandalinas and Mittler, 2022). Abiotic stresses already account for >50% of yield losses worldwide (Akpinar et al., 2013), and climate models predict increasing frequency and severity of combined stress events (Jiang et al., 2025). Field studies confirm this reality: drought rarely occurs alone, often coexisting with heat, while coastal agriculture frequently faces saline flooding (Shabbir et al., 2022).

Combined stresses impose unique physiological conflicts. For example, drought × heat forces opposing stomatal responses, such as closure to conserve water versus opening for transpirational cooling, culminating in reduced photosynthesis and yield (Costa et al., 2021). Similarly, salinity × flooding imposes both ionic toxicity and hypoxia (Chakraborty et al., 2024). At the molecular level, multiomics analyses reveal combination-specific signatures, including transcriptional hubs (*NAC, WRKY, DREB*, etc.), Reactive oxygen species (ROS) signaling networks, and hormonal-gene cascades (Khan et al., 2025; Zandalinas and Mittler, 2022). Such nonlinear outcomes explain why varieties bred under controlled single-stress conditions often fail in farmers' fields.

In this context, plant genetic resources (PGR), including landraces, traditional varieties, breeding lines, as well as crop wild relatives (CWRs), are invaluable. Modern agriculture has narrowed crop diversity, trading resilience for yield and increasing vulnerability (Khoury et al., 2014). Mining PGR enables the rediscovery of lost alleles and the identification of novel adaptive traits. Comparative genomics illustrates this potential: diploid cottons carry Ca^{2+}-signaling and oxidative stress regulators conferring cold–salt tolerance (Xu et al., 2020); sugarcane shows a broad expansion of ABA-related PP2Cs (Wang et al., 2025); and *Brachypodium* networks under drought, salinity, cold, and heat are syntenic with rice, wheat, and maize, facilitating cross-crop translation (Lei et al., 2023). Notably, Tibetan wild barley accumulates high levels of proline and glycine betaine under drought–salinity, highlighting CWRs as reservoirs of MFS resilience (Khan et al., 2025).

The domestication and spread of major staples reflect adaptations to diverse agro-ecologies. Rice diversified into upland (drought-prone) and lowland (submergence/salinity-prone) ecologies; wheat emerged in the Fertile Crescent under fluctuating drought and temperature; maize spread from Mesoamerica into temperate and tropical zones; sorghum evolved stay-green resilience in African drylands; and barley adapted to cold, saline, and arid conditions. These evolutionary legacies underscore the conserved nature of stress-response pathways such as ICE1–CBF–COR for cold and ABA–*SnRK2* for drought–salinity (Chinnusamy et al., 2004; Jiang et al., 2025; Vaughan et al., 2007). Polyploidization events, such as those in wheat and cotton, further broadened adaptive repertoires (Xu et al., 2020).

Multiple Abiotic Stresses in Plants. DOI: https://doi.org/10.1016/B978-0-443-27424-4.00001-X

This chapter synthesizes current understanding of MFS impacts, emphasizing the nonadditive nature of stress interactions and the opportunity offered by PGR. It discusses how comparative genomics and synteny analyses illuminate conserved pathways across crops, and how advanced breeding, genomic selection, and genome editing can translate PGR diversity into climate-smart varieties. By integrating biodiversity with biotechnology, sustainable strategies can be devised to ensure the resilience of agriculture under MFS conditions.

2.2 Complexities of combined stresses on crop plants

Abiotic stresses seldom occur alone; drought, salinity, extreme temperatures, nutrient imbalance, heavy metals, and submergence frequently co-occur and interact nonlinearly across the crop life cycle (Mittler, 2006; Zandalinas and Mittler, 2022). These interactions impose physiological conflicts (e.g., drought × heat: stomatal closure vs transpirational cooling; salinity × flooding: ion toxicity plus hypoxia) that intensify injury beyond single-stress expectations (Chakraborty et al., 2024; Costa et al., 2021) (Fig. 2.1). Omics studies show combination-specific signatures, with large fractions of

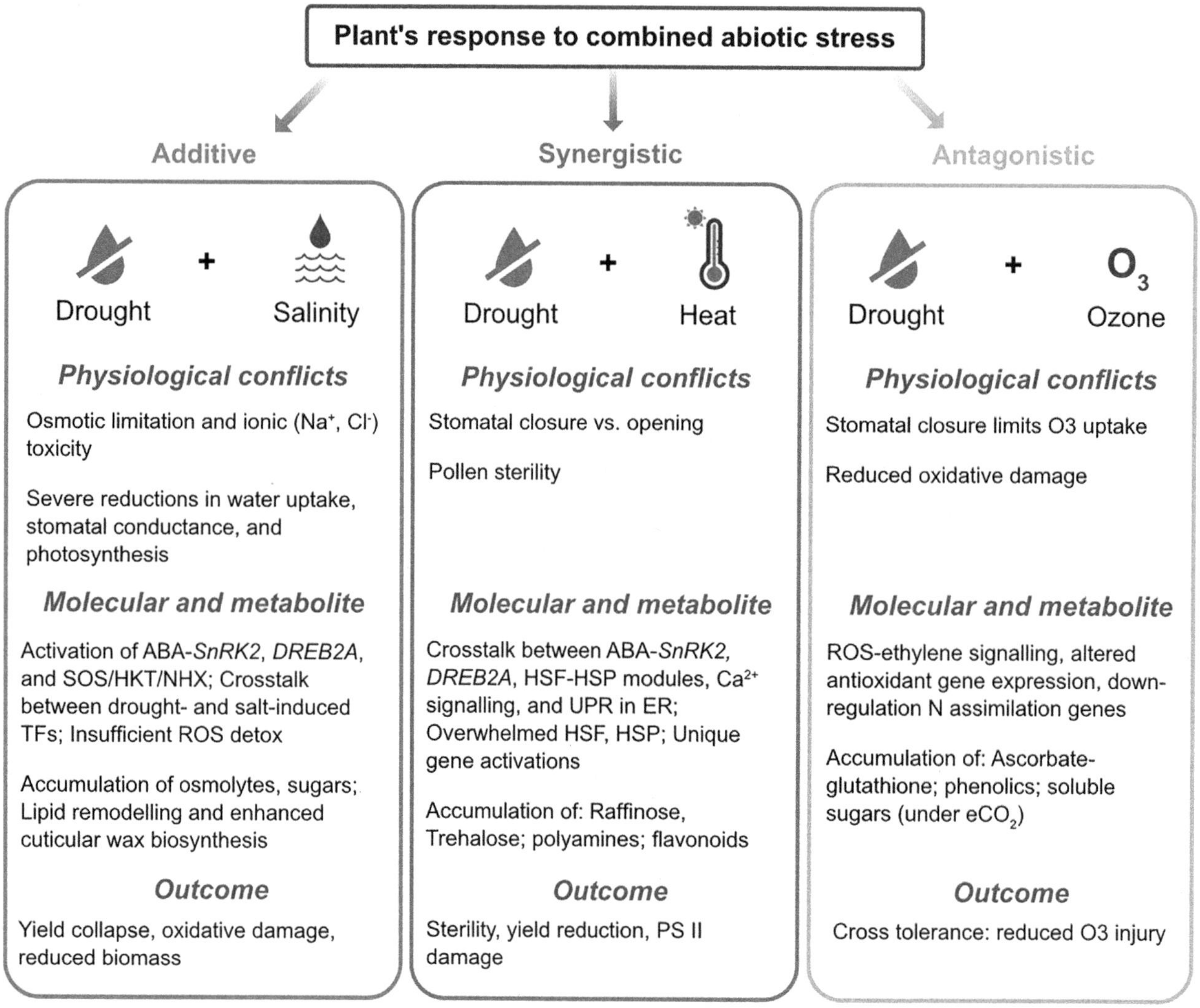

FIGURE 2.1 Conceptual framework of plant responses to combined abiotic stresses. Plants exposed to multiple stresses exhibit nonadditive outcomes: Additive (combined effect equals the sum of individual stresses, e.g., drought + salinity), Synergistic (combined effect exceeds individual impacts, e.g., drought + heat), and Antagonistic (one stress mitigates the other, e.g., drought + ozone). Stress combinations impose distinct physiological conflicts (osmotic stress, ion toxicity, stomatal closure vs opening, pollen sterility) and activate combination-specific molecular/metabolite responses (ABA–SnRK2-AREB/ABF, ion transporters, HSPs, flavonoids, 2-hydroxymelatonin). These interactions ultimately determine crop productivity outcomes, ranging from yield collapse to partial recovery or cross-tolerance.
Source: Adapted from Mittler (2006), Zandalinas and Mittler (2022), Shabbir et al. (2022).

transcripts/metabolites unique to stress pairs, often governed by intersecting ABA, Ca^{2+}, ROS, and heat-shock/endoplasmic reticulum (ER)-stress modules (Rasmussen et al., 2013; Zandalinas and Mittler, 2022). Outcomes depend on intensity, sequence, stage, and genotype; the same pair may be synergistic or antagonistic across species (Khan et al., 2025). Multifactorial triads (e.g., heat × salinity × nutrient limitation) are especially damaging and typify field realities under climate change (Suzuki et al., 2014). This complexity explains why varieties bred under single stresses can fail in farmers' fields and underscores the need to treat stress combinations as distinct targets.

2.2.1 Drought combined with other stresses

Evidence shows that combined stresses (drought–heat, drought–salinity, or drought–chilling) cause greater reductions in plant growth and yield than individual stresses. In wheat, drought and heat during anthesis reduced grain weight by 83%, exceeding reductions from drought (13%–27%) or heat alone (Mahrookashani et al., 2017). Maize drought–heat combinations reduced yields by 40%–70% through tassel blasting and delayed silk emergence (Meseka et al., 2018). Tolerance to single stresses does not ensure resilience to combinations. Physiological conflicts, such as heat-promoting stomatal opening while drought drives closure, cause photoinhibition and reproductive sterility (Zandalinas et al., 2018; Priya et al., 2025). Combined drought and heat disrupt source–sink relations and photosynthesis. Drought restricts leaf expansion and carbohydrate supply, whereas heat impairs pollen viability and grain filling. These stresses cause floral sterility and yield collapse. Combined stress reduces photosystem II efficiency more than individual stresses, as demonstrated in tomato (Zhou et al., 2017). In legumes, drought–heat combinations reduce sucrose synthesis while destabilizing membranes (Priya et al., 2025). Under drought–salinity stress, biomass declines significantly, chlorophyll drops by 25%, and antioxidant defenses often fail. Drought–chilling reduces root hydraulic conductivity, lowering leaf water potential (Hussain et al., 2018). Across stress combinations, oxidative stress emerges as a key mechanism, damaging cellular components.

Stress combinations reprogram molecular signaling unpredictably. Transcriptomic studies show that over 60% of drought–heat-induced genes are unique (Rasmussen et al., 2013). Regulatory modules include ABA-dependent and ABA-independent signaling, with ABA–PP2C–*SnRK2* cascades triggered by water deficit and heat activating HSFs and HSPs. Under combined stress, these pathways intersect with *DREB2A* transcription factor, Ca^{2+} signaling, ROS waves, and unfolded protein responses, creating a complex network (Priya et al., 2025). While antioxidant enzymes (superoxide dismutase [SOD], Catalase [CAT], and ascorbate peroxidase [APX]) are central, their activity is often insufficient under combined stresses. Metabolomic profiling in maize under drought–salinity stress showed unique accumulation of solutes, sugars, and lipids absent in single stresses (Sun et al., 2014). Osmoprotectants, such as proline, glycine betaine, and trehalose, are implicated across stress combinations. Ion transporters HKT1 and the SOS pathway regulate ionic balance under drought–salinity, whereas CBF/DREB factors contribute to drought–chilling responses (Hussain et al., 2018). These findings highlight the importance of studying combined stresses directly, as many responsive elements are absent from single-stress studies.

The interaction of drought with elevated ozone (O_3) or elevated CO_2 (eCO_2) creates complex stress combinations for crops. Stomatal closure in response to drought may provide protection from high O_3 influx to leaves. However, elevated O_3 may prevent drought-induced stomatal closure, leading to depletion of soil water and aggravation of drought stress. Studies in soybean revealed that elevated O_3 consistently reduced soybean Rubisco carboxylation capacity (−17%) and maximum electron transport capacity (−9%). Higher O_3 levels did not alter the relationships between soil moisture, abscisic acid, and g_s (Martin et al., 2025). However, eCO_2 often mitigates these effects by improving water-use efficiency and counteracting oxidative damage. Open-top chamber experiments with soybean showed eCO_2 compensated for O_3-induced yield losses by maintaining leaf area, despite reduced growth under O_3 (Booker et al., 2004). Free-air ozone and CO_2 enrichment studies in chickpea demonstrated that while O_3 shortened growth cycles, eCO_2 enhanced photosynthesis, with combined treatment improving yield by 10% (Singh et al., 2021b). Tomato genotypes under combined stress and eCO_2 showed reduced photosynthetic inhibition, though genotype-specific differences existed (Zhou et al., 2020). Studies in grassland species confirmed eCO_2 reduced drought-induced oxidative injury through enhanced ascorbate–glutathione cycling (Farfan-Vignolo and Asard, 2012). While O_3 and drought intensify crop damage, eCO_2 provides protection by enhancing water-use efficiency, though this mitigation varies by species and may diminish under extreme stress.

2.2.2 Submergence combined with other stresses

Flooding and submergence impose multiple stresses, including oxygen deprivation, reduced photosynthesis, nutrient restriction, and energy crises, often compounded when combined with salinity, drought, heat, or chilling. Plant responses to such stress combinations reflect unique physiological and molecular reprogramming (Tamang and Fukao, 2015), with major implications for rice and cereals in flood-prone coastal areas threatened by climate change (Jagadish et al., 2012). When submergence coincides with salinity in coastal deltas, plants face conflicting demands: quiescence conserves energy, whereas salt tolerance requires active ion transport for Na^+ exclusion (Chakraborty et al., 2024). Studies showed that combined waterlogging and salinity reduced growth and biomass more than individual stresses (Mahmood et al., 2021). Experiments with IR64 and IR64-*SUB1* revealed that combined submergence and osmotic stress lowered growth and increased oxidative damage, though *SUB1A* conferred higher survival (Chi et al., 2023). *SUB1* genotypes survive flooding and recover better from postsubmergence dehydration (Fukao et al., 2011). Environmental studies confirm that prolonged submergence with saline intrusion alters wetland plant communities, favoring specialized halophytes (Jordine et al., 2024). Combined submergence stresses intensify constraints on gas exchange and ion balance. Hypoxia suppresses respiration while salinity adds osmotic stress. Chlorophyll fluorescence shows sharper declines under saline submergence than individual stresses (Pradhan et al., 2019). During recovery, desubmergence causes oxygen influx and ROS bursts, with higher survival in plants that manage ROS detoxification (Tamang and Fukao, 2015).

Wild halophytes show contrasting strategies. *Porteresia coarctata* tolerates high salinity and submergence through ion homeostasis and carbohydrate metabolism (Garg et al., 2014). *Salicornia europaea* thrives under tidal flooding and high NaCl by combining salt compartmentation and oxygen transport. Molecular studies reveal crosstalk between submergence and stress pathways. *SUB1A* introgressed into rice varieties suppresses ethylene and gibberellin responses during flooding, conserves carbohydrates, and enhances post-flood recovery through ABA responsiveness and ROS-scavenging enzymes (Fukao et al., 2011). *SNORKEL1/2* activate gibberellin-mediated elongation escape, beneficial in deepwater but costly under saline floods (Tamang and Fukao, 2015). Transcriptome profiling of *Porteresia coarctata* identified >15,000 genes responsive to salinity and submergence, including AP2/ERFs, NACs, MYBs, and ion transporters (Garg et al., 2014). Submergence with osmotic stress alters the expression of ethylene response factor (ERF)-VIIs, *ADH1*, and *LDH1*, indicating shifts in glycolysis pathways (Chi et al., 2023). Under waterlogging-salinity, ROS accumulation exceeds detoxification capacity unless genotypes maintain high SOD, CAT, and APX activity. The N-degron oxygen-sensing pathway influences both hypoxia and salinity responses. Compatible solutes accumulate under saline flooding, while silicon deposition reduces Na^+ uptake (Gao et al., 2007). Adaptation involves both conserved and stress-specific molecular signatures. *SUB1A* introgression revolutionized flood-prone rice cultivation. Combined with *Saltol* QTL, partial tolerance to saline submergence exists, though single-trait tolerance does not guarantee combined stress resilience (Pradhan et al., 2019). Future breeding requires pyramiding *SUB1* with ion transport genes, antioxidant regulators, and recovery networks. Wild relatives provide genetic reservoirs for these traits.

2.2.3 Cold combined with other stresses

Plants in temperate and high-altitude agroecosystems face episodes of low temperature combined with drought, salinity, or nutrient limitation. These combinations are detrimental as cold limits enzymatic kinetics and membrane fluidity, while other stresses add osmotic, ionic, or oxidative burdens. Field observations in wheat, rice, maize, and legumes show that combined cold stress reduces yields more severely than individual stresses, particularly during germination or reproduction (Ejaz et al., 2023). In wheat, grain weight declines more under cold-drought interaction than single stresses, with spring wheat showing higher vulnerability. Chilling with drought delays germination and accelerates reproductive abortion in chickpea and soybean (Swigonska et al., 2014). Cold and drought stresses restrict water relations and photosynthetic capacity. Cold reduces root hydraulic conductivity, worsening drought-induced turgor loss (Hussain et al., 2018). Combined stresses impair Photosystem II efficiency and chlorophyll content, reducing CO_2 assimilation. In soybean radicles, cold-osmotic stress increases phenolic compounds and alters antioxidant capacity (Swigonska et al., 2014). Cold-salinity interactions disrupt ion homeostasis through reduced K^+ uptake and elevated Na^+/K^+ ratios (Liu et al., 2024).

Molecular responses to combined cold and other stresses reveal shared and unique regulatory modules. Transcriptomic studies show that over half of genes activated under cold-drought or cold-salinity treatments differ from individual stresses (Liu et al., 2024; Nakashima et al., 2014). ABA-dependent and ABA-independent pathways play central roles. The ABRE–AREB/ABF module mediates osmotic adjustment, while DREB/CBF transcription factors drive cold and dehydration responses (Roychoudhury et al., 2013). Genes with both ABRE and DRE/CRT elements integrate ABA and cold signaling. In alfalfa under cold and saline-alkali stress, flavonoid biosynthesis, regulated by *MsMYB12–MsFLS13*, determines stress resilience through ROS scavenging (Liu et al., 2024). In tobacco and tomato, 2-hydroxymelatonin induced under cold-drought stress provides antioxidant protection (Back and Lee, 2019). Priming mechanisms enhance combined cold tolerance, as heat or cold pre-exposure improves cross-tolerance via ROS, nitric oxide, and methylglyoxal signaling. Nanoparticle-mediated seed priming enhances maize tolerance to drought, cold, and salinity by activating mitogen-activated protein kinase (MAPK) and hormone signaling cascades (Chen et al., 2023).

2.2.4 Nutrient imbalance combined with other stresses

Soil nutrient imbalances, particularly phosphorus (P) and nitrogen (N) deficiencies, and aluminum (Al) toxicity in acidic soils, often co-occur with drought, chilling, salinity, or heat. Their interaction creates nonadditive impacts as impaired nutrient uptake affects root growth while concurrent stresses impose osmotic, ionic, or oxidative burdens. Drought × nutrient imbalance interactions are well documented. Phosphorus deficiency restricts ATP and nucleic acid synthesis, limiting root expansion and stomatal regulation. Combined with drought, impaired P acquisition worsens declines in water-use efficiency and photosynthetic rate (Khan et al., 2023). In wheat, combined P starvation and drought caused greater reductions in root nutrient content than either stress alone, though some genotypes showed compensatory root growth improving P uptake (Zhang et al., 2022). Similar effects occur in bamboo, maize, and rice, where P supplementation partially restored photosynthetic function under drought (Wu and Bose, 2024). Phosphate starvation responses (PSRs) interact with hormonal networks through transcription factors *PHR1*, MYB, and WRKY, along with *miR399* and PHT transporters, which are co-regulated by drought and salinity cues (Shukla et al., 2021). Field trials showed PSR activation precedes ABA responses under mild drought (Nagatoshi et al., 2023). Nitrogen deficiency compounds drought and heat stress. In potato, low N restricted carbon transport under heat-drought stress (Demirel et al., 2020). Low soil pH worsens N and P imbalances while increasing oxidative stress. In wheat, acidic soils triggered higher malondialdehyde accumulation, with cultivar-specific antioxidant responses (Toth et al., 2020). Aluminum toxicity in acid soils inhibits root elongation, reducing water and nutrient extraction capacity. When drought co-occurs, restricted root growth magnifies water deficit in legumes and cereals. Studies showed that combined Al × drought stress more strongly inhibited root growth, though some *Phaseolus coccineus* accessions showed that drought partially alleviated Al uptake (Butare et al., 2011; Yang et al., 2012).

Cold × nutrient stress interactions have been elucidated through transcriptomic studies showing Pi starvation with chilling triggers photosynthetic gene upregulation, altered Fe accumulation, and STOP1-ALMT1 activation in roots (Gao et al., 2022). In wheat, P application mitigated chilling effects by maintaining root absorption and spike development (Xu et al., 2022). These findings show the role of P nutrition in buffering oxidative stress under cold stress. Genetic loci such as *Pup1/PSTOL1* demonstrate nutrient efficiency genes' multistress resilience. *PSTOL1* introgression in rice and sorghum enhanced root growth and yield under low-P conditions while stabilizing drought performance (Hufnagel et al., 2014). In legumes, MATE transporters confer Al resistance, while ABA-ethylene crosstalk modulates Al × drought tolerance (Yang et al., 2012). The rhizosphere microbiome mediates combined nutrient and abiotic stresses. Arbuscular mycorrhizal fungi (AMF) and Pi-solubilizing bacteria enhance nutrient uptake and stress tolerance. AMF supply up to 70%–90% of plant P while conferring drought resistance (Hartman and Tringe, 2019). Microbes producing ACC-deaminase reduce ethylene under flooding × nutrient limitation, supporting root growth (Ali and Kim, 2018).

2.3 Geographical spread of crop species and adaptation to different environments

The domestication of crop plants approximately 9000–11,000 years ago signified the advent of agriculture, with multiple centers emerging across Asia, the Fertile Crescent, Africa, and the Americas (Gross and Olsen, 2010; Meyer et al., 2012). These centers were interconnected through migration and exchange, and local climates and soils influenced the domestication trajectories of key species. Asian rice was independently domesticated in the Yangtze and Ganges basins, with wild relatives (*Oryza rufipogon* and *O. nivara*) contributing alleles for drought, submergence, and nutrient stress tolerance (Meyer and Purugganan, 2013; Gross, 2014). Wheat (*Triticum* spp.) originated in the Fertile Crescent through the hybridization of diploid and tetraploid progenitors, while maize (*Zea mays* ssp. *mays*) was derived from teosinte in Mexico and disseminated throughout the Americas, adapting to environments ranging from the Andean highlands to northern prairies (Purugganan and Fuller, 2009; Doebley et al., 2006) (Fig. 2.1). Similar pathways influenced the domestication of barley, sorghum, and millets, with centers of origin across the Near East, Africa, and Asia (Gutaker and Purugganan, 2023). Domestication was driven by selection on developmental traits, often through loss-of-function or regulatory mutations (Fuller et al., 2023). Notable examples include the *tb1* Hopscotch transposon influencing maize architecture (Liu et al., 2016), *PROG1* and *qSH1* alleles affecting rice growth habit and shattering (Konishi et al., 2006), and the *Rc* 14-bp deletion resulting in white pericarps (Sweeney et al., 2006). Structural changes also played a role, as evidenced by maize *Sh1–5.1* rearrangements (Alam and Purugganan, 2024). Collectively, these examples illustrate how selection targeted a limited number of loci with significant phenotypic effects.

The geographical dispersal of crops followed human migration, trade, and cultural exchange. Yet, adaptation to new ecological zones was equally critical. Gutaker and Purugganan (2023) describe three major phases: (1) Early diffusion (11,000–4000 yBP): Rice, wheat, and barley spread locally beyond centers of domestication, often accompanied by wild relatives. (2) Ancient trade networks (4000–550 yBP): The Silk Road and Indian Ocean routes facilitated crop exchanges across Asia, Africa, and Europe. For example, japonica rice spread from China to Southeast Asia, where it adapted to tropical lowlands. (3) Post-Columbian exchanges (last 550 years): The Columbian Exchange dispersed maize, potato, cassava, and tomato globally, introducing unprecedented ecological challenges and opportunities. As crops dispersed, they encountered new abiotic stresses such as rice adapted to upland rainfed ecosystems in South Asia (drought-prone) and coastal deltas (salinity, submergence). Notably, *aus* rice landraces in Eastern India evolved under drought-prone, low-input systems, accumulating unique alleles for phosphorus-deficiency tolerance and anaerobic germination (Cortes, 2024; Sar et al., 2024). Wheat diversified into spring and winter types to cope with seasonal cold and drought. Maize is adapted to photoperiodic variation, highland cold, and nutrient-limited soils. Barley colonized environments from the Mediterranean drought to the Arctic cold. Sorghum and pearl millet diversified in the arid Sahel, with selection for early maturity and heat tolerance. This spatial diversification illustrates how abiotic stress factors acted as evolutionary filters, shaping landrace diversity and conferring resilience across ecologies (Govindaraj et al., 2015). The environmental heterogeneity, along with the varied cultural preferences of the farming communities, has resulted in today's crop diversity in the form of landraces, traditional cultivars, and farmers' varieties.

2.4 Genetic bottlenecks and erosion underscore the need for plant genetic resource conservation

Domestication, however, imposed a genetic bottleneck. Selection for yield, grain size, and harvestability reduced allelic richness, leading to genetic erosion (Fu, 2015). This narrowing is evident in crops such as potato during the Irish famine and maize in the USA during the Southern corn leaf blight epidemic, where genetic uniformity increased vulnerability (Govindaraj et al., 2015). Modern plant breeding, while enhancing yield, often exacerbates this erosion. For example, high-yielding rice and wheat varieties from the Green Revolution replaced diverse landraces, causing erosion in farmers' fields throughout Asia (Eliazer Nelson et al., 2019). However, different studies show that diversity has not been uniformly lost; introgression from landraces and wild relatives into breeding pools has, in some cases, expanded diversity (Kushanov et al., 2024).

The importance of genetic diversity for adaptation to abiotic stress is now unequivocal. It was opined that abiotic stress tolerance is a polygenic trait, underpinned by thousands of small-effect alleles whose value often becomes apparent only under harsh or fluctuating environments (Cortes, 2024). Standing variation in wild relatives and traditional cultivars thus provides a "hidden reservoir" of alleles for drought, salinity, and heat tolerance. Mining this diversity through GWAS, transcriptomics, and comparative genomics has identified genes such as *SUB1A* in rice for submergence tolerance, *PSTOL1* for phosphorus uptake, and *HKT1;5* in sorghum for salt exclusion. Therefore, domestication and subsequent crop spread not only narrowed but also redistributed adaptive diversity across continents (Gross and Olsen, 2010).

Recent advances in genomics, gene banks, and prebreeding are enabling the reintroduction of lost diversity. Global networks such as CGIAR and national gene banks conserve >7 million accessions worldwide, representing landraces, wild relatives, and underutilized crops (Kushanov et al., 2024). Efficient characterization of these resources through high-throughput sequencing, phenomics, and molecular markers is allowing breeders to identify and deploy resilience traits. Evaluation of accessions conserved in the International Rice Genebank located at IRRI, Philippines, as well as in national gene banks, has resulted in the identification of many potential donors for specific stress conditions (Torres et al., 2013; Kaur et al., 2024). The challenge lies in integrating this diversity into elite germplasm without yield penalties, a goal potentially achievable through genomic selection, pangenomics, and gene editing.

From an abiotic stress perspective, the spread of crops across climatic gradients provides lessons for modern resilience breeding. The adaptation of barley to both Mediterranean drought and Arctic cold, or maize to both tropical and temperate zones, exemplifies the plasticity embedded in crop genomes (Cortes, 2024; Fu, 2015). These historical adaptations show that crop evolution has always been a story of managing multiple stresses simultaneously, whether through natural selection, farmer-driven selection, or modern breeding. Therefore understanding domestication, evolution, and distribution of staple crops is more than a historical exercise; it is central to harnessing PGR for climate-resilient agriculture. Biodiversity conservation serves not only as a safeguard against genetic erosion but also as a proactive strategy to secure alleles for stress adaptation.

Comparative genomic and functional studies reveal that both conserved and lineage-specific mechanisms underlie crop adaptation to abiotic stresses. A striking example of conservation is the ABA–PP2C-*SnRK2* signaling cascade, a central regulator of drought, salinity, and cold tolerance. This module, in which ABA perception leads to inhibition of PP2C phosphatases and subsequent activation of *SnRK2* kinases, is deeply conserved across cereals, including rice, maize, wheat, and barley (Wang et al., 2025). The pathway regulates stomatal closure, osmolyte biosynthesis, and expression of stress-inducible transcription factors such as DREB and NAC, illustrating how a shared signaling backbone has been repeatedly harnessed in parallel domestication and breeding histories. In contrast, certain divergent adaptations evolved in specific lineages to cope with unique ecological niches. Deepwater rice developed the *SNORKEL1* and *SNORKEL2* genes, encoding ethylene-responsive transcription factors that promote rapid internode elongation during submergence. This contrasts with the *SUB1A*-mediated quiescence strategy in lowland rice, underscoring how different rice ecotypes evolved alternative genetic solutions to the same stress (Jiang et al., 2025). Such divergent modules illustrate the plasticity of plant genomes under selective pressures imposed by heterogeneous environments. Wild relatives also harbor unique adaptations that highlight the role of positive selection in shaping stress tolerance. Comparative transcriptome analyses of diploid wild cottons (*Gossypium* spp.) have revealed signatures of selection in genes involved in Ca^{2+} signaling and ROS detoxification pathways, which confer resilience under combined cold and salt stress (Xu et al., 2020). These findings demonstrate how natural selection in marginal environments enriches alleles that are absent in cultivated gene pools but highly valuable for modern breeding. Together, the interplay of conserved signaling modules and divergent evolutionary innovations contributes to crop resilience. Leveraging both dimensions is critical for designing crops capable of withstanding the MFSs associated with climate change.

Domestication, spread, and evolution of crops underscore one critical lesson: resilience is embedded in diversity. Historical adaptations demonstrate that crops have always faced combinations of stresses such as drought with heat, salinity with flooding, and cold with nutrient deficiencies. Landraces and wild relatives embody the genomic solutions forged over millennia (Dwivedi et al., 2016; Mittler et al., 2025). Modern breeding must now emulate this process with deliberate strategies: (1) Harness pangenomic diversity to capture structural variants lost during domestication, (2) employ comparative genomics and synteny to transfer stress-tolerance loci across species, (3) use genome editing tools to reintroduce alleles lost in domestication bottlenecks, and (4) design multistress phenotyping platforms to identify resilience not visible under single-stress conditions. By reconnecting modern crops with their evolutionary and domestication histories, breeders can build climate-smart ideotypes that withstand the MFSs.

2.5 Leveraging crop diversity and crop wild relatives for trait and gene discovery

The genetic variability necessary for plant improvement largely originates from CWRs. Harlan and de Wet's "gene pool" concept remains foundational, distinguishing the primary gene pool (cross-compatible wild or weedy conspecifics), secondary gene pool (species that require supportive techniques for hybridization), and tertiary gene pool (distant relatives requiring advanced technologies such as embryo rescue or chromosome engineering). Despite reproductive barriers, these pools together represent a vast reservoir of adaptive alleles that crops lost during domestication and intensive selection. CWRs have contributed repeatedly to modern agriculture. For example, resistance to grassy stunt virus in rice was sourced from *Oryza nivara*, rust resistance in wheat from *Aegilops tauschii*, and perennial traits and pest resistance in maize from *Zea diploperennis*. Such contributions highlight their role not only as "insurance" against emerging stresses but as a continual source of resilience traits that elite cultivars lack. Importantly, the diversity within CWRs often reflects long evolutionary histories of adaptation to marginal environments such as saline coastal zones, arid uplands, or cold highlands, making them especially relevant under climate change (Mammadov et al., 2018) (Table 2.1).

Direct use of wild relatives in breeding is often constrained by linkage drag and maladaptive traits. Prebreeding or germplasm enhancement acts as an intermediary step, transferring favorable alleles from exotic materials into intermediate lines that breeders can use more readily. This approach has been central to enriching modern germplasm pools (Dempewolf et al., 2017). Wild relatives also provide striking examples of stress-adaptive traits. For instance wild wheat relatives such as *Thinopyrum elongatum* exhibit enhanced salinity tolerance through mechanisms such as Na^+ sequestration and silica deposition in leaves. In rice, *Oryza coarctata* contributes alleles for salt exclusion, while aus and rayada ecotypes from South Asia harbor traits for drought, early vigor, and anaerobic germination (Basha et al., 2024; Sar et al., 2024). Similarly, in legumes, *Cicer reticulatum* and *C. echinospermum* are being exploited to broaden the genetic base of chickpea, especially for drought and cold resilience (Rani et al., 2019).

The world's gene banks hold more than 7 million accessions of crops and their relatives. Yet, fewer than 2% of these are actively used in breeding (Dwivedi et al., 2016). Limited use arises from inadequate characterization, redundancy in collections, and practical barriers to handling vast numbers of lines. To make exploration feasible, core collections and reference sets have been developed, capturing ~10% of genetic diversity while retaining most allelic richness. India's wheat core set of 22,000 accessions, for instance, has been instrumental in identifying terminal heat tolerance alleles, whereas rice diversity panels have revealed loci for P deficiency tolerance, anaerobic germination, and drought resilience.

Trait discovery hinges on accurate, repeatable, and scalable phenotyping. Traditional manual methods, although useful for initial characterization, are time-consuming, prone to bias, and inadequate for dissecting quantitative traits under field conditions. High-throughput phenotyping (HTP) has therefore become useful. Platforms such as Field Scanalyzer,

TABLE 2.1 Domestication histories, crop wild relatives, and key stress-tolerance traits in major cereals.

Crop	Domestication & progenitors	Priority stresses	Key genes/QTLs (mechanisms)	CWR donors/ taxa	Example outcomes
Wheat	Polyploid from *Triticum urartu*, *Aegilops speltoides/tauschii*	Salinity, drought, heat, and cold	*TmHKT1;5-A*, *Nax1/Nax2* (Na^+ exclusion); *Dro1* homolog (root depth); *P5CS* (proline); *TaNAC67* (TF for multistress)	*T. monococcum*, *Ae. tauschii*, and *Thinopyrum* spp.	Salt-tolerant durum with *TmHKT1;5-A*; editing targets in HKT/SOS/NHX modules
Maize	Domesticated from teosinte (*Zea mays* ssp. *parviglumis*), enriched by *Tripsacum*	Drought × heat, waterlogging, salinity, and cold	*ZmVPP1* (drought); *ZmNAC111*, *ZmNAC48/49/75* (NAC TFs); *AP2/ERF TFs*; *ZmbZIP68*, *ZmICE1* (cold)	*Z. nicaraguensis*, *Z. luxurians*, and *Tripsacum* spp.	Waterlogging tolerance via radial O_2 loss barrier; drought tolerance via NAC TFs
Rice	*Oryza rufipogon* to Asian rice	Submergence, salinity, drought, cold, and low-P	*SUB1A* (quiescence); *SK1/SK2* (escape); *Saltol/OsHKT1;5* (Na^+ regulation); *PSTOL1* (low-P); *OsTPP7* (anaerobic germination); *COLD1* (cold sensing)	*O. coarctata*, *O. glaberrima*, *O. australiensis*, and CC genome species	Mega-varieties with *SUB1A*; pyramiding *SUB1* + *Saltol* + drought QTLs
Sorghum	*S. bicolor* from *S. verticilliflorum*; GP-3 wild taxa	Post-flowering drought, heat, salinity, and cold	*Stg1–Stg4* QTLs (stay-green); *SbHSF* gene (heat); *SbSOS1* (Na^+ transport); salt gland traits	*S. macrospermum*, *S. leiocladum*, and *S. halepense*	Stay-green ideotypes; prebreeding at ICRISAT; salt exclusion via *SbSOS1*; heat-tolerant Sudanese landraces
Barley	*Hordeum spontaneum* to *H. vulgare*	Drought, salinity, cold	*HvHKT1;5*, *HvHKT2;1*, *HvNHX6* (ion balance); *HVA1* (LEA protein); CBF-COR module (cold acclimation); *Ppd-H1*, *HvCEN* (flowering time)	Wild *H. spontaneum* ecotypes	Salinity/cold-tolerant lines; flowering-time tuning for adaptation; LEA-based drought resilience

LemnaTec, and PhenoMobile integrate RGB, hyperspectral, LiDAR, and thermal sensors to capture canopy temperature, biomass, plant height, chlorophyll fluorescence, and water status across thousands of plots (Singh et al., 2021a). These aspects are discussed in the next section.

Advances in next-generation sequencing (NGS) have revolutionized the exploration of genetic variation in crop germplasm. Techniques, such as genotyping-by-sequencing (GBS), whole-genome resequencing (WGRS), and RAD-seq, produce dense SNP datasets that underpin GWAS and linkage mapping. GWAS, in particular, leverages natural diversity and historical recombination to resolve minor-effect loci that traditional biparental populations may miss. Numerous QTLs and genes for abiotic stress tolerance have been identified through these approaches. Complementary techniques, such as QTL-Seq and MutMap, have accelerated discovery in segregating populations. For example, QTL-Seq was used to rapidly identify major-effect loci for salt tolerance in chickpea, while MutMap pinpointed causal mutations in rice for anaerobic germination tolerance. Haplotype-based association mapping is further resolving allelic series at known stress-responsive loci, guiding haplotype-led breeding strategies (Vasupalli et al., 2024).

The identification of single genes is increasingly being complemented by network-level insights arising from omics studies (Raza et al., 2025). Examples include the DREB/CBF family for dehydration and cold, HSF-HSP cascades for heat, antioxidant enzymes for ROS detoxification, and ion transporters for salinity and Al toxicity tolerance. Notably, many of these networks are conserved across cereals and legumes, opening opportunities for translational breeding. Single-cell and tissue-specific omics are now revealing where stress responses originate. For instance, tapetal cell failure under drought × heat explains sterility in cereals, while root endodermal barriers govern ion balance under salinity × drought. Understanding these cell-type–specific vulnerabilities provides direct editing or selection targets (Wang et al., 2025).

2.5.1 Breeding for combined stress tolerance by utilizing plant genetic resources

Breeding for combined tolerance has long been constrained by the fact that selection under single stresses rarely ensures resilience under multifactorial field conditions. Studies in wheat and maize demonstrated that tolerance to drought alone or heat alone did not guarantee yield stability when both stresses occurred simultaneously, because stress combinations trigger unique physiological conflicts (Cairns et al., 2013; Mahrookashani et al., 2017; Meseka et al., 2018). Similar findings extend to legumes and rice, where genotypes resilient to one stress succumb under combinations. Nonetheless, donor germplasm has been identified across crops, including maize hybrids maintaining ear number under drought × heat, rice landraces tolerant to drought × salinity, and wheat lines sustaining kernel weight under drought × heat (Mahrookashani et al., 2017). A broader strategic shift is also underway: frameworks for "breeding future varieties" argue that MFSs must be considered at plant, plot, and landscape scales, and that selection targets should extend beyond the plant to the holobiont (host + microbiome) to stabilize performance under fluctuating environments (Salse et al., 2024).

The advent of HTP has transformed how breeders evaluate complex traits under simultaneous stresses. Standardized protocols, such as those for combined flooding × salinity screening in rice (Chakraborty et al., 2024), now allow systematic and repeatable assessment of genotypes under co-imposed constraints. HTP platforms integrate imaging, chlorophyll fluorescence, hyperspectral reflectance, and thermal sensing to capture stress signatures before visible symptoms appear (Singh et al., 2021a). These approaches address a critical bottleneck: the subjectivity and low throughput of manual scoring in field stress nurseries. Machine learning and deep learning algorithms further enable automated extraction of stress severity indices at leaf, canopy, and field scales, greatly enhancing precision and scalability (Raza et al., 2025; Singh et al., 2021b). For example, convolutional neural networks have achieved >90% accuracy in classifying abiotic stress signatures in soybean and maize datasets, offering a path toward robust, large-scale screening (Singh et al., 2021b).

Integrating phenomics with genomics has created powerful pipelines for trait dissection under combined stresses. Multiomics approaches, including panomics that integrate genomics, transcriptomics, metabolomics, proteomics, and phenomics, are unraveling regulatory hubs such as *DREB2A*, HSF-HSP modules, antioxidant genes, and ion transporters that underpin cross-stress tolerance (Raza et al., 2025). These insights are accelerating genomic selection models, where HTP-derived indices (e.g., canopy temperature depression, NDVI, and fluorescence efficiency) serve as predictors in genomic prediction frameworks. Such integration improves prediction accuracy for yield and resilience under complex environments, thereby increasing genetic gain per unit time. Coupled with advances in clustered regularly interspaced short palindromic repeats (CRISPR)-based editing of regulatory nodes (e.g., ERF-VIIs for flooding, CBF/DREB for cold, *PSTOL1* and SPX for nutrient deficiency), these pipelines could enable the stacking of multiple tolerance traits.

Participatory breeding and field phenotyping under managed stress environments remain crucial complements. Networks such as multienvironment trials of IRRI and living labs in farmer fields expose breeding lines to realistic combinations of stresses. These are increasingly supported by drone-based phenomics, sensor networks, and AI-driven stress modeling to bridge the gap between controlled and field conditions (Pandey and Senthil-Kumar, 2024; Salse et al., 2024).

Image and sensor platforms (RGB, thermal, hyperspectral, and LiDAR) now operate from leaf to canopy to field scales, whereas machine learning enables the ICQP (Identification, Classification, Quantification, and Prediction) cycle to screen large populations for co-stress traits (e.g., canopy temperature under drought × heat; fluorescence signatures under salinity × waterlogging) (Singh et al., 2016, 2021a). Field-mimic designs are equally essential: recent guidance urges the use of stress-interaction matrices to predefine stress pairs or triads, establish co-imposition timing and order, and prioritize realistic stress sequences observed in farmers' fields (Pandey and Senthil-Kumar, 2024). Recent recommendations also call for single-cell panomics and cell/tissue-specific phenotyping to resolve which tissues determine whole-plant failure under co-stresses, for example, tapetal cells in drought × heat sterility or endodermal barriers in salinity × drought (Raza et al., 2025). The "quantitative stress-sensing" concept further posits that plants integrate MFS through a dual system involving specific stress sensors alongside a general redox/ROS-based quantitative sensor. This provides a mechanistic rationale for why stress combinations reconfigure signaling load and why ROS homeostasis is a high-value breeding/editing target (Dietz and Vogelsang, 2024). Meanwhile, the incorporation of CWRs and landraces continues to enrich the gene pool with novel alleles conferring multistress tolerance. Panomics-driven identification of beneficial haplotypes is guiding introgression into elite backgrounds (Dwivedi et al., 2016; Raza et al., 2025; Shabbir et al., 2022).

Overall, to build climate-smart crops, future breeding must integrate: (1) Diverse germplasm utilization from landraces and wild relatives; (2) advanced genomic tools, including genomic selection and CRISPR editing of key regulatory networks; (3) AI-driven high-throughput phenotyping for noninvasive, repeated measurements across environments; (4) systems-level breeding frameworks that combine genetic, phenotypic, and modeling data; (5) stress priming and memory to harness epigenomic signatures from long-term exposure to MFS; and (6) model-assisted variety placement using climate and agroecological predictions to define 2050 stress niches and tailor breeding targets (Salse et al., 2024) (Fig. 2.2). By uniting PGR, phenomics, omics, and modeling, breeders can finally overcome the limitations of single-stress pipelines and develop varieties resilient to the MFS realities of modern agriculture.

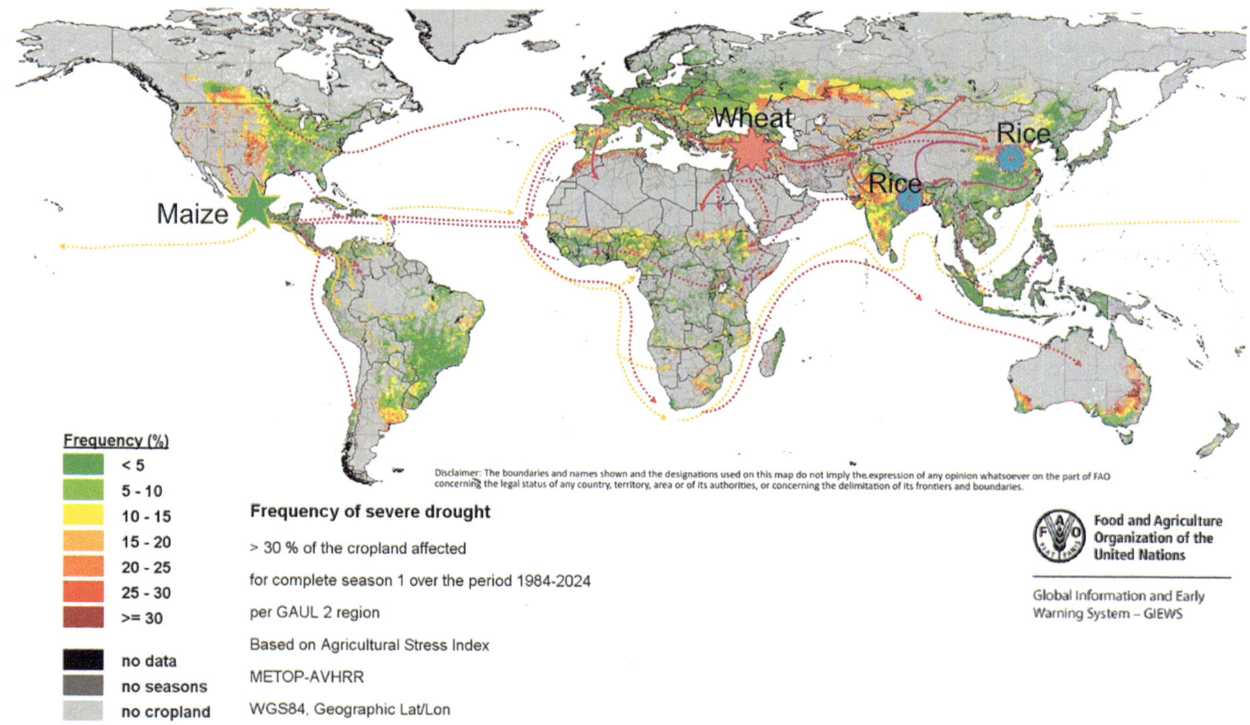

FIGURE 2.2 World map showing the approximate centers of domestication of three staple crops: rice (Yangtze-Ganges basins), wheat (Fertile Crescent), and maize (Mexico). Arrows (*red*, wheat; *purple*, rice; *yellow*, maize) indicate approximate dispersal routes in three epochs: during early crop diffusion, pre-Columbian movement, and post-Columbian exchange. Map lines delineate study areas and do not necessarily depict accepted national boundaries.

Adapted from Gutaker R.M., M.D. Purugganan, Adaptation and the geographic spread of crop species. Annual Review of Plant Biology. 75 (1) (2023), 679–706, Available from: http://www.annualreviews.org/journal/arplant, https://doi.org/10.1146/annurev-arplant-060223-030954. The world map showing the historic drought frequency was accessed from FAO Earth Observation (https://www.fao.org/giews/earthobservation/), and the layout was modified using Affinity Design software.

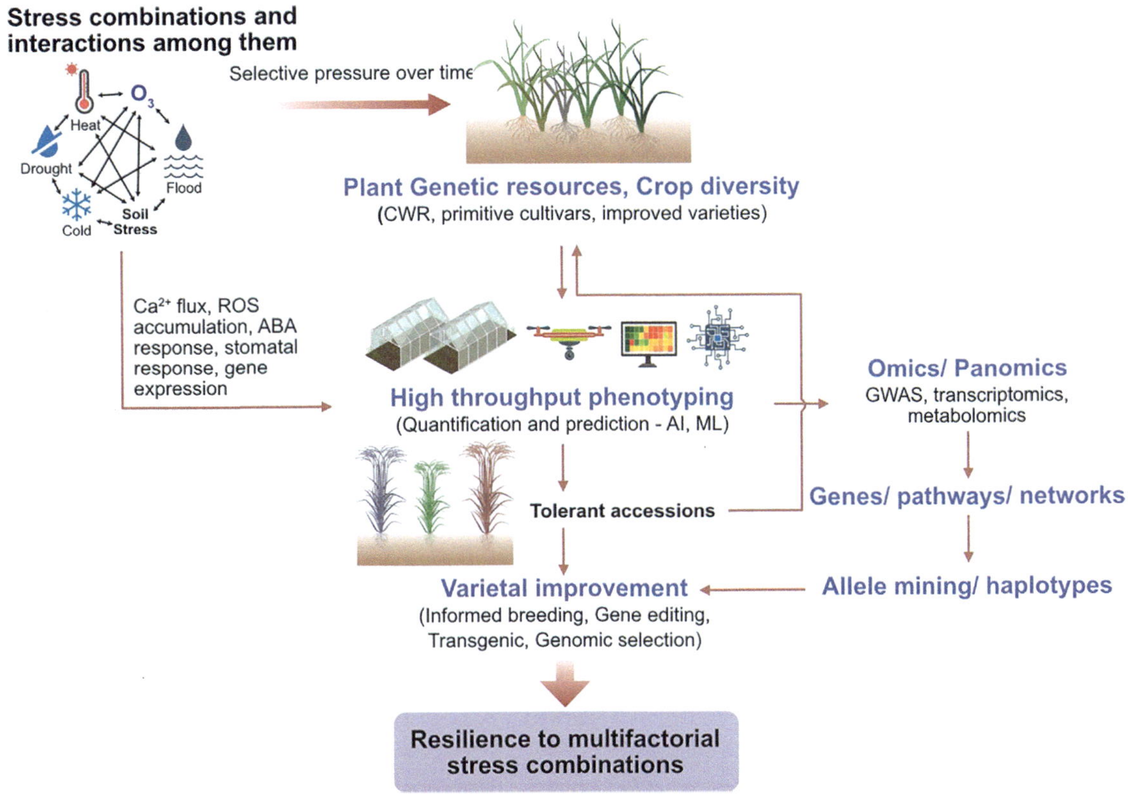

FIGURE 2.3 Breeding and phenomics pipeline for combined stress tolerance.

Conceptual schematic showing how MFSs imposed on crops drive the generation of adaptive traits and novel alleles in PGRs) and CWRs. These can be harnessed through phenomics platforms that mimic combined stress conditions. Integration of phenomics with panomics enables genomic selection, gene editing, and model-assisted variety placement, while agronomic practices further complement genetic strategies to develop high-yielding, climate-smart, multistress-resilient crops (Fig. 2.3).

2.6 Concluding remarks

Crop production in the coming decades will increasingly be shaped by MFS, which act in complex and unpredictable ways. Lessons from the past two decades show that tolerance to individual stresses seldom translates into resilience under stress combinations, highlighting the urgent need for research and breeding pipelines that explicitly target stress interactions. PGRs, including landraces, CWRs, and traditional cultivars, remain the bedrock of resilience, offering unique alleles and adaptive traits that modern varieties often lack. Their systematic exploration, coupled with high-throughput phenotyping and multiomics dissection, is already revealing combination-specific molecular signatures and cross-stress regulatory hubs. Future progress depends on integration rather than isolation: integrating genetic diversity with phenomics and modeling; integrating genomic innovations such as CRISPR with participatory breeding; and integrating crop improvement with complementary agronomy, soil health, and microbiome management. Importantly, strategies must look beyond yield protection to resilience across heterogeneous field environments, guided by climate forecasts and agroecological models. In conclusion, sustainable management of MFS requires not only advanced technologies but also responsible stewardship of biodiversity. By conserving, characterizing, and effectively deploying PGR, while embracing systems-level innovations, agriculture can move toward climate-smart crop varieties that thrive not in the idealized single-stress environments, but in the unpredictable, stress-prone landscapes where food security will be decided.

Glossary of abbreviations

MFS Multifactorial Stresses

NAC	NAM, ATAF, and CUC (a plant-specific transcription factor family)
WRKY	WRKY transcription factors (named after the conserved WRKYGQK domain)
DREB	Dehydration-Responsive Element Binding proteins
ROS	Reactive Oxygen Species
PGR	Plant Genetic Resources
CWR	Crop Wild Relatives
ABA	Abscisic Acid
PP2C	Protein Phosphatase 2C
ICE1	Inducer of CBF Expression 1
CBF	C-Repeat Binding Factor (also known as *DREB1*)
COR	Cold-Responsive GENES/Proteins
SnRK2	*SNF1-Related Protein Kinase 2*
ASI	Anthesis–Silking Interval
HSF	Heat-Shock Factor
HSP	Heat-Shock Protein
TF	Transcription Factor
UPR	Unfolded Protein Response
GB	Glycine Betaine
RFO	Raffinose Family Oligosaccharides
HKT	High-Affinity K^+ Transporter
SOS	Salt Overly Sensitive (SOS1=Na^+/H^+ antiporter pathway)
NHX	Na^+/H^+ Exchanger (vacuolar)
WUE	Water-Use Efficiency
ERF	Ethylene Response Factor
SOD	Superoxide Dismutase
CAT	Catalase
APX	Ascorbate Peroxidase
PSII	Photosystem II
ABRE	ABA-Responsive Element
AREB/ABF	ABRE-Binding Factor (ABA-responsive bZIP TFs)
MAPK	Mitogen-Activated Protein Kinase
PSR	Phosphate Starvation Response
PHR1	*Phosphate Starvation Response 1* (transcription factor)
PHT	Phosphate Transporter (high-affinity *PHT1* family)
FAOCE	Free-Air Ozone and CO_2 Enrichment
PSTOL1	*Phosphorus Starvation Tolerance 1* gene (in rice, also known as Pup1 QTL)
Pup1	*Phosphorus uptake 1* locus (QTL associated with PSTOL1)
AMF	Arbuscular Mycorrhizal Fungi
ACC	1-Aminocyclopropane-1-Carboxylic Acid (ethylene precursor)
GWAS	Genome-Wide Association Study
CGIAR	Consultative Group on International Agricultural Research
IRRI	International Rice Research Institute
HTP	High-Throughput Phenotyping
NGS	Next-Generation Sequencing
GBS	Genotyping-by-Sequencing
WGRS	Whole-Genome Resequencing
AI	Artificial Intelligence
ML	Machine Learning
CRISPR	Clustered Regularly Interspaced Short Palindromic Repeats (genome editing system)

References

Akpinar, B.A., Lucas, S.J., Budak, H., 2013. Genomics approaches for crop improvement against abiotic stress. Scientific World Journal 2013, 361921.

Alam, O., Purugganan, M.D., 2024. Domestication and the evolution of crops: variable syndromes, complex genetic architectures, and ecological entanglements. The Plant Cell 36 (5), 1227–1241. Available at https://doi.org/10.1093/plcell/koae013.

Ali, S., Kim, W.C., 2018. Plant growth promotion under water: decrease of waterlogging-induced ACC and ethylene levels by ACC deaminase-producing bacteria. Frontiers in Microbiology 9 (May). Available at https://doi.org/10.3389/fmicb.2018.01096.

Back, K., Lee, H.-J., 2019. 2-Hydroxymelatonin confers tolerance against combined cold and drought stress in tobacco, tomato, and cucumber as a potent anti-stress compound in the evolution of land plants. Melatonin Research 2, 35–46.

Basha, F.T.M., Sar, P., Bhowmick, P.K., Mahato, A., Bisht, D.S., Iquebal, M.A., et al., 2024. Genome-wide association study identified QTLs and genes underlying early seedling vigour in aus rice (Oryza sativa L. Molecular Genetics and Genomics 299 (1). Available at https://doi.org/10.1007/s00438-024-02204-8.

Booker, F.L., Fiscus, E.L., Miller, J.E., 2004. Combined effects of elevated atmospheric carbon dioxide and ozone on soybean whole-plant water use. Environmental Management 33.

Butare, L., Rao, I., Lepoivre, P., Polania, J., Cajiao, C., Cuasquer, J., et al., 2011. New genetic sources of resistance in the genus phaseolus to individual and combined aluminium toxicity and progressive soil drying stresses. Euphytica 181 (3), 385–404. Available at https://doi.org/10.1007/s10681-011-0468-0.

Cairns, J.E., Crossa, J., Zaidi, P.H., Grudloyma, P., Sanchez, C., Luis Araus, J., et al., 2013. Identification of drought, heat, and combined drought and heat tolerant donors in maize. Crop Science 53 (4), 1335–1346. Available at https://doi.org/10.2135/cropsci2012.09.0545.

Chakraborty, K., Mondal, S., Tripathy, S., Jena, P., Bose, L.K., Chattopadhyay, K., et al., 2024. Energy conservation or expense? A possible dilemma under combined stresses of salinity and submergence in rice. Journal of Experimental Botany. Available at https://doi.org/10.1093/jxb/erae499.

Chen, S., Liu, H., Yangzong, Z., Gardea-Torresdey, J.L., White, J.C., Zhao, L., 2023. Seed priming with reactive oxygen species-generating nanoparticles enhanced maize tolerance to multiple abiotic stresses. Environmental Science and Technology 57 (48), 19932–19941. Available at https://doi.org/10.1021/acs.est.3c07339.

Chi, H.Y., Ou, S.L., Wang, M.C., Yang, C.Y., 2023. Physiological responses and ethylene-response AP2/ERF factor expression in Indica rice seedlings subjected to submergence and osmotic stress. BMC Plant Biology 23 (1). Available at https://doi.org/10.1186/s12870-023-04380-y.

Chinnusamy, V., Schumaker, K., Zhu, J.K., 2004. Molecular genetic perspectives on cross-talk and specificity in abiotic stress signalling in plants. Journal of Experimental Botany 55, 225–236.

Cortes, A.J., 2024. Abiotic stress tolerance boosted by genetic diversity in plants. International Journal of Molecular Science 25.

Costa, M.V.J.D., Ramegowda, Y., Ramegowda, V., Karaba, N.N., Sreeman, S.M., Udayakumar, M., 2021. Combined drought and heat stress in rice: responses, phenotyping and strategies to improve tolerance. Rice Science 28 (3), 233–242. Available at https://doi.org/10.1016/j.rsci.2021.04.003.

Demirel, U., Morris, W.L., Ducreux, L.J.M., Yavuz, C., Asim, A., Tindas, I., et al., 2020. Physiological, biochemical, and transcriptional responses to single and combined abiotic stress in stress-tolerant and stress-sensitive potato genotypes. Frontiers in Plant Science 11. Available at https://doi.org/10.3389/fpls.2020.00169.

Dempewolf, H., Baute, G., Anderson, J., Kilian, B., Smith, C., Guarino, L., 2017. Past and future use of wild relatives in crop breeding. Crop Science 57 (3), 1070–1082. Available at https://doi.org/10.2135/cropsci2016.10.0885.

Dietz, K.J., Vogelsang, L., 2024. A general concept of quantitative abiotic stress sensing. Trends in Plant Science 29 (3), 319–328. Available at https://doi.org/10.1016/j.tplants.2023.07.006.

Doebley, J.F., Gaut, B.S., Smith, B.D., 2006. The molecular genetics of crop domestication. Cell 127, 1309–1321.

Dwivedi, S.L., Ceccarelli, S., Blair, M.W., Upadhyaya, H.D., Are, A.K., Ortiz, R., 2016. Landrace germplasm for improving yield and abiotic stress adaptation. Trends in Plant Science 21 (1), 31–42. Available at https://doi.org/10.1016/j.tplants.2015.10.012.

Ejaz, I., Pu, X., Naseer, M.A., Bohoussou, Y.N.', Liu, Y., Farooq, M., et al., 2023. Cold and drought stresses in wheat: a global meta-analysis of 21st century. Journal of Plant Growth Regulation 42 (9), 5379–5395. Available at https://doi.org/10.1007/s00344-023-10960-x.

Eliazer Nelson, A.R.L., Ravichandran, K., Antony, U., 2019. The impact of the green revolution on indigenous crops of India. Journal of Ethnic Foods 6.

Farfan-Vignolo, E.R., Asard, H., 2012. Effect of elevated CO2 and temperature on the oxidative stress response to drought in Lolium perenne L. and Medicago sativa L. Plant Physiology and Biochemistry 59, 55–62. Available at https://doi.org/10.1016/j.plaphy.2012.06.014.

Fu, Y.B., 2015. Understanding crop genetic diversity under modern plant breeding. Theoretical and Applied Genetics 128 (11), 2131–2142. Available at https://doi.org/10.1007/s00122-015-2585-y.

Fukao, T., Yeung, E., Bailey-Serres, J., 2011. The submergence tolerance regulator SUB1A mediates crosstalk between submergence and drought tolerance in rice. The Plant Cell 23, 412–427.

Fuller, D.Q., Denham, T., Allaby, R., 2023. Plant domestication and agricultural ecologies. Current Biology: CB 33, R636–R649.

Gao, J.P., Chao, D.Y., Lin, H.X., 2007. Understanding abiotic stress tolerance mechanisms: recent studies on stress response in rice. Journal of Integrative Plant Biology 49, 742–750.

Gao, X., Dong, J., Rasouli, F., Pouya, A.K., Tahir, A.T., Kang, J., 2022. Transcriptome analysis provides new insights into plants responses under phosphate starvation in association with chilling stress. BMC Plant Biology 22 (1). Available at https://doi.org/10.1186/s12870-021-03381-z.

Garg, R., Verma, M., Agrawal, S., Shankar, R., Majee, M., Jain, M., 2014. Deep transcriptome sequencing of wild halophyte rice, porteresia coarctata, provides novel insights into the salinity and submergence tolerance factors. DNA Research 21 (1), 69–84. Available at https://doi.org/10.1093/dnares/dst042.

Govindaraj, M., Vetriventhan, M., Srinivasan, M., 2015. Importance of genetic diversity assessment in crop plants and its recent advances: an overview of its analytical perspectives. Genetics Research International 2015, 431487.

Gross, B.L.Z.Z., 2014. Archaeological and genetic insights into the origins of domesticated rice. Proceedings of the National Academy of Sciences 111, 6190–6197.

Gross, B.L., Olsen, K.M., 2010. Genetic perspectives on crop domestication. Trends in Plant Science 15 (9), 529–537. Available at https://doi.org/10.1016/j.tplants.2010.05.008.

Gutaker, R.M., Purugganan, M.D., 2023. Adaptation and the geographic spread of crop species. Annual Review of Plant Biology 75 (1), 679–706. Available at https://doi.org/10.1146/annurev-arplant-060223-030954.

Hartman, K., Tringe, S.G., 2019. Interactions between plants and soil shaping the root microbiome under abiotic stress. Biochemical Journal 476 (19), 2705–2724. Available at https://doi.org/10.1042/BCJ20180615.

Hufnagel, B., de Sousa, S.M., Assis, L., Guimaraes, C.T., Leiser, W., Azevedo, G.C., et al., 2014. Duplicate and conquer: Multiple homologs of phosphorus-starvation tolerance1 enhance phosphorus acquisition and sorghum performance on low-phosphorus soils. Plant Physiology 166 (2), 659–677. Available at https://doi.org/10.1104/pp.114.243949.

Hussain, H.A., Hussain, S., Khaliq, A., Ashraf, U., Anjum, S.A., Men, S., et al., 2018. Chilling and drought stresses in crop plants: implications, cross talk, and potential management opportunities. Frontiers in Plant Science 9. Available at https://doi.org/10.3389/fpls.2018.00393.

Jagadish, S.V.K., Septiningsih, E.M., Kohli, A., Thomson, M.J., Ye, C., Redoña, E., et al., 2012. Genetic advances in adapting rice to a rapidly changing climate. Journal of Agronomy and Crop Science 198 (5), 360–373. Available at https://doi.org/10.1111/j.1439-037X.2012.00525.x.

Jiang, Z., van Zanten, M., Sasidharan, R., 2025. Mechanisms of plant acclimation to multiple abiotic stresses. Communications Biology 8, 655.

Jordine, A., Retzlaff, J., Gens, L., Ehrt, B., Fürtauer, L., van Dongen, J.T., 2024. Introducing the halophyte Salicornia europaea to investigate combined impact of salt and tidal submergence conditions. Functional Plant Biology 51 (3). Available at https://doi.org/10.1071/FP23228.

Kaur, V., Payasi, D.K., Choudhary, R., Kumar, V., Yadav, S.K., Singh, D., et al., 2024. Genetic resources for enhancing drought tolerance from National Genebank collection of linseed (Linum usitatissimum L.) in India. Plant and Soil 510, 923–946.

Khan, F., Siddique, A.B., Shabala, S., Zhou, M., Zhao, C., 2023. Phosphorus plays key roles in regulating plants' physiological responses to abiotic stresses. Plants 12 (15). Available at https://doi.org/10.3390/plants12152861.

Khan, N.A., Owens, L., Nuñez, M.A., Khan, A.L., 2025. Complexity of combined abiotic stresses to crop plants. Plant Stress 17. Available at https://doi.org/10.1016/j.stress.2025.100926.

Khoury, C.K., Bjorkman, A.D., Dempewolf, H., Ramirez-Villegas, J., Guarino, L., Jarvis, A., et al., 2014. Increasing homogeneity in global food supplies and the implications for food security. Proceedings of the National Academy of Sciences of the United States of America 111 (11), 4001–4006. Available at https://doi.org/10.1073/pnas.1313490111.

Konishi, S., Izawa, T., Lin, S.Y., Ebana, K., Fukuta, Y., Sasaki, T., et al., 2006. An SNP caused loss of seed shattering during rice domestication. Science (New York, N.Y.) 312 (5778), 1392–1396. Available at https://doi.org/10.1126/science.1126410.

Kushanov, F.N., He, S., Patel, J., 2024. Editorial: Plant biodiversity and genetic resources: their utilization for crop improvement. Front Genetics 15, 1513972.

Lei, L., Gordon, S.P., Liu, L., Sade, N., Lovell, J.T., Rubio Wilhelmi, M.D.M., et al., 2023. The reference genome and abiotic stress responses of the model perennial grass Brachypodium sylvaticum. G3 (Bethesda) 14.

Liu, Z., Cook, J., Melia-Hancock, S., Guill, K., Bottoms, C., Garcia, A., et al., 2016. Expanding maize genetic resources with predomestication alleles: maize–teosinte introgression populations. Plant Genome 9 (1). Available at https://doi.org/10.3835/plantgenome2015.07.0053.

Liu, L., Si, L., Zhang, L., Guo, R., Wang, R., Dong, H., et al., 2024. Metabolomics and transcriptomics analysis revealed the response mechanism of alfalfa to combined cold and saline-alkali stress. The Plant Journal 119 (4), 1900–1919. Available at https://doi.org/10.1111/tpj.16896.

Mahmood, U., Hussain, S., Hussain, S., Ali, B., Ashraf, U., Zamir, S., et al., 2021. Morpho-physio-biochemical and molecular responses of maize hybrids to salinity and waterlogging during stress and recovery phase. Plants 10 (7). Available at https://doi.org/10.3390/plants10071345.

Mahrookashani, A., Siebert, S., Hüging, H., Ewert, F., 2017. Independent and combined effects of high temperature and drought stress around anthesis on wheat. Journal of Agronomy and Crop Science 203 (6), 453–463. Available at https://doi.org/10.1111/jac.12218.

Mammadov, J., Buyyarapu, R., Guttikonda, S.K., Parliament, K., Abdurakhmonov, I.Y., Kumpatla, S.P., 2018. Wild relatives of maize, rice, cotton, and soybean: treasure troves for tolerance to biotic and abiotic stresses. Frontiers in Plant Science 9. Available at https://doi.org/10.3389/fpls.2018.00886.

Martin, D.G., Aspray, E.K., Li, S., Leakey, A.D.B., Ainsworth, E.A., 2025. Drought does not mitigate reductions in soybean photosynthesis and yield caused by elevated ozone. Plant Physiology 199 (1). Available at https://doi.org/10.1093/plphys/kiaf350.

Meseka, S., Menkir, A., Bossey, B., Mengesha, W., 2018. Performance assessment of drought tolerant maize hybrids under combined drought and heat stress. Agronomy 8 (12), 274. Available at https://doi.org/10.3390/agronomy8120274.

Meyer, R.S., DuVal, A.E., Jensen, H.R., 2012. Patterns and processes in crop domestication: an historical review and quantitative analysis of 203 global food crops. The New Phytologist 196, 29–48.

Meyer, R.S., Purugganan, M.D., 2013. Evolution of crop species: genetics of domestication and diversification. Nature Reviews. Genetics 14 (12), 840–852. Available at https://doi.org/10.1038/nrg3605.

Mittler, R., 2006. Abiotic stress, the field environment and stress combination. Trends in Plant Science 11 (1), 15–19. Available at https://doi.org/10.1016/j.tplants.2005.11.002.

Mittler, R., Karlova, R., Bassham, D.C., Lawson, T., 2025. Crops under stress: can we mitigate the impacts of climate change on agriculture and launch the 'Resilience Revolution'? Philosophical Transactions of the Royal Society B: Biological Sciences 380 (1927). Available at https://doi.org/10.1098/rstb.2024.0228.

Nagatoshi, Y., Ikazaki, K., Kobayashi, Y., Mizuno, N., Sugita, R., Takebayashi, Y., et al., 2023. Phosphate starvation response precedes abscisic acid response under progressive mild drought in plants. Nature Communications 14 (1). Available at https://doi.org/10.1038/s41467-023-40773-1.

Nakashima, K., Yamaguchi-Shinozaki, K., Shinozaki, K., 2014. The transcriptional regulatory network in the drought response and its crosstalk in abiotic stress responses including drought, cold, and heat. Frontiers in Plant Science 5, 170.

Pandey, P., Senthil-Kumar, M., 2024. Unmasking complexities of combined stresses for creating climate-smart crops. Trends in Plant Science 29 (11), 1172–1175. Available at https://doi.org/10.1016/j.tplants.2024.07.005.

Pradhan, B., Chakraborty, K., Prusty, N., Deepa, Mukherjee, A.K., Chattopadhyay, K., et al., 2019. Distinction and characterisation of rice genotypes tolerant to combined stresses of salinity and partial submergence, proved by a high-resolution chlorophyll fluorescence imaging system. Functional Plant Biology 46 (3), 248–261. Available at https://doi.org/10.1071/FP18157.

Priya, M., Farooq, M., Siddique, K.H.M., 2025. Enhancing tolerance to combined heat and drought stress in cool-season grain legumes: mechanisms, genetic insights, and future directions. Plant, Cell & Environment.

Purugganan, M.D., Fuller, D.Q., 2009. The nature of selection during plant domestication. Nature 457 (7231), 843–848. Available at https://doi.org/10.1038/nature07895.

Rani, A., Devi, P., Jha, U.C., Sharma, K.D., Siddique, K.H.M., Nayyar, H., 2019. Developing climate-resilient chickpea involving physiological and molecular approaches with a focus on temperature and drought stresses. Frontiers in Plant Science 10, 1759.

Rasmussen, S., Barah, P., Suarez-Rodriguez, M.C., Bressendorff, S., Friis, P., Costantino, P., et al., 2013. Transcriptome responses to combinations of stresses in Arabidopsis. Plant Physiology 161 (4), 1783–1794. Available at https://doi.org/10.1104/pp.112.210773.

Raza, A., Li, Y., Prakash, C.S., Hu, Z., 2025. Panomics to manage combined abiotic stresses in plants. Trends in Plant Science 30 (10), 1079–1084. Available at https://doi.org/10.1016/j.tplants.2025.03.001.

Roychoudhury, A., Paul, S., Basu, S., 2013. Cross-talk between abscisic acid-dependent and abscisic acid-independent pathways during abiotic stress. Plant Cell Reports 32, 985–1006.

Salse, J., Barnard, R.L., Veneault-Fourrey, C., Rouached, H., 2024. Strategies for breeding crops for future environments. Trends in Plant Science 29 (3), 303–318. Available at https://doi.org/10.1016/j.tplants.2023.08.007.

Sar, P., Gupta, S., Behera, M., Chakraborty, K., Ngangkham, U., Verma, B.C., et al., 2024. Exploring genetic diversity within aus rice germplasm: insights into the variations in agro-morphological traits. Rice 17 (1). Available at https://doi.org/10.1186/s12284-024-00700-4.

Shabbir, R., Singhal, R.K., Mishra, U.N., Chauhan, J., Javed, T., Hussain, S., et al., 2022. Combined abiotic stresses: challenges and potential for crop improvement. Agronomy 12 (11). Available at https://doi.org/10.3390/agronomy12112795.

Shukla, D., Waigel, S., Rouchka, E.C., Sandhu, G., Trivedi, P.K., Sahi, S.V., 2021. Genome-wide expression analysis reveals contrasting regulation of phosphate starvation response (PSR) in root and shoot of Arabidopsis and its association with biotic stress. Environmental and Experimental Botany 188. Available at https://doi.org/10.1016/j.envexpbot.2021.104483.

Singh, A., Ganapathysubramanian, B., Singh, A.K., Sarkar, S., 2016. Machine learning for high-throughput stress phenotyping in plants. Trends in Plant Science 21 (2), 110–124. Available at https://doi.org/10.1016/j.tplants.2015.10.015.

Singh, A., Jones, S., Ganapathysubramanian, B., Sarkar, S., Mueller, D., Sandhu, K., et al., 2021a. Challenges and opportunities in machine-augmented plant stress phenotyping. Trends in Plant Science 26, 53–69.

Singh, R., Mukherjee, J., Sehgal, V.K., Krishnan, P., Das, D.K., Dhakar, R.K., et al., 2021b. Interactive effect of elevated tropospheric ozone and carbon dioxide on radiation utilisation, growth and yield of chickpea (Cicer arietinum L. International Journal of Biometeorology 65 (11), 1939–1952. Available at https://doi.org/10.1007/s00484-021-02150-9.

Sun, C., Gao, X., Fu, J., Zhou, J., Wu, X., 2014. Metabolic response of maize (Zea mays L.) plants to combined drought and salt stress. Plant and Soil 388, 99–117.

Suzuki, N., Rivero, R.M., Shulaev, V., Blumwald, E., Mittler, R., 2014. Abiotic and biotic stress combinations. New Phytologist 203 (1), 32–43. Available at https://doi.org/10.1111/nph.12797.

Sweeney, M.T., Thomson, M.J., Pfeil, B.E., McCouch, S., 2006. Caught red-handed: Rc encodes a basic helix-loop-helix protein conditioning red pericarp in rice. The Plant Cell 18 (2), 283–294. Available at https://doi.org/10.1105/tpc.105.038430.

Swigonska, S., Amarowicz, R., Król, A., Mostek, A., Badowiec, A., Weidner, S., 2014. Influence of abiotic stress during soybean germination followed by recovery on the phenolic compounds of radicles and their antioxidant capacity. Acta Societatis Botanicorum Poloniae 83 (3), 209–218. Available at https://doi.org/10.5586/asbp.2014.026.

Tamang, B.G., Fukao, T., 2015. Plant adaptation to multiple stresses during submergence and following desubmergence. International Journal of Molecular Sciences 16 (12), 30164–30180. Available at https://doi.org/10.3390/ijms161226226.

Torres, R.O., McNally, K.L., Cruz, C.V., Serraj, R., Henry, A., 2013. Screening of rice Genebank germplasm for yield and selection of new drought tolerance donors. Field Crops Research 147, 12–22. Available at https://doi.org/10.1016/j.fcr.2013.03.016.

Toth, B., Juhasz, C., Labuschagne, M., Moloi, M.J., 2020. The influence of soil acidity on the physiological responses of two bread wheat cultivars. Plants (Basel) 9.

Vasupalli, N., Bhat, J.A., Jain, P., Sri, T., Islam, M.A., Shivaraj, S.M., et al., 2024. Omics big data for crop improvement: opportunities and challenges. Crop Journal 12 (6), 1517–1532. Available at https://doi.org/10.1016/j.cj.2024.10.007.

Vaughan, D.A., Balazs, E., Heslop-Harrison, J.S., 2007. From crop domestication to super-domestication. Annals of Botany 100, 893–901.

Wang, T., Wang, F., Deng, S., Wang, K., Feng, D., Xu, F., et al., 2025. Single-cell transcriptomes reveal spatiotemporal heat stress response in maize roots. Nature Communications 16 (1). Available at https://doi.org/10.1038/s41467-024-55485-3.

Wu, H., Bose, J., 2024. Abiotic stress tolerance: adaptations, mechanisms, and new techniques. The Crop Journal 12 (5), 1271–1273. Available at https://doi.org/10.1016/j.cj.2024.10.001.

Xu, Y., Magwanga, R.O., Jin, D., Cai, X., Hou, Y., Juyun, Z., et al., 2020. Comparative transcriptome analysis reveals evolutionary divergence and shared network of cold and salt stress response in diploid D-genome cotton. BMC Plant Biology 20 (1). Available at https://doi.org/10.1186/s12870-020-02726-4.

Xu, H., Hassan, M.A., Sun, D., Wu, Z., Jiang, G., Liu, B., et al., 2022. Effects of low temperature stress on source–sink organs in wheat and phosphorus mitigation strategies. Frontiers in Plant Science 13. Available at https://doi.org/10.3389/fpls.2022.807844.

Yang, Z.B., Eticha, D., Albacete, A., Rao, I.M., Roitsch, T., Horst, W.J., 2012. Physiological and molecular analysis of the interaction between aluminium toxicity and drought stress in common bean (Phaseolus vulgaris). Journal of Experimental Botany 63 (8), 3109–3125. Available at https://doi.org/10.1093/jxb/ers038.

Zandalinas, S.I., Mittler, R., Balfagón, D., Arbona, V., Gómez-Cadenas, A., 2018. Plant adaptations to the combination of drought and high temperatures. Physiologia Plantarum 162 (1), 2–12. Available at https://doi.org/10.1111/ppl.12540.

Zandalinas, S.I., Mittler, R., 2022. Plant responses to multifactorial stress combination. New Phytologist 234 (4), 1161–1167. Available at https://doi.org/10.1111/nph.18087.

Zhang, X., Li, C., Lu, W., Wang, X., Ma, B., Fu, K., et al., 2022. Comparative analysis of combined phosphorus and drought stress-responses in two winter wheat. PeerJ 10, e13887. Available at https://doi.org/10.7717/peerj.13887.

Zhou, R., Yu, X., Ottosen, C.O., Rosenqvist, E., Zhao, L., Wang, Y., et al., 2017. Drought stress had a predominant effect over heat stress on three tomato cultivars subjected to combined stress. BMC Plant Biology 17 (1). Available at https://doi.org/10.1186/s12870-017-0974-x.

Zhou, R., Yu, X., Wen, J., Jensen, N.B., Dos Santos, T.M., Wu, Z., et al., 2020. Interactive effects of elevated CO_2 concentration and combined heat and drought stress on tomato photosynthesis. BMC Plant Biology 20 (1). Available at https://doi.org/10.1186/s12870-020-02457-6.

Section B

Physiological and molecular bases of multiple abiotic stress tolerance in plants

Chapter 3

Morphological, biochemical, and anatomical adaptations of plants to multiple abiotic stresses

Debabrata Panda* and Suraj K. Padhi

Department of Biodiversity and Conservation of Natural Resources, Central University of Odisha, Koraput, Odisha, India

* *Corresponding author. e-mail address: dpanda@cuo.ac.in*

3.1 Introduction

Climate change, primarily caused by human activity, poses numerous issues that greatly affect agricultural systems globally (Salam et al., 2022). As the global population is projected to surge to nearly 10 billion by 2050, the demand for food production continues to escalate, necessitating sustainable practices that also consider environmental conservation [Food and Agriculture Organization of the United Nations (FAO), 2017]. Global food demand is projected to rise by 35%–56% from 2010 to 2050, highlighting the critical necessity for enhanced agricultural methods to satisfy increasing food needs (World Bank Group, 2024). Simultaneously, plants are more exposed to several abiotic challenges, including drought, waterlogging, salinity, temperature extremes, and heavy metal toxicity, as a result of changing climatic circumstances. These stresses present substantial risks to crop viability and agricultural output, heightening concerns regarding food security amid shifting climatic conditions and increasing populations (Godfray et al., 2010).

Abiotic stresses are environmental conditions that adversely impact plant growth and metabolism, resulting in physiological disruptions that may impede overall plant development. These stresses can be classified into various categories, each posing distinct obstacles to plant growth. Drought is defined as inadequate water availability, leading to physiological drought, whereas salinity pertains to elevated salt concentrations in soil or water that hinder plant performance. Extreme temperatures, whether elevated or diminished, can cause cellular damage, whereas heavy metals such as cadmium and lead introduce further toxicity concerns that jeopardize plant health (Munns and Tester, 2008). Comprehending these pressures is essential for formulating adaptive solutions in agricultural techniques. Abiotic stresses significantly affect plants, profoundly impacting their growth, development, and overall production. Adverse environmental conditions can impede critical physiological activities, including photosynthesis, nutrient absorption, and shoot and root growth, resulting in diminished biomass and production (Shinozaki and Yamaguchi-Shinozaki, 2007; Flexas et al., 2013). These stresses have an impact that goes beyond a single plant; in certain situations, drought and salinity stress are thought to reduce agricultural production by as much as 50%. Consequently, such decreases compromise food security, especially in regions that are already food-insecure [Intergovernmental Panel on Climate Change (IPCC), 2014].

As global agricultural systems encounter heightened strain from abiotic pressures, the correlation between environmental difficulties and diminished agricultural output becomes increasingly evident. The declining crop yield, intensified by unfavorable climatic conditions, underscores the necessity of mitigating these effects to secure a steady food supply for a burgeoning population (Farooq et al., 2011). Consequently, a comprehensive understanding of the impact of various pressures on agricultural systems is crucial for modifying methods that sustain or improve productivity. Comprehending plant reactions to abiotic stresses is essential for ensuring food availability amid increasing worldwide population issues. As agricultural systems increasingly face vulnerability due to climate change, understanding plant responses is essential to guide breeding programs and agronomic techniques that enhance resilience (Ashraf and Harris, 2013). This chapter aims to explore the morphological, anatomical, biochemical, and molecular adaptations of plants that

Multiple Abiotic Stresses in Plants. DOI: https://doi.org/10.1016/B978-0-443-27424-4.00003-3

enable survival and growth under multiple abiotic stresses. A comprehensive understanding of plant stress responses is crucial, as it directly relates to the capacity to maximize agricultural productivity while ensuring the health of ecosystems, which is increasingly threatened by climate change (Zhang et al., 2022).

3.2 Different types of abiotic stresses

Environmental stresses can be classified as either biotic or abiotic. Abiotic stresses primarily include drought, salinity, severe temperatures, flooding, oxidative stress, nutritional shortages, and heavy metal stress (He et al., 2018) (Fig. 3.1).

3.2.1 Drought stress

Drought stress presents a considerable challenge to most food crops, which are particularly susceptible to soil moisture deficiencies. The degree of its impact on agricultural productivity and physiological and biochemical processes is contingent upon the severity, timing, and length of the drought (Rane et al., 2021). Severe drought circumstances negatively impact crop morphology, physiological processes, and growth duration, but soil moisture is essential during germination due to its role in enzyme activation, which affects the plant sensitivity at this stage.

3.2.2 High and low temperature

Extreme temperatures are a primary factor in numerous abiotic stresses, including drought. Temperature deviations, whether excessive or inadequate, adversely affect plant growth, development, and productivity. Increased temperatures can expedite the crop life cycle, thereby diminishing the production potential of cereals (Djanaguiraman et al., 2020). Heat stress alters the balance of cells by changing the stability of proteins, membranes, RNA molecules, and cytoskeletal structures. It also changes how well enzymes work inside the cell (Hasanuzzaman et al., 2013; Bita and Gerats, 2013). Cold stress arises when plants encounter exceptionally low temperatures, constituting a critical abiotic factor that reduces crop output by affecting both quality and postharvest longevity. It damagess plants by changing the structure of membranes, slowing down protoplasmic streaming, causing electrolyte leakage, and starting plasmolysis, which destroys

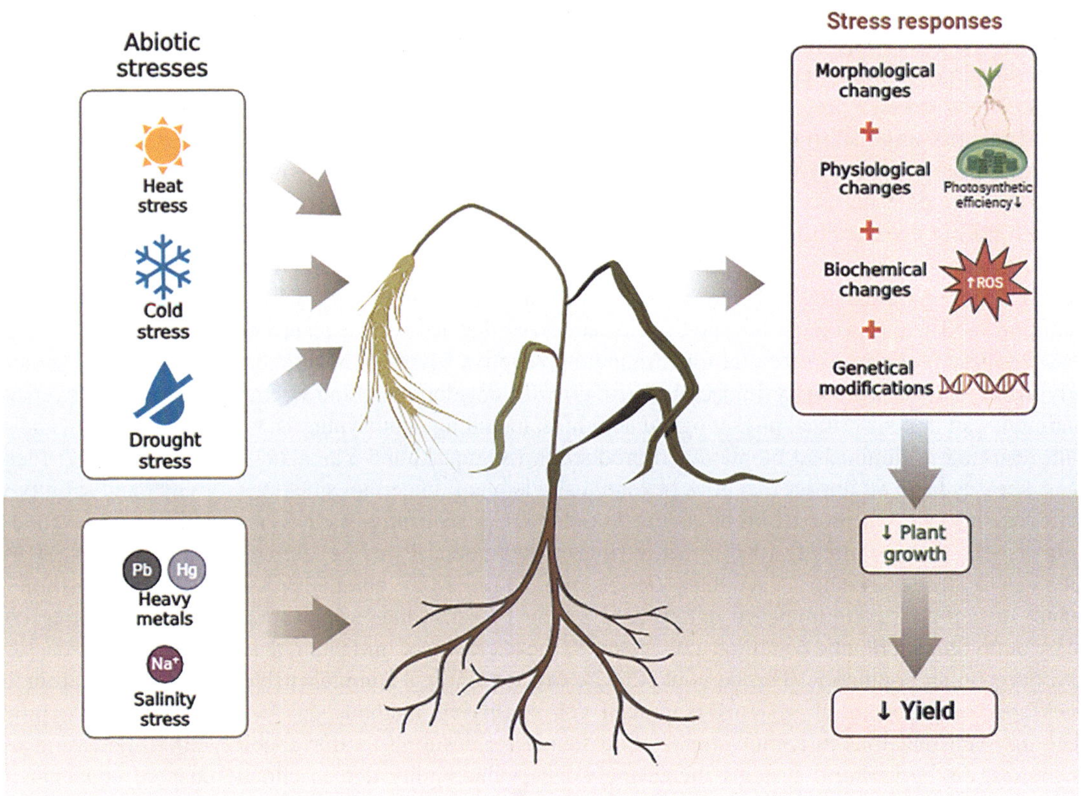

FIGURE 3.1 Different types of abiotic stresses and their possible consequences in plants.

cells. This damage, along with impaired metabolism, leads to stunted growth, irregular fruit ripening, internal discoloration (vascular browning), increased susceptibility to decay, and, in extreme instances, plant mortality (Devasirvatham and Tan, 2018).

3.2.3 Salinity stress

Salinity is a major abiotic stresses that limits crop growth and productivity because it causes too many soluble salts to build up in the soil. This stops plant growth in many irrigated, arid, and semiarid areas around the world (Zaki and Radwan, 2019). Factors such as crop species, cultivar, growth stage, ecological conditions, and the specific composition of soil salts influence the extent of salt damage. Plants' nutritional balance is thrown off by salt stress, which changes ionic ratios such as Na^+/K^+, Na^+/Mg^{2+}, Na^+/Ca^{2+}, $Cl^-/H_2PO_4^-$, and Cl^-/NO^{3-}. These changes impair the functions of cells and metabolic processes (Fahad and Bano, 2012). Salinity stress modifies plant morphology, physiology, and molecular responses. Morphological effects encompass diminished biomass, height, and germination, with severity escalating in accordance with salt concentration and differing by species (Gong et al., 2006; Turan et al., 2009; Li et al., 2017; Munns and Tester, 2008; Soltabayeva et al., 2021). Photosynthesis diminishes under saline stress, evidenced by alterations in stomatal conductance, chlorophyll fluorescence, and chlorophyll concentration, whereas osmotic adjustments transpire through variations in turgor pressure, osmotic pressure, relative water content, and water potential (Munns and Tester, 2008). Salinity disrupts redox homeostasis, leading to the accumulation of reactive oxygen species due to impaired electron transport and diminished water potential (Acosta-Motos et al., 2017). Transcriptomic and proteomic analyses reveal salt-induced reprogramming at the molecular level, with differentially expressed genes acting as potential molecular markers of salinity stress (Gupta and Huang, 2014).

3.2.4 Flooding stress

Waterlogging and submergence markedly diminish yields in food grain crops. Dryland cereals, including maize, wheat, and barley, exhibit significant vulnerability to waterlogging, resulting in yield reductions of up to 20% in irrigated regions and exceeding 40% in rainfed systems (Herzog et al., 2016). In severe instances, waterlogging stress may result in total yield loss, with the degree of damage influenced by crop type, duration of waterlogging, and growth stage. Floods are categorized as either waterlogging, when the water level is shallow and impacts only the root zone, or submergence, when the entire aerial portion of the plant is inundated (Fukao et al., 2019). Flushing water into plants' cells lowers their energy levels (Mustroph, 2018). This makes the soil hypoxic or anoxic, which reduces oxygen availability for root respiration, inhibiting root growth. Morphological and physiological adaptations, such as adventitious root formation and modified shoot growth, are triggered by flooding stress. While total submersion eventually inhibits growth and lowers survival, waterlogging increases root anaerobic respiration, which raises energy consumption and limits growth. However, in certain species, partial submersion encourages internode or petiole elongation to enhance oxygen uptake. Additionally, flooding causes reactive oxygen species (ROS) and malondialdehyde (MDA) to build up (Jia et al., 2019). Plants counteract this by synthesizing proline and glutathione and by activating antioxidant enzymes (Jia et al., 2021). According to Voesenek and Bailey-Serres (2013), oxygen deprivation decreases aerobic respiration, speeds up glycolysis, and, in the case of total submersion, further reduces photosynthesis because of CO_2 and light constraints. Several genes that respond to flooding have been found at the molecular level. These genes include those that control the formation of lysigenous aerenchyma, which improves oxygen diffusion in plants such as wheat, maize, and *Arabidopsis* (Jia et al., 2021).

3.2.5 Metal stress

Heavy metal stress refers to a category of nonbiodegradable inorganic substances defined by an atomic mass greater than 20 and a density surpassing $5\ g\ cm^{-3}$. These metals have toxic effects on cellular and genetic systems, leading to mutagenic impacts on crops through the contamination of irrigated water, soil, drinking water, food chains, and the surrounding environment (Flora et al., 2008; Wuana and Okieimen, 2011). Heavy metals can disrupt a plant's metabolism and normal functions when they are present in large amounts. They do this in a number of ways, including changing the structure of proteins by bonding with sulfhydryl groups (Hall, 2002) and disturbing the functional groups of important cellular molecules (Hossain et al., 2015).

3.2.6 Combined/multiple stresses

Plants experience a range of environmental stresses, such as salinity, drought, cold, and extreme temperatures. The intensities and durations of these stresses may vary, occurring either concurrently or in succession. Drought and heat stress represent two separate abiotic stresses that often coexist in field conditions, and their combined effects result in considerably greater damage to crop growth and productivity than their individual impacts (Mittler, 2006). Some combinations of stress may provide advantages to plants compared to the application of each stress in isolation. Elevated CO_2 levels can alleviate the negative impacts of other stresses, including salinity and high light intensity (Pérez-López et al., 2013). Plants subjected to combined or sequential stresses elicit unique signaling, metabolic, and physiological responses in contrast to individual stresses, although overlapping pathways may also be present (Pareek et al., 2020; Shaar-Moshe et al., 2017). Oxidative stress is a primary outcome, influencing sugar metabolism, growth, and defense (Zhang and Sonnewald, 2017). When plants are exposed to both drought and heat stress, their stomata close, which lowers the water content, membrane stability, photosynthesis, and yield in crops such as potatoes, wheat, and maize (Król, 2013; Hussain et al., 2019; Sattar et al., 2020). This is because stored assimilates increase respiration, which only helps plants get used to the stress (Kadam et al., 2015). Coastal flooding combines salinity and submergence, which causes Na^+ and Cl^- to build up, K^+ to be lost, and ion transport to be limited. Halophytes, on the other hand, adapt by keeping pigments, stabilizing chloroplasts, forming aerenchyma, and using antioxidant defenses (Zheng et al., 2009a, 2009b; Tamang and Fukao, 2015). Cold and salinity separately disrupt the photosynthetic electron transport chain in crops such as sunflower, bean, and maize, yet their synergistic effects are not extensively researched (Liu et al., 2016). Drought and salinity also stop photosynthesis by closing the stomata (Zhang and Sonnewald, 2017). Drought and heat stresses further augment membrane fluidity and permeability, diminishing protein functionality and photosynthesis, while exacerbating photorespiration, resulting in either additive or multiplicative effects (Xu and Zhou, 2005, 2006; Loka et al., 2011; Loka et al., 2020). At the cellular level, combinations of stress stop the flow of electrons, which makes too many reactive oxygen species (ROS) such as $O_2^{\cdot -}$ and H_2O_2, which cause oxidative stress. Plants react by increasing the activity of antioxidant enzymes and scavenging reactive oxygen species (ROS), which is often stronger when two stresses are combined than when each stress is present on its own (Hasanuzzaman et al., 2020).

3.3 Adaptive mechanisms against abiotic stresses

Plants often face conditions that are inadequate for growth and development. Through evolution, specific plants have developed specialized characteristics that allow them to flourish in severe environments, such as deserts, tundra, and swamps. To withstand these adverse conditions, plants have developed morphological and physiological adaptations, as well as signaling pathways that trigger biochemical and molecular mechanisms that enable their survival under diverse stresses (Zhu, 2016) (Fig. 3.2).

3.3.1 Morphophysiological adaptations

Morphophysiological adaptations in plants encompass diverse structural and physiological adjustments that enable their survival and optimal functioning in varying environmental situations. These adaptations are classified into three categories: alterations in root systems, variations in shoots and leaves, and physiological methods.

3.3.1.1 Root system alterations

Modifications in the root system architecture (RSA) are crucial adaptations that allow plants to firmly attach themselves and effectively absorb water and nutrients from the soil. RSA assesses crop responses to abiotic stresses via characteristics including adventitious roots, lateral roots, root hairs, root porosity, and the overall depth and branching of roots. Recent research indicates that specific rice genotypes exhibit compensatory root growth and altered root patterns in response to environmental stresses such as root pruning and water scarcity (Panda et al., 2021). Furthermore, research indicates that particular genetic alterations in rice can facilitate root elongation and lateral root development in drought situations, thereby improving water uptake (Lucob-Agustin et al., 2020). In potatoes, shallow roots restrict drought tolerance and nutrient absorption, whereas basal and stolon roots serve specific roles in water acquisition and tuberization (Porter et al., 1999; Wishart et al., 2013). In sweet potato, adventitious and lateral roots are essential for the formation of storage roots; however, drought, elevated soil temperature, or excessive nitrogen and water inhibit lateral root development and the initiation of storage roots, redirecting biomass to shoots (Villordon et al., 2012; Pardales and Yamauchi, 2003). Stress-specific effects on RSA include deeper roots enhancing water and nitrogen capture during drought (Lynch and Wojciechowski, 2015), shallow and highly branched roots favoring phosphorus uptake under deficiency (Lynch and Brown, 2001), lateral root repression via ABA–auxin signaling (Deak and Malamy, 2005), and aerenchyma formation improving oxygen diffusion under waterlogging (Jia et al., 2021). These examples show how root traits help crops adapt, but they also show how different stresses and environments can have different effects.

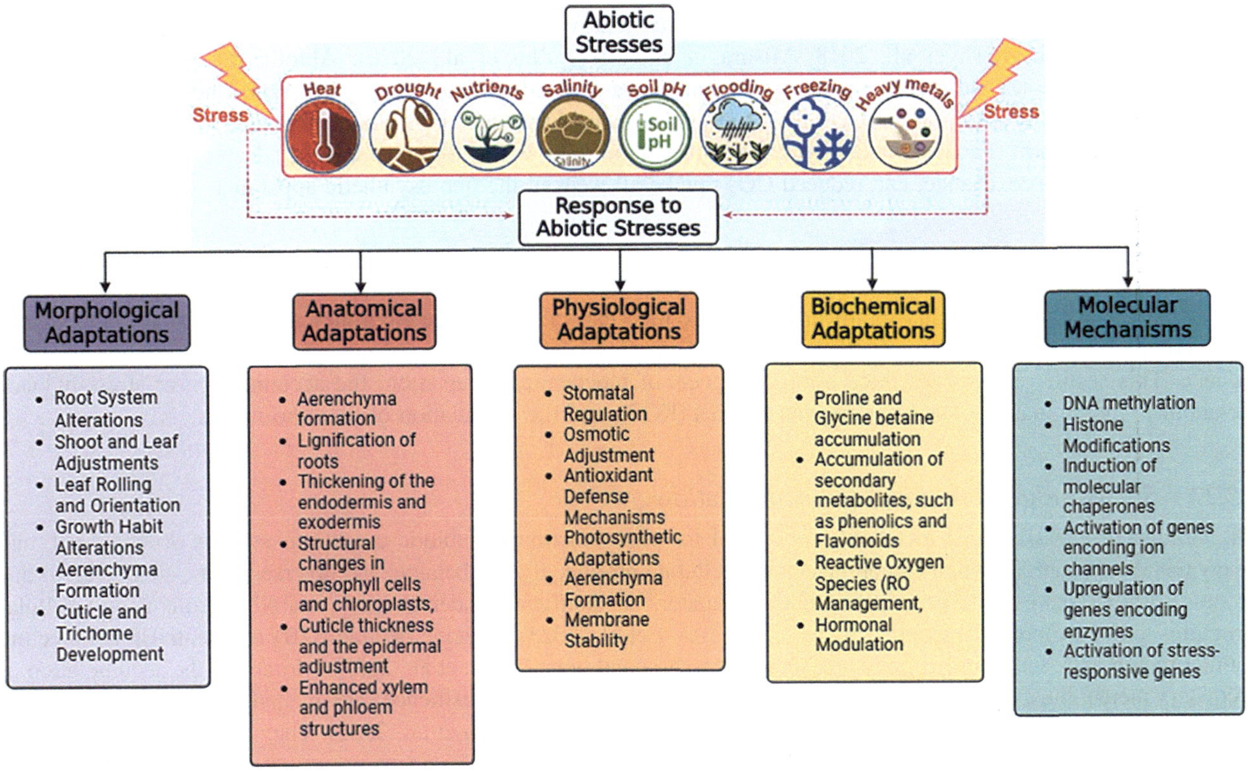

FIGURE 3.2 Morphological, anatomical, physiological, biochemical, and molecular adaptations in plants to different types of abiotic stresses.

3.3.1.2 Shoot and leaf adjustments

Modifications in shoots and leaves are crucial for augmenting photosynthetic efficiency and optimizing water-use efficiency in plants. To enhance light absorption while reducing water loss via transpiration, plants frequently modify their leaf morphology, including adjustments to leaf angle and size. Research indicates that Brachiaria grasses display morpho-anatomical alterations in leaf structure that enhance their tolerance to waterlogging (Lucob-Agustin et al., 2021). The activation of particular genes associated with root and shoot development is essential for adjusting to fluctuations in soil moisture and allowing plants to maximize growth during drought conditions (Suralta et al., 2012). Later research shows that endophyte associations can make plants more resistant to drought by improving the morphophysiological responses of shoots and roots (Lucob-Agustin et al., 2021). In crops, such as populus, soybeans, and many others, water deficit inhibits the leaf growth and expnasion (Zhang et al., 2004; Wullschleger et al., 2005; Farooq et al., 2009). Rubber, cocoa, coconut, and woody plants have been found to exhibit species-specific variation in stomatal behavior, which is frequently used to assess drought tolerance. This behavior is influenced by both internal signals (ABA and leaf water status) and external factors (light, humidity, soil water, and nutrient status) (Willson, 1999; Kozlowski and Pallardy, 1997; Sangsing et al., 2004; Gomes et al., 2008). Additionally, heat stress results in increased cuticle and mesophyll thickness (Natarajan and Kuehny, 2008), as well as leaf scorching, sunburn, senescence, abscission, shoot inhibition, mesophyll damage, and membrane instability (Ismail and Hall, 1999; Vollenweider and Gunthardt-Goerg, 2005; Zhang et al., 2005). By changing mesophyll cells, stomatal frequency, and epidermal cell dimensions, heavy metals also alter leaf anatomy (Tripathi et al., 2012a, 2012b; Vaculík et al., 2012).

3.3.1.3 Physiological mechanisms

Abiotic stress adversely affects numerous physiological processes in plants, eliciting adaptive responses that endure unfavorable conditions. Enhancing crop yields under abiotic stress necessitates optimizing physiological factors and processes before commencing breeding programs (Panda et al., 2021). Lack of water has many negative effects on rice's physiological traits, including less net photosynthesis, transpiration rates, stomatal conductance, water-use efficiency, internal CO_2 concentrations, photosystem II (PSII) activity, relative water content, and membrane stability index (Farooq et al., 2009a, 2009b; Dash et al., 2018; Mishra et al., 2018; Zhu et al., 2020). Abiotic stress impairs standard photosynthetic processes and modifies plant gas exchange characteristics (Zhu et al., 2020). When there is stress, closing the stomata reduces the absorption of CO_2. This causes too many electrons to be made, which makes it easier for reactive oxygen species to form (Farooq et al., 2009a, 2009b; Mishra et al., 2018). Stomatal closure, decreased turgor pressure, compromised leaf gas exchange, and reduced CO_2 uptake all weaken the photosynthetic apparatus.

3.3.2 Biochemical adaptations

Biochemical adaptations in plants significantly enhance their capacity to endure environmental stress. These adaptations entail complex biochemical processes that assist plants in alleviating stress-related damage while preserving metabolic balance. This section emphasizes three primary aspects of biochemical adaptation: the accumulation of stress-induced metabolites, the regulation of reactive oxygen species (ROS), and the modulation of plant hormones.

3.3.2.1 Stress-induced metabolite accumulation

The buildup of stress-induced metabolites is crucial for plant reactions to abiotic conditions such as drought, salt, and heavy metal exposure. Essential osmoprotectants, including proline, glycine betaine, and diverse sugars, are crucial to this process. Proline serves as a principal osmotic regulator, accumulating under stress to stabilize proteins and cellular structures and therefore enhancing stress resilience (Raza et al., 2023). Glycine betaine (GB) is essential for protecting cellular membranes and stabilizing proteins under stressful conditions (Singh et al., 2022). Moreover, the accumulation of secondary metabolites, including phenolics and flavonoids, improves stress resilience. These chemicals exhibit antioxidant capabilities that mitigate oxidative stress by neutralizing free radicals (Chen et al., 2022). They also engage in signaling pathways that trigger defense responses to both biotic and abiotic stresses.

3.3.2.2 Reactive oxygen species management

Reactive oxygen species (ROS) are metabolic by-products that can be detrimental in excessive concentrations, resulting in oxidative stress. Efficient control of reactive oxygen species is essential for preserving cellular equilibrium under stressful circumstances. Plants use both enzymatic and nonenzymatic processes to regulate ROS levels, which is essential for their survival under stress. Augmenting the expression of antioxidant systems can promote tolerance to abiotic stress, providing a way to mitigate oxidative damage and improve stress resistance in plants. Antioxidant enzymes, including superoxide dismutase (SOD), catalase (CAT), and peroxidase (POD), play a crucial role in neutralizing ROS (Mishra and Panda, 2017). Superoxide dismutase (SOD) catalyzes the conversion of superoxide radicals into hydrogen peroxide, which CAT subsequently breaks down into water and oxygen. Studies show that these enzymes' activity goes up a lot during abiotic stress. This shows how important they are for protecting plant cells from oxidative damage and keeping the redox balance (Panda and Sarkar, 2013).

Nonenzymatic antioxidants, such as ascorbate, tocopherol, and glutathione, are crucial in mitigating oxidative stress. Ascorbate (vitamin-C) serves as an effective ROS scavenger and is essential for the regeneration of other antioxidants; tocopherols (vitamin-E) safeguard cell membranes against lipid peroxidation. To make plants more resistant to stress, glutathione, a main nonenzymatic antioxidant, takes part in redox reactions and detoxification processes (Mishra et al., 2023). The synchronized activity of these antioxidants guarantees efficient management of reactive oxygen species and aids plant survival in stressful environments.

3.3.2.3 Hormonal modulation

Hormonal regulation is a vital component of biochemical adaptations, with phytohormones serving critical functions in plant responses to environmental stresses. Abscisic acid, ethylene, jasmonic acid, and salicylic acid are essential hormones that govern stress tolerance mechanisms. Abscisic acid initiates stress responses by regulating stomata closure, retaining water, and altering the expression of stress-related genes (Sah et al., 2016). Ethylene affects multiple plant stress responses, including senescence and fruit ripening, while also improving resistance to stresses such as salt and flooding (Riyazuddin et al., 2020). Jasmonic acid is crucial in defense signaling and the stress response because it modulates the expression of stress-related genes and enhances the production of secondary metabolites. Conversely, salicylic acid facilitates systemic acquired resistance to biotic stress and improves tolerance to abiotic stress (Khan et al., 2015).

3.3.3 Anatomical adaptations

Plants have developed a range of anatomical adaptations to withstand abiotic stresses such as drought, salinity, and extreme temperatures. These structural modifications, observed in roots, leaves, shoots, and vascular systems, are essential for sustaining physiological functions and ensuring survival in challenging environments.

3.3.3.1 Root anatomy modifications

Root adaptations are essential for enhancing a plant's ability to obtain water and nutrients in challenging environmental conditions. These adaptations encompass the formation of aerenchyma tissue, lignification, and the thickening of the endodermis and exodermis.

3.3.3.1.1 Aerenchyma formation and lignification

Aerenchyma is a specialized tissue distinguished by vast air-filled cavities that enhance gas exchange and confer buoyancy in oxygen-poor or submerged environments (Karlova et al., 2021). In soggy or flooded conditions, plants frequently develop aerenchyma to facilitate internal aeration, hence sustaining root respiration and functionality despite restricted oxygen levels. Furthermore, lignification of root tissues improves structural integrity, offering resistance against mechanical stress and pathogen infiltration. The buildup of lignin fortifies the root structure and aids in water regulation by acting as a barrier against excessive water loss (Karlova et al., 2021). By creating a water-resistant barrier around mature xylem tissue, lignin deposition is commonly reported to improve drought tolerance (Xu et al., 2017; Liu et al., 2018; Sharma et al., 2020). One important adaptation to flooding in rice (*Oryza sativa*) is the development of aerenchyma in the roots, which permits effective internal oxygen diffusion in hypoxic environments (Nishiuchi et al., 2012). Similarly, under conditions of nutrient stress and waterlogging, maize (*Zea mays*) produces constitutive and inducible aerenchyma, which lowers the metabolic costs of soil exploration (Zhu et al., 2010; Burton et al., 2013). Additionally, under hypoxia, barley (*Hordeum vulgare*) and wheat (*Triticum aestivum*) exhibit inducible cortical aerenchyma, which improves survival in wet soils (Herzog et al., 2016). Another important stress adaptation has been identified as lignification: in maize, increased lignification of vascular bundles improves mechanical strength and resistance to lodging under stress (Ishfaq et al., 2025), whereas in wheat, drought causes lignin deposition in root xylem, improving water-use efficiency (Moura et al., 2010).

3.3.3.1.2 Thickening of the endodermis and exodermis

The endodermis and exodermis are essential in regulating the passage of water and nutrients into the vascular tissues. Under abiotic conditions such as dryness, these layers frequently thicken due to suberin deposition (Líška et al., 2016). This structural alteration diminishes water permeability, aiding in moisture retention and enhancing the plant's ability to endure arid soil conditions.

3.3.3.2 Leaf and shoot modifications

Modifications in leaves and shoots are crucial for improving photosynthetic efficiency and minimizing water loss in challenging environmental circumstances. These adaptations entail alterations in mesophyll cell and chloroplast architecture, including adjustments to cuticle thickness and epidermal characteristics. Furthermore, in order to reduce transpirational water loss, plants often reduce stomatal size or density during drought (Zhao et al., 2015). Additionally, both glandular and nonglandular leaf trichomes play a role by reflecting excess light, lowering leaf temperature, and preventing water loss, features that have been functionally connected to tomato drought tolerance (Galdon-Armero et al., 2018). Salt bladders, which are specialized epidermal structures in halophytes, serve as external storage sites for Na^+ and Cl^-, preventing ion toxicity in mesophyll tissues and promoting water retention and osmotic balance (Flowers and Colmer, 2008).

3.3.3.2.1 Structural adaptations in mesophyll cells and chloroplasts

Mesophyll cells are essential for photosynthesis, and their structural adaptations are critical for enhancing light absorption and CO_2 assimilation under stressful circumstances. In times of drought or high light exposure, mesophyll cells may increase in density or change form, improving light penetration and gas exchange efficiency. Chloroplasts can similarly undergo structural modifications to optimize light absorption and photosynthetic efficiency despite restricted water availability.

3.3.3.2.2 Cuticle and epidermal modifications

The cuticle serves as a protective barrier that reduces water loss and defends against environmental stresses. In reaction to drought, numerous plants augment their cuticles, thereby lowering transpiration and preserving water. Furthermore, epidermal cells may adjust by augmenting their density or by generating waxy layers, which further mitigate water loss and enhance resistance to desiccation (Munns and Millar, 2023). Cuticular waxes consist of complex mixtures of very-long-chain fatty acids (VLCFAs) embedded in a cutin polyester matrix of C16-C18 hydroxy acids. These VLCFAs include alkanes, aldehydes, primary alcohols, ketones, and wax esters (Yeats and Rose, 2013). Many crops, including wheat, maize, and Arabidopsis, show both an increase in total wax accumulation and a shift towards longer-chain alkanes (C29–C33) under drought stress. These alkanes form hydrophobic barriers and effectively reduce cuticular transpiration (Kosma et al., 2009; Lee, Suh, 2015). When drought and salinity treatments are applied to wheat at the same time, several wax components, such as alkanes, alcohols, fatty acids, aldehydes, and esters, are elevated, improving the barrier's integrity (Tian et al., 2024). Improved drought tolerance is closely associated with these compositional changes, particularly increased alkanes and alcohols, which probably occur by making the cuticle more hydrophobic and reducing nonstomatal water loss.

3.3.3.3 Vascular system adaptations

The vascular system is crucial for the transportation of water and nutrients, and its adaptations are vital for improving stress resilience. Under abiotic stress circumstances, plants often modify the architecture of their xylem and phloem. To enhance water transport efficiency, the xylem may evolve larger vessel widths or a greater quantity of vessels. In woody species, increased lignin deposition typically fortifies the xylem, improving its structural integrity to endure physical stress. Likewise, the phloem experiences structural modifications that facilitate sucrose transfer to regions of active growth, guaranteeing a consistent energy supply in challenging environmental conditions (Cornelis and Hazak, 2022).

3.3.4 Genetic and molecular mechanisms

Understanding the genetic and molecular pathways that govern abiotic stress tolerance in plants is crucial for the creation of stress-resilient crop types. The rising problems of global climate change to agriculture have intensified the demand for crops that can withstand environmental pressures such as drought, salt, and severe temperatures (Raj and Nadarajah, 2022). The genetic basis of stress tolerance encompasses the identification of essential genes, advancements in molecular breeding methods, and the impact of epigenetic alterations, collectively forming a crucial foundation for enhancing agricultural resilience and innovation.

3.3.4.1 Genetic basis of abiotic stress tolerance

3.3.4.1.1 Identification of key genes and quantitative trait loci

A comprehensive study has identified several critical genes and quantitative trait loci (QTLs) associated with abiotic stress tolerance in plants. An examination of 6956 abiotic stress-responsive (ASTR) genes indicated that 73 of these genes colocalize with previously reported QTLs associated with abiotic stress characteristics (Smita et al., 2020). Additionally, genome-wide QTL mapping has revealed large main-effect QTLs linked to drought tolerance, explaining 10.0%–33.9% of phenotypic variance in plants (Pandey et al., 2020). Integrative genetics and genomics methodologies have elucidated reproductive tissue-specific meta-QTLs and candidate genes associated with heat and drought stress, providing enhanced understanding of the genetic intricacies underlying stress tolerance (Raza et al., 2020). Under drought stress, rice grain yield has been consistently linked to QTLs such as *qDTY1.1* and *qDTY3.1* (Vikram et al., 2011). According to a previous report, *qDTL6.05* and *qDTL7.02* in maize improved root characteristics and water-use efficiency, thereby increasing drought tolerance (Semagn et al., 2015). Meta-QTLs for heat tolerance in wheat, such as *qHT2.1* and *qHT7.1,* have been associated with better canopy temperature regulation and grain filling (Acuna Galindo, 2012). *HvNax3* and *HvNax4* are salinity tolerance QTLs in barley that help with sodium exclusion and better ion homeostasis (Shavrukov et al., 2010). These illustrations demonstrate how crop-specific QTLs are important for granting resistance to a variety of stresses and are useful targets for breeding initiatives that use markers.

3.3.4.1.2 Role of transcription factors and regulatory proteins

Transcription factors are essential in regulating plant responses to abiotic stress through the modulation of gene expression. The NAC transcription factor family (NAM, ATAF1/2, and CUC2) has been demonstrated to improve tolerance to drought and salinity by activating stress-responsive genes (Nakashima et al., 2012). Apetala2/ethylene response factor transcription factors similarly modulate the expression of stress-related genes, therefore facilitating various adaptations to abiotic stress (Ma et al., 2024). Moreover, MYB transcription factors are pivotal in regulating pathways associated with stress responses, underscoring their essential role in facilitating plant adaptation to adverse environmental conditions (Li et al., 2019).

3.3.4.2 Advances in molecular breeding and biotechnology

3.3.4.2.1 Role of CRISPR-Cas9 and RNA interference in developing stress-tolerant crops

CRISPR-Cas9 and RNA interference (RNAi) are sophisticated molecular breeding techniques that offer novel approaches for improving crop stress resilience. CRISPR technology facilitates accurate genome modification and has been effectively utilized to develop crops (Table 3.1) with enhanced resilience to drought and saline stress (Rajput et al., 2021). RNAi has demonstrated efficacy in silencing genes that inhibit stress tolerance, therefore enhancing plant resilience in challenging environments (Mushtaq et al., 2018). For instance, by altering stomatal conductance and improving water-use efficiency, CRISPR-Cas9 knockout of the *OsDST* gene in rice markedly increased drought and salinity tolerance (Santosh Kumar et al., 2020). Similarly, by encouraging root growth and lowering transpiration rates, editing the *SlLBD40* gene in tomatoes improved drought resilience (Liu et al., 2020a, 2020b). Plant performance in harsh conditions can be enhanced by RNAi, which has also shown promise in suppressing stress-related genes that impair tolerance (Mushtaq et al., 2018). These developments collectively highlight the potential of RNAi and CRISPR-Cas9 as potent instruments for creating crops that can withstand stress.

3.3.4.2.2 Insights from omics approaches (genomics, proteomics, and metabolomics)

Omics technologies, including genomics, proteomics, and metabolomics, provide significant insights into the molecular pathways that govern plant responses to abiotic stresses. Genomic investigations have revealed regulatory networks and QTLs linked to stress tolerance, enhancing our comprehension of plant responses to environmental adversities (Smita et al., 2020). Proteomic research has clarified the expression profiles of stress-responsive proteins, highlighting the essential functions of enzymes and other proteins associated with stress. Metabolomics enhances these insights by examining metabolites associated with stress adaptation. The increase of metabolites such as proline and sugars during stress is associated with their functions as osmoprotectants and antioxidants. Integrative omics methodologies offer a thorough comprehension of stress tolerance characteristics and improve the accuracy of breeding initiatives focused on creating resilient crop varieties.

3.3.4.3 Epigenetic modifications

3.3.4.3.1 DNA methylation in stress adaptation

DNA methylation is crucial in regulating plant responses to abiotic stresses such as drought, elevated temperatures, and salinity. Research has shown that drought stress causes dynamic changes in DNA methylation patterns, resulting in modifications in gene expression that improve the plant's stress tolerance. Furthermore, temporary alterations in DNA methylation have been associated with stress memory, whereby plants demonstrate enhanced resilience to future stresses due to previous exposure (Rao et al., 2024). This epigenetic mechanism of phenotypic plasticity highlights the potential use of DNA methylation techniques in creating stress-resilient crop varieties (Akhter et al., 2021).

3.3.4.3.2 Histone modifications in stress adaptation

Histone modifications, including acetylation and methylation, are essential regulators of gene expression in plant responses to abiotic stress. Research has revealed markedly elevated histone acetylation levels in genes linked to drought tolerance, hence augmenting their transcriptional activity (Nunez-Vazquez et al., 2022). In Arabidopsis, increased histone acetylation at the promoters of drought-responsive genes has been associated with enhanced stress tolerance. Additionally, some histone modifications, such as H3K27me3, have been associated with gene silencing under stress conditions, facilitating the suppression of nonessential functions to prioritize pathways vital for stress adaptation (Kumar et al., 2021).

3.4 Interactions of multiple stresses

Plants frequently encounter several simultaneous stresses, leading to intricate interactions that may enhance or diminish their impacts. Understanding these relationships is crucial for improving plant resilience in the face of changing environmental circumstances.

TABLE 3.1 Case studies in genetically modified varieties of both model and crop plants designed for enhanced abiotic stress tolerance by using CRISPR genome editing (Rahman et al. 2022).

Sl. no.	Crop plants	Abiotic stress	Genes	Function of genes	Source/year
1.	Rice	DT	*DERF1, PMS3, MSH1, MYB5, SPP*	Involved in the synthesis of amino acids and contributing to drought resistance	Zhang et al. (2014)
2.	Rice	DT	*SRL1, SRL2*	Regulate the rolling of leaves	Liao et al. (2019)
3.	Rice	DT	*ERA1*	Modulates ABA signaling and regulates the response to dehydration	Ogata et al. (2020)
4.	Maize	HerT	*ALS1 ALS2*	Play a role in tolerance to herbicides	Svitashev et al. (2015)
5.	Watermel-on	HerT	*ALS*	Contributes to herbicide resistance	Tian et al. (2018)
6.	Rice	HerT	*ALS*	Involved in herbicide tolerance	Wang et al. (2021a, 2021b, 2021c), Kuang et al., 2020
7.	Rice	HMT	*ARM1*	Regulates the plant's response to arsenic exposure	Wang et al. (2017)
8.	Rice	HMT	*HAK1*	Transports cesium ions	Nieves-Cordones et al. (2017)
9.	Rice	HMT	*Nramp5*	Facilitates the translocation of cadmium	Tang et al. (2017)
10.	Rice	CT	*MYB30*	Acts as a negative regulator of cold stress	Zeng et al. (2020)
11.	Tomato	HT	*AGL6*	Plays a role in fruit development	Klap et al. (2017)
12.	Rice	HT	*PYL1, PYL4 PYL6*	Involved in regulating abscisic acid activity	Miao et al. (2018)
13.	Tomato	HT	*MAPK3*	Functions as a negative regulator of heat stress	Yu et al. (2019)
14.	Rice	HT	*HSA1*	Involved in chloroplast protection and development	Qiu et al. (2018)
15.	Rice	ST	*RR22*	Regulates cytokinin signaling	Zhang et al. (2019)
16.	Rice	ST	*PQT3*	Acts as an E3 ubiquitin ligase	Alfatih et al. (2020)
17.	Rice	ST	*miR535*	Plays a role in regulating salt stress responses	Yue et al. (2020)
18.	Rice	ST	*SOS1*	Functions as a Na^+/H^+ antiporter, facilitating Na^+ transport	Lu et al. (2020)
19.	Rice	ST	*GI*	A component of the circadian clock	Wang et al. (2021a, 2021b, 2021c)
20.	Rice	ST	*bHLH024*	A basic helix–loop–helix transcription factor involved in stress responses and growth	Alam et al. (2022)
21.	Rice	ST	*RAV2*	Regulates various developmental processes	Duan et al. (2016)
22.	Rice	ST	*RR9, RR10*	Act as negative regulators of cytokinin signaling	Wang et al. (2019)
23.	Rice	ST	*DST*	Involved in the development of stomata	Santosh Kumar et al. (2020)

24.	Tomato	ST	*HyPRP1*	Serves as a negative regulator of salt stress	Tran et al. (2021)
25.	Soybean	ST	*AITR*	Regulates ABA signaling	Wang et al. (2021a, 2021b, 2021c)
26.	Rice	ST	*SPL10*	Regulates the development of trichomes	Lan et al. (2019)
27.	Tomato	DT	*GID1a*	Functions as a receptor for gibberellin	Illouz-Eliaz et al. (2020)
28.	Tomato	DT	*LBD40*	Mediates stress response through jasmonic acid	Liu et al. (2020a, 2021b)
29.	Maize	DT	*ARGOS8*	Involved in the ethylene response	Shi et al. (2017)
30.	Maize	DT	*abh2*	Mediates stomatal opening via abscisic acid 80-hydroxylase	Liu et al. (2020a, 2020b)
31.	Rapeseed	DT	*A6.RGA*	A DELLA protein that negatively regulates gibberellin signaling	Wu et al. (2020)
32.	Maize	ST	*STL1*	A dirigent protein that is localized to the Casparian strip	Wang et al. (2022)
33.	Tomato	ST	*ABIC1*	A homeodomain–leucine zipper transcription factor	Ding et al. (2022)

CST: Cold stress tolerance, *DST*: drought stress tolerance, *HerST*: herbicide stress tolerance, *HMT*: heavy metal tolerance, *HST*: heat stress tolerance, *SST*: salinity stress tolerance. Source: Adapted and expanded from Rahman, M. u, Zulfiqar, S., Raza, M.A., Ahmad, N., Zhang, B., 2022. Engineering abiotic stress tolerance in crop plants through CRISPR genome editing. Cells, 11(22), 3590. https://doi.org/10.3390/cells11223590.

3.4.1 Synergistic and antagonistic effects of stress combinations

A synergistic interaction occurs when stresses converge to create an effect that exceeds the cumulative impact of their separate contributions. The concurrent presence of drought and salinity stress significantly impedes plant growth more than the impacts of each stresses individually. Studies demonstrate that simultaneous exposure to salinity and drought markedly diminishes chlorophyll concentrations, photosynthetic efficiency, and biomass production more than individual stresses (Angon et al., 2022). Antagonistic interactions occur when one stresses diminishes the effect of another, resulting in a less severe cumulative impact. Under specific settings, elevated salinity stress might mitigate the adverse impacts of biotic factors, such as herbivory, by diminishing plant vigor. In tomato plants exposed to both drought and herbivory, the simultaneous stresses led to increased tolerance relative to exposure to each stress individually, due to the activation of defense pathways (Vescio et al., 2022).

3.4.2 Mechanistic insights into plant responses to combined stresses

Research has uncovered substantial insights into the systems governing plant responses to concurrent stresses, including drought and heavy metal exposure. Plants subjected to simultaneous stresses frequently demonstrate modifications in their physiological and biochemical processes. Critical reactions involve the activation of antioxidant mechanisms, which are vital for neutralizing reactive oxygen species produced during stress. Moreover, the interaction among plant hormones is significantly altered under these conditions, resulting in hormonal imbalances that might affect growth, development, and stress responses (Gupta et al., 2024).

3.4.2.1 Adaptive crosstalk mechanisms

The ability of plants to respond to simultaneous stresses is crucial for maintaining agricultural productivity and ecological equilibrium. Acquiring mechanistic insights into these responses is essential for comprehending plant adaptation and survival in increasingly harsh environmental situations.

3.4.2.1.1 Integration of morphological responses

Plants utilize many physiological systems to acclimatize to concurrent abiotic stresses. As plants modify their root characteristics to maximize water and nutrient uptake in limiting circumstances, combined stresses frequently cause noticeable changes in RSA. By changing the total root length, surface area, and number of tips, drought and salinity stresses in wheat dramatically change RSA. They also promote the growth of lateral roots and root hairs in saline environments (Tang et al., 2024). Similar to this, different root types in maize exhibit different architectural and anatomical responses to water deficit, with primary, seminal, and lateral roots displaying unique responses (Protto et al., 2024). Furthermore, transcriptomic evidence highlights the additive effects of stress combinations on root morphology by showing that combined drought and heat stress reduce primary root length more severely than individual stresses (Wang et al., 2025). The adaptive importance of root anatomical changes under combined stresses is demonstrated by the way that rice exposed to flooding and salinity stress develops aerenchyma in its roots to enhance oxygen diffusion and maintain root respiration (Tamang and Fukao, 2015).

3.4.2.1.2 Integration of physiological responses

Due to stomatal closure, decreased chlorophyll biosynthesis, and damage to photosystem II, combined stresses physiologically result in more significant decreases in photosynthesis than individual stresses. According to research, the combined effects of salinity and drought worsen water loss and severely impair transpiration and photosynthesis. For instance, when compared to single-stress conditions, maize and soybeans exposed to heat stress and drought exhibit sharp drops in stomatal conductance, relative water content, and chlorophyll fluorescence (Hussain et al., 2018; Sattar et al., 2020). Moreover, when rice experiences flooding and salinity stress, ion homeostasis is disrupted, resulting in the accumulation of Na^+ and Cl^- and a decrease in K^+ concentration. This has a negative impact on photosynthetic capacity and water balance (Zheng et al., 2009a, 2009b). Salinity-induced osmotic stress can trigger stomatal closure, a reaction intensified by drought conditions, resulting in decreased gas exchange and lower photosynthetic efficiency (Zhang et al., 2023).

3.4.2.1.3 Integration of biochemical responses

Alongside physiological alterations, plants initiate metabolic mechanisms to alleviate the effects of these stresses. The accumulation of osmoprotectants, including proline and soluble sugars, has been recognized as a crucial mechanism to mitigate osmotic stress resulting from the combined impacts of salinity and drought (Gupta et al., 2022). The overproduction of ROS is made worse by combined stresses at the biochemical level, which results in oxidative damage to proteins, membranes, and DNA. By increasing the activities of antioxidant defenses such as ascorbate peroxidase (APX), catalase (CAT), and superoxide dismutase (SOD), plants combat this. To stabilize proteins and membranes under stress, barley and tomato exposed to drought and salinity stress, for example, show a significant upregulation of antioxidant enzymes and accumulation of osmolytes such as proline and glycine betaine (Anwar et al., 2018; Hasanuzzaman et al., 2020).

3.4.2.1.4 Molecular mechanisms underlying combined-stress responses

At the molecular level, plants utilize intricate signaling pathways, such as mitogen-activated protein kinases (MAPKs) and heat shock proteins (HSPs), to react to abiotic stress. MAPKs are essential for stress signal transmission, starting cascades that modulate gene expression and improve stress tolerance in response to stimuli such as drought and salinity (He et al., 2020). Likewise, HSPs function as molecular chaperones, protecting proteins against denaturation under stressful circumstances. These proteins enhance cellular stability and resilience by ensuring correct protein folding and functionality (Yurina, 2023). The production of HSPs is meticulously regulated and frequently stimulated by environmental conditions such as severe temperatures and oxidative stress, highlighting their significance in enhancing stress resistance (Ul Haq et al., 2019). Crop-specific insights into these mechanisms are provided by recent studies. By preserving PSII function and reducing ROS accumulation, overexpressing the small HSP *OsHSP17.9* in rice improves tolerance to both high temperatures and salinity (Do et al., 2023). Likewise, overexpression of *ZmMPK4*, a gene that responds to multiple stresses, increases transgenic tobacco's resistance to low temperatures (Zhou et al., 2012). When expressed in rice and Arabidopsis, the mitochondrial sHSP *TaHSP23.9* in wheat provides salt/alkali and thermo-tolerance, demonstrating cross-protection from a single HSP module (Wang et al., 2020). Arabidopsis MPK3/MPK6 directly links MAPK signaling to hormonal rebalancing under ionic stress by inducing targeted degradation of type-B ARR cytokinin response regulators, which in turn promotes salt tolerance for MAPKs (Yan et al., 2021). By maintaining photosynthesis and triggering antioxidant defenses, *StMAPKK5* gain-of-function in potatoes enhances heat (and, in a different study, drought and salinity) tolerance, demonstrating MAPK-mediated protection of core metabolism under combined or sequential challenges (Zhu et al., 2024; Luo et al., 2024).

3.4.2.1.5 Role of abscisic acid in combined-stress responses

Abscisic acid (ABA) is crucial in modulating plant responses to abiotic stress, especially under combined stress conditions. Under drought and salinity conditions, ABA levels rise, inducing stomatal closure to reduce water loss while coordinating systemic stress responses. ABA governs the expression of numerous stress-responsive genes, augmenting the plant's ability to endure various simultaneous stresses (Gupta et al., 2022). Moreover, ABA, ethylene, and salicylic acid are essential for coordinating stress adaptation, including controlling stomatal closure and triggering downstream stress-responsive genes. Therefore, combined stresses also affect hormonal crosstalk (Voesenek & Bailey-Serres, 2013).

3.5 Implications for crop improvement

Crop improvement research seeks to develop plant types with increased tolerance to abiotic stresses, including drought, salinity, and temperature fluctuations. A thorough comprehension of various breeding procedures is crucial for enhancing agricultural output and guaranteeing long-term sustainability in global farming systems.

3.5.1 Breeding for abiotic stress tolerance

Breeding for multiple abiotic stress tolerance employs both conventional and contemporary techniques, including marker-assisted selection, to improve crop varieties. Conventional breeding has been a crucial method in crop enhancement, depending on the selection of superior phenotypes via traditional crossbreeding (Shivhare and Lata, 2017). Marker-assisted breeding enhances this process by employing molecular markers to monitor advantageous features in offspring, therefore increasing the efficacy of breeding programs (Manzoor et al., 2022). This integrated method seeks to increase crop productivity while simultaneously improving resilience to environmental challenges, which is essential due to climate change.

3.5.2 Case studies in model and crop plants

Case studies conducted on both model organisms and crop plants offer valuable information on the mechanisms underlying stress adaptation and the effectiveness of genetically modified varieties developed for improved tolerance to abiotic stresses (Table 3.1).

3.6 Conclusion

The majority of research on how plants react to abiotic stresses has concentrated on specific stresses, such as heat, salinity, or drought, but in natural settings, crops frequently experience several stresses at once. A significant research gap in our current understanding is highlighted by the distinct physiological, biochemical, anatomical, and molecular responses that combined stresses (such as drought–heat, salinity–drought, or flooding–metal toxicity) elicit. Despite the identification of stress-responsive genes, QTLs, and regulatory networks through developments in genomics, transcriptomics, and molecular breeding, there are still few integrative studies that analyze the interplay between signaling pathways under multiple stresses. To identify common and distinct tolerance mechanisms, future studies should focus on multistress systems, emphasizing characteristics such as hormone balance, antioxidant homeostasis, and root system plasticity. Breeding programs must use combined-stress QTL mapping, genomic selection, and gene editing techniques in addition to single-stress tolerance. Additionally, it will be crucial to use artificial intelligence, high-throughput phenotyping, and systems biology to forecast plant performance in challenging field settings. In order to create resilient crop varieties and guarantee global food security in the face of climate variability, it is imperative that combined abiotic stresses be addressed through multidisciplinary research and breeding techniques.

References

Acosta-Motos, J.R., Ortuño, M.F., Bernal-Vicente, A., Diaz-Vivancos, P., Sanchez-Blanco, M.J., Hernandez, J.A., 2017. Plant responses to salt stress: adaptive mechanisms. Agronomy 7 (1), 18.

Acuna Galindo, M.A., 2012. Meta-analysis of wheat QTL regions associated with heat and drought stress. Graduate Theses and Dissertations. Retrieved from https://scholarworks.uark.edu/etd/645.

Akhter, Z., Bi, Z., Ali, K., Sun, C., Fiaz, S., Haider, F.U., et al., 2021. In response to abiotic stress, DNA methylation confers epigenetic changes in plants. Plants 10 (6), 1096.

Alam, M.S., Kong, J., Tao, R., Ahmed, T., Alamin, M., Alotaibi, S.S., et al., 2022. CRISPR/Cas9 mediated knockout of the OsbHLH024 transcription factor improves salt stress resistance in rice (*Oryza sativa* L.). Plants 11 (9), 1184.

Alfatih, A., Wu, J., Jan, S.U., Zhang, Z.S., Xia, J.Q., Xiang, C.B., 2020. Loss of rice PARAQUAT TOLERANCE 3 confers enhanced resistance to abiotic stresses and increases grain yield in field. Plant, Cell & Environment 43 (11), 2743–2754.

Angon, P.B., Tahjib-Ul-Arif, M., Samin, S.I., Habiba, U., Hossain, M.A., Brestic, M., 2022. How do plants respond to combined drought and salinity stress? A systematic review. Plants 11 (21), 2884.

Anwar, A., Liu, Y., Dong, R., Bai, L., Yu, X., Li, Y., 2018. The physiological and molecular mechanism of brassinosteroid in response to stress: a review. Biological Research 51 (1), 46.

Ashraf, M., Harris, P.J.C., 2013. Photosynthesis under stressful environments. Photosynthetica 51 (2), 163–374.

Bita, C.E., Gerats, T., 2013. Plant tolerance to high temperature in a changing environment: scientific fundamentals and production of heat stress-tolerant crops. Frontiers in Plant Science 4, 273.

Burton, A.L., Brown, K.M., Lynch, J.P., 2013. Phenotypic diversity of root anatomical and architectural traits in Zea species. Crop Science 53 (3), 1042–1055.

Chen, D., Mubeen, B., Hasnain, A., Rizwan, M., Adrees, M., Naqvi, S.A.H., et al., 2022. Role of promising secondary metabolites to confer resistance against environmental stresses in crop plants: current scenario and future perspectives. Frontiers in Plant Science 13, 881032.

Cornelis, S., Hazak, O., 2022. Understanding the root xylem plasticity for designing resilient crops. Plant, Cell & Environment 45 (3), 664–676.

Dash, P.K., Rai, R., Rai, V., Pasupalak, S., 2018. Drought-induced signaling in rice: delineating canonical and non-canonical pathways. Frontiers in Chemistry 6, 264.

Deak, K.I., Malamy, J., 2005. Osmotic regulation of root system architecture. The Plant Journal 43 (1), 17–28.

Devasirvatham, V., Tan, D., 2018. Impact of high temperature and drought stresses on chickpea production. Agronomy 8 (8), 145.

Ding, F., Qiang, X., Jia, Z., Li, L., Hu, J., Yin, M., et al., 2022. Knockout of a novel salt responsive gene SlABIG1 enhances salinity tolerance in tomato. Environmental and Experimental Botany 200, 104903.

Djanaguiraman, M., Narayanan, S., Erdayani, E., Prasad, P.V.V., 2020. Effects of high temperature stress during anthesis and grain filling periods on photosynthesis, lipids, and grain yield in wheat. BMC Plant Biology 20 (1), 1–12.

Do, J.M., Kim, H.J., Shin, S.Y., Park, S.I., Kim, J.J., Yoon, H.S., 2023. OsHSP17.9, a small heat shock protein, confers improved productivity and tolerance to high temperature and salinity in a natural paddy field in transgenic rice plants. Agriculture 13 (5), 931.

Duan, Y.B., Li, J., Qin, R.Y., Xu, R.F., Li, H., Yang, Y.C., et al., 2016. Identification of a regulatory element responsible for salt induction of rice OsRAV2 through ex situ and in situ promoter analysis. Plant Molecular Biology 90 (1-2), 49–62.

Fahad, S., Bano, A., 2012. Effect of salicylic acid on physiological and biochemical characterization of maize grown in saline area. Pakistan Journal of Botany 44 (4), 1433–1438.

Farooq, M., Hussain, M., Jabran, K., Cheema, M.A., 2011. Drought stress in plants: a review on morphological, physiological and biochemical aspects. Journal of Agronomy and Crop Science 197 (3), 198–211.

Farooq, M., Kobayashi, N., Wahid, A., Ito, O., Basra, S.M.A., 2009a. Strategies for producing more rice with less water. Advances in Agronomy 101, 351–388.

Farooq, M., Wahid, A., Kobayashi, N., Fujita, D., Basra, S.M.A., 2009b. Plant drought stress: effects, mechanisms and management. Agronomy for Sustainable Development 29 (1), 185–212.

Flexas, J., Bota, J., Loreto, F., Cornic, G., Sharkey, T.D., 2013. Diffusive and metabolic limitations to photosynthesis under drought and salinity in C3 plants. Plant Biology 15 (1), 88–91.

Flora, S.J.S., Mittal, M., Mehta, A., 2008. Heavy metal induced oxidative stress and its possible reversal by chelation therapy. Indian Journal of Medical Research 128 (4), 501–523.

Flowers, T.J., Colmer, T.D., 2008. Salinity tolerance in halophytes. New Phytologist 179 (4), 945–963.

Food and Agriculture Organization of the United Nations (FAO), 2017. Global Food Policy Report. pp. 2016–2017.

Fukao, T., Barrera-Figueroa, B.E., Juntawong, P., Peña-Castro, J.M., 2019. Submergence and waterlogging stress in plants: a review highlighting research opportunities and understudied aspects. Frontiers in Plant Science 10, 340.

Galdon-Armero, J., Fullana-Pericas, M., Mulet, P.A., Conesa, M.A., Martin, C., Galmes, J., 2018. The ratio of trichomes to stomata is associated with water use efficiency in *Solanum lycopersicum* (tomato). The Plant Journal 96 (3), 607–619.

Godfray, H.C., Beddington, J.R., Crute, I.R., Haddad, L., Lawrence, D., Muir, J.F., et al., 2010. Food security: the challenge of feeding 9 billion people. Science 327 (5967), 812–818.

Gomes, F.P., Oliva, M.A., Mielke, M.S., Almeida, A.A.F., Aquino, L.A., 2008. Photosynthetic limitations in leaves of young Brazilian Green Dwarf coconut (*Cocos nucifera* L. 'nana') palm under well-watered conditions or recovering from drought stress. Environmental and Experimental Botany 62 (3), 195–204.

Gong, H.J., Randall, D.P., Flowers, T.J., 2006. Silicon deposition in the root reduces sodium uptake in rice (*Oryza sativa* L.) seedlings by reducing bypass flow. Plant, Cell & Environment 29 (10), 1970–1979.

Gupta, M., Dwivedi, V., Kumar, S., Patel, A., Niazi, P., Yadav, V.K., 2024. Lead toxicity in plants: mechanistic insights into toxicity, physiological responses of plants and mitigation strategies. Plant Signaling & Behavior 19 (1), 2365576.

Gupta, B., Huang, B., 2014. Mechanism of salinity tolerance in plants: physiological, biochemical, and molecular characterization. International Journal of Genomics 2014, 701596.

Gupta, A., Mishra, R., Rai, S., Bano, A., Pathak, N., Fujita, M., et al., 2022. Mechanistic insights of plant growth promoting bacteria mediated drought and salt stress tolerance in plants for sustainable agriculture. International Journal of Molecular Sciences 23 (7), 3741.

Hall, J.L., 2002. Cellular mechanisms for heavy metal detoxification and tolerance. Journal of Experimental Botany 53 (366), 1–11.

Hasanuzzaman, M., Bhuyan, M.H.M.B., Zulfiqar, F., Raza, A., Mohsin, S.M., Mahmud, J.A., et al., 2020. Reactive oxygen species and antioxidant defense in plants under abiotic stress: revisiting the crucial role of a universal defense regulator. Antioxidants 9 (8), 681.

Hasanuzzaman, M., Nahar, K., Alam, M.M., Roychowdhury, R., Fujita, M., 2013. Physiological, biochemical, and molecular mechanisms of heat stress tolerance in plants. International Journal of Molecular Sciences 14 (5), 9643–9684.

Herzog, M., Striker, G.G., Colmer, T.D., Pedersen, O., 2016. Mechanisms of waterlogging tolerance in wheat-a review of root and shoot physiology. Plant, Cell & Environment 39 (5), 1068–1086.

He, M., He, C.Q., Ding, N.Z., 2018. Abiotic stresses: general defenses of land plants and chances for engineering multistress tolerance. Frontiers in Plant Science 9, 1771.

He, X., Wang, C., Wang, H., Li, L., Wang, C., 2020. The function of MAPK cascades in response to various stresses in horticultural plants. Frontiers in Plant Science 11, 952.

Hossain, Z., Mustafa, G., Komatsu, S., 2015. Plant responses to nanoparticle stress. International Journal of Molecular Sciences 16 (11), 26644–26653.

Hussain, H.A., Hussain, S., Khaliq, A., Ashraf, U., Anjum, S.A., Men, S., et al., 2018. Chilling and drought stresses in crop plants: implications, cross talk, and potential management opportunities. Frontiers in Plant Science 9, 393.

Hussain, H.A., Men, S., Hussain, S., Chen, Y., Ali, S., Zhang, S., et al., 2019. Interactive effects of drought and heat stresses on morpho-physiological attributes, yield, nutrient uptake and oxidative status in maize hybrids. Scientific Reports 9 (1), 3890.

Illouz-Eliaz, N., Nissan, I., Nir, I., Ramon, U., Shohat, H., Weiss, D., 2020. Mutations in the tomato gibberellin receptors suppress xylem proliferation and reduce water loss under water-deficit conditions. Journal of Experimental Botany 71 (12), 3603–3612.

Intergovernmental Panel on Climate Change (IPCC), 2014. Climate Change 2014: Impacts, Adaptation, and Vulnerability. Part A: Global and Sectoral Aspects. Cambridge University Press.

Ishfaq, S., Ding, Y., Liang, X., Guo, W., 2025. Advancing lodging resistance in maize: integrating genetic, hormonal, and agronomic insights for sustainable crop productivity. Plant Stress 15, 100777.

Ismail, A.M., Hall, A.E., 1999. Reproductive-stage heat tolerance, leaf membrane thermo stability and plant morphology in cowpea. Crop Science 39 (6), 1762–1768.

Jia, W., Ma, M., Chen, J., Wu, S., 2021. Plant morphological, physiological and anatomical adaption to flooding stress and the underlying molecular mechanisms. International Journal of Molecular Sciences 22 (3), 1088.

Jia, L.T., Qin, X., Lyu, D.G., Qin, S.J., Zhang, P., 2019. ROS production and scavenging in three cherry rootstocks under short-term waterlogging conditions. Scientia Horticulturae 257, 108647.

Kadam, N.N., Yin, X., Bindraban, P.S., Struik, P.C., Jagadish, K.S., 2015. Does morphological and anatomical plasticity during the vegetative stage make wheat more tolerant of water deficit stress than rice? Plant Physiology 167 (4), 1389–1401.

Karlova, R., Boer, D., Hayes, S., Testerink, C., 2021. Root plasticity under abiotic stress. Plant Physiology 187 (3), 1057–1070.

Khan, M.I., Fatma, M., Per, T.S., Anjum, N.A., Khan, N.A., 2015. Salicylic acid-induced abiotic stress tolerance and underlying mechanisms in plants. Frontiers in Plant Science 6, 462.

Klap, C., Yeshayahou, E., Bolger, A.M., Arazi, T., Gupta, S.K., Shabtai, S., et al., 2017. Tomato facultative parthenocarpy results from SlAGAMOUS-LIKE 6 loss of function. Plant Biotechnology Journal 15 (5), 634–647.

Kosma, D.K., Bourdenx, B., Bernard, A., Parsons, E.P., Lü, S., Joubès, J., et al., 2009. The impact of water deficiency on leaf cuticle lipids of Arabidopsis. Plant Physiology 151 (4), 1918–1929.

Kozlowski, T.T., Pallardy, S.G., 1997. Physiology of Woody Plants, 2nd ed. Academic Press.

Król, A., 2013. The growth and water uptake by yellow seed and black seed rape depending on the state of soil compaction. PhD thesis, Bohdan Dobrzanski Institute of Agrophysics, PAS.

Kuang, Y., Li, S., Ren, B., Yan, F., Spetz, C., Li, X., et al., 2020. Base-editing-mediated artificial evolution of OsALS1 in planta to develop novel herbicide-tolerant rice germplasms. Molecular Plant 13 (4), 565–572.

Kumar, V., Thakur, J.K., Prasad, M., 2021. Histone acetylation dynamics regulating plant development and stress responses. Cellular and Molecular Life Sciences 78 (6), 4467–4486.

Lan, T., Zheng, Y., Su, Z., Yu, S., Song, H., Zheng, X., et al., 2019. OsSPL10, a SBP-box gene, plays a dual role in salt tolerance and trichome formation in rice (*Oryza sativa* L.). G3: Genes, Genomes, Genetics 9 (12), 4107–4114.

Lee, S.B., Suh, M.C., 2015. Advances in the understanding of cuticular waxes in Arabidopsis thaliana and crop species. Plant Cell Reports 34 (4), 557–572.

Liao, S., Qin, X., Luo, L., Han, Y., Wang, X., Usman, B., et al., 2019. CRISPR/Cas9-induced mutagenesis of semi-rolled leaf1, 2 confers curled leaf phenotype and drought tolerance by influencing protein expression patterns and ROS scavenging in rice (*Oryza sativa* L.). Agronomy 9 (11), 728.

Liu, X., Huang, M., Fan, B., Buckler, E.S., Zhang, Z., 2016. Iterative usage of fixed and random effect models for powerful and efficient genomewide association studies. PLoS genetics 12 (4), e1005767.

Liu, S., Li, C., Wang, H., Wang, S., Yang, S., Liu, X., et al., 2020a. Mapping regulatory variants controlling gene expression in drought response and tolerance in maize. Genome Biology 21 (1), 163.

Liu, Q., Luo, L., Zheng, L., 2018. Lignins: biosynthesis and biological functions in plants. International Journal of Molecular Sciences 19 (2), 335.

Liu, L., Zhang, J., Xu, J., Li, Y., Guo, L., Wang, Z., et al., 2020b. CRISPR/Cas9 targeted mutagenesis of SlLBD40, a lateral organ boundaries domain transcription factor, enhances drought tolerance in tomato. Plant Science 301, 110683.

Li, N., Han, X., Feng, D., Yuan, D., Huang, L.J., 2019. Signaling crosstalk between salicylic acid and ethylene/jasmonate in plant defense: do we understand what they are whispering? International Journal of Molecular Sciences 20 (3), 671.

Li, N., Wang, X., Ma, B., Du, C., Zheng, L., Wang, Y., 2017. Expression of a Na+/H+ antiporter RtNHX1 from a recretohalophyte Reaumuria trigyna improved salt tolerance of transgenic Arabidopsis thaliana. Journal of Plant Physiology 218, 189–200.

Loka, D.A., Oosterhuis, D.M., Baxevanos, D., Noulas, C., Hu, W., 2020. Single and combined effects of heat and water stress and recovery on cotton (*Gossypium hirsutum* L.) leaf physiology and sucrose metabolism. Plant Physiology and Biochemistry 148, 166–179.

Loka, D.A., Oosterhuis, D.M., Ritchie, G.L., 2011. Water-deficit stress in cotton. In: Oosterhuis, D.M. (Ed.), Stress Physiology in Cotton. The Cotton Foundation, pp. 37–72.

Lucob-Agustin, N., Hasegawa, T., Jinno, K., Suralta, R.R., Niones, J.M., Kano-Nakata, M., et al., 2020. Enhanced root system development responses of a newly identified mutation gene promoting lateral root development to various nitrogen conditions in rice. Journal of International Cooperation for Agricultural Development 18, 48–55.

Lucob-Agustin, N., Kawai, T., Kano-Nakata, M., Suralta, R.R., Niones, J.M., Hasegawa, T., et al., 2021. Morpho-physiological and molecular mechanisms of phenotypic root plasticity for rice adaptation to water stress conditions. Breeding Science 71 (1), 20–29.

Luo, Y., Wang, K., Zhu, L., Zhang, N., Si, H., 2024. StMAPKK5 positively regulates response to drought and salt stress in potato. International Journal of Molecular Sciences 25 (7), 3662.

Lu, Y., Tian, Y., Shen, R., Yao, Q., Wang, M., Chen, M., et al., 2020. Targeted, efficient sequence insertion and replacement in rice. Nature Biotechnology 38 (12), 1402–1407.

Lynch, J.P., Brown, K.M., 2001. Topsoil foraging - an architectural adaptation of plants to low phosphorus availability. Plant and Soil 237 (2), 225–237.

Lynch, J.P., Wojciechowski, T., 2015. Opportunities and challenges in the subsoil: pathways to deeper rooted crops. Journal of Experimental Botany 66 (8), 2199–2210.

Líška, D., Martinka, M., Kohanová, J., Lux, A., 2016. Asymmetrical development of root endodermis and exodermis in reaction to abiotic stresses. Annals of Botany 118 (4), 667–674.

Manzoor, N., Ali, L., Ahmed, T., Noman, M., Adrees, M., Shahid, M.S., et al., 2022. Recent advancements and development in nano-enabled agriculture for improving abiotic stress tolerance in plants. Frontiers in Plant Science 13, 951752.

Ma, Z., Hu, L., Jiang, W., 2024. Understanding AP2/ERF transcription factor responses and tolerance to various abiotic stresses in plants: a comprehensive review. International Journal of Molecular Sciences 25 (10), 893.

Miao, C., Xiao, L., Hua, K., Zou, C., Zhao, Y., Bressan, R.A., et al., 2018. Mutations in a subfamily of abscisic acid receptor genes promote rice growth and productivity. Proceedings of the National Academy of Sciences 115 (23), 6058–6063.

Mishra, S.S., Behera, P.K., Kumar, V., Lenka, S.K., Panda, D., 2018. Physiological characterization and allelic diversity of selected drought tolerant traditional rice (*Oryza sativa* L.) landraces of Koraput, India. Physiology and Molecular Biology of Plants 24 (6), 1035–1046.

Mishra, N., Jiang, C., Chen, L., Paul, A., Chatterjee, A., Shen, G., 2023. Achieving abiotic stress tolerance in plants through antioxidative defense mechanisms. Frontiers in Plant Science 14, 1110622.

Mishra, S.S., Panda, D., 2017. Leaf traits and antioxidant defense for drought tolerance during early growth stage in some popular traditional rice landraces from Koraput, India. Rice Science 24 (4), 207–217.

Mittler, R., 2006. Abiotic stress, the field environment and stress combination. Trends in Plant Science 11 (1), 15–19.

Moura, J.C., Bonine, C.A., de Oliveira Fernandes Viana, J., Dornelas, M.C., Mazzafera, P., 2010. Abiotic and biotic stresses and changes in the lignin content and composition in plants. Journal of Integrative Plant Biology 52 (4), 360–376.

Munns, R., Millar, A.H., 2023. Seven plant capacities to adapt to abiotic stress. Journal of Experimental Botany 74 (15), 4308–4323.

Munns, R., Tester, M., 2008. Mechanisms of salinity tolerance. Annual Review of Plant Biology 59, 651–681.

Mushtaq, M., Bhat, J.A., Mir, Z.A., Sakina, A., Ali, S., Singh, A.K., et al., 2018. CRISPR/Cas approach: a new way of looking at plant–abiotic interactions. Journal of Plant Physiology 224-225, 156–162.

Mustroph, A., 2018. Improving flooding tolerance of crop plants. Agronomy 8 (9), 160.

Nakashima, K., Takasaki, H., Mizoi, J., Shinozaki, K., Yamaguchi-Shinozaki, K., 2012. NAC transcription factors in plant abiotic stress responses. Biochimica et Biophysica Acta (BBA) - Gene Regulatory Mechanisms 1819 (2), 97–103.

Natarajan, S., Kuehny, J.S., 2008. Morphological, physiological, and anatomical characteristics associated with heat preconditioning and heat tolerance in *Salvia splendens*. Journal of the American Society for Horticultural Science 133 (4), 527–534.

Nieves-Cordones, M., Mohamed, S., Tanoi, K., Kobayashi, N.I., Takagi, K., Vernet, A., et al., 2017. Production of low-Cs+ rice plants by inactivation of the K+ transporter OsHAK1 with the CRISPR-Cas system. The Plant Journal 92 (1), 43–56.

Nishiuchi, S., Yamauchi, T., Takahashi, H., Kotula, L., Nakazono, M., 2012. Mechanisms for coping with submergence and waterlogging in rice. Rice 5 (1), 2.

Nunez-Vazquez, R., Desvoyes, B., Gutierrez, C., 2022. Histone variants and modifications during abiotic stress response. Frontiers in Plant Science 13, 984702.

Ogata, T., Ishizaki, T., Fujita, M., Fujita, Y., 2020. CRISPR/Cas9-targeted mutagenesis of OsERA1 confers enhanced responses to abscisic acid and drought stress and increased primary root growth under non-stressed conditions in rice. PLoS One 15 (12), e0243376.

Panda, D., Mishra, S., Behera, P.K., 2021. Drought tolerance in rice: focus on recent mechanisms and approaches. Rice Science 28 (2), 119–132.

Panda, D., Sarkar, R.K., 2013. Characterization of leaf gas exchange and antioxidant defense of rice (*Oryza sativa* L.) cultivars differing in submergence tolerance owing to complete submergence and consequent re aeration. Agricultural Research 2 (4), 301–308.

Pandey, M.K., Gangurde, S.S., Sharma, V., Pattanashetti, S.K., Naidu, G.K., Faye, I., et al., 2020. Improved genetic map identified major QTLs for drought tolerance- and iron deficiency tolerance-related traits in groundnut. Genes 12 (1), 37.

Pardales, J.R., Yamauchi, A., 2003. Regulation of root development in sweetpotato and cassava by soil moisture during their establishment period. Plant and Soil 255 (1), 201–208.

Pareek, A., Joshi, R., Gupta, K.J., Singla-Pareek, S.L., Foyer, C., 2020. Sensing and signalling in plant stress responses: ensuring sustainable food security in an era of climate change. New Phytologist 228 (3), 823–827.

Porter, G.A., Opena, G.B., Bradbury, W.B., McBurnie, J.C., Sisson, J.A., 1999. Soil management and supplemental irrigation effects on potato: I. Soil properties, tuber yield, and quality. Agronomy Journal 91 (3), 416–425.

Protto, V., Bauget, F., Rishmawi, L., Nacry, P., Maurel, C., 2024. Primary, seminal and lateral roots of maize show type-specific growth and hydraulic responses to water deficit. Plant Physiology 194 (4), 2564–2579.

Pérez-López, U., Miranda-Apodaca, J., Muñoz-Rueda, A., Mena-Petite, A., 2013. Lettuce production and antioxidant capacity are differentially modified by salt stress and light intensity under ambient and elevated CO2. Journal of Plant Physiology 170 (15), 1517–1525.

Qiu, Z., Kang, S., He, L., Zhao, J., Zhang, S., Hu, J., et al., 2018. The newly identified heat-stress sensitive albino 1 gene affects chloroplast development in rice. Plant Science 267, 168–179.

Rahman, M.u, Zulfiqar, S., Raza, M.A., Ahmad, N., Zhang, B., 2022. Engineering abiotic stress tolerance in crop plants through CRISPR genome editing. Cells 11 (22), 3590.

Rajput, M., Choudhary, K., Kumar, M., Vivekanand, V., Chawade, A., Ortiz, R., et al., 2021. RNA interference and CRISPR/Cas gene editing for crop improvement: paradigm shift towards sustainable agriculture. Plants 10 (9), 1914. Available at https://doi.org/10.3390/plants10091914.

Raj, S.R.G., Nadarajah, K., 2022. QTL and candidate genes: techniques and advancement in abiotic stress resistance breeding of major cereals. International Journal of Molecular Sciences 24 (1), 6.

Rane, J., Singh, A.K., Kumar, M., Boraiah, K.M., Meena, K.K., Pradhan, A., et al., 2021. The adaptation and tolerance of major cereals and legumes to important abiotic stresses. International Journal of Molecular Sciences 22 (23), 12970.

Rao, X., Yang, S., Lü, S., Yang, P., 2024. DNA methylation dynamics in response to drought stress in crops. Plants 13 (14), 1977.

Raza, A., Charagh, S., Abbas, S., Hassan, M.U., Saeed, F., Haider, S., et al., 2023. Assessment of proline function in higher plants under extreme temperatures. Plant Biology 25 (3), 379–395.

Raza, Q., Riaz, A., Bashir, K., Sabar, M., 2020. Reproductive tissues-specific meta-QTLs and candidate genes for development of heat-tolerant rice cultivars. Plant Molecular Biology 104 (1-2), 97–112.

Riyazuddin, R., Verma, R., Singh, K., Nisha, N., Keisham, M., Bhati, K.K., et al., 2020. Ethylene: a master regulator of salinity stress tolerance in plants. Biomolecules 10 (6), 959.

Sah, S.K., Reddy, K.R., Li, J., 2016. Abscisic acid and abiotic stress tolerance in crop plants. Frontiers in Plant Science 7, 571.

Salam, A., Afridi, M.S., Javed, M.A., et al., 2022. Nano-priming against abiotic stress: a way forward towards sustainable agriculture. Sustainability 14 (22), 14880.

Sangsing, K., Kasemsap, P., Thanisawanyangkura, S., Sangsing, S., Gohet, E., Thaler, P., et al., 2004. Xylem embolism and stomatal regulation in two rubber clones (*Hevea brasiliensis* Muell. Arg.). Trees 18, 109–114.

Santosh Kumar, V.V., Verma, R.K., Yadav, S.K., Yadav, P., Watts, A., Rao, M.V., et al., 2020. CRISPR-Cas9 mediated genome editing of drought and salt tolerance (OsDST) gene in indica mega rice cultivar MTU1010. Physiology and Molecular Biology of Plants 26 (6), 1099–1110.

Sattar, A., Sher, A., Ijaz, M., Ul-Allah, S., Rizwan, M.S., Hussain, M., et al., 2020. Terminal drought and heat stress alter physiological and biochemical attributes in flag leaf of bread wheat. PLoS One 15 (5), e0232974.

Semagn, K., Beyene, Y., Babu, R., Nair, S., Gowda, M., Das, B., et al., 2015. Quantitative trait loci mapping and molecular breeding for developing stress resilient maize for Sub-Saharan Africa. Crop Science 55 (4), 1449–1459.

Shaar-Moshe, L., Blumwald, E., Peleg, Z., 2017. Unique physiological and transcriptional shifts under combinations of salinity, drought, and heat. Plant Physiology 174 (1), 421–434.

Sharma, N.K., Gupta, S.K., Dwivedi, V., Chattopadhyay, D., 2020. Lignin deposition in chickpea root xylem under drought. Plant Signaling & Behavior 15 (6), 1754621.

Shavrukov, Y., Gupta, N.K., Miyazaki, J., Baho, M.N., Chalmers, K.J., Tester, M., et al., 2010. HvNax3–a locus controlling shoot sodium exclusion derived from wild barley (*Hordeum vulgare* ssp. *spontaneum*). Functional & Integrative Genomics 10 (2), 277–291.

Shinozaki, K., Yamaguchi-Shinozaki, K., 2007. Gene networks involved in drought stress response and tolerance. Journal of Experimental Botany 58 (2), 221–227.

Shivhare, R., Lata, C., 2017. Exploration of genetic and genomic resources for abiotic and biotic stress tolerance in pearl millet. Frontiers in Plant Science 7, 2069.

Shi, J., Gao, H., Wang, H., Lafitte, H.R., Archibald, R.L., Yang, M., et al., 2017. ARGOS 8 variants generated by CRISPR-Cas9 improve maize grain yield under field drought stress conditions. Plant Biotechnology Journal 15 (2), 207–216.

Singh, P., Choudhary, K.K., Chaudhary, N., Gupta, S., Sahu, M., Tejaswini, B., et al., 2022. Salt stress resilience in plants mediated through osmolyte accumulation and its crosstalk mechanism with phytohormones. Frontiers in Plant Science 13, 1006617.

Smita, S., Katiyar, A., Lenka, S.K., Dalal, M., Kumar, A., Mahtha, S.K., et al., 2020. Gene network modules associated with abiotic stress response in tolerant rice genotypes identified by transcriptome meta-analysis. Functional & Integrative Genomics 20 (1), 29–49.

Soltabayeva, A., Ongaltay, A., Omondi, J.O., Srivastava, S., 2021. Morphological, physiological and molecular markers for salt-stressed plants. Plants 10 (2), 243.

Suralta, R.R., Lucob, N.B., Perez, L.M., 2012. Shoot and root development in rice (*Oryza sativa* L.) genotypes during progressive drying in soils with varying moisture regimes. Philippine Journal of Crop Science 37 (3), 1–12.

Svitashev, S., Young, J.K., Schwartz, C., Gao, H., Falco, S.C., Cigan, A.M., 2015. Targeted mutagenesis, precise gene editing, and site-specific gene insertion in maize using Cas9 and guide RNA. Plant Physiology 169 (2), 931–945.

Tamang, B.G., Fukao, T., 2015. Plant adaptation to multiple stresses during submergence and following desubmergence. International Journal of Molecular Sciences 16 (12), 30164–30180.

Tang, K., An, C., Li, L., Sun, T., Song, J., Zhao, J., 2024. Effects of drought and salt stress on the root phenotype of wheat seedlings and underlying gene expression analysis. Frontiers in Plant Science 15, 1475500.

Tang, L., Mao, B., Li, Y., Lv, Q., Zhang, L., Chen, C., et al., 2017. Knockout of OsNramp5 using the CRISPR/Cas9 system produces low Cd-accumulating indica rice without compromising yield. Scientific Reports 7 (1), 14438.

Tian, S., Jiang, L., Cui, X., Zhang, J., Guo, S., Li, M., et al., 2018. Engineering herbicide-resistant watermelon variety through CRISPR/Cas9-mediated base-editing. Plant Cell Reports 37 (9), 1353–1356.

Tian, R., Liu, W., Wang, Y., Wang, W., 2024. Cuticular wax in wheat: biosynthesis, genetics, and the stress response. Frontiers in Plant Science 15, 1498505.

Tran, M.T., Doan, D.T.H., Kim, J., Song, Y.J., Sung, Y.W., Das, S., et al., 2021. CRISPR/Cas9-based precise excision of SlHyPRP1 domain(s) to obtain salt stress-tolerant tomato. Plant Cell Reports 40 (6), 999–1011.

Tripathi, D.K., Singh, V.P., Kumar, D., Chauhan, D.K., 2012a. Rice seedlings under cadmium stress: effect of silicon on growth, cadmium uptake, oxidative stress, antioxidant capacity and root and leaf structures. Chemistry and Ecology 28 (3), 281–291.

Tripathi, D.K., Singh, V.P., Kumar, D., Chauhan, D.K., 2012b. Impact of exogenous silicon addition on chromium uptake, growth, mineral elements, oxidative stress, antioxidant capacity, and leaf and root structures in rice seedlings exposed to hexavalent chromium. Acta Physiologiae Plantarum 34 (1), 279–289.

Turan, M.A., Elkarim, A.H.A., Taban, N., Taban, S., 2009. Effect of salt stress on growth, stomatal resistance, proline and chlorophyll concentrations on maize plant. African Journal of Agricultural Research 4 (9), 893–897.

Ul Haq, S., Khan, A., Ali, M., Khattak, A.M., Gai, W.X., Zhang, H.X., et al., 2019. Heat shock proteins: dynamic biomolecules to counter plant biotic and abiotic stresses. International Journal of Molecular Sciences 20 (21), 5321.

Vaculík, M., Landberg, T., Greger, M., Luxová, M., Stoláriková, M., Lux, A., 2012. Silicon modifies root anatomy, and uptake and subcellular distribution of cadmium in young maize plants. Annals of Botany 110 (2), 433–443.

Vescio, R., Caridi, R., Laudani, F., Palmeri, V., Zappalà, L., Badiani, M., et al., 2022. Abiotic and herbivory combined stress in tomato: additive, synergic, and antagonistic effects and within-plant phenotypic plasticity. Life 12 (11), 1804.

Vikram, P., Swamy, B.P., Dixit, S., Ahmed, H.U., Teresa Sta Cruz, M., Singh, A.K., et al., 2011. $qDTY_{1.1}$, a major QTL for rice grain yield under reproductive-stage drought stress with a consistent effect in multiple elite genetic backgrounds. BMC Genetics 12 (1), 89.

Villordon, A., LaBonte, D., Solis, J., Firon, N., 2012. Characterization of lateral root development at the onset of storage root initiation in 'Beauregard' sweetpotato adventitious roots. HortScience 47 (8), 961–968.

Voesenek, L.A., Bailey-Serres, J., 2013. Flooding tolerance: O2 sensing and survival strategies. Current Opinion in Plant Biology 16 (5), 647–653.

Vollenweider, P., Gunthardt-Goerg, M.S., 2005. Diagnosis of abiotic and biotic stress factors using the visible symptoms in foliage. Environmental Pollution 137 (3), 455–465.

Wang, Y., Cao, Y., Liang, X., Zhuang, J., Wang, X., Qin, F., et al., 2022. A dirigent family protein confers variation of Casparian strip thickness and salt tolerance in maize. Nature Communications 13 (1), 2222.

Wang, F.Z., Chen, M.X., Yu, L.J., Xie, L.J., Yuan, L.B., Qi, H., et al., 2017. OsARM1, an R2R3 MYB transcription factor, is involved in regulation of the response to arsenic stress in rice. Frontiers in Plant Science 8, 1868.

Wang, J., Gao, X., Dong, J., Tian, X., Wang, J., Palta, J.A., et al., 2020. Over-expression of the heat-responsive wheat gene TaHSP23.9 in transgenic Arabidopsis conferred tolerance to heat and salt stress. Frontiers in Plant Science 11, 243.

Wang, X., He, Y., Wei, H., Wang, L., 2021a. A clock regulatory module is required for salt tolerance and control of heading date in rice. Plant, Cell & Environment 44 (10), 3283–3301.

Wang, W.C., Lin, T.C., Kieber, J., Tsai, Y.C., 2019. Response regulators 9 and 10 negatively regulate salinity tolerance in rice. Plant and Cell Physiology 60 (6), 2549–2563.

Wang, T., Wang, F., Deng, S., Wang, K., Feng, D., Xu, F., et al., 2025. Single-cell transcriptomes reveal spatiotemporal heat stress response in maize roots. Nature Communications 16 (1), 177.

Wang, T., Xun, H., Wang, W., Ding, X., Tian, H., Hussain, S., et al., 2021b. Mutation of GmAITR genes by CRISPR/Cas9 genome editing results in enhanced salinity stress tolerance in soybean. Frontiers in Plant Science 12, 779598.

Wang, F., Xu, Y., Li, W., Chen, Z., Wang, J., Fan, F., et al., 2021c. Creating a novel herbicide-tolerance OsALS allele using CRISPR/Cas9-mediated gene editing. The Crop Journal 9 (2), 305–312.

Willson, K.C., 1999. Coffee, Cocoa and Tea. CABI Publishing.

Wishart, J., George, T.S., Brown, L.K., Ramsay, G., Bradshaw, J.E., White, P.J., et al., 2013. Measuring variation in potato roots in both field and glasshouse: the search for useful yield predictors and a simple screen for root traits. Plant and Soil 368 (1-2), 231–249.

World Bank Group, 2024. Food Security Update. Retrieved from: https://www.worldbank.org/en/topic/agriculture/brief/food-security-update.

Wuana, R.A., Okieimen, F.E., 2011. Heavy metals in contaminated soils: a review of sources, chemistry, risks and best available strategies for remediation. ISRN Ecology 2011, 1–20.

Wullschleger, S.D., Yin, T.M., DiFazio, S.P., Tschaplinski, T.J., Gunter, L.E., Davis, M.F., et al., 2005. Phenotypic variation in growth and biomass distribution for two advanced-generation pedigrees of hybrid poplar. Canadian Journal of Forest Research 35 (8), 1779–1789.

Wu, J., Yan, G., Duan, Z., Wang, Z., Kang, C., Guo, L., et al., 2020. Roles of the *Brassica napus* DELLA protein BnaA6.RGA, in modulating drought tolerance by interacting with the ABA signaling component BnaA10.ABF2. Frontiers in Plant Science 11, 577.

Xu, C., Fu, X., Liu, R., Guo, L., Ran, L., Li, C., et al., 2017. PtoMYB170 positively regulates lignin deposition during wood formation in poplar and confers drought tolerance in transgenic Arabidopsis. Tree Physiology 37 (12), 1713–1726.

Xu, Z.Z., Zhou, G.S., 2006. Combined effects of water stress and high temperature on photosynthesis, nitrogen metabolism and lipid peroxidation of a perennial grass *Leymus chinensis*. Planta 224 (5), 1080–1090.

Xu, Z.Z., Zhou, G.S., 2005. Effects of water stress and high nocturnal temperature on photosynthesis and nitrogen level of a perennial grass *Leymus chinensis*. Plant and Soil 269 (1-2), 131–139.

Yan, Z., Wang, J., Wang, F., Xie, C., Lv, B., Yu, Z., et al., 2021. MPK3/6-induced degradation of ARR1/10/12 promotes salt tolerance in Arabidopsis. EMBO Reports 22 (10), e52457.

Yeats, T.H., Rose, J.K., 2013. The formation and function of plant cuticles. Plant Physiology 163 (1), 5–20.

Yue, E., Cao, H., Liu, B., 2020. OsmiR535, a potential genetic editing target for drought and salinity stress tolerance in *Oryza sativa*. Plants 9 (10), 1337.

Yurina, N.P., 2023. Heat shock proteins in plant protection from oxidative stress. Molecular Biology 57 (6), 949–964.

Yu, W., Wang, L., Zhao, R., Sheng, J., Zhang, S., Li, R., et al., 2019. Knockout of SlMAPK3 enhances tolerance to heat stress involving ROS homeostasis in tomato plants. BMC Plant Biology 19 (1), 354.

Zaki, R.N., Radwan, T.E.E., 2019. Improving wheat grain yield and its quality under salinity conditions at a newly reclaimed soil by using different organic sources as soil or foliar applications. Journal of Applied Sciences Research 15 (4), 42–55.

Zeng, Y., Wen, J., Zhao, W., Wang, Q., Huang, W., 2020. Rational improvement of rice yield and cold tolerance by editing the three genes OsPIN5b, GS3, and OsMYB30 with the CRISPR–Cas9 system. Frontiers in Plant Science 10, 1663.

Zhang, M., Duan, L., Zhai, Z., Li, J., Tian, X., Wang, B., et al., 2004. Effects of plant growth regulators on water deficit-induced yield loss in soybean. Proceedings of the 4th International Crop Science Congress, Brisbane, 26 September-1 October 2004. pp. 132.

Zhang, J.H., Huang, W.D., Liu, Y.P., Pan, Q.H., 2005. Effects of temperature acclimation pretreatment on the ultrastructure of mesophyll cells in young grape plants (*Vitis vinifera* L. cv. Jingxiu) under cross-temperature stresses. Journal of Integrative Plant Biology 47 (8), 959–970.

Zhang, A., Liu, Y., Wang, F., Li, T., Chen, Z., Kong, D., et al., 2019. Enhanced rice salinity tolerance via CRISPR/Cas9-targeted mutagenesis of the OsRR22 gene. Molecular Breeding 39 (3), 47.

Zhang, H., Sonnewald, U., 2017. Differences and commonalities of plant responses to single and combined stresses. The Plant Journal 90 (5), 839–855.

Zhang, Y., Xu, J., Li, R., Ge, Y., Li, Y., Li, R., 2023. Plants' response to abiotic stress: Mechanisms and strategies. International Journal of Molecular Sciences 24 (13), 10915.

Zhang, H., Zhang, J., Wei, P., Zhang, B., Gou, F., Feng, Z., et al., 2014. The CRISPR/Cas9 system produces specific and homozygous targeted gene editing in rice in one generation. Plant Biotechnology Journal 12 (6), 797–807.

Zhang, H., Zhu, J., Gong, Z., Zhu, J.K., 2022. Abiotic stress responses in plants. Nature Reviews Genetics 23 (2), 104–119.

Zhao, W., Sun, Y., Kjelgren, R., Liu, X., 2015. Response of stomatal density and bound gas exchange in leaves of maize to soil water deficit. Acta Physiologiae Plantarum 37 (8), 1704.

Zheng, C., Jiang, D., Liu, F., Dai, T., Jing, Q., Cao, W., 2009b. Effects of salt and waterlogging stresses and their combination on leaf photosynthesis, chloroplast ATP synthesis, and antioxidant capacity in wheat. Plant Science 176 (4), 575–582.

Zheng, C., Jiang, D., Liu, F., Dai, T., Liu, W., Jing, Q., et al., 2009a. Exogenous nitric oxide improves seed germination in wheat against mitochondrial oxidative damage induced by high salinity. Environmental and Experimental Botany 67 (1), 222–227.

Zhou, Y., Zhang, D., Pan, J., Kong, X., Liu, Y., Sun, L., et al., 2012. Overexpression of a multiple stress-responsive gene, ZmMPK4, enhances tolerance to low temperature in transgenic tobacco. Plant Physiology and Biochemistry 58, 174–181.

Zhu, J., Brown, K.M., Lynch, J.P., 2010. Root cortical aerenchyma improves the drought tolerance of maize (*Zea mays* L.). Plant, Cell & Environment 33 (5), 740–749.

Zhu, X., Li, W., Zhang, N., Jin, H., Duan, H., Chen, Z., et al., 2024. StMAPKK5 responds to heat stress by regulating potato growth, photosynthesis, and antioxidant defenses. Frontiers in Plant Science 15, 1392425.

Zhu, J.K., 2016. Abiotic stress signaling and responses in plants. Cell 167 (2), 313–324.

Zhu, R., Wu, F.Y., Zhou, S., Hu, T., Huang, J., Gao, Y., 2020. Cumulative effects of drought-flood abrupt alternation on the photosynthetic characteristics of rice. Environmental and Experimental Botany 169, 103901.

Chapter 4

Photosynthetic and stomatal regulation to multiple abiotic stresses in plants

Tianyang Liu, Lan Zhu*, and Yali Zhang

Key Laboratory of Oasis Eco-agriculture, Xinjiang Production and Construction Corps/College of Agriculture, Shihezi University, Shihezi, Xinjiang, P.R. China

Corresponding author. e-mail address: zhulan@shzu.edu.cn

4.1 Introduction

Abiotic stresses refer to adverse conditions brought about by nonliving things, such as drought, heat, light level, salinity, and soil pH, which negatively impact plant growth and development. It will affect many important plant activities, such as photosynthesis and stomatal opening. Photosynthesis is vital to plant growth, and stomata are major gas routes. The stomatal opening and closing can directly control CO_2 absorption, and water loss affects photosynthesis. It will have a huge impact on plant survival and distribution (Fig. 4.1).

Under abiotic stress, photosynthesis and stomata make a lot of changes to help plants adapt to their surroundings. Like, for instance, when it is super dry, plants tend to close up tiny air holes on leaves to keep water from going away, but this also means less CO_2 can get into the plant, and that makes the plant's sun-catching machine (photosynthesis) go slower (Fig. 4.2). High-temperature stress causes the leaves to become thicker and have more chlorophyll to make photosynthesis better, and it also produces heat shock proteins (HSPs), which make plants more tolerant to heat. In low light, plants usually increase leaf size and raise chlorophyll concentration to absorb more light and carefully control stomata opening to take in enough but conserve water.

Investigating the regulatory roles of photosynthesis and stomata under abiotic stress can elucidate the mechanisms by which plants adapt to environmental changes, thereby offering potential strategies for enhancing agricultural productivity.

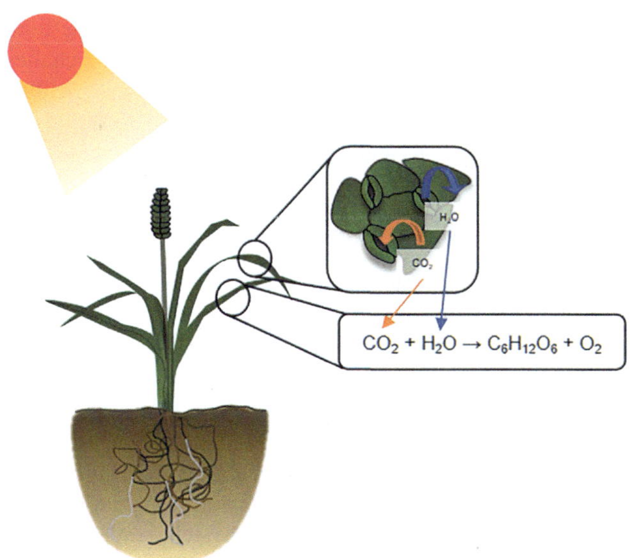

FIGURE 4.1 The relationship between stomatal function and photosynthesis.

Multiple Abiotic Stresses in Plants. DOI: https://doi.org/10.1016/B978-0-443-27424-4.00004-5

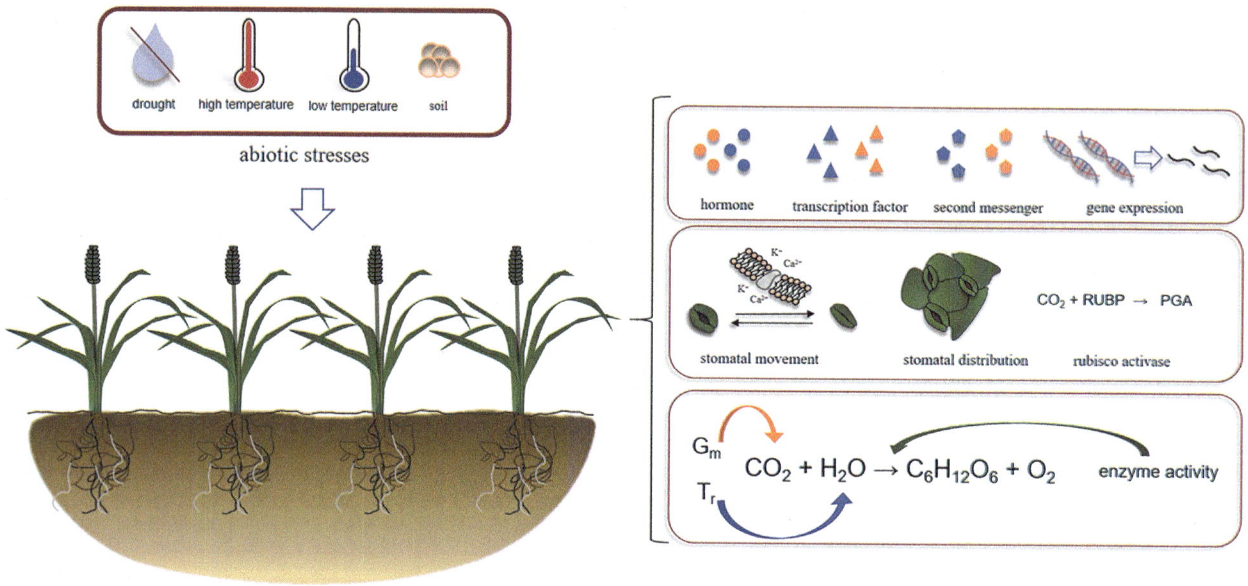

FIGURE 4.2 Photosynthetic and stomatal regulation to abiotic stress in plants.

Improving stomatal responsiveness in plants may increase photosynthetic efficiency and biomass production under fluctuating light conditions. Such advancements could enhance plant resilience to climate change. A thorough understanding of both stomatal and nonstomatal limitations to CO_2 uptake is essential for improving photosynthetic efficiency and crop yield. This knowledge provides new insights into plant stress physiology, ecology, and molecular biology, ultimately contributing to more sustainable agriculture and ecological conservation (Blatt et al., 2024).

4.2 Adaptive changes in stomatal morphology in response to abiotic stress

Leaf stomata play a crucial role in plant physiological processes and mediate interactions between plants and the atmosphere (Wei and Pan, 2003). As the primary sites of gas exchange, stomata regulate the uptake of CO_2 for photosynthesis and the loss of water vapor through transpiration (Zhao et al., 2015). They respond to internal and environmental signals to maintain an optimal balance between carbon acquisition for mesophyll photosynthesis, water use efficiency, and evaporative cooling (Lawson and McElwain, 2016).

Stomata spatial pattern, including density, location, formation, shape, aperture size, and depth, strongly influences stomata function and CO_2 diffusion (Lawson and Matthews, 2020). Alterations in these morphological traits represent an important adaptive strategy through which plants respond to environmental challenges. By regulating stomatal morphology, plants are able to minimize water loss while preserving vital physiological processes under abiotic stressors such as drought, salinity, and high temperatures. The following section will focus on the adaptive changes in stomatal density distribution and stomatal size in response to abiotic stress.

4.2.1 Adjustment of pore density and distribution

Stomatal density plays a significant role in plant evolution and environmental adaptation (Lawson and Matthews, 2020). Through long-term evolution, plants have developed a thickened cuticular layer surrounding their aerial organs, which is mainly accompanied by a general reduction in the number of stomata. However, stomata serve as major structures for gaseous exchange, and the stomatal density of many lineages has shown an increasing trend (Zhao et al., 2015). On leaves, stomata display specific distribution patterns, densities, and placements that are strongly influenced by environmental conditions. By regulating stomatal formation, density, and the differentiation between adaxial and abaxial types, plants precisely regulate gas exchange in response to environmental demands.

In amphistomatous species, the adaxial and abaxial stomata frequently operate independently, demonstrating distinct responses to environmental variables such as light intensity and CO_2 levels (Wall et al., 2022). Plants exhibiting rapid stomatal behavior often feature a more homogeneous distribution across both epidermal surfaces, correlating with enhanced photosynthetic performance (Xiong and Flexas, 2020). Improving both stomatal density and dynamic responsiveness thus represents a promising strategy for augmenting photosynthetic efficiency in crops, with potential benefits for carbon assimilation and water use economy (Haworth et al., 2018).

Stomatal patterning across leaf surfaces exhibits a remarkable degree of spatial organization, reflecting evolutionary adaptations to environmental constraints. Species native to humid habitats typically develop higher stomatal densities on both leaf surfaces and petioles, which facilitates efficient gas exchange. Some plant taxa show plastic responses to abiotic stress by dynamically adjusting stomatal distribution. For example, under drought conditions, certain species reduce the total number of stomata while concentrating them primarily on the abaxial surface. This redistribution helps minimize direct radiative exposure and reduces water loss, while still maintaining a limited capacity for gas exchange.

The distribution of stomata across leaf surfaces often displays distinct regional characteristics, following specific organizational patterns. Plants adapted to humid or wet conditions usually exhibit a high density of stomata on both sides of the leaf as well as on the petiole, thereby supporting enhanced water vapor exchange. Some species improve their environmental adaptability by modulating stomatal distribution in response to abiotic stress. For example, under drought stress, certain plants not only reduce the total number of stomata but also confine them to the abaxial (lower) side of the leaf. This adaptation helps avoid excessive solar exposure and reduces water loss, contributing to the plant's overall resilience.

4.2.2 Variation in stomatal size and its response to stress

Stomatal size plays a crucial role in governing how efficiently plants exchange gases. These microscopic pores control both the uptake of carbon dioxide and the release of water vapor. Differences in stomatal size reflect a significant adaptive response to environmental stresses.

Under drought or high-temperature stress, plants reduce the stomatal aperture to reduce water evaporation. This response is mediated by water status and hormonal signaling pathways, leading to stomatal closure. Such rapid regulation enables plants to conserve water efficiently under extreme environmental conditions. For example, under well-watered conditions, open stomata facilitate evaporative cooling and help regulate leaf temperature; during water deficit, however, stomatal closure reduces water loss at the expense of restricted CO_2 assimilation and potentially higher leaf temperatures (Caine et al., 2019). Additionally, some drought-adapted species exhibit a permanently reduced stomatal density, thereby minimizing water loss. These morphological adjustments arise from a combination of genetic adaptations and physiological responses that enhance survival under water-deficient conditions.

The coordination between photosynthetic rate and stomatal conductance is influenced by the diffusive properties of stomata, which modulate CO_2 intake and water vapor release (Al-Salman et al., 2023). Species exhibiting high stomatal responsiveness can utilize transient light events, such as sunflecks, by rapidly opening stomata to enhance CO_2 assimilation under carbon-limited photosynthetic conditions (Harrison et al., 2020). In contrast, during low-light or dark periods, these plants exhibit swift stomatal closure to reduce nonproductive water loss. Such plasticity in stomatal behavior constitutes an integral adaptive strategy, enabling efficient resource allocation under variable environmental conditions.

Through evolutionary adaptation to abiotic stresses, plants often exhibit directional changes in stomatal morphology. Species from arid regions typically develop smaller stomata and prolonged stomatal closure to reduce water loss (Tulva et al., 2024). In contrast, wetland plants generally possess larger stomata, enabling sustained high photosynthetic rates under conditions of ample water availability. These differences reflect adaptive evolutionary responses shaped by environmental pressures over time.

In summary, stomata play a crucial role in plant adaptation to abiotic stress. By modulating stomatal density, size, morphology, and behavior, plants can reduce water loss and enhance photosynthetic efficiency under adverse conditions. Drought, high temperatures, and salinity trigger stomatal adjustments that help conserve water, maintain carbon assimilation, and support continued growth and reproduction. As climate change exacerbates environmental extremes, understanding the mechanisms behind stomatal plasticity will be essential for developing more resilient crop varieties and promoting ecosystem restoration.

4.3 Physiological and biochemical regulation mechanisms of stomatal behavior and photosynthesis under stress

4.3.1 Signal transmission in stomatal opening and closing

Stomatal movement is regulated by changes in the osmotic potential of guard cells, which is influenced by multiple factors, including hormones, enzymes, and internal water status. Abscisic acid (ABA) serves as a key signaling hormone that induces stomatal closure. ABA binds to specific receptors on the plasma membrane, leading to the activation of protein kinases such as OPEN STOMATA 1 (OST1), a central regulator in ABA-mediated closure. Additionally, the brassinosteroid signaling component CONSTITUTIVE DIFFERENTIAL GROWTH 1 HOMOLOG (CDL1) also activates OST1, further promoting stomatal closure synergistically (Xu et al., 2025).

ABA triggers stomatal closure through the regulation of three major types of ion channels: anion, potassium, and calcium channels. The activation of S-type anion channels induces membrane hyperpolarization. ABA disrupts ionic balance across the guard cell membrane, resulting in potassium efflux, loss of turgor pressure, and subsequent stomatal closure. Furthermore, ABA promotes the opening of calcium-permeable channels, leading to an influx of extracellular Ca^{2+}, and stimulates the production of H_2O_2. Elevated H_2O_2 enhances the activity of calcium channels, amplifying Ca^{2+} signaling.

Second messengers, including inositol trisphosphate (IP_3) and cyclic ADP-ribose, contribute to the rise in cytosolic Ca^{2+} by triggering its release from intracellular stores, thereby modulating stomatal movement. Beyond these mechanisms, ABA also interacts with methyl jasmonate (JA) signaling. JA induces the generation of reactive oxygen species (ROS) and nitric oxide (NO), which in turn regulate Ca^{2+} fluxes—both efflux and influx—across cellular membranes (Daszkowska-Golec and Szarejko, 2013).

4.3.2 Photosynthesis under different stress conditions is also regulated by stomatal regulation

Photosynthesis is a fundamental process through which plants convert light energy into chemical energy, essential for growth and survival. Key enzymes involved include ribulose-1,5-bisphosphate carboxylase/oxygenase (Rubisco) and phosphoenolpyruvate carboxylase (PEPC). Rubisco catalyzes the carboxylation of RuBP in the Calvin cycle, which serves as the primary carbon fixation pathway in most plants. In contrast, PEPC participates in alternative carbon fixation mechanisms, especially in C_4 and CAM plants (Song et al., 2014).

Environmental stresses, such as high temperature, low temperature, and salinity, significantly inhibit the activity and expression of these enzymes, thereby reducing photosynthetic efficiency and impairing the plant's ability to adapt to adverse conditions (Burgess, 2024). For instance the activity of Rubisco declines markedly under both heat and cold stress, negatively affecting carboxylation efficiency and limiting carbon capture. This reduction impedes the synthesis of glyceraldehyde-3-phosphate (G3P), a crucial intermediate for producing sucrose and starch in the Calvin cycle. Furthermore, extreme temperatures can induce conformational changes and even lead to structural denaturation of these enzymes, further compromising their catalytic function. Stresses also harm PSII, a major part of the light-dependent reaction, which is what splits water up and lets out oxygen and gives electrons to the transport train in photosynthesis. PSII impairment impairs electron flow, lowers ATP synthesis, and limits energy availability for the Calvin cycle (Mathur et al., 2014). This damage is also connected with oxidative stress—too much oxidation of chlorophyll results in free radical formation, which encourages degradation of PSII proteins (Björn and McKenzie, 2008).

Extreme temperatures, both high and low, adversely affect the expression of genes encoding key photosynthetic components such as Rubisco, PEPC, and protein subunits of PSII and PSI. Downregulation of these genes impairs chlorophyll synthesis and disrupts the photosynthetic process as a whole.

Salinity stress also inhibits photosynthetic function by directly affecting enzymes, including Rubisco and ATP synthase. The high osmotic pressure induced by salt leads to water loss and ion imbalance, disrupting cellular homeostasis and enzyme activity. Additionally, salt stress downregulates photosynthesis-related genes, further compromising the synthesis and functionality of critical components such as Rubisco and PSII-associated proteins (Pinheiro and Chaves, 2011).

The impact of altered key photosynthetic enzymes on secondary metabolism involves complex regulatory mechanisms, wherein Rubisco activity is modulated by Rubisco activase (RCA) and changes in the chloroplast stromal environment. A decline in Rubisco activity reduces carbon fixation, leading to decreased production of primary photosynthetic assimilates, which in turn influences the biosynthesis of secondary metabolites. Transcriptional regulation also plays a critical role; for instance, cell-specific changes in light-induced gene expression affect photosynthetic efficiency and metabolic flux, resulting in differential accumulation of secondary metabolites. Phytochrome-interacting factors (PIFs) and histone deacetylase 15 (HDA15) interact to coregulate light-responsive genes involved in chlorophyll biosynthesis and photosynthesis, ultimately influencing plant growth and secondary metabolism. Furthermore, damage to PSII impairs electron transport and ATP synthesis, causing photoinhibition and altering the internal redox state and energy supply, thereby disrupting both photosynthetic efficiency and secondary metabolite synthesis.

The production of secondary metabolites is governed by highly specific enzymes, such as those regulating alkaloid synthesis, whose expression and activity can be affected by changes in core photosynthetic enzymes. Strategies to enhance the synthesis of target compounds include increasing the copy number of key transcription factors or modulating the expression of enzymatic genes to coordinate multiple metabolic pathways. This integrated approach represents an emerging direction in the genetic engineering of plant secondary metabolism (Llorente et al., 2017).

4.4 Molecular mechanisms regulating stomatal movement and photosynthesis under stress

4.4.1 Genes associated with stomatal development

Stomatal development is a precisely regulated process involving molecular signaling, asymmetric cell division, cell polarity, environmental cues, and hormonal influences. This developmental sequence, known as the stomatal lineage, begins when multipotent epidermal stem cells differentiate into stomatal lineage cells and transition into meristemoid mother cells (MMCs). The MMC undergoes an asymmetric division, producing a larger stomatal lineage ground cell (SLGC) and a smaller, triangular meristemoid cell (MC). These cells pursue distinct developmental paths: the MC may differentiate directly into a guard mother cell (GMC), which then divides symmetrically to form a stoma, or it may undergo one to three amplifying divisions, generating additional SLGCs and late MCs. The latter eventually differentiate into GMCs and form stomata (Zoulias et al., 2018). SLGCs possess stem cell activity and can either become pavement cells or undergo asymmetric spacing divisions to produce satellite meristemoids (SMCs), which develop into new MCs. This process adheres to the "single cell spacing" rule, ensuring proper stomatal patterning and function (Casson and Gray, 2008).

The initiation and progression of stomatal development are governed by a core regulatory network of basic helix–loop–helix (bHLH) transcription factors, including SPCH SPEECHLESS, MUTE, and FAMA, along with their interacting partners SCRM/ICE1 and SCRM2 (Endo and Torii, 2019; Liu et al., 2015). Stomatal lineage specification begins in protodermal cells, with SPCH acting as a key regulator that promotes the transition into MMC and subsequent asymmetric divisions. Loss-of-function SPCH mutants fail to initiate stomatal lineage and produce only disordered epidermal cells, leading to seedling lethality. Conversely, SPCH overexpression induces excessive asymmetric divisions and sustained meristemoid-like stem cell states. The transition from MC to GMC is controlled by MUTE, which terminates the proliferative meristemoid state and promotes GMC formation. MUTE mutants exhibit arrested meristemoids that continue dividing without differentiating into GMCs, resulting in rosette-like clusters and a complete absence of functional stomata, which also causes seedling death. Overexpression of MUTE, however, drastically increases stomatal density. FAMA regulates the final step of stomatal development, guiding the symmetric division of GMCs into paired guard cells. It also represses SPCH and MUTE to irreversibly exit the proliferative phase. FAMA mutants exhibit uncontrolled symmetric divisions in GMCs, producing caterpillar-like cell files instead of stomata, whereas FAMA overexpression results in unpaired guard cells. SPCH, MUTE, and FAMA physically interact with SCRM and SCRM2 via conserved domains to reinforce stomatal cell fate decisions. Although the development of dumbbell-shaped guard cells (as in grasses) is less fully understood, it is similarly regulated by these bHLH factors (Kanaoka et al., 2008).

Small peptide signaling molecules also play critical roles in stomatal patterning. EPF1 and EPF2 act as negative regulators of stomatal density: EPF2 acts early to restrict MMC formation, while EPF1 influences stomatal spacing and distribution. Overexpression of either peptide reduces stomatal density. In contrast, EPFL9, expressed in mesophyll tissue, promotes stomatal development, and its overexpression leads to stomatal clustering. Receptor kinases such as ERECTA (ER), ERL1 and ERL2 are also involved in regulating stomatal patterning. While single mutants show mild phenotypes, the ER, ERL1 and ERL2 triple mutant exhibits severe stomatal clustering, indicating that these receptors collectively suppress ectopic stomatal formation and help maintain the one-cell spacing rule. ERL1 and ERL2 are particularly important in preserving meristemoid stem cell activity and preventing premature differentiation. TMM acts as a negative regulator of stomatal density and spacing and is expressed specifically in stomatal lineage cells. tmm mutants display stomatal clusters, highlighting their role in enforcing spacing patterns. Similarly, SDD1, a subtilisin-like serine protease, negatively regulates stomatal density. sdd1 mutants exhibit clustered stomata, and their overexpression in species such as maize and tomato reduces stomatal density (Morales-Navarro et al., 2018).

In the final stage of stomatal maturation, the DOF transcription factor SCAP1 (STOMATAL CARPENTER 1) is expressed in guard cells and influences stomatal structure and function by modulating the expression of potassium channels, the MYB60 transcription factor, and pectin methylesterases (Moriwaki et al., 2022). In summary, stomatal development is orchestrated by a tightly coordinated network of transcription factors, peptide signals, and receptor-mediated pathways that together ensure proper stomatal density, spacing, and functionality—each critical for plant growth and adaptation.

4.4.2 Stomatal-related genes regulate their expression in response to stress

Stomata are essential for limiting water loss, and exhibit high sensitivity to water stress, including soil moisture deficit and low atmospheric humidity. The Drought Responsive Element Binding (DREB) gene family helps coordinate this adaptive response by regulating expression of drought-related genes and modulating stomatal behavior, which collectively reduce transpirational water loss and improve tolerance to drought, salinity, and cold. Under water stress, stomatal regulatory genes such as ABA insensitive (ABI) and MYB family genes enhance ABA biosynthesis and strengthen ABA signaling. Accordingly, studies of stomatal responses to water scarcity often emphasize ABA signal transduction within guard cells.

Other ABA-modulated components—including U-box ubiquitin ligases, GHR1, and EAR1 proteins—also contribute to stomatal regulation. When ABA synthesis in roots is low, the peptide signal CLE25 (CLAVATA3/ESR-RELATED 25) acts as a long-distance messenger moving from roots to leaves under soil-drying conditions, triggering local ABA accumulation and stomatal closure. ABA also mediates plant adaptation to saline, alkaline, and heavy metal stressors.

Notably, under high vapor pressure deficit (VPD), stomatal closure persists even in the pyr1pyl1pyl2pyl4pyl5pyl8 hexuple mutant deficient in multiple ABA receptors, indicating the participation of non-ABA signaling in VPD-induced closure (Merilo et al., 2018). The protein kinase OST1, a key component of ABA signaling, contains two regulatory domains: an ABA-box, essential for ABA-dependent activation, and a SnRK2-box, which enables activation independently of ABA under high VPD. Mutants lacking the ABA-box do not respond to ABA but still close stomata under high VPD, illustrating OST1's role in integrating both ABA-dependent and ABA-independent pathways (Yoshida et al., 2006).

Stomatal movement is a key adaptive response of plants to temperature changes, which helps plants retain water and control leaf temperature. At high temperatures, potassium channels in guard cells play a central role in regulating stomatal behavior. In Arabidopsis, several members of the 14-3-3 protein family promote stomatal opening under heat stress. These proteins bind and activate plasma membrane H^+- ATPase, leading to membrane hyperpolarization. This, in turn, triggers the activation of inward rectifying K channels, enhances potassium absorption, and ultimately leads to stomatal opening (Szyroki et al., 2001). In addition, this temperature-induced stomatal opening is closely related to the blue light signaling pathway through thermosensitive photosensitizer receptors.

Under cold stress, the influx of Ca^{2+} into guard cells leads to stomatal closure, even if a slight increase in cytoplasmic Ca^{2+} concentration triggers this reaction. The transcription factor C repeat binding factor (CBF) activates the expression of cold-responsive genes, which helps maintain stomatal function at low temperatures. ICE1 (CBF expression inducer 1) is an upstream regulator of the CBF family, and its protein stability is crucial for cold adaptation. After activation of MKK5 (Mitogen-Activated Protein Kinase Kinase 5), MPK3 (Mitogen-Activated Protein Kinase 3), and MPK6 (Mitogen-Activated Protein Kinase 6) negatively regulate cold signal transduction through phosphorylation of ICE1. This modification promotes ICE1 degradation, followed by inhibition of the CBF pathway and cold tolerance (Zhao et al., 2023).

In addition, some transcription factors (such as members of the NAC family) are upregulated during waterlogging stress. These factors contribute to the regulation of stomatal-related genes, thereby enhancing the adaptive resilience of plants (Saimi et al., 2023). Similarly, under intense-light stress, the levels of Ca^{2+} in guard cells undergo changes, which trigger protective mechanisms to reduce light damage. A common strategy is ABA-induced stomatal closure, which helps to preserve water under strong light.

Under heavy metal stress, the HSP gene family contributes to protein folding and repair, helps alleviate cell damage, and indirectly regulates stomatal function. Concurrently, MATE (multidrug and toxic compound extrusion) transporters mediate the redistribution of heavy metal ions, which in turn influences ion homeostasis and water regulation. Both saline-alkali and heavy metal stresses are known to alter K^+ channel activity. In particular, the GORK (gated outward-rectifying K^+ channel) serves as a critical regulator of stomatal closure. Additionally, ethylene response factors are involved in a range of stress responses and fine-tune stomatal behavior through the transcriptional control of genes associated with stomatal movement.

4.4.3 Regulation of stomatal–photosynthetic pathways under abiotic stress

Abiotic stresses such as drought, salinity, and extreme temperatures threaten plant survival by restricting water availability and impairing photosynthetic function. To adapt to these stresses, plants have developed a finely tuned regulatory system that coordinates stomatal movement with photosynthetic activity. This system incorporates hormonal signals, transcriptional control, messenger molecules, along with epigenetic and posttranscriptional mechanisms, all of which act together to improve water use efficiency and support carbon assimilation under stress.

ABA functions as a primary stress-signaling hormone. Its levels rise under drought or saline conditions, triggering binding to the PYR/PYL/RCAR family of receptors and subsequent activation of the SnRK2 kinase OST1 (Schroeder et al., 2001). OST1 phosphorylates the SLAC1 anion channel, stimulating ion efflux from guard cells, loss of turgor, and stomatal closure. While effective in limiting water loss, this response also reduces CO_2 uptake. To compensate, parallel signaling pathways—such as those involving brassinosteroids—help sustain photosynthetic capacity even when stomata are partially closed (Assmann and Jegla, 2016). Calcium-dependent protein kinases (CDPKs) further modulate stomatal responses by incorporating Ca^{2+} signals initiated by ABA (Kim et al., 2010).

CDPKs further modulate stomatal responses by integrating Ca^{2+} signals activated through ABA-dependent pathways (Kim et al., 2010). Transcription factors play indispensable roles in cellular adaptation to environmental constraints. During stomatal development, three core basic helix–loop–helix transcription factors—SPEECHLESS (SPCH), MUTE, and FAMA—govern cell fate decisions: SPCH initiates asymmetric division and lineage entry, MUTE drives GMC formation, and FAMA ensures terminal differentiation into guard cells (Cai et al., 2017). Under stress conditions, the activity and expression of these developmental regulators are modified, leading to alterations in stomatal density and morphology. Simultaneously, stress-responsive transcription factors from the NAC and WRKY (conserved amino acid sequence Trp-Arg-Lys-Tyr) families participate in reprogramming gene expression related to stomatal function and photosynthetic efficiency. For example, specific WRKY transcription factors enhance drought tolerance by activating ABA-responsive genes while maintaining photosynthetic performance under water deficit (Lawson and Blatt, 2014).

ROS and calcium ions (Ca^{2+}) serve as critical secondary messengers in plant stress responses. Under abiotic stress, ROS accumulation in guard cells and chloroplasts activates MAPK cascades, subsequently modifying ion channel activity and altering transcriptional regulator function (Chen et al., 2015). Simultaneously, oscillatory Ca^{2+} signals activate CDPKs, enabling precise control over ion fluxes. The consequent reduction in CO_2 availability suppresses Rubisco activity and inhibits the Calvin-Benson cycle. In response, plants enhance nonphotochemical quenching (NPQ) to dissipate excess light energy and minimize photo-oxidative damage (Yamori et al., 2014).

Beyond transcriptional regulation, epigenetic mechanisms—including DNA methylation and histone modifications—contribute to long-term adaptation by modulating stress-responsive gene expression in guard cells and mesophyll tissues. Posttranslational modifications such as phosphorylation, ubiquitination, and sumoylation offer an additional regulatory layer, enabling rapid and reversible adjustment of protein activity and stability under fluctuating environmental conditions (Wang et al., 2016).

4.5 Conclusion

In summary, plants deploy an integrated signaling network to combat abiotic stress, incorporating hormonal signals (e.g., ABA and brassinosteroids), developmental transcription factors (SPCH, MUTE, FAMA), stress-responsive regulators (WRKY family proteins), secondary messengers (ROS and Ca^{2+}), and epigenetic and posttranslational mechanisms. This system maintains a dynamic balance between water conservation and carbon assimilation, significantly improving plant survival under adverse conditions. A deeper understanding of these mechanisms provides a foundation for developing stress-resilient crops through targeted breeding strategies.

References

Al-Salman, Y., Ghannoum, O., Cano, F.J., 2023. Midday water use efficiency in sorghum is linked to faster stomatal closure rate, lower stomatal aperture and higher stomatal density. The Plant Journal 115 (6), 1661–1676.

Assmann, S.M., Jegla, T., 2016. Guard cell sensory systems: recent insights on stomatal responses to light, abscisic acid, and CO_2. Current Opinion in Plant Biology 33, 157–167. Available at https://doi.org/10.1016/j.pbi.2016.07.003.

Björn, L.O., McKenzie, R.L., 2008. Ozone Depletion and the Effects of ultraviolet radiation[M]//Photobiology: The Science of Life and Light. Springer New York, New York, NY, pp. 503–530.

Blatt, M.R., Cavanagh, A., Buckley, T.N., 2024. Photosynthesis and the stomatal nexus, past, present and future. Plant, Cell & Environment (9), 47.

Burgess, A.J., 2024. Double trouble: compound effects of heat and drought stress on carbon assimilation. Plant Physiology 194 (3), 1255–1256. Available at https://doi.org/10.1093/plphys/kiad647.

Caine, R.S., Yin, X., Sloan, J., et al., 2019. Rice with reduced stomatal density conserves water and has improved drought tolerance under future climate conditions. New Phytologist 221 (1), 371–384.

Cai, S., Papanatsiou, M., Blatt, M.R., et al., 2017. Speedy grass stomata: emerging molecular and evolutionary features. Molecular Plant 10 (7), 912–914.

Casson, S., Gray, J.E., 2008. Influence of environmental factors on stomatal development. New Phytologist 178 (1), 9–23.

Chen, T.S.U.W.E.I., Kahlen, K., Stützel, H., 2015. Disentangling the contributions of osmotic and ionic effects of salinity on stomatal, mesophyll, biochemical and light limitations to photosynthesis. Plant, Cell & Environment 38 (8), 1528–1542.

Daszkowska-Golec, A., Szarejko, I., 2013. Open or close the gate–stomata action under the control of phytohormones in drought stress conditions. Frontiers in Plant Science 4, 138.

Endo, H., Torii, K.U., 2019. Stomatal development and perspectives toward agricultural improvement. Cold Spring Harbor Perspectives in Biology 11 (5), a034660.

Harrison, E.L., Arce Cubas, L., Gray, J.E., Hepworth, C., 2020. The influence of stomatal morphology and distribution on photosynthetic gas exchange. The Plant Journal 101 (4), 768–779. Available at https://doi.org/10.1111/tpj.14560.

Haworth, M., Scutt, C.P., Douthe, C., Marino, G., Gomes, M.T.G., Loreto, F., et al., 2018. Allocation of the epidermis to stomata relates to stomatal physiological control: stomatal factors involved in the evolutionary diversification of the angiosperms and development of amphistomaty. Environmental and Experimental Botany 151, 55–63. Available at https://doi.org/10.1016/j.envexpbot.2018.04.010.

Kanaoka, M.M., Pillitteri, L.J., Fujii, H., Yoshida, Y., Bogenschutz, N.L., Takabayashi, J., et al., 2008. SCREAM/ICE1 and SCREAM2 specify three cell-state transitional steps leading to Arabidopsis stomatal differentiation. The Plant Cell 20 (7), 1775–1785. Available at https://doi.org/10.1105/tpc.108.060848.

Kim, T.H., Böhmer, M., Hu, H., Nishimura, N., Schroeder, J.I., 2010. Guard cell signal transduction network: advances in understanding abscisic acid, CO_2, and Ca^{2+} signaling. Annual Review of Plant Biology 61, 561–591. Available at https://doi.org/10.1146/annurev-arplant-042809-112226.

Lawson, T., Blatt, M.R., 2014. Stomatal size, speed, and responsiveness impact on photosynthesis and water use efficiency. Plant Physiology 164 (4), 1556–1570. Available at https://doi.org/10.1104/pp.114.237107.

Lawson, T., Matthews, J., 2020. Guard cell metabolism and stomatal function. Annual Review of Plant Biology 71 (1), 273–302.

Lawson, T., McElwain, J.C., 2016. Evolutionary trade-offs in stomatal spacing. New Phytologist 210 (4), 1149–1151.

Liu, Y., Qin, L., Han, L., et al., 2015. Overexpression of maize SDD1 (ZmSDD1) improves drought resistance in Zea mays L. by reducing stomatal density. Plant Cell, Tissue and Organ Culture (PCTOC) 122 (1), 147–159.

Llorente, B., Martinez-Garcia, J.F., Stange, C., Rodriguez-Concepcion, M., 2017. Illuminating colors: regulation of carotenoid biosynthesis and accumulation by light. Current Opinion in Plant Biology 37, 49–55. Available at https://doi.org/10.1016/j.pbi.2017.03.011.

Mathur, S., Agrawal, D., Jajoo, A., 2014. Photosynthesis: response to high temperature stress. Journal of Photochemistry and Photobiology B: Biology 137, 116–126.

Merilo, E., Yarmolinsky, D., Jalakas, P., Parik, H., Tulva, I., Rasulov, B., et al., 2018. Stomatal VPD response: there is more to the story than ABA. Plant Physiology 176 (1), 851–864. Available at https://doi.org/10.1104/pp.17.00912.

Morales-Navarro, S., Pérez-Díaz, R., Ortega, A., de Marcos, A., Mena, M., Fenoll, C., et al., 2018. Overexpression of a SDD1-like gene from wild tomato decreases stomatal density and enhances dehydration avoidance in arabidopsis and cultivated tomato. Frontiers in Plant Science 9. Available at https://doi.org/10.3389/fpls.2018.00940.

Moriwaki, K., Yanagisawa, S., Iba, K., Negi, J., 2022. Two independent cis -acting elements are required for the guard cell-specific expression of SCAP1, which is essential for late stomatal development. The Plant Journal 110 (2), 440–451. Available at https://doi.org/10.1111/tpj.15679.

Pinheiro, C., Chaves, M.M., 2011. Photosynthesis and drought: can we make metabolic connections from available data? Journal of Experimental Botany 62 (3), 869–882.

Saimi, G., Wang, Z., Liusui, Y., Guo, Y., Huang, G., Zhao, H., et al., 2023. The functions of an NAC transcription factor, GhNAC2-A06, in cotton response to drought stress. Plants 12 (21), 3755. Available at https://doi.org/10.3390/plants12213755.

Schroeder, J.I., Allen, G.J., Hugouvieux, V., et al., 2001. Guard cell signal transduction. Annual Review of Plant Biology 52 (1), 627–658.

Song, Y., Chen, Q., Ci, D., et al., 2014. Effects of high temperature on photosynthesis and related gene expression in poplar. BMC Plant Biology 14 (1), 111.

Szyroki, A., Ivashikina, N., Dietrich, P., Roelfsema, M.R.G., Ache, P., Reintanz, B., et al., 2001. KAT1 is not essential for stomatal opening. Proceedings of the National Academy of Sciences 98 (5), 2917–2921. Available at https://doi.org/10.1073/pnas.051616698.

Tulva, I., Koolmeister, K., Hõrak, H., 2024. Low relative air humidity and increased stomatal density independently hamper growth in young Arabidopsis. The Plant Journal 119 (6), 2718–2736.

Wall, S., Vialet-Chabrand, S., Davey, P., Van Rie, J., Galle, A., Cockram, J., et al., 2022. Stomata on the abaxial and adaxial leaf surfaces contribute differently to leaf gas exchange and photosynthesis in wheat. New Phytologist 235 (5), 1743–1756. Available at https://doi.org/10.1111/nph.18257.

Wang, D., Yu, Y., Liu, Z., et al., 2016. Membrane-bound NAC transcription factors in maize and their contribution to the oxidative stress response. Plant Science 250, 30–39. Available at 10.1016/j.plantsci.2016.05.019.

Wei, R., Pan, X., 2003. China Gas exchange characteristics of four desert shrubs. Proceedings of SPIE - The International Society for Optical Engineering 4890 (1), 416–421. Available at https://doi.org/10.1117/12.466258.

Xiong, D., Flexas, J., 2020. From one side to two sides: the effects of stomatal distribution on photosynthesis. New Phytologist 228 (6), 1754–1766.

Xu, X., Liu, H., Praat, M., et al., 2025. Stomatal opening under high temperatures is controlled by the OST1-regulated TOT3–AHA1 module. Nature Plants 11 (1), 105–117.

Yamori, W., Hikosaka, K., Way, D.A., 2014. Temperature response of photosynthesis in C3, C4, and CAM plants: temperature acclimation and temperature adaptation. Photosynthesis Research 119 (1), 101–117.

Yoshida, R., Umezawa, T., Mizoguchi, T., Takahashi, S., Takahashi, F., Shinozaki, K., 2006. The regulatory domain of SRK2E/OST1/SnRK2.6 interacts with ABI1 and integrates abscisic acid (ABA) and osmotic stress signals controlling stomatal closure in Arabidopsis. Journal of Biological Chemistry 281 (8), 5310–5318. Available at https://doi.org/10.1074/jbc.m509820200.

Zhao, Q., Han, R., Cai, K., Yan, H., Li, Y., Qu, G., et al., 2023. Identification and analysis of the CBF gene family in three species of acer under cold stress. International Journal of Molecular Sciences 24 (3), 2088. Available at https://doi.org/10.3390/ijms24032088.

Zhao, W., Sun, Y., Kjelgren, R., Liu, X., 2015. Response of stomatal density and bound gas exchange in leaves of maize to soil water deficit. Acta Physiologiae Plantarum 37, 1704. Available at https://doi.org/10.1007/s11738-014-1704-8.

Zoulias, N., Harrison, E.L., Casson, S.A., Gray, J.E., 2018. Molecular control of stomatal development. Biochemical Journal 475 (2), 441–454. Available at https://doi.org/10.1042/BCJ20170413.

Chapter 5

Plant signaling and signal transduction cascades in response to multiple abiotic stresses

Ruchika Bhagat[1], Pravin Jadhav[1,*], Prashant Kale[1], Umesh Shinde[1], Arpita Mahobia[1], Pranita Thakur[2], Balaji Jadhav[1], Gopika Mote[1], Mangesh Moharil[1], Balwant Munde[1], Aditya Rathod[1], Tarasingh Rathod[1], Rupesh Deshmukh[3], Humira Sonah[3], Shyam Jadhao[1], Shyamsunder Mane[1], Hamroeva Marguba Kamilovna[4,5], Raviprakash Dani[4,5], and Vilas Kharche[1]

[1]*Dr. Panjabrao Deshmukh Krishi Vidyapeeth, Akola, Maharashtra, India,* [2]*School of Biotechnology and Bioinformatics, Dr. D.Y. Patil Deemed to be University, Navi Mumbai, Maharashtra, India,* [3]*Department of Biotechnology, Central University of Haryana, Mahendergarh, Haryana, India,* [4]*Gen Scan Inc. Global Consultancy, Houston, TX, United States,* [5]*Denau Institute of Entrepreneurship and Pedagogy, Surkhandarya Region, Viloyat, Uzbekistan*

[*]*Corresponding author. e-mail address: pvjadhav@pdkv.ac.in*

5.1 Introduction

Being immobile, plants are constantly facing a variety of biotic and abiotic challenges, usually called stress conditions for plants. Plants are facing changing climatic situations, which are witnessed by altered seasons. When considering abiotic stress alone, it refers to nonliving environmental factors that hinder plant growth and overall development. Among these, stresses like drought, salinity, temperature extremes, heavy metal exposure, ultraviolet radiation, and fluctuations in light intensity severely affect plant growth, development, and yield. This is more important when it is related to agricultural production. These stresses disrupt physiological and metabolic processes in a variety of ways, often leading to reduced yields and compromised quality in crops (Zhu, 2016).

Abiotic stresses such as drought, salinity, extreme temperatures, and heavy metals adversely affect plant growth and development by disrupting water balance, ionic homeostasis, enzymatic activity, membrane stability, and nutrient uptake (Mittler et al., 2012). In nature, these stresses often occur simultaneously, and their combination can have additive or synergistic effects, intensifying oxidative damage, osmotic imbalance, and metabolic disruptions, thereby posing a greater threat to plant productivity than individual stresses alone.

Abiotic stresses influence not only specific plant tissues but also disrupt entire agricultural systems, ultimately endangering global food security (Fig. 5.1). Therefore, understanding the mechanisms through which plants sense and adapt to such stresses is crucial for the development of tolerant crop varieties. Over time, plants have evolved complex signaling systems that allow them to recognize environmental cues, transmit these signals within the cell, and trigger suitable adaptive responses.

5.1.1 Perception of abiotic stress

Plants initiate stress perception at the cellular level, where specialized receptors present on the plasma membrane or inside the cell detect external signals. For instance, osmotic stress due to drought or salinity is often detected by osmosensors (Mahajan and Tuteja, 2005).

Multiple Abiotic Stresses in Plants. DOI: https://doi.org/10.1016/B978-0-443-27424-4.00014-8

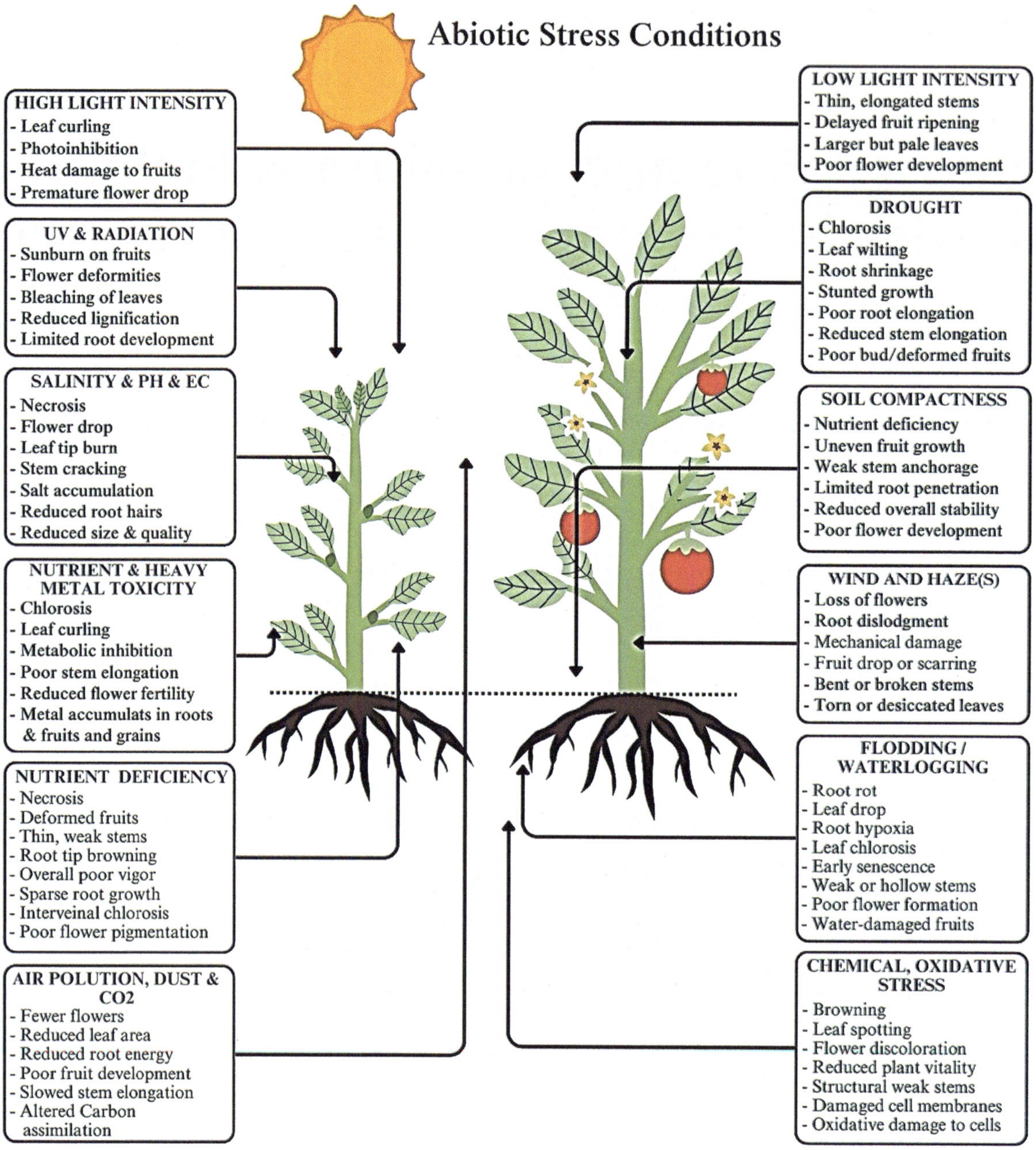

FIGURE 5.1 List of abiotic stresses and their components in the life of plants.

5.1.2 Role of signaling molecules

Key molecules. such as calcium ions (Ca^{2+}), reactive oxygen species (ROS), and phytohormones, such as abscisic acid (ABA), jasmonic acid (JA), and ethylene (ET), regulate stress signaling. These function as secondary messengers, intensifying the initial signals and coordinating subsequent cellular responses.

- Calcium signaling plays a central role by activating calcium-binding proteins such as calmodulins (Sanders et al., 2002).
- ROS act as dual-function molecules, both signaling stress and causing oxidative damage if uncontrolled (Nadarajah, 2020).

5.1.3 Key components of stress signaling pathways

Stress signaling pathways involve a cascade of molecular interactions:

- Receptors/sensors detect the initial stress signal (Nakashima et al., 2012).
- Kinases (e.g., mitogen-activated protein kinases [MAPKs], SNF1-related kinases) convey the signal by means of phosphorylation events (Sinha et al., 2011).
- Transcription factors (TFs) (e.g., DREB, bZIP) regulate the expression of genes associated with stress responses (Nakashima and Yamaguchi-Shinozaki, 2013).

Together, these components work in concert to activate adaptive responses, such as osmoprotectant synthesis, antioxidant defense, and reprogramming of cellular metabolism.

A literature survey on Google Scholar for data spanning the last 30 years, based on the keywords "abiotic stress, signaling, gene, plants," was conducted on December 24, 2024.

Fig. 5.2 highlights a remarkable increase in publications, particularly over the past decade. From a modest 288 publications in 1995–99, the number steadily rose, crossing 1930 in 2000–04 and 6940 in 2005–09. The period 2010–14 marked a significant surge with 18,000 publications. Annual publications continued to climb thereafter, reaching a peak of 20,300 in 2022. While there has been a slight decline in the following years, with 16,700 in 2023 (may be a COVID-19 impact), the data underscore the growing scientific interest in understanding abiotic stress and its signaling mechanisms in plants. This trend reflects the urgency to address agricultural challenges associated with environmental stresses and the increasing complexity of research in this domain.

As climate change intensifies and environmental stresses become more frequent, it is crucial to understand how plants manage these challenges to create resilient agricultural systems. Deciphering the complexities of stress signaling pathways is crucial for advancing crop improvement strategies. By identifying and manipulating key signaling components, researchers can improve plant resilience to abiotic stresses, supporting sustainable agricultural productivity under global environmental pressures. This chapter focuses on dissecting the molecular mechanisms of abiotic stress signaling,

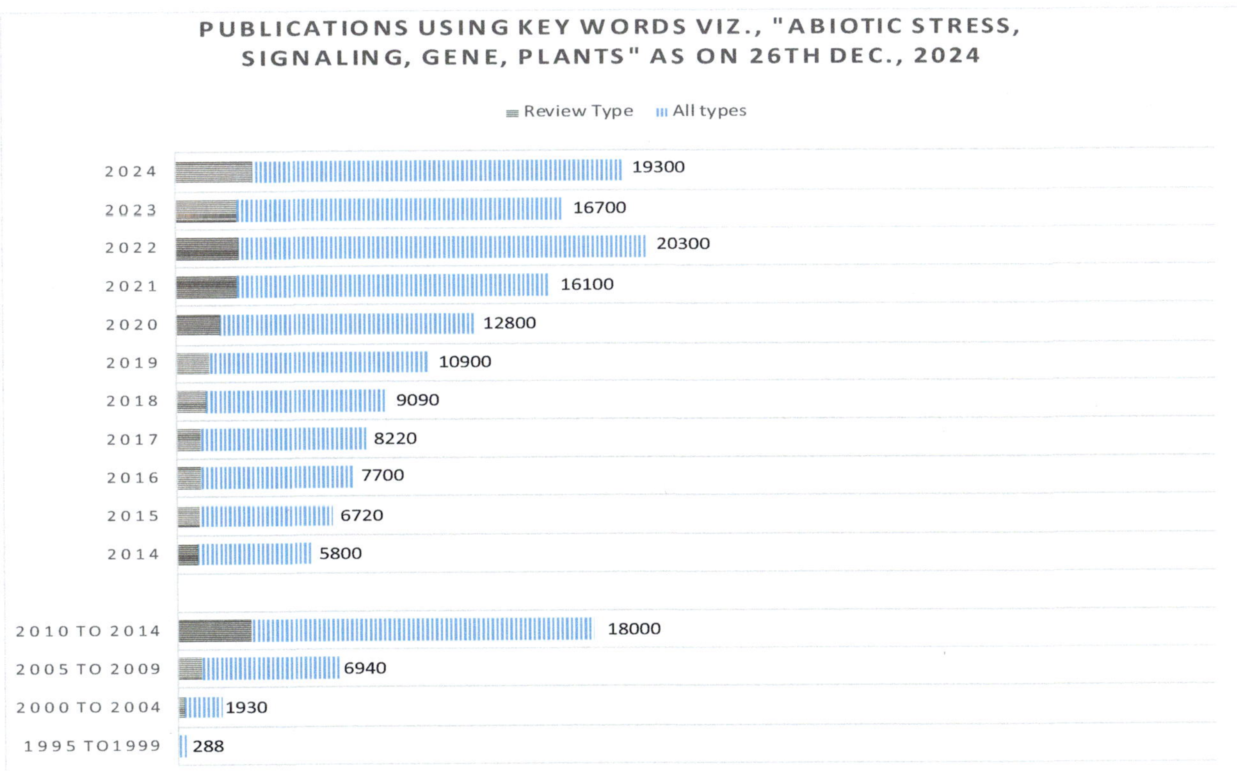

FIGURE 5.2 Graphical representation of literature survey using abiotic stress-related keywords.

representative genes and related pathways, and highlights their potential applications in developing robust, stress-resilient crops.

5.2 General concept of signal transduction

Plants are constantly exposed to diverse environmental factors, including temperature extremes, drought, salinity, nutrient fluctuations, light variations, and biotic stresses. To cope with these challenges, they have evolved a variety of signaling and molecular mechanisms that allow them to perceive these conditions accurately and trigger appropriate defense responses (Hettenhausen et al., 2015). There are four key aspects of plant signaling in response to environmental stimuli: signal detection, signal transduction, gene expression regulation, and physiological responses.

5.2.1 Signal detection

Plants possess specialized receptors and sensor proteins that allow them to recognize a wide range of environmental stimuli. Signal transmission begins when these molecules bind to their corresponding receptors, causing structural changes that activate downstream pathways. For instance, photoreceptors help plants perceive light quality, intensity, and duration, with phytochromes and cryptochromes serving as the main photoreceptors. Likewise, plants can detect temperature fluctuations, sense hormonal signals, and respond to chemical compounds released by pathogens or herbivores (Divekar et al., 2022).

5.2.2 Signal transduction

Once a signal is perceived, it needs to be transmitted throughout the plant to trigger a suitable response. Signal transduction involves a series of molecular events that convey information from the receptor to target molecules. These pathways often include protein kinases, secondary messengers, such as calcium ions and ROS, and TFs, which are activated to regulate gene expression (Divekar et al., 2022).

5.2.3 Control of gene expression

Plants often adjust to environmental changes through changes in gene activity. Signal transduction pathways can turn specific genes on or off, resulting in the production of proteins that aid adaptation. TFs play a central role in this process by binding to specific DNA sequences in gene promoters, thereby modulating the rate at which RNA polymerase initiates transcription (Yang et al., 2012).

5.2.4 Physiological responses

Changes at the molecular level in response to environmental signals eventually produce visible physiological effects in plants. These responses can include alterations in development, growth, defense mechanisms, and metabolism. During drought conditions, plants often conserve water by closing their stomata and producing osmoprotectant compounds (Ghadirnezhad Shiade et al., 2023). Under high salinity conditions, rice plants respond by increasing the production of catalase (CAT), proline, and superoxide dismutase (SOD) to counteract the harmful effects of salt stress. Similarly, wheat plants exposed to light stress have been observed to develop longer spikelets, taller shoots, thinner stems, and a reduction in photosynthetically active radiation (Abdolmaleki et al., 2023).

5.2.5 Types of plant stress signals

Cell-to-cell communication in plants is mediated by various signaling molecules, including small organic compounds, ions, peptides, and physical stimuli. These molecules can be classified into general plant messenger molecules and hormone-based signaling molecules (Ding and Yang, 2022). Plants use various secondary messengers, such as ROS, calcium ions (Ca^{2+}), phosphoinositides, nitric oxide (NO), and cyclic nucleotides, which help initiate downstream signaling cascades in response to multiple environmental stimuli (Jain et al., 2018). Within the group of cyclic nucleotides, cGMP and cAMP serve as vital secondary messengers with key roles in cellular signaling. Their levels are regulated by cyclases, which synthesize them, and phosphodiesterases (PDEs), which break them down. Plant hormones such as auxin (indole-3-acetic acid), ABA, methyl jasmonate, salicylic acid (SA), cytokinin (CK), JA, gibberellins (GA), strigolactones, brassinosteroids, and ET can move over long distances within the plant to enable intercellular communication. These hormones are essential for controlling growth and development, as well as coordinating plant responses to a variety of biotic and abiotic stresses.

Calcium ions (Ca^{2+}) act as crucial secondary messengers, becoming activated when plants encounter specific abiotic stress signals. Unique Ca^{2+} patterns or waves convey these signals through a network of sensors, relay proteins, channels, exchangers, and pumps. These signal transduction pathways are essential for carrying information throughout the plant. These signaling pathways include a variety of components such as receptors, phytochromes, protein kinases, including MAPK cascades, receptor-like kinases (RLKs), SNF1-related protein kinases (SnRKs), and calcium-dependent protein kinases (CDPKs/CPKs), protein phosphatases (e.g., salt overly sensitive 3 [SOS3]/protein kinase), TFs, and stress-responsive genes. Together, these elements play a vital role in controlling plant growth, development, and adaptation to stress. For instance members of the MAPK family contribute to tolerance against drought, salinity, cold, and heat stress (De Zélicourt et al., 2016). Similarly, the RLK guard cell hydrogen peroxide resistance 1 (GHR1) is essential for ABA- and H_2O_2-mediated stomatal closure and contributes to drought tolerance. OPEN STOMATA 1 (OST1), also called SnRK2.6, participates in signaling pathways for both cold and drought stress. SOS2, a Ser/Thr kinase belonging to the SnRK3 family, functions as part of the SOS pathway, which is crucial for plant salt stress responses (Quan et al., 2007).

5.3 Major signaling pathways in plant abiotic stress responses

Plants are constantly exposed to various abiotic stresses, which are the primary harmful elements influencing their development and growth. Abiotic stress is the negative effect of any abiotic factor on a plant in a particular environment, resulting in different reactions, from modifications that impact biological processes such as cell metabolism and gene expression to growth and development (Zhang et al., 2023). Various stresses can elicit distinct reactions. In natural environments, the main factors limiting plant survival and distribution are extreme temperatures, water scarcity, and saline soils. Often, drought conditions occur alongside high temperatures. Understanding how plants perceive, transmit, and respond to stress signals is a major focus in plant biology. Osmotic stress, caused by both salinity and drought, impacts numerous inorganic and organic metabolites and harms plants through ion toxicity, accumulation of ROS, disruption of the plasma membrane, and cell wall damage. Adaptive responses can enhance plant tolerance, whereas nonadaptive responses arise from protein misfolding and deterioration of cell wall structure under stress conditions (Radha et al., 2023). Plants have evolved complex and interconnected signaling networks that enable them to detect environmental stresses and respond in ways that promote survival and adaptation. The molecular response to abiotic stress operates on multiple levels, involving numerous genes, signaling pathways, and metabolic processes, including stress sensing, signal transduction, gene transcription, RNA processing, and protein translation and modification. Numerous cellular functions, including gene expression, ion homeostasis, osmotic regulation, metabolic adaptations, and protective protein synthesis, are regulated by these signaling pathways, which are listed in Fig. 5.3.

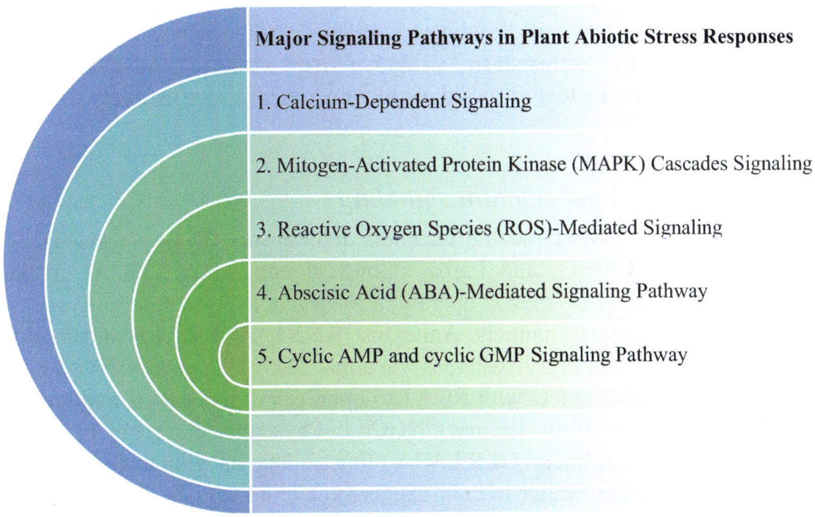

FIGURE 5.3 Signaling pathways in plant abiotic stress responses.

5.3.1 Calcium-dependent signaling pathway

Calcium (Ca^{2+}) is a crucial second messenger in plant stress signaling, mediating responses to drought, salinity, cold, and heat. Its compartmentalization in the cytosol, nucleus, vacuole, mitochondria, and chloroplast ensures precise regulation of stress responses, controlling gene expression, osmotic balance, ion homeostasis, and protein modifications. Stress perception activates calcium signaling, which is tightly regulated by channels, binding proteins, kinases, and TFs, enabling rapid but controlled adaptation.

5.3.1.1 Calcium channels and calcium influx

The first step in Ca^{2+} signaling is the influx of calcium into the cytosol, either from the extracellular environment or internal stores such as the ER and vacuole. Stressors such as osmotic imbalance, heat, or wounding activate stretch-activated, L-type, or cyclic nucleotide-gated channels (CNGCs). For instance CNGCs mediate cold and salinity responses, whereas vacuolar two-pore channels and ER-localized IP3 receptors release stored Ca^{2+}. These stress-induced cytosolic spikes in calcium serve as primary signals that initiate downstream signaling pathways.

5.3.1.2 Calcium-binding proteins

Once calcium enters the cytosol, it binds to specialized proteins that decode spatial and temporal "Ca^{2+} signatures." Calmodulin (CaM) undergoes conformational changes that allow it to regulate TFs, kinases, phosphatases, enzymes, and ion channels. Calcineurin B-like proteins (CBLs) specifically sense stress-induced calcium and interact with CBL-interacting protein kinases (CIPKs) to regulate ion transporters, which are critical for salt and drought tolerance. CDPKs are unique because they contain both calcium-binding and catalytic domains; upon activation, they phosphorylate TFs, enzymes, and ion channels, thereby modulating osmotic regulation, antioxidant defense, and cell wall modification.

5.3.1.3 Calcium-dependent protein kinases

CDPKs serve as central mediators of abiotic stress tolerance. Because they directly sense cytosolic Ca^{2+} changes, CDPKs rapidly regulate ion homeostasis by modulating K^+ and Na^+ transporters, maintain osmotic balance, and induce the biosynthesis of osmoprotectants such as proline. They also phosphorylate stress-responsive TFs that activate genes required for water-deficit management, oxidative stress defense, and salinity tolerance. In this way, CDPKs form an essential link between calcium influx and transcriptional reprogramming during stress.

5.3.1.4 Transcription factors activated by calcium signals

Calcium signaling also activates a wide array of TFs that regulate stress-responsive genes. For example, CREB proteins (CAMP-responsive element binding proteins) regulate genes necessary for stress adaptation, whereas CIB1 controls both developmental and stress-tolerance genes following calcium binding. Similarly, calmodulin-binding transcription activators are triggered via CaM and contribute to ion homeostasis, oxidative stress defense, and growth regulation under stress conditions. Collectively, these transcription factors ensure that calcium signals are effectively translated into precise gene expression patterns.

5.3.1.5 Mechanisms of abiotic stress detection via calcium signaling

Plants detect abiotic stresses through specialized calcium-permeable channels and sensors. During mechanical stress, mechanosensitive channels such as MSLs, MCAs, OSCAs, and Piezos allow calcium influx. OSCA1 in particular mediates hyperosmolality-triggered calcium signaling, while Piezos operate at the vacuole. Under salinity stress, GIPC sphingolipids act as Na^+ sensors, activating calcium influx channels. Annexins (ANN1, ANN4), CNGC10, and RLKs, such as FER, also contribute to salinity-induced calcium entry, whereas vacuolar TPC1 channels propagate systemic calcium waves. In cold stress, the rice COLD1 protein interacts with RGA1 to open calcium channels and enhance cold tolerance. Additionally, CNGCs such as CNGC9, 14, and 16 in rice and CNGCb in Physcomitrella regulate cold-induced calcium influx, whereas *Arabidopsis* GLR1.2/1.3 and tomato GLR3.3/3.5, functioning as glutamate receptors, also participate. Annexin ANN1 has been shown to contribute further to cold-triggered calcium signals. Heat stress is sensed via calcium influx mediated by *Arabidopsis* CNGC2/6 and annexins ANN1/ANN4. In the case of ROS signaling, the receptor HPCA1 is oxidized by H_2O_2, which activates calcium influx channels, whereas another RLK, GHR1, integrates ROS and ABA signaling to regulate stomatal closure. Long-distance ROS-calcium waves are also propagated through mechanosensitive channels such as MSL3 and vacuolar TPC1 (Xu et al., 2022). These examples highlight that each abiotic stress generates a distinct calcium "signature," enabling plants to distinguish between different stress types and mount specific adaptive responses (Fig. 5.4).

5.3.1.6 *Calcium signaling in specific abiotic stress responses*

Drought and Water Deficit: Drought induces cytosolic Ca^{2+} accumulation, activating CDPKs, CBLs, and stress-responsive genes encoding antioxidants, chaperones, and dehydrins. By regulating ion channels in guard cells, calcium signaling also drives stomatal closure (Xu et al., 2022).
Salt Stress (Salinity): Ca^{2+} influx through plasma membrane channels triggers the CBL–CIPK pathway, regulating ion transporters to maintain K^+/Na^+ balance by moving K^+ into the cytoplasm and sequestering Na^+ in vacuoles. Calcium further regulates osmotic adjustment by activating proline and compatible solute biosynthesis (Xu et al., 2022).
Cold Stress: Calcium signaling interacts with CDPKs and calmodulin to activate C-repeat binding factor (CBF) TFs, inducing cold-regulated genes for membrane stabilization and antifreeze proteins.
Heat Stress: Heat stress elevates cytosolic Ca^{2+}, activating heat shock proteins (HSPs) that protect against protein denaturation, as well as heat shock factors (HSFs) that regulate the heat shock response (Xu et al., 2022).

5.3.1.7 *Crosstalk with other pathways*

ABA Signaling: ABA induces Ca^{2+} influx in guard cells, coordinating stomatal closure via calcium-dependent proteins and kinases.
ROS Signaling: Calcium and ROS form feedback loops, with ROS activating Ca^{2+} channels and calcium regulating ROS production.
MAPK Pathway: Calcium signaling activates MAPKs, leading to the expression of stress-responsive genes.

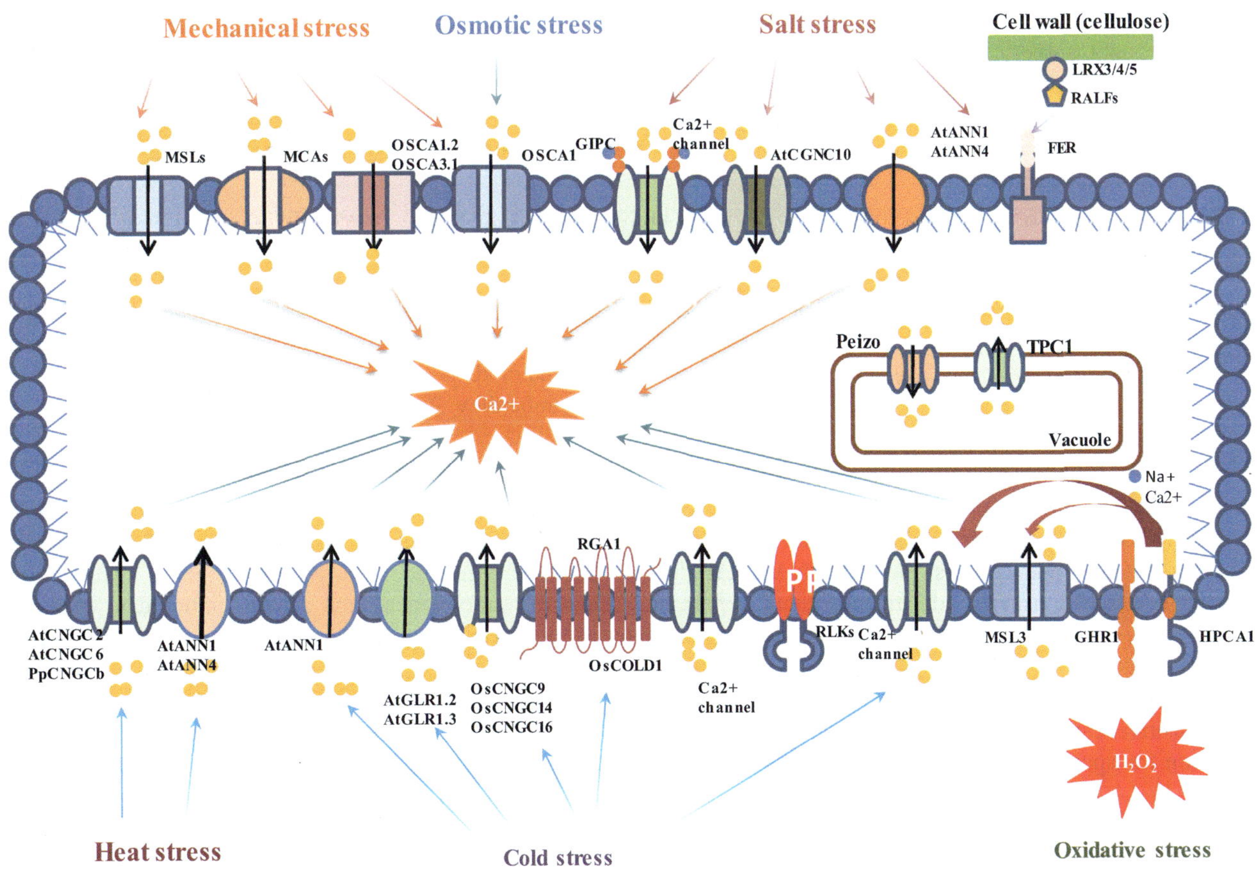

FIGURE 5.4 Calcium signals in abiotic stress.

5.3.2 Mitogen-activated protein kinase cascades

The MAPK cascade is a key signaling pathway involved in plant responses to abiotic stress. This signaling module, which has evolved to be conserved across eukaryotic organisms, connects external stimuli to multiple cellular responses. All eukaryotes contain conserved signaling modules called MAPK cascades, which convert developmental and environmental signals into intracellular reactions. The transmission and integration of stress signals to initiate proper cellular responses depend on MAPK cascades (Jonak et al., 2002). An MAPK cascade consists of at least three tiers: MAP kinase kinases (MAP3Ks or MAPKKKs/MEKKs), MAP kinase kinases (MAP2Ks or MAPKKs/MEKs/MKKs), and MAP kinases (MAPKs or MPKs). According to Dan et al. (2001), MAP3Ks, also known as MAP kinase kinase kinases (or MAP4Ks), are activated by stimuli at the plasma membrane during stress. MAP4Ks can function as adapters, linking core MAPK cascades to upstream signaling events. MAP3Ks are serine/threonine kinases that phosphorylate two amino acids within the S/T-X-S/T motif of the MAP2K activation loop. In turn, MAP2Ks phosphorylate MAPKs on conserved threonine and tyrosine residues within a T-X-Y motif. MAPKs, which are also serine/threonine kinases, can then phosphorylate various substrates, including other kinases and TFs. The formation and stability of a specific MAPK cascade are mediated by scaffold proteins, shared docking domains, and adapter or anchoring proteins. Following signal transmission, the pathway is deactivated by a time-dependent regulator, including MAPK phosphatases (MKPs) (Takekawa et al., 2005) (Table 5.1). MAPKs regulate the activity of various TFs, including WRKY, AP2/ERF, and NAC, to control the expression of stress-responsive genes involved in osmotic adjustment, antioxidant production, and stress-induced senescence. To coordinate plant responses at multiple levels, the MAPK cascade interacts with other signaling pathways, such as those mediated by ABA, ROS, and calcium. By amplifying and diversifying stress signals, MAPK pathways are crucial for enabling plants to simultaneously activate multiple defense mechanisms.

5.3.3 Reactive oxygen species- and nitric oxide-mediated signaling pathway

ROS and NO are vital signaling molecules that regulate plant responses to both biotic (pathogen attack) and abiotic stresses (drought, salinity, heat, cold). Despite their potential to cause oxidative damage at high concentrations, both are short-lived and highly reactive messengers modulating gene expression, protein function, and cellular processes related to growth, development, and stress tolerance. Their major production sites include the cytoplasm, mitochondria, peroxisomes, chloroplasts, and plasma membrane (Fig. 5.5). Tight regulation of ROS/NO production balances stress protection and cellular damage.

The plasma membrane is the primary ROS source, largely through Reduced Nicotinamide Adenine Dinucleotide Phosphate (NADPH) oxidase (RBOH) activity. NADPH oxidases convert NADPH electrons to extracellular oxygen, generating $O_2{\bullet}^-$ and promoting ROS accumulation in the apoplast. Stressful conditions induce NADPH oxidase expression, producing oxidative bursts. Cell wall–associated oxidases and peroxidases also contribute, though their precise roles remain unclear. ROS waves, while essential for systemic acquired acclimation, are often nonspecific and require additional stress-specific signals. Chloroplasts are another major ROS source due to intense electron transport during photosynthesis and oxygen evolution at PSII (Gupta and Igamberdiev, 2015).

5.3.3.1 Reactive oxygen species signaling

ROS include superoxide anion (O_2^-), hydrogen peroxide (H_2O_2), hydroxyl radical (OH•), and singlet oxygen ($^1O_2$). While excessive ROS accumulation leads to oxidative stress, controlled levels serve as secondary messengers during stress signaling. ROS influence TFs (WRKY, NAC, AP2/ERFs, DREB), regulate stress-responsive genes involved in antioxidant defense, osmotic adjustment, protein repair, and cell wall reinforcement, and activate antioxidant enzymes such as SOD, CAT, and peroxidases. H_2O_2 additionally activates MAPK cascades, triggering downstream responses (Apel and Hirt, 2004).

ROS also mediate programmed cell death (PCD), limiting stress damage or restricting pathogen spread. They regulate hypersensitive response (HR), a localized defense involving cell death at infection sites. During drought and salinity, ROS affect ion transport and Ca^{2+} fluxes, supporting Na^+/K^+ homeostasis. They also activate Ca^{2+}-calmodulin pathways and CDPKs. In *Arabidopsis*, H_2O_2 induces AtNDPK2, activating AtMPK3/6 (Laloi et al., 2004).

5.3.3.1.1 Roles in specific abiotic stresses

Drought: ROS triggers stomatal closure and osmotic regulation genes, reducing water loss.
Salinity: ROS regulates ion transport and activates antioxidant defenses to minimize Na^+ toxicity.
Heat: ROS induces HSPs and stabilizes proteins/membranes under high temperature.

TABLE 5.1 MAPKs identified in various plant species and their roles in abiotic stress responses.

Plant	Component of MAPK cascade	Remarks	References
Alfalfa	Medicago mitogen-activated protein kinase (MMK) 2, MMK3, Stress activated mitogen-activated protein kinase (SAMK)	Activated by heavy metal stress	Jonak et al. (2004)
	P44MKK4	Activated by cold and drought	Jonak et al. (1996)
	Oxidative stress-activated MAP triple-kinase1	ROS-induced stress	Nakagami et al. (2004)
	Hordeum vulgare mitogen-activated protein kinase (HAMK)	High-temperature stress	Sangwan et al. (2002)
Arabidopsis	AtMPK3, AtMPK6	Ozone-induced stress	Droillard et al. (2002)
	Arabidopsis thaliana mitogen-activated protein kinase 1 (AtMPK1) , AtMPK4,6	Salt stress, cold stress, dehydration, mechanical stimulation, wounding, and high osmotic stress	Ichimura et al. (2000)
	MAPKK, MKK2	Low-temperature and salinity stress	Teige et al. (2004)
Chorispora bungeana	Cicer (or species-specific) MAP kinase / mitogen-activated protein kinase (CbMAPK) 3	Cold and salinity stress	Zhang et al. (2006)
Cotton	Gossypium hirsutum mitogen-activated protein kinase (GhMAPK)	Mechanical injury, cold, and salt stress	Wang et al. (2007)
	Gossypium hirsutum mitogen-activated protein kinase (GhMPK) 7	Salinity and mechanical injury stress	Shi et al. (2010)
Soybean	MAPK	Activated by light Wound stress	Lee et al. (2001)
Rice	Oryza sativa multiple stress-responsive mitogen-activated protein kinase 2	Wound, ultraviolet radiation, heavy metal, salinity, drought, ozone, and temperature extremes	Agrawal et al. (2002)
	MAPKK4, 6	Cold and salinity stress	Kumar et al. (2008)
	MAPKK1	Salt and drought	
	MAPKK10-2	Cold	
	drought-sensitive mutant (DSM) 1(MAPKKK)	Water-deficit stress	Ning et al. (2010)
Pea	Pisum sativum mitogen-activated protein kinase (PsMPK) 2	Wounding, abscisic acid, and hydrogen peroxide stress	Ortiz-Masia et al. (2008)
Tomato	Tobacco mitogen-activated protein Kinase (tMEK) 2	Wound	Xing et al. (2001)
Potato	Solanum tuberosum mitogen-activated protein kinase (StMPK) 1	Mechanical injury and high-temperature stress	Blanco et al. (2006)
Maize	Zea mays mitogen-activated protein kinase (ZmMPK) 3	Triggered by low temperature, water deficit, ultraviolet radiation, salinity, heavy metals, and mechanical injury	Wang et al. (2010)
	ZmMPK7	Induced by hydrogen peroxide and osmotic stress	Zong et al. (2009)
	Zea mays mitogen-activated protein kinase (ZmMAPK) 5	Activated by hydrogen peroxide, polyethylene glycol, salinity, cadmium chloride, low temperature, mechanical injury, and UV radiation	Ding et al. (2009)
	Zea mays stress-induced mitogen-activated protein kinase (ZmSIMK) 1	Salinity stress	Gu et al. (2010)

MAPK, Mitogen-activated protein kinase.

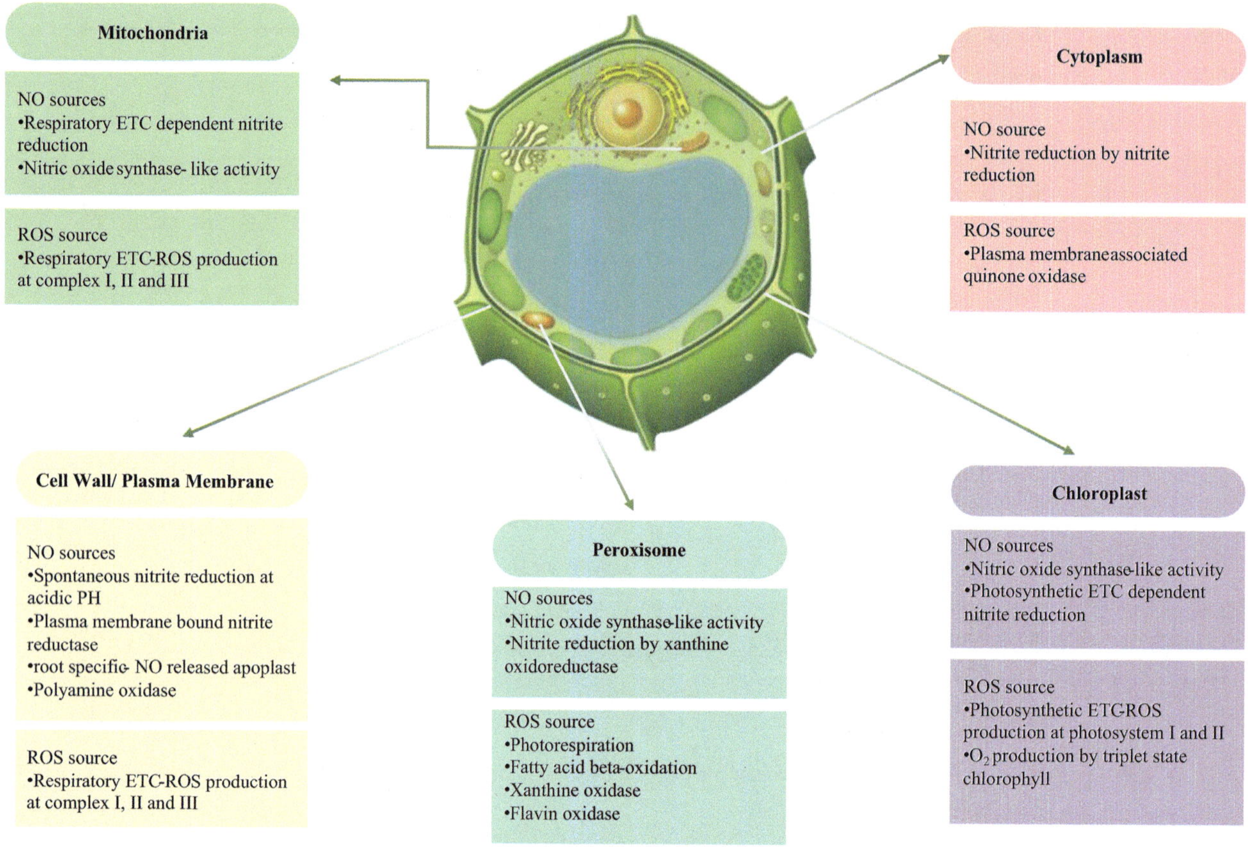

FIGURE 5.5 Main sources of nitric oxide (NO) and reactive oxygen species (ROS) in plant cells.

Cold: ROS activate antioxidant defenses and improve membrane stability during cold stress.

5.3.3.2 Nitric oxide signaling

NO, a gaseous signaling molecule, is endogenously synthesized and also effective when applied exogenously, inducing physiological and molecular adjustments that enhance stress tolerance (Sharma et al., 2025). Plant NO is mainly generated by nitrate reductase (NR), though NOS-like enzymes, nitrite reduction, and oxidative/reductive pathways also contribute. NR, essential for nitrogen assimilation, is a key NO source under specific conditions (e.g., anaerobiosis, high nitrite). NOS-like proteins may be stress-inducible.

NO regulates diverse processes, including defense, hypoxia tolerance (Igamberdiev & Hill, 2004), drought response, floral transition, and lateral root formation. As a signaling molecule, NO modifies TFs (WRKY, NAC, ERFs), regulating genes linked to antioxidant defense, protein stability, and cell wall biosynthesis. NO also interacts closely with ROS, particularly H_2O_2: acting as a ROS scavenger, but also synergizing with ROS to initiate signaling. Together, NO and ROS regulate MAPK cascades, ion homeostasis, antioxidant enzyme activity, and hormone signaling.

NO plays a central role in stomatal regulation. In drought, NO acts with ROS and ABA to close stomata, limiting water loss. In HR, NO induces localized cell death and activates pathogen-response genes. NO also regulates nutrient uptake (K^+, Ca^{2+}) and nitrate assimilation, crucial in nutrient-deficient environments.

5.3.3.3 Roles in specific abiotic stresses

Drought: NO induces ABA-responsive genes to regulate stomata and conserve water.
Salinity: maintains Na^+/K^+ balance, reducing ionic stress.
Heat: NO modulates membrane fluidity, protects proteins, and activates HSPs.
Cold: It enhances antioxidant enzyme expression and promotes cold-regulated proteins.

5.3.3.4 Crosstalk between reactive oxygen species and nitric oxide

ROS and NO act synergistically, integrating stress signaling for coordinated responses. Their interaction involves posttranslational modifications (PTMs) such as S-nitrosylation, tyrosine nitration, and metal nitrosylation (Lamotte et al., 2015). They influence common signaling intermediates, including MAPKs, Ca^{2+}, cGMP, and cADPR. Additionally, oxidative PTMs, GSNO (S-nitrosoglutathione), GSNOR, and truncated hemoglobins regulate ROS/NO balance (Fig. 5.6) (Oz et al., 2015).

5.3.3.4.1 Key interaction mechanisms

Antioxidant Defense: NO scavenges excess ROS (H_2O_2, $O_2{\bullet}^-$, OH•), limiting oxidative damage.
MAPK Activation: Both trigger MAPK cascades, inducing stress-adaptive genes.
Stomatal Regulation: ROS and NO coordinate guard cell signaling, promoting closure under stress.
Hypersensitive Response: Both regulate HR; ROS act as messengers while NO promotes localized cell death.

5.3.4 Abscisic acid-mediated signaling pathway

ABA is present in all plant parts but shows strong basal expression in leaves. Its biosynthesis is markedly upregulated under stress, with protein levels of related genes such as 9-cis-epoxycarotenoid dioxygenase (NCED) increasing alongside transcript abundance. Initially, isoprenoids were thought to derive only from the mevalonate pathway, but a secondary pathway for intrinsically disordered protein (IDP) synthesis was later identified in eubacteria and higher plants (Nambara

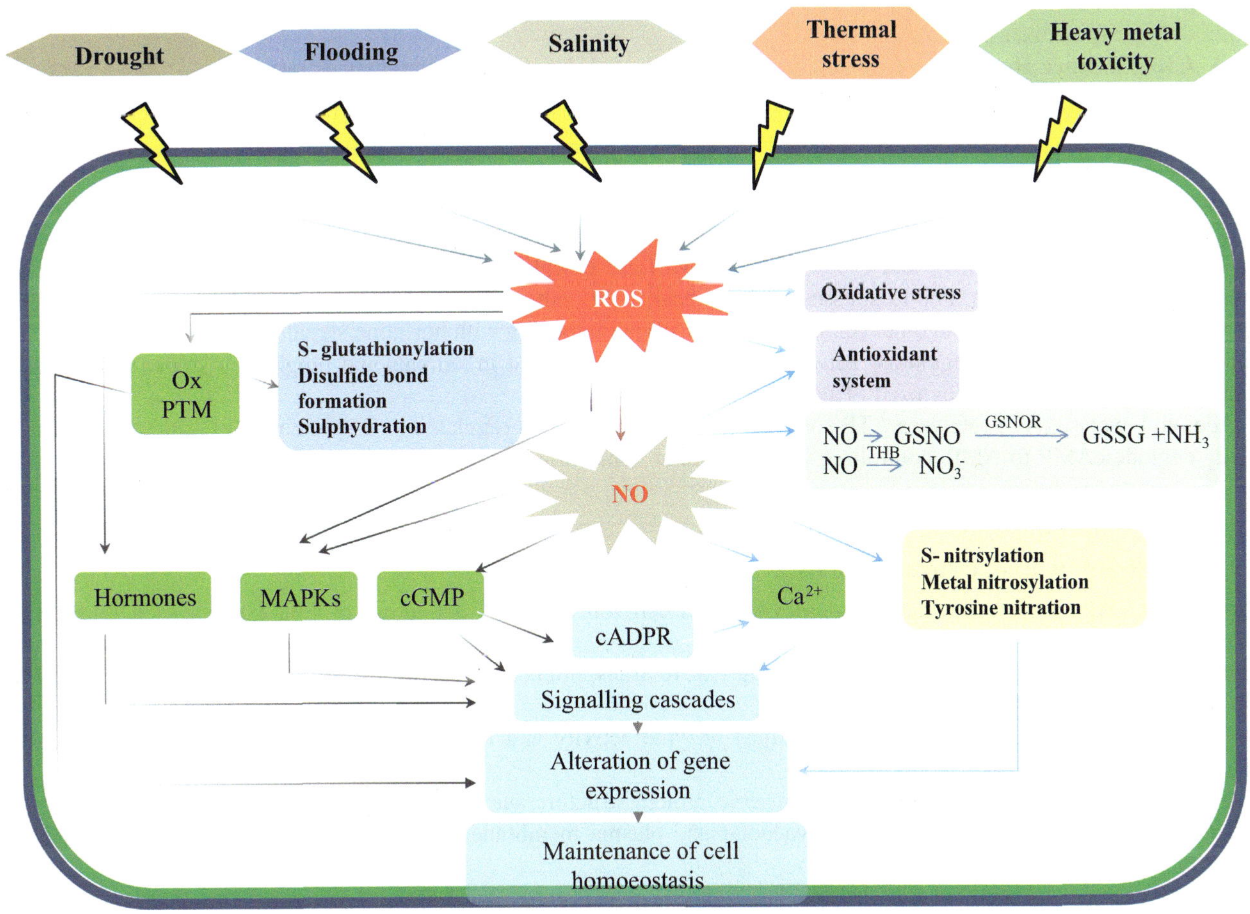

FIGURE 5.6 Schematic representation showing the interplay between NO, ROS, and other signaling molecules in response to abiotic stress. *NO*, Nitric oxide; *ROS*, reactive oxygen species.

and Marion-Poll, 2005). ABA is synthesized from β-carotene through several enzymatic steps and plays key roles in stress adaptation by activating stress-responsive genes, promoting stomatal closure, and conserving water.

ABA perception occurs at the plasma membrane through receptors of the PYR/PYL/RCAR family. Upon ABA binding, these receptors initiate signaling cascades by activating SnRK2 kinases (e.g., SnRK2.2, SnRK2.3). Activated SnRK2s phosphorylate TFs such as ABF/AREB proteins, which bind ABA-responsive elements in target genes. These genes regulate stomatal closure, antioxidant defense, and water conservation. ABA also directly modulates ion fluxes (K^+, Cl^-) in guard cells to induce stomatal closure during drought. Stress-induced de novo ABA biosynthesis is mediated by genes such as ZEP and NCED, which are strongly stress-responsive (Xiong and Zhu, 2003).

5.3.4.1 Abscisic acid receptor and transcription factor crosstalk

Fine-tuning of ABA signaling depends on coordination between receptors (PYR/PYL/RCAR) and TFs (AREBs/ABFs). ABA binding activates SnRK2s, which phosphorylate ABFs, enabling them to trigger stress-responsive genes. These TFs also integrate signals from other pathways, optimizing survival under adverse conditions.

5.3.4.2 Abscisic acid and other signaling pathways

ABA interacts extensively with other hormones to integrate plant responses

ABA–GA Crosstalk: ABA suppresses GA signaling, delaying seed germination under stress and maintaining dormancy until favorable conditions return.
ABA–Ethylene Interaction: Together, they regulate root growth and stomatal closure during drought.
ABA–JA/SA Relationship: ABA and JA often cooperate in balancing abiotic stress responses (drought, salinity) with biotic defense (pathogen infection).
ABA–CK/ET/Other Hormones: Crosstalk ensures coordinated developmental and stress responses.

Through these mechanisms, ABA serves as a central regulator of abiotic stress signaling, linking environmental perception with gene expression, metabolic adjustments, and hormone crosstalk to optimize plant survival.

5.3.5 Cyclic AMP and cyclic GMP pathways in plant abiotic stress signaling

In plants, cyclic AMP (cAMP) and cyclic GMP (cGMP) are essential second messengers that regulate physiological processes and responses to abiotic stresses such as heat, cold, salt, and drought. They control ion channels, stress-responsive gene expression, and other cellular functions, often interacting with hormone signaling pathways, such as ABA and Ca^{2+}, to form complex regulatory networks. Although more studied in animals and fungi, their roles in plant stress tolerance are increasingly recognized (Tuteja and Sopory, 2008).

Adenylyl cyclase (AC) catalyzes ATP conversion to cAMP, which is regulated by G-proteins, Ca^{2+}, and stress signals. PDEs degrade cAMP to AMP, controlling its intracellular concentration.

5.3.5.1 Cyclic AMP signaling in plants

5.3.5.1.1 Roles in abiotic stress responses

Drought: cAMP regulates stomatal closure via guard cell ion channels, activating protein kinases such as PKA to conserve water.
Salt: cAMP mediates salt tolerance by modulating Na^+/K^+ transporters and salt-responsive gene expression, maintaining ion homeostasis.
Cold: cAMP activates protein kinases that adjust enzyme activity and membrane fluidity, regulating cold-responsive genes for acclimation.
Heat: cAMP modulates HSP expression to preserve protein structure and function.
Ion Channel Regulation: cAMP controls vacuolar and plasma membrane K^+ and Ca^{2+} channels, maintaining cellular integrity under osmotic stress.

5.3.5.1.2 Mechanism of action

cAMP activates PKA, which phosphorylates target proteins, promoting stress responses such as stomatal closure, ion homeostasis, and activation of stress-responsive genes, including those involved in antioxidant defense and osmotic regulation.

5.3.5.2 *Cyclic GMP signaling in plants*

cGMP is produced from guanosine triphosphate (GTP) by guanylate cyclases and degraded by PDEs. It exists in soluble and membrane-bound forms and is regulated by Ca^{2+}, hormones, and stress signals.

5.3.5.2.1 Roles in abiotic stress responses

Drought: cGMP mediates ABA-induced stomatal closure and activates TFs controlling osmotic and antioxidant responses.
Salt: It regulates ion channels, transporters, and stress-responsive genes, promoting Na^+ exclusion and ionic balance.
Cold: cGMP modulates protein activity and gene expression for antioxidant defense, membrane stability, and cold acclimation.
Heat: It stabilizes protein structure and influences HSP expression.
Calcium Signaling: cGMP interacts with Ca^{2+} as a secondary messenger in stress pathways, including osmotic stress and stomatal regulation.

5.3.5.2.2 Mechanism of action

cGMP activates cGMP-dependent protein kinases (PKGs), which regulate ion transport, stomatal movement, and stress gene expression. PKGs can also interact with kinases such as SnRK2s, influencing ABA-mediated responses. TFs activated by cGMP control stress proteins, osmotic regulation, and ion homeostasis, while ion channels in vacuolar and plasma membranes adjust fluxes under drought and salinity (Tuteja and Sopory, 2008).

5.3.5.3 *Interaction between cAMP and cGMP signaling*

cAMP and cGMP pathways interact under stress, integrating signaling to regulate gene expression, protein kinases, and ion channels. Both pathways contribute to ABA-mediated responses, particularly stomatal closure during drought and salinity. cGMP activates Ca^{2+} channels and calcium-dependent TFs, whereas cAMP modulates Ca^{2+}-dependent protein kinases. Together, they regulate K^+/Na^+ exchange and Na^+/K^+ pumps, maintaining ion homeostasis and cellular integrity under abiotic stress.

By coordinating ion transport, transcriptional regulation, and kinase activity, cAMP and cGMP serve as central components of plant stress signaling, enhancing adaptation to environmental challenges and linking cyclic nucleotide signaling with hormone and calcium-dependent pathways.

5.4 Crosstalks between abiotic stress pathways

5.4.1 Integrating multiple abiotic stress responses in plants

Numerous abiotic stresses, including drought, salinity, high temperatures, and nutrient deficiencies, continuously challenge plants, significantly affecting growth, productivity, and survival. Plants respond to these stresses through complex signaling networks that detect and integrate multiple stresses to maintain homeostasis and enhance survival. Many abiotic stresses occur simultaneously or in quick succession, necessitating coordinated responses.

Drought stress occurs when plants lose water through transpiration and evaporation. Responses include stomatal closure, accumulation of osmoprotectants (e.g., proline), and activation of ABA-mediated pathways controlling stress-responsive gene expression. Salt stress causes osmotic imbalance and ion toxicity. Plants respond by regulating genes for ion homeostasis and osmotic adjustment, activating ion transporters (Na^+/K^+ pumps), and synthesizing compatible solutes such as glycerol. Hyperosmotic stress may result from drought and high salinity (Munns et al., 2020). Osmoregulation adjusts cellular solute concentrations to lower osmotic potential and improve water retention. Key solutes include inorganic ions (K^+, Na^+, Ca^{2+}) and organic compounds (proline, betaine, soluble sugars (SS), polyols) (Munns et al., 2020). By accumulating proline, SS, and soluble protein (SP), leaves reduce osmotic potential under drought. The proline content increases significantly, but the rate slows as soil drought progresses; SP content first decreases, then rises sharply when relative soil water content drops below 31.38%; and SS content continues to increase steadily. To prevent ion toxicity, plants compartmentalize Na^+ and Cl^- in vacuoles or use organic solutes for osmoregulation. Energy-efficient osmoregulation under high salinity may require modified cell wall structures or water–salt transport pathways (Zhang et al., 2022). In *Arabidopsis*, the SOS pathway regulates ion homeostasis. AtCBL4 (AtSOS3) perceives Ca^{2+} signals in roots and activates AtCIPK24 (AtSOS2), maintaining Na^+ balance.

Heat stress causes protein denaturation, membrane instability, and ROS accumulation. Plants express HSPs to maintain proteostasis. HSPs act as molecular chaperones in mitochondria, cytoplasm, and endoplasmic reticulum and are classified by size: HSP70, sHSP, HSP100, HSP60, and HSP90 (Wang et al., 2004). Heat modifies membrane fluidity and calcium signaling, activating MAPKs that regulate HSP gene expression and enhance heat tolerance. Cold stress damages membranes and reduces enzyme activity. Plants respond by adjusting membrane lipid composition, expressing cold-responsive genes, and producing cryoprotectants (e.g., sugar alcohols). Limited nutrient availability (e.g., nitrogen or phosphorus) triggers adaptive responses, including root architecture modifications, nutrient transporter gene upregulation, and metabolic pathway adjustments. Cold response TFs, such as CBF (AP2 domain), induce downstream COR genes. COR genes and MAPKs modulate cold stress via membrane fluidity, cytoskeleton, and inward calcium flow. Plasma membrane Ca^{2+} channels and regulators serve as cold sensors. *Arabidopsis* AtANN1 mutants show reduced Ca^{2+} influx and lower freezing tolerance (Liu et al., 2021). Accumulation of CBF factors stabilizes phyB photoreceptors, improving freezing resistance. Common elements of cold and heat stress signaling include phospholipid signaling, NO, proteasomal degradation, ROS, membrane fluidity changes, calcium signaling, and MAPK activation.

5.4.2 Mechanisms of integration in stress responses

5.4.2.1 Crosstalk between signaling pathways

Plant stress responses are governed by a complex network of interacting signaling pathways. Central to this network are phytohormones such as ABA, SA, ET, JA, GA, and CKs. ABA is pivotal in abiotic stresses like drought and salinity by inducing stomatal closure and stress-responsive gene expression, while also interacting with ET. Their interaction can be cooperative (e.g., stomatal closure) or antagonistic, as ET may worsen heat stress while ABA promotes tolerance. JA contributes to both abiotic stress tolerance and defense against wounding or herbivory, often in synergy with ABA during drought. While ABA prioritizes water conservation, JA induces defense-related metabolites, resulting in a balance between growth and survival. However, JA can inhibit ABA-mediated tolerance and vice versa, indicating a growth-defense trade-off. Similarly, ABA and GA act antagonistically: GA promotes growth but reduces stress tolerance, whereas ABA enhances tolerance at the expense of growth. SA, a key regulator of defense against biotrophic pathogens, also antagonizes ABA. High ABA levels can suppress SA-mediated immunity during drought, while SA can block ABA-induced drought responses during pathogen attack. ET frequently antagonizes GA, reducing growth under stress to prioritize survival. ET also interacts with CKs, often promoting senescence by triggering CK degradation, although CKs may later assist in recovery. Another layer of integration comes from JA–SA antagonism, where each suppresses the other's signaling, depending on whether the threat is biotic or abiotic. In many plants, SA is more effective against biotrophic pathogens, whereas JA defends against herbivores and necrotrophs. This hormonal crosstalk ensures plants finely tune responses to competing stresses.

At the molecular level, ABA signaling has been well elucidated. ABA perception occurs through receptors such as PYR/PYL/RCAR, which, upon binding ABA, inhibit PP2Cs and release SnRK2 kinases. These kinases phosphorylate downstream proteins, including ion channels (closing stomata), NADPH oxidases (ROS generation), and ABF/AREB TFs (stress gene regulation) (Wang et al., 2013). The PYL–PP2C–SnRK2 module represents a conserved focal point tightly regulated by other proteins (Zhang et al., 2020), highlighting ABA's central role in stress integration.

5.4.2.2 Reactive oxygen species and redox signaling

Abiotic stresses such as drought, salinity, and heat elevate ROS, which act as both damaging agents and crucial secondary messengers. While excess ROS causes oxidative stress, moderate ROS levels integrate signaling pathways to coordinate stress responses. For instance, drought-induced ROS production interacts with ABA signaling, enhancing stomatal closure and stress gene expression. ROS homeostasis depends on antioxidant systems comprising enzymes such as SOD, CAT, and peroxidases, which detoxify ROS and prevent cellular damage. Major ROS-producing organelles include chloroplasts, mitochondria, peroxisomes, and plasma membrane-bound NADPH oxidases (RBOHs). Particularly, RbohD and RbohF generate apoplastic ROS that propagate systemic stress signals, including calcium waves and electrical responses. Beyond local signaling, ROS are involved in retrograde communication from chloroplasts to the nucleus under high-light stress, activating protective transcriptional programs (Zhang et al., 2022). Thus, ROS function as both destructive and integrative signals, with their role depending on cellular redox balance.

5.4.3 Compartmentalization of signals in plant stress responses

Plants respond to biotic (pathogens, herbivory) and abiotic stresses (drought, salinity, temperature extremes) through complex signaling networks. A key feature of these systems is compartmentalization, where signaling molecules are localized to specific cellular or subcellular regions (plasma membrane, cytosol, nucleus, chloroplasts, mitochondria, vacuoles). This spatial organization ensures precise, energy-efficient stress responses while minimizing harmful effects.

5.4.3.1 Hormone compartmentalization in stress responses

Plant hormones such as ABA, ET, JA, SA, CKs, and GA regulate stress signaling, often synthesized or acting in defined compartments.

- Abscisic Acid (ABA): Central in drought, salinity, and cold responses. ABA is synthesized in the cytosol and transported to vacuoles to regulate osmotic adjustment and stomatal closure. In guard cells, it activates ion channels leading to stomatal closure, while downstream SnRK2 kinases in the cytosol regulate stress-responsive genes. Its receptors (PYR/PYL/RCAR) are membrane-localized, allowing precise control.
- Ethylene (ET): Mediates flooding, wounding, and pathogen defense. Synthesized in the cytosol, ET perception occurs via ER-localized receptors, which transmit signals to the cytosol and nucleus. It also interacts with other hormones (e.g., ABA, JA) in vascular tissues and leaves, coordinating herbivore and pathogen defense.
- Jasmonic Acid (JA): Produced in chloroplasts from linolenic acid, JA signaling occurs in the cytosol and nucleus. The bioactive JA-Ile binds the COI1 receptor, leading to degradation of JAZ repressors and activation of defense genes. Spatial regulation ensures JA-triggered defense is energy-efficient and stress-specific.
- Salicylic Acid (SA): Essential for systemic acquired resistance (SAR). Synthesized in the cytosol and transported to the nucleus, SA activates NPR1-dependent defense genes. During pathogen attack, SA accumulates in the apoplast and interacts with ROS in chloroplasts/vacuoles to fine-tune immunity while limiting oxidative damage.
- Cytokinins (CKs): Synthesized in roots and transported via the vascular system, CKs regulate growth, chloroplast development, oxidative stress tolerance, and senescence. They act in the nucleus to control stress-related genes and interact with ABA and ET to balance growth and stress adaptation.

5.4.3.2 Ion compartmentalization in stress responses

Ions, especially Ca^{2+}, Na^+, and K^+, play key roles in abiotic stress signaling, with compartmentalization critical for maintaining cellular homeostasis.

- Calcium (Ca^{2+}): A universal stress signal, Ca^{2+} enters the cytosol via plasma membrane or tonoplast channels. In the cytosol, it binds CaM and CBL proteins to activate stress pathways. Vacuoles act as reservoirs, buffering cytosolic Ca^{2+} and regulating stress-induced fluctuations. Chloroplasts and mitochondria also sequester Ca^{2+}, coordinating oxidative stress management and energy metabolism.
- Sodium (Na^+) and Potassium (K^+): Under salinity stress, excess Na^+ disrupts ionic balance. Plants use Na^+/H^+ antiporters to sequester Na^+ into vacuoles, protecting cytosolic processes. K^+ compartmentalization maintains osmotic balance, ensuring metabolic stability under salt stress.

5.4.3.3 Compartmentalization of reactive oxygen species

ROS such as superoxide (O_2^-), hydrogen peroxide (H_2O_2), and hydroxyl radicals (OH^-) act as both stress signals and damaging agents. Their spatial regulation determines whether they contribute to signaling or oxidative damage.

- ROS are produced in chloroplasts, mitochondria, and peroxisomes during stress. In chloroplasts, light stress drives ROS generation, initiating retrograde signaling to the nucleus. Mitochondria generate ROS during respiration, regulating energy balance under stress.
- Apoplastic ROS from NADPH oxidases (RBOHs) mediate rapid systemic signaling and trigger Ca^{2+} and electrical signals.
- Vacuoles function as storage and degradation sites, restricting ROS damage to vital biomolecules.
- By localizing ROS to specific compartments, plants exploit them as secondary messengers while preventing widespread cytotoxicity.

5.5 Transcriptional regulation in abiotic stress responses

Plants rely on stress-responsive TFs such as MYB, NAC, DREB/CBF, WRKY, bZIP, NF-Y, and WOX to regulate gene expression under abiotic stresses like drought, salinity, heat, and cold (Feng et al., 2023). These TFs link upstream signaling events to downstream stress-inducible genes, operating through ABA-dependent and ABA-independent pathways, often with significant crosstalk and hormonal interactions (Yoon et al., 2020). Gene expression under stress is modulated at transcriptional, posttranscriptional, and translational levels, involving enzymes for osmolyte biosynthesis, detoxification proteins, transporters, and regulatory proteins like TFs. Epigenetic mechanisms, such as histone modifications, DNA methylation, noncoding RNAs, and RNA chemical modifications such as m6A, provide transcriptional memory and fine-tune gene responses to recurring stress (Guo et al., 2018).

Stress-responsive gene networks integrate protective proteins (dehydrins, LEA proteins, HSPs), signaling components (MAPKs, CDPKs, phospholipases), and structural/transport proteins (aquaporins, ion transporters) to maintain cellular homeostasis under stress (Anami et al., 2009). ROS act both as signaling molecules and stress inducers, activating MAPK cascades and TFs to regulate defense responses, often in coordination with ABA and osmolyte pathways such as P5CS1-mediated proline biosynthesis. Computational analyses reveal extensive auto- and cross-regulation among TFs, making them prime targets for genetic engineering to enhance drought, salinity, and temperature tolerance. Overall, the integration of transcriptional, epigenetic, and signaling networks equips plants to perceive, respond, and adapt efficiently to complex abiotic stresses, forming a foundation for developing resilient crops.

5.6 Signaling under specific abiotic stresses

Plants adapt to environmental stresses not only through physiological adjustments but also via highly integrated signaling pathways. These signaling networks perceive external stimuli, transduce them into intracellular messages, and activate stress-responsive genes. Unlike basic descriptions of stress outcomes, understanding the specific pathways underlying these processes provides mechanistic insights essential for crop improvement. Below, the roles of ABA signaling, MAPK cascades, calcium-dependent pathways, ROS networks, and TF regulation are discussed for major abiotic stresses, including drought, salt, heat, cold, and heavy metals, with emphasis on crosstalk and convergence under combined stresses.

5.6.1 Drought stress signaling

5.6.1.1 Abscisic acid–dependent pathways

The ABA signaling pathway is the primary regulator of drought stress responses. Water deficit triggers ABA biosynthesis via NCED, leading to its accumulation in vascular tissues and guard cells. ABA perception occurs through the PYR/PYL/RCAR receptors, which inhibit type 2C protein phosphatases (PP2Cs). This inhibition activates SnRK2 kinases, which phosphorylate ABFs/AREBs TFs, thereby inducing drought-responsive genes, including LEA proteins, aquaporins, and osmoprotectant biosynthesis enzymes (Joshi et al., 2020).

5.6.1.2 Abscisic acid–independent pathways

Besides ABA, DREB2A TFs regulate drought-inducible genes via the ABA-independent pathway. DREB2A binds to the dehydration-responsive element (DRE) motifs in stress gene promoters. Its activation is mediated by MAPK cascades and posttranslational regulation, linking osmotic stress perception with transcriptional reprogramming.

5.6.1.3 Calcium and reactive oxygen species signaling

Drought stress triggers cytosolic Ca^{2+} elevations, sensed by calmodulins (CaMs), CBL–CIPK complexes, and CDPKs. These proteins phosphorylate ion transporters and TFs, contributing to stomatal closure and osmotic balance. Simultaneously, ROS act as secondary messengers, produced via NADPH oxidases (RBOHs). ROS activate MAPK cascades (e.g., MPK3/6), which regulate drought-responsive TFs such as WRKY33 and NAC.

5.6.1.3.1 Integration

Thus, drought stress signaling integrates ABA–SnRK2-ABF pathways, ROS-MAPK cascades, and Ca^{2+}-dependent phosphorylation events, ensuring fine-tuned stomatal regulation and gene expression.

5.6.2 Salt stress signaling

5.6.2.1 *Salt overly sensitive pathway*

Salt stress primarily disrupts ionic balance through excess Na^+. The SOS pathway is central to ionic homeostasis. Cytosolic Na^+ induces Ca^{2+} spikes, activating the Ca^{2+} sensor SOS3/CBL4, which interacts with SOS2 (CIPK24 kinase). The SOS2–SOS3 complex activates SOS1 (plasma membrane Na^+/H^+ antiporter), extruding Na^+ and restoring ion balance (Liu et al., 2024).

5.6.2.2 *Abscisic acid and reactive oxygen species signaling*

Salt stress also induces ABA accumulation, activating SnRK2s and ABFs, similar to drought stress. Concurrently, ROS produced by RBOHs amplify Ca^{2+} signaling, reinforcing SOS pathway activation. ROS also activate MAPKs that regulate ion transporter expression.

5.6.2.3 *Transcriptional regulation*

Salt stress induces TFs such as DREB2, NAC, and WRKYs, which regulate osmoprotectant biosynthesis and ion transporter expression. Epigenetic regulation (histone acetylation and methylation) has also been implicated in modulating salt-responsive genes.

5.6.2.3.1 Integration

Salt tolerance thus relies on a coordinated Ca^{2+}-ROS-SOS module, linked with ABA–SnRK2 and TF networks.

5.6.3 Heat stress signaling

5.6.3.1 *Heat shock transcription factors and heat shock proteins*

Heat stress perception leads to the trimerization of HSFs (e.g., HsfA1a, HsfA2), which bind to heat shock elements (HSEs) in promoters of HSP genes. HSPs (HSP70, HSP90, HSP101) act as molecular chaperones to refold denatured proteins and prevent aggregation (Baniwal et al., 2004)

5.6.3.2 *Mitogen-activated protein kinase and reactive oxygen species signaling*

Heat stress induces ROS bursts, which activate MAPK cascades (MPK3/6). These cascades regulate TFs (e.g., WRKY, ZAT) that promote antioxidant enzyme expression (APX, CAT). The MAPK–ROS module is critical for balancing oxidative stress with protective mechanisms (Qu et al., 2013).

5.6.3.3 *Calcium signaling*

Heat stress elevates cytosolic Ca^{2+} via phospholipase-mediated phosphatidylinositol pathways. Ca^{2+} activates CDPKs and calmodulin, which further regulate HSFs and MAPKs, creating a feedback loop between Ca^{2+} and ROS signaling.

5.6.3.4 *Hormonal crosstalk*

Heat responses are modulated by hormones: ABA regulates stomatal conductance, SA enhances antioxidant defenses, and auxins influence reproductive development. Hormonal integration ensures both vegetative survival and reproductive resilience under heat stress.

5.6.4 Cold stress signaling

5.6.4.1 *Inducer of CBF expression–C-repeat binding factor–cold-responsive pathway*

The ICE1–CBF–COR module is central to cold signaling. Low temperature activates ICE1 (inducer of CBF expression 1), which upregulates CBFs/DREB1s. CBFs bind to DRE/CRT elements of cold-responsive (COR) genes, including dehydrins and LEA proteins, enhancing membrane stability and osmotic adjustment (Zheng et al., 2021)

5.6.4.2 Ca^{2+} and mitogen-activated protein kinase signaling

Cold stress induces Ca^{2+} influx through mechanosensitive channels. Ca^{2+} activates CDPKs and MAPKs, which stabilize ICE1 and regulate CBF activity. ROS produced under cold stress further activate MAPKs, creating a Ca^{2+}–ROS–MAPK network.

5.6.4.3 Osmotic and metabolic regulation

Cold stress enhances osmoprotectant synthesis (proline, sugars, trehalose) via CBF-dependent and independent pathways. Genes such as OsTPP1 in rice regulate trehalose biosynthesis, balancing osmotic and oxidative stress (Guo et al., 2018).

5.6.5 Heavy metal stress signaling

5.6.5.1 Chelation and sequestration

Heavy metals (Cd, Pb, As) are detoxified via phytochelatins and metallothioneins, synthesized in response to metal exposure. These peptides chelate metals and sequester them in vacuoles. Glutathione (GSH) acts as both a chelator and an antioxidant (Kumar & Trivedi, 2016).

5.6.5.2 Reactive oxygen species and mitogen-activated protein kinase signaling

Heavy metal stress induces oxidative bursts, generating ROS as both damaging agents and signals. ROS activate MAPK cascades, which regulate stress-responsive TFs such as WRKY and NAC, leading to antioxidant gene expression (Mhamdi et al., 2010).

5.6.5.3 Hormonal crosstalk

Heavy metals alter hormonal pathways: ET and ABA enhance antioxidant responses, while auxin signaling is often disrupted, affecting root development.

5.6.6 Combined stress signaling

In natural environments, plants rarely face single stresses. Instead, combined stresses (drought + heat, salinity + heavy metals, etc.) trigger unique signaling signatures:

5.6.6.1 Drought + heat

Both activate ABA–SnRK2 pathways, but heat also induces HSF–HSP cascades. ABA and HSFs converge on ROS–MAPK signaling, producing synergistic stomatal regulation.

5.6.6.2 Salt + heavy metal

Both induce ROS bursts, but salt primarily engages SOS signaling, while metals activate chelation pathways. The shared ROS–MAPK module integrates these responses.

5.6.6.3 Cold + drought

Both activate osmoprotectant biosynthesis and ROS scavenging. CBF TFs overlap with DREB2s in regulating stress-inducible genes.

Such overlaps illustrate that Ca^{2+}, ROS, MAPK cascades, ABA signaling, and TF networks are central hubs where multiple stress signals converge.

5.7 Recent advances and future directions

Recent omics approaches (genomics, transcriptomics, proteomics, metabolomics) have uncovered dynamic signaling networks underlying stress responses (Naik et al., 2023). For example, genome-wide studies have revealed stress-specific TF families, whereas phosphoproteomics has identified novel MAPK and CDPK substrates.

Genome editing tools (e.g., CRISPR-Cas9) now enable precise manipulation of stress signaling genes. Engineering ABA receptors, SOS components, HSFs, and CBFs has shown promise in enhancing stress tolerance. Synthetic biology approaches aim to rewire signaling pathways for improved multistress resilience.

Future research must focus on decoding crosstalk between signaling pathways under combined stresses, as this represents the natural field environment. Integrated systems biology approaches will be essential to design crops capable of maintaining yield under climate change-driven challenges.

5.8 Omics-guided strategies for enhancing abiotic stress tolerance in plants

Omics approaches have advanced the understanding of abiotic stress signaling by identifying key genes, proteins, and metabolites involved in plant adaptation. Genomics through GWAS and QTL mapping has revealed stress-responsive genes such as the DREB family (Zhang et al., 2022), while transcriptomics has uncovered TFs such as NAC and bZIP in rice and cotton (Zhang et al., 2022). Proteomics has highlighted proteins such as HSPs and SOD, and metabolomics has identified osmolytes such as proline and glycine betaine (Anzano et al., 2022). Integrative multiomics approaches, supported by tools such as Cytoscape and MapMan, provide systems-level insights into signaling crosstalk under stress (Singhal et al., 2021). Building on these insights, engineering approaches have introduced stress-responsive genes like *TPS1* to improve drought tolerance (Gupta et al., 2020), while TFs such as MYB and WRKY enhance resilience. CRISPR/Cas9 has enabled precise editing of genes such as *OsDST* for drought and salinity tolerance in rice (Santosh Kumar et al., 2020). Further, manipulation of signaling components such as CDPKs has improved calcium-mediated stress tolerance (Parvin et al., 2019). Nonetheless, challenges such as pleiotropy, genotype–environment interactions, and public concerns highlight the need for strategies ensuring stable trait expression across environments.

5.9 Conclusion

Plants are constantly challenged by diverse abiotic stresses, such as drought, salinity, temperature extremes, and heavy metal toxicity, which compromise their growth, development, and productivity. To withstand these adversities, they rely on sophisticated signal transduction pathways that perceive stress stimuli, amplify signals, and activate downstream defense mechanisms. Central to these responses are calcium-dependent pathways, MAPK cascades, ROS and NO networks, ABA-mediated signaling, and cyclic nucleotide signaling (cAMP/cGMP). These pathways act as crucial focal points, enabling plants to maintain cellular homeostasis and reprogram metabolism during stress. Calcium signaling functions as a universal second messenger, rapidly transmitting stress signals and activating calcium-binding proteins, kinases, and TFs to regulate osmotic adjustment, ion balance, and gene expression. MAPK cascades amplify these signals through sequential phosphorylation events, leading to the activation of TFs such as WRKY, NAC, and DREB, which coordinate drought, salinity, and temperature stress responses. ROS and NO serve dual roles as damaging molecules and essential messengers, regulating ion transport, antioxidant defenses, and programmed cell death, while also crosstalking with ABA and MAPK pathways. ABA-mediated signaling plays a pivotal role in drought and salinity tolerance by controlling stomatal closure, inducing osmoprotectant biosynthesis, and activating ABA-responsive TFs. Similarly, cyclic nucleotide signaling integrates with Ca^{2+} and hormone pathways to regulate ion channels and antioxidant defense under multiple stresses. Importantly, these pathways rarely function in isolation. Crosstalk and integration among calcium, ROS, ABA, MAPK, and hormone networks enable plants to mount coordinated and stress-specific responses, particularly under combined stress conditions. This interconnectedness underscores the evolutionary sophistication of plant signaling in balancing growth, survival, and reproduction under challenging environments.

A deeper understanding of these signal transduction mechanisms offers tremendous potential for crop improvement. Targeted manipulation of key signaling components through genomics, proteomics, and genome editing can enhance multistress tolerance and ensure sustainable agricultural productivity in the face of climate change.

References

Abdolmaleki, A.K., Pirdashti, H., Yaghoubian, Y., Abbasian, A., Ghadirnezhad Shiade, S.R., 2023. Endophytic fungi improve growth and yield of wheat (*Triticum aestivum* L.) under limited light conditions. Gesunde Pflanz 75, 1517–1529.

Agrawal, G.K., Rakwal, R., Iwahashi, H., 2002. Isolation of novel rice (*Oryza sativa* L.) multiple stress responsive MAP kinase gene, OsMSRMK2, whose mRNA accumulates rapidly in response to environmental cues. Biochemical and Biophysical Research Communications 294, 1009–1016.

Anami, S., De Block, M., Machuka, J., Van Lijsebettens, M., 2009. Molecular improvement of tropical maize for drought stress tolerance in sub-Saharan Africa. Critical Reviews in Plant Sciences 28, 16–35.

Anzano, A., Bonanomi, G., Mazzoleni, S., Lanzotti, V., 2022. Plant metabolomics in biotic and abiotic stress: a critical overview. Phytochemistry Reviews 21, 503–524.

Apel, K., Hirt, H., 2004. Reactive oxygen species: metabolism, oxidative stress, and signal transduction. Annual Review of Plant Biology 55, 373–399.

Baniwal, S.K., Bharti, K., Chan, K.Y., Fauth, M., Ganguli, A., Kotak, S., et al., 2004. Heat stress response in plants: a complex game with chaperones and more than twenty heat stress transcription factors. Journal of Biosciences 29, 471–487.

Blanco, F.A., Zanetti, M.E., Casalongué, C.A., Daleo, G.R., 2006. Molecular characterization of a potato MAP kinase transcriptionally regulated by multiple environmental stresses. Plant Physiology and Biochemistry PPB / Societe Francaise de Physiologie Vegetale 44, 315–322.

Dan, I., Watanabe, N.M., Kusumi, A., 2001. The Ste20 group kinases as regulators of MAP kinase cascades. Trends in Cell Biology 11, 220–230.

De Zélicourt, A., Colcombet, J., Hirt, H., 2016. The role of MAPK modules and ABA during abiotic stress signaling. Trends in Plant Science 21, 677–685.

Ding, H., Zhang, X., Xu, S., Sun, L., Jiang, M., Zhang, A., et al., 2009. Induction of protection against paraquat-induced oxidative damage by abscisic acid in maize leaves is mediated through mitogen-activated protein kinase. Journal of Integrative Plant Biology 51, 961–972.

Ding, Y., Yang, S., 2022. Surviving and thriving: how plants perceive and respond to temperature stress. Developmental Cell 57, 947–958.

Divekar, P.A., Narayana, S., Divekar, B.A., Kumar, R., Gadratagi, B.G., Ray, A., et al., 2022. Plant secondary metabolites as defense tools against herbivores for sustainable crop protection. International Journal Molecular Science 23, 2690.

Droillard, M.-J., Boudsocq, M., Barbier-Brygoo, H., Laurière, C., 2002. Different protein kinase families are activated by osmotic stresses in Arabidopsis thaliana cell suspensions: involvement of the MAP kinases AtMPK3 and AtMPK6. FEBS Letters 527, 43–50.

Feng, S., Wang, L., Zhou, Y., Wang, Y., Fang, Y., 2023. Research progresses on WOX family genes in regulating plant development and abiotic stress response. Biotechnology Bulletin 39, 1.

Ghadirnezhad Shiade, S.R., Fathi, A., Taghavi Ghasemkheili, F., Amiri, E., Pessarakli, M., 2023. Plants' responses under drought stress conditions: effects of strategic management approaches—a review. Journal of Plant Nutrition 46, 2198–2230.

Gu, L., Liu, Y., Zong, X., Liu, L., Li, D.-P., Li, D.-Q., 2010. Overexpression of maize mitogen-activated protein kinase gene, ZmSIMK1 in Arabidopsis increases tolerance to salt stress. Molecular Biology Reports 37, 4067–4073.

Guo, X., Liu, D., Chong, K., 2018. Cold signaling in plants: insights into mechanisms and regulation. Journal of Integrative Plant Biology 60, 745–756.

Gupta, A., Rico-Medina, A., Caño-Delgado, A.I., 2020. The physiology of plant responses to drought. Science (New York, N.Y.) 368 (80-), 266–269.

Gupta, K.J., Igamberdiev, A.U., 2015. Compartmentalization of reactive oxygen species and nitric oxide production in plant cells: an overview. In: Gupta, K., Igamberdiev, A. (Eds.), Reactive Oxygen and Nitrogen Species Signaling and Communication in Plants, vol 23. Springer, Cham, pp. 1–14. Available at https://doi.org/10.1007/978-3-319-10079-1_1.

Hettenhausen, C., Schuman, M.C., Wu, J., 2015. MAPK signaling: a key element in plant defense response to insects. Insect Science 22, 157–164.

Ichimura, K., Mizoguchi, T., Yoshida, R., Yuasa, T., Shinozaki, K., 2000. Various abiotic stresses rapidly activate Arabidopsis MAP kinases ATMPK4 and ATMPK6. The Plant Journal for Cell and Molecular Biology 24, 655–665.

Igamberdiev, A.U., Hill, R.D., 2004. Nitrate, NO and haemoglobin in plant adaptation to hypoxia: an alternative to classic fermentation pathways. Journal of Experimental Botany 55, 2473–2482.

Jain, M., Nagar, P., Goel, P., Singh, A.K., Kumari, S., Mustafiz, A., 2018. Second messengers: central regulators in plant abiotic stress response. In: Zargar, S., Zargar, M. (Eds.), Abiotic Stress-Mediated Sensing and Signaling in Plants: An Omics Perspective. Springer, Singapore, pp. 47–94. Available at https://doi.org/10.1007/978-981-10-7479-0_2.

Jonak, C., Kiegerl, S., Ligterink, W., Barker, P.J., Huskisson, N.S., Hirt, H., 1996. Stress signaling in plants: a mitogen-activated protein kinase pathway is activated by cold and drought. Proceedings of the National Academy of Sciences of the United States of America 93, 11274–11279.

Jonak, C., Nakagami, H., Hirt, H., 2004. Heavy metal stress. Activation of distinct mitogen-activated protein kinase pathways by copper and cadmium. Plant Physiology 136, 3276–3283.

Jonak, C., Ökrész, Bögre, L., Hirt, H, L., 2002. Complexity, cross talk and integration of plant MAP kinase signalling. Current Opinion in Plant Biology 5, 415–424.

Joshi, R.K., Bharat, S.S., Mishra, R., 2020. Engineering drought tolerance in plants through CRISPR/Cas genome editing. 3 Biotech 10, 400.

Kumar, K., Rao, K.P., Sharma, P., Sinha, A.K., 2008. Differential regulation of rice mitogen activated protein kinase kinase (MKK) by abiotic stress. Plant Physiology and Biochemistry PPB / Societe Francaise de Physiologie Vegetale 46, 891–897.

Kumar, S., Trivedi, P.K., 2016. Heavy metal stress signaling in plants. Plant Metal Interaction. Elsevier, pp. 585–603.

Laloi, C., Apel, K., Danon, A., 2004. Reactive oxygen signalling: the latest news. Current Opinion in Plant Biology 7, 323–328.

Lamotte, O., Bertoldo, J.B., Besson-Bard, A., Rosnoblet, C., Aimé, S., Hichami, S., et al., 2015. Protein S-nitrosylation: specificity and identification strategies in plants. Frontiers in Chemistry. 2, 114.

Lee, S., Hirt, H., Lee, Y., 2001. Phosphatidic acid activates a wound-activated MAPK in glycine max. The Plant Journal for Cell and Molecular Biology 26, 479–486.

Liu, C., Jiang, X., Yuan, Z., 2024. Plant responses and adaptations to salt stress: a review. Horticulturae 10, 1221.

Liu, Q., Ding, Y., Shi, Y., Ma, L., Wang, Y., Song, C., et al., 2021. The calcium transporter ANNEXIN1 mediates cold-induced calcium signaling and freezing tolerance in plants. The EMBO Journal 40, e104559.

Mahajan, S., Tuteja, N., 2005. Cold, salinity and drought stresses: an overview. Archives of Biochemistry and Biophysics 444, 139–158.

Mhamdi, A., Queval, G., Chaouch, S., Vanderauwera, S., Van Breusegem, F., Noctor, G., 2010. Catalase function in plants: a focus on Arabidopsis mutants as stress-mimic models. Journal of Experimental Botany 61, 4197–4220.

Mittler, R., Finka, A., Goloubinoff, P., 2012. How do plants feel the heat? Trends in Biochemical Sciences 37, 118–125.

Munns, R., Passioura, J.B., Colmer, T.D., Byrt, C.S., 2020. Osmotic adjustment and energy limitations to plant growth in saline soil. The New Phytologist 225, 1091–1096.

Nadarajah, K.K., 2020. ROS homeostasis in abiotic stress tolerance in plants. International Journal of Molecular Science 21, 5208.

Naik, B., Kumar, V., Rizwanuddin, S., Chauhan, M., Choudhary, M., Gupta, A.K., et al., 2023. Genomics, proteomics, and metabolomics approaches to improve abiotic stress tolerance in tomato plant. International Journal of Molecular Science 24, 3025.

Nakagami, H., Kiegerl, S., Hirt, H., 2004. OMTK1, a novel MAPKKK, channels oxidative stress signaling through direct MAPK interaction. The Journal of Biological Chemistry 279, 26959–26966.

Nakashima, K., Takasaki, H., Mizoi, J., Shinozaki, K., Yamaguchi-Shinozaki, K., 2012. NAC transcription factors in plant abiotic stress responses. Biochimica et Biophysica Acta (BBA) - Gene Regulatory Mechanisms 1819, 97–103.

Nakashima, K., Yamaguchi-Shinozaki, K., 2013. ABA signaling in stress-response and seed development. Plant Cell Reports 32, 959–970.

Nambara, E., Marion-Poll, A., 2005. Abscisic acid biosynthesis and catabolism. Annual Review of Plant Biology 56, 165–185.

Ning, J., Li, X., Hicks, L.M., Xiong, L., 2010. A Raf-like MAPKKK gene DSM1 mediates drought resistance through reactive oxygen species scavenging in rice. Plant Physiology 152, 876–890.

Ortiz-Masia, D., Perez-Amador, M.A., Carbonell, P., Aniento, F., Carbonell, J., Marcote, M.J., 2008. Characterization of PsMPK2, the first C1 subgroup MAP kinase from pea (Pisum sativum L.). Planta 227, 1333–1342.

Oz, M.T., Eyidogan, F., Yucel, M., Öktem, H.A., 2015. Functional role of nitric oxide under abiotic stress conditions. In: Khan, M., Mobin, M., Mohammad, F., Corpas, F. (Eds.), Nitric Oxide Action Abiotic Stress Responses Plants. Springer, Cham, pp. 21–41. Available at https://doi.org/10.1007/978-3-319-17804-2_2.

Parvin, K., Nahar, K., Hasanuzzaman, M., Bhuyan, M.H.M.B., Fujita, M., 2019. Calcium-mediated growth regulation and abiotic stress tolerance in plants. Plant Abiotic Stress Tolerance: Agronomic, Molecular and Biotechnological Approaches 291–331.

Qu, A.-L., Ding, Y. F., Jiang, Q., Zhu, C., 2013. Molecular mechanisms of the plant heat stress response. Biochemical and Biophysical Research Communications 432, 203–207.

Quan, R., Lin, H., Mendoza, I., Zhang, Y., Cao, W., Yang, Y., et al., 2007. SCABP8/CBL10, a putative calcium sensor, interacts with the protein kinase SOS2 to protect Arabidopsis shoots from salt stress. The Plant Cell 19, 1415–1431.

Radha, B., Sunitha, N.C., Sah, R.P., TP, M.A., Krishna, G.K., Umesh, D.K., et al., 2023. Physiological and molecular implications of multiple abiotic stresses on yield and quality of rice. Frontiers in Plant Science 13, 996514.

Sanders, D., Pelloux, J., Brownlee, C., Harper, J.F., 2002. Calcium at the crossroads of signaling. The Plant Cell 14, S401–S417.

Sangwan, V., Örvar, B.L., Beyerly, J., Hirt, H., Dhindsa, R.S., 2002. Opposite changes in membrane fluidity mimic cold and heat stress activation of distinct plant MAP kinase pathways. The Plant Journal for Cell and Molecular Biology 31, 629–638.

Santosh Kumar, V.V., Verma, R.K., Yadav, S.K., Yadav, P., Watts, A., Rao, M.V., et al., 2020. CRISPR-Cas9 mediated genome editing of drought and salt tolerance (OsDST) gene in indica mega rice cultivar MTU1010. Physiology and Molecular Biology of Plants 26, 1099–1110.

Sharma, L., Parihar, S., Seth, C.S., Shekhawat, G.S., 2025. Abiotic stress tolerance: role of nitric oxide signalling in developing climate-resilient plants. Tree Biology and Biotechnology. Springer, pp. 303–316.

Shi, J., An, H.-L., Zhang, L., Gao, Z., Guo, X.-Q., 2010. GhMPK7, a novel multiple stress-responsive cotton group C MAPK gene, has a role in broad spectrum disease resistance and plant development. Plant Molecular Biology 74, 1–17.

Singhal, R.K., Saha, D., Skalicky, M., Mishra, U.N., Chauhan, J., Behera, L.P., et al., 2021. Crucial cell signaling compounds crosstalk and integrative multi-omics techniques for salinity stress tolerance in plants. Frontiers in Plant Science 12, 670369.

Sinha, A.K., Jaggi, M., Raghuram, B., Tuteja, N., 2011. Mitogen-activated protein kinase signaling in plants under abiotic stress. Plant Signaling & Behavior 6, 196–203.

Takekawa, M., Tatebayashi, K., Saito, H., 2005. Conserved docking site is essential for activation of mammalian MAP kinase kinases by specific MAP kinase kinase kinases. Molecular Cell 18, 295–306.

Teige, M., Scheikl, E., Eulgem, T., Doczi, R., Ichimura, K., Shinozaki, K., et al., 2004. The MKK2 pathway mediates cold and salt stress signaling in Arabidopsis. Molecular Cell 15, 141–152.

Tuteja, N., Sopory, S.K., 2008. Chemical signaling under abiotic stress environment in plants. Plant Signaling & Behavior 3, 525–536.

Wang, J., Ding, H., Zhang, A., Ma, F., Cao, J., Jiang, M., 2010. A novel mitogen-activated protein kinase gene in maize (Zea mays), ZmMPK3, is involved in response to diverse environmental cues. Journal of Integrative Plant Biology 52, 442–452.

Wang, M., Zhang, Y., Wang, J., Wu, X., Guo, X., 2007. A novel MAP kinase gene in cotton (*Gossypium hirsutum* L.), GhMAPK, is involved in response to diverse environmental stresses. BMB Reports 40, 325–332.

Wang, P., Xue, L., Batelli, G., Lee, S., Hou, Y.-J., Van Oosten, M.J., et al., 2013. Quantitative phosphoproteomics identifies SnRK2 protein kinase substrates and reveals the effectors of abscisic acid action. Proceedings of the National Academy of Sciences of the United States of America 110, 11205–11210.

Wang, W., Vinocur, B., Shoseyov, O., Altman, A., 2004. Role of plant heat-shock proteins and molecular chaperones in the abiotic stress response. Trends in Plant Science 9, 244–252.

Xing, T., Malik, K., Martin, T., Miki, B.L., 2001. Activation of tomato PR and wound-related genes by a mutagenized tomato MAP kinase kinase through divergent pathways. Plant Molecular Biology 46, 109–120.

Xiong, L., Zhu, J.-K., 2003. Regulation of abscisic acid biosynthesis. Plant Physiology 133, 29–36.

Xu, T., Niu, J., Jiang, Z., 2022. Sensing mechanisms: calcium signaling mediated abiotic stress in plants. Frontiers in Plant Science 13, 925863.

Yang, C., Fang, X., Wu, X., Mao, Y., Wang, L., Chen, X., 2012. Transcriptional regulation of plant secondary metabolism. Journal of Integrative Plant Biology 54, 703–712.

Yoon, Y., Seo, D.H., Shin, H., Kim, H.J., Kim, C.M., Jang, G., 2020. The role of stress-responsive transcription factors in modulating abiotic stress tolerance in plants. Agronomy 10, 788.

Zhang, H., Liu, S., Ren, T., Niu, M., Liu, X., Liu, C., et al., 2023. Crucial abiotic stress regulatory network of NF-Y transcription factor in plants. International Journal of Molecular Science 24, 4426.

Zhang, H., Zhao, Y., Zhu, J.-K., 2020. Thriving under stress: how plants balance growth and the stress response. Developmental Cell 55, 529–543.

Zhang, H., Zhu, J., Gong, Z., Zhu, J.-K., 2022. Abiotic stress responses in plants. Nature Reviews. Genetics 23, 104–119.

Zhang, T., Liu, Y., Xue, L., Xu, S., Chen, T., Yang, T., et al., 2006. Molecular cloning and characterization of a novel MAP kinase gene in Chorispora bungeana. Plant Physiology and Biochemistry PPB / Societe Francaise de Physiologie Vegetale 44, 78–84.

Zhang, W., Ye, S., Du, Y., Zhao, Q., Du, J., Zhang, Q., 2022. Identification and expression analysis of bZIP members under abiotic stress in mung bean (Vigna radiata). Life (Chicago, Ill. 1978) 12, 938.

Zheng, S., Su, M., Wang, L., Zhang, T., Wang, J., Xie, H., et al., 2021. Small signaling molecules in plant response to cold stress. Journal of Plant Physiology 266, 153534.

Zhu, J.-K., 2016. Abiotic stress signaling and responses in plants. Cell 167, 313–324.

Zong, X., Li, D.-peng, Gu, L., Li, D.-quan, Liu, L., Hu, X., 2009. Abscisic acid and hydrogen peroxide induce a novel maize group C MAP kinase gene, ZmMPK7, which is responsible for the removal of reactive oxygen species. Planta 229, 485–495.

Chapter 6

Regulatory networks and redox homeostasis in plant responses to multiple abiotic stresses

Shubham Joshi[1,2], Anita Kumari[1,2], Jhilmil Nath[1,2], Suman Gusain[1,2], Khushbu Kumari[1,2], Meenakshi Rawat[1,2], Akshita Patial[1,3], Rakhi Chandoliya[1,4], and Rohit Joshi[1,2,*]

[1]*CSIR-Institute of Himalayan Bioresource Technology, Palampur, Himachal Pradesh, India,* [2]*Academy of Scientific and Innovative Research (AcSIR), Ghaziabad, India,* [3]*Department of Molecular Biology & Genetic Engineering, School of Bioengineering and Biosciences, Lovely Professional University, Phagwara, Punjab, India,* [4]*Department of Biotechnology, School of Bioengineering and Biosciences, Lovely Professional University, Phagwara, Punjab, India*

[*]*Corresponding author. e-mail address: rohit.joshi@csir.res.in*

6.1 Introduction

Abiotic stresses such as drought, heat, salinity, and heavy metal accumulation significantly impact plant growth and contribute to reduced agricultural productivity worldwide. According to a report by FAO, around 96.5% of global rural land is affected by these stresses. These challenges are exacerbated by climate change, declining freshwater resources, and the concurrent occurrence of multiple stresses (Hu and Xiong, 2014). To address food security concerns amid an increasing global population and environmental challenges, researchers emphasize the urgent need for a "second green revolution." This effort aims to enhance crop yield and yield stability under suboptimal and adverse conditions by leveraging recent genomic research advancements (You and Chan, 2015). Plants have evolved various cellular responses to cope with stress via activation of numerous stress regulatory genes and synthesis of diverse functional proteins through intricate signal transduction networks to confer stress tolerance (Zhang et al., 2022). Abiotic stresses disrupt the cellular metabolism, leading to the generation of reactive oxygen species (ROS), reactive nitrogen species (RNS), and reactive sulfur species. While potentially damaging when accumulated at high levels due to their reactivity with nucleic acids, proteins, and lipids, these species also serve as crucial signaling molecules. They do so by promoting posttranslational modifications, a process particularly relevant to amino acids like methionine, tyrosine, lysine, and arginine (Raza et al., 2021; Denjalli et al., 2024).

Plants maintain a dynamic balance between ROS development and scavenging, achieved through a network of ROS detoxifying proteins, which helps stabilize ROS as signaling molecules (Mignolet-Spruyt et al., 2016; Gilroy et al., 2016). Each cellular compartment has particular ROS homeostasis mechanisms, and ROS signaling is modulated by factors such as cell type, developmental stage, and stress level. Under abiotic stress conditions, ROS accumulation is detected by specific sensors, triggering stress-specific signaling pathways that support cellular acclimation responses (Fig. 6.1). These ROS signals mediate redox reactions that can modify protein function and structure, impacting transcription factor (TF) binding to DNA and altering transcription (Dietz et al., 2016). This redox system is central to ROS-mediated metabolic regulation and signal transduction, enabling plants to modulate cellular activities under stress (Dietz, 2015; Foyer and Noctor, 2016). ROS, such as superoxide radicals ($O_2^{\bullet-}$), singlet oxygen ($^1O_2$), hydrogen peroxide (H_2O_2), and hydroxyl radicals ($^{\bullet}OH$), are common byproducts of cellular metabolism in aerobic organisms (Raja et al., 2017; Cejudo et al., 2021). Additionally, stress perception activates Reduced Nicotinamide Adenine Dinucleotide Phosphate oxidases at the plasma membrane, which produce signaling ROS through calcium-dependent phosphorylation (RESPIRATORY BURST OXIDASE HOMOLOG) pathways (Gilroy et al., 2014). Recent research highlights the importance of thiol-based redox regulatory networks across cellular compartments, such as nucleoplasm, cytosol, mitochondrial matrix, and stroma. These networks incorporate ferredoxin-dependent thioredoxin (Trx) reductase, NADPH-dependent Trx reductase, and

Multiple Abiotic Stresses in Plants. DOI: https://doi.org/10.1016/B978-0-443-27424-4.00008-2

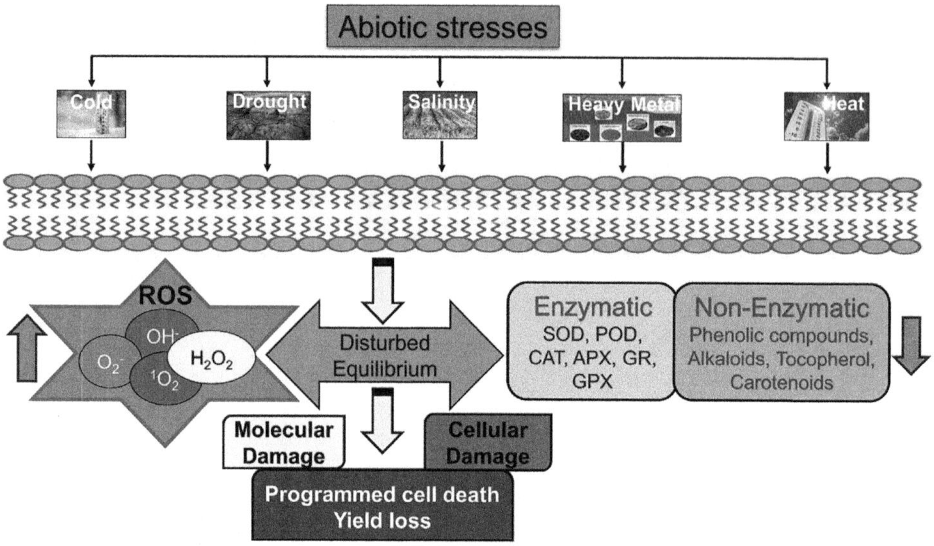

FIGURE 6.1 Regulation of redox homeostasis and response mechanism of multiple abiotic stress mitigation in plants.

NADPH-dependent glutathione reductase (Huang et al., 2023). To connect the dynamics of subcellular redox networks with plant-environment interactions, advanced methodologies are required, including mutant studies, in vivo imaging sensors, proteome analysis, and mathematical modeling. Under abiotic stresses, increased ROS and RNS activate mitogen-activated protein kinase (MAPK), calcium depended protein kinases, and G protein-coupled receptor (GPCR) pathways, which, in turn, modulate defense-related transcripts, phytohormone signaling, and plant immune responses via TFs like ERF, MYC, and AP2 (Denjalli et al., 2024). This redox signaling modulates transcription and translation, ultimately reducing metabolic ROS levels and promoting stress tolerance.

This chapter aims to examine in detail the mechanisms of signal perception and downstream signaling, as well as the physiological and molecular responses plants employ to cope with abiotic stress. Additionally, it explores the interaction of abiotic stress responses, hormonal pathways, and MAPK signaling, highlighting the role of ROS as both a signaling and regulatory molecule in plant stress adaptation.

6.2 Signaling and transcriptional regulation of cold stress tolerance in relation to ROS homeostasis

Plants cope with cold stress by engaging a network of interconnected signaling pathways that involve cold, heat, and ROS responses. Their initial responses involve modifying membrane proteins and activating Ca^{2+} ion channels upon detecting cold stress. Beyond Ca^{2+}, evidence indicates that several signaling molecules, like ROS and nitric oxide (NO), are crucial for modulating responses to low temperature stress. ROS, such as superoxide, hydroxyl radicals, and hydrogen peroxide, are produced as a result of different stressors (Singh et al., 2022). In plant cells, ROS serve dual functions: they activate expression of genes and translation to mitigate stress while also contributing to oxidative damage (Qi et al., 2018). Under low temperature stress, plants tend to build up H_2O_2, which can harm cells when present in excess. To mitigate this, the H_2O_2 scavenging system is activated, converting reduced glutathione to oxidized glutathione (Mishra et al., 2023). Additionally, certain defense genes have promoter regions that feature either antioxidant-responsive elements or sites that bind GSSG, suggesting that redox signaling regulates gene expression during cold stress (Mishra et al., 2023). Identifying COR genes functioning downstream of redox signaling is, therefore a key area for research. Evidence also indicates a connection between calcium ions and ROS, where minimal ROS levels facilitate the entry of Ca^{2+} into the cytoplasm (Rihan et al., 2017). Furthermore, Ca^{2+} participates in regulating ROS generation during environmental stresses like elevated humidity levels and drought (Wang et al., 2016). It is essential to comprehend how Ca^{2+} and ROS function individually and interact to regulate plant responses to cold stress. NO is essential for plants, serving as a key messenger in responding to environmental cues and regulating growth processes (Wilson et al., 2008). In plants, cold stress triggers the production of NO, a response observed across multiple plant species (Zhao et al., 2009). Studies show that NO, generated through nitrate reductase, plays a role in enhancing the ability to withstand cold and freezing temperatures in *Arabidopsis* (Costa-Broseta et al., 2018). However, more recent findings indicate that NO might also act as a negative regulator, reducing constitutive freezing tolerance in the same species (Costa-Broseta et al., 2018). S-nitrosylation, a process regulated by NO, is essential for plant growth, development, and the ability to respond to environmental stressors (Zhan et al., 2018; Puyaubert et al., 2014; Sehrawat et al., 2013).

The dehydration-responsive element (DRE), with the regulatory sequence A/GCCGAC, regulates dehydration- and cold-induced genes in *Arabidopsis thaliana*, known as RD genes (Kidokoro et al., 2022). DREs are found in the promoters of similar genes in rice and soybean (Maruyama et al., 2012; Yamaguchi-Shinozaki and Shinozaki, 2006). Additionally, the C-repeat motif is present in the promoters of cold-responsive COR genes (Joshi et al., 2024a, b). In *Arabidopsis*, DREB1/CBFs are crucial TFs for cold-induced gene expression (Gusain et al., 2025). These AP2/ERF family members bind to DREs, regulating the expression of cold-responsive genes, including *DREB1A*, *DREB1B*, and *DREB1C* (Tiwari et al., 2024). These genes are organized in tandem within the genome in the sequence *DREB1B, DREB1A*, and *DREB1C*. Mutant *Arabidopsis* plants lacking all three genes exhibit significantly reduced tolerance to freezing temperature (Jia et al., 2016). Cold stress rapidly induces the expression of all three *DREB1* genes, initiating a cold-responsive transcriptional cascade that activates numerous cold-inducible genes (Thomashow, 2010). Understanding the regulation of *DREB1* expression under cold stress is therefore essential for unraveling how plants perceive and respond to cold conditions. Numerous TFs control the cold-triggered activation of the *DREB1* gene, highlighting a complex regulatory system (Kidokoro et al., 2020). One such factor, INDUCER OF CBF EXPRESSION 1 (ICE1), was initially discovered through the *ice1-1* mutant. This mutant emerged from a screening aimed at identifying mutations that impact the cold-induced activation of a luciferase reporter gene driven by the *DREB1A* promoter (Chinnusamy et al., 2003). The mutant plants exhibited reduced levels of endogenous *DREB1A* expression and had a reduced ability to withstand freezing temperatures. Over the past 17 years, *ICE1* has played a role as a key positive regulator of *DREB1/CBF* expression and has been extensively supported by research, which has also identified related factors, including different protein kinases and RING finger ubiquitin E3 ligase (Thomashow and Torii, 2020). Previous research found that the suppression of *DREB1A* in *ice1-1* mutants is attributed to gene silencing caused by DNA methylation associated with T-DNA insertion, rather than a mutation in the *ICE1* gene. This suggests that *ICE1* is not directly responsible for regulating *DREB1* expression (Thomashow and Torii, 2020).

6.3 Signaling and transcriptional networks in plant responses to drought stress in relation to ROS homeostasis

Drought is a key stressor that affects plant growth, development, and global productivity. To cope with water scarcity, land plants have evolved sophisticated mechanisms for sensing and responding to drought stress, with ROS playing a pivotal role as key signaling molecules. These responses are intricately coordinated by various signaling molecules (e.g., ROS, Ca^{2+}) and pathways, including calcium (Ca^{2+}) signaling, abscisic acid (ABA) pathways, and MAPK cascades, which collectively regulate plant adaptation under drought conditions. Ca^{2+} signaling acts as a secondary messenger system, amplifying stress signals and modulating various physiological and molecular responses such as ABA signaling and stomatal closure. Drought stress contributes to a surge in ROS levels, which activate membrane-localized hyperpolarization-activated Ca^{2+} channels. This activation elevates free cytosolic calcium ($[Ca^{2+}]_n$) levels (Demidchik et al., 2007). The increase in intracellular Ca^{2+} levels creates a positive feedback loop by further activating RBOH D, amplifying ROS production in the apoplast (Takeda et al., 2008). Additionally, hydrogen peroxide (H_2O_2) binds to HYDROGEN-PEROXIDE-INDUCED Ca^{2+} INCREASES1 (HPCA1), which activates Ca^{2+} channels to enhance Ca^{2+} influx in guard cells, initiating stomatal closure (Wu et al., 2020).

Drought facilitates the synthesis of ABA in plant cells, which is detected by the ABA receptors (PYR1/PYL/RCAR). Upon binding to ABA, the PYLs interact with protein phosphatases (PP2Cs), which negatively regulate ABA signaling. This interaction alleviates PP2C-mediated inhibition, facilitating the induction of protein kinases such as GUARD CELL HYDROGEN PEROXIDE-RESISTANT1 (GHR1) and OPEN STOMATA1 (OST1) (Ma et al., 2009; Hua et al., 2012). These kinases phosphorylate downstream targets, including ALUMINUM-ACTIVATED MALATE TRANSPORTER12 (ALMT12), SLOW ANION CHANNEL-ASSOCIATED1 (SLAC1), and SLAC1 HOMOLOGUE3 (SLAH3). Simultaneously, these kinases also phosphorylate potassium channels, such as KAT1, to prevent the inward flow of K^+ and facilitate stomatal closure (Brandt et al., 2015). NADPH oxidases (Respiratory Burst Oxidase Homolog D (RBOHD) and Homolog F (RBOHF)) mediated ROS generation in guard cells is critical for regulating ABA-mediated drought signaling (Kwak et al., 2003). The activity of RBOH enzymes is tightly regulated through protein phosphorylation. For instance, in *Arabidopsis*, OST1 phosphorylates AtRBOHF at serine residues (Ser13 and Ser174) in the N-terminal region to regulate its activity (Sirichandra et al., 2009). OST1 also forms a complex with BRI1-ASSOCIATED KINASE1 (BAK1) near the cell membrane, further regulating ROS production in guard cells (Shang et al., 2016). Its specific partner and the precise mechanism by which it influences ROS production and stomatal dynamics remain unclear.

Under stress conditions, phospholipase Dα1 (PLDα1) hydrolyzes membrane phospholipids to produce phosphatidic acid (PA). PA plays a critical role in regulating ABA-mediated ROS generation in guard cells and stomatal closure (Zhang et al., 2009). In *pldα1* mutants, ROS generation and stomatal closure are obstructed but can be restored by adding PA, highlighting its essential role (Zhang et al., 2009). Mutations that disrupt the PA-binding domain in RBOHD compromise ABA-induced ROS production (Zhang et al., 2009). PA also interacts with ABI1 to reduce its phosphatase activity and enhance ABA-facilitated stomatal closure (Zhang et al., 2004). Besides these, H_2O_2 is closely associated with the activation of the MAPK cascade, a key pathway in the plant response to drought stress. The MAPK cascade transduces ROS signals to the nucleus, where it controls gene expression by modulating the activity of various TFs (Singh et al., 2019; Joshi et al., 2024a, b). This signaling cascade affects the expression of genes responsible for encoding antioxidant enzymes such as superoxide dismutase (SOD), catalase (CAT), and ascorbate peroxidase (APX) and upregulates genes associated with desiccation tolerance, osmo-protection, and DNA damage repair (Liu and He, 2017). The MAP Kinase pathway is a key signaling pathway for proline synthesis in plants under stress (Meena et al., 2019). A study on potato revealed that overexpression of *StMAPK11* enhanced drought tolerance in plants by increasing proline accumulation and enhancing antioxidant enzyme activity (Zhu et al., 2021). Emerging evidence suggests that ROS signaling can influence epigenetic modifications, impacting chromatin structure and gene expression. ROS-induced changes in histone acetylation, methylation, and phosphorylation can alter chromatin accessibility, facilitating the binding of TFs to promoter regions of stress-inducible genes. These epigenetic modifications provide an additional layer of regulation, enabling plants to adapt to drought stress and potentially "memorize" the stress conditions for enhanced future responses.

6.4 Mechanistic insight into heavy metal stress tolerance in relation to ROS homeostasis in plants

Heavy metal stress presents a substantial challenge to plant health, affecting critical aspects of growth, development, and productivity (Ali et al., 2013). Heavy metals like cadmium (Cd), lead (Pb), mercury (Hg), and arsenic (As) accumulate in the environment due to natural processes and anthropogenic activities, resulting in their uptake by plants. Once inside plant cells, these metals disrupt cellular function by triggering redox imbalance primarily through the excessive production of ROS (Yavaş et al., 2022). While ROS play essential roles in signaling and other physiological processes, their excessive accumulation leads to oxidative damage, which impacts cellular components including lipids, proteins, and nucleic acids (Raza et al., 2020). Consequently, plants have evolved sophisticated regulatory networks that help balance ROS levels, maintain redox homeostasis, and enhance tolerance to heavy metal stress. Heavy metals interfere with various cellular functions upon uptake by plant roots, resulting in a range of physiological disorders. Typical manifestations include decreased seed germination rates, stunted seedling growth, chlorosis (yellowing of leaves), reduced photosynthesis, wilting (loss of turgor), necrosis, and premature senescence. These toxic effects are largely due to heavy metals' ability to catalyze the formation of ROS. The unregulated presence of ROS disrupts cellular membranes, impairs enzyme activities, and induces damage to genetic material, ultimately inhibiting plant growth and productivity (Ghori et al., 2019).

Plants respond to heavy metal stress through complex regulatory mechanisms involving TFs and hormonal signaling. Key TFs such as WRKY TFs (named after the conserved WRKYGQK domain) (WRKY), Myeloblastosis (MYB), and NAM, ATAF1-2 and CUC2 (NAC) play pivotal roles in modulating gene expression related to detoxification and stress responses. These TFs either activate or repress specific target genes in response to metal exposure, regulating antioxidant activity, metal chelation, and transport processes (Zhang and Lu, 2024). Hormones like ethylene, jasmonic acid (JA), and salicylic acid are integral to stress response pathways and interact with TFs and other signaling molecules to initiate a coordinated response to heavy metal stress (Saini et al., 2021). MAPK cascades are also crucial in stress signal transduction, transmitting signals from the cell surface to the nucleus, where they activate stress-responsive genes (Li et al., 2022). Additionally, Ca^{2+} serves as a secondary messenger, with changes in cytosolic Ca^{2+} levels being detected by calcium-binding proteins that activate downstream signaling elements essential for heavy metal tolerance.

In response to oxidative stress induced by heavy metals, plants activate an antioxidant defense system composed of both enzymatic and nonenzymatic components. Key enzymes such as SOD, CAT, and GPx are involved in detoxifying ROS. SOD catalyzes the conversion of superoxide radicals into hydrogen peroxide, which is subsequently broken down by CAT and GPx, thereby minimizing oxidative damage (Fichman and Mittler, 2020). Nonenzymatic antioxidants like glutathione

(GSH), ascorbate, and tocopherols scavenge ROS directly, further contributing to redox homeostasis (Hasanuzzaman et al., 2020). Importantly, ROS are not merely toxic byproducts; they also serve as signaling molecules that modulate stress responses. Controlled ROS production and scavenging are essential for the activation of stress defense. Proteins containing cysteine residues, for instance, undergo redox modifications that allow them to act as sensors and transducers of redox signals, thus influencing protein interactions and overall stress response (Foyer and Kunert, 2024). GSH plays a central role in detoxifying heavy metals and maintaining cellular redox balance. The ratio of reduced GSH to oxidized glutathione (GSSG) serves as a critical indicator of cellular redox status. Additionally, GSH participates in the synthesis of phytochelatins, which are essential for heavy metal sequestration within plant cells (Faizan et al., 2024). The Trx system also contributes to redox regulation by reducing disulfide bonds in target proteins, helping to preserve their functional states under stress conditions (Mata-Pérez and Spoel, 2019).

6.5 Transcriptional reprogramming in response to heat stress in relation to ROS accumulation and scavenging

Extreme heat events, particularly heatwaves, resulting from climate change, present a considerable threat to crop production and worldwide food security. High temperatures negatively impact various stages of plant development, from seed germination to the formation of reproductive structures. Heat stress impedes photosynthesis and other metabolic functions, thereby hindering cellular growth and division (Wang et al., 2018). The ability of plants to cope with elevated temperatures is influenced by the intensity and changes in temperature. Although plants can usually adjust to daily and seasonal temperature changes, extreme summer heatwaves can still harm them, especially when combined with additional stresses like high light or drought. Under stressful circumstances, such as high temperatures, the formation of ROS, which originates from metabolic activities such as mitochondrial electron transport, increases (Choudhury et al., 2017). ROS are key signaling molecules that modulate plant growth and survival under elevated temperatures, even though their reactivity necessitates controlling their concentrations (Huang and Ullah, 2019). ROS are also crucial for heat signaling pathways because they control the synthesis and storage of heat stress response proteins (Driedonks et al., 2015). Additionally, heat stress induces proline accumulation in numerous plants (Cvikrová et al., 2012), and increased proline levels in plants show improved resistance to abiotic stressors (Hanif et al., 2021). Significant alterations occurred in gene expression, which are modulated at both transcriptional and posttranscriptional levels under exposure to high temperatures (Dard et al., 2023). Complex networks of TFs, some of which are redox-responsive, regulate transcription in response to high temperatures (Mittler et al., 2022). Members of the family heat shock factor A (HSFA) are essential for the heat protection response because they directly influence the expression of genes that encode critical heat stress response TFs and make it easier for inhibitory H2A.Z-nucleosomes to be removed from genes and transposable elements (Jacob et al., 2017).

The close relationship between ROS and high temperature responses is highlighted by the activation of several genes involved in ROS metabolism and antioxidant enzymes under heat stress (Babbar et al., 2020). For instance, one of the main genes increased under both brief and transitory moderate heat treatment (BT and TMHT) conditions is the *APX2* gene, which encodes a large H_2O_2-neutralizing enzyme (Dard et al., 2023). *APX2* expression is regulated by *HSFA2* (Schramm et al., 2006). Another HSFA variant, HSFA4, stimulates *APX1* expression, potentially through activation of the H_2O_2-dependent TF ZAT12 (Davletova et al., 2005). Heat shock elements (HSE) in the promoters of several antioxidant-related genes, including *ZAT12*, *APX1*, and *APX2*, demonstrate direct transcriptional activation by *HSF* (Wu et al., 2022). Recent studies have revealed further complexity in *APX2* gene activation during the stepwise acclimation treatment procedure. An initial heat stress (priming) causes HSFA1 and BRI1-Ethyl Methyl Sulfon-SUPPRESSOR (BES1) to accumulate in the nucleus. These proteins interact and bind to HSEs to initiate the transcription of thermotolerance genes (Albertos et al., 2022). Meanwhile, activating H3K4me3 marking replaces the inhibitory H3K27me3 histone marker in the surrounding chromatin in a BES1-dependent mechanism (Oberkofler and Bäurle, 2022). Genes like *APX2* and *HSFA3*, which now have "primed chromatin" (with the activating H3K4me3 marking instead of the inhibiting H3K27me3), exhibit increased expression when subjected to further heat stress, improving heat tolerance. A regulatory pathway necessary for TMHT in *Arabidopsis* is also created by heat stress, which quickly triggers *miR398* and lowers the transcript levels of its target genes encoding SODs, *CSD1* and *CSD2* (Guan et al., 2013).

Many heat stress reactions depend on the build-up of ROS, mainly because it activates TFs. For example, high ROS levels trigger the expression of heat shock response (HSR) genes, including heat shock proteins (HSP) and heat shock factors (HSF) (Willems et al., 2016). A rapid oxidative burst during the first 15 minutes of heat shock is necessary for HSF DNA-binding and the activation of heat-responsive genes, such as *APX2* and HSPs (Volkov et al., 2006). In order to trigger HS target genes, elevated ROS levels activate HSFA4a, which then forms trimers and translocates to the nucleus.

It is thought that conformational changes brought on by cysteine oxidation are essential to this oligomerisation process (Pérez-Salamó et al., 2014). A similar mechanism was proposed for HSFA1a trimerization; however function of specific cysteines involved in stress-induced HSFA1a-DNA interaction remains unclear (Liu et al., 2013). Interestingly, even in the absence of heat shock, changes in chloroplast redox state have been demonstrated to impact HSFA1a binding to its target genes. This discovery further connects the signaling of daily temperature changes to light-dependent regulation of the Calvin-cycle (Dickinson et al., 2018). Cysteine oxidation was also suggested to regulate HSFA8 nuclear translocation (Giesguth et al., 2015). The HS-responsive TF MBF1C, which controls plant systemic acclimation to high light and HS, is likewise moved from the cytosol to the nucleus by elevated ROS levels (Suzuki et al., 2013). Redox control of *bZIP68* is essential for maintaining a balance between development and heat-stress tolerance. The suppression of *HSR* gene transcription is alleviated by an unidentified redox change that H_2O_2 build-up causes on bZIP68 C320 under heat stress (Li et al., 2019).

6.6 Transcriptional reprogramming in response to combined abiotic stresses in plants

Plants integrate multiple signaling pathways to provide a comprehensive defense against combined abiotic stresses (Fig. 6.2). Plants growing in cold regions are exposed to various other abiotic stressors such as UV radiation and high light intensity, which disturb the ROS balance inside the plants. High light intensity in plants causes overflow of electrons from the photosynthetic electron transport system, which results in the generation of ROS, specifically superoxide anion in photosystem I (PSI) and singlet oxygen in photosystem II (PSII) (Hayami et al., 2015). The ROS generation in chloroplasts acts as a retrograde signals, which initiate the expression of nuclear genes involved in chloroplast repair, photoprotection, and antioxidant synthesis. Similarly, plants of cold regions at higher elevations are exposed to UV radiation (UV-B). Plant response to UV-B depends on its intensity, duration, and wavelength. Nondamaging UV-B acts as a priming agent that enhances UV stress tolerance in plants and promotes plants' photomorphogenesis. However, UV-B at high levels causes DNA damage and the generation of ROS and damage to photosynthetic machinery (Chen et al., 2022). The UV-B is perceived by UV RESISTANCE LOCUS (UVR8), a conserved plant-specific photoreceptor. UVR8, along with COP1 (CONSTITUTIVELY PHOTOMORPHOGENIC1), regulates the expression of the UV-B responsive gene for stress tolerance. ELONGATED HYPOCOTYL5 (HY5), a TF

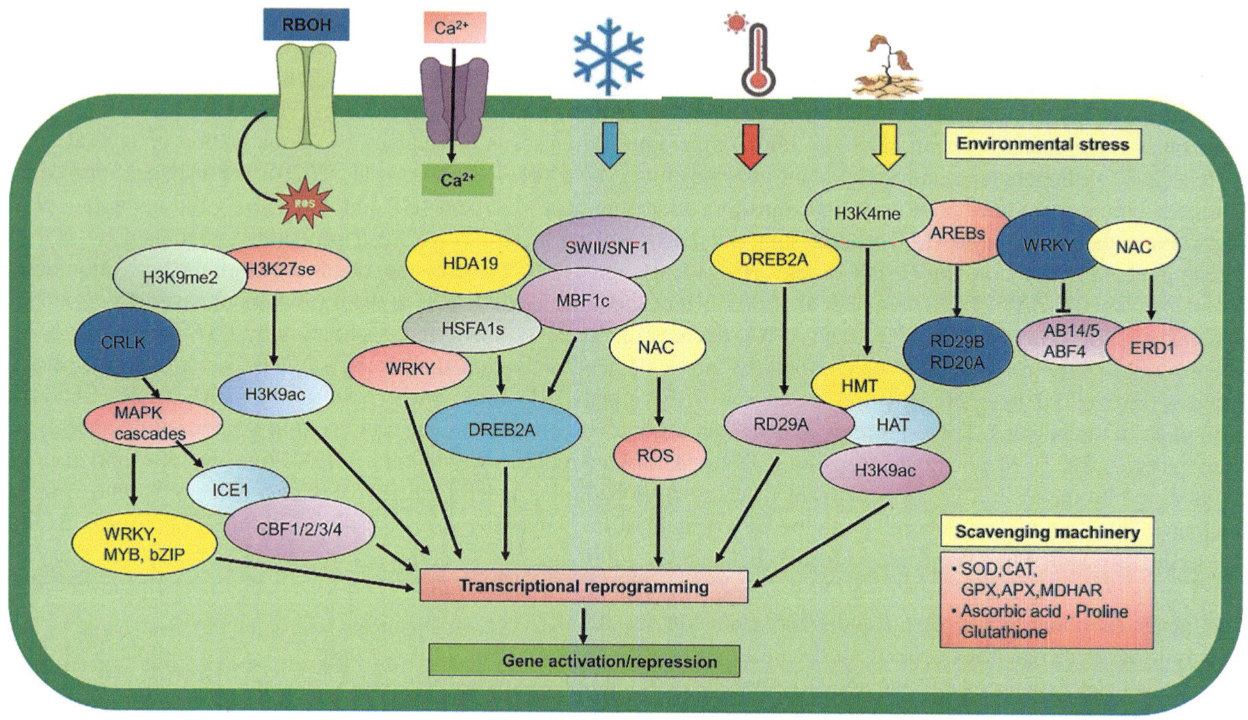

FIGURE 6.2 The overview oftranscriptional reprogramming in response to combined abiotic stress.

that acts downstream of UVR8-COP1, activates antioxidant gene expression and promotes the biosynthesis of flavonoids for UV-B tolerance in plants (Chen et al., 2022).

Plants exposed to drought stress are also subjected to high temperature and salinity stress. These abiotic stressors intensify the impact of drought on plants, such as ionic imbalance, overproduction of ROS, which disrupts cellular homeostasis (Bao et al., 2024). A study on rice showed that heat-salinity interaction has aggravated the effect of salinity stress on the plants. Under combined stress, the level of sodium ion (Na^+) increased in the leaves, disrupting the K^+/Na^+ ratio. This ionic imbalance negatively affected CO_2 assimilation of WUE (water use efficiency) in the plants (Mendes et al., 2024).

Plants integrate multiple signaling pathways to mount a comprehensive defense against heavy metal-induced oxidative stress. These pathways include Ca^{2+} signaling, MAPK cascades, and phytohormone responses. Heavy metals disrupt the electron transport chains in chloroplasts and mitochondria, leading to excessive ROS production. For example, cadmium (Cd) and lead (Pb) are known to compromise membrane integrity, resulting in lipid peroxidation and disturbances in essential metabolic pathways. MAPK cascades play a vital role in orchestrating the antioxidant response. In the model plant *Arabidopsis thaliana*, MAPKs such as MPK3 and MPK6 become activated upon heavy metal exposure, leading to the transcriptional regulation of genes involved in ROS detoxification and metal transport (Zhang et al., 2022). Phytohormones such as ABA and JA are also upregulated, promoting the expression of antioxidant genes and enhancing plant adaptability. ABA, in particular, induces stomatal closure to minimize water loss, further supporting the plant's defense mechanisms under oxidative stress conditions (Singh and Roychoudhury, 2023). Despite the presence of robust regulatory networks, prolonged exposure to heavy metals can overwhelm a plant's antioxidant defense system, resulting in oxidative stress and cellular dysfunction. One of the most detrimental effects is genotoxicity, where ROS accumulation causes DNA damage, leading to chromosomal abnormalities and impaired cell division. The base excision repair pathway, which plays a vital role in repairing oxidative DNA damage, may become inefficient under sustained heavy metal stress, leading to the accumulation of mutations that compromise genomic stability (Dutta et al., 2018). This genetic instability has implications for plant growth, survival, and crop productivity in contaminated soils. The ability of plants to maintain redox homeostasis and activate regulatory networks in response to heavy metal stress is essential for their survival and adaptability. Insights into these molecular mechanisms, particularly in non-model species and hyperaccumulators, offer promising avenues for enhancing phytoremediation efforts and improving crop resilience in metal-contaminated soils (Muthusaravanan et al., 2018). Understanding and leveraging the intricate regulatory networks and redox homeostasis in plants provides a solid foundation for addressing the environmental challenges posed by heavy metal pollution (Kumar et al., 2020).

Redox signaling plays a crucial role in mediating developmental signals induced by high temperatures. For instance, the conflicting effects of light and high ambient temperature (27°C–30°C in *Arabidopsis*) on seed germination are precisely regulated by S-nitrosylation of the HFR1 transcriptional repressor of PIF1 (Ying et al., 2022). To improve seed dormancy expression, high temperatures promote S-nitrosylation of HFR1 at a specific cysteine residue (C164), which activates PIF1 and SOMNUS (*SOM*) gene to alter ABA and GA metabolism. In contrast, light activates the photoreceptor phytochrome B (PhyB) via enhancing S-nitrosoglutathione reductase activity, and prevents S-nitrosylation and degradation of HFR1 caused by high temperatures (Ying et al., 2022). Notably, photo- and thermomorphogenesis are two developmental adaptations to light and high temperatures that are regulated by the PhyB/HFR1/PIF module, suggesting that both pathways may be subject to redox regulation (Zhao and Bao, 2021). According to studies, thiol reduction systems influence hypocotyl elongation brought on by high ambient temperatures by either inactivating certain GRXs (grxS17, grxC3, grxC4) or causing mutations that affect glutathione synthesis (cad2, pad2) (Dard et al., 2023). Members of the rnase three-like (RTL) and DICER-like (DCL) families in *Arabidopsis* have been demonstrated to be redox-controlled by disulfide bond formation, which impacts their RNase III activity. In fact, GRXs can restore RTL1 activity, suggesting that short RNA synthesis is regulated by a redox switch (Seta et al., 2017). As a result, these instances demonstrate a growing connection between epigenetic and redox control.

6.7 Multidimensional regulation of reactive oxygen species in response to combined abiotic stresses

When ROS accumulates in cells under stress conditions, various proteins, including enzymes, receptors, and small molecules that activate, modify, and integrate several stress-related signal transduction pathways, are also changed. These changes influence gene expression and improve plant tolerance towards abiotic stress. Recent advances in understanding these critical processes involve the discovery of specific ROS sensors and regulatory hubs that link ROS signaling with other stress-related pathways. Additionally, AI-driven techniques help to analyze the regulatory mechanisms that are activated by ROS signaling and further expand our knowledge with the discovery of new functions of ROS in cell organelles and during intercellular stress signaling. During basal metabolism, plants convert 1%–2% of the oxygen they consume into ROS at various subcellular locations, such as in the apoplast during oxidation through NADPH oxidases, in mitochondria during respiration, in peroxisomes during photorespiratory processes, and in chloroplasts during photosynthesis (Mittler et al., 2022). Redox signaling between organelles and the nucleus was activated by the accumulation of ROS in these compartments. High levels of ROS trigger an oxidative burst under adverse environmental conditions, which ultimately leads to damage to biomolecules. This damage negatively impacts the growth, development, and survival of the plant. Consequently, for normal cellular functions, it is important to maintain a balance between ROS production and detoxification (Ahammed et al., 2024). The reactivity of atmospheric O_2 is reduced because it is a free molecule with two unpaired parallel spin electrons of the same spin numbers in its ground state. Nevertheless, extra energy from ionizing radiation, electron transport chains, UV-B light, and certain biological activities helps $^3O_2$ to get beyond the spin constraint and turn into ROS. The half-lives of various ROS from their sites of formation inside the cell are determined by their unique chemical reactivity properties (Huang and Ullah, 2019).

Different sensors and receptors to detect the variations in temperature, light, and osmotic pressure are found in plants. For instance, osmotic changes in plants are detected by Ca^{2+}-permeable channels, which include (OSCA1) and (MSL10). Heat stress is detected through cyclic nucleotide-gated channels, and light intensity and quality are detected by (PHYB) and cryptochrome (Mittler et al., 2022; Kang et al., 2023). It is also believed that plants sense light stress by retrograde signaling and temperature changes through the release of ROS and Ca^{2+} from chloroplasts. Stomatal closure is a typical reaction to stress situations like intense light or other abiotic stressors. This lowers intracellular CO_2 levels, which encourages the ROS in chloroplasts. According to Sachdev et al. (2021), the overabundance of ROS in chloroplasts causes photooxidative stress, which damages lipids, compromises membrane proteins linked to the PSII reaction center, and eventually results in cell death. Plants can rapidly modify their metabolome and transcriptome, with changes occurring within seconds to minutes of stress detection. These immediate responses enable the plant to tolerate and eventually recover from the stress by triggering slower, more complex reactions that activate multiple defense and acclimatization mechanisms. Recent studies have shown that ROS are involved in both early and late stress responses, through both active and passive mechanisms. This new understanding reveals that plants detect various stresses quickly through stress-specific receptors, which then activate RBOHs to generate ROS and/or disturbances that alter ROS levels and other stress-related metabolites. Recent studies have emphasized the crucial role of ROS regulation in plant responses to salt stress. In rice, the salt tolerance mechanism is enhanced by (STRK1), which binds to CAT and phosphorylates it at Tyr-210, thereby activating its H_2O_2-dismutation function (Wang et al., 2024). Further, in case of high temperature stress, ROS are quickly formed and accumulate in several regions of plant cells, including the nucleus, mitochondria, plasma membrane, cytosol, and chloroplasts. Plants need to control the production and removal of ROS to avoid damaging concentrations that could harm proteins, lipids, and DNA, and potentially lead to cell injury or death. Several antioxidant enzymes, such as CAT, APX, SOD, POD, and GPX, are vital in regulating ROS levels in plants experiencing high-temperature stress (Hasanuzzaman et al., 2021; Noctor and Foyer, 2016).

In nature, plants frequently experience multiple stresses; for example, a combination of heat, intense light, and drought can trigger various signaling pathways. Research has shown that ROS play a key role in helping plants adapt to these challenging conditions. Specifically, mutants lacking APX1 exhibit increased susceptibility to heat and drought, while mutants deficient in RBOHD are more prone to multiple stressors occurring together (Koussevitzky et al., 2008). Various types of ROS can be formed by different stresses acting on a plant, either at the same time or in sequence. The interaction of these ROS signals may modulate how the plant responds to complex environmental conditions. Additionally, distinct ROS profiles can emerge during a particular stress, like high temperature, during specific developmental phases, such as reproduction. In such cases, the combined ROS levels contribute to the plant's overall tolerance or sensitivity to stress (Wang et al., 2024) (Fig. 6.3).

Receptors and redox sensors activated by ROS detect stress conditions, which lead to the activation and modification of TF machinery that helps plants adapt to diverse environmental conditions. In plants, two primary mechanisms govern transcriptional responses, that is, modifications in signal transduction processes such as phosphorylation, Ca^{2+}-binding, sumoylation, and other pathways triggered by stress or ROS, which alter the function of TFs, and another one is redox regulation, either directly or indirectly, driven by ROS. These mechanisms are closely interconnected, as ROS signaling often interacts with other signaling pathways, such as those involving Ca^{2+} and phosphorylation, through components like RBOHs and aquaporins (AQPs) (Ahammed et al., 2024; You and chan, 2015). Initially thought to be metabolic byproducts that lead to cellular harm, ROS are now recognized as significant signaling molecules, playing a vital role in plant adaptation to various stress conditions and supporting growth and development (Berni et al., 2019) (Table 6.1).

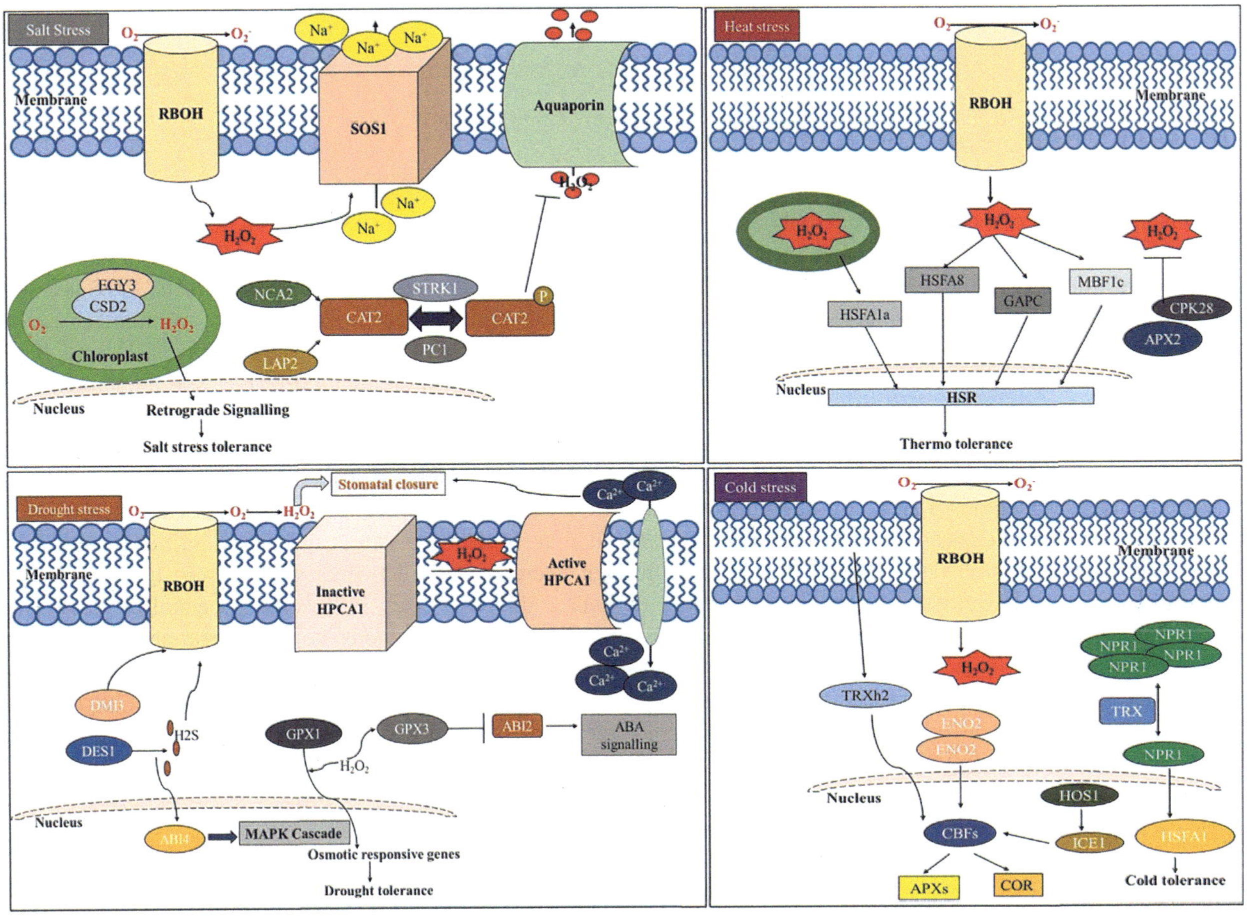

FIGURE 6.3 Multidimensional regulation of reactive oxygen species (ROS) in response to multiple abiotic stresses.

TABLE 6.1 Effect of secondary stresses caused by primary abiotic stresses on plant growth and development.

Abiotic stresses	Secondary stress	Effects	References
Cold stress	Osmotic and oxidative stress	Accumulation of reactive oxidative species, Inhibition of metabolic imbalance, and enzymatic activities.	Esim and Atici (2014), Megha et al. (2018), Gusain et al. (2023)
Heat stress	Oxidative and nutrient stress	Biomass accumulation reduction, an increase in reactive oxidative species	Nievola et al. (2017), Sarkar et al. (2018), Demirel et al., 2020
Drought stress	Heavy metal, oxidative, and osmotic stress	Accumulation of Reactive oxidative stress and ion leakage, rise in leaf temperature, necrosis, decrease in absorption and translocation of mineral nutrients.	Ansari and Lin (2010), Gusain et al. (2024)
Waterlogging stress	Oxidative stress and water, and nutrient deficiency stress	Increased reactive oxidative species, leaf senescence, chlorophyll reduction, decrease in antioxidant content	Bailey-Serres and Voesenek (2008), Ashraf (2012)
Salinity stress	Ionic imbalance	Accumulation of reactive oxidative species restricts the uptake and translocation of mineral nutrients, decreases the ion (K^+, Ca^{2+}, and Mg^{2+}) content, and causes Na^+ toxicity.	Rahman et al. (2016), Joshi et al. (2022)
Light stress	Oxidative stress	Increase in oxidative damage and reactive oxidative species, degradation of pigments, electron transport chain disruption, and an increase in activity of NADPH oxidase	Noshi et al. (2016)
Nutrient imbalance	Oxidative stress	Reduction of antioxidant content, accumulation of reactive oxidative species, solutes, and ion leakage increases, chlorosis, and necrosis.	Bisht et al. (2019), Saleem et al. (2020)

6.8 Conclusions and future prospects

The increasing impact of abiotic stresses on plants, driven by food security challenges and climate change, has made understanding their effects on crop productivity a critical area of research. Efforts have focused on how plants sense stress factors, trigger signaling pathways, and activate downstream defense mechanisms, including ROS, stress-induced gene regulation, hormonal profiling, and metabolite biosynthesis. Initially, as damaging metabolic byproducts, ROS are now recognized as crucial signaling molecules that regulate stress acclimation in plants. This chapter summarizes recent advances in signal transduction, ROS metabolism, and their role in stress tolerance and plant development. Despite these insights, significant questions remain, particularly concerning the spatiotemporal regulation of ROS production and the identification of regulatory proteins that sense ROS fluctuations and mediate crosstalk among diverse abiotic stress signaling pathways.

Extensive research has demonstrated that maintaining ROS equilibrium involves intricate crosstalk between ABA, Ca^{2+}, and various other hormones and signaling molecules. Acting as a signal transducer, ROS activates cascades of genes that enhance abiotic stress tolerance in a ROS-dependent manner. It is now clear that stress responses are not governed by isolated linear signaling pathways; instead, they operate through interconnected networks with shared signaling events, referred to as crosstalk. Given their production in response to diverse abiotic stresses, ROS serve as central players within these complex signaling networks. This pivotal role of ROS arises from their interactions, both upstream and downstream, with other key signaling molecules, including NADPH oxidases, Ca^{2+}, G-proteins, MAPKs, plant hormones such as ABA, SA, JA, and ethylene, and various TFs. Despite significant progress, substantial efforts are needed to reconstruct larger, more comprehensive signaling networks that position ROS at convergent points. Addressing this will be critical for developing biotechnological strategies aimed at enhancing plant yield under progressively increasing adverse climatic conditions.

Author contributions

SJ, JN, AK, SG, KK, and MR participated in the writing of the manuscript; AP prepared the figure; RC prepared the table; RJ conceptualized the idea of the study, reviewed, and edited the final draft of the manuscript. All authors have read and agreed to the final version of the manuscript.

Acknowledgments

The authors thank the Director, CSIR-Institute of Himalayan Bioresource Technology, Palampur, India for infrastructure and support. SJ thanks the Department of Biotechnology, Government of India (File No. DBT/JRF/BET-19/1/2019/AL/212) for providing the Senior Research Fellowship. AK, SG, JN, and KK thank CSIR-UGC (NTA Ref. No. 191620063807; UGC Ref. No. 574/CSIR-UGC NET JUNE 2019; File No. 1029/CSIR-UGC NET JUNE 2019; NTA Ref. No. 191620158661) for providing Senior Research Fellowships. MR thanks the Department of Science and Technology, Government of India (No. DST/INSPIRE Fellowship [IF200333]) for providing the INSPIRE fellowship. The authors also acknowledge the financial support (No. BT/PR45280/NER/95/1918/2022) from DBT and (MLP-207; IHP002401 No. CSPS24/RDSF/IHBT/IHP24/01; GE-Crop MMP025301; MMP25302) from CSIR. This manuscript represents CSIR-IHBT Communication No. 5724.

Conflicts of interest

The authors declare no conflict of interest.

References

Ahammed, G.J., Li, Z., Chen, J., Dong, Y., Qu, K., Guo, T., et al., 2024. Reactive oxygen species signaling in melatonin-mediated plant stress response. Plant Physiology and Biochemistry 207, 108398. Available at https://doi.org/10.1016/j.plaphy.2024.108398.

Albertos, P., Dündar, G., Schenk, P., Carrera, S., Cavelius, P., Sieberer, T., Poppenberger, B., et al., 2022. Transcription factor BES1 interacts with HSFA1 to promote heat stress resistance of plants. EMBO Journal 41 (3), e108664. Available at https://doi.org/10.15252/embj.2021108664.

Ali, H., Khan, E., Sajad, M.A., 2013. Phytoremediation of heavy metals–concepts and applications. Chemosphere 91 (7), 869–881. Available at https://doi.org/10.1016/j.chemosphere.2013.01.075.

Ansari, M.I., Lin, T.P., 2010. Molecular analysis of dehydration in plants. International Research Journal of Plant Science 1, 21–25. Available at http://www.interesjournals.org/IRJPS.

Ashraf, M.A., 2012. Waterlogging stress in plants: a review. African Journal of Agricultural Research 7, 1976–1981. Available at http://www.academicjournals.org/AJAR.

Babbar, R., Karpinska, B., Grover, A., Foyer, C.H., 2020. Heat-Induced Oxidation of the Nuclei and Cytosol. Frontiers in Plant Science 11, 617779. Available at https://doi.org/10.3389/fpls.2020.617779.

Bailey-Serres, J., Voesenek, L.A., 2008. Flooding stress: acclimations and genetic diversity. Annual Review of Plant Biology 59, 313–339. Available at https://doi.org/10.1146/annurev.arplant.59.032607.092752.

Bao, L., Liu, J., Mao, T., Zhao, L., Wang, D., Zhai, Y., 2024. Nanobiotechnology-mediated regulation of reactive oxygen species homeostasis under heat and drought stress in plants. Frontiers in Plant Science 15, 1418515. Available at https://doi.org/10.3389/fpls.2024.1418515.

Berni, R., Luyckx, M., Xu, X., Legay, S., Sergeant, K., Hausman, J.F., et al., 2019. Reactive oxygen species and heavy metal stress in plants: impact on the cell wall and secondary metabolism. Environmental and Experimental Botany 161, 98–106. Available at https://doi.org/10.1016/j.envexpbot.2018.10.017.

Bisht, N., Tiwari, S., Singh, P.C., Niranjan, A., Singh Chauhan, P., 2019. A multifaceted rhizobacterium Paenibacillus lentimorbus alleviates nutrient deficiency-induced stress in Cicer arietinum L. Microbiological Research 223-225, 110–119. Available at https://doi.org/10.1016/j.micres.2019.04.007.

Brandt, B., Munemasa, S., Wang, C., Nguyen, D., Yong, T., Yang, P.G., et al., 2015. Calcium specificity signaling mechanisms in abscisic acid signal transduction in arabidopsis guard cells. eLife 4 (2015), 1–25. Available at https://doi.org/10.7554/eLife.03599.

Cejudo, F.J., Sandalio, L.M., Van Breusegem, F., 2021. Understanding plant responses to stress conditions: redox-based strategies. Journal of Experimental Botany 72 (16), 5785–5788. Available at https://doi.org/10.1093/jxb/erab324.

Chen, Z., Dong, Y., Huang, X., 2022. Plant responses to UV-B radiation: signaling, acclimation and stress tolerance. Stress Biology 2 (1), 51. Available at https://doi.org/10.1007/s44154-022-00076-9.

Chinnusamy, V., Ohta, M., Kanrar, S., Lee, B.H., Hong, X., Agarwal, M., et al., 2003. ICE1: a regulator of cold-induced transcriptome and freezing tolerance in arabidopsis. Genes and Development 17 (8), 1043–1054. Available at https://doi.org/10.1101/gad.1077503.

Choudhury, F.K., Rivero, R.M., Blumwald, E., Mittler, R., 2017. Reactive oxygen species, abiotic stress and stress combination. Plant Journal 90 (5), 856–867. Available at https://doi.org/10.1111/tpj.13299.

Cvikrová, M., Gemperlová, L., Dobrá, J., Martincová, O., Prásil, I.T., Gubis, J., Vanková, R., et al., 2012. Effect of heat stress on polyamine metabolism in proline-over-producing tobacco plants. Plant Science 182, 49–58. Available at https://doi.org/10.1016/j.plantsci.2011.01.016.

Costa-Broseta, Á., Perea-Resa, C., Castillo, M.C., Ruíz, M.F., Salinas, J., León, J., 2018. Nitric Oxide Controls Constitutive Freezing Tolerance in Arabidopsis by Attenuating the Levels of Osmoprotectants, Stress-Related Hormones and Anthocyanins. Scientific Reports 8 (1), 9268. Available at https://doi.org/10.1038/s41598-018-27668-8.

Dard, A., Weiss, A., Bariat, L., Auverlot, J., Fontaine, V., Picault, N., Pontvianne, F., Riondet, C., Reichheld, J.P., et al., 2023. Glutathione-mediated thermomorphogenesis and heat stress responses in Arabidopsis thaliana. Journal of Experimental Botany 74 (8), 2707–2725. Available at https://doi.org/10.1093/jxb/erad042.

Davletova, S., Rizhsky, L., Liang, H., Shengqiang, Z., Oliver, D.J., Coutu, J., Shulaev, V., Schlauch, K., Mittler, R., et al., 2005. Cytosolic ascorbate peroxidase 1 is a central component of the reactive oxygen gene network of Arabidopsis. Plant Cell 17 (1), 268–281. Available at https://doi.org/10.1105/tpc.104.026971.

Demidchik, V., Shabala, S.N., Davies, J.M., 2007. Spatial variation in H2O2 response of Arabidopsis thaliana root epidermal Ca2+ flux and plasma membrane Ca2+ channels. Plant Journal 49 (3), 377–386. Available at https://doi.org/10.1111/j.1365-313X.2006.02971.x.

Demirel, U., Morris, W.L., Ducreux, L.J.M., Yavuz, C., sim A, A., Tindas, I., Campbell, R., Morris, J.A., Verrall, S.R., Hedley, P.E., Gokce, Z.N.O., Caliskan, S., Aksoy, E., Caliskan, M.E., Taylor, M.A., Hancock, R.D., et al., 2020. Physiological, Biochemical, and Transcriptional Responses to Single and Combined Abiotic Stress in Stress-Tolerant and Stress-Sensitive Potato Genotypes. Frontiers in Plant Science 11, 169. Available at https://doi.org/10.3389/fpls.2020.00169.

Denjalli, I., Knieper, M., Uthoff, J., Vogelsang, L., Kumar, V., Seidel, T., Dietz, K.J., et al., 2024. The centrality of redox regulation and sensing of reactive oxygen species in abiotic and biotic stress acclimatization. Journal of Experimental Botany 75 (15), 4494–4511. Available at https://doi.org/10.1093/jxb/erae041.

Dickinson, P.J., Kumar, M., Martinho, C., Yoo, S.J., Lan, H., Artavanis, G., Charoensawan, V., Schöttler, M.A., Bock, R., Jaeger, K.E., Wigge, P.A., et al., 2018. Chloroplast Signaling Gates Thermotolerance in Arabidopsis. Cell Reports 22 (7), 1657–1665. Available at https://doi.org/10.1016/j.celrep.2018.01.054.

Dietz, K.J., 2015. Efficient high light acclimation involves rapid processes at multiple mechanistic levels. Journal of Experimental Botany 66 (9), 2401–2414. Available at https://doi.org/10.1093/jxb/eru505.

Dietz, K.J., Turkan, I., Krieger-Liszkay, A., 2016. Redox- and Reactive Oxygen Species-Dependent Signaling into and out of the Photosynthesizing Chloroplast. Plant Physiology 171 (3), 1541–1550. Available at https://doi.org/10.1104/pp.16.00375.

Driedonks, N., Xu, J., Peters, J.L., Park, S., Rieu, I., 2015. Multi-level interactions between heat shock factors, heat shock proteins, and the redox system regulate acclimation to heat. Frontiers in Plant Science 6, 999. Available at https://doi.org/10.3389/fpls.2015.00999.

Dutta, S., itra M, M., Agarwal, P., Mahapatra, K., De, S., Sett, U., Roy, S., et al., 2018. Oxidative and genotoxic damages in plants in response to heavy metal stress and maintenance of genome stability. Plant Signaling & Behavior 13 (8), e1460048. Available at https://doi.org/10.1080/15592324.2018.1460048.

Esim, N., Atici, O., 2014. Nitric oxide improves chilling tolerance of maize by affecting apoplastic antioxidative enzymes in leaves. Plant Growth Regulation 72, 29–38. Available at https://doi.org/10.1007/s10725-013-9833-4.

Faizan, M., Alam, P., Hussain, A., Karabulut, F., Tonny, S.H., Cheng, S.H., et al., 2024. Phytochelatins: key regulator against heavy metal toxicity in plants. Plant Stress 11. Available at https://doi.org/10.1016/j.stress.2024.100355.

Fichman, Y., Mittler, R., 2020. Rapid systemic signaling during abiotic and biotic stresses: is the ROS wave master of all trades? The Plant Journal 102 (5), 887–896. Available at https://doi.org/10.1111/tpj.14685.

Foyer, C.H., Kunert, K., 2024. The ascorbate-glutathione cycle coming of age. Journal of Experimental Botany 75 (9), 2682–2699. Available at https://doi.org/10.1093/jxb/erae023.

Foyer, C.H., Noctor, G., 2016. Stress-triggered redox signalling: what's in pROSpect? Plant, Cell & Environment 39 (5), 951–964. Available at https://doi.org/10.1111/pce.12621.

Ghori, N.-H., Ghori, T., Hayat, M.Q., Imadi, S.R., Gul, A., Altay, V., et al., 2019. Heavy metal stress and responses in plants. International Journal of Environmental Science and Technology 16 (3), 1807–1828. Available at https://doi.org/10.1007/s13762-019-02215-8.

Giesguth, M., Sahm, A., Simon, S., Dietz, K.J., 2015. Redox-dependent translocation of the heat shock transcription factor AtHSFA8 from the cytosol to the nucleus in Arabidopsis thaliana. FEBS Letters 589 (6), 718–725. Available at https://doi.org/10.1016/j.febslet.2015.01.039.

Gilroy, S, Białasek, M, Suzuki, N, Górecka, M, Devireddy, AR, Karpiński, S, Mittler, R, et al., 2016. ROS, Calcium, and Electric Signals: Key Mediators of Rapid Systemic Signaling in Plants. Plant Physiology 171 (3), 1606–1615. Available at https://doi.org/10.1104/pp.16.00434.

Gilroy, S., Suzuki, N., Miller, G., Choi, W.G., Toyota, M., Devireddy, A.R., Mittler, R., 2014. A tidal wave of signals: calcium and ROS at the forefront of rapid systemic signaling. Trends in Plant Science 19 (10), 623–630. Available at https://doi.org/10.1016/j.tplants.2014.06.013.

Guan, Q., Lu, X., Zeng, H., Zhang, Y., Zhu, J., 2013. Heat stress induction of mi R 398 triggers a regulatory loop that is critical for thermotolerance in Arabidopsis. The Plant Journal 74 (5), 840–851. Available at https://doi.org/10.1111/tpj.12169.

Gusain, S., Joshi, S., Joshi, R., 2023. Sensing, signalling, and regulatory mechanism of cold-stress tolerance in plants. Plant Physiology and Biochemistry 197, 107646. Available at https://doi.org/10.1016/j.plaphy.2023.107646.

Gusain, S., Joshi, S., Kumari, A., Nath, J., Kumari, K., Rawat, M., et al., 2025. Chapter 8 - Cold priming and memory induced acquired tolerance and possible mechanism in plants. Exogenous Priming and Engineering of Plant Metabolic and Regulatory Genes: Stress Mitigation Strategies in Plants. Academic Press, pp. 95–106. Available at https://doi.org/10.1016/B978-0-443-13490-6.00008-4.

Gusain, S., Kumari, K., Joshi, R., 2024. Physiological, hormonal and molecular dynamics of root system architectural response to drought stress signaling in crops. Rhizosphere 31, 100922. Available at https://doi.org/10.1016/j.rhisph.2024.100922.

Hanif, S., Saleem, M.F., Sarwar, M., Irshad, M., Shakoor, A., Wahid, M.A., et al., 2021. Biochemically triggered heat and drought stress tolerance in rice by proline application. Journal of Plant Growth Regulation 40 (1), 305–312. Available at https://doi.org/10.1007/s00344-020-10095-3.

Hasanuzzaman, M., Bhuyan, M.H.M.B., Zulfiqar, F., Raza, A., Mohsin, S.M., Mahmud, J.A., Fujita, M., Fotopoulos, V., et al., 2020. Reactive Oxygen Species and Antioxidant Defense in Plants under Abiotic Stress: Revisiting the Crucial Role of a Universal Defense Regulator. Antioxidants (Basel) 9 (8), 681. Available at https://doi.org/10.3390/antiox9080681.

Hasanuzzaman, M., Raihan, M.R.H., Masud, A.A.C., Rahman, K., Nowroz, F., Rahman, M., et al., 2021. Regulation of reactive oxygen species and antioxidant defense in plants under salinity. International Journal of Molecular Sciences 22 (17). Available at https://doi.org/10.3390/ijms22179326.

Hayami, N., Sakai, Y., Kimura, M., Saito, T., Tokizawa, M., Iuchi, S., Kurihara, Y., atsui M, M., Nomoto, M., Tada, Y., Yamamoto, Y.Y., et al., 2015. The Responses of Arabidopsis Early Light-Induced Protein2 to Ultraviolet B, High Light, and Cold Stress Are Regulated by a Transcriptional Regulatory Unit Composed of Two Elements. Plant Physiology 169 (1), 840–855. Available at https://doi.org/10.1104/pp.15.00398.

Huang, H., Ullah, F., Zhou, D.X., Yi, M., Zhao, Y., 2019. Mechanisms of ROS regulation of plant development and stress responses. Frontiers in Plant Science 10, 800. Available at https://doi.org/10.3389/fpls.2019.00800.

Huang, J., Staes, A., Impens, F., Demichev, V., Van Breusegem, F., Gevaert, K., Willems, P., et al., 2023. CysQuant: Simultaneous quantification of cysteine oxidation and protein abundance using data dependent or independent acquisition mass spectrometry. Redox Biology 67, 102908. Available at https://doi.org/10.1016/j.redox.2023.102908.

Hua, D., Wang, C., He, J., Liao, H., Duan, Y., Zhu, Z., et al., 2012. A plasma membrane receptor kinase, GHR1, mediates abscisic acid- and hydrogen peroxide-regulated stomatal movement in Arabidopsis. The Plant Cell 24 (6), 2546–2561. Available at https://doi.org/10.1105/tpc.112.100107.

Hu, H., Xiong, L., 2014. Genetic Engineering and Breeding of Drought-Resistant Crops. Annual Review of Plant Biology 65 (1), 715–741. Available at https://doi.org/10.1146/annurev-arplant-050213-040000.

Jacob, P, irt H, H, Bendahmane, A, 2017. The heat-shock protein/chaperone network and multiple stress resistance. Plant Biotechnol J 15 (4), 405–414. Available at https://doi.org/10.1111/pbi.12659.

Jia, Y., Ding, Y., Shi, Y., Zhang, X., Gong, Z., Yang, S., 2016. The cbfs triple mutants reveal the essential functions of CBF s in cold acclimation and allow the definition of CBF regulons in Arabidopsis. New Phytologist 212 (2), 345–353. Available at https://doi.org/10.1111/nph.14088.

Joshi, S., Nath, J., Joshi, R., 2024a. Genome-wide identification, characterization, and expression analysis of the MAPK gene family in Nardostachys jatamansi (D. Don) DC. In Vitro Cellular & Developmental Biology-Plant 60, 222–237. Available at https://doi.org/10.1007/s11627-024-10430-3.

Joshi, S., Nath, J., Kumari, A., Gusain, S., Kumari, K., Rawat, M., et al., 2024b. Chapter 4 - Advancement in understanding cold stress tolerance using "omics" tools. Current Omics Advancement in Plant Abiotic Stress Biology. Academic Press, pp. 51–61. Available at https://doi.org/10.1016/B978-0-443-21625-1.00004-X.

Joshi, S., Nath, J., Singh, A.K., Pareek, A., Joshi, R., 2022. Ion transporters and their regulatory signal transduction mechanisms for salinity tolerance in plants. Physiologia Plantarum 174 (3), e13702. Available at https://doi.org/10.1111/ppl.13702.

Kang, X., Zhao, L., Liu, X., 2023. Calcium signaling and the response to heat shock in crop plants. International Journal of Molecular Sciences 25 (1), 324. Available at https://doi.org/10.3390/ijms25010324.

Kidokoro, S., Kim, J.S., Ishikawa, T., Suzuki, T., Shinozaki, K., Yamaguchi-Shinozaki, K., 2020. DREB1A/CBF3 is repressed by transgene-induced DNA methylation in the arabidopsis ice1-1 mutant. The Plant Cell 32 (4), 1035–1048. Available at https://doi.org/10.1105/tpc.19.00532.

Kidokoro, S., Shinozaki, K., Yamaguchi-Shinozaki, K., 2022. Transcriptional regulatory network of plant cold-stress responses. Trends in Plant Science 27 (9), 922–935. Available at https://doi.org/10.1016/j.tplants.2022.01.008.

Koussevitzky, S., Suzuki, N., Huntington, S., Armijo, L., Sha, W., Cortes, D., et al., 2008. Ascorbate peroxidase 1 plays a key role in the response of Arabidopsis thaliana to stress combination. Journal of Biological Chemistry 283 (49), 34197–34203. Available at https://doi.org/10.1074/jbc.m806337200.

Kumar, A., Dubey, A.K., Kumar, V., Ansari, M.A., Narayan, S., Meenakshi, et al., 2020. Over-expression of chickpea glutaredoxin (CaGrx) provides tolerance to heavy metals by reducing metal accumulation and improved physiological and antioxidant defence system. Ecotoxicology and Environmental Safety 192, 110252. Available at https://doi.org/10.1016/j.ecoenv.2020.110252.

Kwak, J.M., Mori, I.C., Pei, Z.M., Leonhard, N., Torres, M.A., Dangl, J.L., et al., 2003. NADPH oxidase AtrbohD and AtrbohF genes function in ROS-dependent ABA signaling in arabidopsis. EMBO Journal 22 (11), 2623–2633. Available at https://doi.org/10.1093/emboj/cdg277.

Liu, Y., He, C., 2017. A review of redox signaling and the control of MAP kinase pathway in plants. Redox Biology 11, 192–204. Available at https://doi.org/10.1016/j.redox.2016.12.009.

Liu, Y., Zhang, C., Chen, J., Guo, L., Li, X., Li, W., et al., 2013. Arabidopsis heat shock factor HsfA1a directly senses heat stress, pH changes, and hydrogen peroxide via the engagement of redox state. Plant Physiology and Biochemistry 64, 92–98. Available at https://doi.org/10.1016/j.plaphy.2012.12.013.

Li, Y., Liu, W., Zhong, H., Zhang, H.L., Xia, Y., 2019. Redox-sensitive bZIP68 plays a role in balancing stress tolerance with growth in Arabidopsis. Plant Journal 100 (4), 768–783. Available at https://doi.org/10.1111/tpj.14476.

Li, S., Han, X., Lu, Z., Qiu, W., Yu, M., Li, H., et al., 2022. MAPK cascades and transcriptional factors: regulation of heavy metal tolerance in plants. International Journal of Molecular Sciences 23 (8), 4463. Available at https://doi.org/10.3390/ijms23084463.

Maruyama, K., Todaka, D., Mizoi, J., Yoshida, T., Kidokoro, S., Matsukura, S., et al., 2012. Identification of cis-acting promoter elements in cold- and dehydration-induced transcriptional pathways in Arabidopsis, rice, and soybean. DNA Research 19 (1), 37–49. Available at https://doi.org/10.1093/dnares/dsr040.

Mata-Pérez, C., Spoel, S.H., 2019. Thioredoxin-mediated redox signalling in plant immunity. Plant Science 279, 27–33. Available at https://doi.org/10.1016/j.plantsci.2018.05.001.

Ma, Y., Szostkiewicz, I., Korte, A., Moes, D., Yang, Y., Christmann, A., et al., 2009. Regulators of PP2C phosphatase activity function as abscisic acid sensors. Science (New York, N.Y.) 324 (5930), 1064–1068. Available at https://doi.org/10.1126/science.1172408.

Meena, M., Divyanshu, K., Kumar, S., Swapnil, P., Zehra, A., Shukla, V., et al., 2019. Regulation of L-proline biosynthesis, signal transduction, transport, accumulation and its vital role in plants during variable environmental conditions. Heliyon 5 (12), e02952. Available at https://doi.org/10.1016/j.heliyon.2019.e02952.

Megha, S., Basu, U., Kav, N.N., 2018. Regulation of low temperature stress in plants by microRNAs. Plant Cell Environment 41 (1), 1–15. Available at https://doi.org/10.1111/pce.12956.

Mendes, K.R., Bonifacio, A., Martins, M.O., Sousa, R.H.V., Monteiro, M.V., Silveira, J.A.G., 2024. Heat shock combined with salinity impairs photosynthesis but stimulates antioxidant defense in rice plants. Environmental and Experimental Botany 225, 105851. Available at https://doi.org/10.1016/j.envexpbot.2024.105851.

Mignolet-Spruyt, L., Xu, E., Idänheimo, N., Hoeberichts, F.A., Mühlenbock, P., Brosché, M., et al., 2016. Spreading the news: subcellular and organellar reactive oxygen species production and signalling. Journal of Experimental Botany 67 (13), 3831–3844. Available at https://doi.org/10.1093/jxb/erw080.

Mishra, N., Jiang, C., Chen, L., Paul, A., Chatterjee, A., Shen, G., 2023. Achieving abiotic stress tolerance in plants through antioxidative defense mechanisms. Frontiers in Plant Science 14, 1110622. Available at https://doi.org/10.3389/fpls.2023.1110622.

Mittler, R., Zandalinas, S.I., Fichman, Y., Van Breusegem, F., 2022. Reactive oxygen species signalling in plant stress responses. Nature Reviews Molecular Cell Biology 23 (10), 663–679. Available at https://doi.org/10.1038/s41580-022-00499-2.

Muthusaravanan, S., Sivarajasekar, N., Vivek, J.S., Paramasivan, T., Naushad, M., Prakashmaran, J., et al., 2018. Phytoremediation of heavy metals: mechanisms, methods and enhancements. Environmental Chemistry Letters 16 (4), 1339–1359. Available at https://doi.org/10.1007/s10311-018-0762-3.

Nievola, C.C., Carvalho, C.P., Carvalho, V., Rodrigues, E., 2017. Rapid responses of plants to temperature changes. Temperature 4 (4), 371–405. Available at https://doi.org/10.1080/23328940.2017.1377812.

Noctor, G., Foyer, C.H., 2016. Intracellular Redox Compartmentation and ROS-Related Communication in Regulation and Signaling. Plant Physiology 171, 1581–1592. Available at https://doi.org/10.1104/pp.16.00346.

Noshi, M., Hatanaka, R., Tanabe, N., Terai, Y., Maruta, T., Shigeoka, S., 2016. Redox regulation of ascorbate and glutathione by a chloroplastic dehydroascorbate reductase is required for high-light stress tolerance in Arabidopsis. Bioscience, Biotechnology, and Biochemistry 80 (5), 870–877. Available at https://doi.org/10.1080/09168451.2015.1135042.

Oberkofler, V., Bäurle, I., 2022. Inducible epigenome editing probes for the role of histone H3K4 methylation in Arabidopsis heat stress memory. Plant Physiology 189 (2), 703–714. Available at https://doi.org/10.1093/plphys/kiac113.

Puyaubert, J., Fares, A., Rézé, N., Peltier, J.B., Baudouin, E., 2014. Identification of endogenously S-nitrosylated proteins in Arabidopsis plantlets: effect of cold stress on cysteine nitrosylation level. Plant Science 215-216, 150–156. Available at https://doi.org/10.1016/j.plantsci.2013.10.014.

Pérez-Salamó, I., Papdi, C., Rigó, G., Zsigmond, L., Vilela, B., Lumbreras, V., et al., 2014. The heat shock factor A4A confers salt tolerance and is regulated by oxidative stress and the mitogen-activated protein kinases MPK3 and MPK6. Plant Physiology 165 (1), 319–334. Available at https://doi.org/10.1104/pp.114.237891.

Qi, J., Song, C., Wang, B., Zhou, J., Kangasjärvi, J., Zhu, J.-K., Gong, Z., 2018. Reactive oxygen species signaling and stomatal movement in plant responses to drought stress and pathogen attack. Journal of Integrative Plant Biology 60, 805–826. Available at https://doi.org/10.1111/jipb.12654.

Rahman, A., Hossain, M.S., Mahmud, J.A., Nahar, K., Hasanuzzaman, M., Fujita, M., 2016. Manganese-induced salt stress tolerance in rice seedlings: regulation of ion homeostasis, antioxidant defense and glyoxalase systems. Physiology and Molecular Biology of Plants 22 (3), 291–306. Available at https://doi.org/10.1007/s12298-016-0371-1.

Raja, V., Majeed, U., Kang, H., Andrabi, K.I., John, R., 2017. Abiotic stress: interplay between ROS, hormones and MAPKs. Environmental and Experimental Botany 137, 142–157. Available at https://doi.org/10.1016/j.envexpbot.2017.02.010.

Raza, A., Habib, M., Kakavand, S.N., Zahid, Z., Zahra, N., Sharif, R., et al., 2020. Phytoremediation of cadmium: physiological, biochemical, and molecular mechanisms. Biology 9 (7), 177. Available at https://doi.org/10.3390/biology9070177.

Raza, A., Charagh, S., Zahid, Z., Mubarik, M.S., Javed, R., Siddiqui, M.H., et al., 2021. Jasmonic acid: a key frontier in conferring abiotic stress tolerance in plants. Plant Cell Reports 40 (8), 1513–1541. Available at https://doi.org/10.1007/s00299-020-02614-z.

Rihan, H.Z., Al-Issawi, M., Fuller, M.P., 2017. Advances in physiological and molecular aspects of plant cold tolerance. Journal of Plant Interactions 12 (1), 143–157. Available at https://doi.org/10.1080/17429145.2017.1308568.

Sachdev, S., Ansari, S.A., Ansari, M.I., Fujita, M., Hasanuzzaman, M., 2021. Abiotic stress and reactive oxygen species: generation, signaling, and defense mechanisms. Antioxidants 10 (2), 1–37. Available at https://doi.org/10.3390/antiox10020277.

Saini, S., Kaur, N., Pati, P.K., 2021. Phytohormones: key players in the modulation of heavy metal stress tolerance in plants. Ecotoxicology and Environmental Safety 223, 112578. Available at https://doi.org/10.1016/j.ecoenv.2021.112578.

Saleem, M.H., Fahad, S., Khan, S.U., Din, M., Ullah, A., Sabagh, A.E., et al., 2020. Copper-induced oxidative stress, initiation of antioxidants and phytoremediation potential of flax (Linum usitatissimum L.) seedlings grown under the mixing of two different soils of China. Environmental Science and Pollution Research 27 (5), 5211–5221. Available at https://doi.org/10.1007/s11356-019-07264-7.

Sarkar, J., Chakraborty, B., Chakraborty, U., 2018. Plant growth promoting rhizobacteria protect wheat plants against temperature stress through antioxidant signalling and reducing chloroplast and membrane injury. Journal of Plant Growth Regulation 37, 1396–1412. Available at https://doi.org/10.1007/s00344-018-9789-8.

Schramm, F., Ganguli, A., Kiehlmann, E., Englich, G., Walch, D., von Koskull-Döring, P., 2006. The heat stress transcription factor HsfA2 serves as a regulatory amplifier of a subset of genes in the heat stress response in Arabidopsis. Plant Molecular Biology 60 (5), 759–772. Available at https://doi.org/10.1007/s11103-005-5750-x.

Sehrawat, A., Abat, J.K., Deswal, R., 2013. RuBisCO depletion improved proteome coverage of cold responsive S-nitrosylated targets in Brassica juncea. Frontiers in Plant Science 4, 342. Available at https://doi.org/10.3389/fpls.2013.00342.

Seta, A., Tabara, M., Nishibori, Y., Hiraguri, A., Ohkama-Ohtsu, N., Yokoyama, T., et al., 2017. Post-translational regulation of the dicing activities of arabidopsis DICER-LIKE 3 and 4 by inorganic phosphate and the redox state. Plant and Cell Physiology 58 (3), 485–495. Available at https://doi.org/10.1093/pcp/pcw226.

Shang, Y., Dai, C., Lee, M.M., Kwak, J.M., Nam, K.H., 2016. BRI1-associated receptor kinase 1 regulates guard cell ABA signaling mediated by open stomata 1 in Arabidopsis. Molecular Plant 9 (3), 447–460. Available at https://doi.org/10.1016/j.molp.2015.12.014.

Singh, A., Kumar, A., Yadav, S., Kumar Singh, I., 2019. Reactive oxygen species-mediated signaling during abiotic stress. Plant Gene 18, 100173. Available at https://doi.org/10.1016/j.plgene.2019.100173.

Singh, A., Mehta, S., Yadav, S., Nagar, G., Ghosh, R., Roy, A., et al., 2022. How to Cope with the challenges of environmental stresses in the era of global climate change: an update on ROS stave off in plants. International Journal of Molecular Sciences 23 (4), 1995. Available at https://doi.org/10.3390/ijms23041995.

Singh, A., Roychoudhury, A., 2023. Abscisic acid in plants under abiotic stress: crosstalk with major phytohormones. Plant Cell Reports 42 (6), 961–974. Available at https://doi.org/10.1007/s00299-023-03013-w.

Sirichandra, C., Gu, D., Hu, H.C., Davanture, M., Lee, S., Djaoui, M., et al., 2009. Phosphorylation of the Arabidopsis AtrbohF NADPH oxidase by OST1 protein kinase. FEBS Letters 583 (18), 2982–2986. Available at https://doi.org/10.1016/j.febslet.2009.08.033.

Suzuki, N., Miller, G., Sejima, H., Harper, J., Mittler, R., 2013. Enhanced seed production under prolonged heat stress conditions in Arabidopsis thaliana plants deficient in cytosolic ascorbate peroxidase 2. Journal of Experimental Botany 64 (1), 253–263. Available at https://doi.org/10.1093/jxb/ers335.

Takeda, S., Gapper, C., Kaya, H., Bell, E., Kuchitsu, K., Dolan, L., 2008. Local positive feedback regulation determines cell shape in root hair cells. Science (New York, N.Y.) 319 (5867), 1241–1244. Available at https://doi.org/10.1126/science.1152505.

Thomashow, M.F., 2010. Molecular basis of plant cold acclimation: insights gained from studying the CBF cold response pathway. Plant Physiology 154 (2), 571–577. Available at https://doi.org/10.1104/pp.110.161794.

Thomashow, M.F., Torii, K.U., 2020. SCREAMing twist on the role of ICE1 in freezing tolerance. The Plant Cell 32 (4), 816–819. Available at https://doi.org/10.1105/tpc.20.00124.

Tiwari, A., Joshi, S., Joshi, R., Bohra, A., 2024. Chapter 2 - Omics advancements in plant abiotic stress. Current Omics Advancement in Plant Abiotic Stress Biology. Academic Press, pp. 23–37. Available at https://doi.org/10.1016/B978-0-443-21625-1.00002-6.

Volkov, R.A., Panchuk, I.I., Mullineaux, P.M., Schöffl, F., 2006. Heat stress-induced H2O2 is required for effective expression of heat shock genes in Arabidopsis. Plant Molecular Biology 61 (4-5), 733–746. Available at https://doi.org/10.1007/s11103-006-0045-4.

Wang, W.H., He, E.M., Guo, Y., Tong, Q.X., Zheng, H.L., 2016. Chloroplast calcium and ROS signaling networks potentially facilitate the primed state for stomatal closure under multiple stresses. Environmental and Experimental Botany 122, 85–93. Available at https://doi.org/10.1016/j.envexpbot.2015.09.008.

Wang, Q.L., Chen, J.H., He, N.Y., Guo, F.Q., 2018. Metabolic reprogramming in chloroplasts under heat stress in plants. International Journal of Molecular Sciences 19 (3), 849. Available at https://doi.org/10.3390/ijms19030849.

Wang, P., Liu, W.C., Han, C., Wang, S., Bai, M.Y., Song, C.P., 2024. Reactive oxygen species: multidimensional regulators of plant adaptation to abiotic stress and development. Journal of Integrative Plant Biology 66 (3), 330–367. Available at https://doi.org/10.1111/jipb.13601.

Willems, P., Mhamdi, A., Stael, S., Storme, V., Kerchev, P., Noctor, G., et al., 2016. The ROS wheel: refining ROS transcriptional footprints. Plant Physiology 171 (3), 1720–1733. Available at https://doi.org/10.1104/pp.16.00420.

Wilson, I.D., Neill, S.J., Hancock, J.T., 2008. Nitric oxide synthesis and signalling in plants. Plant, Cell & Environment 31 (5), 622–631. Available at https://doi.org/10.1111/j.1365-3040.2007.01761.x.

Wu, F., Chi, Y., Jiang, Z., Xu, Y., Xie, L., Huang, F., et al., 2020. Hydrogen peroxide sensor HPCA1 is an LRR receptor kinase in Arabidopsis. Nature 578 (7796), 577–581. Available at https://doi.org/10.1038/s41586-020-2032-3.

Wu, T.Y., Hoh, K.L., Boonyaves, K., Krishnamoorthi, S., Urano, D., 2022. Diversification of heat shock transcription factors expanded thermal stress responses during early plant evolution. The Plant Cell 34 (10), 3557–3576. Available at https://doi.org/10.1093/plcell/koac204.

Yamaguchi-Shinozaki, K., Shinozaki, K., 2006. Transcriptional regulatory networks in cellular responses and tolerance to dehydration and cold stresses. Annual Reviews in Plant Biology 57 (1), 781–803. Available at https://doi.org/10.1146/annurev.arplant.57.032905.105444.

Yavaş, İ., Ali, S., Abbas, Z., Hussain, S., 2022. Heavy metal toxicity in plants: an overview on tolerance mechanisms and management strategies. Turkish Journal of Agriculture - Food Science and Technology 10 (8), 1468–1481. Available at https://doi.org/10.24925/turjaf.v10i8.1468-1481.5220.

Ying, S., Yang, W., Li, P., Hu, Y., Lu, S., Zhou, Y., et al., 2022. Phytochrome B enhances seed germination tolerance to high temperature by reducing S -nitrosylation of HFR1. EMBO Reports 23 (10), EMBR202154371. Available at https://doi.org/10.15252/embr.202154371.

You, J., Chan, Z., 2015. ROS regulation during abiotic stress responses in crop plants. Frontiers in Plant Science 6, 1092. Available at https://doi.org/10.3389/fpls.2015.01092.

Zhang, H., Lu, L., 2024. Transcription factors involved in plant responses to cadmium-induced oxidative stress. Frontiers in Plant Science 15, 1397289. Available at https://doi.org/10.3389/fpls.2024.1397289.

Zhang, W., Qin, C., Zhao, J., Wang, X., 2004. Phospholipase Dα1-derived phosphatidic acid interacts with ABI1 phosphatase 2C and regulates abscisic acid signaling. Proceedings of the National Academy of Sciences 101 (25), 9508–9513. Available at https://doi.org/10.1073/pnas.0402112101.

Zhang, Y., Zhu, H., Zhang, Q., Li, M., Yan, M., Wang, R., et al., 2009. Phospholipase Dα1 and phosphatidic acid regulate nADPH oxidase activity and production of reactive oxygen species in ABA-mediated stomatal closure in Arabidopsis. The Plant Cell 21 (8), 2357–2377. Available at https://doi.org/10.1105/tpc.108.062992.

Zhang, H., Zhu, J., Gong, Z., Zhu, J.K., 2022. Abiotic stress responses in plants. Nature Reviews. Genetics 23 (2), 104–119. Available at https://doi.org/10.1038/s41576-021-00413-0.

Zhan, N., Wang, C., Chen, L., Yang, H., Feng, J., Gong, X., et al., 2018. S-Nitrosylation targets GSNO Reductase for selective autophagy during hypoxia responses in plants. Molecular Cell 71 (1), 142. Available at https://doi.org/10.1016/j.molcel.2018.05.024.

Zhao, H., Bao, Y., 2021. PIF4: integrator of light and temperature cues in plant growth. Plant Science 313, 111086. Available at https://doi.org/10.1016/j.plantsci.2021.111086.

Zhao, M.G., Chen, L., Zhang, L.L., Zhang, W.H., 2009. Nitric reductase-dependent nitric oxide production is involved in cold acclimation and freezing tolerance in Arabidopsis. Plant Physiology 151 (2), 755–767. Available at https://doi.org/10.1104/pp.109.140996.

Zhu, X., Zhang, N., Liu, X., Li, S., Yang, J., Hong, X., et al., 2021. Mitogen-activated protein kinase 11 (MAPK11) maintains growth and photosynthesis of potato plant under drought condition. Plant Cell Reports 40 (3), 491–506. Available at https://doi.org/10.1007/s00299-020-02645-6.

Chapter 7

Modulation of ion transport and uptake dynamics in plants under multiple abiotic stresses

Ibrahim A.A. Mohamed[1,2,3], Muhammad Ikram[4], Ashadu Nyande[2], Ehtisham Hassan Khan[2,5], Ahmad Diyan[2], Zihao Dong[2], Yanhui Li[2], Guangjing Li[2], Jie Wang[1], Zujun Yin[1], and Honghong Wu[6,*]

[1]*National Key Laboratory of Cotton Bio-breeding and Integrated Utilization, Institute of Cotton Research, Chinese Academy of Agricultural Sciences, Anyang, Henan, P.R. China,* [2]*Hubei Hongshan Laboratory, MOA Key Laboratory of Crop Ecophysiology and Farming System in the Middle Reaches of the Yangtze River, The Center of Crop Nanobiotechnology, College of Plant Science and Technology, Huazhong Agricultural University, Wuhan, Hubei, P.R. China,* [3]*Botany Department, Faculty of Agriculture, Fayoum University, Fayoum, Egypt,* [4]*Faculty of Agricultural Sciences and Technology, Institute of Agronomy, Bahauddin Zakariya University, Multan, Punjab, Pakistan,* [5]*Department of Agronomy, Faculty of Crop Production Sciences, The University of Agriculture, Peshawar, Pakistan,* [6]*College of Plant Science and Technology, Huazhong Agricultural University, Wuhan, Hubei, P.R., China*

[*]*Corresponding author. e-mail address: Honghong.Wu@mail.hzau.edu.cn*

7.1 Introduction

Numerous abiotic stressors that crop plants encounter during their lives significantly lower production and jeopardize the world's food security (Savvides et al., 2016). Abiotic stressors that affect crop growth and agricultural productivity globally include water deficit, salt, extremely high temperature, cold, freezing, heavy metal toxicity, flooding, UV radiation, and excessive light (Baweja and Kumar 2020). The negative consequences of these stressors on agricultural productivity are being made worse by human-caused factors, including urbanization and industrialization (Roy et al., 2025) and climate change (Chaudhry and Sidhu 2022), endangering global food security (Sánchez-Bermúdez et al., 2022). The effects of individual stress have been thoroughly researched in recent years, but the capacity of plants to respond to combined stress has received comparatively less attention. Plant metabolism is impacted differently by a mixture of stimuli than by a single stress (Smith et al., 2010; Zhang and Sonnewald 2017; Ju et al., 2021). When many plant species are exposed to external abiotic stressors, an intricate signaling cascade is triggered, which results in a variety of molecular, biochemical, and physiological modifications that help the plants adapt to stressors, including numerous abiotic stresses at once (Kumar et al., 2019). Reactions of plants to abiotic stresses entail the intricate activation of numerous tolerance mechanisms, including the synthesis of ion and water transporters to maintain suitable level of ion homeostasis and relative water content (Barzana et al., 2021; El-Badri et al., 2023), the production of compatible osmolytes (Agami et al., 2016; Mohamed et al., 2020a), the synthesis of specific antioxidant enzymes and compounds that scavenge and quench reactive oxygen species (ROS) (Batool et al., 2022; El-Badri et al., 2022), and the production of chaperones that serve to protect cellular membranes and proteins (Wang et al., 2004; Horváth et al., 2008; Asthir 2015). The presence of upstream regulatory genes that govern ion uptake and transport reactions to different abiotic stresses is suggested via stress signaling pathways coming together to form a common transcription factor (TF) group (Kant et al., 2008). Understanding the mechanisms of uptake and transport of ions in plants exposed to multiple stress conditions at the subcellular level can aid in establishing strategies to improve crop resistance against abiotic stressors.

The transport of ions is crucial for the growth, development, and survival of plants. It involves the movement of ions across cell membranes and tissues and is essential to many physiological functions (Brini and Masmoudi 2012; El-Badri et al., 2023). Membrane transporters (MTs) are proteins embedded in cell membranes, crucial for improving water and nutrient transport efficiency. They help maintain cellular balance by controlling ion movement, xylem loading, and the transfer of sugars from plant leaves (source) to roots, stems, and plant seeds (sink) through the loading of phloem (Gill et al., 2021). The efficiency of MTs to properly utilize plant energy by controlling the transportation of ions and

Multiple Abiotic Stresses in Plants. DOI: https://doi.org/10.1016/B978-0-443-27424-4.00005-7

metabolites in the face of dynamic environmental changes determines the crop yield (Saddhe et al., 2021a; Yadav et al., 2021). Nutrient MTs have a significant part in the cellular processes controlling nutrient acquisition and absorption from the soil and translocating these nutrients to the target plant organs, in addition to their role in removing the excess nutrients from the plant cells (Yadav et al., 2021). The abiotic stressors, including excessive salt and heavy metals, can disturb the complex membrane trafficking system that regulates the function of MTs in the source-to-sink process, thus leading to neglect of ions and sugar homeostasis (Wang 2019).

The capacity of plants to adapt and withstand multiple abiotic stressors depends on their capacity to maintain effective uptake and utilization of ions, maintain suitable levels of ion homeostasis, enhance the antioxidant defense system, and alter the levels of osmoprotectants or metabolites (Mohamed et al., 2020b; Wang et al., 2025). Several known processes allow plants to maximize their usage of ions and nutrients while they are under stress. These processes include alteration in the expression and activity of ions and nutrient transporters, activation of stress-responsive signaling pathways, and adjustment of root architecture that regulate metabolism and allocation of ions and nutrients (Aibara and Miwa 2014; Ahanger and Ahmad 2019). Enhancing agricultural resilience and sustainability requires an understanding of how plants sustain the absorption of ions and nutrient usage efficiency in the face of multiple abiotic stressors. A comprehensive understanding of the mechanics governing these efficiencies under abiotic stressors is still lacking, despite notable advancements in recent years. In this book chapter, our goal is to investigate the different mechanisms underlying the dynamics of the activated signaling in the uptake and transport of important selected ions under multiple abiotic stress circumstances as well as heightening the plant defence mechanism by maintaining ion homeostasis at multilevels and -processes, including sensing molecules, signaling transductions, transcriptional processes, protein translations, metabolic processes, and their alternations. It provides a deciphering for the complex responding networks concerning multiple gene expressions, signaling network pathways, and metabolic processes. By integrating and summarizing these critical aspects, we aim to provide a comprehensive insight that can guide the development of knowledge concerning improving the selected plant ion regulations under combined stress conditions, thereby contributing to the promotion of sustainable farming methods at the global level.

7.2 Potassium uptake and utilization under abiotic stress and its role in stress resistance

7.2.1 Overview of K^+ ion uptake and transport dynamic systems

Potassium ion (K^+) serves as an important macronutrient for plants because it enables adequate growth and developmental processes together with stress tolerance adaptation (Johnson et al., 2022; Mostofa et al., 2022; Hasanuzzaman et al., 2018). K^+ holds fundamental functions in the synthesis of proteins and metabolism of carbohydrates, together with enzyme activation, photosynthesis, synthesis of key metabolites, regulates osmoregulation, and controls stomata function (Hasanuzzaman et al., 2018; Khare et al., 2025) (Sardans and Peñuelas 2021; Li et al., 2022b; Mostofa et al., 2022). Through complex transport systems, plants achieve proper K^+ maintenance under diverse environmental pressures (Wang and Wu 2017). When K^+ homeostasis is maintained, plants can respond properly to abiotic stresses, including water deficit, salinity, and extremely low-temperature stress (Wang et al., 2013b; Mostofa et al., 2022). Momentary abiotic stress conditions, including salt stress and drought, and flood exposure, force plants to use specialized transport systems that help sustain optimal K^+ levels within their cells (Shabala and Pottosin 2014; Bhattacharya 2022). The potassium ion system manages osmotic equilibrium and controls ion stability and stomatal opening and closure while also mediating salt and drought stress conditions (Sardans and Peñuelas 2021; Imtiaz et al., 2023). The movements of stomata regulated by K^+ act to control both water absorption and transpiration rates, especially when plants face drought conditions (Imtiaz et al., 2023). The plant uses K^+ to develop salt stress tolerance by establishing osmotic balance and ion homeostasis, according to Hasanuzzaman et al. (2018). More than 60 enzymes require K^+ participation for activation, while various metabolic processes depend on this mineral element (Pandey 2015). The vital stress-mitigating role of potassium involves its ability to regulate antioxidant mechanisms and control ROS generation (Kumari et al., 2021). The lack of K^+ ions leads to reduced photosynthetic capacity and utilization problems alongside leaf necrosis, which showcases its vital role in plant quality and production (Anil Kumar et al., 2024). Even though K^+ is abundant in soils, available K^+ levels remain constrained because K^+ fertilizers must be soluble (Zörb et al., 2014). Researchers have proposed different methods to increase K^+ use efficiency in plants by expanding roots and improving K^+ transportation to shoot tissue alongside K^+ mobility in the soil layer and by employing molecular breeding techniques (Sardans and Peñuelas 2021). Latest studies show that organic substances released by bacteria and plant root tissues can liberate K^+ from mineral structures, thus making breeding initiatives for K^+ availability accessible genotypes possible (Rawat et al., 2016) (Shah and Wu 2019;

Delgado et al., 2024). Therefore agricultural sustainability in difficult environments depends on maximizing potassium fertilization methods and increasing access to soil potassium reserves.

The evolutionary processes of plants created advanced systems that regulate K^+ uptake as well as cellular and organic movement processes (Sardans and Peñuelas 2021). Roots absorb potassium ions, which the xylem distributes to the shoots before transferring them to the phloem for plant-wide distribution (Weber et al., 2023). The recycling of potassium between plants' root subsystems and aboveground structures allows the transmission of nutrient signal information and generates the energy needed for additional transport operations (Dreyer et al., 2017). The redistribution system for potassium within plants functions more effectively than calcium, so that growing tissue gets a proper nutrient supply. K^+ transport operates through different signaling pathways that respond both to external supply as well as plant stresses and other ions (Karley and White 2009; Szczerba et al., 2009; Awuchi et al., 2020). Therefore the identification of these regulatory mechanisms remains fundamental for enhancing strategies that improve both crop nutrition and nutrient-vitamin content through biofortification.

Under K^+ limitation, the high-affinity K^+ transporter HAK5 together with the low-affinity transporter AKT1 (Arabidopsis K^+ Transporter 1) serve as primary K^+ acquisition elements (Wang and Wu 2017; Li et al., 2018). The AtHAK5 transporter shows decreased expression under salinity conditions, while AtAKT1 increases K^+ loss, resulting in restricted K^+ supply to plants (Nieves-Cordones et al., 2010). It has been demonstrated that the plasma membrane (PM) AtAKT1 in *Arabidopsis thaliana* mediates both low- and high-affinity K^+ acquisition within the root system (Sharma et al., 2013). Plants use transcriptional control systems and posttranslational modifications to maximize their K^+ absorption capabilities and use of K^+ under multistressful environments (Gupta et al., 2021b). The level of K^+ in the cytosol functions as a fundamental controller within plant stress responses since it steers metabolic functions while driving adaptive reactions alongside controlled cellular suicide (Shabala and Pottosin 2014; Talukdar 2019; Sies et al., 2024). Respiratory burst oxidase homolog (RBOH) members have been identified in several plants, including tobacco, rice, maize, and *Arabidopsis,* and they participate in the regulation process of plant growth and development under stress conditions, including drought, heat, salinity, and biotic stresses (Wang et al., 2022, 2023). RBOH members were reported to have an important function in the production of ROS signaling under stress conditions. Recently, OsRBOHA and OsRBOHI have been identified in rice plants, and they have an important role in the ROS-Ca^{2+} signaling in plants (Wang et al., 2023).

Two K^+ efflux channels, a stelar K^+ outward rectifier (SKOR) and a gated outwardly-rectifying K^+ channel (GORK), can interact with SLAH3 and play essential roles in the K^+ translocation from the root to shoot under low K^+ supply in Arabidopsis plants (Drechsler et al., 2015) and as a response to high K^+ levels. SKOR transports K^+ from cells of the root stele into the xylem tissue (vessel element) and from roots to shoots, whereas GORK is expressed in the epidermis and performs the role of K^+ release (Demidchik et al., 2014; Fedorova et al., 2021).

Our knowledge of K^+ signaling improved through modern omics technologies because they allowed us to better understand its roles in stress resistance enhancement from a biological perspective (Saleem et al., 2025). The K^+ signaling and transport regulatory process starts with detecting external K^+ alterations, followed by membrane passage transmission signals until it reaches K^+ channels and transporters (Wang and Wu 2017). The signaling cascades between K^+ and calcium, as well as ROS, result in enhanced plant stress resistance while activating antioxidant system mechanisms (Sewelam et al., 2016; Sachdev et al., 2021). Under drought conditions, H_2S interacts with ABA to promote stomatal closure by activating genes that encode ion channels and downstream genes like *HY1*, which modulate the transmembrane efflux of K^+, Ca^{2+}, and Cl^- ions (Parwez et al., 2022). Plant growth, together with metabolism and stress resistance under biotic and abiotic conditions, benefits from K^+ influence on biochemical and physiological processes. Posttranslational cellular processes using protein interactions and phosphorylation events refine how potassium moves through different plant tissues (Chérel and Gaillard 2019). The molecular study of K^+-induced stress protection combined with optimized fertilizer input methods should enhance plant resistance to climatic weather hazards (Pantha et al., 2023).

7.2.2 Regulation of K^+ ion uptake and homeostasis (hormonal regulation - membrane potential - gene expression) under abiotic stresses

7.2.2.1 Salt stress

Plant development is adversely affected by salinity stress, which changes the ionic composition within the plant cells while also posing osmotic stress. K^+ functions as a vital mediator against salt stress since it helps maintain cell equilibrium while activating enzymes (Ahanger et al., 2017). Salinity-induced oxidative stress becomes manageable for plants through K^+ supplementation because the nutrient enables antioxidant defenses (Mostofa et al., 2022). The accumulation of K^+ as an osmoprotectant together with antioxidant system strengthening allows plants to sustain ion homeostasis (Abbasi et al., 2016). K^+ functions as a fundamental element of plant salt stress tolerance because it enables osmotic regulation as well as enzyme activation and protein synthesis. Plants require various mechanisms to keep salt tolerance functional because this demands a level of high K^+/Na^+ ratio within the cell cytosol, which can be achieved through ion transporters and H^+ pump regulation (Almeida et al., 2017). The K^+ and Na^+ homeostasis functions of H^+-ATPases and SOS1 transporters, as well as high-affinity K transporters (HKTs) and NHXs (Na^+/H^+ exchangers) transporters, are vital (Pardo and Rubio 2011; Almeida et al., 2017; Saibi and Brini 2021). Different K^+ transporters alongside H^+ pumps, including H^+-PPases and H^+-ATPases, operate with SOS1 and NHXs and HKTs to sustain proper ratios of K^+ and Na^+ in the cytosol (Fig. 7.1) (Almeida et al., 2017). The increased levels of Na^+ in saline environments create competition with K^+ for nonselective cation channels (NSCCs), which causes K^+ deficiency that reduces plant growth. When plants are under saline conditions, they use high-affinity K^+ transporter (HAK/HKT) and Arabidopsis K^+ Transporter channels as K^+ transporters to select K^+ absorption along with Na^+ exclusion (Ketehouli et al., 2019). In rice, HAKs, including OsHAK1/5/16/21 participate in K^+ uptake via root under salinity stress and different root zone K^+ levels (Yang et al., 2014; Okada et al., 2018); therefore, this uptake mechanism improves its salinity tolerance. Also, K^+ transporter IbHAK11 was reported to participate in the regulation of high salt stress and K^+ deficiency resistance in sweet potato plants (Zhu et al., 2023). In addition, OsAKT1 is considered an inward-rectifying K^+ channel, and its transcript was found in the root system of rice plants at the early growth stage, and it can be controlled by extracellular protons and Ca^{2+} ions (Fuchs et al., 2005). The same investigations also reported for SsAKT1 in the *Suaeda salsa* plant under salinity conditions (Duan et al., 2015). Furthermore, *AtKUP2/7*, K^+-uptake permease 2/7, also have an important role in the K^+ uptake and transport under salinity stress and low K^+ levels, as their mutant lines have lower K^+ root-shoot transport (Šustr et al., 2020). *WRKY33* and *bHLH122*, two stress-responsive TFs, were found to be direct modulators of *AtKUP2* expression (Rajappa et al., 2020). HKTs are members that belong to the K^+ transporter group. Subsequent research revealed that plants have two classes of HKT transporters: HKT1s, which often mediate highly Na^+ selective transport, and HKT2s, which mediate Na^+/K^+ transport activity (Suzuki et al., 2016).

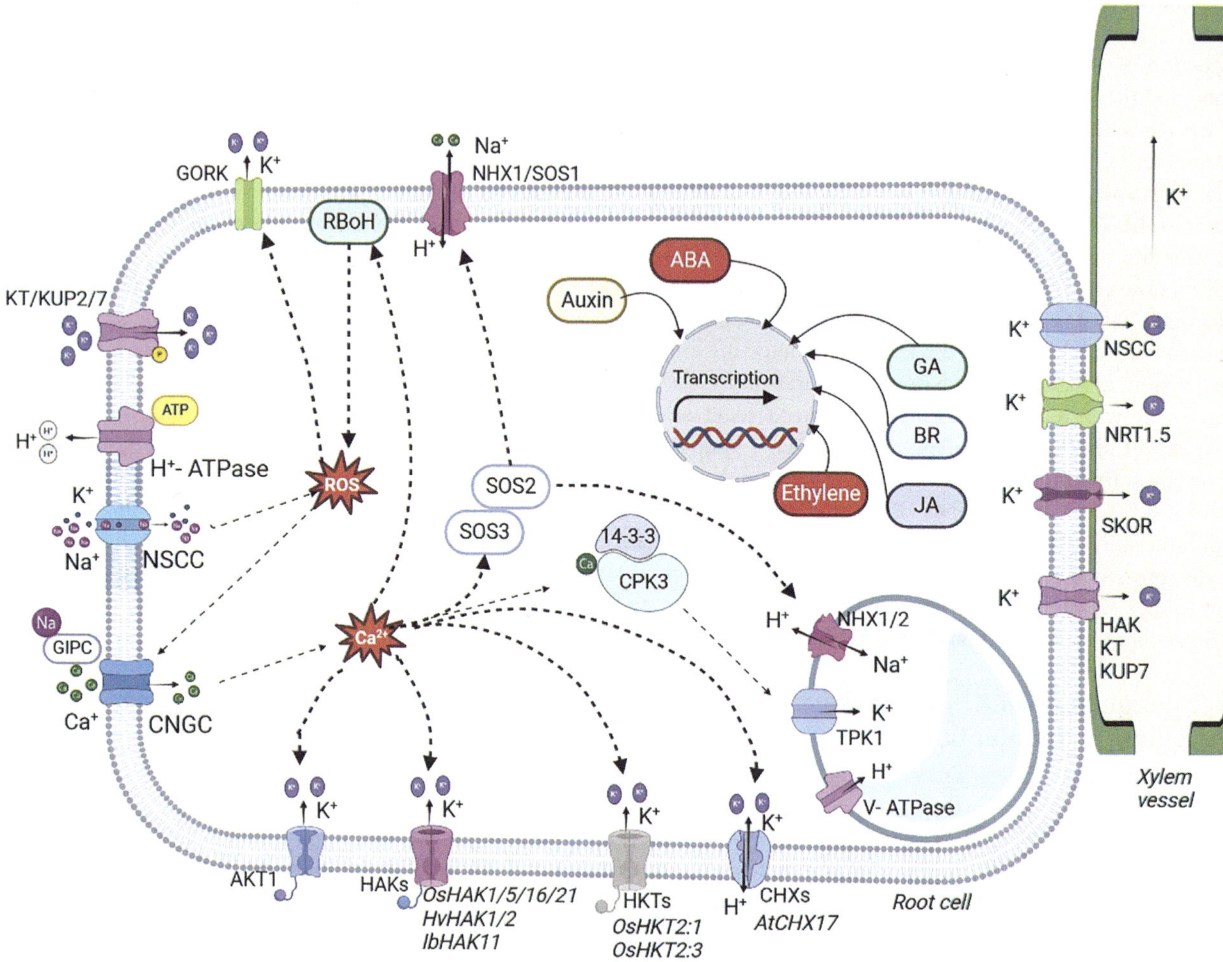

FIGURE 7.1 Illustrating potassium (K^+) ion uptake and transport under salinity stress conditions.

The expression of AtHKT1;1, OsHKT1;5, OsHKT2;1, and OsHKT2;3 was reported in the roots of Arabidopsis or rice, and they have an important role in the regulation of Na^+ and K^+ transport (Hamamoto et al., 2015). The family of cation/H^+ exchanger (CHX) plays a crucial role in regulating the K^+/Na^+ ratio in plants under salinity tolerance. The *KvCHX* gene has been reported to be associated with the *AtCHX17* gene, and its overexpression showed better growth performance under potassium deficiency than wild type and salinity tolerance via enhancing K^+ uptake (Guo et al., 2023). Pathway of Na^+/H^+ antiporter SOS signaling, salt overly sensitive, of plants, leads to Na^+ ion expulsion through the apoplast, which supports a beneficial Na^+/K^+ equilibrium in the cytoplasm. By extruding Na^+ at the root-soil interface, the SOS1, localized in the xylem parenchyma cells, functions to lower the net input of Na^+ (Saibi and Brini 2021). In salt-stressed environments or low K^+ concentrations, several studies have shown that the loss of function of SOS (*sos mutants*) led to salt sensitivity, and strong inhibition in root-cell K^+ permeability (Qi and Spalding 2004). Under salinity stress, phosphatidic acid controls membrane location and SOS2 (CIPK24) activity. This modulation stimulates SOS2-mediated phosphorylation of SOS3-like Ca-binding protein 8 (SCaBP8/CBL10) and activates salt excessively sensitive 1 (SOS1) for Na^+ efflux. This phosphorylation increases K^+ absorption and preserves a balanced cellular Na^+/K^+ ratio by lessening the inhibitory action of SCaBP8 on AKT1 (Li et al., 2023).

In *A. thaliana*, constitutively overexpressed melon SKOR promotes growth and enhances salinity tolerance (Long-Tang et al., 2018). However, the *gorkskor* double mutant showed improvements in the parameters of the salinity tolerance, including lower Na^+ content and higher K^+ content, as well as an increase in primary growth of roots than WT or *skor and gork* single-mutant lines (Hiya et al., 2024). Plant salt tolerance is greatly increased by vacuolar K^+ homeostasis. Tonoplast-localized NHX exchangers (NHX1–4) store incoming Na^+ in the vast central vacuole of roots and shoots. Electrochemical potential gradients for secondary active transport are produced by tonoplast H^+-PPase (V-PPase), tonoplast H^+-ATPase (V-ATPase), PM H^+-PPase (PM-PPase), and PM H^+-ATPase (P-ATPase) (Almeida et al., 2017).

Increased salt tolerance may result from the contribution of two-pore potassium channels (TPKs) to vacuolar K^+ efflux. The 14-3-3 protein GRF6 (GF14-λ) and the Ca^{2+}-dependent protein kinase CPK3 work together to sustain TPK1 vacuolar K^+ efflux activity during extended salt stress. This partnership ensures cellular homeostasis by assisting in the adaptation of the cytoplasmic K^+/Na^+ ratio to salt stress (Shabala and Pottosin 2010; Wang et al., 2013a). To lessen the effects of salinity stress, the voltage-gated K^+ channel is essential. Overexpression of the *KIbB1 gene* encoding a voltage-gated K^+ channel β subunit in sweet potato showed higher K^+ uptake and salinity tolerance than the wild type, while translocation of K^+ from root to shoot was not affected (Zhu et al., 2022).

Under salinity conditions, K^+ regulates stomatal opening as well as controls photosynthesis and allows for efficient water deficit adaptation mechanisms to maintain plant survival (Ahmad et al., 2016a). The treatment with the external K^+ leads to improved salt resistance by boosting various plant attributes, including morphological, physiological, and biochemical markers in plant systems (Abou-Sreea et al., 2022). Plants create osmium-protecting compounds while boosting their antioxidant capabilities to fight salt stress (Abbasi et al., 2016). New research examines salt tolerance through proteomic, metabolomic, genomic, and transcriptomic perspectives, which show that osmoregulation, cell wall modifications, and antioxidative defense represent additional defense strategies. Plants receive improved tools to confront salinity through potassium fertilization, which improves both their inner and outer features and chemical characteristics (Kaya and Ashraf 2024).

Plants' reactions to salt stress are regulated primarily via the functioning of plant hormones. Under saline environments, ABA serves as a crucial hormone to control K^+ movements along with stomatal function (Che et al., 2022). Under salt stress, plants benefit from auxins together with gibberellins and cytokinins (CK), which modify root structures to enhance their water and nutrient extraction capabilities (Egamberdieva et al., 2018) (Bielach et al., 2017; Li et al., 2022d). Jasmonates and salicylic acid (SA), together with these hormones, control root tropisms and both ion homeostasis and ROS signaling in salinity responses (EL Sabagh et al., 2022; Kosakivska et al., 2022). The K^+ transport and accumulation processes controlled by hormones depend on phosphorylation mechanisms as well as protein targeting functions and regulatory partner binding events (Feraille and Doucet 2001; Chérel and Gaillard 2019). The understanding of plant physiology through hormonal relationships helps researchers develop methods for enhancing tolerance in crops while improving agricultural production levels of saline-affected areas (Amir et al., 2019; Yun et al., 2024). Salinity creates major growth limitations for plants and crop yields worldwide, but plants have evolved different survival methods to coexist with salt environments. Plants under salt stress utilize potassium ions both for regulating ion concentrations and osmotic pressure and strengthening their antioxidant defense systems (Ahmad et al., 2019; Che et al., 2022). Both histone modification and DNA methylation use epigenetic mechanisms to regulate salt-responsive gene expression and generate extra adaptive features (Saradadevi et al., 2021; Ma et al., 2024). Proficient crop husbandry requires an integrated use of K^+ mechanistic understanding of hormonal adjustment and epigenetic regulation to develop innovative solutions for salt-tolerant farming systems and improved agricultural production in saline land areas (Aftab and Hakeem 2022; Singh et al., 2024b).

Plants use epigenetic mechanisms as fundamental mechanisms to adapt to salt stress conditions. Histone modifications together with DNA methylation enable the control of expression for salt-stress–responsive gene conditions, including *MYB* TF (Banerjee and Roychoudhury 2017). The stress-responsive TFs become activated under dynamic conditions when DNA methylation combines with histone marks such as H3K4 trimethylation and H3K9 acetylation (Liu et al., 2022). The regulatory system of DNA methylation controls the salt tolerance activity of *GmMYB84* in soybeans because it allows the TF to bind to *GmAKT1* promoter regions (Zhang et al., 2020a). The adaptations of plants to salt stress include contributions from H2A.Z histone variants and noncoding RNAs, as well as more refined responses to salinity (Sharma et al., 2023). The regulatory system formed through the connection of hormone signaling pathways and epigenetic modifications produces protein complexes for the control of stress-responsive gene expressions and improved plant resistance (Kaya et al., 2024). Better strategies for enhancing salt resistance in crops result from understanding these mechanisms (Hao et al., 2021).

7.2.2.2 Drought stress or water deficit

The indispensable function of K^+ in plant resistance to drought happens through boosting water-use efficiency (WUE) while achieving control of different physiological activities (Akhtar et al., 2023). Plants show enhanced growth together with better nutrient intake and better physiological characteristics after receiving K fertilization during drought periods (Rouphael et al., 2012). Plants need potassium ions to develop drought tolerance because of the diverse mechanisms they provide. The potassium ions supplied in cereal crops support water balance, provide membrane protection, enhance osmotic adjustment functions, and stimulate both antioxidant enzyme functions and osmolyte synthesis to develop drought tolerance (Damalas and Koutroubas 2024; Wang et al., 2013b; Hassan et al., 2017; Singh et al., 2024a). The main effect of potassium deficiency on photosynthesis stems from reduced mesophyll conductance, which causes CO_2 assimilation decline and impaired plant-scale WUE with no impact on leaf-scale WUE because stomatal regulation stays unchanged (Kelly and Medlyn 2013). K^+ deficiency activates sophisticated signal pathways that result in both enhanced expression of K^+ transport systems as well as modified root structures that improve K^+ acquisition (Wang and Wu 2010; Chérel and Gaillard 2019). K application shows numerous molecular, biochemical, and physiological mechanisms that enable it to help plants fight drought stress and should be used as a plant resilience-enhancing strategy in drought zones (Hassan et al., 2017; Zia et al., 2021).

The process of K^+ uptake together with its translocation shows sensitivity to drought conditions, while genetic drought tolerance depends on root-based regulator capabilities for K^+ transporters and channels. The main regulatory network controlling K^+ uptake and transport under drought stress is represented in Fig. 7.2 Genes *HAK1*, *HAK5*, *KUP1*, *KUP2*, *AKT1*, and *SKOR* show specific expression patterns under drought conditions, which enable efficient potassium intake and storage (Gierth et al., 2005; Cai et al., 2019). In plants, the main channel for root K^+ absorption is PM inwardly rectifying K^+ channel 1 (AKT1). Suppression of K^+ uptake in barley root epidermal cells and biomass reduction was observed in the silenced *HvAKT1* lines under drought stress, indicating its role in K^+ uptake, mediating ion homeostasis and drought stress tolerance in plants (Feng et al., 2020). Rice plants strangle drought by utilizing *OsHAK1* high-affinity K^+ transporters, which regulate stress-responsive genes while boosting antioxidant enzymes and improving grain yield production when water availability is limited (Chen et al., 2017). Under drought stress, the PM H^+-ATPase functions as a vital regulator of K^+ transporter activity (Zhang et al., 2018b). PM H^+-ATPase activity and higher ROS scavenging ability are major factors associated with potassium retention and thus drought tolerance in tea plants (Zhang et al., 2020b).

Voltage-dependent K^+ channels play an important role in drought stress tolerance as they participate in the regulation of the osmosensing pathway of the stomatal apparatus (Liu and Luan 1998; Liu et al., 2000). Sirichandra et al. (2009) have suggested that K^+ entry via KAT1/KAT2 channels (activated by the OST2 H^+-ATPase) drives stomatal opening, while K^+ efflux through GORK channels (initiated by SLAC1 anion channels) leads to closure. Another suggestion has been reported based on ABA and Ca^{2+} signaling that drought stress increases ABA levels in plant tissues, which then triggers Ca^{2+} oscillations. These oscillations activate both rapid (R-type) and slow (S-type) anion channels, leading to membrane depolarization. As a result, the K^+ inward-rectifying channel (KAT1) in *Arabidopsis thaliana* is inactivated, while the K^+ outward-rectifying channel (GORK) is activated. This process causes turgor loss, ultimately leading to stomatal closure (Jalakas et al., 2021; Tan et al., 2023).

Drought-stress osmotic adjustment strongly depends on KUP-family K^+ transporters. These K^+ transporters, named KUP6, KUP8, and KUP2, are influenced by auxin and ABA, which results in the regulation of root growth and drought defense mechanisms and control of cell expansion (Li et al., 2018). In *Arabidopsis* plants, the KUP potassium transporters control osmotic adjustment and drought responses because they link with ABA signaling components, and their single-gene mutations result in altered cell expansion and potassium efflux while reducing drought survival rates

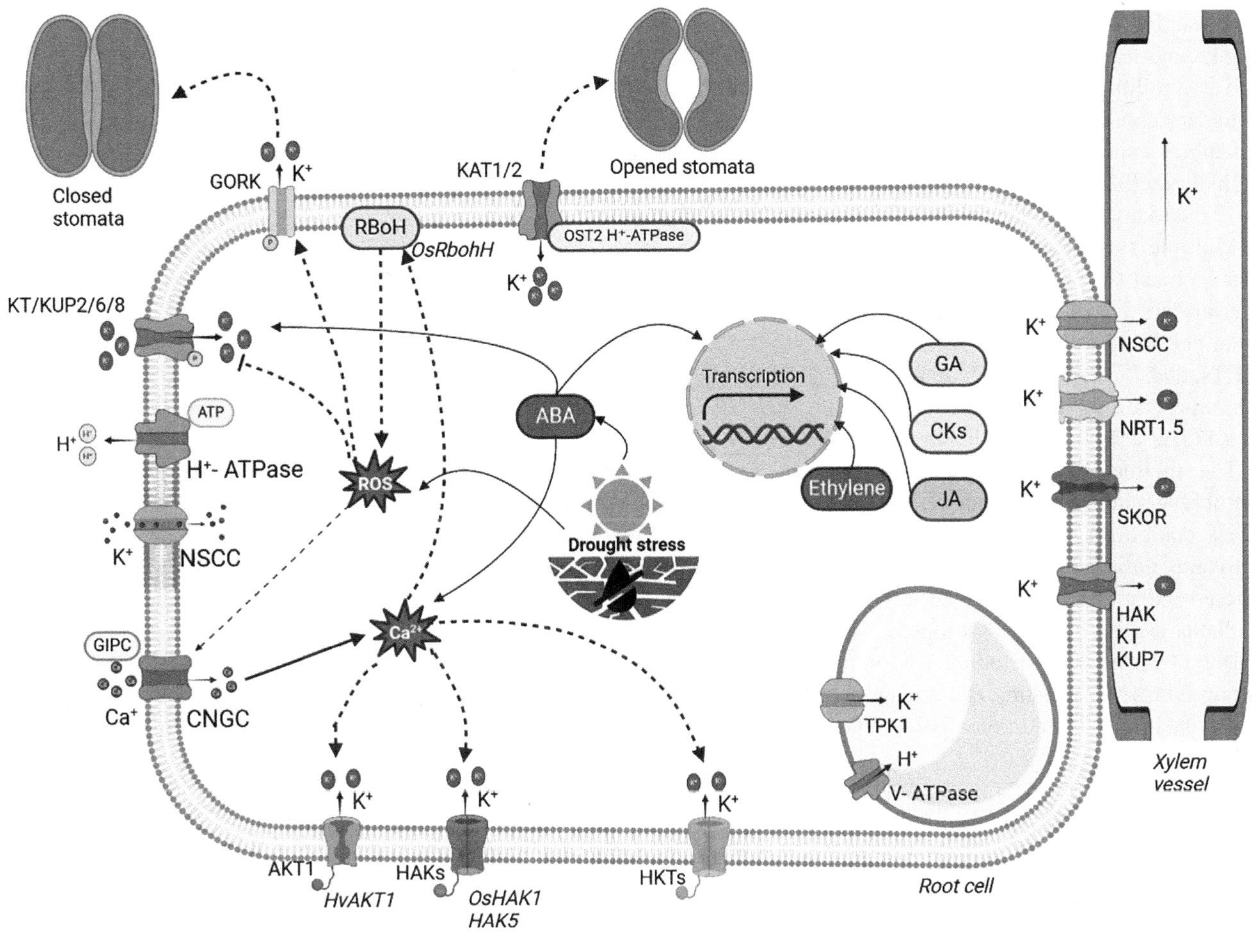

FIGURE 7.2 The regulatory network controlling K+ uptake and transport under drought stress.

(Osakabe et al., 2013). ABA serves as the main plant hormone that controls how plants react to drought stress through hormonal regulation via interactions with jasmonic acid (JA), ethylene, CK, and gibberellins (Parwez et al., 2022). During water deficit conditions, ABA builds up quickly to reduce stomata activity while activating genes that respond to drought (Yari Kamrani et al., 2022; Margay et al., 2024).

Different molecular processes linking K^+ transporters to signaling pathways help control potassium homeostasis as they activate stress-responsive genes (Saddhe et al., 2021b; Mostofa et al., 2022). Therefore the optimal functioning of plant metabolism and performance requires K^+ homeostasis because cytosolic optimum K^+ content may activate plant stress responses, inducing multiple physiological and molecular systems during drought adaptation. K^+-mediated process investigations maintain essential value in developing drought-resistant crops as well as boosting the production efficiency of plant agriculture in stressed water environments. Thus recommending K^+ fertilization methods as a potential solution to increase crop resistance to drought.

7.2.2.3 Heat stress

Heat stress or high-temperature stress is one of the most critical stressors that adversely affects lipid peroxidation and oxidative damage, leading to membrane leakage and altered permeability, resulting in a shortage in plant productivity in several regions in the world due to its effect on the plant at the molecular levels as well as physiological levels. Heat stress in plants can increase the intracellular Ca^{2+} level as a common secondary messenger (Kang et al., 2023). The effectiveness of K^+ in improving antioxidant defense mechanisms along with stabilizing enzymatic functions makes it key to combat the adverse effects of high temperature on plant developmental efficiency and productivity (Hasanuzzaman et al., 2013). K^+ reduces oxidative stress under heat stress through two mechanisms by both inhibiting the activity of NADPH oxidase and preserving electron transport (Belhadj Slimen et al., 2014). The plant requires different channels and transporters to move

K^+ throughout its system for distribution to proper heat stress tolerance (Shabala and Pottosin 2014). The regulations of K^+ uptake and transport under heat stress conditions are presented in Fig. 7.3.

The equilibrium of K^+ in plants depends on multiple transporters and channels that maintain both influx and efflux boundary conditions and internal relocation of K^+ to guarantee metabolic efficiency and plant stability during heat stress (Maathuis and Podar 2011; Rogiers et al., 2017). The maintenance of membrane stability exists through K^+ action in the regulation of H^+-ATPases and aquaporins, which ensures optimal water and ion transport across membranes (Assaha et al., 2017; Sardans and Peñuelas 2021).

Multiple studies indicate that K^+ transporters HAK5 and AKT1 follow hormonal directives for stress homeostasis management (Rubio et al., 2014; Cai et al., 2021). Siddique et al. (2021) have identified several potassium channels and transporters based on genome-wide identification in *Cajanus cajan* under abiotic stress, including heat stress. The identified K^+ channels were 9 shaker K^+ channels, 5 tandem-pore K^+ channels (TPKs), 6 K^+ efflux transporters (KEAs), 2 HKTs, and 17 carrier-like K^+ transporters (KUP/HAK/KTs). Also, Azeem et al. (2022) have found a notable variation in the expression of K^+ transporters, including *VrKUP/HAK/KT*4/5/8.1, *VrTPK3, VrKEA5, VrSKOR, VrAKT2, VrAKT1.1, VrKAT*1/2, and *VrHKT*1.1/1.2 in *Vigna radiata*.

The thermotolerance-promoting actions of K^+ include both the generation of heat-shock proteins (HSPs) and membrane-stability maintenance (Grigorova et al., 2011). HSPs were reported to be expressed under multiple abiotic stress conditions, and in addition to their role in the regulation of cell membrane integrity, they can participate in activating antioxidants, regulating ROS scavenging, and accumulation of metabolic osmolytes (Kumar et al., 2012, 2013). These regulations were reported to be controlled by early activation of PM K^+ and Ca^{2+} channels (Yu et al., 2016).

Plants use K^+ homeostasis together with the hormonal mechanisms to achieve adaptations during heat stress periods. Different hormones, including CK, ethylene, JA, and SA, enhance heat stress tolerance because they influence gene expression while boosting antioxidant defenses (Li et al., 2021; Al-Zahrani et al., 2022; Prerostova and Vankova 2023). Under heat stress conditions, ABA functions to manage stomatal movement through K^+ channel activity control, thus

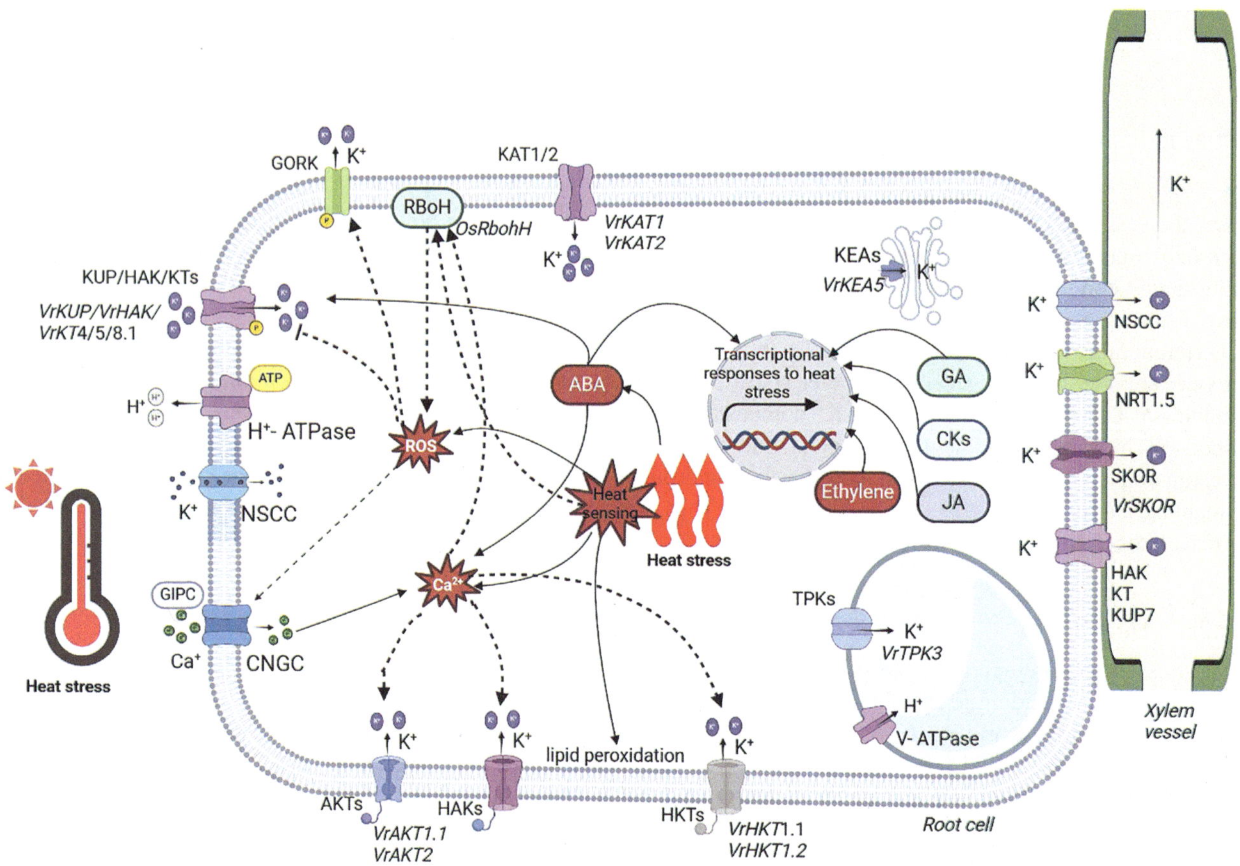

FIGURE 7.3 The regulations of K^+ uptake and transport under heat stress conditions.

maintaining proper water and ion flow (Hasanuzzaman et al., 2018; Chérel and Gaillard 2019). Moreover, the stress tolerance mechanisms of K^+ transport systems depend on interactions between ethylene and JA, which demonstrate the correlated influences of hormones and ions (Shabala and Pottosin 2014). Therefore, the complex network of crosstalk between k^+ signaling and hormonal signaling function provides an improved capacity to handle heat stress challenges.

In conclusion, the elemental composition of K^+ serves as an essential factor for plant high-temperature stress response, so it preserves cell membranes and regulates ion concentrations. Thus, specific research work into K^+-mediated heat stress tolerance mechanisms while studying hormonal regulation enables scientists to generate advanced strategies for improving crop resistance, which ensures farming sustainability during climate change.

7.2.2.4 Waterlogging stress

Waterlogging stress negatively affects plant development and agrarian output because it develops hypoxic environments in root growth areas. Cessation of oxygen availability through waterlogging creates root problems that block K^+ transport and damage cell membranes (Singh et al., 2018). The insufficient oxygen supply during waterlogging harms plant roots because it interrupts nutrient absorption while creating physiological pressure. Molecules of K^+ function as a key factor in reducing waterlogging damage by regulating membrane potential while maintaining osmotic balance and nutrient assimilation (Wang et al., 2013b; Sardans and Peñuelas 2021). The application of external K^+ enhances osmotic regulation mechanisms, enzyme activation, photosynthetic capacity, and mineral uptake in water-logged stressed plants, which indicates its essential role in stress management (Sardans and Peñuelas 2021).

Plants adapt to waterlogging stress through two sets of anatomical and morphological adaptations that include aerenchyma formation in combination with adventitious root development to improve gas exchange and survival potential. Root adaptations are crucial for waterlogging resistance because adventitious roots with aerenchyma perform better than seminal roots when it comes to nutrient absorption in hypoxic circumstances (Huang 2022). Testing plant adaptation systems results in the development of plant varieties that can support crop production even in flood-prone areas. Under waterlogging conditions, root dry mass exhibits a correlation with aerenchyma formation, but shoot dry mass shows a relationship with higher stomatal conductance and Fv/Fm, thus indicating that waterlogging responses between roots and shoots may operate independently from each other (Pang 2023). Plants under waterlogging conditions rely on their ability to keep their membranes at potential while regulating K^+ levels because good oxygen transport between shoots allows H^+-ATPase operation in roots (Zeng et al., 2014; Gill et al., 2018).

The high-affinity K^+ transporter *AtHAK5* is strongly expressed when plants face K^+ deprivation since it facilitates inducible high-affinity K^+ uptake through the roots of *Arabidopsis*, which demonstrates the essential role K^+ transporters play in stress responses (Yang et al., 2020; Cai et al., 2021). Plants use AtHAK5 K^+ transporters along with other K^+ transporters for K^+ homeostasis maintenance under conditions of environmental stress through hormonal signal regulation (Nieves-Cordones et al., 2012; Ankit et al., 2022).

Ethylene, together with ABA hormones, supports the water-logging stress tolerance functions by facilitating aerenchyma formation and adventitious root development to overcome the reoxygenation in plants at post-waterlogging conditions (Bashar 2018). The growth of submerged organs depends on ethylene-driven pathways combined with gibberellin-enhanced pathways, and ethylene-response TFs maintain survival genes under low oxygen conditions. The regulatory hormone ethylene enables both aerenchyma tissue development and the production of new aerial roots to improve oxygen supply to submerged roots (Voesenek and Bailey-Serres 2015). Under waterlogging conditions, ABA controls stomata closing and osmotic changes, which enable plants to preserve water while sustaining cell structures (Yang et al., 2023).

The regulation of voltage-gated K^+-permeable channels to which GORK belongs is important for waterlogged tolerance in plants. Specifically, within the mature root epidermis, the ability to maintain a depolarized state by regulating GORK plays a central role in signaling and retention of K^+ under these conditions (Gill et al., 2018). In comparison to the wild type, the mutant gork1-1 with no functional GORK channels exhibited noticeably more K^+ retention in both the mature and elongation zones during waterlogging stress. Following three days of waterlogging stress, gork1-1 had a K^+ efflux in the elongation zone that was noticeably less than that of the wild type, whereas the wild type maintained an inflow in the mature zone (Wang et al., 2017a).

Waterlogging responses are regulated through TFs, which include family members of *WRKY* and *ERF*. The waterlogging responses of kiwifruit become stronger through *alcohol dehydrogenase* (*ADH*) gene activation under ERF-VII subfamily members *AcERF74* and *AcERF75*, resulting in better alcoholic fermentation (Wang et al., 2021a; Xing et al., 2024). Therefore, enhanced crop waterlogging tolerance can be developed by researchers who apply knowledge of potassium ion maintenance, hormonal modulation, and genetic adaptations.

7.3 Sodium uptake and utilization under abiotic stress

7.3.1 Overview of Na^+ ion uptake and transport dynamic systems

Salinity stress negatively influences the growth performance, survival ability, biomass production, and yield quality of cultivated plants (Mohanavelu et al., 2021; Ondrasek and Rengel 2021). Saline soils have been characterized by elevated concentrations of soluble salts like $CaSO_4$ and Na_2CO_3, with NaCl being the most prevalent (Osman, 2018). The main impacts of salinity on the growth and development of plants include (1) an osmotic effect that causes water deficiency due to the high solute concentration in the growth media, and (2) ion stresses that result in K^+ deficiency caused by imbalanced K^+/Na^+ ratios (Valentovic et al., 2006; Kumari et al., 2021; Zhang et al., 2022a). The alteration of the K^+/Na^+ ratio not only disrupts the various enzymatic activities but also puts a strain on the cell's energy because organic solute synthesis is necessary to make up for the Na^+ exportation for osmotic regulation (Wu 2018). Under salinity stress environments, Na^+ enters through pathways typically used for K^+ uptake, as the hydrated ionic radii of both ions are similar, making it challenging to differentiate between them (Wakeel 2013; Shabala et al., 2015). This inability to discriminate leads to Na^+ toxicity and K^+ deficiency in plants that grow in saline soils. To sustain a lower cytosolic sodium level and a higher K^+/Na^+ ratio, several mechanisms must be in place to: (1) limit Na^+ entry into root system cells, (2) sequester Na^+ into the cell vacuole, and (3) enhance Na^+ expulsion from root system cells (Hauser and Horie 2010; Wakeel 2013). These mechanisms of cellular osmotic regulation and Na^+ detoxification have an important function in improving a plant's ability to withstand salt stress (Arif et al., 2020; Hao et al., 2021).

The high concentration of salts in the soil presents a significant challenge for plants. Sodium and other ions absorbed by the roots are carried to the shoots through the mechanism of transpiration stream force, where they gradually build up over time (Keisham et al., 2018; Tester and Davenport, 2003). As water evaporates, salt concentrations increase in the apoplast and eventually within the cells (Munns 2002). The buildup of ions in the tissues and organs of plants leads to gradual damage. These ion-specific stress effects occur alongside the impacts caused by the high osmolality (Dajic 2006). It remains uncertain whether plants possess specialized transport systems for lower-affinity Na^+ uptake from the soil; therefore, the exact mechanisms behind root Na^+ uptake are not fully understood and are likely to involve transporters from various transport classes and gene families.

This book chapter highlights the advancements made in understanding sodium transport mechanisms, including Na^+ uptake in roots through NSCCs, its efficient sequestration in the vacuole via NHXs, and Na^+ expulsion from roots in response to salt stress through the SOS pathway (Fig. 7.4).

Under salinity stress environments, Na^+ enters the plant root system primarily through **NSCCs, HKTs, AKT1, HAK5, and CHX** channels (Malakar and Chattopadhyay 2021; Wu and Bose 2024). NSCCs are permeable to several cations, such as Na^+, K^+, and Ca^{2+}, and they have an important function in the passive uptake of Na^+ under high levels of Na^+ concentrations (Demidchik and Maathuis 2007; Wu and Hill 2022). NSCCs are found in the PM of cells of the root system, particularly in the epidermis and cortex, where they facilitate the initial ion uptake from the planting media (Demidchik 2014). These nonselective cation channels fall into two categories: voltage-dependent and voltage-independent. The former is primarily in charge of the flow of Na^+ through plant tissues (Tester and Davenport, 2003). Under saline conditions, the high Na^+ level not only can outcompete K^+ for entry through these channels, leading to K^+ deficiency and ionic imbalance (Wu et al., 2021), but also can depolarize the PM, affecting the activities of other ion transporters and channels (Apse and Blumwald 2007), and it can disrupt the K^+ function and homeostasis by its competition with K^+ for binding sites, which is crucial for cellular functions like processes of osmoregulatory and enzymatic activation and also disrupt enzyme activity, protein synthesis, and other cellular processes (Pottosin and Dobrovinskaya 2014). HKTs play a critical role in mediating Na^+ uptake, especially when external Na^+ concentrations are lower or when plants are exposed to moderate salinity. HKTs are a family of transporters that primarily mediate the uptake of K^+ but can also transport Na^+, depending on the specific HKT subtype and plant species (Venkataraman et al., 2021). Cereal crops contain type I and II HKT transporters, whereas *Arabidopsis* only possesses one Na^+-specific HKT (AtHKT1;1) (Huang et al., 2008; Venkataraman et al., 2021). In rice, the OsHKT1;5 transporter is involved in Na^+ exclusion from shoots by retrieving Na^+ from the xylem sap, thereby reducing Na^+ accumulation in leaves (Kobayashi et al., 2017; Khan et al., 2020). In wheat plants, the TaHKT1:5 transporter has a similar function in the exclusion of Na^+, and it is linked to salinity resistance (Ariyarathna et al., 2014).

The family of CHXs (cation/H^+ exchangers) is a group of MTs that perform a crucial function in ion homeostasis, particularly in the exchange of cations (e.g., K^+, Na^+) for protons (H^+) (Sze and Chanroj 2018). These transporters are part of the larger CPA (cation proton antiporter) superfamily, which includes other well-known transporters like NHXs and KEAs (K^+ efflux antiporters) (Brett et al., 2005). CHX proteins participate in various physiological mechanisms, such as ion transport, pH regulation, and stress response (Pardo et al., 2006; Sze and Chanroj 2018). They function as antiporters,

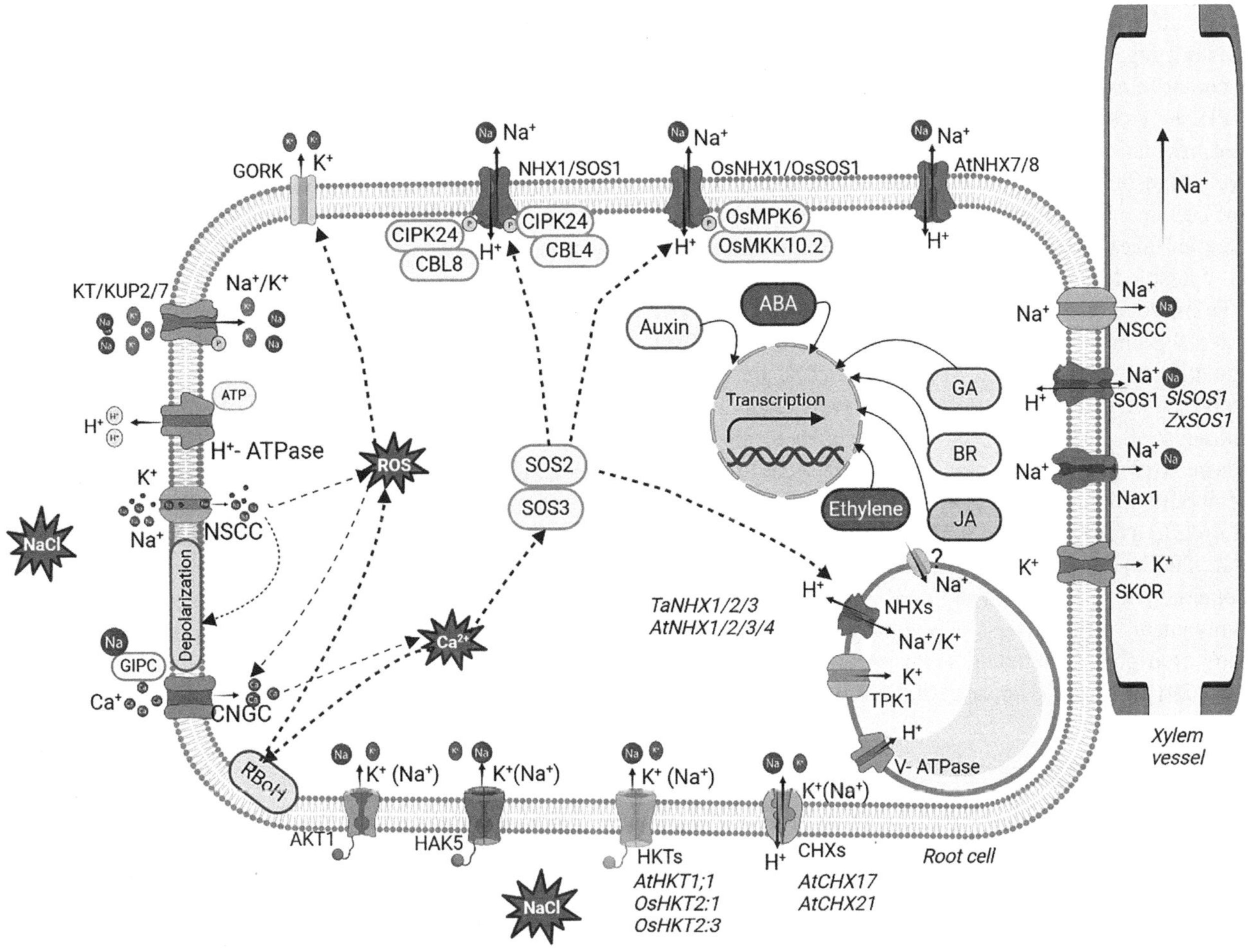

FIGURE 7.4 This overview outlines the mechanisms involved in Na^+ uptake and transport in response to salt stress.

exchanging cations (e.g., K^+, Na^+) for H^+ or protons across membranes, often using the proton gradients produced by H^+ ATPases (Isayenkov et al., 2020). By exchanging cations for H^+, CHX transporters help maintain cellular pH balance (Pittman 2012). As a member of the CHX family in Arabidopsis, CHX21 has been noted to be expressed in the root endodermal cells in salt-stressed plants. Mutants of CHX21 have collected a few Na^+ in their xylem and leaf sap, highlighting that they may be participating in the transfer of Na^+ from the endodermis cells into the xylem tissue (Hall et al., 2006). NHX Transporters, unlike CHX transporters, which are primarily involved in cation/H^+ exchange at the PM or endomembranes, NHX transporters are mainly localized in the vacuolar membrane and mediate Na^+/H^+ antiporter for vacuolar sequestration (Jiang et al., 2010).

Following its passage through the epidermis, Na^+ uptake from the soil is carried radially to the xylem via the apoplastic or symplastic routes (Tester and Davenport, 2003; Keisham et al., 2018). Some of the Na^+ movement via the apoplastic pathway binds to the root apoplast, while the remainder typically reaches the cells of endodermis, where it must cross the PM for continued radial transport (Foster and Miklavcic 2017). Apoplastic barriers, including Casparian strips and suberin lamellae, in the exodermis cells and endodermis layer, regulate the selective movement of ions into and out of the stele. However, in certain cases, Na^+ may pass through the endodermis via a bypass flow, which occurs when the symplastic route is avoided in areas where apoplastic barriers are either underdeveloped or absent (Tester and Davenport, 2003; Krishnamurthy et al., 2009). These regions include young roots, root tips, and sites where lateral roots emerge (Robbins et al., 2014). In rice, bypass flow plays a significant role in Na^+ transport to the shoot, whereas it is not as prominent in *Arabidopsis* (Faiyue 2011; Flam-Shepherd et al., 2018). In *Brassica* plants, it has been recorded that Na^+ can be transported through the symplastic pathway with the control of the endodermis and phi cells (with ingrowth cell walls), which partially block Na^+ transport into the stele via the apoplastic pathway (Fernandez-Garcia et al., 2009).

The compartmentalization of Na^+ ions into the cell vacuole offers an effective way to prevent the effects of Na^+ toxicity in the cell cytosol (Apse and Blumwald 2007). Na^+ transport into the cell vacuole is facilitated by cation/H^+ exchangers, which are powered by the electrochemical gradient mechanism of protons created by enzymes of vacuolar H^+-translocating, including H^+ pyrophosphatase (H^+-PPase) and H^+ ATPases (Sze and Chanroj 2018; Bassil et al., 2019). Vacuolar transport systems include ion channels, $Na^+(K^+)/H^+$ antiporters, Ca^{2+}-ATPase, Ca^{2+}/H^+ exchangers, aquaporins, V-PPase, V-ATPase, and ABC transporters. Their primary contributions include maintaining a high cytosolic K^+/Na^+ ratio, regulating other salinity-responsive processes, sequestering Na^+ and Cl^- into vacuoles, and maintaining a high K^+ level (Mansour 2023). However, additional research and attention are needed because little information is known about how certain vacuolar transporters are regulated and operate under salt stress. Compartmentalization of Na^+ ions in the cell vacuoles takes place in all plant tissues and has an important function in osmotic adjustment and Na^+ detoxification in the cell cytosol (Hasegawa 2013).

To improve resistance to salt stress, plants' vacuolar NHX antiporters have been identified and expressed in heterologous systems. Salinity tolerance is mostly dependent on the MTs encoded by *NHX* genes. These transporters participate in the regulation of Na^+, K^+, and lithium ions (Li^+) movement as well as pH homeostasis (Pu et al., 2021). Vacuolar NHX airporters were reported in several crops, including wheat (Yarra 2019), Arabidopsis (Wang et al., 2007b), and cotton (Akram et al., 2020). In wheat, three vacuolar-localized *NHX* genes, including *TaNHX2 and TaNHX3*, have been identified (Yarra 2019). In Arabidopsis, eight *NHX* gene members were identified, four of them were reported in the vacuolar membrane, that is, *AtNHX1–4*, while *AtNHX7* and *AtNHX8* were reported in the PM (Khan et al., 2018). *AtNHX7*, also referred to as *AtNHX8* and *AtSOS1*, which has been encoded as a Li^+/H^+ antiporter, has been specifically studied (Wang et al., 2007a). Li^+ transport's biological importance is yet unknown; however, *AtNHX4* is found in the vacuole and could cooperate with *AtNHX1* (Yang et al., 2012). Numerous investigations have confirmed the vacuolar transporters' essential involvement in the process of salinity resistance by showing that their activities are enhanced in response to salinity in many crop plants, including switchgrass (Huang et al., 2017), vegetable soybeans (Chen et al., 2014), *Arabidopsis* (Liu et al., 2017), barley (Kargar et al., 2019), and tobacco (Lu et al., 2014).

7.3.2 Long-distance Na^+ transporting

The transport of Na^+ across plant roots and then the movement into xylem vessels occurs via apoplastic and symplastic pathways (Apse and Blumwald 2007). The apoplastic barriers' development, including the suberin lamellae and Casparian band enhances ion selectivity, limiting Na^+ entry into the stele and promoting its efflux (Wei et al., 2021). The deposition of these apoplastic barriers in rice roots decreases Na^+ bypass flow, improving the survival of plants under salinity conditions (Krishnamurthy et al., 2011). The majority of the sodium ions that penetrate the outer root cell portion are probably pumped out again by PM Na^+/H^+ antiporters; the root apex of the *Arabidopsis* protein SOS1 has been shown to perform this role (Hamam 2020). Some of the Na^+ can be sequestered in vacuoles, and the remaining Na^+ can be transported to the shoot via xylem tissue. After the remaining Na^+ ions get to the root stele, they can be transferred into the vessel elements of xylem tissue, and they are involved in the transpiration streams, enabling transport for a long distance from the root system to all parts of the plant shoot system (Nieves-Cordones et al., 2016). Physiological studies have indicated that salinity-tolerant glycophytes and halophytes can activate the movement and transport of Na^+ between the root parts and to the shoot parts, while salinity-susceptible glycophytes seem to restrict Na^+ movement into the transpiration stream to avoid Na^+ buildup in parts of the shoot (Tester and Davenport, 2003). The specific transporter(s) involved in moving Na^+ into or from the vessels of xylem tissue remain unidentified; however, PM Na^+/H^+ antiporters have been suggested to perform this function (Hasegawa 2013). To withstand the salt stress from plants, intracellular Na^+ excretion and vacuolar Na^+ compartmentalization are critical processes that are mediated by the H^+/Na^+ antiporters or exchangers (NHXs) (Sze and Chanroj 2018). It was reported that the transfer of Na^+ over large distances is regulated in plants by the putative PM SOS1 Na^+/H^+ antiporters (Shi et al., 2002). Nevertheless, its precise role may vary depending on the intensity of salt stress and includes extracting Na^+ from the xylem vessel stream under extreme salinity conditions. In tomato plants, PM Na^+/H^+ antiporter SlSOS1 demonstrated the key role that it plays in preserving the homeostasis of ions under a salinity stress environment (Olías et al., 2009). Also, ZxSOS1 in *Zygophyllum xanthoxylum* is crucial for Na^+ distribution and transportation over long distances (Ma et al., 2014). The role of the SOS1 protein in xylem vessel loading of Na^+ is supported by a study in rice, which demonstrated that it controls net Na^+ uptake and distribution, as well as being potentially involved in sodicity stress sensing (El Mahi et al., 2019).

Cotton seedlings were more sensitive to high salt concentrations when *GhNHX1* was silenced, indicating that *GhNHX1* positively controlled cotton's resilience to salt stress (Maryum et al., 2022). Higher plants, such as *Arabidopsis*, barley, and rice, exhibit *HKT* genes that can regulate Na^+ transportation as well as its role in transporting K^+ (Byrt et al., 2007). In rice and *Arabidopsis*, mutations in the *HKT1* gene that cause loss of function lead to an excessive Na^+ accumulation in shoots, making the plant highly sensitive to Na^+ (Hauser and Horie 2010). Heterologous expression studies showed that

OsHKT1;5 acts as a Na^+ transporter, and whole-plant analysis suggested it operates in the parenchyma of root xylem tissue to remove Na^+ from the xylem vessel stream, thus limiting Na^+ buildup in the shoot (Alnayef et al., 2020). In rice, *HKT1;1* and *HKT1;4*, which encode two Na^+ transporters, were shown to be potential *Root Na⁺Content 4* (*RNC4*) candidates. RNA interference revealed decreased expression of both *HKT1;1* and *HKT1;4*, suggesting that *HKT1;1* controls the Na^+ content of shoots and roots and is most likely the causative gene behind *RNC4* (Campbell et al., 2017). *Nax1* and *Nax2* are two identifiedmajor genes for removing Na^+ from leaf blades. *Nax1* and *Nax2* contained lines showed a decrease in the net loading rate of the xylem, causing a decrease in the rate of Na^+ transport from the root system to the shoot system, with lower rates of net intake from the soil or greater rates of translocation in the phloem (James et al., 2006).

Members of the H^+:CHX (monovalent cation exchanger) family are also thought to play a role in Na^+ translocation (Isayenkov and Maathuis 2019). It was shown that AtCHX17/20/21 from Arabidopsis and GmCHX1/19.3 from soybean are involved in the plant's responses to salinity conditions (Qi et al., 2014; Ren et al., 2016; Jia et al., 2017a). Furthermore, cold, osmotic stress, ABA, acidic pH, and Ca^{2+} deficiency all significantly increased the expression of *AtCHX17*, indicating the significance of these transporters in plant adaptation to a variety of abiotic stimuli (Isayenkov et al., 2020). According to the *CHX21* mutant study, seedlings with an *Atchx21* deletion develop more slowly when exposed to 50 and 100 mM NaCl (Hall et al., 2006). According to preliminary research, *CHX21* is most likely encoding a $K^+/Na^+/H^+$ antiporter that contributes to xylem Na^+ (K^+) loading and, as a result, leaf Na^+ buildup. *AtCHX21* is primarily expressed in the root endodermis, and its loss of function decreased Na^+ levels in the xylem sap without impacting Na^+ concentrations in the phloem (Hall et al., 2006). The varying expression levels associated with the tolerant cultivar's increased K^+/Na^+ ratio indicate that CHX11 might play a role in the long-distance transport of Na^+ and/or K^+. Na^+ accumulation in the cell vacuoles could occur either through enhanced uptake or reduced efflux across the vacuolar membrane (Mian et al., 2011; Brini and Masmoudi 2012). The biological roles of CHXs are diverse and may include modification of pH level in the chloroplast stroma, pollen tube development, male fertility, stomata regulation, development of the embryo, and other essential developmental tasks (Isayenkov et al., 2020).

7.4 Regulation of ion uptake and transport under combined abiotic stresses

Research on combined stressors has shown that plants respond differently to each stressor, evoking unique integrated signal networks that may differ from responses to the single stress, especially when the stressors interact antagonistically (Vescio et al., 2022; Jing et al., 2024b). The results of sequential or combined stress can be additive, synergistic, neutral, or occasionally, they might result in new, unforeseen reactions (Anwar et al., 2021). Nonetheless, certain stress combinations also share similar mechanisms and plant responses (Rejeb et al., 2014; Choudhury et al., 2017). These signaling pathways can result in a variety of physiological, morphological, or biochemical responses, contingent on the surrounding stimuli (Rady et al., 2016; Mohamed et al., 2022; Chada et al., 2023; El-Badri et al., 2023). By changing the effectiveness of ion uptake and transport, enzymatic activity, and photosynthetic performance, combined stress results in changes to the main biological processes of plants, causing metabolic imbalance and eventually impacting plant growth and development. Interestingly, some recent research papers have highlighted the complexity of plant responses to multiple simultaneous stressors. As an osmoprotectant, *Arabidopsis*, for instance, accumulates sugar when heat and drought are combined instead of proline, which is normally synthesized under drought alone, as proline may become hazardous when plants are under combined stresses. Approximately 61% of the transcript alterations under double stresses were not anticipated from single-stress responses, according to a transcriptome study conducted across 10 ecotypes of *Arabidopsis* exposed to individual stresses (high temperature, low temperature, high light, salt, and flagellin) and their combinations. This demonstrates how intricately plants react to several stimuli at once.

7.4.1 Effect of combined high temperature and salinity on the ion uptake and transport

The two biggest risks to contemporary agriculture, especially given climate change, are high temperature and salt stress (Rasul et al., 2017). High temperatures cause heat stress and speed up respiration and photosynthetic rates, which result in an excess of ROS that hinders plant development and yield (Moustafa et al., 2014). Several investigations have demonstrated how plants react to heat and salt applied separately, yet in the field, several abiotic stressors frequently happen at the same time. Salinity generates Na^+ toxicity, which can change how root cells absorb macro and micronutrients and has a major negative impact on genes and enzymes, leading to metabolic deficits (Gupta et al., 2021a; Abdelkhalik et al., 2023; Tarfayah et al., 2023). Heat stress causes an increase in the temperature of plant leaves up to 45°C, leading beyond the photosynthetic ideal temperature and increasing respiration rates, and causing a significant vapor pressure difference between the air and plant leaves (Wahid et al., 2007; Mathur et al., 2014; Hassan et al., 2021). Thus it raises the requirement for transpiration from leaves to lower the temperature of leaves, which decreases the water use efficiency. The ion transporters or antiporters found in the PM are inhibited by salt stress (Mansour et al., 2003). Stress tolerance depends on how ion channels change in response to heat and salt exposure (Shabala and Pottosin 2014; Rasul et al., 2017). Signal perception and response are essential for the development of stress resistance at the cellular level. By controlling Na^+ in plant tissues, sodium/potassium antiporters (NHX) and HKT improve salt tolerance (Almeida et al., 2017; El-Badri et al., 2022; Rady et al., 2023). In rice plants, high temperatures and salinity, either separately or in combination, alter their defence mechanisms, survival processes, growth patterns, and reproduction. Osmo-protectants such as proline and antioxidant defense enzymes such as SOD activity rise while photosynthetic efficiency, relative water contents, and protein contents, and plant fresh weight decrease. It was reported that decreasing the heat stress by plant shading can increase salinity tolerance since sun-exposed leaves typically experience higher levels of salinity stress than shaded leaves. During the warmest season, foliar sprays of kaolin clay particle film or artificial shade screens can lower the temperature of citrus leaves and increase water use efficiency (García-Sánchez et al., 2006). Another agronomic strategy to relieve salinity stress impact in tomato crops by spraying the plant leaves with high-quality water has been shown to increase tolerance to soil salinity by reducing the accumulation of harmful ions (Romero-Aranda et al., 2002). In citrus plants, water use has been connected to leaf accumulation in Cl^- ions under heat stress (Syvertsen and Garcia-Sanchez 2014).

Under combined high-temperature and salinity stress, the saline-tolerant rice variety YNU31-2-4 increased *OsHSP18* while downregulating the Na^+/H^+ antiporter *OsNHX*, and the K^+ transporter *OsHKT1;5* (Nahar et al., 2022). Interestingly, the findings of Rivero et al. (2014) have indicated that the combination of heat and salt offers tomato plants a considerable degree of protection from the impacts of salinity, which is in contrast to the anticipated detrimental effect of the stress combination on plant development. They have observed an accumulation of trehalose and glycine betaine, which is linked to ion homeostasis (high K^+ level and lower Na^+/K^+ ratio) and maintained high cell water status and photosynthesis performance under combined stresses as compared with salinity alone. Membrane-bound sensors, such as ion transporters, are activated by saline conditions. This is followed by the activation of second messengers such as Ca^{2+} and signaling cascades, such as TFs, mitogen-activated protein kinases (MAPKs), calcium-dependent protein kinases, heat shock transcription factors (HSFs), SnRK2s, and SOS pathways, which activate response mechanisms (Rasul et al., 2017). These signaling cascades participate in the regulation of ion uptake and transport and activate the production of HSPs, compatible solutes, antioxidants, and ROS scavengers. Response mechanisms (such as TFs, HSFs, chaperones, and osmo-protectants) are activated when heat stress alters membrane characteristics and triggers sensory events (Rasul et al., 2017). It was reported that the TFs, including *WRKY* (Song et al., 2020), *HSFs* (Guo et al., 2016), and signal messengers, have a significant value for plant cellular function in ion homeostasis under salinity and heat stress. Estradiol-dependent induction of HSFA4A plays an important role in conferring salinity stress tolerance and decreasing oxidative stress in *Arabidopsis thaliana* plants. HSFA4A can interact with and activate the MPK3 and MPK6 in plant cells and yeast, leading to *HEAT SHOCK PROTEIN17.6 A* gene transcriptional activation (Pérez-Salamó et al., 2014). Hoang et al. (2012) reported that MPK6 increases *Arabidopsis*'s tolerance to salt via activating and interacting with MYB41. Therefore, the regulation of MPK6 by HSFs under combined salt stress and heat stress suggests that HSFs and MPK6 may function in activating a new signaling pathway in response to the combined salt and heat stresses. In addition, in alfalfa cells, heat shock-activated MAPK (HAMK) was considered one of the first high-temperature responsive MAP kinases (Sangwan et al., 2002; Bharti et al., 2021). The increase of the heat shock protein, HSP70, was seen to coincide with the activation of HAMK in tobacco cells, indicating that HAMK may initiate a heat signaling cascade through *HSP* genes (Saidi et al., 2011). The SOS1 locus is important for the homeostasis of Na^+ and K^+ as it is directly engaged in the movement of sodium ions through PM (El Mahi et al., 2019; Xie et al., 2022). In rice plants, *OsSOS1*, *OsHKT1;5*, and *OsNHX* genes have a pivotal role in plant stress tolerance against salinity and heat conditions (Nahar et al., 2022).

7.4.2 Effect of combined high temperature and drought stress on ion uptake and transport

High temperature or heat stress harms plant development and growth and disrupts several physiological processes, including membrane function, nutrient uptake, respiration, water relations, photosynthesis, and overall plant growth performance and crop quality (Khalil et al., 2018; Peer et al., 2020). Salinity stress causes ionic imbalance, alters physiological traits, decreases water relations, and inhibits photosynthetic machinery performance, decreases reproductive function, and crop yield (El Sabagh et al., 2021; Dos Santos et al., 2022). Combined drought stress and heat stress have more adverse impacts on plants than single stress (Zhou et al., 2017; Handayani and Watanabe 2020). Combined water deficit and high-temperature stress are often complex and have common adverse effects on plants, including a decrease in biomass, membrane damage, denaturation of proteins, inactivation of enzymes specific to respiration and photosynthesis, in addition, a decrease in ion uptake and transport (Silva et al., 2010; Wang et al., 2010; Fahad et al., 2017; Correia et al., 2018). To overcome these adverse conditions, plants have evolved several molecular, physiological, and morphological strategies, including a cascade of signaling, cross-talk, pathways, and specific organelles. ROS, Ca^{2+}, and protein kinases are some of the signaling pathways activated by the combined stresses of drought and heat (Devi et al., 2017; Wang et al., 2017b; Dong et al., 2022). Reduced nutrition availability and absorption may result from heat and water deficit stress-induced ROS production and elevated malondialdehyde concentrations (Fahad et al., 2017; Zhou et al., 2019). Additionally, research in recent years revealed the existence of peroxisomal proteins and their critical function in plant ion transport under multiple environmental stresses, including combined drought and heat stressors (Sousa et al., 2022; Zhang et al., 2022b; Kumar et al., 2024).

Due to the limited research investigations on the impact of combined water deficit and high-temperature stress on ion uptake and transport and their regulatory proteins, it can be concluded that the inhibition in the ion transport could result from the direct heat and drought damages to functional proteins participating in ion uptake and transport in the root cells (Aroca et al., 2012; Dos Santos et al., 2022). More focused heat and drought stress–induced effects on particular ion transport and uptake at the regulatory transcriptional, translational, or posttranslational stages cannot be ruled out, even if the precise underlying mechanism is unclear. Future research should focus specifically on how heat and drought stress affect ion uptake and the transport of genes and proteins.

In recent years, some scientific investigations have shown that the transcriptomic analyses in different plant species including barley (Mahalingam et al., 2022), *Lolium temulentum* (Martin et al., 2021), pearl millet (Jia et al., 2017b), *Sorghum bicolor* (Johnson et al., 2014), wheat (Johnson et al., 2014), and tobacco (Rizhsky et al., 2002) under combined heat and drought stress that revealed both shared and unique gene expression patterns, which differ based on the type of plant, stress duration, and severity. Several signaling pathways and key genes were reported to have an important role in plant growth and development under combined drought and heat stress and participate in stress responses in plants (Xu and Huang 2018; Hosseini et al., 2021; Ju et al., 2021). Interestingly, it was reported that most of the differentially expressed genes altered in response to the application of combined drought and heat stresses were not expressed under single stresses (Hosseini et al., 2021). Plants grown under a combination of high-temperature stress and drought showed increased expressions of at least five distinct transcripts encoding MTs an ABC-type transporter, potassium transporter, Na^{+}/Ca^{2+}-antiporter, aquaporin membrane intrinsic protein, and chloride channel CLC-b (Rizhsky et al., 2004). Johnson et al. (2014) have found that K^{+} transporters of HAK5, AKT1, and AKT2/3 were upregulated (up to eightfold) under combined stress in *Sorghum bicolor* plants.

PM H^{+}ATPases (PM H^{+}ATPase) have an important function in ion uptake regulations and have diverse physiological functions in plants through producing a proton electrochemical gradient, which enables nutrients to enter plant cells and stimulates secondary active transport (Palmgren et al., 2011; Li et al., 2022c; Zeng et al., 2024). Interestingly, a proton gradient (H^{+}) produced by the PM H^{+}-ATPase is necessary for roots to absorb many nutrients (Palmgren 1998; Sondergaard et al., 2004). A cell's PM is one of the main places where PM H^{+} ATPase expression alterations in response to water deficit and high-temperature stresses (Li et al., 2022a; Michalak et al., 2022). High-temperature stress has induced a higher transcriptional level for PM H^{+}ATPase and protein abundance in pea plants, while it has decreased in tomato plants. Previous reports have declared that increased ATPase activity induces heat and drought resistance in plants via enhancing ion uptake and energy status (Ahmed et al., 2018; Zhang et al., 2021; Sousa et al., 2022). By activating the ion-uptake protein found in the root cell's PM, the PM H^{+} ATPase also regulates ion uptake (Zhang et al., 2018a).

Alternative splicing (AS) has an important function in ion homeostasis in several cultivated crops, as it participates in many metabolic regulations in nonstressed conditions and stressed conditions (Dong et al., 2018). The precision and the efficiency of AS during pre-mRNA splicing are pivotal to increasing the functional protein diversity and decreasing the production of harmful proteins (Liu et al., 2024). Liu et al. (2018b) and Ding et al. (2022) have suggested that AS may be important in the combination of high temperature and water deficit stress, as there was a significant increase in the expression of ten spliceosome-related genes.

Moreover, TFs involved in ion homeostasis regulation play important roles in the regulation of stress resistance mechanisms in crops under drought and high-temperature stress. TFs, including bHLH, NAC, MYB, bZIP, WAKY, Arabidopsis transcription activation factor, zinc fingers (ZFs), and AP2/ERF (APETALA 2/ethylene-responsive element binding factor), were reported to participate in ion influx or ion efflux regulations in the cells by opening various membrane ion transporters (Priya et al., 2019; Amin et al., 2021). Two bHLHs, five AP2-EREBPs, three NACs, and five MYBs were identified to have important functions in stress resistance against combined drought and high temperature (Ju et al., 2021). Cyclic nucleotide-gated channels (CNGCs), mostly localized in the PM, have an important function in plant resistance to several environmental stressors such as water deficit and high-temperature stresses (Jha et al., 2016). It was discovered that CGNC16 channels are responsive to drought stress and heat stress, as the *cgnc16* mutant has demonstrated a sensitive trend to $CaCl_2$ exogenous treatment (Liu et al., 2018a).

7.4.3 Combined drought and salinity stress effects on ion uptake and transport

Combined drought and salinity stress can significantly harm ion uptake and transport of plants, leading to complex physiological and biochemical responses. They harm redox hemostasis and cellular energy, accumulate the ABA, ROS, and harmful metabolites, resulting in stomatal closure and a decrease in the transpiration rate, and decreasing photosynthesis, water potential in plant tissues, and growth performance (Abobatta 2020; Ma et al., 2020). These negative effects disrupt the machinery of ion uptake by impairing root function, reducing nutrient availability, and competing the excessive harmful ions with nutrient ions (Dhaliwal et al., 2022; Singhal et al., 2023; Alghamdi 2024; Jing et al., 2024a; Rahim et al., 2024), in addition disrupt machinery of ion transport by altering membrane permeability, interfering the ion loading into vessels of the xylem tissue, limiting phloem flow and redistribution, and disrupting the balance of ions within plant cells and tissues (Ahmad 2014; Agami et al., 2016; Mbarki et al., 2018; Van Zelm et al., 2020).

In the review of literature, there is a lack of research investigations that study the impact of combined water deficit and salinity stresses on ion uptake and transport; therefore, it's a good research area to design research work aimed at deciphering the regulatory networks that control different ion channels and transporters under the combined stress of water deficit and salinity stress. Interestingly, Huda et al. (2013) have reported that *OsACA6*, a P-type IIB Ca^{2+}ATPase pump, has a crucial function in enhancing water deficit and salinity stress resistance in tobacco plants by upregulating stress-responsive genetic elements, TFs related to stress tolerance and signaling components, reducing the accumulation of Na^+ and H_2O_2 levels, and inducing cell membrane stability, photosynthetic efficiency, and osmo-protectant substrates. Under drought stress, the Ca^{2+} efflux mechanism and activity of K^+ uptake and H^+-ATPase pumping could be important signals in barley and tomato seedlings in response to drought stress (Feng et al., 2016; Shabala et al., 2016; Siddiqui et al., 2021). At the seedling stage, the early activation of root hair PM H^+-ATPase in oat (*Avena sativa* L.) genotypes is involved in inducing biosynthesis of major cell osmolytes, which improves the water maintenance system (Gong et al., 2010). The K^+ retention and uptake as a result of rapidly upregulated PM H^+-ATPase, which assists the plant's capacity to sustain a very negative membrane potential to remove Na^+ ions, is essential to enhance *Chenopodium quinoa* and *Atriplex lentiformis* salinity resistance (Bose et al., 2015).

The SOS pathway has a core role in maintaining the dynamics of ion homeostasis and cytoskeleton under stress conditions in plants (Ji et al., 2013; Sathee et al., 2015). The role of *OsSOS2* in combined drought and salinity tolerance was successfully investigated in rice plants (Kumar et al., 2022). Expression of the Na^+/H^+ antiporter gene, *VvNHX1*, has been increased in grapevine rootstocks after 21 days of application of combined water deficit and salinity stresses, indicating its role in ion homeostasis regulations during these two combined stressors (Ullah et al., 2025). In addition, overexpression of *K. caspicaKcNHX1* in *A. thaliana* not only enhanced salinity and drought resistance but also enhanced high-temperature stress at the seedling stage (Wang et al., 2021b).

The high-affinity K^+ transporter HKT1;5, shaker K^+ outward rectifying channel SKOR, high-affinity K^+ transporter HAK5, K^+ channel in *Arabidopsis thaliana* KAT1/2, and inward-rectifier shaker K^+ channel AKT1 are important K^+ channels and transporters that play important roles in ion homeostasis and potassium retention in high salt and drought environments (Ankit et al., 2022; Sonali et al., 2023). Shaker K^+ channels are essential for K^+ movement, distribution, and for plants' reactions to several environmental stresses. In sugar beet plants (*Beta vulgaris* L.), *BvSKOR* overexpression markedly improved transgenic tobacco's resistance to drought and salt stress (Hu et al., 2024).

In rice crops, *OsAKT1* overexpression has shown drought/osmotic stress resistance by improving K^+ concentration in plant tissues (Ahmad et al., 2016b). The *HKT* gene has an important functional role in improving stress resistance in plants, especially in salt- and drought-stressed plants. Overexpression of rice *OsHKT1;5* genes in *Arabidopsis thaliana* showed highly significant tolerance to both combined drought and salinity stresses through increasing K^+ ions and improving the activities of antioxidants (Ullah et al., 2025).

As the stomatal regulation participates in the salinity and drought stress resistance in plants, KAT1 and KAT2 are important transporters in the regulation of potassium transport in the stomatal guard cells and participate in the stress resistance against salinity and drought stress in plants (Obata et al., 2007; Mulet et al., 2023). KAT1 overexpression rescued the stomatal phenotype in the *aks1 aks2-1* double mutants, which lack two bHLH TFs (AKS1 and AKS2) crucial for potassium uptake when the stomata open. AKS TFs promote stomatal opening in wild-type plants by activating genes like *KAT1*, which encode inward-rectifying K^+ channels. However, ABA represses these TFs through phosphorylation, inhibiting their function and thereby reducing stomatal opening (Takahashi et al., 2013). This demonstrates the role of KAT1 and the interplay of TFs and ABA in regulating stomatal behavior and stress tolerance.

It was reported that HAK5 is primarily in charge of K^+ absorption under stress conditions, which cause a decrease in K^+ levels, and 27 genes of *HvHAKs* in barley have been identified to participate in the combined stressors, that is, drought, salinity, and K^+ deficiency (Cai et al., 2021). This information has agronomic ramifications; for instance, it has been shown that HAK5 limits the growth of grafted cucumbers under salt stress (Peng et al., 2023).

7.5 Applications and future directions

In this compilation, we have reviewed the latest developments related to K^+, Na^+ membrane uptake and transport under several abiotic stresses. The encouraging findings so far significantly support the investigation of ion uptake and transport systems that mediate ion homeostasis, including K^+ and Na^+ fluxes within plant systems, and ought to be considered as intriguing goals in biotechnology for the creation of new genotypes that can be able to withstand abiotic stressors. These mechanisms of uptake and systems of transporting usually induce complex signaling networks that have multiple roles within the plant. To further understand combined resistance processes, more scientific investigations are needed to define other ion channels and transporters that participate in the regulation of ion homeostasis under combined stress conditions and evaluate the potential benefits of K^+ and Na^+ at a particular subcellular level, cellular, or tissue. More understanding of the crosstalk between signaling networks controlling the transport and uptake of these ions and plant hormones, including auxins and ABA, during combined abiotic stress conditions is crucial in further research projects. These hormones have an important function in regulating plant cell growth, particularly in the root system, and the transport of K^+, Ca^{2+}, Cl^-, and Na^+. New strategies emerging from such studies could significantly improve agricultural productivity under the challenge of climate change in the 21st century.

References

Abbasi, H., Jamil, M., Haq, A., Ali, S., Ahmad, R., Malik, Z., et al., 2016. Salt stress manifestation on plants, mechanism of salt tolerance and potassium role in alleviating it: a review. Zemdirbyste-Agriculture 103, 229–238.

Abdelkhalik, A., El-Mageed, A., Taia, A., Mohamed, I.A.A., Semida, W.M., Al-Elwany, O.A.A.I., et al., 2023. Soil application of effective microorganisms and nitrogen alleviates salt stress in hot pepper (*Capsicum annum* L.) plants. Frontiers in Plant Science 13, 1079260.

Abobatta, W.F., 2020. Plant responses and tolerance to combined salt and drought stress. In: Hasanuzzaman, M., Tanveer, M. (Eds.), Salt and Drought Stress Tolerance in Plants. Signaling and Communication in Plants. Springer, Cham, pp. 17–52. Available at https://doi.org/10.1007/978-3-030-40277-8_2.

Abou-Sreea, A.I.B., Roby, M.H.H., Mahdy, H.A.A., Abdou, N.M., El-Tahan, A.M., El-Saadony, M.T., et al., 2022. Improvement of selected morphological, physiological, and biochemical parameters of roselle (*Hibiscus sabdariffa* L.) grown under different salinity levels using potassium silicate and Aloe saponaria extract. Plants 11, 497.

Aftab, T., Hakeem, K.R., 2022. Plant Abiotic Stress Physiology: Volume 2: Molecular Advancements. Apple Academic Press.

Agami, R., Medani, R.A., Abd El-Mola, I.A., Taha, R.S., 2016. Exogenous application with plant growth promoting rhizobacteria (PGPR) or proline induces stress tolerance in basil plants (*Ocimum basilicum* L.) exposed to water stress. International Journal of Environmental & Agriculture Research 2, 78–92.

Ahanger, M.A., Ahmad, P., 2019. Chapter 17 - Role of mineral nutrients in abiotic stress tolerance: revisiting the associated signaling mechanisms. In: Khan, M.I.R., Reddy, P.S., Ferrante, A., Khan, N.A. (Eds.), Plant Signaling Molecules: Role and Regulation Under Stressful Environments. Woodhead Publishing, pp. 269–285. Available at https://doi.org/10.1016/B978-0-12-816451-8.00016-2.

Ahanger, M.A., Tomar, N.S., Tittal, M., Argal, S., Agarwal, R.M., 2017. Plant growth under water/salt stress: ROS production; antioxidants and significance of added potassium under such conditions. Physiological and Molecular Plant Pathology 23, 731–744.

Ahmad I., 2014. The role of cation channels in abiotic stress resistance in rice. PhD thesis, University of York.

Ahmad, P., Abdel Latef, A.A., Abd_Allah, E.F., Hashem, A., Sarwat, M., Anjum, N.A., et al., 2016a. Calcium and potassium supplementation enhanced growth, osmolyte secondary metabolite production, and enzymatic antioxidant machinery in cadmium-exposed chickpea (*Cicer arietinum* L.). Frontiers in Plant Science 7, 513.

Ahmad, I., Mian, A., Maathuis, F.J.M., 2016b. Overexpression of the rice AKT1 potassium channel affects potassium nutrition and rice drought tolerance. Journal of Experimental Botany 67, 2689–2698.

Ahmad, W., Ayyub, C.M., Shehzad, M.A., Ziaf, K., Ijaz, M., Sher, A., et al., 2019. Supplemental potassium mediates antioxidant metabolism, physiological processes, and osmoregulation to confer salt stress tolerance in cabbage (*Brassicaoleracea* L.). Horticulture, Environment, and Biotechnology 60, 853–869.

Ahmed, I.M., Nadira, U.A., Qiu, C.-W., Cao, F., Zhang, G., Holford, P., et al., 2018. Tolerance to drought, low pH and Al combined stress in Tibetan wild barley is associated with improvement of ATPase and modulation of antioxidant defense system. International Journal of Molecular Sciences 19, 3553.

Aibara, I., Miwa, K., 2014. Strategies for optimization of mineral nutrient transport in plants: multilevel regulation of nutrient-dependent dynamics of root architecture and transporter activity. Plant & Cell Physiology 55, 2027–2036.

Akhtar, M.N., Akhtar, M.W., Rahi, A.A., ul Haq, T., 2023. Enhancing water use efficiency by using potassium-efficient cotton cultivars based on morphological and biochemical characteristic. Best Crop Management and Processing Practices for Sustainable Cotton Production. IntechOpen. Available at 10.5772/intechopen.112606.

Akram, U., Song, Y., Liang, C., Abid, M.A., Askari, M., Myat, A.A., et al., 2020. Genome-wide characterization and expression analysis of *NHX* gene family under salinity stress in *Gossypium barbadense* and its comparison with *Gossypium hirsutum*. Genes (Basel) 11, 803.

Alghamdi, S.A., 2024. Drought and salinity effects on plant growth: a comprehensive review. SABRAO Journal of Breeding and Genetics 56 (6), 2331–2340. Available at http://doi.org/10.54910/sabrao2024.56.6.14.

Almeida, D.M., Oliveira, M.M., Saibo, N.J.M., 2017. Regulation of Na^+ and K^+ homeostasis in plants: towards improved salt stress tolerance in crop plants. Genetics and Molecular Biology 40, 326–345.

Alnayef, M., Solis, C., Shabala, L., Ogura, T., Chen, Z., Bose, J., et al., 2020. Changes in expression level of *OsHKT1; 5* alters activity of membrane transporters involved in K^+ and Ca^{2+} acquisition and homeostasis in salinized rice roots. International Journal of Molecular Sciences 21, 4882.

Al-Zahrani, H.S., Alharby, H.F., Fahad, S., 2022. Antioxidative defense system, hormones, and metabolite accumulation in different plant parts of two contrasting rice cultivars as influenced by plant growth regulators under heat stress. Frontiers in Plant Science 13, 911846.

Amin, I., Rasool, S., Mir, M.A., Wani, W., Masoodi, K.Z., Ahmad, P., 2021. Ion homeostasis for salinity tolerance in plants: a molecular approach. Physiologia Plantarum 171, 578–594.

Amir, R., Munir, F., Khan, M., Iqbal, T., 2019. Use of plant hormones for the improvement of plant growth and production under salt stress. In: Akhtar, M. (Ed.), Salt Stress, Microbes, and Plant Interactions: Causes and Solution. Springer, Singapore, pp. 59–90. Available at https://doi.org/10.1007/978-981-13-8801-9_3.

Anil Kumar, S., Kaniganti, S., Hima Kumari, P., Sudhakar Reddy, P., Suravajhala, P., P, S., et al., 2024. Functional and biotechnological cues of potassium homeostasis for stress tolerance and plant development. Biotechnology & Genetic Engineering Reviews 40, 3527–3570.

Ankit, A., Kamali, S., Singh, A., 2022. Genomic & structural diversity and functional role of potassium (K^+) transport proteins in plants. International Journal of Biological Macromolecules 208, 844–857.

Anwar, K., Joshi, R., Dhankher, O.P., Singla-Pareek, S.L., Pareek, A., 2021. Elucidating the response of crop plants towards individual, combined and sequentially occurring abiotic stresses. International Journal of Molecular Sciences 22, 6119.

Apse, M.P., Blumwald, E., 2007. Na+ transport in plants. FEBS Letters 581 (12), 2247–2254. Available at https://doi.org/10.1016/j.febslet.2007.04.014.

Arif, Y., Singh, P., Siddiqui, H., Bajguz, A., Hayat, S., 2020. Salinity induced physiological and biochemical changes in plants: an omic approach towards salt stress tolerance. Plant Physiology and Biochemistry: PPB / Societe Francaise de Physiologie Vegetale 156, 64–77.

Ariyarathna, H.A.C.K., Ul-Haq, T., Colmer, T.D., Francki, M.G., 2014. Characterization of the multigene family *TaHKT 2; 1* in bread wheat and the role of gene members in plant Na^+ and K^+ status. BMC Plant Biology 14, 159. Available at 10.1186/1471-2229-14-159.

Aroca, R., Porcel, R., Ruiz-Lozano, J.M., 2012. Regulation of root water uptake under abiotic stress conditions. Journal of Experimental Botany 63, 43–57.

Assaha, D.V.M., Ueda, A., Saneoka, H., Al-Yahyai, R., Yaish, M.W., 2017. The role of Na^+ and K^+ transporters in salt stress adaptation in glycophytes. Frontiers in Physiology 8, 509.

Asthir, B., 2015. Protective mechanisms of heat tolerance in crop plants. Journal of Plant Interactions 10, 202–210.

Awuchi, C.G., Igwe, V.S., Amagwula, I.O., Echeta, C.K., 2020. Health benefits of micronutrients (vitamins and minerals) and their associated deficiency diseases: a systematic review. International Journal of Food Sciences 3, 1–32.

Azeem, F., Ijaz, U., Ali, M.A., Hussain, S., Zubair, M., Manzoor, H., et al., 2022. Genome-wide identification and expression profiling of potassium transport-related genes in vigna radiata under abiotic stresses. Plants 11 (1), 2. Available at https://doi.org/10.3390/plants11010002.

Banerjee, A., Roychoudhury, A., 2017. Epigenetic regulation during salinity and drought stress in plants: histone modifications and DNA methylation. Plant Gene 11, 199–204.

Barzana, G., Rios, J.J., Lopez-Zaplana, A., Nicolas-Espinosa, J., Yepes-Molina, L., Garcia-Ibañez, P., et al., 2021. Interrelations of nutrient and water transporters in plants under abiotic stress. Physiologia Plantarum 171, 595–619.

Bashar, K.K., 2018. Hormone dependent survival mechanisms of plants during post-waterlogging stress. Plant Signaling & Behavior 13, e1529522.

Bassil, E., Zhang, S., Gong, H., Tajima, H., Blumwald, E., 2019. Cation specificity of vacuolar NHX-type cation/H^+ antiporters. Plant Physiology 179, 616–629.

Batool, M., El-Badri, A.M., Wang, Z., Mohamed, I.A.A., Yang, H., Ai, X., et al., 2022. Rapeseed morpho-physio-biochemical responses to drought stress induced by PEG-6000. Agronomy 12, 579.

Baweja, P., Kumar, G., 2020. Abiotic stress in plants: an overview. In: Giri, B., Sharma, M.P. (Eds.), Plant Stress Biology: Strategies and Trends. Springer, Singapore, pp. 1–15. Available at https://doi.org/10.1007/978-981-15-9380-2_1.

Belhadj Slimen, I., Najar, T., Ghram, A., Dabbebi, H., Ben Mrad, M., Abdrabbah, M., 2014. Reactive oxygen species, heat stress and oxidative-induced mitochondrial damage. A review. International Journal of Hyperthermia 30, 513–523.

Bharti, J., Sahil, Mehta, S., Ahmad, S., Singh, B., Padhy, A.K., et al., 2021. Mitogen-activated protein kinase, plants, and heat stress. In: Husen, A. (Ed.), Harsh Environment and Plant Resilience: Molecular and Functional Aspects. Springer, Cham, pp. 323–354. Available at https://doi.org/10.1007/978-3-030-65912-7_13.

Bhattacharya, A., 2022. Low temperature stress and plant-water relationship: a review. Physiological Processes in Plants Under Low Temperature Stress. Springer, Singapore, pp. 107–197. Available at https://doi.org/10.1007/978-981-16-9037-2_2.

Bielach, A., Hrtyan, M., Tognetti, V.B., 2017. Plants under stress: involvement of auxin and cytokinin. International Journal of Molecular Sciences 18, 1427.

Bose, J., Rodrigo-Moreno, A., Lai, D., Xie, Y., Shen, W., Shabala, S., 2015. Rapid regulation of the plasma membrane H^+-ATPase activity is essential to salinity tolerance in two halophyte species, *Atriplex lentiformis* and *Chenopodium quinoa*. Annals of Botany 115, 481–494.

Brett, C.L., Donowitz, M., Rao, R., 2005. Evolutionary origins of eukaryotic sodium/proton exchangers. American Journal of Physiology-Cell Physiology 288, C223–C239.

Brini, F., Masmoudi, K., 2012. Ion transporters and abiotic stress tolerance in plants. International Scholarly Research Notices 2012, 927436.

Byrt, C.S., Platten, J.D., Spielmeyer, W., James, R.A., Lagudah, E.S., Dennis, E.S., et al., 2007. HKT1;5-like cation transporters linked to Na^+ exclusion loci in wheat, Nax2 and Kna1. Plant Physiology 143, 1918–1928.

Cai, K., Gao, H., Wu, X., Zhang, S., Han, Z., Chen, X., et al., 2019. The ability to regulate transmembrane potassium transport in root is critical for drought tolerance in barley. International Journal of Molecular Sciences 20, 4111.

Cai, K., Zeng, F., Wang, J., Zhang, G., 2021. Identification and characterization of *HAK/KUP/KT* potassium transporter gene family in barley and their expression under abiotic stress. BMC Genomics 22, 317.

Campbell, M.T., Bandillo, N., Al Shiblawi, F.R.A., Sharma, S., Liu, K., Du, Q., et al., 2017. Allelic variants of OsHKT1; 1 underlie the divergence between indica and japonica subspecies of rice (*Oryza sativa*) for root sodium content. PLoS Genetics 13, e1006823.

Chada, S., Asiedu, S., Ofoe, R., 2023. An overview of plant morpho-physiology, biochemicals, and metabolic pathways under water stress. Horticulture International Journal 7, 115–125.

Chaudhry, S., Sidhu, G.P.S., 2022. Climate change regulated abiotic stress mechanisms in plants: a comprehensive review. Plant Cell Reports 41, 1–31.

Che, Y., Yao, T., Wang, H., Wang, Z., Zhang, H., Sun, G., et al., 2022. Potassium ion regulates hormone, Ca^{2+} and H_2O_2 signal transduction and antioxidant activities to improve salt stress resistance in tobacco. Plant Physiology and Biochemistry: PPB / Societe Francaise de Physiologie Vegetale 186, 40–51.

Chen, G.-H., Yan, W., Yang, L.F., Gai, J.Y., Zhu, Y.L., 2014. Overexpression of *StNHX1*, a novel vacuolar Na^+/H^+ antiporter gene from *Solanum torvum*, enhances salt tolerance in transgenic vegetable soybean. Horticulture, Environment, and Biotechnology 55, 213–221.

Chen, G., Liu, C., Gao, Z., Zhang, Y., Jiang, H., Zhu, L., et al., 2017. OsHAK1, a high-affinity potassium transporter, positively regulates responses to drought stress in rice. Frontiers in Plant Science 8, 1885.

Choudhury, F.K., Rivero, R.M., Blumwald, E., Mittler, R., 2017. Reactive oxygen species, abiotic stress and stress combination. The Plant Journal: for Cell and Molecular Biology 90, 856–867.

Chérel, I., Gaillard, I., 2019. The complex fine-tuning of K^+ fluxes in plants in relation to osmotic and ionic abiotic stresses. International Journal of Molecular Sciences 20, 715.

Correia, B., Hancock, R.D., Amaral, J., Gomez-Cadenas, A., Valledor, L., Pinto, G., 2018. Combined drought and heat activates protective responses in *Eucalyptus globulus* that are not activated when subjected to drought or heat stress alone. Frontiers in Plant Science 9, 819.

Dajic, Z., 2006. Salt stress. Physiology and Molecular Biology of Stress Tolerance in Plants. Springer, pp. 41–99.

Damalas, C.A., Koutroubas, S.D., 2024. Potassium supply for improvement of cereals growth under drought: a review. Agronomy Journal 116, 3368–3382.

Delgado, A., Quemada, M., Mateos, L., Villalobos, F.J., 2024. Fertilization with phosphorus, potassium, and other nutrients. Principles of Agronomy for Sustainable Agriculture. Springer, pp. 415–437.

Demidchik, V., 2014. Mechanisms and physiological roles of K^+ efflux from root cells. Journal of Plant Physiology 171, 696–707.

Demidchik, V., Maathuis, F.J.M., 2007. Physiological roles of nonselective cation channels in plants: from salt stress to signalling and development. The New Phytologist 175, 387–404.

Demidchik, V., Straltsova, D., Medvedev, S.S., Pozhvanov, G.A., Sokolik, A., Yurin, V., 2014. Stress-induced electrolyte leakage: the role of K^+-permeable channels and involvement in programmed cell death and metabolic adjustment. Journal of Experimental Botany 65, 1259–1270.

Devi, E.L., Kumar, S., Singh, T.B., Sharma, S.K., Beemrote, A., Devi, C.P., et al., 2017. Adaptation strategies and defence mechanisms of plants during environmental stress. In: Ghorbanpour, M., Varma, A. (Eds.), Medicinal Plants and Environmental Challenges. Springer, Cham, pp. 359–413. Available at https://doi.org/10.1007/978-3-319-68717-9_20.

Dhaliwal, S.S., Sharma, V., Shukla, A.K., 2022. Impact of micronutrients in mitigation of abiotic stresses in soils and plants—a progressive step toward crop security and nutritional quality. Advances in Agronomy 173, 1–78.

Ding, Y.Q., Fan, K., Wang, Y., Fang, W.P., Zhu, X.J., Chen, L., et al., 2022. Drought and heat stress-mediated modulation of alternative splicing in the genes involved in biosynthesis of metabolites related to tea quality. Molecular Biology 56, 257–268.

Dong, C., He, F., Berkowitz, O., Liu, J., Cao, P., Tang, M., et al., 2018. Alternative splicing plays a critical role in maintaining mineral nutrient homeostasis in rice (*Oryza sativa*). The Plant Cell 30, 2267–2285.

Dong, Q., Wallrad, L., Almutairi, B.O., Kudla, J., 2022. Ca^{2+} signaling in plant responses to abiotic stresses. Journal of Integrative Plant Biology 64, 287–300.

Drechsler, N., Zheng, Y., Bohner, A., Nobmann, B., von Wirén, N., Kunze, R., et al., 2015. Nitrate-dependent control of shoot K homeostasis by the nitrate transporter1/peptide transporter family member NPF7. 3/NRT1. 5 and the stelar K^+ outward rectifier SKOR in *Arabidopsis*. Plant Physiology 169, 2832–2847.

Dreyer, I., Gomez-Porras, J.L., Riedelsberger, J., 2017. The potassium battery: a mobile energy source for transport processes in plant vascular tissues. The New Phytologist 216, 1049–1053.

Duan, H.-R., Ma, Q., Zhang, J.-L., Hu, J., Bao, A.-K., Wei, L., et al., 2015. The inward-rectifying K^+ channel SsAKT1 is a candidate involved in K^+ uptake in the halophyte *Suaeda salsa* under saline condition. Plant and Soil 395, 173–187.

Egamberdieva, D., Wirth, S., Abd_Allah, E.F., 2018. Plant hormones as key regulators in plant-microbe interactions under salt stress. In: Egamberdieva, D., Ahmad, P. (Eds.), Plant Microbiome: Stress Response. Microorganisms for Sustainability, vol 5. Springer, Singapore, pp. 165–182. Available at https://doi.org/10.1007/978-981-10-5514-0_7.

El-Badri, A.M., Batool, M., Mohamed, I.A.A., Wang, Z., Wang, C., Tabl, K.M., et al., 2022. Mitigation of the salinity stress in rapeseed (*Brassica napus* L.) productivity by exogenous applications of bio-selenium nanoparticles during the early seedling stage. Environmental Pollution (Barking, Essex: 1987) 310, 119815.

El-Badri, A.M., Batool, M., Mohamed, I.A.A., Agami, R., Elrewainy, I.M., Wang, B., et al., 2023. Role of Phytomelatonin in Promoting Ion Homeostasis During Salt Stress. In: Mukherjee, S., Corpas, F.J. (Eds.), Melatonin: Role in Plant Signaling, Growth and Stress Tolerance. Plant in Challenging Environments, vol 4. Springer International Publishing, Cham, pp. 313–342. Available at https://doi.org/10.1007/978-3-031-40173-2_17.

Fahad, S., Bajwa, A.A., Nazir, U., Anjum, S.A., Farooq, A., Zohaib, A., et al., 2017. Crop production under drought and heat stress: plant responses and management options. Frontiers in Plant Science 8, 1147.

Faiyue B., 2011. Bypass flow and sodium transport in rice (*Oryza Sativa* L.). Available at https://hdl.handle.net/10779/uos.23315000.

Fedorova, E.E., Coba De La Penã, T., Lara-Dampier, V., Trifonova, N.A., Kulikova, O., Pueyo, J.J., et al., 2021. Potassium content diminishes in infected cells of *Medicago truncatula* nodules due to the mislocation of channels MtAKT1 and MtSKOR/GORK. Journal of Experimental Botany 72, 1336–1348.

Feng, X., Liu, W., Zeng, F., Chen, Z., Zhang, G., Wu, F., 2016. K^+ uptake, H^+-ATPase pumping activity and Ca^{2+} efflux mechanism are involved in drought tolerance of barley. Environmental and Experimental Botany 129, 57–66.

Feng, X., Liu, W., Cao, F., Wang, Y., Zhang, G., Chen, Z.-H., et al., 2020. Overexpression of *HvAKT 1* improves drought tolerance in barley by regulating root ion homeostasis and ROS and NO signaling. Journal of Experimental Botany 71, 6587–6600.

Feraille, E., Doucet, A., 2001. Sodium-potassium-adenosinetriphosphatase-dependent sodium transport in the kidney: hormonal control. Physiological Reviews 81, 345–418.

Fernandez-Garcia, N., Lopez-Perez, L., Hernandez, M., Olmos, E., 2009. Role of phi cells and the endodermis under salt stress in Brassica oleracea. The New Phytologist 181, 347–360.

Flam-Shepherd, R., Huynh, W.Q., Coskun, D., Hamam, A.M., Britto, D.T., Kronzucker, H.J., 2018. Membrane fluxes, bypass flows, and sodium stress in rice: the influence of silicon. Journal of Experimental Botany 69, 1679–1692.

Foster, K.J., Miklavcic, S.J., 2017. A comprehensive biophysical model of ion and water transport in plant roots. I. Clarifying the roles of endodermal barriers in the salt stress response. Frontiers in Plant Science 8, 1326.

Fuchs, I., Stölzle, S., Ivashikina, N., Hedrich, R., 2005. Rice K^+ uptake channel OsAKT1 is sensitive to salt stress. Planta 221, 212–221.

García-Sánchez, F., Perez-Perez, J.G., Botía, P., Martínez, V., 2006. The response of young mandarin trees grown under saline conditions depends on the rootstock. European Journal of Agronomy 24, 129–139.

Gierth, M., Mäser, P., Schroeder, J.I., 2005. The potassium transporter AtHAK5 functions in K^+ deprivation-induced high-affinity K^+ uptake and AKT1 K^+ channel contribution to K^+ uptake kinetics in *Arabidopsis* roots. Plant Physiology 137, 1105–1114.

Gill, M.B., Zeng, F., Shabala, L., Böhm, J., Zhang, G., Zhou, M., et al., 2018. The ability to regulate voltage-gated K^+-permeable channels in the mature root epidermis is essential for waterlogging tolerance in barley. Journal of Experimental Botany 69, 667–680.

Gill, R.A., Ahmar, S., Ali, B., Saleem, M.H., Khan, M.U., Zhou, W., et al., 2021. The role of membrane transporters in plant growth and development, and abiotic stress tolerance. International Journal of Molecular Sciences 22, 12792.

Gong, D.-S., Xiong, Y.-C., Ma, B.-L., Wang, T.-M., Ge, J.-P., Qin, X.-L., et al., 2010. Early activation of plasma membrane H+-ATPase and its relation to drought adaptation in two contrasting oat (*Avena sativa* L.) genotypes. Environmental and Experimental Botany 69, 1–8.

Grigorova, B., Vaseva, I., Demirevska, K., Feller, U., 2011. Combined drought and heat stress in wheat: changes in some heat shock proteins. Biologia Plantarum 55, 105–111.

Guo, M., Liu, J.-H., Ma, X., Luo, D.-X., Gong, Z.-H., Lu, M.-H., 2016. The plant heat stress transcription factors (HSFs): structure, regulation, and function in response to abiotic stresses. Frontiers in Plant Science 7, 114.

Guo, Y., Zhu, C., Tian, Z., 2023. Overexpression of *KvCHX* enhances salt tolerance in *Arabidopsis thaliana* seedlings. Current Issues in Molecular Biology 45, 9692–9708.

Gupta, S., Schillaci, M., Walker, R., Smith, P.M.C., Watt, M., Roessner, U., 2021a. Alleviation of salinity stress in plants by endophytic plant-fungal symbiosis: current knowledge, perspectives and future directions. Plant and Soil 461, 219–244.

Gupta, A., Shaw, B.P., Sahu, B.B., Munns, R., 2021b. Post-translational regulation of the membrane transporters contributing to salt tolerance in plants. Functional Plant Biology: FPB 48, 1199–1212.

Hall, D., Evans, A.R., Newbury, H.J., Pritchard, J., 2006. Functional analysis of CHX21: a putative sodium transporter in *Arabidopsis*. Journal of Experimental Botany 57, 1201–1210.

Hamam, A., 2020. On the Understanding of Sodium Transport and its Role in Salinity Stress in Plants. University of Toronto, Canada.

Hamamoto, S., Horie, T., Hauser, F., Deinlein, U., Schroeder, J.I., Uozumi, N., 2015. HKT transporters mediate salt stress resistance in plants: from structure and function to the field. Current Opinion in Biotechnology 32, 113–120.

Handayani, T., Watanabe, K., 2020. The combination of drought and heat stress has a greater effect on potato plants than single stresses. Plant, Soil and Environment 66 (4), 175–182. Available at 10.17221/126/2020-PSE.

Hao, S., Wang, Y., Yan, Y., Liu, Y., Wang, J., Chen, S., 2021. A review on plant responses to salt stress and their mechanisms of salt resistance. Horticulturae 7, 132.

Hasanuzzaman, M., Nahar, K., Alam, M.M., Roychowdhury, R., Fujita, M., 2013. Physiological, biochemical, and molecular mechanisms of heat stress tolerance in plants. International Journal of Molecular Sciences 14, 9643–9684.

Hasanuzzaman, M., Bhuyan, M.H.M.B., Nahar, K., Hossain, M.S., Mahmud, J., Al, et al., 2018. Potassium: a vital regulator of plant responses and tolerance to abiotic stresses. Agronomy 8, 31.

Hasegawa, P.M., 2013. Sodium (Na^+) homeostasis and salt tolerance of plants. Environmental and Experimental Botany 92, 19–31.

Hassan, M.U., Aamer, M., Chattha, M.U., Ullah, M.A., Sulaman, S., Nawaz, M., et al., 2017. The role of potassium in plants under drought stress: mini review. Journal of Basic & Applied Sciences 13, 268–271.

Hassan, M.U., Chattha, M.U., Khan, I., Chattha, M.B., Barbanti, L., Aamer, M., et al., 2021. Heat stress in cultivated plants: nature, impact, mechanisms, and mitigation strategies—a review. Plant Biosystems -An International Journal Dealing with all Aspects of Plant Biology 155, 211–234.

Hauser, F., Horie, T., 2010. A conserved primary salt tolerance mechanism mediated by HKT transporters: a mechanism for sodium exclusion and maintenance of high K^+/Na^+ ratio in leaves during salinity stress. Plant, Cell & Environment 33, 552–565.

Hiya, H.J., Nakashima, Y., Takeuchi, A., Nakamura, T., Nakamura, Y., Murata, Y., et al., 2024. Outward-rectifying potassium channels GORK and SKOR function in regulation of root growth under salt stress in *Arabidopsis thaliana*. Journal of Plant Physiology 302, 154322.

Hoang, M.H.T., Nguyen, X.C., Lee, K., Kwon, Y.S., Pham, H.T.T., Park, H.C., et al., 2012. Phosphorylation by AtMPK6 is required for the biological function of AtMYB41 in *Arabidopsis*. Biochemical and Biophysical Research Communications 422, 181–186.

Horváth, I., Multhoff, G., Sonnleitner, A., Vígh, L., 2008. Membrane-associated stress proteins: more than simply chaperones. Biochimica et Biophysica Acta - Biomembranes 1778, 1653–1664.

Hosseini, S.Z., Ismaili, A., Nazarian-Firouzabadi, F., Fallahi, H., Nejad, A.R., Sohrabi, S.S., 2021. Dissecting the molecular responses of lentil to individual and combined drought and heat stresses by comparative transcriptomic analysis. Genomics 113, 693–705.

Hu, Y.-D., Ren, P.-P., Wei, M., Batoko, H., Wu, G.-Q., 2024. Genome-wide identification of *shaker* K^+ *channel* gene family in sugar beet (*Beta vulgaris* L.) and function of *BvSKOR* in response to salt and drought stresses. Environmental and Experimental Botany 228, 106034.

Huang, B., 2022. Mechanisms of plant resistance to waterlogging. In: Basra, A.S., Basra, R.K. (Eds.), Mechanisms of Environmental Stress Resistance in Plants. Routledge, London, pp. 59–81.

Huang, S., Spielmeyer, W., Lagudah, E.S., Munns, R., 2008. Comparative mapping of *HKT* genes in wheat, barley, and rice, key determinants of Na^+ transport, and salt tolerance. Journal of Experimental Botany 59, 927–937.

Huang, Y., Guan, C., Liu, Y., Chen, B., Yuan, S., Cui, X., et al., 2017. Enhanced growth performance and salinity tolerance in transgenic switchgrass via overexpressing vacuolar Na^+ (K^+)/H^+ antiporter gene (*PvNHX1*). Frontiers in Plant Science 8, 458.

Huda, K.M., Banu, M.S., Garg, B., Tula, S., Tuteja, R., Tuteja, N., 2013. OsACA6, a P-type IIB Ca^{2+} ATPase promotes salinity and drought stress tolerance in tobacco by ROS scavenging and enhancing the expression of stress-responsive genes. The Plant Journal: for Cell and Molecular Biology 76, 997–1015.

Imtiaz, H., Mir, A.R., Corpas, F.J., Hayat, S., 2023. Impact of potassium starvation on the uptake, transportation, photosynthesis, and abiotic stress tolerance. Plant Growth Regulation 99, 429–448.

Isayenkov, S.V., Maathuis, F.J.M., 2019. Plant salinity stress: many unanswered questions remain. Frontiers in Plant Science 10.

Isayenkov, S.V., Dabravolski, S.A., Pan, T., Shabala, S., 2020. Phylogenetic diversity and physiological roles of plant monovalent cation/H^+ antiporters. Frontiers in Plant Science 11, 573564.

Jalakas, P., Nuhkat, M., Vahisalu, T., Merilo, E., Brosché, M., Kollist, H., 2021. Combined action of guard cell plasma membrane rapid-and slow-type anion channels in stomatal regulation. Plant Physiology 187, 2126–2133.

James, R.A., Davenport, R.J., Munns, R., 2006. Physiological characterization of two genes for Na^+ exclusion in durum wheat, *Nax1* and *Nax2*. Plant Physiology 142, 1537–1547.

Jha, S.K., Sharma, M., Pandey, G.K., 2016. Role of cyclic nucleotide gated channels in stress management in plants. Current Genomics 17 (4), 315–329. Available at https://doi.org/10.2174/1389202917666160331202125.

Ji, H., Pardo, J.M., Batelli, G., Van Oosten, M.J., Bressan, R.A., Li, X., 2013. The salt overly sensitive (SOS) pathway: established and emerging roles. Molecular Plant 6, 275–286.

Jia, B., Sun, M., DuanMu, H., Ding, X., Liu, B., Zhu, Y., et al., 2017a. *GsCHX19. 3*, a member of cation/H^+ exchanger superfamily from wild soybean contributes to high salinity and carbonate alkaline tolerance. Scientific Reports 7, 9423.

Jia, J., Zhou, J., Shi, W., Cao, X., Luo, J., Polle, A., et al., 2017. Comparative transcriptomic analysis reveals the roles of overlapping heat-/drought-responsive genes in poplars exposed to high temperature and drought. Scientific Reports 7, 43215.

Jiang, X., Leidi, E.O., Pardo, J.M., 2010. How do vacuolar NHX exchangers function in plant salt tolerance? Plant Signaling & Behavior 5, 792–795.

Jing, T., Li, J., He, Y., Shankar, A., Saxena, A., Tiwari, A., et al., 2024. Role of calcium nutrition in plant physiology: advances in research and insights into acidic soil conditions-a comprehensive review. Plant Physiology and Biochemistry: PPB / Societe Francaise de Physiologie Vegetale 108602.

Jing, Z., Liu, N., Zhang, Z., Hou, X., 2024b. Research progress on plant responses to stress combinations in the context of climate change. Plants 13, 469.

Johnson, S.M., Lim, F.-L., Finkler, A., Fromm, H., Slabas, A.R., Knight, M.R., 2014. Transcriptomic analysis of *Sorghum bicolor* responding to combined heat and drought stress. BMC Genomics 15, 1–19.

Johnson, R., Vishwakarma, K., Hossen, M.S., Kumar, V., Shackira, A.M., Puthur, J.T., et al., 2022. Potassium in plants: growth regulation, signaling, and environmental stress tolerance. Plant Physiology and Biochemistry: PPB / Societe Francaise de Physiologie Vegetale 172, 56–69.

Ju, Y., Min, Z., Zhang, Y., Zhang, K., Liu, M., Fang, Y., 2021. Transcriptome profiling provide new insights into the molecular mechanism of grapevine response to heat, drought, and combined stress. Scientia Horticulturae (Amsterdam) 286, 110076.

Kang, X., Zhao, L., Liu, X., 2023. Calcium signaling and the response to heat shock in crop plants. International Journal of Molecular Sciences 25, 324.

Kant, P., Gordon, M., Kant, S., Zolla, G., Davydov, O., Heimer, Y.M., et al., 2008. Functional-genomics-based identification of genes that regulate *Arabidopsis* responses to multiple abiotic stresses. Plant, Cell & Environment 31, 697–714.

Kargar, F., Niazi, A., Fakhrfeshani, M., Malekzadeh, K., 2019. Expression analysis of genes encoding *NHX2* antiporter and subunit A of vacuolar H^+-ATPase pump in salt-resistant and salt-sensitive barley (*Hordeum vulgare* L.) cultivars under salt stress. Russian Journal of Plant Physiology 66, 572–582.

Karley, A.J., White, P.J., 2009. Moving cationic minerals to edible tissues: potassium, magnesium, calcium. Current Opinion in Plant Biology 12, 291–298.

Kaya, C., Ashraf, M., 2024. Foliar fertilization: a potential strategy for improving plant salt tolerance. Critical Reviews in Plant Sciences 43, 94–115.

Kaya, C., Uğurlar, F., Adamakis, I.-D.S., 2024. Epigenetic modifications of hormonal signaling pathways in plant drought response and tolerance for sustainable food security. International Journal of Molecular Sciences 25, 8229.

Keisham, M., Mukherjee, S., Bhatla, S.C., 2018. Mechanisms of sodium transport in plants—progresses and challenges. International Journal of Molecular Sciences 19, 647.

Kelly J.W., Medlyn B., 2013. Productivity and water use of Australian tree species under climate change. Available at https://doi.org/10.25949/19444616.

Ketehouli, T., Idrice Carther, K.F., Noman, M., Wang, F., Li, X., Li, H., 2019. Adaptation of plants to salt stress: characterization of Na^+ and K^+ transporters and role of CBL gene family in regulating salt stress response. Agronomy 9, 687.

Khalil, U., Ali, S., Rizwan, M., Rahman, K.U., Ata-Ul-Karim, S.T., Najeeb, U., et al., Hasanuzzaman, M., Fujita, M., Oku, H., Nahar, K., Hawrylak-Nowak, B., 2018. Role of mineral nutrients in plant growth under extreme temperatures. In: Hasanuzzaman, M., Fujita, M., Oku, H., Nahar, K., Hawrylak-Nowak, B. (Eds.), Plant Nutrients and Abiotic Stress Tolerance. Springer, Singapore, pp. 499–524. Available at https://doi.org/10.1007/978-981-10-9044-8_21.

Khan, I.U., Ali, A., Yun, D.-J., 2018. *Arabidopsis* NHX transporters: Sodium and potassium antiport mythology and sequestration during ionic stress. Journal of Plant Biology 61, 292–300.

Khan, I., Mohamed, S., Regnault, T., Mieulet, D., Guiderdoni, E., Sentenac, H., et al., 2020. Constitutive contribution by the rice OsHKT1; 4 Na^+ transporter to xylem sap desalinization and low Na^+ accumulation in young leaves under low as high external Na^+ conditions. Frontiers in Plant Science 11, 1130.

Khare, N., Khare, P., Singh, S., 2025. Molecular and physiological concepts: macronutrients in crop plant growth and development. In: Husen, A. (Ed.), Agricultural Crop Improvement. CRC Press, Boca Raton, pp. 148–164.

Kobayashi, N.I., Yamaji, N., Yamamoto, H., Okubo, K., Ueno, H., Costa, A., et al., 2017. OsHKT1; 5 mediates Na^+ exclusion in the vasculature to protect leaf blades and reproductive tissues from salt toxicity in rice. The Plant Journal: for Cell and Molecular Biology 91, 657–670.

Kosakivska, I.V., Vedenicheva, N.P., Babenko, L.M., Voytenko, L.V., Romanenko, K.O., Vasyuk, V.A., 2022. Exogenous phytohormones in the regulation of growth and development of cereals under abiotic stresses. Molecular Biology Reports 49, 617–628.

Krishnamurthy, P., Ranathunge, K., Franke, R., Prakash, H.S., Schreiber, L., Mathew, M.K., 2009. The role of root apoplastic transport barriers in salt tolerance of rice (*Oryza sativa* L.). Planta 230, 119–134.

Krishnamurthy, P., Ranathunge, K., Nayak, S., Schreiber, L., Mathew, M.K., 2011. Root apoplastic barriers block Na^+ transport to shoots in rice (*Oryza sativa* L.). Journal of Experimental Botany 62, 4215–4228.

Kumar, R.R., Goswami, S., Sharma, S.K., Singh, K., Gadpayle, K.A., Kumar, N., et al., 2012. Protection against heat stress in wheat involves change in cell membrane stability, antioxidant enzymes, osmolyte, H_2O_2 and transcript of heat shock protein. International Journal of Plant Physiology and Biochemistry 4, 83–91.

Kumar, R.R., Goswami, S., Sharma, S.K., Singh, K., Gadpayle, K.A., Singh, S.D., et al., 2013. Differential expression of heat shock protein and alteration in osmolyte accumulation under heat stress in wheat. Journal of Plant Biochemistry and Biotechnology 22, 16–26.

Kumar, V., Joshi, S., Pant, N.C., Sangwan, P., Yadav, A.N., Saxena, A., et al., 2019. Molecular approaches for combating multiple abiotic stresses in crops of arid and semi-arid region. In: Singh, S.P., Upadhyay, S.K., Pandey, A., Kumar, S. (Eds.), Molecular Approaches in Plant Biology and Environmental Challenges. Springer Singapore, Singapore, pp. 149–170.

Kumar, G., Basu, S., Singla-Pareek, S.L., Pareek, A., 2022. Unraveling the contribution of *OsSOS2* in conferring salinity and drought tolerance in a high-yielding rice. Physiologia Plantarum 174, e13638.

Kumar, P., Paul, D., Jhajhriya, S., Kumar, R., Dutta, S., Siwach, P., et al., 2024. Understanding heat-shock proteins' abundance and pivotal function under multiple abiotic stresses. Journal of Plant Biochemistry and Biotechnology 33, 492–513. Available at https://doi.org/10.1007/s13562-024-00932-x.

Kumari, S., Chhillar, H., Chopra, P., Khanna, R.R., Khan, M.I.R., 2021. Potassium: A track to develop salinity tolerant plants. Plant Physiology and Biochemistry: PPB / Societe Francaise de Physiologie Vegetale 167, 1011–1023.

Li, W., Xu, G., Alli, A., Yu, L., 2018. Plant HAK/KUP/KT K^+ transporters: function and regulation. Seminars in Cell & Developmental Biology. Elsevier, pp. 133–141.

Li, N., Euring, D., Cha, J.Y., Lin, Z., Lu, M., Huang, L.-J., et al., 2021. Plant hormone-mediated regulation of heat tolerance in response to global climate change. Frontiers in Plant Science 11, 627969.

Li, J., Guo, Y., Yang, Y., 2022a. The molecular mechanism of plasma membrane H^+-ATPases in plant responses to abiotic stress. Journal of Genetics and Genomics=Yi Chuan xue bao 49, 715–725.

Li, Y., Hu, J., Qi, J., Zhao, F., Liu, J., Chen, L., et al., 2022b. Improvement of leaf K^+ retention is a shared mechanism behind CeO_2 and Mn_3O_4 nanoparticles improved rapeseed salt tolerance. Stress Biol 2, 1–15.

Li, Y., Zeng, H., Xu, F., Yan, F., Xu, W., 2022c. H+-ATPases in plant growth and stress responses. Annual Review of Plant Biology 73, 495–521.

Li, L., Zheng, Q., Jiang, W., Xiao, N., Zeng, F., Chen, G., et al., 2022d. Molecular regulation and evolution of cytokinin signaling in plant abiotic stresses. Plant & Cell Physiology 63, 1787–1805.

Li, J., Shen, L., Han, X., He, G., Fan, W., Li, Y., et al., 2023. Phosphatidic acid–regulated SOS2 controls sodium and potassium homeostasis in *Arabidopsis* under salt stress. The EMBO Journal 42, e112401.

Liu, K., Luan, S., 1998. Voltage-dependent K^+ channels as targets of osmosensing in guard cells. The Plant Cell 10, 1957–1970.

Liu, K., Fu, H., Bei, Q., Luan, S., 2000. Inward potassium channel in guard cells as a target for polyamine regulation of stomatal movements. Plant Physiology 124, 1315–1326.

Liu, X., Cai, S., Wang, G., Wang, F., Dong, F., Mak, M., et al., 2017. Halophytic NHXs confer salt tolerance by altering cytosolic and vacuolar K^+ and Na^+ in Arabidopsis root cell. Plant Growth Regulation 82, 333–351.

Liu, J., Niu, Y., Zhang, J., Zhou, Y., Ma, Z., Huang, X., 2018a. Ca^{2+} channels and Ca^{2+} signals involved in abiotic stress responses in plant cells: recent advances. Plant Cell, Tissue Organ Culture 132, 413–424.

Liu, Z., Qin, J., Tian, X., Xu, S., Wang, Y., Li, H., et al., 2018b. Global profiling of alternative splicing landscape responsive to drought, heat and their combination in wheat (*Triticum aestivum* L.). Plant Biotechnology Journal 16, 714–726.

Liu, Y., Wang, J., Liu, B., Xu, Z., 2022. Dynamic regulation of DNA methylation and histone modifications in response to abiotic stresses in plants. Journal of Integrative Plant Biology 64, 2252–2274.

Liu, Y., Do, S., Huynh, H., Li, J.-X., Liu, Y.-G., Du, Z.-Y., et al., 2024. Importance of pre-mRNA splicing and its study tools in plants. Adv. Biotechnology 2, 4.

Long-Tang, H., Li-Na, Z., Li-Wei, G., Anne-Aliénor, V., Hervé, S., Yi-Dong, Z., 2018. Constitutive expression of CmSKOR, an outward K^+ channel gene from melon, in *Arabidopsis thaliana* involved in saline tolerance. Plant Science (Shannon, Ireland) 274, 492–502.

Lu, W., Guo, C., Li, X., Duan, W., Ma, C., Zhao, M., et al., 2014. Overexpression of TaNHX3, a vacuolar Na^+/H^+ antiporter gene in wheat, enhances salt stress tolerance in tobacco by improving related physiological processes. Plant Physiology and Biochemistry: PPB / Societe Francaise de Physiologie Vegetale 76, 17–28.

Ma, Q., Li, Y.-X., Yuan, H.-J., Hu, J., Wei, L., Bao, A.-K., et al., 2014. ZxSOS1 is essential for long-distance transport and spatial distribution of Na^+ and K^+ in the xerophyte Zygophyllum xanthoxylum. Plant and Soil 374, 661–676.

Ma, Y., Dias, M.C., Freitas, H., 2020. Drought and salinity stress responses and microbe-induced tolerance in plants. Frontiers in Plant Science 11, 591911.

Ma, L., Xing, L., Li, Z., Jiang, D., 2024. Epigenetic control of plant abiotic stress responses. Journal of Genetics and Genomics=Yi Chuan xue bao 52 (2), 129–144. Available at 10.1016/j.jgg.2024.09.008.

Maathuis, F.J.M., Podar, D., 2011. Uptake, distribution, and physiological functions of potassium, calcium, and magnesium. In: Hawkesford, M.J., Barraclough, P. (Eds.), The Molecular and Physiological Basis of Nutrient Use Efficiency in Crops. Wiley, pp. 265–293. Available at https://doi.org/10.1002/9780470960707.ch13.

Mahalingam, R., Duhan, N., Kaundal, R., Smertenko, A., Nazarov, T., Bregitzer, P., 2022. Heat and drought induced transcriptomic changes in barley varieties with contrasting stress response phenotypes. Frontiers in Plant Science 13, 1066421.

El Mahi, H., Pérez-Hormaeche, J., De Luca, A., Villalta, I., Espartero, J., Gámez-Arjona, F., et al., 2019. A critical role of sodium flux via the plasma membrane Na^+/H^+ exchanger SOS1 in the salt tolerance of rice. Plant Physiology 180, 1046–1065.

Malakar, P., Chattopadhyay, D., 2021. Adaptation of plants to salt stress: the role of the ion transporters. Journal of Plant Biochemistry and Biotechnology 30, 668–683.

Mansour, M.M.F., 2023. Role of vacuolar membrane transport systems in plant salinity tolerance. Journal of Plant Growth Regulation 42, 1364–1401.

Mansour, M.M.F., Salama, K.H.A., Al-Mutawa, M.M., 2003. Transport proteins and salt tolerance in plants. Plant Science (Shannon, Ireland) 164, 891–900.

Margay, A.R., Mehmood, A., Bashir, L., 2024. Review on hormonal regulation of drought stress response in plants. International Journal of Plant & Soil Science 36, 902–916.

Martin, R.C., Kronmiller, B.A., Dombrowski, J.E., 2021. Transcriptome analysis of *Lolium temulentum* exposed to a combination of drought and heat stress. Plants 10, 2247.

Maryum, Z., Luqman, T., Nadeem, S., Khan, S.M.U.D., Wang, B., Ditta, A., et al., 2022. An overview of salinity stress, mechanism of salinity tolerance and strategies for its management in cotton. Frontiers in Plant Science 13, 907937.

Mathur, S., Agrawal, D., Jajoo, A., 2014. Photosynthesis: response to high temperature stress. Journal of Photochemistry and Photobiology B: Biology 137, 116–126.

Mbarki, S., Sytar, O., Cerda, A., Zivcak, M., Rastogi, A., He, X., et al., 2018. Strategies to mitigate the salt stress effects on photosynthetic apparatus and productivity of crop plants. In: Kumar, V., Wani, S., Suprasanna, P., Tran, L.S. (Eds.), Salinity Responses and Tolerance in Plants, Volume 1. Springer, Cham, pp. 85–136. Available at https://doi.org/10.1007/978-3-319-75671-4_4.

Mian, A.A., Senadheera, P., Maathuis, F.J.M., 2011. Improving crop salt tolerance: anion and cation transporters as genetic engineering targets. Plant Stress 5, 64.

Michalak, A., Wdowikowska, A., Janicka, M., 2022. Plant plasma membrane proton pump: one protein with multiple functions. Cells 11, 4052.

Mohamed, I.A.A., Shalby, N., Bai, C., Qin, M., Agami, R.A., Jie, K., et al., 2020. Stomatal and photosynthetic traits are associated with investigating sodium chloride tolerance of *Brassica napus* L. Cultivars. Plants 9 (1), 62. Available at https://doi.org/10.3390/plants9010062.

Mohamed, I.A.A., Shalby, N., El-Badri, A.M.A., Saleem, M.H., Khan, M.N., Nawaz, M.A., et al., 2020b. Stomata and xylem vessels traits improved by melatonin application contribute to enhancing salt tolerance and fatty acid composition of *Brassica napus* L. plants. Agronomy 10, 1186.

Mohanavelu, A., Naganna, S.R., Al-Ansari, N., 2021. Irrigation induced salinity and sodicity hazards on soil and groundwater: an overview of its causes, impacts and mitigation strategies. Agriculture 11, 983.

Mohamed, I.A.A., Shalby, N., El-Badri, A.M., Batool, M., Wang, C., Wang, Z., et al., 2022. RNA-seq analysis revealed key genes associated with salt tolerance in rapeseed germination through carbohydrate metabolism, hormone, and MAPK signaling pathways. Industrial Crops and Products 176, 114262.

Mostofa, M.G., Rahman, M.M., Ghosh, T.K., Kabir, A.H., Abdelrahman, M., Khan, M.A.R., et al., 2022. Potassium in plant physiological adaptation to abiotic stresses. Plant Physiology and Biochemistry: PPB / Societe Francaise de Physiologie Vegetale 186, 279–289.

Moustafa, K., AbuQamar, S., Jarrar, M., Al-Rajab, A.J., Trémouillaux-Guiller, J., 2014. MAPK cascades and major abiotic stresses. Plant Cell Reports 33, 1217–1225.

Mulet, J.M., Porcel, R., Yenush, L., 2023. Modulation of potassium transport to increase abiotic stress tolerance in plants. Journal of Experimental Botany 74, 5989–6005.

Munns, R., 2002. Comparative physiology of salt and water stress. Plant, Cell & Environment 25, 239–250.

Nahar, L., Aycan, M., Hanamata, S., Baslam, M., Mitsui, T., 2022. Impact of single and combined salinity and high-temperature stresses on agro-physiological, biochemical, and transcriptional responses in rice and stress-release. Plants 11, 501.

Nieves-Cordones, M., Alemán, F., Fon, M., Martínez, V., Rubio, F., 2012. K^+ nutrition, uptake, and its role in environmental stress in plants. In: Ahmad, P., Prasad, M. (Eds.), Environmental Adaptations and Stress Tolerance of Plants in the Era of Climate Change. Springer, New York, NY, pp. 85–112. Available at https://doi.org/10.1007/978-1-4614-0815-4_4.

Nieves-Cordones, M., Al Shiblawi, F.R., Sentenac, H., 2016. Roles and transport of sodium and potassium in plants. In: Sigel, A., Sigel, H./, Sigel, R. (Eds.), The Alkali Metal Ions: Their Role for Life. Metal Ions in Life Sciences, vol 16. Springer, Cham, pp. 291–324. Available at https://doi.org/10.1007/978-3-319-21756-7_9.

Nieves-Cordones, M., Alemán, F., Martínez, V., Rubio, F., 2010. The Arabidopsis thaliana HAK5 K^+ transporter is required for plant growth and K^+ acquisition from low K^+ solutions under saline conditions. Molecular Plant 3, 326–333.

Obata, T., Kitamoto, H.K., Nakamura, A., Fukuda, A., Tanaka, Y., 2007. Rice shaker potassium channel *OsKAT1* confers tolerance to salinity stress on yeast and rice cells. Plant Physiology 144, 1978–1985.

Okada, T., Yamane, S., Yamaguchi, M., Kato, K., Shinmyo, A., Tsunemitsu, Y., et al., 2018. Characterization of rice KT/HAK/KUP potassium transporters and K^+ uptake by HAK1 from *Oryza sativa*. Plant Biotechnology 35, 101–111.

Olías, R., Eljakaoui, Z., Pardo, J.M., Belver, A., 2009. The Na^+/H^+ exchanger SOS1 controls extrusion and distribution of Na^+ in tomato plants under salinity conditions. Plant Signaling & Behavior 4, 973–976.

Ondrasek, G., Rengel, Z., 2021. Environmental salinization processes: detection, implications & solutions. The Science of the Total Environment 754, 142432.

Osakabe, Y., Arinaga, N., Umezawa, T., Katsura, S., Nagamachi, K., Tanaka, H., et al., 2013. Osmotic stress responses and plant growth controlled by potassium transporters in *Arabidopsis*. The Plant Cell 25, 609–624.

Osman, K.T., 2018. Saline and sodic soils. In: Osman, K.T. (Ed.), Management of Soil Problems. Springer, Cham, pp. 255–298. Available at https://doi.org/10.1007/978-3-319-75527-4_10.

Palmgren, M.G., 1998. *Proton Gradients and Plant Growth: Role of the Plasma Membrane H^+-ATPase*. Academic Press.

Palmgren, M.G., Bækgaard, L., López-Marqués, R.L., Fuglsang, A.T., 2011. Plasma membrane ATPases. In: Murphy, A., Schulz, B., Peer, W. (Eds.), The Plant Plasma Membrane. Plant Cell Monographs, vol 19. Springer, Berlin, Heidelberg, pp. 177–192. Available at https://doi.org/10.1007/978-3-642-13431-9_7.

Pandey, R., 2015. Mineral nutrition of plants. In: Bahadur, B., Venkat Rajam, M., Sahijram, L., Krishnamurthy, K. (Eds.), Plant Biology and Biotechnology. Volume I, Plant Diversity, Organization, Function and Improvement. Springer, New Delhi, pp. 499–538. Available at https://doi.org/10.1007/978-81-322-2286-6_20.

Pang J., 2023. Physiological responses to waterlogging in barley (*Hordeum vulgare* L.). University of Tasmania. Available at https://doi.org/10.25959/23514678.

Pantha, P., Oh, D.-H., Longstreth, D., Dassanayake, M., 2023. Living with high potassium: balance between nutrient acquisition and K-induced salt stress signaling. Plant Physiology 191, 1102–1121.

Pardo, J.M., Cubero, B., Leidi, E.O., Quintero, F.J., 2006. Alkali cation exchangers: roles in cellular homeostasis and stress tolerance. Journal of Experimental Botany 57, 1181–1199.

Pardo, J.M., Rubio, F., 2011. Na^+ and K^+ transporters in plant signaling. In: Geisler, M., Venema, K. (Eds.), Transporters and Pumps in Plant Signaling. Signaling and Communication in Plants, vol 7. Springer, Berlin, Heidelberg, pp. 65–98. Available at https://doi.org/10.1007/978-3-642-14369-4_3.

Parwez, R., Aftab, T., Gill, S.S., Naeem, M., 2022. Abscisic acid signaling and crosstalk with phytohormones in regulation of environmental stress responses. Environmental and Experimental Botany 199, 104885.

Peer, L.A., Dar, Z.A., Lone, A.A., Bhat, M.Y., Ahamad, N., 2020. High temperature triggered plant responses from whole plant to cellular level. Plant Physiology Reports 25, 611–626.

Peng, Y., Cao, H., Peng, Z., Zhou, L., Sohail, H., Cui, L., et al., 2023. Transcriptomic and functional characterization reveals CsHAK5; 3 as a key player in K^+ homeostasis in grafted cucumbers under saline conditions. Plant Science (Shannon, Ireland) 326, 111509.

Pittman, J.K., 2012. Multiple transport pathways for mediating intracellular pH homeostasis: the contribution of H^+/ion exchangers. Frontiers in Plant Science 3, 11.

Pottosin, I., Dobrovinskaya, O., 2014. Non-selective cation channels in plasma and vacuolar membranes and their contribution to K^+ transport. Journal of Plant Physiology 171, 732–742.

Prerostova, S., Vankova, R., 2023. Phytohormone-mediated regulation of heat stress response in plants. Plant Hormones and Climate Change. Springer, pp. 167–206.

Priya, M., Dhanker, O.P., Siddique, K.H.M., HanumanthaRao, B., Nair, R.M., Pandey, S., et al., 2019. Drought and heat stress-related proteins: an update about their functional relevance in imparting stress tolerance in agricultural crops. TAG. Theoretical and Applied Genetics. Theoretische und Angewandte Genetik 132, 1607–1638. Available at https://doi.org/10.1007/s00122-019-03331-2.

Pu, L., Lin, R., Zou, T., Wang, Z., Zhang, M., Jian, S., 2021. Genome-wide identification, primary functional characterization of the NHX gene family in *Canavalia rosea*, and their possible roles for adaptation to tropical coral reefs. Genes (Basel) 13, 33.

Pérez-Salamó, I., Papdi, C., Rigó, G., Zsigmond, L., Vilela, B., Lumbreras, V., et al., 2014. The heat shock factor A4A confers salt tolerance and is regulated by oxidative stress and the mitogen-activated protein kinases MPK3 and MPK6. Plant Physiology 165, 319–334.

Qi, Z., Spalding, E.P., 2004. Protection of plasma membrane K^+ transport by the salt overly sensitive1 Na^+-H^+ antiporter during salinity stress. Plant Physiology 136, 2548–2555.

Qi, X., Li, M.-W., Xie, M., Liu, X., Ni, M., Shao, G., et al., 2014. Identification of a novel salt tolerance gene in wild soybean by whole-genome sequencing. Nature Communications 5, 4340.

Rady, M.M., El-yazal, M.A.S., Anwar, H., Taie, A., Ahmed, S.M.A., 2016. Response of Triticum aestivum (L.) plants grown under cadmium stress to polyamines pretreatments. Plant 4 (5), 29–36. Available at https://doi.org/10.11648/j.plant.20160405.11.

Rady, M.M., Mossa, A.-T.H., Youssof, A.M.A., Osman, A.S., Ahmed, S.M.A., Mohamed, I.A.A., 2023. Exploring the reinforcing effect of nano-potassium on the antioxidant defense system reflecting the increased yield and quality of salt-stressed squash plants. Scientia Horticulturae 308, 111609.

Rahim, H.U., Ali, W., Uddin, M., Ahmad, S., Khan, K., Bibi, H., et al., 2024. Abiotic stresses in soils, their effects on plants, and mitigation strategies: a literature review. Chemistry and Ecology 41 (4), 552–585. Available at https://doi.org/10.1080/02757540.2024.2439830.

Rajappa, S., Krishnamurthy, P., Kumar, P.P., 2020. Regulation of *AtKUP2* expression by *bHLH* and *WRKY* transcription factors helps to confer increased salt tolerance to *Arabidopsis thaliana* plants. Frontiers in Plant Science 11, 1311.

Rasul, I., Nadeem, H., Siddique, M.H., Atif, R.M., Ali, M.A., Umer, A., et al., 2017. Plants sensory-response mechanisms for salinity and heat stress. Journal of Animal and Plant Sciences 27 (2).

Rawat, J., Sanwal, P., Saxena, J., 2016. Potassium and its role in sustainable agriculture. Potassium Solubilizing Microorganisms for Sustainable Agriculture. Springer, pp. 235–253.

Rejeb, I., Pastor, V., Mauch-Mani, B., 2014. Plant responses to simultaneous biotic and abiotic stress: molecular mechanisms. Plants 3 (4), 458–475. Available at https://doi.org/10.3390/plants3040458.

Ren, S., Lyle, C., Jiang, G., Penumala, A., 2016. Soybean salt tolerance 1 (GmST1) reduces ROS production, enhances ABA sensitivity, and abiotic stress tolerance in *Arabidopsis thaliana*. Frontiers in Plant Science 7, 445.

Rivero, R.M., Mestre, T.C., Mittler, R.O.N., Rubio, F., Garcia-Sanchez, F., Martinez, V., 2014. The combined effect of salinity and heat reveals a specific physiological, biochemical and molecular response in tomato plants. Plant, Cell & Environment 37, 1059–1073.

Rizhsky, L., Liang, H., Mittler, R., 2002. The combined effect of drought stress and heat shock on gene expression in tobacco. Plant Physiology 130, 1143–1151.

Rizhsky, L., Liang, H., Shuman, J., Shulaev, V., Davletova, S., Mittler, R., 2004. When defense pathways collide. The response of *Arabidopsis* to a combination of drought and heat stress. Plant Physiology 134, 1683–1696.

Robbins, N.E., Trontin, C., Duan, L., Dinneny, J.R., 2014. Beyond the barrier: communication in the root through the endodermis. Plant Physiology 166, 551–559.

Rogiers, S.Y., Coetzee, Z.A., Walker, R.R., Deloire, A., Tyerman, S.D., 2017. Potassium in the grape (*Vitis vinifera* L.) berry: transport and function. Frontiers in Plant Science 8, 1629.

Romero-Aranda, R., Soria, T., Cuartero, J., 2002. Greenhouse mist improves yield of tomato plants grown under saline conditions. Journal of the American Society for Horticultural Science. 127, 644–648.

Rouphael, Y., Cardarelli, M., Schwarz, D., Franken, P., Colla, G., 2012. Effects of drought on nutrient uptake and assimilation in vegetable crops. In: Aroca, R. (Ed.), Plant Responses to Drought Stress: From Morphological to Molecular Features. Springer, Berlin, Heidelberg, pp. 171–195. Available at https://doi.org/10.1007/978-3-642-32653-0_7.

Roy, T.K., Islam, M.S., Mahiddin, N.A., Hossain, S.A., Biswas, T., Antu, U.B., et al., 2025. Application of nanoparticles (NPs) to ameliorate abiotic stress in economically important crop species: a potential review. Journal of Crop Health 77, 1–20.

Rubio, F., Fon, M., Ródenas, R., Nieves-Cordones, M., Alemán, F., Rivero, R.M., et al., 2014. A low K^+ signal is required for functional high-affinity K^+ uptake through HAK5 transporters. Physiologia Plantarum 152, 558–570.

El Sabagh, A., Islam, M.S., Skalicky, M., Ali Raza, M., Singh, K., Anwar Hossain, M., et al., 2021. Salinity stress in wheat (*Triticum aestivum* L.) in the changing climate: adaptation and management strategies. Frontiers in Agronomy 3, 661932.

EL Sabagh, A., Islam, M.S., Hossain, A., Iqbal, M.A., Mubeen, M., Waleed, M., et al., 2022. Phytohormones as growth regulators during abiotic stress tolerance in plants. Frontiers in Agronomy 4, 765068.

Sachdev, S., Ansari, S.A., Ansari, M.I., Fujita, M., Hasanuzzaman, M., 2021. Abiotic stress and reactive oxygen species: generation, signaling, and defense mechanisms. Antioxidants 10, 277.

Saddhe, A.A., Manuka, R., Penna, S., 2021a. Plant sugars: homeostasis and transport under abiotic stress in plants. Physiologia Plantarum 171, 739–755.

Saddhe, A.A., Mishra, A.K., Kumar, K., 2021b. Molecular insights into the role of plant transporters in salt stress response. Physiologia Plantarum 173, 1481–1494.

Saibi, W., Brini, F., 2021. Ion transporters and their molecular regulation mechanism in plants. Journal of Plant Science and Phytopathology 5, 28–43.

Saidi, Y., Finka, A., Goloubinoff, P., 2011. Heat perception and signalling in plants: a tortuous path to thermotolerance. The New Phytologist 190, 556–565.

Saleem, M.H., Noreen, S., Ishaq, I., Saleem, A., Khan, K.A., Ercisli, S., et al., 2025. Omics technologies: unraveling abiotic stress tolerance mechanisms for sustainable crop improvement. Journal of Plant Growth Regulation 44, 4165–4187. Available at https://doi.org/10.1007/s00344-025-11674-y.

Sangwan, V., Örvar, B.L., Beyerly, J., Hirt, H., Dhindsa, R.S., 2002. Opposite changes in membrane fluidity mimic cold and heat stress activation of distinct plant MAP kinase pathways. The Plant Journal: For Cell and Molecular Biology 31, 629–638.

Dos Santos, T.B., Ribas, A.F., de Souza, S.G.H., Budzinski, I.G.F., Domingues, D.S., 2022. Physiological responses to drought, salinity, and heat stress in plants: a review. Stresses 2, 113–135.

Saradadevi, G.P., Das, D., Mangrauthia, S.K., Mohapatra, S., Chikkaputtaiah, C., Roorkiwal, M., et al., 2021. Genetic, epigenetic, genomic and microbial approaches to enhance salt tolerance of plants: a comprehensive review. Biology (Basel) 10, 1255.

Sardans, J., Peñuelas, J., 2021. Potassium control of plant functions: ecological and agricultural implications. Plants 10, 419.

Sathee, L., Sairam, R.K., Chinnusamy, V., Jha, S.K., 2015. Differential transcript abundance of salt overly sensitive (SOS) pathway genes is a determinant of salinity stress tolerance of wheat. Acta Physiologiae Plantarum / Polish Academy of Sciences, Committee of Plant Physiology Genetics and Breeding 37, 1–10.

Savvides, A., Ali, S., Tester, M., Fotopoulos, V., 2016. Chemical priming of plants against multiple abiotic stresses: mission possible? Trends in Plant Science 21, 329–340.

Sewelam, N., Kazan, K., Schenk, P.M., 2016. Global plant stress signaling: reactive oxygen species at the cross-road. Frontiers in Plant Science 7, 187.

Shabala, S., Pottosin, I.I., 2010. Potassium and potassium-permeable channels in plant salt tolerance. Ion Channels Plant Stress Responses 87–110.

Shabala, S., Pottosin, I., 2014. Regulation of potassium transport in plants under hostile conditions: implications for abiotic and biotic stress tolerance. Physiologia Plantarum 151, 257–279.

Shabala, S., Wu, H., Bose, J., 2015. Salt stress sensing and early signalling events in plant roots: current knowledge and hypothesis. Plant Science (Shannon, Ireland) 241, 109–119.

Shabala, L., Zhang, J., Pottosin, I., Bose, J., Zhu, M., Fuglsang, A.T., et al., 2016. Cell-type-specific H^+-ATPase activity in root tissues enables K^+ retention and mediates acclimation of barley (*Hordeum vulgare*) to salinity stress. Plant Physiology 172, 2445–2458.

Shah, F., Wu, W., 2019. Soil and crop management strategies to ensure higher crop productivity within sustainable environments. Sustainability 11, 1485.

Sharma, T., Dreyer, I., Riedelsberger, J., 2013. The role of K^+ channels in uptake and redistribution of potassium in the model plant *Arabidopsis thaliana*. Frontiers in Plant Science 4, 224.

Sharma, M., Sidhu, A.K., Samota, M.K., Gupta, M., Koli, P., Choudhary, M., 2023. Post-translational modifications in histones and their role in abiotic stress tolerance in plants. Proteomes 11, 38.

Shi, H., Quintero, F.J., Pardo, J.M., Zhu, J.-K., 2002. The putative plasma membrane Na^+/H^+ antiporter SOS1 controls long-distance Na^+ transport in plants. The Plant Cell 14, 465–477.

Siddique, M.H., Babar, N.I., Zameer, R., Muzammil, S., Nahid, N., Ijaz, U., et al., 2021. Genome-wide identification, genomic organization, and characterization of potassium transport-related genes in *Cajanus cajan* and their role in abiotic stress. Plants 10 (11), 2238. Available at https://doi.org/10.3390/plants10112238.

Siddiqui, M.H., Khan, M.N., Mukherjee, S., Alamri, S., Basahi, R.A., Al-Amri, A.A., et al., 2021. Hydrogen sulfide (H2S) and potassium (K^+) synergistically induce drought stress tolerance through regulation of H^+-ATPase activity, sugar metabolism, and antioxidative defense in tomato seedlings. Plant Cell Reports 40, 1543–1564.

Sies, H., Mailloux, R.J., Jakob, U., 2024. Fundamentals of redox regulation in biology. Nature Reviews. Molecular Cell Biology 25, 701–719.

Silva, E.N., Ferreira-Silva, S.L., de Vasconcelos Fontenele, A., Ribeiro, R.V., Viégas, R.A., Silveira, J.A.G., 2010. Photosynthetic changes and protective mechanisms against oxidative damage subjected to isolated and combined drought and heat stresses in Jatropha curcas plants. Journal of Plant Physiology 167, 1157–1164.

Singh, A.K., Vijai, P., Srivastava, J.P., 2018. Plants under waterlogged conditions: an overview. In: Gupta, S.K., Goyal, M.R., Singh, A. (Eds.), Engineering Practices for Management of Soil Salinity. Apple Academic Press, New York, pp. 335–376.

Singh, A., Agrawal, S., Rajput, V.D., Ghazaryan, K., Yesayan, A., Minkina, T., et al., 2024. Nanoparticles in revolutionizing crop production and agriculture to address salinity stress challenges for a sustainable future. Discover Applied Sciences 6, 317.

Singh, A., Rajput, V.D., Varshney, A., Sharma, R., Ghazaryan, K., Minkina, T., et al., 2024. Revolutionizing crop production: nanoscale wonders-current applications, advances, and future frontiers. Egyptian Journal of Soil Science 64, 221–258.

Singhal, R.K., Fahad, S., Kumar, P., Choyal, P., Javed, T., Jinger, D., et al., 2023. Beneficial elements: new players in improving nutrient use efficiency and abiotic stress tolerance. Plant Growth Regulation 100, 237–265.

Sirichandra, C., Wasilewska, A., Vlad, F., Valon, C., Leung, J., 2009. The guard cell as a single-cell model towards understanding drought tolerance and abscisic acid action. Journal of Experimental Botany 60, 1439–1463.

Smith, S.E., Facelli, E., Pope, S., Andrew Smith, F., 2010. Plant performance in stressful environments: interpreting new and established knowledge of the roles of arbuscular mycorrhizas. Plant and Soil 326, 3–20.

Sonali, Huda, S., Jayasena, V., Ahmed, T., Chen, Z.H., 2023. Potassium transport and use efficiency for sustainable fertigation in protected cropping. Journal of Sustainable Agriculture and Environment 2 (3), 346–356. Available at https://doi.org/10.1002/sae2.12065.

Sondergaard, T.E., Schulz, A., Palmgren, M.G., 2004. Energization of transport processes in plants. Roles of the plasma membrane H^+-ATPase. Plant Physiology 136, 2475–2482.

Song, Y., Li, J., Sui, Y., Han, G., Zhang, Y., Guo, S., et al., 2020. The sweet sorghum *SbWRKY50* is negatively involved in salt response by regulating ion homeostasis. Plant Molecular Biology 102, 603–614.

Sousa, B., Rodrigues, F., Soares, C., Martins, M., Azenha, M., Lino-Neto, T., et al., 2022. Impact of combined heat and salt stresses on tomato plants—insights into nutrient uptake and redox homeostasis. Antioxidants 11, 478.

Suzuki, K., Costa, A., Nakayama, H., Katsuhara, M., Shinmyo, A., Horie, T., 2016. OsHKT2; 2/1-mediated Na^+ influx over K^+ uptake in roots potentially increases toxic Na^+ accumulation in a salt-tolerant landrace of rice Nona Bokra upon salinity stress. Journal of Plant Research 129, 67–77.

Syvertsen, J.P., Garcia-Sanchez, F., 2014. Multiple abiotic stresses occurring with salinity stress in citrus. Environmental and Experimental Botany 103, 128–137.

Szczerba, M.W., Britto, D.T., Kronzucker, H.J., 2009. K^+ transport in plants: physiology and molecular biology. Journal of Plant Physiology 166, 447–466.

Sze, H., Chanroj, S., 2018. Plant endomembrane dynamics: studies of K^+/H^+ antiporters provide insights on the effects of pH and ion homeostasis. Plant Physiology 177, 875–895.

Sánchez-Bermúdez, M., Del Pozo, J.C., Pernas, M., 2022. Effects of combined abiotic stresses related to climate change on root growth in crops. Frontiers in Plant Science 13, 918537.

Šustr, M., Doksanská, T., Doležalová, B., Soukup, A., Tylová, E., 2020. 134Cs uptake and growth at various cs+ and k^+ levels in arabidopsis *AtKUP7* mutants. Plants 9, 1525.

Takahashi, Y., Ebisu, Y., Kinoshita, T., Doi, M., Okuma, E., Murata, Y., et al., 2013. *bHLH* transcription factors that facilitate K^+ uptake during stomatal opening are repressed by abscisic acid through phosphorylation. Science Signaling 6, ra48.

Talukdar, D., 2019. Regulated suicide for survival: toward programmed cell death during reactive species mediated-oxidative stress of plant cells. In: Hasanuzzaman, M., Fotopoulos, V., Nahar, K., Fujita, M. (Eds.), Reactive Oxygen, Nitrogen and Sulfur Species in Plants: Production, Metabolism, Signaling and Defense Mechanisms. Wiley, pp. 1–38. Available at https://doi.org/10.1002/9781119468677.ch1.

Tan, Y.-Q., Yang, Y., Shen, X., Zhu, M., Shen, J., Zhang, W., et al., 2023. Multiple cyclic nucleotide-gated channels function as ABA-activated Ca^{2+} channels required for ABA-induced stomatal closure in *Arabidopsis*. The Plant Cell 35, 239–259.

Tarfayah, D., Ahmed, S., Rady, M., Mohamed, I., 2023. Alleviating saline-calcareous stress in *Atriplex nummularia* seedlings by foliar spraying with silymarin-enriched bee-honey solution. Labyrinth: Fayoum Journal of Science and Interdisciplinary Studies 1, 11–20.

Tester, M., Davenport, R., 2003. Na+ tolerance and Na+ transport in higher plants. Annals of Botany 91, 503–527.

Ullah, M.A., Abdullah-Zawawi, M.-R., Razalli II, Sukiran, N.L., Uddin, M.I., Zainal, Z., 2025. Overexpression of rice high-affinity potassium transporter gene *OsHKT1; 5* improves salinity and drought tolerance in *Arabidopsis*. Molecular Biology Reports 52, 1–12.

Valentovic, P., Luxova, M., Kolarovic, L., Gasparikova, O., 2006. Effect of osmotic stress on compatible solutes content, membrane stability and water relations in two maize cultivars. Plant Soil Environment 52, 184.

Venkataraman, G., Shabala, S., Véry, A.-A., Hariharan, G.N., Somasundaram, S., Pulipati, S., et al., 2021. To exclude or to accumulate? Revealing the role of the sodium HKT1; 5 transporter in plant adaptive responses to varying soil salinity. Plant Physiology and Biochemistry: PPB / Societe Francaise de Physiologie Vegetale 169, 333–342.

Vescio, R., Caridi, R., Laudani, F., Palmeri, V., Zappalà, L., Badiani, M., et al., 2022. Abiotic and herbivory combined stress in tomato: additive, synergic and antagonistic effects and within-plant phenotypic plasticity. Life (Chicago, Ill.: 1978) 12, 1804.

Voesenek, L.A.C.J., Bailey-Serres, J., 2015. Flood adaptive traits and processes: an overview. The New Phytologist 206, 57–73.

Wahid, A., Gelani, S., Ashraf, M., Foolad, M.R., 2007. Heat tolerance in plants: an overview. Environmental and Experimental Botany 61, 199–223.

Wakeel, A., 2013. Potassium–sodium interactions in soil and plant under saline-sodic conditions. Journal of Plant Nutrition and Soil Science 176, 344–354.

Wang, J., 2019. Functional Studies of Arabidopsis Annexins AnnAt1 and AnnAt2 in Primary Root Growth. The University of Texas at Austin. Available at http://dx.doi.org/10.26153/tsw/5673.

Wang, Y., Wu, W.H., 2010. Plant sensing and signaling in response to K^+-deficiency. Molecular Plant 3, 280–287.

Wang, Y., Wu, W.-H., 2017. Regulation of potassium transport and signaling in plants. Current Opinion in Plant Biology 39, 123–128.

Wang, W., Vinocur, B., Shoseyov, O., Altman, A., 2004. Role of plant heat-shock proteins and molecular chaperones in the abiotic stress response. Trends in Plant Science 9, 244–252.

Wang, M., Gu, D., Liu, T., Wang, Z., Guo, X., Hou, W., et al., 2007a. Overexpression of a putative maize calcineurin B-like protein in *Arabidopsis* confers salt tolerance. Plant Molecular Biology 65, 733–746.

Wang, W., Li, Y., Zhang, Y., Yang, C., Zheng, N., Xie, Q., 2007. Comparative expression analysis of three genes from the Arabidopsis vacuolar Na^+/H^+ antiporter (*AtNHX*) family in relation to abiotic stresses. Chinese Science Bulletin 52, 1754–1763.

Wang, G.P., Hui, Z., Li, F., Zhao, M.R., Zhang, J., Wang, W., 2010. Improvement of heat and drought photosynthetic tolerance in wheat by overaccumulation of glycinebetaine. Plant Biotechnology Reports 4, 213–222. Available at https://doi.org/10.1007/s11816-010-0139-y.

Wang, F., Deng, S., Ding, M., Sun, J., Wang, M., Zhu, H., et al., 2013a. Overexpression of a poplar two-pore K^+ channel enhances salinity tolerance in tobacco cells. Plant Cell, Tissue Organ Culture 112, 19–31.

Wang, M., Zheng, Q., Shen, Q., Guo, S., 2013. The critical role of potassium in plant stress response. International Journal of Molecular Science 14, 7370–7390.

Wang, F., Chen, Z.-H., Liu, X., Colmer, T.D., Shabala, L., Salih, A., et al., 2017a. Revealing the roles of GORK channels and NADPH oxidase in acclimation to hypoxia in *Arabidopsis*. Journal of Experimental Botany 68, 3191–3204.

Wang, X., Xu, C., Cai, X., Wang, Q., Dai, S., 2017. Heat-responsive photosynthetic and signaling pathways in plants: insight from proteomics. International Journal of Molecular Sciences 18, 2191.

Wang, L., Dossa, K., You, J., Zhang, Y., Li, D., Zhou, R., et al., 2021a. High-resolution temporal transcriptome sequencing unravels *ERF* and *WRKY* as the master players in the regulatory networks underlying sesame responses to waterlogging and recovery. Genomics 113, 276–290.

Wang, Y., Guo, Y., Li, F., Liu, Y., Jin, S., 2021b. Overexpression of *KcNHX1* gene confers tolerance to multiple abiotic stresses in *Arabidopsis thaliana*. Journal of Plant Research 134, 613–623.

Wang, Q., Ni, L., Cui, Z., Jiang, J., Chen, C., Jiang, M., 2022. The NADPH oxidase *OsRbohA* increases salt tolerance by modulating K^+ homeostasis in rice. Crop Journal 10, 1611–1622.

Wang, H.-Q., Zhao, X.Y., Xuan, W., Wang, P., Zhao, F.J., 2023. Rice roots avoid asymmetric heavy metal and salinity stress via an RBOH-ROS-auxin signaling cascade. Molecular Plant 16, 1678–1694.

Wang, L., Ju, C., Han, C., Yu, Z., Bai, M., Wang, C., 2025. The interaction of nutrient uptake with biotic and abiotic stresses in plants. Journal of Integrative Plant Biology 67 (3), 455–487. Available at https://doi.org/10.1111/jipb.13827.

Weber, J.-N., Minner-Meinen, R., Kaufholdt, D., 2023. The mechanisms of molybdate distribution and homeostasis with special focus on the model plant *Arabidopsis thaliana*. Molecules (Basel, Switzerland) 29, 40.

Wei, X., Liu, L., Lu, C., Yuan, F., Han, G., Wang, B., 2021. *SbCASP4* improves salt exclusion by enhancing the root apoplastic barrier. Planta 254, 81. Available at https://doi.org/10.1007/s00425-021-03731-z.

Wu, H., 2018. Plant salt tolerance and Na^+ sensing and transport. Crop Journal 6, 215–225.

Wu, H., Hill, C., 2022. Editorial: new insights into salinity sensing, signaling and adaptation in plants, volume II. Frontiers in Plant Science 13, 1072658. Available at 10.3389/fpls.2022.1072658.

Wu, H., Bose, J., 2024. Abiotic stress tolerance: adaptations, mechanisms, and new techniques. Crop Journal 12 (5), 1271–1273. Available at https://doi.org/10.1016/j.cj.2024.10.001.

Wu, H., Hill, C.B., Stefano, G., Bose, J., 2021. New insights into salinity sensing, signaling and adaptation in plants. Frontiers in Plant Science 11, 604139.

Xie, Q., Zhou, Y., Jiang, X., 2022. Structure, function, and regulation of the plasma membrane Na^+/H^+ antiporter salt overly sensitive 1 in plants. Frontiers in Plant Science 13, 866265.

Xing, M., Wang, W., Zhang, C., Xi, D., Wang, M., Yin, X., et al., 2024. Identification and functional analyses of the transcription factors *AcWRKY117* and *AcWRKY29* involved in waterlogging response in kiwifruit plant. Scientia Horticulture (Amsterdam) 324, 112568.

Xu, Y., Huang, B., 2018. Comparative transcriptomic analysis reveals common molecular factors responsive to heat and drought stress in *Agrostis stolonifera*. Scientific Reports 8, 15181.

Yadav, B., Jogawat, A., Lal, S.K., Lakra, N., Mehta, S., Shabek, N., et al., 2021. Plant mineral transport systems and the potential for crop improvement. Planta 253, 1–30.

Yang, D.-H., Song, L.-Y., Hu, J., Yin, W.-B., Li, Z.-G., Chen, Y.-H., et al., 2012. Enhanced tolerance to NaCl and LiCl stresses by over-expressing *Caragana korshinskii* sodium/proton exchanger 1 (*CkNHX1*) and the hydrophilic C terminus is required for the activity of *CkNHX1* in *Atsos3-1* mutant and yeast. Biochemical and Biophysical Research Communications 417, 732–737.

Yang, T., Zhang, S., Hu, Y., Wu, F., Hu, Q., Chen, G., et al., 2014. The role of a potassium transporter *OsHAK5* in potassium acquisition and transport from roots to shoots in rice at low potassium supply levels. Plant Physiology 166, 945–959.

Yang, T., Lu, X., Wang, Y., Xie, Y., Ma, J., Cheng, X., et al., 2020. HAK/KUP/KT family potassium transporter genes are involved in potassium deficiency and stress responses in tea plants (*Camellia sinensis* L.): expression and functional analysis. BMC Genomics 21, 1–18.

Yang, L., Li, N., Liu, Y., Miao, P., Liu, J., Wang, Z., 2023. Updates and prospects: morphological, physiological, and molecular regulation in crop response to waterlogging stress. Agronomy 13, 2599.

Yari Kamrani, Y., Shomali, A., Aliniaeifard, S., Lastochkina, O., Moosavi-Nezhad, M., Hajinajaf, N., et al., 2022. Regulatory role of circadian clocks on ABA production and signaling, stomatal responses, and water-use efficiency under water-deficit conditions. Cells 11, 1154.

Yarra, R., 2019. The wheat NHX gene family: potential role in improving salinity stress tolerance of plants. Plant Gene 18, 100178.

Yu, Y., Li, X., Sun, J., Zhang, X., Xu, T., Zhang, J., et al., 2016. Heat shock responses in Populus euphratica cell cultures: important role of crosstalk among hydrogen peroxide, calcium and potassium. Plant Cell, Tissue Organ Culture 125, 215–230.

Yun, P., Kaya, C., Shabala, S., 2024. Hormonal and epigenetic regulation of root responses to salinity stress. Crop Journal 12, 1309–1320.

Van Zelm, E., Zhang, Y., Testerink, C., 2020. Salt tolerance mechanisms of plants. Annual Review of Plant Biology 71, 403–433.

Zeng, F., Konnerup, D., Shabala, L., Zhou, M., Colmer, T.D., Zhang, G., et al., 2014. Linking oxygen availability with membrane potential maintenance and K^+ retention of barley roots: implications for waterlogging stress tolerance. Plant, Cell & Environment 37, 2325–2338.

Zeng, H., Chen, H., Zhang, M., Ding, M., Xu, F., Yan, F., et al., 2024. Plasma membrane H^+-ATPases in mineral nutrition and crop improvement. Trends in Plant Science 29 (9), 978–994. Available at https://doi.org/10.1016/j.tplants.2024.02.010.

Zhang, H., Sonnewald, U., 2017. Differences and commonalities of plant responses to single and combined stresses. The Plant Journal: For Cell and Molecular Biology 90, 839–855.

Zhang, X., Chen, L., Wu, H., Liu, L., Wan, X., 2018a. Root plasma membrane H^+-ATPase is involved in low pH-inhibited nitrogen accumulation in tea plants (*Camellia sinensis* L.). Plant Growth Regulation 86, 423–432.

Zhang, X., Wu, H., Chen, L., Liu, L., Wan, X., 2018. Maintenance of mesophyll potassium and regulation of plasma membrane H^+-ATPase are associated with physiological responses of tea plants to drought and subsequent rehydration. Crop Journal 6, 611–620.

Zhang, W., Wang, N., Yang, J., Guo, H., Liu, Z., Zheng, X., et al., 2020. The salt-induced transcription factor *GmMYB84* confers salinity tolerance in soybean. Plant Science (Shannon, Ireland) 291, 110326.

Zhang, X., Wu, H., Chen, J., Chen, L., Chang, N., Ge, G., et al., 2020. Higher ROS scavenging ability and plasma membrane H^+-ATPase activity are associated with potassium retention in drought tolerant tea plants. Journal of Plant Nutrition and Soil Science 183, 406–415.

Zhang, M., Wang, Y., Chen, X., Xu, F., Ding, M., Ye, W., et al., 2021. Plasma membrane H^+-ATPase overexpression increases rice yield via simultaneous enhancement of nutrient uptake and photosynthesis. Nature Communications 12, 735.

Zhang, Y., Yu, S., Li, Z., Chang, T., Xu, Q., Xu, H., et al., 2022. Effects of excessive nitrogen fertilizer and soil moisture deficiency on antioxidant enzyme system and osmotic adjustment in tomato seedlings. International Journal of Agricultural and Biological Engineering 15, 127–134.

Zhang, H., Zhu, J., Gong, Z., Zhu, J.K., 2022b. Abiotic stress responses in plants. Nature Reviews. Genetics 23, 104–119.

Zhou, R., Yu, X., Ottosen, C.-O., Rosenqvist, E., Zhao, L., Wang, Y., et al., 2017. Drought stress had a predominant effect over heat stress on three tomato cultivars subjected to combined stress. BMC Plant Biology 17, 1–13.

Zhou, R., Kong, L., Yu, X., Ottosen, C.-O., Zhao, T., Jiang, F., et al., 2019. Oxidative damage and antioxidant mechanism in tomatoes responding to drought and heat stress. Acta Physiologiae Plantarum / Polish Academy of Sciences, Committee of Plant Physiology Genetics and Breeding 41, 1–11.

Zhu, H., Yang, X., Li, Q., Guo, J., Ma, T., Liu, S., et al., 2022. The sweetpotato voltage-gated K^+ channel β subunit, KIbB1, positively regulates low-K^+ and high-salinity tolerance by maintaining ion homeostasis. Genes (Basel) 13, 1100.

Zhu, H., Guo, J., Ma, T., Liu, S., Zhou, Y., Yang, X., et al., 2023. The sweet potato K^+ transporter *IbHAK11* regulates K^+ deficiency and high salinity stress tolerance by maintaining positive ion homeostasis. Plants 12.

Zia, R., Nawaz, M.S., Siddique, M.J., Hakim, S., Imran, A., 2021. Plant survival under drought stress: implications, adaptive responses, and integrated rhizosphere management strategy for stress mitigation. Microbiological Research 242, 126626.

Zörb, C., Senbayram, M., Peiter, E., 2014. Potassium in agriculture–status and perspectives. Journal of Plant Physiology 171, 656–669.

Chapter 8

Transcription factors and their role in multiple abiotic stress tolerance in plants

Abida Anjum, Nuram Mubina, Tamima Sultana, Abida Lotif Tamanna, Shajneen Jahan Shoily, and Tahmina Islam*

Plant Breeding and Biotechnology Laboratory, Department of Botany, University of Dhaka, Dhaka, Bangladesh

Corresponding author. e-mail address: t.islam@du.ac.bd

8.1 Introduction

The environment heavily relies on the contribution of various plant species. The global demand for food production, biomass, CO_2, and O_2 supply, raw materials for industries, demand for proteins (60%–70%), and calories (70%–80%). have increased proportionally to the current population of the world (Mall et al., 2017). The human population is not the only variable at play when it comes to the rising demand for these elements. The Earth is shared with countless other species that depend on plants for these exact needs. Due to the ever-changing nature of the atmosphere, currently, even more so due to anthropogenic effects on global climate change, plants are being exposed to various environmental adversities within a very short time frame. To elucidate the effects of environmental adversities on their physiology and morphology, plants have developed adaptive stress-response systems through multiple regulatory pathways, which overlap in gene expression and are interconnected through various elements (Durrant et al., 2000; Schenk et al., 2000). All cellular signaling is ultimately regulated at multiple levels of gene expression. Stress signals are perceived by the cellular membranes and then intracellularly carried by various molecules and elements like phosphatases, kinases, and reactive oxygen species (ROS), to the transcription factors (TFs) or proteins. These factors are responsible for triggering stress signals and stress mitigation through the regulation of downstream stress genes by precise DNA binding to conserved sequences, acting as activators or repressors of the target gene. These incredibly important actions indicate that the extremely specific DNA binding of the TFs, at the promoters of target genes, is a sensitive function that can throw off the biochemical and physiological balance of the plant, as well as cellular homeostasis, even though it is a minor inconvenience.

Approximately 7% of the coding sequences of the entire plant genome are dedicated to various TFs (Udvardi et al., 2007). In the model species *Arabidopsis*, about 1600 TFs account for a total of 6% of their total protein-coding genes (26,000). In the case of rice, the percentage is 4.6% (Riechmann, 2007). Although the number of TFr families identified in plant species is nearly 60, the major TF families that occupy a huge percentage of various stress-responsive mechanisms are myeloblastosis (MYB,) WRKY transcription factors (named after the conserved WRKYGQK domain) (WRKY), Apetala2/ethylene response factor (AP2/ERF), No apical meristem, Arabidopsis transcription activation factor 1–2, and cup-shaped cotyledon (NAC), and basic region/leucine zipper (bZIP). The role of these major families in the adaptive stress response of plants, which has shaped their biology for the last few thousand years as an answer to drought, salinity, heat, cold, nutrient, heavy metal, pathogenic, viral, and waterlogging stresses, may be key to manipulating crop improvement, securing global food supply, and maintaining ecological balance. After the recent surge of genome editing tools and their application in plant biotechnology, it has become more evident to extensively study the involvement of various TFs in specific stress pathways and their molecular mechanisms, identifying their roles as positive or negative regulators of a stress signaling pathway. Isolating these factors and tweaking their functions at the base pair level through site-directed mutagenesis could be the key to creating super-crops, which can be climate-resilient, novel varieties. In this chapter, the fundamentals of what transcription factors are, their elements, structural and evolutionary characteristics of some major plant transcription factors, and recent investigations on their emerging role as regulators of abiotic stress and

Multiple Abiotic Stresses in Plants. DOI: https://doi.org/10.1016/B978-0-443-27424-4.00012-4

their potential to be utilized in novel genome editing technologies have been discussed to further emphasize the importance of novel research being carried out on each member of economically significant crops.

8.2 Transcription factors and their structural elements

Plant TFs continue to play a vital role in the regulation of gene expression through their interactive nature with various cis-regulatory elements of a gene. TFs are regulatory proteins that interact with DNA through unique folding patterns. The general constitution of a transcription factor in higher plants is usually a modular structure composed of four functional domains for the execution of effective DNA binding and regulation. Transcription factors have a DNA-binding domain (DBD), an oligomerization domain, a transcription regulation domain (TRD), and a nuclear localization signal (NLS) (Goff et al., 1992). These domains and other conserved sequences were the basis of the classification of around 56 plant TF types under nine super-classes (Blanc-Mathieu et al., 2024). The number of TFs and their distribution throughout the genome of an organism are variable. To date, 1600 TFs belonging to 28 gene families have been identified in *Arabidopsis*. The accurate structure of a TF has been examined by nuclear magnetic resonance spectroscopy and X-ray crystallography, and to understand the specifics of the function of their domains, functional and mutational analyses are essential (Patikoglou and Burley, 1997). The mechanism by which the DBD works is simple. DBDs often consist of highly conserved sequences that determine their binding specificity with DNA bases. These domains are basic, containing amino acid residues that regulate contact with cis-acting elements. Each TF may have only a single type of DBD and oligomerization unit (with some exceptions), but could occur in multiple copies.

The three-dimensional folding of proteins influences the binding efficiency to some extent. Changes in their folding may cause loss of binding specificity. The increase in binding accuracy may also depend on other amino acid residues aside from the DBD domain, which can interact with deoxyribose moieties and phosphate bonds non-specifically. Classification of the plant TFs is based on their DBD or their conserved regions. For example, zinc finger factors have a zinc ion-organized fold in their structure, basic domains bind using the basic alpha-helix, helix-turn-helix domains insert a recognition alpha-helix into the major groove of DNA, followed by a turn, and often two additional perpendicularly packed alpha-helices are found near it. The beta-hairpin type domains interact with DNA by forming a hairpin structure with their beta-folded sheet and inserting it inside the DNA major groove (Huang et al., 1996). There are large numbers of well-defined and functional DBDs that remain uncharacterized. Characterization of TFs varies not only depending on their DBDs but also on the length of their oligomerization domain. For the TF molecules to function, the formation of heterooligomers or homo-oligomers might be necessary; the oligomerization domain facilitates the promoter binding by increasing affinity. These domains, like the DBDs, also have highly conserved regions and are stabilized through beta sheet-coiled coil hydrophobic interaction (Katagiri et al., 1992, Guiltinan and Miller, 1994).

The TRDs present in these factors function as either repressors or activators of target genes. This is achieved through competitive binding, dimerization of TFs, or the action of individual amino acid residues on the target sequence. TFs are exclusive proteins that enter the nucleus upon their formation and perform activities on DNA. These possess functional domains for nuclear localization known as NLS. The number, type, and organization of NLS can vary within the same TF family. The core peptide composition of these domains is clusters of basic residues like lysine (K) and arginine (R). The copy number or NLS domains can be multiple and monopartite (single or multiple continuous) or bipartite (single/ double or multiple bipartite) in nature. For them to be translocated through the nuclear pores, some TFs often need dimerization with other transport molecules. These are hypothesized to be members who lack the presence of an NLS. The characterization of plant TFs and their division/subdivision into different groups and families is categorized by analyzing the structures, sequences, arrangement, quantity, and evolution of these fundamental domains (Liu et al., 1999).

8.3 Major plant transcription factors

8.3.1 Myeloblastosis family

The myeloblastosis (MYB) TFs family is one of the largest and most abundant TF families conserved across eukaryotes. The MYB family portrays a wide variety of functions in plants and outnumbers other large families. This superfamily has the largest number of members in *Arabidopsis*, with about 197 members (Katiyar et al., 2012; Riechmann, 2007). Genes of this family have also been identified in species like rice (155), maize (157), soybean (244), grape (134), and apple (229). The v-MYB gene of avian myeloblastosis virus was the first member of the MYB gene identified in eukaryotes (Klempnauer et al., 1982). Later, the C1 MYB gene of maize involved in anthocyanin biosynthesis (*ZmMYBC1*) was identified as the first MYB member in plants (Paz-Ares et al., 1987). The characteristic DBD of the MYB family, the

MYB domain, is highly conserved in plants, yeasts, and animals, and is spread across the higher plants in a large number. The physical structure and constitution of the MYB domain play an important part in the classification of this superfamily. The MYB domain consists of about four imperfect repeat arrays of amino acids, each forming a helix-turn-helix structure having about 50–53 amino acids. The structure intercalates with the major groove of DNA to efficiently bind with the cis-acting elements of the target gene (Yanhui et al., 2006). Apart from the MYB domain, MYB proteins show little sequence similarity. The arrangement of the repeats and their number in the binding domain were used to classify the MYB factors into different classes: R1R2R3-MYB, R1R2-MYB, and R1-MYB (Dubos et al., 2010). Two evolutionary schemes of the MYB proteins have been proposed, the first one being the "loss" model of MYB factors, proposed by Lipsick (1996) and supplemented by Jin and Martin (1999). The model predicts the origin of the MYB factors about 1 billion years ago. Plant-specific MYB members are thought to have diverged from the earliest members of MYB, creating a separate lineage. It describes the R1/2 single repeat as the ancestral base from which the other repeats emerged through several amplifications and replications. The loss of the R1/2 domain would have originated the R2R3 class MYB members. The MYB members containing four repeats are a part of the smallest class of the family, while the R2R3-MYB class is the largest (Miyake et al., 2003; Ito, 2005; Hernández et al., 2007).

Like all TFs, MYB factors can act as either transcriptional activators (*AtMYB3, AtMYB65, AtMYB103* genes in *Arabidopsis*, regulating anther production and development process) or repressors (*AtMYB4* repressing cinnamate-4-hydroxylase (C4H) encoding gene, *AtMYB32* repressing the *COMT* gene). The repressor MYB factors are comparatively new, with members of the R2R3 MYB being the majority. MYB family members are largely known to be ABA-dependent mediators, as their signal perception of abiotic stresses is highly ABA-induced. Members of MYB TFs have been extensively studied and are involved in cellular differentiation, response to hormone levels, secondary metabolism (flavonoid pathway), stress cycles like nutrients or light, cold, biological clock maintenance, chromosomal structure maintenance, and regulation of gene expression (Du et al., 2009; Ma et al., 2023).

8.3.2 WRKY transcription factors (named after the conserved WRKYGQK domain) motif family

The WRKY TF superfamily is one of the largest plant-specific TFs families. Members of this family show significant upregulation and respond to both biotic and abiotic stresses. The WRKY family has several members that participate in effector-triggered immunity (ETI) and activate defense-related genes. The family nomenclature was based on the presence of a highly conserved amino acid sequence motif at the N-terminus: WRKYGQK (Schmutz et al., 2010). The other motif present at the C-terminus, the Zn-binding motif composed of Cys and His residues, is crucial for its DNA-binding activity, recognizing the W-box element (TTGACT/C) in the cis-regulatory region of the target gene. The specificity of DNA binding of WRKY factors is attributed largely to the conserved C-terminus sequences; however, truncated WRKY proteins with only the C-terminus or N-terminus region show a fourfold decrease in their DNA binding capacity (Eulgem et al., 1999). The number of members in the *Arabidopsis* WRKY family is reported to be about 74 (Ulker and Somssich 2004). Based on the number of WRKY domains and their associated Zn-binding motif, the family is classified into three major groups, where group 1 has members with two WRKY domains; group 2 has members with one WRKY domain and the same kind of Zn-binding domain; group 3 has one WRKY domain with a different kind of Zn-binding domain (Eulgem et al., 2000). The group 3 WRKY members also seem to contain a calmodulin-binding domain, indicating their involvement in the CAM pathway and Ca^{2+} flux regulation. Group 2 of the family is not monophyletic and is further classified, based on the presence of many additional functional domains aside from the WRKY domain, into five types (group 2a–group 2e). Studies suggest that this group could still be evolving, as its Zn-binding domain shows a variable composition. The overall family shows elasticity in their structural configuration as the signature DNA binding sequence "WRKY" shows occasional amino acid substitutions, like "WRRY," "WKRY," and "WSKY" (Jiang et al., 2017).

The expanded role of the WRKY proteins and their multidimensional functions, ranging from biotic and abiotic stresses, indicate that during evolutionary processes, a chimeric WRKY may have gathered other functional motifs like B3, ZF_SBP and NAC, ATP_GRASP (Chen et al., 2019; Phukan et al., 2016). The evolution of the WRKY factors indicates that climate change may have been a huge factor in the expansion of the family, as the earliest members of the family seen in amoeba (*Dictyostelium purpureum*) and green algae (*Ostreococcus lucimarinus*) had only a single WRKY domain, which evolved into a double WRKY domain in higher taxa of green algae (charophytes). The increase in the number of these domains is surely attributed to the replication of the single-copy domain during gene duplication events (Rinerson et al., 2015). Like other transcription factors, the WRKY members can act as both activators and repressors and are part of ABA-dependent signal perception. Members of WRKY are regulators of other members of the same family when it comes to pathogen-triggered immune responses. The WRKY TFs play a tremendous role in the downstream signaling pathways during pathogen attack by delaying physiological processes like senescence or aiding in acquiring systemic resistance, which is modulated by components like salicylic acid, jasmonic acid, or nitric oxide. WRKY genes can play regulatory roles in salt, heat, drought tolerance (*TaWRKY1, TaWRKY19, TcWRKY53, AtWRKY6*), putrescine biosynthesis (*FcWRKY70*), hormone signal (*GbWRKY2, SiWRKY066/082*), cold tolerance (*OsWRKY71, CsWRKY46*), UV radiation and wounding damage (*OsWRKY89, AtWRKY36, ScWRKY045, AtWRKY6,11,15,22,33,40,53*), sugar starvation (*AtWRKY45* and *65*), and senescence (*AtWRKY6, OsWRKY42, BrWRKY65*) (Goyal et al., 2023). Overexpression of *McWRKY57*-like in *Arabidopsis* (Bai et al., 2023) and *TaWRKY2* in transgenic wheat (Gao et al., 2018) significantly enhances drought tolerance in both species. One of the most well-researched WRKY TFs in *Arabidopsis* is WRKY57, which plays a crucial role in the regulation of disease resistance, leaf senescence, and drought response. By altering the ABA and auxin signaling pathways, it improves drought tolerance (Jiang et al., 2012, 2016), encourages leaf senescence caused by jasmonic acid (Jiang et al., 2014), and affects the expression of jasmonate ZIM-domain genes, which control resistance to *Botrytis cinerea* (Jiang and Yu, 2016). The multifaceted role of WRKY TFs in plant stress responses can be utilized in genome editing systems by knocking out negative regulators, activating beneficial genes, and transgenic approaches to overexpression in high-yielding cultivars to improve abiotic stress tolerance.

8.3.3 APETALA2/ethylene responsive factor family

The AP2/ERF TFs family is the largest among the plant-specific TF families. An essential function of the AP2/ERF TF family is to control how plants react to a variety of abiotic stressors, including drought, salt, and submersion. Complex cross-regulatory networks that improve stress tolerance are formed by these TFs working with hormone signaling pathways. This family is subdivided into five distinct groups based on the number and type of DBDs present in the members. The AP2/ERF family of the *Arabidopsis* species has 145 members. Each member of this family has a conserved AP2/ERF domain. The five distinct groups of this family have diverse types and several conserved DBDs. The AP2 (APETALA2) subfamily is largely involved in floral development and has two AP2/ERF domains. The ERF (Ethylene Responsive Factor) subfamily has only a single AP2/ERF domain and shows involvement in response to biotic stress (Ohme-Takagi and Shinshi, 1995). The RAV subfamily has an AP2/ERF domain and additionally contains a B3 DBD, which is a commonly seen DBD in other plant-specific TF families like ABI3/VP1 (Kagaya et al., 1999; Sakuma et al., 2002). Members of the RAV subfamily regulate leaf senescence (Woo et al., 2010). The dehydration-responsive element binding (DREB) protein subfamily has regions outside the AP2/ERF domain, which contribute to DNA binding specificity (Canella et al., 2010). This subfamily is famously known for its significant role in plant cold stress responses, regulating the function of cold-inducible genes.

The DREB and ERF subfamilies recognize similar sequences that differ slightly from each other. The ERF factors recognize the GCC box sequence 5'-AGCCGCC-3' and the DRE factors recognize the (DRE) sequence 5'-A/GCCGAC-3'. The soloist subfamily contains a single AP2/ERF domain but has low homology to the other members of the family. The characteristic domain of the family, the AP2/ERF domain, is composed of around 60–70 amino acid residues, and these fold into one alpha-helix and two beta folds (Mizoi et al., 2012; Nakano et al., 2006). The evolutionary expansion of this family is striking, as the sequence homologous to the DBD of the family members has been identified in cyanobacteria, ciliates, and bacteriophages (Wuitschick et al., 2004). The 3D folding pattern of the *Arabidopsis* AtERF1 protein has striking similarity with the λ-integrase and the Tn*916*-integrase, both being of bacterial origin (Wojciak et al., 1999; Fadeev et al., 2009). The members of the family are so structurally similar to each other that they might show significant genetic redundancy. The AP2/ERF family plays a significant role in abiotic stress responses, in the case of both hormone-mediated signaling pathways and hormone-dependent signaling pathways. For example, *OsERF71, ZmDRE889, and ZmDREB39* in the case of ABA-mediated osmotic stress response, and *OsDREB2B, ZmEREB20, TmERF5, TmERF14*, and 36 in gibberellic acid-mediated drought, salt, and cold stress. *OsERF096* in IAA-mediated cold stress signaling. *OsERF19* is involved in salt stress without hormone-mediated signaling, *ERF95 and ERF97* enhance heat tolerance during heat stress, *PalERF2* in phosphate stress, and *StAP2075, StAP2126, and StAP2077* in Cd (heavy metal) stress (Ma et al., 2024a, b). *OsRPH1*, another AP2/ERF TF, is a negative regulator of plant height (Ma et al., 2020). Owing to their versatility, AP2/ERF TFs are desirable candidates for genetic enhancement, allowing for the creation of crops with high yields and resilience to stress.

8.3.4 Basic region/leucine zipper family

The bZIP TF family is one of the families conserved across all eukaryotes, from yeast to plants, humans, and other animals. Their functions in plants are extensive. Many members of the bZIP TF family are involved in important physiological processes, such as ABA signaling, defense against pathogen attack, morphological development, other hormone signaling, metabolic activities, flower and seed maturation, and senescence. The naming of the bZIP family is attributed to its structural formation. bZIPs are dimeric TFs that determine their binding specificity to target DNA and contribute to their functional flexibility. These proteins have two fundamental regions dedicated to DNA binding and dimerization. The basic region (BR) of the bZIP proteins lies close to the N-terminal end, while a leucine zipper motif in the C-terminal constitutes its dimerization domain (Vinson et al., 2002). The amino acid leucine provides a much more stable structure to the protein than other amino acids. The basic region is fundamental in binding in a parallel orientation to the major groove of DNA, forming an α-helix. The dimerization and the DNA-binding specificity it confers heavily depend on the leucine zipper (LZ) motif, which is extremely conserved, composed of five repetitions of heptads (seven amino acids sequentially arranged and forming around two alpha helices). The presence of leucine in these heptads is often fixed at two definite positions, often at the 7th position. Each amino acid in the heptad is designated with letters "a" to "g" in specific nomenclature. Depending on their functional diversities and sequence similarity, they can be subdivided into 10 separate groups (Groups A to I and S in *Arabidopsis*) (Jakoby et al., 2002; Nijhawan et al., 2008). The binding specificity of bZIP proteins depends on their dimerization. bZIP proteins preferentially bind to palindromic sequences containing an ACGT core. For example, bZIPs recognize sequences of C-box- GACGTC; G-box- CACGTG; ABRE ABA-responsive element- CCACGTGG; and TGA factors- TGACGT (Liao et al., 2008). To bind to specific DNA grooves, the C-terminus LZ regions wrap around each other in two alpha helices and split slightly at the N-N-terminus region, which gives them a distinguished "Y" like structure. The monomers of each dimer determine sequence binding further, which has been a major topic of study for scientists for many decades. Plant bZIPs mostly bind to the C or G box sequences (Izawa et al., 1993). Apart from their two most important regions, the region between BR and LZ is known as the hinge region, which also contributes to binding specificity to some extent (Niu et al., 1999). bZIP proteins can form both hetero-and homodimers; thus, the combination and permutation of different monomers coming together to form a dimer dedicated to specific binding has extended the function of this family beyond only stress signaling and response. As the function of bZIPs is widespread and intermingling between different pathways, their activity and binding are further regulated by the action of phosphatases. Phosphorylation allows the regulation of bZIPs at multiple levels by alteration of LZ, phosphate group addition in BR, or other regions of the protein (Lee et al., 2010; Deppmann et al., 2003). bZIPs are often found in complex networks of pathways; thus, phosphorylation allows uninterrupted regulation of bZIP activity on multiple genes. Some upstream kinases may contribute to this by simultaneously upregulating and downregulating the activities of specific bZIPs at the same time (Nieva et al., 2005). Such detailed regulation of transcriptional activation allows diverse roles of bZIP proteins in plant physiology, such as pathogen response regulation (*AtbZIP10*) (Kaminaka et al., 2006), production of seed storage protein (*AtbZIP25*) (Onate et al., 1999), negative regulation of ABA, cold, or salinity (*GmbZIP44,62*) (Liao et al., 2008), fiber elongation in cotton (Jiang et al., 2004), and expression of zein gene in maize (*EMBP-1a*) (Nijhawan et al., 2008). Plant responses to abiotic stressors are significantly influenced by bZIP TFs, especially through the ABA-dependent signaling pathway. Master regulators AREB1, AREB2, and ABF3 coordinate ABRE-dependent ABA signaling during drought stress and need ABA to fully activate (Yoshida et al., 2010). These TFs are important targets for agricultural genetic improvement because they control stress-responsive gene expression in addition to aiding in drought tolerance (Jakoby et al., 2002; Kang et al., 2002). Given the rising incidence of climate-induced stresses, using bZIP TFs through genome editing and genetic engineering holds great promise for creating crops that are resilient to climate change. In transgenic *Arabidopsis*, for example, overexpression of the wheat gene *TabZIP60* has been demonstrated to improve resistance to cold, drought, and salt stress (Baillo et al., 2019). Similarly, maize with its *bZIP68* gene disrupted by CRISPR-Cas9 had a higher tolerance to cold (Li et al., 2022). Furthermore, transcriptome and proteomic research have revealed novel functions for bZIPs, expanding their possible uses in enhancing disease resistance and nutrient-use efficiency (Yang et al., 2019; Alves et al., 2015).

8.3.5 No apical meristem, Arabidopsis transcription activation factor 1–2, and Cup-shaped cotyledon family

The NAC TF family is one of the largest plant-specific TF families. These TFs play crucial roles in plant development and responses to multiple abiotic stresses, including drought, salinity, and extreme temperatures. NAC TFs also show an interaction between biotic and abiotic stress responses, improving abiotic stress tolerance and aiding in pathogen defense (Nuruzzaman et al., 2013). Their adaptive relevance in diverse terrestrial habitats is highlighted by the evolutionary diversity and extension of the NAC TF family from early land plants to flowering plants (Kim et al., 2007) and Pu (Li et al., 2024). Members of the NAC family have shown diverse roles in the physiological development of plants, as well as in biotic and abiotic stress responses. The number of NAC members in *Arabidopsis* is 117 (Ooka et al., 2003) and 151 in rice (Nuruzzaman et al., 2010). In 1996, *No Apical Meristem* (*NAM*) was the first gene containing the NAC domain to be discovered in *Petunia*. The naming of the "NAC" TFswas done by combining the first letters of the first three genes

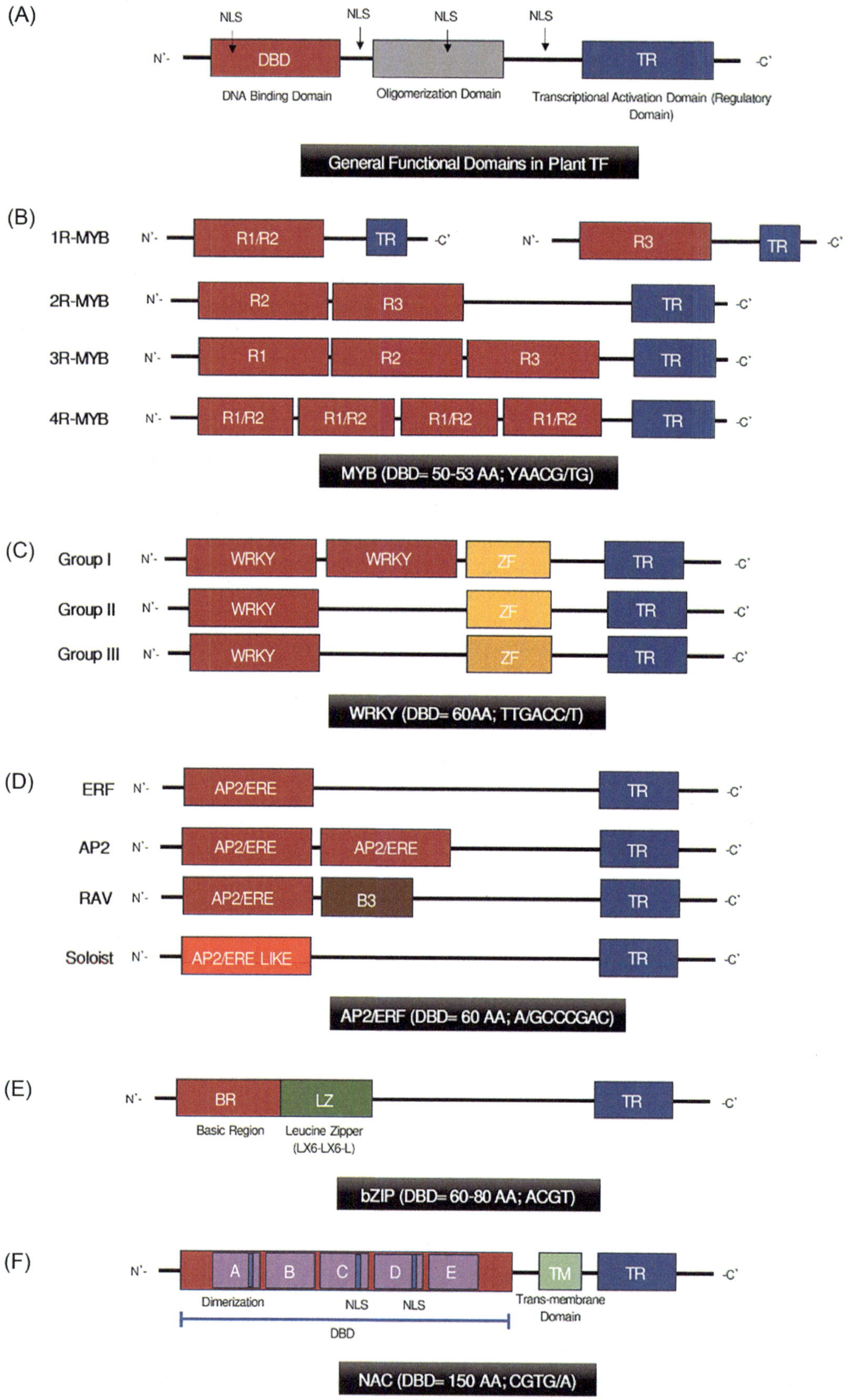
(A)
NLS
NLS
NLS
NLS
N'-
DBD
TR
-C'
DNA Binding Domain
Oligomerization Domain
Transcriptional Activation Domain (Regulatory Domain)
General Functional Domains in Plant TF
(B)
1R-MYB
R1/R2
TR
R3
2R-MYB
R2
R3
3R-MYB
R1
4R-MYB
MYB (DBD= 50-53 AA; YAACG/TG)
(C)
Group I
Group II
Group III
WRKY
ZF
WRKY (DBD= 60AA; TTGACC/T)
(D)
ERF
AP2
RAV
Soloist
AP2/ERE
B3
AP2/ERE LIKE
AP2/ERF (DBD= 60 AA; A/GCCCGAC)
(E)
BR
LZ
Basic Region
Leucine Zipper (LX6-LX6-L)
bZIP (DBD= 60-80 AA; ACGT)
(F)
A
B
C
D
E
TM
Dimerization
NLS
NLS
DBD
Trans-membrane Domain
NAC (DBD= 150 AA; CGTG/A)

identified, containing the NAC domain, namely, the *NAM* (*No Apical Meristem*), the *ATAF1–2*, and the *CUC2* (*Cup-shaped Cotyledon*) genes (Souer et al., 1996). To study the structure and domain constitution of NAC family members, the *Arabidopsis* NAC protein, ANAC's crystal structure, was the first NAC protein to be determined under X-ray crystallography. NAC proteins have a unique composition, which makes most of the members of this family in a particular plant species functionally redundant. They contain their characteristic DBD in the highly conserved N-terminal region, and the C-terminal region of these TFs is highly diverse. The N-terminal DBD is subdivided into five subdomains, based on their functional contribution. The five subdomains (A-E) consist of the NAC domain, which is roughly 150 amino acids long. A, C, and D subdomains are known to be highly conserved in comparison to the B and E subdomains, which are only slightly variable between species. These five contribute to the specific binding of the domain to the DNA sequence CGT(A/G), with a few exceptions. The amino acids Lys79, Lys123, Lys126, Arg85, Arg88, and Val119-Ser183 contribute largely to the specific binding of the proteins, on the other hand, the variable structure of the B and E subdomain, along with transactivating properties of the C-domain contributes to functional diversity of the NAC proteins (Ernst et al., 2004; Olsen et al., 2005; Chen et al., 2011). The amino acid Arg88 is conserved across all NAC proteins to date (Puranik et al., 2012). NAC proteins can form homo or heterodimers to function, where the subdomain A plays a role in their dimerization. In the subdomain D, an NLS resides, which mediates nuclear shuttling through lysine residues (Lee et al., 2010). The C-terminal region may be variable, but a few amino acids, namely, serine, proline, glutamine, and acid-rich residues, characterize it. There are 13 other variable domains present across different subgroups of the NAC family. Most NAC proteins are localized in the nucleus, but some are found in the cytoplasm and other organelles. NAC members, which contain a certain transmembrane domain, are grouped under NTLs (NAC membrane-bound transcription factors). This domain allows the protein to be present in the membranes in an inactive state, which later becomes active by being transported to the nucleus in response to hormone or stress signals. NAC members that are particularly responsive to stress treatments are known as sress-responsive NACs (SNACs) (Kim et al., 2007; Seo et al., 2008). There are reports of approximately 18 NAC TF subfamilies (Shang et al., 2013) owing to the diverse nature of their domains, but no particular fixed physiological activity could be attributed to any of these subfamilies. The range of physiological activities regulated by NAC proteins in plants is wide. The gene NAC1 has been found to regulate downstream genes like *AIR3* and *DBP* for lateral root production (Xie et al., 2000). The interaction between the *CUC1* and *CUC2* genes aids in the expression of an *STM (Shoot Meristemless)* gene in *Arabidopsis*, which is required for embryonic shoot apical meristem formation (Aida et al., 1997). NAC TFs also play a significant role in biotic and abiotic stress control. The genes ANAC019, 055,072, and several other ANACs are induced by wounding, ROS, ABA-dependent stress signaling pathways, salinity, drought, and dehydration (Fujita et al., 2004). The capacity of NAC TFs to control multiple stress responses at once makes them ideal targets for genome-editing technologies, transgenic techniques, and advanced breeding programs that aim to create high-yielding, stress-resilient crops with increased stability and productivity in the face of climate change (Fig. 8.1).

8.4 Common abiotic stressors in plants

Plants, as sessile organisms, face a range of ongoing challenges. They depend on their internal mechanisms to mitigate the impact generated by external conditions. Environmental stress has the potential to inflict physical damage on the plants because of prolonged unfavorable conditions. The negative impact of environmental factors on plants in controlled environments is referred to as abiotic stress. Plants encounter considerable obstacles from abiotic stress, including various nonliving factors that adversely affect their growth, development, and yield. Factors such as drought, salinity, temperature extremes (both heat and cold), heavy metal toxicity, and irradiation pose substantial challenges to global agricultural productivity and the well being of natural ecosystems (Hasanuzzaman et al., 2020). Approximately 10% of the global

FIGURE 8.1 **Simplified schematic representation of structural elements of various plant transcription factors.** (A) The typical functional domains in a transcription factor- A DNA binding domain at the N-terminal (DBD), a typical oligomerization domain, and a transcriptional regulation domain (TR). Nuclear Localization signals (NLS) are present in varied numbers and positions. (B) The four types of MYB transcription factors with MYB domain repeats at the N' terminal and a TR at the C' terminal. (C) The structure of three groups of WRKY transcription factors with the WRKY (WRKYGQK) domain as the main DNA binding domain, and two different types of Zinc Finger Domain (ZF). (D) Four sub-families of the AP2/ERF superfamily, each with a varied number of the AP2/ERE repeat. The RAV (Related to ABI3/VP1) family has a distinct B3 domain along with the main DBD and the soloist family contains a AP2/ERE like domain as their main DBD. E) Structure of the bZIP transcription factor family, their DBD consists of a basic region and a leucine zipper. (F) The five sub-domains (A-E) of the NAC family DBD. Blue boxes within purple box depict their unique function along with DNA binding activities. The transmembrane domain (TM) is depicted in green colored box. The target sequences and total amino acid length of the distinct DNA binding domains of the transcription factor families are mentioned in black colored boxes.

irrigated land area is affected by drought and salinity, according to the Food and Agriculture Organization Report 2024. Abiotic stress reduces water accessibility to plant roots by elevating the concentration of water-soluble salts in the soil, leading to heightened osmotic pressure surrounding the roots. Physiological changes encompass a reduction in osmotic potential, alterations in water potential, and the creation of nutritional imbalances, among other factors. These stressors can impact plants during their three critical stages: vegetative development, preanthesis, and the terminal phase.

8.4.1 Drought stress

Drought is a climatic condition characterized by a considerable decrease in water availability for a prolonged period over substantial areas of land. Drought arises in plants when the absorption of water becomes challenging due to extreme temperatures and high soil salinity, by reducing soil water potential (Hasanuzzaman et al., 2020). These factors make it difficult for the roots to absorb soil water. The intensity, duration, and extent of the affected region are the three main aspects of drought (Tsakiris and Tigkas, 2007). With these dimensions, drought has severe impacts on plants in various ways, influencing their growth, metabolism, and survivability.

The mechanical effect includes the loss of water from the inside of the cell, causing a reduction in cell volume, which ultimately leads to plasmolysis or cell death. An excessive amount of ion accumulation occurs in the water loss condition of cells, in response to metabolic activity, damaging the cell by membrane disruption and protein denaturation (Kalefetoğlu and Ekmekci, 2005). In response to drought, plants usually experience stomatal closure, which confines the abundance of CO_2. On the other hand, chlorophyll absorbs excessive amounts of light energy that cannot be used because of the CO_2 limitation through photosynthesis. This high amount of energy catalyzes the formation of ROS, including superoxide (O_2^-), hydrogen peroxide (H_2O_2), and hydroxyl radicals (OH^-) (Kaur et al., 2021). This compound finally decreases the reaction efficiency of the photosynthetic process, potentially leading to cell death. Drought stress leads to a notable increase in nitrogen levels and a significant decrease in phosphorus levels, while potassium levels remain unaffected (Samarah, 2005). It significantly affects the components of chlorophyll a, b, and carotenoids, as well as impairing photosystem I and photosystem II (Fu and Huang, 2001). For example, the vegetative growth of Solanaceae species is negatively affected by drought stress, which lowers plant height, leaf surface area, and photosynthetic efficiency. This ultimately results in a lower crop yield and poor quality of tomato, potato, eggplant, and pepper (Pang et al., 2024).

8.4.2 Salinity stress

Salt stress is a major abiotic stressor that adversely influences plants at all stages of development, including germination, seedling, and vegetative stages. Soil salinity refers to the presence of a high concentration of soluble salts in the root zone, particularly in arid and semi-arid lands (Hailu and Mehari, 2021). The occurrence of this salinity is frequently a result of natural processes influenced by the existing climatic conditions, elevated levels of evapotranspiration, and a lack of adequate leaching water. While typically associated with arid environments, soil affected by salt can also be present in regions where the quality of irrigation water is subpar (Parida and Das, 2005).

The primary indication of salt stress is plant growth inhibition, which can ultimately result in cell death under severe salt toxicity. Abscisic acid is produced in response to salt stress and is transported to guard cells, causing stomatal closure, which leads to a reduction in the rate of photosynthesis and the onset of oxidative stress. Cell growth is eventually suppressed, and cells die because of this stress. Salt stress has a major impact on metabolic processes and nutrient absorption. For instance, salt stress has several growth-inhibiting effects on rice plants, such as decreased rates of relative growth, dry matter accumulation, leaf growth, leaf cell enlargement, and net CO_2 absorption (Riaz et al., 2019).

The competition between sodium ions (Na^+) and chloride ions (Cl^-) with vital nutrient minerals, such as potassium (K^+), calcium (Ca^{2+}), and nitrate (NO_3^{2-}) ions, causes ion toxicity under saline conditions, thereby lowering nutrient levels in different plant species. Excess Na^+ ions can inhibit the positive effects of K^+ ions and obstruct the functioning of enzymes found in the cytoplasm (Ahanger et al., 2017). Potassium is an essential nutrient mineral for plants, and the negative impacts of Na^+ ion toxicity are inversely associated with the K^+ ion concentration in plant systems; therefore a lack of potassium results in harmful effects on plant growth. Furthermore, higher amounts of Na^+ in saline media inhibit the uptake and transport of Ca^{2+} and might induce Ca^{2+} deficiency in plants because of the higher Na^+/Ca^{2+} ratio (Garg, 1998).

8.4.3 Extreme temperatures stress (cold stress)

Cold is one of the major abiotic stresses that induces substantial alterations in crop biomass and yield. The optimum temperature for plant growth, development, and acclimatization varies from plant to plant. Plants are categorized into different groups according to their endurance to low temperatures: chilling sensitive (0°C–15°C), chilling resistant (tolerates low temperatures until ice formation begins), and frost resistant (very low temperatures), among others. For example, symptoms of damage are evident in maize, soybean, cotton, and tomato at temperatures below 10°C–15°C, as these crops are chilling sensitive (Lynch, 1990). Conversely, barley, wheat, and rye exhibit strong adaptation to frigid temperatures (Pecchioni et al., 2013).

At the seedling stage, cold stress results in decreased germination, stunted growth, reduced leaf area, and chlorosis. Cold stress specifically impacts cellular function by altering protoplasmic streaming, cytomembrane structure and lipid composition, distribution of Ca^{2+} ions, leakage of electrolytes and amino acids, as well as modification to the photosynthetic machinery and electron flow (Thomashow, 1999). The cell membrane of cold-sensitive plants is packed with saturated fatty acids, which limit plant development and production. Cold stress is linked to compromised cell organelle integrity resulting from disrupted compartmentalization, protein assembly, photosynthesis, and metabolic processes (Hannah et al., 2006). Brief exposure to cold induces temporary physiological changes; however, extended exposure may result in necrosis. Cold exposure during anthesis in rice, as well as in the legume crops, like black gram and green gram, led to floral sterility.

8.4.4 Extreme temperature stress (heat stress)

Temperatures exceeding the typical optimum are perceived as heat stress by all living organisms. Heat stress caused by rising temperatures is a serious threat to sustainable crop production worldwide. Plant survival, growth, and yield are negatively affected by heat stress, which causes noticeable alterations at the cell, tissue, and organ levels (Hassan et al., 2021). It occurs with varied degrees of intensity and duration and at different chronological scales.

Two of the most heat-sensitive phases in a plant's life cycle are the early seedling stage and the reproductive period, independent of the plant's functional type (Wahid et al., 2007). While high temperature increases the pace of sexual development, it shortens the time required to include photosynthesis in fruit or seed development (Menzel, 2023). Heat stress mostly influences photosynthesis by two different channels: carbon metabolism in the stroma and light-dependent processes in the thylakoids. The most important effect is found in photosystem II (PSII), which is quite temperature-sensitive and can be either totally or partially inactivated at high temperatures (Yamamoto, 2016). During vital D_1 and D_2 protein denaturation, the damage reaches the oxygen-evolving complex, causing interrupted electron transport. Brief exposure to elevated temperatures during seed filling might result in accelerated filling, consequently diminishing quality and production (Sehgal et al., 2018). Because of increased water loss from transpiration or evaporation, high temperatures can lead to drought stress. Additionally, this temperature can affect seed germination, plant growth, and development, as well as prevent plants from emerging. Internal damage includes serious protein damage, inhibition of certain protein production, inactivation of essential enzymes, damage to membranes, and disruption of the cell division process (Wahid et al., 2007).

Heat is the main limiting factor for dryland crops, such as sorghum, millets, and wheat, which face challenging environmental circumstances during growth and developmental stages (Pfleiderer et al., 2019; Zampieri et al., 2017). For high temperatures (40°C during the day and 35°C at night) for 15 days, the overall rice grain production falls by about 14%, while the wheat grain yield and number of grains per panicle are expected to drop by 8.4% and 7.5%, respectively, for each degree Celsius increase in temperature (Li et al., 2020).

8.4.5 Heavy metal stress

The phrase "heavy metals" denotes metallic elements that are characterized by a density greater than that of water, which can be toxic or harmful even in low concentrations, largely influenced by their chemical properties (Hawkes, 1997). Heavy metals, including zinc (Zn), cadmium (Cd), copper (Cu), mercury (Hg), chromium (Cr), nickel (Ni), lead (Pb), iron (Fe), and cobalt (Co), represent significant environmental pollutants, predominantly located in anthropogenic regions or soils due to industrial waste, fertilizer application, smelting, and sewage disposal (Herawati et al., 2000). The impact of heavy metal toxicity on plants is influenced by various factors, including the species of the plant, the specific metal involved, its concentration, chemical form, as well as the composition and pH of the soil (Gill, 2014). While many of these metals are essential minerals for plant health, their presence in excessive amounts can lead to harmful effects. Certain metals, including Cd, Hg, and Pb, exhibit strong toxicity toward metal-sensitive enzymes, leading to inhibited growth and mortality in organisms (Nieboer and Richardson, 1980).

Metal stressors adversely affect enzymatic activity, disrupt metabolic reactions, and contribute to oxidative stress. Zn toxicity may lead to stunted growth, leaf aging, and signs of chlorosis, especially in younger leaves. It leads to the accumulation of Cu and Mn in both roots and shoots (Nagajyoti et al., 2010), while conversely causing a deficiency in phosphorus (Lee et al., 1996). An excessive concentration of Zn ions displaces Mg^{2+} in RuBisCo, which eventually prevents it from fixing carbon.

High levels of Cd accumulation interfere with Fe (III) reductase, resulting in Fe deficiency and affecting nitrate reductase, which diminishes nitrate absorption in plants (Ghori et al., 2019). The inhibition of CO_2 fixation-related enzymes and decreased chlorophyll synthesis caused by Cd eventually affects photosynthesis. Elevated concentrations of Cd inflict harm on plants at the chromosomal level, resulting in fragmentation or aberration of chromosomes. The breakdown of mitochondria is another significant effect of Cd, which severely hinders plant respiration. In the context of Cd toxicity, Cd replaces calcium in calmodulin, which plays a crucial role in cell signaling. This displacement leads to the inhibition of calmodulin-dependent phosphodiesterase activity (Rivetta et al., 1997).

Cu is crucial for ATP synthesis and carbon assimilation in plants; however, excess Cu can lead to oxidative stress and significant damage to membranes and macromolecules (Yadav, 2010). For example, Excessive Cu negatively affects the germination, seedling length, and root quantity of *Solanum melongena*(Neelima and Reddy, 2002).

Hg functions as a deleterious element for plants, potentially infiltrating aquatic systems and binding to water channel proteins, thereby impeding the flow of water to plants. It also influences mitochondrial and chloroplast functions by disrupting the electron transport chain and triggering oxidative stress (Nagajyoti et al., 2010). The stability of genomic templates in plants is also influenced by Hg treatment, which disrupts cellular functions and obstructs normal growth and development (Malar et al., 2015).

The effects of Cr toxicity on plant physiology exhibit similarities, yet there is a notable distinction in the reduction of amylase enzyme activity and the transport of sugars from the axes of the embryo. It enhances the synthesis of metabolites, such as glutathione and ascorbic acid, which adversely affect plants (Shanker et al., 2005). It additionally includes chlorosis, impedes growth, and leads to the wilting of plants (Öztürk et al., 2015).

Ni is essential for sustaining stress tolerance, reducing the cellular redox state, and enhancing nitrogen use efficiency (Vatansever et al., 2017). The elevated levels present in the soil influence plants, manifesting symptoms such as necrosis, chlorosis, nutrient deficiency, and dysfunction of the cell membrane. The reduction in water content within monocot and dicot plants frequently serves as a significant marker for the presence of Ni toxicity (Yadav, 2010).

Pb alters cell membrane permeability, induces hormonal changes, inhibits various enzymes containing sulfhydryl groups, decreases water content, and disrupts mineral nutrition. This process inhibits the Calvin cycle and the electron transport chain, while simultaneously leading to a deficiency of carbon dioxide due to the closure of stomatal pores (Sharma and Dubey, 2005). Lead toxicity is frequently observed in water hyacinths, with significant accumulation of this metal occurring in the roots, petioles, and leaf tissues, ultimately leading to the inhibition of plant growth (Malar et al., 2014).

These heavy metals can infiltrate the nucleus, potentially inducing genotoxicity in plants through pro-mutagenic damage. This encompasses the molecular cross-linking of DNA and proteins, DNA strand breakage, depurination, and modifications to DNA bases. These elements have a detrimental effect on various crops such as sugarcane, rice, wheat, potato, tomato, cotton, and other cash crops. For instance, the practice of rice cultivation is fundamentally dependent on water; however, the presence of water tainted with hazardous heavy metals such as arsenic, cadmium, lead, and mercury poses significant carcinogenic risks to human health and is generally hazardous to all forms of life (Jallad, 2015).

8.4.6 Ultraviolet radiation

The overuse of halogenated compounds has resulted in the depletion of the ozone layer, which has led to an increase in the amount of ultraviolet (UV) radiation that reaches the surface of the Earth. The effects of UV light on plant growth are severe and include a reduction in growth, a drop in photosynthesis, and a reduction in biomass. According to Liu et al. (2004), the radiation sensitivity of plants varies depending on the species or variety of plants, growth stages and conditions, and the intensity of UV light. Through the inhibition of cell division and metabolic changes, exposure to UV-B light, which is one of the many types of UV light, causes a reduction in plant growth, reduced leaf formation, and stem branching. It has been found that the oxidation of tubulins that occurs under the influence of UV rays causes a reduction in cell division in chickpea and cucumber. Additionally, UV radiation can impede the transcription of histone proteins, which in turn inhibits cell division (Hollósy 2002). There was a higher rate of DNA alterations caused by UV-B and UV-C rays in early growth stages and pollen grains, according to research conducted by Hollósy (2002). According to Dai et al. (1997), the harm caused by UV-B radiation may be linked to the disruption of active oxygen metabolism in rice plants. This occurs because of the breakdown of defense mechanisms.

TABLE 8.1 Role of different transcriptional factor gene families in abiotic stress tolerance in plants.

Family	Gene	Species	Inducible by	References
bHLH	*RrUNE12*	*Rosa roxburghii*	Salt	Yang et al. (2025)
	ZmbHLH36	*Zea mays*	Salt and drought	Dai et al. (2024)
	AhbHLH121	*Arachis hypogaea*		Zhao et al. (2024a)
WRKY	*VhWRKY44*	*Arabidopsis thaliana*	Cold and Salt	Zhang et al. (2024)
	IlWRKY22	*Iris laevigata*	Salt and drought	Fan et al. (2024)
	HbWRKY82	*Hevea brasiliensis*	Dehydration and salt	Kang et al. (2021)
	PeWRKY48	*Passiflora edulis*	Drought, salt, and cold	Ma et al. (2024)
	TaWRKY46	*Triticum aestivum*	Osmotic	Li et al. (2020)
	AhWRKY75	*Arachis hypogaea*	Salt	Zhu et al. (2021)
	TaWRKY55	*Triticum aestivum*	Salt and alkali	Wang et al. (2024a, 2024b, 2024c)
NAC	*FvNAC29*	*Fragaria vesca*	Salt and cold	Li et al. (2024)
	SlNAC12	*Solanum lycopersicum*	Salt	Chen et al. (2025)
	GmNAC20	*Glycine max*	Salt and cold	Yarra and Wei, (2021)
	HaNAC3	*Haloxylon ammodendron*	Temperature, salt, and drought	Li et al. (2022)
	ZmSNAC13	*Zea mays*	Salt and drought	Luo et al. (2022)
MYB	*IbMYB330*	*Ipomoea batatas*	Salt and drought	Wang et al. (2024a, 2024b, 2024c)
	FaMYB63	*Fragaria ananassa*	Salt	Wang et al. (2024a, 2024b, 2024c)
	HbMYB44	*Hevea brasiliensis*	Wound and drought	Qin et al. (2022)
	OsMYB59	*Oryza sativa*	Nitrate	Islam et al. (2021)
bZIP	*ZmbZIP76*	*Zea mays*	Osmotic	He et al. (2024)
	ChbZIP1	*Chlorella sp. BLD*	Alkali	Qu et al. (2021)
HD2	*AtHD2D*	*Arabidopsis thaliana*	Salt, drought, osmotic, and metal ion	Chu et al. (2024)
ERF	*OsERF19*	*Oryza sativa*	Salt	Huang et al. (2021)
NF-Y	*TaNF-YB2*	*Triticum aestivum*	Drought	Zhao et al. (2024a)
CCT	*SlCCT6*	*Solanum lycopersicum*	Drought	Huang et al. (2024)
HHO	*MtHHO3*	*Medicago truncatula*	Salt and osmotic	Wang et al. (2024a, 2024b, 2024c)
HD-Zip	*VvHDZ4*	*Vitis vinifera*	Drought	Li et al. (2021)
Dof	*MtDof32*	*Medicago truncatula*	Salt and osmotic	Guo et al. (2021)
TCP	*BpTCP3*	*Betula platyphylla*	Salt	Ren et al. (2021)
SPL	*CpSPL5CpSP-L8*	*Codonopsis pilosula*	Salt	Li et al. (2025)

8.5 Roles of transcription factors in abiotic stress responses

It is anticipated that abiotic stressors such as oxidative stress, drought, salt, and extreme temperatures will worsen, posing serious threats to human health, agricultural output, and ecological equilibrium. By disrupting ion and osmotic equilibrium, reducing photosynthesis, raising cellular energy expenditure, and producing redox imbalances, these stresses have an impact on the growth and survival of plants on an individual basis. Plant signaling is essential for controlling responses to drought and other abiotic stressors, according to physiological and molecular investigations (Zhu, 2016). By regulating downstream stress-responsive genes, TFs have been identified as essential regulators of certain abiotic stress responses (Joshi et al., 2016). They participate in pathways that alter how plants react to environmental stressors and function as molecular switches for transcriptional control (Manna et al., 2021). Enhancing plant stress tolerance requires an understanding of the mechanisms behind TF-mediated regulation.

8.5.1 The roles of transcription factors in the abiotic stress responses

Recent studies on TFs have demonstrated their significant involvement in stress adaptation within plants (Table 8.1). Major TF families—*bHLH, WRKY, NAC, MYB, AP2/ERF, bZIP*, and many others contribute differently, regulating various signaling pathways related to stress response within plants. By encouraging the production of *AhPOD, AhChAT*, and *AhSOD*, overexpression of the TF *AhbHLH121* within the *BHLH* family increases the activity of antioxidant enzymes under stress and improves ROS scavenging (Zhao et al., 2024b). Transgenic *Arabidopsis* plants overexpressing *ZmbHLH36* are more resistant to salt and drought stress (Dai et al., 2024). By raising the level of AsA and oxidative stress, the *BHLH* TF *RrUNE12* helps to mitigate salt stress by activating AsA production in *Rosa roxburghii* fruit. *RrUNE12* and *RrGGP2* responses to exogenous NaCl treatment as opposed to therapy without NaCl addition. By increasing the expression of *RrGGP2* and genes linked to antioxidants, *RrUNE12* overexpression in fruit calli reduced the damage caused by salt stress (Yang et al., 2025).

Arabidopsis thaliana showing tolerance to cold and salt stress is enhanced by *VhWRKY44*, a *WRKY* TF derived from grapes (Zhang et al., 2024). By favorably controlling the antioxidant system and photosynthesis, *IlWRKY22* improves the resistance of *Nicotiana tabacum* to salt and drought (Fan et al., 2024). Through ABA-dependent signaling and increased antioxidant capacity, *WRKY45* plays a crucial role in regulating excessive salinity and osmotic stress. Overexpression of *WRKY45* may enhance the tolerance of *Arabidopsis* to salinity and osmotic stress, along with significantly higher root length, chlorophyll, and proline content in comparison to the WT plants after salt or PEG treatment. (Zhou et al., 2025). *HbWRKY82* functions as a transcriptional regulator in ET- and ABA-mediated leaf senescence and abiotic stress responses by participating in the ET and ROS signaling pathways (Kang et al., 2021). When exposed to cold stress, *PeWRKY48* is highly expressed, and transgenic yeast and *Arabidopsis* show improved resistance to drought, salt, and cold (Ma et al., 2024). Through both ABA-dependent and ABA-independent pathways, *TaWRKY46* functions as an osmotic stress-positive regulator in wheat, greatly enhancing osmotic tolerance (Li et al., 2020). By increasing the effectiveness of the ROS scavenging system and photosynthesis under stress, *AhWRKY75*-carrying transgenic peanut lines were linked to salt tolerance (Zhu et al., 2021).

The plant TF *TaPLATZ2* is involved in wheat response to saline-alkaline stress by directly suppressing the expression of *TaHA2/TaSOS3*. In addition, by upregulating *TaPLATZ2, TaWRKY55* negatively controls the ability of wheat to withstand salt and alkali stress (Wei et al., 2025). By controlling the expression of *AtRD29a, AtCCA1, AtP5CS1*, and *AtSnRK2.4*, overexpression of *FvNAC29* increases plant resilience to cold and salt stress (Li et al., 2024). By enhancing ion homeostasis, antioxidant enzyme systems, and flavonoid accumulation, *SlNAC12* improves tomato resistance to salt stress. In comparison to the wild type (WT), transgenic tomato lines overexpressing *SlNAC12* exhibited enhanced tolerance to salinity, as opposed to the reduced inhibitory effect of growth metrics under salt stress (Chen et al., 2025). By encouraging proline accumulation and lateral root development, *GmNAC20* overexpression in rice plants increases resilience to cold and salinity stress (Yarra and Wei, 2021). *HaNAC3* is involved in reactions to treatments with phytohormones, especially ABA, high temperatures, high salt, and simulated drought. (Li et al., 2022). By reducing water loss, extending effective photosynthesis, and enhancing cell membrane stability during drought conditions, *ZmSNAC13* overexpression enhanced *Arabidopsis*'s resilience to salt and drought stressors (Luo et al., 2022). Through the c-repeat binding factor system, which mediates plant responses to balance growth and freezing stress tolerance, *NAC056* gives freezing tolerance (Xu et al., 2024).

Under osmotic, salt, and oxidative stress conditions, *IbMYB330* expression is highly stimulated, which results in increased expression of genes linked to stress, including SOD, POD, APX, and P5CS (Wang et al., 2024a, 2024b, 2024c). By directly attaching to the promoter of the SOS1 gene and triggering its expression, *FaMYB63* overexpression improves salt tolerance in *Arabidopsis* (Wang et al., 2024a, 2024b, 2024c). Rubber plants have significant levels of *HbMYB44* expression in response to a variety of abiotic stressors, such as wounding and drought (Qin et al., 2022). Under K^+ stress, *OsMYB59* is upregulated in BRRI dhan56 but downregulated in BRRI dhan48 and BRRI dhan71. In contrast to control plants, *OsMYB59* is elevated in all three kinds under nitrate stress (Islam et al., 2021). *OsERF19* overexpression increases *OsLEA3, OsNHX1, OsHKT6, and OsOTS1*, enhancing rice's ability to withstand salt (Huang et al., 2021). As a nuclear TF *ZmbZIP76* improves ABA levels, osmotic stress resistance, and ROS scavenging (He et al., 2024). *ChbZIP1* overexpression in *Arabidopsis* increased alkali resistance, indicating that *ChbZIP1* might be essential for the alkali stress response. These results imply that *ChbZIP1* may enhance plant response to alkali stress by using the active oxygen detoxification pathway (Qu et al., 2021). By encouraging the growth of lateral roots, *AtHD2D* helps plants deal with abiotic stress. ROS buildup in roots is encouraged by overexpression of *AtHD2D*, suggesting that *AtHD2D* also plays a role in the development of lateral roots mediated by ROS (Chu et al., 2024). *TaNF-YB2* roles in controlling drought adaptation via influencing stomatal mobility, osmolyte production, and ROS balance were validated by transgene analysis (Zhao et al., 2024a). By boosting antioxidant enzyme activity and triggering stress-related genes, *SlCCT6* overexpression improved tomato drought resistance, while *SlCCT6* deletion reduced drought resistance (Huang et al., 2024).

The GARP family TF, *MtHHO3*, is involved in downregulating the expression of stress-associated genes and ABA-responsive pathways. In *Arabidopsis thaliana*, overexpression of *MtHHO3* led to enhanced resistance to ABA inhibition but also hypersensitivity to salt and osmotic stress. Moreover, overexpression of *MtHHO3* in *Medicago truncatula* led to a reduction in salinity tolerance, whereas loss-of-function mutants mthho3-1 and mthho3-2 exhibited greater resistance to salt stress in comparison to plants of the WT (Wang et al., 2024b). The tomato, *Solanum lycopersicum* L. "Alisa Craig," showed higher *VvHDZ4* gene expression, which was found to contribute to improved drought stress tolerance (Li et al., 2021). *MtDof32* transgenic plants showed enhanced resistance to salt and osmotic stress in *Medicago truncatula* (Guo et al., 2021). The *BpTCP3* gene plays a positive regulatory role in salt stress. *BpTCP3* overexpressed lines showed lower levels of hydrogen peroxide, malondialdehyde, and salt damage index than *Betula platyphylla* WT plants, but higher relative/absolute high growth under salt stress. Conversely, the *BpTCP3*-repressed lines displayed indications of salt stress sensitivity. These results imply that the *BpTCP3* TF improves *Betula platyphylla*'s salt tolerance by reducing ROS damage (Ren et al., 2021). *Codonopsis pilosula* attributing tolerance to salt is decreased by *CpSPL5* and *CpSPL8* through inhibition of the SOS pathway (Li et al., 2025) (Fig. 8.2).

8.6 The potential of transcription factors as targets for genome editing

Various TFs that inhibit resistance against stress by acting on genes as a negative regulator of expression or stress signal can be potential targets of knockout experiments through genome editing (Table 8.1). The *ArabidopsisAtERF71/HRE2* is an AP2/ERF TF and regulates responses to hypoxia and osmotic stress. Both positive and negative cis-regulatory elements are present in the promoter sequence of this gene. The transcription factor *HAT22/ABIG1* negatively regulates *AtERF71/HRE2* by binding to its promoter, thus inhibiting its expression under certain stress conditions (Seok et al., 2022). The C2H2-type zinc finger TF *MdZAT10* negatively regulates drought tolerance in apple (Yang et al., 2021). The tomato AP2/ERF TF *SlERF.B1* increases sensitivity to both salt and drought stresses, indicating a negative regulatory role in abiotic stress response. Overexpression of *SlERF.B1* leads to increased sensitivity to these stresses (Wang et al., 2022). The *PalWRKY77* TF negatively regulates responses to salt tolerance and influences abscisic acid signaling in Populus. It acts as a repressor of multiple genes associated with salt tolerance (Jiang et al., 2021). The MADS-box factor *AGL16* is implicated in the plant's response to salt stress, functioning by downregulating salt-responsive genes, and acts as a negative regulator when overexpressed in *Arabidopsis thaliana* (Zhao et al., 2021). In rice, the transcription factor *OsWRKY5* enhances drought tolerance through abscisic acid signaling pathways. Inactivation of *OsWRKY5* leads to improved drought tolerance, indicating its role as a negative regulator of this trait (Li et al., 2022). The R2R3-MYB transcription factor *FtMYB22* negatively regulates both salt and drought stress responses through an ABA-dependent pathway. Its function as a negative regulator suggests that it may inhibit pathways that would otherwise promote stress tolerance (Zhao et al., 2022). The novel *OsERF106MZ*, ERF-X-type TF reduces shoot growth and tolerance to salinity stress in rice. Its overexpression leads to reduced growth and salinity tolerance (Chen et al., 2021). *ZmHsf08* is a heat shock TF that negatively regulates responses to both salt and drought stress in maize. Its overexpression results in decreased stress tolerance, making it a negative regulator of these responses (Wang et al., 2021). The *OsbHLH024* transcription factor improves salt stress resistance when knocked out via CRISPR-Cas9 technology in rice. Its normal function appears to inhibit salt resistance (Alam et al., 2022). *OsbZIP*5 TF negatively regulates drought tolerance in rice. Its activity is linked to reduced drought resilience (Tong et al., 2021). *PtrMYB3* is an R2R3-MYB TF from *Poncirus trifoliata* that negatively regulates salt tolerance and hydrogen peroxide scavenging (Wei et al., 2021). The WRKY TF

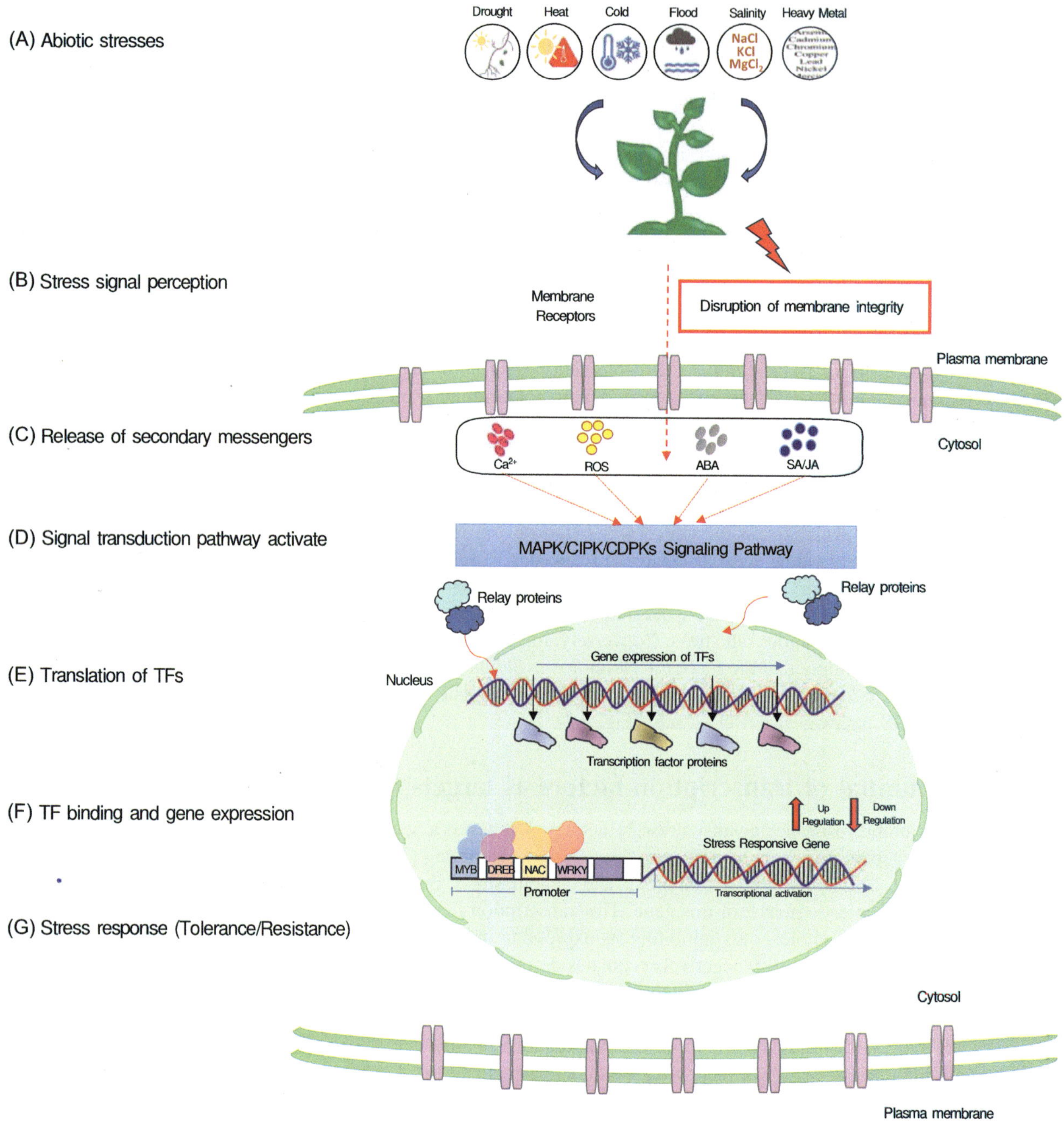

FIGURE 8.2 Schematic representation of abiotic stress signal mechanism in plants. (A) Plants subjected to various different kinds of abiotic stress (Drought, Salinity, Heat, Cold, Flood, Waterlogging, Heavy metals etc.) (B) Signal perception- by the plant's cellular membrane receptors through disruption of membrane integrity and change in osmolarity. (C) The secondary messengers activated- increased level of Ca^{2+} in plants, and release of ROS (Reactive Oxygen Species) by membrane bound NADPH (Reduced Nicotinamide Adenine Dinucleotide Phosphate), increased level of hormones like ABA (Abscisic Acid, for drought and salinity), SA (Salicylic Acid, against pathogen) and JA (Jasmonic Acid, for mechanical stresses) which (D) Entice the activation of MAPK (Mitogen Activated Protein Kinases) and CDPK (Calcium Depended Protein Kinases) signaling pathways. (E) Activation of various transcription factor cassettes by the activity of relay proteins from the signal transduction. (F) Transcription factors binding to their downstream target genes related to adaptive stress response (bZIP, NAC, MYB for ABA depended, salinity, drought, heat etc. stresses; AP2/ERF for cold, drought, dehydration, WRKY for chiefly pathogen responses). (G) Activation or suppression of target genes as a response to stress signals. The target gene's expression is either upregulated (If it has a positive regulatory role to stress) or down regulated (If it acts as a negative regulator) by the activity of the transcription factors on their cis-acting elements.

GmWRKY46 negatively regulates phosphorus tolerance, primarily by modifying root morphology in soybean. It limits phosphorus uptake efficiency, acting as a negative regulator of phosphorus tolerance (Li et al., 2022). The AP2/ERF TF *TaERF-6-3A* identified in wheat, is involved in responses to drought and salinity stresses. Functional characterization studies have implied that the potential inhibition of stress responses is mediated through this gene (Yu et al., 2022) (Table 8.2).

Gene editing of *ZmGA20ox3* improves plant architecture and drought tolerance in maize by manipulating gibberellin biosynthesis pathways (Li et al., 2024). Gene editing of *SlGT30* improves both drought resistance and fruit yield through endoreduplication processes in tomato plants. Editing alleviates this suppression for better performance (Lv et al., 2025). The *SlHyPRP1* and *SlDEA1* genes are involved in imparting genetic tolerance to multiple stress factors through CRISPR-Cas9-based genome editing techniques in tomatoes. Their normal functions may inhibit effective responses against various stresses; editing enhances overall resilience (Saikia et al., 2024). *ZmHDT103* gene negatively regulates drought stress tolerance in maize seedlings by influencing physiological responses under water-limited conditions. Its activity likely reduces the ability of plants to withstand drought conditions effectively, making it a target for enhancing resilience through genetic modifications (Wang et al., 2024a, 2024b, 2024c). Targeted mutations in *OsDSG1* enhance salt tolerance by modifying signaling pathways associated with osmotic balance and ion homeostasis in rice plants (Ly et al., 2024).

TABLE 8.2 Transcription factors as potential targets of genome editing for conferring abiotic stress resistance in various plant species.

Gene	Species	Negative regulator of function	References
AtERF71/HRE2	*Arabidopsis*	Salt stress	Seok et al. (2022).
MdZAT10	Apple	Drought stress	Yang et al. (2021).
SlERF.B1	Tomato	Salt stress	Wang et al. (2022).
PalWRKY77	*Populus*	Salt stress	Jiang et al. (2021).
AGL16	*Arabidopsis*	Salt stress	Zhao et al. (2021)
OsWRKY5	Rice	Drought stress	Li et al. (2022).
FtMYB22	Buckwheat	Salt and drought stress	Zhao et al. (2022).
OsERF106MZ	Rice	Salt stress	Chen et al. (2021).
ZmHsf08	Maize	Drought and salt stress	Wang et al. (2021).
OsbHLH024	Rice	Salt stress	Alam et al. (2022).
OsbZIP5	Rice	Drought stress	Tong et al. (2021).
PtrMYB3	*Poncirus trifoliata*	Salt stress	Wei et al., (2021).
GmWRKY46	Soybean	Phosphorus tolerance	Li et al. (2022).
TaERF-6-3A	wheat	Drought and salt	Yu et al. (2022).
ZmGA20ox3	Maize	Drought	Li et al. (2024).
SlGT30	Tomato	Drought	Ly et al. (2024).
SlHyPRP1 and SlDEA1	Tomato	Multiple stress	Saikia et al. (2024).
ZmHDT103	Maize	Drought	Wang et al. (2024a, 2024b, 2024c)
OsDSG1	Rice	Salt	Ly et al. (2024).
OsNF-YC5	Rice	Salt	Yan et al. (2024).
StDMR6-1	Potato	Multiple stress	Karlsson et al. (2024).
ZmASR1	Maize	Drought	Yan et al. (2024).

OsNF-YC5 is a member of the NF-Y family that negatively regulates salt tolerance by modulating the expression of genes related to osmotic balance and ion transport in rice plants (Yan et al., 2024). Genome editing of *StDMR6–1* results in potato plants being less affected by various environmental stresses through enhanced signaling pathways for stress responses (Karlsson et al., 2024). *ZmASR1* negatively regulates drought stress tolerance by influencing physiological processes related to water retention and use efficiency in maize plants. Targeting this gene can enhance resilience under drought conditions (Yan et al., 2024).

8.7 Significance of transcription factors in biotechnology

TFs and their roles are complex and diverse, and need much investigation to be properly utilized. Extensive studies on the mechanism of action of these factors, their respective pathways, their controlled traits, and the subsequent gene network they are a part of, reveal significant potential for application in the biotechnological sector.

Globally, plant species are threatened by many recurrent issues. As plants are sessile, they are vulnerable to environmental adversities and global climate change. Plants are the primary source of raw materials and foods and are used in many industries. Global food security and the economy depend heavily on the cultivation and yield of many plant species. As these are being threatened now, biotechnological advances must solve the problems hindering the growth, yield, physiology, and behavior of many species of plants. Many studies have been conducted on TFs and their involvement in pathways that may be modulated for generating biofortified crops and high-yielding varieties, increasing biomass and aiding biofuel production, altered flowering fruiting patterns in different species, altered photosynthetic capacity and metabolically engineered plants, climate resilient crops, etc. Due to adverse changes in the global climate, regions that were favorable to salinity-tolerant crops have started to become more saline or additionally exposed to high temperatures. Desert areas have started to go green while green lands have started to become barren. Many dry areas are now becoming waterlogged, moderate temperatures are jumping to high temperatures, and nutrients in certain plots of soil are becoming altered due to excessive or uncontrolled fertilizer usage. To remedy all of these, resilient varieties may be needed. Thus specific TFs that are involved in certain pathways can be modulated in a way to create these resilient plant species. Functional validation of TF members and confirming whether their role in a certain physiological process is positive or negative helps biotechnologists decide the method of genetic engineering that could be applied to modulate said genes.

A well-known member of the AP2/ERF superfamily, the rice *Sub1A* gene, an ERF-type TF which inhibits excessive shoot elongation, conserves energy for survival under submerged conditions and modulates ethylene signaling pathways to give submergence tolerance (Xu et al., 2006). *OsDREB2B* links growth control and stress response by modifying gibberellin metabolism, which in turn negatively influences plant height in rice (Ma et al., 2022). *OsRPH1*, another AP2/ERF TF, is a negative regulator of plant height (Ma et al., 2020). Traits of submergence and plant height could be key studies to the understanding of waterlogging tolerance or submergence tolerance of important cash crops like rice in rural coastal areas, considering the increasing water levels of the ocean due to melting of the icebergs in the polar region, which have started to threaten the use of coastal lands in bay countries. Overexpression of *McWRKY57*-like in *Arabidopsis* (Jiang et al., 2012, 2016) and *TaWRKY2* in transgenic wheat (Gao et al., 2018) significantly enhanced drought tolerance in both species. In transgenic *Arabidopsis*, for example, overexpression of the wheat gene *TabZIP60* has been demonstrated to improve resistance to cold, drought, and salt stress (Baillo et al., 2019). Similarly, maize that had its *bZIP68* gene disrupted by CRISPR-Cas9 had a higher tolerance to cold (Li et al., 2022). Genes mentioned previously, like the *AhbHLH121* (Zhao et al., 2024b), *TaWRKY46* (Li et al., 2020), and *AtHD2D* (Chu et al., 2024), are involved in stress response through ROS scavenging and photosynthesis. The manipulation of TFs that are involved in ROS scavenging, can be key to plants' ion homeostasis and stress response under adverse conditions. The TFs that are involved in photosynthetic control can be utilized in greater industries for creating novel varieties that are high-yielding or have high chlorophyll utilization, resulting in increased biomass. The gene *RrUNE12* (Yang et al., 2025) reduces salt stress in *Rosa roxburghii*, which results in better-quality fruits. Fruit-yielding plants, which are prone to salt stress and other abiotic stresses and are hindered in fruit production, can be a target for engineering to yield better quality fruits, even under stress conditions. *SlNAC12* (Chen et al., 2025) is another such gene that increases salinity tolerance in tomato plants. *ChbZIP1* (Qu et al., 2021), *TaPLATZ2*, and *TaWRKY55* (Wei et al., 2025) have been linked to salt and alkali stress. Altered lands or any accidental change in soil alkalinity may render a plot of land useless, and resilient species or varieties can withstand such alterations and increase productivity in these cases. As far as transgenic and genetically modified organism (GMO) development is concerned, biotechnology has now advanced to the level of genome editing, creating novel genotypes with no requirement for transgenic or GMO development. The TFs mentioned as potential targets of genome editing need to be functionally investigated further through knock-out experiments. The new genotype's capability to serve as a resilient variety may open many possibilities for the plant's stress response and immunity. Future agricultural biotechnology will benefit greatly from these TFs' capacity to integrate various stress signals, which will provide sustainable food production despite environmental obstacles.

8.8 Conclusion

Plant TFs have increasingly become more important to plant biotechnological studies due to severe adversities faced by plant species because of global climate change. Ecological diversity, food security, and various industries face risks if the production of well-known and economically important plant species is threatened. This chapter focuses on the very basic structure and elements of TFs, as well as describes the structure, types, members, domains, and role of major plant TF families like WRKY, AP2/ERF, MYB, bZIP, and NAC. Understanding which member of a TF family is related to a typical physiological role would require proper knowledge of the plant stress response. Common abiotic stressors that threaten various plant species globally are mentioned here for a better understanding of stresses and their effects on plants. Promising members of some TF families summarized in the tables of this chapter could be of great use to biotechnological experiments in understanding the molecular mechanism of genes in a stress signaling pathway, key genes present within a gene network, expression pattern of certain TFsunder different stresses, modulation of their expression through genetic engineering or genome editing, and developing resilient varieties through these methods. The rapid change in the Earth's climatic conditions calls for the rapid evolution of important plant species. Such a feat is achievable through the proper application of molecular biological studies and the utilization of modern technologies like genome editing on the key drivers of physiological response in plants, which are the TFs.

References

Ahanger, M.A., Tomar, N.S., Tittal, M., Argal, S., Agarwal, R.M., 2017. Plant growth under water/salt stress: ROS production; antioxidants and significance of added potassium under such conditions. Physiology and Molecular Biology of Plants 23 (4), 731–744. Available at https://doi.org/10.1007/s12298-017-0462-7.

Aida, M., Ishida, T., Fukaki, H., Fujisawa, H., Tasaka, M., 1997. Genes involved in organ separation in Arabidopsis: an analysis of the cup-shaped cotyledon mutant. The Plant Cell 9 (6), 841–857.

Alam, M.S., Kong, J., Tao, R., Ahmed, T., Alamin, M., Alotaibi, S.S., et al., 2022. CRISPR/Cas9 mediated knockout of the OsbHLH024 transcription factor improves salt stress resistance in rice (Oryza sativa L.). Plants 11 (9). Available at https://doi.org/10.3390/plants11091184.

Alves, M.S., Soares, Z.G., Vidigal, P.M.P., Barros, E.G., Poddanosqui, A.M.P., Aoyagi, L.N., et al., 2015. Differential expression of four soybean bZIP genes during Phakopsora pachyrhizi infection. Functional and Integrative Genomics 15 (6), 685–696. Available at https://doi.org/10.1007/s10142-015-0445-0

Baillo, E.H., Kimotho, R.N., Zhang, Z., Xu, P., 2019. Transcription factors associated with abiotic and biotic stress tolerance and their potential for crops improvement. Genes 10 (10). Available at https://doi.org/10.3390/genes10100771.

Bai, Y., Zhang, T., Zheng, X., Li, B., Qi, X., Xu, Y., et al., 2023. Overexpression of a WRKY transcription factor McWRKY57-like from Mentha canadensis L. enhances drought tolerance in transgenic Arabidopsis. BMC Plant Biology 23 (1). Available at https://doi.org/10.1186/s12870-023-04213-y.

Blanc-Mathieu, R., Dumas, R., Turchi, L., Lucas, J., Parcy, F., 2024. Plant-TFClass: a structural classification for plant transcription factors. Trends in Plant Science 29 (1), 40–51. Available at https://doi.org/10.1016/j.tplants.2023.06.023.

Canella, D., Gilmour, S.J., Kuhn, L.A., Thomashow, M.F., 2010. DNA binding by the Arabidopsis CBF1 transcription factor requires the PKKP/RAGRxKFxETRHP signature sequence. Biochimica et Biophysica Acta - Gene Regulatory Mechanisms 1799 (5-6), 454–462. Available at https://doi.org/10.1016/j.bbagrm.2009.11.017.

Chen, Q., Wang, Q., Xiong, L., Lou, Z., 2011. A structural view of the conserved domain of rice stress-responsive NAC1. Protein & Cell 2 (1), 55–63. Available at https://doi.org/10.1007/s13238-011-1010-9.

Chen, X., Li, C., Wang, H., Guo, Z., 2019. WRKY transcription factors: evolution, binding, and action. Phytopathology Research 1 (1). Available at https://doi.org/10.1186/s42483-019-0022-x.

Chen, H.C., Chien, T.C., Chen, T.Y., Chiang, M.H., Lai, M.H., Chang, M.C., 2021. Overexpression of a Novel ERF-X-type transcription factor, OsERF106MZ, reduces shoot growth and tolerance to salinity stress in rice. Rice 14, 82. Available at https://doi.org/10.1186/s12284-021-00525-5.

Chen, S., Zhang, W., Zhang, Q., Li, B., Zhang, M., Qin, J., et al., 2025. SlNAC12, a novel NAC-type transcription factor, confers salt stress tolerance in tomato. Plant Cell Reports 44 (1). Available at https://doi.org/10.1007/s00299-024-03400-x.

Chu, Y., Duan, R., Song, H., Zhang, W., Zhou, Y., Ma, Y., et al., 2024. AtHD2D is involved in regulating lateral root development and participates in abiotic stress response in Arabidopsis. Journal of Plant Physiology 297, 154242. Available at https://doi.org/10.1016/j.jplph.2024.154242.

Dai, Q., Yan, B., Huang, S., Liu, X., Peng, S., Miranda, M.L.L., et al., 1997. Response of oxidative stress defense systems in rice (Oryza sativa) leaves with supplemental UV-B radiation. Physiologia Plantarum 101 (2), 301–308. Available at https://doi.org/10.1034/j.1399-3054.1997.1010207.x.

Dai, Z., Zhao, K., Zheng, D., Guo, S., Zou, H., Wu, Z., et al., 2024. Heterologous expression of the maize transcription factor ZmbHLH36 enhances abiotic stress tolerance in Arabidopsis. aBIOTECH 5 (3), 339–350. Available at https://doi.org/10.1007/s42994-024-00159-3.

Deppmann, C.D., Thornton, T.M., Utama, F.E., Taparowsky, E.J., 2003. Phosphorylation of BATF regulates DNA binding: a novel mechanism for AP-1 (activator protein-1) regulation. Biochemical Journal 374 (2), 423–431.

Dubos, C., Stracke, R., Grotewold, E., Weisshaar, B., Martin, C., Lepiniec, L., 2010. MYB transcription factors in Arabidopsis. Trends in Plant Science 15 (10), 573–581. Available at https://doi.org/10.1016/j.tplants.2010.06.005.

Durrant, W.E., Rowland, O., Piedras, P., Hammond-Kosack, K.E., Jones, J.D.G., 2000. cDNA-AFLP reveals a striking overlap in race-specific resistance and wound response gene expression profiles. The Plant Cell 12 (6), 963–977. Available at https://doi.org/10.1105/tpc.12.6.963.

Du, H., Zhang, L., Liu, L., Tang, X.F., Yang, W.J., Wu, Y.M., et al., 2009. Biochemical and molecular characterization of plant MYB transcription factor family. Biochemistry (Moscow) 74 (1), 1–11. Available at https://doi.org/10.1134/S0006297909010015.

Ernst, H.A., Olsen, A.N., Skriver, K., Larsen, S., Lo Leggio, L., 2004. Structure of the conserved domain of ANAC, a member of the NAC family of transcription factors. EMBO Reports 5 (3), 297–303. Available at https://doi.org/10.1038/sj.embor.7400093.

Eulgem, T., Rushton, P.J., Schmelzer, E., Hahlbrock, K., Somssich, I.E., 1999. Early nuclear events in plant defence signalling: rapid gene activation by WRKY transcription factors. EMBO Journal 18 (17), 4689–4699. Available at https://doi.org/10.1093/emboj/18.17.4689.

Eulgem, T., Rushton, P.J., Robatzek, S., Somssich, I.E., 2000. The WRKY superfamily of plant transcription factors. Trends in Plant Science 5 (5), 199–206. Available at https://doi.org/10.1016/S1360-1385(00)01600-9.

Fadeev, E.A., Sam, M.D., Clubb, R.T., 2009. NMR structure of the amino-terminal domain of the lambda integrase protein in complex with DNA: immobilization of a flexible tail facilitates beta-sheet recognition of the major groove. Journal of Molecular Biology 388 (4), 682–690.

Fan, L., Niu, Z., Shi, G., Song, Z., Yang, Q., Zhou, S., et al., 2024. WRKY22 transcription factor from iris laevigata regulates flowering time and resistance to salt and drought. Plants 13 (9), 1191. Available at https://doi.org/10.3390/plants13091191.

Fujita, M., Fujita, Y., Maruyama, K., Seki, M., Hiratsu, K., Ohme-Takagi, M., et al., 2004. A dehydration-induced NAC protein, RD26, is involved in a novel ABA-dependent stress-signaling pathway. Plant Journal 39 (6), 863–876. Available at https://doi.org/10.1111/j.1365-313X.2004.02171.x.

Fu, J., Huang, B., 2001. Involvement of antioxidants and lipid peroxidation in the adaptation of two cool-season grasses to localized drought stress. Environmental and Experimental Botany 45 (2), 105–114. Available at https://doi.org/10.1016/s0098-8472(00)00084-8.

Gao, H., Wang, Y., Xu, P., Zhang, Z., 2018. Overexpression of a WRKY transcription factor TaWRKY2 enhances drought stress tolerance in transgenic wheat. Frontiers in Plant Science 9. Available at https://doi.org/10.3389/fpls.2018.00997.

Garg, B.K., 1998. Role of calcium in plants under salt stress. Arid Zone Research Association of India. India Annals of Arid Zone 37 (2), 107–118.

Ghori, N.-H., Ghori, T., Hayat, M.Q., Imadi, S.R., Gul, A., Altay, V., et al., 2019. Heavy metal stress and responses in plants. International Journal of Environmental Science and Technology 16 (3), 1807–1828. Available at https://doi.org/10.1007/s13762-019-02215-8.

Gill, M., 2014. Heavy metal stress in plants: a review. International Journal of Advanced Research 2 (6), 1043–1055.

Goff, S.A., Cone, K.C., Chandler, V.L., 1992. Functional analysis of the transcriptional activator encoded by the maize B gene: evidence for a direct functional interaction between two classes of regulatory proteins. Genes & Development 6 (5), 864–875.

Goyal, P., Devi, R., Verma, B., Hussain, S., Arora, P., Tabassum, R., et al., 2023. WRKY transcription factors: evolution, regulation, and functional diversity in plants. Protoplasma 260 (2), 331–348. Available at https://doi.org/10.1007/s00709-022-01794-7.

Guiltinan, M.J., Miller, L., 1994. Molecular characterization of the DNA-binding and dimerization domains of the bZIP transcription factor, EmBP-1. Plant Molecular Biology 26 (4), 1041–1053.

Guo, T., Wang, S., Zhang, T., Xu, L., Li, Y., Chao, Y., et al., 2021. Expression of the Medicago truncatula MtDof32 transcription factor regulates plant growth and enhances abiotic stress tolerances in transgenic Arabidopsis. Environmental and Experimental Botany 183, 104339. Available at https://doi.org/10.1016/j.envexpbot.2020.104339.

Hailu, B., Mehari, H., 2021. Impacts of soil salinity/sodicity on soil-water relations and plant growth in dry land areas: a review. Journal of Natural Science Research 12 (3), 1–10.

Hannah, M.A., Wiese, D., Freund, S., Fiehn, O., Heyer, A.G., Hincha, D.K., 2006. Natural genetic variation of freezing tolerance in arabidopsis. Plant Physiology 142 (1), 98–112. Available at https://doi.org/10.1104/pp.106.081141.

Hasanuzzaman, M., Carvalho Minhoto Teixeira Filho, M., Fujita, M., Assis Rodrigues Nogueira, T., 2020. Sustainable Crop Production. IntechOpen. Available at https://doi.org/10.5772/intechopen.83521.

Hassan, M.U., Chattha, M.U., Khan, I., Chattha, M.B., Barbanti, L., Aamer, M., et al., 2021. Heat stress in cultivated plants: nature, impact, mechanisms, and mitigation strategies—a review. Plant Biosystems - An International Journal Dealing with All Aspects of Plant Biology 155 (2), 211–234. Available at https://doi.org/10.1080/11263504.2020.1727987.

Hawkes, S.J., 1997. What is a" heavy metal"? Journal of Chemical Education 74 (11), 1374.

Herawati, N., Suzuki, S., Hayashi, K., Rivai, I.F., Koyama, H., 2000. Cadmium, copper, and zinc levels in rice and soil of Japan, Indonesia, and China by soil type. Bulletin of Environmental Contamination and Toxicology 64 (1), 33–39. Available at https://doi.org/10.1007/s001289910006.

Hernández, G., Ramírez, M., Valdés-López, O., Tesfaye, M., Graham, M.A., Czechowski, T., et al., 2007. Phosphorus stress in common bean: root transcript and metabolic responses. Plant Physiology 144 (2), 752–767.

He, L., Wu, Z., Wang, X., Zhao, C., Cheng, D., Du, C., et al., 2024. A novel maize F-bZIP member, ZmbZIP76, functions as a positive regulator in ABA-mediated abiotic stress tolerance by binding to ACGT-containing elements. Plant Science 341, 111952. Available at https://doi.org/10.1016/j.plantsci.2023.111952.

Hollósy, F., 2002. Effects of ultraviolet radiation on plant cells. Micron (Oxford, England: 1993) 33 (2), 179–197.

Huang, H., Tudor, M., Su, T., Zhang, Y., Hu, Y., Ma, H., 1996. DNA binding properties of two Arabidopsis MADS domain proteins: binding consensus and dimer formation. The Plant Cell 8 (1), 81–94. Available at https://doi.org/10.1105/tpc.8.1.81.

Huang, S., Ma, Z., Hu, L., Huang, K., Zhang, M., Zhang, S., et al., 2021. Involvement of rice transcription factor OsERF19 in response to ABA and salt stress responses. Plant Physiology and Biochemistry 167, 22–30. Available at https://doi.org/10.1016/j.plaphy.2021.07.027.

Huang, S., Yang, X., Li, W., Xu, Z., Xie, Y., Meng, X., et al., 2024. Genome-wide analysis of the CCT gene family and functional characterization of SlCCT6 in response to drought stress in tomato. International Journal of Biological Macromolecules 280 (Pt 3), 135906. Available at https://doi.org/10.1016/j.ijbiomac.2024.135906.

Islam, M.Q., Hasan, M.N., Hoque, H., Jewel, N.A., Bhuiyan, M.F.H., Prodhan, S.H., 2021. Characterization of transcription factor MYB59 and expression profiling in response to low K+ and NO3− in indica rice (Oryza sativa L.). Journal of Genetic Engineering and Biotechnology 19 (1). Available at https://doi.org/10.1186/s43141-021-00248-6.

Ito, M., 2005. Conservation and diversification of three-repeat Myb transcription factors in plants. Journal of Plant Research 118 (1), 61–69. Available at https://doi.org/10.1007/s10265-005-0192-8.

Izawa, T., Foster, R., Chua, N.H., 1993. Plant bZIP protein DNA binding specificity. Journal of Molecular Biology 230 (4), 1131–1144.

Jakoby, M., Weisshaar, B., Dröge-Laser, W., Vicente-Carbajosa, J., Tiedemann, J., Kroj, T., et al., 2002. bZIP transcription factors in Arabidopsis. Trends in Plant Science 7 (3), 106–111. Available at https://doi.org/10.1016/s1360-1385(01)02223-3.

Jallad, K.N., 2015. Heavy metal exposure from ingesting rice and its related potential hazardous health risks to humans. Environmental Science and Pollution Research 22 (20), 15449–15458. Available at https://doi.org/10.1007/s11356-015-4753-7.

Jiang, Y., Yu, D., 2016. The WRKY57 transcription factor affects the expression of jasmonate ZIM-domain genes transcriptionally to compromise botrytis cinerea resistance. Plant Physiology 171 (4), 2771–2782. Available at https://doi.org/10.1104/pp.16.00747.

Jiang, J.X., Guo, W.Z., Zhang, T.Z., 2004. Cloning and expression analysis of a bZIP cDNA in Gossypium hirsutum L. Yi chuan xue bao = Acta genetica Sinica 31 (6), 616–621.

Jiang, Y., Liang, G., Yu, D., 2012. Activated expression of WRKY57 confers drought tolerance in Arabidopsis. Molecular Plant 5 (6), 1375–1388. Available at https://doi.org/10.1093/mp/sss080.

Jiang, Y., Liang, G., Yang, S., Yu, D., 2014. Arabidopsis WRKY57 functions as a node of convergence for jasmonic acid– and auxin-mediated signaling in jasmonic acid–induced leaf senescence. The Plant Cell 26 (1), 230–245. Available at https://doi.org/10.1105/tpc.113.117838.

Jiang, Y., Qiu, Y., Hu, Y., Yu, D., 2016. Heterologous expression of AtWRKY57 confers drought tolerance in Oryza sativa. Frontiers in Plant Science 7, 145. Available at https://doi.org/10.3389/fpls.2016.00145.

Jiang, C., Shen, Q.J., Wang, B., He, B., Xiao, S., Chen, L., et al., 2017. Transcriptome analysis of WRKY gene family in Oryza officinalis Wall ex Watt and WRKY genes involved in responses to Xanthomonas oryzae pv. oryzae stress. PLoS One 12 (11), e0188742. Available at https://doi.org/10.1371/journal.pone.0188742.

Jiang, Y., Tong, S., Chen, N., Liu, B., Bai, Q., Chen, Y., et al., 2021. The PalWRKY77 transcription factor negatively regulates salt tolerance and abscisic acid signaling in Populus. The Plant Journal 105 (5), 1258–1273. Available at https://doi.org/10.1111/tpj.15109.

Jin, H., Martin, C., 1999. Multifunctionality and diversity within the plant MYB-gene family. Plant Molecular Biology 41 (5), 577–585. Available at https://doi.org/10.1023/a:1006319732410.

Joshi, R., Wani, S.H., Singh, B., Bohra, A., Dar, Z.A., Lone, A.A., et al., 2016. Transcription factors and plants response to drought stress: current understanding and future directions. Frontiers in Plant Science 7 (2016). Available at https://doi.org/10.3389/fpls.2016.01029.

Kagaya, Y., Ohmiya, K., Hattori, T., 1999. RAV1, a novel DNA-binding protein, binds to bipartite recognition sequence through two distinct DNA-binding domains uniquely found in higher plants. Nucleic Acids Research 27 (2), 470–478. Available at https://doi.org/10.1093/nar/27.2.470.

Kalefetoğlu, T., Ekmekci, Y., 2005. The effects of drought on plants and tolerance mechanisms. Gazi University Journal of Science 18 (4), 723–740.

Kaminaka, H., Näke, C., Epple, P., Dittgen, J., Schütze, K., Chaban, C., et al., 2006. bZIP10-LSD1 antagonism modulates basal defense and cell death in Arabidopsis following infection. EMBO Journal 25 (18), 4400–4411. Available at https://doi.org/10.1038/sj.emboj.7601312.

Kang, J.Y., Choi, H.I., Im, M.Y., Soo, Y.K., 2002. Arabidopsis basic leucine zipper proteins that mediate stress-responsive abscisic acid signaling. Plant Cell 14 (2), 343–357. Available at https://doi.org/10.1105/tpc.010362.

Kang, G., Yan, D., Chen, X., Yang, L., Zeng, R., 2021. HbWRKY82, a novel IIc WRKY transcription factor from Hevea brasiliensis associated with abiotic stress tolerance and leaf senescence in Arabidopsis. Physiologia Plantarum 171 (1), 151–160. Available at https://doi.org/10.1111/ppl.13238.

Karlsson, M., Kieu, N.P., Lenman, M., Marttila, S., Resjö, S., Zahid, M.A., et al., 2024. CRISPR/Cas9 genome editing of potato StDMR6-1 results in plants less affected by different stress conditions. Horticulture Research 11 (7). Available at https://doi.org/10.1093/hr/uhae130.

Katagiri, F., Seipel, K., Chua, N.H., 1992. Identification of a novel dimer stabilization region in a plant bZIP transcription activator. Molecular and Cellular Biology 12 (11), 4809–4816.

Katiyar, A., Smita, S., Lenka, S.K., Rajwanshi, R., Chinnusamy, V., Bansal, K.C., 2012. Genome-wide classification and expression analysis of MYB transcription factor families in rice and Arabidopsis. BMC Genomics 13 (1). Available at https://doi.org/10.1186/1471-2164-13-544India.

Kaur, S., Samiksha, P., Kaur, J., Thakur, S., Sharma, N., Pandit, K., et al., 2021. Reactive oxygen species metabolism and antioxidant defense in plants under stress. Environmental Stress Physiology of Plants and Crop Productivity. Bentham Science Publishers, Sharjah, UAE., pp. 107–123. Available at 10.2174/9781681087900121010011.

Kim, H.S., Park, B.O., Yoo, J.H., Jung, M.S., Lee, S.M., Han, H.J., et al., 2007. Identification of a calmodulin-binding NAC protein as a transcriptional repressor in Arabidopsis. Journal of Biological Chemistry 282 (50), 36292–36302. Available at https://doi.org/10.1074/jbc.m705217200.

Klempnauer, K.H., Gonda, T.J., Bishop, J.M., 1982. Nucleotide sequence of the retroviral leukemia gene v-myb and its cellular progenitor c-myb: the architecture of a transduced oncogene. Cell 31 (2), 453–463.

Lee, C.W., Choi, J.-M., Pak, C.-H., 1996. Micrnutrient toxicity in seed geranium (pelargonium x hortom Bailey). Journal of American Society for Horticultural Science 121 (1), 77–82. Available at 10.21273/JASHS.121.1.77.

Lee, S., Shuman, J.D., Guszczynski, T., Sakchaisri, K., Sebastian, T., Copeland, T.D., et al., 2010. RSK-mediated phosphorylation in the C/EBPβ leucine zipper regulates DNA binding, dimerization, and growth arrest activity. Molecular and Cellular Biology 30 (11), 2621–2635. Available at https://doi.org/10.1128/MCB.00782-09.

Liao, Y., Zou, H.F., Wei, W., Hao, Y.J., Tian, A.G., Huang, J., et al., 2008. Soybean GmbZIP44, GmbZIP62 and GmbZIP78 genes function as negative regulator of ABA signaling and confer salt and freezing tolerance in transgenic Arabidopsis. Planta 228 (2), 225–240. Available at https://doi.org/10.1007/s00425-008-0731-3.

Lipsick, J.S., 1996. One billion years of myb. Oncogene 13 (2), 223–235.

Liu, L., White, M.J., MacRae, T.H., 1999. Transcription factors and their genes in higher plants: functional domains, evolution and regulation. European Journal of Biochemistry 262 (2), 247–257. Available at https://doi.org/10.1046/j.1432-1327.1999.00349.x.

Liu, Q., Callaghan, T.V., Zuo, Y., 2004. Effects of elevated solar UV-B radiation from ozone depletion on terrestrial ecosystems. Journal of Mountain Science 1 (3), 276–288.

Li, X., Tang, Y., Zhou, C., Zhang, L., Lv, J., 2020. A wheat WRKY transcription factor TaWRKY46 enhances tolerance to osmotic stress in transgenic Arabidopsis plants. International Journal of Molecular Sciences 21 (4), 1321. Available at https://doi.org/10.3390/ijms21041321.

Li, G., Quan, R., Cheng, S., Hou, X., Hu, H., 2021. An HD-Zip transcription factor, VvHDZ4, in grapes (Vitis vinifera L.) confers enhanced drought tolerance in transgenic tomato. Journal of Berry Research 11 (2), 217–229. Available at https://doi.org/10.3233/jbr-200632.

Li, Z., Fu, D., Wang, X., Zeng, R., Zhang, X., Tian, J., et al., 2022. The transcription factor bZIP68 negatively regulates cold tolerance in maize. The Plant Cell 34 (8), 2833–2851. Available at https://doi.org/10.1093/plcell/koac137.

Li, W., Li, H., Wei, Y., Han, J., Wang, Y., Li, X., et al., 2024. Overexpression of a Fragaria vesca NAM, ATAF, and CUC (NAC) transcription factor gene (FvNAC29) increases salt and cold tolerance in Arabidopsis thaliana. International Journal of Molecular Sciences 25 (7), 4088. Available at https://doi.org/10.3390/ijms25074088.

Li, Q., Yang, Q., Dong, S., Fu, F., Xin, Y., Kang, H., et al., 2025. Transcription factors Cp SPL5 and Cp SPL8 negatively regulate salt tolerance in Codonopsis pilosula by inhibiting SOS pathway. The Plant Journal 121 (2). Available at https://doi.org/10.1111/tpj.17205.

Luo, P., Chen, Y., Rong, K., Lu, Y., Wang, N., Xu, Z., et al., 2022. ZmSNAC13, a maize NAC transcription factor conferring enhanced resistance to multiple abiotic stresses in transgenic Arabidopsis. Plant Physiology and Biochemistry 170, 160–170. Available at https://doi.org/10.1016/j.plaphy.2021.11.032.

Lv, H., Wang, X., Dong, X., Gao, M., Dong, D., Li, C., et al., 2025. CRISPR/Cas9 edited SlGT30 improved both drought resistance and fruit yield through endoreduplication. Plant, Cell and Environment 48 (4), 2581–2595. Available at https://doi.org/10.1111/pce.14927 Available from.

Lynch, D.V., 1990. CHAPTER 2 - Chilling injury in plants: the relevance of membrane lipids. In: Katterman, F (Ed.), Environmental Injury to Plants. Academic Press, pp. 17–34. Available at https://doi.org/10.1016/B978-0-12-401350-6.50006-0.

Ly, L.K., Ho, T.M., Bui, T.P., Nguyen, L.T., Phan, Q., Le, N.T., et al., 2024. CRISPR/Cas9 targeted mutations of OsDSG1 gene enhanced salt tolerance in rice. Functional and Integrative Genomics 24 (2). Available at https://doi.org/10.1007/s10142-024-01347-6.

Malar, S., Manikandan, R., Favas, P.J.C., Vikram Sahi, S., Venkatachalam, P., 2014. Effect of lead on phytotoxicity, growth, biochemical alterations and its role on genomic template stability in Sesbania grandiflora: a potential plant for phytoremediation. Ecotoxicology and Environmental Safety 108, 249–257. Available at https://doi.org/10.1016/j.ecoenv.2014.05.018.

Malar, S., Sahi, S.V., Favas, P.J.C., Venkatachalam, P., 2015. Mercury heavy-metal-induced physiochemical changes and genotoxic alterations in water hyacinths [Eichhornia crassipes (Mart.)]. Environmental Science and Pollution Research 22 (6), 4597–4608. Available at https://doi.org/10.1007/s11356-014-3576-2.

Mall, R.K., Gupta, A., Sonkar, G., 2017. Effect of climate change on agricultural crops. Current Developments in Biotechnology and Bioengineering. Elsevier, pp. 23–46.

Manna, M., Thakur, T., Chirom, O., Mandlik, R., Deshmukh, R., Salvi, P., 2021. Transcription factors as key molecular target to strengthen the drought stress tolerance in plants. Physiologia Plantarum 172 (2), 847–868. Available at https://doi.org/10.1111/ppl.13268.

Ma, Z., Wu, T., Huang, K., Jin, Y.M., Li, Z., Chen, M., et al., 2020. A novel AP2/ERF transcription factor, OsRPH1, negatively regulates plant height in rice. Frontiers in Plant Science 11, 709. Available at https://doi.org/10.3389/fpls.2020.00709.

Ma, Z., Jin, Y., Wu, T., Hu, L., Zhang, Y., Jiang, W., et al., 2022. OsDREB2B, an Ap2/ERF transcription factor, negatively regulates plant height by conferring GA metabolism in rice. Frontiers in Plant Science 13, 1007811. Available at 10.3389/fpls.2022.1007811.

Ma, R., Liu, B., Geng, X., Ding, X., Yan, N., Sun, X., et al., 2023. Biological function and stress response mechanism of MYB transcription factor family genes. Journal of Plant Growth Regulation 42 (1), 83–95. Available at https://doi.org/10.1007/s00344-021-10557-2.

Ma, Z., Hu, L., Jiang, W., 2024a. Understanding AP2/ERF transcription factor responses and tolerance to various abiotic stresses in plants: a comprehensive review. International Journal of Molecular Sciences 25 (2), 893. Available at https://doi.org/10.3390/ijms25020893.

Ma, F., Zhou, H., Yang, H., Huang, D., Xing, W., Wu, B., et al., 2024. WRKY transcription factors in passion fruit analysis reveals key PeWRKYs involved in abiotic stress and flavonoid biosynthesis. International Journal of Biological Macromolecules 256 (Pt 1), 128063. Available at https://doi.org/10.1016/j.ijbiomac.2023.128063.

Menzel, C.M., 2023. A review of fruit development in strawberry: high temperatures accelerate flower development and decrease the size of the flowers and fruit. Journal of Horticultural Science and Biotechnology 98 (4), 409–431. Available at https://doi.org/10.1080/14620316.2023.2166599.

Miyake, K., Ito, T., Senda, M., Ishikawa, R., Harada, T., Niizeki, M., et al., 2003. Isolation of a subfamily of genes for R2R3-MYB transcription factors showing up-regulated expression under nitrogen nutrient-limited conditions. Plant Molecular Biology 53 (1/2), 237–245. Available at https://doi.org/10.1023/B:PLAN.0000009296.91149.34.

Mizoi, J., Shinozaki, K., Yamaguchi-Shinozaki, K., 2012. AP2/ERF family transcription factors in plant abiotic stress responses. Biochimica et Biophysica Acta (BBA)-Gene Regulatory Mechanisms 1819 (2), 86–96.

Nagajyoti, P.C., Lee, K.D., Sreekanth, T.V.M., 2010. Heavy metals, occurrence and toxicity for plants: a review. Environmental Chemistry Letters 8 (3), 199–216.

Nakano, T., Suzuki, K., Fujimura, T., Shinshi, H., 2006. Genome-wide analysis of the ERF gene family in Arabidopsis and rice. Plant Physiology 140 (2), 411–432. Available at https://doi.org/10.1104/pp.105.073783.

Neelima, P., Reddy, K.J., 2002. Interaction of copper and cadmium with seedling growth and biochemical responses in Solanum melongena. Nature, Environment and Pollution Technology 1 (3), 285–290.

Nieboer, E., Richardson, D.H.S., 1980. The replacement of the nondescript term 'heavy metals' by a biologically and chemically significant classification of metal ions. Environmental Pollution Series B, Chemical and Physical 1 (1), 3–26. Available at https://doi.org/10.1016/0143-148x(80)90017-8.

Nieva, C., Busk, P.K., Domínguez-Puigjaner, E., Lumbreras, V., Testillano, P.S., Risueño, M.C., et al., 2005. Isolation and functional characterisation of two new bZIP maize regulators of the ABA responsive gene rab28. Plant Molecular Biology 58 (6), 899–914. Available at https://doi.org/10.1007/s11103-005-8407-x.

Nijhawan, A., Jain, M., Tyagi, A.K., Khurana, J.P., 2008. Genomic survey and gene expression analysis of the basic leucine zipper transcription factor family in rice. Plant Physiology 146 (2), 333–350. Available at https://doi.org/10.1104/pp.107.112821.

Niu, X., Renshaw-Gegg, L., Miller, L., Guiltinan, M.J., 1999. Bipartite determinants of DNA-binding specificity of plant basic leucine zipper proteins. Plant Molecular Biology 41 (1), 1–13. Available at https://doi.org/10.1023/A:1006206011502.

Nuruzzaman, M., Manimekalai, R., Sharoni, A.M., Satoh, K., Kondoh, H., Ooka, H., et al., 2010. Genome-wide analysis of NAC transcription factor family in rice. Gene 465 (1-2), 30–44. Available at https://doi.org/10.1016/j.gene.2010.06.008.

Nuruzzaman, M., Sharoni, A.M., Kikuchi, S., 2013. Roles of NAC transcription factors in the regulation of biotic and abiotic stress responses in plants. Frontiers in Microbiology 4, 248.

Ohme-Takagi, M., Shinshi, H., 1995. Ethylene-inducible DNA binding proteins that interact with an ethylene-responsive element. The Plant Cell 7 (2), 173–182. Available at https://doi.org/10.1105/tpc.7.2.173.

Olsen, A.N., Ernst, H.A., Leggio, L.L., Skriver, K., 2005. NAC transcription factors: structurally distinct, functionally diverse. Trends in Plant Science 10 (2), 79–87. Available at https://doi.org/10.1016/j.tplants.2004.12.010.

Onate, L., Vicente-Carbajosa, J., Lara, P., Dıaz, I., Carbonero, P., 1999. Barley BLZ2, a seed-specific bZIP protein that interacts with BLZ1 in vivo and activates transcription from the GCN4-like motif of B-hordein promoters in barley endosperm. Journal of Biological Chemistry 274 (14), 9175–9182.

Ooka, H., Satoh, K., Doi, K., Nagata, T., Otomo, Y., Murakami, K., et al., 2003. Comprehensive analysis of NAC family genes in *Oryza sativa* and *Arabidopsis thaliana*. DNA Research 10 (6), 239–247. Available at https://doi.org/10.1093/dnares/10.6.239.

Öztürk, M., Ashraf, M., Aksoy, A., Ahmad, M.S.A., Hakeem, K.R., 2015. Plants, Pollutants and Remediation. Springer Netherlands.

Pang, X., Chen, J., Li, L., Huang, W., Liu, J., 2024. Deciphering drought resilience in solanaceae crops: unraveling molecular and genetic mechanisms. Biology 13 (12), 1076. Available at https://doi.org/10.3390/biology13121076.

Parida, A.K., Das, A.B., 2005. Salt tolerance and salinity effects on plants: a review. Ecotoxicology and Environmental Safety 60 (3), 324–349. Available at https://doi.org/10.1016/j.ecoenv.2004.06.010.

Patikoglou, G., Burley, S.K., 1997. Eukaryotic transcription factor-DNA complexes. Annual Review of Biophysics and Biomolecular Structure 26 (1), 289–325. Available at https://doi.org/10.1146/annurev.biophys.26.1.289.

Paz-Ares, J., Ghosal, D., Wienand, U., Peterson, P.A., Saedler, H., 1987. The regulatory c1 locus of Zea mays encodes a protein with homology to myb proto-oncogene products and with structural similarities to transcriptional activators. The EMBO Journal 6 (12), 3553–3558. Available at https://doi.org/10.1002/j.1460-2075.1987.tb02684.x.

Pecchioni, N., Kosová, K., Vítámvás, P., Prášil, I.T., Milc, J.A., Francia, E., et al., 2013. Genomics of low-temperature tolerance for an increased sustainability of wheat and barley production. Genomics of Plant Genetic Resources: Volume 2. Crop productivity, Food Security and Nutritional Quality. Springer Netherlands, Dordrecht, pp. 149–183.

Pfleiderer, P., Schleussner, C.F., Kornhuber, K., Coumou, D., 2019. Summer weather becomes more persistent in a 2 C world. Nature Climate Change 9 (9), 666–671.

Phukan, U.J., Jeena, G.S., Shukla, R.K., 2016. WRKY transcription factors: molecular regulation and stress responses in plants. Frontiers in Plant Science 7, 760. Available at https://doi.org/10.3389/fpls.2016.00760.

Puranik, S., Sahu, P.P., Srivastava, P.S., Prasad, M., 2012. NAC proteins: regulation and role in stress tolerance. Trends in Plant Science 17 (6), 369–381. Available at https://doi.org/10.1016/j.tplants.2012.02.004.

Qin, B., Fan, S.L., Yu, H.Y., Lu, Y.X., Wang, L.F., 2022. HbMYB44, a rubber tree MYB transcription factor with versatile functions in modulating multiple phytohormone signaling and abiotic stress responses. Frontiers in Plant Science 13. Available at https://doi.org/10.3389/fpls.2022.893896.

Qu, D., Show, P.L., Miao, X., 2021. Transcription factor ChbZIP1 from alkaliphilic microalgae Chlorella sp. BLD enhancing alkaline tolerance in transgenic Arabidopsis thaliana. International Journal of Molecular Sciences 22 (5), 2387.

Ren, L., Li, F., Jiang, J., Li, H., 2021. BpTCP3 transcription factor improves salt tolerance of betula platyphylla by reducing reactive oxygen species damage. Forests 12 (12), 1633. Available at https://doi.org/10.3390/f12121633.

Riaz, M., Arif, M.S., Ashraf, M.A., Mahmood, R., Yasmeen, T., Shakoor, M.B., et al., 2019. Chapter 7 - A comprehensive review on rice responses and tolerance to salt stress. In: Hasanuzzaman, M., Fujita, M., Nahar, K., Biswas, J.K. (Eds.), Advances in Rice Research for Abiotic Stress Tolerance. Woodhead Publishing, pp. 133–158. Available at https://doi.org/10.1016/B978-0-12-814332-2.00007-1.

Riechmann, J.L., 2007. Transcription factors of Arabidopsis and rice: a genomic perspective. In: Grasser, K.D. (Ed.), Annual Plant Reviews Volume 29: Regulation of Transcription in Plants. Wiley, pp. 28–53. Available at https://doi.org/10.1002/9780470988886.ch2.

Rinerson, C.I., Rabara, R.C., Tripathi, P., Shen, Q.J., Rushton, P.J., 2015. The evolution of WRKY transcription factors. BMC Plant Biology 15 (1). Available at https://doi.org/10.1186/s12870-015-0456-y.

Rivetta, A., Negrini, N., Cocucci, M., 1997. Involvement of Ca2+-calmodulin in Cd2+ toxicity during the early phases of radish (*Raphanus sativus* L.) seed germination. Plant, Cell & Environment 20 (5), 600–608.

Saikia, B., S, R., Debbarma, J., Maharana, J., Sastry, G.N., Chikkaputtaiah, C., 2024. CRISPR/Cas9-based genome editing and functional analysis of SlHyPRP1 and SlDEA1 genes of Solanum lycopersicum L. in imparting genetic tolerance to multiple stress factors. Frontiers in Plant Science 15. Available at https://doi.org/10.3389/fpls.2024.1304381.

Sakuma, Y., Liu, Q., Dubouzet, J.G., Abe, H., Yamaguchi-Shinozaki, K., Shinozaki, K., 2002. DNA-binding specificity of the ERF/AP2 domain of Arabidopsis DREBs, transcription factors involved in dehydration- and cold-inducible gene expression. Biochemical and Biophysical Research Communications 290 (3), 998–1009. Available at https://doi.org/10.1006/bbrc.2001.6299.

Samarah, N.H., 2005. Effects of drought stress on growth and yield of barley. Agronomy for Sustainable Development 25 (1), 145–149. Available at https://doi.org/10.1051/agro:2004064.

Schenk, P.M., Kazan, K., Wilson, I., Anderson, J.P., Richmond, T., Somerville, S.C., et al., 2000. Coordinated plant defense responses in Arabidopsis revealed by microarray analysis. Proceedings of the National Academy of Sciences of the United States of America 97 (21), 11655–11660. Available at https://doi.org/10.1073/pnas.97.21.11655.

Schmutz, J., Cannon, S.B., Schlueter, J., Ma, J., Mitros, T., Nelson, W., et al., 2010. Genome sequence of the paleopolyploid soybean. Nature 463 (7278), 178–183. Available at https://doi.org/10.1038/nature08670.

Sehgal, A., Sita, K., Siddique, K.H.M., Kumar, R., Bhogireddy, S., Varshney, R.K., et al., 2018. Drought or/and heat-stress effects on seed filling in food crops: impacts on functional biochemistry, seed yields, and nutritional quality. Frontiers in Plant Science 871. Available at https://doi.org/10.3389/fpls.2018.01705.

Seok, H.Y., Tran, H.T., Lee, S.Y., Moon, Y.H., 2022. AtERF71/HRE2, an Arabidopsis AP2/ERF Transcription factor gene, contains both positive and negative cis-regulatory elements in its promoter region involved in hypoxia and salt stress responses. International Journal of Molecular Sciences 23 (10). Available at https://doi.org/10.3390/ijms23105310.

Seo, P.J., Kim, S.G., Park, C.M., 2008. Membrane-bound transcription factors in plants. Trends in Plant Science 13 (10), 550–556.

Shang, H., Li, W., Zou, C., Yuan, Y., 2013. Analyses of the NAC transcription factor gene family in Gossypium raimondii Ulbr.: chromosomal location, structure, phylogeny, and expression patterns. Journal of Integrative Plant Biology 55 (7), 663–676. Available at https://doi.org/10.1111/jipb.12085.

Shanker, A., Cervantes, C., Lozatavera, H., Avudainayagam, S., 2005. Chromium toxicity in plants. Environment International 31 (5), 739–753. Available at https://doi.org/10.1016/j.envint.2005.02.003.

Sharma, P., Dubey, R.S., 2005. Lead toxicity in plants. Brazilian Journal of Plant Physiology 17 (1), 35–52. Available at https://doi.org/10.1590/s1677-04202005000100004.

Souer, E., van Houwelingen, A., Kloos, D., Mol, J., Koes, R., 1996. The no apical meristem gene of petunia is required for pattern formation in embryos and flowers and is expressed at meristem and primordia boundaries. Cell 85 (2), 159–170. Available at https://doi.org/10.1016/s0092-8674(00)81093-4.

Thomashow, M.F., 1999. Plant cold acclimation: freezing tolerance genes and regulatory mechanisms. Annual Review of Plant Biology 50, 571–599. Available at https://doi.org/10.1146/annurev.arplant.50.1.571.

Tong, Y., Wang, C., Zhao, L.L., Lian, J., Liu, X.M., Zhao, B.C., 2021. Transcription factor OsbZIP5 negatively regulates drought-tolerance in rice. Chinese Journal of Biochemistry and Molecular Biology 37 (6), 798–810. Available at https://doi.org/10.13865/j.cnki.cjbmb.2021.03.1427.

Tsakiris, G., Tigkas, D., 2007. Drought risk in agriculture in Mediterranean regions. Case study: Eastern Crete. Methods and Tools for Drought Analysis and Management. Springer Netherlands, Dordrecht, pp. 399–414.

Udvardi, M.K., Kakar, K., Wandrey, M., Montanari, O., Murray, J., Andriankaja, A., et al., 2007. Legume transcription factors: global regulators of plant development and response to the environment. Plant Physiology 144 (2), 538–549. Available at https://doi.org/10.1104/pp.107.098061.

Ülker, B., Somssich, I.E, 2004. WRKY transcription factors: from DNA binding towards biological function. Current Opinion in Plant Biology 7, 491–498. Available at 10.1016/j.pbi.2004.07.012.

Vatansever, R., Ozyigit, I.I., Filiz, E., 2017. Essential and beneficial trace elements in plants, and their transport in roots: a review. Applied Biochemistry and Biotechnology 181 (1), 464–482.

Vinson, C., Myakishev, M., Acharya, A., Mir, A.A., Moll, J.R., Bonovich, M., 2002. Classification of human B-ZIP proteins based on dimerization properties. Molecular and Cellular Biology 22 (18), 6321–6335. Available at https://doi.org/10.1128/MCB.22.18.6321-6335.2002.

Wahid, A., Gelani, S., Ashraf, M., Foolad, M., 2007. Heat tolerance in plants: an overview. Environmental and Experimental Botany 61 (3), 199–223. Available at https://doi.org/10.1016/j.envexpbot.2007.05.011.

Wang, J., Chen, L., Long, Y., Si, W., Cheng, B., Jiang, H., 2021. A novel heat shock transcription factor (ZmHsf08) negatively regulates salt and drought stress responses in maize. International Journal of Molecular Sciences 22 (21), 11922. Available at https://doi.org/10.3390/ijms222111922.

Wang, Y., Xia, D., Li, W., Cao, X., Ma, F., Wang, Q., et al., 2022. Overexpression of a tomato AP2/ERF transcription factor SlERF.B1 increases sensitivity to salt and drought stresses. Scientia Horticulturae 304, 111332. Available at https://doi.org/10.1016/j.scienta.2022.111332.

Wang, S., Jiang, R., Feng, J., Zou, H., Han, X., Xie, X., et al., 2024a. Overexpression of transcription factor FaMYB63 enhances salt tolerance by directly binding to the SOS1 promoter in *Arabidopsis thaliana*. Plant Molecular Biology 114 (2), 32.

Wang, X., Wei, C., Huang, H., Kang, J., Long, R., Chen, L., et al., 2024b. The GARP family transcription factor MtHHO3 negatively regulates salt tolerance in Medicago truncatula. Plant Physiology and Biochemistry 209, 108542.

Wang, D., Zeng, Y., Yang, X., Nie, S., 2024c. Characterization of DREB family genes in Lotus japonicus and LjDREB2B overexpression increased drought tolerance in transgenic Arabidopsis. BMC Plant Biology 24 (1), 497. Available at https://doi.org/10.1186/s12870-024-05225-y.

Wei, T., Guo, D., Liu, J., 2021. PtrMYB3, a R2R3-MYB transcription factor from Poncirus trifoliata, negatively regulates salt tolerance and hydrogen peroxide scavenging. Antioxidants 10 (9), 1388. Available at https://doi.org/10.3390/antiox10091388.

Wei, L., Ren, X., Qin, L., Zhang, R., Cui, M., Xia, G., et al., 2025. TaWRKY55–TaPLATZ2 module negatively regulate saline–alkali stress tolerance in wheat. Journal of Integrative Plant Biology 67 (1), 19–34. Available at https://doi.org/10.1111/jipb.13793.

Wojciak, J.M., Connolly, K.M., Clubb, R.T., 1999. NMR structure of the Tn916 integrase–DNA complex. Nature Structural Biology 6 (4), 366–373. Available at https://doi.org/10.1038/9369.

Woo, H.R., Kim, J.H., Kim, J., Kim, J., Lee, U., Song, I.J., et al., 2010. The RAV1 transcription factor positively regulates leaf senescence in Arabidopsis. Journal of Experimental Botany 61 (14), 3947–3957. Available at https://doi.org/10.1093/jxb/erq206.

Wuitschick, J.D., Lindstrom, P.R., Meyer, A.E., Karrer, K.M., 2004. Homing endonucleases encoded by germ line-limited genes in Tetrahymena thermophila have Apetela2 DNA binding domains. Eukaryotic Cell 3 (3), 685–694. Available at https://doi.org/10.1128/EC.3.3.685-694.2004.

Xie, Q., Frugis, G., Colgan, D., Chua, N.-H., 2000. Arabidopsis NAC1 transduces auxin signal downstream of TIR1 to promote lateral root development. Genes & Development 14 (23), 3024–3036. Available at https://doi.org/10.1101/gad.852200.

Xu, K., Xu, X., Fukao, T., Canlas, P., Maghirang-Rodriguez, R., Heuer, S., et al., 2006. Sub1A is an ethylene-response-factor-like gene that confers submergence tolerance to rice. Nature 442 (7103), 705–708. Available at https://doi.org/10.1038/nature04920.

Xu, P., Ma, W., Feng, H., Cai, W., 2024. The NAC056 transcription factor confers freezing tolerance by positively regulating expression of CBFs and NIA1 in Arabidopsis. Plant Communications 5 (7), 100923. Available at https://doi.org/10.1016/j.xplc.2024.100923.

Yadav, S.K., 2010. Heavy metals toxicity in plants: an overview on the role of glutathione and phytochelatins in heavy metal stress tolerance of plants. South African Journal of Botany 76 (2), 167–179. Available at https://doi.org/10.1016/j.sajb.2009.10.007.

Yamamoto, Y., 2016. Quality control of photosystem II: the mechanisms for avoidance and tolerance of light and heat stresses are closely linked to membrane fluidity of the thylakoids. Frontiers in Plant Science 7 (2016). Available at https://doi.org/10.3389/fpls.2016.01136.

Yang, J., Wang, M., Li, W., He, X., Teng, W., Ma, W., et al., 2019. Reducing expression of a nitrate-responsive bZIP transcription factor increases grain yield and N use in wheat. Plant Biotechnology Journal 17 (9), 1823–1833. Available at https://doi.org/10.1111/pbi.13103.

Yang, K., Li, C.Y., An, J.P., Wang, D.R., Wang, X., Wang, C.K., et al., 2021. The C2H2-type zinc finger transcription factor MdZAT10 negatively regulates drought tolerance in apple. Plant Physiology and Biochemistry 167, 390–399. Available at https://doi.org/10.1016/j.plaphy.2021.08.014.

Yang, Z., Lin, L., Lu, M., Ma, W., An, H., 2025. A bHLH transcription factor RrUNE12 regulates salt tolerance and promotes ascorbate synthesis. Plant Cell Reports 44 (2). Available at https://doi.org/10.1007/s00299-025-03428-7.

Yanhui, C., Xiaoyuan, Y., Kun, H., Meihua, L., Jigang, L., Zhaofeng, G., et al., 2006. The MYB transcription factor superfamily of arabidopsis: expression analysis and phylogenetic comparison with the rice MYB family. Plant Molecular Biology 60 (1), 107–124. Available at https://doi.org/10.1007/s11103-005-2910-y.

Yan, X., Han, M., Li, S., Liang, Z., Ouyang, J., Wang, X., et al., 2024. A member of NF-Y family, OsNF-YC5 negatively regulates salt tolerance in rice. Gene 892, 147869. Available at https://doi.org/10.1016/j.gene.2023.147869.

Yarra, R., Wei, W., 2021. The NAC-type transcription factor GmNAC20 improves cold, salinity tolerance, and lateral root formation in transgenic rice plants. Functional & Integrative Genomics 21 (3), 473–487. Available at https://doi.org/10.1007/s10142-021-00790-z.

Yoshida, T., Fujita, Y., Sayama, H., Kidokoro, S., Maruyama, K., Mizoi, J., et al., 2010. AREB1, AREB2, and ABF3 are master transcription factors that cooperatively regulate ABRE-dependent ABA signaling involved in drought stress tolerance and require ABA for full activation. The Plant Journal 61 (4), 672–685. Available at https://doi.org/10.1111/j.1365-313x.2009.04092.x.

Yu, Y., Yu, M., Zhang, S., Song, T., Zhang, M., Zhou, H., et al., 2022. Transcriptomic identification of wheat AP2/ERF transcription factors and functional characterization of TaERF-6-3A in response to drought and salinity stresses. International Journal of Molecular Sciences 23 (6), 3272. Available at https://doi.org/10.3390/ijms23063272.

Zampieri, M., Ceglar, A., Dentener, F., Toreti, A., 2017. Wheat yield loss attributable to heat waves, drought and water excess at the global, national and subnational scales. Environmental Research Letters 12 (6), 064008.

Zhang, L., Xing, L., Dai, J., Li, Z., Zhang, A., Wang, T., et al., 2024. Overexpression of a grape WRKY transcription factor VhWRKY44 improves the resistance to cold and salt of *Arabidopsis thaliana*. International Journal of Molecular Sciences 25 (13), 7437. Available at https://doi.org/10.3390/ijms25137437.

Zhao, P.X., Zhang, J., Chen, S.Y., Wu, J., Xia, J.Q., Sun, L.Q., et al., 2021. Arabidopsis MADS-box factor AGL16 is a negative regulator of plant response to salt stress by downregulating salt-responsive genes. New Phytologist 232 (6), 2418–2439. Available at https://doi.org/10.1111/nph.17760.

Zhao, H., Yao, P., Zhao, J., Wu, H., Wang, S., Chen, Y., et al., 2022. A novel R2R3-MYB transcription factor FtMYB22 negatively regulates salt and drought stress through ABA-dependent pathway. International Journal of Molecular Sciences 23 (23), 14549. Available at https://doi.org/10.3390/ijms232314549.

Zhao, Y.J., Ma, C.Y., Zheng, M.J., Yao, Y.R., Lv, L.H., Zhang, L.H., et al., 2024a. Transcription factor TaNF-YB2 interacts with partners TaNF-YA7/YC7 and transcriptionally activates distinct stress-defensive genes to modulate drought tolerance in T. aestivum. BMC Plant Biology 24 (1). Available at https://doi.org/10.1186/s12870-024-05420-x.

Zhao, X., Wang, Q., Yan, C., Sun, Q., Wang, J., Li, C., et al., 2024b. The bHLH transcription factor AhbHLH121 improves salt tolerance in peanut. International Journal of Biological Macromolecules 256, 128492. Available at https://doi.org/10.1016/j.ijbi.128492.

Zhou, X., Sun, Z., Huang, Y., He, D., Lu, L., Wei, M., et al., 2025. WRKY45 positively regulates salinity and osmotic stress responses in Arabidopsis. Plant Physiology and Biochemistry 219, 109408. Available at https://doi.org/10.1016/j.plaphy.2024.109408.

Zhu, J.K., 2016. Abiotic stress signaling and responses in plants. Cell 167 (2), 313–324.

Zhu, H., Jiang, Y., Guo, Y., Huang, J., Zhou, M., Tang, Y., et al., 2021. A novel salt inducible WRKY transcription factor gene, AhWRKY75, confers salt tolerance in transgenic peanut. Plant Physiology and Biochemistry 160, 175–183. Available at https://doi.org/10.1016/j.plaphy.2021.01.014.

Further reading

Akhtar, M., Jaiswal, A., Taj, G., Jaiswal, J.P., Qureshi, M.I., Singh, N.K., 2012. DREB1/CBF transcription factors: their structure, function and role in abiotic stress tolerance in plants. Journal of Genetics 91 (3), 385–395.

Allen, D.J., Nogués, S., Baker, N.R., 1998. Ozone depletion and increased UV-B radiation: is there a real threat to photosynthesis? Journal of Experimental Botany 49 (328), 1775–1788.

Chai, M., Cheng, H., Yan, M., Priyadarshani, S.V.G.N., Zhang, M., He, Q., et al., 2020. Identification and expression analysis of the DREB transcription factor family in pineapple (*Ananas comosus* (L.) Merr.). PeerJ 8, e9006.

Hopkins, L., Bond, M.A., Tobin, A.K., 2002. Ultraviolet-B radiation reduces the rates of cell division and elongation in the primary leaf of wheat (*Triticum aestivum* L. cv Maris Huntsman). Plant Cell & Environment 25 (5), 617–624.

Huang, Y., Li, M.Y., Wu, P., Xu, Z.S., Que, F., Wang, F., et al., 2016. Members of WRKY Group III transcription factors are important in TYLCV defense signaling pathway in tomato (*Solanum lycopersicum*). BMC Genomics 17 (1), 788.

Kierzek, E., Christensen, S.M., Eickbush, T.H., Kierzek, R., Turner, D.H., Moss, W.N., 2009. Secondary structures for 5′ regions of R2 retrotransposon RNAs reveal a novel conserved pseudoknot and regions that evolve under different constraints. Journal of Molecular Biology 390 (3), 428–442.

Lee, S., Seo, P.J., Lee, H.J., Park, C.M., 2012. A NAC transcription factor NTL4 promotes reactive oxygen species production during drought-induced leaf senescence in Arabidopsis. The Plant Journal 70 (5), 831–844.

Lim, C., Kang, K., Shim, Y., Yoo, S.C., Paek, N.C., 2022. Inactivating transcription factor OsWRKY5 enhances drought tolerance through abscisic acid signaling pathways. Plant Physiology 188 (4), 1900–1916. Available at https://doi.org/10.1093/plphys/kiab587.

Liu, X., Ma, Q., Yu, H., Li, Y., Zhou, L., He, Q., et al., 2020. Responses of plant biomass and yield component in rice, wheat, and maize to climatic warming: a meta-analysis. Planta 252 (5), 90.

Liu, X., Yang, Y., Wang, R., Cui, R., Xu, H., Sun, C., et al., 2022a. GmWRKY46, a WRKY transcription factor, negatively regulates phosphorus tolerance primarily through modifying root morphology in soybean. Plant Science 315, 111148. Available at https://doi.org/10.1016/j.plantsci.2022.111148.

Liu, X., Zong, X., Wu, X., Liu, H., Han, J., Yao, Z., et al., 2022b. Ectopic expression of NAC transcription factor HaNAC3 from Haloxylon ammodendron increased abiotic stress resistance in tobacco. Planta 256 (6), 105.

Liu, Y., Chen, Z., Zhang, C., Guo, J., Liu, Q., Yin, Y., et al., 2024. Gene editing of ZmGA20ox3 improves plant architecture and drought tolerance in maize. Plant Cell Reports 43 (1), 18.

Marquez, C.P., Pritham, E.J., 2010. Phantom, a new subclass of Mutator DNA transposons found in insect viruses & widely distributed animals. Genetics 185 (4), 1507–1517. Available at https://doi.org/10.1534/genetics.110.118708.

Puglia, G.D., Prjibelski, A.D., Vitale, D., Bushmanova, E., Schmid, K.J., Raccuia, S.A., 2020. Hybrid transcriptome sequencing approach improved assembly and gene annotation in *Cynara cardunculus* (L.). BMC Genomics 21 (1), 317. Available at https://doi.org/10.1186/s12864-020-6670-5.

Sadau, S.B., Liu, Z., Ninkuu, V., Guan, L., Sun, X., 2024. DREB transcription factors are crucial regulators of abiotic stress responses in Gossypium spp. Plant Stress 11, 100350. Available at https://doi.org/10.1016/j.stress.2024.100350.

Shoaib, M., Sharma, N., Kotthoff, L., Lindauer, M., Kant, S., 2023. AutoML: advanced tool for mining multivariate plant traits. Trends in Plant Science 28 (12), 1451–1452.

Wang, X., Liu, J., Luo, L., Yuan, Z., 2023. Overexpression of GmNAC12 from soybean enhances salinity tolerance in transgenic Arabidopsis. Plant Molecular Biology Reporter 41 (4), 504–516. Available at https://doi.org/10.1007/s11105-023-01476-6.

Yang, Y., Li, A., Liu, Y., Shu, J., Wang, J., Guo, Y., et al., 2024. ZmASR1 negatively regulates drought stress tolerance in maize. Plant Physiology and Biochemistry 211, 108684. Available at https://doi.org/10.1016/j.plaphy.2023.11.009.

Zhang, X., Long, Y., Chen, X., Zhang, B., Xin, Y., Li, L., et al., 2021. A NAC transcription factor OsNAC3 positively regulates ABA response and salt tolerance in rice. BMC Plant Biology 21 (1), 546.

Chapter 9

Role of phytohormones and plant growth regulators in multiple abiotic stress tolerance in plants

Mohammad Shah Jahan[1,*], Abu Bakar Siddique[2], Md Arif Hussain[3], and Chinmay Saha[4]

[1]*Key Laboratory of Stress Physiology and Molecular Biology, Department of Horticulture, Sher-e-Bangla Agricultural University, Dhaka, Bangladesh,* [2]*Tasmanian Institute of Agriculture, University of Tasmania, Launceston, Tasmania, Australia,* [3]*Department of Biochemistry and Molecular Biology, Sher-e-Bangla Agricultural University, Dhaka, Bangladesh,* [4]*Department of Agricultural Chemistry, Sher-e-Bangla Agricultural University, Dhaka, Bangladesh*

[*]*Corresponding author. email address: shahjahanhort@sau.edu.bd*

9.1 Introduction

Global climate change is the foremost threat to crop productivity and efficiency, adversely affecting food security. Agricultural production and food security around the world are being affected by an increase in extreme weather events like heatwaves, intense precipitation, drought periods, flooding, and changes in freezing patterns. A result of human actions, for example, deforestation and industrialization, is a constant rise in the global temperature, CO_2, and O_3 levels, in addition to a decline in the accessibility of water for agriculture (Kabir et al., 2023). Field conditions frequently expose plants to several abiotic stresses, for example, drought, extreme temperatures, waterlogging, salinity, and metal toxicity (Hussain et al., 2024; Shabbir et al., 2021). As a result of these atmospheric conditions, crop performance, grain yield, and nutrient levels are adversely affected. For example, in wheat, the concentrations of protein, Zn, and Fe are lower when atmospheric CO_2 is elevated (546–586 ppm) (Myers et al., 2014). Likewise, 14 days of waterlogging at vegetative and ear emergence stages affected grain storage protein patterns and sulfur uptake in wheat (Hussain et al., 2022).

Feeding an increasing population amid production losses triggered by climate change is a major challenge, and mitigating its effects on the environment is another challenge. Over the past few decades, research has primarily focused on understanding plant responses to individual abiotic stresses (Hussain et al., 2024; Manzoor et al., 2023). However, field conditions tend to be much more complex, and abiotic stresses are usually cumulative rather than individual (Galani et al., 2022; Sánchez-Bermúdez et al., 2022). Moreover, plants can be disturbed by several types of abiotic stresses at diverse stages of their development. Therefore, the combined effects of abiotic stresses have been of concern to researchers in recent years. While plants counter abiotic stresses, certain metabolic modifications are common to many of them. According to He et al. (2018), plants deploy several general metabolites as defenses against abiotic stresses, including the cuticle, oxylipin precursors, membrane modulators, reactive species scavengers, unsaturated fatty acids, molecular chaperones (which stabilize subcellular structures and proteins), and compatible solutes. Additionally, to evade the intracellular water cost affected by salinity and drought, plants gather osmoprotectants. However, to reduce the water potential and promote osmotic adjustment, plants gather these metabolites inside the cell, which gives rise to cellular impairment (Hussain et al., 2024). In cellular metabolism, reactive oxygen species (ROS) are naturally occurring compounds that act as secondary messengers (Hussain et al., 2023; Manzoor et al., 2023). Oxidative stress triggered by abiotic stressors, for example, drought, salinity, waterlogging, nutrient deficiency, temperature stress, and heavy metals, causes DNA, carbohydrates, and proteins to be damaged (Hussain et al., 2024; Møller et al., 2007). Several crops have been exposed to abiotic stress, resulting in the gathering of ROS, including wheat, rice, soybean, rapeseed, tomato, and many others (Kamal and Komatsu, 2015; Khan et al., 2019a, 2019b; Hussain et al., 2023; Kushwaha et al., 2019).

Multiple Abiotic Stresses in Plants. DOI: https://doi.org/10.1016/B978-0-443-27424-4.00013-6

Increased ROS and H_2O_2, in turn, induce the activation of Ca^{2+} channels, which further increases cytosolic Ca^{2+} intensities triggered by abiotic stress (Ranty et al., 2016).

Many plant hormones (PHs) are linked to abiotic stress feedback, and are currently classified into nine types in plants (Abdel Latef et al., 2021; Su et al., 2017): auxins, ethylene (ET), jasmonic acid (JA), abscisic acid (ABA), brassinosteroids (BRs), strigolactones (SL), salicylates (SA), cytokinins (CKs), and gibberellins (GAs). However, of these PHs, SA, ABA, JA, and ET are among those PHs that have been demonstrated to perform a critical role in modulating cellular functions to adapt to environmental stresses at the molecular level. Melatonin, a widely known plant hormone, is engaged in numerous physiological activities, viz., photosynthesis, seed germination, rooting, and defense against abiotic stresses (Afsana et al., 2017; Wasternack, 2014; Yadav et al., 2021). Furthermore, ABA biosynthesis pathway genes are activated and act in root growth under different abiotic conditions (Tuteja, 2014). Likewise, JA and ET also participate in the amelioration of many abiotic stresses like drought, salinity, and thermal stress (Habibi et al., 2019; Ali et al., 2020; Shinohara et al., 2017). Additionally, Sánchez-Bermúdez et al. (2022) showed that resistance to salinity, chilling, and heavy metal stress in barley, rice, and tomatoes can be induced with exogenous SA application. Moreover, Pospíšilová et al. (2016) revealed that overexpression of the gene responsible for cytokinin dehydrogenase in roots results in greater drought tolerance in barley. Similarly, Zhang et al. (2021) revealed that in strawberry seedlings, exogenous application of auxin protects root growth from salt stress. Thus, similar to separate stresses, PHs contribute significantly to combined abiotic stresses. According to the latest proteomic and transcriptomic analyses, plants have unique signaling pathways prompted by combined stress conditions (Atkinson et al., 2013). The hormone signaling of plants is crucial to the direction of these tailored reactions to combined stresses since hormone signaling can be controlled adaptively based on abiotic stresses (Suzuki, 2016). There is also evidence that natural stress can simultaneously affect hormone synthesis and signaling in opposing ways at the same time. Thus, acclimation to various combinations of stresses is dependent on precise hormonal regulation and signaling (Suzuki, 2016). Thus this chapter aims to investigate the physiological, biochemical, and molecular responses of plants to combined abiotic stresses, focusing on the role of PHs and signaling pathways in modulating stress tolerance and improving crop performance under complex field conditions.

9.2 Effects of combined (multiple) abiotic stresses on plant performance

The incidence of combined abiotic stresses at any stage of the life stages of a plant has an extreme influence on plant growth and development, ion balance, physiological routes, and molecular mechanisms (Fig. 9.1) (Vaezi et al., 2024). These combined stresses can have complex and often more severe impacts on plant performance compared to individual stress factors (Habti et al., 2020). However, under field circumstances, salinity-waterlogging, drought-heat, and salinity-drought stress recurrently coincide worldwide (Paul et al., 2019; Mahmood et al., 2021). The combined occurrence of

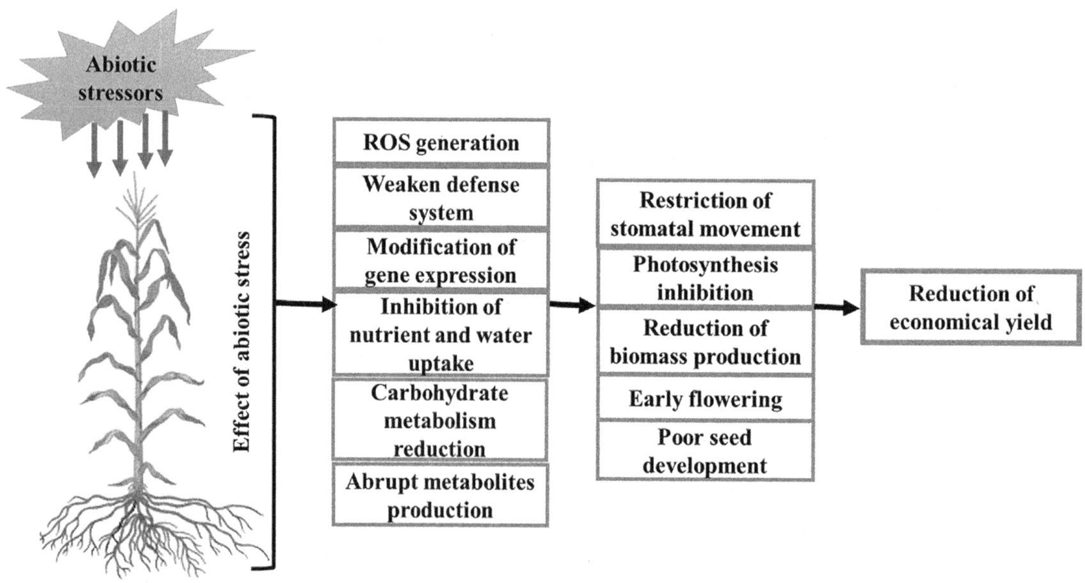

FIGURE 9.1 Detrimental effects of combined abiotic stress on plants.
Created in BioRender, https://BioRender.com/j40d740.

these stresses makes crop performance vulnerable. For example, in wheat, Vaezi et al. (2024) conveyed that combined heat and drought condensed photosynthetic parameters, decline in chlorophyll content, growth, and ultimately yield penalty are much higher compared with individual stresses. Similarly, maize under salinity and waterlogging shows reduced shoot and root length and biomass, leading to poor development and lower productivity (Mahmood et al., 2021). However, if plants experience combined stresses throughout the vegetative and reproductive growth stages, the reduction of growth and yield is almost double compared with other stages. For example, Paul et al. (2019) stated that salinity and drought in critical growth phases in wheat reduced yield by 18% on average under salinity, 50% by drought, and 83% when salinity was merged with drought stress. Moreover, the combined stress severely disrupts ion uptake. For example, in rice, Angon et al. (2022) observed that excessive sodium ion accumulation during merged drought and salinity stress in rice plants impairs ion homeostasis, contributing to diminished plant height and grain yield. Similarly, salinity combined with heat stress in tomato alters nutrient uptake, disrupting overall nutrient homeostasis and further impairing growth (Rivero et al., 2013). Combined stresses negatively impact physiological processes, including photosynthesis, respiration, and water use efficiency (Vile et al., 2011). In cotton, joint heat and drought stress-induced oxidative stress, which affects the antioxidant defense system as well as photosynthetic pigments, ultimately increases lipid peroxidation (Zafar et al., 2023). Similarly, drought and heat stress in Arabidopsis increase respiration rates, decrease photosynthetic efficiency, and lead to differential partitioning of metabolites between shoots and roots (Vile et al., 2011). Paul et al. (2019) described that the combined stresses of drought and salinity at the vegetative stage of wheat reduced photosynthetic efficiency, resulting in a drastic decrease in growth and development. At the molecular level, combined abiotic stress triggers significant fluctuations in hormonal regulation and gene expression.

For example, soybean under heat and water stress activates heat shock proteins (HSPs), which have a significant impact on performing cellular tasks under stress (Katam et al., 2020). Moreover, drought and salinity in tomatoes alter quantities of GAs and ABA expression, which hampers seed germination and growth (Biglia et al., 2024). The overall combined effects of multiple stresses on plant performance are summarized in Table 9.1.

9.3 Tolerance mechanisms of plants to withstand combined abiotic stresses

Owing to their sessile nature, plants cannot transition from unfavorable to favorable environmental conditions. The necessity of mechanisms to evade adverse conditions has triggered the evolution of distinctive responses to environmental contests. Plant species universally employ adaptive mechanisms to contend with environmental stressors. Upon detecting external stressors, these entities undergo activation and subsequently elicit the appropriate stress-related responses. This process is achieved through the conduction of signals from the cell exterior or in-cell receptors to the nuclear transcriptional apparatus via multiple signaling pathways. Signaling cascades act as intermediaries that bridge the detection of stress-inducing environments by eliciting suitable physiological and biochemical adaptations. Plant stress induces a multifaceted response, encompassing alterations in genetic expression, metabolic processes, growth dynamics, and agricultural productivity (Huang et al., 2012). However, plants have developed several defense mechanisms to survive abiotic stress conditions, including drought, waterlogging, salinity, high temperatures, and heavy metals (Hussain et al., 2024; Amjad et al., 2024). Stress-induced signal transduction is initiated during the first response to hostile environmental situations. It contains multiple second messengers, viz. Ca^{2+}, ROS, nitric oxide (NO), and phospholipids, along with several types of protein kinases. Signals are sent and amplified by these secondary messengers after a plant cell detects stress (Zhu, 2016). Certain features are common to different stresses in terms of how they affect plants and how they are perceived. As an example, plant cells are usually subjected to oxidative stress when exposed to these abiotic stresses (Zhang et al., 2022). To neutralize oxidative damage triggered by ROS produced during stresses, plants strengthen their antioxidant systems, which include enzymes such as peroxidases (PODs), catalase (CAT), and superoxide dismutase (SOD) (Hussain et al., 2023; Saharan et al., 2022). Furthermore, they transiently increase intracellular Ca^{2+} concentrations as well. Due to its properties, Ca is considered a universal second messenger in primary stress signals since there are several Ca^{2+} transporters or osmotic channels that can specifically regulate concentration alterations and thus act as an ubiquitous signaling molecule (Dong et al., 2022). For example, under salt stress, Hrabak et al. (2003) reported that the salt excessively sensitive (SOS) pathway perceives a specific cytosolic Ca^{2+} signal, while SOS3 interacts with calcium-dependent protein kinases and activates SOS2. Such SOS3-SOS2 complexes play a vital role in those abiotic stresses, particularly involving ion transport protein activity (Zhang et al., 2022). To respond to the adverse consequences of salinity, temperature fluctuations, and osmotic stress, plants use their inherent genetic mechanisms and implement a range of adaptive and biological approaches to preserve viability and enhance productivity. The regulatory genes involved in abiotic stress reactions may operate redundantly or cumulatively, eliciting activation of both mutual and separate downstream targets (Ahanger et al., 2017). The metabolic adaptations associated with these various stresses cause the

TABLE 9.1 Effects of combined abiotic stresses on plant development.

Species name	Types of stress	Stress duration	Stress effects	References
Wheat	Drought + Salinity	Short term-2 weeks Moderate-4 weeks Long term-Entire growth cycle	• Leaf and shoot area decrease • Biomass and yield loss • Decline in photosynthetic efficiency	Paul et al. (2019)
	Heat + Drought	Short term-14 days Moderate-21 days Long-term-Entire growth cycle	• Reduce plant growth and productivity. • Affecting photosynthetic parameters, final grain weight, and overall yield	Vaezi et al. (2024)
	Drought + Waterlogging	Germination and early stage of seedling: 7–10 days. Vegetative stage: 14–21 days. Reproductive stage: 7–14 days Grain filling stage: 14–21 days.	• Adversely affect physiological processes (water use efficiency and photosynthesis) • Reduction in crop yield	Habti et al. (2020)
Maize	Salinity + Waterlogging	Both stages-22 Days	• Reduce shoot-root length • Reduce root-shoot fresh and dry weight • Reduce the number of leaves and leaf diameter per plant • Reduce leaf carotenoid and chlorophyll content	Mahmood et al. (2021)
	Drought +Heat	Reproductive stage (Tasseling to pollination): 30–50 days Grain filling and maturity stage: 50–80 days	• Weak crop establishment • Interruption in silking • Tassel blasting, leaf firing • Reduction of kernel size • Pollen dehiscence and silk emergence interval increase	Nasser et al. (2020)
	Drought + Salinity	All stages: 21 days	• Decline in photosynthetic efficiency • Adversely affect growth and yield	Yao et al. (2016)
Arabidopsis	Drought + Heat	Germination and seedling stage: 0–10 days Vegetative stage: 10–30 days Reproductive stage: 30–45 days Postreproductive stage: 45–60 days	• Modulation in canopy temperature • Increasing respiration • Increasing cellular aldoses and ketoses • Distinction portioning of metabolites between the root and shoot	Vile et al. (2011)
	Light + Drought	Germination and Seedling stage: 0–10 days Vegetative stage: 10–30 days Reproductive stage: 30–45 days Postreproductive Stage: 45–60 days	• Increasing abscisic acid (ABA) and altered nitrate signaling • Decreasing stomatal conductance	Giraud et al. (2008)
	Drought + Salinity	30 days	• Disrupt the ion balance and decrease photosynthetic efficiency • Higher levels of oxidative stress	Jakab et al. (2005)

Soybean	Heat + Water Stress	20 days	• Activation/Suppression of a specific gene • Accumulation of various chaperone proteins, especially heat shock proteins	Katam et al. (2020)
	Flooding + Salinity	13 days	• Disrupt photosynthesis and respiration • Reducing plant vigor and seed quality	Alam et al. (2011)
	Drought + Salinity	Both stages-10–14 days	• Shift in the levels of abscisic acid and gibberellins • Hampering seed germination and growth	Begum et al. (2022)
Rice	High Temperature + Drought	10 days of flowering	• Spikelet sterility • Abnormal anther dehiscence • Impaired pollination and pollen germination	Sethuraman et al. (2024)
	Drought + Salinity	15 days	• Affect plant height, biomass accumulation, and grain production • Intensifies oxidative stress • Excessive accumulation of sodium ions and disruption of the ion balance	Angon et al. (2022)
	Cd + Zn stress	Seedling stage-10–14 days; vegetative stage-20–30 days	• Negative regulators of auxin signaling in plants • Reduces chlorophyll content and photosynthetic performance	Zhao et al. (2014)
Cotton	Drought + Heat	10 days at the flowering stage	• Decrease in plant vigor and reproductive success • Decrease the activities of catalase • Higher H_2O_2 accumulation and oxidative stress-induced lipid peroxidation	Zafar et al. (2023)
	Drought + Salinity	10 days at the flowering stage	• Lowering in plant growth and lint yield • Substandard fiber properties • Induces cellular water loss • Drop in cytosolic and vacuolar volumes in the cell	Abdelraheem et al. (2018)
	Salinity + Heavy stress	10 days	• Reduced growth and Biomass • Decreased chlorophyll content and photosynthetic efficiency	Ibrahim et al. (2018)
Tomato	Salinity + Heat	14 days	• Accelerate leaf senescence and reduction in yield and fruit quality • Disrupt photosynthesis and exacerbate the negative effect of heat stress on carbon dioxide uptake • Disrupt nutrient homeostasis	Rivero et al. (2013)
	Drought + Salinity	10 days	• Exhibit smaller and darker leaves • Decline in plant development, chlorophyll concentration, and photosynthesis	Biglia et al. (2024)
	Salinity + Heavy metals	14 days	• Decrease in plant height, dry weight, and biomass • Adversely affect the photosynthetic process	Amjad et al. (2024)

synthesis and gathering of diverse organic compounds, for example, proline, sugars, betaines, and polyols (Suprasanna et al., 2016).

However, plant growth regulators (PGRs) are supposed to manage abiotic stresses during the plant's development cycle. Thus hormones in plants permit PGRs to detect and respond to hostile environmental stresses. (Khan et al., 2024a,b) described in a review that ABA helps plants deal with salt and drought stress, while ET combats salinity, waterlogging, cold, and drought. Similarly, mechanically damaged and drought-stressed plants can recover with the help of JA. The "stress hormone," ABA, plays an essential role in enhancing a plant's resistance to stress by regulating multiple processes (Verma et al., 2016). In response to stress, ABA causes stomatal closure, which lowers transpiration rates, thereby reducing water loss. Additionally, ABA allows plants to recover from water scarcity by facilitating root elongation and gradually increasing hydraulic conductivity (Daszkowska-Golec, 2016). Similarly, Habibpourmehraban et al. (2023) demonstrated that pretreatment with ABA enhanced the tolerance of the rice genotype (IAC1131) to multiple abiotic stresses following a 4-day exposure to combined drought, salinity, and extreme temperature conditions. The exogenous application of ABA significantly mitigated leaf damage under combined stress, as reflected at the proteome level. Interestingly, ABA pretreatment had minimal effects on the proteome profile of control plants, suggesting that its role is more stress-inducible than constitutive. These findings highlight the priming potential of ABA in improving rice seedlings' tolerance to combined abiotic stresses, primarily by modulating ABA-dependent stress-responsive mechanisms and signaling pathways. Further, GA can mediate signaling through the breakdown of repressor proteins. Thus plant tissues generate GA, either endogenously or exogenously, to mitigate the negative consequences of stress and improve plant resilience (Silverstone et al., 2001). For example, Devireddy et al. (2021a, 2021b) stated that under water shortage circumstances, *GA2ox6* ectopic expression decreased endogenous GA levels, which improved rice seed yield and drought tolerance. Furthermore, JAs are well-recognized stress hormones derived from lipids and play an important role in responding to stress from biotic and abiotic sources. However, under several stress conditions, JAs can upsurge antioxidant capability and produce osmoprotectants, for example, proline and glycine betaine, which help conserve plant integrity (Boutahiri et al., 2024; Goossens et al., 2016). Similarly, SA improved physiological parameters and antioxidant defense systems in many crops. As Cao et al. (2024) revealed that the use of SA under combined high temperature and water stress increases leaf transpiration, enhances root growth, and accumulates osmoprotectants, as well as increases antioxidant enzyme activity.

9.4 Phytohormones and growth regulators: key contributors to alleviating the combined abiotic stresses

Abiotic stresses in single forms, such as salinity, drought, extreme temperatures (both low and high), nutrient deficiencies, heavy metal toxicity, waterlogging, and in different combined forms, negatively impact all aspects of growth and development by disrupting cellular, physiological, and molecular activities. The effect is visualized as poor plant growth and death in many cases. However, plants adapt to such unfavorable environmental conditions by initiating changes at the physiological, anatomical, and molecular levels, thereby mitigating the damage. Many of these processes are regulated by PHs or phytohormones, including ABA, ET, CK, auxins, GAs, SA, JA, and BRs, melatonin (MT), and PGRs, such as polyamines (PAs) (Fig. 9.2). Phytohormones contribute to the adaptation processes by contributing to the changes at morphological, physiological, and molecular levels (Zheng et al., 2023). This section highlights the individual roles of these phytohormones and growth regulators in the alleviation of single or combined abiotic stress-induced damage in plants.

ABA is a key phytohormone that plays a pivotal role in stress signaling, guiding subsequent stress responses, and enabling plant adaptation to various abiotic stresses. It ensures the balance between stress adaptation and growth by regulating their underlying physiological and metabolic processes. For instance, ABA regulates stomatal closure during drought stress, enhancing water use efficiency and reducing water loss through transpiration. Abiotic stresses often lead to the overproduction of ROS, which can cause cell death. ABA enhances ROS detoxification by upregulating the activity of enzymes such as SOD, POD, CAT, ascorbate peroxidase (APX), and glutathione reductase (You and Chan, 2015). ABA also promotes the accumulation of osmolytes and osmoprotectants such as proline, glycine betaine, and soluble sugars, which maintain cellular turgidity, water balance, and protein stability under drought and other abiotic stresses. Additionally, ABA enhances the activity of genes, contributing to water scarcity and low temperature stresses. In addition to contributing to single abiotic stresses, ABA has been reported to play crucial roles in combined stresses. For instance, transcriptomics profiling of plants exposed to a combination of salt and heat stress was linked to ABA, validated by mutant and metabolic analysis (Suzuki, 2016). This study suggested that ABA is crucial in plants' adaptation to

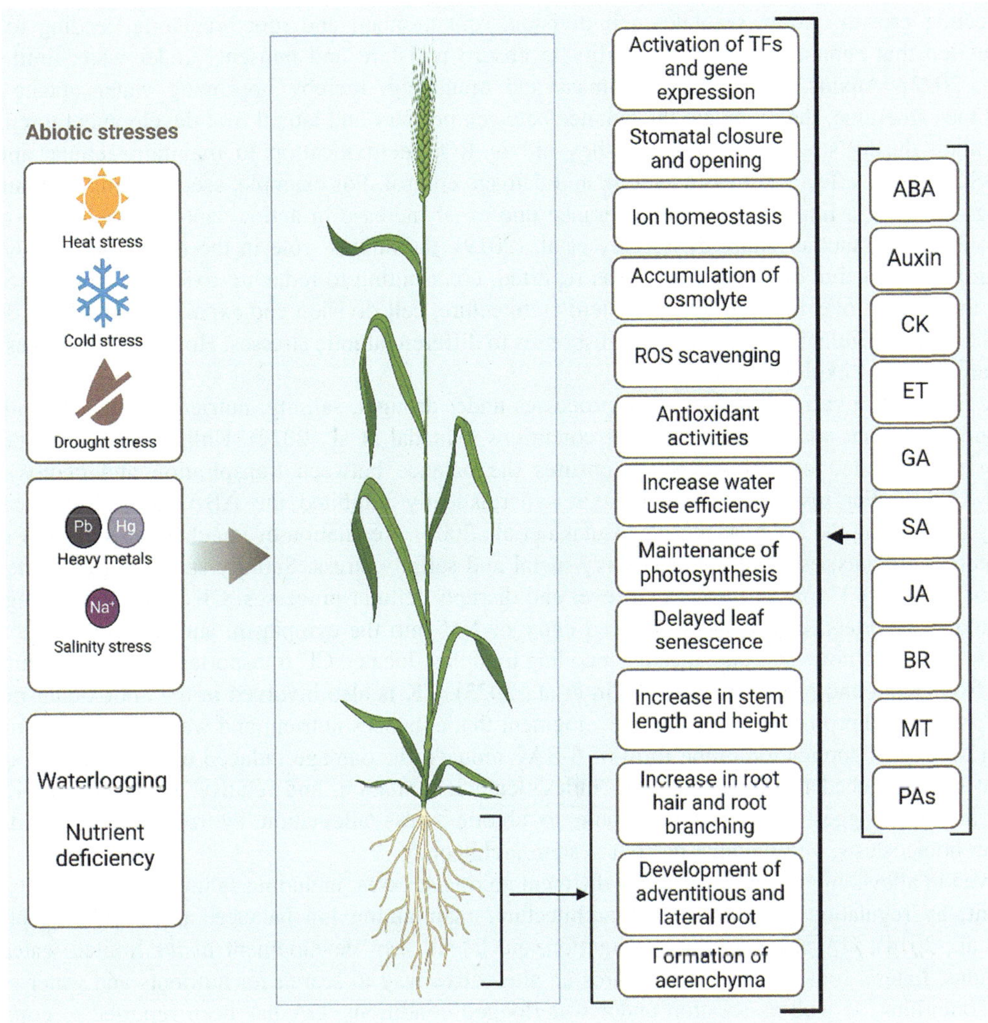

FIGURE 9.2 Schematic diagram showing the phytohormonal contributions in plants' adaptation to different abiotic stresses. Created in BioRender. https://BioRender.com/j40d740.

combined salt and heat stress. In summary, ABA contributes to stomatal regulation, water use efficiency, ROS detoxification, and osmolyte accumulation, effectively mitigating the detrimental effects of abiotic stresses on plants.

ET participates in various physiological processes and plays a major role in stress perception, signal transduction, and stress adaptation. It acts as a signaling molecule to induce ROS production and activate stress regulatory networks. ET involves the activation of signaling cascades and the prevention of protein denaturation by HSPs, particularly under high temperature and water scarcity conditions (Sharma et al., 2023). Additionally, it contributes to osmotic adjustment by promoting the synthesis of compatible solutes like glycine betaine and proline, which maintain cellular redox potential and support metabolism under stress (Hartman et al., 2021). ET has been reported to modulate morphological and anatomical adjustments, such as root system architecture. For instance, under salinity and drought conditions, ET modulates root system architecture, enhancing water and nutrient uptake. For instance, ET mediates the development of adventitious roots during submergence, enabling plants to access aerial oxygen (Steffens, 2014). ET also facilitates the formation of aerenchyma during waterlogging stress by promoting ROS generation and programmed cell death (PCD), thereby maintaining oxygen storage and cellular respiration (Pan et al., 2021). ET also involves shaping the plant responses towards combined abiotic stresses. For instance, a proteomic analysis of two wheat lines under combined drought and nutrient-deficient stress suggested the ET regulation on plant responses (Yan et al., 2023). However, the underlying molecular signaling cascades initiated by ET are not fully understood and need further studies.

Auxins are pivotal in regulating the root-associated responses, particularly under salt and water scarcity conditions. The root system is significantly impacted by both of these abiotic stresses, resulting in disruption in water and nutrient uptake, ultimately affecting growth. Auxin regulates cell division, root number, and root branching, leading to greater root biomass production that enhances the plant's ability to absorb moisture and nutrients under water-limited conditions (Verma et al., 2022). Auxins enhance root biomass and branching, thereby improving water uptake under stress conditions. At the same time, they regulate the balance between primary and lateral root development to ensure efficient resource allocation during stress. Additionally, they aid in ROS detoxification to maintain cellular integrity while regulating ROS levels for effective stress signaling and damage control. For example, seed priming with auxin (IAA) in maize seedlings resulted in improved salinity tolerance due to an increase in antioxidant-mediated ROS detoxification (Ellouzi et al., 2024). In another study by Sergiev et al. (2019), the auxins' role in the reduction of H_2O_2 and higher activity of antioxidative activities in pea plants was reported, contributing to reducing oxidative damage (Sergiev et al., 2019). Due to the crucial roles in shaping root system architecture, cell division and expansion, and ROS detoxification, auxin acts as the central regulator in shaping plant responses to different abiotic stresses. However, their roles in combined abiotic stress are yet to be exploited.

Cytokinin is involved in various physiological processes under drought, salinity, nutrient deficiency, and temperature stress, contributing to plant adaptation to extreme conditions (Mandal et al., 2022). Unlike ABA, CKs delay stomatal closure during water-limited conditions, which ensures the balance between transpiration and photosynthesis rates (Tanaka et al., 2006). For instance, CK and auxin synergistically inhibited the ABA-induced stomatal closure in Arabidopsis by modulating the ET biosynthesis (Tanaka et al., 2006). Regulation of ion channels is one of the important adaptation features in plants, especially under heavy metal and salinity stress. Salinity stress causes higher uptake and accumulation of Na^+ and Cl^- ions at the cellular level and disrupts cellular processes. CK is reported to regulate several ion channels and transporters, contribute to reduced entry of Na^+ into the cytoplasm, and reduce cellular toxicity. For instance, CK signaling promotes salt tolerance by encoding tonoplast-located Cl^- transporters, resulting in higher levels of Cl^- exclusion from shoot and root cortex cells (Yin et al., 2023). CK is also involved in the antioxidant-mediated ROS detoxification process and promotes lateral root development that enhances nutrient and water uptake, as summarized by Mughal et al. (2024). CK application in the form of 6-BAP reduced the damage induced by combined drought and heat stress in wheat by enhanched membrane stability index, leaf pigmentation, and relative water content (Kumari et al., 2018). These findings suggest that CKs contribute to abiotic stress alleviation by regulating ROS detoxification, maintaining ion homeostasis, and modulating root system architecture.

GA is involved in alleviating damage caused by different abiotic stresses, including salinity, water scarcity, and limited nutrient content, by regulating the root system architecture, maintaining ion balance, and regulating photosynthesis (Colebrook et al., 2014). GA enhances root elongation and lateral root development under limited water supply and salinity conditions. Lateral root development ensures an alternative way to search for nutrients and water sources under water-limited conditions, as well as aeration under waterlogged conditions. GA has been reported to contribute to the development of lateral roots in the Populus plant (Gou et al., 2010). Synergistic association of GA with auxin has also been reported to develop lateral roots in cucumber (Cai et al., 2022). GA also contributes to ion homeostasis by regulating ion channels and promoting the accumulation of osmoprotectants such as proline and soluble sugars. For instance, exogenous GA application has been shown to enhance salt stress tolerance by modulating ROS levels and ion balance, thereby reducing oxidative damage (Shahzad et al., 2021). Heat stress causes chlorophyll degradation, protein folding, and denaturation, and affects proper growth. GA ensures efficient photosynthesis by preventing chlorophyll degradation, reducing oxidative damage, and increasing membrane stability. For example, the exogenous application of GA enhances heat stress tolerance in tomato by reducing the heat stress-mediated chlorophyll degradation and enhancing the antioxidative-mediated ROS degradation (Guo et al., 2022). These findings suggest GA as one of the crucial hormones in regulating the plant responses to different abiotic stresses.

In addition to ABA, ET, GA, CK, and auxin, other phytohormones such as SA, JA, and BRs also play important roles in alleviating abiotic stress-induced damage in plants. Similar to ABA, SA induces stomatal closure under drought, thereby reducing water loss. For instance, SA signaling in guard cells integrates with ABA pathways to mediate stomatal closure in Arabidopsis (Prodhan et al., 2018). Moreover, exogenous SA application has been shown to mitigate the adverse effects of heat stress in tomato seedlings by enhancing photosynthetic efficiency and antioxidant enzyme activities (Jahan et al., 2019). Additionally, SA is involved in the accumulation of compatible solutes, osmo-protectants, regulation of ion transporters, and antioxidant-mediated ROS detoxification under diverse abiotic stress conditions, as summarized by Shaukat et al. (2022).

BRs are steroidal phytohormones involved in diverse plant growth and developmental processes, and their roles in abiotic stress alleviation have recently gained increasing attention (Hafeez et al., 2021). Under drought stress, BRs enhance antioxidant defense by stimulating the biosynthesis of SOD, CAT, POD, and APX. The BR application also promotes proline accumulation, contributing to osmotic adjustment and improved yield and oil content. Furthermore, foliar application of BRs in Capsicum under cold stress improved photosynthetic activity, increased photosynthetic pigment levels, and reduced ROS production. Collectively, these findings highlight BRs as key phytohormones that modulate multiple physiological processes to mitigate abiotic stress-induced damage in plants.

MT is recognized as a key regulator of abiotic stress tolerance due to its protective roles under drought, salinity, and temperature stress. MT contributes to stress resilience by enhancing ROS detoxification, activating antioxidant enzymes, regulating photosynthetic processes, and modulating transcription factors such as WRKY and NAC (Zeng et al., 2022; Colombage et al., 2023). Abiotic stresses often trigger oxidative damage through excessive ROS accumulation; however, MT application strengthens antioxidant capacity, reduces ROS levels, and prevents cellular damage (Wang et al., 2024). For instance, soaking barley seeds in MT solution before germination reduced malondialdehyde (MDA) content, improved seedling establishment and growth, and increased chlorophyll content under cold stress (Chang et al., 2021). Similarly, root-zone application of MT in soybean under water-limited conditions enhanced stress tolerance by preventing chlorophyll degradation, maintaining photosynthetic efficiency, increasing antioxidant enzyme activities, and reducing ROS accumulation (Imran et al., 2021). In addition to contributing to single abiotic stress, MT has been exploited to mitigate the negative damage caused by combined abiotic stresses. For instance, exogenous MT application reduced the oxidative stress damage caused by combined salinity and water-limited conditions by enhancing the root growth, chlorophyll biosynthesis, and reducing ROS level (Khan et al., 2024a, 2024b). Collectively, these findings highlight MT as a major regulator of abiotic stress tolerance, functioning through antioxidative defense, photosynthetic stability, and transcriptional reprogramming in plants.

PAs play crucial roles in diverse cellular processes, including gene expression, cell proliferation, enzymatic activities, and membrane stability (Gill and Tuteja, 2014). Recently, PAs have gained attention for their involvement in enhancing tolerance to abiotic stresses such as temperature, salinity, and hypoxia (Shao et al., 2022). Under these stress conditions, PAs help protect membranes, promote the accumulation of osmolytes like proline, and facilitate antioxidative-mediated ROS detoxification (Shao et al., 2022). For example, exogenous application of PAs in cucumber reduced chlorophyll degradation, improved photosynthesis, and minimized thylakoid membrane damage, thereby conferring chilling stress tolerance (He et al., 2002). Similarly, water stress often damages the photosynthetic apparatus, lowering photosynthesis and growth; however, maintaining higher water content under drought helps sustain photosynthesis. Exogenous PAs have been reported to enhance antioxidative activities and transcriptional regulation, increase accumulation of phenolic compounds and osmolytes, and reduce damage to photosynthetic organs under drought stress (Kotakis et al., 2014; Li et al., 2016). Collectively, these studies highlight PAs as key contributors to plant tolerance against major abiotic stresses

Due to their significant contribution toward abiotic stress tolerance, phytohormones and their derived growth regulators have been used exogenously to alleviate the abiotic stress-induced damage in different plants, as summarized in Table 9.2

9.5 Potential functions of phytohormones and growth regulators in plants in response to combined abiotic stresses

Plants exhibit diverse responses to abiotic stresses, with some being common across stress types and others being stress-specific. Common responses include the modulation of root system architecture, stomatal closure, osmotic adjustment, antioxidant production, osmolyte and Osmo protectant accumulation, activation of transcription factors (TFs), and stress-responsive gene expression (Han et al., 2023). However, plants have several stress-specific responses, including ionic homeostasis under salinity stress, aerenchyma and adventitious root formation under waterlogging stress, heavy metal chelation and sequestration in vacuoles under heavy metal stress, and specific gene expression for unique metabolites against particular stresses. This section summarizes the roles of phytohormones and growth regulators in mediating some of the common and stress-specific responses observed in plant species.

Most abiotic stresses cause water scarcity, necessitating efficient water-use strategies. Plants minimize water loss by closing stomata during stress. Stomatal closure involves complex physiological processes, primarily regulated by phytohormones (Agurla et al., 2018). While ABA is the primary hormone involved,ET, JA, CK, SA, auxins, and GA also contribute (Aslam et al., 2022). ABA is synthesized in roots under drought conditions, transported to leaves, and binds to guard cell receptors, triggering a signaling cascade that increases cytoplasmic calcium (Ca^{2+}) (Hsu et al., 2021). This activates protein kinases and ion channels, causing ion efflux (e.g., K^+, Cl^-) from guard cells, reducing osmotic pressure, and leading to stomatal closure. In addition to ABA, ET promotes ion channel regulation during waterlogging, while JA and SA enhance ROS production in guard cells, acting as secondary messengers in stomatal regulation (Postiglione and Muday, 2020). CKs enhance ABA sensitivity, while GA often antagonizes ABA. Collectively, these hormones coordinate to optimize water conservation under stress by regulating stomatal closure.

TABLE 9.2 Representative studies describing the contribution of phytohormones and PGRs in alleviating the single and combined abiotic stress-induced damages in plants.

Phytohormones/growth regulator	Abiotic stress	Plant species	Alleviation mechanism/effect	References
ABA	Drought	*Camellia oleifera*	Higher accumulation of osmolytes, enhanced membrane stability, and higher antioxidative enzymatic activities	Yang et al. (2024)
ABA	Combined salt and heat stress	*Arabidopsis*	Regulation of Na^+/K^+	Suzuki (2016)
ABA	High temperature and drought	*Medicago sativa L*	Protecting membrane damage, maintaining ion balance, and regulating photosynthesis	Khan et al. (2024a, 2024b)
Auxin (IAA)	Salinity	*Vicia faba* L.	Lower MDA content with higher SOD, CAT, POD, and APA activities, higher accumulation of osmolytes, and ionic homeostasis	Abdel Latef et al. (2021)
GA (GA_3)	Salinity	*Zea mays* L., *Pisum sativum* Var. Abyssinicum A. Braun and *Lathyrus sativus* L.	Increased germination percentage, higher root and shoot length, and total weight of crops.	Tsegay and Andargie (2018)
CK (meta-topolin-9-(tetrahydropyran-2-yl)purine (mT9THP))	High temperature	*Oryza sativa*	Hormonal dynamics, gene expression, and volatile emission, higher activity of heat shock protein, and SOD	Prerostova et al. (2023)
CK (6-BAP)	Drought and high temperature	*Tritichum aestivum L.*	Reduced membrane damage, enhanced photosynthesis, pigment stability, and higher water content	Kumari et al. (2018)
CK (6-BAP)	Heat stress	*Tritichum aestivum L.*	Higher grain filling rate due to endosperm cell division and number	Yang et al. (2016)
Ethylene (Ethephon)	Waterlogging	*Glycine max* L	Enhanced photosynthetic activity, higher amino acid contents, higher root surface area, and higher activity of GSH	Kim et al. (2018)
Ethylene	Drought and low nitrogen	*Tritichum aestivum L.*	Higher survival rate by regulating proteins related to ethylene biosynthesis and signal transduction	Yan et al. (2023)
JA (Methyl Jasmonate)	Hight temperature	*Lolium perenne* L	Higher antioxidant activity, maintenance of relative water content, decreased photosynthetic pigment damage, and lowered ROS accumulation.	Su et al. (2021)
SA	High temperature	*Solanum lycopersicum* L	Enhanced photosynthesis, higher water use efficiency, reduced electrolyte leakage, higher antioxidative activities, and reduced oxidative damage due to lower ROS accumulation	Jahan et al. (2019)

BRs (24-epibrassinolide)	Lead (heavy metal)	*Brassica juncea* L.	Higher antioxidative activities and higher seedling germination	Soares et al. (2020)
MT	Salinity and drought	*Oryza sativa*	Promotion of root growth, biomass accumulation, and an increased level of photosynthetic pigment	Khan et al. (2024a,b)
MT	Low temperature	*Hordeum vulgare* L.var. *nudum*	Reduced ROS accumulation, higher accumulation of antioxidants, and restored cellular redox homeostasis	
PAs (Spermidine)	Low temperature	*Cucumis sativus* L.	Higher chlorophyll content maintained a normal photosynthesis rate and enhanced antioxidative activities	He et al. (2002)

CAT, catalase; *POD*, peroxidase; *SA*, salicylic acid; *SOD*, superoxide dismutase; *MDA*, malondialdehyde.

Abiotic stress disrupts cellular water potential and affects the cellular osmotic and ionic balance due to the higher accumulation of ROS (Nadarajah, 2020). Higher accumulation of ROS results in membrane damage and causes cell death. Besides, salinity stress results in higher accumulation of Na^+ and Cl^- ion uptake and transport to the cytoplasm, causing ionic imbalance and affecting the cellular processes. To neutralize the ROS activities and control the osmotic balance, plants produce different osmolytes, including proline, glycine betaine, and soluble sugars. Also, plants activate the enzymatic and non-enzymatic antioxidants, which react with ROS and neutralize its impact (Hasanuzzaman et al., 2020). Osmotic adjustment, a common response to stresses like drought, salinity, and extreme temperatures, involves accumulating osmolytes (e.g., proline, glycine betaine, and soluble sugars) and regulation of ion homeostasis, both of which are regulated by major phytohormones such as ABA, ET, and JA. Phytohormones regulate both ROS biosynthesis and the ROS detoxification process. Phytohormones like ABA, ET, JA, SA, and CK upregulate antioxidant enzymes (e.g., SOD, POD, CAT, and APX) to balance the ROS production. ET, JA, and SA synergize with ABA to enhance enzymatic and no-enzymatic antioxidant defenses, while CK, auxins, and GA fine-tune ROS homeostasis to support growth and survival (Devireddy et al., 2021a, 2021b). Phytohormones are also involved in the accumulation of different osmolytes, which assist in maintaining the redox balance at the cellular level. For example, the application of SA on salt-treated lentils exhibited higher accumulation of proline and glycine betaine, resulting in lower salinity damage (Misra and Saxena, 2009). Exogenous application of JA improved cadmium stress by modulating the antioxidative activities and higher accumulation of osmolytes such as proline and glycine betaine (Ahmad et al., 2017). Exogenous application of ABA, ET, and CK also promotes osmolyte accumulation under different abiotic stress conditions, as summarized by Singh et al. (2022). These non-toxic compounds provide protective benefits against ROS and enhance plant tolerance to abiotic stresses.

Upon exposure to unfavorable conditions, stress stimuli are received by membrane-based kinases, which are transduced and transferred by different primary and secondary messengers and activate the TFs, followed by stress-responsive gene expression (Akpinar et al., 2012). Activation of stress-related TFs and gene expression is a common response irrespective of stress type. Several TF families, including WRKY, NAC TFs, DREB/CBFs, MYB TFs, AP2/ERF TFs, and bZIP, have been reported to regulate the stress-responsive gene expression (Javed et al., 2020; Hoang et al., 2017). These TFs are involved in gene expression, which regulates the osmolyte synthesis, ion transporters, and antioxidative defense. Phytohormones interact with TFs to regulate gene expression, which controls the stress responses, acclimation, and growth under abiotic stress conditions (Ghorbel and Brini, 2021). NAC TFs are one of the major TFs in stress regulatory networks, which respond to different abiotic stresses and enhance multiple stress tolerance. In a study by Han et al. (2023), ET treatment enhanced the *PtNAC3* in conifer plants, and overexpression of *PtNAC3* leads to enhanced abiotic stress tolerance. *bZIP* TFs are involved in drought resistance-related gene expression. Abiotic stress-induced ABA biosynthesis enhances the expression of bZIP transcription factors and mediates phosphorylation of AREB1/2, which in turn regulates downstream stress-responsive genes under drought conditions (Furihata et al., 2006; Yoshida et al., 2010). These findings highlight the critical role of phytohormonal signaling and transcriptional regulation in orchestrating plant responses to abiotic stresses.

Plants exhibit stress-specific responses and vary depending on the stress type, duration, and intensity. Stress-specific stresses include the development of aerenchyma and adventitious roots under waterlogging, and chelation, sequestration, and compartmentalization under heavy metal stress. Phytohormones are involved in shaping and fine-tuning stress-specific responses. As an example, the development of aerenchyma is dependent on the activities of ET. Under waterlogged conditions, ET accumulation triggers ROS bursts via RBOH gene expression, leading to PCD and aerenchyma development. Aerenchyma stores oxygen, ensuring cellular respiration and root growth. ET and ROS-mediated aerenchyma formation has been observed in sunflower (Ni et al., 2019), cotton (Pan et al., 2022), rice and maize (Yamauchi et al., 2016), etc., which provided tolerance to waterlogged conditions and improved growth, yield components, and grain quality as reported in barley (Manik et al., 2022).

A common plant response to heavy metal stress is the chelation, sequestration, and compartmentalization of metals and metalloids, which helps prevent cellular phytotoxicity (Hossain et al., 2023; Sandhi et al., 2023). In addition to enhancing the antioxidative-mediated defense, phytohormones are involved in specific responses towards heavy metal stress and tolerance (Rahman et al., 2023). Phytohormones are involved in cell wall remodeling at the root zone and promote the biosynthesis of chelators such as phytochelatins and metallothioneins, which reduce their accumulation and bioavailability (Saini et al., 2021). For example, exogenous application of IBA resulted in higher Cd^{2+} fixation in the cell wall and increased the endogenous auxin with higher binding to the cell wall, resulting in amelioration of Cd stress (Šípošová et al., 2023). ABA is involved in limiting the heavy metal transport from root to shoot by closing the stomatal regulations, slowing transpiration, and thus reducing the metal ion concentration at the leaf (Hu et al., 2020). ABA promotes the synthesis of chelators like phytochelatins and metallothioneins, reducing heavy metal bioavailability (Bano et al., 2023).

ET and ABA regulate metal ion transporters for vacuolar sequestration, while auxin-mediated cell wall remodeling enhances metal deposition in root zones, minimizing toxicity (Saini et al., 2021).

9.6 Interactions of phytohormone signaling pathways to mitigate combined abiotic stresses

Combined abiotic stress sensing by various receptor molecules embedded in the cell membrane and other organelles within the cells serves as the primary trigger to stimulate the signaling pathways (Khan et al., 2020). The convergence of hormonal and signal transduction pathways within cellular systems is conceptualized as the interplay between phytohormones and signaling molecules (Fig. 9.3). The entity under consideration may function as both an upstream and downstream modulator, potentially activated by phytohormones and secondary messengers within the phosphorylation signal (Jogawat et al., 2021). Phytohormones act as initiators of mitogen-activated protein kinase (MAPK) cascades, which synthesize and modulate signaling crosstalk within plant systems. MAPK cascades play a pivotal role in mediating the transmission of various environmental cues. The plant contains a plethora of MAPK components, which are linked to various stresses (De Zélicourt et al., 2016). The activation of various MAPK cascades, notably MEKK1/MKK2/MPK4 and MPK6 cascades, is triggered in plants by diverse stress combinations. Phytohormone-mediated regulation of specific gene expression serves a protective role to mitigate oxidative damage caused by stressors (Sytar et al., 2019). Intricate signaling and transduction cascades exert a dual influence, orchestrating gene expression regulation while simultaneously affecting the cellular biosynthesis of various hormones (Wani et al., 2016). PGRs exhibiting phytohormone-like properties, participate in intricate interactions with other PGRs, resulting in substantial alterations in plant phenological patterns.

The principal nexus for manipulating crop responses to multiple stressors lies within the ET and ABA signaling pathways (Danquah et al., 2014). ET-induced activation of *ABI1* and *ABI2* serves as a negative regulatory mechanism in ABA signaling. Drought stress elicits protective mechanisms and is corroborated by the synergistic interactions of ABA and JA under various combined stress responses. Furthermore, the sophisticated interplay among ET, JA, and SA defense signaling cascades facilitates the selection of an optimal defence strategy (Jogawat et al., 2021). Additionally, the critical hormonal crosstalk between GA and JA exerts a significant influence on the regulation of growth–defense trade-offs. JA exerts an inhibitory effect on GA-mediated growth responses with the interplay of JAZs and DELLAs. Moreover, the

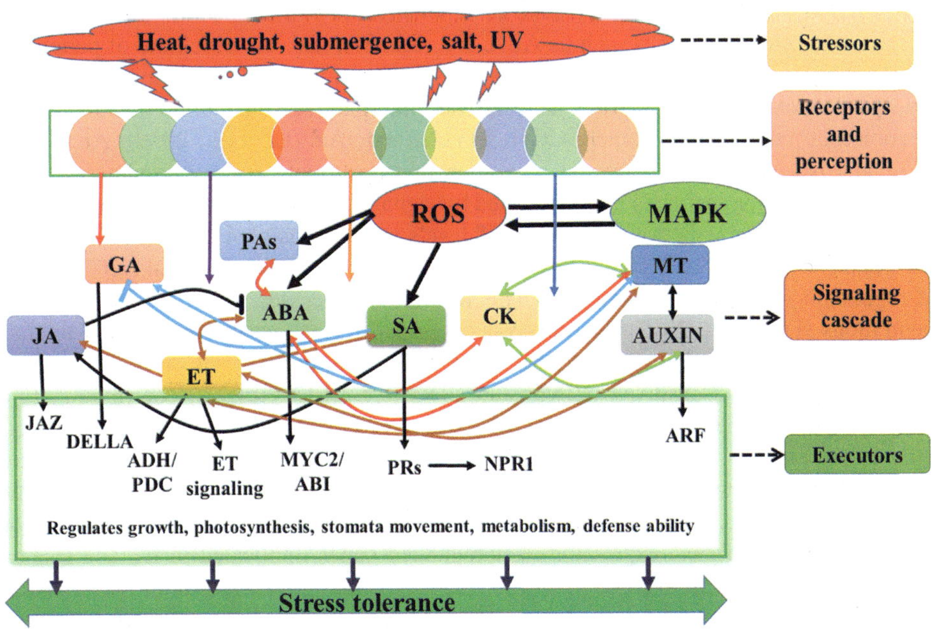

FIGURE 9.3 Cross-talk of phytohormones and PGR during diverse abiotic stresses. Environmental stressors provoked plants to initiate perception or reception, followed by a series of signaling events, culminating in defensive mechanisms and/or adaptive modifications, manifested through physiological changes, specific molecular components, transcription factors, and so on. *ABA*, abscisic acid; *ARF*, auxin response factor; *ADH*, alcohol dehydrogenase; *CK*, cytokinin; *ET*, ethylene; *GA*, gibberellic acid; *JA*, jasmonic acid; *MAPK*, mitogen-activated protein kinase; *MT*, melatonin; *PAs*, polyamines; *PDC*, pyruvate decarboxylase; *PR*, pathogensis related; *ROS*, reactive oxygen species; *SA*, salicylic acid.

interaction of ET and JA limits the processes of leaf cell cycle and hastens it by disrupting the function of growth-promoting factors such as auxins (Khan et al., 2020).

ABA is a pivotal phytohormone that plays a key function in adaptive mechanisms of plants during drought conditions, orchestrating various physiological responses to mitigate stress-induced damage (Jogawat et al., 2021). ABA regulation potentially encompasses dual regulatory mechanisms: coordination of its biosynthetic processes and fine-tuning of gene expression patterns. Studies have shown that JA modulates the expression of 9-cis-epoxy carotenoid dioxygenase (*NCED*) and ABA 8-hydroxylation (*CYP707A*) genes and plays an important role in ameliorating stresses (Xing et al., 2019). Drought exerts a profound effect on stomatal closure, a crucial morphological trait that is extensively regulated in plants. In addition, ABA potentially interacts with JA and NO to promote stomatal closure in response to drought conditions. Concurrently, water scarcity modulates the ET, CK, and auxin signaling pathways associated with gene expression. The transcription family of auxin response factors (ARFs) plays a fundamental role in plant development and the mitigation of combined abiotic stressors as well. ARF either promotes various transcription factor functions or suppresses auxin-responsive genes by direct interactions with their promoter regions (Bouzroud et al., 2018).

Research findings elucidated that ARF triggers both carotenoid biosynthesis and the enhancement of combined stress responses (drought and salt stress) in sweet potatoes (Kang et al., 2018). Dopamine treatment stimulates ET production while concurrently suppressing the oxidative processes of indole-3-acetic acid (IAA) in apples during NaCl stress (Liu et al., 2020). The DELLA protein functions negatively in the GA signaling network. Evidence suggests that DELLA directly connects with BZRI/BES1 within the GRAS domain, potentially suppressing BZR1 DNA-binding capacity and consequently interfering with BR-regulated signaling processes in plants (Unterholzner et al., 2015). Auxins bolster drought resilience in *Arabidopsis* by balancing ROS metabolism and ABA-responsive gene expression modification (Shi et al., 2014). In contrast to ABA, CKs function and play a vital role in the thermal regulation of leaves during water deficit stress. This process involves the accumulation of multiple compounds, such as carbohydrates, organic acids, and amino acids, which collectively enhance the plant's resilience to osmotic stress and facilitate plant performance (Gujjar and Supaibulwatana, 2019). The synergistic functions of salicylic acid and putrescine (Put) elevate drought tolerance in *Cicer arietinum* (chickpea) by the augmentation of different elements like protein, sugar, and phenolic compounds as well (Khan et al., 2019a, 2019b).

The activation of MYC2, MYC3, and MYC4 transcription factors is hindered by the interplay between ABA and JA. Research has demonstrated that ET and ABA exhibit antagonistic relationships, whereas ET and JA interact synergistically. In contrast, ABA manifests antagonistic or negative interactions with both JA and ET during combined stress responses in tobacco (Khan et al., 2020). GA is a potential phytohormone that has elucidated the intrinsic relationship among the other phytohormones such as ABA, ET, GA, SA, and JA, which are of paramount importance in overcoming combined stresses (Wani et al., 2016). ET exerts a substantial influence on stomatal closure, which facilitates H_2O_2 synthesis and production (Wu et al., 2014). The intricate interplay between auxin and JA signaling networks is facilitated by key molecular components, namely COI1, MYC2, and JAZ (Li et al., 2019). ABA engages in intricate interactions with JA and NO to facilitate stomatal regulation during drought stress. Simultaneously, ABA modulates drought-responsive gene expression, which is interconnected by signaling pathways involving CKs, auxins, and ET (Wani et al., 2016). The synergistic interplay between elevated ET levels and various phytohormones, such as ABA, IAA, CK, and GA, facilitates plant adaptation to waterlogged stress conditions (Shimamura et al., 2014). MT and dopamine contribute to the regulation of nitrogen assimilation and metabolic processes in *Malus hupehensis* exposed to low ammonium concentrations (Du et al., 2022). Overexpression of *StEREBP1* results in elevated tolerance to both low temperature and salt stress conditions (Per et al., 2018). The ET signaling pathway encompasses additional downstream components, notably the ethylene-insensitive3 (EIN3) and EIN3-like (EIL) proteins, contributing to regulatory roles during combined abiotic stress conditions in plants (Jyoti et al., 2021). PAs are ubiquitous molecules that act as the mitigators of diverse abiotic stresses and facilitate normal plant development. Putrescine exhibited a stimulatory effect on ABA synthesis gene expression while concurrently suppressing ET, JA, and gibberellin production-related gene expression. Conversely, spermidine elicited the opposite response in these genetic regulatory pathways. In contrast, spermine exhibited a dual effect on gene expression: it upregulated genes implicated in ET and jasmonate production, while downregulating those involved in gibberellin and ABA biosynthesis (Bitrián et al., 2012). Experimental evidence has revealed the differential effects of PAs in the context of hormone-signaling cascades. Specifically, spermidine upregulated genes associated with salicylic acid signaling pathways. Conversely, spermine was correlated with genes involved in auxin and cytokinin signaling. Putrescine exhibits a range of effects on jasmonate signaling, ranging from neutral to mildly positive. PAs exert no discernible influence on either the synthesis or signaling mechanisms of BRs (Anwar et al., 2015). MT performs multidimensional functions in terms of modulating plant physiological, biochemical, and molecular processes. As a universal antioxidant protector, MT ameliorates the negative impact of environmental stressors (Jahan et al., 2021a; Altaf et al., 2022). MT functioned as a stress signaling molecule in plants, eliciting responses that facilitate plant acclimation to adverse extreme weather conditions (Jahan et al., 2021b; Arnao and Hernández-Ruiz, 2019). MT interacts with various signaling molecules in plants, including auxins, CK, GA, and ABA, and participates in enhancing diverse stress tolerance (heat and waterlogging) and other physiological activities (Jahan et al., 2021b; Jahan et al., 2024). MT treatment enhanced the germination quality of melon seeds under combined autotoxicity and saline-alkalinity stress through regulating starch metabolism, elevating antioxidant enzyme activity, and increasing MT biosynthesis gene expression (Liu et al., 2023). MT treatment efficiently alleviates the harmful effects of combined drought and high temperature stress by elevating antioxidant activity, which acts as the safeguard of the photosystem from oxidative damage (Annadurai et al., 2023). Mahnoor et al. demonstrated that the combined application of MT and *Bacillus licheniformis* mitigates arsenic and cadmium stress in spinach without compromising any physiological attributes. Another study conducted by Mumithrakamatchi et al. (2024) noted that MT confers resistance to combination drought and high-temperature stressors in tomatoes via osmotic regulation and ABA accumulation.

9.7 Conclusion and the way forward

The intricate relationship between phytohormones, PGRs, and multiple and/or combined abiotic stresses is a critical focus of contemporary research because of their significant implications for agricultural productivity and stress resistance. Despite the considerable progress achieved in this domain, numerous knowledge gaps persist, necessitating further comprehensive investigation and analysis. Although a broad understanding exists regarding the functions of phytohormones in regulating plant responses to mitigate multiple stressors, the precise in-depth cellular mechanisms and signaling pathways are yet to be fully delineated, and exclusive investigation is warranted to delineate the nuanced interplay between diverse phytohormones and their interactions with stress-related signaling cascades. It is well documented that individual phytohormones are effective in mitigating specific adverse stress effects; however, it would be more advantageous if further research were conducted on multiple stresses with specific hormones or combinations thereof. The spatial allocation of phytohormones and growth regulators within plant tissues under stressful conditions remains poorly understood. The investigation of temporal variations and spatial distribution patterns of phytohormones within diverse plant tissues may provide critical insights into the fundamental mechanisms governing stress adaptation processes. Subsequent investigations should prioritize the unification of multiomics-based approaches, genetic modification, and traditional plant breeding strategies to bolster crop resilience towards environmental stressors. Advanced cutting-edge genetic engineering techniques, such as CRISPR-Cas9, might be applied to rectify specific genes associated with phytohormone signaling to ameliorate combined stresses. The adoption of climate-smart agricultural strategies incorporating phytohormones and PGRs offers the potential to mitigate combined stress effects on crop yields. Epigenetic regulation of phytohormones and PGRs, as well as their underlying interplay under combined stresses, refers to a promising avenue for expanding our understanding of mechanisms to enhance plant stress tolerance. In summary, additional investigation will be focused on unraveling the sophisticated regulatory mechanisms of PHs individually or combined for the development of robust crop varieties and safeguarding global food security amidst climate change and environmental stressors.

References

Abdel Latef, A.A.H., Tahjib-Ul-Arif, M., Rhaman, M.S., 2021. Exogenous auxin-mediated salt stress alleviation in faba bean (*Vicia faba* L.). Agronomy 11 (3), 547. Available at https://doi.org/10.3390/agronomy11030547.

Abdelraheem, A., Esmaeili, N., O'Connell, M., Zhang, J., 2018. Progress and perspective on drought and salt stress tolerance in cotton. Industrial Crops and Products 130, 118–129. Available at https://doi.org/10.1016/j.indcrop.2018.12.070.

Afsana, N., Islam, M., Hossain, E., Nizam, R., Monalesa, N., Hussain, A., et al., 2017. Response of tomato (Solanum lycopersicum L.) to salicylic acid and calcium. Journal of Applied Life Sciences International 15 (2), 1–7. Available at https://doi.org/10.9734/jalsi/2017/37408.

Agurla, S., Gahir, S., Munemasa, S., Murata, Y., Raghavendra, A.S., 2018. Mechanism of stomatal closure in plants exposed to drought and cold stress. In: Iwaya-Inoue, M., Sakurai, M., Uemura, M. (Eds.), Survival Strategies in Extreme Cold and Desiccation. Advances in Experimental Medicine and Biology, vol 1081. Springer, Singapore, pp. 215–232. Available at https://doi.org/10.1007/978-981-13-1244-1_12.

Ahanger, M.A., Akram, N.A., Ashraf, M., Alyemeni, M.N., Wijaya, L., Ahmad, P., 2017. Plant responses to environmental stresses-From gene to biotechnology. AoB Plants 9 (4). Available at https://doi.org/10.1093/aobpla/plx025.

Ahmad, P., Alyemeni, M.N., Wijaya, L., Alam, P., Ahanger, M.A., Alamri, S.A., 2017. Jasmonic acid alleviates negative impacts of cadmium stress by modifying osmolytes and antioxidants in faba bean (Vicia faba L.). Archives of Agronomy and Soil Science 63 (13), 1889–1899. Available at https://doi.org/10.1080/03650340.2017.1313406.

Akpinar, B.A., Avsar, B., Lucas, S.J., Budak, H., 2012. Plant abiotic stress signaling. Plant Signaling & Behavior 7 (11), 1450–1455. Available at https://doi.org/10.4161/psb.21894.

Alam, I., Sharmin, S.A., Kim, K.H., Kim, Y.G., Lee, J.J., Bahk, J.D., et al., 2011. Comparative proteomic approach to identify proteins involved in flooding combined with salinity stress in soybean. Plant and Soil 346 (1), 45–62. Available at https://doi.org/10.1007/s11104-011-0792-0.

Ali, A.Y.A., Ibrahim, M.E.H., Zhou, G., Nimir, N.E.A., Jiao, X., Zhu, G., et al., 2020. Exogenous jasmonic acid and humic acid increased salinity tolerance of sorghum. Agronomy Journal 112 (2), 871–884. Available at https://doi.org/10.1002/agj2.20072.

Altaf, M.A., Shahid, R., Ren, M.X., Naz, S., Altaf, M.M., Khan, L.U., et al., 2022. Melatonin improves drought stress tolerance of tomato by modulation plant growth, root architecture, photosynthesis, and antioxidant defense system. Antioxidants 11 (2). Available at https://doi.org/10.3390/antiox11020309.

Amjad, M., Kousar, R., Naeem, M.A., Imran, M., Nadeem, M., Abbas, G., et al., 2024. An interplay of salt and Ni stress on contrasting tomato (*Solanum lycopersicum* L.) genotypes: a physiological and biochemical insight. International Journal of Phytoremediation 27, 711–723. Available at https://doi.org/10.1080/15226514.2024.2438772.

Angon, P.B., Tahjib-Ul-Arif, M., Samin, S.I., Habiba, U., Hossain, M.A., Brestic, M., 2022. How do plants respond to combined drought and salinity stress?—a systematic review. Plants 11 (21). Available at https://doi.org/10.3390/plants11212884.

Annadurai, M.K.K., Alagarsamy, S., Karuppasami, K.M., Ramakrishnan, S., Subramanian, M., Venugopal, P.R.B., et al., 2023. Melatonin decreases negative effects of combined drought and high temperature stresses through enhanced antioxidant defense system in tomato leaves. Horticulturae 9 (6). Available at https://doi.org/10.3390/horticulturae9060673.

Anwar, R., Mattoo, A.K., Handa, A.K., 2015. Polyamine interactions with plant hormones: crosstalk at several levels. Polyamines: a universal molecular nexus for growth, survival, and specialized metabolism. In: Kusano, T., Suzuki, H. (Eds.), Polyamines. Springer, Tokyo, pp. 267–302. Available at https://doi.org/10.1007/978-4-431-55212-3_22.

Arnao, M.B., Hernández-Ruiz, J., 2019. Melatonin: a new plant hormone and/or a plant master regulator? Trends in Plant Science 24 (1), 38–48. Available at https://doi.org/10.1016/j.tplants.2018.10.010.

Atkinson, N.J., Lilley, C.J., Urwin, P.E., 2013. Identification of genes involved in the response of Arabidopsis to simultaneous biotic and abiotic stresses. Plant Physiology 162 (4), 2028–2041. Available at https://doi.org/10.1104/pp.113.222372.

Bano, A., Singh, K., Singh, S.P., Sharma, P., 2023. Abscisic acid: metabolism, signaling, and crosstalk with other phytohormones under heavy metal stress. Stresses 3 (4), 665–686. Available at https://doi.org/10.3390/stresses3040046.

Begum, N., Hasanuzzaman, M., Li, Y., Akhtar, K., Zhang, C., Zhao, T., 2022. Seed germination behavior, growth, physiology and antioxidant metabolism of four contrasting cultivars under combined drought and salinity in soybean. Antioxidants 11 (3), 498. Available at https://doi.org/10.3390/antiox11030498.

Biglia, A., Gresta, F., Patono, D.L., Comba, L., Lovisolo, C., Gay, P., et al., 2024. Identification of drought-salinity combined stress in tomato plants by vegetation indices. Journal of Agricultural Engineering 55 (4). Available at https://doi.org/10.4081/jae.2024.1599.

Bitrián, M., Zarza, X., Altabella, T., Tiburcio, A.F., Alcázar, R., 2012. Polyamines under abiotic stress: metabolic crossroads and hormonal cross-talks in plants. Metabolites 2 (3), 516–528. Available at https://doi.org/10.3390/metabo2030516.

Boutahiri, S., Benrkia, R., Tembeni, B., Idowu, O.E., Olatunji, O.J., 2024. Effect of biostimulants on the chemical profile of food crops under normal and abiotic stress conditions. Current Plant Biology 40. Available at https://doi.org/10.1016/j.cpb.2024.100410.

Bouzroud, S., Gouiaa, S., Hu, N., Bernadac, A., Mila, I., Bendaou, N., et al., 2018. Auxin response factors (ARFs) are potential mediators of auxin action in tomato response to biotic and abiotic stress (Solanum lycopersicum). PLoS One 13 (2), e0193517. Available at https://doi.org/10.1371/journal.pone.0193517.

Cai, B., Wang, T., Sun, H., Liu, C., Chu, J., Ren, Z., et al., 2022. Gibberellins regulate lateral root development that is associated with auxin and cell wall metabolisms in cucumber. Plant Science 317, 110995. Available at https://doi.org/10.1016/j.plantsci.2021.110995.

Cao, G., Huang, H., Yang, Y., Xie, B., Tang, L., 2024. Analysis of drought and heat stress response genes in rice using co-expression network and differentially expressed gene analyses. PeerJ 12 (4), e17255. Available at https://doi.org/10.7717/peerj.17255.

Chang, T., Zhao, Y., He, H., Xi, Q., Fu, J., Zhao, Y., 2021. Exogenous melatonin improves growth in hulless barley seedlings under cold stress by influencing the expression rhythms of circadian clock genes. PeerJ 9, e10740. Available at https://doi.org/10.7717/peerj.10740.

Colebrook, E.H., Thomas, S.G., Phillips, A.L., Hedden, P., 2014. The role of gibberellin signalling in plant responses to abiotic stress. Journal of Experimental Biology 217 (1), 67–75. Available at https://doi.org/10.1242/jeb.089938.

Colombage, R., Singh, M.B., Bhalla, P.L., 2023. Melatonin and abiotic stress tolerance in crop plants. International Journal of Molecular Sciences 24 (8), 7447. Available at https://doi.org/10.3390/ijms24087447.

Danquah, A., de Zelicourt, A., Colcombet, J., Hirt, H., 2014. The role of ABA and MAPK signaling pathways in plant abiotic stress responses. Biotechnology Advances 32 (1), 40–52. Available at https://doi.org/10.1016/j.biotechadv.2013.09.006.

Daszkowska-Golec, A., 2016. The role of abscisic acid in drought stress: How ABA helps plants to cope with drought stress. In: Hossain, M., Wani, S., Bhattacharjee, S., Burritt, D., Tran, L.S. (Eds.), Drought Stress Tolerance in Plants, Vol 2: Molecular and Genetic Perspectives. Springer International Publishing, Poland, pp. 123–151. Available at 10.1007/978-3-319-32423-4_5.

Devireddy, A.R., Tschaplinski, T.J., Tuskan, G.A., Muchero, W., Chen, J., 2021a. Role of reactive oxygen species and hormones in plant responses to temperature changes. International Journal of Molecular Sciences 22 (16), 8843. Available at https://doi.org/10.3390/ijms22168843.

Devireddy, A.R., Zandalinas, S.I., Fichman, Y., Mittler, R., 2021b. Integration of reactive oxygen species and hormone signaling during abiotic stress. The Plant Journal 105 (2), 459–476. Available at https://doi.org/10.1111/tpj.15010.

Dong, Q., Wallrad, L., Almutairi, B.O., Kudla, J., 2022. Ca2+ signaling in plant responses to abiotic stresses. Journal of Integrative Plant Biology 64 (2), 287–300. Available at https://doi.org/10.1111/jipb.13228.

Du, P., Yin, B., Cao, Y., Han, R., Ji, J., He, X., et al., 2022. Beneficial effects of exogenous melatonin and dopamine on low nitrate stress in *Malus hupehensis*. Frontiers in Plant Science 12, 807472. Available at https://doi.org/10.3389/fpls.2021.807472.

Ellouzi, H., Ben Slimene Debez, I., Amraoui, S., Rabhi, M., Hanana, M., Alyami, N.M., et al., 2024. Effect of seed priming with auxin on ROS detoxification and carbohydrate metabolism and their relationship with germination and early seedling establishment in salt stressed maize. BMC Plant Biology 24 (1). Available at https://doi.org/10.1186/s12870-024-05413-w.

Furihata, T., Maruyama, K., Fujita, Y., Umezawa, T., Yoshida, R., Shinozaki, K., et al., 2006. Abscisic acid-dependent multisite phosphorylation regulates the activity of a transcription activator AREB1. Proceedings of the National Academy of Sciences 103 (6), 1988–1993. Available at https://doi.org/10.1073/pnas.0505667103.

Galani, Y.J.H., Hansen, E.M.Ø., Droutsas, I., Holmes, M., Challinor, A.J., Mikkelsen, T.N., et al., 2022. Effects of combined abiotic stresses on nutrient content of European wheat and implications for nutritional security under climate change. Scientific Reports 12 (1). Available at https://doi.org/10.1038/s41598-022-09538-6.

Ghorbel, M., Brini, F., 2021. Hormone mediated cell signaling in plants under changing environmental stress. Plant Gene 28, 100335. Available at https://doi.org/10.1016/j.plgene.2021.100335.

Gill, S.S., Tuteja, N., 2014. Polyamines and abiotic stress tolerance in plants. Plant Signaling & Behavior 5 (1), 26–33. Available at https://doi.org/10.4161/psb.5.1.10291.

Giraud, E., Ho, L.H.M., Clifton, R., Carroll, A., Estavillo, G., Tan, Y.F., et al., 2008. The absence of alternative oxidase1a in Arabidopsis results in acute sensitivity to combined light and drought stress. Plant Physiology 147 (2), 595–610. Available at https://doi.org/10.1104/pp.107.115121.

Goossens, J., Fernández-Calvo, P., Schweizer, F., Goossens, A., 2016. Jasmonates: signal transduction components and their roles in environmental stress responses. Plant Molecular Biology 91 (6), 673–689. Available at https://doi.org/10.1007/s11103-016-0480-9.

Gou, J., Strauss, S.H., Tsai, C.J., Fang, K., Chen, Y., Jiang, X., et al., 2010. Gibberellins regulate lateral root formation in Populus through interactions with auxin and other hormones. The Plant Cell 22 (3), 623–639. Available at https://doi.org/10.1105/tpc.109.073239.

Gujjar, R.S., Supaibulwatana, K., 2019. The mode of cytokinin functions assisting plant adaptations to osmotic stresses. Plants 8 (12). Available at https://doi.org/10.3390/plants8120542.

Guo, T., Gull, S., Ali, M.M., Yousef, A.F., Ercisli, S., Kalaji, H.M., et al., 2022. Heat stress mitigation in tomato (Solanum lycopersicum L.) through foliar application of gibberellic acid. Scientific Reports 12 (1). Available at https://doi.org/10.1038/s41598-022-15590-z.

Habibi, F., Ramezanian, A., Rahemi, M., Eshghi, S., Guillén, F., Serrano, M., et al., 2019. Postharvest treatments with γ -aminobutyric acid, methyl jasmonate, or methyl salicylate enhance chilling tolerance of blood orange fruit at prolonged cold storage. Journal of the Science of Food and Agriculture 99 (14), 6408–6417. Available at https://doi.org/10.1002/jsfa.9920.

Habibpourmehraban, F., Wu, Y., Masoomi-Aladizgeh, F., Amirkhani, A., Atwell, B.J., Haynes, P.A., 2023. Pre-treatment of rice plants with ABA makes them more tolerant to multiple abiotic stress. International Journal of Molecular Sciences 24 (11). Available at https://doi.org/10.3390/ijms24119628.

Habti, A.E., Fleury, D., Jewell, N., Garnett, T., Tricker, P.J., 2020. Tolerance of combined drought and heat stress is associated with transpiration maintenance and water-soluble carbohydrates in wheat grains. Frontiers in Plant Science 11. Available at https://doi.org/10.3389/fpls.2020.568693.

Hafeez, M.B., Zahra, N., Zahra, K., Raza, A., Batool, A., Shaukat, K., et al., 2021. Brassinosteroids: molecular and physiological responses in plant growth and abiotic stresses. Plant Stress 2, 100029. Available at https://doi.org/10.1016/j.stress.2021.100029.

Han, F., Wang, P., Chen, X., Zhao, H., Zhu, Q., Song, Y., et al., 2023. An ethylene-induced NAC transcription factor acts as a multiple abiotic stress responsor in conifer. Horticulture Research 10 (8). Available at https://doi.org/10.1093/hr/uhad130.

Hartman, S., Sasidharan, R., Voesenek, L.A., 2021. The role of ethylene in metabolic acclimations to low oxygen. New Phytologist 229 (1), 64–70. Available at https://doi.org/10.1111/nph.16378.

Hasanuzzaman, M., Bhuyan, M.H.M.B., Zulfiqar, F., Raza, A., Mohsin, S.M., Mahmud, J.A., et al., 2020. Reactive oxygen species and antioxidant defense in plants under abiotic stress: revisiting the crucial role of a universal defense regulator. Antioxidants 9 (8), 1–52. Available at https://doi.org/10.3390/antiox9080681.

He, M., He, C., Ding, N., 2018. Abiotic stresses: general defenses of land plants and chances for engineering multistress tolerance. Frontiers in Plant Science 9. Available at https://doi.org/10.3389/fpls.2018.01771.

He, L., Nada, K., Tachibana, S., 2002. Effects of spermidine pretreatment through the roots on growth and photosynthesis of chilled cucumber plants (*Cucumis sativus* L.). Journal of the Japanese Society for Horticultural Science 71 (4), 490–498. Available at https://doi.org/10.2503/jjshs.71.490.

Hoang, X.L.T., Nhi, D.N.H., Thu, N.B.A., Thao, N.P., Tran, L.S.P., 2017. Transcription factors and their roles in signal transduction in plants under abiotic stresses. Current Genomics 18 (6), 483–497. Available at https://doi.org/10.2174/1389202918666170227150057.

Hossain, M.A., Tahjib-Ul-Arif, M., Jahin, S.A., Siddique, A.B., Mim, M.H., Sharif-Ar-Raffi, Haque Bhuiyan, M.J., et al., 2023. Genetic engineering for heavy metal/metalloid stress tolerance in plants. Heavy Metal Toxicity and Tolerance in Plants: A Biological, Omics, and Genetic Engineering Approach. Wiley, Bangladesh, pp. 573–592. Available at 10.1002/9781119906506.ch26.

Hrabak, E.M., Chan, C.W., Gribskov, M., Harper, J.F., Choi, J.H., Halford, N., et al., 2003. The Arabidopsis CDPK-SNRK superfamily of protein kinases. Plant Physiology 132 (2), 666–680. Available at https://doi.org/10.1104/pp.102.011999.

Hsu, P.K., Dubeaux, G., Takahashi, Y., Schroeder, J.I., 2021. Signaling mechanisms in abscisic acid-mediated stomatal closure. The Plant Journal 105 (2), 307–321. Available at https://doi.org/10.1111/tpj.15067.

Hu, B., Deng, F., Chen, G., Chen, X., Gao, W., Long, L., et al., 2020. Evolution of abscisic acid signaling for stress responses to toxic metals and metalloids. Frontiers in Plant Science 11. Available at https://doi.org/10.3389/fpls.2020.00909.

Huang, G.T., Ma, S.L., Bai, L.P., Zhang, L., Ma, H., Jia, P., et al., 2012. Signal transduction during cold, salt, and drought stresses in plants. Molecular Biology Reports 39 (2), 969–987. Available at https://doi.org/10.1007/s11033-011-0823-1.

Hussain, M.A., Naeem, A., Sulieman, S., Pitann, B., Mühling, K.H., 2022. Sulfur uptake and distribution, grain yield, and quality of hybrid and inbred winter wheat (Triticum aestivum L.) varieties under early and late waterlogging. Journal of Plant Nutrition and Soil Science 185 (5), 622–631. Available at https://doi.org/10.1002/jpln.202200149.

Hussain, M.A., Naeem, A., Pitann, B., Mühling, K.H., 2023. High sulfur (S) supplementation imparts waterlogging tolerance to oilseed rape (Brassica napus L.) through upregulating S metabolism and antioxidant pathways. Journal of Plant Growth Regulation 42 (12), 7591–7605. Available at https://doi.org/10.1007/s00344-023-11034-8.

Hussain, M.A., Naeem, A., Pitann, B., Mühling, K.H., 2024. Foliar-applied sulfur is supplemental to soil application in alleviating waterlogging stress through the mitigation of oxidative stress and ion homeostasis in rapeseed (Brassica napus L.). Plant Stress 13. Available at https://doi.org/10.1016/j.stress.2024.100529.

Ibrahim, W., Qiu, C., Zhang, C., Cao, F., Shuijin, Z., Wu, F., 2018. Comparative physiological analysis in the tolerance to salinity and drought individual and combination in two cotton genotypes with contrasting salt tolerance. Physiologia Plantarum 165 (2), 155–168. Available at https://doi.org/10.1111/ppl.12791.

Imran, M., Latif Khan, A., Shahzad, R., Aaqil Khan, M., Bilal, S., Khan, A., et al., 2021. Exogenous melatonin induces drought stress tolerance by promoting plant growth and antioxidant defence system of soybean plants. AoB Plants 13 (4). Available at https://doi.org/10.1093/aobpla/plab026.

Jahan, M.S., Wang, Y., Shu, S., Zhong, M., Chen, Z., Wu, J., et al., 2019. Exogenous salicylic acid increases the heat tolerance in Tomato (*Solanum lycopersicum* L) by enhancing photosynthesis efficiency and improving antioxidant defense system through scavenging of reactive oxygen species. Scientia Horticulturae 247, 421–429. Available at https://doi.org/10.1016/j.scienta.2018.12.047.

Jahan, M.S., Guo, S., Sun, J., Shu, S., Wang, Y., Abou El-Yazied, A., et al., 2021a. Melatonin-mediated photosynthetic performance of tomato seedlings under high-temperature stress. Plant Physiology and Biochemistry 167, 309–320. Available at https://doi.org/10.1016/j.plaphy.2021.08.002.

Jahan, M.S., Shu, S., Wang, Y., Hasan, M.M., El-Yazied, A.A., Alabdallah, N.M., et al., 2021. Melatonin pretreatment confers heat tolerance and repression of heat-induced senescence in tomato through the modulation of ABA-and GA-mediated pathways. Frontiers in Plant Science 12, 650955. Available at https://doi.org/10.3389/fpls.2021.650955.

Jahan, M.S., Huang, Y., Qin, N.M., Wu, H.Y., Zhou, X.B., 2024. Exogenous application of melatonin attenuates waterlogging stress through adopting quiescence adaptation technique in tomato seedlings. Environmental and Experimental Botany 225. Available at https://doi.org/10.1016/j.envexpbot.2024.105855.

Jakab, G., Ton, J., Flors, V., Zimmerli, L., Métraux, J.P., Mauch-Mani, B., 2005. Enhancing Arabidopsis salt and drought stress tolerance by chemical priming for its abscisic acid responses. Plant Physiology 139 (1), 267–274. Available at https://doi.org/10.1104/pp.105.065698.

Javed, T., Shabbir, R., Ali, A., Afzal, I., Zaheer, U., Gao, S.J., 2020. Transcription factors in plant stress responses: challenges and potential for sugarcane improvement. Plants 9 (4). Available at https://doi.org/10.3390/plants9040491.

Jogawat, A., Yadav, B., Chhaya, Lakra, N., Singh, A.K., Narayan, O.P., 2021. Crosstalk between phytohormones and secondary metabolites in the drought stress tolerance of crop plants: a review. Physiologia Plantarum 172 (2), 1106–1132. Available at https://doi.org/10.1111/ppl.13328.

Jyoti, S.D., Azim, J.B., Robin, A.H.K., 2021. Genome-wide characterization and expression profiling of EIN3/EIL family genes in *Zea mays*. Plant Gene 25, 100270. Available at https://doi.org/10.1016/j.plgene.2020.100270.

Kabir, M., Habiba, U.E., Khan, W., Shah, A., Rahim, S., De los Rios-Escalante, P.R., et al., 2023. Climate change due to increasing concentration of carbon dioxide and its impacts on environment in 21st century; a mini review. Journal of King Saud University - Science 35 (5), 102693. Available at https://doi.org/10.1016/j.jksus.2023.102693.

Kamal, A.H.M., Komatsu, S., 2015. Involvement of reactive oxygen species and mitochondrial proteins in biophoton emission in roots of soybean plants under flooding stress. Journal of Proteome Research 14 (5), 2219–2236. Available at https://doi.org/10.1021/acs.jproteome.5b00007.

Kang, C., He, S., Zhai, H., Li, R., Zhao, N., Liu, Q., 2018. A sweetpotato auxin response factor gene (IbARF5) is involved in carotenoid biosynthesis and salt and drought tolerance in transgenic Arabidopsis. Frontiers in Plant Science 9. Available at https://doi.org/10.3389/fpls.2018.01307.

Katam, R., Shokri, S., Murthy, N., Singh, S.K., Suravajhala, P., Khan, M.N., et al., 2020. Proteomics, physiological, and biochemical analysis of cross tolerance mechanisms in response to heat and water stresses in soybean. PLoS One 15 (6). Available at https://doi.org/10.1371/journal.pone.0233905.

Khan, N., Bano, A., Babar, M.A., 2019a. Metabolic and physiological changes induced by plant growth regulators and plant growth promoting rhizobacteria and their impact on drought tolerance in *Cicer arietinum* L. PLoS One 14 (3), e0213040. Available at https://doi.org/10.1371/journal.pone.0213040.

Khan, M.Y., Prakash, V., Yadav, V., Chauhan, D.K., Prasad, S.M., Ramawat, N., et al., 2019b. Regulation of cadmium toxicity in roots of tomato by indole acetic acid with special emphasis on reactive oxygen species production and their scavenging. Plant Physiology and Biochemistry 142, 193–201. Available at https://doi.org/10.1016/j.plaphy.2019.05.006.

Khan, N., Bano, A., Ali, S., Babar, M.A., 2020. Crosstalk amongst phytohormones from planta and PGPR under biotic and abiotic stresses. Plant Growth Regulation 90 (2), 189–203. Available at https://doi.org/10.1007/s10725-020-00571-x.

Khan, Z., Jan, R., Asif, S., Farooq, M., Jang, Y., Kim, E., et al., 2024a. Exogenous melatonin induces salt and drought stress tolerance in rice by promoting plant growth and defense system. Scientific Reports 14, 1214. Available at https://doi.org/10.1038/s41598-024-51369-0.

Khan, S., Singh, R., Kaur, H., Kumar, A., Vashishth, A., Shahwan, M., et al., 2024b. Plant growth regulator-mediated response under abiotic stress: a review. Journal of Applied Biology and Biotechnology 12 (2), 13–21. Available at https://doi.org/10.7324/JABB.2024.157299.

Kim, Y., Seo, C.W., Khan, A.L., Mun, B.G., Shahzad, R., Ko, J.W., et al., 2018. Exo-ethylene application mitigates waterlogging stress in soybean (Glycine max L.). BMC Plant Biology 18 (1). Available at https://doi.org/10.1186/s12870-018-1457-4.

Kotakis, C., Theodoropoulou, E., Tassis, K., Oustamanolakis, C., Ioannidis, N.E., Kotzabasis, K., 2014. Putrescine, a fast-acting switch for tolerance against osmotic stress. Journal of Plant Physiology 171 (2), 48–51. Available at https://doi.org/10.1016/j.jplph.2013.09.015.

Kumari, S., Kumar, S., Prakash, P., 2018. Exogenous application of cytokinin (6-BAP) ameliorates the adverse effect of combined drought and high temperature stress in wheat seedling. Journal of Pharmacognosy and Phytochemistry 7 (1), 1176–1180.

Kushwaha, B.K., Singh, S., Tripathi, D.K., Sharma, S., Prasad, S.M., Chauhan, D.K., et al., 2019. New adventitious root formation and primary root biomass accumulation are regulated by nitric oxide and reactive oxygen species in rice seedlings under arsenate stress. Journal of Hazardous Materials 361, 134–140. Available at https://doi.org/10.1016/j.jhazmat.2018.08.035.

Li, Z., Zhang, Y., Zhang, X., Peng, Y., Merewitz, E., Ma, X., et al., 2016. The alterations of endogenous polyamines and phytohormones induced by exogenous application of spermidine regulate antioxidant metabolism, metallothionein and relevant genes conferring drought tolerance in white clover. Environmental and Experimental Botany 124, 22–38. Available at https://doi.org/10.1016/j.envexpbot.2015.12.004.

Li, Y., Yang, X., Li, X., 2019. Role of jasmonate signaling pathway in resistance to dehydration stress in Arabidopsis. Acta Physiologiae Plantarum 41, 1–12. Available at https://doi.org/10.1007/s11738-019-2897-7.

Liu, Q., Gao, T., Liu, W., Liu, Y., Zhao, Y., Liu, Y., et al., 2020. Functions of dopamine in plants: a review. Plant Signaling & Behavior 15 (12), 1827782. Available at https://doi.org/10.1080/15592324.2020.1827782.

Liu, Y., Li, Z., Zhong, C., Zhang, Y., Wang-Pruski, G., Zhang, Z., et al., 2023. Alleviating effect of melatonin on melon seed germination under autotoxicity and saline-alkali combined stress. Journal of Plant Growth Regulation 42 (4), 2474–2485. Available at https://doi.org/10.1007/s00344-022-10720-3.

Mahmood, U., Hussain, S., Hussain, S., Ali, B., Ashraf, U., Zamir, S., et al., 2021. Morpho-physio-biochemical and molecular responses of maize hybrids to salinity and waterlogging during stress and recovery phase. Plants 10 (7). Available at https://doi.org/10.3390/plants10071345.

Mandal, S., Ghorai, M., Anand, U., Samanta, D., Kant, N., Mishra, T., et al., 2022. Cytokinin and abiotic stress tolerance -what has been accomplished and the way forward? Frontiers in Genetics 13. Available at https://doi.org/10.3389/fgene.2022.943025.

Manik, S.M.N., Quamruzzaman, M., Livermore, M., Zhao, C., Johnson, P., Hunt, I., et al., 2022. Impacts of barley root cortical aerenchyma on growth, physiology, yield components, and grain quality under field waterlogging conditions. Field Crops Research 279, 108461. Available at https://doi.org/10.1016/j.fcr.2022.108461.

Manzoor, M., Abdalla, M.A., Hussain, M.A., Mühling, K.H., 2023. Silicon-selenium interplay imparts cadmium resistance in wheat through an up-regulating antioxidant system. International Journal of Molecular Sciences 25 (1), 387. Available at https://doi.org/10.3390/ijms25010387.

Misra, N., Saxena, P., 2009. Effect of salicylic acid on proline metabolism in lentil grown under salinity stress. Plant Science 177 (3), 181–189. Available at https://doi.org/10.1016/j.plantsci.2009.05.007.

Mughal, N., Shoaib, N., Chen, J., li, Y., He, Y., Fu, M., et al., 2024. Adaptive roles of cytokinins in enhancing plant resilience and yield against environmental stressors. Chemosphere 364, 143189. Available at https://doi.org/10.1016/j.chemosphere.2024.143189.

Mumithrakamatchi, A.K., Alagarswamy, S., Anitha, K., Djanaguiraman, M., Kalarani, M.K., Swarnapriya, R., et al., 2024. Melatonin imparts tolerance to combined drought and high-temperature stresses in tomato through osmotic adjustment and ABA accumulation. Frontiers in Plant Science 15. Available at https://doi.org/10.3389/fpls.2024.1382914.

Myers, S.S., Zanobetti, A., Kloog, I., Huybers, P., Leakey, A.D.B., Bloom, A.J., et al., 2014. Increasing CO2 threatens human nutrition. Nature 510 (7503), 139–142. Available at https://doi.org/10.1038/nature13179.

Møller, I.M., Jensen, P.E., Hansson, A., 2007. Oxidative modifications to cellular components in plants. Annual Review of Plant Biology 58 (1), 459–481. Available at https://doi.org/10.1146/annurev.arplant.58.032806.103946.

Nadarajah, K.K., 2020. ROS homeostasis in abiotic stress tolerance in plants. International Journal of Molecular Sciences 21 (15), 5208. Available at https://doi.org/10.3390/ijms21155208.

Nasser, L.M., Badu-Apraku, B., Gracen, V.E., Mafouasson, H.N.A., 2020. Combining ability of early-maturing yellow maize inbreds under combined drought and heat stress and well-watered environments. Agronomy 10 (10), 1585. Available at https://doi.org/10.3390/agronomy10101585.

Ni, X.L., Gui, M.Y., Tan, L.L., Zhu, Q., Liu, W.Z., Li, C.X., 2019. Programmed cell death and aerenchyma formation in water-logged sunflower stems and its promotion by ethylene and ROS. Frontiers in Plant Science 9. Available at https://doi.org/10.3389/fpls.2018.01928.

Pan, J., Sharif, R., Xu, X., Chen, X., 2021. Mechanisms of waterlogging tolerance in plants: research progress and prospects. Frontiers in Plant Science 11. Available at https://doi.org/10.3389/fpls.2020.627331.

Pan, R., Buitrago, S., Feng, X., Hu, A., Zhou, M., Zhang, W., 2022. Ethylene regulates aerenchyma formation in cotton under hypoxia stress by inducing the accumulation of reactive oxygen species. Environmental and Experimental Botany 197, 104826. Available at https://doi.org/10.1016/j.envexpbot.2022.104826.

Paul, K., Pauk, J., Kondic-Spika, A., Grausgruber, H., Allahverdiyev, T., Sass, L., et al., 2019. Co-occurrence of mild salinity and drought synergistically enhances biomass and grain retardation in wheat. Frontiers in Plant Science 10. Available at https://doi.org/10.3389/fpls.2019.00501.

Per, T.S., Khan, M.I.R., Anjum, N.A., Masood, A., Hussain, S.J., Khan, N.A., 2018. Jasmonates in plants under abiotic stresses: crosstalk with other phytohormones matters. Environmental and Experimental Botany 145, 104–120. Available at https://doi.org/10.1016/j.envexpbot.2017.11.004.

Pospíšilová, H., Jiskrová, E., Vojta, P., Mrízová, K., Kokáš, F., Čudejková, M.M., et al., 2016. Transgenic barley overexpressing a cytokinin dehydrogenase gene shows greater tolerance to drought stress. New Biotechnology 33 (5), 692–705. Available at https://doi.org/10.1016/j.nbt.2015.12.005.

Postiglione, A.E., Muday, G.K., 2020. The role of ROS homeostasis in ABA-induced guard cell signaling. Frontiers in Plant Science 11. Available at https://doi.org/10.3389/fpls.2020.00968.

Prerostova, S., Rezek, J., Jarosova, J., Lacek, J., Dobrev, P., Marsik, P., et al., 2023. Cytokinins act synergistically with heat acclimation to enhance rice thermotolerance affecting hormonal dynamics, gene expression and volatile emission. Plant Physiology and Biochemistry 198, 107683. Available at https://doi.org/10.1016/j.plaphy.2023.107683.

Prodhan, M.Y., Munemasa, S., Nahar, M.N.E.N., Nakamura, Y., Murata, Y., 2018. Guard cell salicylic acid signaling is integrated into abscisic acid signaling via the Ca2+/CPK-dependent pathway. Plant Physiology 178 (1), 441–450. Available at https://doi.org/10.1104/pp.18.00321.

Rahman, S.U., Li, Y., Hussain, S., Hussain, B., Riaz, L., Ashraf, M.N., et al., 2023. Role of phytohormones in heavy metal tolerance in plants: a review. Ecological Indicators 146, 109844. Available at https://doi.org/10.1016/j.ecolind.2022.109844.

Ranty, B., Aldon, D., Cotelle, V., Galaud, J.-P., Thuleau, P., Mazars, C., 2016. Calcium sensors as key hubs in plant responses to biotic and abiotic stresses. Frontiers in Plant Science 7. Available at https://doi.org/10.3389/fpls.2016.00327.

Rivero, R.M., Mestre, T.C., Mittler, R., Rubio, F., Garcia-sanchez, F., Martinez, V., 2013. The combined effect of salinity and heat reveals a specific physiological, biochemical and molecular response in tomato plants. Plant, Cell & Environment 37 (5), 1059–1073. Available at https://doi.org/10.1111/pce.12199.

Saharan, B.S., Brar, B., Duhan, J.S., Kumar, R., Marwaha, S., Rajput, V.D., et al., 2022. Molecular and physiological mechanisms to mitigate abiotic stress conditions in plants. Life (Chicago, Ill.: 1978) 12 (10). Available at https://doi.org/10.3390/life12101634.

Saini, S., Kaur, N., Pati, P.K., 2021. Phytohormones: key players in the modulation of heavy metal stress tolerance in plants. Ecotoxicology and Environmental Safety 223, 112578. Available at https://doi.org/10.1016/j.ecoenv.2021.112578.

Sandhi, A., Siddique, A.B., Vithanage, M., 2023. Mercury toxicity: plant response and tolerance. In: Hossain, M.A., Zakir Hossain, A.K.M., Bourgerie, S., Fujita, M., Dhankher, O.P., Haris, P. (Eds.), Heavy Metal Toxicity and Tolerance in Plants: A Biological, Omics, and Genetic Engineering Approach. Wiley, pp. 349–371. Available at https://doi.org/10.1002/9781119906506.ch17.

Sergiev, I., Todorova, D., Shopova, E., Jankauskiene, J., Jankovska-Bortkevič, E., Jurkonienė, S., 2019. Exogenous auxin type compounds amend PEG-induced physiological responses of pea plants. Scientia Horticulturae 248, 200–205. Available at https://doi.org/10.1016/j.scienta.2019.01.015.

Sethuraman, G., Zain, N.A.M., Osman, N., Ismail, M.R., Mispan, M.S., Hanafiah, N.M., et al., 2024. Individual and concurrent effects of heat and drought stress on the growth and yield of two Malaysian rice cultivars. Chilean Journal of Agricultural Research 84 (5), 709–717. Available at https://doi.org/10.4067/s0718-58392024000500709.

Shabbir, R., Javed, T., Afzal, I., Sabagh, A.E., Ali, A., Vicente, O., et al., 2021. Modern biotechnologies: innovative and sustainable approaches for the improvement of sugarcane tolerance to environmental stresses. Agronomy 11 (6), 1042. Available at https://doi.org/10.3390/agronomy11061042.

Shahzad, K., Hussain, S., Arfan, M., Hussain, S., Waraich, E.A., Zamir, S., et al., 2021. Exogenously applied gibberellic acid enhances growth and salinity stress tolerance of maize through modulating the morpho-physiological, biochemical and molecular attributes. Biomolecules 11 (7). Available at https://doi.org/10.3390/biom11071005.

Shao, J., Huang, K., Batool, M., Idrees, F., Afzal, R., Haroon, M., et al., 2022. Versatile roles of polyamines in improving abiotic stress tolerance of plants. Frontiers in Plant Science 13. Available at https://doi.org/10.3389/fpls.2022.1003155.

Sharma, M., Negi, S., Kumar, P., Srivastava, D.K., Choudhary, M.K., Irfan, M., 2023. Fruit ripening under heat stress: the intriguing role of ethylene-mediated signaling. Plant Science 335. Available at https://doi.org/10.1016/j.plantsci.2023.111820.

Shaukat, K., Zahra, N., Hafeez, M.B., Naseer, R., Batool, A., Batool, H., et al., 2022. Chapter 2 - Role of salicylic acid–induced abiotic stress tolerance and underlying mechanisms in plants. In: Aftab, T., Naeem, M. (Eds.), Emerging Plant Growth Regulators in Agriculture: Roles in Stress Tolerance. Academic Press, pp. 73–98. Available at https://doi.org/10.1016/B978-0-323-91005-7.00008-4.

Shi, H., Chen, L., Ye, T., Liu, X., Ding, K., Chan, Z., 2014. Modulation of auxin content in Arabidopsis confers improved drought stress resistance. Plant Physiology and Biochemistry 82, 209–217. Available at https://doi.org/10.1016/j.plaphy.2014.06.008.

Shimamura, S., Yoshioka, T., Yamamoto, R., Hiraga, S., Nakamura, T., Shimada, S., et al., 2014. Role of abscisic acid in flood-induced secondary aerenchyma formation in soybean (Glycine max) hypocotyls. Plant Production Science 17 (2), 131–137. Available at https://doi.org/10.1626/pps.17.131.

Shinohara, T., Martin, E., Leskovar, D., 2017. Ethylene regulators influence germination and root growth of globe artichoke seedlings exposed to heat stress conditions. Seed Science and Technology 45 (1), 167–178. Available at https://doi.org/10.15258/sst.2017.45.1.07.

Silverstone, A.L., Jung, H.S., Dill, A., Kawaide, H., Kamiya, Y., Sun, T.P., 2001. Repressing a repressor: gibberellin-induced rapid reduction of the RGA protein in Arabidopsis. The Plant Cell 13 (7), 1555–1565. Available at https://doi.org/10.1105/tpc.13.7.1555.

Singh, P., Choudhary, K.K., Chaudhary, N., Gupta, S., Sahu, M., Tejaswini, B., et al., 2022. Salt stress resilience in plants mediated through osmolyte accumulation and its crosstalk mechanism with phytohormones. Frontiers in Plant Science 13. Available at https://doi.org/10.3389/fpls.2022.1006617.

Šípošová, K., Labancová, E., Hačkuličová, D., Kollárová, K., Vivodová, Z., 2023. The changes in the maize root cell walls after exogenous application of auxin in the presence of cadmium. Environmental Science and Pollution Research 30 (37), 87102–87117. Available at https://doi.org/10.1007/s11356-023-28029-3.

Soares, T.F.S.N., Dias, D.C.Fd.S., Oliveira, A.M.S., Ribeiro, D.M., Dias, L.Ad.S., 2020. Exogenous brassinosteroids increase lead stress tolerance in seed germination and seedling growth of Brassica juncea L. Ecotoxicology and Environmental Safety 193. Available at https://doi.org/10.1016/j.ecoenv.2020.110296.

Steffens, B., 2014. The role of ethylene and ROS in salinity, heavy metal, and flooding responses in rice. Frontiers in Plant Science 5, 685. Available at https://doi.org/10.3389/fpls.2014.00685.

Su, Y., Xia, S., Wang, R., Xiao, L., 2017. 13 - Phytohormonal quantification based on biological principles. In: Li, J., Li, C., Smith, S.M. (Eds.), Hormone Metabolism and Signaling in Plants. Academic Press, pp. 431–470. Available at https://doi.org/10.1016/B978-0-12-811562-6.00013-X.

Su, Y., Huang, Y., Dong, X., Wang, R., Tang, M., Cai, J., et al., 2021. Exogenous methyl jasmonate improves heat tolerance of perennial ryegrass through alteration of osmotic adjustment, antioxidant defense, and expression of jasmonic acid-responsive genes. Frontiers in Plant Science 12. Available at https://doi.org/10.3389/fpls.2021.664519.

Suprasanna, P., Nikalje, G.C., Rai, A.N., 2016. Osmolyte accumulation and implications in plant abiotic stress tolerance. Osmolytes and Plants Acclimation to Changing Environment: Emerging Omics Technologies. Springer India, New Delhi, pp. 1–12. Available at https://doi.org/10.1007/978-81-322-2616-1_1.

Suzuki, N., 2016. Hormone signaling pathways under stress combinations. Plant Signaling & Behavior 11 (11), e1247139. Available at https://doi.org/10.1080/15592324.2016.1247139.

Sytar, O., Kumari, P., Yadav, S., Brestic, M., Rastogi, A., 2019. Phytohormone priming: regulator for heavy metal stress in plants. Journal of Plant Growth Regulation 38 (2), 739–752. Available at https://doi.org/10.1007/s00344-018-9886-8.

Sánchez-Bermúdez, M., Del Pozo, J.C., Pernas, M., 2022. Effects of combined abiotic stresses related to climate change on root growth in crops. Frontiers in Plant Science 13. Available at https://doi.org/10.3389/fpls.2022.918537.

Tanaka, Y., Sano, T., Tamaoki, M., Nakajima, N., Kondo, N., 2006. S. Hasezawa, Cytokinin and auxin inhibit abscisic acid-induced stomatal closure by enhancing ethylene production in Arabidopsis. Journal of Experimental Botany 57 (10), 2259–2266. Available at https://doi.org/10.1093/jxb/erj193.

Tsegay, B.A., Andargie, M., 2018. Seed priming with gibberellic acid (GA3) alleviates salinity induced inhibition of germination and seedling growth of Zea mays L., Pisum sativum Var. abyssinicum A. Braun and Lathyrus sativus L. Journal of Crop Science and Biotechnology 21 (3), 261–267. Available at https://doi.org/10.1007/s12892-018-0043-0.

Tuteja, N., 2014. Abscisic acid and abiotic stress signaling. Plant Signaling & Behavior 2 (3), 135–138. Available at https://doi.org/10.4161/psb.2.3.4156.

Unterholzner, S.J., Rozhon, W., Papacek, M., Ciomas, J., Lange, T., Kugler, K.G., et al., 2015. Brassinosteroids are master regulators of gibberellin biosynthesis in Arabidopsis. The Plant Cell 27 (8), 2261–2272. Available at https://doi.org/10.1105/tpc.15.00433.

Vaezi, B., Arzani, A., Roberts, T.H., 2024. How do drought, heat stress, and their combination impact stem reserve mobilization in wheat genotypes? Agronomy 14 (8), 1867. Available at https://doi.org/10.3390/agronomy14081867.

Verma, V., Ravindran, P., Kumar, P.P., 2016. Plant hormone-mediated regulation of stress responses. BMC Plant Biology 16 (1). Available at https://doi.org/10.1186/s12870-016-0771-y.

Verma, S., Negi, N.P., Pareek, S., Mudgal, G., Kumar, D., 2022. Auxin response factors in plant adaptation to drought and salinity stress. Physiologia Plantarum 174 (3), e13714. Available at https://doi.org/10.1111/ppl.13714.

Vile, D., Pervent, M., Belluau, M., Vasseur, F., Bresson, J., Muller, B., et al., 2011. Arabidopsis growth under prolonged high temperature and water deficit: independent or interactive effects? Plant, Cell & Environment 35 (4), 702–718. Available at https://doi.org/10.1111/j.1365-3040.2011.02445.x.

Wang, X., Cao, M., Li, H., Liu, Y., Fan, S., Zhang, N., et al., 2024. Strategies and prospects for melatonin to alleviate abiotic stress in horticultural plants. Horticultural Plant Journal 10 (3), 601–614. Available at https://doi.org/10.1016/j.hpj.2023.03.011.

Wani, S.H., Kumar, V., Shriram, V., Sah, S.K., 2016. Phytohormones and their metabolic engineering for abiotic stress tolerance in crop plants. Crop Journal 4 (3), 162–176. Available at https://doi.org/10.1016/j.cj.2016.01.010.

Wasternack, C., 2014. Action of jasmonates in plant stress responses and development - applied aspects. Biotechnology Advances 32 (1), 31–39. Available at https://doi.org/10.1016/j.biotechadv.2013.09.009.

Wu, D., Ji, J., Wang, G., Guan, C., Jin, C., 2014. LchERF, a novel ethylene-responsive transcription factor from Lycium chinense, confers salt tolerance in transgenic tobacco. Plant Cell Reports 33 (12), 2033–2045. Available at https://doi.org/10.1007/s00299-014-1678-4.

Xing, Q., Zhang, X., Li, Y., Shao, Q., Cao, S., Wang, F., et al., 2019. The lipoxygenase CmLOX13 from oriental melon enhanced severe drought tolerance via regulating ABA accumulation and stomatal closure in Arabidopsis. Environmental and Experimental Botany 167, 103815. Available at https://doi.org/10.1016/j.envexpbot.2019.103815.

Yadav, B., Jogawat, A., Gnanasekaran, P., Kumari, P., Lakra, N., Lal, S.K., et al., 2021. An overview of recent advancement in phytohormones-mediated stress management and drought tolerance in crop plants. Plant Gene 25, 100264. Available at https://doi.org/10.1016/j.plgene.2020.100264.

Yamauchi, T., Tanaka, A., Mori, H., Takamure, I., Kato, K., Nakazono, M., 2016. Ethylene-dependent aerenchyma formation in adventitious roots is regulated differently in rice and maize. Plant, Cell & Environment 39 (10), 2145–2157. Available at https://doi.org/10.1111/pce.12766.

Yan, J., Wang, D., Kang, S., He, Z., Li, X., Tang, W., et al., 2023. Ethylene regulates combined drought and low nitrogen tolerance in wheat: proteomic analysis. Agronomy 13 (7), 1950. Available at https://doi.org/10.3390/agronomy13071950.

Yang, D., Li, Y., Shi, Y., Cui, Z., Luo, Y., Zheng, M., et al., 2016. Exogenous cytokinins increase grain yield of winter wheat cultivars by improving stay-green characteristics under heat stress. PLoS One 11 (5), e0155437. Available at https://doi.org/10.1371/journal.pone.0155437.

Yang, D., Chen, Y., Wang, R., He, Y., Ma, X., Shen, J., et al., 2024. Effects of exogenous abscisic acid on the physiological and biochemical responses of Camellia oleifera seedlings under drought stress. Plants 13 (2), 225. Available at https://doi.org/10.3390/plants13020225.

Yao, H.M., Li, Y.S., Zhang, T.Z., Zhao, J., Wang, C., Wang, H.N., et al., 2016. Effects of combined drought and salinity stress on germination and physiological characteristics of maize (Zea mays). Chinese Journal of Applied Ecology 27 (7), 2301–2307. Available at https://doi.org/10.13287/j.1001-9332.201607.023.

Yin, P., Liang, X., Zhao, H., Xu, Z., Chen, L., Yang, X., et al., 2023. Cytokinin signaling promotes salt tolerance by modulating shoot chloride exclusion in maize. Molecular Plant 16 (6), 1031–1047. Available at https://doi.org/10.1016/j.molp.2023.04.011.

Yoshida, T., Fujita, Y., Sayama, H., Kidokoro, S., Maruyama, K., Mizoi, J., et al., 2010. AREB1, AREB2, and ABF3 are master transcription factors that cooperatively regulate ABRE-dependent ABA signaling involved in drought stress tolerance and require ABA for full activation. The Plant Journal 61 (4), 672–685. Available at https://doi.org/10.1111/j.1365-313x.2009.04092.x.

You, J., Chan, Z., 2015. ROS regulation during abiotic stress responses in crop plants. Frontiers in Plant Science 6. Available at https://doi.org/10.3389/fpls.2015.01092.

Zafar, M.M., Chattha, W.S., Khan, A.I., Zafar, S., Subhan, M., Saleem, H., et al., 2023. Drought and heat stress on cotton genotypes suggested agro-physiological and biochemical features for climate resilience. Frontiers in Plant Science 14. Available at https://doi.org/10.3389/fpls.2023.1265700.

De Zélicourt, A., Colcombet, J., Hirt, H., 2016. The role of MAPK modules and ABA during abiotic stress signaling. Trends in Plant Science 21 (8), 677–685. Available at https://doi.org/10.1016/j.tplants.2016.04.004.

Zeng, W., Mostafa, S., Lu, Z., Jin, B., 2022. Melatonin-mediated abiotic stress tolerance in plants. Frontiers in Plant Science 13. Available at https://doi.org/10.3389/fpls.2022.847175.

Zhang, R., Xu, C., Bao, Z., Xiao, R., Chen, X., Xiao, W., et al., 2021. Auxin alters sodium ion accumulation and nutrient accumulation by playing protective role in salinity challenged strawberry. Plant Physiology and Biochemistry 164, 1–9. Available at https://doi.org/10.1016/j.plaphy.2021.04.008.

Zhang, H., Zhu, J., Gong, Z., Zhu, J.K., 2022. Abiotic stress responses in plants. Nature Reviews. Genetics 23 (2), 104–119. Available at https://doi.org/10.1038/s41576-021-00413-0.

Zhao, F.Y., Han, M.M., Zhang, S.Y., Ren, J., Hu, F., Wang, X., 2014. MAPKs as a cross point in H2O2 and auxin signaling under combined cadmium and zinc stress in rice roots. Russian Journal of Plant Physiology 61 (5), 608–618. Available at https://doi.org/10.1134/s1021443714040232.

Zheng, Y., Wang, X., Cui, X., Wang, K., Wang, Y., He, Y., 2023. Phytohormones regulate the abiotic stress: an overview of physiological, biochemical, and molecular responses in horticultural crops. Frontiers in Plant Science 13. Available at https://doi.org/10.3389/fpls.2022.1095363.

Zhu, J.K., 2016. Abiotic stress signaling and responses in plants. Cell 167 (2), 313–324. Available at https://doi.org/10.1016/j.cell.2016.08.029.

Further reading

An, Y., Zhou, P., Liang, J., 2014. Effects of exogenous application of abscisic acid on membrane stability, osmotic adjustment, photosynthesis and hormonal status of two lucerne (*Medicago sativa* L.) genotypes under high temperature stress and drought stress. Crop & Pasture Science 65 (3), 274–286.

Asif, M., Ahmad, R., Pervez, A., Al Farraj, D.A., Elshikh, M.S., Shahzad, M., et al., 2023. Combination of melatonin and plant growth promoting rhizobacteria improved the growth of *Spinacia oleracea* L. under the arsenic and cadmium stresses. Physiological and Molecular Plant Pathology 127, 102097. Available at https://doi.org/10.1016/j.pmpp.2023.102097.

Brestic, M., Zivcak, M., Kunderlikova, K., Allakhverdiev, S.I., 2016. High temperature specifically affects the photoprotective responses of chlorophyll b-deficient wheat mutant lines. Photosynthesis Research 130 (1–3), 251–266. Available at https://doi.org/10.1007/s11120-016-0249-7.

Chatzistathis, T., Zoukidis, K., Vasilikiotis, C., Apostolidis, A., Giannakoula, A.E., Bountla, A., et al., 2024. Plant-growth-promoting rhizobacteria and arbuscular mycorrhizal fungi may improve soil fertility and the growth, nutrient uptake, and physiological performance of batavia lettuce (*Lactuca sativa* L. var. *longifolia*) Plants. Horticulturae 10 (5), 449. Available at https://doi.org/10.3390/horticulturae10050449.

Kang, S., Hoque, M.I.U., Woo, J., Lee, I., 2023. Mitigation of salinity stress on soybean seedlings using indole acetic acid-producing *Acinetobacter pittii* YNA40. Agriculture 13 (5), 1021. Available at https://doi.org/10.3390/agriculture13051021.

Khan, A., Pan, X., Najeeb, U., Tan, D.K.Y., Fahad, S., Zahoor, R., et al., 2018. Coping with drought: stress and adaptive mechanisms, and management through cultural and molecular alternatives in cotton as vital constituents for plant stress resilience and fitness. Biological Research 51 (1). Available at https://doi.org/10.1186/s40659-018-0198-z.

Llanes, A., Palchetti, M.V., Vilo, C., Ibañez, C., 2021. Molecular control to salt tolerance mechanisms of woody plants: recent achievements and perspectives. Annals of Forest Science 78 (4). Available at https://doi.org/10.1007/s13595-021-01107-7.

Mehari, T.G., Hou, Y., Xu, Y., Umer, M.J., Shiraku, M.L., Wang, Y., et al., 2022. Overexpression of cotton *GhNAC072* gene enhances drought and salt stress tolerance in transgenic *Arabidopsis*. BMC Genomics 23 (1). Available at https://doi.org/10.1186/s12864-022-08876-z.

Muhammad Aslam, M., Waseem, M., Jakada, B.H., Okal, E.J., Lei, Z., Saqib, H.S.A., et al., 2022. Mechanisms of abscisic acid-mediated drought stress responses in plants. International Journal of Molecular Sciences 23 (3), 1084. Available at https://doi.org/10.3390/ijms23031084.

Ndlovu, E., Van Staden, J., Maphosa, M., 2021. Morpho-physiological effects of moisture, heat and combined stresses on *Sorghum bicolor* [Moench (L.)] and its acclimation mechanisms. Plant Stress 2, 100018. Available at https://doi.org/10.1016/j.stress.2021.100018.

Nezhadahmadi, A., Prodhan, Z.H., Faruq, G., 2013. Drought tolerance in wheat. The Scientific World Journal 2013 (1). Available at https://doi.org/10.1155/2013/610721.

Sharma, A., Shahzad, B., Kumar, V., Kohli, S.K., Sidhu, G.P.S., Bali, A.S., et al., 2019. Phytohormones regulate accumulation of osmolytes under abiotic stress. Biomolecules 9 (7), 285. Available at https://doi.org/10.3390/biom9070285.

Suzuki, N., Bassil, E., Hamilton, J.S., Inupakutika, M.A., Zandalinas, S.I., Tripathy, D., et al., 2016. ABA is required for plant acclimation to a combination of salt and heat stress. PLoS One 11 (1), e0147625. Available at https://doi.org/10.1371/journal.pone.0147625.

Todaka, D., Takahashi, F., Yamaguchi-Shinozaki, K., Shinozaki, K., 2019. ABA-responsive gene expression in response to drought stress: cellular regulation and long-distance signaling. Advances in Botanical Research, Vol. 92. Academic Press, pp. 83–113. Available at https://doi.org/10.1016/bs.abr.2019.05.001.

Yanqing, W., Jiao, L., Lu, Z., Hao, W., Yiming, Z., Irshad, A., et al., 2024. Abiotic stress responses in crop plants: a multi-scale approach. Journal of Integrative Agriculture. Available at https://doi.org/10.1016/j.jia.2024.09.003.

Zhang, Y., Xu, J., Li, R., Ge, Y., Li, Y., Li, R., 2023. Plants' response to abiotic stress: mechanisms and strategies. International Journal of Molecular Sciences 24 (13), 10915. Available at https://doi.org/10.3390/ijms241310915.

Chapter 10

Secondary metabolites, metallothioneins, and phytochelatins in combating multiple abiotic stresses in plants

Shrushti Joshi, Suraj Patil, and Vinay Kumar*

Department of Biotechnology, Modern College of Arts, Science and Commission, Savitribai, Phule Pune University, Ganeshkhind, Pune, Maharashtra, India

**Corresponding author. e-mail address: vinay.kumar@moderncollegegk.org*

10.1 Introduction

Plants are exposed to abiotic stresses, including drought, salinity, heavy metal (HM) toxicity, and extreme temperatures, which significantly affect their growth, metabolism, and productivity. Plants, being sessile organisms that cannot escape unfavorable environmental conditions, are highly dependent on external stimuli to be in optimal conditions for survival. They rely on intricate mechanisms, such as biochemical, molecular, and physiological, to mitigate the adverse effects of stress. The differing abiotic stress conditions have differing effects on the plants. Abiotic stress significantly contributes to global crop losses, with salinity alone affecting over one billion hectares of irrigated agricultural land across 100 countries worldwide (Singh, 2021). Drought leads to water deficiency, reduced cell turgor, and impaired photosynthetic efficiency (Oguz et al., 2022). Under water-deficient conditions, plants close their stomata to reduce transpiration losses, which in turn limits carbon dioxide uptake, ultimately reducing biomass accumulation. Extended drought exposure induces oxidative stress, resulting in an overproduction of reactive oxygen species (ROS) like hydrogen peroxide and superoxide radicals. These ROS cause lipid peroxidation, protein denaturation, and DNA damage (Oguz et al., 2022). Similarly, soil salinity disrupts ion homeostasis and induces osmotic stress, limiting nutrient uptake and reducing overall plant vigor. High levels of sodium and chloride ions interfere with metabolic functions, leading to physiological damage, growth retardation, and yield losses (Balasubramaniam et al., 2023). This is particularly observed in crops like rice, wheat, and maize.

Conversely, HMtoxicity from environmental pollution caused by industrial processes, mining, and the overuse of chemical fertilizers and pesticides also hampers plant growth and development in that region (Thakur et al., 2022). Metals such as cadmium (Cd), lead, arsenic (As), mercury, and copper disrupt enzymatic activities, interfere with essential physiological processes, and generate oxidative stress by inducing excessive ROS production. Since these metals have a nondegradable nature, they accumulate in plant tissues and pose a serious threat to the food chain and ecosystem health (Thakur et al., 2022). Extreme temperatures, heat, and cold cause severe damage by disrupting membrane integrity, protein stability, and enzymatic functions (Ding & Yang, 2022). High temperatures accelerate cellular respiration, leading to energy imbalances and reduced photosynthetic efficiency. Whereas cold stress impairs membrane fluidity, causing ice crystal formation that mechanically injures plant tissues.

To mitigate the harmful impacts of abiotic stresses, plants activate defense strategies such as producing secondary metabolites (SMs), metallothioneins (MTs), and phytochelatins (PCs). SMs—like flavonoids, alkaloids, terpenoids, and phenolics—are vital in stress adaptation because they act as antioxidants, osmoprotectants, and signaling agents (Salam et al., 2023). Flavonoids are well-known for their ROS-scavenging properties, protecting plant cells from oxidative damage by neutralizing free radicals (Hajam et al., 2023). Alkaloids contribute to stress adaptation by modulating plant hormonal responses and improving metabolic stability under stress conditions (Salam et al., 2023). Terpenoids, crucial for

Multiple Abiotic Stresses in Plants. DOI: https://doi.org/10.1016/B978-0-443-27424-4.00009-4

membrane stabilization, help protect plants from extreme temperatures by preventing lipid peroxidation and preserving cell membrane integrity (Liga et al., 2023). These bioactive compounds not only enhance stress tolerance but also contribute to plant-pathogen interactions, providing additional protection against biotic stress.

MTs are involved particularly in response to HM toxicity. These small, cysteine-rich proteins function as metal chelators, binding toxic metal ions and preventing their interference with essential cellular components (Kumar et al., 2021). MTs are essential for maintaining metal ion balance and detoxification by binding metals like Cd, lead, and mercury into cellular compartments, which decreases their availability and toxicity. In addition to metal binding, MTs also contribute to oxidative stress mitigation by acting as ROS scavengers, thereby preventing lipid peroxidation and protein oxidation (Kumar et al., 2021). Similarly, PCs, small peptides made from glutathione (GSH), are crucial for detoxifying HMs (Faizan et al., 2024). These peptides form stable complexes with metal ions, facilitating their sequestration into vacuoles, thereby preventing their accumulation in metabolically active cellular regions. The ability of PC to chelate metals such as As and Cd makes them essential in plants growing in contaminated environments, enabling them to survive in metal-polluted soils (Faizan et al., 2024). Both MTs and PC are essential for metal detoxification, improving plant resilience against metal stress and minimizing the risk of damage to cellular functions caused by metals (Raychaudhuri et al., 2021).

This chapter offers a detailed overview of how plants defend themselves against various abiotic stresses by producing SMs, MTs, and other mechanisms such as PC. The key objectives include understanding the physiological and biochemical consequences of abiotic stress, highlighting the specific roles of these protective compounds in stress adaptation, and examining the latest research on plant stress biology. Further, this chapter will discuss the biotechnological applications of these mechanisms, including the development of stress-resistant crops, bioremediation strategies, and sustainable agricultural practices. By analyzing these aspects, this discussion aims to contribute to the ongoing efforts to enhance plant resilience, improve crop productivity, and develop innovative solutions to address the challenges posed by abiotic stresses in a changing global environment. A deeper understanding of plant stress responses allows researchers to develop innovative biotechnological approaches, including transgenic plants with enhanced stress tolerance, plant-based bioremediation strategies for metal-contaminated soils, and the application of stress-protective compounds in agriculture.

10.2 Role of secondary metabolites, metallothioneins, and phytochelatins

To cope with these stresses, plants have developed various biochemical defenses, such as synthesizing SMs, MTs, and PC. These SMs are crucial in stress adaptation, functioning as antioxidants, signaling molecules, and antimicrobial agents (Salam et al., 2023). MTs are proteins that bind metals, managing metal balance, detoxifying excess ions, and shielding against oxidative stress (Kumar et al., 2021). PCs, on the other hand, function as metal chelators that sequester HMs and facilitate their detoxification (Faizan et al., 2024). This chapter explores the roles of these three categories of stress-responsive biomolecules and their importance in enhancing plant tolerance to abiotic stressors (Fig. 10.1).

10.2.1 Secondary metabolites in plant stress response

SMs are specialized compounds in plants that do not participate in primary metabolic activities but are vital for stress adaptation (Salam et al., 2023). Secondary metabolite biosynthesis mainly derives from precursors generated in the Krebs cycle and the shikimate pathway. These pathways supply essential primary metabolites that act as foundational building blocks for SMs, which are distinguished from primary metabolites by their chemical structure, roles, and distribution within plants. (Jamwal et al., 2018). Although the biosynthetic pathways of primary metabolites are conserved across plant species,SMs experience various modifications, including glycosylation, methylation, hydroxylation, acylation, oxidation, phosphorylation, and prenylation, that create structural diversity (Jan et al., 2021). SMs are classified into three main groups based on their biosynthetic origin: phenolic compounds from the shikimate pathway, terpenes produced via the mevalonic pathway, and nitrogen-containing compounds generated through the tricarboxylic acid cycle (Jan et al., 2021). These metabolites are categorized into phenolics, terpenoids, alkaloids, and glucosinolates, each with distinct functions in stress mitigation.

Phenolic compounds, like flavonoids and lignins, act as antioxidants that shield plants from oxidative stress by neutralizing ROS produced under challenging environmental conditions (Hajam et al., 2023). These compounds are crucial in defending plants against temperature-related stress. Research shows that cold or chilling stress leads to increased accumulation of anthocyanins in plant tissues, along with the heightened expression of key flavonoid biosynthesis genes such as PAL, C4H, 4CL, CHI, dihydroflavonol-4-reductase (DFR), and anthocyanidin synthase (ANS) (Rao and Zheng, 2025). In *Brassica rapa*, cold tolerance has been strongly linked to the expression of anthocyanin biosynthesis genes,

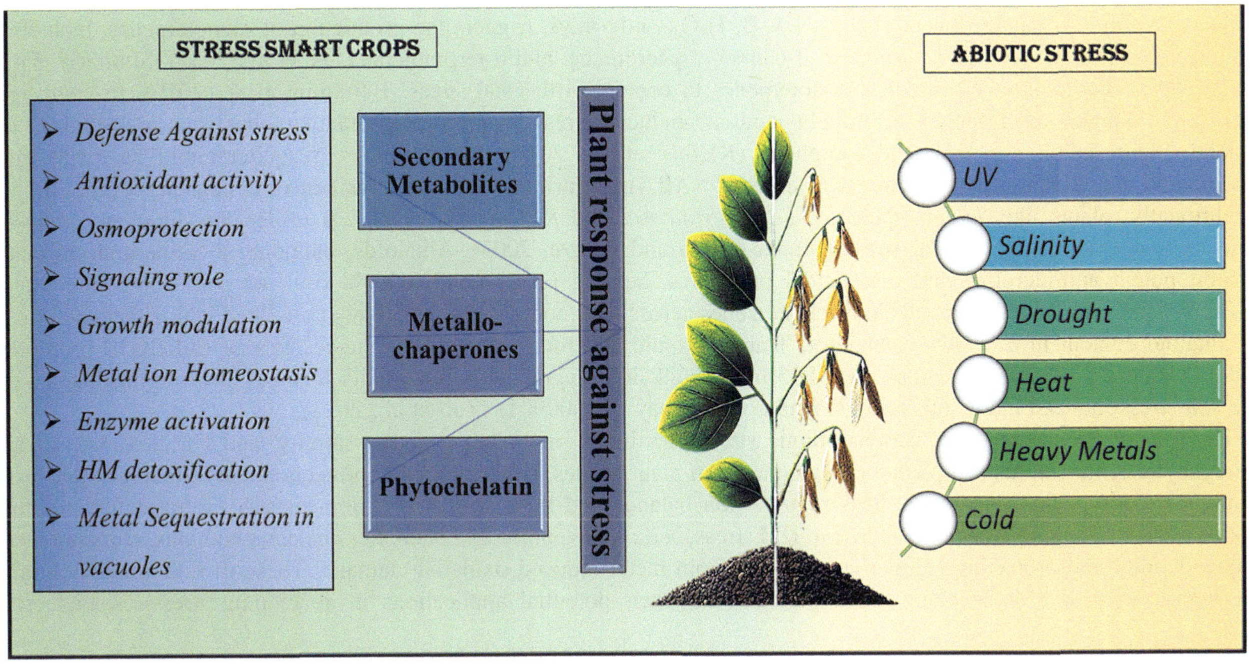

FIGURE 10.1 Roles of secondary metabolites, metallochaperones, and phytochelatin in multiple abiotic stress tolerance.

particularly DFR and ANS (Ahmed et al., 2020). Similarly, in *Malus sieversii*, low temperatures enhance anthocyanin accumulation through the regulatory influence of the MdMYBPA1 transcription factor (TF) (Wang et al., 2018). Recent research has emphasized how different light wavelengths influence the biosynthesis and accumulation of polyphenols in plants (Kołton et al., 2022). A study by Park et al. (2020) demonstrated that blue light promotes polyphenol production in *Brassica napus* and *Brassica oleracea*. Likewise, in *Camellia japonica* callus cultures, the highest levels of phenols and flavonoids were found under red-blue or blue-green light combinations (Jang et al., 2020). *Zea mays* exhibited increased total phenolic and flavonoid levels under salinity stress (Salama et al., 2015). In *Oryza sativa*, salinity improved grain nutritional quality by enhancing antioxidant activity, anthocyanins, and total phenolic content (Chunthaburee et al., 2015). The responses are governed by light-sensitive TFs that activate specific molecular pathways. For instance, in *Fagopyrum tataricum*, the R2R3-MYB TFs FtMYB6 exhibits light-induced promoter activity, enhancing flavanol biosynthesis by upregulating FtF3H and FtFLS1 while downregulating Ft4CL (Yao et al., 2020).

Flavonoids act as UV protectants, shielding plant tissues from radiation-induced damage, while lignins reinforce the cell wall, enhancing mechanical strength and resistance against dehydration and pathogen invasion (Liga et al., 2023). Exposure to UV-B radiation has been shown to enhance flavonoid accumulation and antioxidant capacity in various plant species. In *Medicago sativa*, UV-B radiation increases flavonoid content, thereby strengthening the plant's antioxidant defense system (Gao et al., 2019). Similarly, in *Kalanchoe pinnata*, UV-B exposure leads to a rise in total flavonoid and quercitrin levels, both of which contribute to antioxidant protection (Nascimento et al., 2015). Under combined UV-B stress and drought conditions, a transgenic line of *Populus tremula* × *P. tremuloides* with elevated proanthocyanidin content exhibited reduced hydrogen peroxide accumulation, indicating improved oxidative stress resistance (Gourlay et al., 2021). In response to salinity stress, *Zea mays* demonstrated enhanced plant performance through increased antioxidant activity, while in *Arabidopsis thaliana*, the expression of *CrUGT87A1*, a UDP-sugar glycosyltransferase (*UGT*) gene, improved salt tolerance by promoting flavonoid accumulation and boosting antioxidant capacity (Khalil et al., 2022; Zhang et al., 2021). In *Solanum nigrum*, increasing salinity levels boosted the expression of flavonoid genes, resulting in higher production of quercetin 3-d-glucoside and lutein. It also caused overexpression of carotenoid-related genes like phytoene synthase 2 and lycopene cyclase (Ben Abdallah et al., 2016).

Terpenoids, such as carotenoids and essential oils, also contribute to stress tolerance by modulating oxidative stress responses and deterring herbivores. *Nicotiana attenuata* synthesizes sesquiterpenes such as (E)-β-Farnesene and (E)-α-Bergamotene in response to tropospheric ozone and ROS (Palmer-Young et al., 2015; Loreto et al., 2001). In *Oryza sativa*, exposure to high irradiance levels, such as UV-B, H_2O_2, and γ-rays, triggers the production of monoterpenes, including limonene, sabinene, myrcene, α-terpinene, β-ocimene, γ-terpinene, and α-terpinolene. (Lee et al., 2015). Similarly, *Vitis vinifera* cv. Chardonnay accumulates monoterpenes to cope with thermal stress (Bertamini et al., 2019). *Pseudotsuga menziesii* (Douglas fir) responds to drought- and salt-induced stress by producing monoterpenes such as β-pinene, α-pinene, β-citronellol, 3-carene, and camphene (Kleiber et al., 2017). In *Zea mays*, sesquiterpenoid zealexins and diterpenoid kaulalexins, along with abscisic acid (ABA), contribute to stress resilience (Vaughan et al., 2015). Additionally, plants like *Salvia officinalis*, *Salvia fruticosa*, and *Rosmarinus officinalis* utilize the diterpene carnosic acid to counteract environmental stress (Munné-Bosch and Alegre, 2003). Alkaloids, including nicotine and caffeine, exhibit potent antimicrobial and insecticidal properties, helping plants combat herbivory and pathogenic infections. WGCNA analysis identified six alkaloid-related genes across five modules that positively correlated with either salt stress or alkaloid content in *Eucommia ulmoides*. Transcriptomic and metabolomic data suggested that EuRBG10 promotes indole alkaloid production, whereas EuAMO5 inhibits alkaloid degradation, potentially forming the core mechanism of the salt stress-induced indole alkaloid biosynthesis pathway (map00901) (Zuo et al., 2022).

The role of SMs extends to various abiotic stress conditions, including drought, salinity, and HM toxicity. Under drought stress, flavonoids and phenolics accumulate in plant tissues, enhancing osmoprotection and reducing water loss. Similarly, under saline conditions, SMs regulate ion balance and maintain cellular homeostasis by mitigating sodium toxicity and scavenging ROS. In response to HM stress, certain flavonoids and phenolics chelate metal ions, reducing their bioavailability and protecting cellular components from metal-induced oxidative damage. These diverse roles highlight the significance of SMs in plant stress tolerance and their potential applications in developing stress-resistant crop varieties.

10.2.2 Metallothioneins: metal-binding proteins in abiotic stress tolerance

MTs are a group of small, cysteine-rich proteins essential for metal regulation, detoxification, and combating oxidative stress in plants. They are found throughout various plant tissues and are categorized into four types based on their cysteine-rich regions and sequence similarity (Kumar et al., 2021). Type 1 MTs mainly bind zinc (Zn) and copper (Cu), with high expression in plant roots. Type 2 MTs regulate Cu homeostasis and are mainly present in leaf tissues. Type 3 MTs are located in seeds and fruit tissues, aiding in metal sequestration and storage. Meanwhile, Type 4 MTs play a role in the development of reproductive tissues (Kumar et al., 2021).

The primary function of MTs in abiotic stress tolerance is HMsequestration, where they bind and neutralize excess metal ions such as Cd, Zn, and Cu, reducing their toxic effects. These proteins also play a crucial role in mitigating oxidative stress, which is often triggered by HM accumulation. Metal-induced ROS generation can damage cellular membranes, proteins, and DNA, but MTs help alleviate this oxidative stress by directly scavenging free radicals and chelating toxic metal ions (Kumar et al., 2021). Additionally, MTs regulate essential metal homeostasis by maintaining physiological levels of Zn and Cu, which are necessary for various enzymatic and metabolic activities.

The expression of MT genes often increases in response to metal stress, drought, and oxidative damage, highlighting their role in helping plants adapt. Multiple studies have shown that overexpressing MT genes in transgenic plants improves their ability to tolerate and accumulate metals, making these plants suitable for phytoremediation. The study examines the functions of OsMT-I-Id, a type-I metallothionein in rice, by exploring its involvement in metal accumulation and cellular protection. Using HM-responsive yeast mutants—cup1Δ (DTY4) for Cu, ycf1Δ for , and acr3Δ for trivalent As—the effects of OsMT-I-Id expression were examined. The results showed that yeast cells expressing OsMT-I-Id had increased tolerance and accumulated significantly more Cu, As, and Cd compared to control cells (Gautam et al., 2023). Another study identified 20 *GmaMTP* genes in soybeans, examining their phylogeny, structures, and expression under HM stress. These genes are grouped into three CDF categories (Mn-, Zn-, and Fe/Zn-CDFs), with their expansion primarily driven by segmental duplication. Expression patterns differed between leaves and roots, with *GmaMTP1.1, GmaMTP1.2, GmaMTP3.1, GmaMTP3.2, GmaMTP4.1, and GmaMTP4.3* exhibiting strong responses to metal treatments (El-Sappah et al., 2023). Similarly, Nine *ZmMT* genes were identified in the maize genome, with eight containing a conserved metallothio_2 domain and *ZmMT4* featuring a Metallothio_PEC domain. *ZmMT* genes exhibited tissue-specific expression across 23 developmental stages. They responded differently to Cu, Cd, and Pb stress, revealing their role in plant growth, development, and stress resistance to HMs (Gao et al., 2022).

Another study centered on creating transgenic chickpeas that overexpress the CarMT1 gene to evaluate its function in drought and HM stress resistance. These transgenic plants showed a 22-fold increase in CarMT1 transcript levels during drought conditions, along with enhanced physiological and biochemical traits (Kumar et al., 2022). Twelve *Metallothionein* (*MT*) genes were identified in *B. papyrifera* and were observed to play a key role in HM stress response and ABA signaling, with their Cys-rich domains being essential for metal tolerance and chelation (Xu et al., 2024). The ability of MTs to bind metals and mitigate oxidative stress underscores their importance in plant abiotic stress tolerance and highlights their potential applications in improving plant resilience in metal-contaminated environments.

10.2.3 Phytochelatins: peptide-based metal chelators

PCs are peptides produced through enzymatic processes that serve as essential metal chelators in plants, helping to defend against HM toxicity. They are formed from GSH by the enzyme phytochelatin synthase (PCS), which becomes active when plants encounter metal stress (Faizan et al., 2024). PCs generally consist of the repeating unit (γ-Glu-Cys)_n-Gly, with n ranging from 2 to 11 depending on the plant species and metal exposure levels. Their main function is to detoxify HMs by forming metal-PC complexes. These complexes bind toxic metal ions like Cd, lead (Pb), mercury (Hg), and As, preventing these metals from interacting with essential cellular components and thereby mitigating metal toxicity damage (Faizan et al., 2024). PC biosynthesis is triggered by a broad range of HM ions and anionic species, initiating with the formation of GSH–CdS complexes, which are later replaced by PC–CdS complexes (Lesniewska et al., 2004). These complexes exhibit size variations, possibly due to different intracellular sulfide concentrations at the time of their formation. Studies have also confirmed the involvement of platinum-group elements (PGEs) such as Pt(II), Rh(III), and Pd (II) in PC formation in specific plant tissues (Lesniewska et al., 2004). Among inorganic ions, Cd(II) is a particularly potent inducer of PC synthesis across multiple studies. Once formed, these metal-PC complexes are transported into vacuoles by ATP-binding cassette transporters, effectively compartmentalizing and sequestering the metals away from sensitive cellular structures.

In addition to metal detoxification, PCs also play a protective role against oxidative stress caused by HM accumulation (Raychaudhuri et al., 2021). The presence of excess metals induces the overproduction of ROS, leading to oxidative damage. PCs, along with GSH, help mitigate this oxidative stress by neutralizing ROS and maintaining redox balance in plant cells. Due to their ability to sequester toxic metals, PCs play a significant role in phytoremediation strategies, where plants with enhanced PC production are used to clean metal-contaminated soils (Raychaudhuri et al., 2021). Several transgenic approaches have been explored to enhance PC synthesis in plants, leading to improved metal tolerance and accumulation.

A study identified and characterized the *BnPCS1* gene from ramie (*Boehmeria nivea*), revealing its conserved catalytic regions and stress-related promoter elements. Overexpression of *BnPCS1* enhanced root growth, fresh weight, and Cd accumulation while reducing oxidative stress, suggesting its potential role in phytoremediation of Cd-contaminated soils (Zhu et al., 2021). Another study investigated the role of phytochelatin (PC) and GSH in detoxifying Cd stress in corn plants. It was observed that Cd stress induced higher GSH (72–235 nmol g^{-1} F.W.) and PC (79–505 nmol g^{-1} F.W.) levels, with a strong positive correlation (0.67–0.99). However, prolonged Cd exposure led to a decline in GSH and PC levels, especially at 40 μM Cd. These findings suggest that PC provides transient Cd detoxification, with its levels dependent on Cd concentration and cellular thresholds (Zare et al., 2022). The roles of Caffeic Acid O-Methyltransferase (COMT) and PCS in melatonin-induced cadmium (Cd) stress resistance in tomato plants were investigated using TRV-COMT, TRV-PCS, and TRV-COMT-PCS lines. (Xing et al., 2023). Cd stress reduced chlorophyll content and CO_2 assimilation while increasing Cd, H_2O_2, and MDA accumulation, particularly in PCS-deficient plants. It was observed that melatonin treatment enhanced endogenous melatonin and PC levels, alleviated oxidative stress, and improved antioxidant capacity by maintaining GSH:GSSG and ASA:DHA ratios. Additionally, melatonin regulates PC synthesis, promoting osmotic balance and nutrient absorption (Xing et al., 2023). The effectiveness of PCs in chelating and detoxifying HMs highlights their indispensable role in plant survival under metal stress conditions and their potential applications in environmental bioremediation.

10.3 Molecular and genetic regulation of defense mechanisms

10.3.1 Transcription factors

TFs are crucial in controlling the production of SMs, MTs, and PCs. These proteins attach to specific DNA sequences, either turning genes on or off depending on environmental cues. Among the most studied families are MYB, WRKY, bZIP, NAC, and AP2/ERF, which play roles in plant stress responses by managing important adaptive processes. MYB TFs specifically regulate the synthesis of phenylpropanoids, flavonoids, and lignin, aiding in ROS detoxification, UV protection, and metal chelation (Mariyam et al., 2025). WRKY TFs modulate defense signaling and biotic as well as abiotic stress resistance by regulating stress-responsive genes (Li et al., 2024). bZIP TFs control ABA-mediated responses, photomorphogenesis, and secondary metabolism, enhancing drought and salinity tolerance (Han et al., 2023). NAC TFs contribute to root and shoot development, cell wall remodeling, and senescence under stress conditions (Xiong et al., 2025). AP2/ERF TFs regulate multiple signal transduction pathways, such as ethylene and jasmonic acid (JA) pathways, which are crucial for abiotic stress adaptation and pathogen defense (Nie and Wang, 2023). Together, these TF families orchestrate complex regulatory networks to enhance plant resilience. For example, *VvMYBF1* and MYB111in *Arabidopsis thaliana* have been shown to upregulate flavonoid biosynthesis during drought and salinity stress, enhancing antioxidant defenses (Wang et al., 2019a, b; Li et al., 2019). Similarly, WRKY TFs like *WRKY46, WRKY54, WRKY70,* and *WRKY54* are involved in activating flavonoid biosynthesis pathways under salinity, drought, cold, and HM stress (Zhang et al., 2023) (Fig. 10.2).

The bZIP TFs family is another essential regulator of plant defense mechanisms. In barley, HvbZIP21 is activated during drought stress by improving ROS scavenging activity (Pan et al., 2022). bZIP is also known to regulate ABA signaling and enhance drought stress tolerance, as demonstrated by overexpressing OsbZIP42 in rice, which exhibited upregulation of ABA-responsive LEA3 and Rab16 genes and increased sensitivity to ABA (Joo et al., 2019). bZIP TFs often work in conjunction with MYB and WRKY TFs, forming intricate regulatory networks that ensure balanced metabolic responses under multiple stress conditions. Further, NAC TFs are also known for their role in plant development and stress response, such as *NtNAC028* from tobacco, which was shown to have negative regulation under salt, drought, and ABA stress. Knockout transgenes of the same were further shown to have increased tolerance to stress, such as delayed leaf senescence, lower accumulation of ROS and H_2O_2 content (Wen et al., 2022). Similarly, AP2/ERF TFs particularly *ERF1*, are present on the upstream of ethylene and jasmonate, which are involved in stress conditions. In Arabidopsis, Overexpressed lines of ERF1 showed enhanced tolerance to heat, salinity, and drought stress (Cheng et al., 2013). Overall, all these TFs are integrated into broader signaling networks mediated by ROS and phytohormones, ensuring a coordinated and adaptive response to environmental challenges. ROS, despite being potentially harmful at high concentrations, act as crucial signaling molecules by activating various signaling cascades, such as activating mitogen-activated protein kinase. While phytohormonal signaling pathways involving ABA, salicylic acid (SA), and JA further integrate these molecular signals.

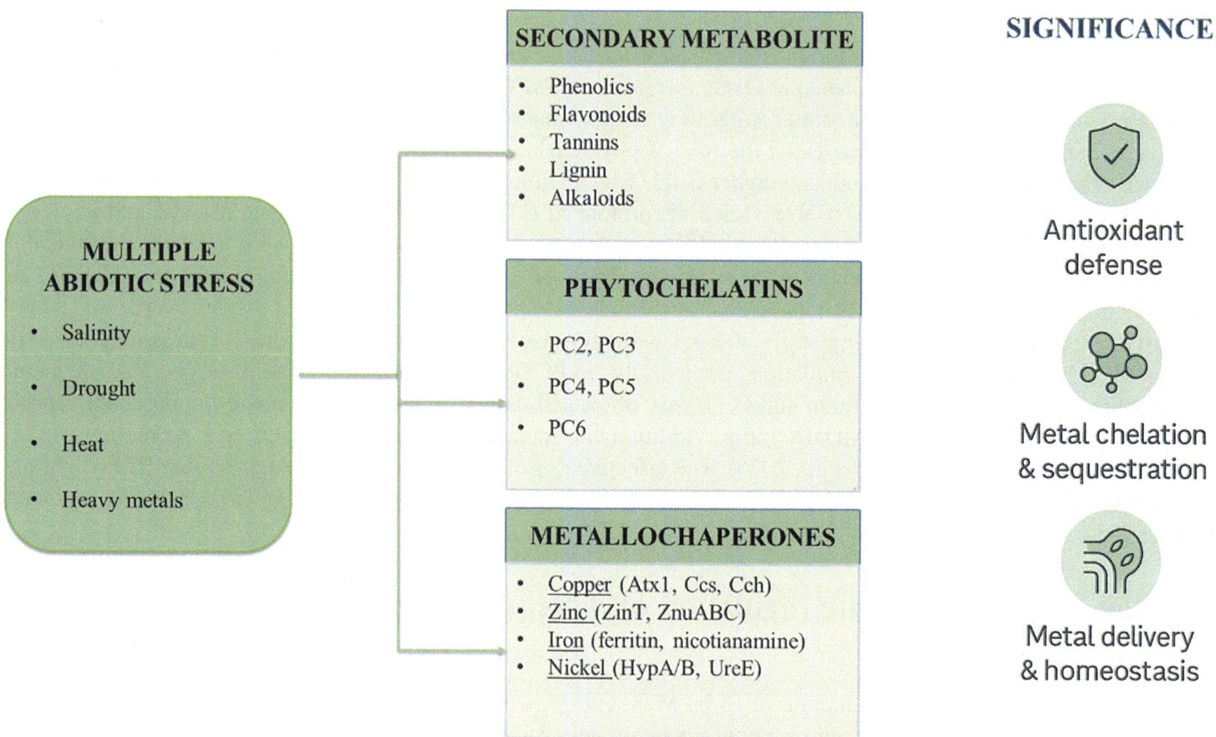

FIGURE 10.2 Interplay of secondary mtabolites, phytochelatins, and metallochaperones under abiotic stress. The interaction of multiple abiotic stresses influences the coordinated involvement of secondary metabolites (e.g., flavonoids), phytochelatins (PC1, PC2, PC3, PC4), and metallochaperones, including those involved in copper (Atx1, Ccs, Cch), zinc (ZinT, ZnuABC), iron (ferritin, nicotianamine), and nickel (HypA/B, UreE) homeostasis.

10.3.2 Epigenetic and posttranscriptional regulation

Besides transcriptional regulation, plants also use epigenetic and posttranscriptional strategies to refine their stress responses. Epigenetic changes, including DNA methylation, histone modifications, and chromatin remodeling, offer a flexible and reversible way to control gene activity without changing the DNA sequence itself. For instance, DNA methylation at cytosine bases, especially within promoter regions, can either turn genes on or off, depending on the specific methylation pattern. Production of SMs and polyphenols is also governed by DNA methylation. For example, in the medicinal plant *Catharanthus roseus*, enzymes involved in the secondary metabolite pathway for Monoterpene Indole Alkaloid were found to be tissue-specific, differentially methylated (Dugé de Bernonville et al., 2020). Under HM stress, many plant species exhibit hypomethylation in stress-responsive genes, promoting their upregulation. For example, in rice (*Oryza sativa*), methylation of HM transportation is cultivar-specific when exposed to Cu, Cd, and Cr metal stresses (Miao et al., 2022). In contrast, drought and salinity stress can induce both hypermethylation and hypomethylation in different genes, allowing plants to fine-tune energy allocation. For instance, in *Tibetan hulless barley*, drought stress induces genome-wide demethylation of drought-responsive genes, facilitating their rapid activation (Jiabu et al., 2023). Meanwhile, under salinity stress, genes that are not immediately required are often hypermethylated, effectively repressing unnecessary cellular activities to conserve energy.

Apart from DNA methylation, histone modifications play a crucial role in stress responses by altering chromatin structure and accessibility to transcriptional machinery. Histone acetylation, controlled by histone acetyltransferases, generally promotes gene expression, while histone deacetylases repress gene activity. For example, in rice, global reduction of H3K14 acetylation during cold stress was observed without changing the status of H3K9ac and H3K27ac (Roy et al., 2014). In addition to acetylation, histone methylation plays a critical role in abiotic stress responses. Histone H3K4-tri-methylation (H3K4me3) analysis in switchgrass under heat, drought, and their combined stress showed peaks corresponding to enzymes involved in secondary metabolite pathways, transport, and signal receptor proteins, as well as stress-responsive genes (Ayyappan et al., 2024). These modifications provide plants with the ability to modulate their stress responses dynamically.

Beyond epigenetic modifications, posttranscriptional mechanisms such as RNA interference (RNAi) contribute significantly to plant stress adaptation. RNAi is mediated by microRNAs (miRNAs) and small interfering RNAs, which fine-tune gene expression by degrading target mRNAs or inhibiting translation. Multiple transgenic studies have been reported by targeting miRNAs to analyze abiotic stress tolerance under drought, heat, salinity, HM exposure, light, and various combined stress conditions. For example, miR408, miR414c, miR156, miR163, and miR828 were overexpressed under various stresses such as drought, salinity, and cold by different authors to demonstrate the importance for stress tolerance (Hajyzadeh et al., 2015; Wang et al., 2019a, b; Visentin et al., 2020; Li et al., 2021a, b). These examples highlight how RNAi mechanisms add another layer of complexity to plant stress tolerance strategies.

In addition to RNAi, another important posttranscriptional mechanism is alternative splicing (AS), which allows a single gene to produce multiple protein isoforms with distinct functions. This mechanism plays a crucial role in stress responses by expanding the functional diversity of proteins involved in stress adaptation. For example, in *Triticum aestivum*, AS of the DREB2 TF produced three isoforms, Wdreb2α, Wdreb2β, and Wdreb2γ, resulting in the differential production of an active isoform under cold, drought, salt, and exogenous ABA stress. Thus expression of DREB2 was under control by transcriptional level and AS (Egawa et al., 2006). In maize (*Zea mays*), heat stress induces AS of multiple genes, leading to the production of an isoform that enhances thermotolerance. Further heat stress also induced AS of genes that are regulators of alternative RNA splicing mechanisms (Li et al., 2021a, b). These findings demonstrate how AS increases plant resilience by generating protein diversity in response to environmental cues.

10.3.3 Advances in genetic engineering to enhance stress tolerance via manipulation of these pathways

Advances in genetic engineering have significantly expanded our ability to manipulate these molecular pathways to improve plant stress tolerance. CRISPR-Cas9 genome editing, in particular, has revolutionized plant biotechnology by enabling precise modifications of genes involved in secondary metabolite production, MT synthesis, and PC biosynthesis. In rice, CRISPR-mediated knockout of negative regulators like *OsbZIP53* showed enhanced H_2O_2 accumulation and modulation of the SA pathway (Lijuan et al., 2024). Similarly, Cas9 mutant of *MpPCS* enzyme from PCS in *M. polymorpha* results in increased phytochelatin production, thereby improving detoxification of Cd^{2+} (Li et al., 2022). Overexpression of stress-responsive genes offers another promising strategy. For example, transgenic Arabidopsis overexpressing *Brassica juncea* BjMT2 exhibits greater Cd accumulation and reduced oxidative damage due to enhanced

TABLE 10.1 Summary of studies on gene manipulation of secondary metabolites, phytochelatins, and metallothioneins for understanding plant responses to multiple abiotic stresses.

Sr.	Gene	Plant	Stress	Details	References
1	OsMT1e-P	Tobacco	Salinity, Drought, ABA	Overexpression of Metallothioneins gene confers stress tolerance by scavenging ROS	Kumar et al. (2012)
2	OsMTs and AtMTs	Rice & *Arabidopsis*	Heavy metal(Cd, Pb, Cu), salt stress, H_2O_2, Cold, ABA	Differential expression of OsMT1a-c, g, OsMT2a, AtMT1a, AtMT1c, and AtMT2a under various combined stresses in rice and Arabidopsis, respectively.	Kim and Kang (2018)
3	GhMT3a	Tobacco and yeast	Salt, drought, and low-temperature stresses, H_2O_2, Heavy metal	Overexpression of GhMT3a in tobacco demonstrated higher tolerance to multiple abiotic stress by scavenging ROS production and efficient ROS signaling	Xue et al. (2009)
4	BjMT2	*Arabidopsis*	Copper and cadmium	Enhanced tolerance under Cu^{2+} and Cd^{2+} but it also showed a decrease in root growth under control conditions.	Zhigang et al. (2006)
5	OsMT1a	Rice	Zinc and Drought	Enhanced tolerance to stress by increased CAT, POX, and APX activity and expression of the zinc finger transcription factors.	Yang et al. (2009)
6	phytochelatin synthase (alr0975)	*Anabaena* sp. PCC 7120	UV-B, NaCl, heat, $CuCl_2$, carbofuran, and $CdCl_2$	Overexpressed lines showed enhanced growth rate, nitrogenase activity, expression of proteins involved in transport, protein folding, and carbon fixation.	Chaurasia et al. (2017)
7	OsPCS2a and OsPCS2b	Rice and yeast	Cadmium and arsenic	Transgenic lines demonstrated the role of OsPCS2 in accumulation and tolerance to metal uptake. The grain content of Cd and As in transgenic lines was reduced compared to the control by 51% and 35%, respectively.	Das et al. (2017)
8	AtABCC3	*Arabidopsis*	Cadmium	abcc3 mutants showed increased Cd sensitivity, while those overexpressing AtABCC3 (AtABCC3ox) exhibit enhanced Cd tolerance, with Cd accumulating in vacuoles. Further AtABCC3 function depends on phytocheletins.	Brunetti et al. (2015)
9	AtMYB12	*Arabidopsis*	Salt and drought	Overexpression of AtMYB12 enhanced salt and drought tolerance in transgenic Arabidopsis by elevating flavonoid and ABA levels. Tolerance was supplemented by increased ABA, proline, SOD, and POD, while reducing H_2O_2 and MDA levels.	Wang et al. (2016)
10	*CtCYP82G24*	*Arabidopsis*	Osmotic stress	CtCYP82G24 from *Carthamus tinctorius* enhances flavonoid biosynthesis and osmotic stress tolerance in transgenic *Arabidopsis*. Transgenic plants showed enhanced expression levels of several flavonoid biosynthetic genes and metabolites.	Ahmad et al. (2019)

MT synthesis (Zhigang et al., 2006). Overexpression of AtMYB12 in *Arabidopsis* has also been shown to elevate flavonoid production, resulting in improved drought resistance (Yang et al. 2009).

RNAi technology has also been employed to manipulate stress-related gene expression. Silencing genes that negatively regulate secondary metabolite production or MT synthesis can significantly enhance stress tolerance. In rice, RNAi-mediated suppression of *OsNAC5* TFs resulted decrease in salinity and ABA stress tolerance. This suggests that *OsNAC5* plays an important to stress tolerance and regulation of metabolites (Song et al., 2011). Synthetic biology approaches further extend the potential of genetic engineering by introducing novel genetic circuits that optimize stress responses (Table 10.1).

The applications of these genetic engineering techniques hold immense potential for agriculture, particularly in the development of stress-resilient crops capable of thriving in adverse environments. Additionally, crops engineered to produce higher levels of antioxidant SMs may exhibit greater resistance to drought, salinity, heat stress, and HM contamination. Future research focusing on the integration of these molecular mechanisms will further our understanding of plant stress adaptation and facilitate the development of next-generation crops tailored to withstand evolving environmental challenges.

10.4 Conclusion

Plants face various biotic and abiotic stresses, often simultaneously, in their natural habitats. To survive these challenges, they have developed complex biochemical defenses such as the production of SMs, MTs, and PCs. These SMs aid stress tolerance by neutralizing ROS, strengthening structural integrity, adjusting hormonal responses, and stabilizing metabolism. MTs function as metal-binding proteins that regulate metal homeostasis and detoxification while alleviating oxidative stress. Similarly, PC play a crucial role in HM detoxification by forming stable metal-PC complexes and sequestering toxic ions in vacuoles.

The biosynthesis of these protective compounds is tightly regulated by key TFs such as MYB, WRKY, bZIP, NAC, and AP2/ERF, highlighting their importance in plant adaptation to extreme conditions. Advances in genetic engineering, including transgenic approaches and CRISPR-Cas9-mediated gene editing, offer promising opportunities to manipulate these pathways for enhanced stress resilience in crops. Future research should focus on integrating multiomics approaches, synthetic biology, and advanced breeding strategies to fully harness the potential of these regulatory mechanisms for sustainable agriculture and the development of climate-resilient crops.

References

Ahmad, N., Jianyu, L., Xu, T., Noman, M., Jameel, A., Na, Y., et al., 2019. Overexpression of a novel cytochrome P450 promotes flavonoid biosynthesis and osmotic stress tolerance in transgenic Arabidopsis. Genes 10 (10), 756. Available at https://doi.org/10.3390/genes10100756.

Ahmed, A., Tariq, A., Habib, S., 2020. Interactive biology of auxins and phenolics in plant environment. In: Lone, R., Shuab, R., Kamili, A. (Eds.), Plant Phenolics in Sustainable Agriculture. Springer, Singapore. Available at https://doi.org/10.1007/978-981-15-4890-1_5.

Ayyappan, V., Sripathi, V.R., Xie, S., et al., 2024. Genome-wide profiling of histone (H3) lysine 4 (K4) tri-methylation (me3) under drought, heat, and combined stresses in switchgrass. BMC Genomics 25, 223. Available at https://doi.org/10.1186/s12864-024-10068-w.

Balasubramaniam, T., Shen, G., Esmaeili, N., Zhang, H., 2023. Plants' Response Mechanisms to Salinity Stress. Plants 12, 2253. Available at 10.3390/plants12122253.

Ben Abdallah, S., Aung, B., Amyot, L., Lalin, I., Lachâal, M., Karray-Bouraoui, N., et al., 2016. Salt stress (NaCl) affects plant growth and branch pathways of carotenoid and flavonoid biosyntheses in Solanum nigrum. Acta Physiologiae Plantarum 38 (3), 72. Available at https://doi.org/10.1007/s11738-016-2096-8.

Bertamini, M., Grando, M.S., Zocca, P., Pedrotti, M., Lorenzi, S., Cappellin, L., et al., 2019. Linking monoterpenes and abiotic stress resistance in grapevines. BIO Web of Conferences 13, 01003. Available at https://doi.org/10.1051/bioconf/20191301003.

Brunetti, P., Zanella, L., De Paolis, A., Di Litta, D., Cecchetti, V., Falasca, G., et al., 2015. Cadmium-inducible expression of the ABC-type transporter AtABCC3 increases phytochelatin-mediated cadmium tolerance in Arabidopsis. Journal of Experimental Botany 66 (13), 3815–3829. Available at https://doi.org/10.1093/jxb/erv185.

Chaurasia, N., Mishra, Y., Chatterjee, A., Rai, R., Yadav, S., Rai, L.C., 2017. Overexpression of phytochelatin synthase (pcs) enhances abiotic stress tolerance by altering the proteome of transformed Anabaena sp. PCC 7120. Protoplasma 254, 1715–1724. Available at https://doi.org/10.1007/s00709-016-1059-7.

Cheng, M.C., Liao, P.M., Kuo, W.W., Lin, T.P., 2013. The Arabidopsis ETHYLENE RESPONSE FACTOR1 regulates abiotic stress-responsive gene expression by binding to different cis-acting elements in response to different stress signals. Plant Physiology 162 (3), 1566–1582. Available at https://doi.org/10.1104/pp.113.221911.

Chunthaburee, S., Sanitchon, J., Pattanagul, W., Theerakulpisut, P., 2015. Effects of salt stress after late booting stage on yield and antioxidant capacity in pigmented rice grains and alleviation of the salt-induced yield reduction by exogenous spermidine. Plant Production Science 18 (1), 32–42. Available at https://doi.org/10.1626/pps.18.32.

Das, N., Bhattacharya, S., Bhattacharyya, S., et al., 2017. Identification of alternatively spliced transcripts of rice phytochelatin synthase 2 gene OsPCS2 involved in mitigation of cadmium and arsenic stresses. Plant Molecular Biology 94, 167–183. Available at https://doi.org/10.1007/s11103-017-0600-1.

Ding, Y., Yang, S., 2022. Surviving and thriving: How plants perceive and respond to temperature stress. Developmental Cell 57, 947–958. Available at 10.1016/j.devcel.2022.03.010.

Dugé de Bernonville, T., Maury, S., Delaunay, A., Daviaud, C., Chaparro, C., Tost, J., et al., 2020. Developmental methylome of the medicinal plant *Catharanthus roseus* unravels the tissue-specific control of the monoterpene indole alkaloid pathway by DNA methylation. International Journal of Molecular Sciences 21 (17), 6028. Available at https://doi.org/10.3390/ijms21176028.

Egawa, C., Kobayashi, F., Ishibashi, M., Nakamura, T., Nakamura, C., Takumi, S., 2006. Differential regulation of transcript accumulation and alternative splicing of a DREB2 homolog under abiotic stress conditions in common wheat. Genes & Genetic Systems 81 (2), 77–91. Available at https://doi.org/10.1266/ggs.81.77.

El-Sappah, A.H., Abbas, M., Rather, S.A., Wani, S.H., Soaud, N., Noor, Z., et al., 2023. Genome-wide identification and expression analysis of metal tolerance protein (MTP) gene family in soybean (Glycine max) under heavy metal stress. Molecular Biology Reports 50 (4), 2975–2990. Available at https://doi.org/10.1007/s11033-022-08100-x.

Faizan, M., Alam, P., Hussain, A., Karabulut, F., Tonny, S.H., Cheng, S.H., et al., 2024. Phytochelatins: key regulator against heavy metal toxicity in plants. Plant Stress 11. Available at https://doi.org/10.1016/j.stress.2024.100355.

Gao, C., Gao, K., Yang, H., Ju, T., Zhu, J., Tang, Z., et al., 2022. Genome-wide analysis of metallothionein gene family in maize to reveal its role in development and stress resistance to heavy metal. Biological Research 55 (1). Available at https://doi.org/10.1186/s40659-021-00368-w.

Gao, L., Liu, Y., Wang, X., Li, Y., Han, R., 2019. Lower levels of UV-B light trigger the adaptive responses by inducing plant antioxidant metabolism and flavonoid biosynthesis in Medicago sativa seedlings. Functional Plant Biology 46 (10), 896. Available at https://doi.org/10.1071/fp19007.

Gautam, N., Tiwari, M., Kidwai, M., Dutta, P., Chakrabarty, D., 2023. Functional characterization of rice metallothionein OsMT-I-Id: insights into metal binding and heavy metal tolerance mechanisms. Journal of Hazardous Materials 458, 131815. Available at https://doi.org/10.1016/j.jhazmat.2023.131815.

Gourlay, G., Hawkins, B.J., Albert, A., Schnitzler, J., Peter Constabel, C., 2021. Condensed tannins as antioxidants that protect poplar against oxidative stress from drought and UV-B. Plant, Cell & Environment 45 (2), 362–377. Available at 10.1111/pce.14242.

Hajam, Y.A., Lone, R., Kumar, R., 2023. Role of plant phenolics against reactive oxygen species (ROS) induced oxidative stress and biochemical alterations. In: Lone, R., Khan, S., Mohammed Al-Sadi, A. (Eds.), Plant Phenolics in Abiotic Stress Management. Springer, Singapore. Available at https://doi.org/10.1007/978-981-19-6426-8_7.

Hajyzadeh, M., Turktas, M., Khawar, K.M., Unver, T., 2015. miR408 overexpression causes increased drought tolerance in chickpea. Gene 555, 186–193. Available at https://doi.org/10.1016/j.gene.2014.11.002.

Han, H., Wang, C., Yang, X., et al., 2023. Role of bZIP transcription factors in the regulation of plant secondary metabolism. Planta 258, 13. Available at https://doi.org/10.1007/s00425-023-04174-4.

Jamwal, K., Bhattacharya, S., Puri, S.K., 2018. Plant growth regulator mediated consequences of secondary metabolites in medicinal plants. Journal of Applied Research on Medicinal and Aromatic Plants 9, 26–38. Available at https://doi.org/10.1016/J.JARMAP.2017.12.003.

Jan, R., Asaf, S., Numan, M., Lubna, Kim, K.-M., 2021. Plant secondary metabolite biosynthesis and transcriptional regulation in response to biotic and abiotic stress conditions. Agronomy 11 (5), 968. Available at https://doi.org/10.3390/agronomy11050968.

Jang, E.B., Ho, T.T., Park, S.Y., 2020. Effect of light quality and tissue origin on phenolic compound accumulation and antioxidant activity in *Camellia japonica* calli. In Vitro Cellular & Developmental Biology - Plant 56, 567–577. Available at https://doi.org/10.1007/s11627-020-10121-9.

Jiabu, D., Yu, M., Xu, Q., et al., 2023. Genome-wide DNA methylation dynamics during drought responsiveness in tibetan Hulless barley. Journal of Plant Growth Regulation 42, 4391–4401. Available at https://doi.org/10.1007/s00344-022-10903-y.

Joo, J., Lee, Y.H., Song, S.I., 2019. OsbZIP42 is a positive regulator of ABA signaling and confers drought tolerance to rice. Planta 249, 1521–1533. Available at https://doi.org/10.1007/s00425-019-03104-7.

Khalil, R., Yusuf, M., Bassuony, F., Haroun, S., Gamal, A., 2022. Alpha-tocopherol reinforce selenium efficiency to ameliorates salt stress in maize plants through carbon metabolism, enhanced photosynthetic pigments and ion uptake. South African Journal of Botany 144, 1–9. Available at https://doi.org/10.1016/j.sajb.2021.08.033.

Kim, Y.-O., Kang, H., 2018. Comparative expression analysis of genes encoding metallothioneins in response to heavy metals and abiotic stresses in rice (Oryza sativa) and Arabidopsis thaliana. Bioscience, Biotechnology, and Biochemistry 82 (9), 1656–1665. Available at https://doi.org/10.1080/09168451.2018.1486177.

Kleiber, A., Duan, Q., Jansen, K., Verena Junker, L., Kammerer, B., Rennenberg, H., et al., 2017. Drought effects on root and needle terpenoid content of a coastal and an interior Douglas fir provenance. Tree Physiology 37 (12), 1648–1658. Available at https://doi.org/10.1093/treephys/tpx113.

Kołton, A., Długosz-Grochowska, O., Wojciechowska, R., Czaja, M., 2022. Biosynthesis regulation of folates and phenols in plants. Scientia Horticulturae 291, 110561. Available at https://doi.org/10.1016/j.scienta.2021.110561.

Kumar, G., Kushwaha, H.R., Panjabi-Sabharwal, V., et al., 2012. Clustered metallothionein genes are co-regulated in rice and ectopic expression of OsMT1e-Pconfers multiple abiotic stress tolerance in tobacco via ROS scavenging. BMC Plant Biology 12, 107. Available at https://doi.org/10.1186/1471-2229-12-107.

Kumar, S., Yadav, A., Kumar, A., Verma, R., Lal, S., Srivastava, S., et al., 2021. Plant metallothioneins as regulators of environmental stress responses. International Journal of Plant And Environment 7 (01), 27–38. Available at https://doi.org/10.18811/ijpen.v7i01.3.

Kumar, S., Yadav, A., Verma, R., Kumar Dubey, A., Narayan, S., Pandey, A., et al., 2022. Metallothionein (MT1): a molecular stress marker in chickpea enhances drought and heavy metal stress adaptive efficacy in transgenic plants. Environmental and Experimental Botany 199, 104871. Available at https://doi.org/10.1016/j.envexpbot.2022.104871.

Lee, G.W., Lee, S., Chung, M.S., et al., 2015. Rice terpene synthase 20 (OsTPS20) plays an important role in producing terpene volatiles in response to abiotic stresses. Protoplasma 252, 997–1007. Available at https://doi.org/10.1007/s00709-014-0735-8.

Lesniewska, B.A., Messerschmidt, J., Jakubowski, N., Hulanicki, A., 2004. Bioaccumulation of platinum group elements and characterization of their species in Lolium multiflorum by size-exclusion chromatography coupled with ICP-MS. Science of the Total Environment 322 (1-3), 95–108. Available at https://doi.org/10.1016/j.scitotenv.2003.09.019.

Li, B., Fan, R., Guo, S., Wang, P., Zhu, X., et al., 2019. The Arabidopsis MYB transcription factor, MYB111 modulates salt responses by regulating flavonoid biosynthesis. Environmental and Experimental Botany 166, 103807. Available at https://doi.org/10.1016/j.envexpbot.2019.103807.

Li, M., Leso, M., Buti, M., Bellini, E., Bertoldi, D., Saba, A., et al., 2022. Phytochelatin synthase de-regulation in Marchantia polymorpha indicates cadmium detoxification as its primary ancestral function in land plants and provides a novel visual bioindicator for detection of this metal. Journal of Hazardous Materials 440, 129844. Available at https://doi.org/10.1016/j.jhazmat.2022.129844.

Li, S., Khoso, M.A., Xu, H., Zhang, C., Liu, Z., Wagan, S., et al., 2024. WRKY transcription factors (TFs) as key regulators of plant resilience to environmental stresses: current perspective. Agronomy 14 (10), 2421. Available at https://doi.org/10.3390/agronomy14102421.

Li, T., Lian, H., Li, H., et al., 2021. HY5 regulates light-responsive transcription of microRNA163 to promote primary root elongation in Arabidopsis seedlings. Journal of Integrative Plant Biology 63, 1437–1450. Available at https://doi.org/10.1111/jipb.13099.

Li, Z., Tang, J., Bassham, D.C., Howell, S.H., 2021b. Daily temperature cycles promote alternative splicing of RNAs encoding SR45a, a splicing regulator in maize. Plant Physiology 186 (2), 1318–1335. Available at https://doi.org/10.1093/plphys/kiab110.

Liga, S., Paul, C., Péter, F., 2023. Flavonoids: overview of biosynthesis, biological activity, and current extraction techniques. Plants 12 (14), 2732. Available at https://doi.org/10.3390/plants12142732.

Lijuan, W., Cong, H., Huimei, W., Yuchang, H., Hai, L., Lei, W., et al., 2024. OsbZIP53 negatively regulates immunity response by involving in reactive oxygen species and salicylic acid metabolism in rice. Rice Science 31 (2), 190–202. Available at https://doi.org/10.1016/j.rsci.2023.12.002.

Loreto, F., Mannozzi, M., Maris, C., Nascetti, P., Ferranti, F., Pasqualini, S., 2001. Ozone quenching properties of isoprene and its antioxidant role in leaves. Plant Physiology 126 (3), 993–1000. Available at https://doi.org/10.1104/pp.126.3.993.

Mariyam, S., Kumar, V., Roychoudhury, A., et al., 2025. Functional diversification and mechanistic insights of myb transcription factors in mediating plant growth and development, secondary metabolism, and stress responses. Journal of Plant Growth Regulation 44, 1465–1484. Available at https://doi.org/10.1007/s00344-025-11622-w.

Miao, Y., Cong, W., Mu, J., Fu, T., Zhuang, T., Yan, Y., et al., 2022. Various potentially toxic element tolerances in different rice genotypes correlate with distinct physiological responses and alterations in DNA methylation. Chemosphere 292, 133462. Available at https://doi.org/10.1016/j.chemosphere.2021.133462.

Munné-Bosch, S., Alegre, L., 2003. Drought-induced changes in the redox state of α-tocopherol, ascorbate, and the diterpene carnosic acid in chloroplasts of labiatae species differing in carnosic acid contents. Plant Physiology 131 (4), 1816–1825. Available at https://doi.org/10.1104/pp.102.019265.

Nascimento, L.B.D.S., Leal-Costa, M.V., Menezes, E.A., Lopes, V.R., Muzitano, M.F., Costa, S.S., et al., 2015. Ultraviolet-B radiation effects on phenolic profile and flavonoid content of Kalanchoe pinnata. Journal of Photochemistry and Photobiology B: Biology 148, 73–81. Available at https://doi.org/10.1016/j.jphotobiol.2015.03.011.

Nie, S., Wang, D., 2023. AP2/ERF transcription factors for tolerance to both biotic and abiotic stress factors in plants. Tropical Plant Biology 16, 105–112. Available at https://doi.org/10.1007/s12042-023-09339-9.

Oguz, M.C., Aycan, M., Oguz, E., Poyraz, I., Yildiz, M., 2022. Drought Stress Tolerance in Plants: Interplay of Molecular, Biochemical and Physiological Responses in Important Development Stages. Physiologia 2, 180–197. Available at 10.3390/physiologia2040015.

Palmer-Young, E.C., Veit, D., Gershenzon, J., Schuman, M.C., 2015. The sesquiterpenes(E)-β-farnesene and (E)-α-bergamotene quench ozone but fail to protect the wild tobacco Nicotiana attenuata from ozone, UVB, and drought stresses. PLoS One 10 (6). Available at https://doi.org/10.1371/journal.pone.0127296.

Pan, R., Buitrago, S., Feng, Z., Abou-Elwafa, S.F., Xu, L., Li, C., et al., 2022. *HvbZIP21*, a novel transcription factor from wild barley confers drought tolerance by modulating ROS scavenging. Frontiers in Plant Science 13, 878459. Available at https://doi.org/10.3389/fpls.2022.878459.

Park, W.T., Yeo, S.K., Sathasivam, R., Park, J.S., Kim, J.K., Park, S.U., 2020. Influence of light-emitting diodes on phenylpropanoid biosynthetic gene expression and phenylpropanoid accumulation in Agastache rugosa. Applied Biological Chemistry 63 (1). Available at https://doi.org/10.1186/s13765-020-00510-4.

Rao, M.J., Zheng, B., 2025. The role of polyphenols in abiotic stress tolerance and their antioxidant properties to scavenge reactive oxygen species and free radicals. Antioxidants 14 (1), 74. Available at https://doi.org/10.3390/antiox14010074.

Raychaudhuri, S.S., Pramanick, P., Talukder, P., Basak, A., 2021. Polyamines, metallothioneins, and phytochelatins—natural defense of plants to mitigate heavy metals. Studies in Natural Products Chemistry 69, 227–261. Available at https://doi.org/10.1016/B978-0-12-819487-4.00006-9.

Roy, D., Paul, A., Roy, A., Ghosh, R., Ganguly, P., Chaudhuri, S., 2014. Differential acetylation of histone H3 at the regulatory region of OsDREB1b promoter facilitates chromatin remodelling and transcription activation during cold stress. PLoS One 9, e100343. Available at https://doi.org/10.1371/journal.pone.0100343.

Salam, U., Ullah, S., Tang, Z.H., Elateeq, A.A., Khan, Y., Khan, J., et al., 2023. Plant metabolomics: an overview of the role of primary and secondary metabolites against different environmental stress factors. Life (Chicago, Ill.: 1978) 13 (3). Available at https://doi.org/10.3390/life13030706.

Salama, Z.A., Gaafar, A.A., El-Fouly, M.M., 2015. Genotypic variations in phenolic, flavonoids and their antioxidant activities in maize plants treated with Zn (II) HEDTA grown in salinized media. Agricultural Sciences 6 (3), 397.

Singh, A., 2021. Soil salinity: A global threat to sustainable development. Soil Use and Management 38, 39–67. Available at 10.1111/sum.12772.

Song, S.Y., Chen, Y., Chen, J., et al., 2011. Physiological mechanisms underlying OsNAC5-dependent tolerance of rice plants to abiotic stress. Planta 234, 331–345. Available at https://doi.org/10.1007/s00425-011-1403-2.

Thakur, M., Praveen, S., Divte, P.R., Mitra, R., Kumar, M., Gupta, C.K., Kalidindi, U., Bansal, R., Roy, S., Anand, A., Singh, B., 2022. Metal tolerance in plants: Molecular and physicochemical interface determines the "not so heavy effect" of heavy metals. Chemosphere 287, 131957. Available at 10.1016/j.chemosphere.2021.131957.

Vaughan, M.M., Christensen, S., Schmelz, E.A., Huffaker, A., Mcauslane, H.J., Alborn, H.T., et al., 2015. Accumulation of terpenoid phytoalexins in maize roots is associated with drought tolerance. Plant Cell and Environment 38 (11), 2195–2207. Available at https://doi.org/10.1111/pce.12482.

Visentin, I., Pagliarani, C., Deva, E., Caracci, A., Turečková, V., Novák, O., et al., 2020. A novel strigolactone-miR156 module controls stomatal behaviour during drought recovery. Plant, Cell & Environment 43 (7), 1613–1624. Available at https://doi.org/10.1111/pce.13758.

Wang, F., Kong, W., Wong, G., et al., 2016. AtMYB12 regulates flavonoids accumulation and abiotic stress tolerance in transgenic *Arabidopsis thaliana*. Molecular Genetics and Genomics: MGG 291, 1545–1559. Available at https://doi.org/10.1007/s00438-016-1203-2.

Wang, J., Wang, F., Jin, C., Tong, Y., Wang, T., 2019b. A R2R3-MYB transcription factor VvMYBF1 from grapevine (*Vitis vinifera* L.) regulates flavonoids accumulation and abiotic stress tolerance in transgenic *Arabidopsis*. The Journal of Horticultural Science and Biotechnology 95 (2), 147–161. Available at https://doi.org/10.1080/14620316.2019.1665480.

Wang, N., Qu, C., Jiang, S., Chen, Z., Xu, H., Fang, H., et al., 2018. The proanthocyanidin-specific transcription factor MdMYBPA1 initiates anthocyanin synthesis under low-temperature conditions in red-fleshed apples. Plant Journal 96 (1), 39–55. Available at https://doi.org/10.1111/tpj.14013.

Wang, W., Liu, D., Chen, D., Cheng, Y., Zhang, X., Song, L., et al., 2019a. MicroRNA414c affects salt tolerance of cotton by regulating reactive oxygen species metabolism under salinity stress. RNA Biology 16 (3), 362–375. Available at https://doi.org/10.1080/15476286.2019.1574163.

Wen, L., Liu, T., Deng, Z., Zhang, Z., Wang, Q., Wang, W., et al., 2022. Characterization of NAC transcription factor NtNAC028 as a regulator of leaf senescence and stress responses. Frontiers in Plant Science 13, 941026. Available at https://doi.org/10.3389/fpls.2022.941026.

Xing, Q., Hasan, M.K., Li, Z., Yang, T., Jin, W., Qi, Z., et al., 2023. Melatonin-induced plant adaptation to cadmium stress involves enhanced phytochelatin synthesis and nutrient homeostasis in Solanum lycopersicum L. Journal of Hazardous Materials 456, 131670. Available at https://doi.org/10.1016/j.jhazmat.2023.131670.

Xiong, H., He, H., Chang, Y., Miao, B., Liu, Z., et al., 2025. Multiple roles of NAC transcription factors in plant development and stress responses. Journal of Integrative Plant Biology. Available at https://doi.org/10.1111/jipb.13854.

Xu, Z., Yang, S., Li, C., Xie, M., He, Y., Chen, S., et al., 2024. Characterization of metallothionein genes from Broussonetia papyrifera: metal binding and heavy metal tolerance mechanisms. BMC Genomics 25 (1). Available at https://doi.org/10.1186/s12864-024-10477-x.

Xue, T., Li, X., Zhu, W., Wu, C., Yang, G., Zheng, C., 2009. Cotton metallothionein GhMT3a, a reactive oxygen species scavenger, increased tolerance against abiotic stress in transgenic tobacco and yeast. Journal of Experimental Botany 60 (1), 339–349. Available at https://doi.org/10.1093/jxb/ern291.

Yang, Z., Wu, Y., Li, Y., et al., 2009. OsMT1a, a type 1 metallothionein, plays the pivotal role in zinc homeostasis and drought tolerance in rice. Plant Molecular Biology 70, 219–229. Available at https://doi.org/10.1007/s11103-009-9466-1.

Yao, P., Huang, Y., Dong, Q., Wan, M., Wang, A., Chen, Y., et al., 2020. FtMYB6, a light-induced SG7 R2R3-MYB transcription factor, promotes flavonol biosynthesis in tartary buckwheat (Fagopyrum tataricum). Journal of Agricultural and Food Chemistry 68 (47), 13685–13696. Available at https://doi.org/10.1021/acs.jafc.0c03037.

Zare, L., Ronaghi, A., Ghasemi, R., Zarei, M., Sepehri, M., 2022. Alterations in glutathione, phytochelatin and micronutrients of corn plants exposed to cadmium stress at different time periods. Communications in Soil Science and Plant Analysis 54 (9), 1185–1197. Available at https://doi.org/10.1080/00103624.2022.2138909.

Zhang, J., Zhao, H., Chen, L., Lin, J., Wang, Z., Pan, J., et al., 2023. Multifaceted roles of WRKY transcription factors in abiotic stress and flavonoid biosynthesis. Frontiers in Plant Science 14, 1303667. Available at https://doi.org/10.3389/fpls.2023.1303667.

Zhang, K., Sun, Y., Li, M., Long, R., 2021. CrUGT87A1, a UDP-sugar glycosyltransferases (UGTs) gene from Carex rigescens, increases salt tolerance by accumulating flavonoids for antioxidation in Arabidopsis thaliana. Plant Physiology and Biochemistry 159, 28–36. Available at https://doi.org/10.1016/j.plaphy.2020.12.006.

Zhigang, A., Cuijie, L., Yuangang, Z., Yejie, D., Wachter, A., Gromes, R., et al., 2006. Expression of BjMT2, a metallothionein 2 from Brassica juncea, increases copper and cadmium tolerance in Escherichia coli and Arabidopsis thaliana, but inhibits root elongation in Arabidopsis thaliana seedlings. Journal of Experimental Botany 57 (14), 3575–3582. Available at https://doi.org/10.1093/jxb/erl102.

Zhu, S., Shi, W., Jie, Y., 2021. Overexpression of *BnPCS1*, a novel phytochelatin synthase gene from ramie (*Boehmeria nivea*), enhanced Cd tolerance, accumulation, and translocation in *Arabidopsis thaliana*. Frontiers in Plant Science 12, 639189. Available at https://doi.org/10.3389/fpls.2021.639189.

Zuo, Y., Li, B., Guan, S., Jia, J., Xu, X., Zhang, Z., et al., 2022. EuRBG10 involved in indole alkaloids biosynthesis in Eucommia ulmoides induced by drought and salt stresses. Journal of Plant Physiology 278, 153813. Available at https://doi.org/10.1016/j.jplph.2022.153813.

Section C

Approaches for improving multiple stress tolerance in plants

Chapter 11

Conventional and marker-assisted breeding approaches to improve multiple abiotic stress tolerance in plants

Manu Priya[1], Uday Chand Jha[2,*], Harsh Nayyar[3], and Giverson Mupambi[1]
[1]*Umass Cranberry Research Station, East Wareham, MA, United States,* [2]*ICAR - Indian Institute of Pulses Research (IIPR), Kanpur, Uttar Pradesh, India,* [3]*Department of Botany, Panjab University, Chandigarh, Panjab, India*
[*]*Corresponding author. e-mail address: uday.jha@icar.org.in*

11.1 Introduction

The growth and productivity of crop plants have been greatly affected by global climate change, which is a direct result of increased greenhouse gas emissions occurring at an accelerated rate (Nazir et al., 2024). Extreme weather events, such as heat episodes, drought, salinity, and cold, pose a significant and immediate threat to different crops and, consequently, to food security worldwide (Raza et al., 2019). Addressing these challenges requires strategies that enhance crop resilience under adverse environmental conditions (Fig. 11.1). These abiotic stresses impair plant metabolic and physiological functions, leading to considerable yield losses (Ahuja et al., 2010). Plant scientists have long forecasted that elevated temperatures and extended periods of drought will have a greater impact on agriculture compared to low temperatures and salinity (Piao et al., 2019; O'neill et al., 2017; Dahal et al., 2019). Moreover, an increase in the average world temperature may additionally deplete the soil moisture content and accelerate water loss from plants through an enhanced rate of transpiration (Choudhury et al., 2017). Hence adaptation of crop plants to future changing conditions is crucial for meeting the food demands of rapidly growing worldwide populations (Sato et al., 2024). This is an emerging issue that necessitates the collaboration of multiple disciplines to understand stress tolerance processes and develop a strategy, techniques, and specific characteristics to identify varieties that are resistant/tolerant to multiple stress environments (Meena et al., 2017). Approximately 10,000 years ago, plant breeding became a fundamental method for plant domestication by utilizing wild relatives for selection of desirable features through a sustained selection process across multiple generations for crop enhancement (Purugganan and Fuller, 2009; Rai et al., 2021). Numerous significant crops cultivated globally are established through selective breeding processes (Meena et al., 2017). However, in the past 150 years, crop development has been significantly transformed due to modern advanced genetics (Derbyshire et al., 2022). Innovative methodologies established in this decade, including genomic selection, high-throughput phenotyping (HTP), and contemporary speed breeding, have demonstrated the capacity to expedite plant breeding (Becker and Angulo, 2019; Devasirvatham et al., 2016). Various techniques have been employed to reduce the length of plant reproductive cycles, and superior crop varieties have been identified through hybridization and an ongoing screening process (Purugganan and Fuller, 2009; Becker and Angulo, 2019). An essential aspect of basic plant breeding is the integration of genetic diversity through intercrossing agronomic cultivars with wild relatives or landraces, followed by the selection of genotypes exhibiting desirable traits (Devasirvatham et al., 2016; Lavarenne et al., 2018). Despite considerable innovation, conventional breeding methods for crop enhancement are constrained by genetic drag, genetic erosion, hybridization bottlenecks, and a labor-intensive selection process (Abberton et al., 2016; Rai et al., 2021). The development of a crop variety with certain qualities may need 10–20 years, rendering it a complex and time-intensive process (Fischer et al., 2014). Conversely, modern breeding methods like genetic engineering have made substantial advancements to address the limitations of traditional breeding procedures (Rai et al., 2021). Genetic engineering and molecular techniques have contributed to the development of crops with beneficial traits through gene transformation (Singh et al., 2021). Additional techniques, such as large-scale sequencing, genomics, quick gene isolation, and high-throughput molecular markers, have been suggested to enhance the breeding of commercially important crop species

Multiple Abiotic Stresses in Plants. DOI: https://doi.org/10.1016/B978-0-443-27424-4.00007-0

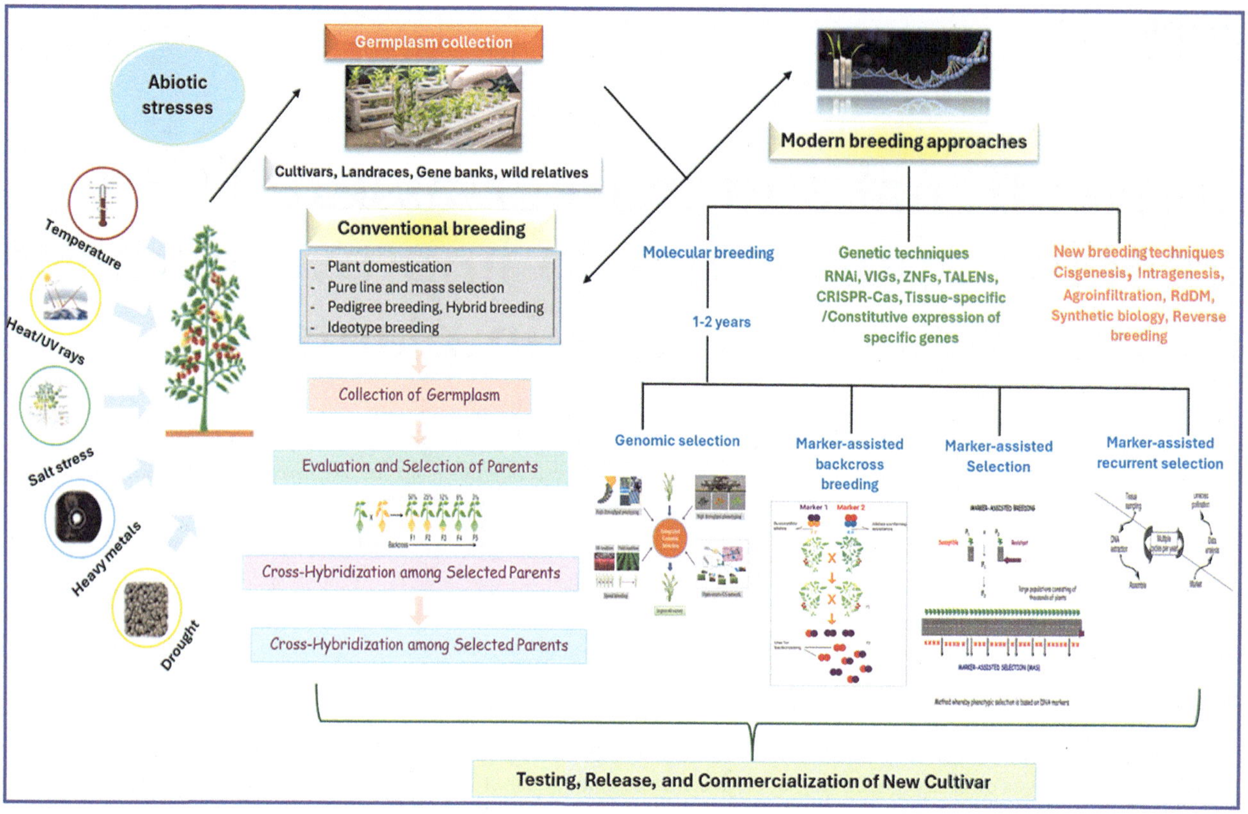

FIGURE 11.1 Conventional and advanced breeding approaches for abiotic stress tolerance in crop plants.

(Devasirvatham et al., 2016; Lavarenne et al., 2018). The primary objective is to cultivate crop types that possess improved resistance to abiotic stressors, while still preserving their high yield potential and nutritional value to ensure the stability of food production in the face of progressively adverse climatic situations (Piao et al., 2019). Considering the above facts, plant breeding has seen a significant change in its approach over time, with the adoption of advanced scientific technologies that are both sustainable and constantly improving (Slafer et al., 2015; Abberton et al., 2016; Rai et al., 2021). Over the past few decades, extensive endeavors have been undertaken to improve different types of crops in challenging agro-climatic conditions (Lateef, 2015). However, there is a need for advanced techniques, such as marker-assisted selection (MAS) combined with a precise phenotyping strategy, to significantly increase agricultural production and productivity (Nair and Pandey, 2024). Molecular breeding tools, such as MAS, provide a method for crop improvement by precisely selecting genotypes that can tolerate harsh conditions (Wani et al., 2018; Younis et al., 2020; Derbyshire et al., 2022). Utilizing molecular markers, such as single-nucleotide polymorphisms (SNPs) and simple sequence repeats (SSRs), allows breeders to more efficiently identify and incorporate stress-tolerance genes (Nair and Pandey, 2024). Further, MAS has expedited the advancement of climate-resilient crops by facilitating the detection and integration of genes of interest for stress tolerance into high-quality cultivars (Joshi et al., 2024). Moreover, progress in genomic and transgenic techniques has broadened the possibilities for enhancing crops' ability to withstand climate challenges (Sivia et al., 2024). Genomic methods, such as genome-wide association studies (GWAS), quantitative trait locus (QTL) mapping, and genomic selection, provide a more comprehensive insight into the genetic structure that underlies stress tolerance (Kumari et al., 2024). These techniques allow breeders to uncover crucial genes, regulatory networks, and molecular markers linked to tolerance of abiotic stresses (Sivia et al., 2024). Considering these facts, conventional and molecular breeding methods, together with genomic and transgenic technologies, provide valuable approaches to increase stress tolerance, mitigate abiotic stressors, and enhance nutritional quality (Kamboj et al., 2020). The continuous progress in these areas presents great potential for ensuring worldwide food security, considering a shifting environment (Slafer et al., 2015; Abberton et al., 2016). Nevertheless, there are still various obstacles that need to be addressed, such as legislative frameworks, public acceptance, and intellectual property rights (Wani et al., 2018). Ongoing study and cooperation among breeders, scientists, policymakers, and stakeholders are essential for fully harnessing the capabilities of breeding and engineering crops that can withstand climate challenges (Jattan et al., 2023; Rao et al., 2016). This chapter will cover marker-assisted breeding and procedures that can be used alongside conventional breeding technologies to generate varieties and hybrids tolerant to abiotic stresses.

11.2 Impact of multiple abiotic stresses in plants

Abiotic stresses such as extreme temperatures, drought, flooding, salinity, and heavy metal contamination are among the most severe constraints to crop growth, productivity, and quality (Kumar, 2020). The inherent ability of plants to withstand these adverse conditions constitutes the foundation of stress resilience and is critical for sustaining stable yields (Sachdev et al., 2021). Therefore accelerating the breeding of stress-tolerant genotypes is essential, with particular focus on traits that enhance tolerance while preserving grain yield, which remains the central objective of most crop improvement programs (Fahad et al., 2017). In addition to direct physiological limitations, abiotic stresses induce secondary oxidative and osmotic imbalances that compromise cellular membranes and impair overall plant performance (Hasanuzzaman et al., 2020; Sachdev et al., 2021).

Different stresses disrupt plant growth through distinct yet overlapping mechanisms. For example, heat stress inhibits photosynthesis by reducing Rubisco activity, decreasing pollen viability, and shortening grain-filling periods, which collectively result in poor seed set and reduced grain weight in crop plants (Crafts-Brandner and Salvucci, 2002; Sita et al., 2018). Conversely, chilling stress hampers germination, root elongation, and membrane stability, while inducing electrolyte leakage and oxidative injury in sensitive crops such as rice, maize, and grapevine (Subramanian et al., 2016; Londo et al., 2018). Drought remains one of the most pervasive constraints, limiting leaf expansion, root activity, and photosynthesis, with yield losses reaching up to 90% in cereals and 40%–60% in legumes during reproductive stages (Araus et al., 2002; Nam et al., 2001). Similarly, salinity imposes osmotic stress and ion toxicity, reducing chlorophyll levels, stunting growth, and lowering yields in crops like maize, cotton, and chickpea (Parida and Das, 2005; Kaya et al., 2018). Heavy metal contamination has comparable impacts, suppressing germination and biomass accumulation while activating stress-response pathways such as proline biosynthesis and antioxidant defenses (Shankar, 2020; Jalmi et al., 2018).

Despite their diverse origins, these abiotic stresses converge on common physiological consequences, including oxidative stress, metabolic disruption, and reproductive impairment, all of which threaten crop productivity and nutritional quality worldwide (Fahad et al., 2017; Shavrukov et al., 2017). Understanding these shared pathways is therefore crucial for breeding, as it provides the foundation for identifying robust traits and molecular markers that can confer multistress tolerance. Integrating conventional breeding with molecular tools and MAS represents a promising route toward developing climate-resilient cultivars capable of maintaining yield under increasingly variable and extreme environments. At present, a major objective of agricultural research is the creation of climate-smart cultivars and the elucidation of their physiological and molecular responses to stress. Advances in genomic technologies, HTP, and precision genotyping are accelerating this process, offering new opportunities to mitigate the inevitable impacts of climate change on global agricultural production (Scheben et al., 2016; Abhinandan et al., 2018; De Pinto et al., 2020).

11.3 Breeding for multiple abiotic stress tolerance in crop plants

11.3.1 Conventional breeding

Traditional breeding has long been employed to develop crop varieties capable of withstanding climatic stress (Kaiser et al., 2020). This approach relies on the deliberate selection and crossing of plants carrying favorable traits, leading to progeny with improved adaptive performance (Lamichhane and Thapa, 2022). Among the principal strategies, selection is the most direct, involving the identification and propagation of individuals that express superior tolerance to drought, heat, salinity, or pathogens in specific environments. Through repeated selection cycles across generations, these beneficial traits can be fixed within breeding populations (Krishna et al., 2023) (Fig. 11.1). Hybridization is another widely used method, where genetically distinct parents are crossed to combine desirable attributes from different lineages. This process generates hybrids that often exhibit heterosis, resulting in enhanced resilience and productivity under stress conditions (Singh et al., 2021; Okada and Whitford, 2019). In parallel, mutation breeding has expanded the available genetic variation by employing chemical or physical mutagens to induce novel alleles. Some of these mutations confer stress tolerance and, once identified, can be integrated into breeding programs (Bhoi et al., 2022; Pandit et al., 2021).

These conventional breeding techniques have contributed significantly to the development of climate-resilient cultivars, enabling farmers to mitigate the negative impacts of abiotic stress factors (Gaba et al., 2021) (Fig. 11.1). The effectiveness of such approaches depends on the genetic variability present in germplasm collections, wild relatives, and landraces, with evaluation typically carried out in field or controlled trials (Qaim, 2020; Gaba et al., 2021). Despite their successes, traditional breeding remains time-consuming, labor-intensive, and expensive (Kaiser et al., 2020). The method also suffers from linkage drag, as beneficial genes are often coinherited with undesirable alleles (Krishna et al., 2023), and multiple cycles may be required to achieve stable improvement (Pandey et al., 2022). In crops with limited genetic diversity, progress can be especially constrained (Andronic, 2022). These limitations highlight the need for integrating more advanced approaches, including molecular and genetic engineering tools, to accelerate the development of stress-tolerant cultivars for climate adaptation (Begna, 2021).

11.3.1.1 Hybridization and selection

Hybridization remains one of the most effective conventional breeding strategies for improving crop tolerance to abiotic stresses (Cassells and Doyle, 2003). It involves the crossing of genetically distinct parents to produce segregating offspring, which are subsequently evaluated under stress conditions, with superior plants advanced for further breeding (Ahloowalia et al., 2004). This approach allows the combination of advantageous traits, such as higher yield potential and resilience, into a single genotype (Cassells and Doyle, 2003) (Fig. 11.1). The efficiency of hybridization is largely determined by parental choice and the robustness of the selection process (Okada and Whitford, 2019). Parents with complementary traits and sufficient genetic diversity are particularly valuable, as they increase the likelihood of generating high-performing recombinants (Singh et al., 2021). Selection methods such as pedigree selection, bulk selection, and single seed descent are routinely applied to identify promising lines under stress environments (Singh et al., 2021). Several studies illustrate the success of hybridization. For instance, Interspecific hybrids were successfully produced from an incompatible cross between *Brassica juncea* and *Brassica tournefortii* via in vitro culture of hybrid embryos (Kumar et al., 2001). The F_2 open-pollinated and BC_1 progenies exhibited significant diversity in morphological traits, leading to the selection of promising plants characterized by reduced aphid infection, drought tolerance, and high yield (Kumar et al., 2001). In a similar way, Li et al. (2019) produced a complete amphiploid (13G819) between emmer wheat and Sea wheatgrass (SWG). Cytogenetic analysis revealed SWG to be an allotetraploid (Li et al., 2019), and the amphiploid displayed superior tolerance to waterlogging, manganese toxicity, salinity, and low nitrogen, and potentially to heat stress, when compared with its wheat parent. These examples demonstrate how hybridization can serve as a powerful tool to create climate-resilient hybrids (Goulet et al., 2017). While conventional hybridization has proven highly valuable, it often remains insufficient to address the increasing demand for stress-tolerant crops (Ahmar et al., 2020). Therefore modern tools such as double haploid production, marker-assisted backcrossing, GWAS, somatic hybridization, and transgenic technologies are now being integrated in parallel with classical approaches to accelerate progress (Okada and Whitford, 2019).

11.3.1.2 Mutation breeding

Mutation breeding utilizes physical or chemical mutagens to provoke random mutations in the DNA of a crop species (Das et al., 2014). The mutants are subsequently evaluated for advantageous characteristics, including improved abiotic stress resilience, and the chosen specimens are utilized as progenitors in breeding initiatives (Hayashi et al., 2007). Mutation breeding has significantly contributed to the development of high-yielding and stress-tolerant crop varieties (Cassells and Doyle, 2003; Parry et al., 2009). Efforts to enhance abiotic stress tolerance in significant crop varieties via mutation breeding have achieved moderate success. Rice seeds irradiated with C or Ne ions have produced a mutant variety exhibiting high salt tolerance (Hayashi et al., 2007). In rice, ion-beam irradiation generated mutants with high salt tolerance (Hayashi et al., 2007), while in another study, the variety Zhefu 802 was derived from gamma irradiation of Simei no. 2 and showed improved cold tolerance (Ahloowalia et al., 2004). Similarly, gamma irradiation produced Golden Promise barley in Europe, a salt-tolerant variety widely used in breeding programs (Kharkwaland Shu, 2009). In Pakistan, cold-tolerant Basmati 370 and the heat-tolerant cotton variety NIAB-78 were released through induced mutagenesis (Awan, 1991; Ahloowalia et al., 2004; Haq, 2009). Other notable examples include barley mutants Bastion, Secret, and Taran with cold tolerance (Shevtsov et al., 2003), Balder J barley with drought tolerance (Kharkwaland Shu, 2009), and Luther and Pennrad barley with lodging resistance (Kharkwaland Shu, 2009). In another study, mutagenesis produced Barberton-B-30-3 groundnut with drought tolerance (Kharkwaland Shu, 2009). The first tomato mutant variety, Maybel, released in Cuba, also showed improved drought performance (Kharkwaland Shu, 2009). Moreover, induced mutagenesis has improved the *Azolla-Anabaena* symbiotic system, generating strains tolerant to salinity, aluminum toxicity, and herbicides, which can serve as biofertilizers for flooded rice (Novak and Brunner, 1992). Tulmann Neto et al. (2001) further demonstrated the role of gamma rays in generating variation in wheat cultivar IAC-24 with improved resilience. Collectively, these studies demonstrate that induced mutagenesis has produced a wide range of climate-resilient cultivars across different crops and regions (Das et al., 2014).

11.3.2 Conventional breeding for heat tolerance

Enhancing crop thermotolerance through conventional breeding represents a key strategy to alleviate the detrimental effects of elevated temperatures on agricultural productivity (Rai et al., 2021). Breeding programs are generally implemented in agro-climatic zones that closely resemble the target production environments, ensuring that selected genotypes are evaluated under conditions analogous to their eventual cultivation sites (Jha et al., 2014). Consequently, the evaluation and selection of breeding lines intended for warmer environments are typically conducted under elevated temperature regimes (Mickelbart et al., 2015). In such regions, traits associated with thermotolerance are often retained through passive selection by local farmers and breeders (Rai et al., 2021). As a result, cultivars originating from hot climates generally display greater inherent heat tolerance compared with those adapted to cooler regions, demonstrating the effectiveness of this long-term selection strategy (Tonsor et al., 2008; Yamamoto et al., 2014). Traditional breeding has also been deliberately employed to generate novel heat-resilient genotypes, where targeted selection, recurrent breeding, and introgression from landraces or wild relatives have facilitated the development of cultivars capable of sustaining growth and yield under high-temperature stress (Rai and Pandey-Rai, 2021). For instance, in Brassica oleracea, an early-maturing broccoli cultivar was developed that maintains head quality by avoiding the heat-sensitive floral initiation stage, thereby mitigating the detrimental effects of elevated temperatures later in the season (Farnham and Bjorkman, 2011). In a similar way, novel cowpea varieties demonstrated higher average grain yields when grown under high temperatures and prolonged photo periods during reproductive development (Ehlers and Hall, 1998). Recurrent selection has also been successfully applied in wheat, where the introgression of *Triticum tauschii* alleles into elite lines improved grain filling and resulted in larger kernels in BC_1F_6 progeny (Gororo et al., 2002). In parallel, potato breeding achieved substantial genetic gains through three cycles of recurrent selection, producing cultivars with enhanced thermotolerance and yield increases of nearly 38% under heat stress (Benites and Pinto, 2011).

Despite such achievements, the genetic and physiological mechanisms underpinning thermotolerance in these conventionally bred cultivars remain poorly characterized (Rai et al., 2021). This lack of mechanistic understanding hinders the development of molecular markers or physiological indices that could streamline germplasm screening and accelerate the breeding of heat-resilient cultivars (Janni et al., 2020). Another limitation arises from the heavy reliance on advanced elite germplasm that has already been extensively utilized in regional breeding programs, which restricts the genetic diversity available for further improvement (Fahad et al., 2017). Consequently, the potential for achieving substantial gains in heat tolerance is constrained by narrow genetic bases (Ladizinsky et al., 1985; Paran and Van Der Knaap, 2007).

Nevertheless, conventional breeding has contributed significantly to the identification of genotypes capable of sustaining growth and yield under elevated temperatures (Jagadish et al., 2008). Screening-based approaches for heat tolerance have proven effective across multiple crop species, with breeding programs in heat-prone environments facilitating the development of cultivars with thermotolerant characteristics (Janni et al., 2020). Understanding the physiological and genetic foundations of heat tolerance remains crucial for guiding future selection efforts (Shanmugavel et al., 2020). For example, evaluation of 64 soybean cultivars identified line 45A46 as highly tolerant to heat stress, based on morphological and physiological performance (Alsajri et al., 2019). In soybeans, conventional strategies such as mass selection, pure-line selection, hybridization, and backcrossing are commonly employed to enhance thermotolerance (Singh and Nelson, 2015). Hybridization in particular has been highly effective, producing numerous heat-tolerant cultivars (Singh and Nelson, 2015; Mammadov et al., 2018). Notable examples include the MG5 line TN09–239, developed through four cycles of backcrossing (Boehm et al., 2017a, 2017b), and PI587982A, a Chinese plant introduction conserved in the USDA-GRIN collection, which has exhibited extended tolerance to heat stress during the seed-filling stage (Smith et al., 2008a, 2008b; Gillman et al., 2021).

11.3.3 Conventional breeding for cold tolerance

Conventional breeding has been fundamental in the development of cold-tolerant cultivars across multiple crop species (Jha et al., 2017). A notable example is the EP80 x Puenteareas maize population, which exhibited strong low-temperature tolerance in both controlled and field trials (Rodríguez Graña et al., 2007). Similarly, Swiss maize landraces displayed superior early vigor when exposed to cold stress, confirming their potential as sources of tolerance (Peter et al., 2006, 2009). In parallel, rice breeding has extensively incorporated cold-tolerant traits from landraces and crop wild relatives into high-yielding yet cold-susceptible varieties, such as Guichao 2 (Liu et al., 2003), Towada (Xu et al., 2008; Zhou et al., 2012a, 2012b), Hitomebore (Shirasawa et al., 2012), HJX74 (Zhang et al., 2004), Norin-PL11 (Kuroki et al., 2007), Norin-PL8 (Saito et al., 1995; Dai et al., 2004), and Milyang 23 (Oh et al., 2004). Similar backcross strategies have also been applied successfully in barley (Casao et al., 2011) and tomato (Vallejos and Tanksley, 1983).

Genetic resources remain indispensable for breeding because they serve as reservoirs of allelic variation and harbor loci associated with stress adaptation (Mickelbart et al., 2015). Important sources of cold tolerance have been documented in japonica rice varieties such as Silewah (Satake and Toriyama, 1979), Koshihikari (Sasaki et al., 1981), Chhomrang (Sthapit and Witcombe, 1998), and Jumli Marshi (Lindlöf et al., 2015). Consistent with this, japonica rice grown in temperate regions generally shows greater cold tolerance than indica types (Glaszmann et al., 1990; Mackill and Lei, 1997). However, a few indica cultivars, such as BR-IRGA 410 and IRGA 416, have also demonstrated resilience to low temperatures (de Cruz and Milach, 2004). Screening based on diverse stress parameters has led to the identification of tolerant rice genotypes, including HSC55, M103, and Jyoudeki, which exhibited reduced spikelet sterility under cold stress (Farrell et al., 2006; Ye et al., 2009). Likewise, NERICA rice genotypes showed higher filled grain ratios during the reproductive phase, ranging between 51.9% and 57.9% (Wainaina et al., 2015), while 12 USDA minicore genotypes were identified based on reduced shoot and root growth under low temperatures (Moraes de Freitas et al., 2016).

Introgression from wild relatives has further expanded the genetic base for cold tolerance. For instance, the genotype Norin PL8, which carries alleles from the landrace Silewah, displayed tolerance during the booting stage (Liu et al., 2003). *Oryza rufipogon* Griff., a wild progenitor of rice, exhibited tolerance during both flowering and booting stages, underlining the potential of crop wild relatives as donors of adaptive traits (Liu et al., 2003). In another study, Chinese rice varieties B55, Bangjiemang, and Lijiangheigu, as well as the Hungarian cultivar HSC55, demonstrated tolerance at multiple developmental stages under field conditions (Ye et al., 2009) (Table 11.1).

Cold tolerance has also been advanced in other cereals and vegetables. In barley, the cultivar OAC Elmira showed high tolerance following traditional breeding efforts (Falk et al., 2001). In wheat, the semi-winter genotype Yannong 19 exhibited enhanced photosynthetic performance under cold stress (Zhang et al., 2022), while significant variation in thousand-kernel weight and yield was observed among more than 600 durum wheat varieties under similar conditions (Mohammadi et al., 2015). The winter wheat cultivar Norstar, carrying the combined alleles Vrn-A1w + Fr-A2T + Fr-B2WT, provides a strong example of frost tolerance across vegetative stages (Eagles et al., 2016). In another comparative study, winter barley genotypes displayed greater tolerance than facultative lines, with Franklin, Amagi Nijo, and Haruna Nijo exhibiting frost resistance at post-heading stages in southern Australia (Kolar et al., 1991; Frederiks et al., 2011).

Together, these examples demonstrate that while conventional breeding has substantially contributed to cold tolerance in diverse crops, its efficiency can be further enhanced through integration with genomic tools that allow precise identification and deployment of stress-resilient alleles (Jha et al., 2017).

11.3.4 Conventional breeding for drought tolerance

Traditional breeding has historically played a critical role in enhancing drought tolerance, a trait essential for maintaining crop productivity under water-limited conditions (Oladosu et al., 2019). While early breeding efforts focused primarily on yield under drought, breeding objectives have gradually shifted toward indirect selection, where secondary traits, such as phenological, morphological, or physiological attributes, are emphasized as indicators of crop performance in stress environments (Rauf et al., 2016; Ali et al., 2017). Conventional breeding enables the exploitation of natural genetic variation for drought tolerance, either within cultivated species or from compatible relatives, which can subsequently be integrated into elite cultivars through diverse mating schemes (Maqbool et al., 2017).

Over the past century, international breeding programs have made notable advances in developing drought-tolerant lines of major food crops (Oladosu et al., 2019). One of the most prominent efforts began in the 1970s at the International Maize and Wheat Improvement Center (CIMMYT), where recurrent selection strategies were applied to maize. These programs achieved consistent yield gains of 59–233 kg ha^{-1} per cycle under drought stress, while simultaneously addressing other limiting factors such as poor nitrogen availability and prevalent diseases (Bänziger et al., 2004). Similarly, a collaborative breeding effort involving the Crops Research Institute (Ghana), IITA (Nigeria), CIMMYT (Mexico), and Sasakawa Global led to the release of the drought-tolerant maize cultivar Obatanpa GH in 2006 (Bänziger et al., 2004).

Genetic diversity remains a prerequisite for the successful breeding of drought-tolerant crops (Ortiz, 2002). In cases where elite germplasm is limited, crop wild relatives serve as valuable reservoirs of adaptive alleles (Mir et al., 2012). CIMMYT exemplified this approach by generating synthetic hexaploid wheat through crosses between durum wheat (*Triticum turgidum* var. durum) and the wild diploid *Aegilops tauschii*. These hybrids, which incorporate the complete D-genome from *A. tauschii*, introduced novel variation for both abiotic and biotic stress tolerance (Valkoun, 2001). Similarly, the International Crops Research Institute for the Semi-Arid Tropics (ICRISAT, India) and the International Center for Agricultural Research in the Dry Areas (ICARDA, Syria) have prioritized dryland cereals (sorghum, millet, and barley) and pulses (lentil, chickpea, pigeonpea, and faba bean) to strengthen food security in arid and semi-arid regions (Valkoun, 2001). ICARDA recently released a barley cultivar with superior drought tolerance, developed through the

TABLE 11.1 Abiotic stress-tolerant cultivars/lines of different crops developed through conventional breeding at different centers and research institutes.

Stress type	Crop	Cultivar/line	Breeding method	Research institute/center involved	References
Heat	Rice	M9962	Crossed with the Phisanulok 2 (PSL2) cultivar	Rice Science Center & Rice Gene Discovery Unit,	Malumpong et al. (2020)
			(the recurrent parent) in a backcross breeding program	Kasetsart University, Thailand	
			Cross of Xieqingzao B / N22 // Xieqingzao		
		'Xieqingzao B	introgression breeding via transfer of HT from 'N22' to '	Key Laboratory of Crop Physiological Ecology and Genetic Breeding,	Liao et al. (2011)
			Xieqingzao B' ine by developing BC1F8 lines	Jiangxi Agricultural University, Nanchang, China	
	Potato	TA209	Backcross inbred lines originated from a cross between	The Robert H. Smith Faculty of Agriculture,	Brog et al. (2019)
			S. pimpinellifolium M82 and S. lycopersicum LA1589	Food and Environment at the Hebrew University of Jerusalem	
	Cotton	Paymaster 404	Backcross breeding with 7456 of G. barbadense	USDA-ARS, Jornada Experimental Range, Las Cruces	Rodriguez-Garay and Barrow (1988)
	Soybean	TN09–239 is an MG5	Backcrossing with 5601 T cultivar	Knoxville, TN, at the East Tennessee Research and Education Center	Boehm et al. (2017a, 2017b)
		PI587982A	Field trials and greenhouse screening	USDA-ARS-CG&PR, Stoneville, MS, USA	Smith et al. (2008a, 2008b)
	Broccoli	Marathon, Gypsy, Arcadia,	Hybrids developed through recurrent selection	U.S. Department of Agriculture, Agricultural Research Service,	Farnham and Bjorkman (2011)
		Packman, USVL HY1,		U.S. Vegetable Laboratory, Charleston, SC, USA	
		USVL, HY5, USVL, HY6			
	Cowpea	California Blackeye No. 27	Backcross breeding	Bean/Cowpea Collaborative Research Support Program (CRSP)	Hall et al. (2003)
			Hybrids developed through recurrent selection	Universidade Federal de Viçosa, in Coimbra-MG	
	Chickpea	ICCV 92944	Screening through field trials	ICRISAT, India	Gaur et al. (2012)
		ICC1205 and ICC15614	Screening through field trials	Plant Breeding Institute, Faculty of Agriculture and Environment,	Devasirvatham et al. (2012, 2013)
	Wheat	PBW550	Backcrossing with Aegilops tauschii	University of Sydney, Cobbitty, Australia	Sehgal et al. (2011)
	Maize	YH-1898, KJ. Surabhi	Hybridization and selection	1Maize & Millets Research Institute Yusafwala-Sahiwal,	Rahman et al. (2015)

(Continued)

TABLE 11.1 (Continued)

Stress type	Crop	Cultivar/line	Breeding method	Research institute/center involved	References
		FH-793 ND-6339, NK -64017		Punjab Pakistan.	
	Tomato	CL1131-0-043-0-6	Hybridization and selection	Asian Vegetable Research and Development Center (AVRDC),	Abdul-Baki (1991)
		CL5915-153D-3-3-0-NSI		Taiwan	
Drought	Chickpea	FLIP87	Developed by crossing IL C3843 with FLIP87	Research in the Dry Areas (ICARDA)	Singh et al. (1997)
	Common	SEA 5	Developed from the interracial	International Center for Tropical	Singh et al. (2001)
	bean		double cross population BAT 477/San	Agriculture (CIAT), Cali, Colombia	
			Cristobal 83/Guanajuato 31/Rio Tibagi		
		SEA 13	Derived from double cross	International Center for Tropical	Singh et al. (2001)
			population BAT 477San Cristobal	Agriculture (CIAT), Cali, Colombia	
			83/BAT 93/Jalo EEP 558		
		A 195	Developed from the single cross Red	Centro Internacional de Agricultura	Singh (2007)
			Kloud ×ICA 10009	Tropical (CIAT), Palmira, Colombia	
		Line CO46348	The complete pedigree of CO46348	Colorado Agricultural Experiment	Brick et al. (2008)
			was unknown, however it was derived	Station in cooperation with the	
			from a single cross with the	University of Idaho and USDA-ARS	
			pinto cultivar Othello		
	Soybean	R01–416F and	Both lines were originated after	Arkansas Agricultural Experiment	Chen et al. (2007)
		R01–581F	selection from F9 population.	Station, USA	
			developed from a cross between		
			Jackson and KS4895		
	Peanut	ICGV 87354	Derived from a cross between	International	Reddy et al. (2001)
			Argentine and PI259747 and	Crops Research Institute for the	
			developed through nine	Semi-Arid Tropics (ICRISAT), India	
			generations of bulk selection		
	Barley	Lenetah	Developed using a pedigree selection	Agricultural Research Service,	Obert et al. (2008)
			procedure with all early generation	Aberdeen, ID, in cooperation	
			population and selected from the	with the Idaho Agricultural	
			cross, 94Ab12981 × 91Ab3148	Experimental Station	

		Giza 2000	The pedigree breeding method was used for development and it was originated from the cross between the Egyptian local cultivar Giza 121 and the line 356/13/1 Giza 117/Bahteem 52//Giza 118/FAO 86	Barley Research Department, Agricultural Research Center at Giza, Egypt	Noaman et al. (2007)
		Giza 121, Line 366/13/1	Plant selection within superior F4 populations	Sakha Research Station Northern Delta Region, Egypt	Noaman et al. (2007)
		Giza 132	Derived from an F3 population. The pedigree method of breeding was used and Giza 132 originated from the cross Rihane-05//As46/Aths/3/Aths/Lignee 686	Barley Research Department, Agricultural Research Center at Giza, Egypt, and International Center for Agricultural Research in the Dry Areas (ICARDA), Aleppo Syria	Noaman et al. (2007)
	Wheat	Willow Creek	Through breeding in single replication observation (SROB) nurseries	Montana Agricultural Experiment Station, Sydney	Cash et al. (2009)
		Ripper	Developed by using a modified bulk breeding procedure	Colorado Agricultural Experiment Station, USA	Haley et al. (2007)
		NE01643	A bulk breeding procedure was used and approximately 50% of F3 population was visually selected on the basis of agronomic appearance	Nebraska Agricultural Experiment Station and the USDA-ARS	Baenziger et al. (2008)
		Prairie Red	Derived from the crosses and backcrosses of CO850034/PI372129/5 TAM 107	Colorado Agricultural Experiment Station, USA	Quick et al. (2001)
		Jinmai 50	Developed from the cross, Pingyang181 ×Qingfeng1	Wheat Breeding Innovation Group (WBIG) in the Cotton Research Institute of Shanxi Agri. Sci. Academy, Yuncheng, China and Testing and Appraising Committee of Crop Cultivars of Shanxi	Xinglai et al. (2006)

(Continued)

TABLE 11.1 (Continued)

Stress type	Crop	Cultivar/line	Breeding method	Research institute/center involved	References
				Province (TACCCSP)	
Cold stress	Rice	Norin PL 8	Introgression breeding from tolerant cultivar Silewah	Hokkaido National Agricultural Experiment Station	Saito et al. (1995)
		B55, Bangjiemang,	Hybridization and selection	Yunnan Academy of Agricultural Sciences	Ye et al. (2009)
		Lijiangheigu,			
		and the Hungarian			
	Barley	OAC Elmira			Falk et al. (1996)
Salt stress	Wheat	S24, KRL1–4	Hybridization and selection	Department of Plant Sciences, College of Agriculture,	Ashraf and O'leary (1996)
				University of Arizona, Tucson	Hollington (2000)
	Rice	Bicol	Hybridization and selection	Plant Breeding, Genetics, and Biochemistry Division, Philippines	Senadhira et al. (2002)
				International Rice Research Institute	
		KDML105	Marker-assisted backcrossing	Rice Gene Discovery Unit, National, Thailand	Punyawaew et al. (2016)
				Center for Genetic Engineering and Biotechnology	
				(BIOTEC, NSTDA),	
	Tomato	Walter	Crossing and Selection	Department of Land, Air and Water Resources,	Rush and Epstein (1981)
				University of California, Davis, USA	
		F1 hybrid M82 ×	Crossing and Selection	Department of Field & Vegetable Crops, Faculty of Agriculture,	
		L. peruvianum LA-716		Hebrew University of Jerusalem	Saranga et al. (1993)
Heavy	Rice	Parag 401	Hybridization of Prabhavati and Basmati rice		Patil et al. (1997)
metals	Wheat	Strongfield	Hybridization and selection	SPARC, AAFC, Saskatchewan, Canada	Clarke et al. (2005)
		Beniswelf-7, Gimmiza-11	Hybridization and selection	Department of Agronomy, Faculty of Agriculture,	Mourad et al. (2021)
				Assiut University, Egypt	
				USDA-ARS, United States	

crossing of a Palestinian landrace with a wild barley line. This variety showed a 50% yield advantage over conventional cultivars under arid conditions (Carter et al., 2019).

Parallel efforts in rice improvement illustrate the versatility of conventional breeding. In Indonesia, mutation breeding generated a "super green" rice line that combined high yield with drought tolerance under low-input conditions (Efendi et al., 2017). In Malaysia, the drought-resilient lines MR219-9 and MR219-4 were derived from the widely cultivated variety MR219 through conventional selection (Abdul et al., 2012). These cases highlight how both induced variation and recurrent selection approaches have been integrated into rice breeding pipelines.

Collectively, a wide array of drought-tolerant cultivars and breeding lines, documented in journals such as Crop Science, testify to the success of conventional strategies in strengthening abiotic stress resilience (Rauf et al., 2016). These efforts have contributed substantially to sustaining crop yields in marginal environments, while simultaneously improving quality traits and resource-use efficiency (Ali et al., 2017). However, conventional breeding for drought tolerance is hindered by several challenges, particularly the simultaneous presence of multiple abiotic stresses such as elevated temperatures, intense solar radiation, and nutrient imbalances, which interact in complex ways and obscure the accurate expression and selection of drought-resilient traits (Mohammadi, 2017). These limitations underscore the need to integrate molecular and genomic tools with traditional approaches to accelerate genetic gains for drought tolerance.

11.3.5 Conventional breeding for salt tolerance

Despite significant advancements in agricultural production and quality through conventional breeding in the 20th century, there has been very little effort to enhance crop tolerance to abiotic challenges, particularly salinity stress (Ashraf and Akram, 2009). This may be mostly attributable to plant breeders prioritizing enhancements in crop production and quality above the development of stress tolerance (Schubert et al., 2009). Despite limited emphasis on stress tolerance in early breeding programs, plant breeders have conducted extensive efforts over the past century to exploit genetic variation at intraspecific, interspecific, and intergeneric levels for the development of salt-tolerant cultivars (Afzal et al., 2023). Through these approaches, several salt-tolerant lines across diverse crops have been successfully generated using conventional methods (Jattan et al., 2023). Importantly, most of these lines were tested under natural field conditions, ensuring their adaptability and practical relevance (Ashraf and Akram, 2009). For example, multiple alfalfa (*Medicago sativa L.*) lines, including AZ-90NDC-ST (Johnson et al., 1991), AZ-Germ Salt II (Dobrenz et al., 1993), AZ-97MEC and AZ-97MEC-ST (Al-Doss and Smith, 1998), and ZS-9491 and ZS-9592 (Dobrenz, 1999), were evaluated directly under saline environments. In tomato, attempts were made to introgress salt tolerance from *Lycopersicon pimpinellifolium* into the cultivated species *L. esculentum* (Asins et al., 1993); however, the resulting F_1 hybrids largely resembled the sensitive parent, highlighting the complexity of transferring this trait.

Although breeding for salinity tolerance has been attempted in several crops, most progress has been achieved in rice and wheat. Between 2007 and 2014, 27 salt-tolerant rice cultivars were released in Bangladesh, India, and the Philippines (Ismail and Horie, 2017), while 14 wheat cultivars with improved salt tolerance were developed in South Asia, Egypt, and China (Naeem et al., 2020). Among these, the bread wheat (*Triticum aestivum L.*) cultivars S24 (Ashraf and O'leary, 1996) and KRL1–4 (Hollington et al., 2002) were specifically evaluated on naturally salt-affected soils. Nearly all conventional methods, including introduction, selection, hybridization, mutation, and shuttle breeding, have been applied in salinity improvement (Schubert et al., 2009). For instance, early salt-tolerant rice cultivars such as Damodar (CSR 1), Dasal (CSR 2), and Getu (CSR 3) were developed via pure-line selection from traditional Sundarbans landraces (Arzani and Ashraf, 2016). Subsequent breeding cycles produced improved CSR varieties (e.g., CSR 10, CSR 13, CSR 23, CSR 27, CSR 30, CSR 36) through pedigree-based recombination, modified bulk pedigree, and shuttle breeding techniques (Reddy et al., 2014). Advances also include the discovery of the *Saltol* QTL, which regulates Na^+/K^+ homeostasis in rice seedlings (Thomson et al., 2010), and its subsequent deployment via marker-assisted backcrossing into elite genetic backgrounds (Hasan et al., 2022). Mutation breeding has also been applied, as demonstrated by the EMS-derived salt-tolerant rice variety *Kaijin* (Takagi et al., 2015).

These examples highlight both the achievements and limitations of conventional breeding for salinity tolerance. Because this approach relies heavily on available genetic variation, its success often depends on the presence of stress-tolerant germplasm (Singer et al., 2021). Crop wild relatives, in particular, remain valuable genetic reservoirs for broadening the base of tolerance (Turan et al., 2012). Looking forward, it is increasingly evident that conventional breeding alone cannot deliver sufficient genetic gains for complex, multigenic traits such as salinity tolerance. A more integrated strategy combining genomics, molecular markers, transcriptomics, proteomics, and metabolomics studies will be essential to identify and deploy key loci underlying salt tolerance in order to develop resilient, high-yielding cultivars (Fita et al., 2015).

11.3.6 Conventional breeding for heavy metal tolerance

Plant breeding facilitates the incorporation of adaptive traits that enhance a crop's capacity to withstand adverse environmental conditions, including both biotic and abiotic stress factors (Farooq et al., 2016). Heavy metal toxicity constitutes a substantial limitation on plant growth and development, as discussed earlier (Das and Jayalekshmy, 2015). Creating crops with enhanced tolerance to heavy metals may serve as an effective approach to address the persistent issue of heavy metal toxicity in agriculture (Shaw, 2022). The high-quality grain variety Parag 401 was created through hybridization of Prabhavati and Basmati using anther culture to confer resilience to Fe chlorosis in rice (Patil et al., 1997). Tolerant brown rice varieties (BR-14, BR-26, and BRRI-29) have been produced and commercialized in Bangladesh (Khan et al., 2016). Further, 16 rice types have been assessed in a hydroponic setting for copper tolerance (Lin et al., 2017). Four cultivars, namely Xiushui 123, Xianghu 301, Xiushui 134, and Jiahe 991, exhibited improved root dehydrogenase activity, root length, and dry weight under copper stress and have been suggested for local cultivation (Lin et al., 2017). In a separate study, the rice variety BRRI-29, developed by the Bangladesh Rice Research Institute (BRRI), demonstrated superior performance across multiple traits, recording the highest biomass (0.071 g), maximum germination rate (82%), and longest root length (90 cm) among the evaluated genotypes (Halim et al., 2014). A durum wheat cultivar named Strongfield is highly favored in Canada due to high-grain protein concentration, elevated yield, and less accumulation of cadmium in grains (Clarke, 2008) (Table 11.1). Similarly, Pirsabak 2004, a lead-tolerant wheat variety, was identified based on epicotyl length, root length, and germination rate (Shafiq et al., 2019). Further, based on observed kernel features, 103 spring wheat genotypes were evaluated for lead and tin tolerance by Mourad et al. (2021). Among all genotypes, Benisweif-7 exhibited significant resistance to Pb, while the Gimmiza-11 genotype showed significant tolerance to Sn and may be utilized as a parent in various breeding initiatives (Mourad et al., 2021) (Table 11.1). The selection and breeding of cultivars with enhanced genetic diversity have been proposed as an effective strategy to mitigate metal accumulation in food crops (Grant et al., 2008). Notably, breeding programs in China, the United States, and Canada have reported success in developing low-accumulating varieties, including rice, hybrid sunflower (*Helianthus laetiflorus*), and durum wheat (*Triticum durum*) (Jena and Mackill, 2008; Miller et al., 2006). The breeding program established in North Dakota has yielded lines with cadmium content significantly lower than the average for sunflower (Miller et al., 2006). The utilization of these lines as progenitors resulted in hybrids exhibiting a 50% reduction in kernel cadmium content. A breeding effort was initiated in Canada in 1991 to diminish the cadmium concentration in Canadian durum wheat (Clarke et al., 2006). The crossing procedure generated near-isogenic lines with high and low grain cadmium concentrations from five durum wheat crosses (Clarke et al., 2006).

The selection of cultivars through plant breeding offers a more pragmatic and viable option for cultivating crops in soils with elevated bioavailability of heavy metals (Farooq et al., 2016). However, traditional breeding methods have shown partial success in the development of heavy metal-tolerant varieties as this approach is expensive, requires significant work, and demands considerable time (Grant et al., 2008).

11.4 Marker-assisted breeding

The field of crop development has been transformed using molecular breeding techniques, which allow breeders to choose to target desired traits at the molecular level (Wani et al., 2018) (Fig. 11.1). These strategies utilize the understanding of plant genomes and the discovery of genes linked to stress tolerance (Joshi et al., 2024). Notable progress in molecular breeding involves the utilization of molecular markers (Wani et al., 2018). These markers are distinct areas in the genome that can be readily identified and areas associated with specific qualities of interest (Kumari et al., 2024). MAS enables breeders to quickly discover and choose plants with desired stress-tolerance traits (Rai and Pandey-Rai, 2021; Sivia et al., 2024). Breeders can use the identification of genetic markers linked to stress tolerance to make reliable predictions about the performance of possible progeny (Tyagi et al., 2024). This allows them to pick the most promising individuals for further breeding. Molecular breeding involves the application of genetic resources, including wild crop relatives and landraces (Kumari et al., 2024). These genetic resources contain valuable genes that can enhance the ability of various crop plants to tolerate stress (Wani et al., 2018). By utilizing both traditional breeding methods and molecular markers, breeders can transfer stress-tolerance genes from these genetic resources into superior cultivars (Sivia et al., 2024; Joshi et al., 2024) (Fig. 11.1). This increases the range of genetic diversity that may be used in breeding programs, which in turn improves the possibility of generating crops that are adaptable to climate change (Henkrar and Udupa, 2020). MAS is a molecular technique that involves selecting parental breeding lines based on a selection index derived from genotypic data (Kumari et al., 2024). MAS has addressed the challenges faced in traditional plant breeding by enhancing the effectiveness and accuracy of selection (Collard and Mackill, 2008) (Fig. 11.1). Conventional breeding procedures necessitate multiple or repetitive breeding cycles to amalgamate desired features in a certain genotype or lineages (Babu et al., 2004). On the other hand, MAS could combine desired features in a single genotype more effectively and accurately, requiring fewer selection cycles and resulting in fewer unintended losses (Babu et al., 2004). This leads to a reduction in time and resources while increasing the production capacity of crops. Marker-assisted selection (MAS) offers a potential strategy to enhance agricultural productivity and mitigate damage caused by both biotic and abiotic stressors (Rai and Pandey-Rai, 2021). Using MAS, breeding programs have observed a twofold increase in the rate of genetic improvement compared to phenotypic selection (Younis et al., 2020). This improvement has been observed in numerous variables, including yield, tolerance to biotic and abiotic stresses, and qualitative attributes (Joshi et al., 2024). For MAS to be practical and successful, there must be a strong correlation and/or close linkage between the molecular markers and the gene of interest (Younis et al., 2020). Additionally, the markers used should be stable, repeatable, and easy to assess, as stated by Yu et al. (2004). MAS has proven to be highly efficient at identifying, monitoring, preserving, merging, and stacking various beneficial genes, such as those responsible for disease resistance (O'Boyle et al., 2007). A list of various crops developed through marker-assisted breeding is included in Table 11.1.

11.4.1 Molecular markers

SSRs or microsatellites are the markers most utilized in plant breeding. SSR polymorphism is a common occurrence in eukaryotes and indicates variation in the quantity of repeated units in a certain area of the genome (Kumari et al., 2024). In addition to SSRs, other methods such as restriction fragment length polymorphism (RFLP) and random amplified polymorphic DNA (RAPD) can be utilized (Younis et al., 2020). A recently developed efficient marker is an SNP that represents a single nucleotide variation among two DNA sequences or individuals (Chugh et al., 2022). They represent the most basic and definitive type of molecular markers, as a single nucleotide base is the smallest unit of heredity (Nadeem et al., 2018).

11.4.2 Marker-assisted selection for abiotic stress tolerance

Abiotic stress tolerance is a multifaceted characteristic regulated by numerous genes/QTLs and is influenced by significant environmental factors (Miklas et al., 2006). The identification and research of abiotic stress resilient/tolerant genes is challenging and complex, both from a physiological and genetic perspective (Liu et al., 2007). Certain quantitative traits can be precisely analyzed using molecular marker technologies through QTL mapping, allowing for the focused selection of individual loci in MAS. QTL mapping investigations are necessary before using MAS to combine multiple desirable QTLs for resistance to multiple abiotic stresses (Witcombe et al., 2008).

11.4.2.1 MAS for heat tolerance

Conventional breeding approaches have historically been employed to study the inheritance of heat tolerance in plants (Wahid et al., 2007; Farooq et al., 2011), but the discovery of DNA markers and the development of high-resolution genotyping assays have considerably advanced the identification of QTLs and genomic regions associated with thermotolerance in major crops (Pinto et al., 2010; Bonneau et al., 2013). In chickpea, composite interval mapping identified 77 QTLs, including 37 major and 40 minor, associated with 12 out of 13 heat-responsive traits, while significant loci were also detected for physiological parameters such as chlorophyll content, nitrogen balance index, NDVI, and cell membrane stability (Jha et al., 2021). Candidate genes within these regions included heat shock proteins (HSP), heat shock transcription factors, and regulators of flowering and pollen development, highlighting the complex genetic architecture of heat tolerance (Jha et al., 2021). More recently, an integrated genomics-assisted strategy using QTL-seq, haplotyping, and association mapping revealed the critical role of natural alleles of *CaHSFA5* in regulating ROS homeostasis and thermotolerance in chickpea (Mohanty et al., 2024). Meta-analysis of 138 reported QTLs further identified six meta-QTLs on CaLG01 and CaLG06 harboring genes such as *CaPRK3*, *CaFPF1*, *CaFLC*, *CaHsfsA5*, and *CaLRXs*, which are involved in flowering, pollen germination, and yield stability under high temperatures (Kumar et al., 2023).

Parallel studies in cereals have reported similar progress. In wheat, 15 consistent QTL clusters for heat response were mapped, with candidate genes including *TaWRKY74-B* and *TaSnRK3.15-B*, which function as regulators of thermotolerance at the seedling stage (Fu et al., 2024). Earlier mapping with RFLP markers detected QTLs on chromosomes 1, 4, and 7 associated with tolerance during grain filling, explaining 8.9–17.3% of the phenotypic variance (Chang-lan et al., 2005), while additional loci linked to spikelet fertility were reported on chromosomes 1, 2, 3, 8, 10, and 11 (Jagadish et al., 2010). In rice, the dominant locus *OsHTAS* from genotype HT54 was shown to confer tolerance up to 48°C during the seedling and grain-filling phases (Wei et al., 2013). Other significant loci, including *qHTSF1.1* and *qHTSF4.1* on chromosomes 1 and 4, were detected in IR64 x N22 populations (Ye et al., 2012), and introgression line studies identified further QTLs such as *qHTS1-1* and *qHTS3* derived from *Oryza rufipogon* (Lei et al., 2013; Lie et al., 2013).

Several agronomic and physiological traits related to yield stability have also been genetically dissected under terminal heat stress. In wheat, QTLs controlling stay-green and senescence traits were located on chromosomes 7A, 6A, 6B, 2A, 3A, and 3B (Vijayalakshmi et al., 2010), while heat susceptibility index and temperature depression traits were used to detect 14 loci associated with heat resilience (Esten Mason et al., 2011; Bennett et al., 2012; Bonneau et al., 2013). Other studies reported major QTLs on 2B, 7B, and 7D explaining up to 20% variation (Paliwal et al., 2012), a locus on 4A accounting for 17% variance in yield under heat (Pinto et al., 2010), and SNP markers capable of distinguishing tolerant from sensitive wheat cultivars (Garg et al., 2012).

Beyond wheat and rice, QTL mapping has contributed to understanding heat response in other crops. In maize, loci regulating pollen germination, pollen tube growth, and membrane stability under stress were identified in RIL populations (Ottaviano et al., 1991; Frova and Sari-Gorla, 1994). In sorghum, the stay-green QTLs (*Stg1–4*) were localized on linkage groups A, D, and J, with evidence that *Stg1* and *Stg2* co-localize with HSP and photosynthetic genes, thereby linking drought and heat tolerance mechanisms (Xu et al., 2000). In cowpea, SNP-based analysis identified five genomic regions associated with thermotolerance in an IT82E-18 × CB27 RIL population (Lucas et al., 2013a). In tomato, several QTLs related to fruit set under heat were mapped (Grilli et al., 2007; Xiang-yang et al., 2008), while RAPD and AFLP marker-based studies in *Brassica campestris* identified five loci associated with heat resilience (Shuancang et al., 2003).

Taken together, these findings illustrate the progress of QTL mapping and molecular dissection of thermotolerance across multiple crops. The identification of stable loci, candidate genes, and diagnostic markers demonstrates the potential of MAS to accelerate the introgression of heat-tolerant alleles into elite germplasm (Jha et al., 2014). Furthermore, fine-mapping and cloning of QTLs, combined with advanced molecular breeding strategies, are expected to provide robust tools for deploying thermotolerance in breeding pipelines and thereby enhancing crop resilience to global warming (Driedonks et al., 2016).

11.4.2.2 Marker-assisted selection for cold tolerance

Low-temperature stress is regarded as the primary abiotic factor restricting plant growth and potential agricultural development (Adhikari et al., 2022). Adapting crops to temperature constraints is a crucial breeding goal, as it influences production stability in environmentally sustainable agricultural techniques (Frascaroli, 2018). Traditional breeding techniques have achieved minimal success in enhancing the cold tolerance of significant crop species due to the intricacy of stress tolerance characteristics, insufficient genetic variability, and the absence of effective selection criteria (Adhikari et al., 2022). Contemporary breakthroughs in physiology, genetics, and DNA technology have significantly enhanced the ability of breeders to forecast the breeding value of optimal genotypes through the utilization of physiological, genetic, and molecular data (Adhikari et al., 2022). The approach to choosing cold-tolerant crops more successfully will encompass efficient genotyping and dependable phenotyping together with appropriate statistical models (Takahashi et al., 2013; Shinada et al., 2014). Rice crops are highly susceptible to low temperatures and experience significant growth retardation from seed germination to seed filling till maturity (Mori et al., 2011; Takahashi et al., 2013). The RAPD marker OPT8.511 has been discovered to be strongly associated with cold tolerance in rice (Kim et al., 2000). In their study, Chen and Li (2005) established a strong association between an SSR marker, RM202, and a QTL named *Qsct-11* in the rice line "Lemont." This association is relevant for the application of MAS in enhancing cold tolerance (Chen and Li, 2005). Three QTLs that regulate cold tolerance at the fertilization stage were discovered in backcrossed inbred lines obtained through a hybrid between the Suisei (weak CTF) and Eikei 88223 (high CTF) rice cultivars (Shinada et al., 2014). Twelve QTLs following treatment at 4 °C were identified for the seedling survival rate (SR) in rice (Li et al., 2021). Among these, seven QTLs (*qSR2–1, qSR3–4, qSR3–5, qSR3–6, qSR3–7, qSR4*, and *qSR7*) were reported for the first time, while five (*qSR2–2, qSR3–1, qSR3–2, qSR3–3*, and *qSR9*) were colocalized with those from prior studies (Li et al., 2021). In another similar study, a significant QTL (*qCTS6*) regulating the SR at low temperatures of 9°C for 10 days was identified in rice within a 2.60-Mb area on chromosome 6 by Yang et al. (2021). The QTLs discovered for cold resilience can be effectively employed in rice breeding programs and MAS for tolerant rice genotypes (Pan et al., 2023). Similarly, in winter wheat, cold tolerance is a crucial agronomic characteristic that determines the capability of the crop to withstand temperatures below the freezing point (Bolouri et al., 2023). Revilla et al. (2016) identified 47 flint inbred lines that had six advantageous QTLs for winter hardiness. Using B73× Mo17 (IBM) Syn4 as a reference, Hu et al. (2016) identified six QTLs that regulate the rate of low temperature germination in maize on chromosomes 9, 7, 6, 5, and 4through RIL populations examination for low temperature germination ability. Six QTLs regulating primary root length at low temperature were identified on chromosomes 9, 6, 5, and 4 (Hu et al., 2016). Notably, four sets of QTL were discovered to be situated at the same chromosomal location and together regulated the pace of germination and the length of main roots when subjected to cold stress (Hu et al., 2016). A significant effector QTL, *qCT6.1*, regulating cold tolerance was consistently identified on chromosome 6 by the cold injury index in cucumber (Li et al., 2022). Sixteen QTLs associated with two cold-tolerance indices pertinent to overwintering success were identified across the two populations of winter rapeseed (*Brassica napus*) (Liu et al., 2022). Two significant QTLs, *qOWRTA07 and qOWRLA07*, accounting for over 10% of the phenotypic variation, were identified in overlapping regions containing gene *BnaA07G0198300ZS* for overwintering rate (Liu et al., 2022). In alfalfa, three QTLs, *dft1, dft2, dft3, dft4*, and *dft5*, were identified for cold tolerance (Adhikari et al., 2022). It has been proposed that cold tolerance/resistance is usually a quantitative trait regulated by multiple genes (Liu et al., 2022). Similarly, in lentil QTL analysis, utilizing a 130-point linkage map identified a significant QTL on linkage group 6, accounting for 20.45% of the phenotypic variation in leaf area (Khoshro and Lotfi, 2021). The QTL for leaf area was located in the same region as one of the QTL for winter hardiness, which was identified between leaf area and winter hardiness (Khoshro and Lotfi, 2021). The results demonstrated a correlation between winter hardiness and leaf area, offering insights relevant to lentil breeding (Khoshro and Lotfi, 2021). A chromosome segment substitution line population derived from wild soybean ZYD00006 and cultivated soybean SN14 was utilized to map three QTLs for cold tolerance in soybean (Zheng et al., 2023). Further, five candidate genes were identified through gene annotation, GO enrichment analysis, and protein function prediction (Zheng et al., 2023). Among those genes, *Glyma.09G162700* was found to be a potential candidate gene associated with cold stress tolerance during the germination time of soybean (Zheng et al., 2023).

11.4.2.3 Marker-assisted selection for drought tolerance

Drought remains the primary constraint on agricultural yield in rain-fed systems, particularly in the semi-arid tropics (Sandhu et al., 2019). The major challenge is to reduce variability in production across seasons and improve food security for resource-poor farmers in these vast regions (Wani et al., 2020). Drought tolerance is typically defined as the capacity of a crop to maintain stable yields in environments prone to water scarcity (Serraj et al., 2005). Despite decades of breeding efforts, the identification of robust traits conferring drought resilience and their successful introgression into high-yielding genetic backgrounds has met with only limited success (Barik et al., 2020). More recently, systems analysis using simulation models has been proposed to assess the potential benefits of incorporating drought tolerance traits in targeted environments (Asati et al., 2022). Complementarily, QTL mapping has proven useful for dissecting the genetic and physiological basis of traits linked to drought resilience (Sandhu et al., 2019).

In rice, multiple drought-associated QTLs have been consistently mapped across diverse environments and genetic backgrounds (Barik et al., 2020). A recent study reported nine QTLs distributed across chromosomes 2, 6, 7, 8, 9, and 10, associated with traits such as plant height (*qPH10.1*), effective panicle number (*qEPN6.1*), panicle length (*qPL9.1*), thousand-grain weight (*qTGW2.1*, *qTGW6.1*, *qTGW8.1*), leaf length (*qLL7.1*), leaf width (*qLW7.1*), and leaf area (*qLA7.1*) (Yi et al., 2023). Another analysis identified three QTLs related to physiological attributes, including relative chlorophyll content (*qRCC1.1*), chlorophyll a (*qCHLa1.1*), and proline concentration (*qPRO3.1*), with *qPRO3.1* highlighted as a novel locus explaining 78.19% of phenotypic variation (Barik et al., 2020). Additional QTLs for plant height (*QPH-4-1*, *QPH-8-1*) and for grain yield (*QGY-6-1*, *QGY-7-1*) have been identified, with the latter representing stable loci of potential value for MAS under climate-induced water stress (Paul et al., 2023).

In upland cotton, a QTL hotspot for drought tolerance was identified on chromosome 8, harboring loci for chlorophyll stability index and other stress-associated traits (Shukla et al., 2021). In wheat, two meta-QTLs, MQTL4 and MQTL12, were found to play a central role in drought resilience and marker-assisted breeding (Kumar et al., 2020). Maize mapping using the B73 RefGen_v2 physical map revealed four key chromosomal regions containing multiple QTLs, such as *qWS-GY3-1*, *qWS-ESP3-1*, *qWW-GY3-1*, and *qWW-ASI3-1*, which together explained up to 8.94% of phenotypic variance and originated from the tolerant donor Han 21 (Hu et al., 2021). In chickpea, populations derived from interspecific crosses between drought-tolerant and sensitive accessions have been extensively studied. For instance, 187 F8 recombinant inbred lines from *Cicer arietinum* (GPF 2) × *C. reticulatum* (ILWC 292) were screened for 21 morphological, physiological, and root traits under both rainfed and irrigated conditions, enabling the discovery of new drought-linked loci (Kushwah et al., 2022). Similarly, QTLs *qRWC1.1* and *qYLD7.1* were mapped in RILs derived from a cross between Pusa 362 and SBD 377 using SNP genotyping, demonstrating the effectiveness of high-resolution approaches for trait dissection (Yadava et al., 2023).

In soybean, QTLs associated with plant height (*qPH2*, *qPH6*, *qPH7*, *qPH17*, *qPH19-1*, *qPH19-2*, *qPH19-3*) and seed weight (*qSWPP2*, *qSWPP6*, *qSWPP13*, *qSWPP17*, *qSWPP19*) were identified across multiple chromosomes (Ren et al., 2020). Pigeonpea, recognized for its inherent drought resilience, has also been investigated at the genomic scale. Analysis of 73% of the pigeonpea genome revealed drought-specific genes (Varshney et al., 2012; Choudhary et al., 2015), and interspecific populations generated from crosses such as ICPL 87 × ICPW 94 and ICPL 8755 × ICPL 227 have been used to map drought-related loci (Pazhamala et al., 2015). In tomato, QTLs linked to seed germination under water deficit were mapped to chromosomes 12, 9, 8, and 1 in backcross populations involving *L. esculentum and L. pimpinellifolium* (Foolad et al., 2003).

Collectively, these studies demonstrate that QTL mapping has provided a foundation for understanding the genetic control of drought adaptation across cereals, legumes, oilseeds, and vegetables. While progress has been significant, ongoing efforts aim to refine these discoveries, validate loci across environments, and integrate them into marker-assisted and genomic selection pipelines to accelerate the development of drought-resilient cultivars (Wani et al., 2020).

11.4.2.4 Marker-assisted selection for salt tolerance

Over the past two decades, extensive research across a wide range of crop species has sought to identify QTLs that regulate plant responses to salinity stress at multiple developmental stages, including germination, seedling establishment, vegetative growth, and reproduction (Ashraf and Foolad, 2013). Salt tolerance is now recognized as a highly complex trait governed by networks of genes that act at both physiological and molecular levels (Fan et al., 2016). With the advent of molecular markers and MAS, a large number of investigations have aimed to dissect salinity-related traits and map genomic regions controlling ion homeostasis, growth maintenance, and stress-induced biochemical adjustments in crop plants (Alpuerto et al., 2009). Despite these advances, however, the translation of marker discoveries into the release of salt-tolerant cultivars through MAS has been limited, highlighting the challenges of deploying this technology in practical breeding pipelines (Witcombe et al., 2008).

A considerable number of QTLs associated with salt stress responses have been mapped across cereals, legumes, and vegetables, and these loci explain variation in growth-related, physiological, and biochemical traits expressed under salinity (Fan et al., 2016). For example, in soybean, two major loci, *qST6* on chromosome 6 controlling ion toxicity, and *qST10* on chromosome 10 influencing physiological performance under salinity, were reported as key determinants of tolerance (Cho et al., 2021). In sorghum, six loci (*qTB6, qSFW9, qJW9, qBrix2, qBrix10, qSTI-Brix9*) were characterized as major-effect QTLs, each explaining more than 10% of the phenotypic variance and thus representing valuable targets for MAS-based improvement (Wang et al., 2020). Similarly, Pundir et al. (2021) identified *qGY-2* for grain yield on chromosome 2, as well as loci for stress tolerance and salt susceptibility indices, the latter explaining up to 45% of the phenotypic variation in sorghum. In zombi pea, newly discovered loci *qSaltol_3.1* and *qSaltol_7.1* have been highlighted as unique contributors to salinity tolerance (Laosatit et al., 2024).

In rice, genome-wide association studies identified 20 SNPs significantly correlated with the Na^+/K^+ ratio, a central physiological determinant of salt tolerance (Kumar et al., 2015). The well-characterized *Saltol* locus remains one of the most widely studied genomic regions for salinity tolerance in rice, regulating both reduced Na^+ accumulation and a favorable K^+/Na^+ balance during the seedling stage (Ali et al., 2013; Krishnamurthy et al., 2020). Comparable QTL discoveries have also been reported in barley, where a salinity tolerance locus was found linked to a DArT marker on chromosome 4 (Fan et al., 2016). Evidence from wheat, barley, and rice further indicates that seedling-stage tolerance to salinity is underpinned by both additive and epistatic QTL effects, underscoring the polygenic and environment-dependent nature of this trait (Genc et al., 2010; Wang et al., 2012; Nguyen et al., 2013; Ahmadi-Ochtapeh et al., 2015). Five QTLs associated with salt tolerance are located on the H chromosomes of barley at locations 1, 2, 5, 6, and 7. These QTLs contribute to about fifty percent of the variation observed during the seedling stage (Zhou et al., 2012a, 2012b). Furthermore, biparental population mapping and other mapping methods for linkage were extensively investigated to identify several genomic areas in barley that include potential genes associated with tolerance (Nguyen et al., 2013). Although significant progress has been made in identifying the primary QTL for tolerance to salt, their identification in molecular breeding is not fully understood. Nevertheless, it is feasible to integrate substantial adaptive characteristics in the genomic age with a comparatively lesser QTL impact (Mirdar Mansuri et al., 2020). Nevertheless, the use of the integrative metaanalysis approach resulted in the identification of numerous fascinating genes associated with salt resilience, including crucial genes/gene families delineated by substantial evidence.

11.4.2.5 Marker-assisted selection for heavy metal toxicity

Heavy metals exert toxic effects on plants at physiological, biochemical, and molecular levels, either directly or indirectly through the generation of reactive oxygen species and free radicals (Haider et al., 2023). To mitigate these effects, different approaches have been adopted, including soil amendments with fertilizers, selection of naturally tolerant genotypes, and transgenic interventions (Mansoor et al., 2023). Advances in plant genetics and genomics have highlighted that elucidating the molecular basis of heavy metal tolerance, along with mapping QTLs, provides a pathway for breeding cultivars with improved tolerance or reduced metal accumulation (Gantait and Mitra, 2019). Such QTLs can serve as valuable tools for MAS to generate low-accumulating or metal-tolerant crop varieties (Gupta and Praveen, 2019).

In rice, GWAS identified 35 QTLs strongly correlated with cadmium (Cd) accumulation, among which *qCd1–3* showed the highest significance and was linked to the candidate gene *OsABCB24*, whose upregulation was associated with reduced Cd accumulation (Pan et al., 2020). Another QTL, *qCDSL1*, was identified at the seedling stage and shown to contribute to cadmium tolerance (Shilin et al., 2021). In barley, two major QTLs explained up to 47.24% and 38.59% of phenotypic variance for Cd tolerance, while 16 candidate genes related to Cd detoxification were also reported (Derakhshani et al., 2020). Further investigations in barley revealed QTLs linked to manganese stress, including *QSur.yf.1H*, *QSur.yf.3H*, *QSur.yf.4H*, and *QSur.yf.6H*, which controlled survival under Mn^{2+} stress, and *QLC.yf.3H* and *QLC.yf.6H*, associated with leaf chlorosis (Huang et al., 2018).

For arsenic (As) tolerance, marker-trait association studies in rice identified nine QTLs, including *qAsS9.1*, *qAsS9.2*, *qAsS6*, *qAsS5.1*, *qAsS5.2*, and *qAsS2* for shoot accumulation, *qAsR8.1* and *qAsR8.2* for root-specific traits, and *qRChlo1* linked to relative chlorophyll content (Murugaiyan et al., 2019). These intervals harbored 25 candidate genes with regulatory roles in As tolerance. Other rice studies reported four QTLs: *qAsTSL8*, *qAsTRL8*, *qAsTRSB8*, and *qAsTSL12* that contributed significantly to arsenic resilience (Syed et al., 2016). In addition, the *Alt12.1* locus was associated with aluminum (Al) resistance in rice, where *ART1* was identified as the major candidate gene (Arbelaez et al., 2017).

In wheat, large-scale mapping studies identified 79 QTLs for Al toxicity, with 22 loci explaining 4.38%–12.24% of phenotypic variance (Farokhzadeh et al., 2020b). Subsequent investigations reported 40 loci linked to Al stress tolerance and 58 QTLs related to spikelet survival under Al toxicity, demonstrating the complex genetic basis of Al resilience in wheat (Farokhzadeh et al., 2020a, 2020b). In common bean, eight QTLs associated with Al resistance were identified for traits such as root length, shoot dry weight, and root dry weight, with phenotypic variance ranging from 7.6% to 14.7% (Njobvu et al., 2020). In rice, QTLs associated with the accumulation of Fe, Zn, Hg, Cu, and As were reported, emphasizing the multimetal response of the crop (Lu et al., 2020). In tomato, 103 QTLs were mapped, including loci for Fe accumulation in fruits, leaves, and yield components, with candidate genes encoding xylem sap proteins implicated in modulating iron homeostasis and fruit quality (Asins et al., 2020). Collectively, these findings confirm that QTL discovery for heavy metal tolerance has expanded across cereals, legumes, and vegetables, providing crucial genetic resources for marker-assisted breeding to mitigate heavy metal toxicity in crops.

11.5 Combined abiotic stress tolerance through marker-assisted selection

In natural field environments, crops are rarely exposed to a single stress factor but often experience combinations such as drought with heat or salinity with waterlogging. These combined stresses impose synergistic damage on crop productivity by disrupting photosynthesis, oxidative balance, and reproductive processes (Shabbir et al., 2022; Zandalinas et al., 2022; Khan et al., 2025). While conventional breeding has made progress against individual stress factors, the polygenic and complex nature of combined stress demands more precise strategies (Derbyshire et al., 2022). MAS provides such an avenue by enabling the identification, pyramiding, and deployment of QTL associated with tolerance to multiple stress factors (Kumari et al., 2024; Joshi et al., 2024).

In wheat, several studies have identified genomic regions conferring combined stress tolerance. Tahmasebi et al. (2016) reported a stable QTL on chromosome 7D-b that explained up to 19.6% of yield variation across drought, heat, and combined stress trials, making it an excellent candidate for MAS due to its stability and contribution to grain yield. More recently, Shamloo-Dashtpagerdi et al. (2024) integrated meta-QTL and transcriptomic analyses to identify 118 MQTL and 690 shared meta-DEGs under drought and salinity, highlighting ERD15 as a central regulator of shared stress responses. Similarly, Manjunath et al. (2024) mapped 35 QTL across 14 chromosomes under combined drought and heat stress in a recombinant inbred population, identifying eight stable QTL and hotspots on 6B and 5B linked to TGW, NDVI, and plant height. Taria et al. (2025) further highlighted stem density as an adaptive trait by identifying ninc QTL, of which seven were major and expressed specifically under combined drought and heat, linked to carbohydrate remobilization during grain filling. Together, these studies provide robust genomic targets that can be exploited through MAS and genomic selection to improve stress-resilient wheat cultivars.

Barley also offers strong examples of MQTL-based integration across stresses. Zhang et al. (2017) identified MQTL2H.1, which aggregated 21 major QTL for drought, salinity, waterlogging, and combined salinity–waterlogging tolerance. MQTL1H.4 was particularly notable as it contained both the salinity-tolerance gene HvNax4 and drought-related QTL, allowing simultaneous improvement for dual stresses. Fan et al. (2015) further demonstrated that QTL for proline accumulation under drought (QPC-D.TxFr.3H) and salinity (QPC-S.TxFr.3H) co-localized on chromosome 3H, pointing to a shared genetic basis for osmotic adjustment. These MQTLs offer promising opportunities for pyramiding tolerance to multiple stresses via MAS, with clear physiological relevance.

In maize, Sheoran et al. (2022) summarized 542 QTL from 33 studies and refined them into 32 meta-QTL across diverse backgrounds. Several loci, including MQTL2.1, 5.1, 7.1, and 9.1, were associated with traits contributing to combined abiotic stress tolerance. Candidate genes in these regions included transcription factor families such as ERF,

NAM, ATAF1-2 and CUC2, bZIP, WRKY, and HSP, highlighting regulatory hubs that can be targeted in MAS pipelines. Importantly, these MQTL were validated across multiple genetic backgrounds and environments, making them highly reliable for breeding climate-resilient maize cultivars.

Rice breeding has demonstrated the power of QTL pyramiding in achieving combined stress tolerance. Dixit et al. (2017) combined three drought QTL (qDTY3.1, qDTY6.1, qDTY6.2) with SUB1 for submergence tolerance in TDK1 through marker-assisted backcrossing coupled with phenotypic selection. The resulting NILs maintained stable yield under both drought and submergence, with qDTY3.1 consistently showing the largest effect. Similarly, marker-assisted introgression of drought and submergence QTLs into Swarna produced lines resilient to both stresses (Sandhu et al., 2019). With the availability of large-effect drought QTLs and ongoing identification of heat-responsive loci, rice breeding is now moving toward pyramiding drought and heat QTLs, an approach that has shown significant potential for improving performance under combined stress (Shanmugavadivel et al., 2019; Prakash et al., 2019).

In legumes, MQTL studies are rapidly advancing and provide new opportunities for MAS. In pea, a meta-analysis of 594 QTL identified 17 MQTL, of which seven were validated through GWAS and found to be associated with drought, heat, and yield traits (Jha et al., 2025). Specific SNPs within MQTL regions, such as M-QTL-2.1 and 2.2, were linked to combined drought and heat tolerance (Tafesse et al., 2021). These refined MQTL regions reduce confidence intervals and increase precision, offering breeders reliable targets for improving pea resilience under combined stress. In chickpea, Panigrahi et al. (2024) compiled 1512 QTL for drought, heat, cold, and salinity tolerance, refining them into 59 MQTL across eight chromosomes, with 23 validated by GWAS. Several MQTL, such as CaMQAST1.1, CaMQAST4.4, and CaMQAST7.7, encompassed stress-responsive genes, including HSP and hormone regulators, making them valuable for MAS. Sharma et al. (2025) further refined 712 QTL into 54 MQTL with narrow confidence intervals, identifying 11 "breeder's MQTL" that explained >10% phenotypic variance and overlapped with GWAS hits. Candidate gene mining revealed 49 stress-related genes, while 23 ortho-MQTL regions shared with pea and pigeon pea suggested translational potential across legumes. These advances make chickpea one of the most promising models for pyramiding tolerance to multiple stresses using MAS and genomic selection.

Taken together, these findings demonstrate that MAS, especially when combined with meta-QTL analysis and genomic selection, provides powerful opportunities to pyramid QTL for multiple abiotic stresses. However, most current studies focus on individual stress factors, and reports on marker-assisted breeding for *combined* stress tolerance are still limited. While promising QTL and MQTL regions have been identified in wheat, rice, maize, barley, and legumes, their direct deployment into breeding pipelines remains scarce. Future efforts must therefore prioritize stress combinations that occur naturally in the field, supported by dedicated MAS studies to validate and pyramid QTL for simultaneous drought, heat, salinity, and waterlogging tolerance. Exploiting the inherent variation in crop germplasm under such combined conditions will be critical to identify novel resistance sources and accelerate the development of climate-resilient cultivars adapted to complex stress environments.

11.6 Conclusion and future prospects

Abiotic stresses pose a significant threat to global food security, with projections indicating an escalation of these impacts in the future. Numerous research findings indicate that abiotic stress conditions have a detrimental impact on the growth, development, physiology, and yield of plants. Traditional plant breeding has significantly contributed to enhancing crop quality and output, as well as improving tolerance to abiotic stresses such as heat, drought, salinity, and heavy metal toxicity. However, producing varieties or hybrids that are resilient to abiotic stress still requires substantial innovation and refinement. Meeting the increasing global food demand with limited natural resources will necessitate a dual approach that integrates conventional and molecular breeding strategies. Conventional selection, hybridization, and mutagenesis remain indispensable for capturing natural genetic variation, whereas molecular tools, such as MAS, provide the precision required to introgress stress-tolerant traits while maintaining desirable agronomic performance. Combining these methods offers a powerful pathway to generate genotypes that sustain productivity under adverse conditions.

Future breeding programs must be guided by comprehensive knowledge of genomics, genetic diversity, germplasm management, and advanced statistical approaches, alongside strong experimental design and HTP capabilities. Expanding the exploration of wild relatives, landraces, and underutilized crops with unique allelic combinations will be essential for broadening the genetic base of stress tolerance. At the same time, genomic selection and GWAS, coupled with rapid genotyping technologies, provide powerful frameworks for accelerating the identification and deployment of beneficial loci in breeding pipelines. Looking forward, the development of climate-resilient "sustainable crops" will depend on integrating traditional breeding with next-generation tools such as CRISPR/Cas9-mediated genome editing, genomic selection, and multiomics-driven trait discovery. These approaches, when combined with precision agriculture and sustainable management practices, hold the potential to mitigate the adverse effects of climate change and ensure food and nutritional security for the growing global population.

References

Abberton, M., Batley, J., Bentley, A., Bryant, J., Cai, H., Cockram, J., et al., 2016. Global agricultural intensification during climate change: a role for genomics. Plant Biotechnology Journal 14 (4), 1095–1098.

Abdul, R.H., Kamarudin, S., Bhuiyan, M.A. R., Narimah, M.K., Wickneswari, R., Abdullah, M.Z., et al., 2012. Evaluation and characterization of advanced rice mutant line of rice (Oryza sativa), MR219–4 and MR219–9 under drought condition. In R and D Seminar 2012: Research and Development Seminar 2012, Malaysia, pp. 1–21.

Abdul-Baki, A.A., 1991. Tolerance of tomato cultivars and selected germplasm to heat. Stress 116 (6), 1113–1116.

Abhinandan, K., Skori, L., Stanic, M., Hickerson, N.M., Jamshed, M., Samuel, M.A., 2018. Abiotic stress signaling in wheat–an inclusive overview of hormonal interactions during abiotic stress responses in wheat. Frontiers in Plant Science 9, 734.

Adhikari, L., Baral, R., Paudel, D., Min, D., Makaju, S.O., Poudel, H.P., et al., 2022. Cold stress in plants: strategies to improve cold tolerance in forage species. Plant Stress 4, 100081.

Afzal, M., Alghamdi, S.S., Khan, M.A., Al-Faifi, S.A., Rahman, M.H.U., 2023. Transcriptomic analysis reveals candidate genes associated with salinity Stress tolerance during the early vegetative stage in fababean genotype, Hassawi-2. Scientific Reports 13 (1), 21223.

Ahloowalia, B.S., Maluszynski, M., Nichterlein, K., 2004. Global impact of mutation-derived varieties. Euphytica 135, 187–204.

Ahmadi-Ochtapeh, H., Soltanloo, H., Ramezanpour, S.S., Naghavi, M.R., Nikkhah, H.R., Yoosefi Rad, S., 2015. QTL mapping for salt tolerance in barley at seedling growth stage. Biologia Plantarum 59, 283–290.

Ahmar, S., Gill, R.A., Jung, K.H., Faheem, A., Qasim, M.U., Mubeen, M., et al., 2020. Conventional and molecular techniques from simple breeding to speed breeding in crop plants: recent advances and future outlook. International Journal of Molecular Sciences 21 (7), 2590.

Ahuja, I., de Vos, R.C., Bones, A.M., Hall, R.D., 2010. Plant molecular stress responses face climate change. Trends in Plant Science 15 (12), 664–674.

Al-Doss, A.A., Smith, S.E., 1998. Registration of AZ-97MEC and AZ-97MEC-ST very nondormant alfalfa germplasm pools with increased shoot weight and differential response to saline irrigation. Crop Science 38, 568.

Ali, Z., Khan, D., Ahmed, N., 2013. Physiological parameters of salt tolerance in three cultivars of Sorghum bicolor (L.) Moench. at seedling stage under single salt (NaCl) salinity. International Journal of Biology and Biotechnology 10 (1), 125–142.

Ali, F., Bano, A., Fazal, A., 2017. Recent methods of drought stress tolerance in plants. Plant Growth Regulation 82, 363–375.

Alpuerto, V.L.E.B., Norton, G.W., Alwang, J., Ismail, A.M., 2009. Economic impact analysis of marker-assisted breeding for tolerance to salinity and phosphorous deficiency in rice. Applied Economic Perspectives and Policy 31 (4), 779–792.

Alsajri, F.A., Singh, B., Wijewardana, C., Irby, J.T., Gao, W., Reddy, K.R., 2019. Evaluating soybean cultivars for low-and high-temperature tolerance during the seedling growth stage. Agronomy 9 (1), 13.

Andronic, L., 2022. Conventional plant breeding and biotechnological tools for a resilient system of plant improvement. In Biotehnologiiavansate–realizărişi perspective, pp. 123-125. 10.53040/abap6.2022.41

Araus, J.L., Slafer, G.A., Reynolds, M.P., Royo, C., 2002. Plant breeding and drought in C3 cereals: what should we breed for? Annals of Botany 89 (7), 925–940.

Arbelaez, J.D., Maron, L.G., Jobe, T.O., Piñeros, M.A., Famoso, A.N., Rebelo, A.R., et al., 2017. Aluminum resistance transcription factor 1 (ART 1) contributes to natural variation in aluminum resistance in diverse genetic backgrounds of rice (O. sativa). Plant Direct 1 (4), e00014.

Arzani, A., Ashraf, M., 2016. Smart engineering of genetic resources for enhanced salinity tolerance in crop plants. Critical Reviews in Plant Sciences 35 (3), 146–189.

Asati, R., Tripathi, M.K., Tiwari, S., Yadav, R.K., Tripathi, N., 2022. Molecular breeding and drought tolerance in chickpea. Life 12 (11), 1846.

Ashraf, M., Akram, N.A., 2009. Improving salinity tolerance of plants through conventional breeding and genetic engineering: an analytical comparison. Biotechnology Advances 27 (6), 744–752.

Ashraf, M., Foolad, M.R., 2013. Crop breeding for salt tolerance in the era of molecular markers and marker-assisted selection. Plant Breeding 132 (1), 10–20.

Ashraf, M., O'leary, J.W., 1996. Responses of some newly developed salt-tolerant genotypes of spring wheat to salt Stress: 1. Yield components and ion distribution. Journal of Agronomy and Crop Science 176 (2), 91–101.

Asins, M.J., Breto, M.P., Carbonell, E.A., 1993. Salt tolerance in Lycopersicon species. II. Genetic effects and a search for associated traits. Theoretical and applied Genetics 86, 769–774.

Asins, M.J., Raga, M.V., Torrent, D., Roca, D., Carbonell, E.A., 2020. QTL and candidate gene analyses of rootstock-mediated tomato fruit yield and quality traits under low iron Stress. Euphytica 216, 1–19.

Awan, M.A., 1991. Use of induced mutations for crop improvement in Pakistan. Plant Mutation Breeding for Crop Improvement 1, 67–72.

Babu, R., Nair, S.K., Prasanna, B.M., Gupta, H.S., 2004. Integrating marker-assisted selection in crop breeding–prospects and challenges. Current Science 607–619.

Baenziger, P.S., Beecher, B., Graybosch, R.A., Ibrahim, A.M.H., Baltensperger, D.D., Nelson, L.A., et al., 2008. Registration of 'NE01643'wheat. Journal of Plant Registrations 2 (1), 36–42.

Bänziger, M., Setimela, P.S., Hodson, D., Vivek, B., 2004, September. Breeding for improved drought tolerance in maize adapted to southern Africa. In New Directions for a Diverse Planet: Proceedings of the 4th International Crop Science Congress, Brisbane, Australia (Vol. 26).

Barik, S.R., Pandit, E., Mohanty, S.P., Nayak, D.K., Pradhan, S.K., 2020. Genetic mapping of physiological traits associated with terminal stage drought tolerance in rice. BMC Genetics 21, 1–12.

Becker, M., Angulo, C., 2019. The evolution of lowland rice-based production systems in Asia: historic trends, determinants of change, future perspective. Advances in Agronomy 157, 293–327.

Begna, T., 2021. Global role of plant breeding in tackling climate change. International Journal of Agricultural Science and Food Technology 7, 223–229.

Benites, F.R.G., Pinto, C.A.B.P., 2011. Genetic gains for heat tolerance in potato in three cycles of recurrent selection. Crop Breeding and Applied Biotechnology 11, 133–140.

Bennett, D., Reynolds, M., Mullan, D., Izanloo, A., Kuchel, H., Langridge, P., et al., 2012. Detection of two major grain yield QTL in bread wheat (*Triticum aestivum* L.) under heat, drought and high yield potential environments. Theoretical and Applied Genetics 125, 1473–1485.

Bhoi, A., Yadu, B., Chandra, J., Keshavkant, S., 2022. Mutagenesis: a coherent technique to develop biotic stress resistant plants. Plant Stress 3, 100053.

Boehm Jr, J.D., Ibba, M.I., Kiszonas, A.M., See, D.R., Skinner, D.Z., Morris, C.F., 2017a. Identification of genotyping-by-sequencing sequence tags associated with milling performance and end-use quality traits in hard red spring wheat (*Triticum aestivum* L.). Journal of Cereal Science 77, 73–83.

Boehm Jr, J.D., Walker, F.R., Bhandari, H.S., Kopsell, D., Pantalone, V.R., 2017b. Seed inorganic phosphorus stability and agronomic performance of two low-phytate soybean lines evaluated across six southeastern US environments. Crop Science 57 (5), 2555–2563.

Bolouri, P., Haliloğlu, K., Mohammadi, S.A., Türkoğlu, A., İlhan, E., Niedbała, G., et al., 2023. Identification of novel QTLs associated with frost tolerance in winter wheat (*Triticum Aestivum* L.). Plants 12 (8), 1641.

Bonneau, J., Taylor, J., Parent, B., Bennett, D., Reynolds, M., Feuillet, C., et al., 2013. Multi-environment analysis and improved mapping of a yield-related QTL on chromosome 3B of wheat. Theoretical and Applied Genetics 126, 747–761.

Brick, M.A., Ogg, J.B., Singh, S.P., Schwartz, H.F., Johnson, J.J., Pastor-Corrales, M.A., 2008. Registration of drought-tolerant, rust-resistant, high-yielding pinto bean germplasm line CO46348. Journal of Plant Registrations 2 (2), 120–124.

Brog, Y.M., Osorio, S., Yichie, Y., Alseekh, S., Bensal, E., Kochevenko, A., et al., 2019. A Solanum neorickii introgression population providing a powerful complement to the extensively characterized *Solanum pennellii* population. The Plant Journal 97 (2), 391–403.

Carter, A.Y., Hawes, M.C., Ottman, M.J., 2019. Drought-tolerant barley: I. Field observations of growth and development. Agronomy 9 (5), 221.

Casao, M.C., Igartua, E., Karsai, I., Bhat, P.R., Cuadrado, N., Gracia, M.P., et al., 2011. Introgression of an intermediate VRNH1 allele in barley (Hordeum vulgare L.) leads to reduced vernalization requirement without affecting freezing tolerance. Molecular Breeding 28, 475–484.

Cash, S.D., Bruckner, P.L., Wichman, D.M., Kephart, K.D., Berg, J.E., Hybner, R., et al., 2009. Registration of 'Willow Creek' forage wheat. Journal of Plant Registrations 3 (2), 185–190.

Cassells, A.C., Doyle, B.M., 2003. Genetic engineering and mutation breeding for tolerance to abiotic and biotic stresses: science, Technology and safety. Bulgarian Journal of Plant Physiology 52–82.

Chang-lan, Z.H.U., Ying-hui, X.I.A.O., Chun-ming, W.A.N.G., Ling, J.I.A.N.G., Hu-Qu, Z., Jian-min, W.A.N., 2005. Mapping QTL for heat-tolerance at grain filling stage in rice. Rice Science 12 (1), 33.

Chen, P., Sneller, C.H., Purcell, L.C., Sinclair, T.R., King, C.A., Ishibashi, T., 2007. Registration of soybean germplasm lines R01–416F and R01–581F for improved yield and nitrogen fixation under drought Stress. Journal of Plant Registrations 1, 166–167.

Chen Wei, C.W., Li Wei, L.W., 2005. Mapping of QTL conferring cold tolerance at the early seedling stage in rice by molecular markers. Plant Science 23, 116–120.

Cho, K.H., Kim, M.Y., Kwon, H., Yang, X., Lee, S.H., 2021. Novel QTL identification and candidate gene analysis for enhancing salt tolerance in soybean (*Glycine max* (L.) Merr.). Plant Science 313, 111085.

Choudhary, A.K., Sultana, R., Ontagodi, T., Singh, I.S., Bhatt, B.P., 2015, December. Recent advances in breeding pigeonpea [Cajanus cajan (L.) Millsp.]. In Proc. Natl Conf Global Res Initiatives for Sustainable Agriculture and Allied Sciences, pp. 12-13.

Choudhury, F.K., Rivero, R.M., Blumwald, E., Mittler, R., 2017. Reactive oxygen species, abiotic stress and stress combination. The Plant Journal 90 (5), 856–867.

Chugh, V., Kaur, D., Purwar, S., Kaushik, P., Sharma, V., Kumar, H., et al., 2022. Applications of molecular markers for developing abiotic-stress-resilient oilseed crops. Life 13 (1), 88.

Clarke, J.M., 2008. Genetic improvement of Canadian durum wheat for sustainable production systems. Options Méditerranéennes. Série A: SéminairesMéditerranéens (CIHEAM) 81.

Clarke, F.R., Clarke, J.M., DePauw, R.M., Fernandez, M.R., Fox, S., Gilbert, J., et al., 2005. Strategic approach to mitigating weather induced defects of wheat quality. Euphytica 143, 285–290.

Clarke, J.M., McCaig, T.N., DePauw, R.M., Knox, R.E., 2006. Registration ofStrongfield'durum wheat. Crop Science 46 (5), 2306.

Collard, B.C., Mackill, D.J., 2008. Marker-assisted selection: an approach for precision plant breeding in the twenty-first century. Philosophical Transactions of the Royal Society B: Biological Sciences 363 (1491), 557–572.

Crafts-Brandner, S.J., Salvucci, M.E., 2002. Sensitivity of photosynthesis in a C4 plant, maize, to heat stress. Plant Physiology 129 (4), 1773–1780.

Cruz, R.P.D., Milach, S.C.K., 2004. Cold tolerance at the germination stage of rice: methods of evaluation and characterization of genotypes. Scientia Agricola 61, 1–8.

Dahal, K., Li, X.Q., Tai, H., Creelman, A., Bizimungu, B., 2019. Improving potato stress tolerance and tuber yield under a climate change scenario–a current overview. Frontiers in Plant Science 10, 563.

Dai, L., Lin, X., Ye, C., Ise, K., Saito, K., Kato, A., et al., 2004. Identification of quantitative trait loci controlling cold tolerance at the reproductive stage in Yunnan landrace of rice, Kunmingxiaobaigu. Breeding Science 54 (3), 253–258.

Das, R., Jayalekshmy, V.G., 2015. Mechanism of heavy metal tolerance and improvement of tolerance in crop plants. Journal of Global Biosciences 4 (7), 2678–2698.

Das, P., Mishra, M., Lakra, N., Singla-Pareek, S.L., Pareek, A., 2014. Mutation breeding: a powerful approach for obtaining abiotic stress tolerant crops and upgrading food security for human nutrition. Mutagenesis: Exploring Novel Genes and Pathways. Wageningen Academic Publishers, pp. 615–621.

de Freitas, G.P.M., Basu, S., Ramegowda, V., Braga, E.B., Pereira, A., 2016. Comparative analysis of gene expression in response to cold stress in diverse rice genotypes. Biochemical and Biophysical Research Communications 471 (1), 253–259.

De Pinto, A., Cenacchi, N., Kwon, H.Y., Koo, J., Dunston, S., 2020. Climate smart agriculture and global food-crop production. PLoS One 15 (4), e0231764.

Derakhshani, B., Jafary, H., Maleki Zanjani, B., Hasanpur, K., Mishina, K., Tanaka, T., et al., 2020. Combined QTL mapping and RNA-Seq profiling reveals candidate genes associated with cadmium tolerance in barley. PLoS One 15 (4), e0230820.

Derbyshire, M.C., Batley, J., Edwards, D., 2022. Use of multiple 'omics techniques to accelerate the breeding of abiotic stress tolerant crops. Current Plant Biology 32, 100262.

Devasirvatham, V., Tan, D.K.Y., Gaur, P.M., Raju, T.N., Trethowan, R.M., 2012. High temperature tolerance in chickpea and its implications for plant improvement. Crop and Pasture Science 63 (5), 419–428.

Devasirvatham, V., Gaur, P.M., Mallikarjuna, N., Raju, T.N., Trethowan, R.M., Tan, D.K., 2013. Reproductive biology of chickpea response to heat stress in the field is associated with the performance in controlled environments. Field Crops Research 142, 9–19.

Devasirvatham, V., Tan, D.K., Trethowan, R.M., 2016. Breeding strategies for enhanced plant tolerance to heat stress. Advances in Plant Breeding Strategies: Agronomic, Abiotic and Biotic Stress Traits 447–469.

Dixit, S., Singh, A., Sandhu, N., et al., 2017. Combining drought and submergence tolerance in rice: marker-assisted breeding and QTL combination effects. Molecular Breeding 37, 143.

Dobrenz, A.K., 1999. Salt-Tolerant Alfalfa. Agripro Seeds. Inc, Shawnee Mission, KS, US.

Dobrenz, A.K., Smith, S.E., Poteet, D., Miller, W.B., 1993. Carbohydrates in alfalfa seed developed for salt tolerance during germination. Agronomy Journal 85 (4), 834–836.

Driedonks, N., Rieu, I., Vriezen, W.H., 2016. Breeding for plant heat tolerance at vegetative and reproductive stages. Plant Reproduction 29, 67–79.

Eagles, H.A., Wilson, J., Cane, K., Vallance, N., Eastwood, R.F., Kuchel, H., et al., 2016. Frost-tolerance genes Fr-A2 and Fr-B2 in Australian wheat and their effects on days to heading and grain yield in lower rainfall environments in southern Australia. Crop and Pasture Science 67 (2), 119–127.

Efendi, B., Sabaruddin, Z., Lukman, H., 2017. Mutation with gamma rays irradiation to assemble green super rice tolerant to drought stress and high yield rice (*Oryza sativa* L.). International Journal of Advanced Science, Engineering and Technology 5, 1–5.

Ehlers, J.D., Hall, A.E., 1998. Heat tolerance of contrasting cowpea lines in short and long days. Field Crops Research 55 (1-2), 11–21.

Esten Mason, R., Mondal, S., Beecher, F.W., Hays, D.B., 2011. Genetic loci linking improved heat tolerance in wheat (Triticum aestivum L.) to lower leaf and spike temperatures under controlled conditions. Euphytica 180, 181–194.

Fahad, S., Bajwa, A.A., Nazir, U., Anjum, S.A., Farooq, A., Zohaib, A., et al., 2017. Crop production under drought and heat stress: plant responses and management options. Frontiers in Plant Science 8, 1147.

Falk, D.E., Meatherall, G., Rossnagel, B.G., 1996. Codac barley. Canadian Journal of Plant Science 76 (4), 799–800.

Falk, A., Jansson, C., Schulman, A.H., Rasmussen, S.K., 2001. Barley mutagenesis. Progress in Botany: Genetics Physiology Systematics Ecology 34–50.

Fan, Y., Shabala, S., Ma, Y., Xu, R., Zhou, M., 2015. Using QTL mapping to investigate the relationships between abiotic stress tolerance (drought and salinity) and agronomic and physiological traits. BMC Genomics 16 (1), 43.

Fan, Y., Zhou, G., Shabala, S., Chen, Z.H., Cai, S., Li, C., et al., 2016. Genome-wide association study reveals a new QTL for salinity tolerance in barley (*Hordeum vulgare* L.). Frontiers in Plant Science 7, 946.

Farnham, M.W., Bjorkman, T., 2011. Breeding vegetables adapted to high temperatures: a case study with broccoli. HortScience 46 (8), 1093–1097.

Farokhzadeh, S., Fakheri, B.A., Mahdi Nezhad, N., Tahmasebi, S., Mirsoleimani, A., 2020a. Genetic dissection of spike-related traits in wheat (*Triticum aestivum* L.) under aluminum stress. Genetic Resources and Crop Evolution 67 (5), 1221–1243.

Farokhzadeh, S., Fakheri, B.A., Nezhad, N.M., Tahmasebi, S., Mirsoleimani, A., Lynne McIntyre, C.L., 2020b. Genetic control of some plant growth characteristics of bread wheat (*Triticum aestivum* L.) under aluminum stress. Genes & Genomics 42, 245–261.

Farooq, M., Bramley, H., Palta, J.A., Siddique, K.H., 2011. Heat stress in wheat during reproductive and grain-filling phases. Critical Reviews in Plant Sciences 30 (6), 491–507.

Farooq, M.A., Ali, B., Gill, R.A., Islam, F., Cui, P., Zhou, W., 2016. Breeding oil crops for sustainable production: heavy metal tolerance. Breeding Oilseed Crops for Sustainable Production. Academic Press, pp. 19–31.

Farrell, T.C., Fox, K.M., Williams, R.L., Fukai, S., 2006. Genotypic variation for cold tolerance during reproductive development in rice: screening with cold air and cold water. Field Crops Research 98 (2–3), 178–194.

Fischer, T.C., Malnoy, M., Hofmann, T., Schwab, W., Palmieri, L., Wehrens, R., et al., 2014. F 1 hybrid of cultivated apple (Malus× domestica) and European pear (Pyrus communis) with fertile F 2 offspring. Molecular Breeding 34, 817–828.

Fita, A., Rodríguez-Burruezo, A., Boscaiu, M., Prohens, J., Vicente, O., 2015. Breeding and domesticating crops adapted to drought and salinity: a new paradigm for increasing food production. Frontiers in Plant Science 6, 978.

Foolad, M.R., Subbiah, P., Kramer, C., Hargrave, G., Lin, G.Y., 2003. Genetic relationships among cold, salt and drought tolerance during seed germination in an interspecific cross of tomato. Euphytica 130, 199–206.

Frascaroli, E., 2018. Breeding cold-tolerant crops. In: Wani, S., Herath, V. (Eds.), Cold Tolerance in Plants: Physiological, Molecular and Genetic Perspectives. Springer, Cham, pp. 159–177. Available at https://doi.org/10.1007/978-3-030-01415-5_9.

Frederiks, T.M., Christopher, J.T., Fletcher, S.E.H., Borrell, A.K., 2011. Post head-emergence frost resistance of barley genotypes in the northern grain region of Australia. Crop and Pasture Science 62 (9), 736–745.

Frova, C., Sari-Gorla, M., 1994. Quantitative trait loci (QTLs) for pollen thermotolerance detected in maize. Molecular and General Genetics MGG 245, 424–430.

Fu, C., Zhou, Y., Liu, A., Chen, R., Yin, L., Li, C., et al., 2024. Genome-wide association study for seedling heat tolerance under two temperature conditions in bread wheat (Triticum aestivum L.). BMC Plant Biology 24 (1), 430.

Gaba, Y., Pareek, A., Singla-Pareek, S.L., 2021. Raising climate-resilient crops: journey from the conventional breeding to new breeding approaches. Current Genomics 22 (6), 450–467.

Gantait, S., Mitra, M., 2019. Applications of synthetic seed technology for propagation, storage, and conservation of orchid germplasms. In: Faisal, M., Alatar, A. (Eds.), Synthetic Seeds: Germplasm Regeneration, Preservation and Prospects. Springer, Cham, pp. 301–321. Available at https://doi.org/10.1007/978-3-030-24631-0_13.

Garg, D., Sareen, S., Dalal, S., Tiwari, R., Singh, R., 2012. Heat shock protein-based SNP marker for terminal heat stress in wheat (*'Triticum aestivum* 'L.). Australian Journal of Crop Science 6 (11), 1516–1521.

Gaur, P.M., Jukanti, A.K., Varshney, R.K., 2012. Impact of genomic technologies on chickpea breeding strategies. Agronomy 2 (3), 199–221.

Genc, Y., Oldach, K., Verbyla, A.P., Lott, G., Hassan, M., Tester, M., et al., 2010. Sodium exclusion QTL associated with improved seedling growth in bread wheat under salinity stress. Theoretical and Applied Genetics 121, 877–894.

Gillman, J.D., Chebrolu, K., Smith, J.R., 2021. Quantitative trait locus mapping for resistance to heat-induced seed degradation and low seed phytic acid in soybean. Crop Science 61 (3), 2023–2035.

Glaszmann, J.C., Kaw, R.N., Khush, G.S., 1990. Genetic divergence among cold tolerant rices (*Oryza sativa* L.). Euphytica 45, 95–104.

Gororo, N.N., Eagles, H.A., Eastwood, R.F., Nicolas, M.E., Flood, R.G., 2002. Use of *Triticum tauschii* to improve yield of wheat in low-yielding environments. Euphytica 123, 241–254.

Goulet, B.E., Roda, F., Hopkins, R., 2017. Hybridization in plants: old ideas, new techniques. Plant Physiology 173 (1), 65–78.

Grant, C.A., Clarke, J.M., Duguid, S., Chaney, R.L., 2008. Selection and breeding of plant cultivars to minimize cadmium accumulation. Science of the Total Environment 390 (2-3), 301–310.

Grilli, G.V.G., Braz, L.T., Lemos, E.G.M., 2007. QTL identification for tolerance to fruit set in tomato by fAFLP markers. Crop Breeding and Applied Biotechnology 7 (3), 234–241. Available at 10.12702/1984-7033.v07n03a02.

Gupta, M., Praveen, A., 2019. Quantitative Trait Loci mapping of heavy metal accumulation and resistance in crop plants. Genomics Assisted Breeding of Crops for Abiotic Stress Tolerance 2, 21–30.

Haider, Z., Ahmad, I., Zia, S., Gan, Y., 2023. Recent developments in rice molecular breeding for tolerance to heavy metal toxicity. Agriculture 13 (5), 944.

Haley, S.D., Johnson, J.J., Peairs, F.B., Quick, J.S., Stromberger, J.A., Clayshulte, S.R., et al., 2007. Registration of 'Ripper'wheat. Journal of Plant Registrations 1 (1), 1–6.

Halim, M.A., Ghosh, M., Nigar, M., Hossain, F., Akhter, N., 2014. Screening of arsenic tolerance in rice at germination and early seedling stage as influenced by sodium arsenate. Jahangirnagar University Journal of Biological Sciences 3 (1), 17–26.

Hall, A.E., Cisse, N., Thiaw, S., Elawad, H.O., Ehlers, J.D., Ismail, A.M., et al., 2003. Development of cowpea cultivars and germplasm by the Bean/Cowpea CRSP. Field Crops Research 82 (2-3), 103–134.

Haq, M.A., 2009. Development of Mutant Varieties of Crop Plants at NIAB and the Impact on Agricultural Production in Pakistan. Food and Agricultural Organization of the United Nations, Rome (Italy), pp. 61. Available at https://inis.iaea.org/records/eg92c-qrc75.

Hasan, M.J., Kulsum, M.U., Sarker, U., Matin, M.Q.I., Shahin, N.H., Kabir, M.S., et al., 2022. Assessment of GGE, AMMI, regression, and its deviation model to identify stable rice hybrids in Bangladesh. Plants 11 (18), 2336.

Hayashi, S., Kamoshita, A., Yamagishi, J., Kotchasatit, A., Jongdee, B., 2007. Genotypic differences in grain yield of transplanted and direct seeded rainfed lowland rice (Oryza sativa L.) in northeastern Thailand. Field Crops Research 102 (1), 9–21.

Henkrar, F., Udupa, S., 2020. Marker assisted selection in plant breeding. Moroccan Journal of Agricultural Sciences 1 (5).

Hollington, P.A., Akhtar, J., Aragüés, R., Hussain, Z., Mahar, A.R., Quarrie, S.A., et al., 2002. Recent advances in the development of salinity and waterlogging tolerant bread wheats. In: Ahmad, R., Malik, K.A. (Eds.), Prospects for Saline Agriculture. Tasks for vegetation science, vol 37. Springer, Dordrecht, pp. 83–99. Available at https://doi.org/10.1007/978-94-017-0067-2_9.

Hollington, P.A., 2000. Technological breakthroughs in screening/breeding wheat varieties for salt tolerance. In Proceedings of the National Conference 'Salinity Management in Agriculture, pp. 273–289.

Hu, S., Lübberstedt, T., Zhao, G., Lee, M., 2016. QTL mapping of low-temperature germination ability in the maize IBM Syn4 RIL population. PLoS One 11 (3), e0152795.

Hu, X., Wang, G., Du, X., Zhang, H., Xu, Z., Wang, J., et al., 2021. QTL analysis across multiple environments reveals promising chromosome regions associated with yield-related traits in maize under drought conditions. The Crop Journal 9 (4), 759–766.

Huang, X., Fan, Y., Shabala, L., Rengel, Z., Shabala, S., Zhou, M.X., 2018. A major QTL controlling the tolerance to manganese toxicity in barley (*Hordeum vulgare* L.). Molecular Breeding 38, 1–9.

Ismail, A.M., Horie, T., 2017. Genomics, physiology, and molecular breeding approaches for improving salt tolerance. Annual Review of Plant Biology 68 (1), 405–434.

Jagadish, S.V.K., Craufurd, P.Q., Wheeler, T.R., 2008. Phenotyping parents of mapping populations of rice for heat tolerance during anthesis. Crop Science 48 (3), 1140–1146.

Jagadish, S.V.K., Cairns, J., Lafitte, R., Wheeler, T.R., Price, A.H., Craufurd, P.Q., 2010. Genetic analysis of heat tolerance at anthesis in rice. Crop Science 50 (5), 1633–1641.

Jalmi, S.K., Bhagat, P.K., Verma, D., Noryang, S., Tayyeba, S., Singh, K., et al., 2018. Traversing the links between heavy metal stress and plant signaling. Frontiers in Plant Science 9, 12.

Janni, M., Gullì, M., Maestri, E., Marmiroli, M., Valliyodan, B., Nguyen, H.T., et al., 2020. Molecular and genetic bases of heat stress responses in crop plants and breeding for increased resilience and productivity. Journal of Experimental Botany 71 (13), 3780–3802.

Jattan, M., Kumari, N., Rani, B., Phogat, D.S., Kumari, S., Ravika, et al., 2023. Conventional breeding and advance approaches to mitigate drought and salt stress in crop plants. Salinity and Drought Tolerance in Plants: Physiological Perspectives. Springer Nature Singapore, Singapore, pp. 121–135.

Jena, K.K., Mackill, D.J., 2008. Molecular markers and their use in marker-assisted selection in rice. Crop Science 48 (4), 1266–1276.

Jha, U.C., Bohra, A., Singh, N.P., 2014. Heat stress in crop plants: its nature, impacts and integrated breeding strategies to improve heat tolerance. Plant Breeding 133 (6), 679–701.

Jha, U.C., Nayyar, H., Palakurthi, R., Jha, R., Valluri, V., Bajaj, P., et al., 2021. Major QTLs and potential candidate genes for heat stress tolerance identified in chickpea (*Cicer arietinum* L.). Frontiers in Plant Science 12, 655103.

Jha, S., Quaiyum, Z., Prasad, B.D., Sahni, S., Mandyal, S.S., Singh, A., et al., 2025. Meta-QTL analysis, candidate gene identification, and validation by GWAS for agronomic traits aND STRESS tolerance in pea (*Pisum sativum* L.). Plant Biotechnology Reports 19, 481–494. Available at https://doi.org/10.1007/s11816-025-00995-9.

Johnson, D.W., Smith, S.E., Dobrenz, A.K., 1991. Registration of AZ-90NDC-ST non-dormant alfalfa germplasm with improved forage yield in saline environments. Crop Science 31, 1098–1099.

Joshi, A., Kumar, A., Adhikari, S., Saroj, R., 2024. Marker-assisted selection: strategy for biotic and abiotic stress tolerance in plants. Climate-Resilient Agriculture. Apple Academic Press, pp. 137–191.

Kaiser, N., Douches, D., Dhingra, A., Glenn, K.C., Herzig, P.R., Stowe, E.C., et al., 2020. The role of conventional Plant breeding in ensuring safe levels of naturally occurring toxins in food crops. Trends in Food Science & Technology 100, 51–66.

Kaya, C., Akram, N.A., Ashraf, M., Sonmez, O., 2018. Exogenous application of humic acid mitigates salinity stress in maize (*Zea mays* L.) plants by improving some key physico-biochemical attributes. Cereal Research Communications 46, 67–78.

Khan, S., Javed, M.A., Jahan, N., Manan, F.A., 2016. A short review on the development of salt tolerant cultivars in rice. International Journal of Public Health Science 5 (2), 201–212.

Khan, N.A., Owens, L., Nuñez, M.A., Khan, A.L., 2025. Complexity of combined abiotic stresses to crop plants. Plant Stress 17, 100926.

Kharkwal, M.C., Shu, Q.Y., 2009. The role of induced mutations in world food security. Induced Plant Mutations in the Genomics Era. Food and Agriculture Organization of the United Nations, Rome, pp. 33–38.

Khoshro, H.H., Lotfi, R., 2021. Advanced breeding approaches for cold-tolerant chickpea and lentil in dryland areas. Legumes Research-Volume 1. IntechOpen.

Kim, K.M., Sohn, J.K., Chung, I.K., 2000. Analysis of OPT8511 RAPD fragments closely linked withcold sensitivity at seedling stage in rice (*Oryza sativa* L.). Molecules and Cells 10 (4), 382–385.

Kolar, S.C., Hayes, P.M., Chen, T.H.H., Linderman, R.G., 1991. Genotypic variation for cold tolerance in winter and facultative barley. Crop Science 31 (5), 1149–1152.

Krishna, T.P.A., Veeramuthu, D., Maharajan, T., Soosaimanickam, M., 2023. The era of plant breeding: conventional breeding to genomics-assisted breeding for crop improvement. Current Genomics 24 (1), 24–35.

Krishnamurthy, S.L., Pundir, P., Warraich, A.S., Rathor, S., Lokeshkumar, B.M., Singh, N.K., et al., 2020. Introgressedsaltol QTL lines improves the salinity tolerance in rice at seedling stage. Frontiers in Plant Science 11, 833.

Kumar, S., 2020. Abiotic stresses and their effects on plant growth, yield and nutritional quality of agricultural produce. International Journal of Food Science and Agriculture 4 (4), 367–378.

Kumar, R., Chowdhury, J.B., Jain, R.K., 2001. Interspecific hybridization in Brassica juncea and Brassica toumefortii through embryo rescue and their evaluation for biotic and abiotic stress tolerance. Indian Journal of Experimental Biology 39, 911–915.

Kumar, I.S., Rao, S.R., Vardhini, B.V., 2015. Role of phytohormones during salt stress tolerance in plants. Current Trends in Biotechnology and Pharmacy 9 (4), 334–343.

Kumar, A., Saripalli, G., Jan, I., Kumar, K., Sharma, P.K., Balyan, H.S., et al., 2020. Meta-QTL analysis and identification of candidate genes for drought tolerance in bread wheat (Triticum aestivum L.). Physiology and Molecular Biology of Plants 26, 1713–1725.

Kumar, R., Sharma, V.K., Rangari, S.K., Jha, U.C., Sahu, A., Paul, P.J., et al., 2023. High confidence QTLs and key genes identified using Meta-QTL analysis for enhancing heat tolerance in chickpea (*Cicer arietinum* L.). Frontiers in Plant Science 14, 1274759.

Kumari, M., Dubey, A.K., Kumar, R., Kumar, A., 2024. Marker-assisted selection in plant breeding for stress tolerance. Improving Stress Resilience in Plants. Academic Press, pp. 371–387.

Kuroki, M., Saito, K., Matsuba, S., Yokogami, N., Shimizu, H., Ando, I., et al., 2007. A quantitative trait locus for cold tolerance at the booting stage on rice chromosome 8. Theoretical and Applied Genetics 115, 593–600.

Kushwah, A., Bhatia, D., Barmukh, R., Singh, I., Singh, G., Bindra, S., et al., 2022. Genetic mapping of QTLs for drought tolerance in chickpea (*Cicer arietinum* L.). Frontiers in Genetics 13, 953898.

Ladizinsky, G., Cohen, D., Muehlbauer, F.J., 1985. Hybridization in the genus Lens by means of embryo culture. Theoretical and Applied Genetics 70, 97–101.

Lamichhane, S., Thapa, S., 2022. Advances from conventional to modern plant breeding methodologies. Plant Breeding and Biotechnology 10 (1), 1–14.

Laosatit, K., Amkul, K., Wang, L., Somta, P., 2024. Identification of novel QTLs for salt tolerance in zombi pea (Vigna vexillata). Euphytica 220 (7), 1–12.

Lateef, D.D., 2015. DNA marker technologies in plants and applications for crop improvements. Journal of Biosciences and Medicines 3 (5), 7–18.

Lavarenne, J., Guyomarc'h, S., Sallaud, C., Gantet, P., Lucas, M., 2018. The spring of Systems Biology-driven breeding. Trends in Plant Science 23 (8), 706–720.

Lei, D., Tan, L., Liu, F., Chen, L., Sun, C., 2013. Identification of heat-sensitive QTL derived from common wild rice (*Oryza rufipogon* Griff.). Plant Science 201, 121–127.

Li, W., Zhang, Q., Wang, S., Langham, M.A., Singh, D., Bowden, R.L., et al., 2019. Development and characterization of wheat–sea wheatgrass (*Thinopyrumjunceiforme*) amphiploids for biotic stress resistance and abiotic stress tolerance. Theoretical and Applied Genetics 132, 163–175.

Li, C., Liu, J., Bian, J., Jin, T., Zou, B., Liu, S., et al., 2021. Identification of cold tolerance QTLs at the bud burst stage in 211 rice landraces by GWAS. BMC Plant Biology 21, 1–11.

Li, C., Dong, S., Beckles, D.M., Liu, X., Liu, D., Miao, H., et al., 2022. Identification of QTLs controlling cold tolerance in cucumber (*Cucumis sativus* L.) seedlings. Scientia Horticulturae 305, 111383.

Liao, J.L., Zhang, H.Y., Shao, X.L., Zhong, P.A., Huang, Y.J., 2011. Identification for heat tolerance in backcross recombinant lines and screening of backcross introgression lines with heat tolerance at milky stage in rice. Rice Science 18 (4), 279–286.

Lin, S., Yang, T., Zhu, S., Wang, J., Ni, W., 2017. A method for screening copper-tolerant rice (Oryza sativa L.) cultivars based on hydroponic experiments and cluster analysis. International Journal of Phytoremediation 19 (12), 1093–1099.

Lindlöf, A., Chawade, A., Sikora, P., Olsson, O., 2015. Comparative transcriptomics of Sijung and Jumli Marshi rice during early chilling stress imply multiple protective mechanisms. PloS One 10 (5), e0125385.

Liu, F., Sun, C., Tan, L., Fu, Y., Li, D., Wang, X., 2003. Identification and mapping of quantitative trait loci controlling cold tolerance of Chinese common wild rice (*O. rufipogon* Griff.) at booting to flowering stages. Chinese Science Bulletin 48, 2068–2071.

Liu, S., Banik, M., Yu, K., Park, S.J., Poysa, V., Guan, Y., 2007. Marker-assisted selection (MAS) in major cereal and legume crop breeding: current progress and future directions. International Journal of Plant Breeding 1 (2), 75–78.

Liu, Z., Dong, X., Cao, X., Xu, C., Wei, J., Zhen, G., et al., 2022. QTL mapping for cold tolerance and higher overwintering survival rate in winter rapeseed (*Brassica napus*). Journal of Plant Physiology 275, 153735.

Londo, J.P., Kovaleski, A.P., Lillis, J.A., 2018. Divergence in the transcriptional landscape between low temperature and freeze shock in cultivated grapevine (*Vitis vinifera*). Horticulture Research 5, 10. Available at https://doi.org/10.1038/s41438-018-0020-7.

Lucas, M.R., Ehlers, J.D., Huynh, B.L., Diop, N.N., Roberts, P.A., Close, T.J., 2013a. Markers for breeding heat-tolerant cowpea. Molecular Breeding 31, 529–536.

Mackill, D.J., Lei, X., 1997. Genetic variation for traits related to temperate adaptation of rice cultivars. Crop Science 37 (4), 1340–1346.

Malumpong, C., Buadchee, R., Thammasamisorn, B., Moung-Ngam, P., Wasuri, B., Saensuk, C., et al., 2020. Backcross breeding for improvement of heat tolerance at reproductive phase in Thai rice (*Oryza sativa* L.) varieties. The Journal of Agricultural Science 158 (6), 496–510.

Mammadov, J., Buyyarapu, R., Guttikonda, S.K., Parliament, K., Abdurakhmonov, I.Y., Kumpatla, S.P., 2018. Wild relatives of maize, rice, cotton, and soybean: treasure troves for tolerance to biotic and abiotic stresses. Frontiers in Plant Science 9, 886.

Manjunath, K.K., Krishna, H., Devate, N.B., Sunilkumar, V.P., Patil, S.P., Chauhan, D., et al., 2024. QTL mapping: insights into genomic regions governing component traits of yield under combined heat and drought Stress in wheat. Frontiers in Genetics 14, 1282240.

Mansoor, S., Ali, A., Kour, N., Bornhorst, J., AlHarbi, K., Rinklebe, J., et al., 2023. Heavy metal induced oxidative stress mitigation and ROS scavenging in plants. Plants 12 (16), 3003.

Maqbool, M.A., Aslam, M., Ali, H., 2017. Breeding for improved drought tolerance in Chickpea (*Cicer arietinum* L.). Plant Breeding 136 (3), 300–318.

Meena, K.K., Sorty, A.M., Bitla, U.M., Choudhary, K., Gupta, P., Pareek, A., et al., 2017. Abiotic stress responses and microbe-mediated mitigation in plants: the omics strategies. Frontiers in Plant Science 8, 172.

Mickelbart, M.V., Hasegawa, P.M., Bailey-Serres, J., 2015. Genetic mechanisms of abiotic stress tolerance that translate to crop yield stability. Nature Reviews Genetics 16 (4), 237–251.

Miklas, P.N., Kelly, J.D., Beebe, S.E., Blair, M.W., 2006. Common bean breeding for resistance against biotic and abiotic stresses: from classical to MAS breeding. Euphytica 147, 105–131.

Miller, J.F., Green, C.E., Li, Y.M., Chaney, R.L., 2006. Registration of three low cadmium (HA 448, HA 449, and RHA 450) confection sunflower genetic stocks. Crop Science 46 (1), 489–491.

Mir, R.R., Zaman-Allah, M., Sreenivasulu, N., Trethowan, R., Varshney, R.K., 2012. Integrated genomics, physiology and breeding approaches for improving drought tolerance in crops. Theoretical and Applied Genetics 125, 625–645.

Mirdar Mansuri, R., Shobbar, Z.S., Babaeian Jelodar, N., Ghaffari, M., Mohammadi, S.M., Daryani, P., 2020. Salt tolerance involved candidate genes in rice: an integrative meta-analysis approach. BMC Plant Biology 20, 1–14.

Mohammadi, R., Amri, A., Ahmadi, H., Jafarzadeh, J., 2015. Characterization of tetraploid wheat landraces for cold tolerance and agronomic traits under rainfed conditions of Iran. The Journal of Agricultural Science 153 (4), 631–645.

Mori, M., Onishi, K., Tokizono, Y., Shinada, H., Yoshimura, T., Numao, Y., et al., 2011. Detection of a novel quantitative trait locus for cold tolerance at the booting stage derived from a tropical japonica rice variety Silewah. Breeding Science 61 (1), 61–68.

Mourad, A.M., Amin, A.E.E.A.Z., Dawood, M.F., 2021. Genetic variation in kernel traits under lead and tin stresses in spring wheat diverse collection. Environmental and Experimental Botany 192, 104646.

Murugaiyan, V., Ali, J., Mahender, A., Aslam, U.M., Jewel, Z.A., Pang, Y., et al., 2019. Mapping of genomic regions associated with arsenic toxicity stress in a backcross breeding population of rice (*Oryza sativa* L.). Rice 12, 1–14.

Nadeem, M.A., Nawaz, M.A., Shahid, M.Q., Doğan, Y., Comertpay, G., Yıldız, M., et al., 2018. DNA molecular markers in plant breeding: current status and recent advancements in genomic selection and genome editing. Biotechnology and Biotechnological Equipment 32 (2), 261–285.

Naeem, M., Iqbal, M., Shakeel, A., Ul-Allah, S., Hussain, M., Rehman, A., et al., 2020. Genetic basis of ion exclusion in salinity stressed wheat: implications in improving crop yield. Plant Growth Regulation 92, 479–496.

Nair, R.J., Pandey, M.K., 2024. Role of molecular markers in crop breeding: a review. Agricultural Reviews 45 (1).

Nam, N.H., Chauhan, Y.S., Johansen, C., 2001. Effect of timing of drought stress on growth and grain yield of extra-short-duration pigeonpea lines. The Journal of Agricultural Science 136 (2), 179–189.

Nazir, M.J., Li, G., Nazir, M.M., Zulfiqar, F., Siddique, K.H., Iqbal, B., et al., 2024. Harnessing soil carbon sequestration to address climate change challenges in agriculture. Soil and Tillage Research 237, 105959.

Nguyen, V.L., Ribot, S.A., Dolstra, O., Niks, R.E., Visser, R.G., van der Linden, C.G., 2013. Identification of quantitative trait loci for ion homeostasis and salt tolerance in barley (Hordeum vulgare L.). Molecular Breeding 31, 137–152.

Njobvu, J., Hamabwe, S.M., Munyinda, K., Kelly, J.D., Kamfwa, K., 2020. Quantitative trait loci mapping of resistance to aluminum toxicity in common bean. Crop Science 60 (3), 1294–1302.

Noaman, M.M., Ahmed, I.A., El-Sayed, A.A., Abo-El-Enin, R.A., El-Gamal, A.S., El-Sherbiny, A.M., et al., 2007. Registration of'Giza 2000'drought-tolerant six-rowed barley for rainfed and new reclaimed areas in Egypt. Crop Science 47, 440.

Novak, F.J., Brunner, H., 1992. Plant breeding: induced mutation technology for crop improvement. IAEA Bulletin 4, 25–33.

Obert, D.E., Evans, C.P., Wesenberg, D.M., Windes, J.M., Erickson, C.A., Jackson, E.W., et al., 2008. Registration of 'Lenetah'spring barley. Journal of Plant Registrations 2 (2), 85–87.

O'Boyle, P.D., Kelly, J.D., Kirk, W.W., 2007. Use of marker-assisted selection to breed for resistance to common bacterial blight in common bean. Journal of the American Society for Horticultural Science 132 (3), 381–386.

Oh, C.S., Choi, Y.H., Lee, S.J., Yoon, D.B., Moon, H.P., Ahn, S.N., 2004. Mapping of quantitative trait loci for cold tolerance in weedy rice. Breeding Science 54 (4), 373–380.

Okada, T., Whitford, R., 2019. Hybrid wheat and abiotic stress. Genomics Assisted Breeding of Crops for Abiotic Stress Tolerance 2, 211–224.

Oladosu, Y., Rafii, M.Y., Samuel, C., Fatai, A., Magaji, U., Kareem, I., et al., 2019. Drought resistance in rice from conventional to molecular breeding: a review. International Journal of Molecular Sciences 20 (14), 3519.

O'neill, B.C., Oppenheimer, M., Warren, R., Hallegatte, S., Kopp, R.E., Pörtner, H.O., et al., 2017. IPCC reasons for concern regarding climate change risks. Nature Climate Change 7 (1), 28–37.

Ortiz, R., 2002. Development of drought-resistant and water-stress tolerant crops through traditional breeding. Proceedings of the 8th JIRCAS International Symposium. pp. 11–21.

Ottaviano, E., Sari Gorla, M., Pe, E., Frova, C., 1991. Molecular markers (RFLPs and HSPs) for the genetic dissection of thermotolerance in maize. Theoretical and Applied Genetics 81, 713–719.

Paliwal, R., Röder, M.S., Kumar, U., Srivastava, J.P., Joshi, A.K., 2012. QTL mapping of terminal heat tolerance in hexaploid wheat (*T. aestivum* L.). Theoretical and Applied Genetics 125, 561–575.

Pan, X., Li, Y., Liu, W., Liu, S., Min, J., Xiong, H., et al., 2020. QTL mapping and candidate gene analysis of cadmium accumulation in polished rice by genome-wide association study. Scientific Reports 10 (1), 11791.

Pan, Y.H., Nong, B.X., Chen, L., Yang, X.H., Xia, X.Z., Zhang, Z.Q., et al., 2023. QTL mapping and identification of candidate genes for cold tolerance at the germination stage in wild rice. Genes and Genomics 45 (7), 867–885.

Pandey, S., Singh, A., Parida, S.K., Prasad, M., 2022. Combining speed breeding with traditional and genomics-assisted breeding for crop improvement. Plant Breeding 141 (3), 301–313.

Pandit, R., Bhusal, B., Regmi, R., Neupane, P., Bhattarai, K., Maharjan, B., et al., 2021. Mutation breeding for crop improvement: a review. Reviews in Food and Agriculture 2 (1), 31–35.

Panigrahi, S., Kumar, U., Swami, S., Singh, Y., Balyan, P., Singh, K.P., et al., 2024. Meta QTL analysis for dissecting abiotic stress tolerance in chickpea. BMC Genomics 25 (1), 439.

Paran, I., Van Der Knaap, E., 2007. Genetic and molecular regulation of fruit and plant domestication traits in tomato and pepper. Journal of Experimental Botany 58 (14), 3841–3852.

Parida, A.K., Das, A.B., 2005. Salt tolerance and salinity effects on plants: a review. Ecotoxicology and Environmental Safety 60 (3), 324–349.

Parry, M.A., Madgwick, P.J., Bayon, C., Tearall, K., Hernandez-Lopez, A., Baudo, M., et al., 2009. Mutation discovery for crop improvement. Journal of Experimental Botany 60 (10), 2817–2825.

Patil, V.D., Nerkar, Y.S., Misal, M.B., &Harkal, S.R., 1997. Parag 401, a semidwarf rice variety developed through anther culture. International Rice Research Newsletter 22 (2), 19.

Paul, T., Debnath, S., Das, S.P., Natarajan, S., Perveen, K., Alshaikh, N.A., et al., 2023. Identification of major and stable QTLs conferring drought tolerance in rice RIL populations. Current Research in Biotechnology 5, 100125.

Pazhamala, L., Saxena, R.K., Singh, V.K., Sameerkumar, C.V., Kumar, V., Sinha, P., et al., 2015. Genomics-assisted breeding for boosting crop improvement in pigeonpea (*Cajanus cajan*). Frontiers in Plant Science 6, 50.

Peter, R., Eschholz, T.W., Stamp, P., &Liedgens, M., 2006. Swiss maize landraces-early vigour adaptation to cool conditions. Acta AgronomicaHungarica 54 (3), 329–336.

Peter, R., Eschholz, T.W., Stamp, P., &Liedgens, M., 2009. Swiss Flint maize landraces-a rich pool of variability for early vigour in cool environments. Field Crops Research 110 (2), 157–166.

Piao, S., Liu, Q., Chen, A., Janssens, I.A., Fu, Y., Dai, J., et al., 2019. Plant phenology and global climate change: current progresses and challenges. Global Change Biology 25 (6), 1922–1940.

Pinto, R.S., Reynolds, M.P., Mathews, K.L., McIntyre, C.L., Olivares-Villegas, J.J., Chapman, S.C., 2010. Heat and drought adaptive QTL in a wheat population designed to minimize confounding agronomic effects. Theoretical and Applied Genetics 121, 1001–1021.

Prakash, C., Sevanthi, A.M., Shanmugavadivel, P.S., 2019. Use of QTLs in developing abiotic stress tolerance in rice. Advances in Rice Research for Abiotic Stress Tolerance. Woodhead Publishing, pp. 869–893.

Pundir, P., Devi, A., Krishnamurthy, S.L., Sharma, P.C., Vinaykumar, N.M., 2021. QTLs in salt rice variety CSR10 reveals salinity tolerance at reproductive stage. Acta Physiologiae Plantarum 43 (2), 35.

Punyawaew, K., Suriya-Arunroj, D., Siangliw, M., Thida, M., Lanceras-Siangliw, J., Fukai, S., et al., 2016. Thai jasmine rice cultivar KDML105 carrying Saltol QTL exhibiting salinity tolerance at seedling stage. Molecular Breeding 36, 1–13.

Purugganan, M.D., Fuller, D.Q., 2009. The nature of selection during plant domestication. Nature 457 (7231), 843–848.

Qaim, M., 2020. Role of new plant breeding technologies for food security and sustainable agricultural development. Applied Economic Perspectives and Policy 42 (2), 129–150.

Quick, J.S., Stromberger, J.A., Clayshulte, S., Clifford, B., Johnson, J.J., Peairs, F.B., et al., 2001. Registration of prairie red' wheat. Crop Science 41 (4), 1362.

Rahman, S.U., Arif, M., Hussain, K., Arshad, M., Hussain, S., Mukhtar, T., et al., 2015. Breeding for heat stress tolerance of maize in Pakistan. Journal of Environmental and Agricultural Sciences 5, 27–33.

Rai, S.K., Pandey-Rai, S., 2021. Marker-assisted breeding for abiotic stress tolerance in horticultural crops. Stress Tolerance in Horticultural Crops. Woodhead Publishing, pp. 63–74.

Rai, K.K., Kumar, A., Rai, A., Rai, V.P., Rai, A.C., 2021. Conventional breeding approaches for abiotic stress management in horticultural crops. Stress Tolerance in Horticultural Crops. Woodhead Publishing, pp. 21–32.

Rao, G.J., Reddy, J.N., Variar, M., Mahender, A., 2016. Molecular breeding to improve plant resistance to abiotic stresses. Advances in Plant Breeding Strategies: Agronomic, Abiotic and Biotic Stress Traits 283–326.

Rauf, S., Al-Khayri, J.M., Zaharieva, M., Monneveux, P., Khalil, F., 2016. Breeding strategies to enhance drought tolerance in crops. Advances in Plant Breeding Strategies: Agronomic, Abiotic and Biotic Stress Traits 397–445.

Raza, A., Razzaq, A., Mehmood, S.S., Zou, X., Zhang, X., Lv, Y., et al., 2019. Impact of climate change on crops adaptation and strategies to tackle its outcome: a review. Plants 8 (2), 34.

Reddy, L.J., Nigam, S.N., RAO, N.R., Reddy, N.S., 2001. Registration of ICGV 87354 peanut germplasm with drought tolerance and rust resistance. Crop Science 41 (1), 274 274.

Reddy, A.M., Francies, R.M., Rasool, S.N., Reddy, V.R.P., 2014. Breeding for tolerance to stress triggered by salinity in rice. International Journal of Applied Biology and Pharmaceutical Technology 5, 167–176.

Ren, H., Han, J., Wang, X., Zhang, B., Yu, L., Gao, H., et al., 2020. QTL mapping of drought tolerance traits in soybean with SLAF sequencing. The Crop Journal 8 (6), 977–989.

Revilla, P., Rodríguez, V.M., Ordás, A., Rincent, R., Charcosset, A., Giauffret, C., et al., 2016. Association mapping for cold tolerance in two large maize inbred panels. BMC Plant Biology 16, 1–10.

Rodriguez-Garay, B., Barrow, J.R., 1988. Pollen selection for heat tolerance in cotton. Crop Science 28 (5), 857–859.

Rodríguez Graña, V.M., Butrón Gómez, A.M., Sandoya Miranda, G., Ordás Pérez, A., Revilla Temiño, P., 2007. Combining maize base germplasm for cold tolerance breeding. Crop Science 47, 1467–1474.

Rush, D.W., Epstein, E., 1981. Comparative studies on the sodium, potassium, and chloride relations of a wild halophytic and a domestic salt-sensitive tomato species. Plant Physiology 68 (6), 1308–1313.

Sachdev, S., Ansari, S.A., Ansari, M.I., Fujita, M., Hasanuzzaman, M., 2021. Abiotic stress and reactive oxygen species: generation, signaling, and defense mechanisms. Antioxidants 10 (2), 277.

Saito, K., Miura, K., Nagano, K., Hayano-Saito, Y., Saiito, A., Araki, H., et al., 1995. Chromosomal location of quantitative trait loci for cool tolerance at the booting stage in rice variety'Norin-PL8' Japanese Journal of Breeding 45 (3), 337–340.

Sandhu, N., Dixit, S., Swamy, B.P.M., Raman, A., Kumar, S., Singh, S.P., et al., 2019. Marker assisted breeding to develop multiple stress tolerant varieties for flood and drought prone areas. Rice 12 (1), 8.

Saranga, Y., Zamir, D., Marani, A., Rudich, J., 1993. Breeding tomatoes for salt tolerance: variations in ion concentrations associated with response to salinity. Journal of the American Society for Horticultural Science 118 (3), 405–408.

Sasaki, O., Yamazaki, K., kawaka, S.I., 1981. The relationship between the diameters and the structures of lateral roots in rice plants. Japanese Journal of Crop Science 50 (4), 476–480.

Satake, T., Toriyama, K., 1979. Two extremely cool-tolerant varieties. International Rice Research Newsletter 4, 9–10.

Sato, H., Mizoi, J., Shinozaki, K., Yamaguchi-Shinozaki, K., 2024. Complex Plant responses to drought and heat Stress under climate change. The Plant Journal 117 (6), 1873–1892.

Scheben, A., Yuan, Y., Edwards, D., 2016. Advances in genomics for adapting crops to climate change. Current Plant Biology 6, 2–10.

Schubert, S., Neubert, A., Schierholt, A., Sümer, A., Zörb, C., 2009. Development of salt-resistant maize hybrids: the combination of physiological strategies using conventional breeding methods. Plant Science 177 (3), 196–202.

Sehgal, S.K., Kaur, S., Gupta, S., Sharma, A., Kaur, R., Bains, N.S., 2011. A direct hybridization approach to gene transfer from Aegilops tauschii Coss. to *Triticum aestivum* L. Plant Breeding 130 (1), 98–100.

Senadhira, D., Zapata-Arias, F.J., Gregorio, G.B., Alejar, M.S., De La Cruz, H.C., Padolina, T.F., et al., 2002. Development of the first salt-tolerant rice cultivar through indica/indica anther culture. Field Crops Research 76 (2-3), 103–110.

Serraj, R., Hash, C.T., Rizvi, S.M.H., Sharma, A., Yadav, R.S., Bidinger, F.R., 2005. Recent advances in marker-assisted selection for drought tolerance in pearl millet. Plant Production Science 8 (3), 334–337.

Shabbir, R., Singhal, R.K., Mishra, U.N., Chauhan, J., Javed, T., Hussain, S., et al., 2022. Combined abiotic stresses: challenges and potential for crop improvement. Agronomy 12 (11), 2795.

Shafiq, S., Zeb, Q., Ali, A., Sajjad, Y., Nazir, R., Widemann, E., et al., 2019. Lead, cadmium and zinc phytotoxicity alter DNA methylation levels to confer heavy metal tolerance in wheat. International Journal of Molecular Sciences 20 (19), 4676.

Shamloo-Dashtpagerdi, R., Tanin, M.J., Aliakbari, M., Saini, D.K., 2024. Unveiling the role of the ERD15 gene in wheat's tolerance to combined drought and salinity stress: a meta-analysis of QTL and RNA-Seq data. Physiologia Plantarum 176 (5), e14570.

Shankar, S.R., 2020. Effect of heavy metals on growth and development of cultivated plants with reference to cadmium, chromium and lead–a review. Journal of Stress Physiology and Biochemistry 16 (3), 84–102.

Shanmugavel, P., Rajaprakasam, S., Chockalingam, V., Ramasamy, G., Thiyagarajan, K., Marimuthu, R., 2020. Breeding mechanisms for high temperature tolerance in crop plants. Plant Breeding-Current and Future Views. IntechOpen, pp. 1–12.

Sharma, S., Sharma, J., Bindra, S., et al., 2025. Meta-analysis of QTLs for drought, heat, and salinity tolerance identifies high-confidence MQTLs and candidate genes in chickpea (*Cicer arietinum* L.). Journal of Bioscience 50, 19.

Shavrukov, Y., Kurishbayev, A., Jatayev, S., Shvidchenko, V., Zotova, L., Koekemoer, F., et al., 2017. Early flowering as a drought escape mechanism in plants: how can it aid wheat production? Frontiers in Plant Science 8, 1950.

Shaw, A.J., 2022. The evolution of heavy metal tolerance in plants: adaptations, limits, and costs. Genetics and Ecotoxicology. CRC Press, pp. 9–30.

Sheoran, S., Gupta, M., Kumari, S., Kumar, S., Rakshit, S., 2022. Meta-QTL analysis and candidate genes identification for various abiotic stresses in maize (Zea mays L.) and their implications in breeding programs. Molecular Breeding 42 (5), 26.

Shilin, D., Chaolei, L., Lianguang, S., Shenglong, Y., Anpeng, Z., Hongzhen, J., et al., 2021. Identification of QTLs for cadmium tolerance during seedling stage and validation of qCDSL1 in rice. Rice Science 28 (1), 81–88.

Shinada, H., Iwata, N., Sato, T., Fujino, K., 2014. QTL pyramiding for improving ofcold tolerance at fertilization stage in rice. Breeding Science 63 (5), 483–488.

Shirasawa, S., Endo, T., Nakagomi, K., Yamaguchi, M., Nishio, T., 2012. Delimitation of a QTL region controlling cold tolerance at booting stage of a cultivar,'Lijiangxintuanheigu', in rice, Oryza sativa L. Theoretical and Applied Genetics 124, 937–946.

Shuancang, Y., Yongiian, W., Xiaoying, Z., 2003. Mapping and analysis QTL controlling heat tolerance in Brassica campestrisL. ssp. pekinensis. Acta Horticulturae Sinica 30 (4), 417.

Shukla, R.P., Tiwari, G.J., Joshi, B., Song-Beng, K., Tamta, S., Boopathi, N.M., et al., 2021. GBS-SNP and SSR based genetic mapping and QTL analysis for drought tolerance in upland cotton. Physiology and Molecular Biology of Plants 27 (8), 1731–1745.

Singer, S.D., Laurie, J.D., Bilichak, A., Kumar, S., Singh, J., 2021. Genetic variation and unintended risk in the context of old and new breeding techniques. Critical Reviews in Plant Sciences 40 (1), 68–108.

Singh, S.P., 2007. Drought resistance in the race Durango dry bean landraces and cultivars. Agronomy Journal 99 (5), 1219–1225.

Singh, R.J., Nelson, R.L., 2015. Intersubgeneric hybridization between Glycine max and G. tomentella: production of F 1, amphidiploid, BC 1, BC 2, BC 3, and fertile soybean plants. Theoretical and Applied Genetics 128, 1117–1136.

Singh, K.B., Omar, M., Saxena, M.C., Johansen, C., 1997. Screening for drought resistance in spring chickpea in the Mediterranean region. Journal of Agronomy and Crop Science 178 (4), 227–235.

Singh, S.P., Terán, H., Gutierrez, J.A., 2001. Registration of SEA 5 and SEA 13 drought tolerant dry bean germplasm. Crop Science 41 (1), 276.

Singh, M., Nara, U., Kumar, A., Choudhary, A., Singh, H., Thapa, S., 2021. Salinity tolerance mechanisms and their breeding implications. Journal of Genetic Engineering and Biotechnology 19 (1), 173.

Sita, K., Sehgal, A., Bhandari, K., Kumar, J., Kumar, S., Singh, S., et al., 2018. Impact of heat stress during seed filling on seed quality and seed yield in lentil (*Lens culinaris*Medikus) genotypes. Journal of the Science of Food and Agriculture 98 (13), 5134–5141.

Sivia, S.S., Singh, G., Rani, R., 2024. Marker-assisted selection approaches for abiotic stress tolerance in crop plants. Smart Breeding. Apple Academic Press, pp. 261–286.

Slafer, G.A., Kantolic, A.G., Appendino, M.L., Tranquilli, G., Miralles, D.J., Savin, R., 2015. Genetic and environmental effects on crop development determining adaptation and yield. Crop Physiology. Academic Press, pp. 285–319.

Smith, J.R., Mengistu, A., Nelson, R.L., Paris, R.L., 2008a. Identification of soybean accessions with high germinability in high-temperature environments. Crop Science 48 (6), 2279–2288.

Smith, J.S.C., Hussain, T., Jones, E.S., Graham, G., Podlich, D., Wall, S., et al., 2008b. Use of doubled haploids in maize breeding: implications for intellectual property protection and genetic diversity in hybrid crops. Molecular Breeding 22, 51–59.

Sthapit, B.R., Witcombe, J.R., 1998. Inheritance of tolerance to chilling stress in rice during germination and plumule greening. Crop Science 38 (3), 660–665.

Subramanian, P., Kim, K., Krishnamoorthy, R., Mageswari, A., Selvakumar, G., Sa, T., 2016. Cold stress tolerance in psychrotolerant soil bacteria and their conferred chilling resistance in tomato (*Solanum lycopersicum* Mill.) under low temperatures. PloS one 11 (8), e0161592.

Syed, M.A., Iftekharuddaula, K.M., Mian, M.K., Rasul, M.G., Rahmam, G.M., Panaullah, G.M., et al., 2016. Main effect QTLs associated with arsenic phyto-toxicity tolerance at seedling stage in rice (*Oryza sativa* L.). Euphytica 209, 805–814.

Tahmasebi, S., Heidari, B., Pakniyat, H., McIntyre, C.L., 2016. Mapping QTLs associated with agronomic and physiological traits under terminal drought and heat Stress conditions in wheat (*Triticum aestivum* L.). Genome 60 (1), 26–45.

Takagi, H., Tamiru, M., Abe, A., Yoshida, K., Uemura, A., Yaegashi, H., et al., 2015. MutMap accelerates breeding of a salt-tolerant rice cultivar. Nature Biotechnology 33 (5), 445–449.

Takahashi, D., Li, B., Nakayama, T., Kawamura, Y., Uemura, M., 2013. Plant plasma membrane proteomics for improving cold tolerance. Frontiers in Plant Science 4, 90.

Taria, S., Arora, A., Krishna, H., Manjunath, K.K., Kumar, S., Singh, B., et al., 2025. Mapping of the QTLs governing stem-specific weight for stem reserve mobilisation in wheat (*Triticum aestivum* L.) under combined heat and drought Stress. Plant Physiology Reports 30 (2), 411–425.

Thomson, M.J., de Ocampo, M., Egdane, J., Rahman, M.A., Sajise, A.G., Adorada, D.L., et al., 2010. Characterizing the Saltol quantitative trait locus for salinity tolerance in rice. Rice 3, 148–160.

Tonsor, S.J., Scott, C., Boumaza, I., Liss, T.R., Brodsky, J.L., Vierling, E., 2008. Heat shock protein 101 effects in A. thaliana: genetic variation, fitness and pleiotropy in controlled temperature conditions. Molecular Ecology 17 (6), 1614–1626.

Tulmann Neto, A., Alves, M.C., Camargo, C.E.D.O., Castro, J.L.D., Ferreira Filho, W.P., 2001. New wheat genotypes tolerant to aluminum toxicity obtained by mutation induction. Pesquisa Agropecuária Brasileira 36, 61–70.

Turan, S., Cornish, K., Kumar, S., 2012. Salinity tolerance in plants: breeding and genetic engineering. Australian Journal of Crop Science 6 (9), 1337–1348.

Tyagi, A., Mir, Z.A., Almalki, M.A., Deshmukh, R., Ali, S., 2024. Genomics-assisted breeding: a powerful breeding approach for improving plant growth and stress resilience. Agronomy 14 (6), 1128.

Valkoun, J.J., 2001. Wheat pre-breeding using wild progenitors. Euphytica 119, 17–23.

Vallejos, C.E., Tanksley, S.D., 1983. Segregation of isozyme markers and cold tolerance in an interspecific backcross of tomato. Theoretical and Applied Genetics 66, 241–247.

Varshney, R.K., Chen, W., Li, Y., Bharti, A.K., Saxena, R.K., Schlueter, J.A., et al., 2012. Draft genome sequence of pigeonpea (*Cajanus cajan*), an orphan legume crop of resource-poor farmers. Nature Biotechnology 30 (1), 83.

Vijayalakshmi, K., Fritz, A.K., Paulsen, G.M., Bai, G., Pandravada, S., Gill, B.S., 2010. Modeling and mapping QTL for senescence-related traits in winter wheat under high temperature. Molecular Breeding 26, 163–175.

Wahid, A., Gelani, S., Ashraf, M., Foolad, M.R., 2007. Heat tolerance in plants: an overview. Environmental and Experimental Botany 61 (3), 199–223.

Wainaina, C.M., Inukai, Y., Masinde, P.W., Ateka, E.M., Murage, H., Kano-Nakata, M., et al., 2015. Evaluation of cold tolerance in NERICA s compared with Japanese standard rice varieties at the reproductive stage. Journal of Agronomy and Crop Science 201 (6), 461–472.

Wang, Z., Cheng, J., Chen, Z., Huang, J., Bao, Y., Wang, J., et al., 2012. Identification of QTLs with main, epistatic and QTL× environment interaction effects for salt tolerance in rice seedlings under different salinity conditions. Theoretical and Applied Genetics 125, 807–815.

Wang, H., Wang, R., Liu, B., Yang, Y., Qin, L., Chen, E., et al., 2020. QTL analysis of salt tolerance in Sorghum bicolor during whole-Plant growth stages. Plant Breeding 139 (3), 455–465.

Wani, S.H., Choudhary, M., Kumar, P., Akram, N.A., Surekha, C., Ahmad, P., et al., 2018. Marker-assisted breeding for abiotic stress tolerance in crop plants. Biotechnologies of Crop Improvement: Genomic Approaches 3, 1–23.

Wani, S.H., Choudhary, J.R., Choudhary, M., Rana, M., Gosal, S.S., 2020. Recent advances in genomics assisted breeding for drought stress tolerance in major cereals. Journal of Cereal Research 12 (1), 1–12.

Wei, H., Liu, J., Wang, Y., Huang, N., Zhang, X., Wang, L., et al., 2013. A dominant major locus in chromosome 9 of rice (*Oryza sativa* L.) confers tolerance to 48 C high temperature at seedling stage. Journal of Heredity 104 (2), 287–294.

Witcombe, J.R., Hollington, P.A., Howarth, C.J., Reader, S., Steele, K.A., 2008. Breeding for abiotic stresses for sustainable agriculture. Philosophical Transactions of the Royal Society B: Biological Sciences 363 (1492), 703–716.

Xinglai, P., Sangang, X., Qianying, P., Yinhong, S., 2006. Registration of 'Jinmai 50' wheat. Crop Science 46 (2), 983.

Xu, W., Subudhi, P.K., Crasta, O.R., Rosenow, D.T., Mullet, J.E., Nguyen, H.T., 2000. Molecular mapping of QTLs conferring stay-green in grain sorghum (*Sorghum bicolor* L. Moench). Genome 43 (3), 461–469.

Xu, L.M., Zhou, L., Zeng, Y.W., Wang, F.M., Zhang, H.L., Shen, S.Q., et al., 2008. Identification and mapping of quantitative trait loci for cold tolerance at the booting stage in a japonica rice near-isogenic line. Plant Science 174 (3), 340–347.

Xu Xiang Yang, X.X., Wang DongMei, W.D., Kang LiGong, K.L., Li JingFu, L.J., 2008. Selection of SSR and RAPD markers related to tomato heat tolerance. Acta Horticulturae Sinica 35, 47–52.

Yadava, Y.K., Chaudhary, P., Yadav, S., Rizvi, A.H., Kumar, T., Srivastava, R., et al., 2023. Genetic mapping of quantitative trait loci associated with drought tolerance in chickpea (*Cicer arietinum* L.). Scientific Reports 13 (1), 17623.

Yamamoto, T., Uga, Y., Yano, M., 2014. Genomics-assisted allele mining and its integration into rice breeding. Genomics of Plant Genetic Resources: Crop Productivity, Food Security and Nutritional Quality 2, 251–265.

Yang, L., Wang, J., Han, Z., Lei, L., Liu, H.L., Zheng, H., et al., 2021. Combining QTL-seq and linkage mapping to fine map a candidate gene in qCTS6 for cold tolerance at the seedling stage in rice. BMC Plant Biology 21 (1), 278.

Ye, C., Fukai, S., Godwin, I., Reinke, R., Snell, P., Schiller, J., et al., 2009. Cold tolerance in rice varieties at different growth stages. Crop and Pasture Science 60 (4), 328–338.

Ye, C., Argayoso, M.A., Redoña, E.D., Sierra, S.N., Laza, M.A., Dilla, C.J., et al., 2012. Mapping QTL for heat tolerance at flowering stage in rice using SNP markers. Plant Breeding 131 (1), 33–41.

Yi, Y., Hassan, M.A., Cheng, X., Li, Y., Liu, H., Fang, W., et al., 2023. QTL mapping and analysis for drought tolerance in rice by genome-wide association study. Frontiers in Plant Science 14, 1223782.

Younis, A., Ramzan, F., Ramzan, Y., Zulfiqar, F., Ahsan, M., Lim, K.B., 2020. Molecular markers improve abiotic stress tolerance in crops: a review. Plants 9 (10), 1374.

Yu, J.K., La Rota, M., Kantety, R.V., Sorrells, M.E., 2004. EST derived SSR markers for comparative mapping in wheat and rice. Molecular Genetics and Genomics 271, 742–751.

Zandalinas, S.I., Balfagón, D., Gómez-Cadenas, A., Mittler, R., 2022. Plant responses to climate change: metabolic changes under combined abiotic stresses. Journal of Experimental Botany 73 (11), 3339–3354.

Zhang, G.Q., Zeng, R.Z., Zhang, Z.M., Ding, X.H., Li, W.T., Liu, G.M., et al., 2004. 35. The construction of a library of single segment substitution lines in rice (*Oryza sativa* L.). Rice Genetic Newsletter 121, 85–87.

Zhang, X., Shabala, S., Koutoulis, A., et al., 2017. Meta-analysis of major QTL for abiotic stress tolerance in barley and implications for barley breeding. Planta 245, 283–295. Available at https://doi.org/10.1007/s00425–016–2605–4.

Zhang, Y., Liu, L., Chen, X., Li, J., 2022. Effects of low-temperature stress during the anther differentiation period on winter wheat photosynthetic performance and spike-setting characteristics. Plants 11 (3), 389.

Zheng, L., Xie, J., Sun, X., Zheng, Y., Meng, F., Fan, X., et al., 2023. QTL mapping and candidate gene analysis of low-temperature tolerance at the germination stage of soybean. Plant Breeding 142 (6), 758–768.

Zhou, G., Johnson, P., Ryan, P.R., Delhaize, E., Zhou, M., 2012a. Quantitative trait loci for salinity tolerance in barley (*Hordeum vulgare* L.). Molecular Breeding 29, 427–436.

Zhou, L., Zeng, Y., Hu, G., Pan, Y., Yang, S., You, A., et al., 2012b. Characterization and identification of cold tolerant near-isogenic lines in rice. Breeding Science 62 (2), 196–201.

Chapter 12

Genomic breeding empowered by next-generation sequencing and generative artificial intelligence for multiple abiotic stress tolerance in plants

Vanavichit A.[1,*], Poncheewin W.[2], Wanchana S.[2], Ruenjaichol V.[2], Ruengphayak S.[1], and Arikit S.[1]

[1]*Rice Science Center, Kasetsart University, Kamphang Saen, Nakhon Pathom, Thailand,* [2]*National Center for Genetic Engineering and Biotechnology (BIOTEC), National Science and Technology Development Agency (NSTDA), Khlong Luang, Pathum Thani, Thailand*

[*]*Corresponding author. e-mail address: rdiavv@ku.ac.th*

12.1 Introduction

Plants face diverse environmental stressors that affect their growth, development, and productivity. These stresses can be broadly classified into biotic factors, caused by living organisms such as bacteria, fungi, insects, and viruses, and abiotic factors, including drought, salinity, extreme temperatures, and nutrient deficiencies (Khraiwesh et al., 2012; Neelapu et al., 2015). Among these, abiotic stresses significantly threaten global food security by significantly reducing crop yields. Plants have evolved molecular and physiological mechanisms to mitigate these challenges, including the production of stress-related proteins, transporter molecules, antioxidants, and phytohormones (Neelapu et al., 2015; Surekha et al., 2015).

Abiotic stresses such as drought, salinity, extreme temperatures, and nutrient deficiencies represent critical challenges that significantly constrain agricultural productivity on a global scale. The simultaneous occurrence of these stresses often exacerbates their adverse effects, making the development of crops with enhanced stress tolerance increasingly complex. This complexity stems from the intricate interplay between the genetic, biochemical, and physiological mechanisms. These stressors limit plant growth, disrupt metabolic processes, and drastically reduce crop yields, threatening food security and the sustainability of farming systems worldwide. Unpredictable climate change has magnified the need to develop crops resilient to harsh environmental conditions. Therefore improving crop stress tolerance must be emphasized in modern agricultural research (Shinozaki and Yamaguchi-Shinozaki, 2000; Carmo et al., 2012).

Although classical quantitative trait loci (QTL) discovery, marker-assisted selection (MAS), and gene-pyramided breeding have been instrumental, they often fail to address the multifaceted nature of these challenges and meet the rising demand for climate-resilient agricultural systems because they rely on time-consuming and resource-intensive processes.

12.2 Constraints toward gene finding underlying tolerance to abiotic stresses

New rice varieties that are tolerant to stress, responsive to production inputs, and highly nutritious are the most desirable for the next century in the face of climate change. Integrating genetic markers and biotechnology into breeding programs is revolutionizing the development of crops that can withstand the increasingly erratic challenges posed by climate change (Karavolias et al., 2021). As conventional breeding methods still play a vital role in generating new cultivars in crops, bioinformatics that can improve breeding efficiency will significantly affect breeding efficiency. One of the constraints is the use of loosely linked molecular markers for the target QTL. If breeders identify functional or tightly linked markers, MAS and gene pyramiding can be more efficient. Therefore, gene identification is key to success in breeding.

Multiple Abiotic Stresses in Plants. DOI: https://doi.org/10.1016/B978-0-443-27424-4.00011-2

Abiotic stress tolerance manifests as a QTL complexed with the phenotype/genotype interaction, rendering the direct molecular cloning approach ineffective. Genome mapping requires statistical methods to localize QTLs without prior knowledge of gene function. Breeders initially mapped the region where the candidate genes were located and narrowed down to critical genomic regions for cloning and sequencing. After sequencing, large clones or a set of overlapping clones may contain the target candidate genes. Tremendous successes in positional cloning have been demonstrated in human and plant diseases (Bent et al., 1994; Grant et al., 1995; Martin et al., 1993; Song et al., 1995).

Over the last decade, gene mining on numerous QTLs has revealed that specific sections of plant chromosomes include sites that regulate the expression of other QTLs. Breeders target genomic regions utilizing numerous quantitative features, including QTL mapping, genome-wide association studies (GWAS), and QTL-seq analysis. Approximately 8000 QTLs influence various complex properties of rice (http://www.gramene.org/qtl/index.html). QTL mapping aims to elucidate the genes responsible for polygenic characteristics and acquire a better understanding of their physiology and biochemistry (Korstanje and Paigen, 2002). In recent decades, researchers have used candidate gene techniques to characterize and clone QTL. This approach complements map-based cloning to identify genes within QTL intervals (Pflieger et al., 2001).

Breeders select candidate genes based on data gathered through classical genetics and molecular techniques (Borevitz and Chory, 2004). Currently, numerous publicly valid biological datasets, particularly high-throughput technologies, have dramatically expanded the amount of biological information available to aid in identifying gene functions. Most biological data are freely accessible via public domain resources. Scientists use these databases to conduct bioinformatic analyses and data integration systems to identify the most promising candidate genes and to elucidate the roles of rice genes. Such bioinformatics skills that can handle big data are not in most breeders' type (Table 12.1).

In drought stress, key genes such as *ZmDREB1A* and *SlDREB2* activate pathways that enhance water retention and osmotic regulation. Genes related to abscisic acid biosynthesis are vital for stress signaling. During heat stress, *HSP70* and *SlHSP17* prevent protein aggregation and stabilize cellular structures at high temperatures. Heat shock transcription

TABLE 12.1 A summary of key genes and findings related to drought, heat, and salinity stress tolerance in maize and tomato.

Stress type	Key genes/mechanisms	Key findings	References
Drought stress	– *ZmDREB1A* (maize)	– ZmDREB1A is involved in drought stress tolerance by enhancing the expression of stress-responsive genes.	Kizis and Pagès. (2002), Liu et al. (2013)
	– *SlDREB2* (tomato)	– SlDREB2 plays a key role in improving drought tolerance by activating genes related to water retention and osmotic regulation.	Hichri et al. (2016)
	– ABA biosynthesis genes (maize & tomato)	– Abscisic acid (ABA)-related genes are essential for drought stress signaling pathways in both maize and tomato.	Tuberosa and Salvi. (2006)
Heat stress	– *HSP70* (maize)	– HSP70 helps in the stabilization of proteins under heat stress by preventing aggregation and assisting in protein folding.	Hahn et al. (2011); Zai et al. (2015)
	– *SlHSP17* (tomato)	– SlHSP17 functions as a molecular chaperone under heat stress in tomato, protecting cellular integrity.	Zhang et al. (2020)
	– Heat shock transcription factors (HSFs)	– HSFs regulate the expression of heat shock proteins (HSPs) under heat stress, facilitating heat tolerance.	Guo et al. (2016); Li et al. (2024)
Salinity stress	– SOS1 (*maize* & tomato)	– SOS1 (Salt Overly Sensitive 1) is involved in Na^{+} ion extrusion, crucial for maintaining cellular ionic balance under salinity stress.	Munns and Tester. (2008)
	– NHX1 (maize & *tomato*)	– NHX1 (Na^{+}/H^{+} antiporter) plays a significant role in vacuolar Na^{+} sequestration, contributing to salt tolerance.	Flowers and Yeo. (1986)

factors regulate heat shock protein expression, which is vital for heat tolerance. The *SOS1* and *NHX1* transporters help maintain ion balance by regulating Na^+ and K^+ under saline conditions, while salt-responsive transcription factors, such as SlGSTs, manage gene expression to improve salt tolerance. These findings provide crucial insights into the genetic basis of stress tolerance, enabling breeders and researchers to develop improved varieties of maize and tomatoes capable of withstanding drought, heat, and salinity stress.

12.3 Constraints of marker-assisted selection and gene pyramiding

Single-MAS is straightforward compared to gene pyramiding for improving resistance to multiple abiotic stresses. When several QTLs are involved in abiotic stresses, breeders typically implement gene-pyramiding projects by introgressing several donors, each providing a target QTL with flanking markers into a preferred genetic background. Conventional gene pyramiding is time-consuming and tedious when many target QTLs are involved. Introducing multiple parental backgrounds caused even more complications in recovering the recurrent parental genomic background (Luo and Yin, 2013; Singh et al., 2013) and minimizing undesirable donor genomic background caused by linkage drag.

High background recovery poses a significant constraint on genomic breeding of self-pollinated crops and inbred-line development aimed at single-cross hybrid breeding programs. The cost of background recovery contributed significantly to the gene pyramiding and MAS programs. Transferring a single resistance QTL using MAS requires at least three to four backcrosses to guarantee a high recovery of the recurrent parent phenotype without background selection (Joshi and Nayak, 2010; Suh et al., 2013). For gene pyramiding programs integrating multiple QTLs, selecting donors' genomic background costs even more and possibly lengthens the project considerably. Without background screening, low background recovery has been reported for wheat introgression with multiple stripe rust resistance (Randhawa et al., 2009). With phenotypic selection, background recovery is slightly improved (Korinsak et al., 2011; Singh et al., 2013; Sundaram et al., 2008). Indeed, genome-wide marker scanning during backcrossing has been suggested to be the best method for improving targeted background recovery (Rajpurohit et al., 2011; Suh et al., 2013).

The number of molecular markers used for genome-wide marker scanning, which ultimately determines cost versus precision, has varied from 44 to 51 SSRs (Siangliw et al., 2003; Tomita, 2009; Wongsaprom et al., 2010; Yi et al., 2009), 67 to 72 SSRs (Singh et al., 2013; Win et al., 2012), 84 to 97 SSRs (Jantaboon et al., 2011; Siangliw et al., 2007), and 107 to 205 SSRs (Korinsak et al., 2011; Rajpurohit et al., 2011; Suh et al., 2013). Nonetheless, most MAS projects utilize molecular markers during the final selection stage. One successful approach was reported using pseudo-backcrossing in rice, which can minimize the donor's genomic background and considerably improve recurrent genomic background recovery (Ruengphayak et al., 2015).

In conventional backcrossing, many portions of the donor genome are instantly inserted into both the carrier and non-carrier recipient chromosomes during the early cycles. Later, depending on the number of backcrossings and genomic background screenings, donor genomic segments are gradually replaced by sequences from the recurrent parent at varying rates. Nonetheless, linkage drag, which causes long-lasting donor genome segments, makes MAS and gene pyramiding inefficient. Reports have revealed that the degree of linkage drag is independent of the size of the target genes/QTLs, but depends on the nature of the donor's genomic segments (Ruengphayak et al., 2015). Persistent linkage drags have been reported to transfer bacterial leaf blight (BB) genes to *indica* and *japonica* rice varieties (Rajpurohit et al., 2011; Suh et al., 2013). One possible cause of persistent linkage drag has been reported in transferring the *Xa21* gene derived from a wild rice species, *Oryza longistaminata*, through several cycles of backcrossing with *indica* rice (Song et al., 1995). Recombination suppression in specific genomic segments may cause a high degree of linkage drag among genes inherited from the wild species. Breeders have implemented additional rounds of backcrossing using flanking molecular markers to overcome persistent linkage drag (Rajpurohit et al., 2011; Tomita, 2009). In tomatoes, pyramiding of resistance genes against multiple diseases, such as bacterial wilt and Verticillium wilt, can lead to unintended adverse effects, such as lower fruit yield (Hanson et al., 2016). In cross-pollinated crops, such as corn, pyramiding resistance to European corn borers and Southern corn leaf rust either reduces the overall fitness of the plant due to linkage drag or masks the full expression of pyramided traits due to heterosis (Duvick and Cassman, 1999; López-Malvar et al., 2021).

12.4 Genome-wide association studies: seeking quantitative trait loci directly on natural genetic variation

The reproductive barrier between cultivated varieties and wild relatives is an essential constraint for exploiting the germplasm for QTL mapping, map-based cloning, and selective breeding. Efforts to enhance abiotic stress tolerance have relied on traditional breeding and transgenic technologies, but these methods often fall short owing to the polygenic nature of stress tolerance (Falconer and Mackay, 1996; Kearsey, 1998). Genome-wide association studies (GWAS) have emerged as powerful tools for dissecting the genetic basis of abiotic stress tolerance and identifying genes linked to adaptive traits. Unlike conventional QTL mapping, which relies on controlled crosses, GWAS leverages natural genetic variation in diverse populations, allowing the identification of multiple alleles that influence stress tolerance traits (Bush and Moore, 2012). Advances in high-throughput sequencing and genotyping technologies have further enabled large-scale single-nucleotide polymorphism (SNP) identification, making GWAS an essential tool in modern crop improvement programs (DiStefano and Taverna, 2011; Kumar et al., 2015).

12.5 Advancements in genome-wide association study models

Recent advancements in GWAS models have improved the accuracy and efficiency of identifying the genetic loci associated with complex traits in plants. Traditional single-locus GWAS models, such as the General Linear Model and Mixed Linear Model, are widely used, but often suffer from false positives and limited power in detecting multiple causal variants (Yoosefzadeh-Najafabadi et al., 2022a). To overcome these challenges, multilocus GWAS (ML-GWAS) models, including Fixed and Random Model Circulating Probability Unification (FarmCPU), Bayesian information, and Linkage-disequilibrium Iteratively Nested Keyway (BLINK), have been developed to improve statistical power by analyzing multiple SNPs simultaneously (Enyew et al., 2022).

Haplotype-based GWAS has also gained attention because of its ability to capture linkage disequilibrium blocks, improving resolution in identifying candidate genes (Bhat et al., 2021). More recently, machine learning (ML)-driven GWAS have emerged, integrating artificial intelligence (AI) algorithms such as random forests, support vector machines (SVM), and deep learning networks to enhance SNP-trait associations (Yoosefzadeh-Najafabadi et al., 2022b). These AI-based approaches facilitate the detection of gene-gene interactions (epistasis) and improve the prediction accuracy for polygenic traits, such as abiotic stress tolerance. Additionally, multiomics GWAS, which combines transcriptomics, metabolomics, and epigenomics, has revolutionized gene discovery by providing a more comprehensive understanding of plant stress responses (Knoch et al., 2024). These advancements have significantly accelerated molecular breeding, enabling the development of climate-resilient crops with enhanced tolerance to drought, salinity, and extreme temperature.

12.6 Applications of genome-wide association studies for abiotic stress tolerance in plants

GWAS have been widely applied to study plant responses to abiotic stresses, such as drought, salinity, and extreme temperatures. Researchers have identified key genetic loci and candidate genes that contribute to plant adaptation to stress conditions using this approach. GWAS enables researchers to identify genetic markers associated with abiotic stress tolerance by analyzing large populations of genetically diverse plants. This approach has become instrumental in uncovering the genetic basis of complex traits and in facilitating the development of stress-resilient crop varieties (Korte and Farlow, 2013).

12.7 Drought tolerance

Drought stress significantly reduces crop productivity by affecting water uptake and metabolism, necessitating GWAS to identify genetic markers associated with drought tolerance in various crops, including sweet sorghum, where 354 accessions and 6186 SNPs led to the discovery of 79 significant loci and two candidate genes homologous to transcription factors in *Arabidopsis* and maize (Xin et al., 2022); soybean, where a GWAS on 240 accessions using 34,817 SNPs identified 23 and 27 quantitative trait nucleotides linked to germination rate (GR), root length (RL), and whole seedling length, with four key candidate genes (*Glyma08g156800*, *Glyma08g160000*, *Glyma08g162700*, and *Glyma13g249600*) playing roles in drought tolerance (Aleem et al., 2024); faba bean, where a high-density SNP genotyping array analysis of 100 accessions under drought conditions revealed 29 SNP markers and novel genes, including uncharacterized proteins and stress-inducible signal transducers that contribute to drought resilience (Gutiérrez et al., 2023); wheat, where a GWAS using a 90 K SNP array on 125 accessions under well-watered and drought stress conditions identified 53 SNPs significantly associated with stress susceptibility and tolerance indices, particularly on chromosomes 2A and 3B, with 44 unique candidate genes involved in plant growth and stress responses (Nouraei et al., 2024). Drought, particularly terminal drought, severely affects wheat growth and yield, making it crucial to understand the genetic mechanisms underlying drought resistance. A GWAS using high-throughput phenotyping identified 5320 drought-related SNPs and 27 SNP clusters, with *TaPP2C6* emerging as a key negative regulator of drought response. Knockout lines of *TaPP2C6* showed enhanced drought resistance without yield loss, and a haplotype analysis confirmed its potential for wheat breeding, while a machine learning-based prediction model using 24 image-based traits further advanced strategies for developing drought-resistant, high-yielding wheat varieties (Zhang et al., 2024); maize, where a GWAS on 379 inbred lines using 97,862 SNPs identified 15 genetic variants and candidate genes linked to metabolism, transcriptional regulation, programmed cell death, autophagy, and cell growth, offering new insights for breeding drought-resistant varieties (Chen et al., 2023); and rice, where a GWAS on 507 accessions analyzed 51 image-based traits (i-traits) under drought conditions, identifying 470 loci, including *OsPP15*, whose role in drought resistance was validated through genetic transformation experiments, demonstrating the potential of i-traits for dissecting complex traits and enhancing drought tolerance in rice (Guo et al., 2018).

12.8 Salt tolerance

Salt stress disrupts the plant water balance and ion homeostasis, leading to reduced growth and yield. GWAS has been employed to identify the genetic factors that contribute to salt tolerance in various crop species. Kumar et al. (2015) used a custom 6000-SNP array to genotype 220 rice accessions and identified 20 SNPs associated with Na^+/K^+ ratio and 44 SNPs linked to other salinity stress traits. The *Saltol* QTL on chromosome 1, known for seedling-stage salinity tolerance, was also found to influence the Na^+/K^+ ratio during the reproductive stage, along with new QTLs on chromosomes 4, 6, and 7. These findings highlight *indica* rice accessions as valuable sources of salt tolerance genes and demonstrate the efficiency of SNP-based GWAS in identifying the genomic regions regulating salinity tolerance in rice. Improving salt tolerance in direct-seeding rice at the germination stage is crucial to ensure seedling establishment in saline soils. Shi et al. (2017) conducted a GWAS on 478 rice accessions, identifying 11 loci containing 22 SNPs associated with salt tolerance, with major loci explaining 20.2% and 26.4% of the phenotypic variation for vigor index and mean germination time, respectively. Key association regions were found on chromosomes 1 and 2, including a region near the Na^+/K^+ regulatory loci and nitrate transporter genes (*OsNRT2.1* and *OsNRT2.2*) that influence gene expression under salt stress. These findings provide valuable insights into gene cloning and breeding of salt-tolerant direct-seeding rice varieties through genomic and MAS. Dilnur et al. (2019) analyzed 215 Asiatic cotton accessions using GWAS. They identified nine SNP-rich regions linked to salt tolerance traits, with 40 candidate genes associated with salt tolerance. Notably, two SNP-rich regions on chromosome 7 were linked to key salinity-related traits, providing valuable insights for future molecular breeding to develop salt-tolerant cotton cultivars. Recently, Raggi et al. (2024) screened 192 common bean genotypes for salt tolerance at the seedling stage and identified seven significant genomic regions associated with salt stress tolerance through GWAS. The detected candidate genes showed strong homology with known salt-tolerance genes in Arabidopsis, offering valuable insights for breeding salt-tolerant common bean varieties to mitigate salinity-induced yield losses.

12.9 Thermal tolerance

Climate variability poses a significant threat to global crop production, highlighting the urgent need to understand genetic variation to develop temperature-resilient breeding lines. Extreme temperatures adversely affect plant metabolism, reproduction, and yield, and heat stress during fertilization is particularly problematic in tropical regions under climate change scenarios. GWAS have identified genetic factors linked to heat and cold tolerance across various crop species. Lafarge et al. (2017) analyzed 167 *indica* rice landraces and varieties under 38°C during anthesis. They identified 14 loci significantly associated with spikelet sterility (SPKST), linked to abiotic stress responses, gametophyte development, and heat shock proteins. Notably, favorable alleles were scattered across *Oryzasativa* genetic groups but were rarely combined in single accessions, with heat-tolerant donors such as N22 and specific Indian and Taiwanese accessions serving as valuable genetic resources for breeding heat-resilient rice. Besides heat, cold stress also significantly impacts plant growth and development. A GWAS study on 127 rice accessions, including 57 Korean landraces, identified four QTLs on chromosomes 1, 4, and 5 associated with cold tolerance at the seedling stage, with two QTLs overlapping with previously reported loci linked to cold tolerance and seedling vigor, explaining 16.5%–18.5% of variance (Ham et al., 2021; Shimoyama et al., 2020) also analyzed chilling tolerance traits in 354 rice accessions from Rice Diversity Panel 1 (RDP1), identifying 500 QTLs covering 25% of the rice genome. These loci are associated with genes that regulate carbohydrate biosynthesis, transmembrane transport, and stress responses. Further analysis highlighted Golgi-mediated subcellular processes and intracellular carbohydrate regulation as the key mechanisms in chilling tolerance. Chopra et al. (2017) investigated the sorghum association panel. They identified 30 SNPs linked to cold stress tolerance and 12 SNPs associated with heat stress, with key genes involved in anthocyanin expression, carbohydrate metabolism, and ion transport. Coexpression network analysis revealed complex gene interactions, including upregulated expression of the

Sb06g025040 transcription factor in cold-tolerant lines. Collectively, these findings offer valuable genetic insights into temperature stress resilience, paving the way for marker-assisted breeding of heat- and cold-tolerant cultivars of rice and other cereal crops.

12.10 Flood tolerance

Flooding stress poses a significant challenge to crop production; however, certain plant species, such as rice, have developed unique adaptive mechanisms to tolerate submergence. Rice is particularly notable for its ability to germinate under flooded conditions through rapid coleoptile elongation, a trait with considerable variation among *indica* rice varieties. Zhang et al. (2017b) screened 432 *indica* rice accessions for flooded coleoptile length (FCL) and identifying 11 SNPs associated with FCL and nine linked to the flooding tolerance index, with five SNPs overlapping between both traits. This study further highlighted *LOC_Os06g03520* as a key candidate gene, showing a remarkable 507-fold induction under anoxia, making it a strong target for breeding flood-tolerant rice varieties. Similar to rice, wild soybeans possess genetic variation that enhances flooding tolerance in cultivated soybeans. Tran et al. (2024) conducted a GWAS on 163 wild soybean accessions and identified six and three tolerant accessions at the early vegetative and germination stages, respectively. The study detected 16 and 20 SNPs linked to submergence-related traits across different chromosomes, identifying 10 and 4 putative genes, respectively, at the two growth stages. These findings suggest that specific genes contribute to submergence stress tolerance in wild soybeans, thereby offering valuable genetic resources for breeding flood-resistant soybean varieties.

Seed flooding stress presents a major environmental challenge that severely limits soybean germination; however, its genetic basis remains largely unknown. Sharmin et al. (2021) conducted a GWAS on 243 soybean accessions using 34,718 SNPs to identify loci associated with seed-flooding tolerance at the germination stage. This study uncovered several SNPs linked to key traits, such as electrical conductivity (EC), GR, shoot length, and RL, with two major SNPs, *Gm_08_11971416* and *Gm_08_46239716*, consistently associated with EC and GR across different environments. Further analysis predicted eight candidate genes and three hub genes within the SNP regions. This study provides critical genetic resources for future functional validation and soybean breeding programs to improve seed flooding tolerance. These studies highlight the importance of genetic variation in flooding tolerance across rice and soybean and offer valuable insights for breeding climate-resilient crop varieties.

12.11 Heavy metal toxicity tolerance

Heavy metal toxicity, particularly from excessive iron (Fe) and zinc (Zn), poses a significant challenge to rice production by affecting both yield and grain quality. Understanding the genetic and physiological mechanisms underlying rice tolerance to heavy metal stress is crucial for developing resilient rice varieties through molecular breeding. A GWAS in rice has identified multiple loci and candidate genes associated with Fe and Zn toxicity tolerance, shedding light on the key biochemical and physiological mechanisms underlying stress adaptation. Matthus et al. (2015) discovered 18 loci linked to iron toxicity tolerance, including a locus on chromosome 1 containing glutathione-S-transferases, which were strongly expressed under iron stress and exhibited sequence polymorphisms associated with tolerance. This study emphasizes that maintaining foliar redox homeostasis is crucial for iron toxicity tolerance. Similarly, Zhang et al. (2017a) identified 29 and 31 QTL associated with Fe and Zn toxicity tolerance at the seedling stage, respectively, affecting various traits such as shoot height, RL, and ion concentrations. Notably, five QTLs were shared between the Fe and Zn stress conditions, and 22 candidate genes were identified through haplotype analysis, revealing a partial genetic overlap between Fe and Zn toxicity tolerance. In another GWAS conducted on a highly diverse Thai *indica* rice population, Kaewcheenchai et al. (2021) identified 16 significant SNPs associated with iron toxicity tolerance. Among them, SNP:2_21262165 showed a strong potential for Fe toxicity tolerance, making it a promising candidate for SNP marker development. Additionally, *RAR1* was found to play a crucial role in cellular redox homeostasis by interacting with the stress-responsive protein, Hsp90. Further research by Theerawitaya et al. (2022) explored agro-morphological and physiological traits across different growth stages in a diverse rice panel and identified 29 QTLs linked to Fe toxicity tolerance, including key colocalized QTLs on chromosomes 5, 6, and 11. Among the candidate genes, *LOC_Os01g72370* (*OsIRO2*) and *LOC_Os05g16670* were associated with iron homeostasis, providing valuable genetic resources for developing Fe toxicity-tolerant rice varieties. More recently, Miao et al. (2024) conducted a GWAS on 551 rice accessions and identified 29 QTLs related to Fe toxicity tolerance, with five key QTLs and candidate genes, *LOC_Os05g40160* and *LOC_Os12g36890*, which play significant roles in Fe homeostasis and antioxidant response. Collectively, these studies provide valuable insights into the genetic and molecular mechanisms of Fe and Zn toxicity tolerance in rice, offering essential genetic resources for functional characterization and marker-assisted breeding to improve resilience against heavy metal stress. These findings provide genetic targets for developing crops resilient to climate variability.

12.12 Pangenomics: a game-changer in mining into natural genetic variation

Understanding natural genetic variation can be explored more efficiently by resequencing multiple varieties and germplasm using new whole-genome sequencing (WGS) technologies. Conventional whole-genome resequencing assembled short-read sequences using a single-reference genome may handicap the discovery of new genes underlying natural genetic variation. Recently, the advent of single-molecule genome sequencing technologies led by Oxford Nanopore sequencing, combined with specialized algorithms, has unleashed gapless, telomere-to-telomere genome sequences (T2T) for any complex genome. Single-molecule sequencing technology has enabled the *de novo* assembly of multiple accessions, particularly among less-studied germplasm, including chickpea (Varshney et al., 2021), foxtail millet (He et al., 2023), and pearl millet (Yan et al., 2023). Pangenome takes comprehensive computational methods to assemble whole-genome sequences from multiple germplasm within species or genera into representative pan-references.

Pangenomics redefines how different individuals within a species vary at the genomic scale. It is a game-changer in gene discovery and precision breeding. Unlike current resequencing projects, pangenomics considers whole-genome sequences from multiple germplasm rather than a single variety. Assembling multiple germplasm is now possible at a reasonable cost with the advent of the 3rd generation sequencing. The T2T-based pangenome assembled a long-read sequence with the power to solve highly repetitive telomeric regions. Panomics, combining OMICs with a single-cell sorter, mass spectroscopy, and plant phenomics, allows breeders to thoroughly investigate natural genetic variation contributing to gene mining, significant plant development, stress tolerance, metabolomes, agronomics, and domestication (Tao et al., 2019). The combination of long-read genome sequencing and rigorous genome assembly has been close to capturing the entire gene content of species and has shed new light on the landscape of crop diversity and improvement. Advances in pangenomics have been reported for soybean (Ge et al., 2022; Li et al., 2014), rice (Schatz et al., 2014; Wang et al., 2018, 2023; Zhao et al., 2018), maize (Hirsch et al., 2014; Yang et al., 2019), wheat (Montenegro et al., 2017), *Brachypodiumdistachyon* (Gordon et al., 2017), *Brassicarapa* (Li et al., 2014), *B. oleracea* (Golicz et al., 2016; Bayer et al., 2016), and *B. napus* (Hurgobin et al., 2018).

Whole-genome de novo assembly without a reference genome remains a challenge for large-scale resequencing of germplasm. A highly repetitive repertoire, high heterozygosity, and polyploidy do not align straightforwardly with the genome, resulting in fragmented assemblies (Kaur et al., 2024). Approaches to constructing pangenomes have been reported (Golicz et al., 2016). Three common assembly strategies are comparative *de novo* assembly (Gordon et al., 2017; Li et al., 2014; Schatz et al., 2014; Zhao et al., 2018), iterative assembly (Golicz et al., 2016; Hurgobin et al., 2018; Montenegro et al., 2017), and the map-to-pan approach (Wang et al., 2018). Additional improved methodologies included reference-based resequencing (Kaur et al., 2024), graph-pangenomics (He et al., 2023; Wang et al., 2023; Yan et al., 2023), pan-transcriptomics (Tang et al., 2016; Wang et al., 2024), and metagenomics (Conesa et al., 2005).

12.13 Gene mining pangenomes

The pangenome concept was first discovered in the comparative genomics of *Streptococcus agalactiae* (Tettelin et al., 2005) and plants (Morgante et al., 2007). The first pangenome in plants was reported in maize (Hirsch et al., 2014) and polyploid wheat (Montenegro et al., 2017). The pangenome represents the entire set of genes within a species, including core (100% shared), dispensable (1%–99% shared), softcore (95%–98%), shell (5%–94%), and cloud (1%–5%) genes (Gao et al., 2019; Gordon et al., 2017; Liu et al., 2022; Montenegro et al., 2017; Morgante et al., 2007; Yang et al., 2022). Genes represented by a single variety are called private or unique genes (Yan et al., 2023; Yang et al., 2022). Pangenomics represents the genetic diversity within a species. Pangenomics may represent global genetic diversity from multiple related species (Cochetel et al., 2023; Hoopes et al., 2022).

Pangenomes refer to genes as the whole content of the species as a pan-reference. Genomic variations include SNP, small insertions and deletions, and major structural variations (SV). SVs include presence or absence variations, copy number variations, chromosomal rearrangements, inversions, and translocations (Liu et al., 2022; Shang et al., 2022; Tao et al., 2019).

The most extensive multipanomics study has been reported in rice using 5105 accessions (Ma et al., 2022; Qin et al., 2021; Shang et al., 2022; Wang et al., 2018, 2023; Wu et al., 2023; Zhang et al., 2022; Zhao et al., 2018; Zhou et al., 2020b, 2023). Plant pangenomes are considered open in the initial phase and closed when no new genes are added to each additional genome, thereby reaching an asymptote phase. Examples include maize (Hirsch et al., 2014), peppers (Ou et al.,

2018), rice (Wang et al., 2018), sunflower (Hübner et al., 2019), tomato (Gao et al., 2019), soybean (Torkamaneh et al., 2021), sorghum (Ruperao et al., 2021), chickpea (Varshney et al., 2021), pea (Yang et al., 2022), and mungbean (Liu et al., 2022) are considered closed pangenomes. The genome ranges included among these closed pangenomes were 118 (pea)-3366 (chickpea), with several germplasm types to maximize genetic diversity.

The inclusion of wild relatives can expand a simple pangenome into a superpangenome and dramatically enhance the power of gene discovery. A super pan-genome was reported in rice by assembling 251 accessions of cultivated and wild rice species (Shang et al., 2022). They uncovered inter- and intra-specific diversity of LRR genes, enhancers, and key domestication genes in Asian and African rice. The pan-genome approach enhances gene discovery and allele mining associated with resistance to biotic and abiotic stresses, agronomic traits, and nutrient density.

12.14 Pangenomes impact genomic breeding

A pangenome reference will decrease the bias of using a single-genome reference for genomic breeding and gene editing (Della Coletta et al., 2021). Insights into the unique sources of natural genetic variation among crops could improve the accuracy of MAS. Pangenomes from several crops have revealed that dispensable genes, which vary between 8% and 61% (Tao et al., 2019), may determine genotype specificity.

Core genes are conserved, whereas dispensable genes involve agronomically important QTLs, such as resistance to multiple abiotic stresses, biomass, and grain yield (Golicz et al., 2016; Hirsch et al., 2014; Montenegro et al., 2017; Yao et al., 2015). The expression of dispensable genes is lower with tight regulation, and their functions are largely unknown (Contreras-Moreira et al., 2017). Evolutionarily, dispensable genes evolve more rapidly than core genes and are often enriched with nonsynonymous mutations and transposable elements (Gordon et al., 2017). Therefore, dispensable genes may determine genetic diversity among plant varieties and germplasm.

Many SVs, often associated with QTLs, are not linked with SNPs in rice pangenomes, which cannot be helpful in association studies, QTL mapping, and MAS (Qin et al., 2021; Yan et al., 2023; Yang et al., 2022; Zhao et al., 2020). Additionally, all homologs and haplotypes can be accurately assembled using long-read sequencing, even in polyploid species (Chen et al., 2020; Long et al., 2022). These haplotype-aware assemblies are helpful for QTL mapping, GWAS, MAS, and gene editing (Chen et al., 2020). Allele-specific sequences are critical for designing transgene-free gene editing systems that accurately target specific haplotypes (Long et al., 2022). Additionally, a pangenome can provide accurate structural variants and functional information to avoid off-target effects in CRISPR-Cas gene editing (Tay Fernandez et al., 2022). Transitioning from employing a single-reference genome to a pan-reference genome can improve variant calling accuracy and facilitate gene-trait linkage and QTL mapping for genomic breeding (Zhou et al., 2023).

With the additional pangenome sequence under investigation, scientists still believe the overall number of dispensable genes may be underestimated. Pangenomes have improved the accuracy of variant calling, QTL analyses, and GWAS for essential traits in tomato (Zhou et al., 2022), pearl millet (Yan et al., 2023), and pea (Zhao et al., 2020a). While most molecular markers have been developed primarily from WGS of a single or a few model plant varieties, a significant proportion of molecular markers have yet to be uncovered from dispensable genomes. One good example is the discovery of 8000 genes absent from the Nipponbare reference genome (Yao et al., 2015). These genomic regions may be unaccounted for in the background genome recovery in MAS/pyramiding. When functional SVs are missing from the reference genome, this affects QTL and GWAS analyses, thereby reducing the effectiveness of MAS and gene editing.

Pangenome resources provide valuable insights into the significance of core, dispensable, and novel genes in plant breeders (Hirsch et al., 2014; Shang et al., 2022; Tay Fernandez et al., 2022; Torkamaneh et al., 2021; Zhang et al., 2021). Many crop online portals have been published for RPAN (Sun et al., 2017) and RiceSuperPIRdb (Guangdong Laboratory for Lingnan Modern Agriculture. (2021). *RiceSuperPIRdb* http://www.ricesuperpir.com/, no date) in rice, MaizeGDB (Lawrence et al., 2004) and ZEAMAP (Gui et al., 2020) in maize, The Wheat Panache Web Portal (Bayer et al., 2022) and Wheat Pangenome (Montenegro et al., 2017) in wheat, PepperPan (Ou et al., 2018) in peppers, and PanSoy (Torkamaneh et al., 2021) in soybean, allowing plant biologists and breeding community access and visualization in an easy-to-read format.

12.15 Artificial intelligence–driven genomic breeding

Recent advancements in AI, particularly ML, deep learning (DL), and data mining techniques, have introduced innovative approaches to better understand, predict, and address these issues, offering solutions that were previously unattainable (Eftekhari et al., 2024).

Multiomics integration: Multiomics technology encompasses genomics, transcriptomics, proteomics, and metabolomics, all of which are essential for understanding the genetic basis of complex traits (Yang et al., 2021). AIplays a pivotal role in harmonizing and analyzing these diverse datasets, particularly in managing the high-dimensionality of omics data (Yan and Wang, 2023), thereby enabling the identification of gene-trait associations and underlying biological mechanisms (Mahmood et al., 2022). For example, the integration of AI with omics technologies has enhanced our understanding of plant defense mechanisms against biotic and abiotic stresses (Murmu et al., 2024), and the identification of salt-tolerant genes in smooth cordgrass (*Spartina alterniflora*) (Gao et al., 2023; Huang et al., 2024; Yan et al., 2023).

Genomic prediction: Targeted genes can be predicted mainly through genomic selection (GS), which has revolutionized modern breeding by leveraging genomic data to accurately predict the genetic potential of individuals, thereby offering unprecedented precision in identifying desirable traits (Crossa et al., 2017). AI-driven ML models, such as SVMs, random forests, and DL techniques, have significantly enhanced the efficiency of GS by analyzing large-scale genomic datasets and capturing complex, nonlinear relationships between genetic markers and traits (Lourenço et al., 2024; Montesinos-López et al., 2021). For instance, AI has been applied in rice breeding to predict susceptibility to rice blast disease, achieving improved accuracy compared to conventional methods (Huang et al., 2019). Similarly, DL techniques have been used to predict wheat resistance to leaf and stripe rust, outperforming traditional linear approaches by capturing genotype-by-environment interactions (Montesinos-López et al., 2019). In maize, ensemble ML models have been employed to predict resistance to Septoria and Gray Leaf Spot diseases, integrating environmental variables to further enhance the prediction (Pérez-Rodríguez et al., 2020). Additionally, AI has been utilized in breeding program improvements through simulation-based approaches (Hamazaki and Iwata, 2024). These studies collectively illustrate the transformative role of AI in genomic prediction, enabling breeders to accelerate the development of high-yielding, stress-tolerant crop varieties, while addressing critical challenges related to food security and climate change.

Phenotyping and image analysis: These nondestructive detection methods have transformed high-throughput technologies, including hyperspectral imaging, LiDAR, and drones, which facilitate the collection of vast and detailed datasets for evaluating various plant traits, such as height, biomass, leaf area, and disease severity (Abebe et al., 2023; Das Choudhury et al., 2019). These innovations enable efficient, accurate analysis of plant characteristics, allowing researchers to conduct in-depth phenotypic evaluations with minimal manual intervention. AI-driven tools, particularly convolutional neural networks (CNNs), also enhance the speed and precision of phenotyping by detecting early stress responses and nutrient deficiencies in crops, which are critical factors for managing crop health (Jiang and Li, 2020). Furthermore, AI-based methods, such as CNNs, have automated trait detection, including glume pubescence in wheat, significantly improving breeding efficiency. This AI-powered approach overcomes the traditional phenotyping constraints and is a reliable tool for selecting stress-resistant cultivars (Artemenko et al., 2023). Integrating AI with automated phenotyping platforms like multispectral and thermal cameras also enhances precision in field trials, accelerating breeding programs by ensuring consistent and accurate trait measurements (Walsh et al., 2024). Additionally, hyperspectral imaging combined with machine learning has been effectively used to study salt stress in okra (*Abelmoschus esculentus* L.), demonstrating the power of these technologies in identifying environmental stress responses (Feng et al., 2020). A 23-layered CNN has also been successfully used to measure nitrogen deficiency-induced stress in plant shoots, providing further evidence for the capabilities of AI in detecting subtle changes in plant health (Azimi et al., 2021).

Challenges and limitations: AI adoption for plant breeding is challenging, including the need for high-quality, standardized datasets, variability in data collection methods, and insufficient open-access datasets presenting significant obstacles. These challenges hinder model development, particularly for smaller institutions that lack the resources to generate comprehensive datasets. Biological complexity, including gene-environment interactions and the integration of management practices, poses significant barriers as AI models often struggle with overfitting and lack interpretability. Moreover, many breeders in developing countries lack comprehensive computational infrastructure and expertise, which constrains the adoption of AI. Integrating AI into established workflows also presents difficulties, with ethical concerns regarding data ownership and privacy, adding further complexity. These issues are exacerbated by proprietary models of plant breeding, which may deepen inequities in access to agricultural innovations. To overcome these hurdles, enhanced collaboration between AI experts and breeders, strategic investments in infrastructure, and the development of user-friendly AI tools tailored to specific contexts are critical (Atapattu et al., 2024; Williamson et al., 2023).

Future opportunities: AI for plant breeding is expanding, with several promising advancements on the horizon. The development of explainable AI is a key area of progress, with techniques such as attention mechanisms and feature-importance analysis offering greater transparency and validation for AI-driven decisions. For example, AI models have the potential to predict the outcome of CRISPR-Cas9 gene editing. Another promising technology is digital twin modeling, which allows breeders to simulate crop growth under various environmental conditions and accurately predict crop performance (Escribà-Gelonch et al., 2024; Pylianidis et al., 2021). Additionally, the integration of edge AI, which

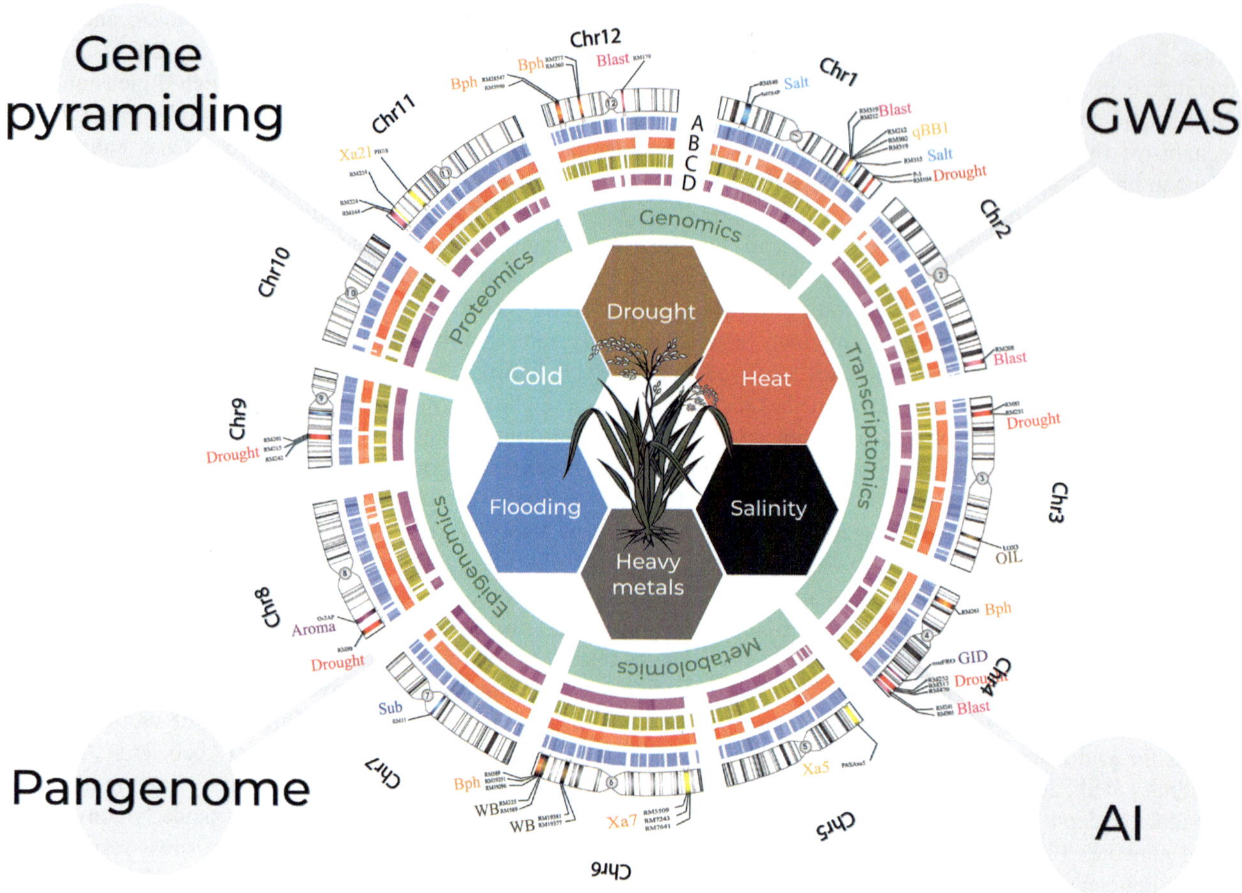

FIGURE 12.1 Integrated gene mining from QTL mapping, GWAS, pangenomes, OMICS, and AI-assisted genomic prediction.

processes data locally in the field using sensors, facilitates real-time phenotyping and enhances farmers' decision-making. AI can also drive sustainability by identifying breeding strategies that optimize resource use, reduce water consumption, and improve nutrient efficiency, ultimately contributing to climate-resilient farming practices. These advancements could address global challenges, such as climate change and food security, ushering in a new era of AI-driven precision agriculture (Fig. 12.1).

References

Abebe, A.M., Kim, Y., Kim, J., Kim, S.L., Baek, J., 2023. Image-based high-throughput phenotyping in horticultural crops. Plants 12 (10). Available at https://doi.org/10.3390/plants12102061.

Aleem, M., Razzaq, M.K., Aleem, M., Yan, W., Sharif, I., Siddiqui, M.H., et al., 2024. Genome-wide association study provides new insight into the underlying mechanism of drought tolerance during seed germination stage in soybean. Scientific Reports 14 (1). Available at https://doi.org/10.1038/s41598-024-71357-8.

Artemenko, N.V., Genaev, M.A., Epifanov, R.U.I., Komyshev, E.G., Kruchinina, Y.V., Koval, V.S., et al., 2023. Image-based classification of wheat spikes by glume pubescence using convolutional neural networks. Federation Frontiers in Plant Science 14. Available at https://doi.org/10.3389/fpls.2023.1336192.

Atapattu, A.J., Perera, L.K., Nuwarapaksha, T.D., Udumann, S.S., Dissanayaka, N.S., 2024. Challenges in achieving artificial intelligence in agriculture. Artificial Intelligence Techniques in Smart Agriculture. Springer Nature, Sri Lanka, pp. 7–34.

Azimi, S., Kaur, T., Gandhi, T.K., 2021. A deep learning approach to measure stress level in plants due to nitrogen deficiency. Measurement 173, 108650. Available at https://doi.org/10.1016/j.measurement.2020.108650.

Bayer, P.E., Petereit, J., Durant, É., Monat, C., Rouard, M., Hu, H., et al., 2022. Wheat Panache: a pangenome graph database representing presence–absence variation across sixteen bread wheat genomes. Plant Genome 15 (3). Available at https://doi.org/10.1002/tpg2.20221.

Bent, A.F., Kunkel, B.N., Dahlbeck, D., Brown, K.L., Schmidt, R., Giraudat, J., et al., 1994. RPS2 of Arabidopsis thaliana: a leucine-rich repeat class of plant disease resistance genes. Science (New York, N.Y.) 265 (5180), 1856–1860. Available at https://doi.org/10.1126/science.8091210.

Bhat, J.A., Yu, D., Bohra, A., Ganie, S.A., Varshney, R.K., 2021. Features and applications of haplotypes in crop breeding. Communications Biology 4 (1). Available at https://doi.org/10.1038/s42003-021-02782-y.

Borevitz, J.O., Chory, J., 2004. Genomics tools for QTL analysis and gene discovery. Current Opinion in Plant Biology 7 (2), 132–136. Available at https://doi.org/10.1016/j.pbi.2004.01.011.

Bush, W.S., Moore, J.H., 2012. Chapter 11: Genome-wide association studies. PLoS Computational Biology 8 (12). Available at https://doi.org/10.1371/journal.pcbi.1002822.

Carmo, B., Zanelato, F., de Almei, A.-A.F., Rehem, B.C., Almeida, A.-A.F.D., Bertolde, F.Z., 2012. Regulation of gene expression in response to abiotic stress in plants. In: Bubulya, P. (Ed.), Cell Metabolism - Cell Homeostasis and Stress Response. IntechOpen, pp. 38. Available at https://doi.org/10.5772/26636.

Chen, H., Zeng, Y., Yang, Y., Huang, L., Tang, B., Zhang, H., et al., 2020. Allele-aware chromosome-level genome assembly and efficient transgene-free genome editing for the autotetraploid cultivated alfalfa. Nature Communications 11 (1). Available at https://doi.org/10.1038/s41467-020-16338-x.

Chen, S., Dang, D., Liu, Y., Ji, S., Zheng, H., Zhao, C., et al., 2023. Genome-wide association study presents insights into the genetic architecture of drought tolerance in maize seedlings under field water-deficit conditions. Frontiers in Plant Science 14. Available at https://doi.org/10.3389/fpls.2023.1165582.

Chopra, R., Burow, G., Burke, J.J., Gladman, N., Xin, Z., 2017. Genome-wide association analysis of seedling traits in diverse Sorghum germplasm under thermal stress. BMC Plant Biology 17 (1). Available at https://doi.org/10.1186/s12870-016-0966-2.

Cochetel, N., Minio, A., Guarracino, A., Garcia, J.F., Figueroa-Balderas, R., Massonnet, M., et al., 2023. A super-pangenome of the North American wild grape species. Genome Biology 24 (1). Available at https://doi.org/10.1186/s13059-023-03133-2.

Conesa, A., Götz, S., García-Gómez, J.M., Terol, J., Talón, M., Robles, M., 2005. Blast2GO: a universal tool for annotation, visualization and analysis in functional genomics research. Bioinformatics (Oxford, England) 21 (18), 3674–3676. Available at https://doi.org/10.1093/bioinformatics/bti610.

Contreras-Moreira, B., Cantalapiedra, C.P., García-Pereira, M.J., Gordon, S.P., Vogel, J.P., Igartua, E., et al., 2017. Analysis of plant pan-genomes and transcriptomes with GET_HOMOLOGUES-EST, a clustering solution for sequences of the same species. Frontiers in Plant Science 8. Available at https://doi.org/10.3389/fpls.2017.00184.

Crossa, J., Pérez-Rodríguez, P., Cuevas, J., Montesinos-López, O., Jarquín, D., de los Campos, G., et al., 2017. Genomic selection in plant breeding: methods, models, and perspectives. Trends in Plant Science 22 (11), 961–975. Available at https://doi.org/10.1016/j.tplants.2017.08.011.

Das Choudhury, S., Samal, A., Awada, T., 2019. Leveraging image analysis for high-throughput plant phenotyping. Frontiers in Plant Science 10. Available at https://doi.org/10.3389/fpls.2019.00508.

Della Coletta, R., Qiu, Y., Ou, S., Hufford, M.B., Hirsch, C.N., 2021. How the pan-genome is changing crop genomics and improvement. Genome Biology 22 (1). Available at https://doi.org/10.1186/s13059-020-02224-8.

DiStefano, J.K., Taverna, D.M., 2011. Technological issues and experimental design of gene association studies. In: DiStefano, J.K. (Ed.), Disease Gene Identification: Methods and Protocols. Humana Press, pp. 3–16. Available at https://doi.org/10.1007/978-1-61737-954-3_1.

Dilnur, T., Peng, Z., Pan, Z., Palanga, K.K., Jia, Y., Gong, W., et al., 2019. Association analysis of salt tolerance in asiatic cotton (gossypium arboretum) with SNP markers. International Journal of Molecular Sciences 20 (9). Available at https://doi.org/10.3390/ijms20092168.

Duvick, D.N., Cassman, K.G., 1999. Post–green revolution trends in yield potential of temperate maize in the North-Central United States. Crop Science 39 (6), 1622–1630. Available at https://doi.org/10.2135/cropsci1999.3961622x.

Eftekhari, M., Ma, C., Orlov, Y.L., 2024. Editorial: Applications of artificial intelligence, machine learning, and deep learning in plant breeding. Frontiers in Plant Science 15. Available at https://doi.org/10.3389/fpls.2024.1420938.

Enyew, M., Feyissa, T., Carlsson, A.S., Tesfaye, K., Hammenhag, C., Seyoum, A., et al., 2022. Genome-wide analyses using multi-locus models revealed marker-trait associations for major agronomic traits in Sorghum bicolor. Frontiers in Plant Science 13. Available at https://doi.org/10.3389/fpls.2022.999692.

Escribà-Gelonch, M., Liang, S., van Schalkwyk, P., Fisk, I., Long, N.V.D., Hessel, V., 2024. Digital twins in agriculture: orchestration and applications. Journal of Agricultural and Food Chemistry 72 (19), 10737–10752. Available at https://doi.org/10.1021/acs.jafc.4c01934.

Falconer, D.S., Mackay, T.F.C., 1996. Introduction to Quantitative Genetics, 4th ed. Benjamin Cummings, England.

Feng, X., Zhan, Y., Wang, Q., Yang, X., Yu, C., Wang, H., et al., 2020. Hyperspectral imaging combined with machine learning as a tool to obtain high-throughput plant salt-stress phenotyping. Plant Journal 101 (6), 1448–1461. Available at https://doi.org/10.1111/tpj.14597.

Flowers, T.J., Yeo, A.R., 1986. Ion relations of plants under drought and salinity. Australian Journal of Plant Physiology 13 (1), 75–91. Available at https://doi.org/10.1071/PP9860075.

Gao, L., Gonda, I., Sun, H., Ma, Q., Bao, K., Tieman, D.M., et al., 2019. The tomato pan-genome uncovers new genes and a rare allele regulating fruit flavor. Nature Genetics 51 (6), 1044–1051. Available at https://doi.org/10.1038/s41588-019-0410-2.

Gao, S., Chen, S., Yang, M., Wu, J., Chen, S., Li, H., 2023. Mining salt stress-related genes in Spartina alterniflora via analyzing co-evolution signal across 365 plant species using phylogenetic profiling. aBIOTECH 4 (4), 291–302. Available at https://doi.org/10.1007/s42994-023-00125-5.

Ge, F., Qu, J., Liu, P., Pan, L., Zou, C., Yuan, G., et al., 2022. Genome assembly of the maize inbred line A188 provides a new reference genome for functional genomics. Crop Journal 10 (1), 47–55. Available at https://doi.org/10.1016/j.cj.2021.08.002.

Golicz, A.A., Bayer, P.E., Barker, G.C., Edger, P.P., Kim, H.R., Martinez, P.A., et al., 2016. The pangenome of an agronomically important crop plant Brassica oleracea. Nature Communications 7. Available at https://doi.org/10.1038/ncomms13390.

Golicz, A.A., Batley, J., Edwards, D., 2016. Towards plant pangenomics. Plant Biotechnology Journal 14 (4), 1099–1105. Available at https://doi.org/10.1111/pbi.12499.

Gordon, S.P., Contreras-Moreira, B., Woods, D.P., Des Marais, D.L., Burgess, D., Shu, S., et al., 2017. Extensive gene content variation in the Brachypodium distachyon pan-genome correlates with population structure. Nature Communications 8 (1). Available at https://doi.org/10.1038/s41467-017-02292-8.

Grant, G., More, L.J., McKenzie, N.H., Dorward, P.M., Buchan, W.C., Telek, L., et al., 1995. Nutritional and haemagglutination properties of several tropical seeds. The Journal of Agricultural Science 124 (3), 437–445. Available at https://doi.org/10.1017/s0021859600073391.

Gui, S., Yang, L., Li, J., Luo, J., Xu, X., Yuan, J., et al., 2020. ZEAMAP, a comprehensive database adapted to the maize multi-omics era. iScience 23 (6). Available at https://doi.org/10.1016/j.isci.2020.101241.

Guo, M., Liu, J.H., Ma, X., Luo, D.X., Gong, Z.H., Lu, M.H., 2016. The plant heat stress transcription factors (HSFS): structure, regulation, and function in response to abiotic stresses. Frontiers in Plant Science 7, 2016. Available at https://doi.org/10.3389/fpls.2016.00114.

Guangdong Laboratory for Lingnan Modern Agriculture, 2021. RiceSuperPIRdb.

Guo, Z., Yang, W., Chang, Y., Ma, X., Tu, H., Xiong, F., et al., 2018. Genome-wide association studies of image traits reveal genetic architecture of drought resistance in rice. Molecular Plant 11 (6), 789–805. Available at https://doi.org/10.1016/j.molp.2018.03.018.

Gutiérrez, N., Pégard, M., Balko, C., Torres, A.M., 2023. Genome-wide association analysis for drought tolerance and associated traits in faba bean (Vicia faba L. Frontiers in Plant Science 14. Available at https://doi.org/10.3389/fpls.2023.1091875.

Hahn, A., Bublak, D., Schleiff, E., Scharf, K.D., 2011. Crosstalk between Hsp90 and Hsp70 chaperones and heat stress transcription factors in tomato. The Plant Cell 23 (2), 741–755. Available at https://doi.org/10.1105/tpc.110.076018.

Ham, T.-H., Kwon, Y., Lee, Y., Choi, J., Lee, J., 2021. Genome-wide association study reveals the genetic basis of cold tolerance in rice at the seedling stage. Agriculture 11 (4), 318. Available at https://doi.org/10.3390/agriculture11040318.

Hamazaki, K., Iwata, H., 2024. AI-assisted selection of mating pairs through simulation-based optimized progeny allocation strategies in plant breeding. Frontiers in Plant Science 15. Available at https://doi.org/10.3389/fpls.2024.1361894.

Hanson, P., Lu, S.F., Wang, J.F., Chen, W., Kenyon, L., Tan, C.W., et al., 2016. Conventional and molecular marker-assisted selection and pyramiding of genes for multiple disease resistance in tomato. Scientia Horticulturae 201, 346–354. Available at https://doi.org/10.1016/j.scienta.2016.02.020.

He, Q., Tang, S., Zhi, H., Chen, J., Zhang, J., Liang, H., et al., 2023. A graph-based genome and pan-genome variation of the model plant Setaria. Nature Genetics 55 (7), 1232–1242. Available at https://doi.org/10.1038/s41588-023-01423-w.

Hichri, I., Muhovski, Y., Clippe, A., Žižková, E., Dobrev, P.I., Motyka, V., et al., 2016. SlDREB2, a tomato dehydration-responsive element-binding 2 transcription factor, mediates salt stress tolerance in tomato and A rabidopsis. Plant, Cell & Environment 39 (1), 62–79. Available at https://doi.org/10.1111/pce.12591.

Hirsch, C.N., Foerster, J.M., Johnson, J.M., Sekhon, R.S., Muttoni, G., Vaillancourt, B., et al., 2014. Insights into the maize pan-genome and pan-transcriptome. The Plant Cell 26 (1), 121–135. Available at https://doi.org/10.1105/tpc.113.119982.

Hoopes, G., Meng, X., Hamilton, J.P., Achakkagari, S.R., de Alves Freitas Guesdes, F., Bolger, M.E., et al., 2022. Phased, chromosome-scale genome assemblies of tetraploid potato reveal a complex genome, transcriptome, and predicted proteome landscape underpinning genetic diversity. Molecular Plant 15 (3), 520–536. Available at https://doi.org/10.1016/j.molp.2022.01.003.

Huang, M., Balimponya, E.G., Mgonja, E.M., McHale, L.K., Luzi-Kihupi, A., Wang, G.L., et al., 2019. Use of genomic selection in breeding rice (Oryza sativa L.) for resistance to rice blast (Magnaporthe oryzae). Molecular Breeding 39 (8). Available at https://doi.org/10.1007/s11032-019-1023-2.

Huang, Z., Chen, S., He, K., Yu, T., Fu, J., Gao, S., et al., 2024. Exploring salt tolerance mechanisms using machine learning for transcriptomic insights: case study in Spartina alterniflora. Horticulture Research 11 (5). Available at https://doi.org/10.1093/hr/uhae082.

Hurgobin, B., Golicz, A.A., Bayer, P.E., Chan, C.K.K., Tirnaz, S., Dolatabadian, A., et al., 2018. Homoeologous exchange is a major cause of gene presence/absence variation in the amphidiploid Brassica napus. Plant Biotechnology Journal 16 (7), 1265–1274. Available at https://doi.org/10.1111/pbi.12867.

Hübner, S., Bercovich, N., Todesco, M., Mandel, J.R., Odenheimer, J., Ziegler, E., et al., 2019. Sunflower pan-genome analysis shows that hybridization altered gene content and disease resistance. Nature Plants 5 (1), 54–62. Available at https://doi.org/10.1038/s41477-018-0329-0.

Jantaboon, J., Siangliw, M., Im-mark, S., Jamboonsri, W., Vanavichit, A., Toojinda, T., 2011. Ideotype breeding for submergence tolerance and cooking quality by marker-assisted selection in rice. Field Crops Research 123 (3), 206–213. Available at https://doi.org/10.1016/j.fcr.2011.05.001.

Jiang, Y., Li, C., 2020. Convolutional neural networks for image-based high-throughput plant phenotyping: a review. Plant Phenomics 2020, 4152816. Available at https://doi.org/10.34133/2020/4152816.

Joshi, R., Nayak, S., 2010. Gene pyramiding-a broad spectrum technique for developing durable stress resistance in crops. Biotechnology and Molecular Biology Review 5 (3), 51–60.

Kaewcheenchai, R., Vejchasarn, P., Hanada, K., Shirai, K., Jantasuriyarat, C., Juntawong, P., 2021. Genome-wide association study of local thai indica rice seedlings exposed to excessive iron. Plants 10 (4), 798. Available at https://doi.org/10.3390/plants10040798.

Karavolias, N.G., Horner, W., Abugu, M.N., Evanega, S.N., 2021. Application of gene editing for climate change in agriculture. Frontiers in Sustainable Food Systems 5. Available at https://doi.org/10.3389/fsufs.2021.685801.

Kaur, H., Shannon, L.M., Samac, D.A., 2024. A stepwise guide for pangenome development in crop plants: an alfalfa (Medicago sativa) case study. BMC Genomics 25 (1). Available at https://doi.org/10.1186/s12864-024-10931-w.

Kearsey, M.J., 1998. The principles of QTL analysis (a minimal mathematics approach). Journal of Experimental Botany 49 (327), 1619–1623. Available at https://doi.org/10.1093/jxb/49.327.1619.

Khraiwesh, B., Zhu, J.K., Zhu, J., 2012. Role of miRNAs and siRNAs in biotic and abiotic stress responses of plants. Biochimica et Biophysica Acta - Gene Regulatory Mechanisms 1819 (2), 137–148. Available at https://doi.org/10.1016/j.bbagrm.2011.05.001.

Kizis, D., Pagès, M., 2002. Maize DRE-binding proteins DBF1 and DBF2 are involved in rab17 regulation through the drought-responsive element in an ABA-dependent pathway. Plant Journal 30 (6), 679–689. Available at https://doi.org/10.1046/j.1365-313X.2002.01325.x.

Knoch, D., Meyer, R.C., Heuermann, M.C., Riewe, D., Peleke, F.F., Szymański, J., et al., 2024. Integrated multi-omics analyses and genome-wide association studies reveal prime candidate genes of metabolic and vegetative growth variation in canola. Plant Journal 117 (3), 713–728. Available at https://doi.org/10.1111/tpj.16524.

Korinsak, S., Sirithunya, P., Meakwatanakarn, P., Sarkarung, S., Vanavichit, A., Toojinda, T., 2011. Changing allele frequencies associated with specific resistance genes to leaf blast in backcross introgression lines of Khao Dawk Mali 105 developed from a conventional selection program. Field Crops Research 122 (1), 32–39. Available at https://doi.org/10.1016/j.fcr.2011.02.005.

Korstanje, R., Paigen, B., 2002. From QTL to gene: the harvest begins. Nature Genetics 31 (3), 235–236. Available at https://doi.org/10.1038/ng0702-235.

Korte, A., Farlow, A., 2013. The advantages and limitations of trait analysis with GWAS: a review. Plant Methods 9 (1), 29. Available at https://doi.org/10.1186/1746-4811-9-29.

Kumar, V., Singh, A., Mithra, S.V.A., Krishnamurthy, S.L., Parida, S.K., Jain, S., et al., 2015. Genome-wide association mapping of salinity tolerance in rice (Oryza sativa). DNA Research 22 (2), 133–145. Available at https://doi.org/10.1093/dnares/dsu046.

Lafarge, T., Bueno, C., Frouin, J., Jacquin, L., Courtois, B., Ahmadi, N., et al., 2017. Genome-wide association analysis for heat tolerance at flowering detected a large set of genes involved in adaptation to thermal and other stresses. PLoS One 12 (2), e0171254. Available at https://doi.org/10.1371/journal.pone.0171254.

Lawrence, C.J., Dong, Q., Polacco, M.L., Seigfried, T.E., Brendel, V., 2004. MaizeGDB, the community database for maize genetics and genomics. Nucleic Acids Research 32, D393.

Li, Y.H., Zhou, G., Ma, J., Jiang, W., Jin, L.G., Zhang, Z., et al., 2014. De novo assembly of soybean wild relatives for pan-genome analysis of diversity and agronomic traits. Nature Biotechnology 32 (10), 1045–1052. Available at https://doi.org/10.1038/nbt.2979.

Li, Z., Li, Z., Ji, Y., Wang, C., Wang, S., Shi, Y., et al., 2024. The heat shock factor 20-HSF4-cellulose synthase A2 module regulates heat stress tolerance in maize. The Plant Cell 36 (7), 2652–2667. Available at https://doi.org/10.1093/plcell/koae106.

Liu, C., Wang, Y., Peng, J., Fan, B., Xu, D., Wu, J., et al., 2022. High-quality genome assembly and pan-genome studies facilitate genetic discovery in mung bean and its improvement. Plant Communications 3 (6). Available at https://doi.org/10.1016/j.xplc.2022.100352.

Liu, S., Wang, X., Wang, H., Xin, H., Yang, X., Yan, J., et al., 2013. Genome-wide analysis of ZmDREB genes and their association with natural variation in drought tolerance at seedling stage of Zea mays L. PLoS Genetics 9 (9). Available at https://doi.org/10.1371/journal.pgen.1003790.

Long, R., Zhang, F., Zhang, Z., Li, M., Chen, L., Wang, X., et al., 2022. Genome assembly of alfalfa cultivar zhongmu-4 and identification of SNPs associated with agronomic traits. Genomics, Proteomics and Bioinformatics 20 (1), 14–28. Available at https://doi.org/10.1016/j.gpb.2022.01.002.

Lourenço, V.M., Ogutu, J.O., Rodrigues, R.A.P., Posekany, A., Piepho, H.-P., 2024. Genomic prediction using machine learning: a comparison of the performance of regularized regression, ensemble, instance-based and deep learning methods on synthetic and empirical data. BMC Genomics 25 (1). Available at https://doi.org/10.1186/s12864-023-09933-x.

Luo, Y., Yin, Z., 2013. Marker-assisted breeding of Thai fragrance rice for semi-dwarf phenotype, submergence tolerance and disease resistance to rice blast and bacterial blight. Molecular Breeding 32 (3), 709–721. Available at https://doi.org/10.1007/s11032-013-9904-2.

López-Malvar, A., Reséndiz, Z., Santiago, R., Jiménez-Galindo, J.C., Malvar, R.A., 2021. Relationships between stalk resistance and corn borers, agronomic traits, and cell wall hydroxycinnamates in a set of recombinant inbred lines from a Maize MAGIC population. Agronomy 11 (6). Available at https://doi.org/10.3390/agronomy11061132.

Ma, J., Wei, H., Yu, X., Lv, Y., Zhang, Y., Qian, Q., et al., 2022. Compared analysis with a high-quality genome of weedy rice reveals the evolutionary game of de-domestication. Frontiers in Plant Science 13. Available at https://doi.org/10.3389/fpls.2022.1065449.

Mahmood, U., Li, X., Fan, Y., Chang, W., Niu, Y., Li, J., et al., 2022. Multi-omics revolution to promote plant breeding efficiency. Frontiers in Plant Science 13. Available at https://doi.org/10.3389/fpls.2022.1062952.

Martin, G.B., Brommonschenkel, S.H., Chunwongse, J., Frary, A., Ganal, M.W., Spivey, R., et al., 1993. Map-based cloning of a protein kinase gene conferring disease resistance in tomato. Science (New York, N.Y.) 262 (5138), 1432–1436. Available at https://doi.org/10.1126/science.7902614.

Matthus, E., Wu, L.B., Ueda, Y., Höller, S., Becker, M., Frei, M., 2015. Loci, genes, and mechanisms associated with tolerance to ferrous iron toxicity in rice (Oryza sativa L. Theoretical and Applied Genetics 128 (10), 2085–2098. Available at https://doi.org/10.1007/s00122-015-2569-y.

Miao, S., Lu, J., Zhang, G., Jiang, J., Li, P., Qian, Y., et al., 2024. Candidate genes and favorable haplotypes associated with iron toxicity tolerance in rice. International Journal of Molecular Sciences 25 (13). Available at https://doi.org/10.3390/ijms25136970.

Montenegro, J.D., Golicz, A.A., Bayer, P.E., Hurgobin, B., Lee, H.T., Chan, C.K.K., et al., 2017. The pangenome of hexaploid bread wheat. Plant Journal 90 (5), 1007–1013. Available at https://doi.org/10.1111/tpj.13515.

Montesinos-López, O.A., Martín-Vallejo, J., Crossa, J., Gianola, D., Hernández-Suárez, C.M., Montesinos-López, A., et al., 2019. A benchmarking between deep learning, support vector machine and Bayesian threshold best linear unbiased prediction for predicting ordinal traits in plant breeding. G3: Genes, Genomes, Genetics 9 (2), 601–618. Available at https://doi.org/10.1534/g3.118.200998.

Montesinos-López, O.A., Montesinos-López, A., Pérez-Rodríguez, P., Barrón-López, J.A., Martini, J.W.R., Fajardo-Flores, S.B., et al., 2021. A review of deep learning applications for genomic selection. BMC Genomics 22 (1). Available at https://doi.org/10.1186/s12864-020-07319-x.

Morgante, M., Depaoli, E., Radovic, S., 2007. Transposable elements and the plant pan-genomes. Current Opinion in Plant Biology 10 (2), 149–155. Available at https://doi.org/10.1016/j.pbi.2007.02.001.

Munns, R., Tester, M., 2008. Mechanisms of Salinity Tolerance. Annual Review of Plant Biology 59 (1), 651–681. Available at https://doi.org/10.1146/annurev.arplant.59.032607.092911.

Murmu, S., Sinha, D., Chaurasia, H., Sharma, S., Das, R., Jha, G.K., et al., 2024. A review of artificial intelligence-assisted omics techniques in plant defense: current trends and future directions. Frontiers in Plant Science 15. Available at https://doi.org/10.3389/fpls.2024.1292054.

Neelapu, N.R.R., Kgk, D., Surekha, C., 2015. Transgenic plants for higher antioxidant contents and salt stress tolerance. In: Wani, S.H., Hossain, M.A. (Eds.), Managing Salt Tolerance in Plants: Molecular and Genomic Perspectives. CRC Press, Boca Raton, FL, pp. 391–406. Available at https://doi.org/10.1201/b19246-22.

Nouraei, S., Mia, M.S., Liu, H., Turner, N.C., Yan, G., 2024. Genome-wide association study of drought tolerance in wheat (Triticum aestivum L.) identifies SNP markers and candidate genes. Molecular Genetics and Genomics 299 (1). Available at https://doi.org/10.1007/s00438-024-02104-x.

Ou, L., Li, D., Lv, J., Chen, W., Zhang, Z., Li, X., et al., 2018. Pan-genome of cultivated pepper (Capsicum) and its use in gene presence–absence variation analyses. New Phytologist 220 (2), 360–363. Available at https://doi.org/10.1111/nph.15413.

Pflieger, S., Lefebvre, V., Causse, M., 2001. The candidate gene approach in plant genetics: a review. Molecular Breeding 7 (4), 275–291. Available at https://doi.org/10.1023/a:1011605013259.

Pylianidis, C., Osinga, S., Athanasiadis, I.N., 2021. Introducing digital twins to agriculture. Computers and Electronics in Agriculture 184, 105942. Available at https://doi.org/10.1016/j.compag.2020.105942.

Pérez-Rodríguez, P., Flores-Galarza, S., Vaquera-Huerta, H., del Valle-Paniagua, D.H., Montesinos-López, O.A., Crossa, J., 2020. Genome-based prediction of Bayesian linear and non-linear regression models for ordinal data. Plant Genome 13 (2). Available at https://doi.org/10.1002/tpg2.20021.

Qin, P., Lu, H., Du, H., Wang, H., Chen, W., Chen, Z., et al., 2021. Pan-genome analysis of 33 genetically diverse rice accessions reveals hidden genomic variations. Cell 184 (13), 3542. Available at https://doi.org/10.1016/j.cell.2021.04.046.

Raggi, L., Caproni, L., Ciancaleoni, S., D'Amato, R., Businelli, D., Negri, V., 2024. Investigating the genetic basis of salt-tolerance in common bean: a genome-wide association study at the early vegetative stage. Scientific Reports 14 (1). Available at https://doi.org/10.1038/s41598-024-55403-z.

Rajpurohit, D., Kumar, R., Kumar, M., Paul, P., Awasthi, A., Osman Basha, P., et al., 2011. Pyramiding of two bacterial blight resistance and a semidwarfing gene in Type 3 Basmati using marker-assisted selection. Euphytica 178 (1), 111–126. Available at https://doi.org/10.1007/s10681-010-0279-8.

Randhawa, H.S., Mutti, J.S., Kidwell, K., Morris, C.F., Chen, X., Gill, K.S., 2009. Rapid and targeted introgression of genes into popular wheat cultivars using marker-assisted background selection. PLoS One 4 (6). Available at https://doi.org/10.1371/journal.pone.0005752.

Ruengphayak, S., Chaichumpoo, E., Phromphan, S., Kamolsukyunyong, W., Sukhaket, W., Phuvanartnarubal, E., et al., 2015. Pseudo-backcrossing design for rapidly pyramiding multiple traits into a preferential rice variety. Rice 8 (1). Available at https://doi.org/10.1186/s12284-014-0035-0.

Ruperao, P., Thirunavukkarasu, N., Gandham, P., Selvanayagam, S., Govindaraj, M., Nebie, B., et al., 2021. Sorghum pan-genome explores the functional utility for genomic-assisted breeding to accelerate the genetic gain. Frontiers in Plant Science 12. Available at https://doi.org/10.3389/fpls.2021.666342.

Schatz, M.C., Maron, L.G., Stein, J.C., Hernandez Wences, A., Gurtowski, J., Biggers, E., et al., 2014. Whole genome de novo assemblies of three divergent strains of rice, Oryza sativa, document novel gene space of aus and indica. Genome Biology 15 (11), 506. Available at https://doi.org/10.1186/s13059-014-0506-z.

Shang, L., Li, X., He, H., Yuan, Q., Song, Y., Wei, Z., et al., 2022. A super pan-genomic landscape of rice. Cell Research 32 (10), 878–896. Available at https://doi.org/10.1038/s41422-022-00685-z.

Sharmin, R.A., Karikari, B., Chang, F., Al Amin, G.M., Bhuiyan, M.R., Hina, A., et al., 2021. Genome-wide association study uncovers major genetic loci associated with seed flooding tolerance in soybean. BMC Plant Biology 21 (1). Available at https://doi.org/10.1186/s12870-021-03268-z.

Shi, Y., Gao, L., Wu, Z., Zhang, X., Wang, M., Zhang, C., et al., 2017. Genome-wide association study of salt tolerance at the seed germination stage in rice. BMC Plant Biology 17 (1). Available at https://doi.org/10.1186/s12870-017-1044-0.

Shimoyama, N., Johnson, M., Beaumont, A., Schläppi, M., 2020. Multiple cold tolerance trait phenotyping reveals shared quantitative trait loci in Oryza sativa. Rice 13 (1). Available at https://doi.org/10.1186/s12284-020-00414-3.

Shinozaki, K., Yamaguchi-Shinozaki, K., 2000. Molecular responses to dehydration and low temperature: differences and cross-talk between two stress signaling pathways. Current Opinion in Plant Biology 3 (3), 217–223. Available at https://doi.org/10.1016/s1369-5266(00)80068-0.

Siangliw, J.L., Jongdee, B., Pantuwan, G., Toojinda, T., 2007. Developing KDML105 backcross introgression lines using marker-assisted selection for QTLs associated with drought tolerance in rice. ScienceAsia 33 (2), 207–214. Available at https://doi.org/10.2306/scienceasia1513-1874.2007.33.207.

Siangliw, M., Toojinda, T., Tragoonrung, S., Vanavichit, A., 2003. Thai jasmine rice carrying QTLch9 (SubQTL) is submergence tolerant. Annals of Botany 91, 255–261. Available at https://doi.org/10.1093/aob/mcf123.

Singh, V.K., Singh, A., Singh, S.P., Ellur, R.K., Singh, D., Gopala Krishnan, S., et al., 2013. Marker-assisted simultaneous but stepwise backcross breeding for pyramiding blast resistance genes Piz5 and Pi54 into an elite Basmati rice restorer line 'PRR78' Plant Breeding 132 (5), 486–495. Available at https://doi.org/10.1111/pbr.12077.

Song, W.Y., Wang, G.L., Chen, L.L., Kim, H.S., Pi, L.Y., Holsten, T., et al., 1995. A receptor kinase-like protein encoded by the rice disease resistance gene, Xa21. Science (New York, N.Y.) 270 (5243), 1804. Available at https://doi.org/10.1126/science.270.5243.1804.

Suh, J.P., Jeung, J.U., Noh, T.H., Cho, Y.C., Park, S.H., Park, H.S., et al., 2013. Development of breeding lines with three pyramided resistance genes that confer broad-spectrum bacterial blight resistance and their molecular analysis in rice. Rice 6 (1), 1–11. Available at https://doi.org/10.1186/1939-8433-6-5.

Sun, C., Hu, Z., Zheng, T., Lu, K., Zhao, Y., Wang, W., et al., 2017. RPAN: Rice pan-genome browser for ~3000 rice genomes. Nucleic Acids Research 45 (2), 597–605. Available at https://doi.org/10.1093/nar/gkw958.

Sundaram, R.M., Vishnupriya, M.R., Biradar, S.K., Laha, G.S., Reddy, G.A., Rani, N.S., et al., 2008. Marker assisted introgression of bacterial blight resistance in Samba Mahsuri, an elite indica rice variety. Euphytica 160 (3), 411–422. Available at https://doi.org/10.1007/s10681-007-9564-6.

Surekha, C., Aruna, L., Hossain, M.A., Wani, S., Neelapu, N.R.R., 2015. Present status and future prospects of transgenic approaches for salt tolerance in plants/crop plants. In: Wani, S.H., Hossain, M.A. (Eds.), Managing Salt Tolerance in Plants, pp. 329–352. Available at https://doi.org/10.1201/b19246-19.

Tang, Y., Liu, X., Wang, J., Li, M., Wang, Q., Tian, F., et al., 2016. GAPIT version 2: an enhanced integrated tool for genomic association and prediction. Plant Genome 9 (2). Available at https://doi.org/10.3835/plantgenome2015.11.0120.

Tao, Y., Zhao, X., Mace, E., Henry, R., Jordan, D., 2019. Exploring and exploiting pan-genomics for crop improvement. Molecular Plant 12 (2), 156–169. Available at https://doi.org/10.1016/j.molp.2018.12.016.

Tay Fernandez, C.G., Nestor, B.J., Danilevicz, M.F., Marsh, J.I., Petereit, J., Bayer, P.E., et al., 2022. Expanding gene-editing potential in crop improvement with pangenomes. International Journal of Molecular Sciences 23 (4). Available at https://doi.org/10.3390/ijms23042276.

Tettelin, H., Masignani, V., Cieslewicz, M.J., Donati, C., Medini, D., Ward, N.L., et al., 2005. Genome analysis of multiple pathogenic isolates of Streptococcus agalactiae: implications for the microbial "pan-genome" Proceedings of the National Academy of Sciences of the United States of America 102 (39), 13950–13955. Available at https://doi.org/10.1073/pnas.0506758102.

Theerawitaya, C., Wanchana, S., Ruanjaichon, V., Tisaram, R., Samphumphuang, T., Sotesaritkul, T., et al., 2022. Determination of traits responding to iron toxicity stress at different stages and genome-wide association analysis for iron toxicity tolerance in rice (Oryza sativa L. Frontiers in Plant Science 13. Available at https://doi.org/10.3389/fpls.2022.994560.

Tomita, M., 2009. Introgression of green revolution sd1 gene into isogenic genome of rice super cultivar Koshihikari to create novel semidwarf cultivar 'Hikarishinseiki' (Koshihikari-sd1). Field Crops Research 114 (2), 173–181. Available at https://doi.org/10.1016/j.fcr.2009.05.004.

Torkamaneh, D., Lemay, M.A., Belzile, F., 2021. The pan-genome of the cultivated soybean (PanSoy) reveals an extraordinarily conserved gene content. Plant Biotechnology Journal 19 (9), 1852–1862. Available at https://doi.org/10.1111/pbi.13600.

Tran, H.A., Jo, H., Nguyen, T.C., Lee, J.D., Seo, H.S., Song, J.T., 2024. Genome-wide association analysis for submergence tolerance at the early vegetative and germination stages in wild soybean (Glycine soja). Agriculture (Switzerland) 14 (9). Available at https://doi.org/10.3390/agriculture14091627.

Tuberosa, R., Salvi, S., 2006. Genomics-based approaches to improve drought tolerance of crops. Trends in Plant Science 11 (8), 405–412. Available at https://doi.org/10.1016/j.tplants.2006.06.003.

Varshney, R.K., Roorkiwal, M., Sun, S., Bajaj, P., Chitikineni, A., Thudi, M., et al., 2021. A chickpea genetic variation map based on the sequencing of 3,366 genomes. Nature 599 (7886), 622–627. Available at https://doi.org/10.1038/s41586-021-04066-1.

Walsh, J.J., Mangina, E., Negrão, S., 2024. Advancements in imaging sensors and AI for plant stress detection: a systematic literature review. Plant Phenomics 6, 0153. Available at https://doi.org/10.34133/plantphenomics.0153.

Wang, W., Mauleon, R., Hu, Z., Chebotarov, D., Tai, S., Wu, Z., et al., 2018. Genomic variation in 3,010 diverse accessions of Asian cultivated rice. Nature 557 (7703), 43–49. Available at https://doi.org/10.1038/s41586-018-0063-9.

Wang, Y., Li, F., Zhang, F., Wu, L., Xu, N., Sun, Q., et al., 2023. Time-ordering japonica/geng genomes analysis indicates the importance of large structural variants in rice breeding. Plant Biotechnology Journal 21 (1), 202–218. Available at https://doi.org/10.1111/pbi.13938.

Wang, Z., Yin, M., Han, J., Wang, X., Chang, J., Ren, Z., et al., 2024. Pan-genome-wide identification and transcriptome-wide analysis of ZIP genes in cucumber. Agriculture (Switzerland) 14 (1). Available at https://doi.org/10.3390/agriculture14010133.

Williamson, H.F., Brettschneider, J., Caccamo, M., Davey, R.P., Goble, C., Kersey, P.J., et al., 2023. Data management challenges for artificial intelligence in plant and agricultural research. F1000Research 10. Available at https://doi.org/10.12688/f1000research.52204.2.

Win, K.M., Korinsak, S., Jantaboon, J., Siangliw, M., Lanceras-Siangliw, J., Sirithunya, P., et al., 2012. Breeding the Thai jasmine rice variety KDML105 for non-age-related broad-spectrum resistance to bacterial blight disease based on combined marker-assisted and phenotypic selection. Field Crops Research 137, 186–194. Available at https://doi.org/10.1016/j.fcr.2012.09.007.

Wongsaprom, C., Sirithunya, P., Vanavichit, A., Pantuwan, G., Jongdee, B., Sidhiwong, N., et al., 2010. Two introgressed quantitative trait loci confer a broad-spectrum resistance to blast disease in the genetic background of the cultivar RD6 a Thai glutinous jasmine rice. Field Crops Research 119 (2-3), 245–251. Available at https://doi.org/10.1016/j.fcr.2010.07.013.

Wu, D., Xie, L., Sun, Y., Huang, Y., Jia, L., Dong, C., et al., 2023. A syntelog-based pan-genome provides insights into rice domestication and de-domestication. Genome Biology 24 (1). Available at https://doi.org/10.1186/s13059-023-03017-5.

Xin, Y., Gao, L., Hu, W., Gao, Q., Yang, B., Zhou, J., et al., 2022. Genome-wide association study based on plant height and drought-tolerance indices reveals two candidate drought-tolerance genes in sweet sorghum. Sustainability 14 (21), 14339. Available at https://doi.org/10.3390/su142114339.

Yan, H., Sun, M., Zhang, Z., Jin, Y., Zhang, A., Lin, C., et al., 2023. Pangenomic analysis identifies structural variation associated with heat tolerance in pearl millet. Nature Genetics 55 (3), 507–518. Available at https://doi.org/10.1038/s41588-023-01302-4.

Yan, J., Wang, X., 2023. Machine learning bridges omics sciences and plant breeding. Trends in Plant Science 28 (2), 199–210. Available at https://doi.org/10.1016/j.tplants.2022.08.018.

Yang, N., Liu, J., Gao, Q., Gui, S., Chen, L., Yang, L., et al., 2019. Genome assembly of a tropical maize inbred line provides insights into structural variation and crop improvement. Nature Genetics 51 (6), 1052–1059. Available at https://doi.org/10.1038/s41588-019-0427-6.

Yang, T., Liu, R., Luo, Y., Hu, S., Wang, D., Wang, C., et al., 2022. Improved pea reference genome and pan-genome highlight genomic features and evolutionary characteristics. Nature Genetics 54 (10), 1553–1563. Available at https://doi.org/10.1038/s41588-022-01172-2.

Yang, Y., Saand, M.A., Huang, L., Abdelaal, W.B., Zhang, J., Wu, Y., et al., 2021. Applications of multi-omics technologies for crop improvement. Frontiers in Plant Science 12. Available at https://doi.org/10.3389/fpls.2021.563953.

Yao, W., Li, G., Zhao, H., Wang, G., Lian, X., Xie, W., 2015. Exploring the rice dispensable genome using a metagenome-like assembly strategy. Genome Biology 16 (1). Available at https://doi.org/10.1186/s13059-015-0757-3.

Yi, M., Nwe, K.T., Vanavichit, A., Chai-arree, W., Toojinda, T., 2009. Marker assisted backcross breeding to improve cooking quality traits in Myanmar rice cultivar Manawthukha. Field Crops Research 113 (2), 178–186. Available at https://doi.org/10.1016/j.fcr.2009.05.006.

Yoosefzadeh-Najafabadi, M., Eskandari, M., Belzile, F., Torkamaneh, D., 2022a. Genome-wide association study statistical models: a review. In: Torkamaneh, D., Belzile, F. (Eds.), Genome-Wide Association Studies. Springer US, pp. 43–62. Available at https://doi.org/10.1007/978-1-0716-2237-7_4.

Yoosefzadeh-Najafabadi, M., Eskandari, M., Torabi, S., Torkamaneh, D., Tulpan, D., Rajcan, I., 2022b. Machine-learning-based genome-wide association studies for uncovering QTL underlying soybean yield and its components. International Journal of Molecular Sciences 23 (10), 5538. Available at https://doi.org/10.3390/ijms23105538.

Zai, W.S., Miao, L.X., Xiong, Z.L., Zhang, H.L., Ma, Y.R., Li, Y.L., et al., 2015. Comprehensive identification and expression analysis of Hsp90s gene family in Solanum lycopersicum. Genetics and Molecular Research 14 (3), 7811–7820. Available at https://doi.org/10.4238/2015.july.14.7.

Zhang, F., Xue, H., Dong, X., Li, M., Zheng, X., Li, Z., et al., 2022. Long-read sequencing of 111 rice genomes reveals significantly larger pan-genomes. Genome Research 32 (5), 853–863. Available at https://doi.org/10.1101/gr.276015.121.

Zhang, J., Chen, K., Pang, Y., Naveed, S.A., Zhao, X., Wang, X., et al., 2017a. QTL mapping and candidate gene analysis of ferrous iron and zinc toxicity tolerance at seedling stage in rice by genome-wide association study. BMC Genomics 18 (1). Available at https://doi.org/10.1186/s12864-017-4221-5.

Zhang, M., Lu, Q., Wu, W., Niu, X., Wang, C., Feng, Y., et al., 2017b. Association mapping reveals novel genetic loci contributing to flooding tolerance during germination in indica rice. Frontiers in Plant Science 8. Available at https://doi.org/10.3389/fpls.2017.00678.

Zhang, N., Zhao, H., Shi, J., Wu, Y., Jiang, J., 2020. Functional characterization of class I SlHSP17.7 gene responsible for tomato cold-stress tolerance. Plant Science 298. Available at https://doi.org/10.1016/j.plantsci.2020.110568.

Zhang, X., Liu, T., Wang, J., Wang, P., Qiu, Y., Zhao, W., et al., 2021. Pan-genome of Raphanus highlights genetic variation and introgression among domesticated, wild, and weedy radishes. Molecular Plant 14 (12), 2032–2055. Available at https://doi.org/10.1016/j.molp.2021.08.005.

Zhang, Z., Qu, Y., Ma, F., Lv, Q., Zhu, X., Guo, G., et al., 2024. Integrating high-throughput phenotyping and genome-wide association studies for enhanced drought resistance and yield prediction in wheat. New Phytologist 243 (5), 1758–1775. Available at https://doi.org/10.1111/nph.19942.

Zhao, J., Bayer, P.E., Ruperao, P., Saxena, R.K., Khan, A.W., Golicz, A.A., et al., 2020. Trait associations in the pangenome of pigeon pea (Cajanus cajan). Plant Biotechnology Journal 18 (9), 1946–1954. Available at https://doi.org/10.1111/pbi.13354.

Zhao, Q., Feng, Q., Lu, H., Li, Y., Wang, A., Tian, Q., et al., 2018. Pan-genome analysis highlights the extent of genomic variation in cultivated and wild rice. Nature Genetics 50 (2), 278–284. Available at https://doi.org/10.1038/s41588-018-0041-z.

Zhao, Y., Li, J., Zhao, R., Xu, K., Xiao, Y., Zhang, S., et al., 2020a. Genome-wide association study reveals the genetic basis of cold tolerance in wheat. Molecular Breeding 40 (4). Available at https://doi.org/10.1007/s11032-020-01115-x.

Zhou, Y., Chebotarov, D., Kudrna, D., Llaca, V., Lee, S., Rajasekar, S., et al., 2020b. A platinum standard pan-genome resource that represents the population structure of Asian rice. Scientific Data 7 (1). Available at https://doi.org/10.1038/s41597-020-0438-2.

Zhou, Y., Zhang, Z., Bao, Z., Li, H., Lyu, Y., Zan, Y., et al., 2022. Graph pangenome captures missing heritability and empowers tomato breeding. Nature 606 (7914), 527–534. Available at https://doi.org/10.1038/s41586-022-04808-9.

Zhou, Y., Yu, Z., Chebotarov, D., Chougule, K., Lu, Z., Rivera, L.F., et al., 2023. Pan-genome inversion index reveals evolutionary insights into the subpopulation structure of Asian rice. Nature Communications 14 (1). Available at https://doi.org/10.1038/s41467-023-37004-y.

Chapter 13

Role of phenomics and precision phenotyping in breeding strategies to develop multiple abiotic stress-tolerant crops

H.G.K. Chanumolu[1], Siddartha Kumar Neteti[1], Mahesh Kumar[1,*], Biswabiplab Singh[1], Elangovan A.[1], Sudhir Kumar[1], Madhurima Das[1], Jagadish Rane[2], and Viswanathan C.[1]

[1]*Division of Plant Physiology, ICAR-Indian Agricultural Research Institute, Pusa Campus, New Delhi, India,* [2]*ICAR-CIAH, Bikaner, Rajasthan, India*

[*]*Corresponding author. e-mail address: mahesh@iari.res.in*

13.1 Introduction

When environmental conditions are unfavorable for growth, plants experience stress. Abiotic stresses such as drought, waterlogging, salinity, high temperatures, nutrient imbalances, and high light intensity significantly limit crop performance by affecting physiological and metabolic processes (Deshmukh et al., 2023). These stresses, occurring individually or in combination, pose serious threats to global agriculture, especially under the current scenario of climate change and growing population pressure (Kopecká et al., 2023; Zandalinas et al., 2021).

There is a need to protect the crops from abiotic and biotic stresses that are likely to be amplified due to climate change. This can be achieved by developing climate-resilient crops through advanced crop improvement programs. Developing crops tolerant to multiple abiotic stresses is not merely important; it is essential for ensuring global food security under a climate change situation (Doggalli et al., 2024; Zhou et al., 2022). A holistic approach that considers positive and negative effects is crucial for developing resilient cultivars (Singh et al., 2021). Phenomics, which is equipped to assure greater precision in large-scale phenotyping, typically featured by high-throughput and accurate measurements, is revolutionizing crop improvement efforts (Zhao et al., 2019). They provide crucial tools for identifying and characterizing traits linked to abiotic stress tolerance and thus accelerating the plant breeding process (Visakh et al., 2024). The integration of phenomics-mediated precision phenotyping with genomics and other omics technologies offers a powerful approach to explore the genetic basis of complex traits and to develop superior crop varieties (Fernie and Gutierrez-Marcos, 2019; Zhao et al., 2019). In this chapter, we discuss the concept and techniques of plant phenomics that are utilized in phenotyping crop plants in breeding programs, both in controlled and field environments, with a focus on the identification of potential quantitative trait loci (QTLs)/ genes responsible for stress tolerance.

13.2 Phenomics: concept and technologies

Plant phenomics is a transdisciplinary field integrating physics, biology, statistics, computer science, metrology, and related disciplines, emerging as a new biological field widely applied in plant sciences (Ninomiya et al., 2019). It is defined as a comprehensive study of phenotypes of organisms on an organism-wide scale (Houle et al., 2010). This field is all about gathering and analyzing multidimensional phenotypic data in a systematic way, which is pivotal to finding the associations between genetic information and observable traits/phenotypes. Enhancing abiotic stress tolerance in crops using phenomics as a tool can be a potential game-changer in agricultural research. Phenomics as a tool can directly be involved in:

Multiple Abiotic Stresses in Plants. DOI: https://doi.org/10.1016/B978-0-443-27424-4.00010-0

1. **Dissecting complex traits**: Phenomics allows for the simplification and quantification of complex traits like drought resistance. Digital features, such as stay-green and leaf-rolling, can be quantified and linked to genetic loci that affect stress yield (Guo et al., 2018).
2. **Root architecture**: Understanding root system architecture is crucial for improving nutrient and water uptake. The phenomics method may reveal major novel genes in comparison to genes identified through conventional methods, such as DRO1, which controls the depth and angle of root growth, hence enhancing drought resistance (Uga et al., 2013).
3. **Data analysis with deep learning**: Phenomics takes advantage of deep learning (DL) to look at large datasets to assess stress effects. To identify stress tolerance loci, DL can use data from different sensors (such as thermal and hyperspectral) and the environments (Yang et al., 2020).
4. **Climate resilience:** Phenomics can assist in the development of crops that are more resistant to climate stresses, including abiotic (e.g., drought, flood, and extreme temperatures) and biotic (e.g., pests and diseases), by examining the morphological, physiological, and biochemical characteristics of plants (Maheswari et al., 2024). It facilitates the early and efficient identification of germplasm with specific characteristics of interest, thereby increasing the value of breeding programs.

This multidisciplinary approach helps in characterizing the plant not only in a single environment but also in multiple environments, that is, under combined abiotic stresses. It enables researchers to develop phenotyping protocols for multifactoral stresses in crops such as rice to evaluate multiple stress tolerance in crop plants (Kota et al., 2023). Furthermore, it offers wide applicability by allowing precise, noninvasive characterization, high-throughput screening at different crop growth stages, rather than being limited to single-point measurements. Consequently, it can monitor plant responses to multiple stresses temporally throughout their lifecycle. Ultimately, this accelerates the crop improvement programs by shortening breeding cycles to develop climate-resilient crop varieties.

13.2.1 High-throughput phenotyping platforms

High-throughput phenotyping (HTP) has been used in genetic studies within a species, allowing for the collection of huge sets of data on the morphology and physiology of plants (Guo et al., 2021). Using spectral reflection information to find out a plant's phenotype depends on the qualities of the light that comes through the canopy after it has been reflected, transmitted, and absorbed by the plant's tissues. Imaging techniques are highly useful for finding plant optical properties, especially those that cannot be seen with the human eye. HTP involves the integration of imaging platforms and sensors to characterize the plant phenotypes across various conditions in both controlled and field environments. Different imaging platforms and sensors are listed in Table 13.1, which play a key role in HTP (Li et al., 2014).

13.2.2 Integration of omics data

An in-depth understanding of the molecular mechanisms governing plant phenotypes and their interactions is made possible by the integration of omics data. For a variety of crops, multiomics analysis has been crucial in identifying molecular processes, development and growth, and tolerance to stress. Numerous omics methodologies, such as transcriptomics, proteomics, metabolomics, ionomics, phenomics, and genomics, have used high-throughput techniques to analyze gene networks, molecular processes of genes, and functional analysis in crop research (Yang et al., 2021). Machine learning techniques and the integration of multi-omics data have emerged as potent techniques for the analysis and interpretation of omics data. This has enabled the development of prediction models that can accelerate the breeding process and reduce the time and money required for traditional breeding procedures. Integrating omics data also allows for the development of new varieties of crops that are more resistant to diseases, have higher stress tolerance levels, and produce higher yields. It also helps in identifying new targets for crop improvement (Chao et al., 2023).

13.3 Precision phenotyping: techniques and applications

13.3.1 Controlled environment phenotyping

The ability to effectively control environmental parameters (such as temperature, humidity, and light intensity) while phenotyping in a growth chamber or greenhouse aids in precisely capturing how plants interact with that particular environment. It offers off-season phenotyping that is nondestructive, high-throughput, and extremely repeatable (Kumari et al., 2024). For instance, controlled environment phenotyping enables screening for drought tolerance by applying standardized water deficit conditions during critical growth stages in crops such as soybean, allowing for precise assessment of each genotype's response (Pardo et al., 2014).

Also, combining controlled environment phenotyping with AI and IoT technologies led to the creation of numerous high-throughput plant phenotyping (HTPP) platforms, such as WPScanConveyor, LemnaTec Scanalyzer, Lab Scanalyzer HTS, PhenoCenter, PlantEye F500, PHENOSPEX, PHENOPSIS, and others (Kumari et al., 2024). The sensor-to-plant method (Fig. 13.1) has the plant stay in one place while the imaging setup moves to each plant. The plant-to-sensor

TABLE 13.1 Different imaging platforms and sensors used in high-throughput phenotyping.

Imaging platforms/sensors	Key role	Representation
RGB imaging	Enables measurement of various parameters, including architecture, shoot/root growth, biomass, leaf area, leaf rolling, senescence, growth rates, tillering, early vigor, plant height, leaf greenness, convex hull, and compactness.	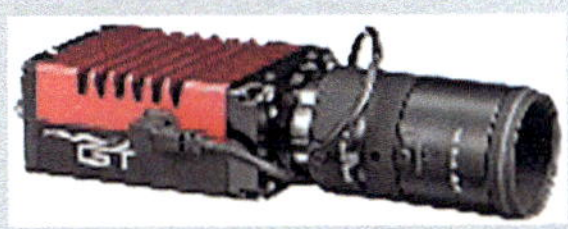 Part of your document could not be imported as it contained an unsupported element. For more information about unsupported elements, please visit the Support Center
NIR imaging	NIR shoot imaging systems are employed to measure water content in plants, leaf thickness, and leaf area index. NIR root imaging technologies are also used to study how roots grow and how they take moisture from the soil.	
IR (thermal) imaging	Allows measurement of tissue temperature, which is primarily influenced by evapotranspiration. IR thermal imaging is valuable for inferring stomatal conductance and assessing plant health under biotic and abiotic stress conditions.	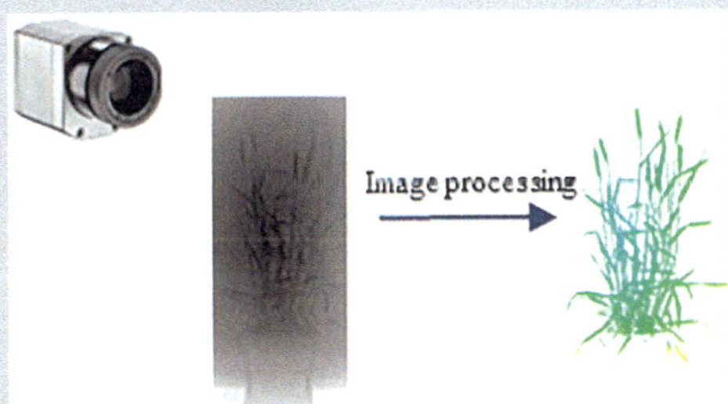
Fluorescence imaging	Measuring chlorophyll fluorescence, parameters such as Fv/Fm, photochemical, and nonphotochemical quenching are useful indicators of resource availability and environmental stresses.	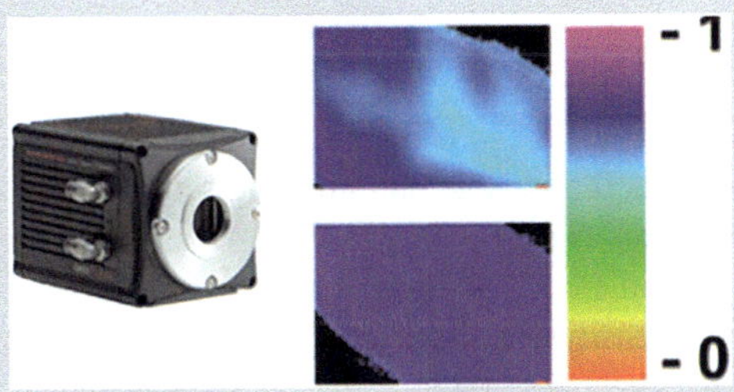
Hyperspectral imaging	Hundreds of spectral channels are used to collect spectral and temporal information about reflected light from the imaged surface. Chlorophyll concentration, RWC, mineral status, chemical composition, photochemical reflectance-based index, genotype barcoding, and other factors can be evaluated using a variety of spectral indices.	

(*Continued*)

TABLE 13.1 (Continued)

Imaging platforms/sensors	Key role	Representation
Soil moisture sensors	Measure the soil moisture content. These sensors are key instruments in precision agriculture, helping to optimize irrigation, increase crop yield, and conserve water resources.	
Weather stations	Provide detailed weather data that allows the researchers to monitor the plant growth precisely under different environmental scenarios.	

method (Fig. 13.2) has the sensors and imaging setup stay in one place while the plants move to the image station one by one. The ICAR Indian Agricultural Research Institute, located in New Delhi, India, has implemented the LamnaTec Scanalyzer, which is a Plant-to-Sensor platform. Moreover, utilizing open-source tools for HTPP in controlled growth chambers, which are easily accessible, could provide accurate and remote monitoring of plant responses to abiotic stresses (Kim et al., 2023; Lee et al., 2023).

13.3.2 Field phenotyping

In general, the QTL and candidate genes identified under controlled environment phenotyping do not perform as well under real-field environments. Also, interpreting field data is challenging due to the highly variable environmental conditions, which are mostly uncontrollable in nature (Araus and Cairns, 2014). For example, compared to the field, the volume of soil available to roots in a pot is substantially smaller, limiting the amount of water and nutrients that plants may absorb. It is difficult to reproduce the soil environment under controlled conditions, despite the fact that it is critical for plant growth and development. Thus, it is sometimes necessary to phenotype the crop plants under field conditions to get a better understanding of variations in phenotypes due to real field conditions. High-throughput field phenotyping can

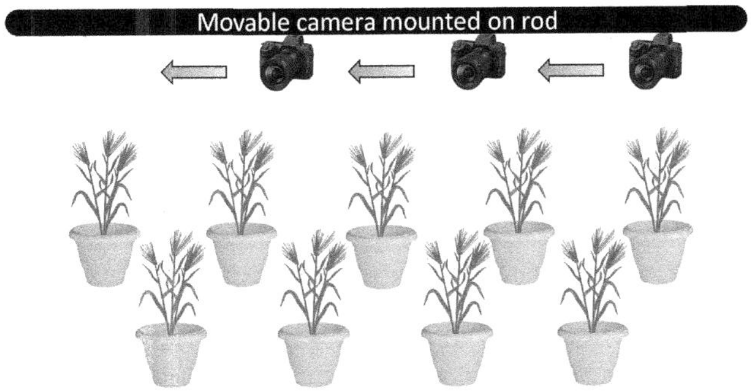

FIGURE 13.1 Sensor-to-plant–based phenotyping.

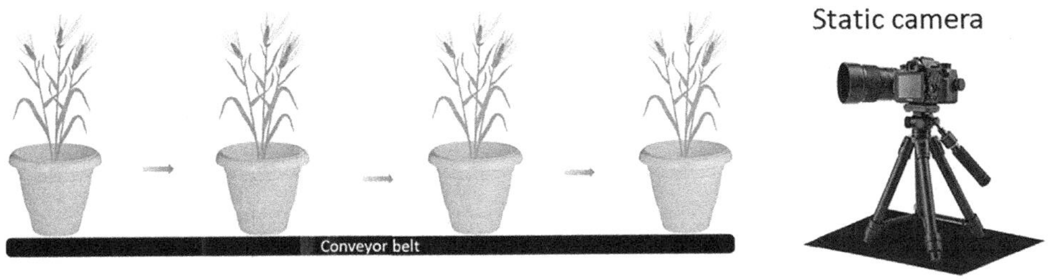

FIGURE 13.2 Plant-to-sensor–based phenotyping.

be delivered through a variety of platforms, including drone and satellite-based remote sensing, ground-based platforms, and proximal sensing, across a range of scales (Fig. 13.3; Fig. 3).

UAV/ drones have become a common tool for the HTP of crops. UAV or drone-based platforms provide a technology that uses established protocols and noninvasive imaging techniques to gather quantitative plant data on hundreds and thousands of crop genotypes in a field (Gano et al., 2024). These are composed of wings to fly (rotary, fixed, and hybrid) and sensors to phenotype (RGB, hyperspectral, multispectral, LiDAR, and thermal sensors, etc.). UAVs are advantageous for trials with a smaller area for each genotype, like wheat nursery trials, because they can fly at lower altitudes and speeds, allowing for greater resolution photos (Atkinson et al., 2018; Link et al., 2013). Drones can be utilized in many places because they are less expensive than other aerial devices and require smaller landing/take-off sites.

Due to the fact that drones have a limited flight time and payload capacity, the types and quantities of sensors that they are able to carry are restricted (Yang et al., 2017; Atkinson et al., 2018). Weather is another factor that restricts drone flying; optimal weather conditions allow for the most successful flights. Moreover, rapidly emerging HTP techniques using UAV-based sensors require the integration of machine learning algorithms (Angel and McCabe, 2022; Gano et al., 2024). The use of new machine learning algorithms in various phenotype extraction techniques has greatly expedited crop improvement (Castro et al., 2020) and precision agriculture (Gašparović et al., 2020) in recent years.

Recent studies show that satellite imagery can substitute the UAV or drone-based phenotyping for yield estimation in crops like maize (Shrestha et al., 2025). Satellites are the most cost-effective source of data (Atkinson et al., 2018), but their use in plant field phenotyping research has been restricted due to their lower resolution, fixed sensor configurations, and infrequent flyovers. However, multispectral imaging at 0.5 m resolution with nearly daily acquisition is now possible worldwide, especially with the development of high-resolution satellites like the SkySat constellation (Pinto et al., 2023).

Fig. 13.3 Aerial phenotyping tools: UAV and satellite.

To acquire accurate phenotyping data of the crops from the field for crop improvement programmes, ground-based platforms and proximal sensing tools are utilized. Traditional tractor-based field-based ground phenotyping systems have given way to more creative options like lightweight pushcarts and autonomous ground robots (Xu and Li, 2022). Because of their portability and ability to carry a number of sensors, tractors were at first seen as the best option. However, their heavy weight can lead to soil compaction, which diminishes their utility for collecting data often. Pushcarts and motorized carts, on the other hand, offer a portable, affordable, and more manageable option; however, they also need to be operated manually, which may adversely impact crops (Bai et al., 2016; Crain et al., 2016).

Although all these phenotyping systems have their own pros and cons, they allow researchers to relate the genotype to phenotype, enabling the identification of desirable traits for developing resilient and high-yielding crops. By utilizing

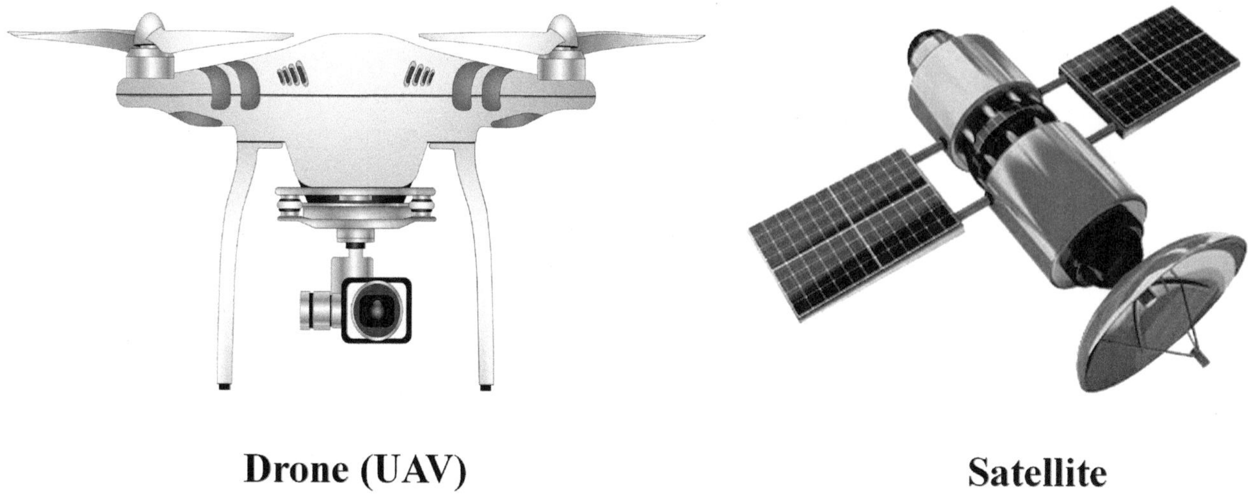

FIGURE 13.3 Multiscale plant phenotyping approaches using ground-based and aerial sensing platforms for Field Phenotyping.

sensors, including RGB cameras, multispectral, and thermal sensors, researchers can monitor plant health, density, and physiological traits throughout the growing season (Gano et al., 2024; Atkinson et al., 2018). They also contribute significantly to precision phenotyping by providing detailed data on plant health and growth stages, which aids in more efficient resource use, including water and fertilizers.

13.4 Trait identification and analysis

The process of breeding crops for abiotic stress tolerance, especially drought tolerance, can be challenging due to the complicated nature of plant stress response. Exploring naturally existing variations in abiotic stress tolerance in crop varieties, landraces, and wild relatives, as well as understanding the traits that enhance tolerance, is therefore a more effective strategy (Roy et al., 2011). To understand the genetic basis of complex traits, many analyses are employed to relate genetic and phenotypic data. It assists in uncovering chromosomal areas that are associated with particular traits (Skelly et al., 2019). By comparing trait values with the genetic variation in a notable mapping population developed from the contrasting parents, the genetic loci determining these traits can subsequently be determined (Roy et al., 2011).

13.4.1 Identification of key traits associated with abiotic stress tolerance

Drought, salinity, high temperatures, and nutritional imbalances are all important factors that might slow down plant growth and productivity (Mittler, 2006). Enhancing crop tolerance to abiotic stresses by breeding and biotechnological methods requires an understanding of related physiological processes and their molecular regulation (Varshney et al., 2018). Many of the physiological characteristics associated with abiotic stress tolerance are outlined in this section, along with the ways they help plants adapt to adverse environments.

Osmotic adjustment enables the cell to lower its osmotic potential, which elevates the gradient for turgor maintenance and water influx. Osmotically active substances, including proline, glycine betaine, and soluble carbohydrates, accumulate to keep cells healthy and turgid (Turner et al., 2001). Changes in cell wall elasticity and/or osmotic adjustment can result in improved tissue water status. For prolonged environmental stress, like drought, this is necessary to maintain physiological activity. Weather factors that affect transpiration and assimilation, such as vapor pressure deficit (VPD), pan evaporation, and relative humidity, have a major impact on WUE (Abbate et al., 2004). Under water-limited conditions, WUE typically increases, most likely as a result of stomatal closure lowering transpiration loss, particularly at midday when VPD is at its maximum. Due to decreased transpiration but maintained biomass development, cereals like wheat have shown an increase in WUE under abiotic stresses such as drought (Barraclough et al., 1989).

Plants protect themselves from oxidative stress by antioxidant metabolites such as ascorbic acid (AsA), glutathione (GSH), tocopherol, and carotene. Since heat-acclimated turf grass produced more AsA and GSH, it produced reduced reactive oxygen species (ROS) (Hasanuzzaman et al., 2013). Heat-tolerant cultivars exhibited higher glutathione Stransferase, ascorbate peroxidase, and catalase activity, which contributes to ROS detoxification (Hasanuzzaman et al., 2013). Small fine root diameters, long specific root length, and high root length density (RLD) are root traits that are associated with an environmental stress such as drought, especially in deeper soil layers where water is available (Comas et al., 2013). Under nutrient-deficient conditions, improved soil exploration and resource acquisition are made possible by lateral root proliferation and enhanced root branching (Bi et al., 2020). Steeper root angles are typically linked to greater drought adaptation, and root distribution patterns and root angles exhibit considerable plasticity in response to different abiotic stresses (Uga et al., 2013). Under stress, the anatomical traits of roots show exceptional plasticity. Aerenchyma tissue formation lowers metabolic expenses while maintaining the oxygen supply during flooding stress (Yamauchi et al., 2013). To allow the plant to adjust to stress, leaf morphological traits are vital. In order to minimize loss of water while maintaining photosynthetic capacity, reduced leaf size and increased leaf thickness are typical reactions to water scarcity (Xing et al., 2021). Leaf rolling, folding, and leaf anatomical traits like increased cuticular thickness and wax deposition enhance water retention capacity in water deficit stress conditions, providing tolerance to drought (Liu et al., 2024; Mondal et al., 2014).

Under a variety of environmental stresses, stay-green traits are an essential adaptation strategy that allows plants to maintain their photosynthetic capacity and delay senescence. Under abiotic stress circumstances, stay-green traits provide numerous benefits. As reported by Christopher et al. (2018), functional stay-green phenotypes have been associated with increased water use efficiency (WUE). Since the stay-green trait maintains the integrity of the photosynthetic apparatus, it also helps to tolerate heat stress better (Chen et al., 2015). One essential characteristic of plant response to abiotic stressors, particularly salinity, is ion homeostasis. Under stress, plants maintain cellular equilibrium and assure proper metabolic functioning through processes like ion exclusion, vacuolar sequestration, and selective ion transport (Munns and Tester, 2008).

13.4.2 Genome-wide association studies

Genome-wide association studies (GWAS), which take advantage of earlier recombination events, have been useful in identifying the genomic regions that regulate specific traits in a variety of crop species. Identification of the genes linked to abiotic stress tolerance has also been made easier by GWAS (Kumar et al., 2022) (Table 13.2).

The candidate gene *LOC_Os08g05520*, associated with the leaf rolling trait, confers abiotic stress tolerance, such as drought in rice, by avoiding water loss from the leaf surfaces (Liu et al., 2024). In addition, two candidate genes, *LOC_Os07g41180* and *LOC_Os07g41200*, located on chromosome 7, were identified in rice GWAS studies for the trait Flag leaf width (Zhang et al., 2015). Furthermore, chromosome 9 showed significant associations with rice root architectural traits, where *DRO1* (*LOC_Os09g26840*) has been identified as a key regulator of root angle (Uga et al., 2013, 2018). Similarly, two candidate genes, *GRMZM2G007555* and *GRMZM2G098494*, which are involved in the regulation of ion homeostasis and are certainly related to salt tolerance, are located on the same chromosome as a significant QTL for a plant-height-based salt tolerance index (Luo et al., 2017). In maize, the candidate gene *ZmRAP2.7 (GRMZM2G700665)*, associated with multiple traits such as seed vigor (Guo et al., 2024), flowering time (Liang et al., 2018), brace root development (Li et al., 2019), plays a wide range of roles for improving growth under stress conditions. Further, six candidate genes associated with the trait biomass were identified in soybean GWAS studies under drought stress, involved in different functions such as transcriptional regulation, signal transduction, metabolite biosynthesis, transport, and cell cycle control. The QTL *qBM-DS17.4* is linked to the candidate gene *Glyma.17G057100*, which encodes a WRKY transcription factor 11, implicating its potential role in regulating biomass under drought stress (Li et al., 2023). Studies on wheat revealed that the candidate gene *TaHXK3-2A (TraesCS2A02G005200)*, which encodes a hexokinase, has been reported to be associated with stomatal index (Li et al., 2022). Moreover, two candidate genes, *Sobic.001G018300* and *Sobic.001G018600*, which encode for cytochrome P450 proteins, were associated with stay green traits and play an important role in the biosynthesis of hormones, accumulation of metabolites, and regulation of chlorophyll content, thereby contributing to drought tolerance mechanisms (Tsehaye et al., 2024). Collectively, these candidate genes highlight the diverse roles essential for abiotic stress tolerance in different crops.

13.4.3 Quantitative trait loci mapping

An efficient approach for analyzing the genetic makeup of complex traits, such as abiotic stress tolerance, is QTL mapping. Many candidate genes/QTLs are identified through QTL mapping (Table 13.3). For many crop species to be able to tolerate water deficiencies, osmotic adjustment (OA) is crucial. Several QTLs for relative water content were identified in rice by Nayyeripasand et al. (2021), located on chromosomes 1 and 11. These include *qRWCs1.1* (*Os01g0228300*), qRWCs11.1 and qRWCs11.2 (*Os11g0216000*), which are associated with genes involved in abiotic stress tolerance,

TABLE 13.2 Candidate genes responsible for a phenotype identified through GWAS.

Crop	Identified candidate gene	Trait	References
Rice	*LOC_Os08g05520*	Leaf rolling	Liu et al. (2024)
Rice	*LOC_Os07g41180*	Flag Leaf Width	Zhang et al. (2015)
Rice	*LOC_Os07g41200*	Flag Leaf Width	Zhang et al. (2015)
Rice	*DRO1* (*LOC_Os09g26840*)	Root angle	Uga et al. (2013)
Maize	*GRMZM2G007555* and *GRMZM2G098494*	plant-height-based salt tolerance index	Luo et al. (2017)
Maize	*ZmRAP2.7 (GRMZM2G700665)*	Pleotropic (seed vigor, flowering time, brace root development)	Guo et al. (2024), Liang et al. (2018), Li et al. (2019)
Soybean	*Glyma.17G057100*	Biomass	Li et al. (2023)
Wheat	*TaHXK3-2A (TraesCS2A02G005200)*	Stomatal index	Li et al. (2022)
Sorghu-m	*Sobic.001G018300* and *Sobic.001G018600*	Stay-green	Tsehaye et al. (2024)

GWAS, genome-wide association studies.

TABLE 13.3 Candidate genes responsible for a phenotype identified through QTL mapping.

Crop	Identified QTL/gene	Trait	References
Rice	*qRWCs1.1 (Os01g0228300)*, qRWCs11.1 and qRWCs11.2 (*Os11g0216000)*	Relative water content (RWC)	Nayyeripasand et al. (2021)
Rice	*qLRS1.1*	Leaf rolling	Trijatmiko et al. (2014)
Rice	*qLRS-8-1 (LOC_Os08g05520)*	Leaf rolling	Liu et al. (2024)
Rice	*qCCAJ-9 and qCCAI-9*	Delayed senescence	Abbate et al. (2004); Abdelkhalik et al. (2005)
Wheat	*QOA.ubo-2B.2, QOA.ubo-6A.1* and *QOA.ubo-6A.2*	Osmotic adjustment	Condorell et al. (2022)
Wheat	*rlT73*	Leaf rolling	Huang et al. (2024)
Wheat	QTLs on chr 4A, 4B, and 4D	Stay-green	Christopher et al. (2018)
Wheat	*QWax.tam08-1B*	Cuticular wax	Mondal et al. (2014)
Maize	*ZmNC1*	Ion homeostasis	Zhang et al. (2017)
Chickpe-a	*qrsr-01, qrld-01* and *qrdw-01*	Root traits	Kushwah et al. (2022)

QTL, quantitative trait loci.

particularly salinity stress. Three major QTLs for OA were identified on chromosomes 2B *(QOA.ubo-2B.2)* and 6 A *(QOA.ubo-6A.1 and QOA.ubo-6A.2)*, suggesting their potential role in conferring drought tolerance (Condorell et al., 2022).

Understanding the genetic makeup underlying abiotic stress adaptation is aided by mapping these traits. Plants adopt morphological adaptations like leaf folding and rolling to reduce transpiration and respond to drought. According to Trijatmiko et al. (2014), the QTL *qLRS1.1* found on chromosome 1 in rice has been linked to leaf rolling and has remained constant across several populations, suggesting that it may be useful in breeding programs. Further, the QTL *qLRS-8-1* contains *OsMYB103L* (*LOC_Os08g05520*), an R2R3-MYB transcription factor that regulates cellulose synthesis and leaf rolling, thereby enhancing drought tolerance in rice (Liu et al., 2024). The candidate region of *rlT73* harbors one candidate gene, *TraesCS1B03G0515400*, which has been shown to play a role in stress tolerance by controlling the leaf rolling trait in wheat, and is located on chromosome 1B (Huang et al., 2024).

Moreover, an important trait linked to lowering the leaf temperatures under heat stress. The QTL QWax.tam08-1B, associated with cuticular wax production, is located on chromosome 1B in wheat and contributes to effective heat stress tolerance (Mondal et al., 2014). Under stress, root traits are essential for absorbing water and acquiring nutrients. Important loci for root traits in chickpeas have been found by QTL mapping studies, most notably *qrsr-01, qrld-01*, and *qrdw-01* on chromosomes 2, 7, and 4, respectively, which are connected to root traits such as root-shoot ratio, RLD, and root dry weight (Kushwah et al., 2022). Stress tolerance relies on ion homeostasis; in maize, a major QTL, *ZmNC1*, for leaf Na^+ concentration was identified on chromosome 3 with *ZmHKT1*, a plasma membrane class I HKT transporter, proposed as the candidate gene that controls the transport of ions across membranes and confers salinity tolerance (Zhang et al., 2017). When drought occurs, the ability to stay green is essential to maintain photosynthesis. Recent studies have identified QTLs on chromosomes 4A, 4B, and 4D of wheat that are constitutively linked to five stay-green traits under multiple environments (Christopher et al., 2018). Notably, rice was shown to have the QTLs *qCCAJ-9 and qCCAI-9* on chromosome 9, which were linked to delayed leaf senescence (Abdelkhalik et al., 2005).

13.4.4 Integration of phenotypic and genotypic data

A key element of current plant research is the integration of genotypic and phenotypic data, which improves our knowledge of trait variations and the genetic mechanisms driving them. The systematic measurement of observable traits in organisms is known as phenotyping. Phenotypic data in plant studies contain a wide range of traits, including rate of growth, leaf morphology, and stress responses in various environments. HTP techniques have become more popular, using nondestructive data collection methods like imaging and sensors (Rahaman et al., 2015). This vast amount of trait data makes it easier to do statistical analyses that link certain genetic variants to observable traits. The unPAK project, which gathers phenotypic information for large mutant datasets in *Arabidopsis*, is one important effort. Understanding plant adaptation to climate change is greatly aided by this kind of data, which enables extensive statistical analyses to connect phenotypes with genomic traits (Rutter et al., 2019). As advances in genomics have developed, it has developed the ability to evaluate genomic data. Through the use of approaches like GWAS, genotyping can uncover certain genetic variations linked to phenotypes of interest. Databases such as solQTL enable users to display QTL data associated with certain genomic areas for efficient QTL mapping and genetic study (Tecle et al., 2010). Researchers can rank genes according to their functional value by combining genotypic data with phenotypic traits (Alexander et al., 2017). Plant breeding efforts are further supported by tools such as the Variant Ranker, which allow genomic data to be ranked based on their expected influence on traits (Alexander et al., 2017).

13.4.5 Bridging the phenotype and genotype gap

Complex traits can only be understood by delving into the interplay between genotype and phenotype. Recent research has proven that integrated approaches that use genotypic and phenotypic data are successful. Studies can more accurately forecast quantitative trait phenotypes from genetic data by utilizing computational biology and bioinformatics (Ober et al., 2012). This prediction power has the potential to improve plant breeding results and deepen our understanding of genetic processes. The development of deep phenotyping methods has made it possible to evaluate phenotypic traits in a more sophisticated manner. The studies can identify superior genotypes for particular traits by conducting in-depth analyses of each component of the phenotype (Robinson, 2012). The quality and accessibility of combined phenotypic and genotypic data must be maintained by efficient data management and curation techniques (Jung et al., 2011). Furthermore, in order to facilitate the improved analysis of these intricate datasets using cutting-edge machine learning and data mining approaches, computational tools must be developed.

13.5 Aligning phenomics with emerging breeding strategies for multiple abiotic stress tolerance

Abiotic stresses that cause significant losses to agricultural production include extreme temperatures, drought, waterlogging, and salinity; all of which are the focus of extensive research. Selection for tolerance to a single stress often fails when plants encounter multiple stresses simultaneously. This failure is due to the fundamental fact that plant responses to combined stresses are unique and cannot be predicted by simply studying responses to individual stresses applied separately (Mittler, 2006). When seedlings were subjected to six different stresses individually at low levels, each stress had minimal impact on plant growth and survival. However, when these same stresses were applied in combination, the accumulated effect was devastating to the plants (Zandalinas et al., 2021). This indicates that selection protocols focused exclusively on individual stress tolerance may not produce varieties able to withstand the complex stress combinations plants encounter in natural environments. Multiple stress interactions in plants represent a vast and complex field that extends beyond simple additive effects. These interactions can be synergistic, where the combined effect is greater than the sum of individual effects, or antagonistic, where one stress mitigates the impact of another (Pandey et al., 2023). For example, rice (*Oryza sativa*) is often subjected to cooccurring stresses like drought and high salinity, which can exacerbate the negative effects on plant physiology and yield (Shabbir et al., 2022).

Further, when heat and drought stress are combined, they have a greater impact on plant development and productivity than when they occur alone (Barnabás et al., 2007). Plant reproductive stages, particularly flowering and anthesis, are especially vulnerable to these stresses (Barnabás et al., 2007). This can lead to impaired pollen development, pollen sterility, and fertilization failure because pollen and ovule functionality is compromised (Prasad et al., 2015). In many crops, including cereals like wheat and maize (Prasad et al., 2011; Barnabás et al., 2007) and legumes like field pea (Sadras et al., 2013) and lentil (Sita et al., 2018), the combination of heat and drought has a significant impact on the reproductive stage.

So, it is very important to think about breeding methods that will make plants more tolerant to various types of abiotic stresses in crops. It needs new ideas that combine traditional breeding practices with cutting-edge genetic technologies (Varshney et al., 2021). Conventional breeding has played a crucial role in developing stress-tolerant varieties, utilizing the existing genetic diversity within crop species (Sandhu et al., 2021). Pyramiding genes associated with multiple abiotic stress tolerances has been identified as a promising method. For instance, studies have demonstrated the effectiveness of pyramiding QTLs that control tolerance against drought, salinity, and submergence in crops like rice, thereby enhancing productivity in marginal environments (Muthu et al., 2020; Zhao et al., 2023). A comprehensive study by Schmidt et al. (2020) investigated combined drought and heat stress tolerance in 315 diverse spring bread wheat accessions. The study revealed that combined drought and heat stress severely impacted grain weight (over 92% reduction), single grain weight, and grain number. The key finding was the identification of a novel QTL on chromosome 6A, validated in field conditions, which consistently increased grain weight and harvest index under combined stress, offering promising avenues for breeding climate-resilient wheat varieties.

Traditional breeding methods, although effective in some scenarios, encounter considerable constraints in developing resistance to multiple stresses. It is predominantly dependent on phenotypic selection, which is intrinsically constrained by environmental conditions that affect trait expression (Krishna et al., 2023). The environment has significant impacts on phenotypic traits, which makes it difficult to select plants with a real genetic advantage for stress tolerance. To speed up crop improvement for combined stress tolerance, new methodologies are urgently needed to overcome the shortcomings of traditional breeding techniques, especially phenotyping. Plant phenomics with HTP technologies serves as a powerful tool for bridging the genotype-phenotype gap.

Plant phenomics enhances the understanding and identification of stress resistance in plants by capturing dynamic, nondestructive measurements of key traits under actual environmental conditions. Effectively integrating precision phenotyping and phenomics into breeding pipelines is essential for accurately assessing plant responses to abiotic stresses throughout the plant life cycle. The development of machine learning models to predict phenotypic outcomes based on genetic and environmental factors further enhances breeding efficiency (Meher et al., 2022). Such data-driven approaches enable breeders to efficiently select candidate varieties that exhibit resilience to multiple stresses and meet yield and quality standards.

By understanding how root architectures and other morphological traits influence plant responses to environmental challenges, breeders can develop varieties that not only survive but thrive under adverse conditions (Bonfiglioli et al., 2024). For example, the root system's plasticity plays a crucial role in drought tolerance, facilitating better crop performance under limited water availability (Bonfiglioli et al., 2024). Phenomics has been successfully used to precisely phenotype plants' underground root systems, and further research in this area can help identify the causal genes and suitable environmental target population (Takahashi and Pradal, 2021). The continuous monitoring of plant responses over time is crucial to capture the complex morphophysiological adjustments that plants undergo in different environments. Another strategy is the identification of genes, pathways, and networks involved in multiple stress responses through GWAS in combination with omics technologies such as phenomics, proteomics, ionomics, and transcriptomics (Zandalinas and Mittler, 2022; Zandalinas et al., 2021). It is possible to use these in breeding programs in order to develop superior cultivars that are tolerant to many types of abiotic stress.

Furthermore, the exploitation of genetic variation within crop wild relatives (CWR) has emerged as a critical route for minimizing the effects of abiotic pressures. This is a significant development. As CWR harbor unique genetic traits that confer resilience, they serve as invaluable resources for breeding programs aimed at improving stress tolerance. Studies have shown that traits such as heat stress resilience and drought tolerance can be successfully introgressed into cultivated varieties from CWR (Uga et al., 2013; van Zonneveld et al., 2019; Souter et al., 2017). HTP can be utilized for phenotyping of a large number of CWR, which can be helpful in the identification of novel traits and genomic locations that can be used for introgression into elite lines. Furthermore, advancements in molecular markers have facilitated the identification and incorporation of genes responsible for abiotic stress resilience from wild relatives into cultivated varieties (van Zonneveld et al., 2019; Souter et al., 2017).

Revolutionizing crop improvement, new breeding techniques, especially genome editing tools like Clustered Regularly Interspaced Short Palindromic Repeats (genome editing system) (CRISPR)/Cas9, enable precise alterations to the plant's genetic makeup. Tolerance to abiotic stresses can be improved through the targeted change of genes associated with stress responses made possible by CRISPR technology (Li et al., 2022; Chen et al., 2021). For example, the knockout of transcription repressor genes in *Arabidopsis* improved drought and salinity tolerance without affecting other trait behavior, indicating a possibility of developing robust crop varieties (Chen et al., 2021). CRISPR may have wider uses in agriculture as it has also been demonstrated to improve drought tolerance in crops like tobacco (Li et al., 2022). These genetic engineering technologies enable crops to be improved in a variety of ways, addressing genotype-specific susceptibilities to abiotic stressors.

By adopting an integrated approach, breeders can improve crops that are not only high-yielding but also robust in the face of multiple concurrent abiotic stresses (Table 13.4). Such strategic advancements are crucial for ensuring food security in a rapidly changing world, where agricultural systems must balance productivity with sustainability.

13.5.1 Case studies and applications

The phenomics tools for precision phenotyping have enabled more accurate and efficient screening of large populations for complex traits in crops (Visakh et al., 2024). Studies have shown that, especially when breeding complex traits, phenomics selection can successfully complement genomic selection. For example, a case study on cereal crops showed how field phenomics made it possible to evaluate important characteristics like biomass and radiation use efficiency (Furbank et al., 2019). Advances in thermal and multispectral imaging have transformed the capacity to gather precise phenotypic data, especially considering the conventional hurdles connected with quantifying complex features (Deery and Jones, 2021). By utilizing machine learning algorithms with phenomics data, breeders were able to hasten the selection process for high-yielding cultivars compared to conventional approaches (Su et al., 2024). HTP speeds up the breeding cycle while maintaining a high level of trait selection accuracy. The US Corn Belt's efforts to produce drought-tolerant maize hybrids are a noteworthy illustration of phenomics in action (Cooper et al., 2014).

Breeding efficiency and selection intensity were increased by the integration of molecular technologies, which made it possible to preselect superior genotypes before thorough phenotyping (Masuka et al., 2017; Zhu et al., 2021) revealed that the wide range of phenomics selection across a range of agronomic variables highlights its role in *triticale*. Their results highlight how phenomics prediction techniques can outperform genomic ones (Zhu et al., 2021, 2022). Researchers can more effectively identify promising germplasm by linking genetic information with phenotypic features by using phenomics databases, as demonstrated in the Pulse Crop Database (Singh et al., 2024; Haghighattalab et al., 2016), which showed how to employ unmanned aerial systems to phenotype huge wheat breeding nurseries in a high-throughput manner. By developing the use of drones to record specific plant metrics like canopy cover and growth factors, their research greatly accelerated the data collection process of breeding projects.

According to Patil and Ram (2024), the importance of phenomics for wheat heat tolerance evaluation has prompted targeted breeding efforts. In order to find and create heat-resistant wheat varieties, their work focused on combining precision phenotyping methods with cutting-edge molecular genetics. The importance of resilient cultivars in guaranteeing food security is highlighted by this work, which is essential for tackling the effects of climate change on wheat production systems (Patil and Ram, 2024; Cooper et al., 2014) described wheat lines that could tolerate heat and drought, and their work showed that some cultivars could thrive significantly even in harsh environments, which served as a model for breeding initiatives aimed at creating stress-tolerant cultivars. Masuka et al. (2017) emphasized the importance of developing wheat cultivars that are resistant to salinity and water stress. Their study highlights the increasing need to incorporate HTP tools capable of precisely evaluating how different wheat cultivars react to these stresses. The significance of nitrogen-use efficiency in breeding programs was reported by Furbank et al. (2019), especially in the context of nitrogen deficit, which frequently coexists with drought stress.

Crossa et al. (2013) reported using genomic prediction to improve genetic gains across a variety of conditions in International Maize and Wheat Improvement Center's wheat and maize breeding programs. They identified high-yielding

TABLE 13.4 Phenomics approaches employed in different combinations of stresses.

Multiple stresses	Phenomic approach	Phenomic indicators	Study focus	References
Drought, heat	Field-based high-throughput phenotyping (UAV imaging), GWAS, integration with SNPs genomics	Flowering date, plant height, and temporal reflectance	Prediction of flowering and plant height in maize under stress; gene discovery	Adak et al. (2023)
Biotic (bacterial/fungal diseases), Abiotic (chemical injury, nutrient deficiency)	High-throughput image-based phenotyping, convolutional neural network, and artificial intelligence	Visual foliar symptoms	Automated, explainable plant stress phenotyping in soybean	Ghosal et al. (2018)
Drought, extreme temperature, salinity, acidity, light, submergence, anaerobiosis, nutrient starvation	Multiomics (genomics, transcriptomics, proteomics, metabolomics, phenomics)	Photosynthesis, root growth, and hormone levels	Abiotic stress responses and microbe-mediated mitigation	Meena et al. (2017)
Drought, herbivory (combined and individual)	Multiomics, GWAS, QTL mapping, CRISPR, RNA interference	Leaf area, root growth, hormone levels, epigenetic marks	Integration of ecophysiology and omics for drought and herbivory	Shafi et al. (2024)
Drought, salinity, disease, pest	High-throughput phenotyping (ground/aerial imaging, remote sensing), machine learning /deep learning, root phenotyping, spectral indices	Root/shoot traits, leaf area, biomass	Using machine learning and high-throughput phenotyping to study plant stress	Gill et al. (2022)
Salinity, drought, nitrogen deficiency	Imaging (red-green-blue, multispectral, hyperspectral), unmanned aerial vehicle, machine learning/deep learning	Physiological traits	Image-based technologies for crop adaptation to abiotic stress	Al-Tamimi et al. (2022)

GWAS, genome-wide association studies; *QTL*, quantitative trait loci.

lines that were adaptable to a variety of environmental conditions by using genotype and phenotype data. A more robust selection of candidate lines was the result, proving that genomic selection provides significant benefits over conventional phenotypic evaluation techniques by enabling more effective use of resources (Crossa et al., 2013). In order to improve the selection process, breeders have to bridge the gap in validated trait-linked markers for abiotic stress in maize, as highlighted by Su et al. (2024). Breeders of maize can greatly increase genetic gains and manage the environmental constraints limiting yields in tropical locations by combining molecular technology with phenomics assessments (Zhu et al., 2021). Recent research on breeding strategies for *Oryza sativa* emphasizes how critical it is to address salinity resistance. Rice growth and production are significantly impacted by soil salinity (Ullah et al., 2022; Parmley et al., 2019), who advocated for the creation of salt-tolerant cultivars, highlighted how well-characterized salinity tolerance traits may be screened for by combining conventional breeding techniques with cutting-edge phenomics technologies.

13.6 Future perspectives and challenges

1. Future research in phenomics and precision phenotyping should focus on integrating multiple sensors and imaging technologies, developing more robust data analysis methods, and improving the automation of phenotyping processes.
2. Improving the accuracy and efficiency of trait prediction by developing more sophisticated statistical models and machine learning algorithms to integrate phenotypic and genotypic data.
3. The scope of phenotyping should expand in the future through incorporating more complex traits, such as root architecture, anatomy, and plant-microbe interactions, into phenotyping studies.
4. In order to speed up the development of novel crop varieties and enable collaborative research, it is vital to improve data exchange and accessibility.

A number of obstacles persist, even if phenomics and precision phenotyping have come a long way. These include the high cost and specialized expertise required for some HTP platforms, which can limit their accessibility, particularly in developing countries. Analyzing high-dimensional datasets generated by HTP platforms requires sophisticated statistical methods and computational resources. The lack of standardized protocols can hinder data comparability and reproducibility across studies. Understanding genotype-by-environment interactions remains a significant challenge in breeding for multiple abiotic stress tolerance. Addressing these challenges through continued research and development, along with collaboration and data sharing, will be crucial for unlocking the full potential of phenomics and precision phenotyping in developing multiple abiotic stress-tolerant crops and ensuring global food security. A paradigm shift in crop improvement has occurred with the incorporation of these technologies into breeding pipelines, opening the door for the creation of high-yielding, climate-resilient cultivars that can endure the difficulties posed by a changing climate.

References

Abbate, P.E., Dardanelli, J.L., Cantarero, M.G., Maturano, M., Melchiori, R.J.M., Suero, E.E., 2004. Climatic and water availability effects on water-use efficiency in wheat. Crop Science 44 (2), 474–483. Available at https://doi.org/10.2135/cropsci2004.4740.

Abdelkhalik, A.F., Shishido, R., Nomura, K., Ikehashi, H., 2005. QTL-based analysis of leaf senescence in an indica/japonica hybrid in rice (Oryza sativa L. Theoretical and Applied Genetics 110 (7), 1226–1235. Available at https://doi.org/10.1007/s00122-005-1955-2.

Adak, A., Kang, M., Anderson, S.L., Murray, S.C., Jarquin, D., Wong, R.K.W., et al., 2023. Phenomic data-driven biological prediction of maize through field-based high-throughput phenotyping integration with genomic data. Journal of Experimental Botany 74 (17), 5307–5326. Available at https://doi.org/10.1093/jxb/erad216.

Al-Tamimi, N., Langan, P., Bernád, V., Walsh, J., Mangina, E., Negrão, S., 2022. Capturing crop adaptation to abiotic stress using image-based technologies. Open Biology 12 (6). Available at https://doi.org/10.1098/rsob.210353.

Alexander, J., Mantzaris, D., Georgitsi, M., Drineas, P., Paschou, P., 2017. Variant Ranker: a web-tool to rank genomic data according to functional significance. BMC Bioinformatics 18 (1). Available at https://doi.org/10.1186/s12859-017-1752-3.

Angel, Y., McCabe, M.F., 2022. Machine learning strategies for the retrieval of leaf-chlorophyll dynamics: model choice, sequential versus retraining learning, and hyperspectral predictors. Frontiers in Plant Science 13. Available at https://doi.org/10.3389/fpls.2022.722442.

Araus, J.L., Cairns, J.E., 2014. Field high-throughput phenotyping: the new crop breeding frontier. Trends in Plant Science 19 (1), 52–61. Available at https://doi.org/10.1016/j.tplants.2013.09.008.

Atkinson, J.A., Jackson, R.J., Bentley, A.R., Ober, E., Wells, D.M., 2018. Field phenotyping for the future. Annual Plant Reviews Online 1 (3), 719–736. Available at https://doi.org/10.1002/9781119312994.apr0651.

Bai, G., Ge, Y., Hussain, W., Baenziger, P.S., Graef, G., 2016. A multi-sensor system for high throughput field phenotyping in soybean and wheat breeding. Computers and Electronics in Agriculture 128, 181–192. Available at https://doi.org/10.1016/j.compag.2016.08.021.

Barnabás, B., Jäger, K., Fehér, A., 2007. The effect of drought and heat stress on reproductive processes in cereals. Plant, Cell & Environment 31 (1), 11–38. Available at https://doi.org/10.1111/j.1365-3040.2007.01727.x.

Barraclough, P.B., Kuhlmann, H., Weir, A.H., 1989. The effects of prolonged drought and nitrogen fertilizer on root and shoot growth and water uptake by winter wheat. Journal of Agronomy and Crop Science 163 (5), 352–360. Available at https://doi.org/10.1111/j.1439-037x.1989.tb00778.x.

Bi, J., Liu, X., Liu, S., Wang, Y., Liu, M., 2020. Microstructural and physiological responses to cadmium stress under different nitrogen forms in two contrasting Populus clones. Environmental and Experimental Botany 169, 103897. Available at https://doi.org/10.1016/j.envexpbot.2019.103897.

Bonfiglioli, L., Urbanavičiūtė, I., Pagnotta, M.A., 2024. Durum wheat (*Triticum turgidum* L. var. *durum*) root system response to drought and salt stresses and genetic characterization for root-related traits. Frontiers in Plant Science 15. Available at https://doi.org/10.3389/fpls.2024.1362917.

Castro, W., Marcato Junior, J., Polidoro, C., Osco, L.P., Gonçalves, W., Rodrigues, L., et al., 2020. Deep learning applied to phenotyping of biomass in forages with UAV-based RGB imagery. Sensors 20 (17), 4802. Available at https://doi.org/10.3390/s20174802.

Chao, H., Zhang, S., Hu, Y., Ni, Q., Xin, S., Zhao, L., et al., 2023. Integrating omics databases for enhanced crop breeding. Journal of Integrative Bioinformatics 20 (4). Available at https://doi.org/10.1515/jib-2023-0012.

Chen, Y., Cothren, J.T., Chen, D.H., Ibrahim, A.M.H., Lombardini, L., 2015. Ethylene-inhibiting compound 1-MCP delays leaf senescence in cotton plants under abiotic stress conditions. Journal of Integrative Agriculture 14 (7), 1321–1331. Available at https://doi.org/10.1016/S2095-3119(14)60999-0.

Chen, S., Zhang, N., Zhou, G., Hussain, S., Ahmed, S., Tian, H., et al., 2021. Knockout of the entire family of AITR genes in Arabidopsis leads to enhanced drought and salinity tolerance without fitness costs. BMC Plant Biology 21 (1). Available at https://doi.org/10.1186/s12870-021-02907-9.

Christopher, M., Chenu, K., Jennings, R., Fletcher, S., Butler, D., Borrell, A., et al., 2018. QTL for stay-green traits in wheat in well-watered and water-limited environments. Field Crops Research 217, 32–44. Available at https://doi.org/10.1016/j.fcr.2017.11.003.

Comas, L.H., Becker, S.R., Cruz, V.M.V., Byrne, P.F., Dierig, D.A., 2013. Root traits contributing to plant productivity under drought. Frontiers in Plant Science 4. Available at https://doi.org/10.3389/fpls.2013.00442.

Condorell, G.E., Newcomb, M., Groli, E.L., Maccaferri, M., Forestan, C., Babaeian, E., et al., 2022. Genome wide association study uncovers the QTLome for osmotic adjustment and related drought adaptive traits in durum wheat. Genes 13 (2). Available at https://doi.org/10.3390/genes13020293.

Cooper, M., Gho, C., Leafgren, R., Tang, T., Messina, C., 2014. Breeding drought-tolerant maize hybrids for the US corn-belt: discovery to product. Journal of Experimental Botany 65 (21), 6191–6204. Available at https://doi.org/10.1093/jxb/eru064.

Crain, J.L., Wei, Y., Barker, J., Thompson, S.M., Alderman, P.D., Reynolds, M., et al., 2016. Development and deployment of a portable field phenotyping platform. Crop Science 56 (3), 965–975. Available at https://doi.org/10.2135/cropsci2015.05.0290.

Crossa, J., Pérez, P., Hickey, J., Burgueño, J., Ornella, L., Cerón-Rojas, J., et al., 2013. Genomic prediction in CIMMYT maize and wheat breeding programs. Heredity 112 (1), 48–60. Available at https://doi.org/10.1038/hdy.2013.16.

Deery, D.M., Jones, H.G., 2021. Field phenomics: will it enable crop improvement? Plant Phenomics 2021. Available at https://doi.org/10.34133/2021/9871989.

Deshmukh, H., Sonone, M., Hegde, V., Tamilselvan, A., Pradhan, A., Rane, J., et al., 2023. Advances in phenomics and its implications for crop improvement under multiple stress conditions through conventional and genomic approaches. Multiple Abiotic Stress Tolerances in Higher Plants: Addressing the Growing Challenges. CRC Press, India, pp. 25–40.

Doggalli, G., Monya, D., Kumar, M.B., Manjunath, P., Choudhuri, S., Das, J., et al., 2024. Breeding techniques and approaches for developing abiotic stress-tolerant crop cultivars: a comprehensive review. Plant Cell Biotechnology And Molecular Biology 25 (7-8), 101–125. Available at https://doi.org/10.56557/pcbmb/2024/v25i7-88745.

Fernie, A.R., Gutierrez-Marcos, J., 2019. From genome to phenome: genome-wide association studies and other approaches that bridge the genotype to phenotpe gap. Plant Journal 97 (1), 5–7. Available at https://doi.org/10.1111/tpj.14219.

Furbank, R.T., Jimenez-Berni, J.A., George-Jaeggli, B., Potgieter, A.B., Deery, D.M., 2019. Field crop phenomics: enabling breeding for radiation use efficiency and biomass in cereal crops. New Phytologist 223 (4), 1714–1727. Available at https://doi.org/10.1111/nph.15817.

Gano, B., Bhadra, S., Vilbig, J.M., Ahmed, N., Sagan, V., Shakoor, N., 2024. Drone-based imaging sensors, techniques, and applications in plant phenotyping for crop breeding: a comprehensive review. Plant Phenome Journal 7 (1). Available at https://doi.org/10.1002/ppj2.20100.

Gašparović, M., Zrinjski, M., Barković, Đ., Radočaj, D., 2020. An automatic method for weed mapping in oat fields based on UAV imagery. Computers and Electronics in Agriculture 173, 105385. Available at https://doi.org/10.1016/j.compag.2020.105385.

Ghosal, S., Blystone, D., Singh, A.K., Ganapathysubramanian, B., Singh, A., Sarkar, S., 2018. An explainable deep machine vision framework for plant stress phenotyping. Proceedings of the National Academy of Sciences 115 (18), 4613–4618. Available at https://doi.org/10.1073/pnas.1716999115.

Gill, T., Gill, S.K., Saini, D.K., Chopra, Y., de Koff, J.P., Sandhu, K.S., 2022. A comprehensive review of high throughput phenotyping and machine learning for plant stress phenotyping. Phenomics 2 (3), 156–183. Available at https://doi.org/10.1007/s43657-022-00048-z.

Guo, S., Ai, J., Zheng, N., Hu, H., Xu, Z., Chen, Q., et al., 2024. A genome-wide association study uncovers a ZmRap2.7-ZCN9/ZCN10 module to regulate ABA signalling and seed vigour in maize. Plant Biotechnology Journal 22 (9), 2472–2487. Available at https://doi.org/10.1111/pbi.14362.

Guo, H., Ayalew, H., Seethepalli, A., Dhakal, K., Griffiths, M., Ma, X.F., et al., 2021. Functional phenomics and genetics of the root economics space in winter wheat using high-throughput phenotyping of respiration and architecture. New Phytologist 232 (1), 98–112. Available at https://doi.org/10.1111/nph.17329.

Guo, Z., Yang, W., Chang, Y., Ma, X., Tu, H., Xiong, F., et al., 2018. Genome-wide association studies of image traits reveal genetic architecture of drought resistance in rice. Molecular Plant 11 (6), 789–805. Available at https://doi.org/10.1016/j.molp.2018.03.018.

Haghighattalab, A., González Pérez, L., Mondal, S., Singh, D., Schinstock, D., Rutkoski, J., et al., 2016. Application of unmanned aerial systems for high throughput phenotyping of large wheat breeding nurseries. Plant Methods 12 (1). Available at https://doi.org/10.1186/s13007-016-0134-6.

Hasanuzzaman, M., Nahar, K., Alam, M., Roychowdhury, R., Fujita, M., 2013. Physiological, biochemical, and molecular mechanisms of heat stress tolerance in plants. International Journal of Molecular Sciences 14 (5), 9643–9684. Available at https://doi.org/10.3390/ijms14059643.

Houle, D., Govindaraju, D.R., Omholt, S., 2010. Phenomics: the next challenge. Nature Reviews. Genetics 11 (12), 855–866. Available at https://doi.org/10.1038/nrg2897.

Huang, L., Gan, M., Zhao, W., Hu, Y., Du, L., Li, Y., et al., 2024. Characterization and mapping of a rolling leaf mutant allele rlT73 on chromosome 1BL of wheat. International Journal of Molecular Sciences 25 (7), 4103. Available at https://doi.org/10.3390/ijms25074103.

Jung, S., Menda, N., Redmond, S., Buels, R.M., Friesen, M., Bendana, Y., et al., 2011. The Chado natural diversity module: a new generic database schema for large-scale phenotyping and genotyping data. Database 2011 (0), bar051. Available at https://doi.org/10.1093/database/bar051.

Kim, J.Y., Abdel-Haleem, H., Luo, Z., Szczepanek, A., 2023. Open-source electronics for plant phenotyping and irrigation in controlled environment. Smart Agricultural Technology 3, 100093. Available at https://doi.org/10.1016/j.atech.2022.100093.

Kopecká, R., Kameniarová, M., Černý, M., Brzobohatý, B., Novák, J., 2023. Abiotic stress in crop production. International Journal of Molecular Sciences 24 (7), 6603. Available at https://doi.org/10.3390/ijms24076603.

Kota, S., Vispo, N.A., Quintana, M.R., Cabral, C.L.U., Centeno, C.A., Egdane, J., et al., 2023. Development of a phenotyping protocol for combined drought and salinity stress at seedling stage in rice. Frontiers in Plant Science 14. Available at https://doi.org/10.3389/fpls.2023.1173012.

Krishna, T.P.A., Veeramuthu, D., Maharajan, T., Soosaimanickam, M., 2023. The era of plant breeding: conventional breeding to genomics-assisted breeding for crop improvement. Current Genomics 24 (1), 24–35. Available at https://doi.org/10.2174/1389202924666230517115912.

Kumari, P., Bhatt, A., Meena, V.K., Adhikari, S., Dhar, N., Chawda, H., et al., 2024. Plant phenomics: the force behind tomorrow's crop phenotyping tools. Journal of Plant Growth Regulation. Available at https://doi.org/10.1007/s00344-024-11450-4.

Kumar, P., Choudhary, M., Halder, T., Prakash, N.R., Singh, V., V.T., V., et al., 2022. Salinity stress tolerance and omics approaches: revisiting the progress and achievements in major cereal crops. Heredity 128 (6), 497–518. Available at https://doi.org/10.1038/s41437-022-00516-2.

Kushwah, A., Bhatia, D., Barmukh, R., Singh, I., Singh, G., Bindra, S., et al., 2022. Genetic mapping of QTLs for drought tolerance in chickpea (Cicer arietinum L. Frontiers in Genetics 13. Available at https://doi.org/10.3389/fgene.2022.953898.

Lee, T.S., Jung, D.H., Kim, J.Y., Lee, J.Y., Park, J.E., Kim, H.S., et al., 2023. Development of a low-cost plant growth chamber for improved phenotyping research. Journal of Biosystems Engineering 48 (3), 355–363. Available at https://doi.org/10.1007/s42853-023-00197-7.

Liang, Y., Liu, Q., Wang, X., Huang, C., Xu, G., Hey, S., et al., 2018. ZmMADS69 functions as a flowering activator through the ZmRap2.7-ZCN8 regulatory module and contributes to maize flowering time adaptation. New Phytologist 221 (4), 2335–2347. Available at https://doi.org/10.1111/nph.15512.

Link, J., Senner, D., Claupein, W., 2013. Developing and evaluating an aerial sensor platform (ASP) to collect multispectral data for deriving management decisions in precision farming. Computers and Electronics in Agriculture 94, 20–28. Available at https://doi.org/10.1016/j.compag.2013.03.003.

Liu, T., Li, S., Du, H., Cui, J., Xu, S., Wang, J., et al., 2024. The identification of drought tolerance candidate genes in Oryza sativa L. ssp. Japonica seedlings through genome-wide association study and linkage mapping. Agriculture 14 (4), 603. Available at https://doi.org/10.3390/agriculture14040603.

Li, S., Cao, Y., Wang, C., Yan, C., Sun, X., Zhang, L., et al., 2023. Genome-wide association mapping for yield-related traits in soybean (Glycine max) under well-watered and drought-stressed conditions. Frontiers in Plant Science 14. Available at https://doi.org/10.3389/fpls.2023.1265574.

Li, J., Chen, F., Li, Y., Li, P., Wang, Y., Mi, G., et al., 2019. ZmRAP2.7, an AP2 transcription factor, is involved in maize brace roots development. Frontiers in Plant Science 10. Available at https://doi.org/10.3389/fpls.2019.00820.

Li, G., Ma, Y., Wang, X., Cheng, N., Meng, D., Chen, S., et al., 2022. CRISPR/Cas9 gene editing of NtAITRs, a family of transcription repressor genes, leads to enhanced drought tolerance in tobacco. International Journal of Molecular Sciences 23 (23), 15268. Available at https://doi.org/10.3390/ijms232315268.

Li, S., Yu, S., Zhang, Y., Zhu, D., Li, F., Chen, B., et al., 2022. Genome-wide association study revealed TaHXK3-2A as a candidate gene controlling stomatal index in wheat seedlings. Plant Cell and Environment 45 (8), 2306–2323. Available at https://doi.org/10.1111/pce.14342.

Li, L., Zhang, Q., Huang, D., 2014. A review of imaging techniques for plant phenotyping. Sensors 14 (11), 20078–20111. Available at https://doi.org/10.3390/s141120078.

Luo, M., Zhao, Y., Zhang, R., Xing, J., Duan, M., Li, J., et al., 2017. Mapping of a major QTL for salt tolerance of mature field-grown maize plants based on SNP markers. BMC Plant Biology 17 (1). Available at https://doi.org/10.1186/s12870-017-1090-7.

Maheswari, M., Singh, M.P., Rane, J., Chinnusamy, V., 2024. Unlocking crop potential—advancing plant phenomics for climate-smart agriculture. Plant Physiology Reports 29 (4), 697–698. Available at https://doi.org/10.1007/s40502-024-00845-w.

Masuka, B., Atlin, G.N., Olsen, M., Magorokosho, C., Labuschagne, M., Crossa, J., et al., 2017. Gains in maize genetic improvement in eastern and southern Africa: I. CIMMYT hybrid breeding pipeline. Crop Science 57 (1), 168–179. Available at https://doi.org/10.2135/cropsci2016.05.0343.

Meena, K.K., Sorty, A.M., Bitla, U.M., Choudhary, K., Gupta, P., Pareek, A., et al., 2017. Abiotic stress responses and microbe-mediated mitigation in plants: the omics strategies. Frontiers in Plant Science 8. Available at https://doi.org/10.3389/fpls.2017.00172.

Meher, P.K., Sahu, T.K., Gupta, A., Kumar, A., Rustgi, S., 2022. ASRpro: A machine-learning computational model for identifying proteins associated with multiple abiotic stress in plants. The Plant Genome 17 (1). Available at https://doi.org/10.1002/tpg2.20259.

Mittler, R., 2006. Abiotic stress, the field environment and stress combination. Trends in Plant Science 11 (1), 15–19. Available at https://doi.org/10.1016/j.tplants.2005.11.002.

Mondal, S., Mason, R.E., Huggins, T., Hays, D.B., 2014. QTL on wheat (*Triticum aestivum* L.) chromosomes 1B, 3D and 5A are associated with constitutive production of leaf cuticular wax and may contribute to lower leaf temperatures under heat stress. Euphytica 201 (1), 123–130. Available at https://doi.org/10.1007/s10681-014-1193-2.

Munns, R., Tester, M., 2008. Mechanisms of salinity tolerance. Annual Review of Plant Biology 59 (1), 651–681. Available at https://doi.org/10.1146/annurev.arplant.59.032607.092911.

Muthu, V., Abbai, R., Nallathambi, J., Rahman, H., Ramasamy, S., Kambale, R., et al., 2020. Pyramiding QTLs controlling tolerance against drought, salinity, and submergence in rice through marker assisted breeding. PLoS One 15 (1), e0227421. Available at https://doi.org/10.1371/journal.pone.0227421.

Nayyeripasand, L., Garoosi, G.A., Ahmadikhah, A., 2021. Genome-wide association study (GWAS) to identify salt-tolerance QTLs carrying novel candidate genes in rice during early vegetative stage. Rice 14 (1). Available at https://doi.org/10.1186/s12284-020-00433-0.

Ninomiya, S., Baret, F., Cheng, Z.-M.(M.), 2019. Plant phenomics: emerging transdisciplinary science. Plant Phenomics 2019, 2765120. Available at https://doi.org/10.34133/2019/2765120.

Ober, U., Ayroles, J.F., Stone, E.A., Richards, S., Zhu, D., Gibbs, R.A., et al., 2012. Using whole-genome sequence data to predict quantitative trait phenotypes in Drosophila melanogaster. PLoS Genetics 8 (5). Available at https://doi.org/10.1371/journal.pgen.1002685.

Pandey, P., Patil, M., Priya, P., Senthil-Kumar, M., 2023. When two negatives make a positive: the favorable impact of the combination of abiotic stress and pathogen infection on plants. Journal of Experimental Botany 75 (3), 674–688. Available at https://doi.org/10.1093/jxb/erad413.

Pardo, E.M., Vellicce, G.R., Aguirrezabal, L., Pereyra Irujo, G., Rocha, C.M.L., García, M.G., et al., 2014. Drought tolerance screening under controlled conditions predicts ranking of water-limited yield of field-grown soybean genotypes. Journal of Agronomy and Crop Science 201 (2), 95–104. Available at https://doi.org/10.1111/jac.12106.

Parmley, K.A., Higgins, R.H., Ganapathysubramanian, B., Sarkar, S., Singh, A.K., 2019. Machine learning approach for prescriptive plant breeding. Scientific Reports 9 (1). Available at https://doi.org/10.1038/s41598-019-53451-4.

Patil, A., Ram, M., 2024. Breeding for heat stress tolerance in wheat: a review. Journal of Advances in Biology & Biotechnology 27 (7), 227–237. Available at https://doi.org/10.9734/jabb/2024/v27i7983.

Pinto, F., Zaman-Allah, M., Reynolds, M., Schulthess, U., 2023. Satellite imagery for high-throughput phenotyping in breeding plots. Frontiers in Plant Science 14. Available at https://doi.org/10.3389/fpls.2023.1114670.

Prasad, P.V.V., Pisipati, S.R., Momčilović, I., Ristic, Z., 2011. Independent and combined effects of high temperature and drought stress during grain filling on plant yield and chloroplast EF-Tu expression in spring wheat. Journal of Agronomy and Crop Science 197 (6), 430–441. Available at https://doi.org/10.1111/j.1439-037x.2011.00477.x.

Prasad, P.V.V., Staggenborg, S.A., Ristic, Z., 2015. Impacts of drought and/or heat stress on physiological, developmental, growth, and yield processes of crop plants. In Advances in Agricultural Systems Modeling. American Society of Agronomy and Soil Science Society of America, pp. 301–355. Available at https://doi.org/10.2134/advagricsystmodel1.c11.

Rahaman, M.M., Chen, D., Gillani, Z., Klukas, C., Chen, M., 2015. Advanced phenotyping and phenotype data analysis for the study of plant growth and development. Frontiers in Plant Science 6. Available at https://doi.org/10.3389/fpls.2015.00619.

Robinson, P.N., 2012. Deep phenotyping for precision medicine. Human Mutation 33 (5), 777–780. Available at https://doi.org/10.1002/humu.22080.

Roy, S.J., Tucker, E.J., Tester, M., 2011. Genetic analysis of abiotic stress tolerance in crops. Current Opinion in Plant Biology 14 (3), 232–239.

Rutter, M.T., Murren, C.J., Callahan, H.S., Bisner, A.M., Leebens-Mack, J., Wolyniak, M.J., et al., 2019. Distributed phenomics with the unPAK project reveals the effects of mutations. Plant Journal 100 (1), 199–211. Available at https://doi.org/10.1111/tpj.14427.

Sadras, V.O., Lake, L., Leonforte, A., McMurray, L.S., Paull, J.G., 2013. Screening field pea for adaptation to water and heat stress: Associations between yield, crop growth rate and seed abortion. Field Crops Research 150, 63–73. Available at https://doi.org/10.1016/j.fcr.2013.05.023.

Sandhu, N., Yadav, S., Catolos, M., Cruz, M.T.S., Kumar, A., 2021. Developing climate-resilient, direct-seeded, adapted multiple-stress-tolerant rice applying genomics-assisted breeding. Frontiers in Plant Science 12. Available at https://doi.org/10.3389/fpls.2021.637488.

Schmidt, J., Tricker, P.J., Eckermann, P., Kalambettu, P., Garcia, M., Fleury, D., 2020. Novel alleles for combined drought and heat stress tolerance in wheat. Frontiers in Plant Science 10. Available at https://doi.org/10.3389/fpls.2019.01800.

Shabbir, R., Singhal, R.K., Mishra, U.N., Chauhan, J., Javed, T., Hussain, S., et al., 2022. Combined abiotic stresses: challenges and potential for crop improvement. Agronomy 12 (11). Available at https://doi.org/10.3390/agronomy12112795.

Shafi, I., Gautam, M., Kariyat, R., 2024. Integrating ecophysiology and omics to unlock crop response to drought and herbivory stress. Frontiers in Plant Science 15. Available at https://doi.org/10.3389/fpls.2024.1500773.

Shrestha, N., Powadi, A., Davis, J., Ayanlade, T.T., Liu, H., Tross, M.C., et al., 2025. Plot-level satellite imagery can substitute for UAVs in assessing maize phenotypes across multistate field trials. Plants People Planet 7 (4), 1011–1026. Available at https://doi.org/10.1002/ppp3.10613.

Singh, A., Jones, S., Ganapathysubramanian, B., Sarkar, S., Mueller, D., Sandhu, K., et al., 2021. Challenges and opportunities in machine-augmented plant stress phenotyping. Trends in Plant Science 26 (1), 53–69. Available at https://doi.org/10.1016/j.tplants.2020.07.010.

Singh, N., Ujinwal, M., Singh, N.K., 2024. Ppomicsdb: a multi-omics database for genetic and molecular breeding applications in Pigeonpea. Legume Science 6 (2). Available at https://doi.org/10.1002/leg3.220.

Sita, K., Sehgal, A., Bhandari, K., Kumar, J., Kumar, S., Singh, S., et al., 2018. Impact of heat stress during seed filling on seed quality and seed yield in lentil (Lens culinaris Medikus) genotypes. Journal of the Science of Food and Agriculture 98 (13), 5134–5141. Available at https://doi.org/10.1002/jsfa.9054.

Skelly, D.A., Raghupathy, N., Robledo, R.F., Graber, J.H., Chesler, E.J., 2019. Reference trait analysis reveals correlations between gene expression and quantitative traits in disjoint samples. Genetics 212 (3), 919–929. Available at https://doi.org/10.1534/genetics.118.301865.

Souter, J.R., Gurusamy, V., Porch, T.G., Bett, K.E., 2017. Successful introgression of abiotic stress tolerance from wild tepary bean to common bean. Crop Science 57 (3), 1160–1171. Available at https://doi.org/10.2135/cropsci2016.10.0851.

Su, H., Wang, Y., Fu, W., Feng, B., Wang, W., Chen, T., et al., 2024. Improving rice yield and quality through high-throughput phenomics, linear regression, and machine learning neural network models. Research Square. Available at https://doi.org/10.21203/rs.3.rs-4866351/v1.

Takahashi, H., Pradal, C., 2021. Root phenotyping: important and minimum information required for root modeling in crop plants. Breeding Science 71 (1), 109–116. Available at https://doi.org/10.1270/jsbbs.20126.

Tecle, I.Y., Menda, N., Buels, R.M., van der Knaap, E., Mueller, L.A., 2010. solQTL: a tool for QTL analysis, visualization and linking to genomes at SGN database. BMC Bioinformatics 11, 525. Available at https://doi.org/10.1186/1471-2105-11-525.

Trijatmiko, K.R., Supriyanta, Prasetiyono, J., Thomson, M.J., Vera Cruz, C.M., Moeljopawiro, S., et al., 2014. Meta-analysis of quantitative trait loci for grain yield and component traits under reproductive-stage drought stress in an upland rice population. Molecular Breeding 34 (2), 283–295. Available at https://doi.org/10.1007/s11032-013-0012-0.

Tsehaye, Y., Menamo, T.M., Abay, F., Tadesse, T., Bantte, K., 2024. Multi-locus genome-wide association study for grain yield and drought tolerance indices in sorghum accessions. Plant Genome 17 (4). Available at https://doi.org/10.1002/tpg2.20505.

Turner, N.C., Wright, G.C., Siddique, K.H.M., 2001. Adaptation of grain legumes (pulses) to water-limited environments. Advances in Agronomy. Elsevier, pp. 193–231. Available at https://doi.org/10.1016/s0065-2113(01)71015-2.

Uga, Y., Assaranurak, I., Kitomi, Y., Larson, B.G., Craft, E.J., Shaff, J.E., et al., 2018. Genomic regions responsible for seminal and crown root lengths identified by 2D & 3D root system image analysis. BMC Genomics 19 (1). Available at https://doi.org/10.1186/s12864-018-4639-4.

Uga, Y., Sugimoto, K., Ogawa, S., Rane, J., Ishitani, M., Hara, N., et al., 2013. Control of root system architecture by DEEPER ROOTING 1 increases rice yield under drought conditions. Nature Genetics 45 (9), 1097–1102. Available at https://doi.org/10.1038/ng.2725.

Ullah, M.A., Abdullah-zawawi, M.R., Zainal-abidin, R.A., Sukiran, N.L., Uddin, M.I., Zainal, Z., 2022. A review of integrative omic approaches for understanding rice salt response mechanisms. Plants 11 (11). Available at https://doi.org/10.3390/plants11111430.

Varshney, R.K., Bohra, A., Roorkiwal, M., Barmukh, R., Cowling, W.A., Chitikineni, A., et al., 2021. Fast-forward breeding for a food-secure world. Trends in Genetics 37 (12), 1124–1136. Available at https://doi.org/10.1016/j.tig.2021.08.002.

Varshney, R.K., Tuberosa, R., Tardieu, F., 2018. Progress in understanding drought tolerance: from alleles to cropping systems. Journal of Experimental Botany 69 (13), 3175–3179. Available at https://doi.org/10.1093/jxb/ery187.

van Zonneveld, M., Rakha, M., Tan, Sy, Chou, Yy, Chang, Ch, Yen, J., et al., 2019. Understanding patterns of abiotic and biotic stress resilience to unleash the potential of crop wild relatives for climate-smart legume breeding. bioRxiv. Available at https://doi.org/10.1101/596072.

Visakh, R., Anand, S., Reddy, S., Jha, U., Sah, R., Beena, R., 2024. Precision phenotyping in crop science: from plant traits to gene discovery for climate-smart agriculture. Plant Breeding. Available at https://doi.org/10.1111/pbr.13228.

Xing, K., Niinemets, Ü., Rengel, Z., Onoda, Y., Xia, J., Chen, H.Y.H., et al., 2021. Global patterns of leaf construction traits and their covariation along climate and soil environmental gradients. New Phytologist 232 (4), 1648–1660. Available at https://doi.org/10.1111/nph.17686.

Xu, R., Li, C., 2022. A review of high-throughput field phenotyping systems: Focusing on ground robots. Plant Phenomics 2022, 9760269. Available at https://doi.org/10.34133/2022/9760269.

Yamauchi, T., Shimamura, S., Nakazono, M., Mochizuki, T., 2013. Aerenchyma formation in crop species: a review. Field Crops Research 152, 8–16. Available at https://doi.org/10.1016/j.fcr.2012.12.008.

Yang, W., Feng, H., Zhang, X., Zhang, J., Doonan, J.H., Batchelor, W.D., et al., 2020. Crop phenomics and high-throughput phenotyping: past decades, current challenges, and future perspectives. Molecular Plant 13 (2), 187–214. Available at https://doi.org/10.1016/j.molp.2020.01.008.

Yang, G., Liu, J., Zhao, C., Li, Z., Huang, Y., Yu, H., et al., 2017. Unmanned aerial vehicle remote sensing for field-based crop phenotyping: current status and perspectives. Frontiers in Plant Science 8. Available at https://doi.org/10.3389/fpls.2017.01111.

Yang, Y., Saand, M.A., Huang, L., Abdelaal, W.B., Zhang, J., Wu, Y., et al., 2021. Applications of multi-omics technologies for crop improvement. Frontiers in Plant Science 12. Available at https://doi.org/10.3389/fpls.2021.563953.

Zandalinas, S.I., Mittler, R., 2022. Plant responses to multifactorial stress combination. New Phytologist 234 (4), 1161–1167. Available at https://doi.org/10.1111/nph.18087.

Zandalinas, S.I., Sengupta, S., Fritschi, F.B., Azad, R.K., Nechushtai, R., Mittler, R., 2021. The impact of multifactorial stress combination on plant growth and survival. New Phytologist 230 (3), 1034–1048. Available at https://doi.org/10.1111/nph.17232.

Zhang, M., Cao, Y., Wang, Z., Wang, Z., Shi, J., Liang, X., et al., 2017. A retrotransposon in an HKT1 family sodium transporter causes variation of leaf Na+ exclusion and salt tolerance in maize. New Phytologist 217 (3), 1161–1176. Available at https://doi.org/10.1111/nph.14882.

Zhang, B., Ye, W., Ren, D., Tian, P., Peng, Y., Gao, Y., et al., 2015. Genetic analysis of flag leaf size and candidate genes determination of a major QTL for flag leaf width in rice. Rice 8 (1). Available at https://doi.org/10.1186/s12284-014-0039-9.

Zhao, G., Liu, Y., Li, L., Che, R., Douglass, M., Benza, K., et al., 2023. Gene pyramiding for boosted plant growth and broad abiotic stress tolerance. Plant Biotechnology Journal 22 (3), 678–697. Available at https://doi.org/10.1111/pbi.14216.

Zhao, C., Zhang, Y., Du, J., Guo, X., Wen, W., Gu, S., et al., 2019. Crop phenomics: current status and perspectives. Frontiers in Plant Science 10. Available at https://doi.org/10.3389/fpls.2019.00714.

Zhou, R., Jiang, F., Niu, L., Song, X., Yu, L., Yang, Y., et al., 2022. Increase crop resilience to heat stress using omic strategies. Frontiers in Plant Science 13. Available at https://doi.org/10.3389/fpls.2022.891861.

Zhu, X., Leiser, W.L., Hahn, V., Würschum, T., 2021. Phenomic selection is competitive with genomic selection for breeding of complex traits. Plant Phenome Journal 4 (1). Available at https://doi.org/10.1002/ppj2.20027.

Zhu, X., Maurer, H.P., Jenz, M., Hahn, V., Ruckelshausen, A., Leiser, W.L., et al., 2022. The performance of phenomic selection depends on the genetic architecture of the target trait. Theoretical and Applied Genetics 135 (2), 653–665. Available at https://doi.org/10.1007/s00122-021-03997-7, https://link.springer.de/link/service/journals/00122/index.htm.

Chapter 14

Use of nanobiotechnology to improve multiple abiotic stress tolerance in plants

Mohammad Faizan[1,**], Mohammad Nauman Khan[2,3,*,**], Honghong Wu[4], Lixiao Nie[3], and Pravej Alam[5]

[1]*Botany Section, School of Sciences, Maulana Azad National Urdu University, Hyderabad, Telangana, India,* [2]*National Key Laboratory for Germplasm Innovation & Utilization of Horticultural Crops, MOA Key Laboratory of Crop Ecophysiology and Farming System in the Middle Reaches of the Yangtze River, The Center of Crop Nanobiotechnology, Huazhong Agricultural University, Wuhan, Hubei, P.R. China,* [3]*School of Breeding and Multiplication (Sanya Institute of Breeding and Multiplication), Hainan University, Sanya, Hainan, P.R. China,* [4]*College of Plant Science and Technology, Huazhong Agricultural University, Wuhan, Hubei, P.R. China,* [5]*Department of Biology, College of Science and Humanities, Prince Sattam bin Abdulaziz University, Alkharj, Riyadh Province, Saudi Arabia*

[*]*Corresponding author. e-mail address: 184268@hainanu.edu.cn*

14.1 Introduction

Drought, salt, severe temperatures, and heavy metal toxicity are all abiotic factors that make it hard for plants to thrive and farms to be productive all over the world. This is especially true when you think about how climate change and environmental degradation affect things (Faizan et al., 2022; Bhat et al., 2022). These pressures make it tougher for plants to grow and develop by messing up how cells divide, grow longer, and become different from each other. This means less biomass and yield. Drought stress, for instance, makes it harder to get water, which lowers turgor pressure and stops leaves from growing and roots from getting longer. Salinity stress causes osmotic and ionic toxicity, which slows growth and makes leaves turn yellow. Heavy metals such as cadmium (Cd) and arsenic (As), conversely, inhibit plants' nutrient absorption and use, hence damaging roots and impeding shoot growth (Finnegan and Chen, 2012). As the climate changes around the world, plants are more prone to experiencing abiotic stress. Photosynthesis works a lot better when these things are in place. For example, when stomata close because of salt stress or dryness, CO_2 absorption goes down, and the photosynthetic electron transport chain stops working (Chaves et al., 2009). When chloroplasts are stressed, such as when it's hot or salty, their structure falls apart, the chlorophyll pigments fall apart, and important enzymes like Rubisco work more slowly. Heat stress changes the shape of thylakoid proteins and stops the generation of ATP. Cold stress, on the other hand, makes membranes less flexible, which affects how plants breathe and create food. Stressful situations like dry air and high salt levels can make breathing harder, which is even worse because it might cause mitochondria to stop working. This makes reactive oxygen species (ROS) that damage proteins, lipids, and nucleic acids. Water and salt stress also mess with metabolic processes by making it harder for the body to take in and use important nutrients like nitrogen and phosphorus, stopping enzymes like nitrate reductase, and slowing down the production of proteins and amino acids. The downregulation of enzymes such as phosphoenolpyruvate carboxylase impacts primary carbon metabolism by decelerating the transit through the TCA cycle and energy generation pathways (Flexas et al., 2004).

Stressful situations might also change secondary metabolism. For example, when the body is under stress, it creates more secondary metabolites like flavonoids and phenolics. But too much stress can hurt the body more than its defenses (Ruelland et al., 2009). One of the worst things that can happen when there is abiotic stress is that too many ROS are generated. These are superoxide anions (O^{2-}), hydrogen peroxide (H_2O_2), and hydroxyl radicals (OH^-). If you do not clear these up right away, they can harm DNA, proteins, and lipids (Gill and Tuteja, 2010). Plants have a very complicated way of keeping free radicals from hurting them. It has both enzymatic parts, such as superoxide dismutase (SOD), catalase

** These authors contribute equally to this work.

Multiple Abiotic Stresses in Plants. DOI: https://doi.org/10.1016/B978-0-443-27424-4.00006-9

(CAT), peroxidases (POD), ascorbate peroxidase (APX), and glutathione reductase (GR), and nonenzymatic parts, including ascorbate, glutathione, carotenoids, and flavonoids. But when stress lasts for a long time or is particularly high, this antioxidant system is often too much for the body to handle. Drought and salinity stress, for example, have been demonstrated to lower CAT and POD activity in plants that are sensitive to them. This elevates the levels of H_2O_2 and malondialdehyde (MDA) even higher. Cd and other heavy metals can prohibit antioxidant enzymes from working by attaching to sulfhydryl groups or taking the place of important cofactors. As stress increases ROS levels and decreases the activities of APX and GR. This messes up the cycle of glutathione and ascorbate and the balance of redox. Stress from heavy metals also makes the plant use up nonenzymatic antioxidants like glutathione, which makes it much harder for the plant to keep its cells in balance. This is because glutathione is used to make phytochelatin, which gets rid of metals.

All of these physiological, metabolic, and oxidative issues work together to slow down plant growth, make it harder for plants to photosynthesize, make it harder for plants to take in nutrients, and make plants more sensitive to stress from the environment. In order to make crops that can manage abiotic challenges, we need to learn about these complex stress-response pathways. Progress in molecular biology, genomics, and biotechnology is facilitating the identification of stress-responsive genes and metabolic pathways that can be modified to improve stress tolerance. Furthermore, the external application of substances such as silicon, salicylic acid, and plant growth-promoting rhizobacteria, along with recent advancements in nanotechnology, is being investigated as an effective strategy to alleviate the impacts of abiotic stress by enhancing antioxidant responses and physiological resilience in plants (Hasanuzzaman et al., 2013). These kinds of combined tactics could help plants survive bad weather and make sure there is enough food even when abiotic needs are growing. To meet the growing need for food around the world, researchers are aiming to increase crop yields and food production, lessen the negative effects of abiotic stresses, and achieve sustainability and food security. Scientists obviously need to keep thinking about innovative approaches to solve these major problems that affect everyone.

There are a lot of complicated and connected reasons why farmers all around the world are having a harder difficulty cultivating food. These include biotic and abiotic pressures, climate change, the loss of natural resources like soil and water, and the need for additional energy. These problems demand long-term, low-cost solutions so that more people can get enough food. The current agricultural paradigm needs to focus on getting more crops from fewer resources while also using fewer synthetic fertilizers and pesticides, which can harm nearby ecosystems (Tilman et al., 2011). We need to quickly move away from old-fashioned farming methods and toward more inventive and environmentally friendly ones that leverage new technologies. One of the greatest options is to use high-tech farming instruments like precision farming, smart irrigation systems, and new ways for plants to get nutrients. Additionally, due to increasing abiotic stresses such as drought, salinity, heavy metal toxicity, and severe temperatures (Zhang et al., 2022), it is essential to develop and utilize crop types that exhibit greater resilience to these conditions. In this context, nanotechnology has become a game-changer in plant biotechnology and agricultural research.

Nanotechnology is the study of materials and technologies that are smaller than 100 nanometers (nm). It encompasses the design, synthesis, and application of various materials and technologies. Nanomaterials have unique physical and chemical qualities that make them useful for new things, such as spreading nutrients, keeping pests away, and lowering plant stress (Dubchak et al., 2010). Some of these traits are a higher surface-to-volume ratio, better responsiveness, and the ability to change how they work. Nanofertilizers and nanoformulations can utilize nutrients more effectively while simultaneously reducing pollution and toxicity (Khan et al., 2020). Recent research indicates that specifically tailored nanoparticles (NPs) can penetrate plant tissues due to their diminutive size, hence enhancing physiological responses in plants under stress situations. Studies have shown that they can make seeds germinate faster, boost photosynthesis, turn on plant defense systems, and change how much water plants take in Faizan et al. (2023a). NPs also help with controlled release and targeted distribution of agrochemicals. This makes better use of resources and reduces negative effects. Natural NPs, including nanoclays, magnetosomes, soil-derived organo-mineral complexes, and protein-based NPs like ferritin, are essential in biological systems and provide biocompatible alternatives for agricultural applications (Raliya et al., 2016). It is also well known that NPs can help the body's defenses against oxidative stress when there is abiotic stress. They can help reduce oxidative stress by improving the function of antioxidant enzymes such as SOD, CAT, and POD. This keeps plants from hurting their cells (Zuverza-Mena et al., 2017).

The objective of this study was to elucidate the mechanistic insights into the application of nanobiotechnology for alleviating abiotic stress toxicity and enhancing stress tolerance in plants. This was achieved by integrating the latest advancements in plant responses to various abiotic stresses, including salinity, drought, heavy metal exposure, and heat stress, through nanobiotechnological approaches.

14.2 Abiotic stress in crops and the current scenario

The process of global climate change is still going on, but it has sped up a lot in the last hundred years. Climate change caused by global warming is making food security a very real threat. Anthropogenic activities have led to the accumulation of greenhouse gases, resulting in global warming and an increase in global temperatures (Fahad et al., 2017) by 0.9°C during the 19th century. Strong sun radiation during the day also elevates body temperature by making things like evapotranspiration happen. The forecasted 4.5°C rise in global temperature by the end of the 21st century poses a substantial threat to agricultural production (Giordano et al., 2021). Predictions say that by 2050, 45% of the world's maize-producing areas will probably have an annual mean episode lasting five days during which the maximum temperature will reach above 35°C during the reproductive stage. This is essential since a 1°C rise in the average seasonal temperature can lower the commercial output of maize crops by 3%–13% (Waqas et al., 2021). The exponential rise in global temperature has led to more climate change events, such as waterlogging, drought, irregular rainfall patterns, the melting of glaciers, powerful typhoons and hurricanes, changes to animal habitats, the movement of plant habitats northward, and other extreme events. As climate change gets worse and abiotic stress events happen more often, agricultural productivity is expected to go down.

Plants are living organisms that do not move around and are always under stress from other living things and the environment. These stresses have a huge effect on how plants grow, which in turn affects how much food they make (Kesawat et al., 2023). There are two main types of stress that plants go through: abiotic and biotic stress. Most people agree that UV radiation, high temperatures, heavy metals, salt, and drought are all major abiotic stressors. A lot of studies have shown that stress can significantly reduce many growth indicators, including plant height, leaf area, and branch number (Kesawat et al., 2020). This event leads to oxidative stress, which harms cell parts and worsens some biological processes, like generating proline, MDA, and other electrolytes. This indicates that it messes up redox homeostasis, which makes it hard for cells to do redox activities (Chen et al., 2017). One method to decrease the harmful effects of stress is to grow plants that can handle it. You can reach this goal by carefully picking cultivars that can handle stress.

One of the worst things that can happen to crops is abiotic stress. They are responsible for 50%–70% of the output losses in important food crops. Too many ROS oxidize lipids, proteins, and DNA, which damages them. Plants grow and develop poorly because of this (Table 14.1). Dryness and saltiness, for instance, cause osmotic stress and ionic toxicity, which make it harder for plants to take in water and close their stomata. This makes it harder for plants to take in carbon (Farooq et al., 2009). Cd, As, and lead (Pb) are other heavy metals that can similarly change how enzymes work and how nutrients move around. This makes cells poisonous and stops them from developing (Shahid et al., 2017). When it's hot,

TABLE 14.1 Effect of abiotic stresses on crop plants.

Abiotic stress	Effects on plants	Key impacted parameters	References
Drought	Reduces water availability; causes oxidative stress and impaired photosynthesis	↓ Relative water content, ↓ chlorophyll, ↑ proline, ↑ MDA, ↓ biomass	Faizan et al. (2023b)
Salinity	Disrupts ion homeostasis and osmotic balance; oxidative damage	↑ Na^+ uptake, ↓ K^+/Na^+ ratio, ↓ growth, ↑ electrolyte leakage, ↑ antioxidant enzyme activity	
Heavy metals	Induce ROS production and damage to cellular structures; inhibit photosynthesis and nutrient uptake	↑ ROS, ↑ MDA, ↓ chlorophyll, ↓ root/shoot length, ↑ antioxidant defense	Jiang et al. (2012)
Heat stress	Inhibits enzymatic activity, reduces membrane stability, and photosynthetic efficiency	↓ chlorophyll, ↓ pollen viability, ↑ HSPs, ↑ oxidative stress	Djanaguiraman et al. (2018)
Cold stress	Disrupts membrane fluidity and metabolic processes	↓ germination rate, ↓ photosynthesis, ↑ sugar and proline content, ↑ ABA levels	Khan et al. (2020)

HSPs, heat shock proteins; *MDA*, malondialdehyde.

proteins are not as stable, photosynthesis does not work as effectively, and membranes are not as flexible. This has a big impact on how plants create seeds and how healthy the food they grow.

Things are becoming worse because the soil is getting worse, it's harder to access water, and people want more food. It requires a lot of genes to develop plants that can endure abiotic stress. This is why traditional methods of growing crops have not worked particularly well. More and more people want to use new technologies to grow crops in climates that are changing. These technologies include genome editing (like CRISPR-Cas9), omics-based approaches, plant-microbe interactions, and nanobiotechnology. Nanotechnology has demonstrated potential by facilitating the targeted delivery of nutrients and stress-relieving compounds, improving photosynthesis, and stimulating antioxidant defense mechanisms to mitigate abiotic stresses (Wang et al., 2022).

14.3 Nanobiotechnology for sustainable agriculture

Nanotechnology has transformed farming by changing how we grow food, how long it lasts, and how healthy the plants are. This novel technology takes advantage of the unique physical and chemical properties of materials that are between 1 and 100 nm in size. These materials are very different from the bigger ones because they have more surface area, are more reactive, and can get through biological membranes. There are many uses for it in farming, and new ones are being identified all the time. Some of these goals are to watch how plants grow, find diseases, and make food better and easier to get. Nanotechnology is also very important for keeping plants safe from damage caused by the environment, which makes farming systems more resilient (Afzal et al., 2024; Faizan et al., 2024a). Using nanofertilizers is one of the best new ideas since they work far better than regular fertilizers. Farming is more sustainable when it makes better use of nutrients, pollutes the environment less, and releases nutrients for longer. Plants can get the nutrients they need more easily using nanofertilizers since they are smaller and dissolve faster. This means you do not need as much fertilizer, which is good for the environment and saves you money. Research indicates that nano-fertilizers significantly enhance both innovative and sustainable agricultural practices, contributing to worldwide initiatives aimed at establishing enduring systems for food and energy production (Faisal et al., 2024a). Carmen et al. highlighted the extraordinary characteristics of nanomaterials, including quantum effects and high surface-area-to-volume ratios, which make them beneficial for promoting crop growth and alleviating stress. Nanomaterials can be used in farming for a lot of different things because of how they work. For instance, they can spread agrochemicals, make soil more fertile, and shield plants from both living and nonliving stressors. NPs, for instance, can be a smart way to get nutrients and insecticides to where they need to go. When the pH, temperature, or humidity changes, they can release what they hold.

NPs made of metals like zinc (ZnO), copper (CuO), silver (Ag), and titanium dioxide (TiO_2) are some of the most important new things in this sector. These NPs can help plants grow by giving them micronutrients, killing pathogens, and helping them thrive. These particles are so tiny that they can enter plant cells through stomata or root pores. This makes it easier for nutrients and other critical things to get to certain places (Faisal et al., 2024b). These NPs can also assist in keeping nutrients from getting away or evaporating, which is a huge problem with normal techniques of fertilization. This is because they send nutrients to certain places in a controlled way. Khan et al. posited, through their experimental research and literature review, that NPs could considerably boost plant resilience to unfavorable environmental conditions. They suggested that NPs make plant defense mechanisms work better and help plants make genes that respond to stress, which makes plants more resistant to environmental stressors. Some people say that NPs are particularly significant for modifying signaling pathways, especially those that are related to calcium (Ca^{2+}) signaling. Plants use calcium ions as secondary messengers to relay signals to one another. Plants may be able to sense and respond to abiotic conditions such as drought, salinity, and heavy metal toxicity due to the influence of Ca^{2+} channels and Ca^{2+}-binding proteins on the transmembrane transport of Ca^{2+} (Thapa et al., 2011). When NPs come into contact with plants, the calcium levels in the plant cells change a lot. This starts a chain reaction of actions, including the activation of transcription factors, the production of protective proteins, and the formation of enzymes that break down ROS. These chemical systems work together to assist living things in dealing with stress from things that are not alive. For instance, NPs have been demonstrated to help plants handle drought stress better than regular fertilizers. For example, ZnO NPs helped soybeans grow and sprout better, even when they were stressed out because they did not have enough water. The NPs probably helped the seeds sprout faster by making it easier for them to take in water and speeding up the work of enzymes. Sorghum plants that were given ZnO NPs took in more nitrogen and grew more biomass (Dogan et al., 2024). The NPs made it easier for roots to grow, absorb nutrients, and carry out other physiological activities, which led to this improvement. A different study discovered that adding ZnO NPs to the culture media helped banana plantlets grow better in vitro. The treatment improved somatic embryogenesis, made plants more able to handle stress, and sped up the formation of healthy plantlets. This research shows that ZnO NPs can help plants grow and deal with stress.

Plants are always under stress because of things like salt, drought, severe temperatures, and heavy metals in their environment. When plants are under stress, they have a harder time accomplishing things like photosynthesis, breathing, taking in nutrients, and keeping their water levels stable. This makes things bigger and less helpful. Plants have made incredibly complex defense systems that work on the molecular, cellular, and physiological levels to deal with these kinds of problems. Some of these strategies function by turning on genes that respond to stress, generating osmoprotectants like proline and glycine betaine, and making antioxidant defense systems stronger. NPs can make the plant's defenses stronger by changing these biological pathways. To understand how NPs affect plants, you need to know how many there are and how strong they are. NPs can help plants develop and make them stronger when they are present in small numbers. But when there are a lot of them, they can hurt plants by producing oxidative damage and making cells work less well than they could. Laware and Raskar (2014) asserted that NPs enhance the effectiveness of critical antioxidant enzymes. Some of these enzymes are SOD, CAT, and POD. When plants are stressed, they create ROS, which these enzymes break down. NPs make these enzymes work better, which helps cells stay in balance and protects plants from oxidative damage. Plants alter in a lot of ways when they are under abiotic stress, such as at the molecular, biochemical, physiological, and morphological levels. These alterations help the plant survive by allowing it to handle bad weather. Researchers have been exploring genetic and biotechnological ways to help plants adapt better. They produced genetically modified vegetables, for instance, that can endure stress. Adding nanotechnology to these designs would be a lot better since it modifies how plants take in nutrients and makes their stress-response circuits work better.

NPs not only help seeds sprout and seedlings grow stronger, but they also modify how glucose is formed, how nitrogen is used, and how proteins are made. These changes to the plant's physiology help it get bigger, make more food, and build more biomass. Researchers have also found that some NPs, such as silicon dioxide (SiO_2) and TiO_2, can help plants photosynthesize better when they are under stress by making chloroplasts more stable and enabling them to take in more light. Nanostructured materials, including hydrogels containing NPs, can help with water stress by making the soil store more water and making the roots more permeable to water. These things help keep the plant's water supply steady when it does not rain. This helps the plant's body deal with the stress of not getting enough water. NPs can also help nutrients flow around in the plant system, which speeds up and makes the process of moving important parts to different tissues more efficient. Because they move in different ways, nutrients can get to the plant's growth tips, reproductive organs, and other important regions. This helps the plant grow better and produce more. Research indicates that NPs may function as elicitors, enhancing systemic acquired resistance and stimulating the synthesis of phytoalexins, which confer significant resistance to diseases in plants. Nanotechnology is a useful tool for figuring out how to farm in a way that can handle climate change, since it helps plants grow and makes them more resistant to disease.

14.4 Role of nanobiotechnology in plants under abiotic stress conditions

14.4.1 Plants under stress from drought

Drought stress is one of the most important factors that are not alive that impacts the growth of plants and the amount of crops grown across the world. Plants' shape, function, metabolism, and gene expression all change when they do not get enough water for a long time. This changes the amount and quality of the crops. When there is not enough water, it becomes harder to transfer food, photosynthesis is not as effective, cellular homeostasis is disrupted, and oxidative stress occurs because too many ROS are generated. Changes in plants caused by stress can harm their health and food security in areas where droughts are expected to happen (Faizan et al., 2023b). Plants have adapted in various ways to cope with drought. Some of the changes to the structure include roots that go deeper and smaller leaf areas. Some things happen in their body, such as controlling the stomata and adjusting the osmotic pressure. Biochemical processes also create osmolytes and turn on enzymes that keep cells from being hurt. Nanotechnology is being used in a new way to assist plants in dealing with drought stress by using customized NPs. NPs are expected to change how plants work at the cellular and molecular levels, which makes them a better approach to relieve stress than other ways. Linh et al. investigated the impact of metal-based NPs, particularly iron (Fe), Cu, cobalt (Co), and ZnO, on soybean (Glycine max) plants experiencing drought conditions. Their research showed that using these NPs, specifically Fe NPs, made some physiological traits that are related to drought tolerance much better. The drought tolerance index, relative water content, and biomass retention all improved a lot. You can find out if a plant can live without enough water by doing these experiments. The study's findings indicate that NPs may assist plants in maintaining their water equilibrium, minimizing biomass loss during arid periods, and enhancing overall stress resilience. Using SiO_2 NPs has shown a lot of potential for helping different kinds of crops survive dry spells. For instance, using varying concentrations of SiO_2 NPs on sorghum (*Sorghum bicolor*) improved its biochemistry and physiology. Plants that had been treated exhibited decreased levels of

MDA, which is clear evidence of damage to membranes and lipid peroxidation. This suggests that oxidation did not bother them as much. The amounts of proline, soluble carbohydrates, carotenoids, and chlorophyll also went up. The photosynthetic efficiency and membrane stability also got better (as seen by less electrolyte leakage). Plants can better handle drought stress with these reactions because they improve osmotic adjustment, maintain robust membranes, and aid energy metabolism. Research has shown that NPs change how genes that respond to drought are expressed at the molecular level. Quantitative PCR research on plants exposed to NPs during drought conditions has revealed significant activation of various stress-related genes. After being treated with NPs, soybeans have been revealed to have considerably greater levels of many critical transcription factors and regulatory genes, including *GmRD20A*, *GmDREB2*, *GmERD1*, *GmFDL19*, *GmNAC11*, *GmWRKY27*, *GmMYB118*, and *GmMYB174*. These genes protect cells in a multitude of ways, such as by transmitting signals, responding to dehydration, and signaling stress-related genes on what to do next. Scientists have proven that these genes are genetic markers of drought tolerance in model organisms like soybeans and *Arabidopsis thaliana*. Huang et al. (2012) contend that they serve as preliminary reactions to stress resulting from dehydration. The interaction between nanomaterials and plant molecular pathways is complex, as evidenced by the tissue-, gene-, and NP-specific responses of these genes to treatment. This specificity highlights the importance of choosing the right kinds and amounts of NPs to get the desired stress-reduction effects while keeping toxicity to a minimum.

14.4.2 Plants under stress from salinity

Salinity stress is one of the most prevalent and severe abiotic challenges facing modern agriculture, significantly impairing plant growth, development, and productivity. It is estimated that more than 20% of cultivated land worldwide is currently affected by salinity, a figure that continues to rise due to factors such as climate change, poor irrigation practices, seawater intrusion, and unsustainable land use (Faizan et al., 2024b). Salinity stress disrupts critical physiological and biochemical processes in plants by inducing osmotic stress, ion toxicity, nutrient imbalance, oxidative damage, and reduced water uptake, ultimately affecting plant health and crop yield. At the physiological level, salinity stress causes excessive accumulation of toxic ions like Na^+ and Cl^-, which interfere with essential cellular functions. This ionic stress leads to impaired photosynthesis, protein degradation, and inhibition of enzyme activity. To combat these effects, plants have developed complex defense mechanisms, including osmotic adjustment, activation of antioxidant enzymes, and regulation of ion transporters. However, these mechanisms are often insufficient under prolonged or high salinity conditions. Therefore innovative strategies are needed to enhance plant resilience to salt stress, and nanotechnology has emerged as a promising approach in this regard.

The application of Si-NPs has been found to play a significant role in improving plant tolerance to salinity stress. Si-NPs possess high surface area, unique physicochemical properties, and the ability to interact with plant metabolic pathways at the cellular level. Their application enhances nutrient uptake, fortifies the antioxidant defense system, and improves water-use efficiency. In basil (*Ocimum basilicum*), the foliar or soil application of silicon-based fertilizers and NPs under saline conditions resulted in marked improvements in plant morphological, physiological, and biochemical traits. Treated basil plants exhibited enhanced growth parameters, including shoot length, root biomass, and leaf area, along with increased chlorophyll content and proline accumulation, a critical osmolyte that stabilizes proteins and membranes under osmotic stress. The observed improvements are largely attributed to the role of Si-NPs in reducing Na^+ uptake and increasing the selectivity of K^+ and Ca^{2+} ions, thus mitigating ionic toxicity. In addition, Si-NPs help stabilize cellular membranes and protect chloroplast structures, leading to enhanced photosynthetic efficiency and delayed senescence. The mechanical and chemical interactions between Si-NPs and plant cells mediated by their catalytic reactivity, nanoscale dimensions, and high surface energy facilitate their entry into plant tissues and enable their involvement in redox regulation and stress signaling pathways. Beyond silicon-based materials, other metallic NPs have also demonstrated considerable potential in enhancing salt stress tolerance in plants. For instance, Ag-NPs have been shown to activate antioxidant defense systems. Jiang et al. (2012) reported that Spirodela polyrhiza, a fast-growing aquatic plant, exhibited increased activity of key antioxidant enzymes such as SOD, CAT, and POD following exposure to 6 nm Ag-NPs at a concentration of 5 mg/L. These enzymes are crucial for scavenging ROS, which accumulate under salinity-induced oxidative stress. By enhancing antioxidant capacity, Ag-NPs help maintain cellular redox homeostasis and protect vital biomolecules from oxidative damage. Similarly, gold nanoparticles (Au-NPs) have also been employed to combat salinity stress. Das and Das (2019) found that the application of Au-NPs to Brassica juncea seedlings significantly increased the activities of GR, guaiacol peroxidase, and CAT. In parallel, elevated levels of H_2O_2 and proline were observed in treated plants. While H_2O_2 is a harmful ROS at high concentrations, its controlled accumulation under NP treatment may act as a signaling molecule, triggering defense gene activation and stress acclimation. The combined upregulation of enzymatic and nonenzymatic antioxidants contributes to enhanced resilience in plants under saline conditions. Taken together, these findings underscore the promise of nanotechnology in managing the adverse effects of salinity stress on agricultural crops. By leveraging the unique properties of engineered nanomaterials, such as their ability to penetrate biological membranes, catalyze redox reactions, and interact with macromolecules, researchers and agronomists can develop sustainable and efficient strategies for stress mitigation. However, careful consideration must be given to the type, concentration, and application method of NPs, as excessive or inappropriate use may lead to phytotoxicity or environmental concerns (Khan et al., 2009).

14.4.3 Plants under stress from heavy metals

The accumulation of heavy metals such as Cd, Pb, mercury (Hg), As, and Fe in the soil constitutes a significant environmental concern in modern agriculture. This buildup is largely caused by human activities, including mining, rapid industrialization, urbanization, and the widespread use of chemical pesticides and fertilizers. Heavy metals in agricultural soils at levels that are too high can hurt crop production, growth, and development by messing with plant physiological systems and diminishing soil fertility and quality. Heavy metal contamination can also cause toxic elements to build up in edible sections of plants. These elements can then enter the food chain and offer serious health concerns to humans and animals, either directly through eating them or indirectly through contaminated water and soil. Heavy metals interfere with important physiological and metabolic functions in plants. Some of the bad things they do are stop seeds from germinating, lower photosynthetic activity, mess up nitrogen intake, and hurt root and shoot systems. Heavy metal toxicity leads to an oxidative burst in cells, characterized by an abundance of ROS, including hydroxyl radicals, superoxide anions, and H_2O_2. These ROS damage lipids, proteins, and DNA, which leads to cell death, lipid peroxidation, and a stop in growth. Heavy metals also cause water imbalance, protein denaturation, and enzyme inactivation, all of which make it harder for plants to survive. When there is a lot of heavy metal stress, structural changes happen a lot, like damage to the plasma membrane, breakdown of mitochondria, and changes in chloroplasts. At the molecular level, these metals can mess up protein synthesis, gene expression, and signal transduction pathways. Plants have reacted by creating a multitude of ways to protect themselves and get rid of toxins. Some of these are activating antioxidant defense systems like SOD, CAT, and GPX; putting metals into vacuoles; and using organic acids, amino acids, and phytochelatins to bind heavy metals. To mitigate heavy metal toxicity, external interventions must be established, as natural defense systems sometimes prove inadequate under elevated or protracted exposure (Theocharis et al., 2012).

Recent advances in nanotechnology have made it possible to find promising ways to lower the stress that heavy metals cause in plants. Engineered NPs can interact with heavy metals to make them less harmful and less available to plants by adsorbing, changing their redox state, precipitating, and chelating. Magnetic FeO_4 NPs, in particular, have attracted attention because they have a large surface area, are magnetic, and like heavy metal ions. Konate et al. (2017) demonstrated that FeO_4 NPs significantly mitigated the detrimental effects of Pb, Zn, Cd, and Cu on wheat seedlings by reducing metal absorption and oxidative damage. These NPs helped keep metals in the soil, which made it less likely for them to get to the aerial sections of plants. ZnO NPs have also shown potential in making *Leucaena leucocephala* (Lam.) seedlings less toxic to Cd and Pb. Venkatachalam et al. (2017) say that ZnO NPs decreased MDA levels, which means there was less oxidative stress and lipid peroxidation. Moreover, the application of ZnO NPs enhanced shoot biomass, root length, and chlorophyll content, signifying their role in restoring normal physiological functions under metal stress. It has also been shown that titanium dioxide NPs make Cd less poisonous. Lian et al. found that spraying TiO_2 NPs on maize (*Zea mays*) leaves and roots reduced the amount of Cd that built up in those parts of the plant. There was also less oxidative damage and more activity of antioxidant enzymes, which led to improved plant development in the end. TiO_2 NPs and Cd ions probably come together to make complexes that make it harder for plants to take in the ions and make them more resistant to stress. In another investigation, SiO_2 NPs were found to be effective at lowering the toxicity of aluminum in maize. According to de Sousa et al. (2019), SiO_2 NPs helped maize plants that were stressed by Al by making their roots grow longer, making more chlorophyll, and boosting their antioxidant activity. Silicon is known to help minimize metal stress by strengthening cell walls, regulating ion transporters, and boosting antioxidant defenses. When used as NPs, it is far more bioavailable and effective, which makes it easier to fix damage caused by metals. All things considered, our results suggest that NPs are useful tools for making heavy metals less harmful in farming systems. NPs enhance plants' resilience to adverse environmental conditions by increasing physiological characteristics, augmenting antioxidant defenses, and regulating metal absorption. Their utilization also provides cost-effective and environmentally friendly methods for rehabilitating contaminated soils, positioning them as a viable component of sustainable agriculture and phytoremediation strategies (Srivastava et al., 2005).

14.4.4 Plants under stress from heat

Temperature is one of the most critical environmental variables influencing plant growth, development, metabolism, and overall productivity. Both high and low temperature fluctuations are known to disrupt normal physiological and biochemical functions in plants. In recent decades, climate change has accelerated the frequency and intensity of extreme temperature events, posing a significant threat to global agricultural output. Heat stress, in particular, has become a major concern, especially in tropical and subtropical regions where crops are already grown at or near their thermal maximum. Temperature extremes can result in poor germination, delayed flowering, abnormal fruit set, and reduced yield. As such, understanding plant responses to thermal stress and developing strategies to enhance thermotolerance are among the top priorities in sustainable agriculture. Plants exposed to high temperatures exhibit numerous morphological, physiological, and molecular changes, including protein denaturation, increased membrane fluidity, and enhanced generation of ROS. These ROS can cause lipid peroxidation, oxidative damage to cellular components, and disruption of normal cellular homeostasis. High temperatures also negatively impact photosynthesis, primarily by inhibiting photosystem II activity and reducing chlorophyll content. Moreover, reproductive processes such as pollen viability, fertilization, and seed formation are particularly sensitive to heat stress, making crops vulnerable during critical stages like flowering and grain filling. To cope with temperature-induced stress, plants have evolved several defense mechanisms, including heat shock protein (HSP) synthesis, osmolyte accumulation (e.g., proline, glycine betaine), and activation of antioxidant enzymes such as SOD, CAT, and APX. However, under extreme or prolonged heat conditions, these natural responses may not suffice. In recent years, the application of engineered nanomaterials has emerged as a promising strategy to enhance plant resilience to temperature stress. Among these, Se NPs have gained significant attention due to their multifunctional roles in enhancing stress tolerance. Selenium is a micronutrient involved in antioxidant defense, hormone regulation, and redox homeostasis. When applied in nanoform, selenium's bioavailability and reactivity are greatly enhanced, allowing for targeted interactions at the cellular and molecular levels. Djanaguiraman et al. (2018) demonstrated that the application of Se NPs to Sorghum bicolor under heat stress conditions significantly improved physiological parameters, including fatty acid unsaturation, photosynthetic efficiency, and pollen germination. Moreover, Se NPs reduced oxidative stress markers such as MDA, thereby protecting membrane integrity and supporting growth under elevated temperatures. Similarly, Haghighi et al. reported that foliar application of Se NPs in tomato (*Solanum lycopersicum*) plants under heat stress enhanced plant biomass, chlorophyll content, leaf water potential, and stomatal conductance. These improvements contributed to better water use efficiency and overall plant vigor during stress periods. The enhanced antioxidant activity observed in Se NP-treated plants suggests their involvement in ROS detoxification, thus minimizing heat-induced cellular damage.

In addition to selenium, Ag NPs have also shown beneficial effects under heat stress. Iqbal et al. (2019) found that Ag NPs enhanced morphological traits in wheat (*Triticum aestivum*), including plant height, root-shoot length, and biomass accumulation, when subjected to high temperatures. The treatment was associated with improved cell membrane stability and higher levels of antioxidant enzymes, indicating a role for Ag NPs in stress amelioration. Due to their small size and surface reactivity, Ag NPs can interact with plant tissues at the nano-bio interface, potentially triggering signaling pathways involved in stress resistance. Several studies have explored the mechanisms by which NPs confer thermotolerance. These mechanisms include improved antioxidant enzyme activity, osmolyte accumulation, hormonal modulation, and gene expression related to HSPs and metabolic regulation. For instance, NP-treated plants often show upregulation of genes encoding for HSPs and antioxidant enzymes, which help maintain protein integrity and redox balance under heat stress. Although the potential of nanomaterials in enhancing heat tolerance is increasingly evident across various crop species, further research is necessary to optimize NP formulations, concentrations, and delivery methods. It is also essential to evaluate their long-term environmental effects and biosafety before widespread agricultural application.

14.5 Conclusion

Nanotechnology in agriculture has created new strategies to assist plants in dealing with abiotic stressors, which are big challenges for farming expansion all over the world. The research reviewed demonstrates that NPs substantially improve plant growth, physiology, and stress tolerance in unfavorable climatic conditions, attributable to their unique physicochemical characteristics, such as minuscule size, large surface area, and catalytic reactivity. Nanomaterials like ZnO, SiO_2, Se, Ag, TiO_2, Fe_3O_4, and Au NPs have been shown to change how plants respond to stressors like drought, salt, heavy metal toxicity, and extreme temperatures. They do this by making antioxidant enzymes work harder, helping cells take in more water, building up osmolytes, keeping cell membranes healthy, and changing how stress-responsive genes are turned on. NPs have been shown to help plants like soybeans and sorghum deal with drought stress by increasing the number of stress-related genes, the amount of chlorophyll, the amount of water, and the rates of germination. NPs such as silicon and silver have enhanced plants' resilience to salt stress by modulating their physiological parameters and augmenting their antioxidant systems. When crops, including wheat, maize, and Leucaena, are exposed to heavy metals, metal oxide and magnetic NPs have been shown to protect them against oxidative damage and limit their absorption of metals. Se and Ag NPs have also helped with temperature stress, especially heat, by making photosynthesis work better, keeping membranes strong, and improving development. Nanotechnology is a new and long-lasting way to make farming more productive as environmental problems get worse. But more research is needed to figure out how safe NPs are for the ecosystem and how they will affect the environment in the long run. Future investigations should focus on the molecular signaling pathways influenced by NPs to improve their application in precision agriculture and climate-resilient crop production.

References

Afzal, S., Singh, N.K., Lal, A.F., Sohrab, S., Singh, N., Gupta, P.S., et al., 2024. Nanostructure and plant uptake: assessing the ecological footprint and root-to-leaf dynamics. Plant Nano Biology 10. Available at https://doi.org/10.1016/j.plana.2024.100122.

Bhat, J.A., Faizan, M., Bhat, M.A., Huang, F., Yu, D., Ahmad, A., et al., 2022. Defense interplay of the zinc-oxide nanoparticles and melatonin in alleviating the arsenic stress in soybean (Glycine max L. Chemosphere 288. Available at https://doi.org/10.1016/j.chemosphere.2021.132471.

Chaves, M.M., Flexas, J., Pinheiro, C., 2009. Photosynthesis under drought and salt stress: Regulation mechanisms from whole plant to cell. Annals of Botany 103, 551–560.

Chen, Q., Zhao, X., Lei, D., Hu, S., Shen, Z., Shen, W., et al., 2017. Hydrogen-rich water pretreatment alters photosynthetic gas exchange, chlorophyll fluorescence, and antioxidant activities in heat-stressed cucumber leaves. Plant Growth Regulation 83 (1), 69–82. Available at https://doi.org/10.1007/s10725-017-0284-1.

Das, A., Das, B., 2019. Nanotechnology a potential tool to mitigate abiotic stress in crop plants. Abiotic and Biotic Stress in Plants 1–13.

de Sousa, A., Saleh, A.M., Habeeb, T.H., Hassan, Y.M., Zrieq, R., Wadaan, M.A.M., et al., 2019. Silicon dioxide nanoparticles ameliorate the phytotoxic hazards of aluminum in maize grown on acidic soil. Science of The Total Environment 693, 133636. Available at https://doi.org/10.1016/j.scitotenv.2019.133636.

Djanaguiraman, M., Belliraj, N., Bossmann, S.H., Prasad, P.V.V., 2018. High-temperature stress alleviation by selenium nanoparticle treatment in grain sorghum. ACS Omega 3 (3), 2479–2491. Available at https://doi.org/10.1021/acsomega.7b01934.

Dogan, Y., Alam, P., Sultan, H., Sharma, R., Soysal, S., Baran, M.F., et al., 2024. Zinc oxide nanoparticles for sustainable agriculture: a tool to combat salinity stress in rice (*Oryza sativa*) by modulating the nutritional profile and redox homeostasis mechanisms. Journal of Agriculture and Food Research 19, 101598.

Dubchak, S., Ogar, A., Mietelski, J.W., Turnau, K., 2010. Influence of silver and titanium nanoparticles on arbuscular mycorrhiza colonization and accumulation of radiocaesium in Helianthus annuus. Spanish Journal of Agricultural Research 8, 103–108. Available at https://doi.org/10.5424/sjar/201008s1-1228.

Fahad, S., Bajwa, A.A., Nazir, U., Anjum, S.A., Farooq, A., Zohaib, A., et al., 2017. Crop production under drought and heat stress: plant responses and management options. Frontiers in Plant Science 8. Available at https://doi.org/10.3389/fpls.2017.01147.

Faisal, M., Faizan, M., Soysal, S., Alatar, A.A., 2024a. Synergistic application of melatonin and silicon oxide nanoparticles modulates reactive oxygen species generation and the antioxidant defense system: a strategy for cadmium tolerance in rice. Frontiers in Plant Science 15. Available at https://doi.org/10.3389/fpls.2024.1484600.

Faisal, M., Faizan, M., Alatar, A.A., 2024. Metallic allies in drought resilience: unveiling the influence of silver and zinc oxide nanoparticles on enhancing tomato (*Solanum lycopersicum*) resistance through oxidative stress regulation. Plant Physiology and Biochemistry 212, 108722.

Faizan, M., Bhat, J.A., El-Serehy, H.A., Moustakas, M., Ahmad, P., 2022. Magnesium oxide nanoparticles (MgO-NPs) alleviate arsenic toxicity in soybean by modulating photosynthetic function, nutrient uptake and antioxidant potential. Metals 12 (12). Available at https://doi.org/10.3390/met12122030.

Faizan, M., Ahmed, S.M., Ahmad, L., Chen, C., Yu, F., 2023a. Nanotechnology in agriculture. In: In: Raina, A., Wani, M.R., Laskar, R.A., Tomlekova, N. (Eds.), Advanced Crop Improvement Volume 1 Springer Nature, Switzerland.

Faizan, M., Alam, P., Rajput, V.D., Shareen, Kaur, K., Faraz, A., et al., 2023b. Potential role of tocopherol in protecting crop plants against abiotic stresses. Physiology and Molecular Biology of Plants 29 (10), 1563–1575.

Faizan, M., Alam, P., Kumari, A., Suresh, G., Sharma, P., Karabulut, F., et al., 2024. Unraveling the nano-biochar mediated regulation of heavy metal stress tolerance for sustaining plant health. Plant Stress 14, 100615.

Faizan, M., Ashoka, H., Karabulut, F., Alam, P., Ahmed, S.M., Khan, I., et al., 2024. Annexins: a family of calcium binding proteins with variety of roles in plant development and abiotic stress tolerance. Plant Stress 14, 100573.

Farooq, M., Wahid, A., Kobayashi, N., Fujita, D., Basra, S.M.A., 2009. Plant drought stress: effects, mechanisms and management. Agronomy for Sustainable Development 29 (1), 185–212. Available at https://doi.org/10.1051/agro:2008021.

Finnegan, P.M., Chen, W., 2012. Arsenic toxicity: the effects on plant metabolism. Frontiers in Physiology 3. Available at https://doi.org/10.3389/fphys.2012.00182.

Flexas, J., Bota, J., Loreto, F., Cornic, G., Sharkey, T.D., 2004. Diffusive and metabolic limitations to photosynthesis under drought and salinity in C 3 plants. Plant Biology 6 (3), 269–279. Available at https://doi.org/10.1055/s-2004-820867.

Gill, S.S., Tuteja, N., 2010. Reactive oxygen species and antioxidant machinery in abiotic stress tolerance in crop plants. Plant Physiology and Biochemistry 48 (12), 909–930. Available at https://doi.org/10.1016/j.plaphy.2010.08.016.

Giordano, M., Petropoulos, S.A., Rouphael, Y., 2021. Response and defence mechanisms of vegetable crops against drought, heat and salinity stress. Agriculture 11, 463.

Hasanuzzaman, M., Nahar, K., Alam, M.M., Bhowmik, P.C., Hossain, M.A., Rahman, M.M., et al., 2013. Potential use of halophytes to remediate saline soils. BioMed Research International 2013, 589341.

Huang, G.T., Ma, S.L., Bai, L.P., Zhang, L., Ma, H., Jia, P., et al., 2012. Signal transduction during cold, salt, and drought stresses in plants. Molecular Biology Reports 39 (2), 969–987. Available at https://doi.org/10.1007/s11033-011-0823-1.

Iqbal, M., Raja, N.I., Mashwani, Z.U.R., Wattoo, F.H., Hussain, M., Ejaz, M., 2019. Assessment of green synthesized silver nanoparticles in wheat seedlings at the anatomical level in relation to their uptake, translocation, and accumulation. Iranian Journal of Science and Technology, Transaction A: Science 43 (4), 1551–1561. Available at https://doi.org/10.1007/s40995-018-0639-0.

Jiang, H.S., Li, M., Chang, F.Y., Li, W., Yin, L.Y., 2012. Physiological analysis of silver nanoparticles and AgNO3 toxicity to Spirodela polyrhiza. Environmental Toxicology and Chemistry 31 (8), 1880–1886. Available at https://doi.org/10.1002/etc.1899.

Kesawat, M.S., Shivaraj, M.S., Kim, D.K., Kumar, M., Hahn, B.S., Deshmukh, R., 2020. Metalloids and their role in the biological system. Metalloids in Plants: Advances and Future Prospects. wiley, South Korea, pp. 1–17. Available at https://doi.org/10.1002/9781119487210.ch1.

Kesawat, M.S., Satheesh, N., Kherawat, B.S., Kumar, A., Kim, H.U., Chung, S.M., et al., 2023. Regulation of reactive oxygen species during salt stress in plants and their crosstalk with other signaling molecules—current perspectives and future directions. Plants 12 (4). Available at https://doi.org/10.3390/plants12040864.

Khan, M.H., Singh, S., Nazar, R., 2009. Photosynthetic adaptation of plants to salinity stress. Handbook of Photosynthesis, 2nd ed. CRC Press, pp. 687–707.

Khan, Z.H., Rizvi, T.F., Jamil, M., 2020. Nanofertilizers: a novel way to increase crop productivity. Frontiers in Plant Science 11, 590.

Konate, A., He, X., Zhang, Z., Ma, Y., Zhang, P., Alugongo, G., et al., 2017. Magnetic (Fe3O4) nanoparticles reduce heavy metals uptake and mitigate their toxicity in wheat seedling. Sustainability 9 (5), 790. Available at https://doi.org/10.3390/su9050790.

Laware, S.L., Raskar, S., 2014. Effect of titanium dioxide nanoparticles on hydrolytic and antioxidant enzymes during seed germination in onion. International Journal of Current Microbiology and Applied Sciences 3 (7), 749–760.

Raliya, R., Nair, R., Chavalmane, S., Wang, W.N., Biswas, P., 2016. Mechanistic evaluation of translocation and physiological impact of titanium dioxide and zinc oxide nanoparticles on the tomato (Solanum lycopersicum L.) plant. Metallomics 8 (5), 582–590.

Ruelland, E., Vaultier, M.N., Zachowski, A., Hurry, V., 2009. Chapter 2 cold signalling and cold acclimation in plants. Advances in Botanical Research 49 (C), 35–150. Available at https://doi.org/10.1016/S0065-2296(08)00602-2.

Shahid, M., Dumat, C., Khalid, S., Niazi, N.K., Antunes, P.M.C., 2017. Cadmium bioavailability, uptake, toxicity and detoxification in soil-plant system. Reviews of Environmental Contamination and Toxicology Volume 241. Springer, Cham, pp. 73–137. Available at https://doi.org/10.1007/398_2016_8.

Srivastava, M., Ma, L.Q., Singh, N., Singh, S., 2005. Antioxidant responses of hyper-accumulator and sensitive fern species to arsenic. Journal of Experimental Botany 56 (415), 1335–1342. Available at https://doi.org/10.1093/jxb/eri134.

Thapa, G., Dey, M., Sahoo, L., Panda, S.K., 2011. An insight into the drought stress induced alterations in plants. Biologia Plantarum 55 (4), 603–613. Available at https://doi.org/10.1007/s10535-011-0158-8.

Theocharis, A., Clement, C., Barka, E.A., 2012. Physiological and molecular changes in plants grown at low temperatures. Planta 235, 1091–1105.

Tilman, D., Balzer, C., Hill, J., Befort, B.L., 2011. Global food demand and the sustainable intensification of agriculture. Proceedings of the National Academy of Sciences 108 (50), 20260–20264. Available at https://doi.org/10.1073/pnas.1116437108.

Venkatachalam, P., Jayaraj, M., Manikandan, R., Geetha, N., Rene, E.R., Sharma, N.C., et al., 2017. Zinc oxide nanoparticles (ZnONPs) alleviate heavy metal-induced toxicity in Leucaena leucocephala seedlings: a physiochemical analysis. Plant Physiology and Biochemistry 110, 59–69. Available at https://doi.org/10.1016/j.plaphy.2016.08.022.

Wang, P., Lombi, E., Zhao, F.J., Kopittke, P.M., 2022. Nanotechnology: a new opportunity in plant sciences. Trends in Plant Science 27 (7), 728–739.

Waqas, M.A., Wang, X., Zafar, S.A., Noor, M.A., Hussain, H.A., Azher Nawaz, M., et al., 2021. Thermal stresses in maize: effects and management strategies. Plants 10 (2), 1–23. Available at https://doi.org/10.3390/plants10020293.

Zhang, H., Lang, Z., Zhu, J.K., 2022. Dynamics and function of DNA methylation in plants. Nature Reviews. Molecular Cell Biology 23 (10), 694–710.

Zuverza-Mena, N., Martínez-Fernández, D., Du, W., Hernandez-Viezcas, J.A., Bonilla-Bird, N., López-Moreno, M.L., et al., 2017. Exposure of engineered nanomaterials to plants: Insights into the physiological and biochemical responses-a review. Plant Physiology and Biochemistry 110, 236–264. Available at https://doi.org/10.1016/j.plaphy.2016.05.037.

Chapter 15

Transgenic and genome editing-based approaches for improving multiple abiotic stress tolerance in plants

Tanmoy Halder[1,**], Sonali Panda[1,2,**], Chandana Ghosh[1], M.J. Baig[1], and Kutubuddin A. Molla[1,*]
[1]*ICAR-Central Rice Research Institute, Cuttack, Odisha, India,* [2]*Ravenshaw University, Cuttack, Odisha, India*
[*]*Corresponding author. e mail address: kutubuddin.molla@icar.org.in*

15.1 Introduction

Agriculture stands at a crossroads today, challenged by the immense task of feeding a growing global population while coping with the intensifying severity of environmental stresses. The impact of climate change has converted many fertile lands into drought-stricken wastelands and rendered fertile soils saline, drastically affecting food production worldwide. Unpredictable temperature fluctuations and heavy metal contamination further exacerbate this crisis, threatening agricultural sustainability and food security.

Abiotic stresses such as drought, salinity, heat, cold, and heavy metal toxicity undermine plant development and agricultural performance, accounting for more than 50% of global losses in major crop yields each year (Yadav et al., 2020). Crops commonly encounter multiple environmental stresses and their sensing and responding mechanisms to different factors are thought to be interconnected. Often, crops are simultaneously threatened by multiple environmental stresses such as drought, salinity, extreme temperatures, and pathogen attacks. The plant's response to these combined stresses is complex. Plants have evolved intricate and interconnected signaling networks involving phytohormones (such as abscisic acid, salicylic acid, jasmonic acid, and ethylene), transcription factors (TFs), reactive oxygen species (ROS), and secondary messengers such as calcium ions. These pathways often overlap and cross-communicate, facilitating plants to optimize their responses depending on the nature, duration, and intensity of stressors.

In view of global warming, climate variability, and increased anthropogenic pressures, crops are now more often subjected to combinations of abiotic stresses such as drought, heat, salinity, and flooding. These stressors often occur simultaneously or sequentially, complicating the plant's ability to mount effective defense responses. Therefore, developing crops that can withstand complex stress environments, which are becoming the norm under climate change, is the primary goal today. This is vital to strengthen global food resilience, agricultural sustainability, and economic stability of farming communities while confronted by increasingly unpredictable and extreme weather conditions. To manage these recurring environmental stresses, plants have established numerous defense strategies, including the processes of signal perception, signal transduction via either abscisic acid (ABA)-dependent or ABA-independent pathways, expression of stress-responsive genes, and, in turn, the activation of physiological and metabolic responses, which might be associated with imparting a certain extent of protection to these stresses (Pérez-Clemente et al., 2013; Yamaguchi-Shinozaki and Shinozaki, 2006; Chaves et al., 2003; Xiong et al., 2002).

Over the years, a wide range of stress-responsive genes has been identified in many plants, and traditional breeding practices have been used to create stress-tolerant crop varieties; however, their advancement is restricted because of the polygenic nature of stress responses and the slow rate of genetic improvement. With the progress of molecular biology and plant biotechnology, transgenic and genome-editing-based methods have become efficient tools for enhancing abiotic

** Both authors have contributed equally.

Multiple Abiotic Stresses in Plants. DOI: https://doi.org/10.1016/B978-0-443-27424-4.00020-3

stress resilience in crops. These methods facilitate the incorporation, upregulation, or targeted modification of stress-associated genes, making it possible to introduce or alter desirable traits with greater efficiency and precision than traditional breeding techniques. The significance of climate-resilient crop development is highlighted by the increased worldwide population and the demand for sustainable agriculture. To ensure global food security in the coming decades, it will be essential to develop crops that can withstand a variety of environmental challenges.

This chapter presents a discussion of transgenic and genome-editing technologies used to improve plant stress tolerance, with a focus on existing applications, technological advances, and future directions. It explores how these cutting-edge molecular tools are remaking the future of crop improvement and holding promise for a more sustainable and robust agricultural system.

15.2 Historical perspectives on genetic engineering for stress tolerance

The development of transgenic plants began in May 1983, when the Van Montagu and Schell lab introduced foreign genes into plants using *Agrobacterium*-mediated transformation (Herrera-Estrella et al., 1983). Subsequently, researchers have successfully engineered plants to express a number of genes that offer resistance to abiotic stresses, paving the way for the development of commercial genetically modified crops. Early transgenic approaches primarily focused on the modification of single genes, such as overexpression or gene silencing. However, recent advancements have led to multigene modifications. In modern times, genome editing tools have become popular for fine-tuning stress responses. These approaches are now integrated with systems biology, omics technologies, and high-throughput phenotyping to evolve robust stress-tolerant crops.

15.3 Transgene-based strategies to withstand multiple environmental abiotic stresses

Genome engineering has enabled the generation of transgenic plants with enhanced multiple abiotic stress resistance by introducing genes from self- or nonself-sources. These approaches include increased expressions of stress-responsive proteins and TFs, gene silencing , and engineering hormonal pathways to elevate stress tolerance.

15.3.1 Overexpression of stress-responsive genes

Overexpression of specific stress-responsive genes has been a widely employed strategy to enhance plant tolerance to abiotic stresses. These genes fall into different categories based on their function. These include genes that produce proteins with known structural, enzymatic, or regulatory functions.

15.3.1.1 Stress-responsive structural proteins

Overexpression of stress-responsive structural proteins is a well-proven strategy for inducing stress tolerance in plants. Earlier studies have shown significant improvement in the performance of plants under multiple abiotic stresses by expressing stress-responsive structural proteins (Table 15.1). Late embryogenesis abundant (LEA) proteins, heat shock

TABLE 15.1 Stress-responsive proteins are utilized for developing multiple abiotic stress tolerance.

Gene	Combine abiotic stress	References
SbDhn1 (Sorghum bicolor)	High temperature, osmotic stress	Halder et al. (2017)
MusaDHN1 (Banana)	Drought, high salinity	Shekhawat et al. (2011)
TAS14 (Tomato)	Drought, salinity	Muñoz-Mayor et al. (2012)
DHN5 (Wheat)	Salinity, osmotic stress	Brini et al. (2007)
PMA80 (Wheat)	Drought, salt	Cheng et al. (2002)
HVA1 (Barley)	Water deficit, salt stress	Xu et al. (1996)
HSFA1b (Arabidopsis)	Drought, salt, biotic stress	Bechtold et al. (2013)

protein (HSP) genes, and chaperones are the most commonly used genes. LEA proteins represent a category of high molecular weight proteins that are abundant during late embryogenesis and accumulate during seed desiccation and in response to water stress (Galau et al., 1987). Significant work has been done on a special type of LEA known as dehydrin. Overexpression of *SbDhn1*, a type of LEA protein from *Sorghum bicolor*, improves protection under high temperatures and osmotic stress in transgenic tobacco plants (Halder et al., 2017). A novel *MusaDHN1* was found to confer protection under drought and high-salinity conditions in transgenic banana (Shekhawat et al., 2011). Transgenic tomato plants overexpressing the *TAS14* gene exhibit long-term drought and salinity tolerance compared to the wild-type plants (Muñoz-Mayor et al., 2012). Arabidopsis plants overexpressing *DHN5* from wheat exhibited enhanced salinity and osmotic stress tolerance (Brini et al., 2007). Overexpression of wheat dehydrin *PMA80* in rice resulted in increased drought and salt tolerance (Cheng et al., 2002). The barley LEA protein gene *HVA1* in transgenic rice plants provides multiple stress tolerances. The transgenic rice plants expressing the *HVA1* gene demonstrated significantly improved tolerance to both water deficit and salt stress compared to nontransformed plants. Transgenic plants were able to maintain higher growth rates under stress conditions (Xu et al., 1996).

HSPs and chaperones are found in most prokaryotes and eukaryotes as well as in some viruses (Maaroufi and Tanguay, 2013). The main function of HSPs/chaperones is to act as a buffer to limit misfolding and resolve aggregates of proteins under stress situations. However, most transgenic approaches have focused on heat shock transcription factor (HSF) rather than constitutive overexpression of HSPs , as the latter causes severe growth reduction in overexpression lines. *HSFA1b* is particularly promising owing to its multistress resistance potential. *HSFA1b* overexpression confers resistance to drought, salt, and biotic stress and enhances seed yield (Bechtold et al., 2013). Overexpression of endogenous HSF has been performed in tomato (Mishra et al., 2002) and soybean (Zhu et al., 2006), successfully increasing plant tolerance to stress.

15.3.1.2 Transcription factors

Transcription factors (TFs) regulate the expression of multiple stress-responsive genes and play essential roles in plant adaptation to different abiotic stresses. Upregulation of different TFs has been shown to significantly enhance stress tolerance in plants, positioning them as attractive targets for genetic manipulation. Among these, some TFs are of particular interest because they bind to the promoter of stress-responsive genes and trigger gene expression cascades that collectively enhance plant resilience to various abiotic stresses (Shinozaki and Yamaguchi-Shinozaki, 1997; Winicov and Bastola, 1997). Over the past few decades, many studies have explored the potential of such TFs in improving stress tolerance, revealing that numerous TFs are involved in coordinating plant responses to multiple abiotic stress conditions (Bartels and Sunkar, 2005; Vinocur and Altman, 2005). Most of these fall into several large TF families, such as Dehydration-Responsive Element-Binding proteins (DREB), APETALA2/Ethylene-Responsive Factor, Basic Leucine Zipper proteins, NAC [*No Apical Meristem* (NAM), Arabidopsis transcription activation factor (ATAF), and *Cup shaped Cotyledon* (CUC)], Myeloblastosis (MYB), myelocytomatosis (MYC), Cys2His2 zinc-finger, and WRKY transcription factors (named after the conserved WRKYGQK domain) (WRKY).

It has been reported that the expression of the C-repeat/DRE binding factor 1 (CBF-1) TF in tomato (*Lycopersicon esculentum*) under the control of a CaMV35S promoter improved tolerance to chilling, drought, and salt stress (Hsieh et al., 2002); however, the transgenic plants exhibited a dwarf phenotype and reduction in fruit set and seed number. Similarly, increased tolerance to freezing and drought in *Arabidopsis* was achieved by overexpressing *CBF4*, a close CBF/DREB1 homolog whose expression is rapidly induced during drought stress and ABA treatment (Haake et al., 2002). Expression of *DREB1A* under the control of CaMV35S promoter in transgenic plants exhibited strong constitutive expression of various stress-inducible genes and brings about increased tolerance to freezing, salt, and drought stresses in various plant species (Liu et al., 1998). For instance, overexpression of *DREB1A* has been reported to improve drought- and low-temperature stress tolerance in wheat, groundnut, and tobacco (Kasuga et al., 2004; Pellegrineschi et al., 2004). Similarly, overexpression of *DREB1A* in transgenic *Arabidopsis* plants resulted in freezing and dehydration tolerance (Liu et al., 1998). Another study demonstrated that overexpression of the *OsDREB2A* in rice significantly improves the tolerance to dehydration and salt stress. Transgenic rice lines exhibited improved growth performance under simulated stress conditions compared to wild-type plants (Mallikarjuna et al., 2011).

NAC proteins are a large family of plant-specific TFs that play pivotal roles in diverse biological processes, including plant growth, development, and responses to both abiotic and biotic stresses. The NAC domain was originally defined through conserved sequence motifs identified in Petunia NAM and *Arabidopsis* ATAF1/2 and CUC2 proteins, with the acronym "NAC" derived from the initials of these genes. Members of this TF family are key regulators of plant stress signaling pathways and are widely recognized as contributors to enhanced stress tolerance. It has been reported that *ONAC045*, a novel NAC TF in rice, plays a significant role in enhancing drought and salt tolerance. Transgenic rice plants overexpressing *ONAC045* exhibited significantly increased tolerance to drought and salt at the seedling stage (Zheng et al., 2009). Similarly, transgenic *Arabidopsis* plants expressing *ONAC063* demonstrated improved tolerance to salinity and osmotic stress, particularly during germination. Microarray and real-time polymerase chain reaction (PCR) analyzes revealed that *ONAC063* overexpression led to the upregulation of several salinity-inducible genes, including the amylase gene *AMY1* (Yokotani et al., 2009). In another study, the overexpression of the wheat NAC TF *TaNAC2* in *Arabidopsis* significantly improved tolerance to multiple abiotic stresses, including drought, salt, and cold. The transgenic plants showed better survival rates and physiological responses, such as longer primary roots and increased biomass, than wild-type plants under stress treatment. Data suggested that *TaNAC2* not only enhances stress tolerance but also contributes to better growth under water-deficit conditions (Mao et al., 2012). Transgenic rice overexpressing *SNAC1* showed improved tolerance to drought and salt stress. Under severe drought conditions, transgenic rice showed better spikelet fertility than wild-type rice. After treatment with 200 mM NaCl for 12 days, the transgenic seedlings showed survival rates of over 80% (Hu et al., 2006a). Transgenic rice lines overexpressing *ONAC022* demonstrated increased tolerance to drought and salt stress. Under drought stress conditions, transgenic lines exhibited higher survival rates and better growth than wild-type plants. Similarly, these plants accumulate less sodium in their roots and shoots when grown under saline conditions (Hong et al., 2016).

WRKY TFs represent a major family of plant-specific regulators that control diverse processes, including growth, development, and defense signaling. Transgenic rice plants overexpressing maize *ZmWRKY58* showed enhanced tolerance to drought and salt stress. They had higher survival rates, increased relative water content, and lower levels of malonaldehyde and electrical leakage when subjected to these stress conditions. However, transgenic rice plants exhibited delayed germination and inhibited postgermination growth compared to wild-type plants (Cai et al., 2014). In another study, transgenic wheat plants overexpressing *AtWRKY30* exhibited a significantly better tolerance to drought and heat stress. Transgenic plants show enhanced growth and biomass compared to wild-type plants under stress conditions (El-Esawi et al., 2019). Transgenic tomato plants overexpressing *SlWRKY39* showed notable improvements in stress resistance (Sun et al., 2015).

MYB TFs have also been reported to play crucial roles in plant stress tolerance. Transgenic rice plants overexpressing *OsMYB6* exhibited better performance under drought and salinity stress conditions compared to wild-type plants (Tang et al., 2019). Overexpression of the R2R3 MYB gene *MdSIMYB1* in transgenic tobacco and apple plays a significant role in enhancing multiple abiotic stress tolerance, such as high salinity, drought, and cold (Wang et al., 2014).

Transgenic *Arabidopsis* plants overexpressing soybean *GmbZIP1* exhibited improved tolerance to drought, salt, and cold stress. This was evidenced by their ability to maintain higher relative survival rates under these stress conditions compared to nontransgenic plants (Gao et al., 2011). Transgenic peanut plants expressing *AtHDG11* showed significant improvements in drought and salt stress tolerance. In this study, *AtHDG11* was expressed under a stress-inducible promoter (rd29A) to tightly regulate the expression (Banavath et al., 2018).

Along with different TFs, transcriptional activators can also be potential targets for genetic manipulation. *AtHsfA6a*, a transcriptional activator of stress-responsive genes, which is highly induced by exogenous ABA, NaCl, and drought, can provide stress tolerance. Transgenic plants overexpressing *AtHsfA6a* exhibited drought and salt stress tolerance (Hwang et al., 2014). A list of TF genes, which were overexpressed in plants to enhance tolerance to multiple abiotic stresses, is given in Table 15.2.

15.3.1.3 Osmo-protectants and antioxidant enzymes

Plants produce osmoprotectants and antioxidants to counteract abiotic stresses. Genetic engineering has enabled the overexpression of key genes involved in biosynthesis of osmoprotectants and antioxidants to achieve tolerance to multiple stresses.

Ectopic expression of the mannitol-accumulating *mtlD* gene in transgenic wheat plants improved tolerance to water stress and salinity. Transgenic wheat plants exhibited significantly better growth performance under both water stress and salinity conditions compared to nontransgenic plants (Abebe et al., 2003). Overexpression of the *1-pyrroline-5-carboxylate synthetase (P5CS)* gene in transgenic rice under the regulation of a stress-inducible promoter significantly increased the production of the P5CS enzyme and subsequent proline accumulation in transgenic rice plants. Transgenic rice plants exhibited improved biomass under both salt and water stress conditions compared to nontransformed control plants (Zhu et al., 1998).

Transgenic plants overexpressing the glycine betaine synthesizing gene *codA* (*choline oxidase*) from the soil bacterium *Arthrobacter globiformis* exhibited improved tolerance against salt and cold stress. Glycine betaine acts as an osmoprotectant, helping to stabilize cellular structures and maintain water balance during stress conditions. Transgenic

TABLE 15.2 Transcription factors used for developing multiple abiotic stress tolerance.

Transcription factor	Abiotic stress	References
CBF1 (Tomato)	Chilling, drought, salt	Hsieh et al. (2002)
CBF4 (*Arabidopsis*)	Freezing, drought	Haake et al. (2002)
DREB1A (*Arabidopsis*)	Freezing, salt, drought	Kasuga et al. (2004), Pellegrineschi et al. (2004)
OsDREB2A (Rice)	Dehydration, salt	Mallikarjuna et al. (2011)
ONAC045 (Rice)	Drought, salt	Zheng et al. (2009)
ONAC063 (Rice)	High salinity, osmotic stress	Yokotani et al. (2009)
TaNAC2 (Wheat)	Drought, salt, cold	Mao et al. (2012)
SNAC1 (Rice)	Drought, salt	Hu et al. (2006b)
ONAC022 (Rice)	Drought, salt	Hong et al. (2016)
ZmWRKY58 (Maize)	Drought, salt	Cai et al. (2014)
AtWRKY30 (*Arabidopsis*)	Drought, heat	El-Esawi et al. (2019)
SlWRKY39 (Tomato)	General stress resistance	Sun et al. (2015)
OsMYB6 (Rice)	Drought, salinity	Tang et al. (2019)
MdSIMYB1 (Apple)	High salinity, drought, cold	Wang et al. (2014)
GmbZIP1 (Soybean)	Drought, salt, cold	Gao et al. (2011)
AtHDG11 (*Arabidopsis*)	Drought, salt	Banavath et al. (2018)

plants demonstrated improved recovery rates after salt stress compared to wild-type plants. The transgenic plants were found to maintain photosynthetic activity better than wild-type plants when exposed to low temperatures (Sakamoto and Murata, 1998). Similar results were observed in transgenic tobacco plants overexpressing choline dehydrogenase and betaine aldehyde dehydrogenase, which are involved in the biosynthesis of glycine betaine. The transgenic plants exhibited higher levels of glycine betaine compared to nontransgenic control plants and showed enhanced tolerance to environmental stresses such as salinity and low temperatures (Holmström et al., 2000).

Another interesting finding demonstrated that transgenic rice plants that overexpress the glyoxalase pathway showed enhanced tolerance to various abiotic stresses, including salinity, drought, and heat. This is attributed to the improved detoxification of methylglyoxal, a cytotoxic metabolite that accumulates under stress conditions (Gupta et al., 2018).

The introduction of superoxide dismutase (SOD) genes into plants has been shown to help plants to more efficiently eliminate ROS under environmental stresses such as chilling, ozone, water deficit, and salt stress (Bowler et al., 1991; Foyer et al., 1994; Gupta et al., 1993; McKersie et al., 1999; McKersie et al., 1993; Van Camp et al., 1996; Wang et al., 2005). Transgenic potato plants expressing SOD and ascorbate peroxidase in chloroplasts exhibited enhanced tolerance to oxidative stress and high temperature (Tang et al., 2006). Overexpression of novel durum wheat catalase in *Arabidopsis* resulted in multiple abiotic stress tolerance (Feki et al., 2015).

15.3.2 Gene silencing strategies

RNA interference (RNAi) and antisense RNA technologies have been used to downregulate inhibitors of stress responses, leading to enhanced abiotic stress tolerance. Studies have shown that suppressing proline degradation using antisense technology can lead to increased proline levels, thereby improving the ability of plants to withstand abiotic stresses. Transgenic plants with a suppressed *ProDH* gene exhibited significantly increased accumulation of proline in comparison to wild-type plants. Importantly, these antisense transgenics demonstrated improved ability to cope with both freezing and high salinity conditions, indicating a positive and significant correlation between proline accumulation and stress resistance (Nanjo et al., 1999). In another study, RNAi-mediated downregulation of the *SlHB2* gene in transgenic tomato plants showed improved tolerance to drought and salt stress. These plants exhibited greater chlorophyll content, improved water content, reduced water loss, and lower malondialdehyde accumulation compared to wild-type plants under stress conditions (Hu et al., 2017).

15.3.3 Engineering hormonal pathways

Phytohormones play a vital role in enabling plants to adapt to stress. Genetic engineering has been used to modify hormone pathways to improve stress responses in plants. The study revealed that the overexpression of *CrNCED1* from *Citrus reshni* in tobacco drives the accumulation of higher ABA levels than in wild-type plants. Overexpressed lines showed increased tolerance levels under multiple abiotic stresses, including dehydration, drought, salt, and oxidative stress (Xian et al., 2014). Ectopic expression of rice *OsNCED5* in *Arabidopsis* resulted in improved tolerance to water and salt stress, indicating *OsNCED5* plays a significant role in stress adaptation (Huang et al., 2019). In a similar study, overexpression of *OsNCED3* in transgenic plants resulted in increased ABA content and enhanced multiple abiotic stress tolerance . This suggests that manipulating *OsNCED3* expression could be a strategy to improve crop resilience and growth under stress conditions (Huang et al., 2018). Transgenic *Pelargonium graveolens*, overexpressing the *ACC deaminase* gene from *Achromobacter xylosoxidans*, exhibited enhanced tolerance to both salt and drought stress. This was indicated by elevated relative water content, chlorophyll levels, and improved photosynthetic parameters in comparison to wild-type plants (Singh et al., 2021). By integrating these transgenic approaches, researchers have successfully developed plants demonstrating improved resilience to abiotic stresses, supporting the advancement of climate-resilient and sustainable agricultural practices in response to global climate change.

15.4 Genome editing-based approaches for abiotic stress tolerance

Genome editing technologies hold tremendous potential for improving crop resilience and boosting agricultural productivity. Unlike transgenic approaches, genome editing addresses key limitations and significantly shortens the time required to develop enhanced crop varieties. This technology enables accurate and site-directed modifications within the genome, leading to the creation of beneficial genetic variations. In contrast to transgenic techniques, which involve the random insertion of DNA from related or unrelated species, genome editing utilizes site-directed nucleases to accurately modify specific DNA sequences (Molla and Yang, 2019).

CRISPR-Cas technology has also been proven to be a potent approach for expanding stress resilience in crops to abiotic environmental conditions. By enabling precise genome alterations, CRISPR-Cas allows the targeted editing of genes contributing to plant defense against abiotic stresses such as drought, salinity, heat, and cold. Recent advances have demonstrated successful editing of key regulatory genes, leading to enhanced stress adaptation and sustained yield levels under harsh environmental conditions . As research progresses, CRISPR-Cas holds great promise for creating climate-smart crops essential for sustainable agriculture.

In a recent study, CRISPR-Cas9 facilitated the knockout of all six ABA-induced transcription repressors (AITR) genes in *Arabidopsis*, resulting in enhanced drought and salinity tolerance. They found that the *AITR* mutants showed greater adaptability to drought and salinity conditions, but did not have any fitness costs concerning plant growth, development, and response to pathogen infection. These results suggest that *AITR* genes could contribute to improved tolerance against abiotic stresses in plants without compromising other agronomic traits (Chen et al., 2021).

In another study, drought and salt tolerance *(DST)* gene mutant, generated by CRISPR-Cas9,showed enhanced osmotic and salt stress tolerance. The *DST* mutant exhibited a reduction in stomatal density and increased leaf water maintenance under dehydration conditions, indicating enhanced drought tolerance. At the seedling stage, the *dst* mutant demonstrated medium tolerance to osmotic stress and high tolerance to NaCl stress (Santosh Kumar et al., 2020).

DPA4 and *SOD7* have been identified as negative regulators of seed size. The research indicates that these two transcription repressors play a significant role in regulating not only seed size but also inflorescence architecture, shoot branching, and floral organ development. The CRISPR-Cas generated *dpa4 sod7 aitr256* quintuple mutants not only had increased seed size but also demonstrated enhanced abiotic stress tolerance (Chen et al., 2019). A further study demonstrated that CRISPR mediates knockout of *OsPQT3*, a rice homolog of *Arabidopsis* PARAQUAT TOLERANCE 3 (*AtPQT3*), enhances the plant's resistance to several abiotic stresses, such as oxidative and salt stress. This study indicated that *OsPQT3* plays a crucial role in stress response mechanisms (Alfatih et al., 2020). It has been reported that *OsmiR535* plays a negative role in drought and salinity stress tolerance in rice, indicating that higher levels of *OsmiR535* may reduce the plant's ability to withstand such conditions. The *osmir535* mutant generated via the CRISPR-Cas9 system exhibited improved tolerance toward drought and salinity stress (Yue et al., 2020). Auxin exerts its effects on the transcriptional regulation of target genes mainly via Auxin Response Factors (*ARF*). A CRISPR-Cas-based knockout tomato line for the *ARF4* gene showed improved tolerance to salinity and osmotic stress (Bouzroud et al., 2020). Many emerging small gene editing variants, such as *TnpB* (Karmakar et al., 2024), *IscB* (Kato et al., 2022), and *Tas* (Faure et al., 2025), are likely to facilitate robust multiplex editing of genes regulating various abiotic stress tolerance.

15.5 Commercial success stories in abiotic stress-tolerant crops

The integration of transgenic technologies and genome editing into agricultural systems has yielded commercially available crops. A pioneering example is the development of HB4 transgenic wheat in Argentina, engineered by introducing the *Helianthus annuus Homeobox 4* gene from sunflower. These genetically modified wheat plants exhibit enhanced drought tolerance, demonstrating the ability to maintain or even increase yield under water-limited conditions compared to conventional varieties. Following rigorous safety assessments, HB4 wheat has received regulatory approval for commercial cultivation in Argentina, Brazil, and the United States. HB4 wheat has also received approvals for food/feed consumption in Australia, Colombia, New Zealand, Nigeria, South Africa, Paraguay, and Indonesia.

A landmark achievement in genome editing emerged on May 4, 2025, with the approval of commercial release of the CRISPR-Cas9-edited rice cultivar in India. This *PUSA Rice DST1* variety was developed by disrupting the *DST* gene to achieve simultaneous tolerance to drought and salinity. The transgenic HB4 wheat and *PUSA Rice DST1* exemplify how advanced biotechnologies can bridge the gap between sustainable agriculture and economic viability in a warming world.

15.6 Concluding remarks and future prospects

Expressing single or multiple candidate genes in transgenic crops enhances their capacity to withstand diverse abiotic stressors. These crops hold great potential for deployment in regions vulnerable to specific stresses, provided that they receive regulatory approval. However, stringent regulatory hurdles faced by transgenic crops delay the realization of the benefits of this powerful technology, which enables the utilization of the entire gene pool for crop improvement. The history of safe use of commercially available transgenic crops over the past two decades should now be sufficient to ease the red tape surrounding their approval and unlock their full potential. On the other hand, transgene-free genome-edited crops have been declared equivalent to conventionally bred varieties and have received regulatory clearance for field use in many countries (Mushtaq et al., 2020). Genome editing has been effectively employed to produce plants that are tolerant to various abiotic stresses. Most of these examples use conventional non-homologous end joining (NHEJ)-based editing, which is effective for disrupting negative regulator genes or introducing novel variations in regulatory regions. However, NHEJ-based editing has limited utility in crop improvement because it cannot achieve precise allele swapping.

Advanced technologies such as base editing (Molla et al., 2020, 2021; Molla and Yang, 2019) and prime editing (Molla et al., 2021) allow targeted allele replacement and hold immense promise for crop improvement. Simultaneous base and prime editing at multiple loci can generate favorable alleles that confer tolerance to multiple abiotic stresses. Likewise, combining gene overexpression with genome editing could unlock unprecedented opportunities to engineer crops with complex stress tolerance. However, such engineered crops may still be classified as genetically modified organisms and thus face the same regulatory challenges, potentially hindering their commercialization. If a crop wild relative possesses traits for multiple abiotic stress tolerance, gene editing could facilitate rapid domestication by removing undesirable wild characteristics while retaining the beneficial traits (Pattnaik et al., 2023). Although genome editing tools are widely used worldwide for research and development, intellectual property restrictions limit their deployment in commercial crop development, particularly in addressing food security challenges in the Global South. To overcome this, we must work collectively to enhance accessibility to these genome editing tools in the Global South.

References

Abebe, T., Guenzi, A.C., Martin, B., Cushman, J.C., 2003. Tolerance of mannitol-accumulating transgenic wheat to water stress and salinity. Plant Physiology 131 (4), 1748–1755. Available at https://doi.org/10.1104/pp.102.003616.

Alfatih, A., Wu, J., Jan, S.U., Zhang, Z.S., Xia, J.Q., Xiang, C.B., 2020. Loss of rice PARAQUAT TOLERANCE 3 confers enhanced resistance to abiotic stresses and increases grain yield in field. Plant Cell and Environment 43 (11), 2743–2754. Available at https://doi.org/10.1111/pce.13856.

Banavath, J.N., Chakradhar, T., Pandit, V., Konduru, S., Guduru, K.K., Akila, C.S., et al., 2018. Stress inducible overexpression of AtHDG11 leads to improved drought and salt stress tolerance in peanut (Arachis hypogaea L.). Frontiers in Chemistry 6. Available at https://doi.org/10.3389/fchem.2018.00034.

Bartels, D., Sunkar, R., 2005. Drought and salt tolerance in plants. Critical Reviews in Plant Sciences 24 (1), 23–58. Available at https://doi.org/10.1080/07352680590910410.

Bechtold, U., Albihlal, W.S., Lawson, T., Fryer, M.J., Sparrow, P.A.C., Richard, F., et al., 2013. Arabidopsis heat shock transcription factora1b overexpression enhances water productivity, resistance to drought, and infection. Journal of Experimental Botany 64 (11), 3467–3481. Available at https://doi.org/10.1093/jxb/ert185.

Bouzroud, S., Gasparini, K., Hu, G., Barbosa, M.A.M., Rosa, B.L., Fahr, M., et al., 2020. Down regulation and loss of auxin response factor 4 function using CRISPR/Cas9 alters plant growth, stomatal function and improves tomato tolerance to salinity and osmotic stress. Morocco Genes 11 (3). Available at https://doi.org/10.3390/genes11030272.

Bowler, C., Slooten, L., Vandenbranden, S., De Rycke, R., Botterman, J., Sybesma, C., et al., 1991. Manganese superoxide dismutase can reduce cellular damage mediated by oxygen radicals in transgenic plants. The EMBO Journal 10 (7), 1723–1732. Available at https://doi.org/10.1002/j.1460-2075.1991.tb07696.x.

Brini, F., Hanin, M., Lumbreras, V., Amara, I., Khoudi, H., Hassairi, A., et al., 2007. Overexpression of wheat dehydrin DHN-5 enhances tolerance to salt and osmotic stress in Arabidopsis thaliana. Plant Cell Reports 26 (11), 2017–2026. Available at https://doi.org/10.1007/s00299-007-0412-x.

Cai, R., Zhao, Y., Wang, Y., Lin, Y., Peng, X., Li, Q., et al., 2014. Overexpression of a maize WRKY58 gene enhances drought and salt tolerance in transgenic rice. Plant Cell, Tissue and Organ Culture 119 (3), 565–577. Available at https://doi.org/10.1007/s11240-014-0556-7.

Chaves, M.M., Maroco, J.P., Pereira, J.S., 2003. Understanding plant responses to drought - from genes to the whole plant. Functional Plant Biology 30 (3), 239–264. Available at https://doi.org/10.1071/FP02076,.

Chen, S., Zhang, N., Zhang, Q., Zhou, G., Tian, H., Hussain, S., et al., 2019. Genome editing to integrate seed size and abiotic stress tolerance traits in arabidopsis reveals a role for DPA4 and SOD7 in the regulation of inflorescence architecture. International Journal of Molecular Sciences 20 (11), 2695. Available at https://doi.org/10.3390/ijms20112695.

Chen, S., Zhang, N., Zhou, G., Hussain, S., Ahmed, S., Tian, H., et al., 2021. Knockout of the entire family of AITR genes in Arabidopsis leads to enhanced drought and salinity tolerance without fitness costs. BMC Plant Biology 21 (1). Available at https://doi.org/10.1186/s12870-021-02907-9.

Cheng, Z., Targolli, J., Huang, X., Wu, R., 2002. Wheat LEA genes, PMA80 and PMA1959, enhance dehydration tolerance of transgenic rice (Oryza sativa L.). Molecular Breeding 10 (1-2), 71–82. Available at https://doi.org/10.1023/a:1020329401191.

El-Esawi, M.A., Al-Ghamdi, A.A., Ali, H.M., Ahmad, M., 2019. Overexpression of atWRKY30 transcription factor enhances heat and drought stress tolerance in wheat (Triticum aestivum L.). Egypt Genes 10 (2). Available at https://doi.org/10.3390/genes10020163.

Faure, G., Saito, M., Wilkinson, M.E., Quinones-Olvera, N., Xu, P., Flam-Shepherd, D., et al., 2025. TIGR-Tas: a family of modular RNA-guided DNA-targeting systems in prokaryotes and their viruses. Science (New York, N.Y.) 388 (6746). Available at https://doi.org/10.1126/science.adv9789.

Feki, K., Kamoun, Y., Ben Mahmoud, R., Farhat-Khemakhem, A., Gargouri, A., Brini, F., 2015. Multiple abiotic stress tolerance of the transformants yeast cells and the transgenic Arabidopsis plants expressing a novel durum wheat catalase. Plant Physiology and Biochemistry 97, 420–431. Available at https://doi.org/10.1016/j.plaphy.2015.10.034.

Foyer, C.H., Descourvières, P., Kunert, K.J., 1994. Protection against oxygen radicals: an important defence mechanism studied in transgenic plants. Plant, Cell & Environment 17 (5), 507–523. Available at https://doi.org/10.1111/J.1365-3040.1994.TB00146.X.

Galau, G.A., Bijaisoradat, N., Hughes, D.W., 1987. Accumulation kinetics of cotton late embryogenesis-abundant mRNAs and storage protein mRNAs: coordinate regulation during embryogenesis and the role of abscisic acid. Developmental Biology 123 (1), 198–212. Available at https://doi.org/10.1016/0012-1606(87)90442-8.

Gao, S.Q., Chen, M., Xu, Z.S., Zhao, C.P., Li, L., Xu, H.J., Tang, Y.M., et al., 2011. The soybean GmbZIP1 transcription factor enhances multiple abiotic stress tolerances in transgenic plants. Plant Molecular Biology 75 (6), 537–553. Available at https://doi.org/10.1007/s11103-011-9738-4.

Gupta, A.K., Webb, R.P., Holaday, A.S., Allen, R.D., 1993. Overexpression of superoxide dismutase protects plants from oxidative stress (induction of ascorbate peroxidase in superoxide dismutase-overexpressing plants). Plant Physiology 103 (4), 1067–1073. Available at 10.1104/pp.103.4.1067.

Gupta, B.K., Sahoo, K.K., Ghosh, A., Tripathi, A.K., Anwar, K., Das, P., et al., 2018. Manipulation of glyoxalase pathway confers tolerance to multiple stresses in rice. Plant Cell and Environment 41 (5), 1186–1200. Available at https://doi.org/10.1111/pce.12968.

Haake, V., Cook, D., Riechmann, J.L., Pineda, O., Thomashow, M.F., Zhang, J.Z., 2002. Transcription factor CBF4 is a regulator of drought adaptation in Arabidopsis. Plant Physiology 130 (2), 639–648. Available at https://doi.org/10.1104/pp.006478.

Halder, T., Upadhyaya, G., Ray, S., 2017. YSK2 type dehydrin (SbDHN1) from sorghum bicolor showed improved protection under high temperature and osmotic stress condition. Frontiers in Plant Science 8, 259380. Available at https://doi.org/10.3389/FPLS.2017.00918.

Herrera-Estrella, L., Depicker, A., Van Montagu, M., Schell, J., 1983. Expression of chimaeric genes transferred into plant cells using a Ti-plasmid-derived vector. Nature 303 (5914), 209–213. Available at https://doi.org/10.1038/303209a0.

Holmström, K.O., Somersalo, S., Mandal, A., Palva, T.E., Welin, B., 2000. Improved tolerance to salinity and low temperature in transgenic tobacco producing glycine betaine. Journal of Experimental Botany 51 (343), 177–185. Available at https://doi.org/10.1093/jexbot/51.343.177.

Hong, Y., Zhang, H., Huang, L., Li, D., Song, F., 2016. Overexpression of a stress-responsive NAC transcription factor gene ONAC022 improves drought and salt tolerance in rice. Frontiers in Plant Science 7 (2016). Available at https://doi.org/10.3389/fpls.2016.00004.

Hsieh, T.H., Lee, J.T., Yang, P.T., Chiu, L.H., Charng, Y.Y., Wang, Y.C., et al., 2002. Heterology expression of the Arabidopsis C-repeat/dehydration response element binding factor 1 gene confers elevated tolerance to chilling and oxidative stresses in transgenic tomato. Plant Physiology 129 (3), 1086–1094. Available at https://doi.org/10.1104/pp.003442.

Hu, H., Dai, M., Yao, J., Xiao, B., Li, X., Zhang, Q., et al., 2006b. Overexpressing a NAM, ATAF, and CUC (NAC) transcription factor enhances drought resistance and salt tolerance in rice. Proceedings of the National Academy of Sciences of the United States of America 103 (35), 12987–12992. Available at https://doi.org/10.1073/pnas.0604882103.

Hu, H., Dai, M., Yao, J., Xiao, B., Li, X., Zhang, Q., et al., 2006a. Overexpressing a NAM, ATAF, and CUC (NAC) transcription factor enhances drought resistance and salt tolerance in rice. Proceedings of the National Academy of Sciences of the United States of America 103 (35), 12987–12992. Available at https://doi.org/10.1073/PNAS.0604882103,.

Hu, J., Chen, G., Yin, W., Cui, B., Yu, X., Lu, Y., et al., 2017. Silencing of SlHB2 improves drought, salt stress tolerance, and induces stress-related gene expression in tomato. Journal of Plant Growth Regulation 36 (3), 578–589. Available at https://doi.org/10.1007/s00344-017-9664-z.

Huang, Y., Guo, Y., Liu, Y., Zhang, F., Wang, Z., Wang, H., et al., 2018. 9-cis-Epoxycarotenoid dioxygenase 3 regulates plant growth and enhances multi-abiotic stress tolerance in rice. Frontiers in Plant Science 9. Available at https://doi.org/10.3389/fpls.2018.00162.

Huang, Y., Jiao, Y., Xie, N., Guo, Y., Zhang, F., Xiang, Z., et al., 2019. OsNCED5, a 9-cis-epoxycarotenoid dioxygenase gene, regulates salt and water stress tolerance and leaf senescence in rice. Plant Science 287, 110188. Available at https://doi.org/10.1016/j.plantsci.2019.110188.

Hwang, S.M., Kim, D.W., Woo, M.S., Jeong, H.S., Son, Y.S., Akhter, S., et al., 2014. Functional characterization of ArabidopsisHsfA6a as a heat-shock transcription factor under high salinity and dehydration conditions. Plant, Cell and Environment 37 (5), 1202–1222. Available at https://doi.org/10.1111/pce.12228.

Karmakar, S., Panda, D., Panda, S., Dash, M., Saha, R., Das, P., et al., 2024. A miniature alternative to Cas9 and Cas12: transposon-associated TnpB mediates targeted genome editing in plants. Plant Biotechnology Journal 22 (10), 2950–2953. Available at https://doi.org/10.1111/pbi.14416.

Kasuga, M., Miura, S., Shinozaki, K., Yamaguchi-Shinozaki, K., 2004. A Combination of the Arabidopsis DREB1A gene and stress-inducible rd29A promoter improved drought- and low-temperature stress tolerance in tobacco by gene transfer. Plant and Cell Physiology 45 (3), 346–350. Available at https://doi.org/10.1093/pcp/pch037.

Kato, K., Okazaki, S., Kannan, S., Altae-Tran, H., Esra Demircioglu, F., Isayama, Y., et al., 2022. Structure of the IscB–ωRNA ribonucleoprotein complex, the likely ancestor of CRISPR-Cas9. Nature Communications 13 (1). Available at https://doi.org/10.1038/s41467-022-34378-3.

Liu, Q., Kasuga, M., Sakuma, Y., Abe, H., Miura, S., Yamaguchi-Shinozaki, K., et al., 1998. Two transcription factors, DREB1 and DREB2, with an EREBP/AP2 DNA binding domain separate two cellular signal transduction pathways in drought- and low-temperature-responsive gene expression, respectively, in Arabidopsis. The Plant Cell 10 (8), 1391–1406. Available at https://doi.org/10.1105/tpc.10.8.1391.

Maaroufi, H., Tanguay, R.M., 2013. Analysis and phylogeny of small heat shock proteins from marine viruses and their cyanobacteria host. PLoS One 8 (11). Available at https://doi.org/10.1371/journal.pone.0081207.

Mallikarjuna, G., Mallikarjuna, K., Reddy, M.K., Kaul, T., 2011. Expression of OsDREB2A transcription factor confers enhanced dehydration and salt stress tolerance in rice (Oryza sativa L. Biotechnology Letters 33 (8), 1689–1697. Available at https://doi.org/10.1007/s10529-011-0620-x.

Mao, X., Zhang, H., Qian, X., Li, A., Zhao, G., Jing, R., 2012. TaNAC2, a NAC-type wheat transcription factor conferring enhanced multiple abiotic stress tolerances in Arabidopsis. Journal of Experimental Botany 63 (8), 2933–2946. Available at https://doi.org/10.1093/jxb/err462.

McKersie, B.D., Bowley, S.R., Jones, K.S., 1999. Winter survival of transgenic alfalfa overexpressing superoxide dismutase. Plant Physiology 119 (3), 839–848. Available at https://doi.org/10.1104/PP.119.3.839.

McKersie, B.D., Chen, Y., De Beus, M., Bowley, S.R., Bowler, C., Inzé, D., et al., 1993. Superoxide dismutase enhances tolerance of freezing stress in transgenic alfalfa (Medicago sativa L.). Plant Physiology 103 (4), 1155–1163. Available at https://doi.org/10.1104/pp.103.4.1155.

Mishra, S.K., Tripp, J., Winkelhaus, S., Tschiersch, B., Theres, K., Nover, L., et al., 2002. In the complex family of heat stress transcription factors, HsfA1 has a unique role as master regulator of thermotolerance in tomato. Genes and Development 16 (12), 1555–1567. Available at https://doi.org/10.1101/gad.228802.

Molla, K.A., Yang, Y., 2019. CRISPR/Cas-mediated base editing: technical considerations and practical applications. Trends in Biotechnology 37 (10), 1121–1142. Available at https://doi.org/10.1016/j.tibtech.2019.03.008.

Molla, K.A., Shih, J., Yang, Y., 2020. Single-nucleotide editing for zebra3 and wsl5 phenotypes in rice using CRISPR/Cas9-mediated adenine base editors. ABIOTECH 1 (2), 106–118. Available at https://doi.org/10.1007/S42994-020-00018-X.

Molla, K.A., Sretenovic, S., Bansal, K.C., Qi, Y., 2021. Precise plant genome editing using base editors and prime editors. Nature Plants 7 (9), 1166–1187. Available at https://doi.org/10.1038/s41477-021-00991-1.

Muñoz-Mayor, A., Pineda, B., Garcia-Abellán, J.O., Antón, T., Garcia-Sogo, B., Sanchez-Bel, P., et al., 2012. Overexpression of dehydrin tas14 gene improves the osmotic stress imposed by drought and salinity in tomato. Journal of Plant Physiology 169 (5), 459–468. Available at https://doi.org/10.1016/j.jplph.2011.11.018.

Mushtaq, M., Mukhtar, S., Sakina, A., Dar, A.A., Bhat, R., Deshmukh, R., et al., 2020. Tweaking genome-editing approaches for virus interference in crop plants. Plant Physiology and Biochemistry 147, 242–250. Available at https://doi.org/10.1016/j.plaphy.2019.12.022.

Nanjo, T., Kobayashi, M., Yoshiba, Y., Kakubari, Y., Yamaguchi-Shinozaki, K., Shinozaki, K., 1999. Antisense suppression of proline degradation improves tolerance to freezing and salinity in Arabidopsis thaliana. FEBS Letters 461 (3), 205–210. Available at https://doi.org/10.1016/s0014-5793(99)01451-9.

Pattnaik, D., Avinash, S.P., Panda, S., Bansal, K.C., Chakraborti, M., Kar, M.K., et al., 2023. Accelerating crop domestication through genome editing for sustainable agriculture. Journal of Plant Biochemistry and Biotechnology 32 (4), 688–704. Available at https://doi.org/10.1007/s13562-023-00837-1.

Pellegrineschi, A., Reynolds, M., Pacheco, M., Brito, R.M., Almeraya, R., Yamaguchi-Shinozaki, K., et al., 2004. Stress-induced expression in wheat of the Arabidopsis thaliana DREB1A gene delays water stress symptoms under greenhouse conditions. Genome / National Research Council Canada = Genome / Conseil National de Recherches Canada 47 (3), 493–500. Available at https://doi.org/10.1139/g03-140.

Pérez-Clemente, R.M., Vives, V., Zandalinas, S.I., López-Climent, M.F., Muñoz, V., Gómez-Cadenas, A., 2013. Biotechnological approaches to study plant responses to stress. BioMed Research International 2013. Available at https://doi.org/10.1155/2013/654120.

Sakamoto, A., Murata, A.N., 1998. Metabolic engineering of rice leading to biosynthesis of glycinebetaine and tolerance to salt and cold. Plant Molecular Biology 38 (6), 1011–1019. Available at https://doi.org/10.1023/a:1006095015717.

Santosh Kumar, V.V., Verma, R.K., Yadav, S.K., Yadav, P., Watts, A., Rao, M.V., et al., 2020. CRISPR-Cas9 mediated genome editing of drought and salt tolerance (OsDST) gene in indica mega rice cultivar MTU1010. Physiology and Molecular Biology of Plants 26 (6), 1099–1110. Available at https://doi.org/10.1007/s12298-020-00819-w.

Shekhawat, U.K.S., Srinivas, L., Ganapathi, T.R., 2011. MusaDHN-1, a novel multiple stress-inducible SK 3-type dehydrin gene, contributes affirmatively to drought- and salt-stress tolerance in banana. Planta 234 (5), 915–932. Available at https://doi.org/10.1007/S00425-011-1455-3,.

Shinozaki, K., Yamaguchi-Shinozaki, K., 1997. Gene expression and signal transduction in water-stress response. Plant Physiology 115 (2), 327–334. Available at https://doi.org/10.1104/pp.115.2.327.

Singh, P., Pandey, S.S., Dubey, B.K., Raj, R., Barnawal, D., Chandran, A., et al., 2021. Salt and drought stress tolerance with increased biomass in transgenic Pelargonium graveolens through heterologous expression of ACC deaminase gene from Achromobacter xylosoxidans. Plant Cell, Tissue and Organ Culture 147 (2), 297–311. Available at https://doi.org/10.1007/s11240-021-02124-0.

Sun, X.C., Gao, Y.F., Li, H.R., Yang, S.Z., Liu, Y.S., 2015. Over-expression of SlWRKY39 leads to enhanced resistance to multiple stress factors in tomato. Journal of Plant Biology 58 (1), 52–60. Available at https://doi.org/10.1007/s12374-014-0407-4.

Tang, L., Kwon, S.Y., Kim, S.H., Kim, J.S., Choi, J.S., Cho, K.Y., et al., 2006. Enhanced tolerance of transgenic potato plants expressing both superoxide dismutase and ascorbate peroxidase in chloroplasts against oxidative stress and high temperature. Plant Cell Reports 25 (12), 1380–1386. Available at https://doi.org/10.1007/s00299-006-0199-1.

Tang, Y., Bao, X., Zhi, Y., Wu, Q., Guo, Y., Yin, X., et al., 2019. Overexpression of a MYB family gene, OsMYB6, increases drought and salinity stress tolerance in transgenic rice. Frontiers in Plant Science 10. Available at https://doi.org/10.3389/fpls.2019.00168.

Van Camp, W., Capiau, K., Van Montagu, M., Inzé, D., Slooten, L., 1996. Enhancement of oxidative stress tolerance in transgenic tobacco plants overproducing Fe-superoxide dismutase in chloroplasts. Plant Physiology 112 (4), 1703–1714. Available at https://doi.org/10.1104/pp.112.4.1703.

Vinocur, B., Altman, A., 2005. Recent advances in engineering plant tolerance to abiotic stress: achievements and limitations. Current Opinion in Biotechnology 16 (2), 123–132. Available at https://doi.org/10.1016/j.copbio.2005.02.001.

Wang, F.Z., Wang, Q.B., Kwon, S.Y., Kwak, S.S., Su, W.A., 2005. Enhanced drought tolerance of transgenic rice plants expressing a pea manganese superoxide dismutase. Journal of Plant Physiology 162 (4), 465–472. Available at https://doi.org/10.1016/j.jplph.2004.09.009.

Wang, R.K., Cao, Z.H., Hao, Y.J., 2014. Overexpression of a R2R3 MYB gene MdSIMYB1 increases tolerance to multiple stresses in transgenic tobacco and apples. Physiologia Plantarum 150 (1), 76–87. Available at https://doi.org/10.1111/ppl.12069.

Winicov, I., Bastola, D.R., 1997. Salt tolerance in crop plants: new approaches through tissue culture and gene regulation. Acta Physiologiae Plantarum 19, 435–449 https://doi.org/10.1007/s11738-997-0040-7.

Xian, L., Sun, P., Hu, S., Wu, J., Liu, J.H., 2014. Molecular cloning and characterization of CrNCED1, a gene encoding 9-cis-epoxycarotenoid dioxygenase in Citrus reshni, with functions in tolerance to multiple abiotic stresses. Planta 239 (1), 61–77. Available at https://doi.org/10.1007/s00425-013-1963-4.

Xiong, L., Schumaker, K.S., Zhu, J.K., 2002. Cell signaling during cold, drought, and salt stress. The Plant Cell 14 (suppl_1), S165–S183. Available at https://doi.org/10.1105/TPC.000596.

Xu, D., Duan, X., Wang, B., Hong, B., Ho, T.H.D., Wu, R., 1996. Expression of a late embryogenesis abundant protein gene, HVA1, from barley confers tolerance to water deficit and salt stress in transgenic rice. Plant Physiology 110 (1), 249–257. Available at https://doi.org/10.1104/pp.110.1.249.

Yadav, S., Modi, P., Dave, A., Vijapura, A., Patel, D., Patel, M., 2020. Effect of abiotic stress on crops. In: Hasanuzzaman, M., Carvalho Minhoto Teixeira Filho, M., Fujita, M., Assis Rodrigues Nogueira, T. (Eds.), Sustainable Crop Production. IntechOpen. Available at 10.5772/intechopen.88434.

Yamaguchi-Shinozaki, K., Shinozaki, K., 2006. Transcriptional regulatory networks in cellular responses and tolerance to dehydration and cold stresses. Annual Review of Plant Biology 57 (1), 781–803. Available at https://doi.org/10.1146/annurev.arplant.57.032905.105444.

Yokotani, N., Ichikawa, T., Kondou, Y., Matsui, M., Hirochika, H., Iwabuchi, M., et al., 2009. Tolerance to various environmental stresses conferred by the salt-responsive rice gene ONAC063 in transgenic Arabidopsis. Planta 229 (5), 1065–1075. Available at https://doi.org/10.1007/s00425-009-0895-5.

Yue, E., Cao, H., Liu, B., 2020. *OsmiR535*, a potential genetic editing target for drought and salinity stress tolerance in *Oryza sativa*. Plants (Basel) 9 (10), 1337. Available at https://doi.org/10.3390/plants9101337.

Zheng, X., Chen, B., Lu, G., Han, B., 2009. Overexpression of a NAC transcription factor enhances rice drought and salt tolerance. Biochemical and Biophysical Research Communications 379 (4), 985–989. Available at https://doi.org/10.1016/j.bbrc.2008.12.163.

Zhu, B., Su, J., Chang, M., Pal, D., Verma, S., Fan, Y.-L., et al., 1998. Overexpression of a D 1-pyrroline-5-carboxylate synthetase gene and analysis of tolerance to water-and salt-stress in transgenic rice. Plant Science 139 (11), 41–48. Available at https://doi.org/10.1016/S0168-9452(98)00175-7.

Zhu, B., Ye, C., Lü, H., Chen, X., Chai, G., Chen, J., et al., 2006. Identification and characterization of a novel heat shock transcription factor gene, GmHsfA1, in soybeans (Glycine max). Journal of Plant Research 119 (3), 247–256. Available at https://doi.org/10.1007/s10265-006-0267-1.

Chapter 16

Multiomics-based approaches for developing multiple abiotic stress-tolerant crops

Soham Ray[1,*], Joshitha Vijayan[2], Laxmi Sharma[3], Shivani Nagar[1], Archana Watts[1], Pranjal Yadava[1], Madhurima Das[1], Chandan K. Deb[4], Ranjeet R. Kumar[1], Ankesh Pandey[1], Ashok Kumar[1], and Nazima Nasrullah[1]

[1]*Division of Plant Physiology, ICAR-Indian Agricultural Research Institute, Pusa Campus, New Delhi, India,* [2]*ICAR- National Institute for Plant Biotechnology, New Delhi, Delhi, India,* [3]*ICAR- National Bureau of Plant Genetic Resources, New Delhi, Delhi, India,* [4]*ICAR Indian Agricultural Statistical Research Institute, New Delhi, Delhi, India*

[*]*Corresponding author. e-mail address: Soham@iari.res.in*

16.1 Introduction

The escalating global climate crisis casts a long shadow over agriculture, threatening to disrupt the delicate balance of food production and security. Increasing temperatures, unpredictable rainfall patterns, and more frequent extreme weather events—including droughts, floods, heat waves, and cold spells—pose major challenges to crop growth and development. These abiotic stresses exert multifaceted pressures on plants, impacting physiological processes, disrupting metabolic pathways, and ultimately compromising yield and quality. In this era of unprecedented environmental change, developing crop varieties with improved resilience to multiple abiotic stresses has become paramount. Traditional breeding approaches, while invaluable, often prove insufficient to address the intricate and multidimensional character of abiotic stress tolerance. The complex interactions among genetic, molecular, and physiological factors that regulate plant stress responses necessitate a more comprehensive and integrated approach to crop improvement. Fortunately, the emergence of high-throughput technologies has ushered in a new era in plant science, enabling researchers to delve deeper into the molecular foundations of stress tolerance. This chapter examines the transformative potential of multiomics approaches, integrating genomics, transcriptomics, proteomics, metabolomics, and phenomics, in unraveling the complex genetic and molecular mechanisms that underpin crop resilience to abiotic stresses.

The term "omics" refers to the comprehensive study of a particular class of molecules on a large scale. Integrating data from diverse "omics" platforms enables researchers to achieve a comprehensive understanding of the complex molecular networks underlying plant stress responses. This systems biology approach has the potential to surpass the limitations of studying individual genes or proteins in isolation, providing a more comprehensive view of the interconnectedness of biological processes. To begin with, genomics provides the foundation by analyzing complete genomes, facilitating the identification of stress-responsive genes, genetic variations, and molecular markers for breeding programs. Advanced approaches such as genome-wide association studies (GWAS) further enable the precise mapping of genetic loci associated with specific traits. Complementing this, transcriptomics examines gene expression dynamics, revealing stress-induced molecular pathways through RNA sequencing and identifying key regulatory genes like transcription factors. Vis-À-vis noncoding RNA sequencing focuses on identifying and characterizing RNA transcripts lacking protein-coding potential, including microRNAs, lncRNAs, and circular RNAs. These noncoding RNAs serve essential functions in gene regulation, epigenetic modification, and cellular signaling. Proteomics further adds another layer by focusing on proteins – the cellular workhorses – studying their abundance, modifications, and interactions, which are critical for stress responses. Metabolomics dives into the biochemical realm, profiling metabolites that reflect the physiological state of the cell and

Multiple Abiotic Stresses in Plants. DOI: https://doi.org/10.1016/B978-0-443-27424-4.00021-5

uncovering altered metabolic pathways using advanced spectrometry techniques. Finally, phenomics bridges these molecular insights with observable plant traits, utilizing high-throughput platforms like drones and sensors to measure growth, yield, and water use efficiency under stress conditions. By correlating multiomics data with phenotypes, researchers can identify key genes, pathways, and traits, driving the development of crop varieties capable of withstanding environmental challenges and securing global food production.

16.2 Genomics: unlocking the blueprint of stress tolerance

16.2.1 Genome sequencing and genome databases of plants

A plant genome can be termed as the complete genetic endowment of the plant, stored in the form of DNA, governing its development, growth, and environmental responses. The concept of plant genomics emerged from the amalgamation of molecular biology and genetics to improve crop breeding. Genomics encompasses the study of genome structure, function, evolution, and mapping, offering vital insights into the genetic blueprint of plants. The emergence of high-throughput sequencing, particularly second- and third-generation technologies, has greatly accelerated progress in crop genomics. Second-generation sequencing, commonly referred to as next-generation sequencing, is characterized by massively parallel short read sequencing, including several sequencing technologies in which Illumina sequencing is the most prominent one. It allows for high-throughput, cost-effective generation of short reads, enabling the identification of genetic variations linked to stress tolerance. This technology aids in mapping key traits, accelerating marker-assisted selection, and improving breeding efficiency. On the other hand, third-generation sequencing platforms, such as single-molecule real-time sequencing by Pacific Biosciences and Oxford Nanopore technologies, are characterized by single-molecule sequencing in real time, generating longer read lengths with greater accuracy. Third-generation sequencing thus provides deeper insights into structural variations, epigenetic modifications, and complex genomic regions, which are often missed by short reads. The integration of both sequencing technologies allows for more precise and comprehensive genome assemblies, offering a better understanding of the genetic architecture behind multistress tolerance.

In plant sciences, genomics has played a crucial role in unraveling the complexities of genome structure and function, providing essential tools for precise and targeted crop improvement. The study of plant genomics has been instrumental in identifying genetic variations and understanding the molecular mechanisms behind key agronomic traits such as yield, stress tolerance, and disease resistance. Genomics has particularly advanced the development of crops that can withstand environmental challenges, such as drought, salinity, and temperature extremes. Significant contributions to the genomics of various crop groups have been made across multiple studies, including those on cereals (Walkowiak et al., 2020; Siddiqui et al., 2021; Jackson et al., 2023), oilseeds (Sandhu et al., 2022; Li et al., 2021), pulses (Ojiewo et al., 2019; Jha et al., 2022; Kumar et al., 2020), millets (Krishna et al., 2022), and horticultural crops (Ahmad et al., 2020). These studies provide deeper insights into genetic traits that improve both productivity and resilience. The availability of plant genome databases has significantly enhanced the accessibility and utility of genomic information. Public repositories like the International Nucleotide Sequence Database Collaboration and National Center for Biotechnology Information archive thousands of plant genomes, ranging from model organisms such as *Arabidopsis thaliana* and *Brachypodium distachyon* to major crop species like rice (*Oryza sativa*), maize (*Zea mays*), and wheat (*Triticum aestivum*). Plant genome databases such as Gramene, Ensembl Plants, Plant genome database (GDB), and Phytozome provide essential resources for comparative genomics, enabling researchers to explore synteny, sequence diversity, and structural variations among species.

16.2.2 Utilization of genome sequence information to unravel genetic mechanisms governing stress tolerance

Decoding crop genomes has facilitated the identification of molecular markers, which are essential for trait dissection, selection, and enhancement. GWAS have emerged as an indispensable tool for crop improvement, helping identify key genes and markers for breeding resilient and high-yielding varieties. GWAS statistically identify associations between genotypes and phenotypes linked to specific traits in plants. By genotyping large germplasm sets or populations with numerous genetic markers, GWAS locates genomic regions associated with traits of interest, typically using single-nucleotide polymorphisms (SNPs). These studies utilize linkage disequilibrium (LD) to link variants with traits across populations, aiding in candidate gene identification. GWAS has revealed genetic variants tied to biological pathways and traits such as stress tolerance and productivity in various crops. The GWAS process involves several steps, such as selecting a germplasm set, genotyping, data quality control, statistical analysis, and replication. Genotyping is often done through microarray or whole-genome sequencing. Whole-genome sequencing is preferred for GWAS as it can capture the complete genetic variation available in the dataset. Quality control removes errors and ensures valid genotype-phenotype associations, followed by genotype imputation to predict missing data. Data analysis involves statistical tests to identify genomic regions associated with traits. This is done using linear or logistic regression models, with population structure and kinship factors addressed using mixed linear models (MLM). The results are visualized using Manhattan and quantile–quantile plots, and candidate genes are identified and validated through functional analysis (Fig. 16.1).

The elucidation of underlying mechanisms and the genes governing an important trait that regulates the phenotypic expression under varying environments is crucial for crop improvement under global climate change scenarios. Traits contributing to abiotic stress tolerance are generally quantitative and are regulated by multiple genes. Thus it is important to identify new genomic resources using advanced genomic tools for the crop improvement program. Recent progress in sequencing technologies and high-throughput genotyping has enabled mining of new genomic resources like SNPs from the genomic and transcriptomic databases and assisted in the advancement of high-density SNP array technologies. Moreover, the high-throughput phenotyping platforms also have the added advantage of acquiring precise and large-scale measurements of the phenotype for the trait of interest. The availability of such genomic resources coupled with large-scale phenotyping facilities has allowed the capture of numerous genetic polymorphisms across the genome through GWAS. GWAS is such a genomic tool that identifies the genetic markers that are tightly associated with the quantitative trait locus (QTLs) through the LD between the genetic marker and the mutation. It thus explores the variation in a diversified population of a crop species to identify potential candidate genes and the respective pathway that regulates abiotic stress tolerance. Moreover, the markers developed through this technique can be further explored for crop improvement programs. To date, several studies have been conducted in different crops to identify the putative gene responsible for abiotic stress tolerance using GWAS (Challa and Neelapu, 2018). Under natural field conditions, abiotic

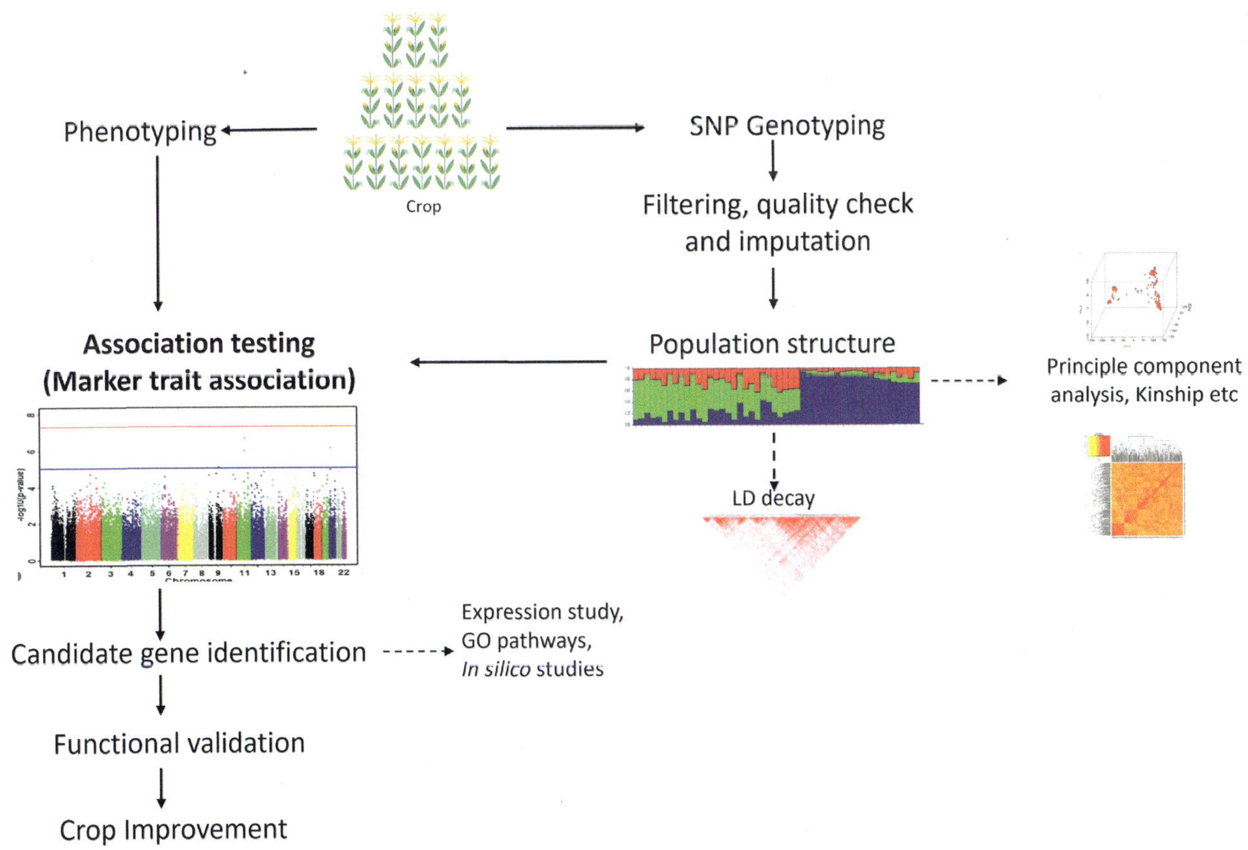

FIGURE 16.1 Schematic representation of the protocol of MGI = GWAS. The diverse panel of accessions/genotypes is phenotyped for the trait of interest. Complete genome sequences for the same genotypes are obtained. The sequence quality is checked and filtered. The diversity of the population and its structure is derived, and the linkage disequilibrium is checked. After confirming the sample quality, the association of the phenotype with the genotype is studied. A group of candidate genes responsible for the trait of interest is derived from the Manhattan plot and Q-Q plot, followed by validation of the genes. The putative gene is targeted for functional validation.

stresses often occur in combination rather than in isolation. For instance, heat stress is frequently accompanied by drought stress, and the two typically manifest together. Therefore it is crucial to identify the genomic regions or genes that confer simultaneous tolerance to multiple stresses using advanced molecular tools. However, reports for GWAS for multiple stress tolerance or tolerance to a combination of stress are very few.

16.2.3 Utilization of genome-wide association studies for studying multistress tolerance

GWAS is an important genetic tool to detect genetic variants for complex traits. It has been successful in capturing the quantitative variation underlying abiotic stress tolerance. Several QTLs and putative genes have been identified for simultaneous tolerance to drought and heat stress. The SNPs identified are highly correlated with the desired trait(s) under drought. A wheat diversity panel of 145 multiple synthetic derivative lines and five check cultivars of *Aegilops tauschii* was evaluated for heat and heat-drought tolerance to identify genes and genomic regions for heat-drought resilience. Marker trait association was observed for grain yield with 20.3% variation on chromosome 3D. Associations for leaf traits—such as canopy temperature, vegetation indices, and carbon isotope composition—were located on chromosomes 2D, 3D, and 5D. Further analysis showed that the *GA20ox* was responsible for grain yield stability, and *CaaX prenyl protease 2* was responsible for canopy temperature at the flowering stage, under heat and heat-drought stresses in wheat (Itam et al., 2022). In maize, GWAS using 7834 SNPs was conducted for phenotypic traits like days to 50% anthesis and silking, husk cover, and grain yield in 162 tropical maize elite cultivars, showing the association of 66, 27, and 24 SNPs for drought and heat and their combined effect (Osuman et al., 2022). They further identified 12 candidate genes containing phytohormone *cis*-acting elements that played a regulatory role in multiple stress tolerance. In another study, a diverse panel of 490 sesame accessions was used to study the stress tolerance index of NaCl-induced salinity and polyethylene glycol-induced drought/osmotic stress. A total of 132 and 120 SNPs for drought and salinity, respectively, were found that pointed to 9 and 15 QTLs and 13 and 27 potential candidate genes, respectively. Out of these, two QTLs were common for both stresses. These candidate genes were related to osmoprotectants, antioxidative enzymes, transcription factors involved in hormonal biosynthesis, ion sequestration, and signal transduction pathways (Li et al., 2018). Apart from the heat drought and salt stress, genetic mechanisms underlying heavy metal stress are also addressed through GWAS. In a diverse panel of 236 *Medicagotruncatula* accessions, GWAS was performed for relative root growth and heavy metal accumulation in leaves under cadmium (Cd) and mercury (Hg) stress. The candidate genes identified from the study relate to oxidative stress response genes, stress-related UDP-glycosyltransferases, P-type ATPase transporters, and ABC transporters (Paape et al., 2022). Although less, GWAS has thus enriched the literature with potential candidate genes underlying various tolerance mechanisms responsible for multiple abiotic stress tolerance. Currently, the focus of research in the community is mainly to multiplex various important traits under an abiotic stress using MLM GWAS analysis so that important trait and genes underlying the stress tolerance mechanism is revealed. However, on its application part, it is necessary to understand that a crop under the field conditions withstands one or more stresses together. However, mimicking the combined stress and its phenotyping is a major challenge. Efforts in these areas will thus allow us to bring out the candidate genes and the genomic regions that will have a major impact on crop improvement for climate resilience.

16.3 Transcriptomics: decoding gene expression under stress

16.3.1 Plant stress perception and the role of transcriptomics

Plants sense environmental stress via diverse receptors that activate secondary messengers, such as calcium and reactive oxygen species (ROS). These signals, in turn, initiate downstream responses mediated through calcium-dependent protein kinases and mitogen-activated protein kinase pathways. These cascades regulate diverse transcriptional responses, inducing transcriptional, posttranscriptional, and translational shifts in stress-related genes (Jeknić et al., 2014). Transcriptomics, the study of the transcriptome, is crucial for elucidating the early responses of plants under stress conditions. The transcriptome represents the complete set of RNA transcripts synthesized within cells at a given time point and under defined environmental conditions. It encompasses all classes of RNA, including messenger RNA, transfer RNA (tRNA), ribosomal RNA, and various noncoding RNAs. Since the transcriptome is a dynamic entity that changes in response to internal and external stimuli, transcriptome study of any plant in a time-course manner after subjecting it to any abiotic stress or a combination of stresses provides a snapshot of genes involved in stress perception and adaptation. Noncoding RNAs also contribute to stress-responsive gene regulation, but their roles and networks remain poorly understood, posing challenges for unraveling these mechanisms (Shin and Shin, 2016). However, much progress has been made in global profiling of the noncoding regulatory RNAs in recent times. Overall, the transcriptome study is a profiling technique of both coding and noncoding RNAs, which can profoundly help in gaining insights into plants' responses to challenges like salinity, drought, heat, cold, etc., individually or in combination. These stresses activate defense mechanisms, and transcriptomic analyzes across tissues and developmental stages help identify functional genomic regions and regulatory pathways involved in stress responses.

16.3.2 Approaches for transcriptome profiling in plants

There are several approaches for studying the transcriptome. Conventional methods such as expressed sequence tags (ESTs), serial analysis of gene expression (SAGE), and microarrays, alongside advanced techniques like RNA-seq, continue to enhance understanding of stress responses. However, characterizing these regulatory networks is critical for developing resilient crops under multiple abiotic stresses. ESTs were foundational in early transcriptomics. ESTs, derived from complementary DNA (cDNA) libraries, provided insights into gene composition. ESTs have played a crucial role in gene discovery and transcriptome analysis, especially in organisms with limited genomic information. However, there are several limitations of EST-based transcriptomics. For example, EST frequencies may not accurately reflect the true abundance of transcripts, especially for low-abundance transcripts. Highly expressed genes tend to be overrepresented in EST libraries. Besides, the single-pass sequencing nature of ESTs introduces sequencing errors, potentially leading to incorrect gene annotations or assembly. Many of these issues were addressed through relatively advanced techniques like microarray and SAGE, along with their variants. While microarrays enabled high-throughput expression profiling of thousands of genes simultaneously with the prior knowledge of the genes being studied, SAGE complemented these methods by quantifying gene expression levels without prior sequence knowledge, offering a snapshot of transcriptomes. However, microarrays and SAGE are limited in exploring regulatory networks across the genome. Specifically, both fail to faithfully profile low-abundant novel transcripts due to their rigid experimental frameworks or limited dynamic range, respectively. RNA sequencing (RNA-Seq), a massively parallel sequencing-based technique, overcomes these limitations by offering broader genome coverage and detecting both known and novel transcripts, including noncoding RNAs. This is by far the most advanced technique for transcriptome analysis, which has successfully been employed in several studies to identify stress-responsive genes and pathways in several crops under various abiotic stresses (Fig. 16.2). The advantages of RNA-Seq over other transcriptome profiling technique are flexibility (not limited by rigid framework like in case of microarray), capability to identify novel and rare transcripts, identification of splice variants etc. since it does not require the a prior knowledge of gene sequence for which the expression profiling has to be performed, it can be easily applied to

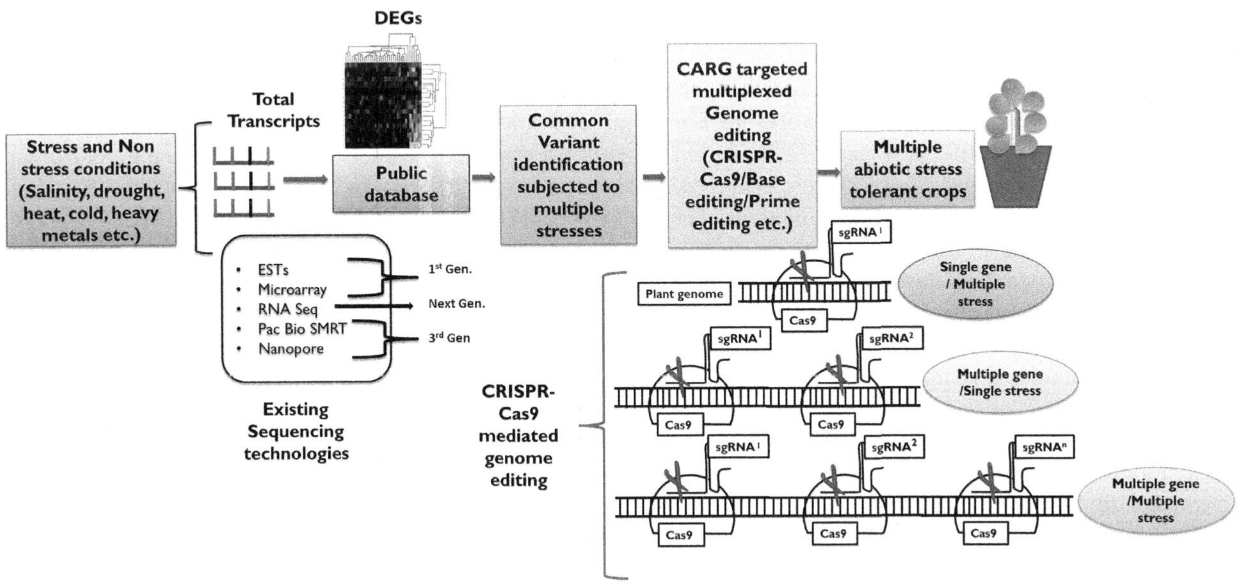

FIGURE 16.2 Schematic representation of a transcriptomic study under stress and nonstress conditions through existing sequencing technologies, followed by the identification of differential gene expression (DEGs). Common variant analysis and identification of core abiotic stress-responsive genes (CARG) are eventually candidate targets to launch genome editing machinery (CRISPR-Cas9/base editing/prime editing, etc.) in order to develop climate-resilient crops against multiple abiotic stresses.

study the transcriptome of orphan crops and wild crop-relatives demonstrating its utility in understanding and optimizing crop resilience.f

16.3.3 Integrative and comparative transcriptome analyses to understand multistress resilience

High-throughput transcriptome analysis has facilitated the identification of intricate stress-induced gene expression networks. This approach has revealed substantial overlap in plant responses to different abiotic stresses, providing insights into shared and unique regulatory mechanisms (Ni et al., 2021). For example, plant responses to drought and salt stress share significant similarities while also exhibiting stress-specific regulatory patterns. In a perennial herbaceous species (*Iris lactea var. chinensis*), Ni et al. (2021) observed that a total of 897 differentially expressed genes (DEGs) exhibited comparable expression patterns during both drought and salt stress conditions. Although many responses overlapped, gene ontology (GO) enrichment analysis identified distinct GO terms uniquely enriched in salt stress or drought responses, suggesting that these related stress conditions elicit both common and specific transcriptional responses in this species. Antagonism between the abiotic stress adaptation process can also be observed in some cases. It can be typically exemplified by one of our studies (Chakraborty et al., 2021), where we have found that the formation of aerenchyma, the key factor for flooding tolerance in rice, is inhibited by the presence of salt during saline water flooding. By comparing the transcriptomes of two rice genotypes, Varshadhan and Rashpanjor, which exhibit contrasting responses to saline water flooding, we were able to infer that ethylene production and signaling, which are crucial for the ROS burst leading to aerenchyma formation, were significantly affected by the salinity component of saline water flooding. Several genes implicated in abiotic stress responses appear to be shared across multiple stress conditions. A comprehensive transcriptome analysis of rapeseed (*Brassica napus*) subjected to abscisic acid, salt, dehydration, and cold stressors at two time points identified 30,908 DEGs (Zhang et al., 2019). Of these, a total of 2568 genes were consistently upregulated and 4376 genes were consistently downregulated (with a twofold expression change) across all stress treatments. Among the DEGs, 225 upregulated and 294 downregulated transcription factors were identified from 35 to 40 distinct families, respectively. Notably, transcription factors belonging to the ERF, bZIP, and MYBR1 families, as well as core abiotic stress-responsive genes such as *ERD15, RAB18*, and *LEA14*, were found to be commonly regulated under all four stress conditions. These findings by Zhang et al. (2019) highlight critical molecular targets for developing stress-resilient rapeseed cultivars. High-throughput expression profiling has significantly advanced the understanding of transcriptional crosstalk among abiotic stress responses. Comparative transcriptome analyzes across related plant species offer potential for uncovering deeper interconnections among stress responses. Gene coexpression analysis of four cotton (*Gossypium*) species exposed to salt and cold stress, a total of 29 coexpression modules exhibiting significant associations were identified (Xu et al., 2020). These coexpressed gene clusters were species-specific, as evidenced by strong correlations between four primary modules and the respective genomes of each species. These studies underscored the evolutionary divergence and conserved regulatory networks governing salt and cold stress responses in related species. In a separate study by Dossa et al. (2019), a total of 72 RNA-Seq datasets encompassing osmotic, salt, waterlogging, and drought stresses in sesame (*Sesamum indicum*) were analyzed, leading to the identification of 543 genes that were differentially expressed consistently across all stress conditions. These genes, collectively referred to as core abiotic stress-responsive genes (CARG), represent fundamental components of the plant's stress response machinery. Weighted gene coexpression network analysis identified three functional modules governed by key transcription factors, which exhibited strong correlations with various abiotic stressors, except for salt stress. Network topology analysis identified several hub genes with pivotal roles in stress response pathways. Overexpression of hub and nonhub genes in *Arabidopsis thaliana* provided experimental validation, demonstrating enhanced tolerance to osmotic, drought, and waterlogging stressors (Dossa et al., 2019).

Meta-analysis of transcriptome data, a statistical method that integrates gene expression data from multiple independent transcriptomic studies, provides a more comprehensive and reliable understanding of gene expression patterns across diverse conditions, species, or experimental setups. This approach enhances the statistical power and generalizability of findings, enabling the identification of consistent biological trends, biomarkers, and key molecular pathways that may not be apparent in individual studies. It is particularly valuable for investigating gene responses to environmental factors or stresses. Meta-analysis of transcriptome data can be very helpful in identifying consistent gene expression patterns, thereby optimizing the use of existing datasets. Recently, Azad et al. (2024) conducted a meta-analysis of microarray datasets from rice seedlings to examine the impacts of multiple abiotic stresses, including drought, salinity, cold, heat, alkalinity, heavy metals, and nutrient deficiencies. The study identified 561 DEGs, of which 422 were upregulated and 139 were downregulated. These results offer valuable insights that may support the development of crop varieties with improved tolerance to multiple stresses.

16.4 Proteomics: understanding the functional machinery

16.4.1 Significance of proteomics in plant biology and crop improvement

Proteomics, the large-scale study of proteins, has emerged as a transformative approach in the field of plant biology, offering deep insights into the molecular mechanisms that drive crop growth, development, stress responses, and productivity. It bridges the gap between genomics and functional biology by validating the roles of genes and elucidating complex metabolic pathways, enabling targeted improvements in traits. It can also assist marker-assisted breeding by identifying protein markers linked to desirable traits, making selection processes more precise. Additionally, it unravels the molecular basis of complex traits like nutrient-use efficiency and stress tolerance, which are governed by multiple proteins and pathways.

Proteomic technologies have become indispensable tools in plant biology, particularly for crop improvement, as they enable comprehensive analysis of proteins, their interactions, and functions. Key proteomic technologies include two-dimensional gel electrophoresis (2-DE), which separates proteins according to their isoelectric points and molecular weights, and mass spectrometry (MS) approaches such as matrix-assisted laser desorption ionization-time of flight (MALDI-TOF) and liquid chromatography (LC)-MS/MS, which enable precise protein identification and quantification. Advances in quantitative proteomics, including isobaric tags for relative and absolute quantification (iTRAQ) and tandem mass tags (TMT), have further facilitated comparative analyzes of protein expression across diverse samples and stress conditions. More recently, label-free proteomics has emerged as a powerful high-throughput alternative, enabling protein quantification without the need for chemical labeling. In addition, posttranslational modification analysis, such as phosphorylation and ubiquitination studies, provides insights into protein regulation and signaling mechanisms. Protein–protein interaction studies, often performed using affinity purification-MS or yeast two-hybrid systems, reveal functional networks crucial for understanding plant stress responses and development. Technologies like plant-specific proteome databases and bioinformatics tools aid in the functional annotation of identified proteins, enhancing the understanding of metabolic pathways and trait regulation. These proteomic approaches continue to drive innovations in plant biology, providing critical insights for the development of stress-resilient, high-yielding crops (Fig. 16.3).

16.4.2 Proteomic insights into plant responses to abiotic stresses

Abiotic stresses trigger extensive physiological, molecular, and biochemical changes in plants. To cope with these stresses, plants modulate gene expression, which, in turn, alters protein abundance. Numerous studies now report that several proteins are responsive to such stresses at pretranscriptional, posttranscriptional, and translational levels. Understanding the functional roles of these stress-inducible proteins is essential for comprehensively elucidating the mechanisms underlying stress tolerance in plants.

Proteomics provides a powerful approach to identify proteins and pathways in crops that are linked to physiological responses under stress conditions. During abiotic stresses such as drought, salinity, heat, and cold, plants commonly upregulate stress-responsive proteins, including heat shock proteins (HSPs), late embryogenesis abundant (LEA) proteins, photosynthesis-associated proteins, and antioxidant enzymes. HSPs function as molecular chaperones, preventing protein denaturation and aggregation during heat stress, whereas LEA proteins contribute to stabilizing cellular structures and membranes during dehydration or freezing (Wang et al., 2004). Antioxidant enzymes such as superoxide dismutase (SOD), catalase (CAT), and peroxidases (POD) are crucial for scavenging ROS generated under stress, thereby reducing oxidative damage (Gill and Tuteja, 2010). However, under multistress scenarios, although these general response proteins and pathways are modulated, they often involve complex interactions and overlaps between or among the multiple signaling pathways, posing a challenge for precise protein identification. MS-based methods, such as label-free quantification and stable isotope labeling selective isotope labelling by/with amino acid in cell culture (SILAC), iTRAQ, TMT, are widely used for their high throughput and sensitivity. Gel-based techniques like 2-DE can also provide valuable information, although they may have limitations in terms of dynamic range and throughput. Other methods, including Western blotting, enzyme linked immunosorbent assay (ELISA), and parallel reaction monitoring (PRM), offer targeted approaches for specific protein analysis. The choice of technique depends on factors such as sample complexity, desired sensitivity, and available resources. Rigorous sample preparation, sophisticated data analysis, and independent validation are crucial for obtaining reliable and meaningful results in quantitative proteomics. Real-life examples of proteomic studies that examine plants' responses to multiple abiotic stresses are limited. Nevertheless, several studies have investigated the intricate interactions among stress-responsive proteins, providing valuable insights into plant adaptation under simultaneous challenges such as drought, salinity, and extreme temperatures. In rice, the combination of drought and heat stress has been shown to upregulate HSPs and antioxidant enzymes, including SOD and CAT, which safeguard

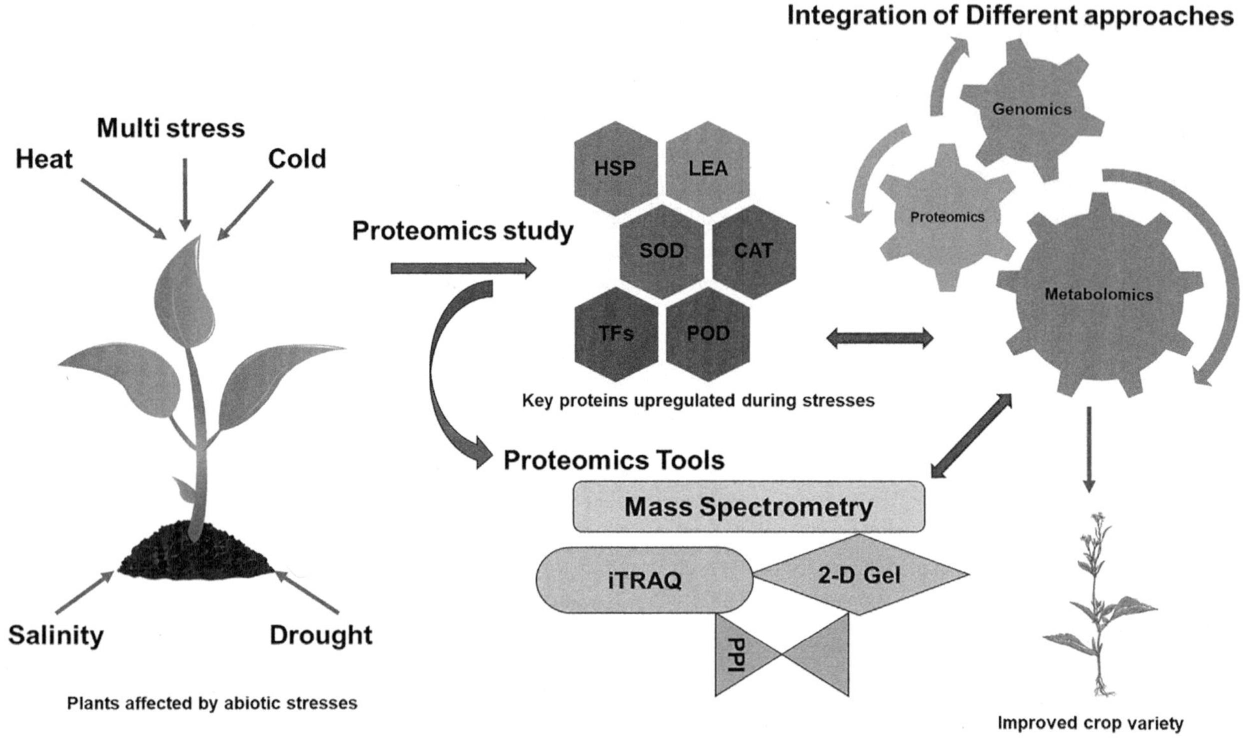

FIGURE 16.3 Schematic representation of a proteomics study to identify key stress-responsive proteins, including heat shock proteins (HSPs), late embryogenesis abundant (LEA) proteins, superoxide dismutase (SOD), catalase (CAT), peroxidase (POD), and transcription factors (TFs). Various proteomics tools, such as mass spectrometry, isobaric tags for relative and absolute quantitation (iTRAQ), two-dimensional gel electrophoresis (2-D Gel), and protein–protein interaction (PPI) analyses, help in identifying the key proteins. The integration of genomics, proteomics, and metabolomics contributes to the development of improved crop varieties with enhanced stress tolerance.

against oxidative damage and support the maintenance of cellular homeostasis (Scafaro et al., 2010). Similarly, in wheat, simultaneous exposure to salinity and heat stress resulted in the induction of LEA proteins and stress-induced chaperones, which stabilize cellular structures and assist in protein folding under stress (Kumar et al., 2017). Proteomic studies in barley under combined salinity and drought stress identified key metabolic enzymes, including adenosine triphosphate (ATP) synthase and glycolytic enzymes, highlighting the role of energy reprogramming in multistress tolerance (Wu et al., 2014). In soybeans, the simultaneous occurrence of cold and drought stress induced differential expression of proteins associated with osmoprotection and ROS scavenging, including dehydrins and glutathione-S-transferases, highlighting the critical role of crosstalk between stress-response pathways (Zhou et al., 2022).

Despite these advancements, identifying proteins uniquely involved in multistress responses remains challenging due to the overlapping nature of stress signaling pathways and the dynamic spatiotemporal regulation of protein expression. High-throughput proteomics, coupled with advanced bioinformatics tools, is essential to disentangle these complexities and identify core proteins and pathways conferring multistress tolerance. Future strategies that integrate proteomics with transcriptomics and metabolomics will facilitate the discovery of both stress-specific and combined responses, thereby supporting the development of crop varieties with improved resilience to multiple abiotic stresses.

16.5 Metabolomics: profiling cellular responses

16.5.1 Utility of metabolomics in predicting plants' responses to environmental cues

Metabolites are the end or intermediate products of biological reactions, with a size range of less than 15 kDa, influencing cellular function and overall organismal processes. The collection of metabolites within a biological system, such as a cell, tissue, or organism, is known as the metabolome. Primary metabolites are fundamental to plant growth and development, playing key roles in cell structure, energy generation, and the storage or transmission of genetic information, and they remain highly conserved across species. In contrast, secondary metabolites are not involved in basic cellular functions but play crucial roles in ecological adaptation and plant defense against biotic and abiotic stresses. These metabolites are unique to certain plant species and contribute to metabolite pool diversity. Metabolomics is the scientific study of metabolites, profiling them in relation to genetic makeup, environmental factors, or a combination of both. Metabolomics, which includes high-throughput profiling of metabolites coupled with comprehensive data analysis, can provide meaningful insights into metabolic perturbations in response to environmental factors such as abiotic stresses (Johnson et al., 2016). Being closest to the phenotype, metabolomics offers valuable insights into the overall expected phenotype of an organism under specific conditions, making it useful for discovering biomarkers associated with specialized states, such as the interaction between multiple abiotic stresses.

16.5.2 Analytical platforms and advances in plant metabolomics

Plant metabolic networks are highly intricate, involving numerous biochemical pathways and the dynamic interplay between primary and specialized metabolites, which are strongly influenced by environmental stresses. To decipher and quantify changes in metabolite composition, precise analytical methods are indispensable. Modern integrated platforms, including nuclear magnetic resonance (NMR) spectroscopy, MS, capillary electrophoresis-MS, LC-MS, gas chromatography-MS, and Fourier transform ion cyclotron resonance MS, are widely employed (Misra, 2021). Advances in MS-imaging technologies—such as MALDI and desorption electrospray ionization—when coupled with high-resolution MS, have enabled in situ metabolome profiling. For liquid-phase separations, ultrahigh-performance liquid chromatography and high-performance liquid chromatography (HPLC) remain powerful approaches, particularly when integrated with MS for comprehensive metabolite analysis (Alseekh et al., 2021; Raza, 2022). Techniques like NMR and HPLC, based on gas chromatography, play a pivotal role in metabolite profiling. Over the past decade, plant metabolomics has evolved from a theoretical concept into a well-established and widely used approach. It is now extensively applied to study metabolites in various crop plants. Given the untargeted nature of many metabolomic methodologies, these studies offer a holistic view of both primary and secondary metabolites. The estimated number of plant metabolites has risen to approximately 200,000, with individual plant species containing between 7000 and 15,000 metabolites, and plant leaves alone hosting 3000–5000 metabolites. This estimate continues to expand as advancements in analytical technologies progress. Selective metabolite profiling extends beyond simple identification to explore the biological characteristics of plants. It involves identifying specialized metabolites that act as markers for physical stimuli, such as abiotic stress, or for physiological processes like growth and development.

16.5.3 Metabolomic insights into plant responses to abiotic and combined stresses

Metabolomic studies in various plants have revealed distinct metabolic responses to different abiotic stress combinations. A comprehensive metabolomics study in *Arabidopsis* by Secomandi et al. (2025) identified varying numbers of differentially accumulated compounds under different stress conditions individually (*viz*., 132 for cold, 71 for heat, 166 for drought, and 259 for salinity) and in combinations (*viz*., 204 compounds for drought × cold, 313 for drought × heat, 244 for salinity × cold, and 274 for salinity × heat). Delving deeper into the effects of these combined stresses revealed interesting patterns. The study noted an enhanced accumulation of flavonoids under drought stress when combined with heat or cold stress, compared to drought stress alone. In contrast, nitrogen-containing alkaloids glucomalcommin, S-magnoflorine, and caffeoylserotonin decreased under the aforementioned combined stresses compared to drought stress alone. This indicates that stress interactions can result in antagonistic, additive, or synergistic effects, depending on the nature and combination of stresses. In wheat seedlings, Shunkao et al. (2023) reported that amino acids, sugars, and their derivatives were the key discriminant metabolites after five days of combined salt stress (150 mM NaCl) and heat stress (42°C for 4 hours) in contrasting cultivars. The enrichment of pathways related to ABC transporters, glucosinolate metabolism, aminoacyl-tRNA biosynthesis, cyanoamino acid metabolism, and galactose metabolism in both cultivars points to a shared metabolic reprogramming strategy for stress mitigation. Using an integrative omics approach, Yue et al. (2024) observed the upregulation of the phenylpropanoid pathway in wheat grown hydroponically, leading to the accumulation of caffeic acid, fumaric acid, and uric acid under cadmium stress, while purine metabolism was overrepresented under salinity stress. Studies in lettuce (*Lactuca sativa* L.) by Qiao et al. (2023) showed that industrial pollutant cadmium and decabromodiphenylethane toxicity enhanced primary metabolic pathways, including amino acid metabolism, carbon metabolism, and ABC transporters. Li et al. (2024) found that combination of cold and saline stress in alfalfa (*Medicago sativa* L.) led to the differential accumulation of various compounds, including sugars raffinose, stachyose, trehalose, sucrose, cellobiose, galactinol, flavonoids (fustin, chrysin, dihydrokaempferol, p-coumaryl alcohol), amino acids (proline, ornithine, lysine, L-citrulline), fatty acids (α-linolenic acid, phosphatidylcholine, linoleic acid, oleic acid), organic acids (succinate, malate), and plant hormones (jasmonic acid, traumatic acid). López-Hidalgo et al. (2023) reported that high-intensity temperature and water stress reprogrammed the metabolic pathways, leading to the prominent accumulation of alkaloids, lignans, and glycosyloxyflavones. This study also reported succinic acid as a marker metabolite linked with poor seed quality in rice. Metabolomics-derived biomarkers enable rapid, reliable, and real-time monitoring of the physiological state of plants under specific environmental conditions. High-throughput metabolic pathway analysis provides a comprehensive understanding of stress-responsive pathways and the stress response mechanisms of specific genotypes. This information can be applied in metabolome-assisted breeding for stress tolerance (Fig. 16.4). This strategy harnesses the potential of metabolomics to identify or characterize metabolites that indicate better tolerance to environmental stresses, such as drought, heat, salinity, cold, or their combinations, and uses this knowledge to develop stress-resilient crop varieties.

16.6 Phenomics: linking genotype to phenotype

16.6.1 The phenotyping bottleneck in modern crop improvement and advances in high-throughput phenotyping technologies

Revolution in the field of genomics by the advent of high-throughput and cost-effective genome sequencing technologies has made a limited impact so far on crop improvement due to the present limitation in generation and storage of high-quality phenotypic data. The bottleneck in plant breeding has shifted from genotyping to phenotyping, where traditional methods are low-throughput and hinder the understanding of genotype-phenotype correlations and environmental interactions. Overcoming this phenotyping bottleneck is critical for progress in modern plant breeding. High throughput phenotyping (HTP) enhances the process of rapid and large-scale phenotyping by automated imaging-mediated phenotypic data collection – paving the way for phenomics. Phenomics is defined as the study of the phenome, encompassing all possible phenotypes influenced by environmental and developmental factors (Zavafer et al., 2023). Phenomics, like other "omics" fields, aims to investigate the entirety of phenotypic possibilities for a given genotype.

A key advancement in HTP is the capacity to nondestructively capture plant traits, allowing time-series measurements that monitor growth and stress responses in individual plants over time. High-throughput image-based phenotyping encompasses technologies capable of imaging interactions among hundreds or more plants per day (Fahlgren et al., 2015). This scale enables the analysis of mutant populations, QTL detection, and the identification of gene-environment interactions. HTP pipelines help in transforming multidimensional images and sensor data into knowledge, which necessitates big data analysis by artificial intelligence (AI) tools, thus helping to accelerate the selection of climate-resilient next-generation crops. Advancements in imaging and sensor technologies enable the acquisition of high-

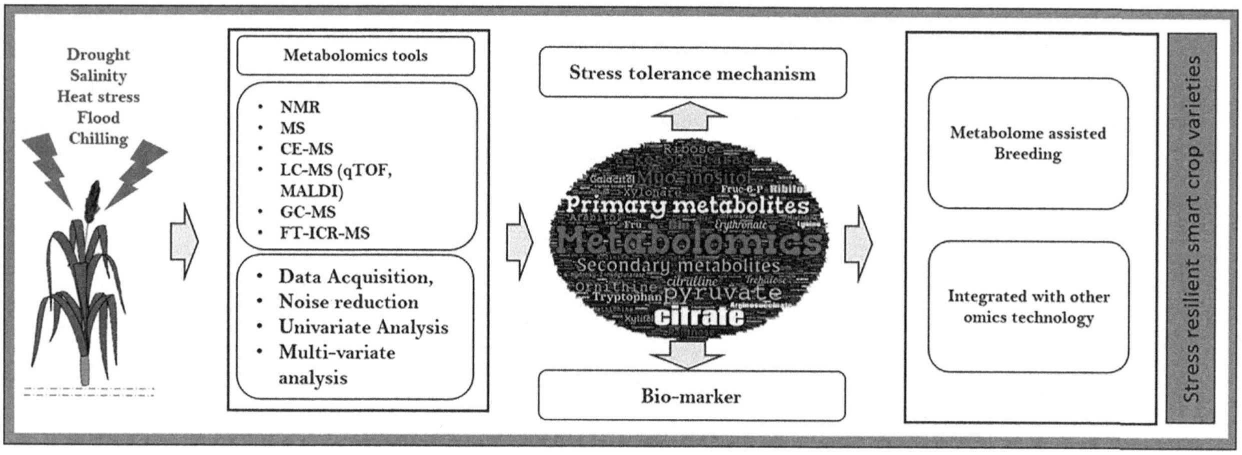

FIGURE 16.4 Schematic representation of metabolomics-led approaches for the development of stress-resilient crop varieties. High-throughput metabolite data generated on NMR and MS-based platforms need massive data processing and analysis to derive useful information, potentially providing stress tolerance mechanisms or unique metabolites (biomarkers) linked with specific stress. Metabolomics information could be integrated with other omics technologies for systematic understanding of stress physiology and for metabolomics-assisted breeding for the development of climate-smart stress resilient varieties.

dimensional phenotypic data, which requires advanced "computational ecosystems" supported by machine learning (ML) or deep learning (DL) tools, which are crucial for deriving meaningful insights from these multidimensional datasets. The integration of AI, including ML/DL, has considerable potential to translate biological knowledge and omics data into precisely designed plant breeding strategies (Fig. 16.5). This is achieved by understanding the mechanisms of trait regulation and identifying target genes, facilitating knowledge-driven molecular design breeding (Varshney et al., 2021).

16.6.2 Potential applications of high-throughput phenotyping technologies in understanding multistress tolerance

As far as we are aware, no specific report exists that has studied the interaction of more than one abiotic stress through plant phenomics. However, phenomics can be applied in such studies in the future. The roadmap involves techniques such as genomic selection, phenomic selection (PS), and Phenome-wide association studies (pheWAS). Genomic selection (GS) predicts breeding values by using dense molecular markers to estimate both the parental average and the Mendelian sampling effect, which represents progeny deviation from this average. Unlike traditional breeding, which relies on pedigree information and multienvironment trials to assess these components, GS quantifies Mendelian sampling without extensive phenotyping. This accelerates breeding cycles, enhances genetic gain, and reduces resource use, especially for traits with low heritability. GS can also improve the accuracy of selection for both complex and simple traits. Integrating phenomics with modern genomics techniques has significantly accelerated genetic gain in numerous plant breeding programs (Masuka et al., 2017; Araus and Cairns, 2014). Multitrait genomic selection models exploit correlations among traits derived from HTP to improve prediction accuracy, as demonstrated in maize (Anche et al., 2020) and wheat (Kaushal et al., 2024). Phenomic selection (PS), proposed by Rincent et al. (2018), follows principles similar to GS but relies on near-infrared (NIR) spectroscopy rather than genotyping. Traditionally, NIR spectroscopy can predict the chemical composition of tissue. In PS, NIR spectra capture genetic similarities between genotypes, enabling predictions of polygenic traits. The spectra, in these cases, serve as proxies for complex traits like grain yield, water stress, or chlorophyll content using vegetation indices such as Normalized Difference Vegetation Index. This cost-effective, high-throughput

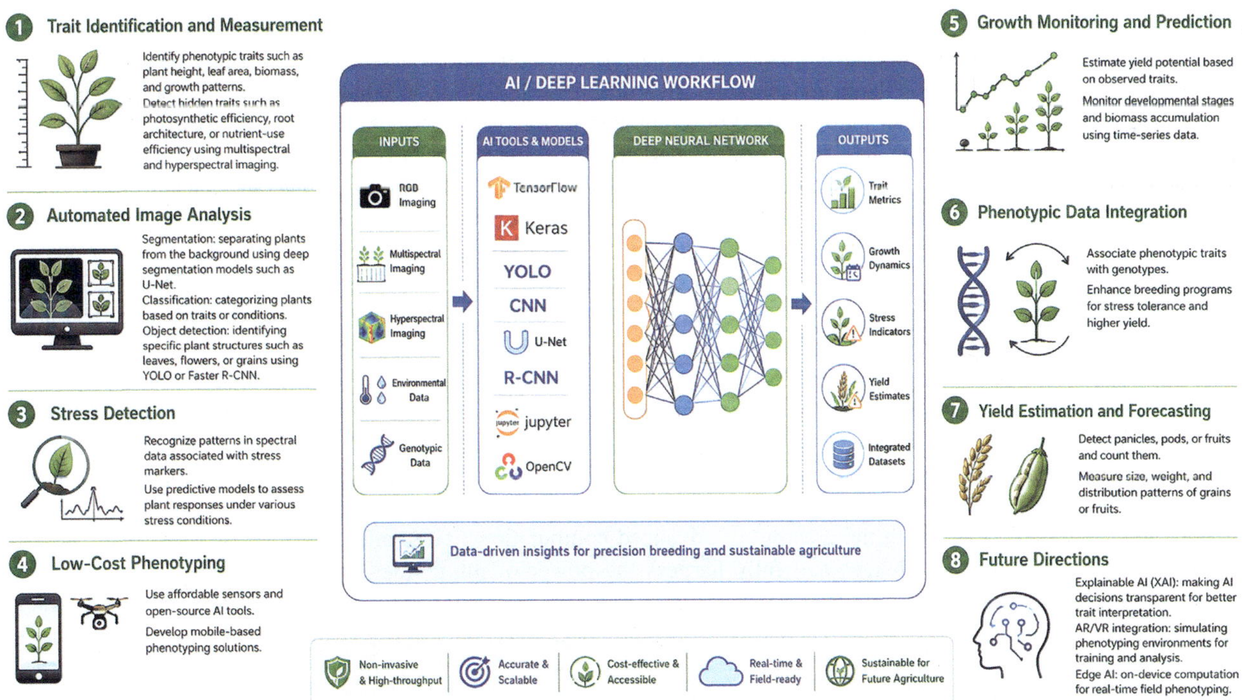

FIGURE 16.5 Schematic representation of AI-driven approaches for plant phenotyping, including automated image analysis (segmentation, classification, and object detection), disease detection, stress detection, and yield estimation. Deep learning models (Convolutional Neural Networks, U-Net, R-CNN, YOLO, etc.) enable trait extraction and growth prediction. Phenotypic data integration supports genotype-trait analysis, while low-cost AI tools enhance mobile-based phenotyping. Future directions include explainable AI, AR/VR simulations, and edge AI for real-time applications.

method captures nonadditive effects and predicts polygenic traits by assessing genetic similarities between genotypes. These indices, linked to photosynthetic activity and yield, enable indirect trait selection, reducing the need for expensive phenotyping. PS has been successfully applied in wheat (Cuevas et al., 2019; Robert et al., 2022) and other species. PheWAS serve as a complement to GWAS by assessing the relationships between genetic variants and a broad spectrum of phenotypic traits, whereas GWAS typically targets only a limited number of traits. By integrating and correlating diverse phenotypic datasets, PheWAS enables a more comprehensive exploration of genotype–phenotype associations (Denny et al., 2010; Pendergrass et al., 2013; Bhat et al., 2015).

16.7 Perspectives and future directions

The true power of multiomics approaches lies in their integration. By combining data from different "omics" platforms, researchers can develop a more comprehensive and systems-level understanding of stress tolerance mechanisms. For example, integrating transcriptomics and proteomics data can provide insights into the posttranslational regulation of gene expression, while integrating metabolomics data along with it can further reveal the metabolic consequences of changes in gene and protein expression. By integrating all of these data sources, researchers can begin to construct a more complete picture of the complex molecular networks that govern plant responses to stress. The rapid advancements in multiomics technologies are transforming our ability to understand and improve plant resilience to abiotic stresses. The knowledge gained through multiomics research can be applied directly to crop improvement efforts. By identifying key genes and pathways associated with stress tolerance, researchers can develop new breeding strategies to improve crop resilience. These advanced strategies may include genomic selection, genome editing, and transgenic approaches. Genomic selection, an advancement of marker-assisted selection, combines advanced computational models with high-throughput genotyping. This approach enhances breeding efficiency by accelerating genetic gains and increasing the precision of crop improvement programs. Genome editing technologies allow targeted modifications of specific genes, enabling the creation of novel alleles and allelic combinations. What was once considered a distant goal—making precise changes at desired loci in the target genome—has now become a reality, facilitating the incorporation of novel traits into crop varieties. Additionally, transgenic approaches, when applied judiciously, provide a powerful means of introducing genes from other organisms into crops, thereby offering new opportunities for engineering stress tolerance.

Today, the "omics" field is experiencing a "data deluge," where the volume of data generated through "omics" innovations exceeds our ability to manage, store, and analyze it effectively. AI tools can potentially provide deeper and less-biased insights from the "big data" generated through different "omics" tools compared to traditional methods (Yang et al., 2021). AI technologies—including ML, DL, computer vision, and robotics—are highly effective for analyzing large-scale datasets and identifying patterns that support the selection of superior plant varieties with desirable traits. Recent progress in AI has been fueled by advancements in ML and DL algorithms, along with enhancements in data processing and computational power. ML is a subset of AI that concentrates on creating algorithms capable of learning from available data and applying the knowledge to uncover hidden patterns and make decisions. DL, a further advanced branch of ML, utilizes deep neural networks and can tackle even more intricate biological problems. The use of ML and DL in plant breeding can accelerate the development of climate-resilient cultivars with enhanced tolerance to multiple stresses, enabling adaptation to both simultaneous and sequential abiotic stress exposures.

While multiomics approaches offer tremendous potential for crop improvement, several challenges remain. One major challenge is the complexity of analyzing and integrating large and diverse datasets. Developing robust computational tools and bioinformatics pipelines is crucial for effectively analyzing and interpreting multiomics data. Another challenge is the cost and accessibility of high-throughput technologies, which can limit the adoption of multiomics approaches in many research settings. Despite these challenges, the future of multiomics research in crop improvement seems bright. Ongoing technological progress, along with the creation of advanced computational tools and bioinformatics pipelines, will help researchers address current challenges and fully harness the power of multiomics approaches. Integrating data across diverse "omics" platforms will provide deeper insights into the molecular mechanisms of stress tolerance, thereby facilitating the development of climate-resilient crops to support global food security in the face of a growing population.

References

Ahmad, R., Anjum, M.A., Balal, R.M., 2020. From markers to genome-based breeding in horticultural crops: an overview. Phyton—International Journal of Experimental Botany 89 (2), 184–199.

Alseekh, S., Aharoni, A., Brotman, Y., Contrepois, K., D'Auria, J., Ewald, J., et al., 2021. Mass spectrometry-based metabolomics: a guide for annotation, quantification and best reporting practices. Nature Research, Germany Nature Methods 18 (7), 747–756. Available at https://doi.org/10.1038/s41592-021-01197-1.

Anche, M.T., Kaczmar, N.S., Morales, N., Clohessy, J.W., Ilut, D.C., Gore, M.A., et al., 2020. Temporal covariance structure of multi-spectral phenotypes and their predictive ability for end-of-season traits in maize. Theoretical and Applied Genetics 133 (10), 2853–2868. Available at https://doi.org/10.1007/s00122-020-03637-6.

Araus, J.L., Cairns, J.E., 2014. Field high-throughput phenotyping: the new crop breeding frontier. Trends in Plant Science 19 (1), 52–61.

Azad, M., Tohidfar, M., Seraj, R.G.M., Mehralian, M., Esmaeilzadeh-Salestani, K., 2024. Identification of responsive genes to multiple abiotic stresses in rice (Oryza sativa): a meta-analysis of transcriptomics data. Scientific Reports 14 (1). Available at https://doi.org/10.1038/s41598-024-54623-7.

Bhat, J.A., Salgotra, R.K., Dar, M.Y., 2015. Phenomics: a challenge for crop improvement in genomic era. Molecular Plant Breeding 6 (22).

Chakraborty, K., Ray, S., Vijayan, J., Molla, K.A., Nagar, R., Jena, P., et al., 2021. Preformed aerenchyma determines the differential tolerance response under partial submergence imposed by fresh and saline water flooding in rice. Physiologia Plantarum 173 (4), 1597–1615. Available at https://doi.org/10.1111/ppl.13536.

Challa, S., Neelapu, N.R., 2018. Genome-wide association studies (GWAS) for abiotic stress tolerance in plants. Biochemical, Physiological and Molecular Avenues for Combating Abiotic Stress Tolerance In Plants. Academic Press, pp. 135–150.

Cuevas, J., Montesinos-LÓpez, O., Juliana, P., GuzmÁn, C., PÉrez-RodrÍguez, P., GonzÁlez-Bucio, J., et al., 2019. Deep kernel for genomic and near infrared predictions in multi-environment breeding trials. G3: Genes|Genomes|Genetics 9 (9), 2913–2924. Available at https://doi.org/10.1534/g3.119.400493.

Denny, J.C., Ritchie, M.D., Basford, M.A., Pulley, J.M., Bastarache, L., Brown-Gentry, K., et al., 2010. PheWAS: demonstrating the feasibility of a phenome-wide scan to discover gene-disease associations. Bioinformatics 26 (9), 1205–1210. Available at https://doi.org/10.1093/bioinformatics/btq126.

Dossa, K., Mmadi, M.A., Zhou, R., Zhang, T., Su, R., Zhang, Y., et al., 2019. Depicting the core transcriptome modulating multiple abiotic stresses responses in sesame (Sesamum indicum L.). International Journal of Molecular Sciences 20 (16), 3930. Available at https://doi.org/10.3390/ijms20163930.

Fahlgren, N., Gehan, M.A., Baxter, I., 2015. Lights, camera, action: high-throughput plant phenotyping is ready for a close-up. Current Opinion in Plant Biology 24, 93–99.

Gill, S.S., Tuteja, N., 2010. Reactive oxygen species and antioxidant machinery in abiotic stress tolerance in crop plants. Plant Physiology and Biochemistry 48 (12), 909–930. Available at https://doi.org/10.1016/j.plaphy.2010.08.016.

Itam, M.O., Mega, R., Gorafi, Y.S.A., Yamasaki, Y., Tahir, I.S.A., Akashi, K., et al., 2022. Genomic analysis for heat and combined heat–drought resilience in bread wheat under field conditions. Theoretical and Applied Genetics 135 (1), 337–350. Available at https://doi.org/10.1007/s00122-021-03969-x.

Jackson, R., Buntjer, J.B., Bentley, A.R., Lage, J., Byrne, E., Burt, C., et al., 2023. Phenomic and genomic prediction of yield on multiple locations in winter wheat. Frontiers in Genetics 14, 1164935.

Jeknić, Z., Pillman, K.A., Dhillon, T., Skinner, J.S., Veisz, O., Cuesta-Marcos, A., et al., 2014. Hv-CBF2A overexpression in barley accelerates COR gene transcript accumulation and acquisition of freezing tolerance during cold acclimation. Plant Molecular Biology 84, 67–82.

Jha, U.C., Nayyar, H., Parida, S.K., Bakır, M., von Wettberg, E.J.B., Siddique, K.H.M., 2022. Progress of genomics-driven approaches for sustaining underutilized legume crops in the post-genomic era. Frontiers in Genetics 13. Available at https://doi.org/10.3389/fgene.2022.831656.

Johnson, C.H., Ivanisevic, J., Siuzdak, G., 2016. Metabolomics: beyond biomarkers and towards mechanisms. Nature Reviews Molecular Cell Biology 17 (7), 451–459.

Kaushal, S., Gill, H.S., Billah, M.M., Khan, S.N., Halder, J., Bernardo, A., et al., 2024. Enhancing the potential of phenomic and genomic prediction in winter wheat breeding using high-throughput phenotyping and deep learning. Frontiers in Plant Science 15. Available at https://doi.org/10.3389/fpls.2024.1410249.

Krishna, T.P.A., Maharajan, T., Ceasar, S.A., 2022. Improvement of millets in the post-genomic era. Physiology and Molecular Biology of Plants 28 (3), 669–685.

Kumar, S., Beena, A.S., Awana, M., Singh, A., 2017. Physiological, biochemical, epigenetic and molecular analyses of wheat (*Triticum aestivum*) genotypes with contrasting salt tolerance. Frontiers in Plant Science 8, 1151.

Kumar, J., Sen Gupta, D., Djalovic, I., 2020. Breeding, genetics, and genomics for tolerance against terminal heat in lentil: current status and future directions. Legume Science 2 (3), e38.

Li, Huang, Hu, Xiao, Lovell, John T., Grabowski, Paul P., Mamidi, Sujan, Chen, Cindy, Amirebrahimi, Mojgan, Kahanda, Indika, Mumey, Brendan, Barry, Kerrie, Kudrna, David, Schmutz, Jeremy, Lachowiec, Jennifer, Lu, Chaofu, 2021. Genetic dissection of natural variation in oilseed traits of camelina by whole-genome resequencing and QTL mapping. The Plant Genome 14. Available at https://doi.org/10.1002/tpg2.20110.

Li, H., Che, R., Zhu, J., Yang, X., Li, J., Fernie, A.R., et al., 2024. Multi-omics-driven advances in the understanding of triacylglycerol biosynthesis in oil seeds. Plant Journal 117 (4), 999–1017. Available at https://doi.org/10.1111/tpj.16545.

Li, D., Dossa, K., Zhang, Y., Wei, X., Wang, L., Zhang, Y., et al., 2018. GWAS uncovers differential genetic bases for drought and salt tolerances in sesame at the germination stage. Genes 9 (2), 87. Available at https://doi.org/10.3390/genes9020087.

LÓpez-Hidalgo, C., Lamelas, L., CaÑal, M.J., Valledor, L., MeijÓn, M., 2023. Untargeted metabolomics revealed essential biochemical rearrangements towards combined heat and drought stress acclimatization in Pinus pinaster. Environmental and Experimental Botany 208, 105261. Available at https://doi.org/10.1016/j.envexpbot.2023.105261.

Masuka, B., Magorokosho, C., Olsen, M., Atlin, G.N., BÄnziger, M., Pixley, K.V., et al., 2017. Gains in maize genetic improvement in eastern and southern Africa: II. CIMMYT open-pollinated variety breeding pipeline. Crop Science 57 (1), 180–191. Available at https://doi.org/10.2135/cropsci2016.05.0408.

Misra, B.B., 2021. New software tools, databases, and resources in metabolomics: updates from 2020. Metabolomics 17 (5). Available at https://doi.org/10.1007/s11306-021-01796-1.

Ni, L., Wang, Z., Guo, J., Pei, X., Liu, L., Li, H., et al., 2021. Full-length transcriptome sequencing and comparative transcriptome analysis to evaluate drought and salt stress in iris lactea var. chinensis. Genes 12 (3), 434. Available at https://doi.org/10.3390/genes12030434.

Ojiewo, C., Monyo, E., Desmae, H., Boukar, O., Mukankusi-Mugisha, C., Thudi, M., et al., 2019. Genomics, genetics and breeding of tropical legumes for better livelihoods of smallholder farmers. Plant Breeding 138 (4), 487–499. Available at https://doi.org/10.1111/pbr.12554.

Osuman, A.S., Badu-Apraku, B., Ifie, B.E., Nelimor, C., Tongoona, P., Obeng-Bio, E., et al., 2022. Combining ability and heterotic patterns of tropical early-maturing maize inbred lines under individual and combined heat and drought environments. Plants 11 (10). Available at https://doi.org/10.3390/plants11101365.

Paape, T., Heiniger, B., Santo Domingo, M., Clear, M.R., Lucas, M.M., Pueyo, J.J., 2022. Genome-wide association study reveals complex genetic architecture of cadmium and mercury accumulation and tolerance traits in Medicago truncatula. Frontiers in Plant Science 12. Available at https://doi.org/10.3389/fpls.2021.806949.

Pendergrass, S.A., Brown-Gentry, K., Dudek, S., Frase, A., Torstenson, E.S., Goodloe, R., et al., 2013. Phenome-wide association study (PheWAS) for detection of pleiotropy within the population architecture using genomics and epidemiology (PAGE) network. PLoS Genetics 9 (1). Available at https://doi.org/10.1371/journal.pgen.1003087.

Qiao, Z., Luo, K., Zhou, S., Fu, M., Shao, X., Gong, K., et al., 2023. Response mechanism of lettuce (Lactuca sativa L.) under combined stress of Cd and DBDPE: an integrated physiological and metabolomics analysis. Science of The Total Environment 887, 164204. Available at https://doi.org/10.1016/j.scitotenv.2023.164204.

Raza, A., 2022. Metabolomics: a systems biology approach for enhancing heat stress tolerance in plants. Plant Cell Reports 41 (3), 741–763. Available at https://doi.org/10.1007/s00299-020-02635-8.

Rincent, R., Charpentier, J.P., Faivre-Rampant, P., Paux, E., Le Gouis, J., Bastien, C., et al., 2018. Phenomic selection is a low-cost and high-throughput method based on indirect predictions: proof of concept on wheat and poplar. G3: Genes, Genomes, Genetics 8 (12), 3961–3972. Available at https://doi.org/10.1534/g3.118.200760.

Robert, P., Goudemand, E., Auzanneau, J., Oury, F.-X., Rolland, B., Heumez, E., et al., 2022. Phenomic selection in wheat breeding: prediction of the genotype-by-environment interaction in multi-environment breeding trials. Theoretical and Applied Genetics 135 (10), 3337–3356. Available at https://doi.org/10.1007/s00122-022-04170-4.

Sandhu, K.S., Merrick, L.F., Sankaran, S., Zhang, Z., Carter, A.H., 2022. Prospectus of genomic selection and phenomics in cereal, legume and oilseed breeding programs. Frontiers in Genetics 12. Available at https://doi.org/10.3389/fgene.2021.829131.

Scafaro, A.P., Haynes, P.A., Atwell, B.J., 2010. Physiological and molecular changes in *Oryza meridionalis* Ng., a heat-tolerant species of wild rice. Journal of Experimental Botany 61 (1), 191–202.

Secomandi, E., De Gregorio, M.A., Castro-CegrÍ, A., Lucini, L., 2025. Biochemical, photosynthetic and metabolomics insights of single and combined effects of salinity, heat, cold and drought in Arabidopsis. Physiologia Plantarum 177 (1). Available at https://doi.org/10.1111/ppl.70062.

Shin, S.Y., Shin, C., 2016. Regulatory non-coding RNAs in plants: potential gene resources for the improvement of agricultural traits. Plant Biotechnology Reports 10 (2), 35–47. Available at https://doi.org/10.1007/s11816-016-0389-4.

Shunkao, S., Theerakulpisut, P., Wanichthanarak, K., Pongdontri, P., Thitisaksakul, M., 2023. Integrative physiological and metabolomics study reveals adaptive strategies of wheat seedlings to salt and heat stress combination. Plant Growth Regulation 100 (1), 181–196. Available at https://doi.org/10.1007/s10725-022-00949-z.

Siddiqui, M.N., LeÓn, J., Naz, A.A., Ballvora, A., 2021. Genetics and genomics of root system variation in adaptation to drought stress in cereal crops. Journal of Experimental Botany 72 (4), 1007–1019. Available at https://doi.org/10.1093/jxb/eraa487.

Varshney, R.K., Bohra, A., Yu, J., Graner, A., Zhang, Q., Sorrells, M.E., 2021. DesigninG Future Crops: Genomics-assisted Breeding Comes Of Age. Trends in Plant Science 26 (6), 631–649. Available at https://doi.org/10.1016/j.tplants.2021.03.010.

Walkowiak, S., Gao, L., Monat, C., Haberer, G., Kassa, M.T., Brinton, J., et al., 2020. Multiple wheat genomes reveal global variation in modern breeding. Nature 588 (7837), 277–283. Available at https://doi.org/10.1038/s41586-020-2961-x.

Wang, W., Vinocur, B., Shoseyov, O., Altman, A., 2004. Role of plant heat-shock proteins and molecular chaperones in the abiotic stress response. Trends in Plant Science 9 (5), 244–252. Available at https://doi.org/10.1016/j.tplants.2004.03.006.

Wu, D., Shen, Q., Qiu, L., Han, Y., Ye, L., Jabeen, Z., et al., 2014. Identification of proteins associated with ion homeostasis and salt tolerance in barley. Proteomics 14 (11), 1381–1392. Available at https://doi.org/10.1002/pmic.201300221.

Xu, Y., Magwanga, R.O., Jin, D., Cai, X., Hou, Y., Juyun, Z., et al., 2020. Comparative transcriptome analysis reveals evolutionary divergence and shared network of cold and salt stress response in diploid D-genome cotton. BMC Plant Biology 20 (1). Available at https://doi.org/10.1186/s12870-020-02726-4.

Yang, Y., Saand, M.A., Huang, L., Abdelaal, W.B., Zhang, J., Wu, Y., et al., 2021. Applications of multi-omics technologies for crop improvement. Frontiers in Plant Science 12, 563953.

Yue, Z., Liu, Y., Zheng, L., Zhang, Q., Wang, Y., Hao, Y., et al., 2024. Integrated transcriptomics, metabolomics and physiological analyses reveal differential response mechanisms of wheat to cadmium and/or salinity stress. Frontiers in Plant Science 15. Available at https://doi.org/10.3389/fpls.2024.1378226.

Zavafer, A., Bates, H., Mancilla, C., Ralph, P.J., 2023. Phenomics: conceptualization and importance for plant physiology. Trends in Plant Science 28 (9), 1004–1013. Available at https://doi.org/10.1016/j.tplants.2023.03.023.

Zhang, Y., Ali, U., Zhang, G., Yu, L., Fang, S., Iqbal, S., et al., 2019. Transcriptome analysis reveals genes commonly responding to multiple abiotic stresses in rapeseed. Molecular Breeding 39 (10-11). Available at https://doi.org/10.1007/s11032-019-1052-x.

Zhou, Y., Li, H., Chen, H., Yang, X., Yu, T., Wang, Y., et al., 2022. Proteomic investigation of molecular mechanisms in response to PEG-induced drought stress in soybean roots. Plants 11 (9), 1173.

Further reading

Liu, L., Si, L., Zhang, L., Guo, R., Wang, R., Dong, H., et al., 2024. Metabolomics and transcriptomics analysis revealed the response mechanism of alfalfa to combined cold and saline-alkali stress. The Plant Journal 119 (4), 1900–1919.

Robert, G., Gosselin, R., 2022. Evaluating the impact of NIR pre-processing methods via multiblock partial least-squares. Analytica Chimica Acta 1189, 339255.

Section D

Management strategies to overcome multiple abiotic stresses in plants

Chapter 17

Management of multiple abiotic stresses through improved agronomic practices

Shanmugam Vijayakumar[1,2,*], B. Raghavendra Goud[3], Pandian Kannan[4], Ashish Srivastava[2], Rajendran Gobinath[1], Virender Kumar[2], and Sudhanshu Singh[2]

[1]*ICAR-Indian Institute of Rice Research, Hyderabad, Telangana, India,* [2]*International Rice Research Institute, Los Banos, Laguna, Philippines,* [3]*ICAR-Central Rice Research Institute, Cuttack, Odisha, India,* [4]*Tamil Nadu Agricultural University, Coimbatore, Tamil Nadu, India*

[*]*Corresponding author. e-mail address: vijayakumar.s@icar.org.in*

17.1 Introduction

Agriculture, the backbone of human civilization, faces unprecedented challenges in the 21st century. Among these, abiotic stress, such as drought, flood, salinity, sodicity, acidity, extreme temperatures, and nutrient deficiencies, has emerged as a formidable obstacle to sustainable food production (Fig. 17.1). The Food and Agriculture Organization (FAO) estimates that abiotic stresses contribute to a 20%–30% reduction in global crop yields annually. For developing countries, where agriculture is a primary livelihood, this equates to billions of dollars in lost income and increased vulnerability to food insecurity. Additionally, the cumulative effect of abiotic stresses on soil degradation further compounds long-term productivity challenges (Chary et al., 2021). Abiotic stress affects crops across all agroecological zones, with varying intensity and frequency.

17.1.1 Moisture stress-induced multiple abiotic stresses in agriculture

Reduced groundwater availability, inefficient irrigation methods, altered rainfall patterns, inadequate drainage facilities, and increased runoff from extreme weather events contribute to moisture stress, which encompasses both deficit and excess moisture conditions in agriculture. Moisture deficit often induces combined drought-salinity stress, combined drought-heat stress, whereas excess moisture often induces combined waterlogging–salinity stress, combined waterlogging-nutrient deficiency, and combined waterlogging-nutrient toxicity stress. This complex multiple abiotic stress affects plant growth, development, and yield more than their individual stress (Khan et al., 2025). Crops under deficit and excess moisture stress also become more susceptible to pests and diseases, further compromising resilience.

According to the World Health Organization, drought or deficit moisture is one of the most pervasive stressors, affecting approximately 55 million people annually and causing substantial yield losses in major staples like wheat, rice, and maize. For instance, drought alone accounts for annual losses of up to 60% in maize production in sub-Saharan Africa (Du and Xiong, 2024). Drought stress frequently occurs in rainfed agriculture, which comprises 80% of global cropland, produces 60% of the world's cereal grains, and supports 100 million farm families. Globally, ~59 M ha (45% of the total rice area) is rainfed and experiences moisture stress at least once during the crop cycle (Rao et al., 2015). Rainfall variability is the primary cause of both drought and reduced crop yields. People dependent on rainfed agriculture are highly vulnerable to both short-term dry spells and long-term droughts. Moreover, changing precipitation patterns and increasing temperatures resulting from climate change exacerbate these challenges for small farmers (WWAP, 2009).

Drought hampers photosynthesis by limiting CO_2 uptake, damaging photosystems, and causing reactive oxygen species (ROS) stress. Plants adapt via nonphotochemical quenching, alternative electron pathways, hormonal regulation, and structural changes, key to developing drought-resistant crops (Qiao et al., 2024). For example, drought stress has been shown to reduce the productivity of wheat, rice, maize, and chickpea by 57%, 22%–92%, 17%–60%, and 45%–69%, respectively (Daryanto et al., 2017). Severe drought during sorghum anthesis can reduce yields by up to 2 t ha^{-1}. Drought

Multiple Abiotic Stresses in Plants. DOI: https://doi.org/10.1016/B978-0-443-27424-4.00015-X

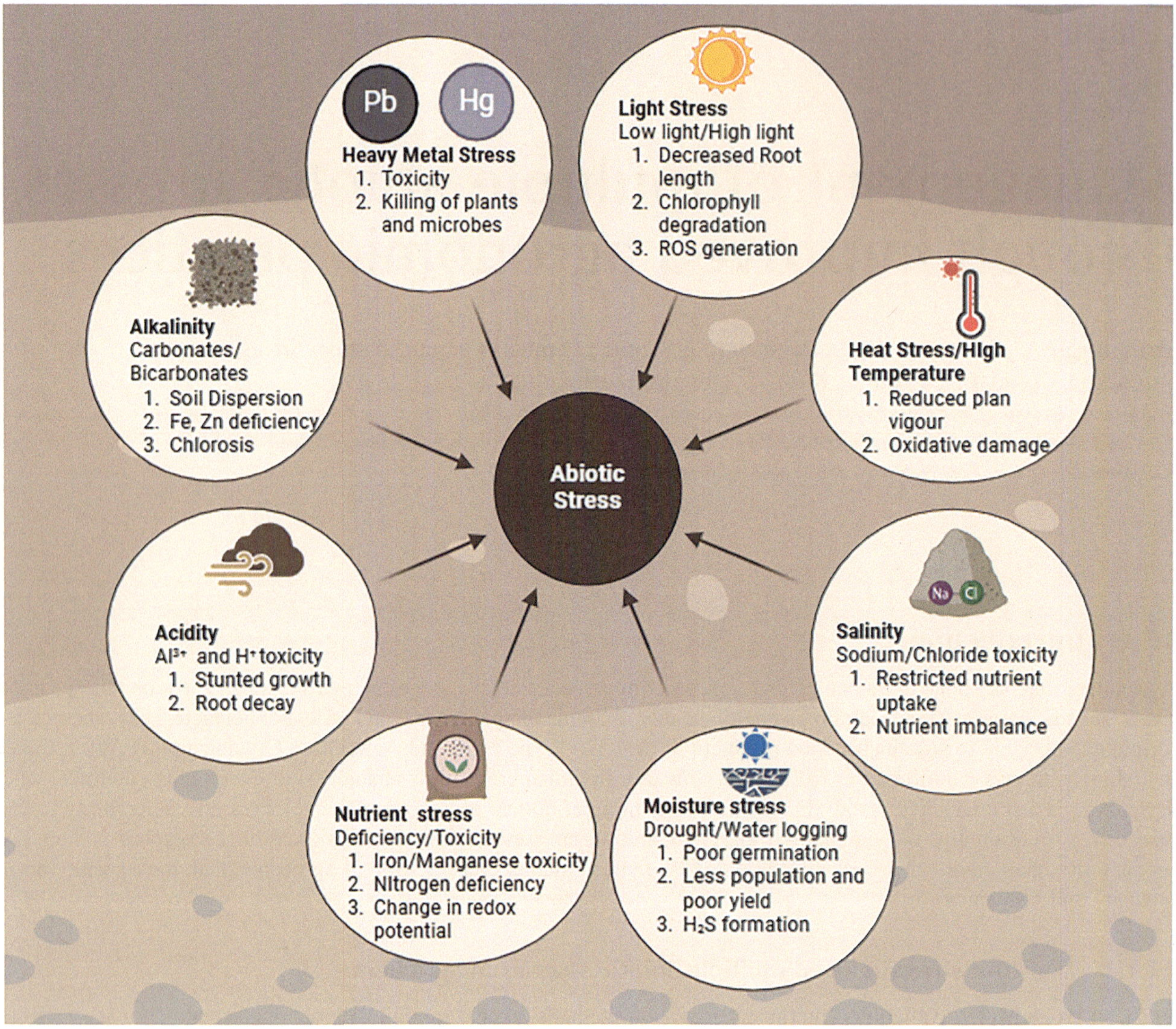

FIGURE 17.1 Major abiotic stresses and their impact on crops.

and heat stress often occur together, particularly in arid and semiarid regions. High temperatures exacerbate the effects of drought by increasing evapotranspiration and reducing soil moisture content. For example, extreme temperatures, along with drought, particularly during flowering and grain-filling stages, reduce wheat yields by 40% in temperate regions. Under drought conditions, plants close their stomata to conserve water, reducing CO_2 uptake and intercellular CO_2 concentrations, which in turn lowers photosynthetic rates (Lawlor and Tezara, 2009). In maize, the combined effects of drought and heat stress resulted in yield reduction, altered flowering time, and reduced kernel size. Similarly, in soybeans, these stresses led to reduced seed filling, lower seed yield, and decreased photosynthetic activity. In chickpea, flower drop, reduced seed yield, and decreased pollen viability were reported under drought and heat conditions. In pepper, the interaction of drought and high temperature caused reduced flowering, enhanced oxidative stress, and accelerated senescence (Khan et al., 2025). Drought also affects biochemical processes within chloroplasts, reducing enzyme activity, such as Rubisco, and damaging the thylakoid membrane (Qiao et al., 2024)

Moisture deficit induces oxidative stress in plants, leading to the production of harmful ROS such as superoxide radicals, hydrogen peroxide, and hydroxyl radicals (Gill and Tuteja, 2010). These ROS damage cellular components, leading to reduced photosynthesis and growth. Plants have antioxidant defense mechanisms to combat ROS, but severe drought can overwhelm these systems. Drought stress significantly reduces crop yield by affecting various yield components like grain number, size, and weight (Chaves et al., 2003). Average rice productivity in drought-prone areas is very low, averaging less than 2.5 t ha^{-1} in rainfed lowlands, even much lower and unstable in flood-prone areas (1.5 t ha^{-1}) as compared to the input-intensive irrigated system (5.0 t ha^{-1}) (Ismail et al., 2013)

Besides drought, excess moisture or submergence stress caused by flooding or waterlogging severely impacts plant growth and productivity by creating anaerobic root-zone conditions, hindering root respiration and nutrient absorption (Kumar et al., 2021). This condition triggers various multiple abiotic stresses, including hypoxia, nutrient imbalances, and toxic accumulation (Fig. 17.2). Waterlogging limits oxygen availability to roots, disrupting cellular respiration and energy production. It also impairs nutrient cycles, increasing nitrogen (N) loss through denitrification, leaching, and runoff while reducing N mineralization rates (Behera et al., 2019). Factors such as extreme precipitation events, degraded soil structure, soil hardpan from heavy machinery, and high clay content exacerbate this challenge. The extent of damage depends on the plant's tolerance to hypoxia, soil properties, and waterlogging duration.

Based on water stagnation duration, the floods are broadly categorized into flash floods and stagnant floods. In India, 12–14 M ha of rice-growing area is prone to flash flooding, resulting in moderate to severe crop losses. The average rice productivity of these areas is <1.5 t ha^{-1} (Dar et al., 2017). Prolonged submergence alters nutrient availability by reducing soil redox potential and increasing percolation, leaching, and denitrification losses, leading to deficiencies. For example, in flooded soils, P availability increases while Zn decreases (Ismail et al., 2007). It also leads to the accumulation of toxic ions like Fe and Mn. Waterlogging delays planting and reduces soil temperature, length of growing season, seed germination, and seedling establishment, thereby reducing plant population density and crop yield (von Haden et al., 2021). It also induces ethylene production, which inhibits root growth and accelerates leaf senescence. Waterlogging delays harvest and increases the risk of postharvest losses. Prolonged waterlogging causes chlorophyll degradation,

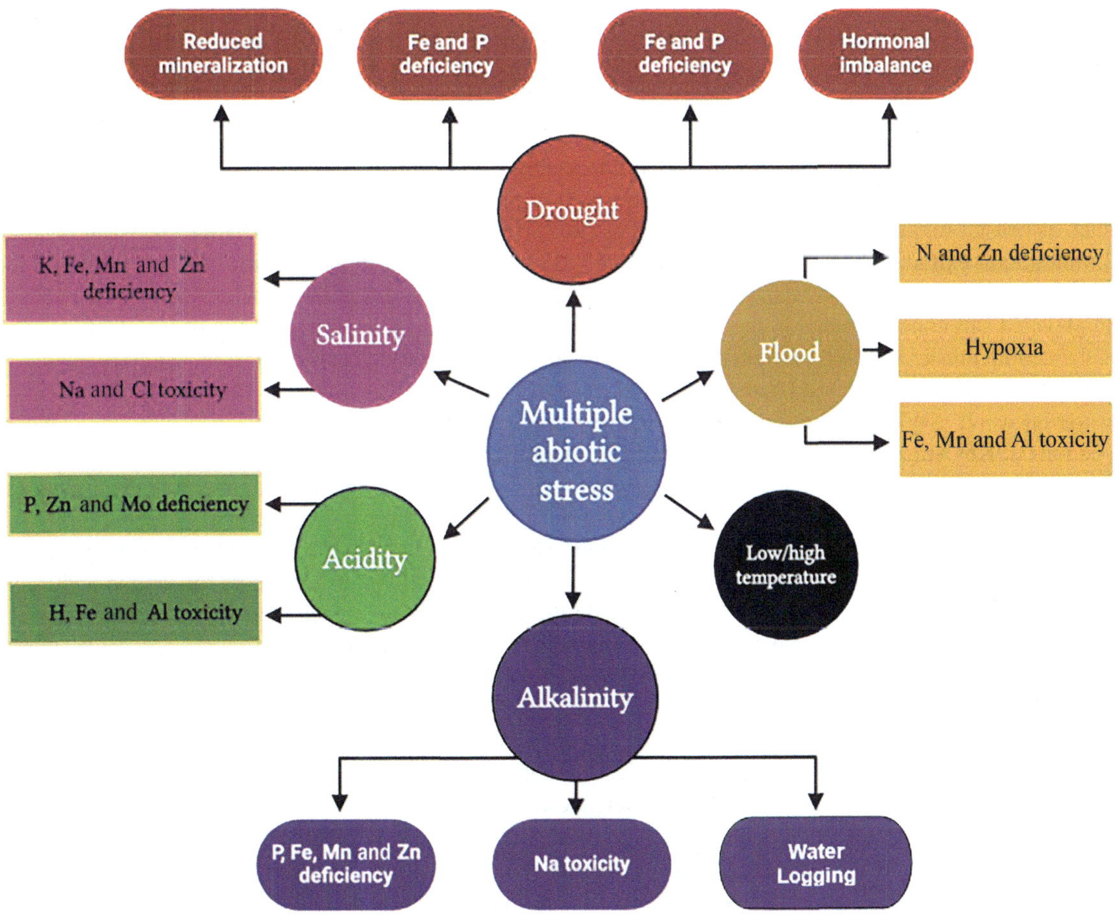

FIGURE 17.2 Multiple abiotic stresses and their impact on crops.

affecting photosynthetic efficiency (Kumar et al., 2021). Understanding the physiological and biochemical mechanisms of moisture stress is crucial for devising effective management strategies.

In combined waterlogging and nutrient stress, plants experience both excessive water in the soil and limited availability or toxicity of essential nutrients. This combined stress often has a more severe impact on plant growth and survival than either stress alone. For example, waterlogging creates oxygen-deficient conditions that impair root growth and may lead to root death, thereby limiting the plant's capacity to absorb and transport water and nutrients. Previous studies with wheat and barley demonstrated that oxygen deficiency results in suboptimal uptake of essential nutrients such as N, P, K, Mn, Cu, and Zn (Steffens et al., 2005). In cotton, waterlogging has been shown to significantly reduce the concentrations of most essential nutrients (Milroy et al., 2009). In barley, combined waterlogging and salinity stress further intensified ionic imbalance, with leaf Na^+ concentrations doubling while leaf K^+ content declined by 40% compared to salinity stress alone (Zeng et al., 2013). In poorly drained soils, oxygen depletion forces microorganisms to use NO_3^- as an electron acceptor, converting it to ammonium (NH_4^+). This can result in ammonium toxicity (Parit et al., 2020)

Continued anaerobic conditions reduce manganese and iron oxides, increasing Mn^{2+} and Fe^{2+} levels to toxicity (Fig. 17.2). Excess Fe disrupts physiological processes, causing oxidative stress and leaf bronzing (Aung and Masuda, 2020). Further reduction forms toxic hydrogen sulfide (H_2S) from sulfate (SO_4^{2-}). Manganese (Mn) toxicity, common in acidic soils, interferes with Fe, Mg, and Ca uptake, while B toxicity is typically observed in irrigated areas with B-rich water. The common nutrient deficiency and toxicity symptoms, such as leaf burning, chlorosis, and yellowing for each of the crop nutrients, are presented in Table 17.1 Fe deficiency and toxicity are the most common among micronutrient-related issues and are closely tied to soil moisture levels.

17.1.2 Salt stress-induced multiple abiotic stresses in agriculture

Another abiotic stress is salinity and sodicity, which arises from the accumulation of salts, particularly sodium chloride, in the root zone, but alkalinity increases soil pH by sodium saturation. Salinity stress may induce combined stress with waterlogging, drought, acidic and alkaline stress, and nutrient deficiency. This can dexterously reduce plant growth and productivity more than the separate conditions. According to the FAO, ~ 932 million hectares (M ha) of the world's land are affected by salinity and alkalinity (FAO, 2024). It impacts over 20% of irrigated land globally, rendering large swathes of agricultural land unproductive (Pitman and Läuchli, 2002). Salinity reduces rice yields by up to 50% in saline-prone areas, such as coastal regions of South Asia (Islam et al., 2021). Soil salinity is measured by the electrical conductivity of the soil's saturation extract (ECe), with levels exceeding 4 dS m^{-1} indicating saline conditions. High salinity affects plant growth by three major mechanisms, *viz.*, osmotic stress, ion toxicity, and nutrient imbalances. Plants are unable to absorb it due to elevated salt concentrations in the soil and reduced water potential, making it difficult for plants to absorb both water and essential nutrients (osmotic stress). This leads to dehydration, reduced cell expansion, stunted growth, and, leads to eventual plant death if prolonged (Acosta-Motos et al., 2017).

Saltwater intrusion (SWI) threatens coastal agriculture, ecosystems, and local economies, driven by climate change and human activities. Key strategies include developing salt-tolerant crops, adopting controlled environment agriculture, and enabling marsh migration to boost biodiversity and protect farmland. Addressing SWI requires multidisciplinary research, supportive policies, and community engagement to ensure sustainable coastal farming (Su et al., 2025). In ion toxicity, the excessive accumulation of ions, particularly sodium ions (Na^+) and chloride (Cl^-), in plant tissues disrupts cellular processes, leading to enzyme dysfunction and membrane instability. Salinity stress hampers the uptake and translocation of vital nutrients like potassium (K^+), calcium (Ca^{2+}), and nitrate (NO_3^-), which are essential for plant metabolic functions by creating nutrient imbalance near the root zone. Salinity stress also induces the overproduction of ROS, which causes oxidative damage to plant cells. This oxidative stress exacerbates the impact of salinity by damaging cellular structures and triggering programmed cell death (Acosta-Motos et al., 2017).

Alkalinity stress, in contrast to salinity, arises from an excess of sodium carbonates ($NaCO_3^-$) or bicarbonates (HCO_3^-) in the soil, resulting in elevated pH levels above 8.5. This high pH creates significant nutrient imbalances, particularly micronutrients such as iron (Fe), zinc (Zn), and manganese (Mn), which are essential for healthy plant growth and development. These deficiencies often manifest as chlorosis, poor root development, and ultimately reduced crop yields. The accumulation of bicarbonate ions in plant roots impairs root elongation, reducing the plant's ability to access both water and nutrients effectively. Additionally, alkalinity disrupts soil structure and negatively impacts soil aggregation by dispersing soil particles, leading to poor water infiltration, root penetration, and increased erosion. Moreover, alkalinity stress also alters soil microbial communities, further negatively impacting nutrient cycling, plant nutrient acquisition, and plant-microbe interactions, leading to stunted plant growth and reduced crop productivity (Msimbira and Smith, 2020). In

TABLE 17.1 Symptoms of nutrient stress in major crops under multiple abiotic stress conditions.

Crop	N	P	K	Zn	Mg	B	S
Rice	Poor tillering, pale yellow-green leaves, yellowing in older leaves, poor yields	Dwarf growth, abnormal bluish-green color, poor yield	Red/brown discoloration, tips, and margins, and burning of leaves	Leaves turned brown under flooding conditions			
Cotton	Yellowish green foliage, dried older leaves, stunted growth		Yellowish rustic appearance in older leaves, scorched and curled downward leaves, dwarf and immature bolls		Purple leaves with green veins appeared in older leaves	Dark rings on leaf petioles and deformed leaves	
Wheat	Reduced plant height, pale green leaves	Purple to brown-color leaves in severe deficiency	Yellow/brown along the leaf tips	Irregular light brown lesions, bordered by a dark brown margin	Striped appearance because of interveinal chlorosis	Abnormally thick stems, twisted awns	Pale green leaves, tissue testing needed to distinguish between N and S deficiency
Maize	Yellow discoloration in older leaves and subsequent necrosis	Dark green leaves, purple tints, reddish discoloration in younger leaves	Narrowed and deformed cob in maize	Pale yellow chlorotic zones, chlorotic yellow or white bud, and reduced internode growth	Interveinal chlorosis and steaky lesions	Yellow and white spots on leaves	Interveinal chlorosis

many cases, soils experience a combination of both salinity and alkalinity stresses, exacerbating the negative impacts on plant health.

Nutrient availability is determined by their ionic forms in the soil solution and by soil chemical properties, including pH (Kabata-Pendias, 2010). Soil pH strongly influences nutrient availability to crop plants by affecting metal solubility (Table 17.2). A nutrient imbalance happens when one nutrient's availability affects another's uptake, leading to deficiencies or toxicities. This can result from natural soil factors like sodicity, salinity, or acidity. For example, acidic soils typically increase micronutrient availability (Fe, Mn), but extreme acidity can cause toxicity (Al, Fe, Mn) and limit phosphorus (P) availability. However, molybdenum (Mo) is an exception, becoming less available under such conditions (Table 17.2).

17.1.3 Multiple abiotic stresses: a global concern for food and nutritional security

Unlike single stressors, which plants may be able to tolerate to some extent, the combined effect of multiple stresses is often synergistic, leading to a much more severe and complex impact on plant health and crop yield. Plants facing combined stresses experience a severe decline in essential physiological processes, such as photosynthesis, water uptake, and nutrient assimilation. This results in stunted growth, reduced biomass, impaired reproductive success, and ultimately, a substantial loss in crop yield. The ripple effects of yield loss due to multiple abiotic stresses extend far beyond agriculture, with declining crop productivity jeopardizing food security for millions, particularly in low-income regions, raising concerns about the affordability and availability of staple crops (Chary et al., 2021).

Additionally, nutritional insecurity escalates as stress-affected crops often exhibit reduced micronutrient content. Multiple abiotic stresses also compromise its nutritional quality. Stressed plants may produce grains or fruits with lower concentrations of essential vitamins, minerals, and proteins, impacting the nutritional value of the food supply. This is a critical issue, particularly in developing countries where the population is highly dependent on a few staple crops. For instance, rice grown under saline conditions shows lower levels of Zn and Fe, key nutrients for combating malnutrition (Negrão et al., 2017). The interplay between abiotic stress and global population growth further complicates the scenario, highlighting the need for transformative interventions.

A plant's response to a combination of heat and drought is different from its response to either stress alone. Traditional breeding methods, which often focus on single-stress tolerance, are proving insufficient to develop resilient crops that can withstand the complex and unpredictable conditions of a changing climate. This chapter critically examines the impact of multiple abiotic stresses on agriculture, encompassing the severity of the issue, its influence on global agricultural yields, and exploring available agronomic management strategies to ensure food and nutritional security for a growing global population.

17.1.4 Agronomic practices for managing multiple abiotic stresses in agriculture

To mitigate these impacts, a variety of agronomic practices have been developed. These practices aim to enhance crop resilience and productivity under stressful conditions. Some key agronomic strategies include improved water management, soil health management, nutrient management, stress-tolerant cultivar selection, tillage practices, etc. (Fig. 17.3). These agronomic practices, when implemented effectively, can significantly improve crop performance under abiotic stress conditions, leading to increased yields, improved quality, and enhanced agricultural sustainability.

17.2 Agronomic management options to mitigate deficit moisture stress-induced multiple abiotic stresses

While droughts cannot be eliminated, their adverse effects on crop yield can be mitigated through several agronomic practices. Management technologies to mitigate drought-induced multiple stresses include water management options, the choice of appropriate stress-tolerant varieties, adjusted cropping systems, crop rotation, mulching, integration of soil organic matter, seed priming, zero and minimum tillage, improved nutrient management, alternate crop establishment methods, better field management, and soil amelioration.

17.2.1 Crop and varietal selection

In drought-prone areas, selecting drought-tolerant crops and short-duration varieties will minimize the impact of drought stress and allow them to escape drought stress. The list of drought-tolerant crop varieties in different crops is presented in Tables 17.3 and 17.4. In general, millets and pulses are more tolerant to drought compared to other cereals, due to their

TABLE 17.2 Essential nutrients concentration and their functions in soil and plants under different pH conditions.

Nutrients	% in plants	Content in soil	Optimal pH range	Function in plants	At low pH	At high pH
Nitrogen (N)	100	0.1%–2%	6.0–7.5	Protein, amino acid production	Reduced nitrification, lower nitrate formation	Reduced ammonium solubility
Phosphorus (P)	6	0.1%	6.0–7.0	Nucleic acid generation, ATP	Forms insoluble compounds with iron and aluminum	Forms insoluble compounds with calcium
Potassium (K)	25	2.0%	5.5–8.0	Ion transport, salt balance in plants	Leaching in acid soils	Reduced availability due to competition with other cations
Calcium (Ca)	12.5	0.5%	6.0–7.5	Cell wall development, cell integrity	Low availability	Precipitate as Carbonates
Magnesium (Mg)	8	0.05%	6.5–8.0	Chlorophyll formation	-	Precipitate as Carbonates
Sulfur (S)	3	0.04%	6.0–7.5	Amino acid development	Formation of sulfides	Sulfates formation
Copper (Cu)	0.01	5–50 ppm	5.0–7.0	Enzyme activation	More soluble and readily available	Insoluble hydroxides and carbonates
Iron (Fe)	0.2	250,000 ppm	4.0–6.5	Chlorophyll and enzyme cofactor	More soluble and readily available	Insoluble hydroxides and carbonates
Manganese (Mn)	0.1	200–300 ppm	5.0–6.5	Enzyme activation	More soluble and readily available	Insoluble hydroxides and carbonates
Zinc (Zn)	0.03	10–30 ppm	5.0–7.0	Carbonic dehydrogenase, enzyme activation	More soluble and readily available	Insoluble hydroxides and carbonates
Boron (B)	0.03–0.08	9.0–80.0 ppm	5.0–7.0	Carbohydrate, auxin metabolism, sugar transport	Low availability	Increases availability
Molybdenum (Mo)	0.1–2.0 ppm	0.25–5.0 ppm	6-5–8.0	Nitrogen assimilation, nitrogenase activity	Low availability	Increases availability

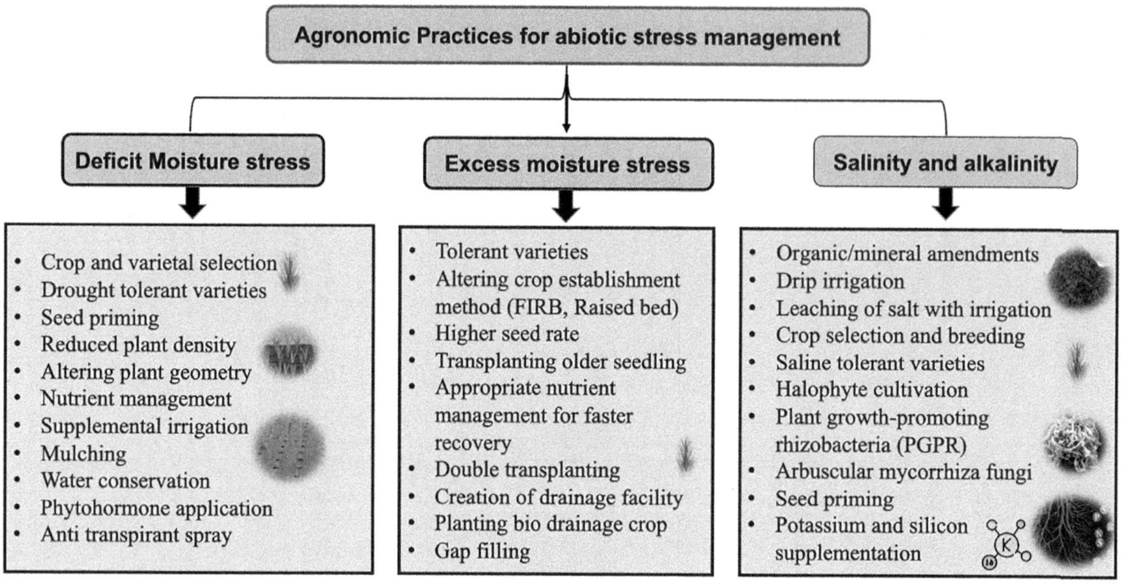

FIGURE 17.3 Various agronomic practices for abiotic stress management.

TABLE 17.3 Short-duration and early-maturing varieties of different crops.

Crop	Varieties
Rice	Vandana, Sahabhagidhan, Satyabhama, Naveen, Satabdi, CR Dhan 201, CR Dhan 202, Phalguni, CR Dhan 40, DRRH-1, DRRH-2, MTU 1010, MTU 1156,
Wheat	HI 1621, HI 1544, HD 3271, PBW 761, N5439, N5749, N8223, NIAW-34, NI 9947, HD 3298, HD 2987, HD 3090, HD 2985, WR 544, PDKV Sardar (AKAW 4210-6), NIAW 34, RS 31-1, Raj 3765, C 306
Sorghum	CSV1, CSV2, CSH1, CSH6, CSH14, CSH17, CSH23, CSH30, CSH33
Maize	DMRH1305, LQMH 1, Pusa Early Hybrid Makka-1 (EH-202892), Pusa Early Hybrid Makka-2 (EH-203492), Pusa Early Hybrid Makka-3 (AH-58), Pusa Extra Early Hybrid Makka-5 (AH-421).
Pearl millet	JKBH 1008, HHB 272, MPMH 21, 86M01, Proagro Tejas, NBH 5767, HHB 234, Bio 70, RHB-177, RHB 154, HHB 197, GHB 757, GHB 719, GHB 538, HHB 67 improved, GHB 577, PB 180, HHB 117, HHB 146, GHB 558, GHB 526, PB 172, PB 106, RHB 121, Nandi 35, CoHCu 8, HHB 94, RHB 90, Pusa 605, Nandi 8, Nandi 32, Pusa 415, MLBH 504, GK 1004, 7686, PAC 903, RHRBH 8924, GHB 183, MLBH 285, RHB 30, JKBH 26, MLBH 267, JBH 1, PABH 3, Pusa 444, RHB 58, RHRBH 8609, GHB 15, GHB 235, AHB 251, MBH 160, HHB 68, Pusa 322, ICMH 312, ICMH 356, Eknath 301, MLBH 104, HHB 67, VBH 4, GHB 181, MBH 136, ICMH 423, HHB 50, HHB 60, GHB 30, Pusa 23, MBH 130, MH 180, MH 182, GHB 32, ICTP 8203, Raj 171, ICMV 221, CZP-1C 923, Pusa Bajri 266, Pusa Composite 334, JBV-2, HC 10, AIMP 92901, Pusa Composite 383, JBV 3, HC 20, CZP 9802, CoCu 9, PPC-6, JBV 4, Pusa Composite 443, MBC 2, X 3, HB 2, HB 5, HB 1, HB 3, NHB 4,
Finger millet	KMR 340, KMR 204, VL 352, GPU 48, GPU 45, GPU 26, BM 9-1, Champavathi (VR 708), PR 230 (Maruthi), VL-146, Suraj (VR 520), KM 65, VL 149, PES 400, Co 13, TRY-1, VL 124,
Foxtail millet	Suryanandi (SiA 3088), SiA 3085, Prathapkangni 1 (SR-51), Sri Lakshmi, Meera (SR 16), PRK 1, PS 4, Gavari (SR 11), TNAU 43, TNAU 186, AK 132 – 1
Barnyard millet	VLMadira 172, Sushrutha (RAU 11), Kanchan
Kodo millet	JK 76
Little millet	BirsaGundli 1
Proso millet	PratapCheena-1 (PR-18), Bhawna, Nagarjuna
Pigeonpea	PRG 176, Pusa Arhar 16, Pusa Arhar 2017-1

TABLE 17.4 Drought-tolerant varieties of different crops.

Crop	Drought-tolerant varieties
Rice	Sahbhagi Dhan, Anjali, Sookha Dhan, Sahod Ulan, Rice CO 53, Vandana, Satyabhama, DRR Dhan 42 (IR 64 Drt 1/ Sukha dhan 3), DRR Dhan 43, DRR Dhan 44, DRR Dhan 46, Birsa Vikas Dhan 203, Birsa Vikas Dhan 111, Rajendra Bhagwati, Jaldi Dhan 6, CR Dhan 40.
Perennial rice	PR23, PR101, PR107, PR24, and PR25
Wheat	Raj 3765, JKW 261, DBW 296, DBW 316, HD 3293, DBW 252, N5439, WH 1142, K7410, K9465, MP3173, HI 1531, HYB 65, A-9–30-1, BRW 3723, Ajanta, K-72, K-78, K65, PBW 527, HI 8627, HD 2888, HPW 349, PBW 644, WH 1080, HD 3043, PBW 396, K9465, K8962, MP 3288, HD 4672, NIAW 1415, HD 2987
Barley	RD 2660, K603
Sorghum	CSV27, CSV17, CSV 8R, CSV 14R, CSV 26, M35-1, CSH 19R, CSV 18, CSH 15R
Maize	LQMH 1, DL 708, RCRMH 2, Pratap Makka 2, Pusa Hybrid Makka 1, HM 4, Pusa Hybrid Makka 5, DHM 121, Buland
Pearl millet	HHB 223, GHB 719, HHB 68, HHB 60, CJ 104, CZP 9802, HC 20, HC 10, HS 1, T 55, Co3, Co2, HHB 67 improved, GHB 757, Dhanshakti, HHB 234, Mandor Bajra Composite 2, HHB-226, RHB-177, Pusa Composite 443
Finger millet	PPR 2700 (Vakula), Saptagiri (PR 2614), Suraj (VR 520), GN 3, RAU 8
Foxtail millet	TNAU 186, AK 132-1
Kodo millet	KK 2, Jawaharkodo 48 (JK 48), KMV 20 (Varugu 1), GK 2
Little millet	JawaharKutki 4 (JK 4), OLM 20
Proso millet	TNAU 143, Co 4
Chickpea	Vijay, Vikas, RSG 14, RSG 888, ICCV 10, Pusa 362, Vijay
Groundnut	Ajaya, Girnar 1, TAG-24, Kadiri 6, ICGV 91114
Soybean	NRC 7, JS 95-60
Sugarcane	Co 98014 (Karan-1), Co 0239, Co 0118, Co 0238, Co 06927, Co 0403, Co 86032
Cotton	HD 324, CICR-1, Raj DH 7, Jawahar Tapti, Pratap Kapi, Suraj, Surabhi, Veena, AK 235
Jute	JBO 1 (Sudhangsu), JRO 204, JRO 524, JRC 80

short growing period and early life cycle completion (plasticity in growth). In addition, the tap root system of pulses helps the plants to extract moisture from deeper soil layers and thrive well under moisture-stress conditions. Therefore, it is recommended to grow these crops in drought-prone areas. Rice varieties such as Swarna, Samba Mahsuri, IR 64, and MTU 1010, originally bred for irrigated environments, are often grown in rainfed conditions due to their high yield potential and farmer preference. However, these varieties are highly susceptible to drought. Selecting suitable rice cultivars is essential to minimize yield losses during adverse weather. Advancements in rice breeding have successfully introduced drought-tolerant quantitative trait loci (QTL) into popular varieties, resulting in improved cultivars like Swarna-Sub1 and Samba-Sub1. In India, the ICAR-National Rice Research Institute, Cuttack, and the ICAR-Indian Institute of Rice Research, Hyderabad, have released drought-tolerant varieties such as CR Dhan 801 and DRR 50, respectively. Nepal has also introduced Bahuguni Dhan 1 and Bahuguni Dhan 2 through its National Rice Research Program in Hardinath. Bangladesh Rice Research Institute (BRRI) and Bangladesh Institute of Nuclear Agriculture (BINA) also introduced drought-tolerant rice varieties like BRRI dhan56, BRRI dhan57, BRRI dhan66, BRRI dhan71, and BINA dhan-17. These varieties have shown promising yields under drought conditions and are being adopted by farmers in drought-prone areas of Bangladesh.

Stress-tolerant rice varieties (STRVs) hold significant potential to enhance agricultural productivity and resilience. STRVs released in India exhibit strong performance under abiotic stress with minimal yield penalties, while maintaining high productivity in normal conditions. Drought is a major constraint in rainfed rice ecosystems, especially in eastern India. The drought-tolerant rice variety Sahbhagi Dhan (IR74371-70-1-1), developed by IRRI and partners, shows high yield, short duration, and wide adaptability. Released as Sahbhagi Dhan (India), Sukha Dhan 3 (Nepal), and BRRI Dhan 56 (Bangladesh), it provides 0.8–1.6 t ha^{-1} yield advantage under drought without a yield penalty in normal years. Its short maturity allows earlier sowing of succeeding crops, boosting cropping intensity (Dar et al., 2020). Perennial rice offers additional advantages, with its underground rhizomes and deep root systems providing greater resilience to environmental stresses like drought, flooding, and high temperatures. Its deep roots access water and nutrients from deeper soil layers, making it more drought-tolerant compared to annual rice varieties (Shanmugam et al., 2025).

17.2.2 Crop management practices

Tillage practices play a critical role in influencing soil moisture and crop yields. Premonsoon tillage captures rainwater effectively, while surface tillage enhances soil roughness, improving water infiltration. In wheat fields, deep plowing increases soil water storage by 7.8% during the fallow season and 13.7% during the growing season compared to no-tillage (Sun et al., 2018). Over 3 years, deep plowing in dryland wheat boosted soil water storage in the top 200 cm by 7.9% compared to no-tillage (Yan et al., 2023). Conservation tillage practices have also improved soil moisture storage by approximately 25% compared to conventional tillage (Ngigi et al., 2006). No-tillage increases soil water storage by 7.1%, 5.4%, and 9.6% in the top 200 cm during dry, normal, and wet years, respectively, compared to plow tillage (Wang et al., 2020). Additionally, no-tillage enhances soil water storage by 3.7% and 4.4% during tasseling and grain filling, respectively, compared to conventional tillage (Zhang et al., 2022).

17.2.3 Seed priming

Seed priming partially hydrates seeds to activate germination-related metabolic processes without radicle emergence. This technique enhances germination rate, uniformity, and percentage, making it an effective short-term strategy for mitigating drought stress in various crops. KNO_3 priming increased shoot dry matter, seed cotton yield, and net photosynthetic rate by 54%, 36%, and 125%, respectively. Similarly, glycine betaine priming improved single boll weight, leaf greenness, and membrane stability by 31%, 11%, and 79%, with both treatments boosting germination rates by over 20% (Khalequzzaman et al., 2023). Priming with 2.5%–5% KNO_3, 3%–3.5% SiO_2, or 1–2.5 mM salicylic acid significantly enhanced rice seedling growth (Ali et al., 2020). Selenium (Se) priming with 15–60 μmo L^{-1} also improved rice emergence and seedling growth compared to hydropriming and controls (Khaliq et al., 2015).

In lentils, hydropriming and KNO_3 priming increased germination rates to 93.3% and 73.3%, respectively, compared to 19.3% in the control under drought stress (-0.3 MPa soil water potential) (Eskandari and Alizadeh-Amraie, 2014). Osmopriming with 1.5% $CaCl_2$ enhanced wheat growth and yield under drought conditions, increasing leaf area by 9%–43%, tissue water status by 2%–47%, osmolyte accumulation by 6%–48%, and grain yield by 14–79% (Tabassum et al., 2018). In rapeseed, priming with distilled water, salicylic acid with 138 mg L^{-1}, gibberellic acid with 300 mg L^{-1}, sodium nitroprusside with 89.4 mg L^{-1}, $CaCl_2$ with 3000 mg L^{-1}, or abscisic acid with 30 mg L^{-1} significantly increased germination by 8.5%–41% for tolerant varieties, while 244% for susceptible varieties under low temperatures and drought stress (Zhu et al., 2021). Si and Zn priming enhanced drought tolerance in *Eruca sativa* by mitigating oxidative stress and increasing antioxidant enzyme activity and osmoprotectants such as soluble sugars and proline (Hussain et al., 2024). In rice, priming with Na-selenite and ZnO nanoparticles has been shown to trigger its seed germination to the maximum and improve the seedling vigor (Adhikary et al., 2022).

17.2.4 Altering sowing/planting time

Adjusting planting dates is critical for mitigating drought stress. Early planting shifts critical growth stages away from dry periods, especially in dry climatic conditions (Korres et al., 2017). On the contrary, late sowing avoids early-season drought but may shorten the growing season. Early planting combined with supplemental irrigation mitigates drought. For instance, supplemental irrigation increased sorghum yields by up to 3 t ha^{-1} in Kobo and 2 t ha^{-1} in Meisso, Ethiopia (Getachew et al., 2021). Adjusting planting time helps crops tolerate salinity stress by aligning with periods of lower soil salt concentrations. Early planting allows crops to establish before salinity peaks, benefiting from natural leaching by rainfall. Conversely, delaying planting provides time for preplanting irrigation or rainfall to flush out salts, reducing salinity before sowing. It also helps avoid peak salt accumulation during hot, dry periods. Additionally, a delay allows soil amendments, such as sulfur or acidifying agents, to take effect, improving soil conditions before planting.

17.2.5 Altering planting density and establishment method

Planting density plays a vital role in crop performance under drought stress. Adjusting spacing optimizes soil moisture use and minimizes the adverse effects of water scarcity. While higher densities increase competition for resources, lower densities improve water use efficiency (WUE) and individual plant growth. Optimal planting density depends on crop type, soil, and climate. In rainfed maize, wider row spacing reduced competition and improved WUE, and found the highest at a density of 67,500 plants ha^{-1}. On the other hand, increasing plant densities to 82,500 and 97,500 plants ha^{-1} reduced WUE by 4.9% and 6.0%, respectively, without increasing the grain yield of maize (Wang et al., 2021). Similarly, in wheat under drought stress, lower planting densities proved optimal by promoting efficient resource use and minimizing competition (Zhu et al., 2021).

Crop establishment methods like direct seeding of rice rather than transplanting provide opportunities to reduce unproductive water losses (deep percolation and evaporation losses from early-season rains) and moderate drought effects by efficiently utilizing the irrigation water. However, a foliar application of 0.5% Fe is recommended for direct-seeded rice, as Fe deficiency is common in this method. The ridge and furrow method conserves water by enabling alternate furrow irrigation, reducing overall irrigation needs. This technique also helps minimize salt accumulation, as water movement leaches salts away from plant roots, lowering the risk of salinity-related crop stress.

17.2.6 Planting geometry

Furrow and bed sowing is more water-efficient than flat sowing. WUE under furrow-irrigated raised bed, and broad bed furrow was 8.3%–22.8% and 4.9%–24.8% higher, respectively, over the flatbed with flooding (Behera and Sharma, 2014). Rice grown on raised beds, whether transplanted or direct-seeded, yielded 5%–6% less than traditionally puddled and transplanted rice. However, these methods reduced irrigation water use by 37%–40%. Total water input for rice on raised beds was 38%–42% lower than that of flooded transplanted rice (Choudhury et al., 2007). In other crops, furrow-irrigated raised beds proved highly water-efficient. Chickpeas grown on 75 cm wide raised beds recorded up to 33.3% water savings with increased WUE (Kumar et al., 2015). Ridge-sown maize also exhibited higher WUE due to its well-developed root system, which enhanced access to soil moisture from deeper layers (Khan et al., 2012).

17.2.7 Nutrient management

Organic manure improves soil water retention, while inorganic fertilizers like P stimulate root growth. N and K fertilization enhances drought tolerance by optimizing plant physiological processes (Cakmak, 2005). K application through soil or foliar spray supports critical plant functions. For instance, a 2% foliar spray of potassium sulfate increased maize photosynthesis by 10% and enhanced root and shoot lengths by 25% and 16%, respectively (Wasaya et al., 2021). In redgram, foliar application of multi-K during flowering and pod formation boosted yields to 1.36 t ha^{-1}, compared to 1.10 t ha^{-1} without the spray. Similarly, Si application mitigated drought stress in aerobic rice (Jinger et al., 2018). In wheat, Si improved water status, dry matter accumulation, and photosynthesis under drought conditions (Jinger et al., 2020). Si also reduced the shoot-to-root ratio, enhancing root growth and maintaining stomatal conductance (Hattori et al., 2005).

In winter oats, Si fertilization increased the photosynthetic rate by 16.8%–149.3%, transpiration by 5.4%–5.6%, and yield by 10.2%. Combined Si and S application further increased yield by 8.0%, with a 5.0% rise in thousand-kernel weight (Kutasy et al., 2021). Potassium sulfate (4%) enhanced durum wheat grain numbers by 21% and test weight by 24.5% under drought conditions (Akhtar et al., 2022). In sunflower, K foliar application (1%) at head formation and achene filling increased biological yield by 19% and 12%, respectively (Raza et al., 2018). In maize, foliar application of a water-soluble nitrogen, phosphorus, potassium, sulphur (NPKS) complex fertilizer (18:18:18:6) at 0.5% + $ZnSO_4$ at 0.5% increased grain yield by 36% (2.96 t ha^{-1}) compared to water spray (2.19 t ha^{-1}) (Chary et al., 2021). The application of ZnO (1.77 g), B_2O_3 (0.80 g), and CuO (0.92 g) significantly improved N, P, and Zn uptake in soybean under drought-induced nutrient stress (Dimkpa et al., 2017).

Nanofertilizers enhance a plant's ability to withstand multiple abiotic stresses by improving nutrient absorption and strengthening its defense systems. Their small size allows for more efficient nutrient uptake, ensuring the plant remains well-nourished even under stressful conditions like drought. Furthermore, certain nanoparticles, such as nano-Zn, nano-Si, and nano-Cu, act as antioxidants, neutralizing harmful ROS and reducing cellular damage (Shoukat et al., 2024). By boosting key processes like photosynthesis and improving water-use efficiency, nanofertilizers increase the plant's overall resilience against combined environmental pressures.

17.2.8 Irrigation management

Efficient irrigation is essential to mitigate drought stress. Irrigation efficiency varies by method, with surface and subsurface drip systems achieving 80%–90%, sprinklers 60%–70%, and surface irrigation 30%–40%. Combining precision farming with variable-rate irrigation lowers drought stress by reducing water losses and enhancing WUE in field crops. Strategies like limiting frequent irrigations, targeting critical growth stages, and adopting modern irrigation techniques further improve drought mitigation (Evans and Sadler, 2008). In rice, alternate wetting and drying (AWD) reduces water consumption by up to 50% in loamy and sandy soils with deeper groundwater, though it can cause yield losses exceeding 20%. Despite this, AWD consistently improves WUE. Saturated soil culture, a specific AWD approach, lowers water input by 23% compared to continuous flooding with a minimal yield reduction of 6% (Wassmann et al., 2009).

Supplemental irrigation is a vital strategy for drought mitigation. For winter wheat, applying 30 mm of water during overwintering and regreening stages increased grain numbers per panicle and thousand-grain weight by 3.4% and 4.7%, respectively, compared to no irrigation (Dai et al., 2023). In rainfed hybrid cotton, two supplemental irrigations increased yield by 30% in shallow red soils and 19% in medium-deep silty clay loam soils (Ambati et al., 2011). One supplemental irrigation during a critical growth stage can boost crop yields by 50%–60% under normal conditions (Das, 2005). For the PRG 176 red-gram variety grown under rainfed conditions, life-saving irrigation during a 17–21-day mid-season dry spell improved yields by 20% (1.20 t ha^{-1}) (Rao et al., 2015). At different locations, supplemental irrigation improved yields by 25% in cotton, 40% in groundnut, and 55% in soybean (Chary et al., 2021). Sprinklers cool the plant canopy, reducing heat stress, and ensure uniform soil moisture for efficient nutrient uptake. Drip irrigation is even more effective as it delivers water directly to the plant's roots in small, controlled amounts, preventing both drought and waterlogging.

17.2.9 Mulching

Mulches mitigate soil salinity and alkalinity by reducing evaporation, which prevents salt accumulation at the surface. By maintaining soil moisture, they keep salts dissolved and facilitate their downward movement, away from the root zone. Organic mulches further enhance soil health by adding organic matter, improving structure, water retention, and nutrient availability. Over time, better soil porosity promotes water infiltration, aiding salt leaching. Additionally, mulches regulate soil temperature, preventing extreme fluctuations that can accelerate evaporation and salt buildup. Organic mulches like rice, sesame, sudan grass, and sorghum straw increased soil moisture by 18.9% and 55.1% at 45 and 75 days after sowing, respectively, compared to unmulched controls (Teame et al., 2017). Straw and plastic mulches improved wheat and maize yields, WUE, and nutrient-use efficiency (NUE) by 20% and 60%, respectively. Straw mulching's effect on wheat was most pronounced under low water input (<250 mm) compared to high water input conditions (>250 mm) (Qin et al., 2015). Plastic, straw, and degradable mulches increased maize yields by 27.31%, 6.98%, and 16.55%, respectively, while improving crop water productivity by 34.08%, 11.48%, and 24.20% (Cai et al., 2022).

Mulching also conserved more water in the soil, with an average of 28 and 20 mm more in the top 100 cm of soil at upland and lowland sites, respectively, during wheat sowing (Zhang et al., 2009). Organic amendments such as compost and manure improve soil organic matter, enhance water-holding capacity, and reduce runoff. Lactobionate, a dairy industry byproduct, increased soil water content by 37% relative to unamended soil across various soil matric potentials (Kallenbach et al., 2019). In almond orchards, the application of composted dairy manure increased soil water retention, especially at low moisture levels (<10%). After two years, this practice improved water retention by 13% between 0 and 100 kPa at 0–10 cm soil depth (Lepsch et al., 2019).

17.2.10 Water conservation

Water harvesting and supplemental irrigation further improve WUE and crop yields. Shading in agroforestry systems, such as coffee, significantly reduces soil evaporation and transpiration. Windbreaks mitigate evapotranspiration by reducing wind speed. A 6-meter-tall poplar hedge in a South African vineyard reduced wind speed at the canopy level by 39.2% during the growing season. This significant wind reduction translated to an 18.4% decrease in reference evapotranspiration (Veste et al., 2020).

Windbreaks also improved the yields of various crops, including groundnuts (33.6%), mungbeans (16.8%), and sesame (9.6%), compared to unsheltered crops (Lay Swe, 2005). Pruning and leaf removal (senescence) reduce plant water demand under drought conditions. Branch removal in jujube trees decreased evapotranspiration by up to 10%, enhancing soil moisture retention. Although this practice slightly reduced fruit yield, it improved WUE, making it a viable drought mitigation strategy (Che et al., 2016). In water-stressed maize, defoliation by 33% and 67% increased yield by 13.5% and 25%, respectively. Similarly, 67% defoliation in cotton doubled the harvestable yield under water stress (Yang and Midmore, 2004).

17.2.11 Application of phytohormones and osmoprotectants

Phytohormones and osmoprotectants are pivotal in enhancing plant resilience to drought stress. Compounds like abscisic acid (ABA), salicylic acid, proline, and glycine betaine play significant roles in mitigating drought impacts by regulating physiological processes and improving stress tolerance (Rabert et al., 2014). For instance, gibberellic acid combined with diluted bee honey has been shown to improve nutrient uptake, photosynthesis, and antioxidant defense in faba bean, resulting in enhanced growth, yield, and WUE under drought conditions (Rady et al., 2021). Among phytohormones, ABA is critical for drought stress tolerance, orchestrating stomatal closure and osmotic adjustment. Other hormones such as auxins, brassinosteroids, cytokinins, ethylene, gibberellins, jasmonic acid, and salicylic acid also contribute to drought adaptation by regulating growth, signaling pathways, and stress-responsive genes. Additionally, plant growth-promoting microbes can boost endogenous phytohormone levels, further improving drought tolerance (Chhaya et al., 2021).

The exogenous application of phytohormones and phytometabolites elevates endogenous levels, enhancing plant defense mechanisms against drought stress. For example, the application of *Casuarina equisetifolia* leaf extract improves rice drought tolerance by modulating the root microbiome. This adaptation is marked by higher chlorophyll and carotenoid content, increased proline levels, and enhanced antioxidant activity. As the extract concentration rises, rice cultivars exhibit greater resilience to drought stress (Purwestri et al., 2025). Despite their potential, challenges such as high costs, complex application techniques, and variable field efficacy hinder the large-scale adoption of phytohormones and osmoprotectants. Addressing these barriers is essential for harnessing their benefits in sustainable agriculture, particularly in drought-prone regions.

Biofertilizers boost nutrient availability by converting nutrients into absorbable forms, ensuring a steady supply even under stressful conditions. They also regulate stress hormone levels, helping plants maintain growth and vigor. Seaweed extracts, on the other hand, provide a direct supply of growth-promoting hormones and antioxidants. These antioxidants directly neutralize harmful ROS, reducing cellular damage and protecting the plant's metabolic functions from stress. Both of these biostimulants ultimately improve the plant's overall health and ability to withstand a combination of environmental pressures.

17.2.12 Antitranspirants spray

Antitranspirants reduce water loss through transpiration, improving crop drought tolerance and WUE, especially when applied at critical growth stages. They have demonstrated significant potential to enhance crop yield under water-limited conditions. For example, $CaCO_3$ application reduced transpiration by 47%–58% in tomato crops, although excessive application hinders photosynthesis and affects plant growth and yield (Patanè et al., 2018). Despite this limitation, these substances are valuable tools for mitigating drought stress. In wheat, applying a film-type antitranspirant before the booting and anthesis stages increased grain yield by 36% and 12%, respectively, compared to untreated plants (Abdullah et al., 2015; Djurovic et al., 2016) reported a 27% increase in tomato yield with kaolin antitranspirant application. Similarly, fulvic acid enhanced relative water content, yield by 24%, and irrigation WUE by 34.8% in tomatoes (Aggag et al., 2015). Antitranspirants have also shown promise in perennial crops like apples. ABA application enhanced leaf water potential by 62%, demonstrating improved water retention and stress tolerance (Tworkoski et al., 2011).

17.3 Agronomic management options to mitigate excess moisture stress-induced multiple abiotic stress

17.3.1 Tolerant varieties

To mitigate yield losses caused by waterlogging and submergence, selecting appropriate crop varieties is crucial. Cultivating early-maturing varieties can minimize crop exposure to prolonged waterlogging. However, it's crucial to recognize significant intercrop and intracrop variations in waterlogging tolerance. Rice exhibits the highest tolerance to submergence among field crops, followed by jute and sugarcane. While maize can tolerate waterlogging in later stages, it is highly susceptible during its early growth phases. In contrast, pulses demonstrate sensitivity to waterlogging throughout their entire growth cycle. Therefore, careful selection of waterlogging-tolerant varieties is paramount for each crop. Table 17.4 provides a list of waterlogging-tolerant cultivars for various crops, including multiple-stress-tolerant rice cultivars. BRRI and BINA also introduced flood-tolerant rice varieties like BRRI dhan51, BRRI dhan52, BRRI dhan79, BINA dhan-11, and BINA dhan-12. These varieties have shown promising yields under flooding conditions and are being adopted by farmers in flood-prone areas of Bangladesh.

In rice, an indica rice variety native to Odisha, FR13A, has been extensively used in the breeding program for submergence tolerance. This variety possesses the Sub1 gene on rice chromosome 9, an ethylene response factor responsible for about 70% tolerance without any effect on the yield and cooking quality (Septiningsih et al., 2015). Through marker-assisted backcrossing Sub1 gene was transferred into the Indian mega variety Swarna for flash flood tolerance. SUB1 is associated with the suppression of the normal elongation response of rice varieties when inundated. This suppression of elongation enhances survival by reducing reserve soluble sugar consumption, allowing the plants to recover faster upon de-submergence (Kato et al., 2014).

Stagnant flooding is an escalating challenge in delta regions like the Ganges Delta due to rising sea levels. Developing rice varieties that can elongate based on the standing water depth (up to 50 cm) is essential. However, these varieties must also exhibit traits such as survival and effective tillering under deep water conditions, along with resistance to lodging. A significant locus on chromosome 12 has been identified to control this elongation response. It contains two ethylene-responsive factor genes, SNORKEL1 and SNORKEL2, which regulate the elongation mechanism (Hattori et al., 2009). These genes function similarly to the SUB1 gene, which is known for submergence tolerance. However, unlike SNORKEL1 and SNORKEL2, the SUB1 gene appears to suppress elongation, emphasizing a distinct role in enabling survival under submerged conditions. The identification and utilization of these genes offer a pathway to developing rice varieties tailored to stagnant flooding scenarios.

17.3.2 Anaerobic germination in rice

Rice germination is highly sensitive to waterlogging-induced hypoxia or anoxia, often referred to as anaerobic germination (AG). Under these conditions, rice seeds may fail to develop roots and shoots, hampering early embryo growth (Angaji et al., 2010). Key QTLs associated with AG tolerance have been identified, leading to the development of breeding lines capable of germinating and growing normally under hypoxic conditions. These AG-tolerant genotypes, such as IR-64 AG1, IR-64 AG2, Ciherang-Sub1-AG1, and PSBRc-AG1, are now available for large-scale field testing. Importantly, the introgression of AG1 and AG2 QTLs has no adverse effects on grain physical or chemical quality. AG cultivars, such as IR 64 and IR 64-Sub1, demonstrated significantly higher germination and survival rates under flooded conditions, recording 120.9% and 98.1% increases, respectively, along with higher grain yields.

AG cultivars maintain higher sugar content in grains, enabling better germination by sustaining energy reserves under flooded conditions. On a physiological level, reduced ascorbate plays a crucial role in germination under hypoxic conditions by limiting malondialdehyde production, a by-product of membrane lipid peroxidation (Mondal et al., 2020). Optimized nutrient management further enhances the performance of AG cultivars. For instance, applying 20% more basal P alongside recommended N-P-K fertilizers increased grain yields by 21%, 16%, and 19% in IR 64-AG, IR 64-Sub1, and IR 64, respectively. These practices, combined with AG-tolerant genotypes, offer a viable approach for improving crop establishment in flooded soils, particularly through direct-seeded rice methods, where waterlogging caused by sudden rains after sowing inhibits seed germination. This strategy not only reduces production costs and water use but also boosts farmers' income (Lal et al., 2018).

17.3.3 Establishment method, spacing, and seed rate

Crop establishment methods significantly influence plant survival and productivity after de-submergence. Direct seeding of rice before the monsoon season proves more resilient to submergence compared to traditional transplanting. Double transplanting, a strategic method for low-lying, flood-prone regions, involves sequential nursery transfers to strengthen seedlings. This technique, particularly effective during the *Aman* season in waterlogged fields, enhances plant vigor and improves rice yields by up to 20% (Vijayakumar et al., 2019). Ridge and furrow sowing is another effective strategy to mitigate submergence stress. Raised ridges with furrows improve drainage, aeration, and root growth, reducing the risk of submergence. This method has demonstrated increased plant height and yield in crops such as maize, cotton, and sunflower compared to flatbed methods under waterlogged conditions.

Raised bed planting reduces soil saturation, promoting healthier root development. Additionally, adopting controlled irrigation minimizes overirrigation and prevents waterlogging, ensuring optimal soil conditions for rice growth. Installing drainage systems improves soil aeration and removes excess water from the root zone. For flood-prone areas, employing higher seeding rates, AG-tolerant rice varieties (e.g., Tulasiphula, Panirohi, BJ 1, Chinamali, ARC 14855, MTU 1140, TKM13, Anna R4), and increased N fertilization can improve crop establishment and yield. Using vigorous seedlings from low-density nurseries and adopting optimal plant spacing (20 × 10 cm, one seedling $hill^{-1}$) further enhances yield potential. Wider spacing (20 × 15 cm) improves light penetration, enhances photosynthesis, and facilitates quick recovery from stress (Bhaduri et al., 2020).

17.3.4 Seedling age and depth of sowing

The age of rice seedlings at transplantation plays a crucial role in their resilience under submerged conditions. Older seedlings, aged 35–45 days, outperform younger ones in systems with poor water control and uneven fields, as taller seedlings are better equipped to avoid flood-induced losses (Bhowmick et al., 2014). Sowing depth also significantly affects seedling emergence under flooding conditions. Seeds sown deeper, at 2 cm, experience greater challenges in establishment compared to shallower sowing depths of 1 cm or less (Chamara et al., 2018). In rice, transplanting 40-day-old seedlings at a seed density of 25 g m^{-2}, combined with balanced nutrient application, resulted in higher benefit–cost ratios across varying conditions (Sarangi et al., 2015).

17.3.5 Nutrient management

Appropriate nutrient management is critical for mitigating the adverse effects of waterlogging and enhancing plant growth, survival, and recovery. A balanced supply of N, P, and K promotes these outcomes. The Si application also strengthens cell walls, enhances lodging resistance, and boosts photosynthetic activity (Jinger et al., 2020). Similarly, K supplementation is particularly effective in waterlogging conditions. Both foliar and soil applications of K were found to be most effective in improving growth, photosynthesis, and nutrient uptake (Vijayakumar et al., 2023). The application of 40 kg K_2O ha^{-1} as basal with a 0.5% K foliar spray at the panicle initiation stage increased yield by 28% compared to the control and by 5.1% over split K_2O application (Dwivedi et al., 2017). Additionally, combining P and N applications after flooding significantly improves outcomes compared to N alone. For example, a urea foliar spray (2%) combined with P fertilization increased grain yields by 87.6%, 53.1%, and 47.6% in IR-20, IR-64 Sub1, and Swarna Sub1 rice varieties, respectively, compared to unfertilized controls whereas N alone reduced yields by 25.8%, 17.8%, and 17.1%, respectively (Gautam et al., 2015).

Using slow-release fertilizers like neem-coated urea, urea super granules, urea mega granules in flooded or calcareous soils to reduce N loss. Postsubmergence N application timing also matters. Applying 20–30 kg N ha^{-1} resulted in 8.4%–15.9% higher yields compared to 10 kg N ha^{-1} in rice. When 30%–35% of plants exhibited at least one green leaf, N application led to 23.6% and 13.2% higher yields compared to applications at 10%–15% or 65%–70% green leaf recovery, respectively, in rice (Behera et al., 2019). Tools like the leaf color chart and site-specific nutrient management improve N application efficiency in deficient soils (Debarati et al., 2023). Overall, in flood-prone conditions, transplanting 40–44-day-old seedlings and optimal fertilization practices, including applying 80-40-40 kg N-P_2O_5-K_2O ha^{-1} + an additional 20-20 kg N-K_2O ha^{-1} at 5–7 days after floodwater recedes, significantly enhance postsubmergence recovery and grain yield of rice crop (Bhowmick et al., 2014). For stagnant flooding in coastal rainfed lowlands, the recommended package includes 50-30-15 kg N-P_2O_5-K_2O ha^{-1}, 5 t FYM ha^{-1}, and transplanting 2 seedlings $hill^{-1}$ of rice at 15 cm x 15 cm spacing. This combination optimizes productivity and economic returns while exceeding 50 kg N ha^{-1}, but did not increase the grain yield of rice and inflated costs (Sarangi et al., 2015).

17.3.6 Sensor-based nutrient application

Sensor-based nutrient management provides soil and crop nutrient levels in real-time for site-specific nutrient applications. These sensors measure parameters like soil moisture, pH, nutrient content, and plant health, enabling precise nutrient application based on crop needs. This approach reduces nutrient wastage, minimizes environmental impact, and enhances crop productivity. Sensors like GreenSeeker, leaf color chart, and Soil Plant Analysis Development (SPAD) value detect N deficiency at early growth stages by estimating plant chlorophyll levels and reducing the overall cost of fertilization by saving N fertilizers (Ladha et al., 2005). Similarly, multispectral remote sensing and machine learning enable real-time N control, facilitating variable rate application and site-specific crop management (Sishodia et al., 2020).

Artificial intelligence (AI) further improves sensor-based nutrient management by analyzing sensor data to identify patterns and provide accurate predictions. AI algorithms optimize nutrient recommendations, predict deficiencies, and suggest corrective actions. Machine learning models integrate data from weather forecasts, soil conditions, and crop growth to dynamically adjust nutrient schedules, enhancing efficiency, reducing costs, and promoting sustainable farming practice (Vijayakumar et al., 2022). Several smartphone-based applications and technologies, such as RiceNxpert (estimating N nutrition status in the rice leaf using digital color image analysis), Farm Calculator (detecting soil nutrient levels), Khamari app (soil based nutrient recommendation for Bangladesh) and the Baikhao app (assessing soil fertility status, identifying soil type, and recommending nutrient inputs), assist farmers in detecting soil nutrient status and receiving tailored recommendations (Vijayakumar et al., 2022).

17.4 Agronomic management options for salinity and alkalinity stress-induced multiple abiotic stress

17.4.1 Soil amendments

17.4.1.1 Mineral amendments

Managing salinity and alkalinity requires targeted amendments. Mineral amendments are particularly effective in reclaiming saline and sodic soils by improving their chemical and physical properties. Gypsum ($CaSO_4 \cdot 2H_2O$) is the most widely used amendment for saline-sodic and sodic soils. It displaces exchangeable Na^+ with Ca^{2+}, enhancing soil structure, infiltration, and drainage. Gypsum application combined with leaching reduced the exchangeable sodium percentage (ESP) by 21%–28% and increased barley yield by 45% in saline-sodic soils (Mbarki et al., 2001). Land reclamation using gypsum and press mud lowered soil sodicity, enhanced crop salt tolerance, and boosted yields by about 35% compared to the control (2.19 t ha^{-1}) (Sheoran et al., 2021). However, its effectiveness is contingent on sufficient water availability for leaching, a critical limitation in arid regions. Elemental sulfur effectively reduces soil pH and enhances nutrient availability in alkaline and calcareous soils. When combined with compost, elemental sulfur decreased soil pH from 8.7 to 7.9, increased P availability, and improved crop yield by 28%. Calcium nitrate [$Ca(NO_3)_2$] is another beneficial amendment, particularly for sodium-affected soils, providing Ca and N. Its foliar application increased Ca uptake, reduced blossom-end rot, and enhanced tomato yield and quality under saline conditions (Sajid et al., 2020).

17.4.1.2 Organic amendments

Organic amendments such as compost, biochar, humic acid, vermicompost, tricho-compost, and farmyard manure improve soil water-holding capacity, aggregation, and nutrient retention, enhancing plant tolerance to salinity and alkalinity. Biochar, derived from the pyrolysis of organic matter, enhances soil porosity, cation exchange capacity, microbial activity, and water retention. Its effectiveness varies based on biomass type and pyrolysis conditions. A meta-analysis by Dai et al. (2023) reported an 18% increase in crop yield in salt-affected soils, with more pronounced effects in sandy soils compared to clayey ones. Different composts, made from decomposed organic materials like crop residues, food waste, and farmyard manure (FYM), improve soil aggregation, infiltration, and organic matter content. During decomposition, it releases organic acids that solubilize $CaCO_3$, aiding sodic soil reclamation.

17.4.1.3 Integrated use of organic and mineral amendments

Combining mineral and organic amendments offers a holistic strategy to improve soil health in salt-affected areas. For example, integrating gypsum with compost or FYM maximizes benefits by addressing both chemical imbalances and physical soil degradation. Nano-materials such as nano-silica, nano-humic acid, and nano-zeolite have shown remarkable potential in improving soil properties and promoting plant growth under saline conditions. For example, nano-silica applications have been found to enhance water retention, reduce sodium uptake, and boost wheat biomass by 23% (Mushtaq et al., 2017).

Combined biochar, humic substances, and gypsum application boosted quinoa root biomass (206% in AZ-51; 176% in AZ-103) and significantly reduced ECe, sodium absorption ratio, and ESP (Alcívar et al., 2018). Incorporating green manure crops like dhaincha (*Sesbania aculeata*) improves soil iron availability. Integrating chemical fertilizers with organic inputs (e.g., farmyard manure, compost) and recommended micronutrient doses enhances soil health and grain quality (Sushma et al., 2024). Split application of vermicompost at the recommended N increased rice yield by 20% over basal application, improved rice grain quality (higher micronutrients and 7% more amylose), and left a positive residual effect on chickpea yield. It also enhanced rice-equivalent yield, benefit–cost ratio, and energy efficiency of the rice–chickpea system compared to chemical fertilizer (Kumar et al., 2018).

17.4.2 Irrigation management

Effective irrigation management is essential for mitigating salinity and alkalinity stresses, as improper practices, such as the overuse of saline water, exacerbate soil salt accumulation. Employing advanced irrigation techniques, such as sprinkler and drip irrigation methods, can optimize water use, minimize salt buildup, and improve crop productivity in salt-affected soils.

17.4.2.1 Precision irrigation techniques

Drip and sprinkler irrigation systems deliver water directly to the root zone, reducing water wastage and salt accumulation. These systems are particularly beneficial in arid and semiarid regions. Leaching, the practice of flushing salts from the root zone with excess water, remains a common approach, but it is water-intensive and less viable in water-scarce areas. To address this, innovations inefficient leaching methods and the use of treated wastewater for irrigation are being explored. Drip irrigation delivers water directly to the root zone, reducing evaporation and preventing salt accumulation on the soil surface. Drip irrigation is particularly effective in arid and semiarid regions, where water scarcity and high evaporation rates exacerbate salinity stress.

Mulched drip irrigation raises shallow (0–40 cm) soil water and fine root biomass, enhancing root absorptive capacity, aboveground growth, and cotton yield. It optimizes water–root relationships and irrigation WUE, but excessive irrigation reduces yield and efficiency (Wang et al., 2021). Similarly, Ramesh et al. (2024) recommended cultivating the short-duration rice variety "TRY (R) 2" under drip irrigation with 90 cm lateral spacing and 30 cm emitter spacing in sodic soils irrigated with saline water to improve grain yields (2.99 t ha^{-1}) and water productivity (3.58 kg ha^{-1} mm^{-1}). Additionally, drip irrigation supports yield stability under deficit irrigation, making it suitable for salt-sensitive crops like peppers. Integrating sensor technologies and automated irrigation systems enables real-time monitoring of soil moisture and salinity, optimizing water use while preventing overirrigation.

17.4.2.2 Other irrigation techniques

Alternate furrow irrigation applies water to every other furrow in successive cycles, reducing waterlogging and minimizing salt accumulation near the root zone. Zhang et al. (2024) reported that partial root-zone irrigation enhances stress resistance over conventional irrigation by reducing transpiration and boosting photosynthesis, proline, and yield. Alternate root-zone irrigation (ARI) balances root growth, delays senescence, and further improves stress resistance and yield. Potassium (K) application strengthens these effects by raising the K/Na ratio, proline, photosynthesis, and yield, with ARI + K being the most effective strategy. Pulse irrigation applies water in short, frequent intervals rather than continuously, ensuring uniform soil moisture distribution and limiting salt accumulation in the root zone. This technique is particularly effective in sandy and loamy soils. By preventing excessive salt accumulation and ensuring effective moisture delivery, pulse irrigation enhanced plant growth and productivity under saline conditions.

17.4.3 Crop selection and breeding

Developing crop varieties that can tolerate salinity and alkalinity is a critical long-term strategy to combat these stresses and enhance agricultural productivity in affected areas. This approach involves both traditional breeding techniques and advanced molecular tools, with varying levels of success and challenges. Traditional breeding has successfully developed salt-tolerant varieties in staple crops like rice, wheat, and barley by selecting genotypes capable of withstanding osmotic stress, ionic toxicity, and nutrient imbalances (Table 17.5). For example, the salt-tolerant rice variety "Saltol"

TABLE 17.5 Crop-wise waterlogging-tolerant cultivars/varieties.

Crop	Deep water/submergence/waterlogging-tolerant variety
Rice	Reeta, CR Dhan 500, CR Dhan 503, MTU 1140, PLA 1100, CR 1009-Sub1 (Savitri-Sub1), Ciherang-Sub1 (BINA dhan 11), Swarna Sub-1 (improved Samba Mahsuri), IR-64 Sub-1, Sambha Mahsuri Sub-1, Varshadhan, Gayatri, Sarla, Pooja, Prateeksha, Durga, JalaMani, CR Dhan 505, CR Dhan 502, Jalnidhi, Neerja, Jaladhi 1, Jaladhi 2, Hemavathi
Submergence + Drought-tolerant rice varieties	CR Dhan 801 (Swarna-Sub1 + Drt *qtl*), DRR 50 (Samba-Sub1 + Drt *qtl*), Bahuguni dhan 1 (Swarna-Sub1 + Drt *qtl*), Bahuguni dhan 2 (Swarna-Sub1 + Drt *qtl*)
Submergence-induced nutrient stress	Swarna Sub1, Jalamani, Ksheera
Wheat	AAI-W9, KRL-19
Maize	HM-5, Seed Tech-2324, HM-10, PMH-2
Sugarcane	Co 98014 (Karan-1), Co 0239, Co 0118, Co 0238, Co 0233, Co 05009
Jute	JRO 7835, JRO 878, JRC 321, JRC 7447, JRC 532, JRC-517, Bidhan Pat-1

demonstrates resilience under saline conditions, retaining 80% of its yield under moderate salinity stress (ECe = 8 dS m^{-1}). Additionally, managing ROS production through antioxidant compounds and stress-responsive gene modulation has emerged as a promising strategy to enhance plant tolerance to salinity (Che et al., 2016). However, breeding for alkalinity tolerance remains particularly challenging due to the intricate interplay between high soil pH and plant physiological responses, requiring further research into alkalinity-specific tolerance mechanisms.

Genetic modification has introduced stress-responsive genes and ion transporters into crops to enhance tolerance to salinity and alkalinity. For example, overexpression of genes like high-affinity potassium transporters and salt overly sensitive 1 (*SOS1*) improves ion homeostasis and salt exclusion. Transgenic crops with these modifications have shown enhanced growth and yield in saline environments. Despite these advancements, the widespread adoption of genetically modified crops is limited by regulatory hurdles, public perception, and the necessity for extensive field trials across diverse agro-climatic conditions. CRISPR/Cas9 technology has revolutionized crop improvement by enabling precise genome edits, particularly for stress-tolerance genes. For instance, editing the *SlHKT1;2* genes in tomatoes improved sodium exclusion and stabilized yields under saline conditions (Wang et al., 2011). This precision breeding approach offers faster and more targeted solutions compared to traditional methods, reducing the time and resources required to develop stress-tolerant varieties.

17.4.4 Saline-tolerant rice varieties

Indigenous rice varieties like Uri dhan, Pokkali, Nona Bokra, Dular, Azgo, and Damodar are renowned for their ability to tolerate high salinity, employing a combination of physiological and biochemical mechanisms. These mechanisms include sodium exclusion (restricting sodium uptake to maintain ionic balance), tissue tolerance (compartmentalizing salts in the apoplast and older tissues to prevent cellular damage), stomatal responsiveness (regulating stomatal closure to reduce water loss during stress), and antioxidant system activation (upregulating antioxidant defenses to mitigate oxidative stress caused by salinity) (Moradi and Ismail, 2007). Over the past decade, significant advancements have been made in identifying QTLs linked to salinity tolerance. The "Saltol" QTL, located on chromosome 1, stands out as a pivotal discovery. Ionic stress from excess Na^+ disrupts K^+ homeostasis, making the Na^+/K^+ balance vital for salinity tolerance. Halophytes outperform glycophytes due to stronger gene expression, protein modifications, and the ability to maintain stable plasma membrane potential. Key regulators like *SOS1*, *Nax1/Nax2*, *haem oxygenase*, and *H^+-ATPase* enhance K^+ uptake and salt tolerance (Assaha et al., 2017). The incorporation of the "Saltol" QTL into locally adapted rice varieties has accelerated the development of salinity-tolerant cultivars. Examples of varieties being enhanced with this QTL include ADT 45, CR 1009, Gayatri, MTU 1010, PR 114, Pusa 44, and Sarjoo 52. These varieties cater to inland saline and sodic areas, offering improved productivity and resilience against salt-induced stresses. BRRI and BINA also introduced salt-tolerant rice varieties like BRRI dhan 53, BRRI dhan 54, BRRI Dhan 67, BRRI dhan 97, BRRI dhan 99, and BINA dhan-10. These varieties have shown promising yields under salinity conditions and are being adopted by farmers in coastal areas of Bangladesh.

17.4.4.1 Seed agreement: a boon to the faster dissemination of stress-tolerant rice varieties

For the faster movement and strengthening of the rice seed system in South Asia, the first seed agreement was signed between India and Bangladesh on February 17, 2013. After one and a half years, Nepal joined the agreement on October 18, 2014. It opens the gate for the quick dissemination of the STRVs across the international boundary. STRVs are evaluated and released in any country without any further evaluation in India, Bangladesh, and Nepal. There was a breakthrough in the seed agreement when Sri Lanka, Cambodia, and Vietnam joined the "Seeds without Borders" agreement at Siem Reap, Cambodia. During 2016–17, India had record rice production, which is 4.54 million tons higher than 2015–16. Out of this, 3.0 million tons are contributed by seven eastern Indian states, where STRVs (mainly Swarna-Sub1 and Sahbhagi dhan) have played a major role.

17.4.5 Halophyte cultivation

Halophytes, naturally salt-tolerant plants, offer a promising avenue for agriculture in highly saline environments. These plants are adapted to thrive in soils with elevated salt concentrations, making them an ideal choice for utilizing marginal lands. For example, *Salicornia europaea* and *Chenopodium quinoa* have been successfully cultivated in soils with electrical conductivity (ECe) exceeding 20 dS m^{-1}, highlighting their potential as a source for food, feed, and biofuel production (Ventura et al., 2015). Phosphorus fertilization (60–70 kg P_2O_5 ha^{-1}) mitigates salinity stress in quinoa by boosting yield, biomass, and essential nutrient uptake. It improved seed yield up to 51% and is an effective strategy for saline soils (Bouras et al., 2022). By integrating halophyte cultivation into agricultural practices, salt-affected lands can be transformed into productive ecosystems, contributing to global food security where traditional crops are unsuitable.

17.4.6 Plant growth-promoting rhizobacteria

Plant growth-promoting rhizobacteria enhance plant growth by improving nutrient availability, uptake, and stimulating root development, and strengthening natural defenses against stress. In saline soils, strains like *Bacillus, Pseudomonas, and Azospirillum* enhance plant growth and yield by boosting root development and nutrient uptake. For example, inoculation of *Bacillus amyloliquefaciens* SQR9 with maize increased its yield by 22% under saline conditions (ECe = 7 dS m^{-1}) by promoting root growth and K uptake (Che et al., 2016).

17.4.7 Arbuscular mycorrhizal fungi

Arbuscular mycorrhizal fungi (AMF) form symbiotic associations with plant roots, facilitating water and nutrient acquisition, notably P and Zn, and enhancing tolerance to salinity through osmolyte accumulation. Inoculating tomato plants with *Rhizophagus irregularis*, combined with gibberellic acid foliar application, boosted tomato fruit yield by 35% under moderate salinity (ECe = 6 dS m^{-1}) (Khalloufi et al., 2017). Indigenous AMF application in wheat showed strong genotype × inoculation and year × genotype interactions, with AMF composition unchanged among genotypes but AMF structure modified by genotype and growth stage (Marrassini et al., 2024).

17.4.8 Seed priming

Seed priming with selected organic and inorganic nutrients has emerged as an effective strategy for improving crop performance in salt-affected soils. Priming rice seeds with 1% $Ca(NO_3)_2$ and 20 mg L^{-1} gibberellic acid significantly increased yield in the NDR 359 rice variety (Khan et al., 2014). The integration of bio-compost prepared from press mud (a sugar industry by product) with a nutrient regimen of 120-40-40 kg ha^{-1} N-P_2O_5-K_2O further boosted the yields of NDR 359 and the salt-tolerant rice variety Usar dhan 3, demonstrating the synergy between organic amendments and chemical fertilizers (Khan et al., 2014).

17.4.9 Potassium and silicon supplementation

K is vital for maintaining ionic balance and osmotic adjustment in plant cells, particularly under saline conditions. Foliar application of potassium sulfate (K_2SO_4) alleviated salinity stress in wheat and increased grain yield by 18% under saline conditions (ECe = 10 dS m^{-1}) (Tuna et al., 2007). Similarly, Si supplementation enhances salt stress tolerance by improving physiological and biochemical processes. In mung beans, it reduced Na accumulation in shoots and increased antioxidant enzyme activity (Ahmad et al., 2019), while Foliar application of silicon (Si) and Se improved tomato growth, with Si boosting yield and Se enhancing fruit quality (sugars, vitamin C, phenols, anthocyanin, lycopene, carotenoids, protein, and reduced nitrates) (Hu et al., 2023). For aerobic rice, the combined application of 120 kg ha^{-1} Si and 90 kg ha^{-1} P_2O_5 significantly increased dry matter accumulation, grain yield, and nutrient uptake in soils with a pH of 7.7 (Jinger et al., 2021).

17.4.10 Customized fertilizers

Customized fertilizers (CF) are multinutrient carriers formulated with macro- and micronutrients to meet specific crop requirements based on location, soil type, and crop stage. CF is particularly suited for nutrient-stressed environments such as acidic, saline, or alkaline soils. Their use improves NUE, which is currently at 40% for N, 20% for P, 50% for K, and 2%–5% for micronutrients, and addresses long-term soil nutrient deficiencies (Dass et al., 2017). Incorporating micronutrients with macronutrients ensures balanced fertilization, enhances NUE, and prevents deficiencies. CF also reduces workloads, saves time and costs, and addresses nonuniform application issues. India's first official CF, "Paras Formula" (10% N, 18% P_2O_5, 25% K_2O, 3% S, and 0.5% Zn), was introduced on November 22, 2010, by Tata Chemicals Fertilizers Limited for wheat cultivation in Western Uttar Pradesh. Another example is the 8% N, 16 P_2O_5, 24 K_2O, 6% S, 0.5% Zn, and 0.15% B formulation for potatoes in Uttar Pradesh's Agra, Aligarh, and Farrukhabad districts. Similarly, a 10% N, 13% P_2O_5, 12% K_2O, 6% S, 2% Zn, 0% B formulation was designed for sweet sorghum in Nanded, Maharashtra. Coromandel Industries Limited quantified the yield loss due to abiotic stress, underscoring its critical impact. Ed developed two approved formulations (15% N, 15 P_2O_5, 15% K_2O, 9S, 0.5 Zn, 0.2 B, and 20% N, 0% P_2O_5, 15% K_2O, 0% S, 0% Zn, 0.2% B) for groundnut and maize in Andhra Pradesh's Anantapur and Warangal districts, respectively. To date, 36 CF formulations have been approved under India's Fertilizer Control Order.

17.4.11 Maintenance of soil pH

As soil pH increases, micronutrient availability decreases significantly, particularly for metal nutrients, which decline by nearly 100 times for every one-unit rise in pH, except for Mo. This occurs because these nutrients sorb onto soil minerals, rendering them unavailable. Conversely, an increase in H^+ ions causes soil acidification, increasing the solubility of heavy metals like Al, Fe, and Mn. These metals can become toxic to plants and microbes, inhibiting microbially-driven processes (Kochian, 1995). Acidity-induced toxicities, such as excess Fe and Mn, can be mitigated by adding organic inputs like organic matter, biochar, and lime (CaO, $CaCO_3$, $CaSiO_3$, $CaCO_3$-$MgCO_3$). Si is another effective agent for mitigating Fe and Mn toxicity. Si applied as a spray or manual broadcasting on the soil surface moderates toxicity by forming Si-Mn and Si-Fe complexes (Che et al., 2016).

For B deficiency, apply borax at 5–10 kg ha^{-1} as a basal dose. In severe cases, use a 0.2% foliar spray of boric acid or 0.05% borax for quick recovery (Vijayakumar et al., 2021). Sulfate-based fertilizers ($ZnSO_4$, $FeSO_4$, $MnSO_4$, $CuSO_4$) are ideal for basal application due to their rapid nutrient delivery. Severe deficiencies can be addressed with foliar sprays, such as 0.5% $ZnSO_4$, 0.25% $FeSO_4$, and 0.5% $MnSO_4$. Combining soil and foliar applications is highly effective. For instance, applying 30 kg ha^{-1} $FeSO_4.7H_2O$ as a basal dose, followed by 1% $FeSO_4.7H_2O$ foliar sprays at 10-day intervals, boosts growth and grain iron content. For multinutrient deficiencies, compound nano fertilizers like Fe-Mn-Zn effectively address deficiencies and improve NUE. On average, yield increases of 10%–25% in cereals, 20%–30% in oilseeds, and 13%–15% in pulses have been observed with nano fertilizer sprays compared to conventional fertilizers (Rathore et al., 2019).

17.4.12 Foliar spray of nano-fertilizers

Unlike conventional fertilizers, which are often applied in large quantities and are prone to significant losses through leaching or runoff, nano-fertilizers are composed of nano-sized particles (typically 1–100 nanometers). This extremely small size gives them unique properties that are highly beneficial for agriculture. Their tiny size allows them to easily penetrate plant cells and be absorbed more efficiently, whether applied to the soil or as a foliar spray on leaves. Foliar nutrition serves as a valuable strategy to address nutrient deficiencies in crops during critical growth stages. It also plays a significant role in reducing the soil limitations on plant development and nutrient uptake (Table 17.6). Foliar application of N and K fertilizers at the reproductive stage, particularly in leguminous crops under stress conditions, significantly slows

TABLE 17.6 Salinity stress-tolerant rice varieties released in India.

Crop	Name of the variety	Ecosystem
Rice	CSR 30, CSR 36, CSR 43, Co-43, TRY 1, TRY 2, Usar dhan 1, Usar dhan 2, Usar dhan 3	Inland salinity
Rice	CR Dhan 405 (Luna Sankhi), CR Dhan 406 (Luna Barial), DRR 39	Coastal salinity
Rice	CSR 46, CSR 60, CSR 56, CARI Dhan 5	Salinity-induced nutrient stress
Rice	CSR 43	Salinity and submergence
Wheat	KRL 19, 210, 283	Waterlogging–salinity-induced nutrient stress
Maize	Jawahar Maize 218, Pragati	Waterlogging-induced nutrient stress
Pigeonpea	Jagriti, Co-5	Salinity-induced nutrient stress
Chickpea	MAL 13, Co-5	Waterlogging and salinity
Greengram	ADT 43	Salinity-induced nutrient stress
Cowpea	Hisar cowpea 46	Salinity-induced nutrient stress
Soybean	Jawahar Soybean 97-52, Ahilya 4, ADT 1	Waterlogging-induced nutrient stress
Sunflower	TNAUSUF-1	Salinity-induced nutrient stress
Groundnut	M-13	Salinity-induced nutrient stress
Linseed	PSL 9	Salinity-induced nutrient stress
Mustard	CS58, 60	Salinity-induced nutrient stress

Note: The value in the bracket indicates the concentration of the foliar spray.

abscisic acid synthesis while accelerating cytokinin production, which promotes nodulation, cell elongation, apical dominance, shoot development, photosynthetic activity, and assimilate translocation to the sink organs and increases stress-tolerant (Swain et al., 2023).

17.5 Future perspectives

While significant progress has been made in managing abiotic stresses, several research gaps remain that need to be addressed to maximize the benefits of different agronomic management strategies (Table 17.7).

- More research is needed to understand the complex interactions between soil properties, plant physiology, and microbial communities under abiotic stress conditions. This knowledge is crucial for developing site-specific management practices. Integrating genomics, proteomics, and metabolomics (omics technologies) could help identify stress-tolerance genes and pathways, accelerating the development of resilient crop varieties.
- Although abiotic stress-tolerant varieties have been developed, there is still limited success in breeding crops with high tolerance to abiotic stress. A significant focus should be given to developing abiotic stress-tolerant varieties through gene-editing technologies like CRISPR/Cas9, as it can enable precise modifications in crop genomes to enhance tolerance to abiotic stresses.
- Water scarcity limits the effectiveness of leaching and irrigation-based solutions. Developing water-saving technologies like precision irrigation and alternative water resources, such as brackish water and wastewater, is essential for long-term sustainability.
- Nanoparticles, nano-fertilizers, and bio-nanocomposites have emerged as innovative materials to enhance soil health and improve plant stress tolerance. When combined with existing amendments like vermicompost, humic acid, biochar, zeolite, and gypsum, these materials provide synergistic benefits, enhancing soil structure, nutrient availability, and water retention. However, technological advancement to make these amendments largely available at lower prices is important.

TABLE 17.7 Foliar application of nutrients to address the nutrient stress.

Nutrients	Compound	Suitability for foliar spray	Effects	Crop
N	Urea	Very good (2%)	Increase yield, N uptake, and grain protein content	Rice (Alim et al., 2023)
	NH_4NO_3	Average		
K	$K_2CO_3{\cdot}2H_2O$ K_2SO_4	Good Very good	Increases K uptake, biotic and abiotic stress resistance	Rice (Sultana et al., 2001)
Ca	$CaCl_2.6H_2O$ $CaNO_3$	Very good Very good	Increase in yield	
Mg	$MgCl.6H_2O$	Very good	Increase yield	
Fe	$FeSO_4$	Very good (0.5%)	Increase yield and Fe concentration in grain	Wheat (El-Fouly et al., 2011)
Mn	$MnSO_4$	Very good (0.25%)	Increase yield and N uptake in grain	
B, Cu	B_2O_3 and CuO	Very good	Increased biomass, uptake of N, Zn, B, and Cu	Soybean (Dimkpa et al., 2017)
Zn	$ZnSO_4$	Very good (0.5%)	Increase in yield and Zn uptake in grain	Rice (Kandil et al., 2022)
Mo	Na_2MoO_4 (NH_4) $6Mo_7O_{24}.4H_2O$	Very good	Improve the Mo availability	-
Si	Silicic acid	Very good	Positive interaction with Fe and Zn uptake and availability	Pavlovic et al. (2021)

- Techniques such as IoT-based technologies, remote sensing, and machine learning enable early detection and mapping of abiotic stress, allowing for targeted interventions. These technologies can also monitor crop health in real time, supporting informed decisions regarding irrigation and nutrient management, thus improving resource efficiency. However, these technologies are still in the research stage. The benefits and adoption feasibility of these technologies are yet to be tested in the farmer's field.
- Cost-benefit analyzes will identify the most economically viable options for developing policies and incentives that will encourage farmers to adopt these improved management practices, ensuring accessibility and affordability, particularly in developing regions.

17.6 Conclusion

As the global food demand keeps increasing, satisfactorily managing the multiple abiotic stresses simultaneously through agronomic management practices is critical for sustaining agricultural productivity and ensuring the food security of ever-growing populations. Understanding the complex interactions between soil, plants, and microbial communities under these stresses will pave the way for management practices that improve soil health and crop resilience under these stresses. A holistic approach, combining soil amendments, tolerant cultivars, organic manures, nutrient management, foliar spray, precision irrigation and drainage, and biological solutions, offers the best prospects for mitigating these stresses. Technological advancements like remote sensing and AI-driven decision support systems enable real-time monitoring and mitigation of stress impacts. Policy interventions must prioritize investment in research, extension services, and infrastructure to facilitate the adoption of these solutions at scale. Collaborative efforts among researchers, policymakers, and farmers will be crucial in building a sustainable and food-secure future.

References

Abdullah, A.S., Aziz, M.M., Siddique, K.H.M., Flower, K.C., 2015. Film antitranspirants increase yield in drought stressed wheat plants by maintaining high grain number. Agricultural Water Management 159, 11–18. Available at https://doi.org/10.1016/j.agwat.2015.05.018.

Acosta-Motos, J.R., Ortuño, M.F., Bernal-Vicente, A., Diaz-Vivancos, P., Sanchez-Blanco, M.J., Hernandez, J.A., 2017. Plant responses to salt stress: adaptive mechanisms. Agronomy 7 (1). Available at https://doi.org/10.3390/agronomy7010018.

Adhikary, S., Biswas, B., Chakraborty, D., Timsina, J., Pal, S., Chandra Tarafdar, J., et al., 2022. Seed priming with selenium and zinc nanoparticles modifies germination, growth, and yield of direct-seeded rice (Oryza sativa L.). Scientific Reports 12 (1). Available at https://doi.org/10.1038/s41598-022-11307-4.

Aggag, A.M., Alzoheiry, A.M., Abdallah, A.E., 2015. Effect of kaolin and fulvic acid antitranspirants on tomato plants grown under different water regimes. Alexandria Science Exchange Journal 36 (April-June), 169–179. Available at https://doi.org/10.21608/ASEJAIQJSAE.2015.2875.

Ahmad, P., Ahanger, M.A., Alam, P., Alyemeni, M.N., Wijaya, L., Ali, S., et al., 2019. Silicon (Si) supplementation alleviates NaCl toxicity in mung bean [Vigna radiata (L.) Wilczek] through the modifications of physio-biochemical attributes and key antioxidant enzymes. Journal of Plant Growth Regulation 38 (1), 70–82. Available at https://doi.org/10.1007/s00344-018-9810-2.

Akhtar, N., Ilyas, N., Arshad, M., Meraj, T.A., Hefft, D.I., Jan, B.L., et al., 2022. The impact of calcium, potassium, and boron application on the growth and yield characteristics of durum wheat under drought conditions. Agronomy 12 (8). Available at https://doi.org/10.3390/agronomy12081917.

Alcívar, M., Zurita-Silva, A., Sandoval, M., Muñoz, C., Schoebitz, M., 2018. Reclamation of saline–sodic soils with combined amendments: impact on quinoa performance and biological soil quality. Sustainability 10 (9), 3083. Available at https://doi.org/10.3390/su10093083.

Alim, M.A., Hossain, S.I., Ditta, A., Hasan, M.K., Islam, M.R., Hafeez, A.S.M.G., et al., 2023. Comparative efficacy of foliar plus soil application of urea versus conventional application methods for enhanced growth, yield, agronomic efficiency, and economic benefits in rice. ACS Omega 8 (39), 35845–35855. Available at https://doi.org/10.1021/acsomega.3c03483.

Ali, L.G., Nulit, R., Ibrahim, M.H., Yien, C.Y.S., 2020. Enhancement of germination and early seedling growth of rice (Oryza sativa) var. FARO44 by seed priming under normal and drought stressed conditions. Journal of Plant Nutrition 43 (11), 1579–1593. Available at https://doi.org/10.1080/01904167.2020.1739298.

Ambati, R.R., Majumdar, G., Reddy, A.R., 2011. Validation of farm pond size for irrigation during drought. Indian Journal of Agronomy 56 (4), 356–364. Available at https://doi.org/10.59797/IJA.V56I4.4701.

Angaji, S.A., Septiningsih, E.M., Mackill, D.J., Ismail, A.M., 2010. QTLs associated with tolerance of flooding during germination in rice (Oryza sativa L.). Euphytica 172 (2), 159–168. Available at https://doi.org/10.1007/s10681-009-0014-5.

Assaha, D.V.M., Ueda, A., Saneoka, H., Al-Yahyai, R., Yaish, M.W., 2017. The role of Na+ and K+ transporters in salt stress adaptation in glycophytes. Frontiers in Physiology 8. Available at https://doi.org/10.3389/fphys.2017.00509.

Aung, M.S., Masuda, H., 2020. How does rice defend against excess iron? physiological and molecular mechanisms. Frontiers in Plant Science 11, 546092. Available at https://doi.org/10.3389/FPLS.2020.01102.

Behera, U.K., Sharma, A.R., 2014. Productivity and water use efficiency of wheat (*Triticum aestivum*) under different resource conservation techniques and irrigation regimes. Cereal Research Communications 42 (3), 439–449. Available at https://doi.org/10.1556/CRC.2013.0072.

Behera, U.K., Singh, S., Sarangi, S.K., Behera, S.K., Bishoyi, B.S., Srivastava, A.K., et al., 2019. Post-submergence nitrogen fertilizer management for enhancing rainfed lowland rice productivity in eastern India. International Journal of Bio-resource and Stress Management 10 (4), 419–428. Available at https://doi.org/10.23910/ijbsm/2019.10.4.2011a.

Bhaduri, D., Chakraborty, K., Nayak, A.K., Shahid, M., Tripathi, R., Behera, R., et al., 2020. Alteration in plant spacing improves submergence tolerance in Sub1 and non-Sub1 rice (cv. IR64) by better light interception and effective carbohydrate utilisation under stress. Functional Plant Biology 47 (10), 891–903. Available at https://doi.org/10.1071/FP19364.

Bhowmick, M.K., Dhara, M.C., Singh, S., Dar, M.H., Singh, U.S., 2014. Improved management options for submergence-tolerant (Sub1) rice genotype in flood-prone rainfed lowlands of West Bengal. Journal of Plant Sciences 05 (1), 14–23. Available at https://doi.org/10.4236/ajps.2014.51003.

Bouras, H., Choukr-Allah, R., Amouaouch, Y., Bouaziz, A., Devkota, K.P., El Mouttaqi, A., et al., 2022. How does quinoa (chenopodium quinoa willd.) respond to phosphorus fertilization and irrigation water salinity? Plants 11 (2), 216. Available at https://doi.org/10.3390/plants11020216.

Cai, W., Gu, X., Du, Y., Chang, T., Lu, S., Zheng, X., et al., 2022. Effects of mulching on water saving, yield increase and emission reduction for maize in China. Agricultural Water Management 274, 107954. Available at https://doi.org/10.1016/j.agwat.2022.107954.

Cakmak, I., 2005. The role of potassium in alleviating detrimental effects of abiotic stresses in plants. Journal of Plant Nutrition and Soil Science 168 (4), 521–530. Available at https://doi.org/10.1002/jpln.200420485.

Chamara, B.S., Marambe, B., Kumar, V., Ismail, A.M., Septiningsih, E.M., Chauhan, B.S., 2018. Optimizing sowing and flooding depth for anaerobic germination-tolerant genotypes to enhance crop establishment, early growth, and weed management in dry-seeded rice (Oryza sativa L.). Frontiers in Plant Science 871. Available at https://doi.org/10.3389/fpls.2018.01654.

Chary, G.R., Gopinath, K.A., Kumar, P.R., Bhaskar, S., Singh, V.K., Deepika, S., et al., 2021. Management of major abiotic stresses in field crops. Journal of Agronomy 66, S258.

Chaves, M.M., Maroco, J.P., Pereira, J.S., 2003. Understanding plant responses to drought - from genes to the whole plant. Functional Plant Biology 30 (3), 239–264. Available at https://doi.org/10.1071/FP02076.

Che, J., Yamaji, N., Shao, J.F., Ma, J.F., Shen, R.F., 2016. Silicon decreases both uptake and root-to-shoot translocation of manganese in rice. Journal of Experimental Botany 67 (5), 1535–1544. Available at https://doi.org/10.1093/jxb/erv545.

Chhaya, B., Yadav, A., Jogawat, P., Gnanasekaran, P., Kumari, N., Lakra, S.K., et al., 2021. An overview of recent advancement in phytohormones-mediated stress management and drought tolerance in crop plants. Plant Gene 25. Available at https://doi.org/10.1016/j.plgene.2020.100264.

Choudhury, B.U., Bouman, B.A.M., Singh, A.K., 2007. Yield and water productivity of rice–wheat on raised beds at New Delhi, India. Field Crops Research 100 (2–3), 229–239. Available at https://doi.org/10.1016/J.FCR.2006.07.009.

Dai, Y., Liao, Z., Lai, Z., Bai, Z., Zhang, F., Li, Z., et al., 2023. Interactive effects of planting pattern, supplementary irrigation and planting density on grain yield, water-nitrogen use efficiency and economic benefit of winter wheat in a semi-humid but drought-prone region of northwest China. Agricultural Water Management 287, 108438. Available at https://doi.org/10.1016/j.agwat.2023.108438.

Daryanto, S., Wang, L., Jacinthe, P.A., 2017. Global synthesis of drought effects on cereal, legume, tuber and root crops production: a review. Agricultural Water Management 179, 18–33. Available at https://doi.org/10.1016/J.AGWAT.2016.04.022.

Dar, M.H., Chakravorty, R., Waza, S.A., Sharma, M., Zaidi, N.W., Singh, A.N., et al., 2017. Transforming rice cultivation in flood prone coastal Odisha to ensure food and economic security. Food Security 9 (4), 711–722. Available at https://doi.org/10.1007/s12571-017-0696-9.

Dar, M.H., Waza, S.A., Shukla, S., Zaidi, N.W., Nayak, S., Hossain, M., et al., 2020. Drought tolerant rice for ensuring food security in eastern India. Sustainability (Switzerland) 12 (6), 1–17. Available at https://doi.org/10.3390/su12062214.

Dass, A., Jinger, D., Kaur, R., Kumar, S.V., Kumari, K., 2017. Addressing multinutrient deficiency in crops using customized fertilizer. Farming 67 (3), 22–25.

Das, H.P., 2005. Agricultural drought mitigation and management of sustained agricultural development in India. In: Sivakumar, M.V., Motha, R.P., Das, H.P. (Eds.), Natural Disasters and Extreme Events in Agriculture. Springer, Berlin Heidelberg, pp. 277–303. Available at http://doi.org/10.1007/3-540-28307-2_16.

Debarati, B., Shanmugam, V., Amrita, D., Bibhash, C.V., Mohanty, S., 2023. Carbon footprint in rice cultivation. In: Rakshit, A., Biswas, A., Sarkar, D., Meena, V.S., Datta, R. (Eds.), Handbook of Energy Management in Agriculture. Springer, Singapore, pp. 3–38. Available at https://doi.org/10.1007/978-981-19-7736-7_1-1.

Dimkpa, C.O., Bindraban, P.S., Fugice, J., Agyin-Birikorang, S., Singh, U., Hellums, D., 2017. Composite micronutrient nanoparticles and salts decrease drought stress in soybean. Agronomy for Sustainable Development 37 (1), 1–13. Available at https://doi.org/10.1007/S13593-016-0412.

Djurovic, N., Cosic, M., Stricevic, R., Savic, S., Domazet, M., 2016. Effect of irrigation regime and application of kaolin on yield, quality and water use efficiency of tomato. Scientia Horticulturae 201, 271–278.

Du, S., Xiong, W., 2024. Weather extremes shock maize production: current approaches and future research directions in Africa. Plants 13 (12), 1585. Available at https://doi.org/10.3390/PLANTS13121585.

Dwivedi, S.K., Kumar, S., Bhakta, N., Singh, S.K., Rao, K.K., Mishra, J.S., et al., 2017. Improvement of submergence tolerance in rice through efficient application of potassium under submergence-prone rainfed ecology of Indo-Gangetic Plain. Functional Plant Biology 44 (9), 907–916. Available at https://doi.org/10.1071/FP17054.

El-Fouly, M.M., Mobarak, Z.M., Salama, Z.A., 2011. Micronutrients (Fe, Mn, Zn) foliar spray for increasing salinity tolerance in wheat Triticum aestivum L. African Journal of Plant Science 5 (5), 314–322.

Eskandari, H., Alizadeh-Amraie, A., 2014. Improvement of lentil germination performance under salt and drought conditions using seed priming treatments. Seed Science and Technology 42 (1), 87–91. Available at https://doi.org/10.15258/SST.2014.42.1.09.

Evans, R.G., Sadler, E.J., 2008. Methods and technologies to improve efficiency of water use. Water Resources Research 44 (7). Available at https://doi.org/10.1029/2007WR006200.

FAO, 2024. Global status of salt-affected soils (Technical report). FAO, Rome, Italy, pp. 1. Available at 10.4060/CD3044EN.

Gautam, P., Lal, B., Raja, R., Tripathi, R., Shahid, M., Baig, M.J., et al., 2015. Effect of simulated flash flooding on rice and its recovery after flooding with nutrient management strategies. Ecological Engineering 77, 250–256. Available at https://doi.org/10.1016/j.ecoleng.2015.01.033.

Getachew, F., Bayabil, H.K., Hoogenboom, G., Teshome, F.T., Zewdu, E., 2021. Irrigation and shifting planting date as climate change adaptation strategies for sorghum. Agricultural Water Management 255. Available at https://doi.org/10.1016/j.agwat.2021.106988.

Gill, S.S., Tuteja, N., 2010. Reactive oxygen species and antioxidant machinery in abiotic stress tolerance in crop plants. Plant Physiology and Biochemistry 48 (12), 909–930. Available at https://doi.org/10.1016/j.plaphy.2010.08.016.

von Haden, A.C., Burnham, M.B., Yang, W.H., DeLucia, E.H., 2021. Comparative establishment and yield of bioenergy sorghum and maize following pre-emergence waterlogging. Agronomy Journal 113 (6), 5602–5611. Available at https://doi.org/10.1002/agj2.20832.

Hattori, T., Inanaga, S., Araki, H., An, P., Morita, S., Luxová, M., et al., 2005. Application of silicon enhanced drought tolerance in Sorghum bicolor. Physiologia Plantarum 123 (4), 459–466. Available at https://doi.org/10.1111/j.1399-3054.2005.00481.x.

Hattori, Y., Nagai, K., Furukawa, S., Song, X.J., Kawano, R., Sakakibara, H., et al., 2009. The ethylene response factors SNORKEL1 and SNORKEL2 allow rice to adapt to deep water. Nature 460 (7258), 1026–1030. Available at https://doi.org/10.1038/nature08258.

Hussain, I., Ayub, A., Nayab, A., Ashraf, M.A., Ashraf, M.A., Hussain, S., et al., 2024. Exogenous application of silicon and zinc attenuates drought tolerance in Eruca sativa L. through increasing chlorophyll pigments, osmoprotectants, and modulating defense mechanisms. Journal of Plant Growth Regulation 43 (9), 3221–3237. Available at https://doi.org/10.1007/s00344-023-11116-7.

Hu, W., Su, Y., Yang, R., Xie, Z., Gong, H., 2023. Effect of foliar application of silicon and selenium on the growth, yield and fruit quality of tomato in the field. Horticulturae 9 (10), 1126. Available at https://doi.org/10.3390/horticulturae9101126.

Islam, M.A., Lobry de Bruyn, L., Warwick, N.W.M., Koech, R., 2021. Salinity-affected threshold yield loss: a signal of adaptation tipping points for salinity management of dry season rice cultivation in the coastal areas of Bangladesh. Journal of Environmental Management 288. Available at https://doi.org/10.1016/j.jenvman.2021.112413.

Ismail, A.M., Heuer, S., Thomson, M.J., Wissuwa, M., 2007. Genetic and genomic approaches to develop rice germplasm for problem soils. Plant Molecular Biology 65 (4), 547–570. Available at https://doi.org/10.1007/s11103-007-9215-2.

Ismail, A.M., Singh, U.S., Singh, S., Dar, M.H., Mackill, D.J., 2013. The contribution of submergence-tolerant (Sub1) rice varieties to food security in flood-prone rainfed lowland areas in Asia. Field Crops Research 152, 83–93. Available at https://doi.org/10.1016/j.fcr.2013.01.007.

Jinger, D., Devi, M.T., Dhar, S., Dass, A., Sharma, V.K., Kumar, V.S., et al., 2020. Silicon application mitigates abiotic stresses in rice: a review. Journal of Agricultural Sciences 90 (11), 2394–3319. Available at https://doi.org/10.56093/ijas.v90i11.108557.

Jinger, D., Dhar, S., Dass, A., Sharma, V.K., Joshi, E., Vijayakumar, S., et al., 2018. Effect of silicon and phosphorus fertilization on growth, productivity and profitability of aerobic rice (Oryza sativa). Journal of Agricultural Sciences 88 (10), 1600–1605.

Jinger, D., Dhar, S., Dass, A., Sharma, V., Paramesh, V., Parihar, M., et al., 2021. Co-fertilization of silicon and phosphorus influenced the dry matter accumulation, grain yield, nutrient uptake, and nutrient-use efficiencies of aerobic rice. Silicon 14, 4683–4697. Available at https://doi.org/10.1007/s12633-021-01239-5.

Kabata-Pendias, A., 2010. Trace Elements in Soils and Plants, 4th ed. CRC Press, pp. 1–520. Available at 10.1201/b10158.

Kallenbach, C.M., Conant, R.T., Calderón, F., Wallenstein, M.D., 2019. A novel soil amendment for enhancing soil moisture retention and soil carbon in drought-prone soils. Geoderma 337, 256–265. Available at https://doi.org/10.1016/j.geoderma.2018.09.027.

Kandil, E.E., El-Banna, A.A.A., Tabl, D.M.M., Mackled, M.I., Ghareeb, R.Y., Al-Huqail, A.A., et al., 2022. Zinc nutrition responses to agronomic and yield traits, kernel quality, and pollen viability in rice (Oryza sativa L.). Frontiers in Plant Science 13. Available at https://doi.org/10.3389/fpls.2022.791066.

Kato, Y., Collard, B.C.Y., Septiningsih, E.M., Ismail, A.M., 2014. Physiological analyses of traits associated with tolerance of long-term partial submergence in rice. AoB PLANTS 6 (0), plu058. Available at https://doi.org/10.1093/aobpla/plu058.

Khalequzzaman, H., Ullah, S.K., Himanshu, N.E.T., Islam, R., Tisarum, S., Cha-um, A., et al., 2023. Seed priming improves germination, yield, and water productivity of cotton under drought stress. Journal of Soil Science and Plant Nutrition 23 (2), 2418–2432. Available at https://doi.org/10.1007/s42729-023-01196-5.

Khaliq, A., Aslam, F., Matloob, A., Hussain, S., Geng, M., Wahid, A., et al., 2015. Seed priming with selenium: consequences for emergence, seedling growth, and biochemical attributes of rice. Biological Trace Element Research 166 (2), 236–244. Available at https://doi.org/10.1007/s12011-015-0260-4.

Khalloufi, M., Martínez-Andújar, C., Lachaâl, M., Karray-Bouraoui, N., Pérez-Alfocea, F., Albacete, A., 2017. The interaction between foliar GA3 application and arbuscular mycorrhizal fungi inoculation improves growth in salinized tomato (Solanum lycopersicum L.) plants by modifying the hormonal balance. Journal of Plant Physiology 214, 134–144. Available at https://doi.org/10.1016/j.jplph.2017.04.012.

Khan, M.B., Yousaf, F., Hussain, M., Haq, M.W., Lee, D.-J., Farooq, M., 2012. Influence of planting methods on root development, crop productivity and water use efficiency in maize hybrids. Chilean Journal of Agricultural Research 72 (4), 556–563. Available at https://doi.org/10.4067/S0718-58392012000400015.

Khan, N.A., Owens, L., Nuñez, M.A., Khan, A.L., 2025. Complexity of combined abiotic stresses to crop plants. Plant Stress 17. Available at https://doi.org/10.1016/j.stress.2025.100926.

Khan, A.H., Singh, A.K., Mubeen, Singh, S., Zaidi, N.W., Singh, U.S., et al., 2014. Response of salt-tolerant rice varieties to biocompost application in sodic soil of eastern Uttar Pradesh. American Journal of Plant Sciences 05 (01), 7–13. Available at https://doi.org/10.4236/ajps.2014.51002.

Kochian, L.V., 1995. Cellular mechanisms of aluminum toxicity and resistance in plants. Annual Review of Plant Physiology and Plant Molecular Biology 46 (1), 237–260. Available at https://doi.org/10.1146/annurev.arplant.46.1.237.

Korres, N.E., Norsworthy, J.K., Burgos, N.R., Oosterhuis, D.M., 2017. Temperature and drought impacts on rice production: an agronomic perspective regarding short- and long-term adaptation measures. Water Resources and Rural Development 9, 12–27. Available at https://doi.org/10.1016/j.wrr.2016.10.001.

Kumar, A., Nayak, A.K., Hanjagi, P.S., Kumari, K., Mohanty, V.S.S., et al., 2021. Submergence stress in rice: adaptive mechanisms, coping strategies and future research needs. Environmental and Experimental Botany 186. Available at https://doi.org/10.1016/j.envexpbot.2021.104448.

Kumar, N., Singh, M.K., Praharaj, C.S., Singh, U., Singh, S.S., 2015. Performance of chickpea under different planting method, seed rate and irrigation level in Indo-Gangetic Plains of India. Journal of Food Legumes 28 (1), 40–44.

Kumar, K.A., Swain, D.K., Bhadoria, P.B.S., 2018. Split application of organic nutrient improved productivity, nutritional quality and economics of rice-chickpea cropping system in lateritic soil. Field Crops Research 223, 125–136. Available at https://doi.org/10.1016/J.FCR.2018.04.007.

Kutasy, E., Buday-Bódi, E., Virág, I.C., Forgács, F., Melash, A.A., Zsombik, L., et al., 2021. Mitigating the negative effect of drought stress in oat (Avena sativa L.) with silicon and sulphur foliar fertilization. Plants 11 (1), 30. Available at https://doi.org/10.3390/PLANTS11010030.

Ladha, J.K., Pathak, H., J. Krupnik, T., Six, J., van Kessel, C., 2005. Efficiency of fertilizer nitrogen in cereal production: retrospects and prospects. Advances in Agronomy 87, 85–156. Available at https://doi.org/10.1016/S0065-2113(05)87003-8.

Lal, B., Gautam, P., Nayak, A.K., Raja, R., Shahid, M., Tripathi, R., et al., 2018. Agronomic manipulations can enhance the productivity of anaerobic tolerant rice sown in flooded soils in rainfed areas. Field Crops Research 220, 105–116. Available at https://doi.org/10.1016/j.fcr.2016.08.026.

Lawlor, D.W., Tezara, W., 2009. Causes of decreased photosynthetic rate and metabolic capacity in water-deficient leaf cells: a critical evaluation of mechanisms and integration of processes. Annals of Botany 103 (4), 561–579. Available at https://doi.org/10.1093/AOB/MCN244.

Lay Swe, K., 2005. Temporary Windbreak Effect on Soil -Water Conservation and Growth and Yield of Certain Crops in the Dry Zone, Myanmar. Agricultural Practices and Meteorology.

Lepsch, H.C., Brown, P.H., Peterson, C.A., Gaudin, A.C.M., Khalsa, S.D.S., 2019. Impact of organic matter amendments on soil and tree water status in a California orchard. Agricultural Water Management 222, 204–212. Available at https://doi.org/10.1016/j.agwat.2019.06.002.

Marrassini, V., Ercoli, L., Piazza, G., Pellegrino, E., 2024. Plant genotype and inoculation with indigenous arbuscular mycorrhizal (AM) fungi modulate wheat productivity and quality of processed products through changes in the frequency of root AM fungal taxa. Field Crops Research 315, 109456. Available at https://doi.org/10.1016/j.fcr.2024.109456.

Mbarki, M., Labidi, N., Mahjoubi, H., Ksontini, M., 2001. Gypsum efficiency in the amelioration of saline sodic/sodic soils. International Journal of Agriculture & Biology 3, 276–280.

Milroy, S.P., Bange, M.P., Thongbai, P., 2009. Cotton leaf nutrient concentrations in response to waterlogging under field conditions. Field Crops Research 113 (3), 246–255. Available at https://doi.org/10.1016/J.FCR.2009.05.012.

Mondal, S., Khan, M.I.R., Entila, F., Dixit, S., Pompe, P.C., Panna Ali, M., et al., 2020. Responses of AG1 and AG2 QTL introgression lines and seed pre-treatment on growth and physiological processes during anaerobic germination of rice under flooding. Scientific Reports 10 (1). Available at https://doi.org/10.1038/s41598-020-67240-x.

Moradi, F., Ismail, A.M., 2007. Responses of photosynthesis, chlorophyll fluorescence and ROS-scavenging systems to salt stress during seedling and reproductive stages in rice. Annals of Botany 99 (6), 1161. Available at https://doi.org/10.1093/AOB/MCM052.

Msimbira, L.A., Smith, D.L., 2020. The roles of plant growth promoting microbes in enhancing plant tolerance to acidity and alkalinity stresses. Frontiers in Sustainable Food Systems 4, 564220. Available at https://doi.org/10.3389/FSUFS.2020.00106.

Mushtaq, A., Jamil, N., Riaz, M., Hornyak, G.L., Ahmed, N., Ahmed, S.S., et al., 2017. Synthesis of silica nanoparticles and their effect on priming of wheat (Triticum aestivum L.) under salinity stress. Biological Forum – An International Journal 9, 150.

Negrão, S., Schmöckel, S.M., Tester, M., 2017. Evaluating physiological responses of plants to salinity stress. Annals of Botany 119 (1), 1–11. Available at https://doi.org/10.1093/AOB/MCW191.

Ngigi, S.N., Rockström, J., Savenije, H.H.G., 2006. Assessment of rainwater retention in agricultural land and crop yield increase due to conservation tillage in Ewaso Ng'iro river basin, Kenya. Physics and Chemistry of the Earth, Parts A/B/C 31 (15–16), 910–918. Available at https://doi.org/10.1016/J.PCE.2006.08.006.

Parit, E.R.K., Bharteey, P.K., Saikia, B.J., Bora, S.S., Maurya, P.K., Tabing, R., 2020. Nitrogen transformation in waterlogged soil in indian rice fields: a review. Journal of Experimental Agriculture International 107–118. Available at https://doi.org/10.9734/JEAI/2020/V42I330488.

Patanè, C., Pellegrino, A., DI Silvestro, I., 2018. Effects of calcium carbonate application on physiology, yield and quality of field-grown tomatoes in a semi-arid Mediterranean climate. Crop and Pasture Science 69 (4), 411–418. Available at https://doi.org/10.1071/CP17424.

Pavlovic, J., Kostic, L., Bosnic, P., Kirkby, E.A., Nikolic, M., 2021. Interactions of silicon with essential and beneficial elements in plants. Frontiers in Plant Science 12. Available at https://doi.org/10.3389/fpls.2021.697592.

Pitman, M.G., Läuchli, A., 2002. Global impact of salinity and agricultural ecosystems. In: Läuchli, A., Lüttge, U. (Eds.), Salinity: Environment - Plants - Molecules. Springer, Dordrecht. https://doi.org/10.1007/0-306-48155-3_1.

Purwestri, Y.A., Nuringtyas, T.R., Wibowo, A.T., Nugrahapraja, H., Salsinha, Y.C.F., Sebastian, A., et al., 2025. Application of osmoprotectant enhance tolerance to drought stress in rice and trigger changes in root microbial composition. Journal of Plant Biochemistry and Biotechnology 34 (2), 413–425. Available at https://doi.org/10.1007/s13562-024-00933-w.

Qiao, M., Hong, C., Jiao, Y., Hou, S., Gao, H., 2024. Impacts of drought on photosynthesis in major food crops and the related mechanisms of plant responses to drought. Plants 13 (13), 1808. Available at https://doi.org/10.3390/plants13131808.

Qin, W., Hu, C., Oenema, O., 2015. Soil mulching significantly enhances yields and water and nitrogen use efficiencies of maize and wheat: a meta-analysis. Scientific Reports 5. Available at https://doi.org/10.1038/srep16210.

Rabert, G.A., Manivannan, P., Somasundaram, R., Panneerselvam, R., 2014. Triazole compounds alters the antioxidant and osmoprotectant status in drought stressed Helianthus annuus L. Plants. Journal of Food and Agriculture 26 (3), 265–276. Available at https://doi.org/10.9755/ejfa.v26i3.17385.

Rady, M.M., Boriek, S.H.K., El-Mageed, T.A.A., El-Yazal, M.A.S., Ali, E.F., Hassan, F.A.S., et al., 2021. Exogenous gibberellic acid or dilute bee honey boosts drought stress tolerance in vicia faba by rebalancing osmoprotectants, antioxidants, nutrients, and phytohormones. Plants 10 (4). Available at https://doi.org/10.3390/plants10040748.

Ramesh, T., Rathika, S., Geetha, S., Sabarinathan, R., Vijayakumar, S., 2024. Improving yield and water productivity of rice in sodic soil with saline water through drip irrigation. Oryza-An International Journal on Rice 61 (1), 65–71. Available at https://doi.org/10.35709/ory.2024.61.1.8.

Rao, C.S., Lal, R., Prasad, J.V.N.S., Gopinath, K.A., Singh, R., Jakkula, V.S., et al., 2015. Potential and challenges of rainfed farming in India. Advances in Agronomy 133, 113–181. Available at https://doi.org/10.1016/bs.agron.2015.05.004.

Rathore, S.S., Shekhawat, K.A., Singh, R.K., Updhyay, P.K., Shekhawat, R., Premi, O.P., 2019. Effect of nano-particles on growth, productivity, profitability of Indian mustard (Brassica juncea) under semi-arid conditions. Journal of Agricultural Sciences 89 (7), 1145–1150. Available at https://doi.org/10.56093/ijas.v89i7.91669.

Raza, M.A.S., Saleem, M.F., Khan, I.H., Hussain, M.B., Shah, G.M., 2018. Amelioration in growth and physiological efficiency of sunflower (Helianthus annuus L.) under drought by potassium application. Communications in Soil Science and Plant Analysis 49 (18), 2291–2300. Available at https://doi.org/10.1080/00103624.2018.1499764.

Sajid, M., Ullah, I., Rab, A., Syed, T., Shah, F.-I.-W., Ahmad, N., et al., 2020. Foliar application of calcium improves growth, yield and quality of tomato cultivars. Pure and Applied Biology 9 (1), 10–19. Available at https://doi.org/10.19045/bspab.2020.90002.

Sarangi, S.K., Maji, B., Singh, S., Burman, D., Mandal, S., Sharma, D.K., et al., 2015. Improved nursery management further enhances the productivity of stress-tolerant rice varieties in coastal rainfed lowlands. Field Crops Research 174, 61–70. Available at https://doi.org/10.1016/j.fcr.2015.01.011.

Septiningsih, E.M., Hidayatun, N., Sanchez, D.L., Nugraha, Y., Carandang, J., Pamplona, A.M., et al., 2015. Accelerating the development of new submergence tolerant rice varieties: the case of Ciherang-Sub1 and PSB Rc18-Sub1. Euphytica 202 (2), 259–268. Available at https://doi.org/10.1007/s10681-014-1287-x.

Shanmugam, V., Tyagi, V.C., Rajendran, G., Chimmili, S.R., Swarnaraj, A.K., Arulanandam, M., et al., 2025. Perennial rice – an alternative to the 'one-sow, one-harvest' rice production: benefits, challenges, and future prospects. Farming System 3 (2). Available at https://doi.org/10.1016/j.farsys.2025.100137.

Sheoran, P., Kumar, A., Sharma, R., Barman, A., Parjapat, K., Singh, R.K., et al., 2021. Managing sodic soils for better productivity and farmers' income by integrating use of salt tolerant rice varieties and matching agronomic practices. Field Crops Research 270. Available at https://doi.org/10.1016/j.fcr.2021.108192.

Shoukat, A., Pitann, B., Hossain, M.S., Saqib, Z.A., Nawaz, A., Mühling, K.H., 2024. Zinc and silicon fertilizers in conventional and nano-forms: mitigating salinity effects in maize (*Zea mays* L.). Journal of Plant Nutrition and Soil Science 187, 678–689. Available at https://doi.org/10.1002/jpln.202300267.

Sishodia, R.P., Ray, R.L., Singh, S.K., 2020. Applications of remote sensing in precision agriculture: a review. Remote Sensing 12 (19), 3136. Available at https://doi.org/10.3390/RS12193136.

Steffens, D., Hütsch, B.W., Eschholz, T., Lošák, T., Schubert, S., 2005. Water logging may inhibit plant growth primarily by nutrient deficiency rather than nutrient toxicity. Plant, Soil and Environment 51 (12), 545–552. Available at https://doi.org/10.17221/3630-pse.

Sultana, N., Ikeda, T., Kashem, M.A., 2001. Effect of foliar spray of nutrient solutions on photosynthesis, dry matter accumulation and yield in seawater-stressed rice. Environmental and Experimental Botany 46 (2), 129–140. Available at https://doi.org/10.1016/S0098-8472(01)00090-9.

Sun, M., Ren, A.X., Gao, Z.Q., Wang, P.R., Xue, F.M.O.L.Z., Lei, M.M., 2018. Long-term evaluation of tillage methods in fallow season for soil water storage, wheat yield and water use efficiency in semiarid southeast of the Loess Plateau. Field Crops Research 218, 24–32. Available at https://doi.org/10.1016/j.fcr.2017.12.021.

Sushma, M., Singh, J.P., Rajpoot, S.K., Bhushan, C., Verma, S.K., Singh, N.K., et al., 2024. Effect of integrated nutrient management on the growth and yield of Yellow Sarson (Brassica rapa var. yellow sarson) under guava (Psidium guajava) based agri-horti system. Indian Journal of Agronomy 69 (3). Available at https://doi.org/10.59797/IJA.V69I3.5535.

Su, Q., Kambale, R.D., Tzeng, J.H., Amy, G.L., Ladner, D.A., Karthikeyan, R., 2025. The growing trend of saltwater intrusion and its impact on coastal agriculture: challenges and opportunities. Science of the Total Environment 966. Available at https://doi.org/10.1016/j.scitotenv.2025.178701.

Swain, R., Sahoo, S., Behera, M., Rout, G.R., 2023. Instigating prevalent abiotic stress resilience in crop by exogenous application of phytohormones and nutrient. Frontiers in Plant Science 14. Available at https://doi.org/10.3389/fpls.2023.1104874.

Tabassum, T., Farooq, M., Ahmad, R., Zohaib, A., Wahid, A., Shahid, M., 2018. Terminal drought and seed priming improves drought tolerance in wheat. Physiology and Molecular Biology of Plants 24 (5), 845–856. Available at https://doi.org/10.1007/s12298-018-0547-y.

Teame, G., Tsegay, A., Abrha, B., 2017. Effect of organic mulching on soil moisture, yield, and yield contributing components of sesame (*Sesamum indicum* L.). International Journal of Agronomy 2017 (1), 4767509. Available at https://doi.org/10.1155/2017/4767509.

Tuna, A.L., Kaya, C., Ashraf, M., Altunlu, H., Yokas, I., Yagmur, B., 2007. The effects of calcium sulphate on growth, membrane stability and nutrient uptake of tomato plants grown under salt stress. Environmental and Experimental Botany 59, 173–178. Available at https://doi.org/10.1016/j.envexpbot.2005.12.007.

Tworkoski, T., Wisniewski, M., Artlip, T., Tworkoski, T., Wisniewski, M., Artlip, T., 2011. Application of BABA and s-ABA for drought resistance in apple. Journal of Applied Horticulture 13 (2), 85–90.

Ventura, Y., Eshel, A., Pasternak, D., Sagi, M., 2015. The development of halophyte-based agriculture: past and present. Annals of Botany 115 (3), 529–540. Available at https://doi.org/10.1093/aob/mcu173.

Veste, M., Littmann, T., Kunneke, A., Toit, B.D., Seifert, T., 2020. Windbreaks as part of climate-smart landscapes reduce evapotranspiration in vineyards, Western Cape Province, South Africa. Plant, Soil and Environment 66 (3), 119–127. Available at https://doi.org/10.17221/616/2019-PSE.

Vijayakumar, S., Choudhary, A.K., Deiveegan, M., Thirumalaikumar, R., Kumar, R.M., 2022. Android based mobile application for rice crop management. Chronicle of Bioresource Management 6 (1), 19–24.

Vijayakumar, S., Kumar, Dinesh, Varatharajan, T., Kaje, Varun V., Deiveegan, M., 2023. Enriching cationic micro-nutrients concentration in basmati rice through potassium fertilization. Journal of Plant Nutrition 46, 3355–3369. Available at https://doi.org/10.1080/01904167.2023.2203183.

Vijayakumar, S., Kumar, D., Ramesh, K., Govindasam, P., Jinger, D., Khanam, R., et al., 2021. Potassium nutrition in rice: a review. Oryza-An International Journal on Rice 58 (3), 341–353. Available at https://doi.org/10.35709/ory.2021.58.3.1.

Vijayakumar, S., Nayak, A.K., Satapathy, B.S., Munda, S., Tripathi, R., Khanam, R., 2019. Double transplanting -a contingent strategy for rice cultivation in low-lying areas. Indian Farming 69 (5), 36–38.

Wang, J., Du, G., Tian, J., Jiang, C., Zhang, Y., Zhang, W., 2021. Mulched drip irrigation increases cotton yield and water use efficiency via improving fine root plasticity. Agricultural Water Management 255, 106992. Available at https://doi.org/10.1016/j.agwat.2021.106992.

Wang, X., Geng, S., Ri, Y.J., Cao, D., Liu, J., Shi, D., et al., 2011. Physiological responses and adaptive strategies of tomato plants to salt and alkali stresses. Scientia Horticulturae 130 (1), 248–255. Available at https://doi.org/10.1016/j.scienta.2011.07.006.

Wang, S., Wang, H., Hafeez, M.B., Zhang, Q., Yu, Q., Wang, R., et al., 2020. No-tillage and subsoiling increased maize yields and soil water storage under varied rainfall distribution: a 9-year site-specific study in a semi-arid environment. Field Crops Research 255, 107867. Available at https://doi.org/10.1016/j.fcr.2020.107867.

Wasaya, A., Affan, M., Yasir, T.A., Atique-Ur-Rehman, Mubeen, K., Ur Rehman, H., Ali, M., et al., 2021. Foliar potassium sulfate application improved photosynthetic characteristics, water relations and seedling growth of drought-stressed maize. Atmosphere 12 (6). Available at https://doi.org/10.3390/atmos12060663.

Wassmann, R., Jagadish, S.V.K., Sumfleth, K., Pathak, H., Howell, G., Ismail, A., et al., 2009. Chapter 3 Regional vulnerability of climate change impacts on asian rice production and scope for adaptation. Advances in Agronomy 102, 91–133. Available at https://doi.org/10.1016/S0065-2113(09)01003-7.

World Water Assessment Programme (WWAP), (2009). Water in a Changing World: The United Nations World Water Development Report 3, Earthscan.

Yang, Z., Midmore, D.J., 2004. Experimental assessment of the impact of defoliation on growth and production of water-stressed maize and cotton plants. Experimental Agriculture 40 (2), 189–199. Available at https://doi.org/10.1017/S0014479703001534.

Yan, Q., Wu, L., Dong, F., Zhang, Q., Yang, F., Yan, S., et al., 2023. Continuous tillage practices improve soil water storage and yields of dryland winter wheat grown for three consecutive years in North China. Archives of Agronomy and Soil Science 69 (3), 446–460. Available at https://doi.org/10.1080/03650340.2021.2004587.

Zeng, F., Shabala, L., Zhou, M., Zhang, G., Shabala, S., 2013. Barley responses to combined waterlogging and salinity stress: separating effects of oxygen deprivation and elemental toxicity. Frontiers in Plant Science 4. Available at https://doi.org/10.3389/fpls.2013.00313.

Zhang, Y., Li, X., Jia, L., Ji, L., Wang, C., Xu, W., et al., 2024. Partial root zone irrigation and K application improves summer maize production and salt resistance in saline soil. Agricultural Water Management 303, 109057. Available at https://doi.org/10.1016/j.agwat.2024.109057.

Zhang, S., Lövdahl, L., Grip, H., Tong, Y., Yang, X., Wang, Q., 2009. Effects of mulching and catch cropping on soil temperature, soil moisture and wheat yield on the Loess Plateau of China. Soil and Tillage Research 102 (1), 78–86. Available at https://doi.org/10.1016/j.still.2008.07.019.

Zhang, Q., Wang, S., Sun, Y., Zhang, Y., Li, H., Liu, P., et al., 2022. Conservation tillage improves soil water storage, spring maize (Zea mays L.) yield and WUE in two types of seasonal rainfall distributions. Soil and Tillage Research 215, 105237. Available at https://doi.org/10.1016/j.still.2021.105237.

Zhu, S.Y., Cheng, Z.G., Tian, T., Gong, D.S., Lv, G.C., Wang, J., et al., 2021. Screening optimum population density in response to soil water availability in dryland wheat: from laboratory to field. Agricultural Water Management 257. Available at https://doi.org/10.1016/j.agwat.2021.107147.

Chapter 18

Adoption of climate smart agricultural practices to mitigate multiple abiotic stresses in crops

Dhiman Burman*, Devika S., Uttam Kumar Mandal, and Tashi Dorjee Lama

Regional Research Station (CSSRI, RRS), ICAR-Central Soil Salinity Research Institute, Canning Town, West Bengal, India

Corresponding author. e-mail address: dhiman.burman@icar.org.in

18.1 Introduction

Agriculture in coastal regions is vital for food security, economic development, and the livelihoods of millions of people. However, these areas face increasing challenges due to abiotic stresses such as rising sea levels, saline water intrusion, water scarcity, temperature fluctuations, and unpredictable weather patterns. Climate change exacerbates these stresses, threatening agricultural productivity and the long-term sustainability of these ecosystems and communities. Coastal environments are essential to agricultural development worldwide because of their rich soils and temperate weather. However, these areas are experiencing growing risks from salt and waterlogging worsened by poor drainage systems, climate change, and rising sea levels. These issues reduce agricultural yields and endanger the livelihoods of millions of farmers who depend on these ecosystems. Among the world's most densely inhabited regions, Asian deltas are especially vulnerable to these problems. These areas are distinguished by high rainfall, low elevations, and substantial sediment deposition and are undergoing widespread salinization, affecting ecosystem health and agriculture (Rahman et al., 2019).

The Bengal Delta in Bangladesh is a critical case study, with around 30 million people facing growing saline levels in water and soil resources. According to Feist et al. (2023), this development poses a hazard to ecosystem services and human well-being. According to Akter et al. (2023), saltwater is expected to spread further inland over the next century, indicating an ongoing trend. A 2009 assessment found that salinization has affected approximately 10,560 km^2 of cultivable land, a 26% increase since 1973 (SRDI 2010). Climate modeling projects a 39% increase in salt accumulation by 2050 (Mainuddin et al., 2021), potentially reducing agricultural production by 25%–50% by 2098 (Clarke et al., 2015). To ensure sustainable agriculture in coastal regions, it is crucial to create crop enhancement strategies that improve tolerance to abiotic stresses.

The multifaceted nature of abiotic stresses in coastal regions makes them particularly challenging to manage. Two or more biotic and abiotic stresses may occur simultaneously in many areas in the world. For example, sea-level rise can lead to saline water encroachment, induce the combined saline and waterlogging stress and nutrient toxicity, reducing soil fertility and limiting freshwater availability for irrigation, imposing drought with salinity and nutrient deficiency. Extreme weather events, such as cyclones and storm surges, further damage crops, infrastructure, and livelihoods, compounding the vulnerability of these areas. These interlinked challenges demand innovative and sustainable results to ensure flexible farming systems. It is anticipated that future climate change scenarios will worsen crop production and plant survival by intensifying the effects of abiotic stressors on plants. Indeed, plants are being exposed to a variety of sometimes conflicting consequences of global warming on the climate, including increased temperatures, a greater vapor pressure deficit, more frequent and severe droughts, unusual frost occurrences, flooding, and more. Most significantly, plants will encounter a wide range of intricate stressors at the same time, which may make it difficult for them to persist in particular environments and occasionally encourage them to spread into other geographic areas.

Exposure of plants to various stress combinations activates many interacting signal transduction pathways (Mittler, 2006). The interplay among these pathways can be neutral, additive, synergistic, or may result in unforeseen reactions

Multiple Abiotic Stresses in Plants. DOI: https://doi.org/10.1016/B978-0-443-27424-4.00019-7

(Prasch and Sonnewald, 2015). Heat and drought exert contrasting influences on leaf physiology and morphology. An initial reaction to heat stress is the opening of stomata to lower the leaf surface temperature through enhanced transpiration. Conversely, water deficiency triggers stomatal closure to mitigate water loss, thus elevating leaf temperature by 2°C–5°C (Prasch and Sonnewald, 2013). Heat stress leads to elongated, thinner leaves with a high specific leaf area and diminished root growth, whereas drought reduces leaf area and enhances root growth to limit water loss and optimize water intake (Vile et al., 2012). Despite these evident disparities, heat and drought similarly influence flowering time, seed abortion, and final output. Several abiotic conditions (drought, salinity, and cold) induce osmotic stress, which is mitigated by the synthesis of suitable osmolytes. Drought and salinity adversely impact photosynthesis by limiting internal carbon dioxide concentration as a result of stomatal closure (Nouri et al., 2015). Plant growth and biomass, as indicated by shoot and root dry weights, were reduced in wheat genotypes subjected to drought and salt stresses. The photosynthetic rate, chlorophyll content, and stomatal conductance were negatively impacted in wheat plants exposed to combined drought and salt stresses (Dugasa et al., 2019). Comparable adverse effects on photosynthesis may also occur with elevated light intensity, resulting from the absorption of substantial photosynthetic energy. Moreover, nutritional deficiency frequently heightens plant susceptibility to various abiotic stresses, and the majority of these stresses typically lead to the generation of reactive oxygen species (Mittler and Blumwald, 2010).

Responses are regulated at the transcriptional, posttranscriptional, translational, and/or posttranslational levels. Most transcriptomic studies have concentrated on identifying common transcriptional regulators or proteins involved in signal transduction pathways. Ectopic expression of these proteins frequently led to enhanced plant performance under specific stress conditions (Century et al., 2008). Recent studies have demonstrated that microRNA-mediated gene regulation plays a significant role in various stress experiments, with modifications in the expression of specific microRNAs leading to the development of transgenic plants exhibiting enhanced performance (Bej and Basak, 2014; Zhang, 2015). MicroRNAs and most transcription factors regulate the expression of various candidate genes and processes; thus, assessing the performance of these plants under field conditions is essential. Currently, diverse innovative methods in transcriptome and proteome analysis, genomic sequencing, and bioinformatics, along with functional studies, will enhance the identification of common regulators. These approaches will improve our understanding of the complex and efficient networks involved in plant responses to various stress conditions, thereby aiding in the enhancement of plant tolerance and productivity.

Climate-smart agriculture (CSA) has emerged as a transformative approach to addressing these challenges. By integrating sustainable practices, advanced technologies, and adaptive strategies, CSA aims to boost agricultural productivity, enhance resilience to climate impacts, and reduce greenhouse gas emissions. Coastal regions can particularly benefit from CSA practices such as adopting salt-tolerant crops, efficient water management techniques, agroforestry systems, conservation agriculture, and climate-adaptive decision-making tools. These practices provide farmers with the tools to withstand environmental stressors while maintaining or improving productivity.

The successful implementation of CSA in coastal regions hinges on several factors, including socioeconomic conditions, institutional frameworks, technological accessibility, and the availability of climate-resilient inputs. Understanding these dynamics is essential for designing effective interventions and scaling up CSA adoption. Moreover, fostering collaboration among stakeholders, such as governments, research institutions, nongovernmental organizations, and farming communities, can accelerate the uptake of CSA practices and embed them in agricultural policies and programs.

This chapter delves into the role of climate-smart agricultural practices in mitigating the impacts of combined multiple abiotic stresses and managing the adverse effects of combined multiple abiotic stresses. It examines the unique challenges faced in these regions, highlights innovative CSA interventions, and explores the socioeconomic and institutional drivers that influence their adoption. This chapter provides actionable insights for policymakers, practitioners, and researchers, drawing on recent research, development, and success stories. By emphasizing the integration of climate-smart strategies into broader sustainable development efforts, it seeks to strengthen the resilience and food security of the coastal regions in an era of climate uncertainty.

18.2 Abiotic stresses and their impact on crop production: relevance to coastal regions

Abiotic stress management is one of agriculture's most pressing concerns. Abiotic stress can constantly limit crop choice and agricultural production across wide areas, and extreme events can result in crop failure. Abiotic stresses harm the livelihoods of individual farmers and their families, as well as national economies and food security.

Abiotic stresses include potential negative consequences such as salinity, drought, flooding, metal toxicity, nutrient deficiency, light, pollutants, nutrient deficiencies, and high and low temperatures (Fig. 18.1). Abiotic stresses, including shade, ultraviolet exposure, photoinhibition, air pollution, wind, hail, and gaseous deprivation, are typically sporadic and localized. Abiotic stress in plants can be caused by a lack of a vital resource or the presence of excessive amounts of a poisonous or antagonistic chemical. In some circumstances, such as water supplies, either too little (drought) or too much (flooding) can cause stress in plants. In actuality, abiotic and biotic stresses are frequently intimately intertwined.

Abiotic variables influence plant growth, development, and productivity by altering physiological, morphological, biochemical, and metabolic processes. Abiotic stress factors such as salinity, drought, heat, and cold significantly impact crop yield (Deryng et al., 2014; Kapazoglou et al., 2018; Loreti et al., 2016). These factors can reduce biomass production and performance by up to 70% (Liliane and Charles, 2020). Abiotic stresses can greatly affect root growth and function by changing how roots take in water and nutrients and how they interact with microorganisms in the rhizosphere (Choi et al., 2021; Koevoets et al., 2016).

Plants subjected to combined stresses, including salinity–waterlogging, salinity–drought, and heat–drought, encounter intricate and frequently synergistic damage that exceeds the impacts of individual stressors. Lettuce exposed to simultaneous salinity and drought experiences more significant reductions in growth parameters, such as plant height, leaf area, and fresh weight, as well as more pronounced declines in nutrient uptake and antioxidant activity, in comparison to either stress condition alone (Abdelkader et al., 2024). Cooccurrence of heat and drought leads to accelerated phenology in crops, including cereals and legumes, resulting in substantial biomass and yield losses, impaired photosynthesis, and decreased grain number and quality, especially when these stresses occur during reproductive stages (Aloui et al., 2023; Benali et al., 2023; Yang et al., 2022). In desert-adapted species such as *Artemisia herba-alba*, the combination of heat and drought induces oxidative damage, reduces photosynthetic efficiency, and prompts significant molecular responses, including the upregulation of HSP70, aquaporins, and osmoprotectants, thereby underscoring the activation of defense mechanisms (Abdulaziz S. Alhaithloul et al., 2020). The findings indicate that combined abiotic stresses elicit unique physiological and molecular responses that are not evident in single-stress studies. Addressing these specific responses, particularly during critical growth stages, is essential for improving crop resilience to climate extremes (Xiong et al., 2025).

The region's climate stressors include tropical cyclones, floods, droughts, saline intrusion, sea level rise, temperature fluctuations, and rainfall. All of these climatic conditions were documented to cause multiple-fold losses to the agricultural industry as a whole and the smallholder farming community in particular. It was also discovered that all of the climatic pressures were connected. Rising sea levels and changes in sea temperature have increased the likelihood of cyclone incidences increasing in severity and frequency, posing additional risks of floods, saline intrusion, temperature, and

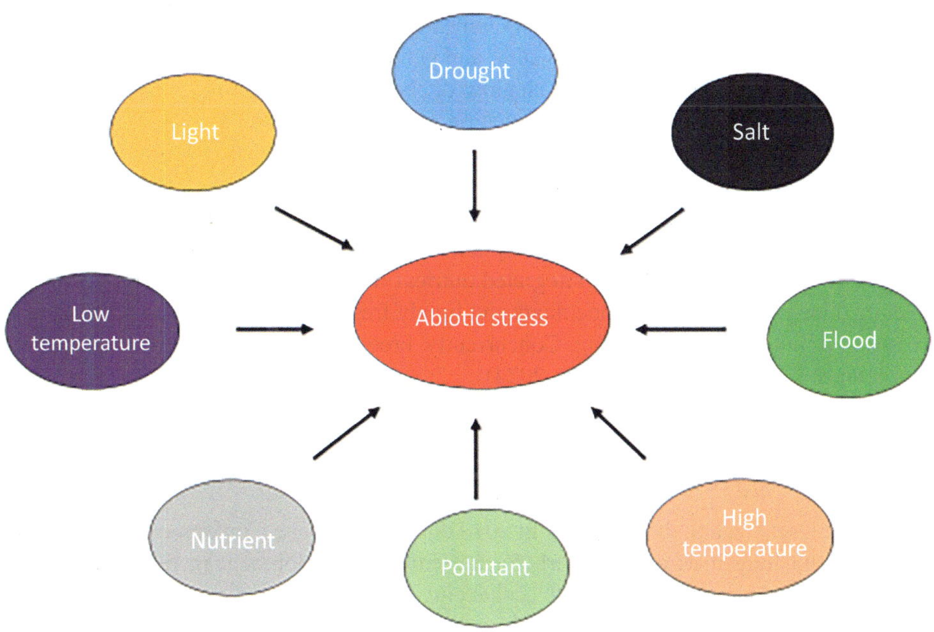

FIGURE 18.1 Common abiotic stresses in plants.

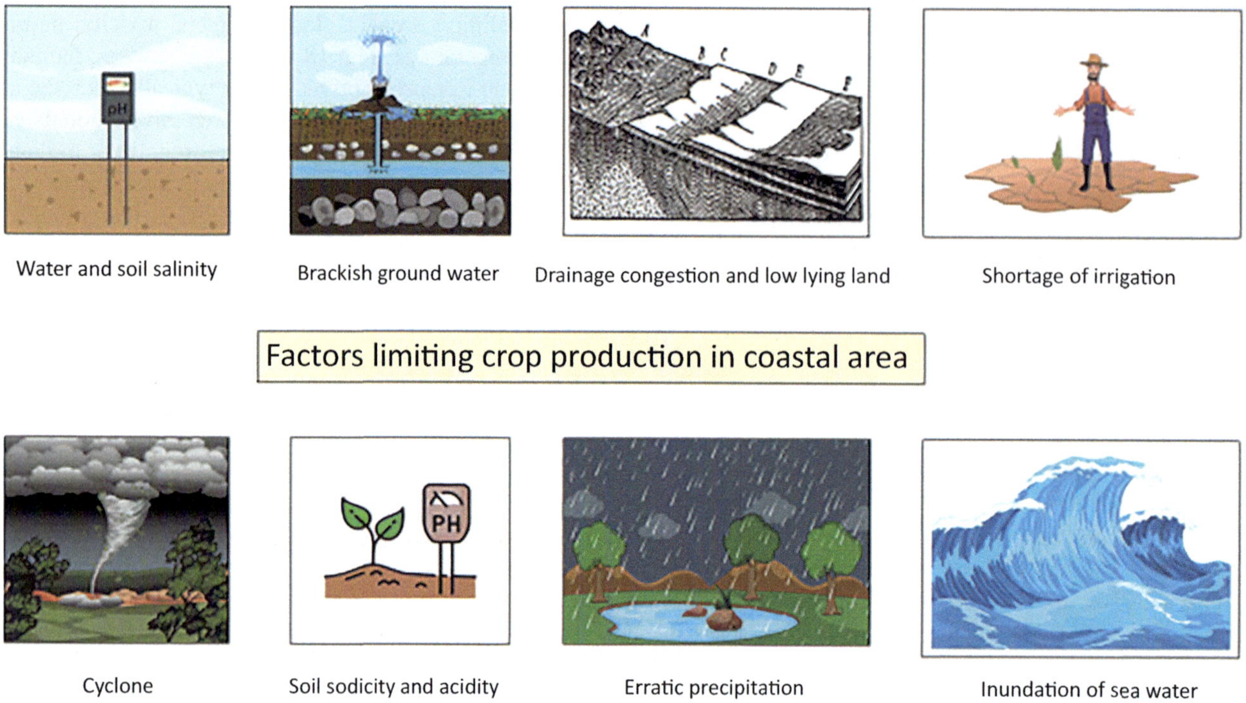

FIGURE 18.2 Impacts of climate change on coastal crop production.

rainfall swings. Kunze (2021) reported that tropical cyclones are the leading cause of coastal disasters. Coastal agriculture is impacted in multiple ways by climate disasters and subsequent climate changes (Fig. 18.2). The coastal region's frequent cyclones and floods have caused saline intrusion in arable lands and freshwater bodies. In coastal areas, salinity can only be alleviated by substantial rainfall over time. According to Sabareshwari and Ramya (2018), salinity increases during dry spells due to fluctuating rainfall and rising temperatures over time. The uneven distribution of rainfall and temperatures during the growing season caused a shift in the crop calendar. In coastal farmlands, the most detrimental combination of stresses is salinity combined with waterlogging, which is influenced by sea-level rise, tidal inundation, inadequate drainage, and storm surges. This combination adversely affects growth and yield more than either stress independently, as it exacerbates ionic and toxic effects alongside hypoxia (Tahjib-Ul-Arif et al., 2023).

Practical management integrates hydrology, soil health, and genetics through land modification and enhanced drainage techniques, such as raised beds, embankments, and shallow ditches, alongside on-farm rainwater harvesting to mitigate salinity and reduce the water table. It employs conservation agriculture practices and organic amendments, including residues and compost/biochar, to enhance soil structure and salt buffering capacity. Additionally, it involves the strategic selection of crops, favoring salt and waterlogging-tolerant cultivars or biosaline crops in the most impacted areas (Sarangi et al., 2024). Recent coastal case studies emphasize adaptive calendars, such as earlier planting to avoid peak inundation, the use of elevated platforms for vegetables, and mixed systems that alternate rice with less sensitive crops to mitigate risk (Akter and Mazumder, 2025). At the field scale, integrated nutrient management, particularly with appropriate sources of potassium and calcium, along with microbial inoculants and timely freshwater flushing or irrigation, effectively alleviates the combined stresses of salinity and waterlogging. Concurrently, breeding efforts continue to focus on physiological traits that support dual tolerance (Palmgren and Shabala, 2024).

18.3 Climate-smart agriculture practices in coastal regions

The Food and Agriculture Organization defined CSA in 2010 during the Hague Conference on Agriculture, Food Security and Climate Change, establishing it as a framework for agricultural development in response to climate change. The definition highlights three fundamental components: enhancing productivity and resilience (adaptation), decreasing or eliminating greenhouse gas emissions (mitigation), and improving food security and development objectives. There are three primary goals for CSA:

1. ***Productivity***: CSA aims to increase overall agricultural productivity and provide greater food security by improving crop and livestock production as well as farm profitability.
2. ***Adaptation***: The goal of CSA is to strengthen agricultural infrastructure to withstand the damaging impacts of climate change. This means taking action to reduce vulnerability to climate-related hazards such as droughts, floods, and excessive heat.
3. ***Mitigation***: One of the main objectives of climatically smart agriculture is to reduce the quantity of greenhouse gases emitted into the atmosphere as a result of farming operations, such as methane emissions from livestock, paddy rice production, and the use of synthetic fertilizers.

This can be achieved through the congruence of water-smart, energy-smart, C-smart, N-smart, weather-smart, and knowledge-smart moves and programs.

18.3.1 Classification of climate-smart agriculture technologies

CSA technologies are categorized into six types based on their capacity to address climate change challenges: weather-smart, water-smart, nutrient-smart, carbon-smart, energy-smart, and knowledge-smart technologies.

18.3.1.1 Weather-smart technology

Weather-smart technologies address the increasing frequency of extreme weather events, like floods and droughts, which adversely impact agricultural productivity and farm income. Solutions such as early warning systems, weather forecasts, agro-advisory services, crop insurance, stress-tolerant crop varieties, adjusted planting schedules, and diversified cropping systems help mitigate risks. These technologies rely on the timely dissemination of climate information, enabling farmers to make informed decisions.

18.3.1.2 Water-smart technology

Agriculture, being the largest consumer of freshwater, is critical under climate stress, and managing water resources efficiently is critical. Water-smart technologies focus on optimizing on-farm and off-farm water use through strategies such as rainwater harvesting, groundwater recharge, watershed management, and efficient irrigation systems like drip irrigation, sprinkler irrigation, and alternate wetting and drying methods. Other examples include practices like the system of rice intensification, contour bunding, mulching, and farm ponds for rainwater storage. Proper drainage systems are also essential for managing excess water.

18.3.1.3 Nutrient-smart technology

High-yielding crop varieties demand effective nutrient management to sustain productivity. Indiscriminate fertilizer use, often due to a lack of knowledge on the optimal timing, quantity, and placement of fertilizers, reduces efficiency, increases greenhouse gas emissions, and causes environmental harm. Nutrient-smart technologies emphasize practices such as integrated nutrient management, site-specific nutrient application, 4R nutrient stewardship, green and brown manuring, organic fertilizers, biofertilizers, and crop rotations with legumes. These approaches optimize nutrient use while minimizing emissions and pollution.

18.3.1.4 Carbon-smart technology

Carbon-smart technologies focus on reducing atmospheric carbon emissions (e.g., CO_2 and CH_4) while enhancing carbon sequestration in soil-plant systems. Practices such as minimum tillage, zero tillage, crop residue management, alternate wetting and drying irrigation practice, and agroforestry help capture carbon in the soil and reduce its release into the atmosphere, thus contributing to climate mitigation.

18.3.1.5 Energy-smart technology

Energy-smart technologies aim to enhance productivity while improving energy efficiency, thereby reducing greenhouse gas emissions. Techniques such as conservation agriculture, zero tillage, and minimum tillage, along with innovations like solar-powered machinery (e.g., solar pumps and sprayers), contribute to sustainable energy use in agriculture.

18.3.1.6 *Knowledge-smart technology*

Knowledge-smart technologies focus on disseminating climate-related knowledge to enhance farmers' resilience and adaptive capacity. This includes practices such as cropping system intensification, integrated farming systems, seed priming, mechanical rice transplanting, integrated weed and pest management, and Information and communication technology (ICT)-based agricultural extension services. Innovations such as climate-smart villages, hydroponics, protected horticulture, seed and fodder banks, and village knowledge centers further support the widespread adoption of CSA practices. CSA provides an integrated framework to address climate challenges while maintaining productivity. Tailored practices ensure sustainable resource use and resilience for coastal regions.

18.3.2 Sustainable agricultural practices for management of multiple abiotic stresses

In India's coastal regions, smallholder farming groups rely on various activities, including livestock husbandry, fisheries, and apiculture, in addition to crop agriculture. Throughout history, the community has valued interdependence among firms. Extension research and advising services are increasingly aligned with integrated farming systems (IFS) (Dutta et al., 2020; Mondal, 2022). The people needed to diversify their revenue sources to withstand the coastal region's rapid climate change. The people needed to diversify their revenue sources to withstand the coastal region's rapid climate change. IFS integrates multiple enterprises, such as crop farming, livestock, fisheries, and poultry, to improve productivity, reduce waste, and increase farmers' net income (Ponnusamy and Devi, 2017). It is also a crucial step toward developing climate-resilient agriculture. Diversifying production within a single agricultural unit optimizes resource utilization and improves tolerance to climate shocks.

Diversified cropping systems, also known as diversified agriculture, are a popular and effective adaptation strategy against climate change (Adam et al., 2018; Bhattacharjee et al., 2023; Dhanya et al., 2022; Jena et al., 2023; Mondal, 2022; Padhan and Madheswaran, 2022; Patel et al., 2020; Rao et al., 2017; Reddy et al., 2022). Diversified cropping systems, such as intercropping, multicropping, mixed cropping, alternative crops, and crop rotation, contribute to farmers' overall development (Davis et al., 2012; Shah et al., 2021). Due to the considerable impact of salinity on coastal regions (DasGupta et al., 2019), traditional farming practices are being used less frequently during nonkharif seasons. To address the issue, research recommended employing sophisticated land modification techniques, such as land shaping and tillage cultivation, as well as shifting to a more diverse cropping system (Burman et al., 2013, 2015; Dutta et al., 2020). Diverse cropping systems combined with integrated agriculture can reduce climate-related hazards, especially in coastal regions. The coastal region's agricultural system mostly includes paddy, jute, cotton, oilseeds, groundnuts, and pulses (Nambi and Bahinipati, 2012; Padhan and Madheswaran, 2022). Diversifying cropping systems to include value-added goods such as vegetables, sunflowers, groundnuts, mungbeans, and pigeonpea leads to higher economic returns and increased cropping intensity. Diversified cropping systems make better use of limited water resources, especially for irrigation during the summer. Rainwater is the primary irrigation source for much of India's coastal regions, particularly in the east. During the summer season, the situation worsens due to high coastal salt and insufficient irrigation for cultivation.

Adaptation strategies in coastal agriculture are essential to address the challenges posed by climate change, including fluctuating precipitation, temperature, and seasonal timing (Dhanya et al., 2022; Nambi and Bahinipati, 2012; Rakshit et al., 2016; Reddy et al., 2022). Adjustments to crop calendars have emerged as a proactive approach, enabling farmers to align agricultural practices with favorable conditions while mitigating adverse effects.

18.3.2.1 *Soil fertility and chemical usage*

Rising salinity in coastal soils has led to a decline infertility and an increase in pest and disease outbreaks, prompting farmers to rely more heavily on chemical inputs. While chemicals can provide immediate benefits, their overuse is not climate-resilient and has been linked to declining soil fertility (Singh, 2000) and rising production costs (Singh and Narayanan, 2015). A more sustainable approach involves judicious application of inputs based on soil testing, deposition of soil amendments, integration of legumes, site-specific application of fertilizers, integrated plant nutrient system, and 4R fertilizer methods (Das and Ghosh, 2020; Reddy et al., 2022; Sarkar et al., 2010; Singh et al., 2019).

18.3.2.2 *Organic and sustainable practices*

Farmers are adopting organic farming methods, including the use of vermicompost, trichocompost, biochar (Das and Ghosh, 2020; Dutta et al., 2020), green manuring (Das et al., 2022; Das and Ghosh, 2020), and biofertilizers (Sarkar et al., 2010). Organic farming shows significant potential for climate-resilient agriculture by reducing greenhouse gas emissions, improving soil nutrients, and lowering cultivation costs (Jouzi et al., 2017). Vermicomposting, in particular, enhances soil's biological, physical, and chemical properties, with studies affirming its environmental benefits (Komakech et al., 2016).

18.3.2.3 Indigenous technical knowledge

Farmers continue to utilize indigenous technical knowledge as part of their adaptation strategies. These include traditional plant protection methods, indigenous insecticides, and ash as a suspension material for chemical fertilizers (Das et al., 2022; Dutta et al., 2020; Nambi and Bahinipati, 2012). Additionally, traditional crop varieties, such as "Dudheswar," "Matla," and "Talmugur," are favored for their genetic diversity and resilience to stress conditions, such as floods, droughts, and salinity (Dhara and Paul, 2016). Abiotic stress-tolerant rice varieties, including "BhutMuri," "Jhumur-Kandi," and "Dular," have also been widely cultivated.

18.3.2.4 Agroforestry, windbreaks and floating agriculture

Agroforestry systems, which incorporate trees into farming landscapes, help prevent soil erosion, act as windbreaks, and improve soil fertility and farm productivity (Das and Ghosh, 2020; Mondal, 2022; Patel et al., 2020; Sarkar et al., 2010). Floating agriculture and "gher" farming, other innovative approaches, enable crop cultivation on flooded lands, enhancing agricultural sustainability in regions prone to prolonged inundation (Mondal, 2022). Windbreaks play a vital role in mitigating wind speeds, with studies indicating they can reduce wind velocity by up to 80%. This reduction has significant benefits, such as enhancing crop yields by as much as 20%, improving air quality, and creating habitats that support local wildlife. To achieve optimal performance, the trees used for windbreaks should meet specific criteria: they need to be tall enough to effectively block wind, have adequate density to slow wind while still permitting sunlight to reach crops, and be robust enough to endure strong wind without structural failure.

18.3.3 Soil and water management practices

Soil and water management are vital for sustaining agriculture in India's coastal regions, where climate change has significantly affected these resources. Frequent tropical cyclones, flooding, and rising sea levels have exacerbated problems like soil salinity, erosion, and freshwater degradation, posing long-term challenges such as water scarcity, soil infertility, and increased pest and disease prevalence (Hoque et al., 2013; Kundu et al., 2020; Paul and Chatterjee, 2019; Thaker et al., 2021).

Rainwater harvesting, mulching, and modern irrigation techniques, particularly sprinkler and drip systems, constitute an effective approach for conserving soil moisture and enhancing water utilization in agriculture. The collection and direction of rainfall into soils or storage enhances shallow soil water storage and groundwater recharge, thereby decreasing dependence on pumped water (González et al., 2025). Surface mulches, whether organic or plastic, reduce soil evaporation, regulate temperature fluctuations, and improve the infiltration and retention of rainfall. This enhancement of plant-available water can lead to increased yields when appropriately aligned with soil and climatic conditions (Ye et al., 2023). Sprinkler systems provide uniform coverage across extensive areas; however, drip irrigation, which directs water to the root zone, generally achieves superior water-use efficiency and crop response. Numerous reviews indicate water savings and yield increases of approximately 20%–30% compared to conventional methods (Guo et al., 2024). Combining measures such as rainwater harvesting for root-zone moisture recharge, surface mulching, and drip irrigation yields synergistic benefits, including enhanced soil moisture retention, reduced evaporation losses, improved crop water productivity, and often significantly better yields (Debebe et al., 2025). For effective management, select mulch materials and irrigation schedules that align with local soil types and seasonal conditions, integrate rooftop or in-field rainwater harvesting where applicable, and prioritize drip irrigation for high-value or water-scarce crops to enhance long-term soil and water sustainability (Muharomah et al., 2025).

18.3.3.1 Adaptation strategies for multiple abiotic stresses

Adaptation strategies for major and combined stresses, such as salinity–drought and salinity–waterlogging, require integrated approaches that operate at genetic, agronomic, and landscape levels. For salinity–drought, key measures include the development of tolerant varieties through marker-assisted breeding and CRISPR-based editing of ion transport and osmotic regulation genes, precision irrigation methods to improve water-use efficiency, and the application of plant growth-promoting microbes and seed priming to enhance root vigor and osmotic balance. In salinity–waterlogging environments, raised-bed planting, improved drainage, and the use of cultivars with aerenchyma formation or adventitious roots help reduce anoxia while maintaining ionic homeostasis. Soil amendments such as gypsum and organic matter improve soil structure, while microbial inoculants and exogenous protectants (e.g., silicon, proline) further alleviate stress impacts. Combining engineering interventions with biotechnological innovations and digital monitoring tools offers a sustainable pathway for managing these overlapping stresses in coastal and irrigated systems (Acharya et al., 2024; Tahjib-Ul-Arif et al., 2023; Tarolli et al., 2024).

18.3.3.1.1 Land modifications

Land shaping and "ail cultivation" are widely practiced adaptation methods to mitigate salinity effects and optimize land use (Burman et al., 2013; Dutta et al., 2020; Sarkar et al., 2010). Land shaping involves modifying land surfaces to reduce soil salinity, harvest rainwater, and support diverse cropping systems, including vegetables, paddy, and aquaculture, thereby increasing cropping intensity and income (Mandal et al., 2019). It also aligns with CSA by enhancing food security, adaptation, and mitigation strategies, such as reducing methane emissions (Bandyopadhyay et al., 2009).

18.3.3.1.2 Rainwater harvesting and soil conservation

Rainwater harvesting is a practical solution for addressing erratic precipitation and salinity issues. It replenishes freshwater supplies, improves soil moisture, and restores ecological balance, making it a widely adopted practice in coastal areas (Das and Ghosh, 2020; Ghosh and Mistri, 2020; Mondal, 2022). Farmers also use ridge-and-furrow techniques (Sorjon technique), mulching, and "land holiday" practices to manage soil salinity and fertility. "Land holiday" involves leaving fields fallow after salinity intrusion until monsoon rains wash away salinity (Nambi and Bahinipati, 2012; Padhan and Madheswaran, 2022).

18.3.3.1.3 Pond management

Ponds play a crucial role in providing freshwater for agriculture and aquaculture in coastal regions. Salinization of ponds is a persistent issue, but farmers address this through measures such as dyke repair, dyke plantation with fruit and wood trees, saline water removal, and lime application for water purification (Sarkar et al., 2010). Maintaining pond water quality is essential to prevent fish diseases and sustain homestead fisheries. The dyke can also be used for "gher" farming, which is widely used in the coastal area of Bangladesh.

18.3.3.1.4 Improved irrigation and drainage systems

The irrigation system is vital for agriculture in coastal regions, but is often disrupted by natural disasters and salinity. Overreliance on groundwater depletes freshwater resources and increases cultivation costs. Improved irrigation methods, such as drip and sprinkler irrigation, help conserve water and enhance soil moisture retention (Goswami et al., 2021; Mondal, 2022). Additionally, upgrading field and regional drainage systems is critical to managing soil and freshwater salinity (Das and Ghosh, 2020; Ghosh and Mistri, 2020).

18.4 Development and deployment of stress-resilient crop varieties

Resilient crop varieties are critical for maintaining agricultural productivity and ensuring food security in the face of climate change. For instance, numerous stress-resilient crop varieties have been developed and implemented worldwide to improve CSA. Sahbhagi Dhan from India, Zhongzhan 3 from China, and drought-tolerant maize hybrids in Africa are significant for drought tolerance. Rice varieties such as CSR-36, CSR-56, CSR-60 from India, Pokkali from Kerala, and the wheat variety Shatabdi from Bangladesh exhibit salinity tolerance. Flood-tolerant rice varieties, including Swarna-Sub1 and IR64-Sub1, are extensively cultivated. In contrast, heat-tolerant varieties comprise HD 2967 and HD 3086 for wheat in India, as well as Awhi for maize in Africa. Furthermore, stress-tolerant rice varieties such as DRR Dhan 44, Bhavani, and Annapoorna exhibit resilience to salinity, drought, and waterlogging. The varieties, developed via conventional breeding, marker-assisted selection, and molecular breeding, result from collaborative efforts among IRRI, CIMMYT, ICAR, ICRISAT, and national breeding programs. Successful adoption of these crops requires investments in research, efficient seed delivery systems, affordable pricing, farmer education, and supportive policies. These efforts collectively strengthen agricultural systems and ensure food security for future generations.

18.4.1 Breeding for drought resistance

Drought is a significant challenge to global food security and agricultural productivity. Breeding drought-resilient crops combines traditional methods, such as backcrossing, pedigree breeding, and phenotypic selection, with advanced techniques like quantitative trait locus mapping, marker-assisted selection, and genomic selection. Cutting-edge technologies, including genome editing and genomic-assisted breeding, are also employed (Blum, 1996; Gao, 2021; Kültz, 2003).

Case studies demonstrate the success of drought-tolerant maize, heat-tolerant wheat, and salt-tolerant rice in maintaining productivity under adverse conditions. Future advancements involve integrating breeding with improved agronomic practices, fostering collaboration among research institutions, seed companies, and farmers, and developing innovative drought-resilience mechanisms. Continued investment in these areas ensures sustainable food production and strengthens global food security (Xu et al., 2020).

18.4.2 Developing flood-tolerant crops

Floods significantly threaten agricultural yields and livelihoods, prompting researchers to develop flood-tolerant crops capable of maintaining productivity during submergence. These crops exhibit morphological adaptations such as aerial roots and elongated internodes, as well as physiological mechanisms such as anaerobic respiration and enhanced antioxidant activity. Breeding strategies include conventional methods, backcrossing flood-tolerant genes into high-yield varieties, marker-assisted selection, and genome editing using CRISPR-Cas9 (Lee et al., 2019).

Examples of flood-tolerant crops include deepwater rice, waterlogging-tolerant soybean, and maize varieties adapted to flash floods. Future efforts aim to harness genetic diversity, develop multistress-tolerant crops, improve evaluation in real field conditions, and involve farmers in breeding programs to align outputs with local needs. Collaborative investments in research and farmer-centric breeding approaches can enhance food security and support livelihoods in flood-prone regions worldwide (Singh et al., 2020).

18.4.3 Use of nanotechnology

Nanotechnology has emerged as a viable approach for the development of stress-resilient crops, providing innovative solutions to alleviate the negative impacts of abiotic stresses, including drought, salinity, heat, cold, and nutrient deficiencies. Nano-fertilizers and nano-encapsulated nutrients enhance nutrient use efficiency through slow and targeted release, which promotes plant growth in stressful environments. Nano-based sensors facilitate real-time monitoring of soil moisture, nutrient status, and stress signals, thereby enabling precise management practices. Nanoparticles, including nano-silicon, nano-zinc, and nano-titanium dioxide, enhance plant antioxidant defense systems, increase photosynthetic efficiency, and mitigate oxidative damage resulting from stress conditions. Furthermore, nanocarriers facilitate the effective delivery of stress-tolerant genes, biomolecules, or agrochemicals, thereby enhancing genetic improvement and crop protection. The integration of nanotechnology in crop management enables farmers to minimize input losses, improve resource efficiency, and develop crops that are more resilient to changing climatic conditions, thereby fostering sustainable agricultural systems.

18.5 Biostimulants and their role in mitigating multiple abiotic stresses

Bio-stimulants are organic or inorganic substances and/or microorganisms that stimulate several processes when applied to plants, resulting in increased growth, productivity, and stress tolerance (Ali et al., 2021; Franzoni et al., 2022). They are available insoluble powder, granular form, or liquid form and can be used as foliar sprays or in soil at the root zone. Bio-stimulants are ecologically beneficial agronomic tools, and the market for these compounds is expected to increase at a 12% annual pace, reaching $5.6 billion by 2026 (Bulgari et al., 2019; Lau et al., 2022). Several reasons have contributed to the growth of the bio-stimulant market, including changes in agricultural and environmental policies, and climate change, which have emphasized the need for more sustainable alternatives to synthetic chemicals (Lau et al., 2022). Several reasons have contributed to the bio-stimulant market's growth, including changes in agricultural and environmental policies and climate change, which have pushed for more sustainable alternatives to synthetic chemicals. Bio-stimulants have low or no toxicity and do not accumulate in the long run (Sangiorgio et al., 2020). Bio-stimulants are categorized into six broad classes based on their raw material source, including seaweed, plant extracts, protein hydrolysates, humic substances, inorganic chemicals, and microorganisms (Franzoni et al., 2021).

18.5.1 Seaweed and plant extracts

The most extensively researched and commercialized bio-stimulants are those derived from seaweeds (Cristiano et al., 2018). This category includes macroscopic and multicellular marine algae from several taxonomic groupings, including red, green, and brown algae. Seaweeds have been utilized as fertilizers in agriculture since ancient times because of their beneficial effects for crops (Franzoni et al., 2022). These algae contain hormones (e.g., auxins and cytokinins), polyphenols (e.g., phloroglucinol and eckol), polysaccharides (e.g., fucoidan, laminaran, carrageenan, and alginates), and kahydrin, a vitamin K1 derivative that promotes growth and activates the plant defense system (Baltazar et al., 2021; Ertani et al., 2018).

Most agricultural products derived from algae are sourced from red algae (*Lithothamnium calcareum, Kappaphycus alvarezii, Jania rubens*, and *Gracilaria edulis), green algae (Ulva lactuca)*, and brown algae *(Ascophyllum nodosum, Ecklonia maxima, Durvillaea potatorum, and Laminaria spp.)* (Van Oosten et al., 2017). Extracts of many plant structures, including leaves, fruits, stems, and flowers, can be used to produce bio-stimulants (Franzoni et al., 2022). These extracts are high in bioactive chemicals, which can activate physiological processes in plants and hence boost their performance (Zulfiqar et al., 2020). Seagrass aqueous extracts reduce the deleterious effects of salt stress on tomato plants by increasing the activity of antioxidant enzymes.

18.5.2 Protein hydrolysates, humic substances, and inorganic compounds

Hydrolyzed proteins are a mixture of amino acids, polypeptides, and oligopeptides formed by the chemical, thermal, and enzymatic hydrolysis of proteins derived from various matrices, such as animal or plant tissue (Cristofano et al., 2021). Protein hydrolysates formed from agricultural and industrial byproducts are gaining popularity since they provide a sustainable and environmentally friendly solution to various wastes while also contributing to a circular economy (Colla et al., 2017). The majority of commercially available protein hydrolysate bio-stimulants are produced from animal proteins, which have a greater nitrogen concentration (9%–16% DM) and a slower release rate. Proline and glycine are the most prevalent amino acids in protein hydrolysate bio-stimulants based on collagen, but glutamic acid is more plentiful in plant-based bio-stimulants (Baroccio et al., 2017). Humic substances are another type of bio-stimulant that contains humic and fulvic acids and is found in lignites, soil, and peat (Van Oosten et al., 2017). Humic substances are supramolecular structures with hydrophilic (-OH and -COOH groups) and hydrophobic portions formed by the decomposition of small molecules by animals, plants, or soil microbes (Franzoni et al., 2022; Popa et al., 2022). Some inorganic substances, such as silicon, selenium, cobalt, sodium, and aluminum, can stimulate plant growth and induce stress responses, but they are not required by all plants (Hoque et al., 2013). They can be present in plants and soils in various forms, including inorganic salts and insoluble compounds such as amorphous silica.

18.5.3 Microorganisms

Bacteria, fungi, and arbuscular mycorrhizal fungi that can be isolated from soils, plants, and organic materials make up the majority of microorganism-based bio-stimulants (Baltazar et al., 2021; Del Buono, 2021). These bio-stimulants can be used directly on soils or seeds (Franzoni et al., 2021). These bio-stimulants interact directly with plants, forming a mutual symbiotic relationship, or indirectly by increasing nutrient availability to plants (Franzoni et al., 2021). Plant growth-promoting bacteria comprise rhizobacteria that colonize the rhizosphere as well as free-living bacteria from the soil (Shah et al., 2021). The most important group is the plant growth-promoting rhizobacteria, which includes the species *Azotobacter, Arthrobacter, Azospirillum, Acinetobacter, Bacillus, Bradyrhizobium, Enterobacter, Pseudomonas, Rhodococcus, Rhizobium, Streptomyces*, and *Ochrobactrum* (Rai et al., 2021).

18.5.4 Mechanism of action of bio-stimulants for mitigating multiple abiotic stresses and promoting plant growth

Bio-stimulants operate as priming agents (Shukla et al., 2019); however, the mechanisms of action of microbial and nonmicrobial bio-stimulants in plants are not yet fully understood. Priming, also known as stress hardening, activates several biochemical and physiological defense mechanisms, increasing plants' ability to defend themselves when subjected to various stresses. These molecular and cellular changes caused by these priming agents in plants are kept as primed memory (Nephali et al., 2020), and once primed, the plants will defend faster and stronger to cope with future shocks.

Bio-stimulants stimulate and control many defensive mechanisms via various action modes (Rai et al., 2021). Bio-stimulants can be applied directly to the leaves (foliar application) or to the soil around the root system. Bio-stimulants derived from proteins and amino acids (e.g., proline and glycine) can penetrate leaf tissues and reach cells directly. Protein hydrolysate-based bio-stimulants penetrate plant cells through membrane pores, incurring energy expenditures (Yakhin et al., 2017). In microorganism-based bio-stimulants, hyphae penetrate tissues and form symbiotic or mycorrhizal relationships. When bio-stimulants reach the leaves and/or roots, they are transported and distributed throughout the plant (Rai et al., 2021).

18.6 Digital tools and technologies for climate-smart agriculture for managing multiple abiotic stresses

18.6.1 Precision agriculture

Precision agriculture (PA) represents an advanced approach to managing agricultural operations by utilizing cutting-edge technologies and information systems that can be used to manage multiple abiotic stresses. This practice involves collecting and analyzing data related to soil conditions, weather, and crop health to enable resource-efficient decision-making, ultimately enhancing yields, profitability, and environmental sustainability. Originating in the 1980s with the use of remote sensing to assess soil organic matter, PA has since evolved with the adoption of modern electromagnetic wavelengths, such as ultraviolet, microwave, and hyperspectral remote sensing. These advancements facilitate detailed analyzes of chemical and biophysical properties, crop stress, and molecular interactions, offering real-time insights into crop and soil management (Mulla, 2013).

The key features of PA include the integration of technologies such as sensors, drones, satellites, yield monitors, geographic information systems (GIS), and Global Positioning System-enabled precision equipment. The benefits of PA extend to increased crop yields, reduced environmental impact, improved profitability, enhanced decision-making, and greater labor efficiency. Despite these advantages, challenges such as high implementation costs, data management complexities, and the need for technological adoption persist.

Technology plays a central role in resource optimization across various sectors, including agriculture, energy, transportation, and manufacturing. In agriculture, PA employs sensors, drones, and data analytics to gather real-time data for soil, crop, and weather monitoring. Other resource-optimizing innovations include vertical farming for reducing land and water use, renewable energy technologies such as solar panels and wind turbines, and smart grids for enhancing energy efficiency (Padarian et al., 2019).

Data-driven decision-making (DDDM) is transforming farming by providing actionable insights that improve yields, profitability, sustainability, and risk management as personalized advice using different interfaces such as apps, software, chatbots, ChatGPT, DeepSeek, etc. By leveraging data from sensors, drones, satellites, weather systems, and farm management software, farmers can optimize operations while minimizing environmental impact. However, DDDM faces challenges, such as high costs, data security concerns, and the need for technological literacy.

18.6.2 Remote sensing and geographic information systems

Remote sensing plays a pivotal role in supporting CSA by providing critical data for sustainable and resilient agricultural practices. The diverse capabilities of remote sensing systems—spanning spatial, spectral, radiometric, and temporal resolutions—offer valuable insights that align closely with CSA objectives.

18.6.2.1 Spatial resolution and precision farming

High spatial resolution data enables detailed mapping of agricultural fields, identifying variations in soil health, crop health, and water availability at a localized level. This allows farmers to adopt precision agriculture techniques, where resources such as fertilizers, water, and pesticides are applied only where needed, reducing waste and environmental impacts. For instance, a 10-meter spatial resolution dataset can help monitor individual crop rows or small-scale irrigation systems.

18.6.2.2 Spectral resolution for crop and soil monitoring

Hyperspectral and multispectral sensors provide detailed spectral data to monitor vegetation health, detect nutrient deficiencies, and assess soil moisture. Farmers can identify early signs of crop stress caused by drought, pests, or diseases by analyzing specific bands related to chlorophyll content or thermal stress. This information supports timely interventions, minimizing yield losses, and improving climate resilience.

18.6.2.3 Radiometric resolution for subtle changes

The high radiometric resolution of remote sensors enables the detection of subtle energy variations emitted by the Earth's surface, such as small temperature changes in soil or water bodies. These measurements are crucial for monitoring microclimatic conditions within agricultural landscapes, helping farmers adjust planting schedules, irrigation systems, and crop choices to mitigate the effects of changing weather patterns.

18.6.2.4 Temporal resolution for seasonal and real-time monitoring

Satellites with high temporal resolution offer frequent updates on agricultural conditions. This capability is essential for tracking crop growth stages, monitoring seasonal changes, and responding to extreme weather events such as droughts or floods. For example, real-time monitoring can guide decisions on harvesting times or applying protective measures during extreme weather events, reducing losses.

18.6.2.5 Climate and weather forecasting with remote sensing

Weather radars, such as Doppler systems, provide critical information on precipitation patterns, intensity, and type (rain, snow, or hail), helping farmers plan irrigation and manage risks associated with extreme weather. Long-term climate data derived from satellite systems also aids in developing strategies for adapting to climate change, such as shifting planting seasons or selecting climate-resilient crops.

18.6.2.6 Sustainability and emissions monitoring

Remote sensing can assess land use changes, monitor deforestation, and measure greenhouse gas emissions from agricultural activities. By identifying practices contributing to emissions, farmers can implement mitigation strategies, such as adopting agroforestry or conservation tillage, to align with CSA goals.

18.6.3 Remote sensing for multiple abiotic stress management

18.6.3.1 Drought monitoring

Remote sensing data, such as multispectral and hyperspectral photography, can detect variations in vegetation moisture content, indicating drought stress. GIS may be used to map drought-prone areas and prioritize the allocation of water resources (Thenkabail et al., 2018).

18.6.3.2 Salinity and soil quality

Remote sensing techniques, along with GIS, can analyze soil salinity and quality across wide areas. This information aids in identifying areas where soil management practices should be modified to reduce stress.

18.6.3.3 Farmer support applications

Farmers frequently require reliable and up-to-date information tailored to their specific location. Here are five apps that give Indian farmers smartphone access to a variety of services, with an emphasis on local and practical information (Table 18.1).

18.6.3.4 Temperature and climate monitoring

Satellite data can be used to collect information about land surface temperature and climatic conditions, as well as to assess vegetation dynamics over time. GIS may then be used to assess temperature patterns and their effects on crop development, allowing farmers to adjust planting schedules and cultivation procedures.

18.6.3.5 Use of artificial intelligence technologies to manage multiple abiotic stresses

Artificial intelligence (AI) technologies are becoming essential in addressing several abiotic stresses in plants, including drought, salinity, heat, cold, and nutrient deficits, which frequently occur concurrently and hinder agricultural productivity. Through the integration of extensive data and sophisticated modeling approaches, AI offers predictive, adaptive, and real-time solutions to improve stress management. In stress prediction and monitoring, AI-driven tools, including remote sensing through drones, satellites, and IoT sensors, consistently monitor soil moisture, temperature, nutrient availability, and plant physiological responses, while machine learning models analyze spectral imagery to identify early stress indicators and anticipate stress events such as droughts or heatwaves. In genomics and breeding, AI expedites the identification of stress-tolerant genes and markers, synthesizes multiomics data to elucidate complicated connections, and directs CRISPR-Cas gene-editing methodologies for the development of resilient crop varieties.

AI-enabled decision support systems enhance smart crop management by optimizing irrigation, nutrient utilization, and resource allocation, while digital crop twins model stress scenarios to inform adaptive approaches. Furthermore, AI-driven phenotyping platforms and image analysis evaluate plant characteristics under simultaneous stress circumstances, aiding breeders in comprehending interaction effects and selecting resilient types. AI synthesises climate and agricultural models

TABLE 18.1 Applications (App) and their uses in agriculture.

Sl. no.	Name of application	Features	Benefits
1	AgroStar: KisanAgridoctor	Personalized agronomy advice, agri products, farmer community	• Access to expert advice and high-quality inputs tailored to local conditions
2	Pusa Krishi	New crop varieties, resource conservation techniques, crop management practices	• Empowers farmers with the latest research • It leads to more efficient and sustainable farming
3	DeHaat	Quality inputs (seeds, fertilizers), crop advisory services, market linkages	• Enhances productivity and profitability • Provides a direct link to markets
4	AgriCentral	Crop care, market view, weather forecasts, community support	• Real-time data and recommendations that improve decision-making and productivity
5	Kisaan Helpline India	Local agricultural, insights pest management, soil health	• Provides solutions for pest management • Soil quality improvement • Local agricultural challenges
6	m-Agri	Facilitating farmers' access to financial services and sourcing agricultural information about input use, practices, and market prices	• The potential to revolutionize agriculture and significantly improve smallholder farmers' livelihoods in Africa
7	WeFarm	Operates in Kenya and Uganda, and adopted the strategy of using crowdsourced knowledge through a message service to give remote farmers access to agricultural information without using Internet facilities	• The application provides a connection to more than 1.1 million users across Kenya and Uganda. This service also allows the farmers to ask questions in any language, and, therefore directly addresses one of the barriers to long-term usage by the farmers. This is where the farmers are not conversant with the language used for the service

to evaluate risks and improve climate resilience, providing tailored suggestions that enable farmers to select appropriate crops and techniques for sustainable agriculture in multistress conditions.

18.7 Institutional and policy frameworks supporting climate-smart agriculture for coastal regions

18.7.1 Climate-smart agriculture in national policy

The greatest barriers to alleviating hunger and poverty in developing nations are socioeconomic divides and inequalities, which are exacerbated by climate change and abiotic stresses. Location-specific and community-based efforts to promote CSA demand further attention. Science contributes to development by providing rigorous evidence that informs policymakers and establishes the science-policy interaction at national, regional, and global levels. Strengthening national capacities in multidisciplinary and participatory research, knowledge development, database building, science-informed policy formulation, strategic planning, and program execution is essential. The scientific method can inform national investment decisions for adaptation and mitigation.

Policies related to agriculture, disaster management, food security, water, and land must be coordinated across multiple levels, including grassroots organizations such as climate-smart villages. Effective implementation necessitates institutional reforms and interministerial collaboration. The integration of CSA into national policies requires strategic investment and financing mechanisms. To address these challenges, the Government of India has introduced several initiatives. In 2008, the National Action Plan on Climate Change (NAPCC) was announced, followed by the launch of the National Mission for Sustainable Agriculture (NMSA) in 2010. The NMSA focuses on key areas such as dryland agriculture, access to information, biotechnology, and risk management. Additionally, certain national missions have a direct influence on the CSA agenda. For instance, the National Mission on Micro Irrigation plays a crucial role in supporting adaptation, mitigation, and sustainable intensification while promoting synergy among CSA interventions.

18.7.2 Policy and programs

18.7.2.1 National action plan on climate change

The national action plan on climate change (NAPCC) provides India's strategy for addressing climate change in a variety of industries, including agriculture. The NAPCC emphasizes the importance of adaptation techniques while recognizing that agriculture is especially sensitive to climate change. It includes missions with specific targets in climate-resilient agriculture and water management, such as the "National Water Mission" and the "National Mission for Sustainable Agriculture (NMSA)" (Government of India, 2008).

18.7.2.2 Pradhan Mantri Krishi Sinchayee Yojana

The Pradhan Mantri Krishi Sinchayee Yojana (PMKSY) increases water use efficiency using a range of approaches such as micro-irrigation, watershed improvement, and rainwater harvesting. PMKSY assists agriculture in adapting to climate change by adopting appropriate water management strategies, particularly in water-stressed areas.

18.7.2.3 Paramparagat Krishi Vikas Yojana

This promotes organic farming methods since they have been shown to improve soil health and biodiversity. Furthermore, by limiting the use of chemical inputs such as pesticides and fertilizers, organic farming aids efforts to minimize the consequences of climate change. Farmers who switch to organic agriculture practices receive cash incentives and training through Paramparagat Krishi Vikas Yojana.

18.7.2.4 Soil health card scheme

This program went into effect in 2015 and seeks to monitor soil health and provide farmers with tailored advice on how to improve soil fertility. This program is critical for adaptation since healthy soils are more resistant to climate change-induced stresses like droughts and floods.

18.7.2.5 Crop insurance program based on weather

Crop insurance program based on weather provides financial protection to farmers against crop losses due to poor weather conditions. By reducing the financial risks associated with climate variability, this initiative encourages farmers to implement risk-reduction methods and climate-resilient practices.

18.7.2.6 National mission for sustainable agriculture

The National Mission for Sustainable Agriculture emphasizes crop diversity, integrated farming systems, and agroforestry as climate-resilient agricultural techniques. Farmers can reduce their vulnerability to the effects of climate change by adopting climate-smart technologies and diversifying their crop varieties (Rao et al., 2024).

18.7.2.7 Pradhan Mantri fasal bima yojana

It was launched in 2016; Pradhan Mantri fasal bima yojana intends to provide farmers with affordable crop insurance. This method makes it easier to recover from crop losses and encourages the adoption of suggested climate-resilient activities, as insurance coverage relies on their implementation (Rao et al., 2024).

18.8 Challenges and future prospects

18.8.1 Challenges in implementing climate-smart agriculture

Implementing CSA presents significant challenges for smallholder farmers, particularly due to high initial costs, limited access to financing, and market uncertainties. Investments in advanced technologies and supporting infrastructure often require substantial capital, which poses a considerable burden for farmers in developing nations. Moreover, the market for climate-smart agricultural products remains unpredictable, with uncertainties surrounding fair pricing and consistent demand.

The impacts of climate change are anticipated to further exacerbate the vulnerabilities of smallholder farmers, potentially resulting in a 28% decline in household incomes. This decline could reduce investments in cash crops and increase the likelihood of economic migration (Choquette-Levy et al., 2021). However, the adoption of risk-transfer mechanisms, coupled with modest cash transfers, has the potential to enhance the uptake of CSA practices, improve community income levels, and mitigate economic inequalities. Despite its benefits, CSA adoption is often hindered by a lack of awareness and inadequate extension services. Many regions suffer from insufficient training and support, leaving farmers ill-equipped to transition to sustainable practices. Additionally, traditional agricultural methods and cultural beliefs can act as barriers to the adoption of innovative technologies and practices.

18.8.2 Opportunities for advancement in technological innovations

Technological developments are fundamentally transforming the agricultural sector, enhancing its productivity, efficiency, and climate resilience. Innovations such as crop breeding, precision agriculture, AI, nanotechnology, and CRISPR-Cas gene-editing technologies are propelling substantial advancements in the creation of high-yielding and climate-resilient crop types. These advancements enhance crop performance and provide improved flexibility to fluctuating weather circumstances. Simultaneously, renewable energy solutions are being incorporated into agricultural systems, diminishing reliance on nonrenewable resources and fostering sustainable farming practices.

As these technologies advance and achieve economies of scale, the related prices are anticipated to decrease, rendering them more affordable and accessible for farmers in various places. Furthermore, the emergence of digital technologies and sophisticated data analytics equips farmers with real-time information and critical insights, enabling them to enhance the utilization of resources such as water, fertilizers, and labor. These techniques assist in forecasting weather patterns and mitigating hazards, allowing farmers to adjust proficiently to climate variability. Collectively, these innovations are establishing the groundwork for a more sustainable and resilient agricultural future. The promotion of CSA can be facilitated through international collaboration, particularly South–South cooperation, where developing nations can share best practices and lessons learned. Financial and technical support from donor agencies and international organizations can further strengthen CSA initiatives. The Global initiatives like The "Blue Carbon Initiative" is the name for multiple efforts to conserve and restore coastal ecosystems like mangroves and seagrass beds, with key initiatives including the International Partnership for Blue Carbon, launched in 2015 at COP21 in Paris. The earlier establishment of a Blue Carbon Scientific Working Group in 2011 to provide scientific support for the initiative. India also launched its own Mangrove Initiative for Shoreline Habitats & Tangible Incomes scheme in recent years to support local communities in mangrove conservation and restoration and it includes Conservation/Restoration Activities: Kaimana Coastal Conservation and Community Development – Indonesia, Tomago Wetland Restoration – Australia; Scientific Activities: Mangrove Sedimentation and Surface Accretion – Indonesia Blue Carbon A-Z: from small projects to policy development – Latin America, Tampa Bay Blue Carbon Assessment – USA, Gulf of Nicoya Blue Carbon Stock Assessment – Costa Rica; Policy Activities: Blue Carbon A-Z: from small projects to policy development – Latin America, Herring River Estuary Restoration Market Feasibility Assessment – USA. International Coral Reef Initiative started in 1994 at the first Conference of the Parties to the Convention on Biological Diversity in the Bahamas, intending to preserve coral reefs and related ecosystems globally. Mangroves for the Future initiative started in December 2006 when former U.S. President Bill Clinton launched it in Phuket, Thailand. Governments also play a critical role by enacting policies and regulations that encourage sustainable agricultural practices, developing robust value chains for climate-smart products, and implementing comprehensive training programs to ensure successful adoption and long-term viability.

Future research should emphasize multidisciplinary collaboration among geneticists, agronomists, soil scientists, and environmentalists to address the complex concerns of abiotic stresses. Integrating climate-smart agricultural practices and crop enhancement initiatives can improve coastal agriculture's resilience to climate change-induced stresses.

18.9 Conclusion

CSA provides a strategy framework for addressing food security, climate change, and environmental sustainability. Through increased production, resource efficiency, and greenhouse gas reduction, CSA strengthens the agricultural industry. Economic hurdles, lack of understanding, and cultural hostility hinder adoption.

Farmers should be encouraged to adopt CSA via government policies, financial incentives, and regulations. Financial organizations offer personalized solutions such as low-interest loans and insurance, while research institutions discover and spread revolutionary technologies. Farmers need training and technical help from extension agencies to undertake CSA. Farmers need continual education, and customers can help by buying sustainable products.

To improve resource use and decision-making, CSA must integrate modern tools like precision farming and data analytics. Addressing climate change requires improving CSA practices, community involvement, and funding for climate-resilient crops and livestock. Improved support systems and social safety nets can protect at-risk groups.

Indian coastal farmers must incorporate local knowledge into adaptive tactics. Understanding farmers' behaviors is crucial to creating client-centered policies that boost social, economic, and environmental resilience. Cyclones, droughts, and salinity intrusion require targeted efforts to mitigate their consequences. Regional climate and agricultural differences necessitate various, location-specific adaptation measures.

Participatory research can help policymakers, local institutions, and agricultural communities create inclusive and actionable CSA policies. These programs are essential for smallholder farmer livelihoods, food security, and climate-resilient agriculture.

References

Abdelkader, M., Suliman, A.A., Salem, S.S., Assiya, A., Voronina, L., Puchkov, M., et al., 2024. Studying the combined impact of salinity and drought stress-simulated conditions on physio-biochemical characteristics of lettuce plant. Horticulturae 10 (11), 1186.

Abdulaziz S. Alhaithloul, H., Abu-Elsaoud, A.M., Soliman, M.H., 2020. Abiotic stress tolerance in crop plants: role of phytohormones. In: Fahad, S., Saud, S., Chen, Y., Wu, C., Wang, D. (Eds.), Abiotic Stress in Plants. IntechOpen, pp. 233. Available at https://doi.org/10.5772/intechopen.93710.

Acharya, B.R., Gill, S.P., Kaundal, A., Sandhu, D., 2024. Strategies for combating plant salinity stress: the potential of plant growth-promoting microorganisms. Frontiers in Plant Science 15, 1406913.

Adam, H.N., Kjosavik, D.J., Shanmugaratnam, N., 2018. Adaptation trajectories and challenges in the Western Ghats: a case study of Attappady, South India. Journal of Rural Studies 61, 1–11.

Akter, R., Mazumder, M.S.U., 2025. Adaptation strategies to waterlogging among coastal farmers in Bangladesh: practices, determinants, and implications for resilient agriculture. Climate Services 39, 100598.

Akter, R., Hasan, N., Reza, F., Asaduzzaman, M., Begum, K., Shammi, M., 2023. Hydrobiology of saline agriculture ecosystem: a review of scenario change in South-West Region of Bangladesh. Hydrobiology 2 (1), 162–180.

Ali, O., Ramsubhag, A., Jayaraman, J., 2021. Biostimulant properties of seaweed extracts in plants: implications towards sustainable crop production. Plants 10, 531.

Aloui, K., Choukri, H., El Haddad, N., Gupta, P., El Bouhmadi, K., Emmrich, P.M., et al., 2023. Impact of heat and drought stress on grasspea and its wild relatives. Plants 12 (19), 3501.

Baltazar, M., Correia, S., Guinan, K.J., Sujeeth, N., Bragança, R., Gonçalves, B., 2021. Recent advances in the molecular effects of biostimulants in plants: an overview. Biomolecules 11 (8), 1096.

Bandyopadhyay, B.K., Burman, D., Sarangi, S.K., Mandal, S., Bal, A.R., 2009. Land shaping techniques to alleviate salinity and water logging problems of mono-cropped coastal land for multi-crop cultivation. Journal of the Indian Society of Coastal Agricultural Research 27 (1), 13–17.

Baroccio, F., Barilaro, N., Tolomei, P., Mascini, M., 2017. Classification of biostimulants origin using amino acids composition of hydrolyzed proteins. Journal of Horticultural Science and Research 1, 30–35.

Bej, S., Basak, J., 2014. MicroRNAs: the potential biomarkers in plant stress response. American Journal of Plant Sciences 05 (05), 748–759. Available at 10.4236/ajps.2014.55089.

Benali, A., El Haddad, N., Patil, S.B., Goyal, A., Hejjaoui, K., El Baouchi, A., et al., 2023. Impact of terminal heat and combined heat-drought stress on plant growth, yield, grain size, and nutritional quality in chickpea (Cicer arietinum L.). Plants 12 (21), 3726.

Bhattacharjee, R., Naskar, M., Sengupta, C., Ghosh, P.K., Datta, D., 2023. Regional environmental change, farmers' perceptions and resilient farming: experiences from the Sagar Island of Indian Sundarbans. Papers in Applied Geography 9 (2), 171–192.

Blum, A., 1996. Crop responses to drought and the interpretation of adaptation. Plant Growth Regulation 20 (2), 135–148.

Bulgari, R., Franzoni, G., Ferrante, A., 2019. Biostimulants application in horticultural crops under abiotic stress conditions. Agronomy 9 (6), 306.

Burman, D., Mandal, S., Bandopadhyay, B.K., Maji, B., Sharma, D.K., Mahanta, K.K., et al., 2015. Unlocking production potential of degraded coastal land through innovative land management practices: a synthesis. Journal of Soil Salinity and Water Quality 7 (1), 12–18.

Burman, D., Bandyopadhyay, B.K., Mandal, S., Mandal, U.K., Mahanta, K.K., Sarangi, S.K., 2013. Land shaping—a unique technology for improving productivity of coastal land. Technical Bulletin: CSSRI/Canning Town/Bulletin/2013/02. https://doi.org/10.13140/RG.2.1.1595.2483.

Century, K., Reuber, T.L., Ratcliffe, O.J., 2008. Regulating the regulators: the future prospects for transcription-factor-based agricultural biotechnology products. Plant Physiology 147 (1), 20–29.

Choi, K., Khan, R., Lee, S.W., 2021. Dissection of plant microbiota and plant-microbiome interactions. Journal of Microbiology 59 (3), 281–291.

Choquette-Levy, N., Wildemeersch, M., Oppenheimer, M., Levin, S.A., 2021. Risk transfer policies and climate-induced immobility among smallholder farmers. Nature Climate Change 11 (12), 1046–1054.

Clarke, D., Williams, S., Jahiruddin, M., Parks, K., Salehin, M., 2015. Projections of on-farm salinity in coastal Bangladesh. Environmental Science: Processes & Impacts 17 (6), 1127–1136.

Colla, G., Hoagland, L., Ruzzi, M., Cardarelli, M., Bonini, P., Canaguier, R., et al., 2017. Biostimulant action of protein hydrolysates: unraveling their effects on plant physiology and microbiome. Frontiers in Plant Science 8, 2202.

Cristiano, G., Pallozzi, E., Conversa, G., Tufarelli, V., De Lucia, B., 2018. Effects of an animal-derived biostimulant on the growth and physiological parameters of potted snapdragon (Antirrhinum majus L.). Frontiers in Plant Science 9, 861.

Cristofano, F., El-Nakhel, C., Rouphael, Y., 2021. Biostimulant substances for sustainable agriculture: origin, operating mechanisms and effects on cucurbits, leafy greens, and nightshade vegetables species. Biomolecules 11 (8), 1103.

Das, U., Ghosh, S., 2020. Factors driving farmers' knowledge on climate change in a climatically vulnerable state of India. Natural Hazards 102 (3), 1419–1434.

Das, U., Ansari, M.A., Ghosh, S., 2022. Effectiveness and upscaling potential of climate smart agriculture interventions: farmers' participatory prioritization and livelihood indicators as its determinants. Agricultural Systems 203, 103515.

DasGupta, R., Shaw, R., Basu, M., 2019. Implication and management of coastal salinity for sustainable community livelihood: case study from the Indian Sundarban Delta. Coastal Management. Academic Press, pp. 251–269.

Davis, A.S., Hill, J.D., Chase, C.A., Johanns, A.M., Liebman, M., 2012. Increasing cropping system diversity balances productivity, profitability and environmental health. PLoS One 7 (10), e47149. Available at https://doi.org/10.1371/journal.pone.0047149.

Debebe, Y., Otterpohl, R., Birhane, E., 2025. Integrating rainwater harvesting and organic soil amendment to enhance crop yield and soil nutrients in agroforestry. Environment, Development and Sustainability 1–19.

Del Buono, D., 2021. Can biostimulants be used to mitigate the effect of anthropogenic climate change on agriculture? It is time to respond. Science of the Total Environment 751, 141763.

Deryng, D., Conway, D., Ramankutty, N., Price, J., Warren, R., 2014. Global crop yield response to extreme heat stress under multiple climate change futures. Environmental Research Letters 9 (3), 034011.

Dhanya, P., Ramachandran, A., Palanivelu, K., 2022. Understanding the local perception, adaptation to climate change and resilience planning among the farmers of semi-arid tracks of South India. Agricultural Research 11 (2), 291–308.

Dhara, S., Paul, A.K., 2016. Status of agriculture-a case study at Patharpratima block of South 24 Parganas district. International Journal of Innovative Science, Engineering & Technology 3 (2), 239–246.

Dugasa, M.T., Cao, F., Ibrahim, W., Wu, F., 2019. Differences in physiological and biochemical characteristics in response to single and combined drought and salinity stresses between wheat genotypes differing in salt tolerance. Physiologia Plantarum 165 (2), 134–143.

Dutta, S., Maiti, S., Garai, S., Abrar, F., Jha, S.K., Bhakat, M., et al., 2020. Analyzing adaptation strategies to climate change followed by the farming community of the Indian Sunderbans using Analytical Hierarchy Process. Journal of Coastal Conservation 24 (5), 61.

Ertani, A., Francioso, O., Tinti, A., Schiavon, M., Pizzeghello, D., Nardi, S., 2018. Evaluation of seaweed extracts from Laminaria and Ascophyllum nodosum spp. as biostimulants in Zea mays L. using a combination of chemical, biochemical and morphological approaches. Frontiers in Plant Science 9, 428.

Feist, S.E., Hoque, M.A., Ahmed, K.M., 2023. Coastal salinity and water management practices in the Bengal Delta: a critical analysis to inform salinisation risk management strategies in Asian Deltas. Earth Systems and Environment 7 (1), 171–187.

Franzoni, G., Bulgari, R., Ferrante, A., 2021. Maceration time affects the efficacy of borage extracts as potential biostimulant on rocket salad. Agronomy 11 (11), 2182.

Franzoni, G., Cocetta, G., Prinsi, B., Ferrante, A., Espen, L., 2022. Biostimulants on crops: their impact under abiotic stress conditions. Horticulturae 8 (3), 189.

Gao, C., 2021. Genome engineering for crop improvement and future agriculture. Cell 184 (6), 1621–1635.

Ghosh, S., Mistri, B., 2020. Coastal agriculture and its challenges: a case study in Gosaba Island, Sundarban, India. Space and Culture 8 (2), 140–154.

González, P.S., Sáez Lazo, R., Vallejos Carrera, C., Fernández Torres, Ó., Bustos-Espinoza, L., et al., 2025. Rainwater harvesting for well recharge and agricultural irrigation: an adaptation strategy to climate change in Central Chile. Sustainability 17 (8), 3549.

Goswami, R., Roy, K., Dutta, S., Ray, K., Sarkar, S., Brahmachari, K., et al., 2021. Multi-faceted impact and outcome of COVID-19 on smallholder agricultural systems: integrating qualitative research and fuzzy cognitive mapping to explore resilient strategies. Agricultural Systems 189, 103051.

Guo, H., Si, D.H., Zhu, H.J., Chen, Z.A., Cao, R., Huang, Y.B., 2024. Boosting CO2 electroreduction over a covalent organic framework in the presence of oxygen. Angewandte Chemie International Edition 63 (14), e202319472.

Hoque, M.A.A., Saika, U., Sarder, B.C., Biswas, K.K., 2013. Environmental and socio-economic impacts of salinity intrusion in the coastal area: a case study on Munshigong Union, Shymnagor, Satkhira. Jahangirnagar University Environmental Bulletin 2, 41–49.

Jena, P.R., Tanti, P.C., Maharjan, K.L., 2023. Determinants of adoption of climate resilient practices and their impact on yield and household income. Journal of Agriculture and Food Research 14, 100659.

Jouzi, Z., Azadi, H., Taheri, F., Zarafshani, K., Gebrehiwot, K., Van Passel, S., et al., 2017. Organic farming and small-scale farmers: main opportunities and challenges. Ecological Economics 132, 144–154.

Kapazoglou, A., Ganopoulos, I., Tani, E., Tsaftaris, A., 2018. Epigenetics, epigenomics and crop improvement. Advances in Botanical Research 86, 287–324. Available at https://doi.org/10.1016/bs.abr.2017.11.007.

Koevoets, I.T., Venema, J.H., Elzenga, J.T.M., Testerink, C., 2016. Roots withstanding their environment: exploiting root system architecture responses to abiotic stress to improve crop tolerance. Frontiers in Plant Science 7, 1335.

Komakech, A.J., Zurbrügg, C., Miito, G.J., Wanyama, J., Vinnerås, B., 2016. Environmental impact from vermicomposting of organic waste in Kampala, Uganda. Journal of Environmental Management 181, 395–402.

Kültz, D., 2003. Evolution of the cellular stress proteome: from monophyletic origin to ubiquitous function. Journal of Experimental Biology 206 (18), 3119–3124.

Kundu, S., Kabir, M.E., Morgan, E.A., Davey, P., Hossain, M., 2020. Building coastal agricultural resilience in Bangladesh: a systematic review of progress, gaps and implications. Climate 8 (9), 98.

Kunze, S., 2021. Unraveling the effects of tropical cyclones on economic sectors worldwide: direct and indirect impacts. Environmental and Resource Economics 78 (4), 545–569.

Lau, S.E., Teo, W.F.A., Teoh, E.Y., Tan, B.C., 2022. Microbiome engineering and plant biostimulants for sustainable crop improvement and mitigation of biotic and abiotic stresses. Discover Food 2 (1), 9.

Lee, J.T., Li, G., Zhang, Y., Yang, Z., Xu, Y., Gao, C., 2019. Genome editing of crops for improved flood tolerance. Plant Biotechnology Journal 17, 1907–1917.

Liliane, T.N., Charles, M.S., 2020. Factors affecting yield of crops. Agronomy-Climate Change & Food Security 9, 9–24.

Loreti, E., van Veen, H., Perata, P., 2016. Plant responses to flooding stress. Current Opinion in Plant Biology 33, 64–71.

Mainuddin, M., Karim, F., Gaydon, D.S., Kirby, J.M., 2021. Impact of climate change and management strategies on water and salt balance of the polders and islands in the Ganges delta. Scientific Reports 11 (1), 7041.

Mandal, U.K., Burman, D., Bhardwaj, A.K., Nayak, D.B., Samui, A., Mullick, S., et al., 2019. Waterlogging and coastal salinity management through land shaping and cropping intensification in climatically vulnerable Indian Sundarbans. Agricultural Water Management 216, 12–26.

Mittler, R., 2006. Abiotic stress, the field environment and stress combination. Trends in Plant Science 11 (1), 15–19.

Mittler, R., Blumwald, E., 2010. Genetic engineering for modern agriculture: challenges and perspectives. Annual Review of Plant Biology 61 (1), 443–462.

Mondal, T.K., 2022. Assessing the scope for promoting climate resilient agriculture in the Indian Sundarban Delta: a SWOT-AHP analysis. Journal of Coastal Conservation 26 (6), 62.

Muharomah, R., Setiawan, B.I., Sands, G.R., Juliana, I.C., Gunawan, T.A., 2025. A review on enhancing water productivities adaptive to the climate change. Journal of Water and Climate Change 16 (3), 860–887.

Mulla, D.J., 2013. Twenty five years of remote sensing in precision agriculture: key advances and remaining knowledge gaps. Biosystems Engineering 114 (4), 358–371.

Nambi, A., Bahinipati, C.S., 2012. Adaptation to climate change and livelihoods: an integrated case study to assess the vulnerability and adaptation options of the fishing and farming communities of selected east coast stretch of Tamil Nadu, India. Asian Journal of Environment and Disaster Management 4 (3), 297–321.

Nephali, L., Piater, L.A., Dubery, I.A., Patterson, V., Huyser, J., Burgess, K., et al., 2020. Biostimulants for plant growth and mitigation of abiotic stresses: a metabolomics perspective. Metabolites 10 (12), 505.

Nouri, M.Z., Moumeni, A., Komatsu, S., 2015. Abiotic stresses: insight into gene regulation and protein expression in photosynthetic pathways of plants. International Journal of Molecular Sciences 16 (9), 20392–20416.

Padarian, J., Minasny, B., McBratney, A.B., 2019. Using deep learning for digital soil mapping. Soil 5 (1), 79–89.

Padhan, N., Madheswaran, S., 2022. Determinants of farm-level adaptation strategies to flood: insights from a farm household-level survey in coastal districts of Odisha. Water Policy 24 (2), 450–469.

Palmgren, M., Shabala, S., 2024. Adapting crops for climate change: regaining lost abiotic stress tolerance in crops. Frontiers in Science 2, 1416023.

Patel, S.K., Mathew, B., Nanda, A., Mohanty, B., Saggurti, N., 2020. Voices of rural people: community-level assessment of effects and resilience to natural disasters in Odisha, India. International Journal of Population Studies 6 (1), 3–15.

Paul, B.K., Chatterjee, S., 2019. Climate change-induced environmental hazards and Aila relief measures undertaken to Sundarbans in Bangladesh and India. The Sundarbans: A Disaster-Prone Eco-Region: Increasing Livelihood Security. Springer International Publishing, Cham, pp. 469–490.

Ponnusamy, K., Devi, M.K., 2017. Impact of integrated farming system approach on doubling farmers' income. Agricultural Economics Research Review 30, 233. Available at https://doi.org/10.5958/0974-0279.2017.00037.4.

Popa, D.G., Lupu, C., Constantinescu-Aruxandei, D., Oancea, F., 2022. Humic substances as microalgal biostimulants—implications for microalgal biotechnology. Marine Drugs 20 (5), 327.

Prasch, C.M., Sonnewald, U., 2013. Simultaneous application of heat, drought, and virus to Arabidopsis plants reveals significant shifts in signaling networks. Plant Physiology 162 (4), 1849–1866.

Prasch, C.M., Sonnewald, U., 2015. Signaling events in plants: stress factors in combination change the picture. Environmental and Experimental Botany 114, 4–14.

Rahman, M.M., Penny, G., Mondal, M.S., Zaman, M.H., Kryston, A., Salehin, M., et al., 2019. Salinization in large river deltas: drivers, impacts and socio-hydrological feedbacks. Water Security 6, 100024.

Rai, N., Rai, S.P., Sarma, B.K., 2021. Prospects for abiotic stress tolerance in crops utilizing phyto-and bio-stimulants. Frontiers in Sustainable Food Systems 5, 754853.

Rakshit, S., Padaria, R.N., Bandyopadhyay, S., 2016. Farmers' adaptation strategies, coping behaviour and barriers to effective adaptation to current climatic risks: a study on Sundarban region. Indian Journal of Extension Education 52 (3–4), 17–20.

Rao, C.A., Raju, B.M.K., Rao, A.V.M., Rao, K.V., Samuel, J., Ramachandran, K., et al., 2017. Assessing vulnerability and adaptation of agriculture to climate change in Andhra Pradesh. Indian Journal of Agricultural Economics 72 (3), 375–384.

Rao, C.S., Baral, K., Chandana, V.M., Jagadesh, M., Karthik, R., 2024. Climate change adaptation and mitigation in Indian agriculture. Journal of Agrometeorology 26, 137–148.

Reddy, K.V., Paramesh, V., Arunachalam, V., Das, B., Ramasundaram, P., Pramanik, M., et al., 2022. Farmers' perception and efficacy of adaptation decisions to climate change. Agronomy 12 (5), 1023.

Sabareshwari, V., Ramya, A., 2018. Coastal saline soils of India: a review. Agricultural Reviews 39 (1), 86–88. Available at https://doi.org/10.18805/ag.R-1768.

Sangiorgio, D., Cellini, A., Donati, I., Pastore, C., Onofrietti, C., Spinelli, F., 2020. Facing climate change: application of microbial biostimulants to mitigate stress in horticultural crops. Agronomy 10 (6), 794.

Sarangi, S.K., Mainuddin, M., Raut, S., Mandal, U.K., Mahanta, K.K., 2024. Coastal salinity management and cropping system intensification through conservation agriculture in the ganges delta. Soil Systems 8 (3), 80.

Sarkar, S., Padaria, R.N., Sivaramane, N., Vijayragavan, K., 2010. Assessment of farmer vulnerability and adaptation to climate change in Sunderbans ecosystem. Indian Journal of Extension Education 46 (3–4), 78–84.

Shah, K.K., Modi, B., Pandey, H.P., Subedi, A., Aryal, G., Pandey, M., et al., 2021. Diversified crop rotation: an approach for sustainable agriculture production. Advances in Agriculture 2021 (1), 8924087.

Shukla, P.S., Mantin, E.G., Adil, M., Bajpai, S., Critchley, A.T., Prithiviraj, B., 2019. Ascophyllum nodosum-based biostimulants: sustainable applications in agriculture for the stimulation of plant growth, stress tolerance, and disease management. Frontiers in Plant Science 10, 462648.

Singh, A.P., Narayanan, K., 2015. Impact of economic growth and population on agrochemical use: evidence from post-liberalization India. Environment, Development and Sustainability 17 (6), 1509–1525.

Singh, P.K., Papageorgiou, K., Chudasama, H., Papageorgiou, E.I., 2019. Evaluating the effectiveness of climate change adaptations in the world's largest mangrove ecosystem. Sustainability 11 (23), 6655.

Singh, R.B., 2000. Environmental consequences of agricultural development: a case study from the Green Revolution state of Haryana, India. Agriculture, Ecosystems & Environment 82 (1-3), 97–103.

Singh, R.P., Chintagunta, A.D., Agarwal, D.K., Kureel, R.S., Kumar, S.J., 2020. Varietal replacement rate: prospects and challenges for global food security. Global Food Security 25, 100324.

SRDI, (Soil Resource Development Institute), 2010. Saline Soils of Bangladesh, Soil Resource Development Institute, SRMAF Project, Ministry of Agriculture, Mrittika Bhaban, Krishikhamar Sarak, Farmgate, Dhaka, Bangladesh.

Tahjib-Ul-Arif, M., Hasan, M.T., Rahman, M.A., Nuruzzaman, M., Rahman, A.S., Hasanuzzaman, M., et al., 2023. Plant response to combined salinity and waterlogging stress: current research progress and future prospects. Plant Stress 7, 100137.

Tarolli, P., Luo, J., Park, E., Barcaccia, G., Masin, R., 2024. Soil salinization in agriculture: mitigation and adaptation strategies combining nature-based solutions and bioengineering. Iscience 27 (2), 108830. Available at https://doi.org/10.1016/j.isci.2024.108830.

Thaker, P.N., Brahmbhatt, N., Shah, K., 2021. A review: impact of soil salinity on ecological, agricultural and socio-economic concerns. International Journal of Advanced Research 9 (07), 979–986.

Thenkabail, P.S., Lyon, J.G., Huete, A., 2018. Fifty years of advances in hyperspectral remote sensing of agriculture and vegetation—summary, insights, and highlights of volume III. Biophysical and Biochemical Characterization and Plant Species Studies, second ed. CRC Press, pp. 303–341. Available at 10.1201/9780429431180-11.

Van Oosten, M.J., Pepe, O., De Pascale, S., Silletti, S., Maggio, A., 2017. The role of biostimulants and bioeffectors as alleviators of abiotic stress in crop plants. Chemical and Biological Technologies in Agriculture 4 (1), 5.

Vile, D., Pervent, M., Belluau, M., Vasseur, F., Bresson, J., Muller, B., et al., 2012. Arabidopsis growth under prolonged high temperature and water deficit: independent or interactive effects? Plant, Cell & Environment 35 (4), 702–718.

Xiong, J., Wang, H., Zhong, Z., Li, S., Qin, P., 2025. Emerging strategies to improve heat stress tolerance in crops. Abiotech 6 (1), 97–115.

Xu, L., Liu, F., Fu, Z., Jin, J., Zhang, H., 2020. Breeding for crop drought resistance: strategies and progress. Plant Methods 16, 1–21.

Yakhin, O.I., Lubyanov, A.A., Yakhin, I.A., Brown, P.H., 2017. Biostimulants in plant science: a global perspective. Frontiers in Plant Science 7, 2049.

Yang, Y., Yu, J., Qian, Q., Shang, L., 2022. Enhancement of heat and drought stress tolerance in rice by genetic manipulation: a systematic review. Rice 15 (1), 67.

Ye, L., Xu, Y., Zhu, G., Zhang, W., Jiao, Y., 2023. Effects of different mulch types on farmland soil moisture in an artificial oasis area. Land 13 (1), 34.

Zhang, B., 2015. MicroRNA: a new target for improving plant tolerance to abiotic stress. Journal of Experimental Botany 66 (7), 1749–1761.

Zulfiqar, F., Casadesús, A., Brockman, H., Munné-Bosch, S., 2020. An overview of plant-based natural biostimulants for sustainable horticulture with a particular focus on moringa leaf extracts. Plant Science 295, 110194.

Chapter 19

Alleviation of multiple abiotic stresses by improving macro- and micronutrient management

Md Arif Hussain[1,2,*], Md. Nazmus Sakib[3], Md. Hasanuzzaman[3], and Karl H. Mühling[2]

[1]*Department of Biochemistry and Molecular Biology, Sher-e-Bangla Agricultural University, Dhaka, Bangladesh,* [2]*Institute of Plant Nutrition and Soil Science, Kiel University, Kiel, Schleswig-Holstein, Germany,* [3]*Department of Agronomy, Sher-e-Bangla Agricultural University, Dhaka, Bangladesh*

[*]*Corresponding author. e-mail address: arifbioc@sau.edu.bd*

19.1 Introduction

The increase in global temperatures projected to increase by approximately +2°C per decade, along with declining freshwater availability, is anticipated to have terrible consequences on the growth of plants and yields of crops (Khan et al., 2025). A discrepancy in a single climatic factor, such as temperature or precipitation, can disrupt interconnected biogeochemical cycles, including those of nitrogen, carbon, and sulfur. Prolonged or sudden episodes of climate variation have a major effect on agricultural output and plant growth. The global population, currently estimated at 7.7 billion, is expected to rise to nearly 10 billion by 2050 (Hussain et al., 2024; Etesami and Jeong, 2018). This growth demands a significant increase in food production, which is increasingly threatened by abiotic stressors. These include salinity, soil acidity, drought, extreme temperatures, waterlogging, and heavy metal toxicity, all of which adversely affect biochemical, physiological, and morphological attributes of plants, ultimately reducing crop productivity (Hossain et al., 2021, 2024). In the field, crops are rarely exposed to isolated stress factors; instead, they frequently face combinations of stressors such as drought–salinity, drought–heat, or salinity–heavy metals, particularly in dry and semiarid regions (Hussain et al., 2022, 2023; Zandalinas et al., 2020; Rivero et al., 2021). These combined stresses exert distinct and often more severe effects on the physiological, morphological, and agronomic reactions of plants. Morphologically, they lead to reduced leaf area, stunted growth, early senescence, and, in some cases, the expansion of adventitious roots (Khan et al., 2025). Physiologically, these stress combinations disrupt cellular water balance, impair photosynthesis, elevate the assembly of reactive oxygen species (ROS), and cause ionic toxicity such as sodium (Na^+) accumulation under salinity stress, thereby compromising plant metabolism and total health (Ramegowda et al., 2024; Zandalinas et al., 2020). Agronomically, such stresses result in delayed flowering, poor seed set, and significant yield losses (Shelake et al., 2022).

Macro- and micronutrient applications play a crucial role in alleviating these adverse effects. For example, nitrogen (urea) use efficiency declines under stress due to reduced uptake and increased loss, but balanced N management is still essential for protein synthesis and chlorophyll maintenance. Potassium helps regulate osmotic balance and stomatal function, whereas the beneficial element silicon enhances antioxidant defense and ion balance (Li et al., 2020; Araus et al., 2020). Zinc supports enzyme stability and reproductive development, and calcium maintains membrane integrity under ionic and oxidative stress (Hanif and Zia, 2023). Many micronutrients contribute to oxidative stress regulation by contributing as essential cofactors for antioxidant enzymes. For example, superoxide dismutase (SOD), a key enzyme in scavenging ROS, requires zinc, copper, or manganese as cofactors, resulting in different isoforms located in various cellular components such as mitochondria and peroxisomes (Khan et al., 2025). Thus, nutrient management under combined stress conditions is vital to sustaining plant resilience, physiological function, and productivity. While substantial improvement has been introduced in considering plant reactions to individual abiotic stresses, research on the combined effects of multiple stressors remains limited and often yields inconsistent findings. Much of the prevailing effort

Multiple Abiotic Stresses in Plants. DOI: https://doi.org/10.1016/B978-0-443-27424-4.00017-3

has been directed under precise conditions, which neglect to reflect the complexity and unpredictability of stress interactions in real field atmospheres (Khan et al., 2025). Moreover, although the significance of macro- and micronutrients in alleviating single-stress conditions is well established, their combined roles in enhancing plant tolerance under simultaneous stresses are still not fully elucidated (Shelake et al., 2022).

There is a prominent gap in comprehensive research exploring how different nutrient application strategies, such as improving urea use efficiency, foliar application of micronutrients, or silicon supplementation, affect the biochemical and physiological resilience of crops under multiple stress scenarios. Bridging these gaps is crucial for formulating effective nutrient management strategies aimed at sustaining crop efficiency in the face of climate variability. Therefore, this chapter aims to synthesize current insights into plant responses to major stress combinations and underscore the integrative role of nutrients in enhancing resilience. Such knowledge is vital for advancing climate-smart agriculture and implementing precision nutrient strategies to improve crop performance under multifactorial stress conditions.

19.2 Effects of combined stresses on different agricultural crops

Combined abiotic stresses exert intricate and often antagonistic effects on plants (Fig. 19.1), resulting in extreme gathering of ROS, disruption of nutrient uptake, loss of cellular homeostasis, and severe oxidative damage (Colmenero-Flores and Rosales, 2013; Cohen et al., 2020). For example, the simultaneous occurrence of heat and drought stress significantly reduced seed weight and the number of grains per year in wheat in comparison with individual stresses (Schmidt et al., 2020). Likewise, in rapeseed, the interaction of salinity and waterlogging resulted in more pronounced damage than either stress alone (Tahjib-Ul-Arif et al., 2023). However, certain stress combinations may exhibit synergistic effects (Fig. 19.2). Rivero et al. (2021) stated that tomato plants exposed to concurrent heat and salinity stress demonstrated enhanced tolerance in contrast to salinity alone, possibly because of the accumulation of osmoprotectants, for example, trehalose and glycine betaine. This response is often attributed to cross-resistance, where exposure to one stress primes plants to tolerate another by triggering shared defense pathways, including antioxidant enzyme activation and osmolyte accumulation (Saeed et al., 2025; Ru et al., 2024; Degenkolbe et al., 2008). The timing of stress exposure also plays a critical role. Ru et al. (2024) found that maize exposed to heat and drought during the reproductive stage experienced more severe physiological impairment than when exposed during the vegetative stage (Fig. 19.2). Interestingly, stress experienced earlier in development induced a memory effect that improved resistance later in the growth cycle.

In this context, nutrient management has arisen as an auspicious methodology to bolster plant resilience (Aslam et al., 2021; Mamnabi et al., 2020). Adequate and balanced application of macro- and micronutrients supports improved water and nutrient uptake, enhances antioxidant shield systems, and regulates stress-responsive gene expression (Hasanuzzaman

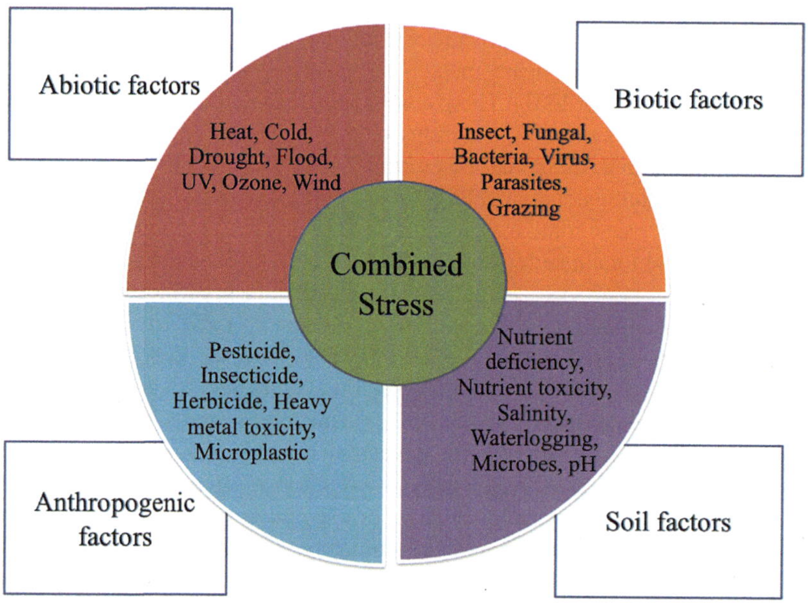

FIGURE 19.1 Global warming and climate change are projected to expose crops to sequential or simultaneous combinations of abiotic stresses.

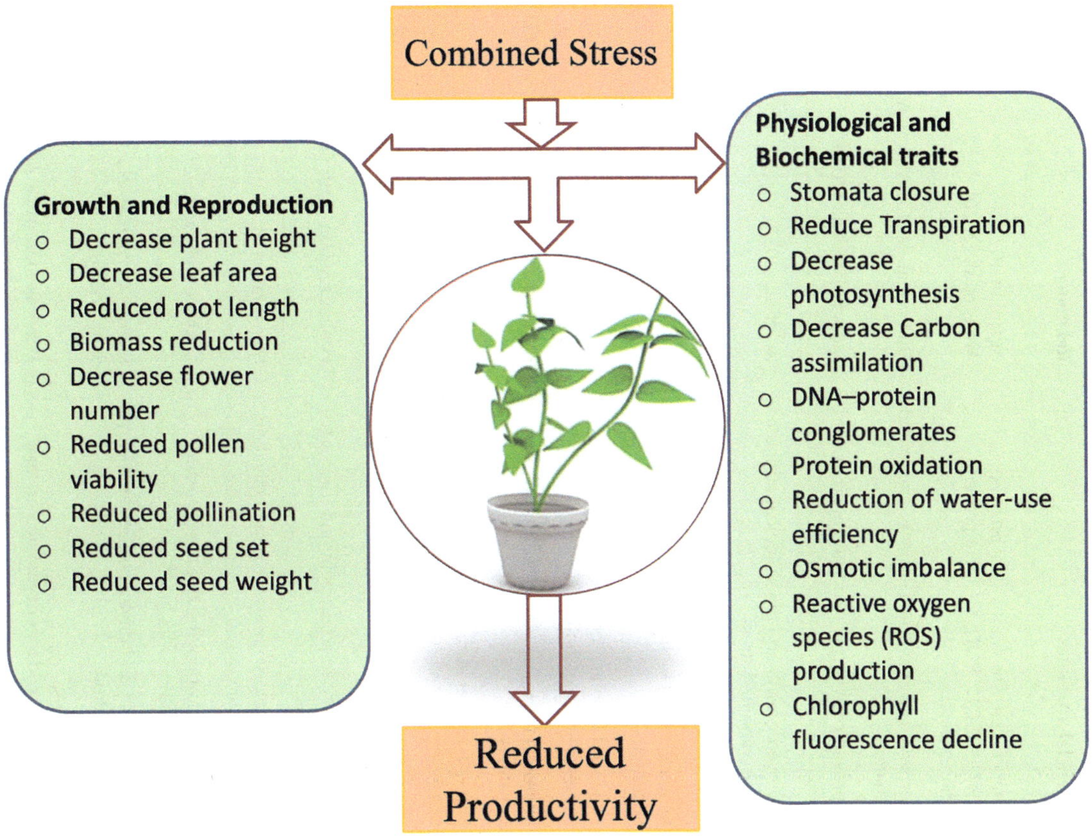

FIGURE 19.2 Combined abiotic stresses pose significant threats to plant growth, development, and overall productivity by disrupting physiological and metabolic processes.

et al., 2018). For instance the joint use of potassium and silicon effectively mitigated the impact of drought and salinity in wheat. Similarly, foliar sprays of zinc and iron enhanced rice resistance to heavy metal stress by improving photosynthetic activity and reducing the uptake of toxic ions (Rizwan et al., 2015). Among the macronutrients, nitrogen (N), phosphorus (P), calcium (Ca), potassium (K), and magnesium (Mg) play vital roles in both metabolism and stress adaptation (De Bang et al., 2020). Recent research has confirmed their significance in mitigating various abiotic stresses. Shafiq et al. (2021) showed that sulfate-based fertilizers improved drought tolerance by enhancing photosynthesis, nutrient assimilation, and antioxidant responses. Furthermore, the joint use of micronutrients such as zinc and selenium or the synergistic application of potassium and silicon has been shown to strengthen plant defenses by decreasing oxidative damage and alleviating osmotic stress.

19.3 Role of macronutrients in alleviating single and combined abiotic stresses

Potassium, nitrogen (N), phosphorus, magnesium (Mg), and sulfur (S) are examples of macronutrients that are vital components that have a major impact on plant growth, development, and resistance to abiotic stress (Hasanuzzaman et al., 2018; De Bang et al., 2020). These nutrients not only support fundamental physiological processes, such as protein synthesis, photosynthesis, respiration, and osmoregulation, but also modulate antioxidant defense systems along with stress-responsive signaling pathways (Table 19.1).

Under individual stress states such as cold, drought, heat, or salinity, adequate macronutrient supply supports plant resilience by maintaining metabolic balance, water status, and membrane integrity. In the case of combined stresses such as salinity with drought, drought with heat, or waterlogging with nutrient toxicity, macronutrients function synergistically to mitigate compounded physiological damage. This is attained employing integrated approaches, along with the

TABLE 19.1 Contribution of macro- and micronutrients in the abatement of single and multiple abiotic stresses in crops.

Sl no.	Nutrient name	Type of stresses	Crop name	Beneficial attributes	References
1	Nitrogen (N)	Drought	Wheat	Enhanced growth, antioxidant enzyme activities (SOD, POD, CAT), improved osmoregulation	Agami et al. (2018), Li et al. (2020), Araus et al. (2020)
		Heat	Rice	Improved tolerance to heat, better grain yield formation	Duan et al. (2013), Waraich et al. (2012)
		Cold	Maize	Improved early growth and tolerance to cold stress	Moradtalab et al. (2020)
		Heat + drought	Wheat	Improved growth and physiological attributes under combined stress	Ru et al. (2024)
2	Phosphorus (P)	Heat + cold	Rice and wheat	Enhanced grain yield and quality	Xu et al. (2022), Faizan et al. (2021)
		Waterlogging	Fodder maize	Increased grain yield under anaerobic soil conditions	Jiménez et al. (2015), Ayub et al. (2002)
		Drought + salinity	Sugar beet and quinoa	Enhanced biomass, enzymatic activities, and stress-related metabolites	Baroowa et al. (2022), Bouras et al. (2022)
		Heavy metals (Cu + Cd)	General crops	Improved tolerance via altered phosphorus metabolism	Liu et al. (2022)
3	Potassium (K)	Waterlogging + heat	Cotton	Improved water status, photosynthetic capacity, and translocation of assimilates	Huang et al. (2023a)
		Drought	General crops	Enhanced proline accumulation and osmoregulation, improved photosynthesis	Zain et al. (2014), Shabala and Lew (2002), Egilla et al. (2005)
		Salinity + heat	General crops	Promoted osmotic adjustment and improved stress response	Hasanuzzaman et al. (2018)
4	Magnesium (Mg)	Drought + salinity	Rice and wheat	Improved photosynthesis, enzyme activation, and osmoregulation	Mengutay et al. (2013)
		Heat + light stress	Maize	Maintains chlorophyll stability and reduces photoinhibition	Cakmak and Yazici (2010)
5	Sulfur (S)	Drought + cadmium stress, waterlogging, salinity	Brassica species and wheat	Enhances GSH and phytochelatin production for detoxification	Khan et al. (2014); Hussain et al. (2023)
		Salt + heat	Wheat	Enhances antioxidant enzyme activity and stress-related metabolite accumulation	Astolfi et al. (2011); Anjum et al. (2015)
6	CuO NPs	General combined stress	Soybean	Yield variations depending on nanoparticle type and dosage	Yusef-Tanha et al. (2020), Kaleem et al. (2024)

		Photosynthetic stress (drought/heat)	Wheat, maize, and tomato	Increased Chl a and b, carotenoids, and photosystem II activity	Singh et al. (2017), Badawy et al. (2021)
7	ZnO NPs	Heavy metal (Cd) + drought	Wheat	Decreased Cd toxicity and improved oxidative stress resistance	Hussain et al. (2018)
		Salinity	Coriander, tomato, and maize	Increased biomass, root/shoot growth, and salt tolerance	Hanif and Zia (2023), Faizan et al. (2021), Shoukat et al. (2024)
		Heat + drought	Cucumber and general crops	Improved antioxidant activity and maintained water balance	Ghani et al. (2022), Thakur et al. (2021)
8	Si & Se	Wheat	Salinity +drought (applied in the form of nanoparticles)	Increase the activity of antioxidant enzymes, photosynthetic attributes, chlorophyll content, transpiration rate, and water relations.	Cunha and De Mello Prado (2023), Cooke and Leishman (2016), Manzoor et al. (2024)
9	Si	Sugarcane and wheat	Salinity + drought + high temperature + heavy metal	Salinity affects plant efficiency and the quality of the crop produced, Lessening of sucrose in sugarcane and cane quality	Verma et al. (2020), Bhardwaj et al. (2022) Etesami and Jeong (2018)
10	Mn	Wheat	Oxidative + drought	Improved the chlorophyll pigments (Chlorophyll a, b) and carotenoids growth	Parveen et al. (2024)

activation of antioxidant enzymes, gathering of osmoprotectants, enhanced nutrient uptake, and regulation of hormonal balance (Fig. 19.2).

19.3.1 Nitrogen

Nitrogen is a vital macronutrient involved in key physiological and metabolic processes, including leaf expansion, biomass accumulation, and nitrogen assimilation pathways (Araus et al., 2020; Kubar et al., 2021; Ru and Hu, 2024; Marschner, 2012). Numerous studies have demonstrated that N deficiency significantly increases plant susceptibility to abiotic stresses (Table 19.1). Likewise, Agami et al. (2018) reported that nitrogen supplementation enhances biomass production and mitigates drought-induced growth reductions. Similarly, Waraich et al. (2012) observed improved plant growth and heat stress tolerance with increased N supply. Duan et al. (2013) noted that higher N use rates at the panicle initiation or flowering stages mitigated high temperature stress and stabilized grain yield. Beyond single-stress conditions, N application also offers benefits under combined abiotic stress scenarios. Under simultaneous drought and heat stresses, Ru et al. (2024) found that exogenous N application improved wheat growth and physiological traits. For example, Moradtalab et al. (2020) demonstrated that in maize, coapplication of ammonium-N, micronutrients (Mn and Zn), and microbial inoculants (*Trichoderma and Bacillus*) enhanced cold stress resistance during initial growth stages. Additionally, Li et al. (2020) showed that adequate N supply alleviated drought-induced physiological destruction by promoting antioxidant enzyme activity and osmoregulation. Gene expression studies further support this, revealing upregulation of stress-responsive isoforms of POD, SOD, APX, and CAT under N-sufficient conditions (Khan et al., 2017; Lin and Lai, 2013; Shafi et al., 2015). Nitrogen improves stress resistance in plants because it is essential for synthesizing amino acids, nucleic acids, proteins, chlorophyll, and many stress-responsive metabolites. Under stress conditions, plants need to quickly produce protective proteins (e.g., heat-shock proteins, antioxidative enzymes), osmolytes (e.g., proline), and signaling molecules (e.g., nitric oxide), all of which are nitrogen-containing. By supporting these biochemical pathways, adequate nitrogen enhances a plant's capacity to maintain cellular function, repair damage, and adapt metabolically to stress (Qaderi et al., 2025).

19.3.2 Phosphorus

Phosphorus plays a vital role in plant energy metabolism, membrane stability, and signal transduction, making it crucial for stress adaptation (Table 19.1). Under abiotic stresses such as waterlogging, drought, salinity, and extreme temperature, P availability significantly affects plant performance (Khan et al., 2023; Baroowa et al., 2022). P fertilization increases root and shoot biomass, promotes enzymatic activity, and supports overall plant vigor (Huang et al., 2023b). Investigations have found that P application improves grain quality and yield in rice and wheat under both heat and cold stress (Xu et al., 2022; Faizan et al., 2021). Under waterlogging situations, P application markedly increased grain yield in maize (Jiménez et al., 2015; Ayub et al., 2002). Furthermore, P supplementation has been linked to enhanced drought and salinity resistance through increased protein and metabolite synthesis (Baroowa et al., 2022). In metal-contaminated soils, P application improved stress resistance by modulating phosphorus metabolism (Liu et al., 2022). Across species, P fertilization promotes growth parameters including shoot/root length, trichome density, and biomass in crops, for example, Echinacea, rapeseed, quinoa, sugar beet, and okra (Mamnabi et al., 2020; Bouras et al., 2022; Attarzadeh et al., 2018). Additionally, P increases photosynthetic capacity through improvements in chlorophyll content, stomatal conductance, PSII efficiency, and water-use efficiency (Jin et al., 2015; Tariq et al., 2019). Enhanced nutrient and water uptake, driven by P fertilization, contributes further to stress resilience (Loudari et al., 2022; Frosi et al., 2017). At the biochemical level, P improves antioxidant defense by enhancing SOD, POD, CAT, and proline content, and sugar accumulation (Jin et al., 2015; Baroowa et al., 2022). Moreover, it reduces ROS generation and modulates hormone biosynthesis under salinity and drought conditions (Bechtaoui et al., 2021; Begum et al., 2020).

19.3.3 Potassium

Potassium is involved in numerous physiological processes such as enzyme stimulation, osmoregulation, photosynthesis, and phloem transport (Huang et al., 2023b). K deficiency disrupts these functions, leading to nonstomatal photosynthesis inhibition and impaired assimilate transport, thereby exacerbating plant sensitivity to abiotic stress (Gerardeaux et al., 2009). Exogenous K application enhances stress resistance by improving germination, stomatal conductance, cation–anion balance, protein synthesis, and osmotic regulation (Hasanuzzaman et al., 2018). Under heat stress, K contributes to osmolyte accumulation and maintains stomatal conductance, acting as both a regulator and osmoprotectant (Azedo-Silva et al., 2004). Foliar K treatments further promote photosynthate transport and dry matter partitioning, supporting yield stability under stress (Egilla et al., 2005; Mesbah, 2009). Under drought, K augments proline synthesis, water uptake, and root surface area, thus improving plant hydration and osmotic balance (Zain et al., 2014; Zia et al., 2020; Römheld and Kirkby, 2010). In cotton, Huang et al. (2023a) demonstrated that K application (150 kg K_2O ha^{-1}) alleviated waterlogging stress by maintaining photosynthetic capacity and enhancing assimilate translocation. Additionally, K supplementation under osmotic stress promoted recovery through the gathering of Cl^-, K^+, and Na^+ in root epidermal cells, contributing to membrane stability and antioxidant defense (Manimaran et al., 2025; Shabala and Lew, 2002).

19.3.4 Magnesium

Magnesium (Mg) is the central atom in chlorophyll and a critical activator of enzymes involved in photosynthesis, nucleic acid metabolism, and energy transfer (Cakmak, 2000). Under stress conditions, Mg maintains photosynthetic efficiency and enhances antioxidative defenses. Mg fertilization has been done to improve Rubisco activity, chlorophyll content, and drought resistance in rice and wheat (Mengutay et al., 2013). Beneath salinity stress, Mg promotes ion homeostasis by refining the K^+/Na^+ ratio and minimizing Na^+ toxicity (Sarraf et al., 2026). Mg-deficient plants exhibit elevated Na^+ uptake, leading to oxidative damage and chlorosis. Adequate Mg reduces this damage by stabilizing membranes and supporting antioxidant enzyme activity. Moreover, Mg stabilizes thylakoid membranes under heat and light stress, thereby protecting PSII from photoinhibition (Cakmak and Yazici, 2010). In multifactorial stress conditions (e.g., drought + salinity), Mg improves proline biosynthesis, enhances ROS scavenging, and supports ATP-dependent ion transport, all of which contribute to sustained growth and water-use efficiency (Mengutay et al., 2013).

19.3.5 Sulfur

Sulfur is a prime example of a nutrient that broadly enhances plant stress resistance, especially under combined abiotic and biotic stress conditions. This is because oxidative stress is a common, nonselective response of nearly all stress factors, emerging as a secondary effect in advanced stages of stress (Hussain et al., 2025). Sulfur is a crucial constituent of amino acids (methionine and cysteine), vitamins, and antioxidant molecules such as glutathione (GSH), making it particularly effective in mitigating oxidative damage (Hussain et al., 2024). It plays a pivotal role in redox regulation, detoxification, and stress signal transduction. S application improves resistance to drought, salinity, heavy metals, and combined abiotic stresses by enhancing antioxidant metabolism and osmotic regulation (Khan et al., 2014). During drought, S enhances GSH biosynthesis and antioxidant enzyme activity, improving water retention and membrane stability. Under salinity stress, S improves ion balance, reduces Na^+ uptake, and limits lipid peroxidation through upregulation of sulfate assimilation pathways (Astolfi et al., 2011). In heavy metal-stressed plants, S promotes phytochelatin and metallothionein synthesis, sequestering toxic metals such as Cd and As (Mazumdar & Chattopadhyay, 2025). Under combined stresses, such as drought + Cd or salt + heat, S enhances both enzymatic (CAT, SOD, APX) and nonenzymatic (GSH, ascorbate) antioxidant systems (Table 19.1). Anjum et al. (2015) showed that S treatment in wheat improved resistance to salt + heat stress via enhanced ROS detoxification, proline accumulation, and improved membrane stability. Similarly, Hussain et al. (2023) reported that under 7D waterlogging in *Brassica napus*, higher S supplementation showed higher resistance and production of GSH compared with the recommended dose of S.

19.4 Role of micronutrients and essential nutrients in alleviating single and combined abiotic stresses

Micronutrients, though needed in minute quantities, are indispensable for plant growth, development, and stress resilience. Acting as cofactors for numerous enzymatic reactions, they regulate antioxidant defense systems, maintain membrane stability, and modulate gene expression, which are key components of crop responses to abiotic stress. This section reviews the roles of several critical micronutrients, i.e., zinc (Zn), copper (Cu), and manganese (Mn), and the beneficial elements selenium (Se) and silicon (Si) in mitigating various single and combined abiotic stresses.

19.4.1 Zinc

Zinc plays essential structural, catalytic, and regulatory roles in plant metabolism. It is crucial for enzyme function, protein synthesis, auxin metabolism, and membrane integrity. Plants utilize Zn in multiple forms—free ions, low molecular weight complexes, or as a structural element of proteins and enzymes (Waraich et al., 2012). Recent research highlights the efficacy of zinc oxide nanoparticles (ZnO NPs) in lessening abiotic stresses. ZnO NPs improve salt and drought resistance by augmenting antioxidant enzyme activity, maintaining chlorophyll content, and reducing heavy metal uptake.

For example, Banerjee et al. (2023), Karmous et al. (2022), and Sun et al. (2023) showed that ZnO NPs reduce cadmium (Cd) bioavailability and toxicity, promoting plant resistance under metal stress (Table 19.1). Similarly, Hussain et al. (2018) stated that Zn supplementation considerably reduced Cd accumulation and improved growth performance in wheat. Under salinity stress, ZnO NPs coated with glycine betaine enhanced root and shoot biomass in coriander (Hanif and Zia, 2023).

In tomatoes, ZnO NP foliar sprays boosted salt resistance and photosynthetic activity (Faizan et al., 2021; Shoukat et al., 2024). Under heat stress, ZnO NPs preserved chlorophyll content by enhancing biosynthesis and reducing chlorophyllase activity, improving growth and biomass (Thakur et al., 2021). Zn application also mitigated drought-induced oxidative damage in wheat and cucumber seedlings by augmenting ROS scavenging and water-use efficiency (Ghani et al., 2022).

19.4.2 Selenium

Although not traditionally categorized as essential for higher plants, selenium has emerged as a potent stress-mitigating element. At appropriate concentrations, Se boosts the action of antioxidant enzymes such as superoxide dismutase (SOD), glutathione peroxidase, and catalase (CAT), and increases the levels of nonenzymatic antioxidants such as glutathione (GSH) and ascorbate (Hasanuzzaman et al., 2020; Manzoor et al., 2024). Se supplementation enhances chlorophyll retention, photosynthetic efficiency, and water balance under drought and salinity stress (Kaur et al., 2014; Feng et al., 2013). Under combined stresses, comprising salinity + drought and oxidative + heavy metal exposure, Se reduces lipid peroxidation and supports redox homeostasis and nutrient uptake (Djanaguiraman et al., 2010; Khan et al., 2014). These benefits collectively increase plant physiological performance and resilience (Table 19.1).

19.4.3 Silicon

Although not universally recognized as essential, silicon is increasingly acknowledged for its beneficial roles under abiotic stress (Cooke and Leishman, 2016; Cunha and De Mello Prado, 2023). Si enhances plant structural integrity by depositing in cell walls and improves resistance (Table 19.1) to drought, salinity, extreme temperature, and heavy metal toxicity (Shoukat et al., 2024; Ahmad et al., 2021; Bhardwaj et al., 2022). Si supplementation boosts water-use efficiency, reduces transpirational loss, and improves antioxidant enzyme activity, comprising CAT, SOD, and POD (Gong et al., 2005; Liang et al., 2005). Si also promotes the gathering of osmolytes such as proline and glycine betaine, helping plants maintain osmotic balance under water-limited conditions (Azedo-Silva et al., 2004). Under combined stresses (e.g., salinity + drought or drought + heavy metals), Si enhances photosynthetic stability, reduces oxidative stress, and modulates stress-regulated gene expression (Debona et al., 2017; Etesami and Jeong, 2018; Rios et al., 2017).

19.4.4 Copper

Copper is a redox-active micronutrient vital for several physiological functions, including electron transport, photosynthesis, and lignin biosynthesis. It serves as a cofactor in enzymes such as Cu/Zn-SOD, which detoxifies ROS, and plastocyanin, which mediates electron transport in photosynthesis (Table 19.1). Emerging research highlights the role of copper-based NPs (CuO NPs) in alleviating abiotic stresses. Their efficacy relies on plant species, soil characteristics, application method, and dosage. Yusef-Tanha et al. (2020) reported differential responses of soybean to various Cu NP formulations, with certain concentrations improving yield under stress. In wheat, CuO NP application (50 ppm) increased total chlorophyll by 23% and carotenoids by 20% (Badawy et al., 2021). Similar improvements in chlorophyll levels were reported in tomato and maize, correlating with enhanced photosystem II action and chloroplast utility (Singh et al., 2017; Toqeer et al., 2020). These effects suggest that CuO NPs strengthen photosynthetic capacity and contribute to greater stress resistance. CuO NPs have shown varying effects on soybean under combined stress conditions, with plant responses largely dependent on the type and concentration of NPs applied. Studies by Yusef-Tanha et al. (2020) and Kaleem et al. (2024) stated that while certain dosages improved stress tolerance and yield, others had neutral or even negative effects, highlighting the importance of optimizing NP formulation and application rate for beneficial outcomes.

19.4.5 Manganese

Manganese is a key component of the oxygen-evolving complex in photosystem II and acts as a cofactor for enzymes involved in antioxidative defense and carbon metabolism. It is particularly important under oxidative and drought stress conditions due to its role in ROS detoxification and pigment synthesis (Hänsch and Mendel, 2009). Parveen et al. (2024) confirmed that Mn utilization considerably enhanced chlorophyll a, chlorophyll b, and carotenoid content in wheat under combined oxidative and drought stress, contributing to improved photosynthetic efficiency and total plant health. Mn's ability to support photosynthetic pigment biosynthesis makes it a critical player in sustaining crop productivity under multifactorial environmental stresses.

19.5 Conclusion and future prospects

In the face of increasingly unpredictable climatic patterns, crops are repeatedly exposed to combinations of abiotic stresses such as drought, salinity, extreme temperatures, waterlogging, and heavy metal toxicity. These combined stresses often exert more severe impacts than individual stresses, largely due to their synergistic or antagonistic interactions, which disrupt plant physiological and biochemical homeostasis. Mounting evidence highlights the substantial potential of macro- and micronutrient management as a sustainable and cost-effective strategy to enhance crop resistance under such multifactorial stress conditions. Macronutrients such as phosphorus, nitrogen, and potassium play vital roles in modulating plant growth, antioxidant defense, osmotic regulation, and photosynthetic efficiency under stress combinations. Similarly, micronutrients such as copper, zinc, and manganese act as critical cofactors for antioxidant enzymes and signaling molecules, reinforcing plant resistance mechanisms. Moreover, beneficial elements such as selenium and silicon contribute to stress resistance. Notably, the application of these nutrients and beneficial elements, including in NP forms, can significantly mitigate the damaging effects of oxidative stress and support key physiological processes even under dual or multiple stress environments. However, nutrient-mediated resistance relies on factors such as dose, form of application, crop species, developmental stage, and the specific stress combination. Excessive or unbalanced nutrient supply may have adverse effects, underscoring the necessity of precise nutrient management approaches. Potential research should focus on unraveling the molecular and genetic pathways by which nutrients confer cross-resistance to combined stresses. Integrating nutrient-based interventions with stress-resilient genotypes and agronomic practices offers a promising avenue for safeguarding crop productivity and food security under climate change.

Looking ahead, future research should focus on deepening our mechanistic understanding of how macro- and micronutrients alleviate plant stress responses, particularly under combined abiotic stress conditions, which are increasingly prevalent due to climate variability. Although nutrient-mediated resistance mechanisms under single-stress conditions have been extensively explored, their roles under multiple or sequential stresses remain poorly characterized, especially in relation to signal integration, metabolic adjustments, and gene regulation. Special emphasis should be placed on identifying optimal nutrient dosages and ratios, timing, and delivery methods such as soil versus foliar application or nanotechnology-based fertilizers, slow-released fertilizers, precision agriculture, biofortification, and microbial approaches across diverse crops and stress scenarios. Moreover, nutrient interactions with phytohormones and beneficial soil microbiota offer a promising frontier for developing holistic strategies that enhance plant resilience. The integration of omics technologies and high-throughput phenotyping can help uncover genotype-specific nutrient responses, enabling the breeding of nutrient-efficient, stress-tolerant cultivars. Simultaneously, more research is needed under field conditions to validate laboratory findings and simulate real-lab environments involving multiple concurrent stresses. Beneficial elements, and may be in future emerging plant nutrients such as silicon and selenium, and their synergistic effects with conventional nutrients, also deserve more attention. Finally, future research must align with the goals of climate-smart agriculture by integrating nutrient management into adaptive cropping systems that consider agroecological variability, environmental sustainability, and global food security.

References

Agami, R.A., Alamri, S.A.M., Abd El-Mageed, T.A., Abousekken, M.S.M., Hashem, M., 2018. Role of exogenous nitrogen supply in alleviating the deficit irrigation stress in wheat plants. Agricultural Water Management 210, 261–270. Available at https://doi.org/10.1016/j.agwat.2018.08.034.

Anjum, N.A., Gill, R., Kaushik, M., Hasanuzzaman, M., Pereira, E., Ahmad, I., Tuteja, N., Gill, S.S., 2015. ATP-sulfurylase, sulfur-compounds, and plant stress tolerance. Frontiers in Plant Science 6. Available at 10.3389/fpls.2015.00210.

Ahmad, R., Lim, J., Park, J.H., et al., 2021. Silicon-mediated resistance in plants under abiotic and biotic stresses: a review. Ecotoxicology and Environmental Safety 211, 111887. Available at https://doi.org/10.1016/j.ecoenv.2021.111887.

Araus, V., Swift, J., Alvarez, J.M., Henry, A., Coruzzi, G.M., 2020. A balancing act: how plants integrate nitrogen and water signals. Journal of Experimental Botany 71 (15), 4442–4451. Available at https://doi.org/10.1093/jxb/eraa054.

Aslam, S., Gul, N., Mir, M.A., Asgher, M., Al-Sulami, N., Abulfaraj, A.A., et al., 2021. Role of jasmonates, calcium, and glutathione in plants to combat abiotic stresses through precise signaling cascade. Frontiers in Plant Science 12. Available at https://doi.org/10.3389/fpls.2021.668029.

Astolfi, S., Zuchi, S., Neumann, G., Cesco, S., di Toppi, L.S., Pinton, R., 2011. Response of barley plants to Fe deficiency and Cd contamination as affected by S starvation. Journal of Experimental Botany 63, 1241–1250. Available at 10.1093/jxb/err344.

Attarzadeh, M., Balouchi, H., Rajaie, M., Dehnavi, M.M., Salehi, A., 2018. Growth and nutrient content of *Echinacea purpurea* as affected by the combination of phosphorus with arbuscular mycorrhizal fungus and Pseudomonas fluorescence under different irrigation regimes. Journal of Environmental Management 231, 182–188. Available at https://doi.org/10.1016/j.jenvman.2018.10.040.

Ayub, M., Nadeem, M.A., Sharar, M.S., Mahmood, N., 2002. Response of maize (Zea mays L.) fodder to different levels of nitrogen and phosphorus. Asian Journal of Plant Sciences 1 (4), 352–354. Available at https://doi.org/10.3923/ajps.2002.352.354.

Azedo-Silva, J., Osório, J., Fonseca, F., Correia, M.J., 2004. Effects of soil drying and subsequent re-watering on the activity of nitrate reductase in roots and leaves of Helianthus annuus. Functional Plant Biology 31 (6), 611. Available at https://doi.org/10.1071/fp04018.

Badawy, A.A., Abdelfattah, N.A.H., Salem, S.S., Awad, M.F., Fouda, A., 2021. Efficacy assessment of biosynthesized copper oxide nanoparticles (Cuo-nps) on stored grain insects and their impacts on morphological and physiological traits of wheat (triticum aestivum l.) plant. Biology 10 (3). Available at https://doi.org/10.3390/biology10030233.

Banerjee, S., Islam, J., Mondal, S., Saha, A., Saha, B., Sen, A., 2023. Proactive attenuation of arsenic-stress by nano-priming: zinc oxide nanoparticles in Vigna mungo (L.) Hepper trigger antioxidant defense response and reduce root-shoot arsenic translocation. Journal of Hazardous Materials 446, 130735. Available at https://doi.org/10.1016/j.jhazmat.2023.130735.

De Bang, T.C., Husted, S., Laursen, K.H., Persson, D.P., Schjoerring, J.K., 2020. The molecular–physiological functions of mineral macronutrients and their consequences for deficiency symptoms in plants. New Phytologist 229 (5), 2446–2469. Available at https://doi.org/10.1111/nph.17074.

Baroowa, B., Paul, S., Gogoi, N., 2022. Role of phosphorus in imparting abiotic stress tolerance to plants. Climate Change and Agriculture: Perspectives, Sustainability and Resilience. Wiley, pp. 239–262. Available at https://doi.org/10.1002/9781119789789.ch11.

Bechtaoui, N., Rabiu, M.K., Raklami, A., Oufdou, K., Hafidi, M., Jemo, M., 2021. Phosphate-dependent regulation of growth and stresses management in plants. Frontiers in Plant Science 12. Available at https://doi.org/10.3389/fpls.2021.679916.

Begum, N., Ahanger, M.A., Zhang, L., 2020. AMF inoculation and phosphorus supplementation alleviates drought-induced growth and photosynthetic decline in *Nicotiana tabacum* by upregulating antioxidant metabolism and osmolyte accumulation. Environmental and Experimental Botany 176, 104088. Available at https://doi.org/10.1016/j.envexpbot.2020.104088.

Bhardwaj, S., Sharma, D., Singh, S., Ramamurthy, P.C., Verma, T., Pujari, M., et al., 2022. Physiological and molecular insights into the role of silicon in improving plant performance under abiotic stresses. Plant and Soil 486 (1–2), 25–43. Available at https://doi.org/10.1007/s11104-022-05395-4.

Bouras, H., Choukr-Allah, R., Amouaouch, Y., Bouaziz, A., Devkota, K.P., El Mouttaqi, A., et al., 2022. How does quinoa (Chenopodium quinoa Willd.) respond to phosphorus fertilization and irrigation water salinity? Plants 11 (2), 216. Available at https://doi.org/10.3390/plants11020216.

Cakmak, I., 2000. Possible roles of zinc in protecting plant cells from damage by reactive oxygen species. New Phytologist 146 (2), 185–205. Available at https://doi.org/10.1046/j.1469-8137.2000.00630.x.

Cakmak, I., Yazici, A.M., 2010. Magnesium: a forgotten element in crop production. Better Crops 95, 23–25.

Cohen, I., Zandalinas, S.I., Huck, C., Fritschi, F.B., Mittler, R., 2020. Meta-analysis of drought and heat stress combination impact on crop yield and yield components. Physiologia Plantarum 171 (1), 66–76. Available at https://doi.org/10.1111/ppl.13203.

Colmenero-Flores, J.M., Rosales, M.A., 2013. Interaction between salt and heat stress: when two wrongs make a right. Plant, Cell & Environment 37 (5), 1042–1045. Available at https://doi.org/10.1111/pce.12229.

Cooke, J., Leishman, M.R., 2016. Consistent alleviation of abiotic stress with silicon addition: a meta-analysis. Functional Ecology 30 (8), 1340–1357. Available at https://doi.org/10.1111/1365-2435.12713.

Cunha, M.L.O., De Mello Prado, R., 2023. Synergy of selenium and silicon to mitigate abiotic stresses: a review. Gesunde Pflanzen 75 (5), 1461–1474. Available at https://doi.org/10.1007/s10343-022-00826-9.

Debona, D., Rodrigues, F.A., Datnoff, L.E., 2017. Silicon's role in abiotic and biotic plant stresses. Annual Review of Phytopathology 55, 85–107. Available at https://doi.org/10.1146/annurev-phyto-080516-035312.

Degenkolbe, T., Thi, P., Do, Z., Zuther, E., Repsilber, D., Walther, D., et al., 2008. Expression profiling of rice cultivars differing in their tolerance to long-term drought stress. Plant Molecular Biology 69 (1–2), 133–153. Available at https://doi.org/10.1007/s11103-008-9412-7.

Djanaguiraman, M., Devi, D.D., Shanker, A.K., et al., 2010. Selenium – an antioxidative protectant in soybean during senescence. Plant and Soil 272 (1–2), 77–86. Available at https://doi.org/10.1007/s11104-004-4039-1.

Duan, H., Fu, L., Ju, C., Liu, L., Yang, J., 2013. Effects of application of nitrogen as panicle-promoting fertilizer on seed setting and grain quality of rice under high temperature stress. Chinese Journal of Rice Science 27, 591–602. Available at https://doi.org/10.3969/j.issn.1001-7216.2013.06.005.

Egilla, J.N., Davies, F.T., Boutton, T.W., 2005. Drought stress influences leaf water content, photosynthesis, and water-use efficiency of *Hibiscus rosa-sinensis* at three potassium concentrations. Photosynthetica 43 (1), 135–140. Available at https://doi.org/10.1007/s11099-005-5140-2.

Etesami, H., Jeong, B.R., 2018. Silicon (Si): review and future prospects on the action mechanisms in alleviating biotic and abiotic stresses in plants. Ecotoxicology and Environmental Safety 147, 881–896. Available at https://doi.org/10.1016/j.ecoenv.2017.09.063.

Faizan, M., Bhat, J.A., Chen, C., Alyemeni, M.N., Wijaya, L., Ahmad, P., et al., 2021. Zinc oxide nanoparticles (ZnO-NPs) induce salt tolerance by improving the antioxidant system and photosynthetic machinery in tomato. Plant Physiology and Biochemistry 161, 122–130. Available at https://doi.org/10.1016/j.plaphy.2021.02.002.

Feng, R., Wei, C., Tu, S., 2013. The roles of selenium in protecting plants against abiotic stresses. Environmental and Experimental Botany 87, 58–68. Available at https://doi.org/10.1016/j.envexpbot.2012.09.002.

Frosi, G., Barros, V.A., Oliveira, M.T., Santos, M., Ramos, D.G., Maia, L.C., Santos, M.G., 2017. Arbuscular mycorrhizal fungi and foliar phosphorus inorganic supply alleviate salt stress effects in physiological attributes, but only arbuscular mycorrhizal fungi increase biomass in woody species of a semiarid environment. Tree Physiology 38, 25–36. Available at 10.1093/treephys/tpx105.

Gerardeaux, E., Jordan-Meille, L., Constantin, J., Pellerin, S., Dingkuhn, M., 2009. Changes in plant morphology and dry matter partitioning caused by potassium deficiency in *Gossypium hirsutum* (L.). Environmental and Experimental Botany 67 (3), 451–459. Available at https://doi.org/10.1016/j.envexpbot.2009.09.008.

Ghani, M.I., Saleem, S., Rather, S.A., Rehmani, M.S., Alamri, S., Rajput, V.D., et al., 2022. Foliar application of zinc oxide nanoparticles: an effective strategy to mitigate drought stress in cucumber seedling by modulating antioxidant defense system and osmolytes accumulation. Chemosphere 289. Available at https://doi.org/10.1016/j.chemosphere.2021.133202.

Gong, H., Zhu, X., Chen, K., Wang, S., Zhang, C., 2005. Silicon alleviates oxidative damage of wheat plants in pots under drought. Plant Science 169 (2), 313–321. Available at https://doi.org/10.1016/j.plantsci.2005.02.023.

Hanif, S., Zia, M., 2023. Glycine betaine capped ZnO NPs eliminate oxidative stress to coriander plants grown under NaCl presence. Plant Physiology and Biochemistry 197, 107651. Available at https://doi.org/10.1016/j.plaphy.2023.107651.

Hasanuzzaman, M., Bhuyan, M.H.M.B., Nahar, K., Hossain, M.S., Mahmud, J.A., Hossen, M.S., et al., 2018. Potassium: a vital regulator of plant responses and tolerance to abiotic stresses. Agronomy 8 (3). Available at https://doi.org/10.3390/agronomy8030031.

Hasanuzzaman, M., Bhuyan, M.H.M.B., Raza, A., Hawrylak-Nowak, B., Matraszek-Gawron, R., Mahmud, J.A., et al., 2020. Selenium in plants: boon or bane? Environmental and Experimental Botany 178. Available at https://doi.org/10.1016/j.envexpbot.2020.104170.

Hossain, A., Skalicky, M., Brestic, M., Maitra, S., Alam, M.A., Syed, M.A., et al., 2021. Consequences and mitigation strategies of abiotic stresses in wheat (Triticum aestivum l.) under the changing climate. Agronomy 11 (2). Available at https://doi.org/10.3390/agronomy11020241.

Huang, L., Li, J., Yang, P., Zeng, X., Chen, Y., Wang, H., 2023a. Potassium application alleviated negative effects of soil waterlogging stress on photosynthesis and dry biomass in cotton. Agronomy 13 (4), 1157. Available at https://doi.org/10.3390/agronomy13041157.

Huang, Q., An, H., Chen, J., Li, X., Shao, G., 2023. Phosphorus application interferes expression of Fe uptake-associated genes to feedback regulate Cd accumulation in rice (Oryza sativa L.) and relieves Cd toxicity via antioxidant defense. Plant Growth Regulation 102, 199–212. Available at https://doi.org/10.1007/s10725-023-00991-5.

Hussain, A., Ali, S., Rizwan, M., Zia ur Rehman, M., Javed, M.R., Imran, M., et al., 2018. Zinc oxide nanoparticles alter the wheat physiological response and reduce the cadmium uptake by plants. Environmental Pollution 242, 1518–1526. Available at https://doi.org/10.1016/j.envpol.2018.08.036.

Hussain, M.A., Naeem, A., Sulieman, S., Pitann, B., Mühling, K.H., 2022. Sulfur uptake and distribution, grain yield, and quality of hybrid and inbred winter wheat (Triticum aestivum L.) varieties under early and late waterlogging. Journal of Plant Nutrition and Soil Science 185 (5), 622–631. Available at https://doi.org/10.1002/jpln.202200149.

Hussain, M.A., Naeem, A., Pitann, B., Mühling, K.H., 2023. High sulfur (S) supplementation imparts waterlogging tolerance to oilseed rape (Brassica napus L.) through upregulating s metabolism and antioxidant pathways. Journal of Plant Growth Regulation 42 (12), 7591–7605. Available at https://doi.org/10.1007/s00344-023-11034-8.

Hussain, M.A., Naeem, A., Pitann, B., Mühling, K.H., 2024. Foliar-applied sulfur is supplemental to soil application in alleviating waterlogging stress through the mitigation of oxidative stress and ion homeostasis in rapeseed (Brassica napus L.). Plant Stress 13. Available at https://doi.org/10.1016/j.stress.2024.100529.

Hussain, M.A., Pitann, B., Mühling, K.H., 2025. Combined effect of melatonin and sulfur on alleviating waterlogging stress in rapeseed. Plant-Environment Interactions 6 (2). Available at https://doi.org/10.1002/pei3.70050.

Hänsch, R., Mendel, R.R., 2009. Physiological functions of mineral micronutrients (Cu, Zn, Mn, Fe, Ni, Mo, B, Cl). Current Opinion in Plant Biology 12 (3), 259–266. Available at https://doi.org/10.1016/j.pbi.2009.05.006.

Jiménez, J.D.L.C., Cardoso, J.A., Arango-Londoño, D., Fischer, G., Rao, I., 2015. Influence of soil fertility on waterlogging tolerance of two Brachiaria grasses. Agronomía Colombiana 33 (1), 20–28. Available at https://doi.org/10.15446/agron.colomb.v33n1.48412.

Jin, J., Lauricella, D., Armstrong, R., Sale, P., Tang, C., 2015. Phosphorus application and elevated CO2 enhance drought tolerance in field pea grown in a phosphorus-deficient vertisol. Annals of Botany 116 (6), 975–985. Available at https://doi.org/10.1093/aob/mcu209.

Kaleem, Z., Xu, W., Ulhassan, Z., Shahbaz, H., He, D., Naeem, S., et al., 2024. Harnessing the potential of copper-based nanoparticles in mitigating abiotic and biotic stresses in crops. Environmental Science and Pollution Research 31 (50), 59727–59748. Available at https://doi.org/10.1007/s11356-024-35174-w.

Karmous, I,, Gammoudi, N., Chaoui, A,, 2022. Assessing the potential role of zinc oxide nanoparticles for mitigating cadmium toxicity in capsicum annuum L. under in vitro conditions. Journal of Plant Growth Regulation 42, 719–734. Available at 10.1007/s00344-022-10579-4.

Kaur, N., Sharma, S., Kaur, S., Nayyar, H., 2014. Selenium in agriculture: a nutrient or contaminant for crops? Archives of Agronomy and Soil Science 60, 1593–1624. Available at 10.1080/03650340.2014.918258.

Khan, M.N., Khan, S., Nazar, R., 2014. Salinity tolerance in plants: revisiting the role of sulfur metabolites. Journal of Plant Biochemistry & Physiology 2 (1), 120. Available at https://doi.org/10.4172/2329-9029.1000120.

Khan, A., Tan, D.K.Y., Afridi, M.Z., Luo, H., Tung, S.A., Ajab, M., et al., 2017. Nitrogen fertility and abiotic stresses management in cotton crop: a review. Environmental Science and Pollution Research 24 (17), 14551–14566. Available at https://doi.org/10.1007/s11356-017-8920-x.

Khan, F., Siddique, A.B., Shabala, S., Zhou, M., Zhao, C., 2023. Phosphorus plays key roles in regulating plants' physiological responses to abiotic stresses. Plants 12 (15), 2861. Available at https://doi.org/10.3390/plants12152861.

Khan, N.A., Owens, L., Nuñez, M.A., Khan, A.L., 2025. Complexity of combined abiotic stresses to crop plants. Plant Stress 17. Available at https://doi.org/10.1016/j.stress.2025.100926.

Kubar, M.S., Feng, M., Sayed, S., Shar, A.H., Rind, N.A., Ullah, H., et al., 2021. Agronomical traits associated with yield and yield components of winter wheat as affected by nitrogen managements. Saudi Journal of Biological Sciences 28 (9), 4852–4858. Available at https://doi.org/10.1016/j.sjbs.2021.07.027.

Loudari, A., Mayane, A., Zeroual, Y., Colinet, G., Oukarroum, A., 2022. Photosynthetic performance and nutrient uptake under salt stress: Differential responses of wheat plants to contrasting phosphorus forms and rates. Front Plant Sci 13, 1038672. Available at 10.3389/fpls.2022.1038672.

Lin, Y.L., Lai, Z.X., 2013. Superoxide dismutase multigene family in longan somatic embryos: A comparison of CuZn-SOD, Fe-SOD, and Mn-SOD gene structure, splicing, phylogeny, and expression. Molecular Breeding 32 (3), 595–615. Available at https://doi.org/10.1007/s11032-013-9892-2.

Liu, W., Gao, J., Wan, X., Li, Q., Fu, Q., Zhu, J., Hu, H., 2022. Effect of phosphorus fertilizer on phytoextraction using *Ricinus communis* L. in Cu and Cd co-contaminated soil. International Journal of Phytoremediation 25, 822–831. Available at 10.1080/15226514.2022.2112144.

Liang, Y.C., Sun, W.C., Si, J., Römheld, V., 2005. Effects of foliar- and root-applied silicon on the enhancement of induced resistance to powdery mildew in *Cucumis sativus*. Plant Pathology 54, 678–685. Available at 10.1111/j.1365-3059.2005.01246.x.

Li, S., Zhou, L., Addo-Danso, S.D., Ding, G., Sun, M., Wu, S., et al., 2020. Nitrogen supply enhances the physiological resistance of Chinese fir plantlets under polyethylene glycol (PEG)-induced drought stress. Scientific Reports 10 (1). Available at https://doi.org/10.1038/s41598-020-64161-7.

Mamnabi, S., Nasrollahzadeh, S., Ghassemi-Golezani, K., Raei, Y., 2020. Improving yield-related physiological characteristics of spring rapeseed by integrated fertilizer management under water deficit conditions. Saudi Journal of Biological Sciences 27 (3), 797–804. Available at https://doi.org/10.1016/j.sjbs.2020.01.008.

Manimaran, G., Duraisamy, S., Subramanium, T., Rangasamy, A., Alagarsamy, S., James, P., et al., 2025. Silicon-driven approaches to salinity stress tolerance: Mechanisms, uptake dynamics, and microbial transformations. Plant Stress 16, 100825. Available at https://doi.org/10.1016/j.stress.2025.100825.

Bose Mazumdar, A., Chattopadhyay, S., 2025. Glutathione pathways and gene networks: central players in crop adaptation to environmental stresses. Plant Growth Regulation 105, 1927–1948. Available at 10.1007/s10725-025-01375-7.

Manzoor, M., Abdalla, M.A., Hussain, M.A., Mühling, K.H., 2024. Silicon-selenium interplay imparts cadmium resistance in wheat through an up-regulating antioxidant system. International Journal of Molecular Sciences 25 (1). Available at https://doi.org/10.3390/ijms25010387.

Marschner, P., 2012. Marschner's Mineral Nutrition of Higher Plants, 3rd ed., Academic Press, pp. 1–651..

Mengutay, M., Ceylan, Y., Kutman, U.B., Cakmak, I., 2013. Adequate magnesium nutrition mitigates adverse effects of heat stress on maize and wheat. Plant and Soil 368 (1-2), 57–72. Available at https://doi.org/10.1007/s11104-013-1761-6.

Mesbah, E.A.E., 2009. Effect of irrigation regimes and foliar spraying of potassium on yield, yield components and water use efficiency of wheat in sandy soils. World Journal of Agricultural Sciences 5 (6), 662–669.

Moradtalab, N., Ahmed, A., Geistlinger, J., Walker, F., Höglinger, B., Ludewig, U., et al., 2020. Synergisms of microbial consortia, N forms, and micronutrients alleviate oxidative damage and stimulate hormonal cold stress adaptations in maize. Frontiers in Plant Science 11. Available at https://doi.org/10.3389/fpls.2020.00396.

Parveen, A., Perveen, S., Tariq, S., Atif, M., Saeed, F., Zafar, S., 2024. Protective role of manganese, proline and rice straw extract in wheat against drought driven oxidative stress. Acta Physiologiae Plantarum 46 (3). Available at https://doi.org/10.1007/s11738-024-03655-7.

Qaderi, M.M., Evans, C.C., Spicer, M.D., 2025. Plant nitrogen assimilation: a climate change perspective. Plants 14 (7), 1025. Available at https://doi.org/10.3390/plants14071025.

Ramegowda, V., Senthil, A., Senthil-Kumar, M., 2024. Stress combinations and their interactions in crop plants. Plant Physiology Reports 29 (1), 1–5. Available at https://doi.org/10.1007/s40502-024-00785-5.

Rios, J.J., Martínez-Ballesta, M.C., Ruiz, J.M., Blasco, B., Carvajal, M., 2017. Silicon-mediated improvement in plant salinity tolerance: the role of aquaporins. Frontiers in Plant Science 8. Available at 10.3389/fpls.2017.00948.

Rivero, R.M., Mittler, R., Blumwald, E., Zandalinas, S.I., 2021. Developing climate-resilient crops: improving plant tolerance to stress combination. The Plant Journal 109 (2), 373–389. Available at https://doi.org/10.1111/tpj.15483.

Rizwan, M., Ali, S., Ibrahim, M., Farid, M., Adrees, M., Bharwana, S.A., et al., 2015. Mechanisms of silicon-mediated alleviation of drought and salt stress in plants: a review. Environmental Science and Pollution Research 22 (20), 15416–15431. Available at https://doi.org/10.1007/s11356-015-5305-x.

Römheld, V., Kirkby, E.A., 2010. Research on potassium in agriculture: needs and prospects. Plant and Soil 335, 155–180. Available at 10.1007/s11104-010-0520-1.

Ru, C., Hu, X., Wang, W., Yan, H., 2024. Impact of nitrogen on photosynthesis, remobilization, yield, and efficiency in winter wheat under heat and drought stress. Agricultural Water Management 302, 109013. Available at https://doi.org/10.1016/j.agwat.2024.109013.

Saeed, K., Hussain, M.A., Abdalla, M.A., Mühling, K.H., 2025. Selenium increases the capacity of antioxidative defense and their accompanying metal cofactors in maize under sulfate salinity. Plant Stress 16. Available at https://doi.org/10.1016/j.stress.2025.100816.

Sarraf, M., Bansal, R., Shackira, A.M., Yadav, V., Zarbakhsh, S., Roychowdhury, R., Chauhan, D.K., Mousavi, H., Hasanuzzaman, M., 2026. Magnesium-mediated stress adaptation in plants: from physio-biochemical insights to climate-resilient agriculture. Frontiers in Plant Science 17. Available at 10.3389/fpls.2026.1715501.

Schmidt, J., Claussen, J., Wörlein, N., Eggert, A., Fleury, D., Garnett, T., Gerth, S., 2020. Drought and heat stress tolerance screening in wheat using computed tomography. Plant Methods 16. Available at 10.1186/s13007-020-00565-w.

Shabala, S.N., Lew, R.R., 2002. Turgor regulation in osmotically stressed arabidopsis epidermal root cells. direct support for the role of inorganic ion uptake as revealed by concurrent flux and cell turgor measurements. Plant Physiology 129, 290–299. Available at 10.1104/pp.020005.

Shafiq, B.A., Nawaz, F., Majeed, S., Aurangzaib, M., Al Mamun, A., Ahsan, M., et al., 2021. Sulfate-based fertilizers regulate nutrient uptake, photosynthetic gas exchange, and enzymatic antioxidants to increase sunflower growth and yield under drought stress. Journal of Soil Science and Plant Nutrition 21 (3), 2229–2241. Available at https://doi.org/10.1007/s42729-021-00516-x.

Shafi, A., Chauhan, R., Gill, T., Swarnkar, M.K., Sreenivasulu, Y., Kumar, S., et al., 2015. Expression of SOD and APX genes positively regulates secondary cell wall biosynthesis and promotes plant growth and yield in Arabidopsis under salt stress. Plant Molecular Biology 87 (6), 615–631. Available at https://doi.org/10.1007/s11103-015-0301-6.

Shelake, R.M., Kadam, U.S., Kumar, R., Pramanik, D., Singh, A.K., Kim, J.Y., 2022. Engineering drought and salinity tolerance traits in crops through CRISPR-mediated genome editing: Targets, tools, challenges, and perspectives. Plant Communications 3 (6). Available at https://doi.org/10.1016/j.xplc.2022.100417.

Shoukat, A., Pitann, B., Hossain, M.S., Saqib, Z.A., Nawaz, A., Mühling, K.H., 2024. Zinc and silicon fertilizers in conventional and nano-forms: Mitigating salinity effects in maize (Zea mays L.). Journal of Plant Nutrition and Soil Science 187 (5), 678–689. Available at https://doi.org/10.1002/jpln.202300267.

Sun, L., Wang, R., Ju, Q., Xing, M., Li, R., Li, W., Li, W., Wang, W., Deng, Y., Xu, J., 2023. Mitigation mechanism of zinc oxide nanoparticles on cadmium toxicity in tomato. Frontiers in Plant Science 14. Available at 10.3389/fpls.2023.1162372.

Singh, A., Singh, N.B., Hussain, I., Singh, H., 2017. Effect of biologically synthesized copper oxide nanoparticles on metabolism and antioxidant activity to the crop plants Solanum lycopersicum and Brassica oleracea var. botrytis. Journal of Biotechnology 262, 11–27. Available at 10.1016/j.jbiotec.2017.09.016.

Tahjib-Ul-Arif, M., Hasan, M.T., Rahman, M.A., Nuruzzaman, M., Rahman, A.M.S., Hasanuzzaman, M., et al., 2023. Plant response to combined salinity and waterlogging stress: current research progress and future prospects. Plant Stress 7. Available at https://doi.org/10.1016/j.stress.2023.100137.

Tariq, A., Pan, K., Olatunji, O.A, Graciano, C., Li, Z., Li, N., Song, D., Sun, F., Wu, X., Dakhil, M.A., Sun, X., Zhang, L., 2019. Impact of phosphorus application on drought resistant responses of *Eucalyptus grandis* seedlings. Physiologia Plantarum 166, 894–908. Available at 10.1111/ppl.12868.

Thakur, S., Asthir, B., Kaur, G., Kalia, A., Sharma, A., 2021. Zinc oxide and titanium dioxide nanoparticles influence heat stress tolerance mediated by antioxidant defense system in wheat. Cereal Research Communications 50, 385–396. Available at 10.1007/s42976-021-00190-w.

Toqeer, I., Raza, A., Naz, M.Y., Ghaffar, A., Hussain, Z., Ghuffar, A., 2020. Synthesis and application of controlled size copper oxide nanoparticles for improving biochemical and growth parameters of maize seedling. Journal of Plant Nutrition 43, 2622–2632. Available at 10.1080/01904167.2020.1793182.

Waraich, E.A., Ahmad, R., Halim, A., Aziz, T., 2012. Alleviation of temperature stress by nutrient management in crop plants: a review. Journal of Soil Science and Plant Nutrition 12 (2), 221–244. Available at https://doi.org/10.4067/s0718-95162012000200003.

Xu, H., Hassan, M.A., Sun, D., Wu, Z., Jiang, G., Liu, B., et al., 2022. Effects of low temperature stress on source–sink organs in wheat and phosphorus mitigation strategies. Frontiers in Plant Science 13. Available at https://doi.org/10.3389/fpls.2022.807844.

Yusefi-Tanha, E., Fallah, S., Rostamnejadi, A., Pokhrel, L.R., 2020. Particle size and concentration dependent toxicity of copper oxide nanoparticles (CuONPs) on seed yield and antioxidant defense system in soil grown soybean (Glycine max cv. Kowsar). Science of The Total Environment 715, 136994. Available at 10.1016/j.scitotenv.2020.136994.

Verma, K.K., Song, X.-P., Li, D.-M., Singh, M., Rajput, V.D., Malviya, M.K., Minkina, T., Singh, R.K., Singh, P., Li, Y.-R., 2020. Interactive role of silicon and plant–rhizobacteria mitigating abiotic stresses: a new approach for sustainable agriculture and climate change. Plants 9, 1055. Available at 10.3390/plants9091055.

Zain, N.A.M., Ismail, M.R., Puteh, A., Mahmood, M., Islam, M.R., 2014. Drought tolerance and ion accumulation of rice following application of additional potassium fertilizer. Communications in Soil Science and Plant Analysis 45 (19–20), 2502–2514. Available at https://doi.org/10.1080/00103624.2014.932374.

Zandalinas, S.I., Mittler, R., Balfagón, D., Arbona, V., Gómez-Cadenas, A., 2020. Plant adaptations to the combination of drought and high temperatures. Physiologia Plantarum 168 (2), 486–500. Available at https://doi.org/10.1111/ppl.12540.

Zia, R., Nawaz, M.S., Siddique, M.J., Hakim, S., Imran, A., 2020. Plant survival under drought stress: implications, adaptive responses, and integrated rhizosphere management strategy for stress mitigation. Microbiological Research 242, 126626. Available at https://doi.org/10.1016/j.micres.2020.126626.

Chapter 20

Role of beneficial elements in mediating tolerance to multiple abiotic stresses in plants

Sheikh Muhammad Masum*, Tanvir Ahmad Sourav, Md. Rakibul Hasan Khan, and Md. Hasanuzzaman

Department of Agronomy, Sher-e-Bangla Agricultural University, Dhaka, Bangladesh

Corresponding author. e-mail address: smmasum607@sau.edu.bd

20.1 Introduction

A limited number of elements are considered beneficial elements (BEs) because they specifically benefit certain plant species (Singhal et al., 2023). BEs are important for certain plant groups and can aid in development, but they aren't crucial for all species. While elements such as Al, Ce, Co, I, La, Na, Se, Si, Ti, and V have significant functions, their roles as both macronutrients and micronutrients remain unclear. These elements promote growth but are not essential for all plant species. Researchers have focused on strategies to mitigate abiotic stresses, finding that plant nutrients play a crucial role in alleviating these stresses (Kumari et al., 2022). BEs boost the antioxidant capabilities of plants by enhancing the activities of enzymes such as superoxide dismutase (SOD), catalase (CAT), and peroxidase (POD) (Wani, 2023). Furthermore, they help regulate phytohormones and interact with beneficial microbes to mitigate plant abiotic stress (Sarraf et al., 2023). Additionally, Upadhyaya and Panda (2013) highlighted the role of these elements in improving stress tolerance and facilitating recovery after stress exposure.

Abiotic stress refers to the detrimental effects of various abiotic factors on plants, including extreme temperatures, water scarcity or excess, high salinity, and toxic heavy metals (An et al., 2020). These stresses impact plant physiology, morphology, and biochemistry, resulting in decreased germination, photosynthesis, and overall growth (Oyebamiji et al., 2024). Plants may encounter multiple stressors, such as concurrent heat and drought, resulting in effects that differ from those experienced when these conditions occur separately. They possess a variety of mechanisms to defend against stress, regulated by a complex network. External stimuli also contribute to these responses. While plants can flourish in changing environments, they need optimal conditions for growth. Adverse circumstances threaten their genetic potential, contributing to food insecurity (Mathivanan, 2021). Furthermore, abiotic stressors can cause physiological and morphological challenges in plants, affecting photosynthesis, leading to oxidative damage, regulating stomata, impacting transpiration rates, influencing chlorophyll content, and resulting in enzyme denaturation (Ma et al., 2023). During the reproductive phase, common issues include floral abscission, pollen grain sterility, distortion of pollen tubes, and ovule abortion (Kumari et al., 2022). Furthermore, various abiotic stresses can potentially decrease crop yields by 50%–70% (Francini and Sebastiani, 2019).

BEs are reported to have the potential to alleviate various forms of stress. Their combined effects are also beneficial for plants, as supported by several researchers. BEs like Si, Co, Na, and Se help plants manage various abiotic stresses by regulating phytohormones and interacting with microorganisms (Sarraf et al., 2023). Together, Se and Si may help plants tackle various abiotic stressors (Luís Oliveira Cunha and de Mello Prado, 2023). Si aids plants in enduring both biotic and abiotic stresses through the hardening of cell walls and the reduction of heavy metal absorption in salt- and drought-prone environments (Seal et al., 2018). Additionally, plants can benefit from Se by mitigating the harmful effects of abiotic stressors such as heavy metals, salt, and drought (Ganguly et al., 2022).

Climate change is likely to heighten both the frequency and severity of abiotic stressors shortly (Mathivanan, 2021). Over the coming decades, fertile agricultural lands face substantial threats of dramatic declines due to increasing

Multiple Abiotic Stresses in Plants. DOI: https://doi.org/10.1016/B978-0-443-27424-4.00022-7

temperatures and recurrent flooding events. Temperature significantly impacts a plant's enzyme activity; elevated temperatures can damage the enzyme's structure in ways that cannot be repaired. Additionally, the prevalence of drought and salinity poses a considerable risk. By 2050, over half of all arable land is expected to face negative impacts. Not only do high temperatures present a challenge, but some crops are not adapted to cold stress, limiting their geographical range to tropical or subtropical regions.

Ahanger and Ahmad (2019) reported that studying the mechanisms of nutrient-induced abiotic stress tolerance through nutrient signaling and crosstalk with other pathways is essential for developing strategies for sustainable agriculture under challenging environmental conditions. Considering this, it is essential to develop strategies that utilize BEs to alleviate abiotic stresses, thereby improving crop resilience and productivity. This chapter centers on utilizing BEs to strengthen plant resilience against abiotic stresses, boost crop performance, and promote sustainable agricultural practices amid a changing climate. Furthermore, this article will offer recommendations for investigating the combined effects of different BEs to alleviate the influences of several abiotic stresses, as well as future research directions with valuable insights.

20.2 Beneficial elements for plants (uptake, translocation, and physiological roles)

Plant behavior depends on essential mineral nutrients, which influence vital processes. BEs promote growth by enhancing resistance to stressors, improving resource use, and reducing negative effects.

20.2.1 Silicon

Silicon (Si) is carried from roots to shoots as silicic acid [$Si(OH)_4$], which at biological pH can pass through the plasma membrane. In contrast to root hairs, lateral roots actively absorb monosilicic acid, which is the main form of Si. Certain transmembrane proteins called Si transporters are required for the absorption mechanism (Marron et al., 2016). Moreover, radial transport from the external solution to the cortical cells, along with xylem loading from the cortical cells, are the two mechanisms that facilitate Si movement uptake. Several crops are reported to accumulate a greater concentration of it. For example, rice had a much greater Si content in the symplast than cucumber and tomato (Mitani, 2005). Si regulates plant ionic homeostasis, nutrient accretion, as well as ethylene (ET)-mediated and jasmonic acid (JA) pathways, together with other phytohormones, particularly in stress conditions (Wani et al., 2016). Moreover, the signaling cascade controlling phytohormone homeostasis involves the merging of Si with the metabolism of reactive oxygen species (ROS) (Tripathi et al., 2021).

20.2.2 Selenium

In agricultural soils, selenium (Se) is most bioavailable as selenate (SeO_4^{2-}). It is prevalent in alkaline soils, whereas acidic soils contain selenite (SeO_3^{2-}). Even so, selenates are transported by sulfate transporters and channels, and selenites are carried via phosphate transport systems. At pH 5, low temperatures and respiratory inhibitors greatly diminished selenite uptake, suggesting a connection between selenite absorption and energy metabolism (Zhang et al., 2006a). In plants, it is vital for developing chloroplasts and cell membranes, which aid in photosynthesis. Furthermore, it plays an active role in managing antioxidants and ROS (Pilon-Smits, 2014). It influences many enzymes, such as CAT, SOD, and glutathione peroxidase in tomatoes (Castillo-Godina et al., 2016). Besides, it has been shown that it has antagonistic and synergistic relationships with other nutrients (Guerrero et al., 2014).

20.2.3 Cobalt

Cobalt (Co) is a crucial micronutrient for plants, essential for synthesizing cobalamin and fixing nitrogen (Banerjee and Roychoudhury, 2021). Once absorbed, it primarily accumulates in the roots before being transported to the shoots and leaves. Subcellularly, it accumulates in vacuoles, cytoplasm, cell walls, chloroplasts, the cortex, endodermis, and vascular tissues. It is absorbed by the roots and moves throughout the plant via the xylem. Its mobility within the plant supports various biochemical processes (Hinton, 2023). Reduced Co levels boost plant tolerance to stress, enhance nitrogen fixation, and improve nutrient absorption and antioxidant defense, leading to greater growth and production (Ulhassan et al., 2023).

20.2.4 Iodine

The primary form of iodine that is typically absorbed by roots is iodide (I^-). It is primarily transported through xylem pathways, and its concentrations decrease from the roots to the upper regions. Proton pumps fuel chloride transporters and ionic channels, allowing plants to absorb iodine as I^-. Iodine is found in soils as I^- and iodate (IO_3^-), with plants more effective in absorbing IO_3^- than I^-. Lettuce and tomato plants use phloemic transport, with low-transpiring parts, such as fruits and seeds, having lower iodine content than leaves (Hong et al., 2008). Soil is the second most important source of iodine due to the activity of living organisms. It can eliminate superoxide, hydroxyl, singlet oxygen, and hydrogen peroxide. Additionally, it increases biomass production, predicts *Arabidopsis thaliana* blooming, etc. It influences the expression of defensive response genes and is found in proteins located in roots and shoots (Kiferle et al., 2021). Although it is not essential for terrestrial plants, many aquatic plants require it for the metabolism of antioxidants (Medrano-Macías et al., 2016). While I^- has beneficial roles in crops, it can also have negative effects, such as reduced biomass in *Brassica oleracea* and an inverse relationship between I^- content and various elements (Mao et al., 2014).

20.2.5 Titanium

Titanium (Ti) exists in several oxidation states, such as Ti^{2+}, Ti^{3+}, and the most stable Ti^{4+}. It is regarded as advantageous for plant development. Ti, often applied as Ti-ascorbate, can directly enter cells in the roots of poaceous plants through YS1 (yellow stripe 1) (Lyu et al., 2017). TiO_2 NPs appear to be size-selectively absorbed by roots (Tripathi et al., 2017). These should have a minimum diameter of 140 nm to pass through the roots of wheat plants, 36 nm or less is necessary to transfer via parenchyma, and must be lower than 36 nm to pass through the Casparian band (Larue et al., 2012a,b). Furthermore, as NPs have been demonstrated to activate membrane receptors and trigger endocytosis, TiO_2 NPs may enter plant cells through this process. Utilizing n-TiO_2 when combined with single-nucleotide polymorphism increased SOD, CAT, and ascorbate peroxidase (APX) activities in salt-stressed barley (Karami and Sepehri, 2018). In addition, Ti-ascorbate was used in a hydroponic experiment led by Hrubý et al. (2002). It increased the amounts of chlorophyll *a* and *b* and raised the activity of nitrate reductase. TiO_2NPs are not only used for the improvement of plants but also used worldwide in the textile, plastic, and cosmetic sectors.

20.2.6 Vanadium

The fifth most abundant element in the Earth's crust is vanadium (V) (Imtiaz et al., 2015). It is chemically hard and malleable. Its mean concentration is reported to be up to 120 mg kg^{-1}, which is comparable to the concentration of Zn in the Earth's crust. Three oxidation states are found in nature: V(III), V(IV), and V(V). Among them, the tetravalent form (V^{4+}) has been reported to be advantageous for plants since it facilitates nitrogen fixation and chlorophyll production and raises potassium absorption and accumulation (Barker and Pilbeam, 2015).

Plant roots mostly take vanadium from the soil using the same transporters that are used to absorb essential plant nutrients (Edelstein and Ben-Hur, 2018; Aihemaiti et al., 2018). Plant energy expenditure is necessary for the active transport process of its absorption (Sun et al., 2019; Kubes et al., 2019). The majority of V in soil solution is found as anions, such as $VO_2(OH)_2^-$, $VO_3(OH)^{2-}$, and $V_4O_{12}^{4-}$(Gustafsson, 2019). The use of phosphorus fertilizer may have an impact on its absorption (Yang et al., 2017). A high pH increases its absorption because an alkaline pH in the soil-plant system improves possible bioavailability (Shaheen et al., 2019; Tian et al., 2015). (Roychoudhury, 2020) noted that the absorption rate peaked at pH 4, and at pH 10, it decreased significantly. Moreover, Qian et al. (2014) found an inverse correlation between total organic carbon and soil pH, suggesting that SOMNUS mineralization may reduce soil pH and affect plant V absorption. It is an essential cofactor in the nitrogenase enzyme, which is present in *Azobacter vinelandii* and catalyzes the transformation of atmospheric dinitrogen into ammonia in a variety of diazotrophic microbes (Akabayov and Akabayov, 2014).

20.2.7 Aluminum

Aluminum (Al) is the third most common metal in the crust of the planet, after Si and oxygen (Buchanan et al., 2015). Al absorption by plants is a complex process that is influenced by soil pH, environmental factors, and plant type (Bojórquez-Quintal et al., 2017). Al ions are absorbed by plant roots, especially in acidic environments. Plants usually uptake soluble Al from acidic soils in the form of Al^{3+}, $Al(OH)^{2+}$, and $Al(OH)_2^+$. Though Al is responsible for the disruption of root systems in some plants, some species have developed strategies to tolerate Al toxicity and often store it in less toxic forms. It has been shown to promote the development of many species, particularly those that are native to acidic soils. Additionally, Al treatment promotes the growth of new roots in tea plants (Sun et al., 2020). In maize, the application of Al triggers the activation of genes that prevent inorganic P deficiency (Maron et al., 2008). According to Bressan et al. (2021), *Vochysia tucanorum* seedlings exposed to 1110 μM Al demonstrated preserved proficient photochemical performance and leaf gas exchange rates, together with improved root development, stem, and leaf biomass.

20.2.8 Sodium

Sodium (Na) is widely present in nature, with seawater containing an unlimited supply. It primarily enters plant roots through nonselective cation channels (NSCCs) and is taken up by vacuoles via Na^+/H^+ exchangers. The negative electrical potential difference (-140 mV) across the plasma membrane creates a substantial Na^+ electrochemical potential gradient, facilitating its passive transport from the surrounding environment into the cytosol. high-affinity K transporter, LCT1 (low-affinity cation transporter), NSCC (including cyclic nucleotide-gated channels and GLRs i.e., glutamate receptors), and other uniporter or ion channel-type transporters are likely the passive entry points into root cells for Na. Besides, the competition between Na^+ and K^+ for absorption by plant roots implies that K^+ transporters also act as gates for its entrance. Since excess Na ions can soak up water, maintain turgor, and elevate osmotic potential, they are potentially harmful to crops (Pardo and Quintero, 2002). However, several reports have demonstrated its several advantageous roles. For example, *Atriplex* spp, *Kochia childsii*, and millet are examples of C_4 plants that need trace amounts of Na^+. It facilitates a Na^+-pyruvate cotransporter's absorption of pyruvate into chloroplasts (Furumoto et al., 2011). Subbarao et al. (2003) reported that under K^+ deprivation, many glycophytic plants respond well to Na^+ fertilization. Na is also a vital plant constituent in nutrition, facilitating glucose conversion into fructose in sugar beet plants. Na is also essential in photosynthesis regulation in some C_4 or CAM (crassulacean acid metabolism) plants, as it regenerates phosphoenolpyruvate from pyruvate, assimilating atmospheric carbon during photosynthesis (Marschner, 2012).

20.2.9 Lanthanum

The metal element lanthanum (La) belongs to the family of rare earth elements (REEs). In agriculture, La is an essential REE since it is used to increase crop yield (Hu et al., 2004). It is readily oxidized. Both the +3 and +2 oxidation states exist, with the +3 state being more stable and dominating (Ma et al., 2023). The absorption of La by plants is significantly influenced by various factors such as soil pH, cation exchange capacity, and the presence of organic acids (Sharma et al., 2024). While some La is moved to the shoots and leaves, most is taken up by the roots (von Tucher and Schmidhalter, 2005). Both roots and aerial organs may absorb it, but the majority of it accumulates in the roots, which affects how other elements are absorbed and accumulated. Uptake reaches a peak value of 48 hours after spraying on leaves (Kastori et al., 2023). Liu et al. (2013) similarly noted that cations can bind to negatively charged free hydroxyl and carboxyl groups present in cell wall components like cellulose, hemicelluloses, and pectin within the apoplast. In rice seedlings, La^{3+} primarily accumulates in the cell walls of the roots. Liu et al. (2012) claim that a signal transduction pathway involving calmodulin is responsible for the import of the La^{3+} ion in rice root cells. A low concentration of La encourages growth, photosynthetic rate, active pigments, and dry matter production to rise (Cui et al., 2019). According to Oliveira et al. (2015), the photosynthetic rate and chlorophyll content of soybean increased when 5 and 10 μM La were applied. Systemic endocytosis is triggered when *Arabidopsis* leaves are treated with La^{3+}, which influences the accumulation of mineral elements, root growth, and leaf expansion (Cheng et al., 2021). Further, LI et al. (2020) observed that enhanced photosynthetic rates, stomatal conductance, and chlorophyll fluorescence characteristics in *Amorphophallus sinensis* were achieved with the foliar spraying of $LaCl_3$ solution. Additionally, $La(NO_3)_3$ was shown to dramatically improve root fresh weight and rooting rate in vitro in *Eriobotrya japonica* rooting medium (Zhang et al., 2013).

20.2.10 Cerium

Cerium (Ce) is one of the most often utilized REEs, which causes it to accumulate widely in the environment. The Earth's crust contains the rare lanthanide element CeO_2 at a trace level of 50 ppm (Xu and Qu, 2014). The trivalent and tetravalent forms of CeO_2 are the most prevalent in bulk states and geochemical environments, with tetravalent states generally being more stable than trivalent ones (Dahle and Arai, 2015). Organisms undergo the reverse conversion from Ce^{4+} to Ce^{3+} as a result of valence shift and electron addition (Farias et al., 2018). Ce oxide nanoparticles (CeO_2 NPs) travel via the apoplastic route through the epidermis, cortex, and endodermis. Yet, the Casparian strip, made of lipophilic hydrocarbons, blocks their entry into the vascular cylinder (Roppolo et al., 2011). Multiple studies have indicated that Ce accumulates in plant roots.(Singh et al., 2019) observed an increase in Ce levels in the roots and shoots of tomato seedlings when exposed to CeO_2 NPs. Additionally, Wojcieszek et al. (2019) found Ce accumulation in radish root tissues. Ce significantly enhanced the electron transport rate of PSII by 11%, improved the effective quantum yield by 4%, and boosted photosynthetic efficiency by 14%. Furthermore, it reduced SOD activity by 7%, increased glutathione (GSH) levels by 12%, and enhanced dehydroascorbate reductase (DHAR) activity by 54% (Agathokleous et al., 2022).

20.3 Mechanisms of beneficial elements in mitigating multiple stresses

Healthy plants have natural defenses that allow them to endure different climate conditions and reduce significant losses. Aluminum, Co, Na, Se, and Si are vital bio-stimulants in plants. They enhance plant productivity and multiple stress tolerance, resulting in phenotypical, physiological, biochemical, and molecular changes. Key processes involve osmotic regulation, ion homeostasis, antioxidant defense, and enhanced photosynthetic efficiency. Besides, BEs often trigger cross-tolerance mechanisms, boosting tolerance to other stresses.

20.3.1 Silicon

Si exists at significant concentrations in the rhizosphere, being the second most abundant element (Sheikh et al., 2018). Monocotyledonous plants typically exhibit higher Si contents compared to dicotyledons (Kaur et al., 2015). It is taken up by plants in the form of $Si(OH)_4^{2+}$, subsequently transformed, and deposited as cell walls composed of hydrated amorphous silica. It is considered responsible for the durability, rigidity, and thickening of cell walls because of its interactions with pectin and polyphenols (Hussain et al., 2021). Moreover, it activates transporters, which are vital for nutrient uptake. This process improves the absorption and movement of essential nutrients, such as nitrogen, phosphorus, and potassium, by altering root architecture and expanding root surface area (Pavlovic et al., 2021).

High Na and chlorine (Cl) levels in salty soils are equally harmful to plant development. Si supplementation is vastly related to the uptake, exclusion, and compartmentalization of Na. Moreover, the alleviation of Cl toxicity can also be lessened through Si intervention, thereby promoting root-shoot transportation. Importantly, an increase in the availability of Si significantly improved water movement in stressed plants, particularly stomatal conductance in plants experiencing salinity stress (Cooke and Carey, 2023). Its use improved the photochemical efficiency of photosystem II and increased water use efficiency in maize treated with 60 mM NaCl (Khan et al., 2016). Besides, it is essential for salt stress reduction because it stimulates phytohormones, which influence the levels of stress hormones, as stated by Dhiman et al. (2021). Exogenous Si can enhance tomato fruit yield and water productivity by 12% to mitigate the negative effects of salt stress (Dou et al., 2023). Fig. 20.1 illustrates the key roles of BEs in alleviating salt stress.

Si effectively mitigates heavy metal stress in plants by downregulating metal transporter genes, such as Nramp5 and HMA2, which are responsible for metal absorption, including Cu and Cd (Tao and Lu, 2022). The addition of Si reduced the inhibitory effect of Ni stress on biomass production by lowering Ni concentrations in the roots and leaves of cotton (*Gossypium hirsutum*) and rapeseed (*Brassica juncea*) (Abd_Allah et al., 2019; Khaliq et al., 2015). It assists in decreasing

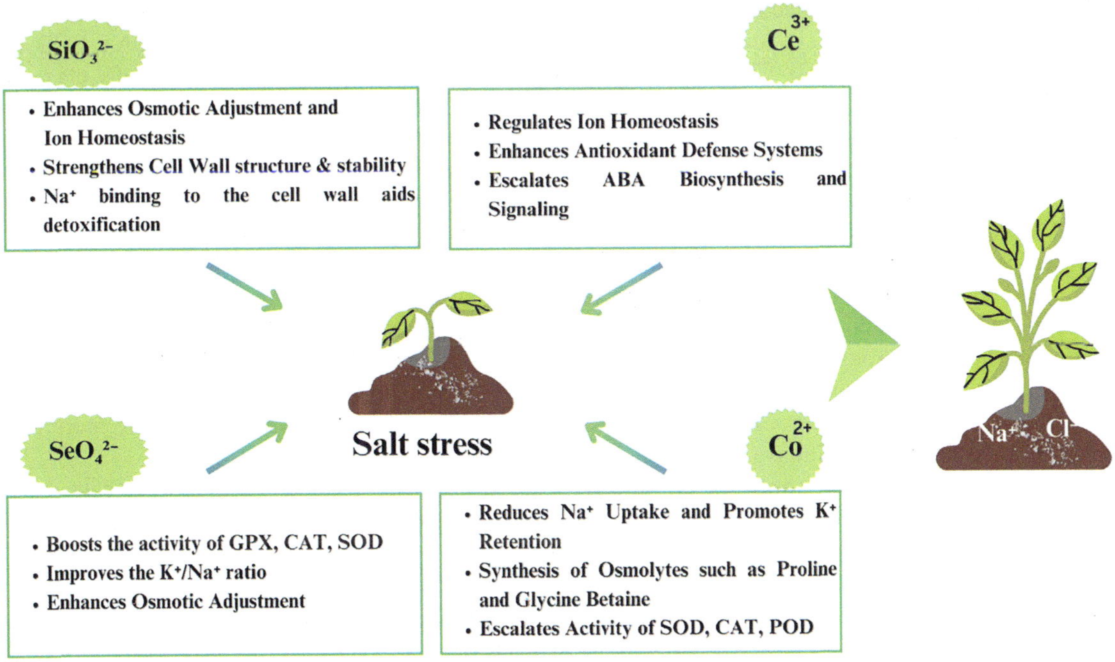

FIGURE 20.1 Roles of beneficial elements in mitigating salt stress.

the toxicity of Cd through vacuolar sequestration, phytochelatin synthesis, and antioxidant system detoxification (Biswas et al., 2023). (Sattar et al., 2022) found that applying it to As-contaminated maize reduced the As content in grains by over 90%. Si treatment improved the expression of related genes, boosting antioxidant molecules such as ascorbic acid (AsA) and GSH, as well as enzymes such as ascorbate POD, GSH reductase, DHAR, and monodehydroascorbate reductase. This significantly reduced oxidative damage in tomato seedlings exposed to V stress (Altaf et al., 2024).

Grain yield and shoot mass in drought-stressed plants are greatly enhanced by their supplementation (Johnson et al., 2022). Likewise, applying 60 ppm of nano-silica topically to wheat under water stress has boosted its dry weight and root length significantly (Boora et al., 2023). (Ahsan et al., 2023) stated that Si supplementation by 40 mg L^{-1} resulted in a satisfactory increase in total chlorophyll, carotenoid concentration, SOD activity, total free proline, and glycine betaine in *Gerbera jamesonii* exposed to water stress. Moreover, Si^{4+} reduces heat stress and electrolyte leakage in leaves at high temperatures (42.5°C) via stabilizing lipids and cell membranes (Rizwan et al., 2015).

The production of phytohormones such as abscisic acid (ABA) and JA is also enhanced, which are crucial for adapting to abiotic stress. Notably, it promotes the accumulation of osmoprotectants, such as proline and glycine betaine, thereby influencing antioxidant defenses, osmoprotection, and stress tolerance, which improves photosynthetic efficiency and minimizes oxidative damage in crops like wheat, maize, and tomato (Khalid et al., 2022). Additionally, Kim et al. (2017) highlighted that a crucial strategy to counteract free radicals and repair damage induced by ROS is the enhancement of an advanced antioxidant system. This system relies on enzymatic antioxidants, such as SOD, POD, CAT, and APX, as well as nonenzymatic antioxidants, including carotenoids, polyphenols, tocopherols, and GSH, to maintain enzymatic homeostasis.

20.3.2 Selenium

Se has been thoroughly researched in plants, with new efforts focusing on utilizing it for biofortification due to its benefits (Malagoli et al., 2015). Plants mainly absorb Se as selenate or selenite. Selenate enters through sulfate transporters because it resembles sulfate ions, while selenite is taken up through phosphate transporters (White, 2018; Trippe and Pilon-Smits, 2021). Since Se can function as an antioxidant in plants, it enhances their tolerance to various types of abiotic stresses (Pilon-Smits, 2014).

It is commonly known that Se has an ameliorative effect on a variety of environmental stressors (Zwolak, 2020). Positive effects on gaseous exchange parameters have been observed under heat stress, salinity, and exposure to cadmium stress (Amirabad et al., 2020). The foliar spray of Se increases plant growth under salinity regimes (Tisarum et al., 2019). Se treatment (100 μmol. L^{-1}) promoted root and shoot development, which in turn increased biomass production of salt-stressed *Brassica rapa* (Hussain et al., 2024). *B. rapa* also showed improvements in growth, APX, CAT, POD, SOD, and proline content (Hussain et al., 2023). Chlorophyll content, shoot, and root lengths are all significantly elevated in salt-stressed *Brassica campestris* when treated with Se–NPs (Sarkar and Kalita, 2022). Additionally, the fresh weights of soybean shoots and roots increased with the use of Na_2SeO_3 (Wang et al., 2023).

LI et al. (2020) noted that its presence effectively reduces the harmful effects of different abiotic stressors at lower concentrations, promoting plant growth and performance. This application is especially beneficial for plants facing drought and saline environments, as it positively influences water use efficiency. This protective effect is due to its capacity to reduce damage to chloroplast structure and improve electron transport rates (Andrade et al., 2018). The application of Se can enhance the photosynthetic rate in plants, suggesting its use through soil incorporation or foliar spraying during drought conditions (Andrade et al., 2018). Applying Se-NPs externally at a concentration of 10 ppm significantly enhances the drought tolerance of grapevine saplings by increasing proline accumulation, strengthening antioxidant defense, and promoting overall growth (Daler et al., 2024). In addition, seedlings treated with foliar Se (8 days before drought stress) showed higher CAT, APX, and SOD, suggesting that previous Se sprays are more effective in overcoming induced drought conditions (Sousa et al., 2023).

In maize, certain antioxidant enzymes like APX, GSH reductase, DHAR, and monodehydroascorbate reductase are stimulated by the application of Se (Yildiztugay et al., 2016). Thus the ROS accumulation in the mesophyll is controlled. Moreover, Cadmium stress is common in solanaceous crops. For example, Alves et al. (2020) applied Se at a concentration of 1 M in SeO_4^{2-} form in the nutrient solution when tomato was subjected to cadmium stress. It led to improved activities of APX and CAT, thereby enhancing the tomatoes' ability to withstand Cd stress. Additionally, various Se forms enhance HM tolerance in rice by promoting antioxidant activities under Cd stress (Xu et al., 2020). It likely enhances V toxicity in rice seedlings, promoting plant growth and composition under metal stress, as shown by balanced nutrient effluxes (AL-Huqail et al., 2024). Tian et al. (2024) confirmed these findings, showing that Se treatment can reduce HM levels in plants growing in both acidic and neutral soils, while also inhibiting the uptake of Hg, Cr, Cd, Pb, As, and Cu.

20.3.3 Cobalt

Co is recognized as a valuable element involved in several physiological processes. While not essential for all plant species, Co is crucial for nitrogen fixation, enzyme activation, and stress tolerance (*The importance of molybdenum and cobalt in plants*, 2024). It is responsible for its usefulness in nitrogen fixation in diazotrophs like *Rhizobium*, *Azotobacter*, *Bacillus*, and *Frankia*, carrying out the process (Soumare et al., 2020). Furthermore, it is an essential part of vitamin B12 or cobalamin, which is necessary for nitrogenase enzyme function. The enzyme is responsible for converting atmospheric N into NH_3. This process is critical for the legumes and the plants that rely on nitrogen-fixing bacteria (Hu et al., 2021).

Numerous studies have highlighted the significance of Co in mitigating salt stress. The crucial role in enhancing tolerance to abiotic stress was showcased through Comediated salt tolerance responses in *Cucumis sativus* (Brengi et al., 2022). It alleviates salt stress by boosting the levels of free proline (Sachdev et al., 2021). The shoot and root weight of radish plants were significantly increased with doses of Co sulfate under salt stress (Inayat et al., 2024). ABA, auxins, and gibberellins in onion leaves are all positively affected by Co rates, which enable plants to tolerate high salt levels of up to 5 dS m^{-1} (Gad et al., 2020). Faiyad and Abd El-Azeiz (2024) obtained the maximum vegetative and yield characteristics of salt-affected *Vicia faba* when Co (10 ppm) and seaweed extract were combined. In *Cucumis sativus*, salt stress reduced H_2O_2 and malondialdehyde levels while increasing protein content and CAT activity (Brengi et al., 2022).

ABA is crucial for regulating transpiration, as it initiates stomatal closure during drought. The use of Co has been shown to increase ABA and ET levels in plants (Kumari et al., 2022). Co sulfate enhances plant growth by regulating stomatal openings, activating antioxidant enzymes, boosting photosynthesis, and influencing gene expression under drought stress (Ali et al., 2023). Collectively, these alterations aid plants to acclimate better in abiotic stress conditions by optimizing their physiological responses (Kumari et al., 2022). By increasing the content of proline, flavonoids, total phenols, and antioxidant enzymes (CAT, POD, and polyphenol oxidase) in rice, Co significantly enhanced drought tolerance and the efficacy of the antioxidant system (Tourky et al., 2023).

Corequisite for the activation of vital antioxidant enzymes that are prominent for counteracting oxidative stress triggered by ROS under abiotic stress conditions. Major antioxidant enzymes like CAT, POD, and SOD are upregulated in the presence of Co, aiding in the neutralization of ROS and safeguarding plant cells from harm (Hareem et al., 2024). It facilitates stomatal opening under stressful conditions, thereby enhancing gas exchange and optimizing the efficiency of photosynthesis, which allows plants to utilize available water resources more effectively (Kumari et al., 2022). Additionally, it affects how other vital nutrients, such as N and P, are absorbed and transported, which is critical for plant growth and development. By facilitating the movement of these macronutrients, Co supports resilience against abiotic stress and makes the plant more resilient under such circumstances (Banerjee and Roychoudhury, 2021).

20.3.4 Iodine

Research indicates that iodine enhances antioxidant levels and resilience against abiotic stressors, such as heavy metals and salt (Leyva et al., 2011; Gupta et al., 2015). Iodine at 20 and 40 μM KIO_3 enhances foliar mass, antioxidant response, and phenolic compound accumulation, even under salt stress (Blasco et al., 2013). $KIO_3 < 40 \times 10^{-6}$ M raises SOD, APX, GSH, AA, and antioxidant potential, which improves lettuce's ability to withstand salt stress (Leyva et al., 2011). Similarly, tomato fruit yield, leaf nitrogen content, and AsA content all increased when 100 μM KIO_3 was applied topically (Fuentes et al., 2022). Kiferle et al. (2022) found that adding it to fertigation reduces salt's negative effects on tomato growth and fruit yield while preserving fruit quality.

Li et al. (2023) summarized that soybean plants exposed to 10 μM KI solution showed an improved tolerance to water shortages, and no phytotoxicity was observed in plants that received adequate irrigation. Additionally, deficiencies in water can be effectively addressed by KI, which enhances the fruit quality of tomato and promotes positive reactions during periods of water deficit (Lima et al., 2023b). Numerous researchers have reported its function in alleviating oxidative stress. The induction of antioxidants is viewed as a crucial element of the adaptive responses that enable plants to withstand stress. Increasing atmospheric O_2 levels during photosynthesis suggest iodine as an early inorganic antioxidant that helps organisms mitigate oxidative damage (Venturi, 2011). Research shows that using iodine as IO_3^- is more advantageous than I^- for synthesizing antioxidant chemicals (Leyva et al., 2011; Blasco et al., 2013). Iodine pretreatment for a brief period significantly increased rapeseed survival in deterioration tests closely related to oxidative stress (Powell et al., 2005). Since iodine inactivates H_2O_2, O_2^-, hydroxyl radicals, and $^1O_2$, its role in marine algae has been established.

20.3.5 Titanium

Ti is known for its ability to alleviate various stresses, including heat, cold, drought, and salinity. The use of Ti oxide nanoparticles has enhanced plant tolerance to temperature stress in chickpea, water stress in wheat, and heavy metal (Cd) stress in soybean (Jaberzadeh et al., 2013; Singh and Lee, 2016; Mohammadi et al., 2014). TiO_2 NPs applied directly increased the amount of chlorophyll in tomato (Raliya et al., 2015). Latef et al. (2018) found that Ti aids in minimizing H_2O_2 accumulation, thereby alleviating harmful impacts from abiotic stresses. Additionally, it plays a role in safeguarding and boosting the resilience of photosynthetic pigments against degradation. Shahandashti et al. (2013) observed that chickpea genotypes under cold stress showed a reduction in MDA content when exposed to low doses of Ti compounds (5 ppm). Similarly, TiO_2 NPs can significantly enhance fenugreek's capacity to produce primary and secondary metabolites, thereby increasing its resistance to cold stress (Babaei et al., 2025). These results are supported by Parvin et al. (2020), who concluded that exposure to TiO_2 NPs enhanced the quantity of photosynthetic pigments in *Linum usitatissimum*.

Khan et al. (2023) investigated the impact of Ti dioxide on cucumbers exposed to drought and nickel stress. They reported that the ascorbate-GSH system, antioxidant enzymes, and the metabolism of nitrogen and carbohydrates were all enhanced when the stressed seedlings received a supplementation of 15 mg L^{-1}. TiO_2 NPs reduced the toxicity caused by Cr(VI) in *Helianthus annuus* by inhibiting the formation of H_2O_2 and MDA, thereby decreasing DNA damage (Kumar et al., 2023). Karvar et al. (2022) found that when maize plants experienced dry conditions, TiO_2 foliar treatment enhanced leaf relative water content (RWC), carotenoids, chlorophylls, proline, soluble protein, and grain yield. Moreover, stevia plants were exposed to different levels of salinity. The application of TiO_2 foliar treatment increased the levels of carotenoids and chlorophyll, the rate of net CO_2 assimilation, and the RWC of the leaves (Sheikhalipour et al., 2021).

20.3.6 Vanadium

Scientists have been searching for new elements and their mechanisms in mitigating the effects of abiotic stressors. Although much information is still unavailable about the roles of vanadium, some researchers have contributed to identifying its beneficial effects. According to Hanus-Fajerska et al. (2021), plants can be primed to withstand abiotic stress by utilizing a stress-protective substance. When vanadium interacts with certain elements, including phosphorus, zinc, molybdenum, and copper, it demonstrates positive impacts on plants. Additionally, it can reduce the toxicity of copper (García-Jiménez et al., 2018). It reduces the adverse impacts or stops the spread of toxic trace metals, such as lead, mercury, and copper (Akoumianaki-Ioannidou et al., 2016).

Recent studies indicate that vanadium (IV) complexes can safeguard plants treated with the pro-oxidant H_2O_2. Treatment with oxidovanadium (IV) complexes enhanced the resilience of *Arabidopsis thaliana* to oxidative stress (Rojek et al., 2019). Furthermore, Ashokan et al. (2024) reinforced this finding, emphasizing that proper nutrient management, which includes micronutrients such as vanadium, can mitigate the negative impacts of abiotic stressors on crop growth and yields.

20.3.7 Sodium

In stressful environments, Na helps plants maintain osmotic balance by producing osmoprotectants, such as proline and glycine betaine. Certain amaranth plants exhibited improved water content due to induced osmotic stress (Yamada et al., 2016). Similar findings were observed in sword-leaf dogbane (*Apocynum venetum*), which improved photosynthetic capability and water content levels (Cui et al., 2019). Another aspect is that Na retrenched Cd concentration during Cd stress in *Carpobrotus rossii* (Cheng et al., 2018). The application of Na sulfide (Na_2S) in Pb and Zn-contaminated soils resulted in minimum accumulation of Pb and Zn in plants, ultimately promoting growth in black locusts (*Robinia pseudoacacia*) (An et al., 2020). An experiment conducted by Bistgani et al. (2019) on *Thymus vulgaris* and *Thymus daenensis* concluded that certain phenolic compounds, such as rutin and quercetin, are synthesized within the plant body as a result of Na exposure. These compounds exhibit synergistic effects in ROS scavenging, ultimately maintaining cellular redox balance (Farahani et al., 2021). However, their antioxidant capacity is higher when these two phenolic compounds are combined (Katiyar et al., 2023). Plant hormones prevent a plant from damage caused by abiotic stress (Arzani, 2008). Sea lavender (*Limonium bicolor*) seedlings exhibited increased indoleacetic acid (IAA) levels in response to NaCl exposure compared to controls (Leng et al., 2019). (Arzani, 2008) states that various phytohormones, including IAA, help address salinity issues.

20.3.8 Aluminum

Al-induced growth stimulation varies among plant species, despite Al being essential for the formation of meristematic cells in roots. It enhances the uptake of macronutrients, such as phosphorus and nitrogen, in low soil pH conditions. While most observed growth-stimulating effects are in tree species that thrive in acidic soils, some studies have also shown similar benefits in certain economically important crops (LI et al., 2020). In maize, a low dose inhibited root growth while significantly enhancing the leaf growth rate, depending on the duration of aluminum exposure (Wang et al., 2015). Additionally, Famoso et al. (2011) observed stimulated root elongation in rice, while Nhan (2013) reported increases in chlorophyll and carotenoid levels.

Although aluminum is toxic, it significantly helps diminish the adverse effects of manganese. Muhammad et al. (2019) noted that higher solubility and availability of aluminum in acidic soils can effectively reduce this toxicity. An experiment by Wang et al. (2015) on rice plants showed that exposure to 200 μM of Mn greatly hindered both shoot and root growth. In contrast, applying 200 μM of aluminum alleviated the manganese-induced toxicity in the rice seedlings. Fluoride in tea plants hinders shoot and root development. However, Al countered F toxicity by creating an $Al^{-}F$ compound that stimulated growth (Yang et al., 2016).

20.3.9 Lanthanum

La enhances the activity of antioxidant enzymes and mitigates the effects of various stressors, including heavy metals, acid rain, UV-B radiation, osmotic stress, and drought (Kastori et al., 2024). (REN et al., 2016) reported a decrease in heavy metal levels (Cd, Pb, Ni, Cu, Zn) in the stems and leaves of Chinese cabbage and rape. Synchronous treatment with La promotes the accumulation of phytochelatin synthetase gene mRNA and phytochelatin content in both leaves and roots, hence improving Cd tolerance in *Lactuca sativa* (He et al., 2005). Besides, when it is applied at the proper concentrations, decreased formation of superoxide and H_2O_2, as well as encouraged their removal by upregulating genes linked to antioxidant enzymes, can be perceived in rice seedling leaves under Cd stress (Wang et al., 2014), additionally, La can increase the amount of soluble sugars and essential oils in flowers and foliage by more than 18% under water stress (Zhu et al., 2018), moreover, it prevents electrolyte leakage by reducing cell membrane permeability and protecting cells from acid rain stress.

Zhong and Chen (2020) suggested that OsGH3.4 formed by La(III) might be capable of suppressing IAA-dependent expansin expression, thereby enhancing rice's resistance to Cu stress. It appears to enhance the capacity of the ROS scavenging system, thereby reducing oxidative damage caused by osmotic stress (Zhang et al., 2006b). The La application effectively lowered oxidative damage, reduced the Na^{+}/K^{+} ratio in both roots and leaves, and enhanced the activities of POD and SOD. According to a study by Wang et al. (2009), La^{3+} can reduce the oxidative damage caused by UV-B radiation in soybean plants by interacting with ROS. Additionally, it prevented chloroplasts from disintegrating (Tong et al., 2024). By lowering the Cu(II) content and increasing seedling resistance to Cu(II) stress, La(III)-amino acid chelate dramatically enhances Cu detoxification in rice tissues at a concentration of 20 mg L^{-1} (Chen and Zhong, 2024).

20.3.10 Cerium

CeO_2 NPs, acting as nanozymes, play a crucial role in mitigating abiotic stressors, including temperature, salinity, drought, and heavy metal exposure (Preetha et al., 2023). The harmful impacts of high-salt environments may be mitigated by applying CeO_2 NPs to stressed plants, according to numerous reports. One such work utilized polyacrylic acid coated with CeO_2 NPs to prime cotton seeds under 200 mM salt stress (An et al., 2020). The results of this study demonstrated that treated seeds exhibited enhanced morphological and physiological characteristics, decreased ROS production, elevated antioxidant levels, and improved ion homeostasis during salt stress. CeO_2 NPs are well-known for their potent oxidizing properties and ability to scavenge ROS at high concentrations in plants (Djanaguiraman et al., 2018; An et al., 2020; Inbaraj and Chen, 2020). Their ability to efficiently scavenge ROS in biological systems stems from the presence of oxygen vacancies occurring during cellular uptake (Gunawan et al., 2019; Niu et al., 2011). They also demonstrated elevated alpha-amylase activity, which lessened the consequences of salt stress (Khan et al., 2021). CeO_2 NPs at 10 mg L^{-1} enhanced photosynthetic efficiency, reduced ROS levels, and maintained Na^{+}/K^{+} equilibrium in salt-stressed maize leaves (Liu et al., 2022). In salt-stressed *Dracocephalum moldavica*, CeO_2 NPs at varying doses enhanced morphological traits, antioxidant levels, and proline content while markedly lowering MDA, H_2O_2, proline, and EL (electrolyte leakage) levels (Mohammadi et al., 2021). They also enhanced spearmint's salt tolerance by boosting chlorophyll and carotenoid levels, alongside increasing the activity of antioxidant enzymes (Milani et al., 2024).

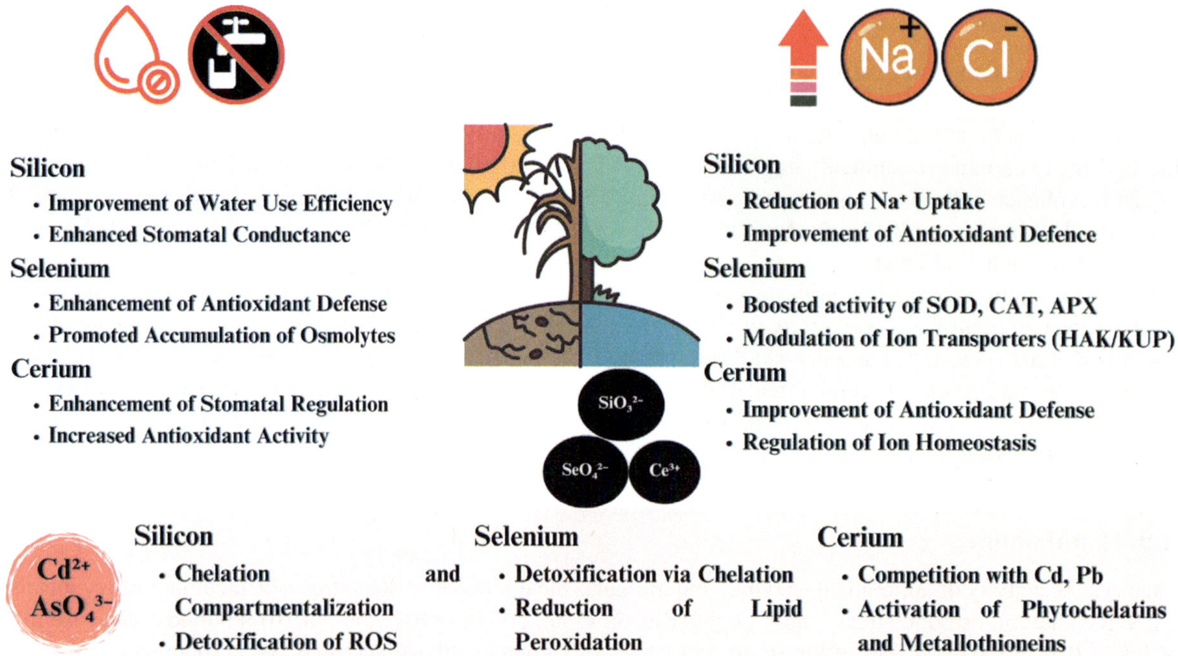

FIGURE 20.2 Role of beneficial elements in alleviating multiple abiotic stresses.

CeO_2 NPs increased photosynthetic indices, decreased ROS activity, and successfully alleviated drought stress in sorghum (Djanaguiraman et al., 2018). Applying Ce^{3+} to common bean (*Phaseolus vulgaris*) enhances its ability to withstand water stress, thereby improving survival, chlorophyll content, and water use efficiency (Salgado et al., 2020). According to the study by Li et al. (2023), an acceptable Ce level (up to 20 mg L^{-1}) aids *Dendrobium nobile* (an ornamental orchid) in accumulating more medicinal components. Thus it lessens the detrimental effects of copper stress. Under chromium stress, CeO_2 NPs dramatically increased enzymatic activity, decreased oxidative stress, and increased biomass yield in sunflower (Ma et al., 2022). They regulate the transfer of Cd from soil to plant tissue, which has a beneficial effect on wheat development in Cd-spiked circumstances (Ayub et al., 2023). Besides, Mukherjee et al. (2023) found that they enhanced plant growth and decreased oxidative stress in Mung bean (*Vigna radiata*) at doses of up to 10 ppm under mercury-stressed circumstances. Fig. 20.2 illustrates how BEs help mitigate various abiotic stresses.

20.4 Role of beneficial elements in alleviating multiple abiotic stresses

Any unfavorable environmental condition that negatively affects the growth and development of plants is considered abiotic stress (Kochhar and Gujral, 2020). Numerous studies have demonstrated the beneficial effects of bioenhancers in mitigating various abiotic stresses (Table 20.1). Recent developments in BEs, such as Co, Na, Se, and Si, enhance both crop yield and quality during abiotic stress by facilitating mineral absorption, modulating transporter functions, and boosting the accumulation of sulfur-containing metabolites. These elements also enhance stress tolerance, influence phytohormone biosynthesis, and are associated with soil bacteria that promote growth (Sarraf et al., 2023). Marschner (2012) specifies that Co, Na, Se, and Si are beneficial for plants, strengthening their defenses against biotic stressors, such as pathogens and herbivores, as well as abiotic stresses, including salt, UV, drought, and cold. Taha et al. (2021) reported that the combined application of Se and Si has been shown to improve growth, strengthen antioxidant defenses, and increase wheat yield under salt stress. Additionally, the epigenome is modified by essential elements, including Al, Co, Na, Se, and Si, which are crucial for epigenetic regulation and enhance plant stress tolerance (Pandey et al., 2023).

20.5 Research gaps and future directions

Research gaps in sustainable agriculture hinder the full use of BEs in helping plants tolerate multiple stresses. These gaps include mechanical interactions, stress-specific pathways, biomarkers, functional genomics, bioremediation integration, bioavailability, plant uptake, climate change effects, bioremediation priming, species-specific responses, and optimizing

TABLE 20.1 Roles of beneficial elements in alleviating multiple abiotic stresses.

Beneficial element	Multiple stresses	Performance	References
Silicon (Si)	Salt	Elevated aquaporin activity enhanced root hydraulic conductivity and reduced osmotic stress in sorghum.	Liu et al. (2015)
		Facilitated the actions of enzymes that metabolize carbohydrates, resulting in higher starch content in cucumber roots.	Zhu et al. (2016)
	Heavy metal	Improved growth and reduced buildup of heavy metals in acidic soil contaminated with multiple metals (Cd, Zn, Cu, and Pb) in rice.	Gu et al. (2011)
		The upregulation of key antioxidant enzyme genes, such as APX, GR, DHAR, and MDHAR, significantly reduced oxidative damage in tomatoes, thereby helping to alleviate vanadium (V) stress.	Altaf et al. (2024)
	Water deficit	Reduced ABA homeostasis and improved stress tolerance through transcriptional control of ABA production and breakdown pathways in barley.	Hosseini et al. (2017)
		Elevated RWC, antioxidant enzymes, and improved morpho-physiological parameters in maize.	Sattar et al. (2020), LI et al. (2020)
	Low temperature	Aided Maize plants retain their water turgor potential and lessen dehydration by forming phytoliths and subcutaneous layers.	Moradtalab et al. (2018)
		Elevated levels of flavonoids, total phenols, and lycopene in tomato seedlings.	Zhang et al. (2025)
Selenium (Se)	Salt	Decreased oxidative damage, lipid peroxidation in cellular membranes, and malondialdehyde buildup in sweet william (*Dianthus barbatus* L.).	Alvan et al. (2025)
		Seed priming with 100 μmol L^{-1} significantly decreased MDA and H_2O_2 contents in Field mustard.	Hussain et al. (2024)
	Heavy metal	Decreased absorption, transfer coefficient, and accumulation in garlic.	NIe et al. (2024)
		Enhanced cucumber biomass and growth under heavy metal stress (Cd, Pb, and Ni) by regulating antioxidant enzyme activities to detoxify ROS-induced damage.	Aroiee et al. (2019)
	Water deficit	Enhanced photosynthesis rates, the level of osmotic substances, antioxidant enzymes, and stress-responsive genes in tobacco.	Han et al. (2022)
		Activated antioxidant defenses reduce oxidative stress effects and improve developmental traits by controlling ROS accumulation in tomato.	Rady et al. (2020), LI et al. (2020)
	Low temperature	Increased their levels of chlorophyll, enzymatic antioxidants, and nonenzymatic antioxidants to combat ROS production in chili.	Nandagopal et al. (2022)
		Decreased lipid peroxidation and increased glutathione peroxidase system activity in plants, as evidenced by lowered MDA and elevated proline levels in strawberry seedlings.	Huang et al. (2018)

(*Continued*)

TABLE 20.1 (Continued)

Beneficial element	Multiple stresses	Performance	References
Cobalt (Co)	Salt	Reduced levels of enzymatic antioxidants, such as catalase and peroxidase, in cabbage.	Alkharpotly et al. (2023)
		Plant cell division and elongation lead to increased length of roots and shoots, along with better fresh and dry weight in rapeseed.	Ali et al. (2018)
	Heavy metal	$CoCl_2$ acted as an ethylene inhibitor in the presence of Cd stress.	Chmielowska-Bąk et al. (2014)
	Water deficit	Increased proline, flavonoid, and total phenol levels, as well as antioxidant enzymes (catalase, peroxidase, and polyphenol oxidase), to enhance the effectiveness of the antioxidant system in rice.	Tourky et al. (2023)
		Adjustment of the stomatal aperture, activation of antioxidant enzymes, enhanced photosynthesis, improved nutrient absorption, and altered gene expression in rapeseed.	Ali et al. (2023)
	Osmotic Stress	It reduces membrane damage and has preventive effects on leaves by mitigating the decline in polyamine content and antioxidative enzyme activity in potato seedlings.	Li et al. (2004)
Iodine (I^-)	Salt	Tomato fruit output was significantly increased by 100 μM KIO_3 without resulting in any reduction in plant biomass.	Fuentes et al. (2022)
		Higher levels of P and K, as well as increased activity of glutathione and ascorbate peroxidase, are found in strawberries.	Macías et al. (2021)
	Water deficit	10 μM KI improved the photosynthetic efficiency, water tolerance, and enzymatic antioxidant protection of soybean plants.	Li et al. (2023)
Titanium (Ti)	Salt	Enhanced germination characteristics, fresh and dry mass, relative water content, membrane stability index, and chlorophyll levels in wheat.	Badshah et al. (2023)
		Increased antioxidant enzyme activity, decreased H_2O_2 levels, and improved essential oil content in Moldavian balm (*Dracocephalum moldavica* L.)	Gohari et al. (2020)
	Water deficit	Achieved the greatest protein and seed oil content levels in linseed.	Aghdam et al. (2016)
		Enhancements were observed in osmolyte levels (sugar and proline), membrane stability index, relative water content, and chlorophyll content in wheat.	Mustafa et al. (2021)
	Heavy metal	Decreased the formation of H_2O_2, MDA, and total chromium accumulation, which in turn reduced DNA damage and the toxicity of Cr(VI) in sunflowers.	Kumar et al. (2023)
		Reduced MDA concentration, decreased antioxidant enzyme activity, and increased levels of carotenoids and chlorophyll, indicating that they mitigated the oxidative stress induced by As and Cd stress in Indian mustard.	Singh et al. (2021)
	Low temperature	Decreased electrolyte leakage and malondialdehyde levels, and had a favorable impact on pigment production as well as enzymatic and nonenzymatic antioxidant activities in fenugreek (*Trigonella foenum-graecum* L.).	Babaei et al. (2025)

(Continued)

TABLE 20.1 (Continued)

Beneficial element	Multiple stresses	Performance	References
Cerium (Ce)	Salt	Increased Chlorophyll contents (a and b), reduced MDA and H_2O_2 in spearmint (*Mentha spicata* L.).	Milani et al. (2024)
		Enhanced plant physiological responses (e.g., chlorophyll fluorescence) were the result of altered root architecture in rapeseed.	Rossi et al. (2017)
	Water deficit	Increased ROS catalytic scavenging can safeguard sorghum grain yield and photosynthesis.	Djanaguiraman et al. (2018)
		Enhanced efficiency of water usage, increased chlorophyll levels, and improved photosynthesis rates in common beans.	Salgado et (LI et al., 2020)
	Heavy metal	In wheat, the bioavailable soil Cd content and its accumulation in grain, root, and shoot tissues decreased, while dry weight in roots and shoots increased.	Ayub et al. (2023)
	UV	Showcased inherent SOD and CAT enzymes that eliminate ROS and alter antioxidant defense genes, improving enzymatic (increased SOD, CAT, and POD activities) and nonenzymatic (raised AsA and GSH levels) antioxidants.	Sun et al. (2025)

bioremediation for economic viability. Understanding the bioavailability and uptake of BEs is crucial for effective crop management and targeted farming strategies. The usefulness of BEs under combined or sequential stresses, which are more similar to actual farming conditions, is not well understood, despite reports that they can mitigate specific stresses such as salt, drought, and heavy metals. Research mainly focuses on individual elements, but plants in natural soils face multiple BEs simultaneously, with their synergistic, additive, or antagonistic interactions still underexplored.

Si, a signaling molecule in plants, influences primary metabolism, yet its precise mechanism remains unclear. Further research is needed to clarify its role in host defense and its effects on microbial gene expression and plant biochemistry. Some BEs (e.g., Al, La, V) show stress-positive effects, but systematic studies on dose-response relationships and toxicity limits are lacking. Safe ranges under various environments are not well established. More investigation into the appropriate forms, doses, and timing of Se application under salt stress is essential to evaluate environmental risks (Nedjimi, 2024). Future studies should explore the physiological, ecological, and molecular processes influencing Se-mediated stress tolerance to develop long-term crop improvement strategies. Research on low Co availability and its effects on plant growth and nutrient uptake remains limited. Similarly, studies on the combined effects of melatonin and Co on plant salt tolerance are scarce. The biochemical mechanisms behind methylglyoxal detoxification, especially the roles of microelements like Ti and Co as stress protectants, are not fully understood. Further research is needed to clarify iodine's role in stress tolerance, particularly regarding its impact on plant pathogens. Future work should examine molecular targets, how protein iodination reduces other biotic and abiotic stressors, and the use of biopolymers to enhance stress resilience. TiO_2 nanoparticles (NPs) are not transported actively by carriers, and their influence on molecular transporters needs more study. The interaction between Ti and Fe is vital for plant metabolism, but claims that it causes phytotoxicity remain unproven. Vascular plants capable of detoxifying various V ion types are still in early stages of metabolomics research, and deeper studies are necessary to understand V compounds' roles in cellular stress responses. The mechanisms of V toxicity are not fully understood, requiring additional research on tolerance and homeostasis across ecosystems. The role of Al in mitigating stress remains unclear, but studying species-specific Al tolerance and molecular techniques will help clarify its function. Current research emphasizes direct application of REEs, but soil immobilization may lower efficiency and cause ecological issues. Future studies could focus on how REEs like La are immobilized in soil and ways to develop mitigation strategies. CeO_2 NPs pose a threat to plants, requiring further investigation into their effects on plant metabolism and food chains. More research should explore their mechanisms of action. It's also unclear how BEs influence microbial populations in the rhizosphere or how they synergize to promote stress tolerance. Much of what is known comes from hydroponic or controlled experiments, and the lack of field validation across different areas limits large-scale use. Concerns about the long-term environmental impacts, such as soil buildup, food chain transfer, and ecological effects, remain.

Deciphering the molecular networks behind BE-mediated stress tolerance will benefit from integrating genomes, transcriptomics, proteomics, metabolomics, and epigenomics. It is essential to establish accurate dose-response curves for each BE under various soil and climate conditions to optimize their benefits while avoiding toxicity. Future research should examine combinations of BEs (e.g., Se + Si, Co + Ti) for improved nutrient use and tolerance to multiple stresses, considering possible synergies with traditional fertilizers. Understanding BE-driven cross-tolerance can be advanced by designing studies that mimic real-world conditions, such as simultaneous heat and drought or salinity with heavy metal stress. To enhance resilience, research should focus on how BEs affect plant-microbe interactions, especially beneficial symbioses like mycorrhizae and nitrogen-fixing bacteria. Developing controlled-release carriers and nano-formulations for BEs could increase bioavailability, reduce toxicity, and enable targeted delivery during stress. Long-term field testing is vital to assess ecological impacts, environmental buildup, and food safety, especially for RREs and nanoparticles. Incorporating BEs into integrated, climate-smart, and sustainable farming practices can boost productivity, quality, and resilience to climate change. Despite progress, understanding nutrient interactions, optimal concentrations, and application timings will help farmers manage abiotic stresses more effectively amid climate challenges. Further investigation into the regulatory mechanisms of BEs in plant growth, resilience, and productivity is needed. Additional research into physiological and biochemical pathways will allow for full potential utilization of these elements, despite ongoing debates about their use.

20.6 Conclusion

Abiotic stress factors, such as salinity, extreme temperatures, drought, heavy metals, etc., impact plant distribution, productivity, and food security. Global warming is expected to worsen these stresses, hindering plant survival, growth, and agricultural yields. BEs such as aluminum, Co, Na, Se, and Si serve as essential bio-stimulants, enhancing productivity and multiple stress tolerance through various physiological and molecular changes. They trigger adaptive responses, increasing nutrient uptake and antioxidant synthesis. However, high doses can inhibit their growth stimulation, making careful application vital for meeting the food demands of a growing population under adverse conditions. BEs significantly enhance tolerance to multiple abiotic stresses by improving nutrient uptake and physiological functions. Although many reports underscore the benefits of BEs, implementation in the field is still ongoing. Some elements may become toxic, with varying effects on plants, necessitating further research to integrate them safely and mitigate potential toxicity. Both BEs and microorganisms can contribute to this goal, alongside consideration of their long-term environmental impacts. Moreover, the exact mechanisms by which BEs improve conditions remain unclear, highlighting a research gap in mitigating multiple abiotic stresses.

References

Abd_Allah, E.F., Hashem, A., Alam, P., Ahmad, P., 2019. Silicon alleviates nickel-induced oxidative stress by regulating antioxidant defense and glyoxalase systems in mustard plants. Journal of Plant Growth Regulation 38 (4), 1260–1273. Available at https://doi.org/10.1007/s00344-019-09931-y.

Agathokleous, E., Zhou, B., Geng, C., Xu, J., Saitanis, C.J., Feng, Z., et al., 2022. Mechanisms of cerium-induced stress in plants: a meta-analysis. Science of the Total Environment 852. Available at https://doi.org/10.1016/j.scitotenv.2022.158352.

Aghdam, M.T.B., Mohammadi, H., Ghorbanpour, M., 2016. Effects of nanoparticulate anatase titanium dioxide on physiological and biochemical performance of *Linum usitatissimum* (Linaceae) under well-watered and drought stress conditions. Brazilian Journal of Botany 39 (1), 139–146. Available at https://doi.org/10.1007/s40415-015-0227-x.

Ahanger, M.A., Ahmad, P., 2019. Role of mineral nutrients in abiotic stress tolerance: revisiting the associated signaling mechanisms. Plant Signaling Molecules: Role and Regulation under Stressful Environments. Elsevier, India, pp. 269–285.

Ahsan, M., Valipour, M., Nawaz, F., Raheel, M., Abbas, H.T., Sajid, M., et al., 2023. Evaluation of silicon supplementation for drought stress under water-deficit conditions: an application of sustainable agriculture. Agronomy 13 (2). Available at https://doi.org/10.3390/agronomy13020599.

Aihemaiti, A., Jiang, J., Li, D.'an, Liu, N., Yang, M., Meng, Y., et al., 2018. The interactions of metal concentrations and soil properties on toxic metal accumulation of native plants in vanadium mining area. Journal of Environmental Management 222, 216–226. Available at https://doi.org/10.1016/j.jenvman.2018.05.081.

Akabayov, S.R., Akabayov, B., 2014. Vanadate in structural biology. Inorganica Chimica Acta 420, 16–23. Available at https://doi.org/10.1016/j.ica.2014.02.010.

Akoumianaki-Ioannidou, A., Barouchas, P.E., Ilia, E., Kyramariou, A., Moustakas, N.K., 2016. Effect of vanadium on dry matter and nutrient concentration in sweet basil (Ocimum basilicum L.). Australian Journal of Crop Science 10 (2), 199–206.

AL-Huqail, A.A., Alghanem, S.M.S., Alhaithloul, H.A.S., Abbas, Z.K., AL-Balawi, S.M., Darwish, D.B.E., et al., 2024. Selenium mitigates vanadium toxicity through enhanced nutrition, photosynthesis, and antioxidant defense in rice (Oryza sativa L.) seedlings. BMC Plant Biology 24 (1). Available at https://doi.org/10.1186/s12870-024-05790-2.

Ali, S., Gill, R.A., Mwamba, T.M., Zhang, N., Lv, M.T., ul Hassan, Z., et al., 2018. Differential cobalt-induced effects on plant growth, ultrastructural modifications, and antioxidative response among four Brassica napus (L.) cultivars. International Journal of Environmental Science and Technology 15 (12), 2685–2700. Available at https://doi.org/10.1007/s13762-017-1629-z.

Ali, S., Gill, R.A., Ulhassan, Z., Zhang, N., Hussain, S., Zhang, K., et al., 2023. Exogenously applied melatonin enhanced the tolerance of Brassica napus against cobalt toxicity by modulating antioxidant defense, osmotic adjustment, and expression of stress response genes. China Ecotoxicology and Environmental Safety 252. Available at https://doi.org/10.1016/j.ecoenv.2023.114624.

Alkharpotly, A.-E.A., Soliman, M.A.E., El-Sherpiny, M.A., Baddour, A.G., Ghazi, D.A., Abdelgawad, Z.A., et al., 2023. Evaluation of cobalt application combined with gypsum and compost as a regulator of cabbage plant tolerance to soil salinity. Open Journal of Ecology 13 (12), 914–930. Available at https://doi.org/10.4236/oje.2023.1312056.

Altaf, M.A., Shahid, R., Naz, S., Ahmad, R., Manzoor, M.A., Alsahli, A.A., et al., 2024. Vanadium toxicity was alleviated by supplementation of silicon in tomato seedlings: upregulating antioxidative enzymes and glyoxalase system. Plant Physiology and Biochemistry 213. Available at https://doi.org/10.1016/j.plaphy.2024.108809.

Alvan, H.A., Jabbarzadeh, Z., Fard, J.R., Noruzi, P., 2025. Selenium foliar application alleviates salinity stress in sweet william (Dianthus barbatus L.) by enhancing growth and reducing oxidative damage. Scientific Reports 15 (1). Available at https://doi.org/10.1038/s41598-025-89463-6.

Alves, L.R., Rossatto, D.R., Rossi, M.L., Martinelli, A.P., Gratão, P.L., 2020. Selenium improves photosynthesis and induces ultrastructural changes but does not alleviate cadmium-stress damages in tomato plants. Protoplasma 257 (2), 597–605. Available at https://doi.org/10.1007/s00709-019-01469-w.

Amirabad, S.A., Behtash, F., Vafaee, Y., 2020. Selenium mitigates cadmium toxicity by preventing oxidative stress and enhancing photosynthesis and micronutrient availability on radish (*Raphanus sativus* L.) cv. Cherry Belle. Environmental Science and Pollution Research International 27 (11), 12476–12490. Available at https://doi.org/10.1007/s11356-020-07751-2.

An, J., Hu, P., Li, F., Wu, H., Shen, Y., White, J.C., et al., 2020. Emerging investigator series: molecular mechanisms of plant salinity stress tolerance improvement by seed priming with cerium oxide nanoparticles. Environmental Science: Nano 7 (8), 2214–2228. Available at https://doi.org/10.1039/d0en00387e.

Andrade, F.R., da Silva, G.N., Guimarães, K.C., Barreto, H.B.F., de Souza, K.R.D., Guilherme, L.R.G., et al., 2018. Selenium protects rice plants from water deficit stress. Ecotoxicology and Environmental Safety 164, 562–570. Available at https://doi.org/10.1016/j.ecoenv.2018.08.022.

Aroiee, H., Shekari, L., Mirshekari, A., 2019. Effects of selenium on damage of heavy metals in germination, growth, and antioxidant activities of cucumber (*Cucumis sativus* L.) seedling. Iranian Journal of Seed Science and Research 6 (2), fa269–fa286. Available at https://doi.org/10.22124/jms.2019.3605.

Arzani, A., 2008. Improving salinity tolerance in crop plants: a biotechnological view. In Vitro Cellular & Developmental Biology - Plant 44 (5), 373–383. Available at https://doi.org/10.1007/s11627-008-9157-7.

Ashokan, A., Mini, V., Rani, B., Anand, S., 2024. Mitigation of abiotic stresses in plants through nutrient management. International Journal of Environment and Climate Change 14 (7), 254–267. Available at https://doi.org/10.9734/ijecc/2024/v14i74267.

Ayub, M.A., Ahmad, H.R., Zia ur Rehman, M., Waraich, E.A., 2023. Cerium oxide nanoparticles alleviates stress in wheat grown on Cd contaminated alkaline soil. Chemosphere 338. Available at https://doi.org/10.1016/j.chemosphere.2023.139561.

Babaei, M.J., Ebrahimi, A., Heidari, P., Azadvari, E., Gharanjik, S., Chaghakaboodi, Z., 2025. Titanium dioxide -mediated regulation of enzymatic and non-enzymatic antioxidants, pigments, and diosgenin content promotes cold stress tolerance in Trigonella foenum-graecum L. Scientific Reports 15 (1). Available at https://doi.org/10.1038/s41598-024-84472-3.

Badshah, I., Mustafa, N., Khan, R., Mashwani, Z.U.R., Raja, N.I., Almutairi, M.H., et al., 2023. Biogenic titanium dioxide nanoparticles ameliorate the effect of salinity stress in wheat crop. Agronomy 13 (2). Available at https://doi.org/10.3390/agronomy13020352.

Banerjee, A., Roychoudhury, A., 2021. Chapter 18 - Beneficial aspects of cobalt uptake in plants exposed to abiotic stresses. In: Aftab, T., Hakeem, K.R. (Eds.), Frontiers in Plant-Soil Interaction: Molecular Insights Into Plant Adaptation. Academic Press, pp. 523–529. Available at http://doi.org/10.1016/b978-0-323-90943-3.00012-2.

Barker, A.V., Pilbeam, D.J., 2015. Handbook of Plant Nutrition, second ed. CRC Press, Boca Raton. Available at https://doi.org/10.1201/b18458.

Bistgani, Z.E., Hashemi, M., DaCosta, M., Craker, L., Maggi, F., Morshedloo, M.R., 2019. Effect of salinity stress on the physiological characteristics, phenolic compounds and antioxidant activity of Thymus vulgaris L. and Thymus daenensis Celak. Industrial Crops and Products 135, 311–320. Available at https://doi.org/10.1016/j.indcrop.2019.04.055.

Biswas, A., Pal, S., Paul, S., 2023. Silicon as a powerful element for mitigation of cadmium stress in rice: a review for global food safety. Plant Stress 10, 100237. Available at https://doi.org/10.1016/j.stress.2023.100237.

Blasco, B., Leyva, R., Romero, L., Ruiz, J.M., 2013. Iodine effects on phenolic metabolism in lettuce plants under salt stress. Journal of Agricultural and Food Chemistry 61 (11), 2591–2596. Available at https://doi.org/10.1021/jf303917n.

Bojórquez-Quintal, E., Escalante-Magaña, C., Echevarría-Machado, I., Martínez-Estévez, M., 2017. Aluminum, a friend or foe of higher plants in acid soils. Frontiers in Plant Science 8. Available at https://doi.org/10.3389/fpls.2017.01767.

Boora, R., Rani, N., Kumari, S., Yashveer, S., Kumari, N., Grewal, S., 2023. Efficacious role of silica nanoparticles in improving growth and yield of wheat under drought stress through stress-gene upregulation. Plant Nano Biology 6, 100051. Available at https://doi.org/10.1016/j.plana.2023.100051.

Brengi, S.H., Khedr, A.E.M., Abouelsaad, I.A., 2022. Effect of melatonin or cobalt on growth, yield, and physiological responses of cucumber (*Cucumis sativus* L.) plants under salt stress. Journal of the Saudi Society of Agricultural Sciences 21 (1), 51–60. Available at https://doi.org/10.1016/j.jssas.2021.06.012.

Bressan, A.C.G., de Oliveira Carvalho Bittencourt, B.M., Silva, G.S., Habermann, G., 2021. Could the absence of aluminum (Al) impair the development of an Al-accumulating woody species from Brazilian savanna? Theoretical and Experimental Plant Physiology 33 (3), 281–292. Available at https://doi.org/10.1007/s40626-021-00216-y.

Buchanan, B.B., Gruissem, W., Jones, R.L., 2015. Biochemistry and Molecular Biology of plPnts. John Wiley & Sons.

Castillo-Godina, R.G., Foroughbakhch-Pournavab, R., Benavides-Mendoza, A., 2016. Effect of selenium on elemental concentration and antioxidant enzymatic activity of tomato plants. Journal of Agricultural Science and Technology 18 (1), 233–244.

Chen, J., Zhong, Y., 2024. Lanthanum(III)-amino acid chelate mitigates copper(II) stress in rice (Oryza sativa). Scientific Reports 14 (1). Available at https://doi.org/10.1038/s41598-024-73771-4.

Cheng, M., Wang, A., Liu, Z., Gendall, A.R., Rochfort, S., Tang, C., 2018. Sodium chloride decreases cadmium accumulation and changes the response of metabolites to cadmium stress in the halophyte Carpobrotus rossii. Annals of Botany 122 (3), 373–385. Available at https://doi.org/10.1093/aob/mcy077.

Cheng, M., Wang, L., Zhou, Q., Chao, D., Nagawa, S., He, D., et al., 2021. Lanthanum(III) triggers AtrbohD- and jasmonic acid-dependent systemic endocytosis in plants. Nature Communications 12 (1). Available at https://doi.org/10.1038/s41467-021-24379-z.

Chmielowska-Bąk, J., Lefèvre, I., Lutts, S., Kulik, A., Deckert, J., 2014. Effect of cobalt chloride on soybean seedlings subjected to cadmium stress. Acta Societatis Botanicorum Poloniae 83 (3), 201–207. Available at https://doi.org/10.5586/asbp.2014.027.

Cooke, J., Carey, J.C., 2023. Stress alters the role of silicon in controlling plant water movement. Functional Ecology 37 (12), 2985–2999. Available at https://doi.org/10.1111/1365-2435.14447.

Cui, W., Kamran, M., Song, Q., Zuo, B., Jia, Z., Han, Q., 2019. Lanthanum chloride improves maize grain yield by promoting photosynthetic characteristics, antioxidants enzymes and endogenous hormone at reproductive stages. Journal of Rare Earths 37 (7), 781–790. Available at https://doi.org/10.1016/j.jre.2018.12.006.

Dahle, J.T., Arai, Y., 2015. Environmental geochemistry of cerium: applications and toxicology of cerium oxide nanoparticles. International Journal of Environmental Research and Public Health 12 (2), 1253–1278. Available at https://doi.org/10.3390/ijerph120201253.

Daler, S., Korkmaz, N., Kılıç, T., Hatterman-Valenti, H., Karadağ, A., Kaya, O., 2024. Modulatory effects of selenium nanoparticles against drought stress in some grapevine rootstock/scion combinations. Chemical and Biological Technologies in Agriculture 11 (1). Available at https://doi.org/10.1186/s40538-024-00609-6.

Dhiman, P., Rajora, N., Bhardwaj, S., Sudhakaran, S.S., Kumar, A., Raturi, G., et al., 2021. Fascinating role of silicon to combat salinity stress in plants: an updated overview. Plant Physiology and Biochemistry 162, 110–123. Available at https://doi.org/10.1016/j.plaphy.2021.02.023.

Djanaguiraman, M., Nair, R., Giraldo, J.P., Prasad, P.V.V., 2018. Cerium oxide nanoparticles decrease drought-induced oxidative damage in sorghum leading to higher photosynthesis and grain yield. ACS Omega 3 (10), 14406–14416. Available at https://doi.org/10.1021/acsomega.8b01894.

Dou, Z., Feng, H., Zhang, H., Abdelghany, A.E., Zhang, F., Li, Z., et al., 2023. Silicon application mitigated the adverse effects of salt stress and deficit irrigation on drip-irrigated greenhouse tomato. Agricultural Water Management 289, 108526. Available at https://doi.org/10.1016/j.agwat.2023.108526.

Edelstein, M., Ben-Hur, M., 2018. Heavy metals and metalloids: sources, risks and strategies to reduce their accumulation in horticultural crops. Scientia Horticulturae 234, 431–444. Available at https://doi.org/10.1016/j.scienta.2017.12.039.

Faiyad, R.M.N., Abd El-Azeiz, E.H., 2024. Mitigation the deleterious effect of salinity on faba bean by cobalt and bio-stimulants. Egyptian Journal of Soil Science 64 (1), 181–192. Available at https://doi.org/10.21608/ejss.2023.238554.1667.

Famoso, A.N., Zhao, K., Clark, R.T., Tung, C.W., Wright, M.H., Bustamante, C., et al., 2011. Genetic architecture of aluminum tolerance in rice (oryza sativa) determined through genome-wide association analysis and qtl mapping. PLoS Genetics 7 (8). Available at https://doi.org/10.1371/journal.pgen.1002221.

Farahani, M., Talebi, M., Etesami, H., 2021. Rutin application alleviates drought stress in Cucumis sativus by improving antioxidant enzyme activities and reducing ROS accumulation. Plant Physiology and Biochemistry 167, 111–121.

Farias, I.A.P., Santos, C.C.L., Sampaio, F.C., 2018. Antimicrobial activity of cerium oxide nanoparticles on opportunistic microorganisms: a systematic review. BioMed Research International 2018, 1923606. Available at https://doi.org/10.1155/2018/1923606.

Francini, A., Sebastiani, L., 2019. Abiotic stress effects on performance of horticultural crops. Horticulturae 5 (4), 67. Available at https://doi.org/10.3390/horticulturae5040067.

Fuentes, J.E.G., Castellanos, B.F.H., Martínez, E.N.R., Ortiz, W.A.N., Mendoza, A.B., Macías, J.M., 2022. Outcomes of foliar iodine application on growth, minerals and antioxidants in tomato plants under salt stress. Folia Horticulturae 34 (1), 27–37. Available at https://doi.org/10.2478/fhort-2022-0003.

Furumoto, T., Yamaguchi, T., Ohshima-Ichie, Y., Nakamura, M., Tsuchida-Iwata, Y., Shimamura, M., et al., 2011. A plastidial sodium-dependent pyruvate transporter. Nature 476 (7361), 472–475. Available at https://doi.org/10.1038/nature10250.

Gad, N., Hassan, N.M.K., El Sayed, S.A.A., 2020. Influence of cobalt on tolerating climatic change (salinity) in onion plant with reference to physiological and chemical approach. Plant Archives 20 (1), 1496–1500.

Ganguly, R., Sarkar, A., Dasgupta, D., Acharya, K., Keswani, C., Popova, V., et al., 2022. Unravelling the efficient applications of zinc and selenium for mitigation of abiotic stresses in plants. Agriculture 12 (10), 1551. Available at https://doi.org/10.3390/agriculture12101551.

García-Jiménez, A., Trejo-Téllez, L.I., Guillén-Sánchez, D., Gómez-Merino, F.C., 2018. Vanadium stimulates pepper plant growth and flowering, increases concentrations of amino acids, sugars and chlorophylls, and modifies nutrient concentrations. PLoS One 13 (8). Available at https://doi.org/10.1371/journal.pone.0201908.

Gohari, G., Mohammadi, A., Akbari, A., Panahirad, S., Dadpour, M.R., Fotopoulos, V., et al., 2020. Titanium dioxide nanoparticles (TiO2 NPs) promote growth and ameliorate salinity stress effects on essential oil profile and biochemical attributes of Dracocephalum moldavica. Scientific Reports 10 (1). Available at https://doi.org/10.1038/s41598-020-57794-1.

Gu, H.H., Qiu, H., Tian, T., Zhan, S.S., Deng, T.H.B., Chaney, R.L., et al., 2011. Mitigation effects of silicon rich amendments on heavy metal accumulation in rice (Oryza sativa L.) planted on multi-metal contaminated acidic soil. Chemosphere 83 (9), 1234–1240. Available at https://doi.org/10.1016/j.chemosphere.2011.03.014.

Guerrero, B., Llugany, M., Palacios, O., Valiente, M., 2014. Dual effects of different selenium species on wheat. Plant Physiology and Biochemistry 83, 300–307. Available at https://doi.org/10.1016/j.plaphy.2014.08.009.

Gunawan, C., Lord, M.S., Lovell, E., Wong, R.J., Jung, M.S., Oscar, D., et al., 2019. Oxygen-vacancy engineering of cerium-oxide nanoparticles for antioxidant activity. ACS Omega 4 (5), 9473–9479. Available at https://doi.org/10.1021/acsomega.9b00521.

Gupta, N., Bajpai, M., Majumdar, R., Mishra, P., 2015. Response of iodine on antioxidant levels of Glycine max L. grown under Cd^{2+} stress. Advances in Biological Research 9 (1), 40–48.

Gustafsson, J.P., 2019. Vanadium geochemistry in the biogeosphere –speciation, solid-solution interactions, and ecotoxicity. Applied Geochemistry 102, 1–25. Available at https://doi.org/10.1016/j.apgeochem.2018.12.027.

Han, D., Tu, S., Dai, Z., Huang, W., Jia, W., Xu, Z., et al., 2022. Comparison of selenite and selenate in alleviation of drought stress in Nicotiana tabacum L. Chemosphere 287 (Pt 2), 132136. Available at https://doi.org/10.1016/j.chemosphere.2021.132136.

Hanus-Fajerska, E., Wiszniewska, A., Kamińska, I., 2021. A dual role of vanadium in environmental systems—beneficial and detrimental effects on terrestrial plants and humans. Plants 10 (6), 1110. Available at https://doi.org/10.3390/plants10061110.

Hareem, M., Danish, S., Obaid, S.A., Ansari, M.J., Datta, R., 2024. Mitigation of drought stress in chili plants (Capsicum annuum L.) using mango fruit waste biochar, fulvic acid and cobalt. Scientific Reports 14 (1). Available at https://doi.org/10.1038/s41598-024-65082-5.

He, Z., Li, J., Zhang, H., Ma, M., 2005. Different effects of calcium and lanthanum on the expression of phytochelatin synthase gene and cadmium absorption in Lactuca sativa. Plant Science 168 (2), 309–318. Available at https://doi.org/10.1016/j.plantsci.2004.07.001.

Hinton, R., 2023. Cobalt's role in plant growth and development, Available from: https://omexcanada.com/blog/cobalts-role-in-plant-growth-and-development/.

Hong, C.L., Weng, H.X., Qin, Y.C., Yan, A.L., Xie, L.L., 2008. Transfer of iodine from soil to vegetables by applying exogenous iodine. Agronomy for Sustainable Development 28 (4), 575–583. Available at https://doi.org/10.1051/agro:2008033.

Hosseini, S.A., Maillard, A., Hajirezaei, M.R., Ali, N., Schwarzenberg, A., Jamois, F., et al., 2017. Induction of barley silicon transporter HvLsi1 and HvLsi2, increased silicon concentration in the shoot and regulated starch and ABA homeostasis under osmotic stress and concomitant potassium deficiency. Frontiers in Plant Science 8. Available at https://doi.org/10.3389/fpls.2017.01359.

Hrubý, M., Cígler, P., Kuzel, S., 2002. Contribution to understanding the mechanism of titanium action in plants. Journal of Plant Nutrition 25 (3), 577–598. Available at https://doi.org/10.1081/PLN-120003383.

Hu, Z., Richter, H., Sparovek, G., Schnug, E., 2004. Physiological and biochemical effects of rare earth elements on plants and their agricultural significance: a review. Journal of Plant Nutrition 27 (1), 183–220. Available at https://doi.org/10.1081/pln-120027555.

Hu, X., Wei, X., Ling, J., Chen, J., 2021. Cobalt: an essential micronutrient for plant growth? Frontiers in Plant Science 12. Available at https://doi.org/10.3389/fpls.2021.768523.

Huang, C., Qin, N., Sun, L., Yu, M., Hu, W., Qi, Z., 2018. Selenium Improves physiological parameters and alleviates oxidative stress in strawberry seedlings under low-temperature stress. International Journal of Molecular Sciences 19 (7), 1913. Available at https://doi.org/10.3390/ijms19071913.

Hussain, S., Shuxian, L., Mumtaz, M., Shafiq, I., Iqbal, N., Brestic, M., et al., 2021. Foliar application of silicon improves stem strength under low light stress by regulating lignin biosynthesis genes in soybean (Glycine max (L.) Merr.). Journal of Hazardous Materials 401, 123256. Available at https://doi.org/10.1016/j.jhazmat.2020.123256.

Hussain, S., Ahmed, S., Yasin, N.A., Akram, W., Sardar, R., Ahmad, A., et al., 2023. In vitro and in silico study of salt stress resilience in Brassica rapa through selenium seed priming. South African Journal of Botany 160, 504–515. Available at https://doi.org/10.1016/j.sajb.2023.07.024.

Hussain, S., Ahmed, S., Akram, W., Ahmad, A., Ahmad Yasin, N., Fu, M., et al., 2024. The potential of selenium to induce salt stress tolerance in Brassica rapa: evaluation of biochemical, physiological and molecular phenomenon. Plant Stress 11, 100331. Available at https://doi.org/10.1016/j.stress.2023.100331.

Imtiaz, M., Rizwan, M.S., Xiong, S., Li, H., Ashraf, M., Shahzad, S.M., et al., 2015. Vanadium, recent advancements and research prospects: a review. Environment International 80, 79–88. Available at https://doi.org/10.1016/j.envint.2015.03.018.

Inayat, H., Mehmood, H., Danish, S., Alharbi, S.A., Ansari, M.J., Datta, R., 2024. Impact of cobalt and proline foliar application for alleviation of salinity stress in radish. BMC Plant Biology 24 (1). Available at https://doi.org/10.1186/s12870-024-04998-6.

Inbaraj, B.S., Chen, B.-H., 2020. An overview on recent in vivo biological application of cerium oxide nanoparticles. Asian Journal of Pharmaceutical Sciences 15 (5), 558–575. Available at https://doi.org/10.1016/j.ajps.2019.10.005.

Jaberzadeh, A., Moaveni, P., Tohidi Moghadam, H.R., Zahedi, H., 2013. Influence of bulk and nanoparticles titanium foliar application on some agronomic traits, seed gluten and starch contents of wheat subjected to water deficit stress. Notulae Botanicae Horti Agrobotanici Cluj-Napoca 41 (1), 201–207. Available at https://doi.org/10.15835/nbha4119093.

Johnson, S.N., Chen, Z.H., Rowe, R.C., Tissue, D.T., 2022. Field application of silicon alleviates drought stress and improves water use efficiency in wheat. Frontiers in Plant Science 13. Available at https://doi.org/10.3389/fpls.2022.1030620.

Karami, A., Sepehri, A., 2018. Nano titanium dioxide and nitric oxide alleviate salt-induced changes in seedling growth, physiological, and photosynthesis attributes of barley. Zemdirbyste-Agriculture 105 (2), 123–132. Available at https://doi.org/10.13080/z-a.2018.105.016.

Karvar, M., Azari, A., Rahimi, A., Maddah-Hosseini, S., Ahmadi-Lahijani, M.J., 2022. Titanium dioxide nanoparticles (TiO2-NPs) enhance drought tolerance and grain yield of sweet corn (Zea mays L.) under deficit irrigation regimes. Acta Physiologiae Plantarum 44 (2). Available at https://doi.org/10.1007/s11738-021-03349-4.

Kastori, R., Putnik-Delić, M., Maksimović, I., 2023. Rare earth elements application in agriculture. Acta Agriculturae Serbica 28 (56), 87–95. Available at https://doi.org/10.5937/AASer2356087K.

Kastori, R., Putnik-Delić, M., Maksimović, I., 2024. Lanthanum and cultivated plants. Biologia Serbica 46, 3–11. Available at https://doi.org/10.5281/zenodo.13925785.

Katiyar, P., Singh, A., Bhattacharya, S., Kumar, A., 2023. Synergistic antioxidant potential of rutin and quercetin against oxidative stress in a cellular model. Molecules 28 (4), 1687.

Kaur, S., Kaur, N., Siddique, K.H.M., Nayyar, H., 2015. Beneficial elements for agricultural crops and their functional relevance in defence against stresses. Archives of Agronomy and Soil Science 62 (7), 905–920. Available at https://doi.org/10.1080/03650340.2015.1101070.

Khalid, M., Rehman, H.M., Ahmed, N., Nawaz, S., Saleem, F., Ahmad, S., et al., 2022. Using exogenous melatonin, glutathione, proline, and glycine betaine treatments to combat abiotic stresses in crops. International Journal of Molecular Sciences 23 (21). Available at https://doi.org/10.3390/ijms232112913.

Khaliq, A., Ali, S., Hameed, A., Farooq, M.A., Farid, M., Shakoor, M.B., et al., 2015. Silicon alleviates nickel toxicity in cotton seedlings through enhancing growth, photosynthesis, and suppressing Ni uptake and oxidative stress. Archives of Agronomy and Soil Science 62 (5), 633–647. Available at https://doi.org/10.1080/03650340.2015.1073263.

Khan, W., Aziz, T., Hussain, I., Ramzani, P.M.A., Reichenauer, T.G., 2016. Silicon: a beneficial nutrient for maize crop to enhance photochemical efficiency of photosystem II under salt stress. Archives of Agronomy and Soil Science 63 (5), 599–611. Available at https://doi.org/10.1080/03650340.2016.1233322.

Khan, M.N., Li, Y., Khan, Z., Chen, L., Liu, J., Hu, J., et al., 2021. Nanoceria seed priming enhanced salt tolerance in rapeseed through modulating ROS homeostasis and α-amylase activities. Journal of Nanobiotechnology 19 (1). Available at https://doi.org/10.1186/s12951-021-01026-9.

Khan, M.N., Siddiqui, M.H., Alhussaen, K.M., El-Alosey, A.R., AlOmrani, M.A.M., Kalaji, H.M., 2023. Titanium dioxide nanoparticles require K+ and hydrogen sulfide to regulate nitrogen and carbohydrate metabolism during adaptive response to drought and nickel stress in cucumber. Environmental Pollution 334. Available at https://doi.org/10.1016/j.envpol.2023.122008.

Kiferle, C., Martinelli, M., Salzano, A.M., Gonzali, S., Beltrami, S., Salvadori, P.A., et al., 2021. Evidences for a nutritional role of iodine in plants. Frontiers in Plant Science 12. Available at https://doi.org/10.3389/fpls.2021.616868.

Kiferle, C., Gonzali, S., Beltrami, S., Martinelli, M., Hora, K., Holwerda, H.T., et al., 2022. Improvement in fruit yield and tolerance to salinity of tomato plants fertigated with micronutrient amounts of iodine. Scientific Reports 12 (1). Available at https://doi.org/10.1038/s41598-022-18301-w.

Kim, Y.H., Khan, A.L., Waqas, M., Lee, I.J., 2017. Silicon regulates antioxidant activities of crop plants under abiotic-induced oxidative stress: a review. Frontiers in Plant Science 8. Available at https://doi.org/10.3389/fpls.2017.00510.

Kochhar, S.L., Gujral, S.K., 2020. Abiotic and biotic stress. Plant Physiology: Theory and Applications. Cambridge University Press, pp. 545–589. Available at https://doi.org/10.1017/9781108486392.021.

Kubes, J., Skalicky, M., Tumova, L., Martin, J., Hejnak, V., Martinkova, J., 2019. Vanadium elicitation of Trifolium pratense L. cell culture and possible pathways of produced isoflavones transport across the plasma membrane. Plant Cell Reports 38 (5), 657–671. Available at https://doi.org/10.1007/s00299-019-02397-y.

Kumar, D., Dhankher, O.P., Tripathi, R.D., Seth, C.S., 2023. Titanium dioxide nanoparticles potentially regulate the mechanism(s) for photosynthetic attributes, genotoxicity, antioxidants defense machinery, and phytochelatins synthesis in relation to hexavalent chromium toxicity in Helianthus annuus L. Journal of Hazardous Materials 454. Available at https://doi.org/10.1016/j.jhazmat.2023.131418.

Kumari, V.V., Banerjee, P., Verma, V.C., Sukumaran, S., Chandran, M.A.S., Gopinath, K.A., et al., 2022. Plant nutrition: an effective way to alleviate abiotic stress in agricultural crops. International Journal of Molecular Sciences 23 (15). Available at https://doi.org/10.3390/ijms23158519.

Larue, C., Laurette, J., Herlin-Boime, N., Khodja, H., Fayard, B., Flank, A.M., et al., 2012a. Accumulation, translocation and impact of TiO2 nanoparticles in wheat (Triticum aestivum spp.): influence of diameter and crystal phase. Science of the Total Environment 431, 197–208. Available at https://doi.org/10.1016/j.scitotenv.2012.04.073.

Larue, C., Veronesi, G., Flank, A.M., Surble, S., Herlin-Boime, N., Carrière, M., 2012b. Comparative uptake and impact of TiO_2 nanoparticles in wheat and rapeseed. Journal of Toxicology and Environmental Health, Part A 75 (13-15), 722–734. Available at https://doi.org/10.1080/15287394.2012.689800.

Latef, A.A.H.A., Srivastava, A.K., El-Sadek, M.S.A., Kordrostami, M., Tran, L.-S.P., 2018. Titanium dioxide nanoparticles improve growth and enhance tolerance of broad bean plants under saline soil conditions. Land Degradation & Development 29 (4), 1065–1073. Available at https://doi.org/10.1002/ldr.2780.

Leng, B., Geng, F., Dong, X., Yuan, F., Wang, B., 2019. Sodium is the critical factor leading to the positive halotropism of the halophyte Limonium bicolor. Plant Biosystems - An International Journal Dealing with all Aspects of Plant Biology 153 (4), 544–551. Available at https://doi.org/10.1080/11263504.2018.1508085.

Leyva, R., Sánchez-Rodríguez, E., Ríos, J.J., Rubio-Wilhelmi, M.M., Romero, L., Ruiz, J.M., et al., 2011. Beneficial effects of exogenous iodine in lettuce plants subjected to salinity stress. Plant Science 181 (2), 195–202. Available at https://doi.org/10.1016/j.plantsci.2011.05.007.

Li, X., Fan, Y., Ma, J., Gao, X., Wang, G., Wu, S., et al., 2023. Cerium improves the physiology and medicinal components of Dendrobium nobile Lindl. under copper stress. Journal of Plant Physiology 280, 153896. Available at https://doi.org/10.1016/j.jplph.2022.153896.

Li, C.Z., Wang, D., Wang, G.X., 2004. The protective effects of cobalt on potato seedling leaves during osmotic stress. Botanical Bulletin of Academia Sinica 46, 119–125.

LI, X.X., YU, B., Dong, Y.Y., Wang, L.S., Zhang, S.L., Shangguan, H.Y., et al., 2020. Lanthanum chloride enhances the photosynthetic characteristics and increases konjac glucomannan contents in Amorphophallus sinensis Belval. Photosynthetica 58 (1), 165–173. Available at https://doi.org/10.32615/ps.2019.184.

Lima, J.S., Andrade, O.V.S., Morais, E.G., Machado, G.G.L., Santos, L.C.D., Andrade, E.S., et al., 2023b. KI increases tomato fruit quality and water deficit tolerance by improving antioxidant enzyme activity and amino acid accumulation: a priming effect or relief during stress? Plants 12 (23), 4023. Available at https://doi.org/10.3390/plants12234023.

Liu, D., Wang, X., Chen, X., Lin, Y., Chen, Z., Xu, H., 2012. Effects of lanthanum on the change of calcium level in the root cells of rice. Communications in Soil Science and Plant Analysis 43 (15), 1994–2003. Available at https://doi.org/10.1080/00103624.2012.693231.

Liu, D., Wang, X., Zhang, X., Gao, Z., 2013. Effects of lanthanum on growth and accumulation in roots of rice seedlings. Plant, Soil and Environment 59 (5), 196–200. Available at https://doi.org/10.17221/760/2012-pse.

Liu, P., Yin, L., Wang, S., Zhang, M., Deng, X., Zhang, S., et al., 2015. Enhanced root hydraulic conductance by aquaporin regulation accounts for silicon alleviated salt-induced osmotic stress in Sorghum bicolor L. Environmental and Experimental Botany 111, 42–51. Available at https://doi.org/10.1016/j.envexpbot.2014.10.006.

Liu, Y., Cao, X., Yue, L., Wang, C., Tao, M., Wang, Z., et al., 2022. Foliar-applied cerium oxide nanomaterials improve maize yield under salinity stress: reactive oxygen species homeostasis and rhizobacteria regulation. Environmental Pollution 299, 118900. Available at https://doi.org/10.1016/j.envpol.2022.118900.

Luís Oliveira Cunha, M., de Mello Prado, R., 2023. Synergy of selenium and silicon to mitigate abiotic stresses: a review. Gesunde Pflanzen 75 (5), 1461–1474. Available at https://doi.org/10.1007/s10343-022-00826-9.

Lyu, S., Wei, X., Chen, J., Wang, C., Wang, X., Pan, D., 2017. Titanium as a beneficial element for crop production. Frontiers in Plant Science 8. Available at https://doi.org/10.3389/fpls.2017.00597.

Ma, J., Alshaya, H., Okla, M.K., Alwasel, Y.A., Chen, F., Adrees, M., et al., 2022. Application of cerium dioxide nanoparticles and chromium-resistant bacteria reduced chromium toxicity in sunflower plants. Frontiers in Plant Science 13. Available at https://doi.org/10.3389/fpls.2022.876119.

Ma, J.F., Zhao, F.J., Rengel, Z., Cakmak, I., 2023. Beneficial elements. Marschner's Mineral Nutrition of Plants. Elsevier, Japan, pp. 387–418.

Macías, J.M., Caltzontzit, M.G.L., Martínez, E.N.R., Ortiz, W.A.N., Mendoza, A.B., Lagunes, P.M., 2021. Enhancement to salt stress tolerance in strawberry plants by iodine products application. Agronomy 11 (3), 602. Available at https://doi.org/10.3390/agronomy11030602.

Malagoli, M., Schiavon, M., Dall'Acqua, S., Pilon-Smits, E.A.H., 2015. Effects of selenium biofortification on crop nutritional quality. Frontiers in Plant Science 6, 1–5. Available at https://doi.org/10.3389/fpls.2015.00280.

Mao, H., Wang, J., Wang, Z., Zan, Y., Lyons, G., Zou, C., 2014. Using agronomic biofortification to boost zinc, selenium, and iodine concentrations of food crops grown on the loess plateau in China. Journal of Soil Science and Plant Nutrition 14 (2), 459–470. Available at https://doi.org/10.4067/S0718-95162014005000036.

Maron, L.G., Kirst, M., Mao, C., Milner, M.J., Menossi, M., Kochian, L.V., 2008. Transcriptional profiling of aluminum toxicity and tolerance responses in maize roots. New Phytologist 179 (1), 116–128. Available at https://doi.org/10.1111/j.1469-8137.2008.02440.x.

Marron, A.O., Ratcliffe, S., Wheeler, G.L., Goldstein, R.E., King, N., Not, F., et al., 2016. The evolution of silicon transport in eukaryotes. Molecular Biology and Evolution 33 (12), 3226–3248. Available at https://doi.org/10.1093/molbev/msw209.

Marschner, P., 2012. Marschner's Mineral Nutrition of Higher Plants, third ed. Academic Press. Available at 10.1016/c2009-0-63043-9.

Mathivanan, S., 2021. Abiotic stress-induced molecular and physiological changes and adaptive mechanisms in plants. Abiotic Stress in Plants. IntechOpen. Available at 10.5772/intechopen.93367.

Medrano-Macías, J., Leija-Martínez, P., González-Morales, S., Juárez-Maldonado, A., Benavides-Mendoza, A., 2016. Use of iodine to biofortify and promote growth and stress tolerance in crops. Frontiers in Plant Science 7. Available at https://doi.org/10.3389/fpls.2016.01146 (2016).

Milani, M.H., Mohammadi, A., Panahirad, S., Farhadi, H., Labib, P., Kulak, M., et al., 2024. Cerium oxide nanoparticles (CeO_2 NPs) enhance salt tolerance in spearmint (*Mentha spicata* L.) by boosting the antioxidant system and increasing essential oil composition. Plants 13 (20), 2934. Available at https://doi.org/10.3390/plants13202934.

Mitani, N., 2005. Uptake system of silicon in different plant species. Journal of Experimental Botany 56 (414), 1255–1261. Available at https://doi.org/10.1093/jxb/eri121.

Mohammadi, R., Maali-Amiri, R., Mantri, N.L., 2014. Effect of TiO2 nanoparticles on oxidative damage and antioxidant defense systems in chickpea seedlings during cold stress. Russian Journal of Plant Physiology 61 (6), 768–775. Available at https://doi.org/10.1134/S1021443714050124.

Mohammadi, M.H.Z., Panahirad, S., Navai, A., Bahrami, M.K., Kulak, M., Gohari, G., 2021. Cerium oxide nanoparticles (CeO2-NPs) improve growth parameters and antioxidant defense system in Moldavian Balm (Dracocephalum moldavica L.) under salinity stress. Plant Stress 1. Available at https://doi.org/10.1016/j.stress.2021.100006.

Moradtalab, N., Weinmann, M., Walker, F., Höglinger, B., Ludewig, U., Neumann, G., 2018. Silicon improves chilling tolerance during early growth of maize by effects on micronutrient homeostasis and hormonal balances. Frontiers in Plant Science 9. Available at https://doi.org/10.3389/fpls.2018.00420.

Muhammad, N., Zvobgo, G., Zhang, G.-P., 2019. A review: the beneficial effects and possible mechanisms of aluminum on plant growth in acidic soil. Journal of Integrative Agriculture 18, 1518–1528. Available at https://doi.org/10.1016/S2095-3119(18)61991-4.

Mukherjee, S., Krishnamoorthy, S.B., Subrayan, R., Goswami, A., Mitra, S., 2023. A brief study on the role of cerium oxide nanoparticles in growth and alleviation of mercury-induced stress in Vigna radiata and soil bacteria Bacillus coagulans. Environmental Science and Pollution Research 30 (29), 73952–73963. Available at https://doi.org/10.1007/s11356-023-27496-y.

Mustafa, H., Ilyas, N., Akhtar, N., Raja, N.I., Zainab, T., Shah, T., et al., 2021. Biosynthesis and characterization of titanium dioxide nanoparticles and its effects along with calcium phosphate on physicochemical attributes of wheat under drought stress. Ecotoxicology and Environmental Safety 223, 112519. Available at https://doi.org/10.1016/j.ecoenv.2021.112519.

Nandagopal, J.G.T., Harinarayanan, U.N.D., Raghavan, S., Girija, S., 2022. Foliar selenium application mitigates low-temperature stress in chilli (Capsicum annuum L.) seedlings. Energy Nexus 6. Available at https://doi.org/10.1016/j.nexus.2022.100079.

Nedjimi, B., 2024. Selenium as a powerful trace element for mitigation of plant salt stress: a review. Journal of Trace Elements and Minerals 8, 100123. Available at https://doi.org/10.1016/j.jtemin.2024.100123.

Nhan, P., 2013. Amelioration of aluminum toxicity on OM4900 rice seedlings by sodium silicate. African Journal of Plant Science 7 (6), 208–212. Available at https://doi.org/10.5897/AJPS11.306.

Nie, X., Luo, D., Ma, H., Wang, L., Yang, C., Tian, X., et al., 2024. Different effects of selenium speciation on selenium absorption, selenium transformation and cadmium antagonism in garlic. Food Chemistry 443, 138460. Available at https://doi.org/10.1016/j.foodchem.2024.138460.

Niu, J., Wang, K., Kolattukudy, P.E., 2011. Cerium oxide nanoparticles inhibit oxidative stress and nuclear factor-κB activation in H9c2 cardiomyocytes exposed to cigarette smoke extract. The Journal of Pharmacology and Experimental Therapeutics 338 (1), 53–61. Available at https://doi.org/10.1124/jpet.111.179978.

Oliveira, C., Ramos, S.J., Siqueira, J.O., Faquin, V., de Castro, E.M., Amaral, D.C., et al., 2015. Bioaccumulation and effects of lanthanum on growth and mitotic index in soybean plants. Ecotoxicology and Environmental Safety 122, 136–144. Available at https://doi.org/10.1016/j.ecoenv.2015.07.020.

Oyebamiji, Y.O., Adigun, B.A., Shamsudin, N.A.A., Ikmal, A.M., Salisu, M.A., Malike, F.A., et al., 2024. Recent advancements in mitigating abiotic stresses in crops. Horticulturae 10 (2). Available at https://doi.org/10.3390/horticulturae10020156.

Pandey, S., Tripathi, D.K., Singh, V.P., Sharma, S., Chauhan, D.K., 2023. Role of beneficial elements in epigenetic regulation of plants in response to abiotic stress factors. Beneficial Chemical Elements of Plants: Recent Developments and Future Prospects. Wiley, pp. 22–37. Available at http://doi.org/10.1002/9781119691419.ch2.

Pardo, J.M., Quintero, F.J., 2002. Plants and sodium ions: keeping company with the enemy. Genome Biology 3 (6), 1017.4.

Parvin, S., Van Geel, M., Yeasmin, T., Verbruggen, E., Honnay, O., 2020. Effects of single and multiple species inocula of arbuscular mycorrhizal fungi on the salinity tolerance of a Bangladeshi rice (*Oryza sativa* L.) cultivar. Mycorrhiza 30, 431–444. Available at 10.1007/s00572-020-00957-9.

Pavlovic, J., Kostic, L., Bosnic, P., Kirkby, E.A., Nikolic, M., 2021. Interactions of silicon with essential and beneficial elements in plants. Frontiers in Plant Science 12. Available at https://doi.org/10.3389/fpls.2021.697592.

Pilon-Smits, E.A.H., 2014. Selenium in Plants. In: Lüttge, U., Beyschlag, W. (Eds.), Progress in Botany. Progress in Botany, vol 76. Springer, Cham, pp. 93–107. Available at http://doi.org/10.1007/978-3-319-08807-5_4.

Powell, A.A., Corbineau, F., Franca-Neto, J., Léchappé, J., Mesterhazy, A., Noli, E., et al., 2005. Towards the future in seed production, evaluation and improvement. Seed Science and Technology 33 (2), 265–281. Available at https://doi.org/10.15258/sst.2005.33.2.01.

Preetha, J.S.Y., Sriram, D., Premasudha, P., Pudake, R.N., Arun, M., 2023. Cerium oxide as a nanozyme for plant abiotic stress tolerance: an overview of the mechanisms. Plant Nano Biology 6. Available at https://doi.org/10.1016/j.plana.2023.100049.

Qian, Y., Gallagher, F.J., Feng, H., Wu, M., Zhu, Q., 2014. Vanadium uptake and translocation in dominant plant species on an urban coastal brownfield site. Science of the Total Environment 476-477, 696–704. Available at https://doi.org/10.1016/j.scitotenv.2014.01.049.

Rady, M.M., Belal, H.E.E., Gadallah, F.M., Semida, W.M., 2020. Selenium application in two methods promotes drought tolerance in *Solanum lycopersicum* plant by inducing the antioxidant defense system. Scientia Horticulturae 266, 109290. Available at https://doi.org/10.1016/j.scienta.2020.109290.

Raliya, R., Nair, R., Chavalmane, S., Wang, W.N., Biswas, P., 2015. Mechanistic evaluation of translocation and physiological impact of titanium dioxide and zinc oxide nanoparticles on the tomato (Solanum lycopersicum L.) plant. Metallomics 7 (12), 1584–1594. Available at https://doi.org/10.1039/c5mt00168d.

Ren, Y., Ren, X., Ma, J., Yan, L., 2016. Effects of mixed rare earth fertilizer on yield and nutrient quality of leafy vegetables during different seasons. Journal of Rare Earths 34 (6), 638–643. Available at https://doi.org/10.1016/s1002-0721(16)60073-x.

Rizwan, M., Ali, S., Ibrahim, M., Farid, M., Adrees, M., Bharwana, S.A., et al., 2015. Mechanisms of silicon-mediated alleviation of drought and salt stress in plants: a review. Environmental Science and Pollution Research 22 (20), 15416–15431. Available at https://doi.org/10.1007/s11356-015-5305-x.

Rojek, J., Kozieradzka-Kiszkurno, M., Kapusta, M., Aksmann, A., Jacewicz, D., Drzezdzon, J., et al., 2019. The effect of vanadium(IV) complexes on development of Arabidopsis thaliana subjected to H2O2-induced stress. Functional Plant Biology 46 (10), 942–961. Available at https://doi.org/10.1071/FP18262.

Roppolo, D., De Rybel, B., Tendon, V.D., Pfister, A., Alassimone, J., Vermeer, J.E.M., et al., 2011. A novel protein family mediates Casparian strip formation in the endodermis. Nature 473 (7347), 381–384. Available at https://doi.org/10.1038/nature10070.

Rossi, L., Zhang, W., Ma, X., 2017. Cerium oxide nanoparticles alter the salt stress tolerance of *Brassica napus* L. by modifying the formation of root apoplastic barriers. Environmental Pollution 229, 132–138. Available at https://doi.org/10.1016/j.envpol.2017.05.083.

Roychoudhury, A., 2020. Vanadium uptake and toxicity in plants. SF Journal of Agriculture & Crop Management 1 (2), 1010.

Sachdev, S., Ansari, S.A., Ansari, M.I., Fujita, M., Hasanuzzaman, M., 2021. Abiotic stress and reactive oxygen species: generation, signaling, and defense mechanisms. Antioxidants 10 (2), 1–37. Available at https://doi.org/10.3390/antiox10020277.

Salgado, O.G.G., Teodoro, J.C., Alvarenga, J.P., de Oliveira, C., de Carvalho, T.S., Domiciano, D., et al., 2020. Cerium alleviates drought-induced stress in Phaseolus vulgaris. Journal of Rare Earths 38 (3), 324–331. Available at https://doi.org/10.1016/j.jre.2019.07.014.

Sarkar, R.D., Kalita, M.C., 2022. Se nanoparticles stabilized with Allamanda cathartica L. flower extract inhibited phytopathogens and promoted mustard growth under salt stress. Heliyon 8 (3). Available at https://doi.org/10.1016/j.heliyon.2022.e09076.

Sarraf, M., Janeeshma, E., Arif, N., Qudrat Ullah Farooqi, M., Kumar, V., Ansari, N.A., et al., 2023. Understanding the role of beneficial elements in developing plant stress resilience: signalling and crosstalk with phytohormones and microbes. Plant Stress 10. Available at https://doi.org/10.1016/j.stress.2023.100224.

Sattar, A., Sher, A., Ijaz, M., Ul-Allah, S., Butt, M., Irfan, M., et al., 2020. Interactive effect of biochar and silicon on improving morpho-physiological and biochemical attributes of maize by reducing drought hazards. Journal of Soil Science and Plant Nutrition 20, 1819–1826. Available at https://doi.org/10.1007/s42729-020-00253-7.

Sattar, A., Sher, A., Abourehab, M.A.S., Ijaz, M., Nawaz, M., Ul-Allah, S., et al., 2022. Application of silicon and biochar alleviates the adversities of arsenic stress in maize by triggering the morpho-physiological and antioxidant defense mechanisms. Frontiers in Environmental Science 10. Available at https://doi.org/10.3389/fenvs.2022.979049.

Seal, P., Das, P., Biswas, A., 2018. Versatile potentiality of silicon in mitigation of biotic and abiotic stresses in plants: a review. American Journal of Plant Sciences 9 (7), 1433–1454. Available at https://doi.org/10.4236/ajps.2018.97105.

Shahandashti, S.S.K., Amiri, R.M., Zeinali, H., Ramezanpour, S.S., 2013. Change in membrane fatty acid compositions and cold-induced responses in chickpea. Molecular Biology Reports 40 (2), 893–903. Available at https://doi.org/10.1007/s11033-012-2130-x.

Shaheen, S.M., Alessi, D.S., Tack, F.M.G., Ok, Y.S., Kim, K.H., Gustafsson, J.P., et al., 2019. Redox chemistry of vanadium in soils and sediments: interactions with colloidal materials, mobilization, speciation, and relevant environmental implications - a review. Advances in Colloid and Interface Science 265, 1–13. Available at https://doi.org/10.1016/j.cis.2019.01.002.

Sharma, P., Jha, A.B., Dubey, R.S., 2024. Addressing lanthanum toxicity in plants: sources, uptake, accumulation, and mitigation strategies. Science of The Total Environment 929, 172560. Available at https://doi.org/10.1016/j.scitotenv.2024.172560.

Sheikh, M.A., Hameed, A., Farooq, T., Ibrahim, M., 2018. Evaluation of sodium silicate as antioxidant activator and growth enhancer in wheat. Agrochimica: International Journal of Plant Chemistry, Soil Science and Plant Nutrition of the University of Pisa 62 (2), 113–125. Available at https://doi.org/10.12871/00021857201822.

Sheikhalipour, M., Esmaielpour, B., Gohari, G., Haghighi, M., Jafari, H., Farhadi, H., et al., 2021. Salt stress mitigation via the foliar application of chitosan-functionalized selenium and anatase titanium dioxide nanoparticles in Stevia (Stevia rebaudiana Bertoni). Molecules 26 (13), 4090. Available at https://doi.org/10.3390/molecules26134090.

Singh, J., Lee, B.K., 2016. Influence of nano-TiO2 particles on the bioaccumulation of Cd in soybean plants (Glycine max): a possible mechanism for the removal of Cd from the contaminated soil. Journal of Environmental Management 170, 88–96. Available at https://doi.org/10.1016/j.jenvman.2016.01.015.

Singh, A., Hussain, I., Singh, N.B., Singh, H., 2019. Uptake, translocation and impact of green synthesized nanoceria on growth and antioxidant enzymes activity of Solanum lycopersicum L. Ecotoxicology and Environmental Safety 182, 109410. Available at https://doi.org/10.1016/j.ecoenv.2019.109410.

Singh, D.K., Ranjana, S., Chatterjee, R.S., 2021. Titanium dioxide nanoparticles (nano-TiO2) mitigate heavy metals stress in Indian mustard (Brassica juncea L.) seedlings. Plant Cell Biotechnology and Molecular Biology 22 (71-72), 131–140.

Singhal, R.K., Fahad, S., Kumar, P., Choyal, P., Javed, T., Jinger, D., et al., 2023. Beneficial elements: new Players in improving nutrient use efficiency and abiotic stress tolerance. Plant Growth Regulation 100 (2), 237–265. Available at https://doi.org/10.1007/s10725-022-00843-8.

Soumare, A., Diedhiou, A.G., Thuita, M., Hafidi, M., Ouhdouch, Y., Gopalakrishnan, S., et al., 2020. Exploiting biological nitrogen fixation: a route towards a sustainable agriculture. Plants 9 (8), 1011. Available at https://doi.org/10.3390/plants9081011.

Sousa, G.Fd, Silva, M.A., Carvalho, M.Rd, Morais, E.G.D., Benevenute, P.A.N., Van Opbergen, G.A.Z., et al., 2023. Foliar selenium application to reduce the induced-drought stress effects in coffee seedlings: induced priming or alleviation effect? Plants 12 (17). Available at https://doi.org/10.3390/plants12173026.

Subbarao, G.V., Ito, O., Berry, W.L., Wheeler, R.M., 2003. Sodium - a functional plant nutrient. Critical Reviews in Plant Sciences 22 (5), 391–416. Available at https://doi.org/10.1080/07352680390243495.

Sun, W., Li, X., Padilla, J., Elbana, T.A., Selim, H.M., 2019. The influence of phosphate on the adsorption-desorption kinetics of vanadium in an acidic soil. Journal of Environmental Quality 48 (3), 686–693. Available at https://doi.org/10.2134/jeq2018.08.0316.

Sun, L., Zhang, M., Liu, X., Mao, Q., Shi, C., Kochian, L.V., et al., 2020. Aluminium is essential for root growth and development of tea plants (Camellia sinensis). Journal of Integrative Plant Biology 62 (7), 984–997. Available at https://doi.org/10.1111/jipb.12942.

Sun, C., Zhao, C., Wang, G., Mao, Q., Han, R., 2025. Cerium oxide nanoparticles alleviate enhanced UV-B radiation-induced stress in wheat seedling roots by regulating reactive oxygen species. Phyton-International Journal of Experimental Botany 94 (2), 455–479. Available at https://doi.org/10.32604/phyton.2025.061462.

Taha, R.S., Seleiman, M.F., Shami, A., Alhammad, B.A., Mahdi, A.H.A., 2021. Integrated application of selenium and silicon enhances growth and anatomical structure, antioxidant defense system and yield of wheat grown in salt-stressed soil. Plants 10 (6). Available at https://doi.org/10.3390/plants10061040.

Tao, J., Lu, L., 2022. Advances in genes-encoding transporters for cadmium uptake, translocation, and accumulation in plants. Toxics 10 (8), 411. Available at https://doi.org/10.3390/toxics10080411.

The importance of molybdenum and cobalt in plants, 2024, Available from: https://www.fertiberia.com/en/the-importance-of-molybdenum-and-cobalt-in-plants/.

Tian, L.Y., Yang, J.Y., Huang, J.H., 2015. Uptake and speciation of vanadium in the rhizosphere soils of rape (*Brassica juncea* L.). Environmental Science and Pollution Research International 22 (12), 9215–9223. Available at https://doi.org/10.1007/s11356-014-4031-0.

Tian, Y., Xie, L., Hao, S., Zhou, X., 2024. Application of selenium to reduce heavy metal(loid)s in plants based on meta-analysis. Chemosphere 364, 143150. Available at https://doi.org/10.1016/j.chemosphere.2024.143150.

Tisarum, R., Theerawitaya, C., Samphumphung, T., Takabe, T., Cha-um, S., 2019. Exogenous foliar application of glycine betaine to alleviate water deficit tolerance in two indica rice genotypes under greenhouse conditions. Agronomy 9 (3), 138. Available at https://doi.org/10.3390/agronomy9030138.

Tong, K., Yan, L., Riaz, M., Gao, G., Yu, H., Lu, M., et al., 2024. Mechanisms of Lanthanum-mediated mitigation of salt stress in soybean (Glycine max L. Physiologia Plantarum 176 (4). Available at https://doi.org/10.1111/ppl.14452.

Tourky, S.M.N., Shukry, W.M., Hossain, M.A., Siddiqui, M.H., Pessarakli, M., Elghareeb, E.M., 2023. Cobalt enhanced the drought-stress tolerance of rice (Oryza sativa L.) by mitigating the oxidative damage and enhancing yield attributes. South African Journal of Botany 159, 191–207. Available at https://doi.org/10.1016/j.sajb.2023.05.035.

Tripathi, D.K., Shweta, Singh, S., Singh, S., Pandey, R., Singh, V.P., et al., 2017. An overview on manufactured nanoparticles in plants: uptake, translocation, accumulation and phytotoxicity. Plant Physiology and Biochemistry 110, 2–12. Available at https://doi.org/10.1016/j.plaphy.2016.07.030.

Tripathi, D.K., Vishwakarma, K., Singh, V.P., Prakash, V., Sharma, S., Muneer, S., et al., 2021. Silicon crosstalk with reactive oxygen species, phytohormones and other signaling molecules. Journal of Hazardous Materials 408. Available at https://doi.org/10.1016/j.jhazmat.2020.124820.

Trippe, R.C., Pilon-Smits, E.A.H., 2021. Selenium transport and metabolism in plants: phytoremediation and biofortification implications. Journal of Hazardous Materials 404. Available at https://doi.org/10.1016/j.jhazmat.2020.124178.

von Tucher, S., Schmidhalter, U., 2005. Lanthanum uptake from soil and nutrient solution and its effects on plant growth. Journal of Plant Nutrition and Soil Science 168 (4), 574–580. Available at https://doi.org/10.1002/jpln.200520506.

Ulhassan, Z., Shah, A.M., Khan, A.R., Azhar, W., Hamid, Y., Zhou, W., 2023. Mechanisms of cobalt uptake, transport, and beneficial aspects in plants. In: Pandey, In. S., Tripathi, D.K., Singh, V.P., Sharma, S., Chauhan, D.K. (Eds.), Beneficial Chemical Elements of Plants. Wiley, pp. 169–181. Available at https://doi.org/10.1002/9781119691419.ch7.

Upadhyaya, H., Panda, S.K., 2013. Abiotic stress responses in tea [Camellia sinensis l (o) kuntze]: an overview. Reviews in Agricultural Science 1, 1–10. Available at https://doi.org/10.7831/ras.1.1.

Venturi, S., 2011. Evolutionary significance of iodine. Current Chemical Biology 5 (3), 155–162. Available at https://doi.org/10.2174/2212796811105030155.

Wang, L., Huang, X., Zhou, Q., 2009. Protective effect of rare earth against oxidative stress under ultraviolet-B radiation. Biological Trace Element Research 128 (1), 82–93. Available at https://doi.org/10.1007/s12011-008-8250-4.

Wang, C.R., Wang, Q.Y., Tian, Y., Zhang, J.F., Li, Z.X., Cao, P., et al., 2014. Lanthanum ions intervened in enzymatic production and elimination of reactive oxygen species in leaves of rice seedlings under cadmium stress. Environmental Toxicology and Chemistry 33 (7), 1656–1664. Available at https://doi.org/10.1002/etc.2610.

Wang, W., Zhao, X.Q., Hu, Z.M., Shao, J.F., Che, J., Chen, R.F., et al., 2015. Aluminium alleviates manganese toxicity to rice by decreasing root symplastic Mn uptake and reducing availability to shoots of Mn stored in roots. Annals of Botany 116 (2), 237–246. Available at https://doi.org/10.1093/aob/mcv090.

Wang, Y., Xu, C., Wuriyanghan, H., Lei, Z., Tang, Y., Zhang, H., et al., 2023. Exogenous selenium endows salt-tolerant and salt-sensitive soybeans with salt tolerance through plant-microbial coactions. Agronomy 13 (9), 2271. Available at https://doi.org/10.3390/agronomy13092271.

Wani, S.H., 2023. Editorial: mechanisms of abiotic stress responses and tolerance in plants: physiological, biochemical and molecular interventions, volume II. Frontiers in Plant Science 14. Available at https://doi.org/10.3389/fpls.2023.1272255.

Wani, S.H., Kumar, V., Shriram, V., Sah, S.K., 2016. Phytohormones and their metabolic engineering for abiotic stress tolerance in crop plants. Crop Journal 4 (3), 162–176. Available at https://doi.org/10.1016/j.cj.2016.01.010.

White, P.J., 2018. Selenium metabolism in plants. Biochimica et Biophysica Acta - General Subjects 1862 (11), 2333–2342. Available at https://doi.org/10.1016/j.bbagen.2018.05.006.

Wojcieszek, J., Jiménez-Lamana, J., Bierła, K., Ruzik, L., Asztemborska, M., Jarosz, M., et al., 2019. Uptake, translocation, size characterization and localization of cerium oxide nanoparticles in radish (Raphanus sativus L.). Science of The Total Environment 683, 284–292. Available at https://doi.org/10.1016/j.scitotenv.2019.05.265.

Xu, C., Qu, X., 2014. Cerium oxide nanoparticle: a remarkably versatile rare earth nanomaterial for biological applications. NPG Asia Materials 6 (3), e90. Available at https://doi.org/10.1038/am.2013.88.

Xu, H., Yan, J., Qin, Y., Xu, J., Shohag, M.J.I., Wei, Y., et al., 2020. Effect of different forms of selenium on the physiological response and the cadmium uptake by rice under cadmium stress. International Journal of Environmental Research and Public Health 17 (19), 6991. Available at https://doi.org/10.3390/ijerph17196991.

Yamada, M., Kuroda, C., Fujiyama, H., 2016. Growth promotion by sodium in amaranthaceous plants. Journal of Plant Nutrition 39 (8), 1186–1193. Available at https://doi.org/10.1080/01904167.2015.1069341.

Yang, Y., Liu, Y., Huang, C.F., de Silva, J., Zhao, F.J., 2016. Aluminium alleviates fluoride toxicity in tea (Camellia sinensis). Plant and Soil 402 (1-2), 179–190. Available at https://doi.org/10.1007/s11104-015-2787-8.

Yang, J., Wang, M., Jia, Y., Gou, M., Zeyer, J., 2017. Toxicity of vanadium in soil on soybean at different growth stages. Environmental Pollution 231 (Pt 1), 48–58. Available at https://doi.org/10.1016/j.envpol.2017.07.075.

Yildiztugay, E., Ozfidan-Konakci, C., Kucukoduk, M., Tekis, S.A., 2016. The impact of selenium application on enzymatic and non-enzymatic antioxidant systems in Zea mays roots treated with combined osmotic and heat stress. Archives of Agronomy and Soil Science 63 (2), 261–275. Available at https://doi.org/10.1080/03650340.2016.1201810.

Zhang, L., Shi, W., Wang, X., 2006a. Difference in selenite absorption between high- and low-selenium rice cultivars and its mechanism. Plant and Soil 282, 183–193. Available at https://doi.org/10.1007/s11104-005-5706-6.

Zhang, L., Yang, T., Gao, Y., Liu, Y., Zhang, T., Xu, S., et al., 2006b. Effect of lanthanum ions (La3+) on ferritin-regulated antioxidant process under PEG stress. Biological Trace Element Research 113 (2), 193–208. Available at https://doi.org/10.1385/BTER:113:2:193.

Zhang, C., Li, Q., Zhang, M., Zhang, N., Li, M., 2013. Effects of rare earth elements on growth and metabolism of medicinal plants. Acta Pharmaceutica Sinica B 3 (1), 20–24. Available at https://doi.org/10.1016/j.apsb.2012.12.005.

Zhang, F., Zhao, Y., Zhang, Y., Shi, Y., Hou, L., Khan, A., et al., 2025. Mechanism of exogenous silicon in enhancing cold stress tolerance in Solanum lycopersicum L. seedlings: insights from resistance and quality indicators. Horticulturae 11 (1), 4. Available at https://doi.org/10.3390/horticulturae11010004.

Zhong, Y., Chen, J., 2020. Effects of Lanthanum(III) on Copper(II) Stressed Rice (Oryza sativa) and Its Molecular Mechanism Revealed by Transcriptome Profiling. Authorea. Available at https://doi.org/10.22541/au.158282420.05199470.

Zhu, Y., Guo, J., Feng, R., Jia, J., Han, W., Gong, H., 2016. The regulatory role of silicon on carbohydrate metabolism in Cucumis sativus L. under salt stress. Plant and Soil 406 (1-2), 231–249. Available at https://doi.org/10.1007/s11104-016-2877-2.

Zhu, L., Song, L., Gao, Y., Qian, J., Zhang, X., Li, S., 2018. Effects of lanthanum on the growth and essential oil components of lavender under osmotic stress. Journal of Rare Earths 36 (8), 891–897. Available at https://doi.org/10.1016/j.jre.2018.04.001.

Zwolak, I., 2020. The role of selenium in arsenic and cadmium toxicity: an updated review of scientific literature. Biological Trace Element Research 193 (1), 44–63. Available at https://doi.org/10.1007/s12011-019-01691-w.

Further reading

Lima, J.S., Andrade, O.V.S., Santos, L.C.D., Morais, E.G., Martins, G.S., Mutz, Y.S., et al., 2023a. Soybean plants exposed to low concentrations of potassium iodide have better tolerance to water deficit through the antioxidant enzymatic system and photosynthesis modulation. Plants 12 (13), 2555. Available at https://doi.org/10.3390/plants12132555.

Liu, Y., Tao, J., Cao, J., Zeng, Y., Li, X., Ma, J., et al., 2020. The beneficial effects of aluminum on plant growth in *Camellia japonica*. Journal of Soil Science and Plant Nutrition 20 (4), 1799–1809. Available at https://doi.org/10.1007/s42729-020-00251-9.

Sarma, B., Kashtoh, H., Tamang, T.L., Bhattacharyya, P.N., Mohanta, Y.K., Baek, K., 2023. Abiotic stress in rice: visiting the physiological response and its tolerance mechanisms. Plants 12 (23), 3948. Available at https://doi.org/10.3390/plants12233948.

Zhang, H., Zhao, Y., Zhu, J.K., 2020a. Thriving under stress: how plants balance growth and the stress response. Developmental Cell 55 (5), 529–543. Available at https://doi.org/10.1016/j.devcel.2020.10.012.

Zhang, X., Lou, X., Zhang, H., Ren, W., Tang, M., 2020. Effects of sodium sulfide application on the growth of *Robinia pseudoacacia*, heavy metal immobilization, and soil microbial activity in Pb–Zn polluted soil. Ecotoxicology and Environmental Safety 197, 110563. Available at https://doi.org/10.1016/j.ecoenv.2020.110563.

Zhang, Y., Xu, J., Li, R., Ge, Y., Li, Y., Li, R., 2023. Plants' response to abiotic stress: mechanisms and strategies. International Journal of Molecular Sciences 24 (13), 10915. Available at https://doi.org/10.3390/ijms241310915.

Chapter 21

Soil management strategies for improving multiple abiotic stress tolerance in plants

Nirmalendu Basak[1,2,*], Priyanka Chandra[2], Parul Sundha[2], Puja Gupta Soni[3], Piu Basak[4], Shreyasi Gupta Choudhury[1], and Arvind Kumar Rai[2]

[1]*ICAR-National Bureau of Soil Survey and Land Use Planning, Regional Centre Kolkata, Salt Lake city Kolkata, West Bengal, India,* [2]*ICAR-Central Soil Salinity Research Institute, Karnal, Haryana, India,* [3]*Forage Section, Department of Genetics and Plant Breeding, Chaudhary Charan Singh Haryana Agricultural University, Hisar, Haryana, India,* [4]*Department of Agricultural Chemistry and Soil Science, Bidhan Chandra Krishi Viswavidyalaya, Nadia, West Bengal, India*

[*]*Corresponding author. e-mail address: nirmalendu.basak@icar.org.in*

21.1 Introduction

The global earth is expected to have a population of 9.3 billion by 2050 (FAO, 2024), and the stresses of biotic and abiotic factors are the major limitations for good harvests in the changing climate scenario. Poor water accessibility and salinisation cover nearly up to half of the arable lands, and which will hamper food production by the year 2050 (Wang et al., 2013). Acidity, salinity/sodicity, waterlogging, nutrient toxicity, deficiency, drought/increasing aridity, and increasing emission of anthropogenic greenhouse gases (GHGs) are major abiotic stresses that threaten crop productivity worldwide (Cook et al., 2014; Guo et al., 2014). Global warming is driven by increased mean temperature and raising evaporation demand, the appearance of frequent and severe drought and extreme phenomena, in the future, and consequently, the decrease of regional precipitation and hamper economic activities (Ji et al., 2024). Ten per cent of arable land in arid and semiarid areas is under the threat of drought, and in some cases, desertification causes salinisation, which is rapidly spreading on a global scale with declining average crop yields (Cook et al., 2014). Land degradation is also caused by geogenic salinity, and irrigation induces salinity in arid and semiarid regions (Basak et al., 2022a).

Water stress to the plant roots reduces the transpiration rate in plants and causes low physiological activities of plants, and showing physiological dryness, and manifests low primary productivity. The presence of undesirable electrolytes in the root zone, irregular and variable distribution of rainfall, further accelerates the water stress for the plant. The occurrence of drought and extreme events in agriculture is serious due to the reduction of water resources and the high food demand for the burgeoning world population. Almost 95% of the fresh biomass of the crop plant is composed of water, which plays a vital role in many physiological processes with many features of plant growth, metabolism, and productivity (Chieb and Gachomo, 2023). The severity, duration, and time course of water limitation and low soil moisture cause drought stress, which frequently encourages stomatal closing that limits the diffusion of carbon dioxide into the leaf, or due to nonstomatal restrictions, which decrease C-assimilation, pathogen attack, and heat stress and other processes of photosynthesis (Ahluwalia et al., 2021).

Nearly 1.4 billion hectares of land are salt-affected, which has appreciable consequences on global food–feed–fiber–bioenergy supply (FAO, 2024). In another way, more than 11% of the earth's land is affected by salt stress, and the extent is increasing day by day. Higher salt content slows down the capacity of plants to uptake water and faces water limitation and accumulates salt in the plant parts, which undesirably disturbs growth by damaging metabolic developments and declining photosynthetic efficiency, partly through stomatal closure (Guo et al., 2025). The increasing electrolyte concentration slows down the water uptake by the root because osmotic pressure around the roots is similar to drought stress (Arzani and Ashraf, 2016), and the toxic effect of salt ions (Na^+ and Cl^-) accumulates in the plant tissues of leaves. Soil salinity represses plant growth because of osmotic stress by ion toxicity of [Na^+/Cl^-]. This salt accumulation is a type of hyperosmotic stress experienced by roots in a saline environment and manifested as 'physiological drought and/or slow and low water absorption' by the crop.

Multiple Abiotic Stresses in Plants. DOI: https://doi.org/10.1016/B978-0-443-27424-4.00018-5

Acid soil pH ranges from 4.5 to 6.5, and sometimes it reaches lower than 4.5. Excess of acid-forming cations, Al^{3+}, Fe^{3+}, Mn^{3+}, causes the deficiency of nutrients Ca, Mg, P, and Mo in low pH soil below 5.5. This soil reaction causes poor soil biological activities, low flourishing of microbial population, stunted and stubby root growth, and poor physical properties of soil, limiting crop yield.

Soil salinity imposes specific ion toxicity, osmotic stress, limits water uptake, generates oxidative stress on plants, and finally causes nutritional deficiency (N, Ca, K, P, Fe, Zn) and deleteriously affects the reproductive phase of the crop. Therefore it is necessary to adopt strategic management techniques and tolerant crop and crop cultivars to reduce the stress caused by drought and salinity. Therefore to succeed over adverse and extreme effects, precise and/or considerable/advanced reclamation processes and methodology need to be adopted to alleviate abiotic stress(s): salinity, sodicity, drought, submergence, nutrition deficiency or toxicity, and promotion of sustainable land uses of stress-affected soils.

21.2 Salinity stress

Salinity stress is considered a detrimental environmental stress that severely affects seed germination, plant growth, and crop productivity due to higher concentrations of ions such as Na^+, Cl^-, Ca^{2+}, SO_4^{2-}, and HCO_3^{2-} (Basak et al., 2022b; Khanh et al., 2024). The causal factors of salinity development are the land topography, precipitation and evapotranspiration, irrigation water quality, and higher use of inorganic fertilizers. Such soils exhibit electrical conductivity (EC) of the soil sample >4.0 dS m^{-1} (EC of soil water saturation paste extract), and the exchangeable sodium content is >15%. Salt stress inhibits the growth of the crops, thereby reducing the fresh and dry weight of roots, stems, and leaves. The morphology of the leaves is altered, diminishing the leaf expansion and stomatal activities. Salinity often results in metabolic imbalances due to higher sodium levels in the plants and oxidative stress by lowering the photosynthesis process and disrupting the plant ion homeostasis (Sharma et al., 2012; Ji et al., 2022). This causes the reduced uptake of the other nutrients such as magnesium, phosphorus, potassium, and nitrogen. The biochemical processes of the plants are altered by oxidative damage to biological molecules, affecting phytohormone synthesis (Mukherjee et al., 2022; Sharma et al., 2012). The functioning of the cell organelles, such as chloroplasts and mitochondria, is affected (Hameed et al., 2021). The intracellular osmotic pressure is increased, and the plant spends high energy to balance this pressure. Thereby, the growth of the major cereal crops such as rice and wheat and other sensitive plants is reduced during the stress conditions (Krishnamurthy et al., 2016).

21.3 Management strategies for combating salinity stress

Salt stress is a prominent environmental stress that distresses plant growth and development. The adoption of different strategies like crop diversification, conservation tillage, use of mulch, and deficit saline irrigation helps to minimize the hazardous effects of salinity in soil (Rai et al., 2022; Soni et al., 2023). Also, the plants adopt various mechanisms such as activating the osmotic stress pathways to acclimate to high saline environments (Zhao et al., 2021). Salt stress generates the intracellular osmotic pressure and causes the buildup of Na^+ to toxic levels. Plants like sorghum (*Sorghum bicolor*) can grow in harsh environments and have evolved tolerance to multiple abiotic stresses compared to other crops (Basak et al., 2020; Rai et al., 2022). Different plant species and varieties vary in their degree of salt tolerance, such as barley is more salt-tolerant than rice and wheat (Verma and Yadava, 1986). Plants develop adaptive and resistive mechanisms by modifying physiological processes, metabolic pathways, and networking at the molecular or gene level. Such adaptations help to develop salt-tolerant plant varieties for salt-affected environments. Further, there is a need for a detailed understanding of the mechanisms involved in the counterreaction of plants to salinity stress at various levels.

21.4 Sodicity stress

Sodicity stress is a type of salt-stress that occurs when soil contains high levels of sodium (Na^+), which negatively impacts plant growth and crop yield. Sodicity refers specifically to the amount of Na^+ present in the soil and the irrigation water (Rai et al., 2021b). The high sodium content creates soil particle dispersal and clogging of soil pores, and reducing the water infiltration in soils. Sodicity also alters plant–water relations and affects the photosynthetic pigments and colouration (Sheoran et al., 2021). Irrigation carrying excessive amounts of Na^+ deteriorates the soil structure due to soil sodicity and thus highly affects crop yield (Yan et al., 2024). The ionic and oxidative stress in the plants leads to an imbalance in the sodium: potassium ratio and affects plant growth.

21.5 Sodicity stress management

Need-based gypsum application in soil and water, or both, is considered the most appropriate and effective measure for sodicity reclamation. However, there are several issues associated with the use of gypsum regarding the purity and availability of this mined material. Looking into the urgency of alternate sources for sodic soil reclamation, researchers now focus upon the use of industrial by products like formulated forms of elemental sulfur, flue gas desulfurization (FGD) gypsum, phosphogypsum, marine gypsum, polymers, nanomaterials and organics like, acidified composts, city waste compost (CWC), sewage sludge and biogas slurry have the potential for replacing gypsum as an amendment in future (Rai et al., 2021; Sundha et al., 2025).

Elemental S is effective in amending the calcareous-sodic soil upon oxidation by the microbes (Thiobacillus) under moist conditions. The sodium salts of carbonate and bicarbonate are converted to sodium sulfate by the biochemical action of elemental S and mobilization of calcium carbonate (Rai et al., 2024). The petroleum industry by-product, Reliance-formulated sulfur application in the sodic soils decreases the soil pH by 0.6–1.5 units, as reported in research carried out at ICAR–Central Soil Salinity Research Institute, Karnal (Sharma et al., 2019). Another product, marine gypsum extracted from the coastal areas of Gujarat and Tamil Nadu, is believed to contain 89.7%–92.6% of $CaSO_4$, $2H_2O$ is one of the emerging sources of alkali soil reclamation (Jha et al., 2022). However, the issues of impurities of other salts such as NaCl, $MgCl_2$, and $MgSO_4$, may interfere with its reclamation potential. The FGD gypsum is the by-product produced during the scrubbing of the sulfur from combustion gases in coal-burning-based power-producing plants. The application of the FGD gypsum in the sodic soils results in a lower soil pH, ESP, and an increase in the yield of corn, alfalfa, sunflower, and rice (Wang et al., 2023). Besides the sodicity reclamation potential of FGD gypsum, this material does not contain toxic elements as reported in the literature (Sundha et al., 2023). Conjunctive application of city waste compost (10 t ha^{-1}), farm wastes, and earthworms with an application of waste decomposers in combination with gypsum reduced the soil pH from 9.80 to 8.84 and exchangeable sodium percent (ESP) from 78.3 to 28.0 under rice–wheat cropping system in the sodic soils (Singh et al., 2017). The municipal solid waste compost (MSWC), when applied with a 25% gypsum requirement (25 GR, gypsum requirement) under high SAR irrigation in a saline-sodic soil amendment, showed higher reclamation efficiency and larger supply of Ca^{2+}, and also increased the availability of N and P (Sundha et al., 2018, 2017). The interactive effect of organic (compost) and inorganic amendments application and the quality of the irrigation water influences dynamics and leaching losses of P during sodic soil reclamation (Sundha et al., 2022). Conjoint addition of gypsum and MSWC increases several inorganic (P_{NaHCO3} and P_{H2O}) and organic P concentration than gypsum alone. The rapid acidification of the compost with elemental S (S°) and S°–oxidizers showed successful results in declining the soil pH within 28 days of incubation (Bedwal et al., 2023, 2025). The integrated composting of degradable wastes with inorganic S-rich materials has an extensive possibility for halting alkalinization through scientific interventions and government-aided support. Enriched biogas slurry produced under the Gobardhan scheme of the Government of India is another researchable material for sodicity reclamation and reduced load of synthetic fertilizers (Mukhtiar et al., 2024). The conjoint application of slurry (25% N requirement) and suboptimal doses (75%) of NPK fertilizers can be a sustainable way out for productivity improvement of diversified rice-based cropping systems (Mohanty et al., 2024). Surface and underground drainage is a feasible, consistent, and suitable problem-solving method for slowing down the water table and washing salts, and it is a deliverable option for managing adequate salt balance in the root zone (Neha et al., 2022). Holed corrugated PVC pipe covered with a sieve mesh constantly customized in a suitable method below the need rooting depth to lower down the saline water table and percolate extra salts and water by gravitational stroke or hydraulic pump (Kamra et al., 2019). Several countries such as the United States of America, Pakistan (Imran et al., 2021), Egypt, and Gulf countries of middle Asia, installed subsurface drainage technology (SSD) to rehabilitate a considerable area of waterlogged saline soil. SSD increases threefold in farmers' profit on crop yields. Soil salinity and sodicity deleteriously affect soil health and crop growth and yield in salt-affected ecologies across the globe. There are different approaches to address the soil salinity and sodicity stress. The broad-scale pilot studies are desirable to evaluate the efficacy and effectiveness of emergent sources that will help cope with the adverse effects of salts present in the soil layers.

21.6 Acidity stress

Soil having a pH less than 5.5 and base saturation (BS) less than 50% is categorized as an acid soil, and amending these categories of soil with lime and liming materials to keep soil pH and BS greater than 6.5% and 80%. The presence of acidic parent materials, leaching of exchangeable bases, accumulation of Fe and Al-hydroxy compounds, hot and humid climatic conditions, heavy precipitation, laterization, podzolization, marshy landform, long-term use of acid-generating fertilizers cause acidification. Acidity is not a favorable soil environment for the growth and development of bacteria, earthworms, and other soil organisms, and it loses the survival environment of useful rhizobacteria, and harmful effect on N–fixers for legumes, and poor microbial activity causes low S availability (Panda et al., 2012).

The nature of acidity quite differs in diverse acidic nature soil, viz., formation of free acids or acidic substances in the humid forests and in the foot hill regions, presence of Fe and Al–hydroxy compounds in Red and lateritic, intensive cultivation in Alluvial zone and indiscriminate use of acid-forming fertilizers, losses of exchangeable bases in Hill, oxidation of pyrites or Fe^{2+} in saline-acid (sulfate), pond, problem differ. Acid soil has the problem of a Ca deficiency or Al toxicity, or low P availability because of high P–P-fixing capacity. Acid soil of different agroecological zones has a deficiency of multinutrients like B, Mo, Zn and Si in Alluvial soil; B, Mo, Zn and Cu deficiency in Red and Lateritic soil; B and Mo deficiency in Hilly soil; additionally, toxicity of Fe normally found in Alluvial and Red and Lateritic; and acid soil of Coastal region is deficient in B. Soil acidity causes the aggravation of soil borne diseases through acid-loving pathogens, e.g., Club root of mustard.

Use of amendments (e.g., lime), low-cost (dolomite, basic slag, press mud, and coral shells) liming materials, cost-intensive, efficient amendments (calcite, slaked lime), agronomic practices, growing of acid-tolerant crops, and follow of proper crop rotation. Results from On-farm and farmers' fields of Jharkhand and Odisha (Eastern Plateau and Hill Region) states of India reveal the imperative need to use organic manures, biofertilizers, and lime in acid soils ($pH < 5.5$) in conjunction with chemical fertilizers for better soil health and higher crop production as against the use of chemical fertilizers alone (Sarkar et al., 2020).

21.7 Drought stress

Drought is among the most impactful abiotic stresses because of low stream flow (hydrologic) and reduced daily ratio of water uptake; or by declines in precipitation, more demand for evapotranspiration, or a combined effect of the two, impairs plant growth, leaf expansion, and productivity on a global scale. The increasing frequency and intensity of drought events, driven by climate change and unpredictable weather patterns, pose a growing threat to food security and agricultural sustainability. Water makes up 80%–95% of a plant's fresh weight; it is vital for many processes such growth, development, and metabolism. Drought stress happens when plants don't get enough water, especially in arid or semiarid areas, because of higher temperatures and faster water loss from the soil and leaves. Drought stress is one of the most severe environmental factors worldwide, significantly reducing photosynthesis rates and crop yields (Alharby and Fahad, 2020; Yahaya and Shimelis, 2022). Drought profoundly impacts soil quality, particularly in agroecosystems, by altering soil structure, depleting nutrient availability, and accelerating soil erosion. Water stress is the leading abiotic factor that hinders plant growth, reduces yield components, and limits overall crop development (Aslam et al., 2020). Water stress is the primary challenge facing dry and semidry regions worldwide. Future drought conditions are expected to worsen due to a projected 1.5°C rise in global average temperature and ongoing climate variations, leading to significant changes in agricultural productivity and farming systems. Although genetic and physiological strategies for improving drought resistance have received significant focus, soil management practices are equally vital for mitigating the effects of water stress on plants. Key soil properties, including texture, structure, organic matter content, and water-holding capacity, play crucial roles in shaping plant responses to drought stress. The drought stress sometimes conjugates with heat stress, which damages reproductive organs or hampers anthesis and sink limitation. This chapter explores soil management strategies designed to enhance plant tolerance to drought conditions. These strategies encompass practices that boost soil water retention, encourage root development, enhance soil fertility, and strengthen soil resilience to environmental stresses. Understanding the intricate relationship between soil health and plant drought resistance is essential for devising effective soil management approaches. Here, the strategies for halting the adverse effects of abiotic stresses on crop production are described (Table 21.1).

21.8 Drought management

Mulching is a common practice for conserving soil moisture, minimizing evaporation, and regulating the temperature of soil. Straw, wood chips, and grass clippings are decomposable organic mulches, and inorganic mulches create a stable microenvironment at the soil surface, shielding it from heat and wind that increase water loss. Additionally, mulches help prevent soil erosion, suppress weeds that compete for water, and boost soil microbial activity, collectively supporting better plant health during drought conditions. Mulching significantly lowers soil surface temperatures and reduces water evaporation, thereby enhancing water availability for plants. Mulching is an effective method for conserving soil moisture, promoting plant growth, boosting agricultural yields, and controlling pests (Al-Bayati and Hamdoon, 2019). It is also a widely used system globally, particularly in arid regions, to protect plant roots from environmental stresses such as drought and heat (Kazemi and Safari, 2018). Using mulch can reduce evapotranspiration losses from soil by 40%–50% compared to unmulched soil (Chai et al., 2016). In dryland agriculture, mulching offers benefits like moisture

TABLE 21.1 Strategies for halting the adverse effects of abiotic stresses on crop production in the environment.

Types and stress management options	Soil property improvement, initial (I) and final (F) soil	Inference
Sodicity stress-integrated use of city waste compost @ 10 Mg ha^{-1} with mineral gypsum 25GR	Heavy texture saline–alkali soil $pH_{1:2}$ 10.7; $EC_{1:2}$ 12.2 dS m^{-1}; ESP 70%; gypsum requirement (100 gypsum requirement, GR) 22.7 Mg ha^{-1} (initial); GR25 gypsum + city waste compost declined soil $pH_{1:2}$ 8.8; EC_e 1.0 dS m^{-1} (after reclamation)	Productive utilization of city waste and reducing dependency on mineral gypsum (Sundha et al., 2020)
Sodicity stress–Flue gas desulfurization (FGD) gypsum-generated from coal-fired plant	Soil pH1:2 9.4; EC1:2 0.86 dS m^{-1}; pHs 8.6; ECe 2.0 dS m^{-1}; SAR_e 40.0; total alkalinity: 4.6 me L^{-1}; CaCO3 1.3% (initial). Soil $pH_{1:2}$ 8.6; $EC_{1:2}$ 1.95 dS m^{-1}; pH_s 8.0; EC_e 5.0 dS m^{-1}; SAR_e 5.0; total alkalinity: 2.6 me L^{-1}; $CaCO_3$ 1.0% (after reclamation).	FGD gypsum is 1.3 times less bulky than mineral gypsum and 2%–11% more efficient, and requires 30% less labor and energy (Sundha et al., 2025, 2023)
Sodicity stress-elemental S incubated with Sphingomonas olei	Soil $pH_{1:2}$ 9.4; EC1:2 0.86 dS m^{-1}; pH_s 8.6; ECe 2.0 dS m^{-1}; SAR_e 40.0; total alkalinity: 4.6 me L^{-1}; $CaCO_3$ 1.3% (initial). Soil $pH_{1:2}$8.6; $EC_{1:2}$ 1.95 dS m^{-1}; pH_s 8.0; EC_e 5.0 dS m^{-1}; SAR_e 5.0; total alkalinity: 2.6 me L^{-1}; $CaCO_3$ 1.0% (after reclamation).	The integration of elemental S and *Sphingomonas olei* increased rice and wheat yield by up to 21.7%–25.4% over the unamended control. Benefit cost ratio is 1.22 to 1.70 with a payback period of 2–3 years (Rai et al., 2024)
Salinity and saline water irrigation-deficit irrigation, mulch, and salt-tolerant cultivars	Soil pHs 8.2; EC_e 16.2 dS m^{-1} (initial) Soil pHs 7.9; EC_e 3.6 dS m^{-1} (after 3 years)	Saline irrigation (60% of crop water requirement of wheat) in the wheat season, reduced tillage and mulch (5 Mg ha^{-1}) increased grain and biomass production, and reduced soil EC, and increased the cropping intensity, reducing dependence on fresh irrigation (Soni et al., 2021)
Acidity stress– basic slag application @ 20% of liming	Soil $pH_{1:2.5}$ (water) 5.2; $pH_{1:2.5}$ (Ca) 4.5; exchange acidity 0.31 $cmol_{(P+)}$ kg^{-1}; $KMnO_4$–N 146.0 kg ha^{-1}; Bray–P 10.3 kg ha^{-1}; NH_4OAc–K 154.4 kg ha^{-1}; DTPA–Zn 1.82 mg kg^{-1} unamended control. Soil pH $_{1:2.5}$ (water) 5.9; pH $_{1:2.5\ (Ca)}$ 5.5; exchange acidity 0.10 $cmol_{(P+)}kg^{-1}$; $KMnO_4$–N 262.6.0 kg ha^{-1}; Bray–P 26.4 kg ha^{-1}; NH_4OAc–K 268.3 k g ha^{-1}; DTPA–Zn 3.1 mg kg^{-1} unamended control (after reclamation).	A greater benefit: cost (B: C) ratio of 4.8 times obtained by adopting basic slag than calcite (1.44) for mustard cultivation. Steel industry by-product basic slag is a cost-effective material that can be used in acidic, Red and lateritic soils for achieving economic productivity (Bhat et al., 2010)
Drought stress-irrigation drought-tolerant cultivars	Under alternate wetting and drying, drought-resistance rice cultivars presented a mean 23% increase in grain yield and an 21% increase in water use efficiency compared to drought-susceptible rice cultivars (Chu et al., 2018)	

conservation, temperature regulation, weed control, improved soil health, and erosion prevention, thereby enhancing water use efficiency and increasing crop yields (Kishore et al., 2022). Plastic mulching has proven to be a promising approach for enhancing water use efficiency, improving drought tolerance, and ultimately boosting crop yields (Lee et al., 2021; Wang et al., 2020).

21.9 Submergence stress

Global agriculture is badly and frequently affected by the occurrence of extreme weather events. Food security is seriously threatened by occurrences like droughts, submergence, and floods that alter the availability of water (FAO, 2017).

Waterbody overflow causes agricultural floods, which are mostly caused by heavy or prolonged rains. The loss of cattle, seed stocks, infrastructure, tools, and agricultural output is only a few of the extensive economic and social effects of these floods. They also enhance vulnerability, worsen food shortages, and spread diseases, which frequently result in a shortage of commodities for everyday economic crises, poverty, and migration (Reed et al., 2022).

Water is a key component, and it plays an interactive bridge between the plant and its environment (Brendel, 2021). Globally, plant distribution and species richness are greatly impacted by variables such as potential evapotranspiration and the number of rainy days in a year, and events are the outcome of the quantity and timing of rainfall (Gutiérrez-Hernández and García, 2021). In addition to rainfall, the depth of groundwater delineates the line separating oxygenated soil from water-saturated aquifers, which is also crucial. It prevents drainage, creates wetland habitats for plants suited to moist soils, and supports aquatic ecosystems during dry spells. Rainfall and groundwater table depth work together to influence soil hydrology, which in turn influences plant physiological adaptations like rooting depth to deal with fluctuating water supply (Bai et al., 2024).

Worldwide occurrence of floods causes roughly two-thirds of the billions of dollars in crop losses and damage between 2006 and 2016 (FAO, 2017). Agricultural recovery following flooding frequently necessitates extensive remediation techniques, including microbial activity revival, soil property restoration, sediment removal, and, if required, replanting. Flooding can be divided into two categories according to the height of the water column: submergence, in which water fully engulfs aerial plant tissues, and waterlogging, in which water covers the roots just superficially. Both kinds impede the flow of oxygen from the atmosphere to plant tissues, resulting in hypoxia (less than 21% oxygen). Predictive models for estimating crop loss from floods are among the agronomic solutions to combat submergence and waterlogging (Fukao et al., 2019). These models support risk management, economic insurance, and a well-informed decision-making system. These systems predict agricultural areas affected, damage, and flooding frequency based on correlations with climate indices. They are vital resources for organizing, controlling, and lessening the effects of floods on agriculture (Ahmed et al., 2013). Flood-tolerant cultivar, breeding tools provide "genetic insurance" to guarantee steady yields in the face of flooding and many stressors. Plants need to trigger precisely controlled changes in gene expression, including epigenetic, transcriptional, and translational processes, in order to sustain and recover (Fukao et al., 2019). Understanding flooding stress, including submergence, waterlogging, hypoxia, and anoxia, has been prioritized significantly in recent years, especially in model plants such as rice and Arabidopsis (Pan et al., 2021). Extensive research on molecular components is essential to identify the flood tolerance mechanism. Group VII ethylene response factors (ERF-VIIs), for example, are important transcription factors that control genes that respond to hypoxia (Manghwar et al., 2024). Flooding tolerance in Arabidopsis is mostly dependent on ERF-VII members, including HRE1, HRE2, RAP2.2, RAP2.3, and RAP2.12. Groups that are constitutively expressed, such as RAP2.2, RAP2.3, and RAP2.12, function as strong activators. Rice can survive for 14–16 days under total submersion due to the ERF-VII gene SUB1A, which is a master regulator of submergence tolerance. Furthermore, ERF-VII genes, SNORKEL1 and SNORKEL2, encourage internode elongation in flood environments, enabling plants to outgrow floodwaters in deep-water rice (Yuan et al., 2023).

21.10 Cold stress

Crop species' productivity, survivability, and ecological distribution can all be negatively impacted by low and nonfreezing temperatures, which can even cause serious harm or death. As plants are sessile, they develop efficient survival mechanisms to cope with adverse environmental situations. Cold acclimation is the process by which plants develop a tolerance to low temperatures (Sanghera et al., 2011). However, different species and populations experience different temperatures that affect plant growth. For instance, many tropical and subtropical plants can suffer cold injury at temperatures below 15°C, whereas Antarctic algae show heat stress symptoms at temperatures over 5°C (Abhinandan and Skori, 2018).

While freezing tolerance indicates that a plant can survive at temperatures below freezing (<0°C), cold tolerance refers to a plant's capacity to tolerate seasonal low, nonfreezing temperatures (0°C–15°C) (Ritonga and Chen, 2020). Despite these differences, previous studies have frequently examined both phenomena simultaneously (Hussain et al., 2018; Adhikari et al., 2022). Plants that can withstand cold temperatures acclimate during the fall and deacclimate during the late winter or early spring when temperatures rise. Freezing-tolerant plants acclimate to the effects of chilling when frost recurs in late winter or early spring, while sensitive plants do not (Aslam et al., 2022; Juurakko et al., 2021). However, the genetic and genomic basis of deacclimation and acclimation is still poorly understood; the latter is receiving more attention as a result of the growing unpredictability of global climates (Hussain et al., 2018; Ritonga and Chen, 2020).

To increase crops' resilience to cold stress, researchers have suggested several agronomic techniques on the basis of their experiments (Juurakko et al., 2021). Perennial crops can become more cold-tolerant through agronomic techniques

(Jungers et al., 2023; Liu et al., 2023). Managing fertilizer is essential to reducing cold stress and increasing production. Adequate K levels are critical for cold stress protection (Wang et al., 2013). Optimum K controls stomatal function, enhances photosynthesis, lowers respiration and energy loss, keeps turgor pressure high, and synthesizes proteins, sugars, enzymes, and antioxidants. Additionally, it improves the movement of nutrients, water, and carbohydrates and lowers the generation of reactive oxygen species. Similarly, phosphorus (P) is essential for numerous biological functions, such as the creation of ATP and nucleic acids. Application of K and P helps store enough root and crown carbohydrates, supporting physiological and biochemical processes related to cold-tolerance (Ambroise et al., 2020; Hassan et al., 2021). Water management also enhances cold tolerance, as during winter, high soil moisture causes more freezing and thawing, which causes heaving, which harms roots and crown buds, exposes plants to the cold, and interferes with the uptake of nutrients and moisture, which can be minimized by the application of mixed crops and proper drainage systems (Luo et al., 2024). Appropriate cutting techniques in forage crops increase winter survival rates, restore root glycogen stores, and boost photosynthetic efficiency (Adhikari et al., 2022).

The physiology of plants is altered by climate fluctuation and temperature instability, leading to the development of tolerant cultivars essential for ensuring global food security (Dhankher and Foyer, 2018; Lahlali et al., 2024). Crop development is being revolutionized by smart breeding techniques that combine genome-editing technologies like CRISPR-Cas9 with high-throughput genomics and phenomics (Ahmad, 2023; Zhang et al., 2025). The emphasis on the potential of genomic data to find agronomic genes for breeding crops that are climate-adaptable is necessary through CRISPR-Cas9 editing of genes such as OsPIN5b (panicle length), GS3 (grain size), and OsMYB30 (cold tolerance). Rice production and cold tolerance were increased. Investigating secondary and tertiary gene pools of wild cultivars of important crops may also provide new genes for the development of climate-resilient cultivars (Shaheen et al., 2023).

21.11 Microbial interactions and abiotic stress management

Crop exposures to abiotic stress are the most important limiting factors for agricultural production worldwide (Rai et al., 2021a). These circumstances include flooding, high temperatures, unbalanced soil pH, nutrient deficits, exposure to hazardous substances such as organic pollutants and toxic metals, and limited water availability brought on by drought or salinity (Rai et al., 2022). These pressures jeopardize the world's food security by affecting plant metabolic pathways and equilibrium. Nonetheless, environments impacted by stress encourage the emergence of an association between plant growth-promoting (PGP) microbes and plants in their rhizosphere, which mitigate these unfavorable circumstances. Abiotic stressors induce large amounts of reactive oxygen species (ROS), including superoxide radicals (O_2^-), hydroxyl radicals (OH^-), hydrogen peroxide (H_2O_2), and singlet oxygen ($^1O_2$) in a plant's cell, which causes oxidative damage and cell dysfunction by interacting with lipids, proteins, and DNA in cells (Chandra et al., 2022a). The rhizosphere's carbon absorption and respiration are also altered by drought-induced moisture stress, which lowers ATP synthesis and increases ROS generation, all of which impair cellular metabolism (Chandra et al., 2022b; Chandra et al., 2021b). Beneficial microbes boost plant resilience by producing osmoprotectants and antioxidants to counteract these impacts (Chandra et al., 2024b). Drought and salinity stressors substantially lower crop productivity in arable land areas of dry and semiarid climates. Salinity affects crops by reducing nodulation and N-fixation in legumes, as well as decreasing germination and vigor. Salt-stress conditions due to high levels of sodium (Na^+) and chloride (Cl^-) in soil obstruct ion uptake, nutrient transport, and water absorption. Furthermore, saline environments hinder the activity of microbial enzymes such as phosphatase, urease, and nitrogenase, which are important components of biogeochemical processes and are essential to plant health. Interactions between microbes and plants, especially in the rhizosphere, through several mechanisms improve plant nutrition and stress-tolerance (Rai et al., 2024). Microbes possess various characteristics that reduce the adverse effects of abiotic stressors through modulating plant biochemical, physiological, and molecular processes. Various mechanisms through which soil microbes improve plant nutrition and stress tolerance include enhancing nutrient availability to plants, and secreting phytohormones, extracellular enzymes, and volatile compounds. PGP bacteria (PGPB) also produce enzymes such as ACC deaminase, activate antioxidant defense mechanisms, and control ROS levels. These processes initiate physiological and biochemical changes in plants, which lead to improved plant stress tolerance. Azospirillum, Azotobacter, Bacillus, Bradyrhizobium, Burkholderia, Enterobacter, Pseudomonas, Rhizobium, Methylobacterium, and Variovorax are a few examples of beneficial PGPB (Fig. 21.1). These microorganisms reduce stress-induced damage, aid in the absorption of nutrients, and release phytohormones (Chandra et al., 2024a; Chandra et al., 2021a).

Utilizing plant–microbe interactions is a crucial strategy for attaining sustainable agriculture and improving global food supply, especially in light of the difficulties presented by decreasing rainfall and rising salt (Besset-Manzoni et al., 2018). Chemical signals, including hormones, volatile organic chemicals, and signals based on phenolic compounds, sugars,

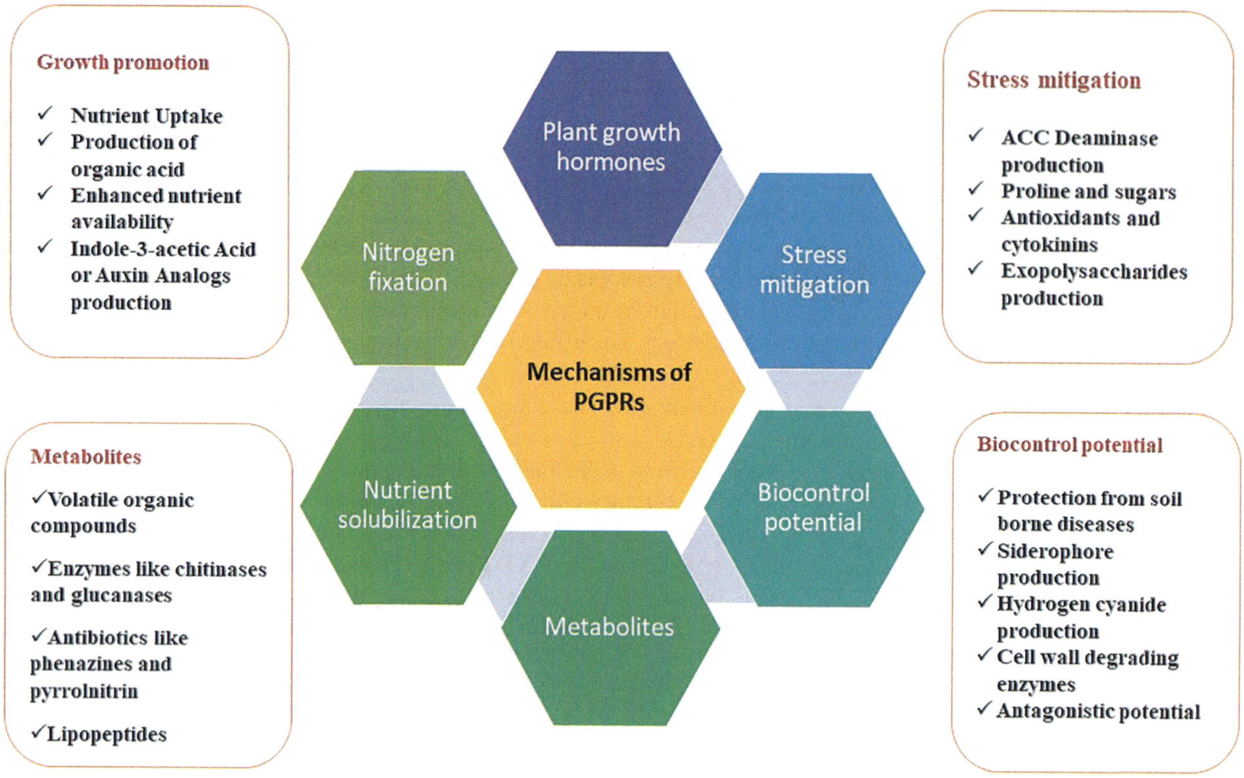

FIGURE 21.1 Different mechanisms of PGPR for performing several biological actions. *PGPR*, plant growth promoting rhizobacteria.

aliphatic and aromatic compounds, are used in plant–microbe communication. Two important signaling molecules that control these relationships are pathogen-associated molecular patterns (PAMPs) and microbe-associated molecular patterns (MAMPs). Crop resilience and productivity under abiotic stress conditions can be minimized by comprehending these pathways and utilizing plant–microbe interactions (Choi and Klessig, 2016).

Interactions between microbes and plants have been dynamic processes, and these relationships can be advantageous or detrimental, which affects the growth and development of plants. Beneficial partnerships can improve plant growth directly (by producing phytohormones and reducing stress) or indirectly (by enhancing nutrient availability) (Chandra et al., 2021). On the other hand, detrimental interactions with pathogens harm plant health. Three categories of beneficial bacterial communities to plants are as follows:

21.12 Phyllospheric bacteria

These bacteria are present on the surfaces of aboveground plants and aid plants through phytoremediation and nitrogen fixation. *Bacillus*, *Methylobacterium*, *Achromobacter*, *Acinetobacter*, and *Pseudomonas* are examples of phyllospheric microbes that have evolved to withstand strong abiotic stressors, such as UV light and high temperatures (35°C–50°C) (Devi et al., 2022).

21.12.1 Endophytic bacteria

Internal plant tissues, such as roots, stems, blooms, fruits, and seeds, are colonized by these bacteria. Among the notable genera are *Klebsiella*, *Herbaspirillum*, *Burkholderia*, *Pantoea*, and *Azoarcus* (Morales-Cedeño et al., 2021).

21.12.2 Rhizospheric bacteria

Plant roots impact the rhizosphere, a section of soil that is home to a variety of microorganisms that aid in nutrient uptake and guard against harmful infections. Beneficial microorganisms such as *Methylobacterium*, *Pseudomonas*, *Serratia*,

Rhizobium, *Paenibacillus*, and *Azospirillum* are found in the rhizosphere microbiome. Rhizodeposits, which include dead cells, mucigel, and root exudates, are what these microorganisms feed on since they promote microbial activity and the synthesis of enzymes (Chandra et al., 2021a).

21.13 Role of key elements in managing abiotic stresses

The role of nutrients in abiotic stress in plants is less investigated. Some available reports showed that different elements exhibit positive as well as negative effects in counteracting abiotic stress, and these are described in the flow diagram (Fig. 21.2) and in the following sections.

21.13.1 Nitrogen

Nitrogen is an important plant nutrient that is necessary for resilience to abiotic stresses, such as soil salinity and drought (Farhan et al., 2024). The amount of stress relief was shown to be controlled by the type of nitrogen supplied, either ammonium (NH_4^+) or nitrate (NO_3^-). Plants fertilized with NO_3^- showed a stronger tolerance to heat and water stress than those fertilized with NH_4^+(Li et al., 2020). Despite these factors in plants, the nodule activity had been severely affected by water stress, the rapid recovery of nitrate reductase activity suggests that the nitrate assimilation process is less sensitive to drought/rehydration cycle when cowpea plants are nodulated in the presence of moderate NO_3^- level. N-supply greatly increased the activities of superoxide dismutase (SOD), lipid peroxidation by triggering proline accumulation, and enzymatic antioxidant activities with respect to peroxidase (POD), catalase (CAT), polyphenol oxidase (PPO), glutamine synthetase (GS), and glutamate synthase (GOGAT) when plantlets were exposed to water stress. These results suggested that N-supply could alleviate the adverse effects of drought stress on plants by enhancing antioxidant defense and improving N-assimilation, while the effects on plant tolerance to drought stress varied with NH_4^+/NO_3^- forms (Abid et al., 2016). Moreover, unused solar energy can occasionally increase photooxidative damage and lipid peroxidation in high-temperature conditions when there is an N deficiency in the plant (Kato et al., 2013). Many crops also exhibit curled small leaves and grains with low protein content (Ashraf et al., 2011). Furthermore, Foliar application of N

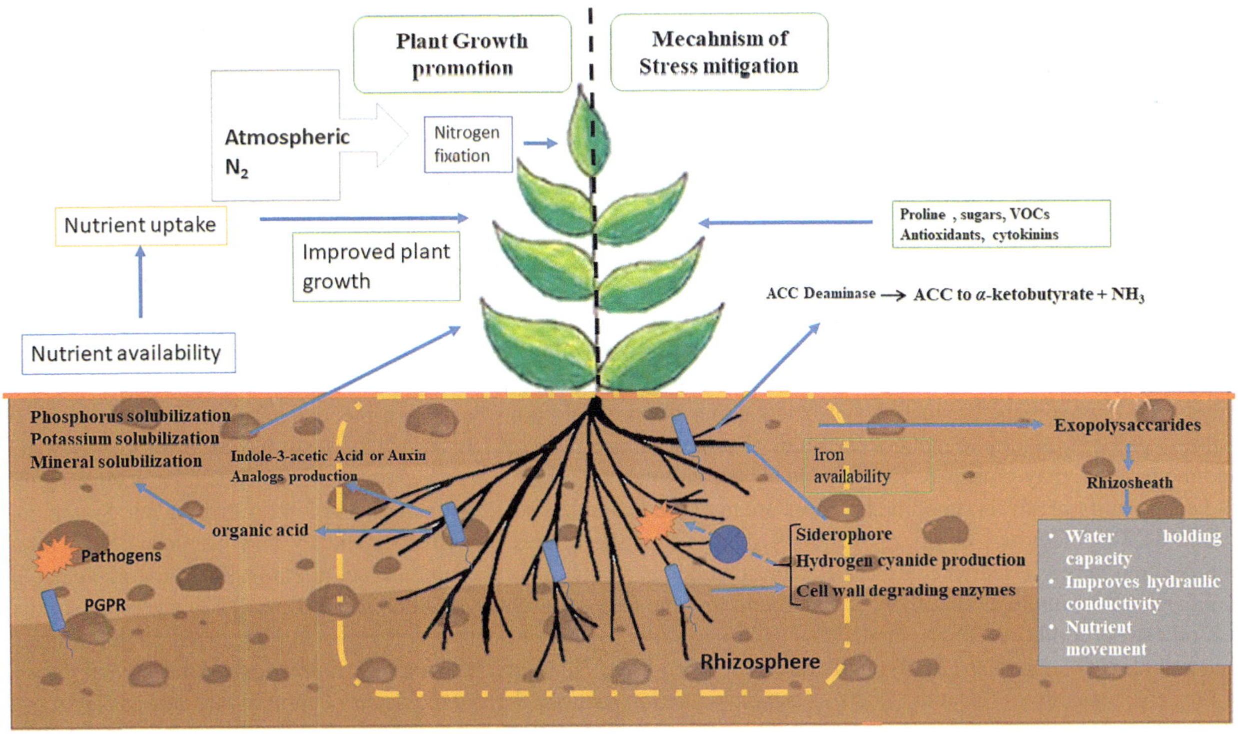

FIGURE 21.2 The mechanisms of uptake and absorption of nutrients in plants.

increases the seed yield and protein content in chickpea (*Cicer arietinum* L), in which terminal drought reduces crop yield (Tarvainen et al., 2017).

21.13.2 Phosphorous

Phosphorus (P) limitation in the soil causes significant challenges for crop productivity, especially when plants are subjected to abiotic stresses such as drought, salinity, and extreme temperatures (Khan et al., 2023). Excess soil moisture and elevated carbon dioxide (CO_2) reduce transpiration rates and alter the cell transporter functions, and thereby slowing down the rate of nutrient transport from the root to the shoot. It helps in the modulation of various morphological, physiological, and biochemical processes inside the plant system, leading to stress-tolerance (Bouras et al., 2022). Apart from this, incorporation of P under various abiotic stresses has a profound impact on tissue formation, carbohydrate and lipid metabolisms, chlorophyll biosynthesis, photosynthesis, production of assimilatory compounds used for energy acquisition, enzymatic functions, and antioxidant metabolism, and osmolyte accumulation, etc. (Begum et al., 2020). P in saline soils could be an alternative for alleviating the negative effects of salinity on plants and may ameliorate salinity tolerance (Dey et al., 2021). On the other hand, abiotic stresses have also been observed to result in P deficiency and hinder the activities of ATPase and phosphatase enzymes in plants, apparently under drought stress (Bouras et al., 2022). However, Al and Cd toxicity in phosphorus (P) deficiency soil are two major limiting factors for plant growth (Chen and Liao, 2016).

21.13.3 Potassium

Potassium (K) is a vital macronutrient for the better growth and physiological development of crop plants under abiotic stress situations, like water shortage, lodging, cold stress, drought, and salinity stress (Hernandez et al., 2012). K has a critical role in mitigating different abiotic stresses. Under moisture stress, K has been shown to have a favorable impact on relative leaf water content and better water use efficiency (Fooladivanda et al., 2014). K supplementation improves the ability of the plants to cope with drought stress through the efficient use of water (Fooladivanda et al., 2014). K fertilizer improves drought stress by enhancing photosynthesis and carbohydrate metabolism in plants (Zahoor and Dong, 2017). However, application of K has been found to regulate osmotic adjustment, cytoplasmic homeostasis, membrane potential, and enzyme activation under salinity stress (Assaha et al., 2017). K regulates antioxidant metabolism and alleviates growth inhibition under water and osmotic stress situations (Chakraborty et al., 2016). Under both normal and stressed conditions, K-deficient plants experienced a significant increase in the synthesis of osmolytes like free proline, amino acids, and sugars, which assumes special significance in growth stress conditions (Fiasconaro et al., 2019). Proline accumulation is one of the major mechanisms of drought tolerance in plants. Stress tolerance through the accumulation of proline has been reported to be intensified through the foliar spray of K (Yadav et al., 2019). The foliar spray of K-containing fertilizer improves the growth, yield, nutrient uptake, levels of antioxidant activities, total polyphenol, flavonoid, chlorophyll, and carotenoid contents by plants under salinity stress conditions (Majeed et al., 2016).

21.13.4 Sulfur

Sulfur is an element that participates in plant systems for controlling growth and development, and in modulating many adaptive responses to different biotic and abiotic stresses (Chen et al., 2015). Recent studies have provided the notion that S not only improves the productivity of plants under normal conditions but also protects them from abiotic stresses like salinity, drought, and toxic metals/metalloids (Hasanuzzaman et al., 2018). Many S-containing chemicals, like cysteine and methionine, play protective functions in abiotic cell mechanisms to improve stress tolerance, cellular acclimatization, and adaptability in adverse conditions. There are multiple examples, such as metabolism of nonproteinogenic S-containing amino acids in legumes (Joshi et al., 2019), analysis of the effects of sulfur deficiency and water stress on protein composition in peas (Henriet et al., 2019), and a detailed examination of the synthesis of organosulfur compounds in onion and garlic (Yoshimoto et al., 2019). It was found that the H_2S avert chlorophyll concentration loss in sweet potato under osmotic stress condition (Zhang et al., 2007), alleviated the negative effects of salt stress and improved photosynthesis in mustard plant (Fatma et al., 2014) and barley (Chen and Liao, 2016) and increase heat-tolerance in tobacco and regrowth ability after heat-stress (Li et al., 2020). S-supplementation reduced the arsenic accumulation in rice leaves (Dixit et al., 2015) and mitigated cadmium and arsenic toxicity in mustard (Sahay et al., 2020). Besides this, it was shown that S dramatically induced the activities of SOD, catalase, and ascorbate POD, and reduced the concentration of lipoxygenase, hydrogen peroxide (H_2O_2), and malondialdehyde (Hasanuzzaman et al., 2018). S can interact with different phytohormones, including auxins, gibberellins, cytokinins, abscisic acid, ethylene, jasmonic acid, and salicylic acid under optimal and limited environmental conditions, and positively regulates plant defense by regulation of S-assimilation, antioxidant metabolism (Hasanuzzaman et al., 2018). Besides this, sodium hydrosulfide improves the heat tolerance of maize, and the acquisition of this heat tolerance may be involved in proline content (Li et al., 2020).

21.13.5 Selenium

Selenium regulates ion balance and maintains cellular structural integrity. These mechanism of plants helps in lowering the multiple abiotic stresses such as salinity, drought, cold, high temperature, heavy-metal toxicity, acid stress, and UV-B irradiation. Besides this, it may interfere with electron transport by affecting the assembly of the photosynthesis complexes (Feng et al., 2013). There are also some other reports where the significant role of Se has been elucidated through modulations in metabolites transport, photosynthetic activity, membrane integrity, and cellular turgor in ameliorating drought, salinity, and heavy-metal stress, aluminum (Cartes et al., 2010), and arsenic (Sahay et al., 2020) in various crops. Severe drought stress decreased chlorophyll and carotenoid contents but increased anthocyanin, proline, POD, ascorbate POD, and catalase activities (CAT) in plants (Ali et al., 2023). The antioxidative activity of Se, higher accumulation of proline and soluble sugar, and decreases in the content of Cl^- in the shoots and facilitate for growth promoting. Increased biosynthesis of the photosynthetic pigments strengthened the metabolic balance to support growth even under salinity (Hasanuzzaman et al., 2018). Se increases the tolerance of the plants to drought-induced oxidative damage by increasing their antioxidant defense and methylglyoxal detoxification systems. Se supplementation improves the recovery of chlorophyll content due to light stress. Se-treated plants were more tolerant to paraquat, and the membrane integrity of the plant improved during oxidative stress. Se altered transcript accumulation of chloroplast SOD and glutathione POD, these results suggest that Se has an antioxidant property or it activates protective mechanisms, which can alleviate oxidative stress in the chloroplasts. Se restrained the accumulation of Pb and Cd (Kumar et al., 2010) in plants and enhanced the absorption of some other nutritional elements, such as Fe, Mn, Cu, Ca, and Mg. Therefore, the application of Se was a practical and reasonable process to improve trace element nutrition in vegetables. Selenium at a lower concentration alleviated Pb toxicity, which improved cell viability, whereas at a high concentration it was prooxidant and enhanced the lipid peroxidation and cell membrane injury (Mroczek-Zdyrska and Wójcik, 2012).

21.13.6 Silicon

Silicon (Si) is another important beneficial trace element for enhancing plant growth and development (Mir et al., 2022) as well as photosynthesis, economic yield, and also withstands soil salinity and drought stress (Cheraghi et al., 2023). Tolerance to salinity and drought stresses, including regulation of plant water relationships, gas exchange, photosynthesis, nutrient balance, oxidative stress reduction, lower-down ionic toxicity, osmoregulation, and root growth, K-uptake, and stimulation of plant hormones. Si is known to enhance plant tolerance against various kinds of biotic and abiotic stresses. However, it has a protective role for plants facing multiple stresses such as salinity and boron (B) deficiency (Farooq et al., 2019) and mercury (Li et al., 2020) toxicity. The application of Si nutrition boosts the antioxidant enzyme production, including SOD, chloramphenicol acetyltransferase, and PODs, while reducing reactive oxygen species generation (Abdel Latef and Tran, 2016; Farooq et al., 2019). In fact, Si alleviates oxidative stresses by the accumulation of protective proteins as well as decreasing the metal-ion toxicity, thereby amplifying the efficiencies of oxidative enzymes, while eliminating oxygen and hydrogen peroxide from the cells and reducing lipid peroxidation (Abdelaal et al., 2020). Due to deposition in the leaf cuticle and epidermal cells in the cell wall, Si lowers the transpiration rate and prevents desiccation of the plant during salt stress (Bybordi, 2015). Si also has a positive role in seed germination under saline conditions (Thorne et al., 2020). Silicon plays a significant detoxifying role under aluminum stress and cadmium stress by stimulating the biosynthesis of proline and phenolic compounds (Ma, 2004).

21.13.7 Zinc

Zinc (Zn) is a micronutrient that facilitates better defense against heat stress under saline conditions by maintaining the membrane integrity inside the plant system (Bashir et al., 2016). Zn is an integral constituent of plant enzymes such as carbonic anhydrase, alkaline phosphatase, alcohol dehydrogenase, RNA polymerase, Cu–Zn SOD, and phospholipase (Cakmak, 2005). Different proteins associated with DNA and RNA synthesis are related to Zn. Zn is an indispensable micronutrient that plays essential roles in crop resistance against drought stress by regulating various physiological and molecular mechanisms. Abiotic stress factors have several harmful effects on different parameters of crop plants. Zn acts as a major contributing factor in maintaining different abiotic stresses by inducing the enzymes involved in chlorophyll synthesis, bringing stability to membrane permeability and integrity, reducing ROS production to keep oxidative stress under control, *etc.* (Ganguly et al., 2022). Zn plays an important role in the mitigation of salinity stress and enhancements in indole–3–acetic acid, whereas reductions in abscisic acid in the shoot and root (Nadeem et al., 2020).

21.13.8 Manganese

Manganese is an essential cofactor of various antioxidants, which is required for the activation of Mn-SOD, Mn-catalase, guaiacol POD, ascorbate POD, glutathione reductase (GR), *etc.*, in the plant system (Demirevska-Kepova et al., 2004). According to Srivastava and Dubey (2011), rice seedlings in excess Mn induce oxidative stress, imbalances the levels of antioxidants, and the antioxidative enzymes play an important role in scavenging ROS and withstanding oxidative stress induced by Mn (Srivastava and Dubey, 2011). Mn induces the relative growth rate of salt-stressed plants, which was attributed to the rise in assimilation rate in salt-stressed plants (Pandya et al., 2005). Sufficient supply of Mn remediates plant manganese deficiency, induces Mn SOD at the transcriptional level to face reactive oxygen species production in plants, and stimulates manganese-dependent proteins to maintain cell integrity. The positive effects of Mn have provided significant evidence to perform further studies incorporating Mn compounds to diminish abiotic stress by promoting Mn–SOD scavenging abilities (Ye et al., 2019). Mn is essential for the activity of some dehydrogenases, decarboxylases, kinases, oxidases, polymerases, PODs, and related enzymes. Mn increases the activity of SOD, GR, and CAT, reduces the salinity-induced oxidative damage, increases membrane stability index and relative water content, respectively (Nadeem et al., 2020). In some reports, it was found that salinity decreased chlorophyll a, chlorophyll b, and carotenoid contents, whereas Mn supplement partially recovered all these photosynthetic pigments by producing reactive oxygen species (ROS) and upregulation of Mn uptake, supported by previous findings (Rahman et al., 2016).

21.13.9 Iron

Iron plays an irreplaceable role in alleviating stress imposed by salinity (Nadeem et al., 2020) and heavy metal stress (Huang et al., 2009). It is one of the chief components of the cell redox systems and also functions as a cofactor regarding various antioxidant enzymes such as CAT, POD, and APX, increasing total soluble sugar and proline content (Singh et al., 2021a). Fe is a component of SOD. The SOD is a scavenger of reactive oxygen species (ROS). In plants, Fe helps in chlorophyll synthesis and is vital for the maintenance of structural integrity as well as functional aspects of the chloroplast. Iron spraying plays a significant role in improving the protein quality and resistance under drought stress. It is important to mention that Fe is applicable in controlling drought effects on root growth (Glinski et al., 1992). In fact, when subjected to Fe deficiency and to increase iron availability in the soil, dicots and nongraminaceous monocots induce some responses involving especially a plasma membrane H^+–ATPase to increase root proton release capacity and a plasma membrane Fe (III) chelate reductase to reduce Fe(III) chelates at the root surface (Rabhi et al., 2007). Besides this, Fe also reduced the uptake of heavy metals and divalent cations (Cd, Mg) by the plant (Shao et al., 2008; Sharma et al., 2004; Singh et al., 2021). Application of Fe enhances the performance of agroforestry trees under abiotic stress conditions, such as salinity (Singh et al., 2021). According to Ghasemi et al. (2014), the adverse effects of salt stress on tomato plants can be alleviated with the application of Fe-amino acids chelates (such as Fe-EDTA and Fe-DTPA), as chelates are appropriate for applications in alkaline and calcareous soils.

21.13.10 Boron

Plants with sufficient boron (B) nutrition have shown an elevated resistance to drought stress due to improved nutritional status and enhanced water uptake from the rhizospheric soil by growing more root hairs and mycorrhizae. Drought sensitivity of crop plants is regulated by B in two different ways. First, it is involved in the ROS detoxification method in chloroplasts and thereby takes a protective role in the prohibition of photo-oxidative losses catalyzed by ROS. The ROS produced within cells can be eliminated by diverse antioxidant systems, through the action of enzymes such as SOD, CAT, ascorbate POD (APX), GR, PPO, and the ascorbate–glutathione cycle activity, and the nonenzymatic antioxidants, ascorbic acid, and proline accumulation were found to be significantly decreased in B-deficient conditions (ASA–GSH) (Simón-Grao et al., 2018; Yan et al., 2018). Second, B is tolerant to drought by protecting against oxidative damage to cell membranes (Ardıc et al., 2009; Hajiboland and Farhanghi, 2011). Besides this, B deficiency in water-deficient conditions could be due to a decreased microbial population that can release B, making it available to plants, and due to decreased mineral dissolution. B deficiency can affect water relations in plants in different ways by altering water uptake in the root, transport through the shoot, and loss of water in the leaf (Wimmer and Eichert, 2013). However, Al-toxicity and B–deficiency are the most common interactions studied in plants in tropical and subtropical areas. Boron application mitigates chilling stress (Ilyas et al., 2021). During chilling stress, the application of B can enhance photosynthetic activity, water uptake, activities of plant antioxidants, and decrease root hydraulic conductance and stomatal dysfunction. This helps to reduce ROS injury caused by temperature stress. Thus B plays an important role in alleviating high-temperature stress (Shahid et al., 2018). Boron nutrition also improves sugar transport in the plant body, which acts as an osmolyte and antifreezing agent (Ilyas et al., 2021). Foliar application of B was found to be more effective in alleviating the deleterious effects of waterlogging and water stress.

21.14 Conclusions

A larger spatiotemporal variability of abiotic stress(s) is the principal constraint for formulating proper management practices. Therefore, frequent and detailed/ or narrow-scale and in-depth monitoring of soil, water, and plant attributes is necessary for the diagnosis and inventory of the correct problem of single or multiple abiotic stresses. Salinity/sodicity or both may be residing on the surface/subsurface soil or below the depth. Very often, this problem may appear because of prolonged irrigation of saline/sodic or poor-quality water. The dynamics of electrolytes in the root zone vary with seasonal temperature, precipitation, evapotranspiration, vegetation cover, and supply of external irrigation and regional drainage. The problem of sodicity may be cured on application of the Ca-based amendments or by mobilizing the calcium carbonate through neutralization of alkalinity on the application of mineral acid or acid-forming agent or afforestation of sodic land by sodicity-tolerant trees. The limited availability, slow rate of reclamation, and associated environmental hazards limit the application of some of the conventional amendments (phosphogypsum, fly ash, pyrites, etc.). The problem of acidity can be solved by amending soil with lime or low-cost liming materials (dolomite, basic slag, press mud, and coral shells), cost-intensive, efficient amendments (calcite, slaked lime), agronomic practices, and growing of acid-tolerant crop/ cultivars. Plastic mulching, plastic film mulching, reduces the net evaporation ratio. Hydrogel and acrylamide–acrylate copolymers absorb and retain water for a longer duration, have proven to be a promising approach for enhancing water use efficiency, improving drought tolerance, and ultimately boosting crop yields. Efficient water management also enhances cold-tolerance during winter because high soil moisture causes more freezing and thawing. Integrated (organic and inorganic fertilizers) and balanced fertilization, application of biostimulants or bioregulators, adoption of 4-R concept of "Nutrition Stewardship" with right fertilizer source, right time—when crops needs, right rate matches amount of fertilizer type as crop needs and right place keep nutrients where crops can use them matches fertilizer type to crop need and application of nano, alternative and new generation fertilizers based on soil and or plant-led strategies can maintain the optimum nutrient content in soil for sustainable crop production.

References

Abdel Latef, A.A., Tran, L.S.P., 2016. Impacts of priming with silicon on the growth and tolerance of maize plants to alkaline stress. Frontiers in Plant Science 7:243. Available at https://doi.org/10.3389/fpls.2016.00243.

Abdelaal, K.A.A., Mazrou, Y.S.A., Hafez, Y.M., 2020. Silicon foliar application mitigates salt stress in sweet pepper plants by enhancing water status, photosynthesis, antioxidant enzyme activity and fruit yield. Plants 9, 733.

Abhinandan, K., Skori, L., Stanic, M., Hickerson, N.M.N., Jamshed, M., Samuel, M.A., 2018. Abiotic stress signaling in wheat – an inclusive overview of hormonal interactions during abiotic stress responses in wheat. Frontiers in Plant Science 9. Available at https://doi.org/10.3389/fpls.2018.00734.

Abid, M., Tian, Z., Ata-Ul-Karim, S.T., Cui, Y., Liu, Y., Zahoor, R., et al., 2016. Nitrogen nutrition improves the potential of wheat (Triticum aestivum L.) to alleviate the effects of drought stress during vegetative growth periods. Frontiers in Plant Science 7. Available at https://doi.org/10.3389/fpls.2016.00981.

Adhikari, L., Baral, R., Paudel, D., Min, D., Makaju, S.O., Poudel, H.P., et al., 2022. Cold stress in plants: strategies to improve cold tolerance in forage species. Plant Stress 4. Available at https://doi.org/10.1016/j.stress.2022.100081.

Ahluwalia, O., Singh, P.C., Bhatia, R., 2021. A review on drought stress in plants: implications, mitigation and the role of plant growth promoting rhizobacteria. Resources, Environment and Sustainability 5, 100032.

Ahmad, M., 2023. Plant breeding advancements with "CRISPR-Cas" genome editing technologies will assist future food security. Frontiers in Plant Science 14. Available at https://doi.org/10.3389/fpls.2023.1133036.

Ahmed, F., Rafii, M.Y., Ismail, M.R., Juraimi, A.S., Rahim, H.A., Asfaliza, R., et al., 2013. Waterlogging tolerance of crops: breeding, mechanism of tolerance, molecular approaches, and future prospects. BioMed Research International 2013, 1–10. Available at https://doi.org/10.1155/2013/963525.

Al-Bayati, H.J.M., Hamdoon, D.N., 2019. Response of eggplant Solanum melongena L. to soil mulching, organic and inorganic fertilizers on vegetative growth traits and yield grown under unheated plastic house. IOP Conference Series: Earth and Environmental Science 388, 012075. Available at 10.1088/1755-1315/388/1/012075.

Alharby, H.F., Fahad, S., 2020. Melatonin application enhances biochar efficiency for drought tolerance in maize varieties: modifications in physio-biochemical machinery. Agronomy Journal 112 (4), 2826–2847. Available at https://doi.org/10.1002/agj2.20263.

Ali, J., Jan, I., Ullah, H., Fahad, S., Saud, S., Adnan, M., et al., 2023. Biochemical response of okra (Abelmoschus esculentus L.) to selenium (Se) under drought stress. Sustainability (Switzerland) 15 (7). Available at https://doi.org/10.3390/su15075694.

Ambroise, V., Legay, S., Guerriero, G., Hausman, J.F., Cuypers, A., Sergeant, K., 2020. The roots of plant frost hardiness and tolerance. Plant and Cell Physiology 61 (1), 3–20. Available at https://doi.org/10.1093/pcp/pcz196.

Ardıc, M., Sekmen, A.H., Tokur, S., Ozdemir, F., Turkan, I., 2009. Antioxidant responses of chickpea plants subjected to boron toxicity. Plant Biology 11 (3), 328–338. Available at https://doi.org/10.1111/j.1438-8677.2008.00132.x.

Arzani, A., Ashraf, M., 2016. Smart engineering of genetic resources for enhanced salinity tolerance in crop plants. Critical Reviews in Plant Sciences 35 (3), 146–189. Available at https://doi.org/10.1080/07352689.2016.1245056.

Ashraf, M.A., Ahmad, M.S.A., Ashref, M., Al-Qurainy, F., Ashraf, M.Y., 2011. Alleviation of waterlogging stress in upland cotton (Gossypium hirsutum L) by exogenous application of potassium in soil and as a foliar spray. Crop and Pasture Science 62 (1), 25–38. Available at https://doi.org/10.1071/CP09225.

Aslam, M.U., Raza, M.A.S., Saleem, M.F., Waqas, M., Iqbal, R., Ahmad, S., et al., 2020. Improving strategic growth stage-based drought tolerance in quinoa by rhizobacterial inoculation. Communications in Soil Science and Plant Analysis 51 (7), 853–868. Available at https://doi.org/10.1080/00103624.2020.1744634.

Aslam, M., Fakher, B., Ashraf, M.A., Cheng, Y., Wang, B., Qin, Y., 2022. Plant low-temperature stress: signaling and response. Agronomy 12 (3), 702. Available at https://doi.org/10.3390/agronomy12030702.

Assaha, D.V.M., Ueda, A., Saneoka, H., Al-Yahyai, R., Yaish, M.W., 2017. The role of Na+ and K+ transporters in salt stress adaptation in glycophytes. Frontiers in Physiology 8. Available at https://doi.org/10.3389/fphys.2017.00509.

Bai, Y., Li, Z., Dai, J., Seki, K., Zhao, Y., Sun, H., et al., 2024. Effects of water table depth variations on canopy transpiration from Mongolian pine plantations in arid ecosystems. Journal of Hydrology 644, 132106. Available at https://doi.org/10.1016/j.jhydrol.2024.132106.

Basak, N., Rai, A.K., Barman, A., Mandal, S., Sundha, P., Bedwal, S., et al., 2022a. Salt affected soils: global perspectives. In: Shit, P.K., Adhikary, P.P., Bhunia, G.S., Sengupta, D. (Eds.), Soil Health and Environmental Sustainability: Application of Geospatial Technology. Springer International Publishing, Cham, pp. 107–129.

Basak, N., Rai, A.K., Sundha, P., Meena, R.L., Bedwal, S., Yadav, R.K., et al., 2022b. Assessing soil quality for rehabilitation of salt-affected agroecosystem: a comprehensive review. Frontiers in Environmental Science 10. Available at https://doi.org/10.3389/fenvs.2022.935785.

Basak, N., Barman, A., Sundha, P., Rai, A.K., 2020. Recent trends in soil salinity appraisal and management. In: Rakshit, A, Ghosh S, SC, Chakraborty, S, Philip, V, Datta, A (Eds.), Soil Analysis: Recent Trends and Applications. Springer, Singapore, pp. 143–161. Available at http://doi.org/10.1007/978-981-15-2039-6_9.

Bashir, K., Rasheed, S., Kobayashi, T., Seki, M., Nishizawa, N.K., 2016. Regulating subcellular metal homeostasis: the key to crop improvement. Frontiers in Plant Science 7 (2016). Available at https://doi.org/10.3389/fpls.2016.01192.

Bedwal, S., Basak, N., Rai, A.K., Sundha, P., Patel, S., Raj, D., et al., 2023. Converting waste to resource: utilisation of elemental S and consortia of halo–alkaliphilic S–oxidisers for preparation of acidified compost from organic wastes. Bioresource Technology Reports 24. Available at https://doi.org/10.1016/j.biteb.2023.101620.

Bedwal, S., Rai, A.K., Basak, N., Chandra, P., Sundha, P., Chaudhary, A., et al., 2025. Acidulated composts govern the soil solution nutrient composition through changes in key functional bacterial community of alkaline soils. Applied Soil Ecology 211, 106164.

Begum, N., Ahanger, M.A., Zhang, L., 2020. AMF inoculation and phosphorus supplementation alleviates drought induced growth and photosynthetic decline in Nicotiana tabacum by up-regulating antioxidant metabolism and osmolyte accumulation. Environmental and Experimental Botany 176, 104088.

Besset-Manzoni, Y., Rieusset, L., Joly, P., Comte, G., Prigent-Combaret, C., 2018. Exploiting rhizosphere microbial cooperation for developing sustainable agriculture strategies. Environmental Science and Pollution Research 25 (30), 29953–29970. Available at https://doi.org/10.1007/s11356-017-1152-2.

Bhat, J.A., Kundu, M.C., Hazra, G.C., Santra, G.H., Mandal, B., 2010. Rehabilitating acid soils for increasing crop productivity through low-cost liming material. Science of the Total Environment 408 (20), 4346–4353. Available at https://doi.org/10.1016/j.scitotenv.2010.07.011.

Bouras, H., Choukr-Allah, R., Mosseddaq, F., Bouaziz, A., Devkota, K.P., Mouttaqi, A.E., et al., 2022. Does phosphorus fertilization increase biomass production and salinity tolerance of blue panicum (Panicum antidotale Retz.) in the salt-affected soils of arid regions? Agronomy 12 (4), 791. Available at https://doi.org/10.3390/agronomy12040791.

Brendel, O., 2021. The relationship between plant growth and water consumption: a history from the classical four elements to modern stable isotopes. Annals of Forest Science 78 (2). Available at https://doi.org/10.1007/s13595-021-01063-2.

Bybordi, A., 2015. Influence of exogenous application of silicon and potassium on physiological responses, yield, and yield components of salt-stressed wheat. Communications in Soil Science and Plant Analysis 46 (1), 109–122. Available at https://doi.org/10.1080/00103624.2014.956936.

Cakmak, I., 2005. The role of potassium in alleviating detrimental effects of abiotic stresses in plants. Journal of Plant Nutrition and Soil Science 168 (4), 521–530. Available at https://doi.org/10.1002/jpln.200420485.

Cartes, P., Jara, A.A., Pinilla, L., Rosas, A., Mora, M.L., 2010. Selenium improves the antioxidant ability against aluminium-induced oxidative stress in ryegrass roots. Annals of Applied Biology 156 (2), 297–307. Available at https://doi.org/10.1111/j.1744-7348.2010.00387.x.

Chai, Q., Gan, Y., Zhao, C., Xu, H.L., Waskom, R.M., Niu, Y., et al., 2016. Regulated deficit irrigation for crop production under drought stress. A review. Agronomy for Sustainable Development 36 (1), 1–21. Available at https://doi.org/10.1007/s13593-015-0338-6.

Chakraborty, K., Bhaduri, D., Meena, H.N., Kalariya, K., 2016. External potassium (K+) application improves salinity tolerance by promoting Na +-exclusion, K+-accumulation and osmotic adjustment in contrasting peanut cultivars. Plant Physiology and Biochemistry 103, 143–153. Available at https://doi.org/10.1016/j.plaphy.2016.02.039.

Chandra, A., Chandra, P., Tripathi, P., 2021. Whole genome sequence insight of two plant growth-promoting bacteria (B. subtilis BS87 and B. megaterium BM89) isolated and characterized from sugarcane rhizosphere depicting better crop yield potentiality. Microbiological Research 247, 126733.

Chandra, P., Khobra, R., Sundha, P., Sharma, R.K., Jasrotia, P., Chandra, A., et al., 2021a. Plant growth promoting Bacillus-based bio formulations improve wheat rhizosphere biological activity, nutrient uptake and growth of the plant. Acta Physiologiae Plantarum 43 (11). Available at https://doi.org/10.1007/s11738-021-03310-5.

Chandra, P., Wunnava, A., Verma, P., Chandra, A., Sharma, R.K., 2021b. Strategies to mitigate the adverse effect of drought stress on crop plants—influences of soil bacteria: a review. Pedosphere 31, 496–509.

Chandra, P., Rai, A.K., Sundha, P., Basak, N., Kaur, H., 2022a. Rhizospheric soil–plant-microbial interactions for abiotic stress mitigation and enhancing crop performance. In: Shit, P.K., Adhikary, P.P., Bhunia, G.S., Sengupta, D. (Eds.), Soil Health and Environmental Sustainability: Application of Geospatial Technology. Springer International Publishing, Cham, pp. 593–614.

Chandra, P., Singh, A., Prajapat, K., Rai, A.K., Yadav, R.K., 2022b. Native arbuscular mycorrhizal fungi improve growth, biomass yield, and phosphorus nutrition of sorghum in saline and sodic soils of the semi–arid region. Environmental and Experimental Botany 201, 104982.

Chandra, P., Khobra, R., Sundha, P., Chandra, A., Sharma, R.K., Singh, G.P., 2024a. Bacillus subtilis strain BS87 as a biocontrol agent against spot blotch disease: effect on growth, nutrient status, and antioxidant enzymes in wheat. Acta Physiologiae Plantarum 46, 35.

Chandra, P., Rai, A.K., Basak, N., Sundha, P., Prajapat, K., Singh, A., et al., 2024b. In vitro P - solubilization activity of halophilic fungi in salt - affected soils and their potential as bio - inoculants. Environmental Engineering Research 29, 230760.

Chen, J., Wang, W.H., Wu, F.H., He, E.M., Liu, X., Shangguan, Z.P., et al., 2015. Hydrogen sulfide enhances salt tolerance through nitric oxide-mediated maintenance of ion homeostasis in barley seedling roots. Scientific Reports 5. Available at https://doi.org/10.1038/srep12516.

Chen, Z.C., Liao, H., 2016. Organic acid anions: an effective defensive weapon for plants against aluminum toxicity and phosphorus deficiency in acidic soils. Journal of Genetics and Genomics 43, 631–638.

Cheraghi, M., Motesharezadeh, B., Mousavi, S.M., Ma, Q., Ahmadabadi, Z., 2023. Silicon (Si): a regulator nutrient for optimum growth of wheat under salinity and drought stresses- a review. Journal of Plant Growth Regulation 42 (9), 5354–5378. Available at https://doi.org/10.1007/s00344-023-10959-1.

Chieb, M., Gachomo, E.W., 2023. The role of plant growth promoting rhizobacteria in plant drought stress responses. BMC Plant Biology 23 (1). Available at https://doi.org/10.1186/s12870-023-04403-8.

Choi, H.W., Klessig, D.F., 2016. DAMPs, MAMPs, and NAMPs in plant innate immunity. BMC Plant Biology 16 (1). Available at https://doi.org/10.1186/s12870-016-0921-2.

Chu, G., Chen, T., Chen, S., Xu, C., Wang, D., Zhang, X., 2018. Agronomic performance of drought-resistance rice cultivars grown under alternate wetting and drying irrigation management in southeast China. The Crop Journal 6 (5), 482–491. Available at https://doi.org/10.1016/j.cj.2018.04.005.

Cook, B.I., Smerdon, J.E., Seager, R., Coats, S., 2014. Global warming and 21st century drying. Climate Dynamics 43 (9-10), 2607–2627. Available at https://doi.org/10.1007/s00382-014-2075-y.

Demirevska-Kepova, K., Simova-Stoilova, L., Stoyanova, Z., Hölzer, R., Feller, U., 2004. Biochemical changes in barley plants after excessive supply of copper and manganese. Environmental and Experimental Botany 52 (3), 253–266. Available at https://doi.org/10.1016/j.envexpbot.2004.02.004.

Devi, R., Kaur, T., Kour, D., Yadav, A., Yadav, A.N., Suman, A., et al., 2022. Minerals solubilizing and mobilizing microbiomes: a sustainable approach for managing minerals' deficiency in agricultural soil. Journal of Applied Microbiology 133 (3), 1245–1272. Available at https://doi.org/10.1111/jam.15627.

Dey, G., Banerjee, P., Sharma, R.K., Maity, J.P., Etesami, H., Shaw, A.K., et al., 2021. Management of phosphorus in salinity-stressed agriculture for sustainable crop production by salt-tolerant phosphate-solubilizing bacteria—a review. Agronomy 11 (8). Available at https://doi.org/10.3390/agronomy11081552.

Dhankher, O.P., Foyer, C.H., 2018. Climate resilient crops for improving global food security and safety. Plant Cell and Environment 41 (5), 877–884. Available at https://doi.org/10.1111/pce.13207.

Dixit, G., Singh, A.P., Kumar, A., Dwivedi, S., Deeba, F., Kumar, S., et al., 2015. Sulfur alleviates arsenic toxicity by reducing its accumulation and modulating proteome, amino acids and thiol metabolism in rice leaves. Scientific Reports 5. Available at https://doi.org/10.1038/srep16205.

FAO, 2017. The Future of Food and Agriculture – Trends and Challenges. Food and Agriculture Organization of the United Nations, Rome. Available at https://openknowledge.fao.org/server/api/core/bitstreams/2e90c833-8e84-46f2-a675-ea2d7afa4e24/content.

FAO, 2024. Global Status of Salt-Affected Soils: Main Report. FAO, Rome, Italy. Available at https://doi.org/10.4060/cd3044en.

Farhan, M., Sathish, M., Kiran, R., Mushtaq, A., Baazeem, A., Hasnain, A., et al., 2024. Plant nitrogen metabolism: balancing resilience to nutritional stress and abiotic challenges. Phyton-International Journal of Experimental Botany 93 (3), 581–609. Available at https://doi.org/10.32604/phyton.2024.046857.

Farooq, M.A., Saqib, Z.A., Akhtar, J., Bakhat, H.F., Pasala, R.K., Dietz, K.J., 2019. Protective role of silicon (Si) against combined stress of salinity and boron (B) toxicity by improving antioxidant enzymes activity in rice. Silicon 11 (4), 2193–2197. Available at https://doi.org/10.1007/s12633-015-9346-z.

Fatma, M., Asgher, M., Masood, A., Khan, N.A., 2014. Excess sulfur supplementation improves photosynthesis and growth in mustard under salt stress through increased production of glutathione. Environmental and Experimental Botany 107, 55–63. Available at https://doi.org/10.1016/j.envexpbot.2014.05.008.

Feng, R., Wei, C., Tu, S., 2013. The roles of selenium in protecting plants against abiotic stresses. Environmental and Experimental Botany 87, 58–68.

Fiasconaro, M.L., Lovato, M.E., Antolín, M.C., Clementi, L.A., Torres, N., Gervasio, S., et al., 2019. Role of proline accumulation on fruit quality of pepper (Capsicum annuum L) grown with a K-rich compost under drought conditions. Scientia Horticulturae 249, 280–288. Available at https://doi.org/10.1016/j.scienta.2019.02.002.

Fooladivanda, Z., Hassanzadehdelouei, M., Zarifinia, N., 2014. Effects of water stress and potassium on quantity traits of two varieties of mung bean (Vigna radiata L). Agronomical Research in Moldavia 47, 107–114.

Fukao, T., Barrera-Figueroa, B.E., Juntawong, P., Peña-Castro, J.M., 2019. Submergence and waterlogging stress in plants: a review highlighting research opportunities and understudied aspects. Frontiers in Plant Science 10. Available at https://doi.org/10.3389/fpls.2019.00340.

Ganguly, R., Sarkar, A., Dasgupta, D., Acharya, K., Keswani, C., Popova, V., et al., 2022. Unravelling the efficient applications of zinc and selenium for mitigation of abiotic stresses in plants. Agriculture 12 (10), 1551. Available at https://doi.org/10.3390/agriculture12101551.

Ghasemi, S., Khoshgoftarmanesh, A.H., Afyuni, M., Hadadzadeh, H., 2014. Iron(II)-amino acid chelates alleviate salt-stress induced oxidative damages on tomato grown in nutrient solution culture. Scientia Horticulturae 165, 91–98. Available at https://doi.org/10.1016/j.scienta.2013.10.037 Available from.

Glinski, D.S., Carrow, R.N., Karnok, K.J., 1992. Iron fertilization effects on shoot/root growth, water use, and drought stress of creeping bentgrass. Agronomy Journal 84, 496–503.

Guo, G., Tan, X., He, Y., Guo, J., Yu, J., Zhuang, W., et al., 2025. Transport of heat, water, and salts in freeze–thaw soils under flood irrigation: experiment and simulation. Journal of Hydrology 652, 132688. Available at https://doi.org/10.1016/j.jhydrol.2025.132688.

Guo, J., Ling, H., Wu, Q., Xu, L., Que, Y., 2014. The choice of reference genes for assessing gene expression in sugarcane under salinity and drought stresses. Scientific Reports 4, 7042. Available at https://doi.org/10.1038/srep07042.

Gutiérrez-Hernández, O., García, L.V., 2021. Relationship between precipitation and species distribution. Precipitation: Earth Surface Responses and Processes. Elsevier, Spain, pp. 239–259. Available at http://doi.org/10.1016/B978-0-12-822699-5.00010-0.

Hajiboland, R., Farhanghi, F., 2011. Effect of low boron supply in turnip plants under drought stress. Biologia Plantarum 55 (4), 775–778. Available at https://doi.org/10.1007/s10535-011-0186-4.

Hameed, A., Ahmed, M.Z., Hussain, T., Aziz, I., Ahmad, N., Gul, B., et al., 2021. Effects of salinity stress on chloroplast structure and function. Cells 10 (8). Available at https://doi.org/10.3390/cells10082023.

Hasanuzzaman, M., Hossain, M.S., Bhuyan, M.H.M.B., Al Mahmud, J., Nahar, K., Fujita, M., 2018. The role of sulfur in plant abiotic stress tolerance: molecular interactions and defense mechanisms. Plant Nutrients and Abiotic Stress Tolerance. Springer, Singapore, pp. 221–252. Available at 10.1007/978-981-10-9044-8_10.

Hassan, M.A., Xiang, C., Farooq, M., Muhammad, N., Yan, Z., Hui, X., et al., 2021. Cold stress in wheat: plant acclimation responses and management strategies. Frontiers in Plant Science 12. Available at https://doi.org/10.3389/fpls.2021.676884.

Henriet, C., Aimé, D., Térézol, M., Kilandamoko, A., Rossin, N., Combes-Soia, L., et al., 2019. Water stress combined with sulfur deficiency in pea affects yield components but mitigates the effect of deficiency on seed globulin composition. Journal of Experimental Botany 70 (16), 4287–4304. Available at https://doi.org/10.1093/jxb/erz114.

Hernandez, M., Fernandez-Garcia, N., Garcia-Garma, J., Rubio-Asensio, J.S., Rubio, F., Olmos, E., 2012. Potassium starvation induces oxidative stress in L. roots. Journal of Plant Physiology 169 (14), 1366–1374. Available at https://doi.org/10.1016/j.jplph.2012.05.015.

Huang, Y., Hu, Y., Liu, Y., 2009. Heavy metal accumulation in iron plaque and growth of rice plants upon exposure to single and combined contamination by copper, cadmium and lead. Acta Ecologica Sinica 29, 320–326.

Hussain, H.A., Hussain, S., Khaliq, A., Ashraf, U., Anjum, S.A., Men, S., et al., 2018. Chilling and drought stresses in crop plants: implications, cross talk, and potential management opportunities. Frontiers in Plant Science 9. Available at https://doi.org/10.3389/fpls.2018.00393.

Ilyas, M., Ayub, G., Imran, Ali Awan, A., Ahmad, M., 2021. Calcium and boron effect on production and quality of autumn potato crop under chilling temperature. Communications in Soil Science and Plant Analysis 52 (4), 375–388. Available at https://doi.org/10.1080/00103624.2020.1854286.

Imran, M.A., Xu, J., Sultan, M., Shamshiri, R.R., Ahmed, N., Javed, Q., et al., 2021. Free Discharge of subsurface drainage effluent: an alternate design of the surface drain system in Pakistan. Sustainability 13 (7), 4080. Available at https://doi.org/10.3390/su13074080.

Jha, S.K., Mishra, V.K., Damodaran, T., Singh, Y.P., Sharma, P.C., 2022. Utilization of marine gypsum as an alternative to mineral gypsum in the reclamation of degraded sodic soil and as a means of enhancing crop productivity. Land Degradation and Development 33 (13), 2276–2288. Available at https://doi.org/10.1002/ldr.4282.

Ji, X., Tang, J., Zhang, J., 2022. Effects of salt stress on the morphology, growth and physiological parameters of Juglansmicrocarpa L. seedlings. Plants (Basel) 11.

Ji, Y., Fu, J., Liu, B., Tan, X., 2024. Can limiting global temperature rise to below 2°C warming prevent the emergence of unprecedented drought? Agricultural and Forest Meteorology 352, 110047. Available at https://doi.org/10.1016/j.agrformet.2024.110047.

Joshi, J., Saboori-Robat, E., Solouki, M., Mohsenpour, M., Marsolais, F., Kopriva, S., 2019. Distribution and possible biosynthetic pathway of non-protein sulfur amino acids in legumes. Journal of Experimental Botany 70 (16), 4115–4121. Available at https://doi.org/10.1093/jxb/erz291.

Jungers, J., Runck, B., Ewing, P.M., Maaz, T., Carlson, C., Neyhart, J., et al., 2023. Adapting perennial grain and oilseed crops for climate resiliency. Crop Science 63 (4), 1701–1721. Available at https://doi.org/10.1002/csc2.20972.

Juurakko, C.L., diCenzo, G.C., Walker, V.K., 2021. Cold acclimation and prospects for cold-resilient crops. Plant Stress 2, 100028.

Kamra, S.K., Kumar, S., Kumar, N., Dagar, J.C., 2019. Engineering and biological approaches for drainage of irrigated lands. In: Dagar, J., Yadav, R., Sharma, P. (Eds.), Research Developments in Saline Agriculture. Springer, Singapore, pp. 537–577. Available at http://doi.org/10.1007/978-981-13-5832-6_18.

Kato, S., Wachi, T., Yoshihira, K., Nakagawa, T., Ishikawa, A., Takagi, D., et al., 2013. Rice (Oryza sativa L) roots have iodate reduction activity in response to iodine. Frontiers in Plant Science 4. Available at https://doi.org/10.3389/fpls.2013.00227.

Kazemi, F., Safari, N., 2018. Effect of mulches on some characteristics of a drought tolerant flowering plant for urban landscaping. Desert 23, 75–84.

Khan, F., Siddique, A.B., Shabala, S., Zhou, M., Zhao, C., 2023. Phosphorus plays key roles in regulating plants' physiological responses to abiotic stresses. Plants 12 (15). Available at https://doi.org/10.3390/plants12152861.

Khanh, P.T., Ngoc, T.T.H., Pramanik, S., 2024. Evaluation of machine learning models for mapping soil salinity in Ben Tre province, Vietnam. Multimedia Tools and Applications 84, 201–220.

Kishore, G., Babu, B.M., Mattaparti, L.D., 2022. Influence of plastic mulching and irrigation levels on soil temperature, moisture and water use efficiency of tomato crop (Solanum lycopersicum). International Journal of Plant & Soil Science 277–282.

Krishnamurthy, S.L., Sharma, P.C., Ravikiran, K.T., Basak, N., Vineeth, T.V., Singh, Y.P., et al., 2016. G×E interaction and stability analysis for salinity and sodicity tolerance in rice at reproductive stage. Journal of Soil Salinity and Water Quality 8, 162–172.

Kumar, P., Tewari, R.K., Sharma, P.N., 2010. Sodium nitroprusside-mediated alleviation of iron deficiency and modulation of antioxidant responses in maize plants. AoB Plants 2010.

Lahlali, R., Taoussi, M., Laasli, S.E., Gachara, G., Ezzouggari, R., Belabess, Z., et al., 2024. Effects of climate change on plant pathogens and host-pathogen interactions. Crop and Environment 3 (3), 159–170. Available at https://doi.org/10.1016/j.crope.2024.05.003.

Lee, J.G., Chae, H.G., Hwang, H.Y., Kim, P.J., Cho, S.R., 2021. Effect of plastic film mulching on maize productivity and nitrogen use efficiency under organic farming in South Korea. Science of the Total Environment 787. Available at https://doi.org/10.1016/j.scitotenv.2021.147503.

Li, S., Zhou, L., Addo-Danso, S.D., Ding, G., Sun, M., Wu, S., et al., 2020. Nitrogen supply enhances the physiological resistance of Chinese fir plantlets under polyethylene glycol (PEG)-induced drought stress. Scientific Reports 10, 7509.

Liu, Y., Song, W., Song, A., Wu, C., Ding, J., Yu, X., et al., 2023. The improvement of agronomic performances in the cold weather conditions for perennial wheatgrass by crossing Thinopyrum intermedium with wheat-Th. intermedium partial amphiploids. Frontiers in Plant Science 14. Available at https://doi.org/10.3389/fpls.2023.1207078.

Luo, C., Wang, R., Hao, K., Jia, X., Zhu, J., Xin, Z., et al., 2024. Effects of seasonal freezing and thawing on soil moisture and salinity in the farmland shelterbelt system in the hetao irrigation district. Agronomy 14 (7), 1425. Available at https://doi.org/10.3390/agronomy14071425.

Ma, J.F., 2004. Role of silicon in enhancing the resistance of plants to biotic and abiotic stresses. Soil Science and Plant Nutrition 50 (1), 11–18. Available at https://doi.org/10.1080/00380768.2004.10408447.

Majeed, S., Akram, M., Latif, M., Ijaz, M., Hussain, M., 2016. Mitigation of drought stress by foliar application of salicylic acid and potassium in mungbean (Vigna radiata L). Legume Research 39 (2), 208–214. Available at https://doi.org/10.18805/lr.v39i2.9532.

Manghwar, H., Hussain, A., Alam, I., Khoso, M.A., Ali, Q., Liu, F., 2024. Waterlogging stress in plants: unraveling the mechanisms and impacts on growth, development, and productivity. Environmental and Experimental Botany 224, 105824. Available at https://doi.org/10.1016/j.envexpbot.2024.105824.

Mir, R.A., Bhat, B.A., Yousuf, H., Islam, S.T., Raza, A., Rizvi, M.A., et al., 2022. Multidimensional role of silicon to activate resilient plant growth and to mitigate abiotic stress. Frontiers in Plant Science 13. Available at https://doi.org/10.3389/fpls.2022.819658.

Mohanty, S., Saha, S., Saha, B., Asif, S.M., Poddar, R., Ray, M., et al., 2024. Substitution of fertilizer-N with biogas slurry in diversified rice-based cropping systems: effect on productivity, carbon footprints, nutrients and energy balance. Field Crops Research 307. Available at https://doi.org/10.1016/j.fcr.2023.109242.

Morales-Cedeño, L.R., Orozco-Mosqueda, M.D.C., Loeza-Lara, P.D., Parra-Cota, F.I., de los Santos-Villalobos, S., Santoyo, G., 2021. Plant growth-promoting bacterial endophytes as biocontrol agents of pre- and post-harvest diseases: fundamentals, methods of application and future perspectives. Microbiological Research 242. Available at https://doi.org/10.1016/j.micres.2020.126612.

Mroczek-Zdyrska, M., Wójcik, M., 2012. The influence of selenium on root growth and oxidative stress induced by lead in Vicia faba L. minor plants. Biological Trace Element Research 147 (1-3), 320–328. Available at https://doi.org/10.1007/s12011-011-9292-6.

Mukherjee, A., Gaurav, A.K., Singh, S., Yadav, S., Bhowmick, S., Abeysinghe, S., et al., 2022. The bioactive potential of phytohormones: a review. Biotechnology Reports 35. Available at https://doi.org/10.1016/j.btre.2022.e00748.

Mukhtiar, A., Mahmood, A., Zia, M.A., Ameen, M., Dong, R., Shoujun, Y., et al., 2024. Role of biogas slurry to reclaim soil properties providing an eco-friendly approach for crop productivity. Bioresource Technology Reports 25. Available at https://doi.org/10.1016/j.biteb.2023.101716.

Nadeem, F., Azhar, M., Anwar-ul-Haq, M., Sabir, M., Samreen, T., Tufail, A., et al., 2020. Comparative response of two rice (Oryza sativa L.) cultivars to applied zinc and manganese for mitigation of salt stress. Journal of Soil Science and Plant Nutrition 20, 2059–2072.

Neha, G., Yadav, R.K., Yadav, A., Kumar, A.K., Rai, M., Sehwag, J., et al., 2022. Desalinization effect of simulated cut-soiler-based residue filled preferential shallow subsurface drainage: improvement of physiology and yield of rain-fed pearl millet in saline semiarid regions. Land Degradation and Development 33 (10), 1758–1770. Available at https://doi.org/10.1002/ldr.4237.

Pan, J., Sharif, R., Xu, X., Chen, X., 2021. Mechanisms of waterlogging tolerance in plants: research progress and prospects. Frontiers in Plant Science 11. Available at https://doi.org/10.3389/fpls.2020.627331.

Panda, N., Sarkar, A.K., Chamuah, G.C., 2012. Soil acidity. In: Goswami, N.N., Rattan, R.K., Dev, G., Narayanasamy, G., Das, D.K., Sanyal, S.K., Pal, D.K., Rao, D.L.N. (Eds.), Fundamental of Soil Science. Indian Society of Soil Science, New Delhi, pp. 317–328.

Pandya, D.H., Mer, R.K., Prajith, P.K., Pandey, A.N., 2005. Effect of salt stress and manganese supply on growth of barley seedlings. Journal of Plant Nutrition 27 (8), 1361–1379. Available at https://doi.org/10.1081/pln-200025835.

Rabhi, M., Barhoumi, Z., Ksouri, R., Abdelly, C., Gharsalli, M., 2007. Interactive effects of salinity and iron deficiency in Medicago ciliaris. Comptes Rendus Biologies 330 (11), 779–788. Available at https://doi.org/10.1016/j.crvi.2007.08.007.

Rahman, A., Hossain, M.S., Mahmud, J.A., Nahar, K., Hasanuzzaman, M., Fujita, M., 2016. Manganese-induced salt stress tolerance in rice seedlings: regulation of ion homeostasis, antioxidant defense and glyoxalase systems. Physiology and Molecular Biology of Plants 22 (3), 291–306. Available at https://doi.org/10.1007/s12298-016-0371-1.

Rai, A.K., Basak, N., Sundha, P., 2021a. Alternate amendments for reclamation of alkali soils. In: Minhas, P.S., Yadav, R.K., Sharma, P.C. (Eds.), Managing Salt-Affected Soil for Sustainable Agriculture. Indian Council of Agricultural Research ICAR, New Delhi, pp. 332–345.

Rai, A.K., Dinkar, A., Basak, N., Dixit, A.K., Das, S.K., Dev, I., et al., 2021. Phosphorus nutrition of oats genotypes in acidic soils: exploiting responsive plant-microbe partnership. Applied Soil Ecology 167. Available at https://doi.org/10.1016/j.apsoil.2021.104094.

Rai, A.K., Basak, N., Sundha, P., Gobinath, R., Dixit, A.K., 2022. Problem soils and their management for crop production. In: Kumar, S., Tripathi, A.K., Palsaniya, D.R., Ghosh, P.K. (Eds.), A Text Book on Recent Advances in Agronomy. Kalyani Publishers, Ludhiana, pp. 4.1–4.14.

Rai, A.K., Basak, N., Chandra, P., Sundha, P., Kaur, H., Bedwal, S., et al., 2024. Occurrence of Sphingomonas olei with elemental S oxidation capability in sodic soil: potential role in sodicity reclamation and plant growth promotion. Ecotoxicology and Environmental Safety 278, 116396.

Rai, A.K., Basak, N., Sundha, P., 2021. Saline and sodic ecosystems in the changing world. In: Rakshit, A., Singh, S.K., Abhilash, P.C., Biswas, A. (Eds.), Soil Science: Fundamentals to Recent Advances. Springer Singapore, Singapore, pp. 175–190.

Rai, A.K., Basak, Soni, N., Kumar, P.G., Sundha, S., Narjary, P., et al., 2022. Bioenergy sorghum as balancing feedback loop for intensification of cropping system in salt-affected soils of the semi–arid region: Energetics, biomass quality and soil properties. European Journal of Agronomy 134, 126452.

Reed, C., Anderson, W., Kruczkiewicz, A., Nakamura, J., Gallo, D., Seager, R., et al., 2022. The impact of flooding on food security across Africa. Proceedings of the National Academy of Sciences 119 (43). Available at https://doi.org/10.1073/pnas.2119399119.

Ritonga, F.N., Chen, S., 2020. Physiological and molecular mechanism involved in cold stress tolerance in plants. Plants 9 (5), 560. Available at https://doi.org/10.3390/plants9050560.

Sahay, S., Khan, E., Praveen, A., Panthri, M., Mirza, Z., Gupta, M., 2020. Sulphur potentiates selenium to alleviate arsenic-induced stress by modulating oxidative stress, accumulation and thiol-ascorbate metabolism in Brassica juncea L. Environmental Science and Pollution Research 27 (11), 11697–11713. Available at https://doi.org/10.1007/s11356-019-07520-w.

Sanghera, G.S., Wani, S.H., Hussain, W., Singh, N.B., 2011. Engineering cold stress tolerance in crop plants. Current Genomics 12 (1), 30–43. Available at https://doi.org/10.2174/138920211794520178.

Sarkar, A.K., Pattanayak, S.K., Singh, S., Mahapatra, P., Kumar, A., Ghosh, G.K., 2020. Integrated nutrient management strategies for acidic soils. Indian Journal of Fertilizers 16, 476–491.

Shaheen, N., Ahmad, S., Alghamdi, S.S., Rehman, H.M., Javed, M.A., Tabassum, J., et al., 2023. CRISPR-Cas system, a possible "Savior" of rice threatened by climate change: an updated review. Rice 16 (1). Available at https://doi.org/10.1186/s12284-023-00652-1.

Shahid, M., Nayak, A.K., Tripathi, R., Katara, J.L., Bihari, P., Lal, B., et al., 2018. Boron application improves yield of rice cultivars under high temperature stress during vegetative and reproductive stages. International Journal of Biometeorology 62 (8), 1375–1387. Available at https://doi.org/10.1007/s00484-018-1537-z.

Shao, G.S., Chen, M.X., Wang, D.Y., Xu, C.M., Mou, R.X., Cao, Z.Y., et al., 2008. Using iron fertilizer to control Cd accumulation in rice plants: a new promising technology. Science in China, Series C: Life Sciences 51 (3), 245–253. Available at https://doi.org/10.1007/s11427-008-0031-y.

Sharma, P.C., Singh, A., Choudhary, M., 2019. Annual Report. ICAR-Central Soil Salinity Research Institute, Karnal, Haryana, India.

Sharma, P., Jha, A.B., Dubey, R.S., Pessarakli, M., 2012. Reactive oxygen species, oxidative damage, and antioxidative defense mechanism in plants under stressful conditions. Journal of Botany 2012, 1–26. Available at https://doi.org/10.1155/2012/217037.

Sharma, S.S., Kaul, S., Metwally, A., Goyal, K.C., Finkemeier, I., Dietz, K.J., 2004. Cadmium toxicity to barley (Hordeum vulgare) as affected by varying Fe nutritional status. Plant Science 166 (5), 1287–1295. Available at https://doi.org/10.1016/j.plantsci.2004.01.006.

Sheoran, P., Kumar, A., Singh, A., Kumar, A., Parjapat, K., Sharma, R., et al., 2021. Pressmud alleviates soil sodicity stress in a rice–wheat rotation: effects on soil properties, physiological adaptation and yield-related traits. Land Degradation and Development 32 (9), 2735–2748. Available at https://doi.org/10.1002/ldr.3953.

Simón-Grao, S., Nieves, M., Martínez-Nicolás, J.J., Cámara-Zapata, J.M., Alfosea-Simón, M., García-Sánchez, F., 2018. Response of three citrus genotypes used as rootstocks grown under boron excess conditions. Ecotoxicology and Environmental Safety 159, 10–19. Available at https://doi.org/10.1016/j.ecoenv.2018.04.042.

Singh, D., Sillu, D., Kumar, A., Agnihotri, S., 2021. Dual nanozyme characteristics of iron oxide nanoparticles alleviate salinity stress and promote the growth of an agroforestry tree, Eucalyptus tereticornis Sm. Environmental Science: Nano 8 (5), 1308–1325. Available at https://doi.org/10.1039/d1en00040c.

Singh, Y.P., Arora, S., Mishra, V.K., Dixit, H., Gupta, R.K., 2017. Composting of municipal solid waste and farm wastes for its use as amendment in sodic soil. Journal of Soil and Water Conservation 16 (2), 172. Available at https://doi.org/10.5958/2455-7145.2017.00025.x.

Soni, P.G., Basak, N., Rai, A.K., Sundha, P., Narjary, B., Kumar, P., et al., 2021. Deficit saline water irrigation under reduced tillage and residue mulch improves soil health in sorghum-wheat cropping system in semi-arid region. Scientific Reports 11 (1). Available at https://doi.org/10.1038/s41598-020-80364-4.

Soni, P.G., Rai, A.K., Basak, N., Sundha, P., Narjary, B., Kumar, P., et al., 2023. Effect of deficit saline irrigation, tillage and rice straw mulch in wheat-sorghum cropping system on yield and nutritive value of rainfed forage sorghum in salt-affected soils. Land Degradation and Development 34 (17), 5635–5646. Available at https://doi.org/10.1002/ldr.4869.

Srivastava, S., Dubey, R.S., 2011. Manganese-excess induces oxidative stress, lowers the pool of antioxidants and elevates activities of key antioxidative enzymes in rice seedlings. Plant Growth Regulation 64 (1), 1–16. Available at https://doi.org/10.1007/s10725-010-9526-1.

Sundha, P., Basak, N., Rai, A., Yadav, R., Sharma, D., Sharma, P., 2017. N and P release pattern in saline-sodic soil amended with gypsum and municipal solid waste compost. Journal of Soil Salinity and Water Quality 9, 145–155.

Sundha, P., Basak, N., Rai, A.K., Yadav, R.K., Sharma, D.K., 2018. Utilization of municipal solid waste compost in reclamation of saline-sodic soil irrigated with poor quality water. Journal of the Indian Society of Soil Science 66 (1), 28–39. Available at https://doi.org/10.5958/0974-0228.2018.00004.X.

Sundha, P., Basak, N., Rai, A.K., Yadav, R.K., Sharma, P.C., Sharma, D.K., 2020. Can conjunctive use of gypsum, city waste composts and marginal quality water rehabilitate saline-sodic soils? Soil and Tillage Research 200. Available at https://doi.org/10.1016/j.still.2020.104608.

Sundha, P., Rai, A.K., Basak, N., Yadav, R.K., Sharma, P.C., 2022. P solubility and release kinetics in the leachate of saline–sodic soil: effect of reclamation strategies and water quality. Soil and Tillage Research 222. Available at https://doi.org/10.1016/j.still.2022.105440.

Sundha, P., Mukhopadhyay, R., Basak, N., Rai, A.K., Bedwal, S., Patel, S., et al., 2023. Characterization of flue gas desulphurized (FGD) gypsum of a coal-fired plant and its relevant risk of associated potential toxic elements in sodic soil reclamation. Scientific Reports 13 (1). Available at https://doi.org/10.1038/s41598-023-45706-y.

Sundha, P., Basak, N., Rai, A.K., Chandra, P., Chaudhary, A., Bedwal, S., et al., 2025. Improving soil properties through various soil amendments under changing climate scenario. Cutting Edge Technologies for Developing Future Crop Plants. Springer Nature, India, pp. 37–54. Available at http://doi.org/10.1007/978-981-96-2508-6_3.

Tarvainen, L., Näsholm, T., Way, D., 2017. Can adjustments in foliar nitrogen-use efficiency reduce drought stress impacts on boreal trees? Tree Physiology 37 (4), 415–417. Available at https://doi.org/10.1093/treephys/tpx003.

Thorne, S.J., Hartley, S.E., Maathuis, F.J.M., 2020. Is silicon a panacea for alleviating drought and salt stress in crops? Frontiers in Plant Science 11. Available at https://doi.org/10.3389/fpls.2020.01221.

Verma, O.P.S., Yadava, R.B.R., 1986. Salt tolerance of some oats (Avena sativa L) varieties at germination and seedling stage. Journal of Agronomy and Crop Science 156 (2), 123–127. Available at https://doi.org/10.1111/j.1439-037x.1986.tb00016.x.

Wang, J., Zhao, A., Ma, F., Liu, J., Xiao, G., Xu, X., 2023. Amendment of saline–alkaline soil with flue-gas desulfurization gypsum in the Yinchuan Plain, Northwest China. Sustainability 15 (11), 8658. Available at https://doi.org/10.3390/su15118658.

Wang, L., Coulter, J.A., Li, L., Luo, Z., Chen, Y., Deng, X., et al., 2020. Plastic mulching reduces nitrogen footprint of food crops in China: a meta-analysis. Science of The Total Environment 748, 141479. Available at https://doi.org/10.1016/j.scitotenv.2020.141479.

Wang, M., Zheng, Q., Shen, Q., Guo, S., 2013. The critical role of potassium in plant stress response. International Journal of Molecular Sciences 14 (4), 7370–7390. Available at https://doi.org/10.3390/ijms14047370.

Wimmer, M.A., Eichert, T., 2013. Review: mechanisms for boron deficiency-mediated changes in plant water relations. Plant Science 203-204, 25–32. Available at https://doi.org/10.1016/j.plantsci.2012.12.012.

Yadav, G.S., Devi, A.G., Das, A., Kandpal, B., Babu, S., Das, R.C., et al., 2019. Foliar application of urea and potassium chloride minimizes terminal moisture stress in lentil (Lens culinaris L) crop. Legume Research-An International Journal 44 (6), 627–633. Available at https://doi.org/10.18805/LR-4148.

Yahaya, M.A., Shimelis, H., 2022. Drought stress in sorghum: mitigation strategies, breeding methods and technologies—a review. Journal of Agronomy and Crop Science 208 (2), 127–142. Available at https://doi.org/10.1111/jac.12573.

Yan, L., Riaz, M., Wu, X., Du, C., Liu, Y., Lv, B., et al., 2018. Boron inhibits aluminum-induced toxicity to citrus by stimulating antioxidant enzyme activity. Journal of Environmental Science and Health, Part C 36 (3), 145–163. Available at https://doi.org/10.1080/10590501.2018.1490513.

Yan, S., Zhang, T., Zhang, B., Feng, H., 2024. A revised saline water quality assessment method considering including Mg2+/Na+ as a new indicator for an arid irrigated area. Journal of Hydrology 639, 131619. Available at https://doi.org/10.1016/j.jhydrol.2024.131619.

Ye, Y., Medina-Velo, I.A., Cota-Ruiz, K., Moreno-Olivas, F., Gardea-Torresdey, J.L., 2019. Can abiotic stresses in plants be alleviated by manganese nanoparticles or compounds? Ecotoxicology and Environmental Safety 184. Available at https://doi.org/10.1016/j.ecoenv.2019.109671.

Yoshimoto, N., Saito, K., Kopriva, S., 2019. S-Alk(en)ylcysteine sulfoxides in the genus Allium: proposed biosynthesis, chemical conversion, and bioactivities. Journal of Experimental Botany 70 (16), 4123–4137. Available at https://doi.org/10.1093/jxb/erz243.

Yuan, L.B., Chen, M.X., Wang, L.N., Sasidharan, R., Voesenek, L.A.C.J., Xiao, S., 2023. Multi-stress resilience in plants recovering from submergence. Plant Biotechnology Journal 21 (3), 466–481. Available at https://doi.org/10.1111/pbi.13944.

Zahoor, R., Dong, H., Abid, M., Zhao, W., Wang, Y., Zhou, Z., 2017. Potassium fertilizer improves drought stress alleviation potential in cotton by enhancing photosynthesis and carbohydrate metabolism. Environmental and Experimental Botany 137, 73–83. Available at https://doi.org/10.1016/j.envexpbot.2017.02.002.

Zhang, L.-X., Li, S.-X., Zhang, H., Liang, Z.-S., 2007. Nitrogen rates and water stress effects on production, lipid peroxidation and antioxidative enzyme activities in two maize (Zea mays L) genotypes. Journal of Agronomy and Crop Science 193 (6), 387–397. Available at https://doi.org/10.1111/j.1439-037x.2007.00276.x.

Zhang, Y., Huang, G., Zhao, Y., Lu, X., Wang, Y., Wang, C., et al., 2025. Revolutionizing crop breeding: next-generation artificial intelligence and big data-driven intelligent design. Engineering 44, 245–255. Available at https://doi.org/10.1016/j.eng.2024.11.034.

Zhao, S., Zhang, Q., Liu, M., Zhou, H., Ma, C., Wang, P., 2021. Regulation of plant responses to salt stress. International Journal of Molecular Sciences 22 (9), 4609. Available at https://doi.org/10.3390/ijms22094609.

Chapter 22

Importance of beneficial microbes in improving multiple abiotic stress tolerance of plants: special reference to plant growth-promoting rhizobacteria and arbuscular mycorrhizal fungi

Rishu Thakur[1,*], Hena Dhar[2], Amit Yadav[3], Badal Mahakalkar[3], and Rupesh Deshmukh[3]

[1]*Menzies School of Health Research, Charles Darwin University, Alice Springs, NT, Australia,* [2]*Faculty of Applied Sciences and Biotechnology, School of Biotechnology, Shoolini University, Solan, Himachal Pradesh, India,* [3]*Department of Biotechnology, Central University of Haryana, Mahendergarh, Haryana, India*

[*]*Corresponding author. e-mail address: rishu.thakur@menzies.edu.au*

22.1 Introduction

Climate change has been associated with an increase in temperatures, more frequent and intense weather events such as floods, cyclones, and changes in rainfall patterns, and a rise in the sea level. These changes have a substantial impact on agricultural production. Agricultural productions are vulnerable to heat stress, and elevated temperatures which may reduce yields or result in complete crop failures (Abebaw, 2025; Waqas et al., 2025). Changes in precipitation patterns affect agriculture and cause disruptions for traditional farming practices. Flooding events can wash away soil and damage crops. Another main challenge is sea-level rise, which is a threat to many agricultural regions along the coastline. Rising sea levels may result in the salinization of freshwater aquifers used for irrigation. Growers still need to alleviate the effects of these abiotic stresses on their produce. It is predicted that soil salinity will be a problem in more than 50% of agricultural land by 2050 (Hassani et al., 2021). To adapt to climate change, it is necessary to develop innovative and sustainable strategies that can enhance crop production and food security. Also, the increased demand for organic food after the recent global pandemic necessitates the need for sustainable and eco-friendly agricultural practices. To meet the consumers' demand, the use of chemical fertilizers further exacerbates the issue of soil and environmental pollution.

Abiotic stresses, including extreme temperatures, drought, salinity, soil conditions, and nutrient availability, cause significant crop losses worldwide (Zhang et al., 2023a). Plants being immobile, environmental factors become detrimental and affect, resulting in various responses, ranging from altered biological functions such as cell metabolism and gene expression to growth, development, and production (Zhang et al., 2023b). According to the FAO report (2007), 96.5% of the world's land surface is subject to some form of environmental constraint, leaving only 3.5% free of any constraints. With the rising temperature, the yield of wheat would decrease by 6.0%, soybean by 3.1%, maize by 7.4%, and rice by 3.2% (Zhao et al., 2017). Heat stress alters plant anatomy at both the cellular and subcellular levels. It induces chloroplast deformation, stromal lamellae enlargement, and vacuole lumpiness, all of which change thylakoid shape and impair photosynthesis and respiration rates. Thermal stress is more damaging during the crop growth's reproductive and gametogenesis stages. Thus, in combination, all of these factors significantly impair growth and productivity (Kumar, 2020).

Abiotic stress is not only responsible for reduced growth and productivity, but also impacts the quality of produce. Studies showed a decrease in the quality of gluten, glutenin, and gliadin under drought and salinity stress in wheat (Kumar, 2020). It is also observed that stress factors can cause both reversible and irreversible changes (Triboi, 2003). The abiotic stress factors usually show the effects in combinations and may not be directly involved. As in the case of flooding, effects can also be seen as increased salinity, altered microbiota in the rhizosphere, even increased disease susceptibility, and reduced drought tolerance for silicon

Multiple Abiotic Stresses in Plants. DOI: https://doi.org/10.1016/B978-0-443-27424-4.00023-9

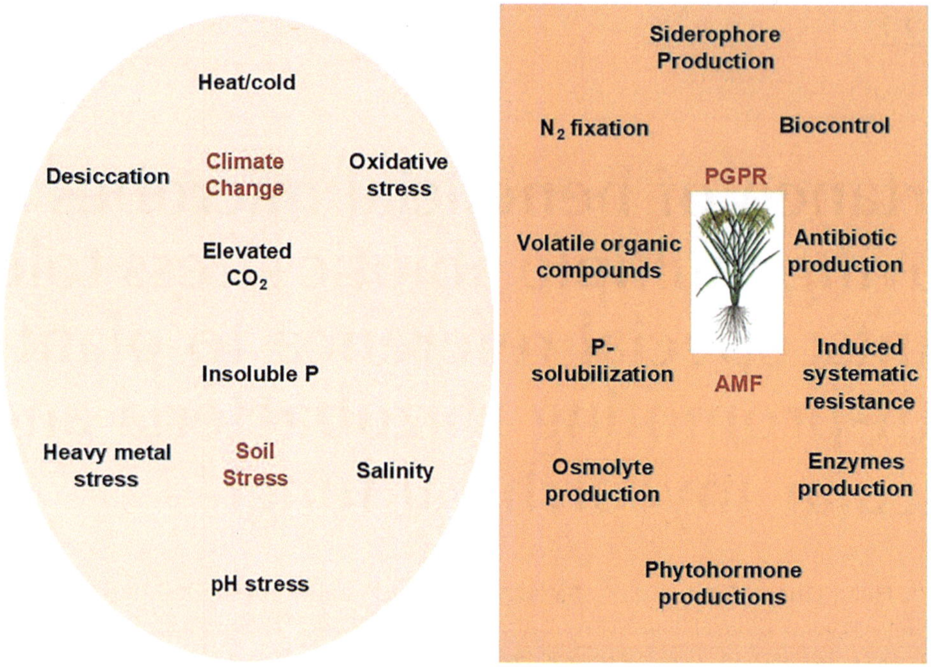

FIG. 22.1 Role of PGPR and AMF in alleviating abiotic stress. *PGPR*, plant growth-promoting rhizobacteria; *AMF*, arbuscular mycorrhizal fungi.

deficiency (Zargar et al., 2019). The gaseous composition of the environment is also a significant abiotic factor. The rise in CO_2 concentrations enhances vegetation productivity, which acts as the major negative feedback to climate warming, although its temporal dynamics over extended periods remain uncertain (Wang et al., 2020). Additionally, studies also indicate that there is a saturation for CFE. Other factors, such as mechanical stress and radiation, and altered gravity, are also significant for plant growth and production to ensure food availability on nonterrestrial systems such as space stations.

Soil is a complex and heterogeneous natural habitat of diverse microbes, such as bacteria, actinomycetes, fungi, algae, etc. Soil microbes are crucial for soil health and physiology, plant growth, and stress tolerance. The rhizosphere, a region adjacent to plant roots, is a rich source of beneficial microbes. These soil bacteria are known as plant growth-promoting rhizobacteria (PGPR). The use of PGPR in agriculture is a promising approach to improve soil quality and crop productivity in the face of climate change. PGPR create a symbiotic relationship in the rhizosphere where they facilitate nutrient uptake, disease resistance, and stress tolerance for plants while obtaining essential nutrients and a habitat from the plant. PGPR are involved in several mechanisms for plant growth promotion and stress tolerance, including phosphate (P) solubilization, siderophore production, auxins and gibberellin production, HCN production, symbiosis with nitrogen-fixers, 1-amino-cyclopropane-1-carboxylic acid (ACC)-deaminase activity, and antagonistic activity against pathogens. Besides growth promotion, PGPR also exerts stress tolerance in plants. PGPR increases the nutrient availability and uptake by increasing nutritional status and helps in relieving the stress posed due to biotic or abiotic factors. PGPR induces systemic resistance in a plant so that plants are more prepared to defend themselves against abiotic stresses. With the rhizosphere colonization mechanism, PGPR competes with pathogenic microorganisms for nutrients and space, reducing disease incidence and severity as well as indirectly benefiting plant stress tolerance. Besides bacteria, fungi such as arbuscular mycorrhizal fungi (AMF) form a symbiotic relationship with plants and help plants in growth promotion and tolerate stress conditions. Fig. 22.1 summarizes the role and impact of PGPR and AMF in abiotic stress alleviation. This chapter aims to summarize the current scenario of the use of PGPR and AMF for tackling abiotic stresses in plants.

22.2 Plant growth-promoting rhizobacteria

22.2.1 Mechanisms by which plant growth-promoting rhizobacteria supports plant growth and stress tolerance

22.2.1.1 Plant growth promotion mechanism

PGPR enhances crop productivity and sustainability in agriculture by influencing plant growth, nutrient acquisition, soil health, and stress-response pathways. Phytohormones and signaling molecules produced by PGPR play a key role in growth promotion. Phytohormones such as indole acetic acid (IAA), cytokinins, abscisic acid (ABA), and gibberellic acid are produced by PGPR. IAA has been widely reported from PGPR belonging to the genera *Bacillus*, *Burkholderia*, *Pantoea*, *Pseudomonas*, and *Rhizobium* (Ngalimat et al., 2021). IAA production has been reported to increase root length, root surface area, and biomass, which improves the ability of the plant to absorb nutrients and grow in nutrient-deficient soils. PGPR produce cytokinins, which are involved in regulating root elongation, activity in the apical meristem, and lateral roots and nodules formation. The accumulation of cytokinins by PGPR promotes plant growth and stress tolerance. Many PGPR, such as *Arthrobacter*, *Azotobacter*, *Bacillus*, and *Pseudomonas*, synthesize cytokinins, which not only increase nutrient availability but also aid in reprogramming the plant's defense response to pathogens (Khoso et al., 2024). PGPR such as *Azospirillumbrasiliense*, *Achromobacter xylosoxidans*, and *Rhizobium* spp. Produced endogenous ABA, which helps the regulation of water loss by inducing stomatal closing under stressful conditions. Interestingly, some PGPR produce ABA to enhance stress tolerance, while others use ABA as a carbon source, leading to a reduced accumulation of ABA in plant tissues.

PGPR can also promote plant growth by reducing plant ethylene levels through the secretion of the enzyme ACC-deaminase, an enzyme that hydrolyzes ACC, a precursor compound of ethylene. Ethylene is a stress-response hormone that also delays growth and promotes fruit ripening and senescence. By degrading ACC, PGPR lowers ethylene levels and allows plants to escape growth inhibition by excess ethylene under stress conditions. This action is especially useful under salt and drought stress, when ethylene production can be elevated. For instance, ACC-deaminase-producing *B. amyloliquefaciens* and *P. oryzihabitans* elongate roots, enhancing plant growth under stress (Gamalero and Glick, 2022).

Besides hormones, PGPR also synthesizes other signaling molecules, including volatile organic compounds (VOC) and exopolysaccharides (EPS) that are important in modulating plant development and enhancing resilience. VOCs are involved in the regulation of plant growth and plant–soil interaction. Moreover, some PGPR strains induce plant priming, such as changing the redox state of a plant, which could increase resistance toward pathogens and environmental stresses. Inoculating with PGPR such as *Pseudomonas* and *Enterobacter* mimics wound healing and induces the plant's defense responses that can respond to stress and activate antioxidant mechanisms more vigorously. This priming effect enables plants to better cope with stresses such as drought, salinity, and pathogen attacks.

Phosphorus (P) is a crucial macronutrient for plant growth. The scarcity of P can cause significant stress in plants. Deficiency of P can lead to stunted growth, darker leaves, poor root development, delayed maturity, and susceptibility to diseases in plants. P generally occurs in insoluble form in the soil. In acidic soils, P is quickly immobilized by metal cations such as Al and Fe, while alkaline P tends to bind with Ca in alkaline soils. PGPR that can solubilize inorganic phosphates are known as phosphate-solubilizing bacteria (PSB) or rhizobacteria. Several mechanisms are involved in the solubilization of P by PSB. PSB produces organic acids, including acetic acid, citric acid, lactic acid, propionic acid, gluconic acid, oxalic acid, succinic acid, and other organic acids. These organic acids are accompanied by a reduction in pH that results in the acidification and release of P. Some PSBs also produce enzymes such as phosphatases, protons, and hydroxyl ions to solubilize the P. Many PSB, such as *Azotobacter*, *Bacillus*, *Bradyrhizobium*, *Burkholderia*, *Kushneria*, *Rhizobium*, *Serratia*, *Salmonella*, and *Thiobacillus,* have been reported to solubilize P under different soil types and different climatic conditions (Alori et al., 2017). Some PGPR fix atmospheric nitrogen and solubilize soil insoluble potassium, release iron for the host plant, and finally produce different phytohormones that promote plant development.

Biotic as well as abiotic stresses induce a variety of physical modifications in plant cells, such as the production of reactive oxygen species (ROS). In plants, PGPR induces the immune response through an induced systemic resistance (ISR). In addition, to cope with oxidative damage, plant cells secrete a diverse array of antioxidant molecules, including carotenoids and phenylpropanoids. PGPR-mediated ISR triggers host plants to resist pathogen invasion by producing defense-related antioxidative enzymes and molecules. Many PSBs have been recorded to undergo P-solubilization under different soil types and different climatic conditions (Gulati et al., 2024; Thakur et al., 2025).

22.3 Stress tolerance mechanisms

IAA also boosts the growth and root development under abiotic stress conditions, such as drought or salinity, thus improving the ability of the plant to absorb nutrients. These microbes utilize different biochemical and physiological strategies, such as sporulation, production of EPS, synthesis of heat-shock proteins (HSPs), and modification of membrane lipid compositions to provide survival and support to the plant chloroplast to tolerate thermal stress. Heat-resistant PGPR are capable of sporulation to resist extreme heat conditions by forming endospores. When environmental conditions are appropriate again, the endospores become active and remain alive in the soil ecosystem. Production of EPS and lipopolysaccharide (LPS) helps in stabilizing the cell environment when exposed to heat stress. Additionally, certain microbes alleviate the negative impacts of heat stress by inducing the expression of heat-responsive genes or by secreting phytohormones that increase the tolerance of the plant to extreme temperatures. Such heat-tolerant strains can act as significant biological agents as better stress relievers in ameliorating heat stress to the plants. These rhizobacteria also influence the rhizosphere chemical milieu to increase nutrient bioavailability by secreting organic acids and enzymes that solubilize phosphorus and other micronutrients. Some PGPR can help plants to tolerate combined stresses such as drought, pathogens, and nutrient deficiency through multiple stress tolerance mechanisms. For instance, ACC-deaminase-producing *B. amyloliquefaciens* MMR04 has been reported to mitigate drought stress as well as combat other abiotic stresses, antioxidant defense mechanism (Murali et al., 2021).

PGPR helps the plants achieve physiological adaptability during stressful conditions by producing phytohormones such as IAA, ABA, cytokinins, gibberellins, and ethylene. PGPR can produce organic acids and metal ions that can alter the bioavailability of heavy metals in the soil. Some PGPR make symbiotic relationships with plant roots, which help in the maintenance of heavy metals within plant tissues. They produce osmoprotectants such as glycine betaine (GB) and proline, which assist in maintaining cellular water content to regulate metabolic processes in plants under osmotic stress. They also release EPS to enhance soil aggregation and root adherence (root-adhering soils), with more water and nutrient retention. Halotolerant PGPR regulate SOS1 genes 1 and HKT, which are involved in ion transport, thus maintaining ionic homeostasis by minimizing toxic Na^+ accumulation and promoting sufficient uptake of essential ions, including K^+, P, Zn^{2+}, and Fe^{3+} (Khan et al., 2019).

22.4 Role of plant growth-promoting rhizobacteria in enhancing multiple abiotic stress tolerance

22.4.1 Thermal stress

One of the biggest threats to agriculture today is thermal stress. The Intergovernmental Panel on Climate Change states that a rise of 1°C in atmospheric temperature results in a decline of 6% in crop yield. A small surge in temperature can impact the overall plant health and different developmental stages, such as plant phenology, gene expression, anatomical features, and the functional and structural orientation. Crops such as wheat, mustard, and peas are vulnerable to heat during crucial periods such as grain filling and the flowering period. Heat stress causes the denaturation of proteins as well as decreased fluidity of membrane lipids. This leads to enzyme inactivation in chloroplasts and mitochondria and inhibition of protein synthesis.

It has been documented that PGPR represent an effective resource to reduce heat stress in plants by several adaptive and protective mechanisms (Tables 22.1– 22.3). Some bacteria able to withstand high temperatures belong to the genera *Rhizobium*, *Azospirillum*, *Pseudomonas*, and *Bacillus*. The foliar application of ACC-deaminase-producing *Brevibacterium linens* RS16 enhanced the photosynthetic performance and modulated stress-related volatile emissions, which helped in alleviating the adverse effects of heat stress in *Eucalyptus grandis* (Chatterjee et al., 2020). Inoculation of *B. linens* RS16 in rice plants maintained higher chlorophyll content and photosynthetic rates during heat stress compared to uninoculated plants (Choi et al., 2022). They also adjust the saturation level of their membrane lipids and increase the concentration of unsaturated lipids to maintain membrane integrity at high temperatures. Another study found that inoculating wheat plants with *Pseudomonas aeruginosa* strain 2CpS1 helped in improving plant growth under heat stress conditions (Dasgupta et al., 2023). Application of *Aeromonasaquariorum* SDT13 and *Pseudomonas composti* SDT3 isolated from *Spartina densiflora* rhizosphere showed effectively enhanced grapevine resilience to heat stress in terms of increased reaction center availability and preserved photosynthetic efficiency, enhanced light-harvesting capabilities, maintained leaf turgidity with lower osmolyte concentrations, and reduction in lipid peroxidation products (Carreiras et al., 2023). The bacteria provided benefits such as increased plant height, biomass, and root length. Heat-resistant PGPR can synthesize HSPs that contribute to the refolding and stabilization of denatured proteins under stress conditions. This mutualistic association illustrates the potential of utilizing microbial solutions for addressing the increasing need to mitigate heat stress effects in agriculture toward the sustainability and resilience of crop production. Several PGPR, including *Bacillus cereus*, *B. pseudomycoides*, *Lelliottia amnigena*, and *L. adecarboxylata*, have been reported to alleviate the adverse effects of drought and heat stress in maize (Notununu et al., 2024). Other studies showed that ACC-deaminase-producing *Klebsiella oxytoca* and *Pseudomonas putida* improved nutrient uptake and alleviated heat and drought (Banerjee et al., 2023; Parasar et al., 2024).

22.4.2 Acidity/alkalinity

Soil pH has a substantial impact on plant growth and development because it affects nutrient availability, microbial activity, and general soil health. Acidic conditions are often associated with nutrient deficiencies. Also, low pH or high pH

TABLE 22.1 PGPR heat and cold stress tolerance mechanism.

PGPR	Mechanism and effects	Associated plant	References
Thermal stress			
B. cereus	Reduced abscisic acid (ABA) and increased salicylic acid (SA) gibberellin; indole-3-acetic acid (IAA); organic acids; HSP; Increased antioxidant content (ascorbic acid, peroxidase, superoxide dismutase, glutathione)	Soyabean	Khan et al. (2020)
B. cereus, Providencia rettgeri, Myroides odoratimimus	P-solubilization; siderophores production; IAA production	Sorghum	Bruno et al. (2020)
B. safensis	Reduced ROS production and photosynthetic damage; maintained chloroplast structure and enhanced antioxidant enzyme activity; produced osmolytes such as proline, glycine betaine; upregulated the expression of HSP; enhanced performance of redox enzymes and antioxidants; low proline accumulation; accumulation of glycine betaine (GB), total soluble sugars (TSS), and phenols	Wheat	Sarkar et al. (2018b), Sarkar *et al.*, (2021)
B. thuringiensis, B. cereus, B. subtilis	Phosphate (P)- solubilization; antimicrobial activity; proline accumulation; temperature-induced proteins	Wheat	Ashraf et al. (2019)
Bacillus spp. *Pseudomonas* spp.	Upregulated of heat-shock proteins (HSP) (HSP1, HSP18, HSP70, HSP101); enhanced antioxidant enzyme activity; reduced malondialdehyde	Maize	Ahmad et al. (2023)
Klebsiella sp	Reduced stress-induced ethylene and regulation of ion transporters	Wheat	Singh *et al.*, (2015)
P. megaterium	Upregulated antioxidant defense system; increased endogenous SA levels and decreased ABA levels; P-solubilization; production of IAA, siderophore, and exopolysaccharide (EPS)	Soyabean	Shaffique et al. (2024)
P. putida	Reduced membrane injury; enhanced antioxidant enzyme activity; increased cellular metabolites such as amino acids, chlorophyll, proline, proteins, starch, and sugars	wheat	Ali et al. (2011)
Pseudomonas spp.	Modified root exudate composition and shift in carbon	Rapeseed, Camelina	Delamare et al. (2023)
Cold stress			
Acinetobacter rhizosphaerae	P, K, Zn solubilization; IAA production; siderophores and hydrolytic enzymes production	Wheat	Kour and Yadav. (2023)
Arthrobacter sp., Flavimonas sp., Flavobacterium sp., Massilia sp., Pedobacter sp., *Pseudomonas* sp.	Reduced membrane damage; antioxidant enzymes; enhanced proline synthesis	Tomato	Subramanian et al. (2016)
B. amyloliquefaciens, Brevibacillus laterosporus	P-solubilization; siderophores production; IAA production Decreased leaf MDA content and electrolyte leakage; increased proline and chlorophyll content; antioxidant enzyme activity; Upregulation of stress-related genes (OsMYB3R-2, OsDIL, OsDREB1A, OsCDPK13)	Rice	Kakar et al. (2016)

(Continued)

TABLE 22.1 (Continued)

PGPR	Mechanism and effects	Associated plant	References
Bacillus spp., *B. velezensis*	Production of IAA, siderophores, and hydrogen cyanide	Wheat	Zubair et al. (2019)
Brevibacterium halotolerans, *B. subtilis*, *Achromobacter xylosoxidans*, *Burkholderia cepacia*	Increased accumulation of proline, starch, and total phenolics, and reduces ethylene	Basil	Singh et al. (2020)
Pseudomonas spp.	Increased cellular metabolites such as amino acids, anthocyanin, chlorophyll, proline, total phenolics, starch, proteins	Wheat	Mishra et al. (2011)

PGPR, plant growth-promoting rhizobacteria.

TABLE 22.2 PGPR drought and salinity stress tolerance mechanism.

PGPR	Mechanism and effects	Associated plant	References
Drought			
Azospirillum sp.	Improved plants' response to water deficit	Maize	García et al. (2017)
B. cereus, B. albus	IAA, siderophore, and EPS production; antioxidant enzyme (catalase)	Dryland plants	Ashry et al. (2022)
B. licheniformis	Enhanced protein and gene expression patterns (Cadhn, VA, sHSP, CaPR-10); production of auxin and ACC deaminase	Pepper	Lim and Kim. (2013)
B. subtilis, B. thuringiensis, B. megaterium	Improved osmoregulation and reduction of oxidative stress; systemic acquired resistance (SAR)	Chickpea	Khan *et al.*, (2018)
B. velezensis, B. amyloliquefaciens	P-solubilization; production of siderophores, IAA, and gibberellic acid	Walnut	Lotfi et al. (2022)
Klebsiella variicola, Raoultella planticola, P. fluorescens	Enhanced accumulation of choline and GB	Maize	Gou et al. (2015)
Ochrobactrum pseudogrignonense, Pseudomonas sp., *B. subtilis*	Enhanced production of ROS scavenging enzymes and cellular osmolytes; downregulated ACC-oxidase gene expression	Black gram and pea	Saikia et al. (2018)
P. azotoformans	Production of EPS, IAA, TCP solubilization; ACC-deaminase activity	Wheat	Ansari *et al.*, (2021)
Pseudomonas sp., *B. lentus, A. brasilens*	Increased antioxidant activity and chlorophyll content	Basil	Heidari and Golpayegani. (2012)
Salinity			
Acinetobacter bereziniae, *Enterobacter ludwigii, Alcaligenes faecalis*	Increased chlorophyll, proline, total soluble sugar content; reduced electrolyte leakage; antioxidant enzyme activities	Pea	Sapre *et al.*, (2022)
Brachybacterium saurashtrense, Brevibacterium casei	Enhanced water content; reduced electrolyte leakage and MDA content; increased Osmoprotectants; auxin production	Peanut	Shukla *et al.*, (2012)

(Continued)

TABLE 22.2 (Continued)

PGPR	Mechanism and effects	Associated plant	References
Brevibacterium frigoritolerans, B. thuringiensis, B. velezensis	Enhanced proline and soluble sugar content and antioxidant enzyme activity; decreased ethylene levels	Wheat	Huang et al. (2022)
Enterobacter sp., *Pseudomonas* sp.	Reduced electrolyte leakage index (ELI); decreased MDA and hydrogen peroxide (H_2O_2); increased antioxidant capacity, proline content, and antioxidant enzyme activities	Canola	Neshat et al. (2022)
Kocuria rhizophila	Enhanced nutrient acquisition and relative water content; reduced Na^+ levels and electrolyte leakage; increased expression of stress-responsive genes (ZmGR1, ZmAPX1, ZmNHX1, ZmNHX2, ZmNHX3, ZmWRKY58, ZmDREB2A)	Maize	Li et al. (2020)
P. fluorescens, B. pumilus, E. aurantiacum	Increased proline content	Wheat	Nawaz et al. (2020)

PGPR, plant growth-promoting rhizobacteria.

conditions make toxic elements soluble, such as aluminum (Al) in acidic soils and manganese (Mn) in alkaline soils, which may restrict root development and alter physiological processes and negatively impact crop yields. In acidic soils, P is quickly immobilized by metal cations such as Al and Fe, while alkaline P tends to bind with Ca in alkaline soils. Scarcity of P can cause significant stress in plants. Plants struggle to carry out photosynthesis and root development in the absence of P. Deficiency of P can lead to stunted growth, darker leaves, poor root development, delayed maturity, and susceptibility to diseases in plants. P is generally present in insoluble form in the soil. Phosphate fertilizers are applied to meet the P demand, but most of the applied P is not available to the plants. Extreme pH levels also impede water uptake by plants and degrade soil structure, compounding the obstacles to plant growth.

PGPR secretes ammonia and other buffering agents to neutralize pH. In acidic soils, PGPR release organic acids that dissolve phosphate that is fixed by Al or iron (Fe), making it available for plants. In alkaline soils, PGPR excrete chelating agents, including siderophores, which bind iron and facilitate plant bioavailability. Rhizobacteria such as *Viridibacillus arenosi* and *Priestia megaterium* are reported to grow under acidic and alkaline conditions (Thakur et al., 2024; Thakur et al., 2017). They also exude acid phosphatases to mobilize P from insoluble calcium phosphates, to alleviate limitations in nutrient availability. PGPR also helps plants in invoking tolerance to stress mechanisms by activating stress-response mechanisms, such as antioxidant defense systems that limit oxidative injury under extreme pH conditions.

22.4.3 Heavy metal

Heavy metal contamination is a significant problem in the agriculture sector. Heavy metals such as Ni, Cd, Hg, Cu, Cr, As, Pb, and Zn are very toxic and harmful to plants. Extensive usage of synthetic fertilizers, urbanization, industrialization, farmyard manure, pesticides, and sewage sludge contributes to the accumulation of heavy metals in soil. Heavy metals are hard to degrade and can accumulate in the soil for a very long time. Soil contaminated by heavy metals results in acidification, biodiversity loss, desertification, ecological unevenness, erosion, salinization, and low soil fertility. Extreme concentrations of heavy metals in the soils have also led to lowered soil microbial activity. High concentrations of heavy metals can increase the plant ethylene production, which can restrict plant growth. Soil factors such as pH and soil types affect the composition of heavy metals in the soil. Some studies also reported multiple stress tolerance by rhizobacteria against acidity/alkalinity, desiccation, salinity, and metal tolerance. For instance, *V. Arenosi* IHB B 7171 exhibited stress tolerance against acidity/alkalinity, desiccation, salinity, and aluminum and iron metal ions with potential as a broad-spectrum PGPR (Thakur et al., 2017). *P. megaterium* IHB B 7164, a PGPR from tea rhizosphere, showed multiple abiotic tolerance against salinity, desiccation, and exposure to calcium, aluminum, and iron salts, corroborated by genomic analysis (Thakur et al., 2024). The role of PGPR in alleviating metal toxicity in plants is well described in many studies. PGPR use different strategies such as acidification, biotransformation, chelation, efflux system, and siderophore

TABLE 22.3 PGPR heavy metals and acidity/alkalinity stress tolerance mechanism.

PGPR	Mechanism and effects	Associated plant	References
Heavy-metal stress			
B. pseudomycoides	Increased superoxide dismutase, ascorbate peroxidase, and catalase activities; reduced electrolyte leakage	Maize	Anbuganesan et al. (2024)
Bacillus gibsonii, B. xiamenensis	Increased antioxidant enzyme activities; IAA production, and ACC-deaminase activity	Sesbania sesban	Zainab et al. (2021)
Ochrobactrum sp., *Bacillus* sp., *Bacillus* sp.	Enhanced amylase and protease activities; reduced superoxide dismutase (SOD) activity and MDA levels; ACC-deaminase activity; siderophore production	Rice	Pandey et al. (2013)
P. azotoformans, S. rubidaea, P. pabuli, B. velezensis	Improved photosynthetic pigments; lower electrolyte leakage; enhanced antioxidant defense systems	Pepper	El-Saadony et al. (2024)
P. aeruginosa	Enhanced antioxidative enzyme activities (SOD, POD, CAT); Increased ascorbic acid and total phenolics; reduced H_2O_2 and MDA content	Wheat	Islam et al. (2014)
P. aeruginosa, Burkholderia gladioli	Enhanced chlorophyll, carotenoid, and xanthophyll content; upregulated metal transporter genes	Lettuce	Khanna et al. (2019)
Pseudomonas sp.	IAA Production; Ammonia Production; P-solubilization	Wheat	Verma *et al.*, (2015)
Pseudomonas sp.	P-solubilization; nitrogen fixation; ammonia, cellulase, chitinase, exopolysaccharides, IAA, hydrogen cyanide, pectinase, and siderophores production	Alfalfa	Tirry et al. (2021)
Acidity/alkalinity			
Azospirillum brasilense, Arthrobacter crystallopoietes	Increased cation exchange capacity (CEC)	Maize	Kari et al. (2021)
Bacillus sp., *Alcaligenes* sp., *Bacillus* sp.	Improved photosynthetic pigments, soluble sugar content, and reduced proline levels; enhanced soil enzymes (dehydrogenase, alkaline phosphatase, beta-glucosidase)	Maize	Dixit et al. (2020)
Bacillus spp., B. humi, B. megaterium, B. drentensis, B. pocheonensis, B. aestuarii, B. arbutinivorans, B. niacini, Brevibacterium casei	Enhanced carbohydrate utilization	Rice	Yadav et al. (2011)
Bacillus spp., *Pseudomonas* spp., *Brevibacillus* sp., *Stenotrophomonas* sp., *Paenibacillus* sp.	Nitrogen fixation; P-solubilization; siderophore production	Grapevine	Karagöz et al. (2012)

PGPR, plant growth-promoting rhizobacteria.

production to remediate heavy-metal contamination. PGPR reduces the uptake of toxic metals by immobilizing metals in the rhizosphere through biofilms, promotes selective uptake of essential nutrients over heavy metals, and enhances the production of antioxidants in plants to combat oxidative stress caused by metal toxicity. Many studies have shown that employing PGPR as bioinoculants can be an effective approach for the bioremediation of heavy-metal-contaminated soils. Inoculation of the crops with PGPR strains tolerant to heavy metals improves root and shoot growth and chlorophyll contents, nutrient uptake, etc. Moreover, they are critical in complexing heavy metals in the soil to decrease their bioaccessibility and subsequent incorporation by plants. This is especially critical for preventing metals from accumulating in the edible parts of plants that are the source of food. PGPR assists in phytoremediation by facilitating the accumulation, stabilization, or transformation of heavy metals in plants. A study on the application of *Bacillus*, *Klebsiella*, *Leifsonia*, and *Enterobacter* on maize showed an enhancement in root and shoot growth under cadmium stress (Ahmad et al., 2016). Siderophore-producing rhizobacteria, including *Pseudomonas aeruginosa*, *P. fluorescens*, and *Ralstonia metallidurans*, showed their ability in phytoextraction, where plants can extract heavy metals from contaminated soils (Braud et al., 2009). They enhance root growth, allowing plants to access larger soil volumes for metal uptake or exclusion. PGPR promotes sustainable agriculture by maintaining soil fertility and structure even in contaminated soils. They support nutrient cycling and improve soil resilience to contamination. The amalgamation of PGPR in phytoremediation strategies offers a sustainable approach to managing metal toxicity in soil and reducing dependence on chemical remediation methods.

22.4.4 Salinity

Salinity is one of the most serious environmental stresses that negatively affects plant growth and productivity by causing osmotic stress, ionic imbalance, and oxidative damage. Osmotic stress limits water intake, inhibiting cell expansion and resulting in a decrease in leaf growth, thus hindering general plant growth. Ionic stress, resulting from the disproportionate build-up of sodium and chloride ions, manifests by disrupting nutrient uptake through the depletion of the requisite elements such as calcium (Ca^{2+}), potassium (K^{+}), nitrogen (N), and nitrate (NO_3^-), culminating in nutrient imbalance and field-upward leaf senescence. ROS leads to oxidative stress by damaging cellular structures such as DNA, RNA, proteins, and lipid membranes and diminishing tracking resources.

Several rhizobacteria have been reported to promote plant growth under salinity stress. Among these are halotolerant PGPR, which can alleviate the detrimental impact of salt stress and enhance plant growth and yields. Halotolerant PGPR are well known for their salt stress alleviation and PGP potential, as well as their effectiveness in halophytic soil reclamation. They colonize the rhizosphere and use a variety of strategies to mitigate the negative effects of salinity. Inoculation of wheat with *B. pumilus* FAB10 significantly improved shoot and root length, biomass, and overall vigor under both normal and saline conditions (Ansari et al., 2019). The application of carotenoid-producing PGPR *Dietzia natronolimnaea* STR1 stimulated the expression of salt-tolerant genes (*TaST*). This application also showed an upregulation in genes encoding antioxidant enzymes such as ascorbate peroxidase, catalase, glutathione peroxidase, glutathione reductase, and peroxidase. This enhancement likely contributes to reduced oxidative damage under salt stress (Bharti et al., 2016). Inoculation with ACC-deaminase-producing *Pseudomonas fluorescens* led to significant improvements in wheat growth under salt stress, including increased germination rate, enhanced seed vigor, longer and heavier coleoptiles and radicles, and higher salinity tolerance index (Safari et al., 2018).

Inoculation with *Enterobacter* sp. P23 led to significant improvements in rice seedling growth under salt stress, including increased root and shoot lengths and overall biomass. The bacterium's ACC-deaminase activity resulted in decreased levels of stress-induced ethylene, a plant hormone associated with growth inhibition under stress conditions (Sarkar et al., 2018a). Inoculation with *Serratia marcescens* significantly improved wheat growth under various salinity levels (150–200 mM NaCl). This enhancement was evident in increased shoot and root lengths, fresh and dry weights, and chlorophyll content (Singh and Jha, 2016).

Halotolerant PGPR also aid in plant protection from oxidative injury via the induction of the plant antioxidant defense system. Moreover, halotolerant PGPR enhances stress tolerance by inducing systemic tolerance (IST), which is accomplished through synthesizing ACC deaminase, which decreases ethylene concentration during stress, and producing VOCs to improve plant resistance against stress. They have developed bioformulations that help in enhancing crop productivity and soil health parameters in saline conditions. This makes halotolerant PGPR act against salinity stress, which helps in sustainable management of salinity stress, leading to improved plant growth for increased agricultural production in degraded soils.

22.4.5 Drought

Drought stress is among the most crucial abiotic threats to global agriculture, with the potential to limit crop yield and thereby jeopardize food security. Drought stress led to changes in metabolic pathways, growth inhibition, reduction in photosynthesis, and reduced yields. PGPR alleviates drought stress mainly by altering the morphology of roots. PGPR can promote root growth, leading to deeper and wider root systems. Such changes loosen soil to allow plants to access water and nutrients more effectively and are associated with a greater ability to withstand periods of water shortages. Plant root-modifying capacity is important for improved drought tolerance, particularly under conditions of water limitation. PGPR assists plants in managing water shortages more efficiently by regulating the expression of drought-resistance genes. By enhancing the plant's ability from the root to withstand the physiological strains of drought, the genetic modulation increases plant resilience. These hormones can modulate the water status of plants via stomatal closure and enhance water-use efficiency. For example, ABA production enhances plant water conservation under drought, and thus it is one of the important agents through which PGPR has improved water-use efficiency in crops. Production of osmolytes and EPS by PGPR is also elaborated in the review. These osmolytes assist in maintaining cellular turgor and safeguarding cellular structures during drought stress, aiding osmotic adjustment and providing cellular protection.

In addition, PGPR enhances the plant antioxidant defense systems that reduce the cellular oxidative damage due to enhanced production of ROS during drought stress. Apart from the above mechanisms, PGPR also solubilize nutrients by producing siderophores that chelate iron. VOCs emitted from PGPR may also enhance plant growth and increase drought tolerance. PGPR can be used successfully in enhancing crop performance under water-deficient conditions via a change in the plant physiology by increasing hormone production, root modulation, and antioxidant reaction.

22.4.6 Cold stress

Cold stress is another major stress in which plants encounter lower temperatures, varying from 0°C to 15°C. Cold stress causes a shift in the metabolic response of plants, which affects the plant cycle and results in loss of crop production. The impact of cold stress varies from structural to molecular levels. The main effect of cold is reduced seed germination. Decline in photosynthesis due to loss of leaf development, weakened photosynthetic machinery, and early leaf aging are also observed under cold stress conditions. Plants generate ROS that harm cellular homeostasis during cold stress. However, plants differ in their tolerance to low temperatures; for instance, temperate region plants are cold-tolerant.

Cold-tolerant PGPR has great agronomical importance. Crop growth cycles are subject to varying cold spells in most parts of the world. A meta-analysis conducted on the PGPR impacts on plants under temperature stress revealed that the inoculation of PGPR increased chlorophyll contents in plants under temperature stress 5°C–14°C (Zhang et al., 2023c). Another study conducted on wheat seedlings showed that *Pseudomonas* strains were effective in mitigating cold stress (Mishra et al., 2011). Cold-stressed plants inoculated with *Kosakonia* sp. CIR2 and *Staphylococcus* sp. CSR1T2 substantially enhanced cold tolerance in rice plants (de Souza et al., 2021). Maize plants inoculated with *Lysinibacillus fusiformis* and *L. sphaericus* showed resistance to osmotic and oxidative stress induced by cold exposure (Jha and Mohamed, 2023). Inoculation of *Streptomyces* sp. in tomato plants enhances the cold stress tolerance by changing the expression of stress tolerance genes (Ma et al., 2023).

22.5 Role of fungi in abiotic stress tolerance

Abiotic stress is the leading cause of crop loss worldwide, cutting average yields for most of the main agricultural plants by more than half (Kopecká et al., 2023). Fungi play a crucial role in helping plants cope with abiotic stresses and may be useful for better yield quantity and quality. Its association is often mutualism or commensalism by symbiotic or free-living interaction. It offers resistance to drought, metals, disease, heat, and herbivory (Singh et al., 2011), while also promoting growth and nutrient solubilization, fixation, and uptake. It has also been demonstrated that some plants are unable to endure habitat-imposed abiotic and biotic challenges in the absence of fungal endophytes. Based on the site of fungal habitat, there are two types of fungal association with plants, namely endophytic and mycorrhizal.

AMF are soil-borne fungi from the phylum *Glomeromycota* that can improve plant nutrient absorption and resilience to a variety of abiotic stressors (Begum et al., 2019). The symbiotic association between plant roots and fungi is observed to be 80% of plant species, suggesting its critical importance (Bhantana et al., 2021). In a recent study on *Oryza sativa*, arbuscular mycorrhizal (AM) plants were found to exhibit higher photochemical efficiency in CO_2 fixation. This enhanced efficiency contributes to improved salt tolerance by protecting the plant's photosystems from damage and optimizing light energy use during photochemical processes (Porcel et al., 2015). Studies have also established that AM also reduces drought stress in *Leymus chinensis* and *Hemarthria altissima*, suggesting that the carbon cycle is not detrimental to the fungal symbiosis and its positive effects (Li et al., 2019).

Similarly, AMF also assists in nutrient availability, as in the case of the essential element phosphorus (P) often lacking in soil. The primary source of phosphatic fertilizer, rock phosphate, is nonrenewable and on the verge of depletion in the next 50–100 years (Gilbert, 2009). Phosphate supplemental techniques using AMF are promising, as in the case of foxtail (Ceasar et al., 2014). AMFs not only enhance nutrient bioavailability and uptake for plants such as Cu, Fe, K, N, P, Mn, S, and Zn (Chen et al., 2017) but also play a critical role in phytoremediation by reducing the harmful effects of Ag, Fe, and Zn (Cao et al., 2016). Fungal association with a virus is also related to thermal stress tolerance. Fungal isolates that have been cured of the virus are unable to transmit heat tolerance; however, heat tolerance is restored when the virus is reintroduced (Márquez et al., 2007). Inoculation with *Piriformospora indica* reduced stress effects, decreased oxidative damage, and improved maize tolerance to hard environments (Hosseini et al., 2018), and potentially even the anchorage in flooding and high winds. Overall, fungi improve plant tolerance to abiotic conditions such as drought, heat, and salinity by increasing nutrient uptake and lowering oxidative damage. Their role in fostering resistance makes them a valuable method for enhancing plant growth and output, and could be a step towards food security in a globally stressed environment, as well as the potential to help overcome dependency on various chemical fertilizers, etc.

22.5.1 Mechanism of abiotic stress tolerance in fungi

Fungi do not simply tolerate and survive the extremes of abiotic factors such as high temperature, drought, heavy metals, nutrient deficiency, salinity, radionuclides, and high concentrations of toxic xenobiotic compounds, microgravity, but are also actively reproducing by adopting different mechanisms (Selbmann et al., 2013). When subjected to abiotic conditions such as drought, salt, severe temperatures, and oxidative damage, fungi develop complex adaptation and survival mechanisms. These adaptive tactics not only maintain their survival but also play a key role in their interaction with plants, especially in symbiotic interactions. Mycorrhizal fungi, for example, improve plant tolerance to abiotic stress by increasing water and nutrient intake while also regulating plant stress-response pathways. It is widely accepted that many environmental stress-related features are the product of genetics in fungi; these specific resistance genes may be useful for improving crop tolerance to single or multiple pressures, thus they provide indirect assistance to plants to cope with the stress (Zhang, 2016). AMF, including *Funneliformis mosseae* and *Glomus constrictum*, have been reported to tolerate salt stress through increasing the levels of enzymes, including catalase, ascorbate peroxidase, and SOD (Evelin et al., 2019; Pandey et al., 2013). These two species are also reported to tolerate low temperatures by increasing the accumulation of soluble sugars and proline (Joudmand et al., 2021). There are mainly two mechanisms: Endophytic Symbiosis and AMF. Endophytic fungi live inside plant tissues and respond to diverse plant stimuli by acting to reduce stress through various mechanisms. Fungi may penetrate plant tissues by a variety of routes, including lenticels, germination radicles, root splits, stomata, and tissue wounds, gaining access to plant sap and allowing the transfers (Lu et al., 2021). The simplest response to stress is ROS scavenging: fungi help plants with their antioxidant defenses. Some endophytic fungi increase plant enzyme levels, lowering damaging ROS accumulation that would otherwise damage plant cells during stress, or by producing the enzyme ACC deaminase. ACC deaminase lowers the amount of ACC, a precursor of the stress hormone ethylene. Lowering ethylene levels can reduce stress-induced plant senescence while also promoting root development, cell division, and elongation (Lu et al., 2021).

AMF mycelia increase root surface area, which improves water and nutrient intake and helps plants deal with thermal stress. They increase antioxidant enzyme activity (APX, CAT, POD, SOD) to reduce ROS damage and stimulate the accumulation of osmolytes (sugars and proline) to maintain cellular osmotic balance. AMF also stabilizes chloroplast function and chlorophyll concentration, which boosts photosynthetic rates. Gene regulation, particularly the upregulation of HSPs and aquaporins, contributes to stress responses and cellular integrity. Together in combinations, these systems make a plant more resistant to heat-induced harm (Jian et al., 2024).

Under high salinity conditions, the nutrient uptake for the plants becomes difficult. In plants with AMF appressorium is generated as a response to the root surface, hyphae are extended into the root cortex, and arbuscules and vesicles are developed to facilitate nutrition transfer. AMF improve water and nutrient uptake, increase osmolyte buildup, and boost water efficiency. This allows plants to better withstand salt stress and improve overall resilience and output (Evelin et al., 2009).

AMF also contribute to heavy-metal remediation by altering plant physiology to increase tolerance to heavy metal stress. AMF cause chelator production in the host plant, which prevents metal transport to the shoot and increases tolerance. AMF's extraradical hyphae effectively bind metals, which are aided by chitin and chitosan in the fungal cell wall. AMF also produces glomalin, a glycoprotein that aids in the immobilization of metals in the mycorrhizosphere, hence improving the plant's ability to sequester and tolerate heavy metals in polluted soil (Riaz et al., 2021).

Overall, fungi considerably improve plant resistance to abiotic stressors such as drought, heat, salt, and heavy metals. This may not be observed in premia facia, such as the disease tolerance by better availability of the nutrients fixed by the fungi, or better flowering and production. Fungi such as mycorrhizal and endophytic fungi assist plants in withstanding hard environments by enhancing nutrient and water intake, scavenging ROS, and stimulating root growth. They also improve heavy metal tolerance and plant antioxidant defenses. Given the growing environmental challenges, fungi-based solutions may be critical for boosting crop resilience and maintaining future food security.

22.6 Conclusion and future directions

PGPR and AMF formulations are well known for exhibiting stress tolerance and growth promotion capabilities under controlled and field conditions. However, these formulations account for only a small segment of global agriculture. The complex soil dynamics and environmental conditions in real-world scenarios could affect the performance of PGPR inoculum. To unlock the full potential of these formulations, solutions that are fit for complex field conditions should be implemented. Selection of strains with multiple PGP attributes and stress tolerance to various biotic and abiotic factors is essential. Achieving effective viability of microbial formulation under real-world scenarios is the most significant challenge. Suitable carriers that can safeguard long-term nutrient supply can enhance their establishment and effectiveness in the soil. Appropriate application methods are important for safeguarding the effective delivery and colonization of microbial formulation in the plant. Combined application of PGPR and AMF can further boost their beneficial outcomes with improved soil health. A coinoculation approach using PGPR and AMF can be a sustainable and adaptive strategy to tackle environmental stresses in the face of climate change. These synergistic interactions can enhance agroecosystem resilience. However, careful optimization is required for consistency and scalability. Continuous monitoring of the plant–microbe interactions through omics techniques such as metaproteomics and metatranscriptomics is also essential to understand the efficacy of PGPR at different stages of the growth cycle.

References

Abebaw, S.E., 2025. A global review of the impacts of climate change and variability on agricultural productivity and farmers' adaptation strategies. Food Science and Nutrition 13 (5). Available at https://doi.org/10.1002/fsn3.70260.

Ahmad, I., Akhtar, M.J., Asghar, H.N., Ghafoor, U., Shahid, M., 2016. Differential effects of plant growth-promoting rhizobacteria on maize growth and cadmium uptake. Journal of Plant Growth Regulation 35 (2), 303–315. Available at https://doi.org/10.1007/s00344-015-9534-5.

Ahmad, M., Imtiaz, M., Nawaz, M.S., Mubeen, F., Sarwar, Y., Hayat, M., et al., 2023. Thermotolerant PGPR consortium B3P modulates physio-biochemical and molecular machinery for enhanced heat tolerance in maize during early vegetative growth. Annals of Microbiology 73 (1). Available at https://doi.org/10.1186/s13213-023-01736-5.

Ali, S.Z., Sandhya, V., Grover, M., Linga, V.R., Bandi, V., 2011. Effect of inoculation with a thermotolerant plant growth promoting Pseudomonas putida strain AKMP7 on growth of wheat (Triticum spp.) under heat stress. Journal of Plant Interactions 6 (4), 239–246. Available at https://doi.org/10.1080/17429145.2010.545147.

Alori, E.T., Glick, B.R., Babalola, O.O., 2017. Microbial phosphorus solubilization and its potential for use in sustainable agriculture. Frontiers in Microbiology 8. Available at https://doi.org/10.3389/fmicb.2017.00971.

Anbuganesan, V., Vishnupradeep, R., Bruno, L.B., Sharmila, K., Freitas, H., Rajkumar, M., 2024. Combined application of biochar and plant growth-promoting rhizobacteria improves heavy metal and drought stress tolerance in Zea mays. Plants 13 (8). Available at https://doi.org/10.3390/plants13081143.

Ansari, F.A., Ahmad, I., Pichtel, J., 2019. Growth stimulation and alleviation of salinity stress to wheat by the biofilm forming Bacillus pumilus strain FAB10. Applied Soil Ecology 143, 45–54. Available at https://doi.org/10.1016/j.apsoil.2019.05.023.

Ansari, F.A., Jabeen, M., Ahmad, I., 2021. Pseudomonas azotoformans FAP5, a novel biofilm-forming PGPR strain, alleviates drought stress in wheat plant. International Journal of Environmental Science and Technology 18 (12), 3855–3870. Available at https://doi.org/10.1007/s13762-020-03045-9.

Ashraf, A., Bano, A., Ali, S.A., Papen, H., 2019. Characterisation of plant growth-promoting rhizobacteria from rhizosphere soil of heat-stressed and unstressed wheat and their use as bio-inoculant. Plant Biology 21 (4), 762–769. Available at https://doi.org/10.1111/plb.12972.

Ashry, N.M., Alaidaroos, B.A., Mohamed, S.A., Badr, O.A.M., El-Saadony, M.T., Esmael, A., 2022. Utilization of drought-tolerant bacterial strains isolated from harsh soils as a plant growth-promoting rhizobacteria (PGPR). Saudi Journal of Biological Sciences 29 (3), 1760–1769. Available at https://doi.org/10.1016/j.sjbs.2021.10.054.

Banerjee, A., Chakraborty, N., Acharya, K., Chakraborty, N., 2023. Role of endophytes, plant growth promoting rhizobacteria, and arbuscular mycorrhizal fungi in stress tolerance. Biology and Biotechnology of Environmental Stress Tolerance in Plants: Sustainable Approaches for Enhancing Environmental Stress Tolerance 3, 57–96.

Begum, N., Qin, C., Ahanger, M.A., Raza, S., Khan, M.I., Ashraf, M., et al., 2019. Role of arbuscular mycorrhizal fungi in plant growth regulation: implications in abiotic stress tolerance. Frontiers in Plant Science 10. Available at https://doi.org/10.3389/fpls.2019.01068.

Bhantana, P., Rana, M.S., Sun, Xc, Moussa, M.G., Saleem, M.H., Syaifudin, M., et al., 2021. Arbuscular mycorrhizal fungi and its major role in plant growth, zinc nutrition, phosphorous regulation and phytoremediation. Symbiosis 84 (1), 19–37. Available at https://doi.org/10.1007/s13199-021-00756-6.

Bharti, N., Pandey, S.S., Barnawal, D., Patel, V.K., Kalra, A., 2016. Plant growth promoting rhizobacteria Dietzia natronolimnaea modulates the expression of stress responsive genes providing protection of wheat from salinity stress. Scientific Reports 6. Available at https://doi.org/10.1038/srep34768.

Braud, A., Jézéquel, K., Bazot, S., Lebeau, T., 2009. Enhanced phytoextraction of an agricultural Cr- and Pb-contaminated soil by bioaugmentation with siderophore-producing bacteria. Chemosphere 74 (2), 280–286. Available at https://doi.org/10.1016/j.chemosphere.2008.09.013.

Bruno, L.B., Karthik, C., Ma, Y., Kadirvelu, K., Freitas, H., Rajkumar, M., 2020. Amelioration of chromium and heat stresses in Sorghum bicolor by Cr6+ reducing-thermotolerant plant growth promoting bacteria. Chemosphere 244, 125521. Available at https://doi.org/10.1016/j.chemosphere.2019.125521.

Cao, J., Feng, Y., Lin, X., Wang, J., 2016. Arbuscular mycorrhizal fungi alleviate the negative effects of iron oxide nanoparticles on bacterial community in rhizospheric soils. Frontiers in Environmental Science 4. Available at https://doi.org/10.3389/fenvs.2016.00010.

Carreiras, J., Cruz-Silva, A., Fonseca, B., Carvalho, R.C., Cunha, J.P., Proença Pereira, J., et al., 2023. Improving grapevine heat stress resilience with marine plant growth-promoting rhizobacteria consortia. Microorganisms 11 (4). Available at https://doi.org/10.3390/microorganisms11040856.

Ceasar, S.A., Hodge, A., Baker, A., Baldwin, S.A., 2014. Phosphate concentration and arbuscular mycorrhizal colonisation influence the growth, yield and expression of twelve PHT1 family phosphate transporters in foxtail millet (Setaria italica). PLoS One 9 (9). Available at https://doi.org/10.1371/journal.pone.0108459.

Chatterjee, P., Kanagendran, A., Samaddar, S., Pazouki, L., Sa, T.-M., Niinemets, Ü., 2020. Influence of Brevibacterium linens RS16 on foliage photosynthetic and volatile emission characteristics upon heat stress in Eucalyptus grandis. Science of The Total Environment 700, 134453. Available at https://doi.org/10.1016/j.scitotenv.2019.134453.

Chen, S., Zhao, H., Zou, C., Li, Y., Chen, Y., Wang, Z., et al., 2017. Combined inoculation with multiple arbuscular mycorrhizal fungi improves growth, nutrient uptake and photosynthesis in cucumber seedlings. Frontiers in Microbiology 8. Available at https://doi.org/10.3389/fmicb.2017.02516.

Choi, J., Roy Choudhury, A., Walitang, D.I., Lee, Y., Sa, T., 2022. ACC deaminase-producing Brevibacterium linens RS16 enhances heat-stress tolerance of rice (Oryza sativa L. Physiologia Plantarum 174 (1). Available at https://doi.org/10.1111/ppl.13584.

Dasgupta, D., Paul, A., Acharya, K., Minkina, T., Mandzhieva, S., Gorovtsov, A.V., et al., 2023. Bioinoculant mediated regulation of signalling cascades in various stress responses in plants. Heliyon 9 (1), e12953. Available at https://doi.org/10.1016/j.heliyon.2023.e12953.

Delamare, J., Brunel-Muguet, S., Boukerb, A.M., Bressan, M., Dumas, L., Firmin, S., et al., 2023. Impact of PGPR inoculation on root morphological traits and root exudation in rapeseed and camelina: interactions with heat stress. Physiologia Plantarum 175 (6). Available at https://doi.org/10.1111/ppl.14058.

Dixit, V.K., Misra, S., Mishra, S.K., Tewari, S.K., Joshi, N., Chauhan, P.S., 2020. Characterization of plant growth-promoting alkalotolerant Alcaligenes and Bacillus strains for mitigating the alkaline stress in Zea mays. Antonie van Leeuwenhoek, International Journal of General and Molecular Microbiology 113 (7), 889–905. Available at https://doi.org/10.1007/s10482-020-01399-1.

El-Saadony, M.T., Desoky, E.S.M., El-Tarabily, K.A., AbuQamar, S.F., Saad, A.M., 2024. Exploiting the role of plant growth promoting rhizobacteria in reducing heavy metal toxicity of pepper (Capsicum annuum L. Environmental Science and Pollution Research 31 (18), 27465–27484. Available at https://doi.org/10.1007/s11356-024-32874-1.

Evelin, H., Kapoor, R., Giri, B., 2009. Arbuscular mycorrhizal fungi in alleviation of salt stress: a review. Annals of Botany 104 (7), 1263–1280. Available at https://doi.org/10.1093/aob/mcp251.

Evelin, H., Devi, T.S., Gupta, S., Kapoor, R., 2019. Mitigation of salinity stress in plants by arbuscular mycorrhizal symbiosis: current understanding and new challenges. Frontiers in Plant Science 10. Available at https://doi.org/10.3389/fpls.2019.00470.

Gamalero, E., Glick, B.R., 2022. Recent advances in bacterial amelioration of plant drought and salt stress. Biology 11 (3). Available at https://doi.org/10.3390/biology11030437.

García, J.E., Maroniche, G., Creus, C., Suárez-Rodríguez, R., Ramirez-Trujillo, J.A., Groppa, M.D., 2017. In vitro PGPR properties and osmotic tolerance of different Azospirillum native strains and their effects on growth of maize under drought stress. Microbiological Research 202, 21–29. Available at https://doi.org/10.1016/j.micres.2017.04.007.

Gilbert, N., 2009. Environment: the disappearing nutrient. Nature 461 (7265), 716–718. Available at https://doi.org/10.1038/461716a.

Gou, W., Tian, L., Ruan, Z., Zheng, P., Chen, F., Zhang, L., et al., 2015. Accumulation of choline and glycinebetaine and drought stress tolerance induced in maize (Zea mays) by three plant growth promoting rhizobacteria (PGPR) strains. Pakistan Journal of Botany 47 (2), 581–586.

Gulati, A., Thakur, R., Vyas, P., Sharma, A., Dhar, H., Pal, M., et al., 2024. Fostering climate-resilient agriculture with ACC-deaminase producing rhizobacterial biostimulants from the cold deserts of the Indian Himalayas. Journal of Environmental Management 371, 123075. Available at https://doi.org/10.1016/j.jenvman.2024.123075.

Hassani, A., Azapagic, A., Shokri, N., 2021. Global predictions of primary soil salinization under changing climate in the 21st century. Nature Communications 12 (1). Available at https://doi.org/10.1038/s41467-021-26907-3.

Heidari, M., Golpayegani, A., 2012. Effects of water stress and inoculation with plant growth promoting rhizobacteria (PGPR) on antioxidant status and photosynthetic pigments in basil (Ocimum basilicum L. Journal of the Saudi Society of Agricultural Sciences 11 (1), 57–61. Available at https://doi.org/10.1016/j.jssas.2011.09.001.

Hosseini, F., Mosaddeghi, M.R., Dexter, A.R., Sepehri, M., 2018. Maize water status and physiological traits as affected by root endophytic fungus Piriformospora indica under combined drought and mechanical stresses. Planta 247 (5), 1229–1245. Available at https://doi.org/10.1007/s00425-018-2861-6.

Huang, Z., Wang, C., Feng, Q., Liou, R.M., Lin, Y.F., Qiao, J., et al., 2022. The mechanisms of sodium chloride stress mitigation by salt-tolerant plant growth promoting rhizobacteria in wheat. Agronomy 12 (3). Available at https://doi.org/10.3390/agronomy12030543.

Islam, F., Yasmeen, T., Ali, Q., Ali, S., Arif, M.S., Hussain, S., et al., 2014. Influence of Pseudomonas aeruginosa as PGPR on oxidative stress tolerance in wheat under Zn stress. Ecotoxicology and Environmental Safety 104 (1), 285–293. Available at https://doi.org/10.1016/j.ecoenv.2014.03.008.

Jha, Y., Mohamed, H.I., 2023. Inoculation with Lysinibacillus fusiformis Strain YJ4 and Lysinibacillus sphaericus Strain YJ5 alleviates the effects of cold stress in maize plants. Gesunde Pflanzen 75 (1), 77–95. Available at https://doi.org/10.1007/s10343-022-00666-7.

Jian, P., Zha, Q., Hui, X., Tong, C., Zhang, D., 2024. Research progress of arbuscular mycorrhizal fungi improving plant resistance to temperature stress. Horticulturae 10 (8), 855. Available at https://doi.org/10.3390/horticulturae10080855.

Joudmand, A., Hajiboland, R., Aliasgharzad, N., 2021. Mitigation of chilling and freezing stresses through colonization with arbuscular mycorrhizal fungi in spring barley. jispp 9 (40), 13--24.

Kakar, K.U., Ren, X.-l, Nawaz, Z., Cui, Z.-Q., Li, B., Xie, G.-L., et al., 2016. A consortium of rhizobacterial strains and biochemical growth elicitors improve cold and drought stress tolerance in rice (Oryza sativa L.). Plant Biology 18 (3), 471–483. Available at https://doi.org/10.1111/plb.12427.

Karagöz, K., Ateş, F., Karagöz, H., Kotan, R., Çakmakçı, R., 2012. Characterization of plant growth-promoting traits of bacteria isolated from the rhizosphere of grapevine grown in alkaline and acidic soils. European Journal of Soil Biology 50, 144–150. Available at https://doi.org/10.1016/j.ejsobi.2012.01.007.

Kari, A., Nagymáté, Z., Romsics, C., Vajna, B., Tóth, E., Lazanyi-Kovács, R., et al., 2021. Evaluating the combined effect of biochar and PGPR inoculants on the bacterial community in acidic sandy soil. Applied Soil Ecology 160, 103856. Available at https://doi.org/10.1016/j.apsoil.2020.103856.

Khan, M.A., Asaf, S., Khan, A.L., Adhikari, A., Jan, R., Ali, S., et al., 2019. Halotolerant rhizobacterial strains mitigate the adverse effects of NaCl Stress in soybean seedlings. BioMed Research International 2019. Available at https://doi.org/10.1155/2019/9530963.

Khan, M.A., Asaf, S., Khan, A.L., Jan, R., Kang, S.M., Kim, K.M., et al., 2020. Thermotolerance effect of plant growth-promoting Bacillus cereus SA1 on soybean during heat stress. BMC Microbiology 20 (1). Available at https://doi.org/10.1186/s12866-020-01822-7.

Khan, N., Bano, A., Zandi, P., 2018. Effects of exogenously applied plant growth regulators in combination with PGPR on the physiology and root growth of chickpea (Cicer arietinum) and their role in drought tolerance. Journal of Plant Interactions 13 (1), 239–247. Available at https://doi.org/10.1080/17429145.2018.1471527.

Khanna, K., Jamwal, V.L., Gandhi, S.G., Ohri, P., Bhardwaj, R., 2019. Metal resistant PGPR lowered Cd uptake and expression of metal transporter genes with improved growth and photosynthetic pigments in Lycopersicon esculentum under metal toxicity. Scientific Reports 9 (1). Available at https://doi.org/10.1038/s41598-019-41899-3.

Khoso, M.A., Wagan, S., Alam, I., Hussain, A., Ali, Q., Saha, S., et al., 2024. Impact of plant growth-promoting rhizobacteria (PGPR) on plant nutrition and root characteristics: current perspective. Plant Stress 11. Available at https://doi.org/10.1016/j.stress.2023.100341.

Kopecká, R., Kameniarová, M., Černý, M., Brzobohatý, B., Novák, J., 2023. Abiotic stress in crop production. International Journal of Molecular Sciences 24 (7), 6603. Available at https://doi.org/10.3390/ijms24076603.

Kour, D., Yadav, A.N., 2023. Alleviation of cold stress in wheat with psychrotrophic phosphorus solubilizing Acinetobacter rhizosphaerae EU-KL44. Brazilian Journal of Microbiology 54 (1), 371–383. Available at https://doi.org/10.1007/s42770-023-00913-7.

Kumar, S., 2020. Abiotic stresses and their effects on plant growth, yield and nutritional quality of agricultural produce. International Journal of Food Science and Agriculture 4 (4), 367–378. Available at https://doi.org/10.26855/ijfsa.2020.12.002.

Li, J., Meng, B., Chai, H., Yang, X., Song, W., Li, S., et al., 2019. Arbuscular Mycorrhizal fungi alleviate drought stress in C3 (Leymus chinensis) and C4 (Hemarthria altissima) grasses via altering antioxidant enzyme activities and photosynthesis. Frontiers in Plant Science 10. Available at https://doi.org/10.3389/fpls.2019.00499.

Li, X., Sun, P., Zhang, Y., Jin, C., Guan, C., 2020. A novel PGPR strain Kocuria rhizophila Y1 enhances salt stress tolerance in maize by regulating phytohormone levels, nutrient acquisition, redox potential, ion homeostasis, photosynthetic capacity and stress-responsive genes expression. Environmental and Experimental Botany 174, 104023. Available at https://doi.org/10.1016/j.envexpbot.2020.104023.

Lim, J.H., Kim, S.D., 2013. Induction of drought stress resistance by multi-functional PGPR Bacillus licheniformis K11 in pepper. Plant Pathology Journal 29 (2), 201–208. Available at https://doi.org/10.5423/PPJ.SI.02.2013.0021.

Lotfi, N., Soleimani, A., Çakmakçı, R., Vahdati, K., Mohammadi, P., 2022. Characterization of plant growth-promoting rhizobacteria (PGPR) in Persian walnut associated with drought stress tolerance. Scientific Reports 12 (1). Available at https://doi.org/10.1038/s41598-022-16852-6.

Lu, H., Wei, T., Lou, H., Shu, X., Chen, Q., 2021. A critical review on communication mechanism within plant-endophytic fungi interactions to cope with biotic and abiotic stresses. Journal of Fungi 7 (9), 719. Available at https://doi.org/10.3390/jof7090719.

Ma, J., Peng, J., Tian, S., He, Y., Zhang, C., Jia, N., et al., 2023. Streptomyces sp. TOR3209 alleviates cold stress in tomato plants. New Zealand Journal of Crop and Horticultural Science 51 (4), 662–682. Available at https://doi.org/10.1080/01140671.2022.2066141.

Márquez, L.M., Redman, R.S., Rodriguez, R.J., Roossinck, M.J., 2007. A virus in a fungus in a plant: three-way symbiosis required for thermal tolerance. Science (New York, N.Y.) 315 (5811), 513–515. Available at https://doi.org/10.1126/science.1136237.

Mishra, P.K., Bisht, S.C., Ruwari, P., Selvakumar, G., Joshi, G.K., Bisht, J.K., et al., 2011. Alleviation of cold stress in inoculated wheat (Triticum aestivum L.) seedlings with psychrotolerant Pseudomonads from NW Himalayas. Archives of Microbiology 193 (7), 497–513. Available at https://doi.org/10.1007/s00203-011-0693-x.

Murali, M., Singh, S.B., Gowtham, H.G., Shilpa, N., Prasad, M., Aiyaz, M., et al., 2021. Induction of drought tolerance in Pennisetum glaucum by ACC deaminase producing PGPR- Bacillus amyloliquefaciens through Antioxidant defense system. Microbiological Research 253, 126891. Available at https://doi.org/10.1016/j.micres.2021.126891.

Nawaz, A., Shahbaz, M., Asadullah, M., Imran, A., Marghoob, M.U., Imtiaz, M., et al., 2020. Potential of salt tolerant PGPR in growth and yield augmentation of wheat (Triticum aestivum L.) under saline conditions. Frontiers in Microbiology 11. Available at https://doi.org/10.3389/fmicb.2020.02019.

Neshat, M., Abbasi, A., Hosseinzadeh, A., Sarikhani, M.R., Dadashi Chavan, D., Rasoulnia, A., 2022. Plant growth promoting bacteria (PGPR) induce antioxidant tolerance against salinity stress through biochemical and physiological mechanisms. Physiology and Molecular Biology of Plants 28 (2), 347–361. Available at https://doi.org/10.1007/s12298-022-01128-0.

Ngalimat, M.S., Hata, E.M., Zulperi, D., Ismail, S.I., Ismail, M.R., Zainudin, N.A.I.M., et al., 2021. Plant growth-promoting bacteria as an emerging tool to manage bacterial rice pathogens. Microorganisms 9 (4). Available at https://doi.org/10.3390/microorganisms9040682.

Notununu, I., Moleleki, L., Roopnarain, A., Adeleke, R., 2024. Enhancing maize drought and heat tolerance: single vs combined plant growth promoting rhizobacterial inoculation. Frontiers in Plant Science 15. Available at https://doi.org/10.3389/fpls.2024.1480718.

Pandey, S., Ghosh, P.K., Ghosh, S., De, T.K., Maiti, T.K., 2013. Role of heavy metal resistant Ochrobactrum sp. and Bacillus spp. strains in bioremediation of a rice cultivar and their PGPR like activities. Journal of Microbiology 51 (1), 11–17. Available at https://doi.org/10.1007/s12275-013-2330-7.

Parasar, B.J., Kashyap, S., Sharma, I., Marme, S.D., Das, P., Agarwala, N., 2024. Microbe mediated alleviation of drought and heat stress in plants-current understanding and future prospects. Discover Plants 1 (1). Available at https://doi.org/10.1007/s44372-024-00022-1.

Porcel, R., Redondo-Gómez, S., Mateos-Naranjo, E., Aroca, R., Garcia, R., Ruiz-Lozano, J.M., 2015. Arbuscular mycorrhizal symbiosis ameliorates the optimum quantum yield of photosystem II and reduces non-photochemical quenching in rice plants subjected to salt stress. Journal of Plant Physiology 185, 75–83. Available at https://doi.org/10.1016/j.jplph.2015.07.006.

Riaz, M., Kamran, M., Fang, Y., Wang, Q., Cao, H., Yang, G., et al., 2021. Arbuscular mycorrhizal fungi-induced mitigation of heavy metal phytotoxicity in metal contaminated soils: a critical review. Journal of Hazardous Materials 402, 123919. Available at https://doi.org/10.1016/j.jhazmat.2020.123919.

Safari, D., Jamali, F., Nooryazdan, H.R., Bayat, F., 2018. Evaluation of ACC deaminase producing Pseudomonas fluorescens strains for their effects on seed germination and early growth of wheat under salt stress. Australian Journal of Crop Science 12 (3), 413–421. Available at https://doi.org/10.21475/ajcs.18.12.03.pne801.

Saikia, J., Sarma, R.K., Dhandia, R., Yadav, A., Bharali, R., Gupta, V.K., et al., 2018. Alleviation of drought stress in pulse crops with ACC deaminase producing rhizobacteria isolated from acidic soil of Northeast India. Scientific Reports 8 (1). Available at https://doi.org/10.1038/s41598-018-21921-w.

Sapre, S., Gontia-Mishra, I., Tiwari, S., 2022. Plant growth-promoting rhizobacteria ameliorates salinity stress in pea (Pisum sativum). Journal of Plant Growth Regulation 41 (2), 647–656. Available at https://doi.org/10.1007/s00344-021-10329-y.

Sarkar, A., Ghosh, P.K., Pramanik, K., Mitra, S., Soren, T., Pandey, S., et al., 2018a. A halotolerant Enterobacter sp. displaying ACC deaminase activity promotes rice seedling growth under salt stress. Research in Microbiology 169 (1), 20–32. Available at https://doi.org/10.1016/j.resmic.2017.08.005.

Sarkar, J., Chakraborty, B., Chakraborty, U., 2018b. Plant growth promoting rhizobacteria protect wheat plants against temperature stress through antioxidant signalling and reducing chloroplast and membrane injury. Journal of Plant Growth Regulation 37 (4), 1396–1412. Available at https://doi.org/10.1007/s00344-018-9789-8.

Sarkar, J., Chakraborty, U., Chakraborty, B., 2021. High-temperature resilience in Bacillus safensis primed wheat plants: a study of dynamic response associated with modulation of antioxidant machinery, differential expression of HSPs and osmolyte biosynthesis. Environmental and Experimental Botany 182, 104315. Available at https://doi.org/10.1016/j.envexpbot.2020.104315.

Selbmann, L., Egidi, E., Isola, D., Onofri, S., Zucconi, L., de Hoog, G.S., et al., 2013. Biodiversity, evolution and adaptation of fungi in extreme environments. Plant Biosystems - An International Journal Dealing with all Aspects of Plant Biology 147 (1), 237–246. Available at https://doi.org/10.1080/11263504.2012.753134.

Shaffique, S., Shah, A.A., Peter, O., Injamum-Ul-Hoque, M., Elansary, H.O., kang, S.M., et al., 2024. The rhizobacterial Priestia megaterium strain SH-19 mitigates the hazardous effects of heat stress via an endogenous secondary metabolite elucidation network and molecular regulation signalling. BMC Plant Biology 24 (1). Available at https://doi.org/10.1186/s12870-024-05534-2.

Shukla, P.S., Agarwal, P.K., Jha, B., 2012. Improved salinity tolerance of Arachis hypogaea (L.) by the interaction of halotolerant plant-growth-promoting rhizobacteria. Journal of Plant Growth Regulation 31 (2), 195–206. Available at https://doi.org/10.1007/s00344-011-9231-y.

Singh, L.P., Gill, S.S., Tuteja, N., 2011. Unraveling the role of fungal symbionts in plant abiotic stress tolerance. Plant Signaling & Behavior 6 (2), 175–191. Available at https://doi.org/10.4161/psb.6.2.14146.

Singh, R.P., Jha, P.N., 2016. The multifarious PGPR Serratia marcescens CDP-13 augments induced systemic resistance and enhanced salinity tolerance of wheat (Triticum aestivum L. PLoS One 11 (6). Available at https://doi.org/10.1371/journal.pone.0155026.

Singh, R.P., Jha, P., Jha, P.N., 2015. The plant-growth-promoting bacterium Klebsiella sp. SBP-8 confers induced systemic tolerance in wheat (Triticum aestivum) under salt stress. Journal of Plant Physiology 184, 57–67. Available at https://doi.org/10.1016/j.jplph.2015.07.002.

Singh, S., Tripathi, A., Singh Chanotiya, C., Barnawal, D., Singh, P., Patel, V.K., et al., 2020. Cold stress alleviation using individual and combined inoculation of ACC deaminase producing microbes in Ocimum sanctum. Environmental Sustainability 3 (3), 289–301. Available at https://doi.org/10.1007/s42398-020-00118-w.

de Souza, E.M., Lamb, T.I., Lamb, T.A., dos Santos Silva, A., da Fré de Carvalho, S., Nyland, V., et al., 2021. Rhizospheric soil from rice paddy presents isolable bacteria able to induce cold tolerance in rice plants. Journal of Soil Science and Plant Nutrition 21 (3), 1993–2006. Available at https://doi.org/10.1007/s42729-021-00496-y.

Subramanian, P., Kim, K., Krishnamoorthy, R., Mageswari, A., Selvakumar, G., Sa, T., et al., 2016. Cold stress tolerance in psychrotolerant soil bacteria and their conferred chilling resistance in tomato (Solanum lycopersicum Mill.) under low temperatures. PLoS One 11 (8), e0161592. Available at https://doi.org/10.1371/journal.pone.0161592.

Thakur, R., Rahi, P., Gulati, A., Gulati, A., 2025. Tea seedlings growth promotion by widely distributed and stress-tolerant PGPR from the acidic soils of the Kangra valley. BMC Microbiology 25 (1). Available at https://doi.org/10.1186/s12866-025-03811-0.

Thakur, R., Sharma, K.C., Gulati, A., Sud, R.K., Gulati, A., 2017. Stress-tolerant Viridibacillus arenosi strain IHB B 7171 from tea rhizosphere as a potential broad-spectrum microbial inoculant. Indian Journal of Microbiology 57 (2), 195–200. Available at https://doi.org/10.1007/s12088-017-0642-8.

Thakur, R., Dhar, H., Swarnkar, M.K., Soni, R., Sharma, K.C., Singh, A.K., et al., 2024. Understanding the molecular mechanism of PGPR strain Priestia megaterium from tea rhizosphere for stress alleviation and crop growth enhancement. Plant Stress 12, 100494. Available at https://doi.org/10.1016/j.stress.2024.100494.

Tirry, N., Kouchou, A., El Omari, B., Ferioun, M., El Ghachtouli, N., 2021. Improved chromium tolerance of Medicago sativa by plant growth-promoting rhizobacteria (PGPR). Journal of Genetic Engineering and Biotechnology 19 (1), 149. Available at https://doi.org/10.1186/s43141-021-00254-8.

Triboi, E., 2003. Environmentally-induced changes in protein composition in developing grains of wheat are related to changes in total protein content. Journal of Experimental Botany 54 (388), 1731–1742. Available at https://doi.org/10.1093/jxb/erg183.

Verma, C., Singh, P., Kumar, R., 2015. Isolation and characterization of heavy metal resistant PGPR and their role in enhancement of growth of wheat plant under metal (cadmium) stress condition. Archives of Applied Science Research 7, 37–43.

Wang, S., Zhang, Y., Ju, W., Chen, J.M., Ciais, P., Cescatti, A., et al., 2020. Recent global decline of CO2 fertilization effects on vegetation photosynthesis. Science (New York, N.Y.) 370 (6522), 1295–1300. Available at https://doi.org/10.1126/science.abb7772.

Waqas, M., Naseem, A., Humphries, U.W., Hlaing, P.T., Shoaib, M., Hashim, S., 2025. A comprehensive review of the impacts of climate change on agriculture in Thailand. Farming System 3 (1). Available at https://doi.org/10.1016/j.farsys.2024.100114.

Yadav, S., Kaushik, R., Saxena, A.K., Arora, D.K., 2011. Diversity and phylogeny of plant growth-promoting bacilli from moderately acidic soil. Journal of Basic Microbiology 51 (1), 98–106. Available at https://doi.org/10.1002/jobm.201000098.

Zainab, N., Amna, Khan, A.A., Azeem, M.A., Ali, B., Wang, T., et al., 2021. Pgpr-mediated plant growth attributes and metal extraction ability of sesbania sesban l. In industrially contaminated soils. Agronomy 11 (9). Available at https://doi.org/10.3390/agronomy11091820.

Zargar, S.M., Mahajan, R., Bhat, J.A., Nazir, M., Deshmukh, R., 2019. Role of silicon in plant stress tolerance: opportunities to achieve a sustainable cropping system. 3 Biotech 9 (3). Available at https://doi.org/10.1007/s13205-019-1613-z.

Zhang, H., Liu, S., Ren, T., Niu, M., Liu, X., Liu, C., et al., 2023a. Crucial abiotic stress regulatory network of NF-Y transcription factor in plants. International Journal of Molecular Sciences 24 (5), 4426. Available at https://doi.org/10.3390/ijms24054426.

Zhang, S.-H., 2016. The genetic basis of abiotic stress resistance in extremophilic fungi: the genes cloning and application. In: Purchase, D. (Ed.), Fungal Applications in Sustainable Environmental Biotechnology. Fungal Biology. Springer, Cham, pp. 29–42. Available at 10.1007/978-3-319-42852-9_2.

Zhang, X., Yang, Z., Wang, L., Yue, Y., Wang, L., Xiulian, Y., 2023. The effects of plant growth-promoting rhizobacteria on plants under temperature stress: a meta-analysis. Rhizosphere 28, 100788. Available at https://doi.org/10.1016/j.rhisph.2023.100788.

Zhang, Y., Xu, J., Li, R., Ge, Y., Li, Y., Li, R., 2023c. Plants' response to abiotic stress: mechanisms and strategies. International Journal of Molecular Sciences 24 (13), 10915. Available at https://doi.org/10.3390/ijms241310915.

Zhao, C., Liu, B., Piao, S., Wang, X., Lobell, D.B., Huang, Y., et al., 2017. Temperature increase reduces global yields of major crops in four independent estimates. Proceedings of the National Academy of Sciences 114 (35), 9326–9331. Available at https://doi.org/10.1073/pnas.1701762114.

Zubair, M., Hanif, A., Farzand, A., Sheikh, T.M.M., Khan, A.R., Suleman, M., et al., 2019. Genetic screening and expression analysis of psychrophilic bacillus spp. reveal their potential to alleviate cold stress and modulate phytohormones in wheat. Microorganisms 7 (9). Available at https://doi.org/10.3390/microorganisms7090337.

Chapter 23

Detection and monitoring of abiotic stresses in crops: recent technological applications

Harmandeep Sharma*, Chibuzor Uche, Arnab Bhowmik, and William Randle

Department of Natural Resources and Environmental Design, North Carolina Agricultural and Technical State University, Greensboro, NC, United States

*Corresponding author. e mail address: hsharma@ncat.edu

23.1 Introduction

The global population is projected to reach 10 billion by 2050, driving an unprecedented demand for food and energy. At the same time, climate change continues to pose significant challenges to agricultural productivity through increased frequency and severity of extreme weather events. Plants are frequently exposed to a variety of stresses that can broadly be classified into two categories: abiotic stress (e.g., drought, heat, cold, salinity, and nutrient deficiency) and biotic stress (e.g., pathogens, pests, and viruses) (Berger et al., 2022; Zhang et al., 2024). Abiotic stress can be further subdivided into physical and chemical categories, encompassing a range of environmental conditions that impede plant growth.

Moreover, plants often face compound stresses, where multiple abiotic stressors occur simultaneously or in succession. For example, heat stress might increase plant transpiration rates, subsequently enhance the uptake of salts, and exacerbate damage from saline-alkaline conditions (Wu et al., 2024). These abiotic stresses impose significant constraints on plant growth, yield, and quality, frequently leading to drastic reductions in crop performance. In some instances, losses in crop yields due to abiotic stresses can vary between 50% and 70% (Francini and Sebastiani, 2019). Given these challenges, this chapter delves into the intricacies of abiotic and compound stresses.

Plants have evolved complex adaptation mechanisms that mediate their stress responses to sustain agricultural productivity under these dynamic and often unsuitable environmental conditions (Zhang et al., 2024). Therefore the ability to rapidly and accurately phenotype plant responses to abiotic stresses is crucial for accelerating plant-breeding programs aimed at developing climate-resilient genotypes.

23.1.1 Impact of abiotic stress on plants

Abiotic stresses disrupt normal physiological processes such as photosynthesis, transpiration, and nutrient uptake, ultimately leading to reduced growth, lower yield, and compromised quality in crops (Wu et al., 2024). Plants respond to these stresses with diverse morphological, physiological, and metabolic changes across various organs. Extremely high temperatures, or heat stress, adversely affect organ vitality, constrain metabolism, and retard growth; prolonged exposure can cause irreversible damage or even plant death. Additionally, high temperatures escalate water evaporation, potentially exceeding water uptake rates, thereby inducing drought-like stress and compounding the plant's physiological strain. Among abiotic stresses, drought is particularly severe and poses a significant threat to crop productivity primarily through dehydration. Observable symptoms of water-deficit stress include morphological changes such as leaf curling, wilting, and drooping of plant branches. Subsequently, leaf stomatal closure reduces transpiration and limits water loss, albeit at the cost of carbon dioxide (CO_2) absorption, followed by alterations in chlorophyll content and reductions in the plant's leaf area index (LAI). These changes collectively disrupt metabolic and osmotic balance.

Chemical analysis has identified 17 essential elements crucial for plant growth and metabolism. Carbon (C), hydrogen (H), and oxygen (O) are primarily sourced from water (H_2O) and atmospheric CO_2. The remaining 14 elements are

Multiple Abiotic Stresses in Plants. DOI: https://doi.org/10.1016/B978-0-443-27424-4.00016-1

predominantly absorbed from the soil. Stress related to mineral nutrients can arise from either excessive concentrations or insufficient availability of these elements. High concentrations of nutrients and mineral ions in soil can hinder water and mineral absorption, significantly impeding plant growth—similar to drought stress but with potentially more severe consequences due to ion toxicity (Pandey et al., 2021). Moreover, excessive levels of specific mineral nutrients can induce deficiencies in other essential nutrients by altering soil pH, thereby adversely affecting plant health. Conversely, deficiencies in mineral nutrients can manifest in various disorders, notably suppressing plant growth and reproduction. For instance nitrogen (N) deficiency leads to specific morphological traits such as reduced leaf area and leaf discoloration, as well as physiological effects such as diminished photosynthesis and overall metabolic activity. Comprehending these intricate and multifaceted responses to individual or combined stresses necessitates the use of advanced tools and methodologies.

Traditional methods for studying plant stress responses include visual inspection, physical and chemical monitoring, and molecular technologies (Li et al., 2023). However, these methods are often labor-intensive, time-consuming, and destructive. Consequently, there is a growing need for high-throughput, nondestructive techniques to enable precise and efficient evaluation of plant traits associated with stress tolerance.

In terms of soil moisture and temperature monitoring, conventional technologies—including both laboratory-based and in situ sensing methods directly connected to the soil—have proven to be highly stable and effective over long periods. However, the landscape of precision agriculture is rapidly changing with the introduction of technologies that eliminate the need for direct contact with the soil. Notably, recent innovations have facilitated the deployment of proximal in-field soil moisture and temperature sensors on unmanned aerial vehicles (UAVs), or drones, and on mobile agricultural equipment (referred to as "on-the-go" sensors), such as tractors or seeders. Additionally, comprehensive monitoring is achievable through satellite and advanced sensor systems (Zhang et al., 2024).

Ground-based sensors continue to play an indispensable role in providing real-time, localized monitoring of abiotic stresses. Over the last decade, significant progress has been made in the fields of soil moisture, temperature, and nutrient monitoring, greatly benefiting precision agriculture. For instance the integration of Internet of Things (IoT) systems has significantly enhanced the accuracy and efficiency of these assessments. Such technological advancements are pivotal in promoting climate-smart agriculture, which aims to optimize input management, thus increasing crop yields, adapting farming practices to climate variability, and improving the efficiency of resource use.

The real-time data provided by these innovative technologies enables farmers to adapt swiftly to environmental changes. This adaptability reduces the reliance on excessive irrigation and fertilizer application, consequently lowering energy consumption and mitigating greenhouse gas emissions. These improvements underscore the vital role of modern agricultural technologies in sustaining and enhancing farming practices in the face of evolving environmental challenges.

23.1.2 Latest imaging technologies

Plant phenotyping involves the quantitative evaluation of plant traits to discern genotypes that exhibit superior tolerance to various stressors. Recent advancements in imaging sensors and technologies have significantly transformed phenotyping by facilitating rapid and noninvasive measurements of plant stress responses (Table 23.1). Imaging technologies encompass a broad spectrum of sensors, including passive panchromatic, multispectral, hyperspectral, ultraviolet (UV), visible (VIS), near-infrared (NIR), mid-infrared, thermal, and medium wave wavebands (Fig. 23.1). Common imaging systems integrate various sensors, such as optical-NIR and radar, deployed on diverse platforms including satellites, airplanes, balloons, and helicopters (Zhang et al., 2024). Among these, NIR, VIS, and TIR sensors are predominantly utilized, with VIS-TIR and VIS-NIR combinations being extensively explored due to their effectiveness in capturing critical data (Berger et al., 2022).

Image-based phenotyping (IBP) applications have been optimized through High-throughput phenotyping (HTP) technology, which leverages developments in computer science, engineering, and big data analytics to accelerate the assessment of plant traits. This technology enables automated and thorough imaging based on predefined objectives, enhancing the efficiency and precision of the phenotyping process (Coppens et al., 2017; Jangra et al., 2021). HTP has proven particularly effective in identifying stress tolerance, especially under conditions like drought stress, across various crops including rice, wheat, and maize (Briglia et al., 2020; Joshi et al., 2021; Kim, 2020, 2020; Kim et al., 2021; Rossi et al., 2022).

However, the effectiveness of HTP can be influenced by several factors, including the environmental conditions of the observation area, the sensors used, and the methodologies employed for data processing and analysis (Jayasinghe et al., 2021; Jangra et al., 2021; Ninomiya, 2022). Notably, the red, blue, and green (RGB) sensor, while sensitive to electromagnetic radiation, can exhibit higher bias compared to other sensors, which may affect its reliability under certain

TABLE 23.1 Summary regarding the use of sensors with respect to specific abiotic stresses.

Sensor	Light used and wavelength range	Abiotic stress measured	Plant measurements	Parameter measured	Citations
Red, blue, and green (RGB)	Visible light (VIS) (380–760 nm)	Drought Heat Nutrient	Leaves Leaves and stem canopy Leaves and flowering stages	Chlorophyll and carotenoid levels Canopy temperature, Relative growth rate, Plant volume Chlorophyll content, Senescence, Plant biomass, Leaf chemical nutrient analysis	Kim and Chung (2021), Chandel et al. (2022), Zhang et al. (2019), Kim et al. (2020), Abdelhakim et al. (2021), Briglia et al. (2020), Zhou et al. (2021)Neilson et al. (2015), Yi et al. (2020), Zermas et al. (2021), Ghazal, Kommineni and Munir (2024), Fevgas et al. (2023), Kim et al. (2018), Abdelhakim et al. (2021), Lazarević et al. (2022), Neilson et al. (2015), Yi et al. (2020), Zermas et al. (2021), Taha et al. (2022), Deichmann et al. (2024)
Multispectral	Visible light (VIS) (380–760 nm), Near-infrared (NIR) (750–900 nm)	Drought Heat Nutrient Water and Nutrient (Compound stress)	Leaves Plant canopy Leaves and Orchard scale	Chlorophyll content, canopy water content, photosynthetic activity, stomatal conductance Chlorophyll content and photosynthetic efficiency, Plant height, Senescence Chlorophyll content, Nutrient analysis, Grain yield Photosynthetic efficiency, Chlorophyll content, and Stomatal conductance	Gerhards et al. (2019), Virlet et al. (2015), Berger et al. (2022), Lazarević et al. (2022), Avgoustaki et al. (2022), Sharma et al. (2024), Kim et al. (2022), Ryu et al. (2022), Spišić et al. (2022), Li et al. (2023), Becker et al. (2020), Sun et al. (2023), Avgoustaki et al. (2022)
Hyperspectral	Visible light (VIS) (380–750 nm), Near-infrared (NIR) (750 – 900 nm), Short infrared (1000–2500 nm)	Drought Heat Nutrient Drought and Heat (Compound stress)	Field plots Leaves, Seed	Water potential, Stomatal conductance, transpiration rate, leaf area index, chlorophyll content Chlorophyll content, Polyphenol content, Leaf metabolites Chlorophyll and nutrient levels Photosynthesis efficiency and Leaf structure	Zhang et al. (2024), Asaari et al. (2018), Faqeerzada et al. (2022), Wang et al. (2022), Laroche-Pinel et al. (2021), Asaari et al. (2022) Faqeerzada et al. (2022), Kang et al. (2024), Park et al. (2021), Melandri et al. (2021), Behmann et al. (2014), Zhang et al. (2024), Mishra et al. (2017), Faqir et al. (2024), Sen et al. (2024), Zhu et al. (2018), Peanusaha et al. (2024), Debnath et al. (2021), Poobalasubramanian et al. (2022)

FIGURE 23.1 Examples of imaging technologies: multispectral sensor installed on a drone (*left*) and a hyperspectral sensor camera in the laboratory (*right*).
Picture Courtesy: SENSE Lab and CAES Ag Communication team at NCA&T State University, Greensboro, NC, USA.

conditions (Al-Tamimi et al., 2022; Jangra et al., 2021). Despite these limitations, the RGB sensor has been effectively employed to study abiotic stress responses, particularly drought, in crops such as rice, wheat, and maize (Laraswati et al., 2021; Jiang et al., 2021; Sakinah et al., 2022; Dodig et al., 2021; Qiu et al., 2019; Zhang et al., 2019; Hairmansis et al., 2014).

Another advanced sensor utilized in visible light imaging is light detection and ranging (LIDAR). LIDAR, an RGB sensor capable of functioning in dark conditions, is valued for its high resolution, sharpness, and accuracy, making it suitable for constructing three-dimensional phenotyping and remote sensing (RS) applications. The primary function of LIDAR involves emitting lasers to assess canopy density and various aspects of plant growth. Despite its advantages, LIDAR is relatively expensive and complex, requiring extensive time and sophisticated analysis to yield valuable data (Al-Tamimi et al., 2022; Kim, 2020; Hama et al., 2022).

Additionally, hyperspectral technology has greatly enhanced its applicability in agriculture, enabling the monitoring and quantification of various plant biochemical and biophysical properties, such as chlorophyll, nitrogen, and water content. This technology also supports crop discrimination, disease and pest monitoring, assessment of environmental stresses, and crop classification (Lu et al., 2020; Sahoo et al., 2015). Hyperspectral imaging (HSI) is a form of RS that utilizes a combination of camera technologies and imaging algorithms to quantify and map nutrient stress in plants, both at the leaf and canopy levels (Ramesh et al., 2023; Sen et al., 2024).

However, each imaging technology has its own limitations. To overcome the limitations of individual technologies, sensor-fusion techniques, which integrate data from multiple sensors, are employed to provide comprehensive insights into plant stress responses. For example, integrating RGB and hyperspectral data enables simultaneous monitoring of both morphological and physiological traits, while combining aerial and terrestrial sensors enhances spatial and temporal resolution, facilitating multiscale analysis of plant responses under both field and controlled conditions.

23.2 Applications of imaging technologies for specific and combined abiotic stresses

This section is organized into three subsections, each dedicated to a commonly occurring abiotic stress: drought, heat, and nutrient stress. In addition to individual stressors, each subsection also includes select examples of combined stresses, illustrating the complex interactions that can exacerbate plant responses under varied environmental conditions. A central focus of these subsections is the incorporation of the latest studies that utilize advanced imaging technologies, including RGB, multispectral, and hyperspectral sensors (Fig. 23.2).

The coverage extends across a diverse array of crops, from field crops to horticultural varieties, highlighting the broad applicability of these technologies. Moreover, the deployment of these imaging technologies is examined across different spatial scales—from detailed assessments in greenhouse settings to expansive field studies conducted at both ground level

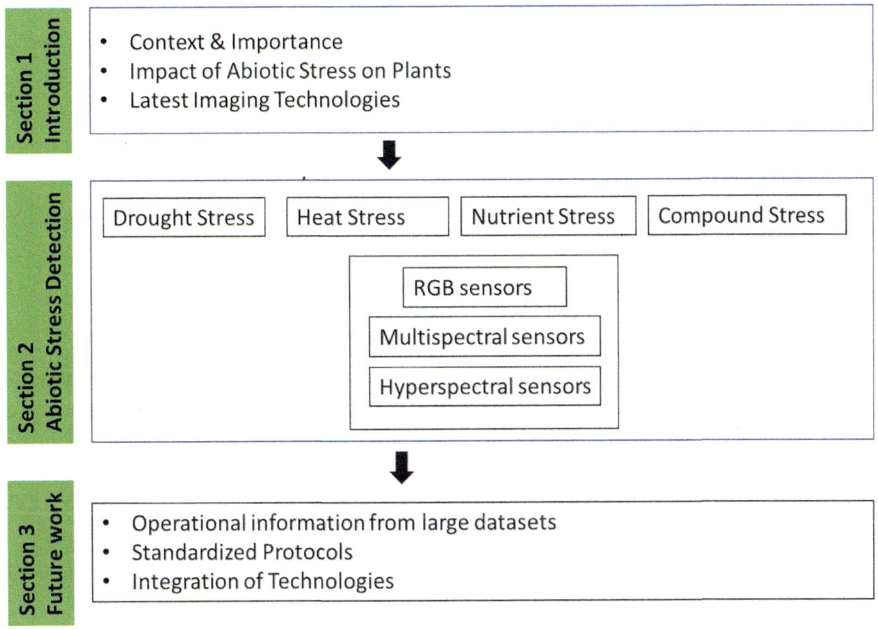

FIGURE 23.2 A stepwise framework illustrating the organization of the book chapter, which encompasses key sections from the introduction to discussions on future developments in the latest imaging technologies for abiotic stress detection.

and from aerial perspectives. This comprehensive approach underscores the versatility of these technological tools in enhancing our understanding of plant responses to abiotic stresses under various agricultural and experimental setups.

23.2.1 Drought stress

Drought represents a significant abiotic stressor within the contemporary climatic context, affecting many regions globally. It is defined by agronomists as a prolonged deficiency of water that detrimentally impacts plant development and survival, consequently diminishing agricultural productivity (Begna, 2021). Either this stress arises from an inadequate water supply to plant roots or excessive water loss through transpiration, rendering plants susceptible to drought conditions (Anjum et al., 2011). The severity of drought effects varies and is influenced by factors including rainfall patterns, soil moisture retention capabilities, and losses through evapotranspiration (Begna, 2021).

Drought conditions impair several physiological processes in plants, such as photosynthesis, assimilate partitioning, and nutrient and water relations, ultimately leading to reduced growth and a significant decline in agricultural yields (Praba et al., 2009). Initial consequences of drought include inhibited seed germination and stunted seedling development, with numerous studies highlighting the adverse effects on these early stages of plant growth (Begna, 2021; Farooq et al., 2009). For example, critical field crops under drought stress exhibit decreased germination rates, early seedling growth, and reductions in root and shoot dry weight, hypocotyl length, and overall vegetative growth (Zeid and Shedeed, 2006). A notable impact of drought was observed in the U.S. in 2012, where maize yields decreased by approximately 21% to 7.7 metric tons per hectare, compared to the preceding 5-year average of 9.7 metric tons per hectare. This reduction in yield mirrored past severe droughts like that of 1988, which saw a reduction of 26%. Moreover, the drought notably compromised grain quality, affecting starch content and protein quality, vital for food, feed, and industrial uses (Boyer et al., 2013).

Plants, predominantly composed of water, which constitutes 80%–90% of their total mass, rely on water for essential functions such as photosynthesis, nutrient uptake, transpiration, and temperature regulation (Im et al., 2018; Qing et al., 2000). Consequently, droughts induce significant alterations in the biological activities of plants and prompt a variety of adaptive and biological responses over both short and long terms (Im et al., 2018).

Drought, often referred to as water deficiency stress, is commonly analyzed using RS techniques, which have become pivotal in studying this prevalent type of crop stress (Damm et al., 2022; Gago et al., 2015; Gerhards et al., 2016; Ihuoma and Madramootoo, 2017; Parkash and Singh, 2020; Virnodkar et al., 2020). One significant aspect of RS in drought assessment is the use of canopy temperature as an indicator of plant water stress (Jones and Vaughan, 2010). Although RS-based surface energy balance (SEB) methods, which estimate actual evapotranspiration indirectly through the SEB equation using thermal data, require extensive meteorological inputs and conductance parameterization, they remain a valuable tool (Tolomio and Casa, 2020). The relationship between canopy temperature and water loss is inversely proportional, and stomatal conductance, which directly correlates with transpiration rates, is an important metric in this analysis (Jones and Vaughan, 2010). Advanced technologies, such as UAVs equipped with RGB, thermal, multispectral, and hyperspectral, are crucial for monitoring water status in agricultural fields, providing essential data for effective water stress management (Baluja et al., 2012; Gago et al., 2015; Kandylakis et al., 2020).

23.2.1.1 Red, blue, and green sensors

The RGB sensors provide the reflected values of RGB bands in the visible spectrum between 380 and 760 nm (Kim and Chung, 2021). RGB imagery detects water stress by analyzing leaf coloration changes that occur due to reductions in chlorophyll and carotenoid levels under water stress conditions (Chandel et al., 2022). These biochemical changes, caused by stomatal closure and subsequent physiological impacts, alter the visible reflectance patterns of plant leaves. RGB sensors capture these reflectance differences, which are then analyzed using deep learning (DL) models such as ResNet-50, GoogLeNet, and convolutional neural networks (CNNs) to classify crops into stressed and nonstressed categories based on their visual characteristics (Chandel et al., 2022; Kamarudin et al., 2021). Additionally, RGB imagery detects plant water stress by analyzing changes in leaf morphology, such as the leaf canopy. The stress-induced reflectance variations are quantified using vegetation indices derived from RGB data. RGB imagery has been used to assess crop water stress using different DL and machine-learning (ML) techniques (Chandel et al., 2021; Goldstein et al., 2018; Chandel et al., 2022) used RGB imagery to detect water stress in winter wheat (*Triticum aestivum* L., cv. HI 1544) by capturing visual indicators such as changes in leaf coloration caused by reduced chlorophyll and carotenoid levels under water stress conditions (75%, 50% and 25% of crop evapotranspiration). These images were taken at critical growth stages and processed using DL models such as ResNet-50, which extracted features like pixel intensity, edges, and object dimensions to classify crops into stressed and nonstressed categories. Although RGB imagery had slightly lower accuracy compared to thermal imagery in the study, it still achieved a classification accuracy of up to 94.6%, demonstrating its effectiveness in identifying water stress through visible changes in the crop canopy (Chandel et al., 2022).

Briglia et al. (2020) conducted a study to assess the utility of RGB and NIR imaging in phenotyping drought stress in grapevines (*Vitis vinifera* L.) at the Leibniz Institute of Vegetable and Ornamental Crops, Germany. Using the LemnaTec 3D Scanalyzer, they evaluated morphological and physiological traits such as leaf area, water consumption, and water potential (Ψ) in 80 grapevines subjected to four irrigation levels (100%, 75%, 50%, and 25% of daily water needs) over 22 days. Images were taken at predawn and midday, and RGB image analysis helped classify leaf colors into categories such as dark green, green, yellow, and brown. The analysis found significant reductions in leaf area, stomatal conductance, and water consumption under drought stress, with a decrease in dark green and an increase in yellow leaves correlating with lower Ψ values ($R^2 = 0.71$ and 0.50, respectively). Additionally, the study modeled leaf area and linked water consumption per unit leaf area to Ψ with high correlation coefficients ($R^2 = 0.78$ and 0.86, respectively). These results demonstrate the effectiveness of RGB imaging in monitoring drought stress and support its use in developing phenotyping protocols and precision irrigation strategies to manage drought in grapevines effectively.

A study conducted in soybean (*Glycine max*) investigated the potential of UAV-based RGB imagery combined with a CNN modeling to estimate the yield of soybean breeding lines under drought stress (Zhou et al., 2021). Conducted in a rain-fed experimental field in Clarkton, Dunklin County, Missouri, USA, the study monitored 972 soybean breeding lines across three maturity groups (MG 3, 4, and 5) under drought conditions. UAV imagery data were collected at three critical growth stages: late vegetative (V6), early reproductive (R1), and late reproductive (R6-R8) stages. RGB images were processed to extract image features, including hue, saturation, and excess green (ExG), representing canopy color and texture. A mixed CNN model was developed using these features along with categorical factors, such as maturity group and drought tolerance. Results indicated that image features collected during the early and late reproductive stages were highly predictive of yield. The mixed CNN model achieved an R^2 of 0.78 and a root mean square error (RMSE) of 391.0 kg/ha, demonstrating the model's accuracy and robustness. RGB imagery provided clear indicators of drought stress, with darker green canopies correlating with higher chlorophyll content and yield. The study concluded that UAV-based RGB imagery, integrated with advanced ML models like CNN, is a promising tool for HTP in drought-stress environments. This approach can significantly improve the selection of drought-tolerant soybean genotypes, ultimately enhancing breeding efficiency and yield prediction accuracy.

23.2.1.2 Multispectral sensors

These sensors detect water stress in plants by analyzing changes in reflectance across the visible (VIS) and NIR spectral bands (Gerhards et al., 2019). In the VIS spectrum, the red band (640–680 nm) shows increased reflectance when chlorophyll content decreases due to water stress. The green band (520–560 nm) typically exhibits reduced reflectance, reflecting changes in chlorophyll and carotenoid levels. The blue band (450–490 nm), though less commonly analyzed, provides insights into changes in pigmentation. In the NIR spectrum (750–900 nm), structural changes in leaf cells and reductions in internal water content due to water stress result in decreased reflectance.

These physiological and structural alterations in plant leaves influence the canopy's overall water content, chlorophyll concentration, and photosynthetic activity, leading to detectable variations in reflectance. Key vegetation indices like the Normalized Difference Vegetation Index (NDVI) and the Normalized Difference Water Index (NDWI) are derived from VIS-NIR data to monitor plant health and water status effectively (Karnieli et al., 2010; Asaari et al., 2018). Additional indices, such as the Enhanced Vegetation Index (EVI), Photochemical Reflectance Index (PRI), and Water Band Index, also contribute to the assessment of water stress (Jinru and Baofeng, 2017). These indices facilitate the early diagnosis of water stress, enabling interventions before visible symptoms manifest (Berger et al., 2022). For example, Virlet et al. (2015) conducted a study to identify genetic factors influencing water stress responses in apple tree hybrids, using multispectral airborne imagery. The research involved 122 hybrids derived from a cross between "Starkrimson" and "Granny Smith," grafted onto M9 rootstock and grown at the INRA Melgueil experimental station in southeast France. Over two growing seasons, these hybrids were exposed to contrasting water regimes—well-watered and water-stressed—and assessed through multispectral imaging across visible, NIR, and thermal infrared (TIR) bands at three annual time points. Key vegetation indices, including NDVI and the Water Deficit Index (WDI), were calculated to evaluate canopy structure, transpiration, and water stress levels.

In addition to RS, in-plant measurements of tree vigor, indicated by trunk cross-sectional area, and fruit production were analyzed. The study employed linear mixed models to assess the impact of genetic and environmental factors on these phenotypic traits, with quantitative trait locus (QTL) mapping used to associate traits with specific genetic markers. The results demonstrated considerable genetic variability in vegetation indices, transpiration rates, and stress responses, with eighteen stable QTLs explaining 15%–20% of the phenotypic variance. Notable findings included QTLs related to NDVI and intracrown heterogeneity, suggesting distinct genetic controls over canopy structure and drought tolerance. Additionally, QTLs for indices like WDI highlighted genetic aspects of transpiration regulation under drought conditions. This comprehensive approach underscored the effectiveness of multispectral imagery for HTP in apple trees, facilitating simultaneous evaluation of genetic and physiological responses to drought. The study provided critical insights into the mechanisms of drought tolerance and their genetic bases, enhancing understanding of adaptive strategies in fruit tree agriculture.

Similarly, another study explored the water stress responses in eight novel *Miscanthus sinensis* hybrids (GRC1–GRC8) through multispectral imaging and image analysis at the University of Zagreb Faculty of Agriculture, Croatia (Lazarević et al., 2022). This study involved 80 plants (10 per hybrid), with half subjected to water stress by withholding irrigation for 21 days—reducing substrate volumetric water content from 45% to 8.5%—followed by a 7-day recovery period. Multispectral imaging captured reflectance at six wavelengths using a CropReporter™ system, and morphological traits were analyzed with a PlantEye F500 multispectral 3D scanner. Key vegetation indices like the NDVI and Plant Senescence Reflectance Index (PSRI) were calculated to assess physiological changes. The findings indicated significant variability in water stress responses among the hybrids. Resilient hybrids, such as GRC4 and GRC5, exhibited smaller reductions in total leaf area and digital biomass, each decreasing by 50%, compared to more sensitive hybrids like GRC2 and GRC3, which showed reductions of 80% in these traits. NDVI decreased by 20% in resilient hybrids versus a 45% decrease in sensitive ones, while PSRI increases were more pronounced in sensitive hybrids, indicating accelerated leaf senescence. The study concluded that multispectral imaging and image analysis are potent tools for evaluating water stress responses in *M. sinensis* hybrids, demonstrating particular promise for drought-tolerant hybrids like GRC4 and GRC5 in breeding programs aimed at enhancing resilience and biomass productivity on marginal lands.

In one of the latest studies, Avgoustaki et al. (2022) employed a CMS-V multispectral camera attached to the robotic arm of an autonomous robot named "Watney," developed by Seasony in Copenhagen, Denmark, to collect multispectral images of basil plants (*Ocimum basilicum*). The study aimed to monitor compound stress, i.e., water and nitrogen stress, in basil grown in an indoor vertical farming setup using multispectral imaging combined with robotic technology. The camera captured images at three different positions per treatment: control (full irrigation and fertilization), water stress (50% reduced irrigation), and nitrogen stress (50% reduced nitrogen). Images were taken three times daily at 10:00, 12:00, and 14:00, with each session beginning with a camera calibration using a dark and a white reference image. The experiment involved three groups of Genovese basil, each subjected to different treatment conditions for 6 days. Multispectral imaging data were analyzed using indices such as the NDVI and the PRI. Concurrently, physiological measurements like photosynthetic efficiency, chlorophyll content, and stomatal conductance were recorded to establish correlations with spectral data. The results demonstrated that NDVI and PRI were effective in detecting changes in reflectance associated with water and nitrogen stress, indicating significant reductions in these indices under stress conditions. Specifically, nitrogen stress correlated with decreased chlorophyll content and photosynthetic efficiency, whereas water stress was linked to lower substrate water content. Despite these stressors, the basil plants showed no visible symptoms, highlighting the utility of spectral indices in detecting subtle physiological changes.

This approach underscores the potential of plant phenotyping through multispectral or HSI as a noninvasive, efficient method for monitoring plant health and stress responses in a controlled environment, supporting the ongoing development of phenomics.

23.2.1.3 Hyperspectral sensors for the detection of drought and combined stresses

HSI is a complex, multidisciplinary field characterized by the simultaneous acquisition of spatial images across many spectrally contiguous bands, with each pixel containing a full spectrum (ElMasry and Sun, 2010). Among various HSI systems, the line-scan or pushbroom approach is the most prevalent, involving an array of components including an objective lens, spectrograph, camera, acquisition system, translation stage, illumination, and computer (Iqbal et al., 2014). This system's spectral range is primarily determined by the camera, spectrograph, and illumination conditions (Gowen et al., 2007). The pushbroom system operates with a spectrograph at its core, characterized by the absence of moving parts, which helps generate a spectrum for each point along the scanned line. Light from the object is focused onto an entrance slit, collimated, dispersed by a grating by wavelength, and then focused onto a two-dimensional detector—either charge-coupled device or complementary metal-oxide-semiconductor. This process produces a two-dimensional image where one dimension represents spectral data (λ) and the other represents spatial data (X) of the scanned line. The second spatial dimension (Y) is produced by moving the object or the optical unit across the scene, generating a comprehensive three-dimensional hyperspectral image (X, Y, λ), where two dimensions provide spatial information, and one represents spectral data (ElMasry et al., 2012).

HSI is based on spectral reflectance spectroscopy and is a noninvasive, fast, high-throughput, and RS method for plant phenotyping. It captures extensive spectral and spatial information, which is compiled into a 3D data matrix or hypercube. This hypercube contains hundreds to thousands of contiguous images across narrow spectral bands ranging from UV to shortwave IR (250–2500 nm) (Li et al., 2020; Lowe et al., 2017). Its nondestructive nature allows for detailed analyses of plant tissues and their biochemical components, estimation of nutrient content, calculating vegetation indices, identifying species, and monitoring plant responses to environmental stresses (Amigo and Grassi, 2020; Abebe et al., 2023; Mishra et al., 2017). In summary, HSI is a critical tool in modern plant phenotyping, leveraging its capability to integrate spectral and spatial data to provide insights into plant structure and physiological conditions (Liu et al., 2020; Sarić et al., 2022). This technology plays a significant role in advancing our understanding of plant biology and enhancing agricultural productivity through precise monitoring and management strategies. In the past half-decade, hyperspectral cameras have become increasingly popular for monitoring soil water content in the 400–1000 nm reflectance range (Zhang et al., 2024).

HSI systems, deployed proximally, extend capabilities by capturing reflectance data that correlates with plant water content. For instance, in the study conducted by Asaari et. al. (2018), HSI systems deployed proximally in an HTP platform (HTPPP) were used to analyze maize plants (*Zea mays* L.) during their early vegetative stages (V5–V7). The system employed a pushbroom visible and NIR (VNIR) hyperspectral camera installed in an enclosed cabin, providing controlled and uniform imaging conditions. Reflectance data were captured across a spectral range of 400–1000 nm, which includes wavelengths sensitive to water absorption and leaf structure properties. By measuring reflectance in specific spectral bands, the system detected subtle changes in spectral signatures associated with variations in plant water content. These changes, particularly in water-sensitive bands, enabled the identification of physiological traits and early detection of drought stress. This proximal setup extended the system's capabilities to monitor plant health noninvasively, offering insights into water-related stress responses in maize plants (Asaari et al., 2018).

Advances in HSI technologies are employed to assess solar-induced fluorescence (SIF) for the prediction of the photosynthesis properties of plants. HSI systems allow for simultaneous and accurate assessment of the SIF emission in plant leaves, which provides a visual representation of physiological changes caused by the photosynthetic activities of the plants in the early stage (Faqeerzada et al., 2022). For instance, water stress in sugar beet was detected based on the SIF using HSI (Wang et al., 2022). This study evaluated the potential of UAV-based sun-induced chlorophyll fluorescence (SIF) indicators for detecting water stress in a sugar beet (*Beta vulgaris* L.) field. (Wang et al., 2022) investigated the response of SIF metrics under conditions of drought (June) and combined drought and heat stress (July) to determine their reliability in identifying plant physiological changes and guiding precision agricultural practices. The study also utilized hyperspectral and thermal data to enhance the interpretation of SIF responses to irrigation treatments. The method involved conducting a field experiment during the 2019 growing season in the Netherlands, where sugar beet plots were subjected to two treatments: irrigation and nonirrigation. UAV-mounted sensors collected SIF data at 687–760 nm wavelengths using the FluorSpec SIF system. Additional hyperspectral data (515–870 nm) were acquired using the Rikola hyperspectral camera, and thermal imagery was captured using the Workswell WIRIS camera. Ground-based measurements of soil moisture, LAI, and chlorophyll content (Cab) were conducted to validate RS observations. SIF metrics such as SIF yield (SIF normalized by absorbed PAR, i.e., photosynthetically active radiation) and fluorescence emission yield (ΦF760) were analyzed to assess physiological plant responses to stress. The results demonstrated that UAV-based SIF metrics, particularly ΦF760 and the SIF ratio (SIFratio), were effective in detecting water stress during June when only drought was present. Significant differences were observed between irrigated and nonirrigated fields, with nonirrigated fields showing decreased SIF metrics due to reduced photosynthetic efficiency. However, under the combined drought and heat stress in July, the response of SIF indicators was less pronounced, with only weak tracking of irrigation effects. The study concluded that UAV-based SIF has potential for detecting water stress at the field level but emphasized the need to account for structural canopy effects and further explore its utility under varying stress conditions. Similarly, studies have been conducted to test the utilization of HSI under combined heat and drought stress conditions (Trachsel et al., 2019; Melandri et al., 2021; Poobalasubramanian et al., 2022) as well as under combined drought and nutrient stress (Okyere et al., 2024; Corti et al., 2017; Camino González, 2019).

Using hyperspectral data, Laroche-Pinel et al. (2021) evaluated the results from several vine plots consisting of three grape varieties with varying growth stages and irrigation methods. Regression methods were used to examine correlations between leaf reflectance and vine water status as determined by stem water potential. The findings revealed the most effective (red [B4], rededge [B6], NIR, B8, and shortwave-infrared [SWIR, B11]) bands of Sentinel-2 satellite imagery, spectral domains, and vegetation indices (NDVI and NDWI) for assessing vine water availability. Water content and NIR directly affected the shortwave-infrared domain, and red-edge bands were indirectly correlated with water status by affecting chlorophyll concentration and cellular morphology. Their study aimed to determine whether vine water status could be monitored over a landscape with the help of multispectral satellite spectral bands, such as Sentinel-2 data.

Poobalasubramanian et al. (2022) conducted a study aimed at detecting early compound stress of heat and water in strawberry plants (*Fragaria ananassa*) using hyperspectral imagery and chlorophyll fluorescence-based indices. A total of 45 strawberry plants were subjected to three experimental stages: controlled conditions, heat stress, and recovery. During the heat-stress stage, plants were exposed to elevated temperatures of 35°C for 16 hours and 30°C for 8 hours daily over 6 days. HSI captured fluorescence reflectance data across wavelengths from 500 to 900 nm. Eight fluorescence-based indices were analyzed for their sensitivity to heat stress, including the pigment-specific normalized difference for chlorophyll A, pigment-specific normalized difference for chlorophyll B, pigment-specific simple ratio for chlorophyll A, pigment-specific simple ratio for chlorophyll B, chlorophyll index at RedEdge, normalized-difference rededge index (NDRE), simple ratio (SR), and rededge vegetation stress index (RVSI). These indices captured changes in chlorophyll fluorescence and reflectance, which are highly sensitive to the physiological impacts of heat stress, such as reduced photosynthetic efficiency and altered leaf structure. ML models, including random forest, gradient boosting, and support vector machines, were employed to classify heat-stress levels based on the extracted indices. The gradient boosting model achieved an accuracy of 91% when selective indices, such as RVSI, PSSRb, and Chl-b, were used. This highlights the importance of these indices as reliable indicators of heat stress. The results demonstrated that spectral properties associated with photosynthetic activity are significantly affected by heat stress, enabling its early detection before visual symptoms appear. In conclusion, the study showcased the effectiveness of hyperspectral imagery combined with fluorescence-based indices and ML models in identifying early heat stress in strawberry plants. This approach provides a noninvasive, accurate method for monitoring plant stress, which can support timely interventions to mitigate heat-induced damage and improve crop management.

In one of the latest studies, Asaari et al. (2022) evaluated the use of HSI for nondestructive analysis of plant physiological traits under drought stress. The research aimed to estimate water potential, stomatal conductance, transpiration rate, and effective quantum yield as indicators of drought response in maize plants. The study was conducted at the PHENOVISION platform, VIB-UGent Center for Plant Systems Biology in Ghent, Belgium. A total of 128 maize plants of the B104 inbred line were used, with 108 plants (54 well-watered and 54 water-stressed) for model training and 20 plants (10 well-watered and 10 water-stressed) for testing. The experiment involved growing maize plants in a controlled greenhouse environment. Drought was induced in the progressive drought group by withholding water for 7 days, reducing the soil water content from 2.4 g H_2O/g dry soil (well-watered) to 1.4 g H_2O/g dry soil (drought).

Hyperspectral images were captured using a VNIR hyperspectral camera (400–1000 nm) over 10 days. The spectral data were processed and normalized, and machine learning (ML) regression algorithms, including PLSR, Kernel ridge regression (KRR), and Gaussian process regression (GPR), were used to develop models for physiological trait estimation. The results showed that GPR outperformed the other algorithms, achieving the highest accuracy with an average normalized root mean squared error (NRMSE) of 6.85% for water potential and R^2 of 0.92. The GPR models accurately detected significant differences between well-watered and drought-stressed plants as early as the third day of drought induction. Spectral bands in the ranges of 500–520 nm, 600–680 nm, and 720–800 nm were particularly effective for water potential prediction. All four physiological traits demonstrated clear trends distinguishing well-watered from drought-stressed plants, with reduced stomatal conductance, transpiration rates, and photosynthetic efficiency observed under drought conditions. The study concluded that HSI combined with GPR is a powerful, noninvasive tool for detecting drought stress and assessing plant physiological traits.

23.2.2 Heat stress

Heat stress refers to the adverse effects experienced by organisms when exposed to high temperatures (Hu et al., 2020). Heat stress represents a significant environmental challenge that adversely affects the physiological and metabolic balance in plants, leading to detrimental impacts such as reduced growth, yield losses, and plant mortality (Hasanuzzaman et al., 2013). Higher temperatures directly affect plant vegetative stages, resource allocation, and, above all, reproductive processes, which can lead to a substantial reduction in yields (Kopecka et al., 2023). Organisms exposed to elevated temperatures experience a surge in phytohormones like abscisic acid (ABA) and indole-3-acetic acid, which play crucial roles in stress response (Sheikh et al., 2024). Additional phytohormones, including salicylic acid, ethylene (ET), and jasmonates (JAs), further contribute to a plant's adaptive mechanisms under stressful conditions (Siddiqi and Husen, 2019).

Research indicates that heat and drought stresses often co-occur, complicating the stress dynamics for plants. While drought stress develops gradually, heat stress can arise abruptly, causing immediate and severe impacts on plant health and productivity, such as a sharp decline in grain yields even after brief episodes of high temperatures (Siebert et al., 2014).

The resilience of different plant species to heat varies significantly. Deryng et al. (2014) employed the PEGASUS global crop model to analyze 72 climate change scenarios, revealing distinct responses among maize, spring wheat, and soybean to extreme heat. Their findings highlighted that spring wheat is particularly vulnerable, whereas soybeans show relative resilience to elevated temperatures. Siebert et al. (2014) demonstrated that surpassing critical temperature thresholds, particularly around the anthesis stage in wheat, leads to notable declines in grain yield and number. The concept of stress thermal time, derived from canopy temperatures, provided a more precise measure of heat stress impacts, indicating yield reductions by up to 22% under field conditions. This approach emphasized that standard weather station measurements often underestimate the real extent of heat stress experienced by plants.

However, recent advancements in sensing technologies, such as TIR sensors and Proximal Hyperspectral Sensing (PHS), offer refined methods for assessing plant stress responses. These technologies enable nondestructive, high-resolution monitoring of both leaf and canopy temperatures, which are crucial for detecting stress-induced physiological changes (Sanaeifar et al., 2023; Zain et al., 2023). Ground-based thermal sensors, particularly effective in controlled environments like greenhouses, highlight reduced transpiration rates indicative of heat and drought stresses.

23.2.2.1 Red, blue, and green sensors

RGB sensors can be employed either individually in controlled environments or on UAVs for detecting heat stress (Abdelhakim et al., 2021; Im et al., 2018). Typically, RGB sensors are combined with TIR sensors to enhance the accuracy of heat stress detection in plants. For instance, Fevgas et al. (2023) utilized a combination of RGB and TIR imaging to detect heat stress in grapevines (*Vitis vinifera*), exploiting their complementary capabilities. TIR imaging measured variations in canopy temperature to identify areas showing elevated temperatures indicative of stress. Conversely, RGB imaging isolated the plant canopy from the soil and background, distinguishing vegetation based on its visual color characteristics. The RGB and TIR images were aligned using metadata, ensuring precise pixel correspondence between the datasets. Elevated temperatures identified in the TIR data were mapped onto the aligned RGB images using pseudocoloring (e.g., yellow) to visualize stressed areas, thereby enhancing the identification of heat-stressed plants by combining the thermal sensitivity of TIR with the spatial and visual information from RGB, resulting in a more intuitive and accurate detection method.

Im et al. (2018) developed and validated crop growth estimation models for Chinese cabbage (*Brassica rapa*) and white radish (*Raphanus sativus* var. *longipinnatus*) using UAV-based RGB imagery. These models quantified biophysical parameters such as leaf length, leaf count, leaf area, and fresh weight, with a particular focus on detecting heat stress through spatial and temporal variations in crop growth. Field experiments conducted over two growing seasons 2015 and 2016 in Muan, Korea, utilized UAVs equipped with RGB cameras to collect imagery at regular intervals, applying radiometric calibration to minimize light condition effects. The vegetation fraction (VF) was determined using the ExG vegetation index, and plant height (PH) was calculated using the structure-from-motion (SfM) algorithm. These metrics served as predictors in multiple linear regression models to estimate biophysical parameters. In 2016, 62 Chinese cabbage plants and 42 white radish plants were grown to validate the regression models, ensuring a diverse range of growth stages were represented. The results showed significant linear relationships between UAV-derived indices and actual crop growth measurements. Coefficients of determination (R^2) ranged from 0.66 to 0.90 for various biophysical parameters. For Chinese cabbage, R^2 values were highest for fresh weight and leaf length (both at 0.94), followed by leaf count (0.90) and leaf width (0.83). For white radish, fresh weight had an R^2 of 0.85, leaf length 0.82, leaf count 0.78, and leaf width 0.66. The lower accuracy for white radish was attributed to challenges in segmenting overlapping leaves, particularly during early growth stages. Heat stress effects were inferred by observing deviations in vegetation fraction and plant height during high-temperature periods. The study concluded that UAV-based RGB imagery is a cost-effective and reliable tool for monitoring crop growth and detecting heat stress. Zhang et al. (2019) presented a study using thermal and high-resolution RGB images captured by a UAV to extract canopy temperature in maize (*Zea mays*). They proposed a method termed the red-green ratio index (RGRI)-Otsu for extracting the canopy temperature, demonstrating a high correlation of 0.94 between the UAV-measured canopy temperatures and the ground-truth temperatures captured by a handheld infrared thermometer.

Abdelhakim et al. (2021) investigated the effects of heat stress on wheat (*Triticum aestivum*) genotypes using RGB imagery in a controlled environment. The objective was to evaluate differences in growth and physiological responses of heat-sensitive and heat-tolerant wheat genotypes under heat-stress conditions and during recovery phases. The experiment was conducted in a climate-controlled growth chamber at the Photon Systems Instruments (PSI) facility in Drásov, Czech Republic, involving four wheat genotypes: LM19, SF1 (heat-sensitive), and LM62, NS3 (heat-tolerant), with a total of 152 plants. A dynamic IBP system monitored plant growth. Plants were grown in 3.1 L pots under optimal conditions until the heat stress treatment, which involved raising the day/night temperature to 36°C/26°C for 11 days. RGB imagery was used to assess changes in plant biomass and growth rate during stress and subsequent recovery phases. Measurements of relative growth rate (RGR) and plant volume were taken daily using RGB cameras. The results demonstrated that heat stress significantly impacted the growth of heat-sensitive genotypes (LM19 and SF1), with notable reductions in RGR and biomass accumulation during the late stress phase. In contrast, heat-tolerant genotypes (LM62 and NS3) exhibited less pronounced reductions in growth and showed better recovery after the heat treatment. RGB-based growth models revealed that NS3 maintained the highest RGR among all genotypes, demonstrating its superior tolerance to heat stress. Heat-sensitive genotypes exhibited higher canopy temperatures, reflecting impaired cooling mechanisms under stress. The phenotyping approach using RGB imagery provided valuable insights into dynamic physiological responses to abiotic stress and demonstrated its utility for selecting robust wheat genotypes for future climate challenges.

In conclusion, the evolving field of plant stress physiology benefits immensely from technological advancements in imaging and sensing. These tools not only enhance our understanding of heat stress impacts but also aid in the development of robust strategies for managing stress in agricultural contexts, thus ensuring sustainable crop production in the face of climatic challenges.

23.2.2.2 Multispectral sensors

Recent studies have demonstrated innovative applications of multispectral imagery combined with ML to assess heat stress in various crops. This approach represents a significant advancement in agricultural technology, offering insights into the physiological changes that crops undergo under heat stress conditions. For example, Sharma et al. (2024) conducted a comprehensive study at the Plant Breeding Centre in Horsham, Victoria, Australia, phenotyping a population of 184 wheat (*Triticum aestivum*) genotypes for heat stress tolerance. They employed a blend of physiological traits, multispectral imagery, and diverse ML algorithms, including random forest classifier (RFC), support vector machine (SVM), linear discriminant analysis (LDA), k-nearest neighbors (KNN), and ResNet50. The integration of these methods enabled effective differentiation of wheat genotypes based on their heat tolerance levels. Particularly, the RFC achieved an overall accuracy of 64% in classifying genotypes into susceptible, moderate, and tolerant categories. This study underscores the potential of using UAV-gathered multispectral data in conjunction with ML to enhance our understanding of how heat stress impacts crop performance.

Kim et al. (2022) explored the potential of snapshot-based VNIR multispectral imaging for early detection of heat stress in Chinese cabbage (*Brassica rapa*). Conducted at the National Institute of Horticultural and Herbal Science in the Republic of Korea, the experiment utilized 72 plants grown under three different temperature regimes. Advanced ML models, such as partial least squares-discriminant analysis (PLS-DA) and least squares support vector machine (LS-SVM), outperformed traditional vegetation indices in detecting heat stress, with the LS-SVM model achieving a classification accuracy of up to 93.6%.

Ryu et al. (2022) assessed the impact of heat stress on paddy rice (*Oryza sativa*) cultivars 'Ilmi' and "Saenuri," using multispectral imagery and vegetation indices like NDVI and PRI. Their research, carried out at Chonnam National University, South Korea, demonstrated that these indices are effective tools for measuring the effects of heat stress, particularly in terms of delayed canopy senescence and reduced photosynthetic efficiency.

Finally, Spišić et al. (2022) investigated the detection of heat stress and water-deficit responses, such as leaf rolling in maize (*Zea mays*), through multispectral imagery at the Agricultural Institute Osijek, Croatia. Utilizing a spectrum of ML algorithms, including single layer perceptron (SLP), multilayer perceptron (MLP), CNNs, and support vector machines (SVM), they found that SVM and CNN models, with F1 scores of approximately 76.9%, effectively monitored heat-induced stress in maize.

These studies collectively highlight the efficacy of integrating multispectral imaging with ML techniques to provide a sophisticated, reliable, and economically feasible method for monitoring and assessing heat stress across different crop types. This integration not only improves our diagnostic capabilities but also enhances our capacity to manage crops under varying climatic stresses effectively.

23.2.2.3 Hyperspectral sensors

HSI has emerged as a powerful nondestructive tool in agricultural research, capable of identifying biochemical markers such as reactive oxygen species (ROS) that accumulate under stress conditions. By analyzing specific reflectance patterns, hyperspectral sensors can detect heat stress in plants before visible symptoms appear, enabling proactive management strategies.

Faqeerzada et al. (2022) utilized line-scan fluorescence hyperspectral imaging (Fl-HSI) in the VIS-NIR range of 400–1000 nm to screen 80 ginseng (*Panax ginseng*) plants across four varieties. This study conducted imaging in a dark room to minimize fluorescence interference from ambient light, assessing plants before and after exposure to 35°C for three days. The study found significant reductions in chlorophyll concentrations, with control plants showing a mean SPAD value of 39.5 compared to 26.8 in heat-stressed plants. Advanced statistical methods like ANOVA and PLSR helped identify critical wavelengths and predict chlorophyll levels with a high determination coefficient ($R^2 = 0.90$), demonstrating a 96% accuracy in distinguishing stressed plants. In contrast, Park et al. (2021) investigated the potential of VIS/NIR and SWIR imaging systems to evaluate heat stress in Korean ginseng (*Panax ginseng*). Their analysis included partial least squares-discriminant analysis (PLS-DA) and identified key wavebands through variable importance in projection (VIP) scores, achieving high validation accuracies in distinguishing between susceptible and resistant varieties of ginseng. This method proved effective in supporting HTP and breeding programs aimed at enhancing heat resistance.

A study conducted on tomato (*Solanum lycopersicum* L.) seedlings used HSI to quantify the impact of low-temperature stress on total polyphenol content (Kang et al., 2024). By subjecting seedlings to varying low temperatures for different durations, the researchers employed Lasso regression to pinpoint spectral bands crucial for predicting polyphenol content. This study demonstrated that cold stress significantly increased polyphenol levels, particularly under more severe conditions, and established that HSI could effectively monitor these biochemical changes, thus paving the way for the development of cost-effective hyperspectral sensors for agricultural applications. Melandri et al. (2021) assessed the relationship between hyperspectral imagery and stress responses in upland cotton (*Gossypium hirsutum* L.), correlating leaf metabolite profiles with spectral data. The study utilized advanced analytical techniques such as GC-MS and LC-MS to analyze metabolites, with hyperspectral reflectance providing a reliable means of detecting stress-induced changes in plant biochemistry. Their findings suggest that HSI, combined with multivariate analysis, offers a robust framework for nondestructively measuring leaf metabolites, which could facilitate the screening of genotypes under varied environmental stresses.

These studies collectively underscore the versatility and precision of HSI in plant science, particularly in detecting and analyzing plant responses to temperature-induced stress. This technology not only enhances our understanding of plant physiology under stress conditions but also supports the development of more resilient crop varieties through informed breeding strategies.

23.2.3 Nutrient stress

Nutrient deficiencies are a major factor contributing to the loss of crop quality and productivity. Poor crop yield is commonly observed under various environmental constraints, particularly when the availability of essential mineral nutrients is limited. Nutrient imbalances in plants significantly impact their overall performance, affecting growth patterns, antioxidant defense mechanisms, and tolerance to both biotic and abiotic stresses. Furthermore, the availability and transport of soil nutrients are heavily influenced by other abiotic stresses, such as drought, since water serves as a vital carrier for nutrient transport to the roots. Drought stress reduces the diffusion and mass flow of key macronutrients and micronutrients, including calcium, silicon, magnesium, and nitrate (Chaudhry and Sidhu, 2022).

However, implementing proper nutrient management practices under stress conditions can improve crop yields. Nutrient management is an effective strategy for mitigating environmental stress and enhancing agricultural productivity. As the climate changes, the detrimental effects of abiotic stress on plants are expected to intensify. While developing stress-tolerant crop varieties is an important adaptation strategy, the application of appropriate plant nutrients can also enhance plants' ability to withstand abiotic stresses. Specifically, nutrients such as nitrogen, potassium, calcium, and magnesium have been shown to increase the concentration of antioxidant enzymes, including superoxide dismutase (SOD), peroxidase (POD), and catalase (CAT), which help reduce reactive oxygen species (ROS) (Kumari et al., 2022). Nutrients like potassium and calcium also improve stomatal regulation and osmotic adjustments by enhancing water uptake (Kumari et al., 2022).

Certain plant nutrients activate specific mechanisms that alleviate particular types of abiotic stress. For instance, boron, an important micronutrient, enhances the activity of the antioxidant system and mitigates the harmful effects of ROS produced under heat stress (Waraich et al., 2011). Similarly, zinc (Zn) plays a crucial role in maintaining membrane permeability, and an optimal supply of Zn helps protect plants from the adverse effects of heat stress (Peck and McDonald, 2010).

Although the use of plant nutrients represents a low-cost and sustainable approach to managing abiotic stresses, much remains to be explored in this area. A deeper understanding of nutrient deficiencies, their optimal concentrations at different phenological stages, and the application of advanced imaging technologies can aid in the precise management of nutrient resources, ultimately improving crop resilience under stress conditions.

23.2.3.1 Red, blue, and green sensors

The utilization of RGB imagery in conjunction with DL models presents a significant advancement in precision agriculture, offering a more objective and error-resistant method for quantifying changes in plant health traditionally assessed by human judgment. This technology facilitates the accurate detection and classification of nutrient deficiencies and stresses in various crop species.

Lazarević et al. (2022) and Neilson et al. (2015) demonstrated how RGB cameras can quantify stress symptoms like senescence due to nitrogen deficiency by counting the number of yellow pixels in leaves. This method reduces subjective error, providing a more reliable assessment of plant health. Furthering this application, Yi et al. (2020) conducted a study at the long-term fertilizer experiment site Dikopshof, near Bonn, Germany, evaluating the capability of RGB imagery and DL models to diagnose nutrient deficiencies in sugar beet (*Beta vulgaris*). The study involved seven nutrient treatment plots, with a dataset named "Deep Nutrient Deficiency for Sugar Beet (DND-SB)" consisting of 5648 RGB images. Using CNNs such as DenseNet-161, which achieved the highest accuracy of 98.4%, the research highlighted the potential of DL to precisely identify deficiencies of nitrogen, phosphorus, potassium, and calcium.

Zermas et al. (2021) developed an automated computer vision methodology using high-resolution RGB imagery and Faster R-CNN with ResNet-50 to detect nitrogen deficiency in corn plants. Conducted over two cultivation seasons in Minnesota, USA, this approach achieved an 82.3% mean average precision, demonstrating its effectiveness in identifying stressed leaves and enabling targeted fertilizer applications. In another study, Ghazal et al. (2024) assessed RGB imagery and ML models for detecting nitrogen stress in maize (*Zea mays*). Using 1200 images captured during the flowering stage, the study successfully classified nitrogen levels using EfficientNetB0, which outperformed other models with a 97% accuracy. This underscores the practicality and precision of CNN models in identifying nutrient stress across various growth stages and conditions.

Taha et al. (2022) utilized RGB imagery and deep CNNs to diagnose nitrogen (N), phosphorus (P), and potassium (K) deficiencies in lettuce (*Lactuca sativa*) grown in aquaponic systems at Zhejiang University, China. Employing Inceptionv3, which achieved a classification accuracy of 96.5%, the study emphasized the potential of RGB imagery and DCNNs for real-time nutrient monitoring in sustainable agriculture systems. In one of the latest studies, Deichmann et al. (2024) explored RGB imagery and DL for detecting macro and micronutrient deficiencies in barley (*Hordeum vulgare*). The research was conducted across diverse experimental setups at the University of Bonn, Germany, including hydroponic, greenhouse, mesocosm, and field experiments. The ResNet-50 model achieved prediction accuracies up to 88.8%, proving the viability of RGB and CNNs for nutrient management in barley cultivation.

In conclusion, the abovementioned studies demonstrate the robustness of RGB imagery combined with advanced ML techniques in enhancing the accuracy and scalability of nutrient deficiency detection in agriculture. This integration not only supports precision agriculture practices but also paves the way for more sustainable farming techniques by optimizing resource use and management.

23.2.3.2 *Multispectral sensors for combined nutrient stresses*

Recent studies have demonstrated the effectiveness of UAV multispectral imaging in diagnosing and managing nutrient deficiencies across various crops, enhancing precision agriculture and sustainable farming practices. For example, Li et al. (2023) conducted a significant study at the Medicinal Botanical Garden of Chengdu University of Traditional Chinese Medicine, focusing on *Ligusticum chuanxiong* Hort., a valuable medicinal plant. The study involved 196 plants categorized into control, nitrogen deficiency (ND), phosphorus deficiency (PD), and potassium deficiency (KD) groups. UAVs captured multispectral imagery at 5 m and 10 m altitudes, and vegetation indices such as NDVI, GNDVI, and OSAVI were analyzed. Logistic regression models at the 5 m altitude achieved remarkable accuracies—100% for nitrogen deficiencies and over 90% for phosphorus and potassium deficiencies. This research highlights the potential of UAV-based multispectral imaging to provide rapid and accurate nutrient diagnostics in medicinal plants.

On the other hand, a study evaluated the capability of multispectral UAV imagery to distinguish between nitrogen and water stress in maize at the University of California's Russell Ranch Sustainable Agriculture Facility (Becker et al., 2020). The study utilized vegetation indices like NDRE, GLI, and BRIN to develop a Combined Index (CI) for assessing stress types, with NDRE showing strong correlations with grain yield (R^2 = 0.50–0.63) and nitrogen uptake (R^2 = 0.71–0.64). This study underlines the efficacy of UAV-based multispectral imaging in identifying and differentiating stress types, thereby supporting targeted interventions in precision agriculture. The use of UAV-based multispectral imagery to predict NPK content in grape leaves throughout various growth stages has also been tested (Peng et al., 2022). This study employed and compared different models, such as partial least squares (PLS), random forest (RF), and extreme learning machine (ELM), with RF and ELM demonstrating effective nutrient prediction with R^2 values above 0.75. This research points to the promising application of UAV imagery for enhancing vineyard nutrient management, ultimately boosting productivity and sustainability.

Sun et al. (2023) applied UAV-based multispectral RS in conjunction with the Diagnosis and Recommendation Integrated System (DRIS) to assess nitrogen, phosphorus, and potassium levels in apple trees in Shaanxi Province, China. The study spanned seven growth stages and utilized ridge regression models that achieved R^2 values up to 0.76, effectively estimating nutrient levels. DRIS methodologies further diagnosed deficiencies, facilitating precise fertilizer management. This approach emphasizes the integration of UAV imagery and advanced diagnostic systems in promoting sustainable orchard management. Collectively, these studies confirm the transformative impact of UAV multispectral imaging technology in agriculture. By enabling precise and timely diagnosis of nutrient deficiencies, this technology not only optimizes resource use but also contributes to the advancement of sustainable farming practices globally.

23.2.3.3 *Hyperspectral sensors for combined stresses of nutrient deficiencies and salinity*

Recent advancements in hyperspectral sensing technology have greatly enhanced the capacity to detect and analyze nutrient deficiencies in various crops, facilitating precise agricultural management and supporting sustainable farming practices. Behmann et al. (2014) and Zhang et al. (2024) have demonstrated the efficacy of UAVs equipped with hyperspectral sensors in detecting nutrient deficiencies across large-scale operations by analyzing plant reflectance in the VIS and NIR spectral regions. This approach provides early diagnosis of fertilizer-related stress by revealing changes in chlorophyll and nutrient content. Ground-based hyperspectral sensors, as discussed by Mishra et al. (2017) and Faqir et al. (2024), complement these aerial assessments by providing detailed spectral data and real-time soil fertility evaluations through portable soil sensors.

Sen et al. (2024) introduced HyperSpectral Photo Evaluation (HSIA) for nutrient stress analysis in plants, emphasizing the growing interest in RS technologies over the past few years (Wang, 2023). These technologies employ vegetation indices such as NDVI and EVI to map nutrient domains and assess plant health, allowing for detailed characterization of nutrient stress and enhancing agricultural management decisions (Ramakrishnan et al., 2023).

A well-established strong correlation between nitrogen levels and chlorophyll content in leaves is the basis of using UAV HSI for quick and accurate estimates of leaf nitrogen in the field (Zhu et al., 2018). Similarly, studies on phosphorus have shifted from direct content monitoring in crops to using hyperspectral responses of soil to predict spatial variability (Kawamura et al., 2021; Lin et al., 2015). For potassium, ground-based hyperspectral data have been effectively used to build prediction models for crop nutrients, as shown in studies by Abdel-Rahman et al. (2017) and Mohamed et al. (2020), achieving high validation accuracies through approaches like PLSR.

Debnath et al. (2021) utilized HSI to identify individual nutrient deficiencies in grapevine leaves at Charles Sturt University, Australia. By extracting features such as reflectance, spectral ratios, and vegetation indices, and employing ML techniques like Support Vector Machines (SVM), the study efficiently distinguished between healthy and nutrient-deficient leaves. This approach not only demonstrated the potential of HSI in precision viticulture but also highlighted its role in optimizing grapevine health and productivity. Similarly, another study also explored various nitrogen retrieval methods using hyperspectral data for grapevine cultivars (*Vitis vinifera*), highlighting that chemometric and ML models like Gaussian Process Regression (GPR) provided the highest accuracy. This study emphasized the potential of hyperspectral sensing in high-resolution vineyard nitrogen management, particularly noting the effectiveness of models in accounting for variations such as leaf age (Peanusaha et al., 2024).

Additionally, HSI has also been utilized to detect salinity stress in both field and specialty crops (Hasana and Miyake, 2017; Lara et al., 2016; Pabuayon et al., 2022) investigated nutrient uptake dynamics under salinity stress by integrating HSIwith conventional ion analysis. Rice recombinant inbred lines with contrasting salinity tolerance were cultivated hydroponically and subjected to salinity stress by increasing the solution conductivity to 9 dS/m (~90 mM NaCl). Daily hyperspectral images were collected across wavelengths ranging from 550 to 1700 nm, generating reflectance spectra indicative of physiological and biochemical changes in the plants. To improve data quality and emphasize treatment-related differences, the spectral datasets were processed through masking, cropping, noise reduction, and wavelength filtering. To validate spectral features as proxies for nutrient uptake, Na^+ and K^+ concentrations were quantified at the end of the experiment using flame photometry on destructively sampled tissues. The imaging and ion datasets were then integrated through PLSR, enabling the prediction of ion content trends over time. This approach successfully captured Na^+ accumulation and K^+ depletion under salinity stress, although predictions for K^+ remained less consistent. Overall, the study demonstrated that HSI offers a high-throughput, nondestructive tool for monitoring nutrient uptake and salinity responses in rice.

Similarly, Lara et al. (2016) applied proximal HSI in combination with physiological measurements to detect nutrient stress induced by salinity in lettuce. Plants were grown hydroponically and exposed to four irrigation treatments: control (0 mM NaCl) and three salinity levels (50, 100, and 150 mM NaCl). After ten days, leaves were harvested and imaged with a hyperspectral vision system covering the 400–1000 nm range. Reflectance spectra extracted from tens of thousands of pixels per leaf were processed using Savitzky–Golay smoothing, spectral derivatives, and Standard Normal Variate normalization. To interpret spectral variation, Principal Component Analysis (PCA) and a newly developed Level Salinity Index (LSI), focusing on the rededge region (~710 nm), were employed. This analysis enabled nondestructive, pixel-level detection of salinity-induced nutrient stress, underscoring the potential of HSI as a rapid and precise monitoring method for crop stress responses.

These studies collectively underscore the significant potential of hyperspectral sensing technologies in agriculture. By providing detailed and accurate assessments of nutrient deficiencies and salinity stress, HSI supports precision agriculture initiatives, enabling farmers and agronomists to make informed decisions that optimize crop health, enhance yield, and promote sustainable agricultural practices.

23.3 Future perspectives and challenges

Despite the promising potential of imaging technologies in mitigating and managing abiotic stresses under changing climatic conditions, several limitations remain that must be addressed in the near future. One of the primary challenges is managing the vast amounts of data generated by imaging systems and the ability to extract and interpret valuable insights from these large datasets. The integration of artificial intelligence (AI), particularly ML and DL, is critical for handling multidimensional and multivariate data automatically recorded by RGB, multispectral, and hyperspectral sensors. However, selecting the appropriate tools for specific applications is equally important. For example, ML is typically employed for the evaluation and prediction of stress, biomass, and yield, while DL is more suitable for the detection, recognition, and counting of objects in large and complex datasets, such as biotic/abiotic stresses and individual plants (Abebe et al., 2023).

The limited use of DL can be attributed to the lack of standardized protocols and the scarcity of labeled datasets. Establishing standard protocols for data collection, processing, and analysis is crucial to ensure reproducibility and comparability across studies (Walsh et al., 2024). Additionally, (Abebe et al., 2023) highlight the absence of technologies capable of capturing root system architecture and predicting the yield of underground parts at the field level, as most current sensors focus on above-ground data. However, integrating below-ground data with above-ground information could uncover new traits that can be measured and compared among crops under abiotic stress conditions.

Advanced computational tools enable automated feature extraction, pattern recognition, and predictive modeling, facilitating data-driven decision-making in agriculture. The combined use of imaging technologies—such as fluorescence, thermal, satellite, and LiDAR imaging—could lead to more sensitive and specific stress detection systems, despite their current limitations when used individually. Big data analysis will simplify the determination of plant tolerance to stress. Furthermore, integrating satellite data with existing imaging technologies is an avenue worth exploring, as it could enhance the accuracy and timeliness of predictions related to environmental stress impacts on crop health, contributing to global food security efforts. LiDAR technology, with its ability to provide detailed structural information, may find enhanced application when paired with AI for 3D phenotyping of plants, improving the understanding of stress manifestations and increasing the accuracy of stress detection and phenotyping in complex environments (Walsh et al., 2024).

While significant progress has been made, the adoption of image-based technologies specifically for combined abiotic stress detection continues to grow. Prospects for detecting combined or multiple abiotic stresses in crops are closely tied to advances in next-generation technologies that can integrate multiple sensing platforms and leverage computational intelligence. Unlike current approaches that often isolate a single-stress signature, future systems are expected to combine HSI, TIR, LiDAR, chlorophyll fluorescence, and even nanosensor-based systems into multimodal frameworks. These integrated technologies will generate complementary datasets, enabling researchers to disentangle overlapping physiological responses caused by drought, salinity, heat, or nutrient deficiencies. Moreover, the growing accessibility of UAVs, satellites, and IoT-based ground sensors offers scalable opportunities to monitor multiple stresses simultaneously across spatial and temporal scales.

The hope for the future lies in moving from single-stress detection toward holistic plant health monitoring, where technologies are capable of capturing the plant's entire stress all at the same time. This shift would not only improve accuracy but also allow proactive stress management in precision agriculture, where interventions can be optimized based on real-time, multistress insights. Innovations such as quantum-based sensors, compact handheld tools that combine multiple sensing capabilities, and the merging of omics data (like genomics, metabolomics, and phenomics) with imaging approaches are showing great potential to make stress detection even more powerful and precise. Ultimately, the convergence of advanced sensor technologies, AI-driven analytics, and systems biology approaches will make it possible to reliably diagnose multiple concurrent stresses, paving the way for resilient crop production under increasingly variable and unpredictable climatic conditions.

References

Abdel-Rahman, E.M., Mutanga, O., Odindi, J., Adam, E., Odindo, A., Ismail, R., 2017. Estimating Swiss chard foliar macro- and micronutrient concentrations under different irrigation water sources using ground-based hyperspectral data and four partial least squares (PLS)-based (PLS1, PLS2, SPLS1 and SPLS2) regression algorithms. Computers and Electronics in Agriculture 132, 21–33. Available at https://doi.org/10.1016/j.compag.2016.11.008.

Abdelhakim, L.O.A., Rosenqvist, E., Wollenweber, B., Spyroglou, I., Ottosen, C.O., Panzarová, K., 2021. Investigating combined drought-and heat stress effects in wheat under controlled conditions by dynamic image-based phenotyping. Agronomy 11 (2), 364. Available at https://doi.org/10.3390/agronomy11020364.

Abebe, A.M., Kim, Y., Kim, J., Kim, S.L., Baek, J., 2023. Image-based high-throughput phenotyping in horticultural crops. Plants 12 (10), 2061. Available at https://doi.org/10.3390/plants12102061.

Al-Tamimi, N., Langan, P., Bernád, V., Walsh, J., Mangina, E., Negrão, S., 2022. Capturing crop adaptation to abiotic stress using image-based technologies. Open Biology 12 (6). Available at https://doi.org/10.1098/rsob.210353.

Amigo, J.M., Grassi, S., 2020. Configuration of hyperspectral and multispectral imaging systems. Data Handling in Science and Technology 32, 17–34. Available at https://doi.org/10.1016/B978-0-444-63977-6.00002-X.

Anjum, S.A., Wang, L., Farooq, M., Khan, I., Xue, L., 2011. Methyl jasmonate-induced alteration in lipid peroxidation, antioxidative defence system and yield in soybean under drought. Journal of Agronomy and Crop Science 197 (4), 296–301. Available at https://doi.org/10.1111/j.1439-037x.2011.00468.x.

Asaari, M.S.M., Mertens, S., Verbraeken, L., Dhondt, S., Inzé, D., Bikram, K., et al., 2022. Non-destructive analysis of plant physiological traits using hyperspectral imaging: a case study on drought stress. Computers and Electronics in Agriculture 195, 106806. Available at https://doi.org/10.1016/j.compag.2022.106806.

Asaari, M.S.M., Mishra, P., Mertens, S., Dhondt, S., Inzé, D., Wuyts, N., et al., 2018. Close-range hyperspectral image analysis for the early detection of stress responses in individual plants in a high-throughput phenotyping platform. ISPRS Journal of Photogrammetry and Remote Sensing 138, 121–138. Available at https://doi.org/10.1016/j.isprsjprs.2018.02.003.

Avgoustaki, D.D., Avgoustakis, I., Miralles, C.C., Sohn, J., Xydis, G., 2022. Autonomous mobile robot with attached multispectral camera to monitor the development of crops and detect nutrient and water deficiencies in vertical farms. Agronomy 12 (11), 2691. Available at https://doi.org/10.3390/agronomy12112691.

Baluja, J., Diago, M.P., Balda, P., Zorer, R., Meggio, F., Morales, F., et al., 2012. Assessment of vineyard water status variability by thermal and multispectral imagery using an unmanned aerial vehicle (UAV). Irrigation Science 30 (6), 511–522. Available at https://doi.org/10.1007/s00271-012-0382-9.

Becker, T., Nelsen, T.S., Leinfelder-Miles, M., Lundy, M.E., 2020. Differentiating between nitrogen and water deficiency in irrigated maize using a UAV-based multi-spectral camera. Agronomy 10 (11), 1671. Available at https://doi.org/10.3390/agronomy10111671.

Begna, T., 2021. Impact of drought stress on crop production and its management options. International Journal of Research in Agronomy 4 (2), 66–74. Available at https://doi.org/10.33545/2618060x.2021.v4.i2a.103.

Behmann, J., Steinrücken, J., Plümer, L., 2014. Detection of early plant stress responses in hyperspectral images. ISPRS Journal of Photogrammetry and Remote Sensing 93, 98–111. Available at https://doi.org/10.1016/j.isprsjprs.2014.03.016.

Berger, K., Machwitz, M., Kycko, M., Kefauver, S.C., Van Wittenberghe, S., Gerhards, M., et al., 2022. Multi-sensor spectral synergies for crop stress detection and monitoring in the optical domain: a review. Remote Sensing of Environment 280. Available at https://doi.org/10.1016/j.rse.2022.113198.

Boyer, J.S., Byrne, P., Cassman, K.G., Cooper, M., Delmer, D., Greene, T., et al., 2013. The U.S. drought of 2012 in perspective: a call to action. Global Food Security 2 (3), 139–143. Available at https://doi.org/10.1016/j.gfs.2013.08.002.

Briglia, N., Williams, K., Wu, D., Li, Y., Tao, S., Corke, F., et al., 2020. Image-based assessment of drought response in grapevines. Frontiers in Plant Science 11. Available at https://doi.org/10.3389/fpls.2020.00595.

Camino González, C.L., 2019. Detection of water and nutritional stress through chlorophyll fluorescence and radiative transfer models from hyperspectral and thermal imagery (PhD Thesis). UCO Press, pp. 191.

Chandel, N.S., Chakraborty, S.K., Rajwade, Y.A., Dubey, K., Tiwari, M.K., Jat, D., 2021. Identifying crop water stress using deep learning models. Neural Computing and Applications 33 (10), 5353–5367. Available at https://doi.org/10.1007/s00521-020-05325-4.

Chandel, N.S., Rajwade, Y.A., Dubey, K., Chandel, A.K., Subeesh, A., Tiwari, M.K., 2022. Water stress identification of winter wheat crop with state-of-the-art AI techniques and high-resolution thermal-RGB imagery. Plants 11 (23), 3344. Available at https://doi.org/10.3390/plants11233344.

Chaudhry, S., Sidhu, G.P.S., 2022. Climate change regulated abiotic stress mechanisms in plants: a comprehensive review. Plant Cell Reports 41, 1–31.

Coppens, F., Wuyts, N., Inzé, D., Dhondt, S., 2017. Unlocking the potential of plant phenotyping data through integration and data-driven approaches. Current Opinion in Systems Biology 4, 58–63. Available at https://doi.org/10.1016/j.coisb.2017.07.002.

Corti, M., Marino Gallina, P., Cavalli, D., Cabassi, G., 2017. Hyperspectral imaging of spinach canopy under combined water and nitrogen stress to estimate biomass, water, and nitrogen content. Biosystems Engineering 158, 38–50. Available at https://doi.org/10.1016/j.biosystemseng.2017.03.006.

Damm, A., Cogliati, S., Colombo, R., Fritsche, L., Genangeli, A., Genesio, L., et al., 2022. Response times of remote sensing measured sun-induced chlorophyll fluorescence, surface temperature and vegetation indices to evolving soil water limitation in a crop canopy. Remote Sensing of Environment 273, 112957. Available at https://doi.org/10.1016/j.rse.2022.112957.

Debnath, S., Paul, M., Rahaman, D.M.M., Debnath, T., Zheng, L., et al., 2021. Identifying individual nutrient deficiencies of grapevine leaves using hyperspectral imaging. Remote Sensing 13 (16), 3317. Available at https://doi.org/10.3390/rs13163317.

Deichmann, M., Yi, J.H., Mihiret, Y.E., Zahra, S., Sauer, C., Hüging, H., et al., 2024. RGB image-based detection of nutrient deficiencies in barley by deep learning methods. agriRxiv. Available at https://doi.org/10.31220/agriRxiv.2024.00271.

Deryng, D., Conway, D., Ramankutty, N., Price, J., Warren, R., 2014. Global crop yield response to extreme heat stress under multiple climate change futures. Environmental Research Letters 9 (3), 034011. Available at https://doi.org/10.1088/1748-9326/9/3/034011.

Dodig, D., Božinović, S., Nikolić, A., Zorić, M., Vančetović, J., Ignjatović-Micić, D., et al., 2021. Dynamics of maize vegetative growth and drought adaptability using image-based phenotyping under controlled conditions. Frontiers in Plant Science 12. Available at https://doi.org/10.3389/fpls.2021.652116.

ElMasry, G., Sun, D.W., Allen, P., 2012. Near-infrared hyperspectral imaging for predicting colour, pH and tenderness of fresh beef. Journal of Food Engineering 110 (1), 127–140.

ElMasry, G., Sun, D.W., 2010. CHAPTER 1 - Principles of hyperspectral imaging technology. In: Sun, D.W. (Ed.), Hyperspectral Imaging for Food Quality Analysis and Control. Academic Press, pp. 3–43. Available at https://doi.org/10.1016/B978-0-12-374753-2.10001-2.

Faqeerzada, M.A., Nabwire, S., Park, E., Wakholi, C., Joshi, R., Cho, B.K., 2022. Fluorescence hyperspectral imaging for early diagnosis of abiotically heat-stressed ginseng plants. SSRN, South Korea SSRN. Available at https://doi.org/10.2139/ssrn.4147603.

Faqir, Y., Qayoom, A., Erasmus, E., Schutte-Smith, M., Visser, H.G., 2024. A review on the application of advanced soil and plant sensors in the agriculture sector. Computers and Electronics in Agriculture 226, 109385. Available at https://doi.org/10.1016/j.compag.2024.109385.

Farooq, M., Wahid, A., Kobayashi, N., Fujita, D., Basra, S.M.A., Lichtfouse, E., Navarrete, M., Debaeke, P., Véronique, S., Alberola, C., 2009. Plant drought stress: effects, mechanisms and management. In: Lichtfouse, E., Navarrete, M., Debaeke, P., Véronique, S., Alberola, C. (Eds.), Sustainable Agriculture. Springer, Dordrecht, pp. 153–188. Available at 10.1007/978-90-481-2666-8_12.

Fevgas, G., Lagkas, T., Argyriou, V., Sarigiannidis, P., 2023. Detection of biotic or abiotic stress in vineyards using thermal and RGB images captured via IoT sensors. IEEE Access 11, 105902–105915. Available at https://doi.org/10.1109/access.2023.3320048.

Francini, A., Sebastiani, L., 2019. Abiotic stress effects on performance of horticultural crops. Horticulturae 5 (4), 67. Available at https://doi.org/10.3390/horticulturae5040067.

Gago, J., Douthe, C., Coopman, R.E., Gallego, P.P., Ribas-Carbo, M., Flexas, J., et al., 2015. UAVs challenge to assess water stress for sustainable agriculture. Agricultural Water Management 153, 9–19. Available at https://doi.org/10.1016/j.agwat.2015.01.020.

Gerhards, M., Rock, G., Schlerf, M., Udelhoven, T., 2016. Water stress detection in potato plants using leaf temperature, emissivity, and reflectance. International Journal of Applied Earth Observation and Geoinformation 53, 27–39. Available at https://doi.org/10.1016/j.jag.2016.08.004.

Gerhards, M., Schlerf, M., Mallick, K., Udelhoven, T., 2019. Challenges and future perspectives of multi-/hyperspectral thermal infrared remote sensing for crop water-stress detection: a review. Remote Sensing 11 (10), 1240. Available at https://doi.org/10.3390/rs11101240.

Ghazal, S., Kommineni, N., Munir, A., 2024. Comparative analysis of machine learning techniques using RGB imaging for nitrogen stress detection in maize. AI 5 (3), 1286–1300. Available at https://doi.org/10.3390/ai5030062.

Goldstein, A., Fink, L., Meitin, A., Bohadana, S., Lutenberg, O., Ravid, G., 2018. Applying machine learning on sensor data for irrigation recommendations: revealing the agronomist's tacit knowledge. Precision Agriculture 19 (3), 421–444.

Gowen, A., Odonnell, C., Cullen, P., Downey, G., Frias, J., 2007. Hyperspectral imaging – an emerging process analytical tool for food quality and safety control. Trends in Food Science & Technology 18 (12), 590–598. Available at https://doi.org/10.1016/j.tifs.2007.06.001.

Hairmansis, A., Berger, B., Tester, M., Roy, S.J., 2014. Image-based phenotyping for non-destructive screening of different salinity tolerance traits in rice. Rice 7 (1), 16.

Hama, A., Matsumoto, Y., Matsuoka, N., 2022. Estimating leaf water content through low-cost LiDAR. Agronomy 12 (5), 1183.

Hasana, R., Miyake, H., 2017. Salinity stress alters nutrient uptake and causes the damage of root and leaf anatomy in maize. KnE Life Sciences 3 (4), 219. Available at https://doi.org/10.18502/kls.v3i4.708.

Hasanuzzaman, M., Nahar, K., Alam, M.M., Roychowdhury, R., Fujita, M., 2013. Physiological, biochemical, and molecular mechanisms of heat stress tolerance in plants. International Journal of Molecular Sciences 14 (5), 9643–9684. Available at https://doi.org/10.3390/ijms14059643.

Hu, S., Ding, Y., Zhu, C., 2020. Sensitivity and responses of chloroplasts to heat stress in plants. Frontiers in Plant Science 11, 375.

Ihuoma, S.O., Madramootoo, C.A., 2017. Recent advances in crop water stress detection. Computers and Electronics in Agriculture 141, 267–275. Available at https://doi.org/10.1016/j.compag.2017.07.026.

Im, H., Lee, S., Naqi, M., Lee, C., Kim, S., 2018. Flexible PI-based plant drought stress sensor for real-time monitoring system in smart farm. Electronics 7 (7), 114. Available at https://doi.org/10.3390/electronics7070114.

Iqbal, A., Sun, D.W., Allen, P., 2014. An overview on principle, techniques and application of hyperspectral imaging with special reference to ham quality evaluation and control. Food Control 46, 242–254.

Jangra, S., Chaudhary, V., Yadav, R.C., Yadav, N.R., 2021. High-throughput phenotyping: a platform to accelerate crop improvement. Phenomics 1 (2), 31–53. Available at https://doi.org/10.1007/s43657-020-00007-6.

Jayasinghe, C., Badenhorst, P., Jacobs, J., Spangenberg, G., Smith, K., 2021. Image-based high-throughput phenotyping for the estimation of persistence of perennial ryegrass (Lolium perenne L.)—a review. Grass and Forage Science 76 (3), 321–339. Available at https://doi.org/10.1111/gfs.12520.

Jiang, Z., Tu, H., Bai, B., Yang, C., Zhao, B., Guo, Z., et al., 2021. Combining UAV-RGB high-throughput field phenotyping and genome-wide association study to reveal genetic variation of rice germplasms in dynamic response to drought stress. New Phytologist 232 (1), 440–455. Available at https://doi.org/10.1111/nph.17580.

Jinru, X., Baofeng, S., 2017. Significant remote sensing vegetation indices: a review of developments and applications. Journal of Sensors 2017, 1353691. Available at https://doi.org/10.1155/2017/1353691.

Jones, H.G., Vaughan, R.A., 2010. Remote Sensing of Vegetation: Principles, Techniques, and Applications. Oxford University Press, Oxford. Available at https://books.google.co.in/books?id=sTmcAQAAQBAJ.

Joshi, S., Thoday-Kennedy, E., Daetwyler, H.D., Hayden, M., Spangenberg, G., Kant, S., et al., 2021. High-throughput phenotyping to dissect genotypic differences in safflower for drought tolerance. PLoS One 16 (7), e0254908. Available at https://doi.org/10.1371/journal.pone.0254908.

Kamarudin, M.H., Ismail, Z.H., Saidi, N.B., 2021. Deep learning sensor fusion in plant water stress assessment: a comprehensive review. Applied Sciences 11 (4), 1403.

Kandylakis, Z., Falagas, A., Karakizi, C., Karantzalos, K., 2020. Water stress estimation in vineyards from aerial SWIR and multispectral UAV Data. Remote Sensing 12 (15), 2499. Available at https://doi.org/10.3390/rs12152499.

Kang, Y.S., Ryu, C.S., Kang, J.G., 2024. Presenting a multispectral image sensor for quantification of total polyphenols in low-temperature stressed tomato seedlings using hyperspectral imaging. Sensors (Basel, Switzerland) 24 (13), 14248220. Available at https://doi.org/10.3390/s24134260.

Karnieli, A., Agam, N., Pinker, R.T., Anderson, M., Imhoff, M.L., Gutman, G.G., et al., 2010. Use of NDVI and land surface temperature for drought assessment: merits and limitations. Journal of Climate 23 (3), 618–633. Available at https://doi.org/10.1175/2009JCLI2900.1.

Kawamura, K., Nishigaki, T., Andriamananjara, A., Rakotonindrina, H., Tsujimoto, Y., Moritsuka, N., et al., 2021. Using a one-dimensional convolutional neural network on visible and near-infrared spectroscopy to improve soil phosphorus prediction in Madagascar. Remote Sensing 13 (8), 1519. Available at https://doi.org/10.3390/rs13081519.

Kim, J.Y., Chung, Y.S., 2021. A short review of RGB sensor applications for accessible high-throughput phenotyping. Journal of Crop Science and Biotechnology 24 (5), 495–499. Available at https://doi.org/10.1007/s12892-021-00104-6.

Kim, S.L., Kim, N., Lee, H., Lee, E., Cheon, K.S., Kim, M., et al., 2020. High-throughput phenotyping platform for analyzing drought tolerance in rice. Planta 252 (3). Available at https://doi.org/10.1007/s00425-020-03436-9.

Kim, M., Lee, C., Hong, S., Kim, S.L., Baek, J., Kim, K.H., 2021. High-throughput phenotyping methods for breeding drought-tolerant crops. International Journal of Molecular Sciences 22 (15), 8266. Available at https://doi.org/10.3390/ijms22158266.

Kim, G., Lee, H., Wi, S.H., Cho, B.-K., 2022. Snapshot-based visible-near infrared multispectral imaging for early screening of heat injury during growth of Chinese cabbage. Applied Sciences 12 (18), 9340. Available at https://doi.org/10.3390/app12189340.

Kim, D.W., Yun, H.S., Jeong, S.J., Kwon, Y.S., Kim, S.G., Lee, W.S., et al., 2018. Modeling and testing of growth status for Chinese cabbage and white radish with UAV-based RGB imagery. Remote Sensing 10 (4), 563. Available at https://doi.org/10.3390/rs10040563.

Kim, J.Y., 2020. Roadmap to high throughput phenotyping for plant breeding. Journal of Biosystems Engineering 45 (1), 43–9355. Available at https://doi.org/10.1007/s42853-020-00043-0.

Kopecka, R., Kameniarova, M., Cerny, M., Brzobohaty, B., Novak, J., 2023. Abiotic stress in crop production. International Journal of Molecular Sciences 24, 6603.

Kumari, V.V., Banerjee, P., Verma, V.C., Sukumaran, S., Chandran, M.A.S., Gopinath, K.A., et al., 2022. Plant nutrition: an effective way to alleviate abiotic stress in agricultural crops. International Journal of Molecular Sciences 23 (15), 8519. Available at https://doi.org/10.3390/ijms23158519.

Laraswati, A.A., Padjung, R., Farid, M., Nasaruddin, N., Anshori, M.F., Nur, A., et al., 2021. Image based-phenotyping and selection index based on multivariate analysis for rice hydroponic screening under drought stress. Plant Breeding and Biotechnology 9 (4), 272–286. Available at https://doi.org/10.9787/PBB.2021.9.4.272.

Lara, M., Diezma, B., Lleó, L., Roger, J., Garrido, Y., Gil, M., et al., 2016. Hyperspectral imaging to evaluate the effect of irrigationwater salinity in lettuce. Applied Sciences 6 (12), 412. Available at https://doi.org/10.3390/app6120412.

Laroche-Pinel, E., Duthoit, S., D. Costard, A., Rousseau, J., Hourdel, J., Vidal-Vigneron, M., et al., 2021. Monitoring vineyard water status using sentinel-2 images: qualitative survey on five wine estates in the south of france. Oeno One 55 (4), 115–127. Available at https://doi.org/10.20870/oeno-one.2021.55.4.4752.

Lazarević, B., Kontek, M., Carović-Stanko, K., Clifton-Brown, J., Al Hassan, M., Trindade, L.M., et al., 2022. Multispectral image analysis detects differences in drought responses in novel seeded Miscanthus sinensis hybrids. GCB Bioenergy 14 (11), 1219–1234. Available at https://doi.org/10.1111/gcbb.12999.

Lin, C., Ma, R., Zhu, Q., Li, J., 2015. Using hyper-spectral indices to detect soil phosphorus concentration for various land use patterns. Environmental Monitoring and Assessment 187 (1). Available at https://doi.org/10.1007/s10661-014-4130-x.

Liu, H., Bruning, B., Garnett, T., Berger, B., 2020. Hyperspectral imaging and 3D technologies for plant phenotyping: from satellite to close-range sensing. Computers and Electronics in Agriculture 175, 105621. Available at https://doi.org/10.1016/j.compag.2020.105621.

Li, W., Wang, K., Han, G., Wang, H., Tan, N., Yan, Z., 2023. Integrated diagnosis and time-series sensitivity evaluation of nutrient deficiencies in medicinal plant (Ligusticum chuanxiong Hort.) based on UAV multispectral sensors. Frontiers in Plant Science 13. Available at https://doi.org/10.3389/fpls.2022.1092610.

Li, Q., Wang, Q., Li, X., 2020. Mixed 2D/3D convolutional network for hyperspectral image super-resolution. Remote Sensing 12 (10), 1660.

Lowe, A., Harrison, N., French, A.P., 2017. Hyperspectral image analysis techniques for the detection and classification of the early onset of plant disease and stress. Plant Methods 13 (1), 80.

Lu, B., Dao, P., Liu, J., He, Y., Shang, J., 2020. Recent advances of hyperspectral imaging technology and applications in agriculture. Remote Sensing 12 (16), 2659. Available at https://doi.org/10.3390/rs12162659.

Melandri, G., Thorp, K.R., Broeckling, C., Thompson, A.L., Hinze, L., Pauli, D., 2021. Assessing drought and heat stress-induced changes in the cotton leaf metabolome and their relationship with hyperspectral reflectance. Frontiers in Plant Science 12. Available at https://doi.org/10.3389/fpls.2021.751868.

Mishra, P., Asaari, M.S.M., Herrero-Langreo, A., Lohumi, S., Diezma, B., Scheunders, P., 2017. Close range hyperspectral imaging of plants: a review. Biosystems Engineering 164, 49–67. Available at https://doi.org/10.1016/j.biosystemseng.2017.09.009.

Mohamed, E.S., El Baroudy, A.A., El-beshbeshy, T., Emam, M., Belal, A.A., Elfadaly, A., et al., 2020. Vis-NIR spectroscopy and satellite landsat-8 OLI data to map soil nutrients in arid conditions: a case study of the Northwest Coast of Egypt. Remote Sensing 12 (22), 3716. Available at https://doi.org/10.3390/rs12223716.

Neilson, E.H., Edwards, A.M., Blomstedt, C.K., Berger, B., Møller, B.L., Gleadow, R.M., 2015. Utilization of a high-throughput shoot imaging system to examine the dynamic phenotypic responses of a C4 cereal crop plant to nitrogen and water deficiency over time. Journal of Experimental Botany 66 (7), 1817–1832. Available at https://doi.org/10.1093/jxb/eru526.

Ninomiya, S., 2022. High-throughput field crop phenotyping: current status and challenges. Breeding Science 72 (1), 3–18. Available at https://doi.org/10.1270/jsbbs.21069.

Okyere, F.G., Cudjoe, D.K., Virlet, N., Castle, M., Riche, A.B., Greche, L., et al., 2024. Hyperspectral imaging for phenotyping plant drought stress and nitrogen interactions using multivariate modeling and machine learning techniques in wheat. Remote Sensing 16 (18), 3446. Available at https://doi.org/10.3390/rs16183446.

Pabuayon, I.C.M., Pabuayon, I.L.B., Singh, R.K., Ritchie, G.L., de los Reyes, B.G., 2022. Applicability of hyperspectral imaging during salinity stress in rice for tracking Na+ and K+ levels in planta. PLoS One 17 (7), e0270931. Available at https://doi.org/10.1371/journal.pone.0270931.

Pandey, R., Vengavasi, K., Hawkesford, M.J., 2021. Plant adaptation to nutrient stress. Plant Physiology Reports 26 (4), 583–586.

Parkash, V., Singh, S., 2020. A review on potential plant-based water stress indicators for vegetable crops. Sustainability 12 (10), 3945. Available at https://doi.org/10.3390/su12103945.

Park, E., Kim, Y.S., Omari, M.K., Suh, H.K., Faqeerzada, M.A., Kim, M.S., et al., 2021. High-throughput phenotyping approach for the evaluation of heat stress in korean ginseng (Panax ginseng meyer) using a hyperspectral reflectance image. Sensors (Basel, Switzerland) 21 (16), 5634. Available at https://doi.org/10.3390/s21165634.

Peanusaha, S., Pourreza, A., Kamiya, Y., Fidelibus, M.W., Chakraborty, M., 2024. Nitrogen retrieval in grapevine (Vitis vinifera L.) leaves by hyperspectral sensing. Remote Sensing of Environment 302, 113966. Available at https://doi.org/10.1016/j.rse.2023.113966.

Peck, A.W., McDonald, G.K., 2010. Adequate zinc nutrition alleviates the adverse effects of heat stress in bread wheat. Plant and Soil 337, 355–374.

Peng, X., Chen, D., Zhou, Z., Zhang, Z., Xu, C., Zha, Q., et al., 2022. Prediction of the nitrogen, phosphorus and potassium contents in grape leaves at different growth stages based on UAV multispectral remote sensing. Remote Sensing 14 (11), 2659. Available at https://doi.org/10.3390/rs14112659.

Poobalasubramanian, M., Park, E.S., Faqeerzada, M.A., Kim, T., Kim, M.S., Baek, I., et al., 2022. Identification of early heat and water stress in strawberry plants using chlorophyll-fluorescence indices extracted via hyperspectral images. Sensors (Basel, Switzerland) 22 (22), 8706. Available at https://doi.org/10.3390/s22228706.

Praba, M.L., Cairns, J.E., Babu, R.C., Lafitte, H.R., 2009. Identification of physiological traits underlying cultivar differences in drought tolerance in rice and wheat. Journal of Agronomy and Crop Science 195 (1), 30–46. Available at https://doi.org/10.1111/j.1439-037x.2008.00341.x.

Qing, X., Bochu, W., Chuanren, D., 2000. Molecular mechanism of water-stress response in plant. Sheng wu hua xue yu Sheng wu wu li jin Zhan 27 (3), 247–250.

Qiu, Q., Sun, N., Bai, H., Wang, N., Fan, Z., Wang, Y., et al., 2019. Field-based high-throughput phenotyping for maize plant using 3D LiDAR point cloud generated with a "Phenomobile" Frontiers in Plant Science 10. Available at https://doi.org/10.3389/fpls.2019.00554.

Ramakrishnan, R., Mohammed, M.A., Mohammed, M.A., Mohammed, V.A., Logeshwaran, J., Maheswaran, S., 2023. An innovation prediction of DNA damage of melanoma skin cancer patients using deep learning. 2023 14th International Conference on Computing Communication and Networking Technologies (ICCCNT), Delhi, India. IEEE, pp. 1–7. Available at 10.1109/ICCCNT56998.2023.10306749.

Ramesh, G., Logeshwaran, J., Kiruthiga, T., Lloret, J., 2023. Prediction of energy production level in large PV plants through AUTO-encoder based neural-network (AUTO-NN) with restricted Boltzmann feature extraction. Future Internet 15 (2), 46. Available at https://doi.org/10.3390/fi15020046.

Rossi, R., Costafreda-Aumedes, S., Summerer, S., Moriondo, M., Leolini, L., Cellini, F., et al., 2022. A comparison of high-throughput imaging methods for quantifying plant growth traits and estimating above-ground biomass accumulation. European Journal of Agronomy 141, 126634. Available at https://doi.org/10.1016/j.eja.2022.126634.

Ryu, J.H., Oh, D., Ko, J., Kim, H.Y., Yeom, J.M., Cho, J., 2022. Remote sensing-based evaluation of heat stress damage on paddy rice using NDVI and PRI measured at leaf and canopy scales. Agronomy 12 (8), 1872. Available at https://doi.org/10.3390/agronomy12081972.

Sahoo, R.N., Ray, S.S., Manjunath, K.R., 2015. Hyperspectral remote sensing of agriculture. Current Science 108 (5), 848–859. Available at https://www.ncfc.gov.in/publications/p2.pdf.

Sakinah, A.I., Musa, Y., Farid, M., Hairmansis, A., Anshori, M.F., Nasaruddin, N., 2022. Rice selection criteria based on morphological and image-based phenotyping under drought- and salinity-stress conditions. Sabrao Journal of Breeding and Genetics 54 (4), 686–699. Available at https://doi.org/10.54910/sabrao2022.54.4.1.

Sanaeifar, A., Yang, C., de la Guardia, M., Zhang, W., Li, X., He, Y., 2023. Proximal hyperspectral sensing of abiotic stresses in plants. Science of The Total Environment 861, 160652. Available at https://doi.org/10.1016/j.scitotenv.2022.160652.

Sarić, R., Nguyen, V.D., Burge, T., Berkowitz, O., Trtílek, M., Whelan, J., et al., 2022. Applications of hyperspectral imaging in plant phenotyping. Trends in Plant Science 27 (3), 301–315. Available at https://doi.org/10.1016/j.tplants.2021.12.003.

Sen, S., Bhambu, P., Satish Kumar, R., 2024. Retracted: remote monitoring of nutrient stress in agricultural crops using hyper spectral image analysis. 2024 15th International Conference on Computing Communication and Networking Technologies (ICCCNT), Kamand, India. IEEE, pp. 1–6. Available at 10.1109/ICCCNT61001.2024.10724757.

Sen, S., Bhambu, P., Satish Kumar, R., 2024. Remote monitoring of nutrient stress in agricultural crops using hyper spectral image analysis. 15th International Conference on Computing Communication and Networking Technologies, ICCCNT 2024. Institute of Electrical and Electronics Engineers Inc, India. Available at 10.1109/ICCCNT61001.2024.10724757.

Sharma, N., Kumar, M., Daetwyler, H.D., Trethowan, R.M., Hayden, M., Kant, S., 2024. Phenotyping for heat stress tolerance in wheat population using physiological traits, multispectral imagery, and machine learning approaches. Plant Stress 14. Available at https://doi.org/10.1016/j.stress.2024.100593.

Sheikh, Z.N., Raina, S., Bhat, U.H., Sharma, N., Saffeullah, P., Mohammad, F., et al., 2024. Deciphering the potential role of nano-based biosensors in growth, development and stress management of crop plants. South African Journal of Botany 167, 542–553. Available at https://doi.org/10.1016/j.sajb.2024.02.035.

Siddiqi, K.S., Husen, A., 2019. Plant response to jasmonates: current developments and their role in changing environment. Bulletin of the National Research Centre 43, 153. Available at https://doi.org/10.1186/s42269-019-0195-6.

Siebert, S., Ewert, F., Eyshi Rezaei, E., Kage, H., Graß, R., 2014. Impact of heat stress on crop yield—on the importance of considering canopy temperature. Environmental Research Letters 9 (4), 044012. Available at https://doi.org/10.1088/1748-9326/9/4/044012.

Spišić, J., Šimić, D., Balen, J., Jambrović, A., Galić, V., 2022. Machine learning in the analysis of multispectral reads in maize canopies responding to increased temperatures and water deficit. Remote Sensing 14 (11), 2596. Available at https://doi.org/10.3390/rs14112596.

Sun, G., Hu, T., Chen, S., Sun, J., Zhang, J., Ye, R., et al., 2023. Using UAV-based multispectral remote sensing imagery combined with DRIS method to diagnose leaf nitrogen nutrition status in a fertigated apple orchard. Precision Agriculture 24 (6), 2522–2548. Available at https://doi.org/10.1007/s11119-023-10051-7.

Taha, M.F., Abdalla, A., Elmasry, G., Gouda, M., Zhou, L., Zhao, N., et al., 2022. Using deep convolutional neural network for image-based diagnosis of nutrient deficiencies in plants grown in aquaponics. Chemosensors 10 (2), 45. Available at https://doi.org/10.3390/chemosensors10020045.

Tolomio, M., Casa, R., 2020. Dynamic crop models and remote sensing irrigation decision support systems: a review of water stress concepts for improved estimation of water requirements. Remote Sensing 12 (23), 3945. Available at https://doi.org/10.3390/rs12233945.

Trachsel, S., Dhliwayo, T., Perez, L.G., Lugo, J.A.M., Trachsel, M., Ali, J., 2019. Estimation of physiological genomic estimated breeding values (PGEBV) combining full hyperspectral and marker data across environments for grain yield under combined heat and drought stress in tropical maize (Zea mays L.). PLoS One 14 (3), e0212200. Available at https://doi.org/10.1371/journal.pone.0212200.

Virlet, N., Costes, E., Martinez, S., Kelner, J.J., Regnard, J.L., 2015. Multispectral airborne imagery in the field reveals genetic determinisms of morphological and transpiration traits of an apple tree hybrid population in response to water deficit. Journal of Experimental Botany 66 (18), 5453–5465. Available at https://doi.org/10.1093/jxb/erv355.

Virnodkar, S.S., Pachghare, V.K., Patil, V.C., Jha, S.K., 2020. Remote sensing and machine learning for crop water stress determination in various crops: a critical review. Precision Agriculture 21 (5), 1121–1155. Available at https://doi.org/10.1007/s11119-020-09711-9.

Walsh, J.J., Mangina, E., Negrão, S., 2024. Advancements in imaging sensors and AI for plant stress detection: a systematic literature review. Plant Phenomics 6, 0153.

Wang, N., Clevers, J.G.P.W., Wieneke, S., Bartholomeus, H., Kooistra, L., 2022. Potential of UAV-based sun-induced chlorophyll fluorescence to detect water stress in sugar beet. Agricultural and Forest Meteorology 323, 109033. Available at https://doi.org/10.1016/j.agrformet.2022.109033.

Wang, Y.D.M., 2023. A survey on multi agent system and its applications in power system engineering. Journal of Computational Intelligence in Materials Science 1, 001–011.

Waraich, E.A., Ahmad, R., Ashraf, M.Y., Saifullah, S., Ahmad, M., 2011. Improving agricultural water use efficiency by nutrient management in crop plants. Acta Agriculturae Scandinavica Section B: Soil and Plant Science 61 (4), 291–304. Available at https://doi.org/10.1080/09064710.2010.491954.

Wu, L., Shao, H., Li, J., Chen, C., Hu, N., Yang, B., et al., 2024. Noninvasive abiotic stress phenotyping of vascular plant in each vegetative organ view. Plant Phenomics 6. Available at https://doi.org/10.34133/plantphenomics.0180.

Yi, J., Krusenbaum, L., Unger, P., Hüging, H., Seidel, S.J., Schaaf, G., et al., 2020. Deep learning for non-invasive diagnosis of nutrient deficiencies in sugar beet using RGB images. Sensors (Basel, Switzerland) 20 (20), 5893. Available at https://doi.org/10.3390/s20205893.

Zain, M., Ma, H., Nuruzzaman, M., Chaudhary, S., Nadeem, M., Shakoor, N., et al., 2023. Nanotechnology based precision agriculture for alleviating biotic and abiotic stress in plants. Plant Stress 10, 100239. Available at https://doi.org/10.1016/j.stress.2023.100239.

Zeid, I.M., Shedeed, Z.A., 2006. Response of alfalfa to putrescine treatment under drought stress. Biologia Plantarum 50 (4), 635–640. Available at https://doi.org/10.1007/s10535-006-0099-9.

Zermas, D., Nelson, H.J., Stanitsas, P., Morellas, V., Mulla, D.J., Papanikolopoulos, N., 2021. A methodology for the detection of nitrogen deficiency in corn fields using high-resolution RGB imagery. IEEE Transactions on Automation Science and Engineering 18 (4), 1879–1891. Available at https://doi.org/10.1109/TASE.2020.3022868.

Zhang, L., Niu, Y., Zhang, H., Han, W., Li, G., Tang, J., et al., 2019. Maize canopy temperature extracted from UAV thermal and RGB imagery and its application in water stress monitoring. Frontiers in Plant Science 10. Available at https://doi.org/10.3389/fpls.2019.01270.

Zhang, K., Yan, F., Liu, P., 2024. The application of hyperspectral imaging for wheat biotic and abiotic stress analysis: a review. Computers and Electronics in Agriculture 221, 109008. Available at https://doi.org/10.1016/j.compag.2024.109008.

Zhou, J., Zhou, J., Ye, H., Ali, M.L., Chen, P., Nguyen, H.T., 2021. Yield estimation of soybean breeding lines under drought stress using unmanned aerial vehicle-based imagery and convolutional neural network. Biosystems Engineering 204, 90–103. Available at https://doi.org/10.1016/j.biosystemseng.2021.01.017.

Zhu, H., Liu, H., Xu, Y., Guijun, Y., 2018. UAV-based hyperspectral analysis and spectral indices constructing for quantitatively monitoring leaf nitrogen content of winter wheat. Applied Optics 57 (27), 7722–7732. Available at https://doi.org/10.1364/AO.57.007722.

Index

Note: Page numbers followed by "*f*" and "*t*" refer to figures and tables, respectively.

D

E

I

J

K

9780443274244